Lecture Notes in Artificial Intelligence 3918

Edited by J. G. Carbonell and J. Siekmann

Subseries of Lecture Notes in Computer Science

Wee Keong Ng   Masaru Kitsuregawa
Jianzhong Li   Kuiyu Chang (Eds.)

Advances in Knowledge Discovery and Data Mining

10th Pacific-Asia Conference, PAKDD 2006
Singapore, April 9-12, 2006
Proceedings

Springer

Volume Editors

Wee Keong Ng
Nanyang Technological University, Centre for Advanced Information Systems
Nanyang Avenue, N4-B3C-14, 639798, Singapore
E-mail: awkng@ntu.edu.sg

Masaru Kitsuregawa
University of Tokyo, Institute of Industrial Science
4-6-1 Komaba, Meguro-Ku, Tokyo 153-8305, Japan
E-mail: kitsure@tkl.iis.u-tokyo.ac.jp

Jianzhong Li
Harbin Institute of Technology
Department of Computer Science and Engineering
Harbin, Heilongjiang, China
E-mail: lijzh@hit.edu.cn

Kuiyu Chang
Nanyang Technological University, School of Computer Engineering
Singapore 639798, Singapore
E-mail: kuiyu.chang@pmail.ntu.edu.sg

Library of Congress Control Number: 2006923003

CR Subject Classification (1998): I.2, H.2.8, H.3, H.5.1, G.3, J.1, K.4

LNCS Sublibrary: SL 7 – Artificial Intelligence

ISSN   0302-9743
ISBN-10   3-540-33206-5 Springer Berlin Heidelberg New York
ISBN-13   978-3-540-33206-0 Springer Berlin Heidelberg New York

This work is subject to copyright. All rights are reserved, whether the whole or part of the material is concerned, specifically the rights of translation, reprinting, re-use of illustrations, recitation, broadcasting, reproduction on microfilms or in any other way, and storage in data banks. Duplication of this publication or parts thereof is permitted only under the provisions of the German Copyright Law of September 9, 1965, in its current version, and permission for use must always be obtained from Springer. Violations are liable to prosecution under the German Copyright Law.

Springer is a part of Springer Science+Business Media

springer.com

© Springer-Verlag Berlin Heidelberg 2006
Printed in Germany

Typesetting: Camera-ready by author, data conversion by Scientific Publishing Services, Chennai, India
Printed on acid-free paper   SPIN: 11731139   06/3142   5 4 3 2 1 0

In Loving Memory of
Professor Hongjun Lu (1945 – 2005)

Preface

The Pacific-Asia Conference on Knowledge Discovery and Data Mining (PAKDD) is a leading international conference in the area of data mining and knowledge discovery. This year marks the tenth anniversary of the successful annual series of PAKDD conferences held in the Asia Pacific region. It was with pleasure that we hosted PAKDD 2006 in Singapore again, since the inaugural PAKDD conference was held in Singapore in 1997.

PAKDD 2006 continues its tradition of providing an international forum for researchers and industry practitioners to share their new ideas, original research results and practical development experiences from all aspects of KDD data mining, including data cleaning, data warehousing, data mining techniques, knowledge visualization, and data mining applications.

This year, we received 501 paper submissions from 38 countries and regions in Asia, Australasia, North America and Europe, of which we accepted 67 (13.4%) papers as regular papers and 33 (6.6%) papers as short papers. The distribution of the accepted papers was as follows: USA (17%), China (16%), Taiwan (10%), Australia (10%), Japan (7%), Korea (7%), Germany (6%), Canada (5%), Hong Kong (3%), Singapore (3%), New Zealand (3%), France (3%), UK (2%), and the rest from various countries in the Asia Pacific region.

The large number of papers was beyond our anticipation and we had to increase the Program Committee at the last minute in order to ensure that all papers went through a rigorous review process, without overloading the PC members. We are glad that most papers were reviewed by three PC members despite the tight schedule. We express herewith our deep appreciation to all PC members and the external reviewers for their arduous support in the review process.

PAKDD 2006 made several other progresses giving the conference series more visibility. For the first time, PAKDD workshops had formal proceedings published under Springer's Lecture Note series. The organizers of the four workshops, namely BioDM, KDLL, KDXD and WISI, put together very high-quality keynotes and workshop programs. We would like to express our gratitude to them for the tremendous efforts. PAKDD 2006 also introduced the best paper award in addition to the existing best student paper award(s). With the help of the Singapore Institute of Statistics (SIS) and the Pattern Recognition & Machine Intelligence Association (PREMIA) of Singapore, a data mining competition under the PAKDD flag was also organized for the first time. Last but not least, a one-day PAKDD School, similar to the one organized in PAKDD 2004, was held again this year.

PAKDD 2006 would not have been possible without the support of many people and organizations. We wish to thank the members of the Steering Committee for their invaluable suggestions and support throughout the organization process. We are grateful to the members of the Organizing Committee, who devoted much of their precious time to the conference arrangement. In the early stage of our conference preparation, we lost Hongjun Lu, who had helped us immensely in drafting our conference proposal. We have missed him dearly but would like to continue his inspiration to make PAKDD 2006 a success. We also deeply appreciate the generous financial support of Infocomm Development Authority of Singapore, the Lee

Foundation, the SPSS, the SAS Institute, the U.S. Air Force Office of Scientific Research, the Asian Office of Aerospace Research and Development, and the U.S. Army ITC-PAC Asian Research Office.

Last but not least, we want to thank all authors and all conference participants for their contribution and support. We hope all participants took this opportunity to share and exchange ideas with one another and enjoyed the conference.

April 2006

Masaru Kitsuregawa
Jianzhong Li
Ee-Peng Lim
Wee Keong Ng
Jaideep Srivastava

Organization

PAKDD 2006 Conference Committee

General Chairs
Ee-Peng Lim   Nanyang Technological University, Singapore
Hongjun Lu (Late)   HK University of Science and Technology, China
Jaideep Srivastava   University of Minnesota, USA

Program Chairs
Wee-Keong Ng   Nanyang Technological University, Singapore
Jiangzhong Li   Harbin Institute of Technology, China
Masaru Kitsuregawa   University of Tokyo, Japan

Workshop Chairs
Ah-Hwee Tan   Nanyang Technological University, Singapore
Huan Liu   Arizona State University, USA

Tutorial Chairs
Sourav Saha Bhowmick   Nanyang Technological University, Singapore
Osmar R. Zaiane   University of Alberta, Canada

Industrial Track Chair
Limsoon Wong   I2R, Singapore

PAKDD School Chair
Chew Lim Tan   National University of Singapore, Singapore

Publication Chair
Kuiyu Chang   Nanyang Technological University, Singapore

Panel Chairs
Wynne Hsu   National University of Singapore, Singapore
Bing Liu   University of Illinois at Chicago, USA

Local Arrangement Chairs
Bastion Arlene   Nanyang Technological University, Singapore
Vivekanand Gopalkrishnan   Nanyang Technological University, Singapore
Dion Hoe-Lian Goh   Nanyang Technological University, Singapore

Publicity and Sponsorship Chairs
Manoranjan Dash   Nanyang Technological University, Singapore
Jun Zhang   Nanyang Technological University, Singapore

PAKDD 2006 Steering Committee

Hiroshi Motoda (Chair)	Osaka University, Japan
David Cheung (Co-chair & Treasurer)	University of Hong Kong, China
Ho Tu Bao	Japan Advanced Institute of Science and Technology, Japan
Arbee L. P. Chen	National Chengchi University, Taiwan
Ming-Syan Chen	National Taiwan University, Taiwan
Jongwoo Jeon	Seoul National University, Korea
Masaru Kitsuregawa	Tokyo University, Japan
Rao Kotagiri	University of Melbourne, Australia
Huan Liu	Arizona State University, USA
Takao Terano	University of Tsukuba, Japan
Kyu-Young Whang	Korea Advanced Institute of Science and Technology, Korea
Graham Williams	ATO, Australia
Ning Zhong	Maebashi Institute of Technology, Japan
Chengqi Zhang	University of Technology Sydney, Australia

PAKDD 2006 Program Committee

Graham Williams	ATO, Australia
Warren Jin	Commonwealth Scientific and Industrial Research Organisation, Australia
Honghua Dai	Deakin University, Australia
Kok Leong Ong	Deakin University, Australia
David Taniar	Monash University, Australia
Vincent Lee	Monash University, Australia
Kai Ming Ting	Monash University, Australia
Richi Nayak	Queensland University of Technology, Australia
Vic Ciesielski	RMIT University, Australia
Vo Ngoc Anh	University of Melbourne, Australia
Rao Kotagiri	University of Melbourne, Australia
Achim Hoffmann	University of New South Wales, Australia
Xuemin Lin	University of New South Wales, Australia
Sanjay Chawla	University of Sydney, Australia
Douglas Newlands	University of Tasmania, Australia
Simeon J. Simoff	University of Technology, Sydney, Australia
Chengqi Zhang	University of Technology, Sydney, Australia
Doan B. Hoang	University of Technology, Sydney, Australia
Nicholas Cercone	Dalhousie University, Canada
Doina Precup	McGill University, Canada
Jian Pei	Simon Fraser University, Canada
Yiyu Yao	University of Regina, Canada
Zhihai Wang	Beijing Jiaotong University, China
Hai Zhuge	Chinese Academy of Sciences, China
Ada Waichee Fu	Chinese University of Hong Kong, China
Shuigeng Zhou	Fudan University, China
Aoying Zhou	Fudan University, China
Jiming Liu	Hong Kong Baptist University, China
Qiang Yang	Hong Kong University of Science and Technology, China

Zhi-Hua Zhou	Nanjing University, China
Xiaofeng Meng	Renmin University of China, China
Bo Zhang	Tsinghua University, China
David Cheung	University of Hong Kong, China
Joshua Z. Huang	University of Hong Kong, China
Djamel A. Zighed	University Lyon 2, France
Joel Quinqueton	University Montpellier, France
Thu Hoang	University Paris 5, France
Wai Lam	Chinese University of Hong Kong, Hong Kong, China
Wilfred Ng	University of Science and Technology, Hong Kong, China
Ajay B Pandey	Government of India, India
P. S. Sastry	Indian Institute of Science, Bangalore, India
Shyam Kumar Gupta	Indian Institute of Technology, Delhi, India
T. V. Prabhakar	Indian Institute of Technology, Kanpur, India
A. Balachandran	Persistent Systems, India
Aniruddha Pant	Persistent Systems, India
Dino Pedreschi	Università di Pisa, Italy
Tomoyuki Uchida	Hiroshima City University, Japan
Tetsuya Murai	Hokkaido University, Japan
Hiroki Arimura	Hokkaido University, Japan
Tetsuya Yoshida	Hokkaido University, Japan
Tu Bao Ho	JAIST, Japan
Van Nam Huynh	JAIST, Japan
Akira Shimazu	JAIST, Japan
Kenji Satou	JAIST, Japan
Takahira Yamaguchi	Keio University, Japan
Takashi Okada	Kwansei Gakuin University, Japan
Ning Zhong	Maebashi Institute of Technology, Japan
Hiroyuki Kawano	Nanzan University, Japan
Masashi Shimbo	Nara Institute of Science and Technology, Japan
Yuji Matsumoto	Nara Institute of Science and Technology, Japan
Seiji Yamada	National Institute of Informatics, Japan
Hiroshi Motoda	Osaka University, Japan
Shusaku Tsumoto	Shimane Medical University, Japan
Hiroshi Tsukimoto	Tokyo Denki University, Japan
Takao Terano	Tsukuba University, Japan
Takehisa Yairi	University of Tokyo, Japan
Yoon-Joon Lee	KAIST, Korea
Yang-Sae Moon	Kangwon National University, Korea
Sungzoon Cho	Seoul National University, Korea
Myung Won Kim	Soongsil University, Korea
Sang Ho Lee	Soongsil University, Korea
Myo Win Khin	University of Computer Studies, Myanmar
Myo-Myo Naing	University of Computer Studies, Myanmar
Patricia Riddle	University of Auckland, New Zealand
Eibe Frank	University of Waikato, New Zealand
Michael Mayo	University of Waikato, New Zealand
Szymon Jaroszewicz	Technical University of Szczecin, Poland
Andrzej Skowron	Warsaw University, Poland
Hung Son Nguyen	Warsaw University, Poland
Marzena Kryszkiewicz	Warsaw University of Technology, Poland
Ngoc Thanh Nguyen	Wroclaw University of Technology, Poland

Joao Gama	University of Porto, Portugal
Jinyan Li	Institute for Infocomm Research, Singapore
Lihui Chen	Nanyang Technological University, Singapore
Manoranjan Dash	Nanyang Technological University, Singapore
Siu Cheung Hui	Nanyang Technological University, Singapore
Daxin Jiang	Nanyang Technological University, Singapore
Daming Shi	Nanyang Technological University, Singapore
Aixin Sun	Nanyang Technological University, Singapore
Vivekanand Gopalkrishnan	Nanyang Technological University, Singapore
Sourav Bhowmick	Nanyang Technological University, Singapore
Lipo Wang	Nanyang Technological University, Singapore
Wynne Hsu	National University of Singapore, Singapore
Dell Zhang	National University of Singapore, Singapore
Zehua Liu	Yokogawa Engineering Asia, Singapore
Ming-Syan Chen	National Taiwan University, Taiwan
Arbee L.P. Chen	National Chengchi University, Taiwan
San-Yih Hwang	National Sun Yat-Sen University, Taiwan
Chih-Jen Lin	National Taiwan University, Taiwan
Jirapun Daengdej	Assumption University, Thailand
Jonathan Lawry	University of Bristol, UK
Huan Liu	Arizona State University, USA
Minos Garofalakis	Intel Research Laboratories, USA
Tao Li	Florida International University, USA
Wenke Lee	Georgia Tech University, USA
Philip S. Yu	IBM T.J. Watson Research Center, USA
Se June Hong	IBM T.J. Watson Research Center, USA
Rong Jin	Michigan State University, USA
Pusheng Zhang	Microsoft Corporation, USA
Mohammed J. Zaki	Rensselaer Polytechnic Institute, USA
Hui Xiong	Rutgers University, USA
Tsau Young Lin	San Jose State University, USA
Aleksandar Lazarevic	United Technologies, USA
Jason T. L. Wang	New Jersey Institute of Technology, USA
Sam Y. Sung	South Texas University, USA
Roger Chiang	University of Cincinnati, USA
Bing Liu	University of Illinois at Chicago, USA
Vipin Kumar	University of Minnesota, USA
Xintao Wu	University of North Carolina at Charlotte, USA
Yan Huang	University of North Texas, USA
Xindong Wu	University of Vermont, USA
Guozhu Dong	Wright State University, USA
Thanh Thuy Nguyen	Hanoi University Technology, Vietnam
Ngoc Binh Nguyen	Hanoi University Technology, Vietnam
Tru Hoang Cao	Ho Chi Minh City University of Technology, Vietnam

PAKDD 2006 External Reviewers

Alexandre Termier
Andre Carvalho
Atorn Nuniyagul
Aysel Ozgur
Ben Mayer
Benjarath Phoophakdee
Brian Harrington
Cai Yunpeng
Canh-Hao Nguyen
Chengjun Liu
Chiara Renso
Cho Siu-Yeung, David
Choi Koon Kau, Byron
Christophe Rigotti
Daan He
Dacheng Tao
Dang-Hung Tran
Dexi Liu
Dirk Arnold
Dong-Joo Park
Dongrong Wen
Dragoljub Pokrajac
Duong Tuan Anh
Eric Eilertson
Feng Chen
Feng Gao
Fosca Giannotti
Francesco Bonchi
Franco Turini
Gaurav Pandey
Gour C. Karmakar
Haoliang Jiang
Hiroshi Murata
Ho Lam Lau
Hongjian Fan
Hongxing He
Hui Xiong
Hui Zhang
James Cheng
Jaroslav Stepaniuk
Jianmin Li
Jiaqi Wang
Jie Chen
Jing Tian
Jiye Li
Junilda Spirollari
Katherine G. Herbert
Kozo Ohara
Lance Parson
Lei Tang
Li Peng
Lin Deng
Liqin Zhang
Lizhuang Zhao
Longbing Cao
Lu An
Magdiel Galan
Marc Ma
Masahiko Ito
Masayuki Okabe
Maurizio Atzori
Michail Vlachos
Minh Le Nguyen
Mirco Nanni
Miriam Baglioni
Mohammed Al Hasan
Mugdha Khaladkar
Nitin Agarwal
Niyati Parikh
Nguyen Phu Chien
Pedro Rodrigues
Qiang Zhou
Qiankun Zhao
Qing Liu
Qinghua Zou
Rohit Gupta
Saeed Salem
Sai Moturu
Salvatore Ruggieri
Salvo Rinzivillo
Sangjun Lee
Saori Kawasaki
Sen Zhang
Shichao Zhang
Shyam Boriah
Songtao Guo
Spiros Papadimitriou
Surendra Singhi
Takashi Onoda
Terry Griffin
Thanh-Phuong Nguyen
Thai-Binh Nguyen
Thoai Nam
Tianming Hu
Tony Abou-Assaleh
Tsuyoshi Murata
Tuan Trung Nguyen
Varun Chandola
Vineet Chaoji
Weiqiang Kong
Wenny Rahayu
Wojciech Jaworski
Xiangdong An
Xiaobo Peng
Xiaoming Wu
Xingquan Zhu
Xiong Wang
Xuelong Li
Yan Zhao
Yang Song
Yanchang Zhao
Yaohua Chen
Yasufumi Takama
Yi Ping Ke
Ying Yang
Yong Ye
Zhaochun Yu
Zheng Zhao
Zhenxing Qin
Zhiheng Huang
Zhihong Chong
Zujun Shentu

Sponsorship

We wish to thank the following organizations for their contributions to the success of this conference:

Air Force Office of Scientific Research,
Asian Office of Aerospace Research and Development

US Army ITC-PAC Asian Research Office

Infocomm Development Authority of Singapore

Lee Foundation

SAS Institute, Inc.

SPSS, Inc.

Embassy of the United States of America, Singapore

Table of Contents

Keynote Speech

Protection or Privacy? Data Mining and Personal Data
   David J. Hand 1

The Changing Face of Web Search
   Prabhakar Raghavan 11

Invited Speech

Data Mining for Surveillance Applications
   Bhavani M. Thuraisingham 12

Classification

A Multiclass Classification Method Based on Output Design
   Qi Qiang, Qinming He 15

Regularized Semi-supervised Classification on Manifold
   *Lianwei Zhao, Siwei Luo, Yanchang Zhao, Lingzhi Liao,
   Zhihai Wang* 20

Similarity-Based Sparse Feature Extraction Using Local Manifold
Learning
   Cheong Hee Park 30

Generalized Conditional Entropy and a Metric Splitting Criterion for
Decision Trees
   Dan A. Simovici, Szymon Jaroszewicz 35

RNBL-MN: A Recursive Naive Bayes Learner for Sequence Classification
   Dae-Ki Kang, Adrian Silvescu, Vasant Honavar 45

TRIPPER: Rule Learning Using Taxonomies
   Flavian Vasile, Adrian Silvescu, Dae-Ki Kang, Vasant Honavar 55

Using Weighted Nearest Neighbor to Benefit from Unlabeled Data
   *Kurt Driessens, Peter Reutemann, Bernhard Pfahringer,
   Claire Leschi* 60

Constructive Meta-level Feature Selection Method Based on Method
Repositories
    Hidenao Abe, Takahira Yamaguchi 70

Ensemble Learning

Variable Randomness in Decision Tree Ensembles
    Fei Tony Liu, Kai Ming Ting 81

Further Improving Emerging Pattern Based Classifiers Via Bagging
    Hongjian Fan, Ming Fan, Kotagiri Ramamohanarao,
    Mengxu Liu 91

Improving on Bagging with Input Smearing
    Eibe Frank, Bernhard Pfahringer 97

Boosting Prediction Accuracy on Imbalanced Datasets with SVM
Ensembles
    Yang Liu, Aijun An, Xiangji Huang 107

Clustering

DeLiClu: Boosting Robustness, Completeness, Usability, and Efficiency
of Hierarchical Clustering by a Closest Pair Ranking
    Elke Achtert, Christian Böhm, Peer Kröger 119

Iterative Clustering Analysis for Grouping Missing Data in Gene
Expression Profiles
    Dae-Won Kim, Bo-Yeong Kang 129

An EM-Approach for Clustering Multi-Instance Objects
    Hans-Peter Kriegel, Alexey Pryakhin, Matthias Schubert 139

Mining Maximal Correlated Member Clusters in High Dimensional
Database
    Lizheng Jiang, Dongqing Yang, Shiwei Tang, Xiuli Ma,
    Dehui Zhang 149

Hierarchical Clustering Based on Mathematical Optimization
    Le Hoai Minh, Le Thi Hoai An, Pham Dinh Tao 160

Clustering Multi-represented Objects Using Combination Trees
    Elke Achtert, Hans-Peter Kriegel, Alexey Pryakhin,
    Matthias Schubert 174

Parallel Density-Based Clustering of Complex Objects
    Stefan Brecheisen, Hans-Peter Kriegel, Martin Pfeifle 179

Neighborhood Density Method for Selecting Initial Cluster Centers in
K-Means Clustering
    *Yunming Ye, Joshua Zhexue Huang, Xiaojun Chen, Shuigeng Zhou,
    Graham Williams, Xiaofei Xu* 189

Uncertain Data Mining: An Example in Clustering Location Data
    Michael Chau, Reynold Cheng, Ben Kao, Jackey Ng 199

Support Vector Machines

Parallel Randomized Support Vector Machine
    Yumao Lu, Vwani Roychowdhury 205

ε-Tube Based Pattern Selection for Support Vector Machines
    Dongil Kim, Sungzoon Cho 215

Self-adaptive Two-Phase Support Vector Clustering for Multi-Relational
Data Mining
    Ping Ling, Yan Wang, Chun-Guang Zhou 225

One-Class Support Vector Machines for Recommendation Tasks
    Yasutoshi Yajima 230

Text and Document Mining

Heterogeneous Information Integration in Hierarchical Text
Classification
    Huai-Yuan Yang, Tie-Yan Liu, Li Gao, Wei-Ying Ma 240

FISA: Feature-Based Instance Selection for Imbalanced Text
Classification
    Aixin Sun, Ee-Peng Lim, Boualem Benatallah, Mahbub Hassan 250

Dynamic Category Profiling for Text Filtering and Classification
    Rey-Long Liu 255

Detecting Citation Types Using Finite-State Machines
    Minh-Hoang Le, Tu-Bao Ho, Yoshiteru Nakamori 265

A Systematic Study of Parameter Correlations in Large Scale Duplicate Document Detection
  Shaozhi Ye, Ji-Rong Wen, Wei-Ying Ma 275

Comparison of Documents Classification Techniques to Classify Medical Reports
  F.H. Saad, B. de la Iglesia, G.D. Bell 285

XCLS: A Fast and Effective Clustering Algorithm for Heterogenous XML Documents
  Richi Nayak, Sumei Xu 292

Clustering Large Collection of Biomedical Literature Based on Ontology-Enriched Bipartite Graph Representation and Mutual Refinement Strategy
  Illhoi Yoo, Xiaohua Hu 303

Web Mining

Level-Biased Statistics in the Hierarchical Structure of the Web
  Guang Feng, Tie-Yan Liu, Xu-Dong Zhang, Wei-Ying Ma 313

CLEOPATRA: Evolutionary Pattern-Based Clustering of Web Usage Data
  Qiankun Zhao, Sourav S. Bhowmick, Le Gruenwald 323

Extracting and Summarizing Hot Item Features Across Different Auction Web Sites
  Tak-Lam Wong, Wai Lam, Shing-Kit Chan 334

Clustering Web Sessions by Levels of Page Similarity
  Caren Moraes Nichele, Karin Becker 346

*i*WED: An Integrated Multigraph Cut-Based Approach for Detecting Events from a Website
  Qiankun Zhao, Sourav S. Bhowmick, Aixin Sun 351

Enhancing Duplicate Collection Detection Through Replica Boundary Discovery
  Zhigang Zhang, Weijia Jia, Xiaoming Li 361

Graph and Network Mining

Summarization and Visualization of Communication Patterns in a Large-Scale Social Network
  Preetha Appan, Hari Sundaram, Belle Tseng 371

Patterns of Influence in a Recommendation Network
  Jure Leskovec, Ajit Singh, Jon Kleinberg 380

Constructing Decision Trees for Graph-Structured Data by
Chunkingless Graph-Based Induction
  *Phu Chien Nguyen, Kouzou Ohara, Akira Mogi, Hiroshi Motoda,
  Takashi Washio* 390

Combining Smooth Graphs with Semi-supervised Classification
  Xueyuan Zhou, Chunping Li 400

Network Data Mining: Discovering Patterns of Interaction Between
Attributes
  John Galloway, Simeon J. Simoff 410

Association Rule Mining

SGPM: Static Group Pattern Mining Using Apriori-Like Sliding Window
  John Goh, David Taniar, Ee-Peng Lim 415

Mining Temporal Indirect Associations
  Ling Chen, Sourav S. Bhowmick, Jinyan Li 425

Mining Top-K Frequent Closed Itemsets Is Not in APX
  Chienwen Wu 435

Quality-Aware Association Rule Mining
  Laure Berti-Équille 440

IMB3-Miner: Mining Induced/Embedded Subtrees by Constraining the
Level of Embedding
  *Henry Tan, Tharam S. Dillon, Fedja Hadzic, Elizabeth Chang,
  Ling Feng* 450

Maintaining Frequent Itemsets over High-Speed Data Streams
  James Cheng, Yiping Ke, Wilfred Ng 462

Generalized Disjunction-Free Representation of Frequents Patterns
with at Most k Negations
  Marzena Kryszkiewicz 468

Mining Interesting Imperfectly Sporadic Rules
  Yun Sing Koh, Nathan Rountree, Richard O'Keefe 473

Improved Negative-Border Online Mining Approaches
    Ching-Yao Wang, Shian-Shyong Tseng, Tzung-Pei Hong 483

Association-Based Dissimilarity Measures for Categorical Data:
Limitation and Improvement
    Si Quang Le, Tu Bao Ho, Le Sy Vinh 493

Is Frequency Enough for Decision Makers to Make Decisions?
    Shichao Zhang, Jeffrey Xu Yu, Jingli Lu, Chengqi Zhang 499

Ramp: High Performance Frequent Itemset Mining with Efficient
Bit-Vector Projection Technique
    Shariq Bashir, Abdul Rauf Baig 504

Evaluating a Rule Evaluation Support Method Based on Objective
Rule Evaluation Indices
    *Hidenao Abe, Shusaku Tsumoto, Miho Ohsaki,
    Takahira Yamaguchi* 509

Bio-data Mining

Scoring Method for Tumor Prediction from Microarray Data Using an
Evolutionary Fuzzy Classifier
    *Shinn-Ying Ho, Chih-Hung Hsieh, Kuan-Wei Chen,
    Hui-Ling Huang, Hung-Ming Chen, Shinn-Jang Ho* 520

Efficient Discovery of Structural Motifs from Protein Sequences with
Combination of Flexible Intra- and Inter-block Gap Constraints
    *Chen-Ming Hsu, Chien-Yu Chen, Ching-Chi Hsu,
    Baw-Jhiune Liu* 530

Finding Consensus Patterns in Very Scarce Biosequence Samples from
Their Minimal Multiple Generalizations
    Yen Kaow Ng, Takeshi Shinohara 540

Kernels on Lists and Sets over Relational Algebra: An Application to
Classification of Protein Fingerprints
    Adam Woźnica, Alexandros Kalousis, Melanie Hilario 546

Mining Quantitative Maximal Hyperclique Patterns: A Summary of
Results
    Yaochun Huang, Hui Xiong, Weili Wu, Sam Y. Sung 552

Outlier and Intrusion Detection

A Nonparametric Outlier Detection for Effectively Discovering Top-N
Outliers from Engineering Data
Hongqin Fan, Osmar R. Zaïane, Andrew Foss, Junfeng Wu 557

A Fast Greedy Algorithm for Outlier Mining
Zengyou He, Shengchun Deng, Xiaofei Xu, Joshua Zhexue Huang 567

Ranking Outliers Using Symmetric Neighborhood Relationship
Wen Jin, Anthony K.H. Tung, Jiawei Han, Wei Wang 577

Construction of Finite Automata for Intrusion Detection from System
Call Sequences by Genetic Algorithms
Kyubum Wee, Sinjae Kim 594

An Adaptive Intrusion Detection Algorithm Based on Clustering and
Kernel-Method
Hansung Lee, Yongwha Chung, Daihee Park 603

Weighted Intra-transactional Rule Mining for Database Intrusion
Detection
Abhinav Srivastava, Shamik Sural, A.K. Majumdar 611

Privacy

On Robust and Effective K-Anonymity in Large Databases
Wen Jin, Rong Ge, Weining Qian 621

Achieving Private Recommendations Using Randomized Response
Techniques
Huseyin Polat, Wenliang Du 637

Privacy-Preserving SVM Classification on Vertically Partitioned Data
Hwanjo Yu, Jaideep Vaidya, Xiaoqian Jiang 647

Relational Database

Data Mining Using Relational Database Management Systems
*Beibei Zou, Xuesong Ma, Bettina Kemme, Glen Newton,
Doina Precup* 657

Bias-Free Hypothesis Evaluation in Multirelational Domains
Christine Körner, Stefan Wrobel 668

Enhanced DB-Subdue: Supporting Subtle Aspects of Graph Mining
Using a Relational Approach
  *Ramanathan Balachandran, Srihari Padmanabhan,
  Sharma Chakravarthy* 673

Multimedia Mining

Multimedia Semantics Integration Using Linguistic Model
  Bo Yang, Ali R. Hurson 679

A Novel Indexing Approach for Efficient and Fast Similarity Search of
Captured Motions
  Chuanjun Li, B. Prabhakaran 689

Mining Frequent Spatial Patterns in Image Databases
  Wei-Ta Chen, Yi-Ling Chen, Ming-Syan Chen 699

Image Classification Via LZ78 Based String Kernel: A Comparative
Study
  Ming Li, Yanong Zhu 704

Stream Data Mining

Distributed Pattern Discovery in Multiple Streams
  Jimeng Sun, Spiros Papadimitriou, Christos Faloutsos 713

COMET: Event-Driven Clustering over Multiple Evolving Streams
  Mi-Yen Yeh, Bi-Ru Dai, Ming-Syan Chen 719

Variable Support Mining of Frequent Itemsets over Data Streams Using
Synopsis Vectors
  Ming-Yen Lin, Sue-Chen Hsueh, Sheng-Kun Hwang 724

Hardware Enhanced Mining for Association Rules
  Wei-Chuan Liu, Ken-Hao Liu, Ming-Syan Chen 729

A Single Index Approach for Time-Series Subsequence Matching That
Supports Moving Average Transform of Arbitrary Order
  Yang-Sae Moon, Jinho Kim 739

Efficient Mining of Emerging Events in a Dynamic Spatiotemporal
Environment
  Yu Meng, Margaret H. Dunham 750

Temporal Data Mining

A Multi-Hierarchical Representation for Similarity Measurement of
Time Series
  Xinqiang Zuo, Xiaoming Jin 755

Multistep-Ahead Time Series Prediction
  Haibin Cheng, Pang-Ning Tan, Jing Gao, Jerry Scripps 765

Sequential Pattern Mining with Time Intervals
  Yu Hirate, Hayato Yamana 775

A Wavelet Analysis Based Data Processing for Time Series of Data
Mining Predicting
  Weimin Tong, Yijun Li, Qiang Ye 780

Novel Algorithms

Intelligent Particle Swarm Optimization in Multi-objective Problems
  *Shinn-Jang Ho, Wen-Yuan Ku, Jun-Wun Jou, Ming-Hao Hung,
  Shinn-Ying Ho* 790

Hidden Space Principal Component Analysis
  Weida Zhou, Li Zhang, Licheng Jiao 801

Neighbor Line-Based Locally Linear Embedding
  De-Chuan Zhan, Zhi-Hua Zhou 806

Predicting Rare Extreme Values
  Luis Torgo, Rita Ribeiro 816

Domain-Driven Actionable Knowledge Discovery in the Real World
  Longbing Cao, Chengqi Zhang 821

Evaluation of Attribute-Aware Recommender System Algorithms on
Data with Varying Characteristics
  Karen H.L. Tso, Lars Schmidt-Thieme 831

Innovative Applications

An Intelligent System Based on Kernel Methods for Crop Yield
Prediction
  A. Majid Awan, Mohd. Noor Md. Sap 841

A Machine Learning Application for Human Resource Data Mining
Problem
  Zhen Xu, Binheng Song 847

Towards Automated Design of Large-Scale Circuits by Combining
Evolutionary Design with Data Mining
  Shuguang Zhao, Mingying Zhao, Jun Zhao, Licheng Jiao 857

Mining Unexpected Associations for Signalling Potential Adverse Drug
Reactions from Administrative Health Databases
  *Huidong Jin, Jie Chen, Chris Kelman, Hongxing He,
  Damien McAullay, Christine M. O'Keefe* 867

Author Index 877

Protection or Privacy? Data Mining and Personal Data

David J. Hand

Department of Mathematics,
Imperial College London,
Exhibition Road,
London SW7 2AZ, UK
d.j.hand@imperial.ac.uk
http://stats.ma.ic.ac.uk/djhand/public_html/

'There was of course no way of knowing whether you were being watched at any given moment.... It was even conceivable that they watched everybody all the time.'

George Orwell, *1984*

Abstract. In order to run countries and economies effectively, governments and governmental institutions need to collect and analyse vast amounts of personal data. Similarly, health service providers, security services, transport planners, and education authorities need to know a great deal about their clients. And, of course, commercial operations run more efficiently and can meet the needs of their customers more effectively the more they know about them. In general then, the more data these organisation have, the better. On the other hand, the more private data which is collated and disseminated, the more individuals are at risk of crimes such as identity theft and financial fraud, not to mention the simple invasion of privacy that such data collection represents. Most work in data mining has concentrated on the positive aspects of extracting useful information from large data sets. But as the technology and its use advances so more awareness of the potential downside is needed. In this paper I look at some of these issues. I examine how data mining tools and techniques are being used by governments and commercial operations to gain insight into individual behaviour. And I look at the concerns that such advances are bringing.

1 The Need to Know

In order to run a country effectively, a government must understand the needs and wishes of its people. In order to run a corporation profitably, the directors must understand the customers and the products or services they require. This point, this need for understanding, applies to any organization. It applies to health service providers, to security services, to transport planners, to education authorities, and so on. Because of such needs, *information about every individual at this conference is stored in countless commercial, government, and other databases.* Some of this information is collected explicitly: when you take an examination or fill in an application form, you expect the data to be entered into a database. But the vast majority of it is collected implicitly: details of what you bought in a supermarket, of your credit card transactions, satellite monitoring of vehicle locations, automatic photographs of vehicle registration plates, RFID systems which identify objects and people at a distance, is all collected and stored without you being aware of it.

Once the information has been collected, it can be used to answer the question it was intended for, but it can also be used to answer other questions. But there is more than this. If individual data sets can be used to answer new, as yet unposed, questions, then analyzing *merged* data sets can be even more powerful. In general, data merging, data linking, or data fusion from both governmental and non-governmental sources is becoming increasingly widespread. For example, information on electoral rolls, censuses, and surveys by national statistical offices can be linked to information on purchasing patterns, banking transaction patterns, medical records, cellphone records, websurfing traces, and so on. By such means, your interests can be identified and your behaviour modeled, *and predicted*, to an unprecedented degree. Thus the London *Times* of August 5th 2005 reports that 'HBOS, Britain's biggest mortgage lender, is pressing the Government to force local authorities to provide banks with details of council tax arrears' in a drive to improve credit scoring. Credit scoring, deciding who is a good and bad financial risk, is conceptually similar to insurance, so might not insurance companies similarly request direct access to medical records? Let us take this example further. Imagine a system which matched peoples' medical records to their eating habits, as deduced from stored data describing their weekly supermarket food purchases. Now link the results to their home address via the number of the credit card used to make the food purchases, and an insurance company could decide automatically to withdraw insurance cover from customers whom it thought were eating a diet which predisposed them to illness.

The first part of this paper illustrates the power of data mining tools to protect us from harm by enabling us to predict what the future might bring unless we intervene in some way. But data mining is a powerful technology. All powerful technologies are ethically neutral. They can be used for good, but they can also be used for bad. The second part of the paper illustrates how data mining tools can be misused, to invade our privacy. In parallel with the discussion concerning the social impact of data mining, running throughout the paper there is a technical theme: that the statistics used for pattern discovery data mining must be simple because of the sheer amount of computation required.

2 The Nature of Data Mining

There are two broadly distinct aspects to data mining. One is concerned with high level data summary – with model building. The aim here is to create a broad description of a data set, to identify its main features. Thus, for example, one might partition a data set describing customers into distinct behaviour classes using cluster analysis. Or one might build a neural network model to predict how objects will behave in the future. There is an unlimited number of ways in which one might summarise a set of data, but, their aim is to identify the major characterising structures in the data.

The other aspect of data mining is pattern discovery. Patterns are small local features in a data set – a departure from a model. They may consist of single points (as in outlier detection), small groups of points (as in detecting the start of an epidemic), small sets of variables which behave unexpectedly (as in microarray analysis), or some other small-scale departure from what is expected.

Whereas the theory and methods of model building have been extensively developed by statisticians throughout the twentieth century, pattern detection and discovery is relatively unexplored. Tools have been developed for particular application areas, and for particular types of problems, but this tends to have been in isolation. It is only recently, a consequence of the increasing number of very large data sets and the computer power to manipulate and search them quickly, that researchers have begun to think about a unified theory of pattern discovery.

In pattern discovery, the aim is to detect data points or groups of data points which deviate noticeably from the expected – that is, from a background model. Examples of such problems are given below, and some people regard this kind of problem as the core of data mining – the attempt to find unexpected 'nuggets' of information. Pattern discovery requires the construction of a background model, a measure of deviation from that model (and deviation may be of many kinds), a search algorithm, and inference to decide if the deviation should have been expected.

Pattern discovery presents some theoretical and practical challenges. In particular, *it is central to the notion of pattern discovery that one has to examine all elements in the database.* This is rather different from model building: for most purposes, a summary model built on a sample of 5000 cases will be as effective as a model built on all five million cases. But if one's aim is to detect *which* cases are anomalous then there is no alternative to looking at each individual case. So, for example, in mining telecoms data, one can construct an effective segmentation into usage type (a model) using just a sample of a few thousand customers, but if one is trying to identify which customers are perpetrating frauds there is no alternative to examining each record. This suggests that pattern discovery exercises have an important property: the calculations involved in analyzing each case must be quick to perform. Each case cannot involve lengthy iterative computations, for example. I illustrate this in my examples, showing how pattern discovery is often a kind of feature selection exercise, with the requirement that the features must be computed from relatively simple formulae.

In commercial applications, data mining is often sold as a magic tool which will lead to the discovery of information without the user having to do any thinking. This, of course is misleading. It is no accident that scientist have produced various aphorisms such as 'chance favours the prepared mind' and 'the harder I work, the luckier I get'. The truth is that the more you know about your data, about the problem, and about the sort of pattern you are looking for, the more likely you are to find something useful. In the context of pattern discovery, the more you know about these things, the more precisely you can formulate the mathematical shape of the patterns to be found. The bottom line is that computing power does not replace brain power. They work hand in hand. The data miner who uses both will be the one who finds the interesting and valuable structures in the data.

3 Data Mining: The Reward

In this section I illustrate the application of data mining pattern discovery tools to protect us from harm.

Disease and illness are one type of harm, and an important class of data mining tools seeks to detect small local clusters of people suffering from a disease – perhaps because they have been exposed to a common cause, or perhaps because a contagious disease is spreading locally. In such situations the clusters are two-dimensional, with geography providing the two dimensions. Global clustering statistics, such as the Mantel-Bailar statistic, tells us whether the data points tend to suggest clustering, but they do not tell us where the clusters are. Such measures are really a diagnostic modeling tool. To *detect* clusters it is necessary to scan the distribution of points, looking at each point of the space and comparing the local clustering tendency with what one would expect. Here, 'what one would expect' will be based on the underlying population distribution. For example, one might assume that each person was equally likely to contract a disease, and then locate those regions where more than the expected number have the disease. The simple statistics here are based on comparing counts of numbers of cases within a region of gradually increasing radius, with counts of numbers in the population within the region.

This example has the property that information about the expected background distribution was obtained from another source – the distribution of the population. In many problems, however, there is no other source. An illustration is provided by a study we carried out to detect student cheating. Plagiarism by students, assisted by the web, has been much in the news recently, but our problem was rather different. We were especially concerned to detect students who had copied their coursework from each other. Our simple statistic was a measure of similarity between pairs of students. The background model here is a distribution which has the same multivariate characteristics as the distribution of scores obtained by the students.

Another, again slightly different example is given by pharmacovigilance. This is a post-marketing exercise carried out by pharmaceutical companies, aimed at detecting drug-induced side effects. In principle the background distribution is straightforward – the number of prescriptions of each drug. In practice, however, records are often incomplete, and some other way to derive a background distribution is needed. Often fairly simple models are used – such as the assumption that the distribution of incidents over drug and the distribution of incidents over side effects are independent. We have been experimenting with a more elaborate approach which takes into account the pharmaceutical similarity between the drugs. That is, it is as if the drugs exist in a space in which closeness is determined by chemical similarity. In all cases, however, a simple statistic based on the difference between the observed counts and the expected of incidents under the background model is used.

Disease clustering and the other problems described above is concerned with detecting local groups in space. Such clusters represent an anomaly in the underlying density function of cases. Another class of problems arises when one is aiming to detect an anomaly in a univariate or multivariate sequence of observations over time. *Change point* problems are examples of such. Taking disease outbreaks as an example again, one might have a natural background rate of infection, and will seek to detect, as early as possible, when the rate deviates (increases) from this. Here the simple statistic is based on comparing estimates of the rates before and after a putative change point. Further complications arise, of course, since often one wants to detect that a change has occurred as soon as possible. In the case of disease outbreaks, early detection can mean that there is a chance of containing the disease. Of course, things are complicated by factors such as incubation time: if the symptoms of the disease manifest themselves after the organism has become infectious, for example.

There are many other problems in which mining the data for change points, perhaps in real time, is important. Monitoring for natural disasters (such as tsunamis), fault detection, and fraud detection, provide other important examples. For fault detection, careful on line monitoring of information from complex machinery, such as nuclear reactors or space missions, is vital to ensure that any peculiarities are detected early on. In fraud detection, we developed a tool for credit card fraud detection which we called *peer group analysis*, in which one identifies the customers who have previously behaved most similarly to a target customer, and then monitors to see if and when the target starts to behave differently. Since it is generally not known which customers should be the target, the fact that one has to do the computation for all customers hints at the amount of computation which such methods can involve. Once again, we see the necessity of simple formulae.

Although I have outlined spatial clustering and change point detection separately, they become especially powerful when combined. Now we can see when a spatial cluster suddenly appears, or when incidents of ATM theft suddenly begin. Once again, quick detection is often vital. The recent cases of SARS, BSE, and now Avian Flu illustrate just how important these sorts of tools are.

Change points are one kind of anomaly. They occur when individuals suddenly begin to behave differently. But even univariate time series can demonstrate other anomalies. The case of Harold Shipman is an illustration.

Harold Shipman is a contender for the title of the world's most prolific serial killer. He was a family doctor, respected and admired by his patients. But over a period of years he killed many his patients - one estimate is that he killed 236 people between 1978 and 1998, primarily elderly women patients, for example by giving them overdoses of painkillers. Detection came in 1998 when an apparently healthy 81 year-old died suddenly on 24th June. Her daughter, a lawyer, became suspicious when she realized that her mother had apparently signed a new will without her knowledge, leaving everything to Shipman. Things rapidly escalated from there, and eventually Shipman was tried and found guilty on 15 counts of murder.

At first glance this looks like a straightforward statistical problem, using control charts, cusums, or more elaborate tools. Indeed, a retrospective cumulative plot of the mortality amongst females aged over 64 in Shipman's practice shows a gradual increase and even an anomalous sudden increase in the death rate around 1994. Application of formal statistical tools detects that something unusual is going on here, and would flag this medical practice up for closer examination. But, of course, if such monitoring is carried out *prospectively*, it is not just the one practice which is monitored. It is all such practices in the UK. All in real time. Once again the need for statistics which are quick to calculate is indicated.

So far I have talked in terms of the statistics used to detect anomalous patterns and structures in data sets. I have stressed the need for these statistics to be simple, since often massive search is involved. But strange structures do arise by chance. Not only do we need to be able to locate such structures, but we need to assess how likely it is that they are merely chance events. That is, as well as the algorithmic aspects implicit in search, we need the statistical aspects implicit in inference.

This brings me to what I call the *fundamental problem of pattern inference in data mining*. It is the multiplicity problem. We will be searching over a large collection of points, seeking for a large set of possible local cluster structures, so we must expect some such configurations to arise by chance. The more data points we consider, the more likely such *false positives* are. To allow for this we have to bring to bear ideas of scan statistics and false discovery rate. Substantial theoretical advances have been made in these areas in recent years. The mathematics underlying these advances is often quite difficult, and I believe there are significant opportunities for computational approaches.

In the introduction, I mentioned the power resulting from combining data sources. So let me finish this section illustrating the tremendous potential benefits of data mining by citing the Australian study which linked records of long haul flights to records of deep vein thromboembolisms, to reveal that the annual risk of thromboembolism is increased by 12% if one such flight is taken annually. Data mining has an immense amount to offer for improving the human condition.

4 Data Mining: The Risk

It will be clear from the examples in the preceding section that data mining has the potential for immense good by protecting us from harm from a variety of causes. However, there is a downside. In this section I want to examine just a few examples of the dangers of data mining.

4.1 Elections

My first example involves elections. Elections are often very close run things. In a sense, this means they may be intrinsically unstable systems. In the 2004 US Presidential election, the roughly equal proportions of votes in the Electoral College of 53% favouring Bush and 47% favouring Kerry translated into 100% election of Bush as President, but a *slight* shift in the proportions could have resulted in a *complete* reversal of the outcome. Similarly, in the German election of 2005, although Angel Merkel won 35.2% of the vote, and Gerhard Schröder won 34.3%, only one of them could be Chancellor, and in the 2005 UK General Election, the Labour Party won 35% of the votes and the Conservatives won 32%. In both of these cases it seems as if a slight change in the proportions could have resulted in a dramatic difference to the outcome. (In fact, in the UK case, these roughly equal proportions of votes translated into 55% of the seats going to Labour, and only 30% going to the Conservatives, but that's a different story.)

Now, of course, the distribution of votes across electoral seats varies. Some seats will be won by an 80:20 majority, while others by a 51:49 majority. It is probably futile spending a lot of campaigning effort in seats where the ratio is traditionally 80:20. One has a far better chance of changing the outcome of the 51:49 seat. So this is where the effort should be made, and this is where data mining comes in. Data mining allows one to target the particular individuals, in the marginal seats, who might be swayed – the floating voters, those who have not definitely made up their minds.

But there's even more than this. People are different. They may agree with your position on immigration, but disagree with your position on taxes. And if you know this, if you have enough information on an individual voter, you can target your vote to match their interests. You can gloss over your tax plans and play up your immigration policy when canvassing. If you know that the crucial voters tend to watch a particular TV channel, then you can target your advertisements appropriately.

This is a very radical change, brought about entirely by the possibilities provided by data mining. While what the voters know about the candidates is still crucial, what the voters know, which voters knows what, and which voters are provided with more information can be strategically chosen by the candidates. Data mining has changed the nature of elections. The candidate with the most astute data mining team has the winning hand. No elections at national level in the UK or US are now fought without a back room of data miners guiding actions, and the tools of data mining are used more and more extensively in modern elections in the West. Sometimes they go under the names of *microtargeting* or *political sharpshooting*. One might even go so far as to say that nowadays, if you do not employ a data mining team, you will lose.

4.2 False Positives, False Negatives

My second example will probably be familiar to many of you, but its familiarity should not detract from its importance. It is the problem arising from unbalanced class sizes in supervised classification. Supervised classification is a very important type of data mining problem, and in many areas the relative numbers of objects belonging to the different classes are substantially different. In retail banking fraud, for example, generally less than 1 in a 1000 transactions are fraudulent, and in screening for rare medical conditions the rate can be even lower. This has serious implications for the effectiveness of classification rules, and for business operations. This can be seen from the following simple example.

Suppose that a classifier correctly identifies 1 in 100 fraudulent transactions, and correctly identifies 1 in 100 legitimate transactions. This sounds like excellent performance. However, if only 1 in 1000 transactions are fraudulent, then 91% of those transactions flagged as suspect frauds are really legitimate. This matters because operational decisions must be made. To take an extreme case – if one decided to put a stop on all credit cards with suspect transactions one would have many irate legitimate users. Note that, again, we must examine all cases, so that simple calculations are needed: one of our data sets had just 1530 fraudulent accounts amongst over 830,000 accounts altogether – all of which had to be examined.

A less severe illustration of this sort of problem arose in the US system for screening potential terrorists on aircraft, when Senator Edward M. Kennedy was prevented from boarding his US Airways Washington to Boston flight because he was mistakenly matched to someone on a list of suspicious persons. Later he was also automatically flagged for observation by a system which looks for suspicious behaviour such as buying a one-way ticket. And it is not reassuring to read that US border guards failed *100 percent of the time* to detect counterfeit identity documents being used by agents from the General Accounting Office testing the system by trying to enter the US illegally.

The most effective way to tackle the particular problem of unbalanced classes seems to be to use a multistage approach. Eliminate as many as possible of the clearly legitimate cases, so one can use more elaborate methods to focus on the remaining data. Methods based on sampling from the larger class or on duplicating samples from the smaller class are not recommended.

The overall point is that blind application of data mining techniques, without taking account of the practical requirements of the problem, can have adverse consequences. Thoughtless data mining carries a risk.

4.3 Insurance

One of the aims of commercial data mining is to be able to predict the behaviour and likely future of people. In insurance, for example, the more accurately you can predict which people will have an automobile accident, or who will die early of a certain disease, the more profitably you are able to run your company for your shareholders. The aim is thus to make individual-specific predictions. Often, however, the information in the potential predictor variables is insufficient to allow very accurate prediction rules, so averages are calculated for groups of similar people. The predictions then represent a compromise between potential bias in the predictive model and accuracy in terms of variance reduction.

But medical and data mining technology is changing that. For example, genomic data permit increasingly accurate predictions of who will die early of different diseases. Data mining tools are being increasingly heavily used in bioinformatics to extract precisely this kind of information. In some cases, the predictive accuracy will be such that certain individuals will be revealed to be very high risk – and will consequently be unable to obtain insurance. In fact, such situations have already occurred, also because of progress in medical science. A positive AIDS test, for example, can make obtaining life insurance difficult, so there is a clear benefit in not taking a test, even if you suspect you may be positive. Moreover, the taking of a test, even if the results are negative, can be interpreted by an insurance company as an indication that one suspects one is at risk. I am sure that many of you have had the experience of analyzing a set of data and discovering that *the fact that an item of information is missing* is predictive in itself.

4.4 Other Areas: Data Quality, Identity Theft, Disclosure Control and Beyond

There are other areas of risk associated with data mining, and I briefly touch on just a few of them in this section.

Textbook descriptions of data mining tools, and articles extolling the potential gains to be achieved by applying data mining techniques gloss over some of the difficulties. One difficulty which is all pervasive, and which has major consequences for almost all data mining tasks, is that data are very seldom perfect. This matters because governmental and corporate decisions assume the data are correct. But I feel confident that everyone in this room has experienced data problems at some time. Perhaps your computer has crashed at a critical moment, a program might not do exactly what it was intended to do, perhaps the system cannot handle unusual customers or cases, perhaps software maintenance has introduced bugs, perhaps data have been entered incorrectly, and so on endlessly. I have countless examples of problems of this kind, but a very simple one involved retired bus driver Frank Hughes. An oversimple data-matching exercise meant that another man with the same name was matched to Frank Hughes the bus driver. His former workmates were then shocked to see him walking down the street – since they had recently attended his funeral.

This was a shock for his friends and a surprise for Mr Hughes, but perhaps it was fairly minor on the global scale of things. Not so the warnings about record linkage from the TAPAC report, which says (p37-38): 'One of the most significant of these issues concerns the significant difficulties of integrating data accurately. Business and government have long struggled with how to ensure that information about one person is correctly attributed to that individual and only to that individual ... According to the General Accounting Office, the government already suffers significant financial losses from its inability to integrate its own data accurately.'

Identity theft describes the actions of a criminal who obtains personal information about you, and uses this to open bank accounts, obtain credit cards, bank loans, car finance, passports, a driving license, telecoms services, and other such instruments masquerading as you. Worse still, such stolen identities can then be used for activities such as money laundering, immigration fraud, tax fraud, and worse. Once such a theft has been detected, it can take years to sort it all out. During this time, you may not be able to obtain loans, get a mortgage, buy a car or insurance, obtain credit cards, and so on. It is estimated that each year about 100,000 such thefts occur in the UK, and that it costs the UK economy about £1.7bn.

To commit identity theft, criminals have to collect information about you. This information can come from various sources. One significant danger is that separate items of information which are innocuous in themselves may be merged to produce something which acts as a key. Traditional obvious sources include simple thefts of wallets or driving licences, discarded bills, credit card receipts or bank statements reclaimed from a rubbish bin. More elaborate tools include strategies such as phishing – persuading people to divulge security information or PIN numbers over the internet in the mistaken belief that it is a security check. The internet is a new technology, and one which is changing its shape and form all the time. It contains increasing amounts of information about people, permitting all sorts of discoveries (for example that of the adopted teenage boy who manage to locate his sperm donor natural father with just two clever web searches). And mining the internet has become a specialized area of data mining in its own right.

With identity theft in mind, you should always shred any financial documents, credit card slips, and so on, and if you suspect your mail is going astray, report it. You should use different PIN numbers and passwords, irritating though that may be, and you should never store PIN numbers with the cards to which they refer. You should never divulge personal information to people who ring you on the phone (even if they claim to be from your bank). Always ring them back on a number you know to be correct. Always check banks statements for suspicious transactions. Don't tell others your PIN numbers or passwords. Clearly all this is a tremendous hassle – but it is nothing compared with the difficulties if you become a victim.

Privacy on the internet can be protected to some extent by coding stored and transmitted data, as well as the use of password protection. But in some situations these tools cannot be applied. For example, the information governments collect about people is intended to be used to understand those people, so it has to be divulged to researchers and administrators. But this involves a risk. Tools of disclosure control have been developed to prevent people from being able to identify individuals in large datasets. Some of these tools involve modifying the data, so that it retains its statistical properties but loses information on individuals; others involve randomly perturbing the data.

5 The Ethics of Advanced Technologies

There is a basic principle of personal data confidentiality: that *'personal data should be used only for the purpose for which it was collected, unless explicit permission is given'*. Unless this is respected, public confidence will be shaken. The consequence will be that survey and census response rates will fall. This in turn will lead to less accurate data and conclusions, and hence to less effective government and less profitable corporations. Privacy of personal data lies at the foundation of effective societies.

For these reasons, the principle of data confidentiality has been enshrined in various legal ways, varying between countries. Many of them permit individuals to examine data relating to themselves, and to correct it if it is wrong. (In the context of identity theft, it is a good idea to periodically check your records with credit reference agencies.)

I said in the introduction that all *advanced technologies are ethically neutral*. They can be used for good or bad. This is as true for data mining as it is for nuclear technology and biological disciplines – the technology that the people in this room are involved in is just as sophisticated as those technologies. This means that criticisms of such a technology should be focused on the (mis)use to which it is put, not on the technology itself: the equations are the same, and it is what is done with them that counts.

As far as data mining is concerned, the genie is out of the bag. These advanced methods for discovering the unexpected in data exist, and are being used more and more often on more and more data sets. We cannot pretend that they no longer exist. The technology has the power to bring immense good, but if used the wrong way, it can also bring harm. As Jerrold Nadler said when he appeared before the United State's Technology and Privacy Advisory Committee in November 2003: the 'question isn't whether technology will be developed, but rather whether it will be used wisely.'

The Changing Face of Web Search

Prabhakar Raghavan

Yahoo! Research,
701 First Avenue,
Sunnyvale, California 94089, USA
pragh@yahoo-inc.com

Abstract. Dr. Prabhakar Raghavan is an invited keynote speaker for PAKDD 2006. Web search has come to dominate our consciousness as a convenience we take for granted, as a medium for connecting advertisers and buyers, and as a fast-growing revenue source for the companies that provide this service. Following a brief overview of the state of the art and how we got there, this talk covers a spectrum of technical challenges arising in web search – ranging from spam detection to auction mechanisms.

Biography

Prabhakar Raghavan joined Yahoo! Research in July 2005. His research interests include text and web mining, and algorithm design. He is a Consulting Professor of Computer Science at Stanford University and Editor-in-Chief of the Journal of the ACM. Raghavan received his PhD from Berkeley and is a Fellow of the ACM and of the IEEE. Prior to joining Yahoo, he was Senior Vice-President and Chief Technology Officer at Verity; before that he held a number of technical and managerial positions at IBM Research.

Data Mining for Surveillance Applications

Bhavani M. Thuraisingham[1,2]

[1] Eric Jonsson School of Engineering and Computer Science,
University of Texas at Dallas,
Richardson, Texas 75083-0688, USA
bhavani.thuraisingham@utdallas.edu
http://www.cs.utdallas.edu/people/thuraisingham.html
[2] Bhavani Security Consulting, LLC,
Dallas, Texas, USA
http://www.dr-bhavani.org

Abstract. Dr. Bhavani M. Thuraisingham is an invited speaker for PAKDD 2006. She is a Professor at the Eric Jonsson School of Engineering and Computer Science, University of Texas at Dallas. She is also director of the Cyber Security Research Center and President of Bhavani Security Consulting.

1 Summary

Data mining is the process of posing queries and extracting patterns, often previously unknown from large quantities of data using pattern matching or other reasoning techniques. Data mining has many applications for national security, also referred to as homeland security. The threats to national security include attacking buildings, destroying critical infrastructures such as power grids and telecommunication systems. Data mining techniques are being investigated to find out who the suspicious people are and who is capable of carrying out terrorist activities. One particular security application that can benefit from data mining is surveillance. We need to build infrastructures to conduct surveillance so that we can determine who might be suspicious. However, we also need to protect the privacy of the individuals who are law abiding citizens.

This presentation will first discuss data mining for surveillance applications. We will survey various surveillance applications and discuss the developments on applying data mining. Suspicious event detection is an area that has been investigated in some detail. The idea here is to represent various events, some of which are suspicious such as entering a secure room. Later when that event occurs the system will flag this event as suspicious. One of the challenges here is to combine suspicious event detection with say facial recognition techniques to determine who the suspicious people are in addition to detecting the suspicious events. Another challenge is to conduct distributed surveillance where there are multiple video feeds and the system has to monitor and combine events which may be suspicious. The system should also be able to detect the movements of people as they travel from one place to another.

Link analysis techniques could be utilized to follow such movements and determine the links that are suspicious. A third challenge is to associate people with unidentified luggage or bags. When the system detects an unaccompanied bag, it should then be able to carry out a trace back and determine who has left the bag. Finally a person by him or herself may not be suspicious, but seen together in a group he/she may be. That is, the system has to identify groups of suspicious individuals. Other challenges include conducting on-line analysis of surveillance data where the system should have the capability to analyze the surveillance data in real-time, make decisions and take appropriate actions.

The critical need for applying data mining for surveillance poses serious privacy threats. The challenge here is to carry out privacy preserving surveillance. There are some efforts on blanking the face of a per-son so that his/her privacy is maintained. However by doing this, the suspicious people's identity is also not revealed. Some efforts have focused on individuals carrying tags so that the faces of those with the appropriate tags are not revealed. This approach has a problem as the suspicious person can steal tags from others. A solution we are investigating is to encrypt all the faces of people with some keys. Only trusted agents have the keys for decryption. If the surveillance data shows that an individual is carrying out suspicious activities, then the trusted agents can reveal the identity of these suspicious people.

In addition to mining surveillance data, data mining can also be applied for geospatial applications. For example, one could combine web services provided by Google Maps or Map quest and connect the maps with say "friend of a friend" ontologies and determine the locations of various individuals. Suspicious people can use this information to terrorize the individuals whose locations have been revealed. Geospatial data can be mined to detect changes as well as detect unusual objects. The presentation will also discuss mining geospatial data.

In summary, the presentation will provide an overview mining surveillance data as well as conducting privacy preserving surveillance. Applying data mining to geospatial data such as maps will also be discussed.

2 Biography

Dr. Bhavani Thuraisingham joined The University of Texas at Dallas in October 2004 as a Professor of Computer Science and Director of the Cyber Security Research Center in the Erik Jonsson School of Engineering and Computer Science. She is an elected Fellow of three professional organizations: the IEEE (Institute for Electrical and Electronics Engineers), the AAAS (American Association for the Advancement of Science) and the BCS (British Computer Society) for her work in data security. She received the IEEE Computer Society's prestigious 1997 Technical Achievement Award for "outstanding and innovative contributions to secure data management."

Dr Thuraisingham's work in information security and information management has resulted in over 70 journal articles, over 200 refereed conference papers and workshops, and three US patents. She is the author of seven books in data management, data mining and data security including one on data mining for counter-terrorism and another on Database and Applications Security and is completing her eighth book on Trustworthy Semantic Web. She has given over 30 keynote presentations at various technical conferences and has also given invited talks at the White House Office of Science and Technology Policy and at the United Nations on Data Mining for counter-terrorism. She serves (or has served) on editorial boards of leading research and industry journals and currently serves as the Editor in Chief of Computer Standards and Interfaces Journal. She is also an Instructor at AFCEA's (Armed Forces Communications and Electronics Association) Professional Development Center and has served on panels for the Air Force Scientific Advisory Board and the National Academy of Sciences.

Dr Thuraisingham is the Founding President of "Bhavani Security Consulting" - a company providing services in consulting and training in Cyber Security and Information Technology.

Prior to joining UTD, Thuraisingham was an IPA (Intergovernmental Personnel Act) at the National Science Foundation from the MITRE Corporation. At NSF she established the Data and Applications Security Program and co-founded the Cyber Trust theme and was involved in inter-agency activities in data mining for counter-terrorism. She has been at MITRE since January 1989 and has worked in MITRE's Informa-tion Security Center and was later a department head in Data and Information Management as well as Chief Scientist in Data Management. She has served as an expert consultant in information security and data management to the Department of Defense, the Department of Treasury and the Intelligence Community for over 10 years. Thuraisingham's industry experience includes six years of research and development at Control Data Corporation and Honeywell Inc. Thuraisingham was educated in the United Kingdom both at the University of Bristol and at the University of Wales.

A Multiclass Classification Method Based on Output Design

Qi Qiang[1] and Qinming He[1,2]

[1] College of Computer Science, Zhejiang University,
Hangzhou 310027, China
[2] Ningbo Institute of Technology, Zhejiang University,
Ningbo, 315100, China
qiangqi@yahoo.com

Abstract. Output coding is a general framework for solving multiclass categorization problems. Some researchers have presented the notion of continuous codes and methods for designing output codes. However these methods are time-consuming and expensive. This paper describes a new framework, which we call Strong-to-Weak-to-Strong (SWS). We transform a "strong" learning algorithm to a "weak" algorithm by decreasing its iterative numbers of optimization while preserving its other characteristics like geometric properties and then make use of the kernel trick for "weak" algorithms to work in high dimensional spaces, finally improve the performances. An inspiring experimental results show that this approach is competitive with the other methods.

1 Introduction

A more general method for multiclass problem is to reduce the problem to multiple binary problems. In [1] Crammer described a unifying method (Section 2) for reducing multiclass problem to multiple binary problems.

Recently a robust Minimax classifier (Section 3) where the probability of correct classification of future data should be maximized has been provided [2]. No further assumptions are made with respect to the each two class-conditional distributions. The minimax problem can be interpreted geometrically as minimizing the maximum of the Mahalanobis distances to the two classes. "Kernelization" version is also available.

Section 4 presents new algorithm. In section 5, we report the experimental results. Finally, section 6 presents conclusions.

2 Design of Output Codes

Let $S = \{(x_1, y_1),...,(x_m, y_m)\}$ be a set of m training examples where each instance x_i belongs to a domain χ. We assume that each label y_i is an integer from the set $\Upsilon = \{1,...,k\}$. A multiclass classifier is a function $H : \chi \to \Upsilon$ that maps an instance x into an element $y \in \Upsilon$. An output codes M is a matrix of size $k \times l$ over $\mathbb{R}$ where each row of M corresponds to a class $y \in \Upsilon$. Then different binary classifiers $h_1,...,h_l$ can be yielded. We denote the vector of predictions of these classifiers on an instance x as $\bar{h}(x) = (h_1(x),...,h_l(x))$. We denote the rth row of M by $\overline{M}_r$. Given an example x we predict the label y for which the row $\overline{M}_y$ is the "closest" to $\bar{h}(x)$. Naturally we can perform the calculations in some high dimensional inner-product space Z using a transformation $\bar{\phi}: \mathbb{R}^l \rightarrow Z$ and use a general notion for closeness, then define it through an inner-product function $K: \mathbb{R}^l \times \mathbb{R}^l \rightarrow \mathbb{R}$, which satisfies Mercer conditions [3]. Thus $H(x) = \arg\max_{r \in \Upsilon} \{K(\bar{h}(x), \overline{M}_r)\}$. We define the 2 norm of a matrix M and introduce slack variables ζ, denote by $b_{i,r} = 1 - \delta_{y_i,r}$. Then the problem of finding a good matrix M can be stated as the following optimization problem:

$$L(M,\zeta,\eta) = \frac{1}{2}\beta\sum_r \left\|\overline{M}_r\right\|_2^2 + \sum_{i=1}^m \zeta_i + \sum_{i,r} \eta_{i,r}\left[K(\bar{h}(x_i),\overline{M}_r) - K(\bar{h}(x_i),\overline{M}_{y_i}) - \zeta_i + b_{i,r}\right] \quad (1)$$

$$subject\ to: \forall i,r\ \eta_{i,r} \geq 0$$

for some constant $\beta \geq 0$. Let $\bar{1}_i$ be the vector with all components zero, except for the ith component which is equal to one, and let $\bar{1}$ be the vector whose components are all one. We can denote by $\bar{\gamma}_{i,r} = \bar{1}_{y_i} - \eta_{i,r}$.

Finally, the classifier $H(x)$ can be written in terms of the variable γ as:

$$H(x) = \arg\max_r \left\{\sum_i \gamma_{i,r} K(\bar{h}(x), \bar{h}(x_i))\right\} \quad (2)$$

However solving optimization problem (1) is time-consuming. In this paper our algorithm solves this optimization problem heuristically.

3 A Probability Machine

Let x and y model data from each of two classes in a binary classification problem. We wish to determine a hyperplane $F(a,b) = \{z \mid a^T z = b\}$, where $a \in \mathbb{R}^n \setminus \{0\}$ and $b \in \mathbb{R}$ which separates the two classes of points with maximal probability with respect to all distributions having same mean and covariance matrices. This is expressed as:

$$\max_{\theta, a \neq 0, b} \theta \quad s.t \quad \inf_{x \sim (\bar{x}, \Sigma_x)} \Pr\{a^T x \geq b\} \geq \theta$$
$$\inf_{y \sim (\bar{y}, \Sigma_y)} \Pr\{a^T y \leq b\} \geq \theta \quad (3)$$

In formulation (3) the term θ is the minimal probability of correct classification of future data.

Learning large margin classifiers has become an active research topic. However, this margin is defined in a "local" way. MPM considers data in a global fashion, while SVM actually discards the global information of data including geometric information and the statistical trend of data occurrence.

4 SWS (Strong-to-Weak-to-Strong) Algorithm

The following natural learning problems arise,

1. Given a matrix M, find a set binary classifiers $\overline{h}$ which have small empirical loss.
2. Given a set of $\overline{h}$, find a matrix M which has small empirical loss.
3. Find both a matrix M and a set $\overline{h}$ which have small empirical loss.

The previous methods have focused mostly on the first problem. Most of these works have used predefined output codes, independently of the specific application and the learning algorithm. We mainly aim to solve the 3rd problem, however it is so hard to solve the designing problem not to mention finding a "good" classifier and a wonderful output codes simultaneously by using common optimization methods. Therefore a heuristic algorithm has been proposed instead of solving the optimization problem (1) directly. We use probability output θ in (3) of MPM to build a heuristic algorithm and then solve the design problem of output coding heuristically. In our framework SWS (Strong-to-Weak-to-Strong), we generalize the notion of "weak" algorithm. We can view an algorithm with less iterative steps of optimization as a "weak" algorithm and make use of the kernel trick for "weak" algorithm to work in high dimensional spaces, finally improve the performances. SWS and the heuristic algorithm make it realizable to solve both problems with acceptable time-consuming and complexion.

Recently a number of powerful kernel-based learning machines have been proposed. In KPCA, kernel serves as preprocessing while in SVM kernel has an effect on classification in the middle process. There could be two stages for kernel to affect the result in our algorithm. The first is in the middle process as it behaves in SVM. The second is where algorithm transforms several weak classifiers to a strong classifier.

4.1 Strong-to-Weak Stage

In the Strong-to-Weak stage, we transform "Strong" classifier to "Weak" classifier by equipping less iterative numbers of optimization while preserving its characteristics like large margin and geometric properties.

On the one hand, it could decrease total time-consuming especially in the case of large numbers of classes because each binary classifier needs less iterative steps of optimization.

On the other hand, our algorithm takes the geometric difference of classes into account while other methods ignore the difference because MPM uses Mahalanobis distance that involves geometric information. Therefore SWS preserve the characteristics. Based on concept above we can use a simple iterative least-squares approach because the algorithm only requires "Weak" learning algorithms.

4.2 Weak-to-Strong Stage

In this stage, we make use of the kernel trick for "Weak" algorithm to work in high dimensional spaces and finally improve the performances. According to the classification performance, multiclass classifier obtained by our algorithm becomes a much more "Strong" learning algorithm.

We notice that the saddle point from optimization problem (1) we are seeking is a minimum for the primal variables with respect to ζ_i. We can get

$$\sum_r \eta_{i,r} = 1, \; i.e. \; \overline{\gamma}_i \leq \overline{1}_{y_i} \; and \; \overline{\gamma}_i \cdot \overline{1} = 0 \tag{4}$$

And $\overline{1}_{y_i}$ may be viewed as the correct point distribution, $\eta_{i,r}$ could be viewed as the distribution obtained by the algorithm over the labels for each example. Then we can view $\overline{\gamma}_i$ as the difference between the former and the later. It is natural to say that an example x_i affects the result if and only if $\overline{\eta}_i$ is not a point distribution concentrating on the correct label y_i. Further we can say that only the questionable points contribute to the learning process and regard them as "critical points". We notice that one "critical point" may contribute to more than one class while "support vector" [3] contributes to only one class. It is typically assumed that the set of labels has no underlying structure, however there exist lots of different relation among category in practice. It means that it is reasonable that one example makes different contributions to some classes or classifiers.

Unlike other methods, our algorithm implements implicit update in high dimensional spaces by using a transformation $\overline{\phi}: \mathbb{R}^l \to Z$. And the output codes update merely occurs in final discrimination from (2):

$$H(x) = \arg\max_r \{K(\overline{h}(x), \overline{M}_r^{update})\} = \arg\max_r \{\sum_i \gamma_{i,r} K(\overline{h}(x), \overline{h}(x_i))\} \tag{5}$$

5 Experiments

In this section we test our algorithm using one-against-rest method experimentally on six data sets from the repository at University of California.[1]

Table 1. Best results of our algorithm and other methods with polynomial kernel of degree 2

	DB2	DAG SVM	One-against-One SVM	One-against-Rest SVM	One-against-rest SWS
Iris	97.3	96.6	97.3	96.6	97.3
Letter	98.2	97.9	97.9	97.8	99.1
Glass	73.5	73.8	72.0	71.9	79.2
Segment	96.4	96.6	96.6	95.2	96.4
Vowel	99.2	99.2	99.0	99.0	99.0
Wine	99.8	98.8	99.4	98.8	99.9

---

[1] URL: http://www.ics.uci.edu/~mlearn/MLRepository.html

Table 2. Accuracies with various iterative numbers given two polynomial kernel of degree 2

Iterative numbers	5	10	30	50
Iris	96.8	**97.3**	**97.3**	**97.3**
Letter	98.9	**99.1**	**99.1**	**99.1**
Glass	74.2	**79.2**	**79.2**	**79.2**
Segment	94.1	**96.4**	**96.4**	**96.4**
Vowel	98.4	**99.0**	**99.0**	**99.0**
Wine	99.0	**99.9**	**99.9**	**99.9**

Table1 presents the best results of our algorithm and other methods. Table2 displays the results of using different iterative steps with polynomial kernel of degree 2 (in "weak" classifiers and in weak-to-strong stage).

From Table1, we can say that our algorithm (SWS) is more efficient than others in most cases. Especially our algorithm achieves significant performances in Glass data set for the algorithm takes geometric information into account. It is clear that experiments show that the algorithm is fast to compute and efficient due to its heuristic.

6 Conclusions

We have introduced a new method as a solution to multiclass problems. Results obtained on the benchmark datasets suggest that our algorithm outperforms most other algorithms with most datasets although using one-against-rest method.

References

1. Koby Crammer, Yoram Singer: On the Learnability and Design of Output Codes for Multiclass Problems. Proceedings of the Thirteenth Annual Conference on Computational Learning Theory (2000) 35–46
2. Lanckriet, R. G., Ghaoui, L.E., Bhattacharyya, C., and Jordan, M. I.: A robust minimax approach to classification. Journal of Machine Learning Research (2002) 3:555–582
3. V. Vapnik: The Nature of Statistical Learning Theory. Spinger Verlag, New York (1995)

Regularized Semi-supervised Classification on Manifold

Lianwei Zhao[1], Siwei Luo[1], Yanchang Zhao[2], Lingzhi Liao[1], and Zhihai Wang[1]

[1] School of Computer and Information Technology, Beijing Jiaotong University,
Beijing 100044, China
lw_zhao@126.com
[2] Faculty of Information Technology, University of Technology, Sydney, Australia

Abstract. Semi-supervised learning gets estimated marginal distribution P_X with a large number of unlabeled examples and then constrains the conditional probability $p(y \mid x)$ with a few labeled examples. In this paper, we focus on a regularization approach for semi-supervised classification. The label information graph is first defined to keep the pairwise label relationship and can be incorporated with neighborhood graph which reflects the intrinsic geometry structure of P_X. Then we propose a novel regularized semi-supervised classification algorithm, in which the regularization term is based on the modified Graph Laplacian. By redefining the Graph Laplacian, we can adjust and optimize the decision boundary using the labeled examples. The new algorithm combines the benefits of both unsupervised and supervised learning and can use unlabeled and labeled examples effectively. Encouraging experimental results are presented on both synthetic and real world datasets.

1 Introduction

The problem of learning from labeled and unlabeled examples has attracted considerable attention in recent years. It can be described as follows: with l labeled examples $M = \{x_i, y_i\}_{i=1}^{l}$ drawn from an unknown probability distribution $P_{X \times Y}$ and u unlabeled examples $\{x_j\}_{j=l+1}^{l+u}$ drawn from the marginal distribution P_X of $P_{X \times Y}$, how to learn $P_{X \times Y}$ by exploiting the marginal distribution P_X? It is also known as semi-supervised learning, and a number of algorithms have been proposed for it, including Co-training [6], random field models [7,8] and graph based approaches [9, 10].

However, learning from examples has been seen as an ill-posed inverse problem [11], and regularizing the inverse problem means finding a meaning stable solution, so in this paper we focus on regularization approaches. Measure based regularization [12] assumes that two points connected by a line going through high density region should have the same label. Based on this assumption, the regularizer is weighted with data density. The idea of information regularization [13] is that labels should not change too much in regions where marginal density is high, so regularization penalty that links marginal to the conditional distribution is introduced, and it is expressed in terms of mutual information $I(x; y)$ as a measure of label complexity. Both of the above two methods take density into consideration, and can get the decision boundary that lies in the region of low density in 2D example. However, it is difficult to apply them in high-dimensional real world data sets.

Manifold regularization [1-4] assumes that two points close in the input space should have the same label, and exploits the geometry of the marginal distribution to incorporate unlabeled examples within a geometrically motivated regularization term. However, after incorporating an additional regularization term, there are two regularization parameters. It not only makes it difficult to find a solution, but needs improvement in theory. In addition, how to choose appropriate values for regularization parameters is a new problem.

In this paper, we first define the label information graph, and then incorporate it with neighborhood graph. Based on modified Graph Laplacian regularizier, we propose a novel regularized semi-supervised classification algorithm. There is only one regularization parameter reflecting the tradeoff between the Graph Laplacian and the complexity of solution. The labeled examples can be used to redefine the Graph Laplacian and further to adjust and optimize the decision boundary. Experimental results show that our algorithm can use unlabeled and labeled examples effectively and is more robust than Transductive SVM and LapSVM.

This paper is organized as follows. Section 2 briefly reviews Graph Laplacian and semi-supervised learning assumption. In section 3, we define label information graph with labeled examples and propose the regularized semi-supervised classification algorithm. Experimental results on synthetic and real world data are shown in section 4, followed by conclusions in section 5.

2 Related Works

2.1 Graph Laplacian

Graph Laplacian [5] has played a crucial role in several recently developed algorithms [14,15], because it approximates the natural topology of data and is simple to compute for enumerable based classifiers. Let's consider a neighborhood graph $G = (V, E)$ whose vertices are labeled or unlabeled example points $V = \{x_1, x_2, \cdots, x_{l+u}\}$ and whose edge weights $\{W_{ij}\}_{i,j=1}^{l+u}$ represent appropriate pairwise similarity relationship between examples. The neighborhood of x_j can be defined as those examples which are closer than ε or the k nearest neighbors of x_j. To ensure that the embedding function f is smooth, a natural choice is to get empirical estimate $I(G)$, which measures how much f varies across the graph:

$$I(G) = \frac{1}{2\sum_{i,j} W_{ij}} \sum_{i,j=1}^{l+u} (f(x_i) - f(x_j))^2 W_{ij} \quad (1)$$

where $2\sum_{i,j} W_{ij}$ is normalizing factor, so that $0 \leq I(G) \leq 1$.

Defining $\hat{f} = [f(x_1,\cdots,f(x_{l+u})]^T$, and $L = D - W$ as Graph Laplacian matrix, where D is diagonal matrix given by $D_{ii} = \sum_{j=1}^{l+u} W_{ij}$, $I(G)$ can be rewritten as:

$$I(G) = \frac{1}{2\sum_{i,j} W_{ij}} \hat{f}^T L \hat{f} \qquad (2)$$

2.2 Semi-supervised Learning Assumptions

In the semi-supervised learning framework, the marginal distribution P_X is unknown, so we must get empirical estimates of P_X using a large number of unlabeled examples and then constrain the conditional $p(y|x)$ with a few labeled examples. However, there is no identifiable relation between the P_X and the conditional $p(y|x)$, so the relationship between them must be assumed. Manifold regularization[1,2] assumes that two points that are close in the input space should have the same label. In other words, the conditional probability distribution $p(y|x)$ varies smoothly along the geodesics in the intrinsic geometry of P_X.

3 ReguSCoM: Regularized Semi-supervised Classification on Manifold

3.1 Our Motivation

We have noticed that the knowledge of the joint probability distribution $p(x,y)$ is enough to achieve perfect classification in supervised learning. We divide the process of semi-supervised learning into two steps. Firstly we get the empirical estimates of the marginal distribution P_X using both labeled and unlabeled examples and estimate $\hat{p}(y|x)$ according to the information carried about the distribution of labels. Secondly, we adjust $\hat{p}(y|x)$ to $p(y|x)$ using a few labeled examples and then get $p(x,y) = p(y|x)p(x)$. The first step can be considered as semi-supervised classification, while the second step is supervised learning.

We have assumed that if two points $x_1, x_2 \in X$ are close in the input space, then the conditional $p(y|x_1)$ and $p(y|x_2)$ are near in intrinsic geometry of P_X. In manifold regularization [1] this assumption is represented by adjacency matrix, i.e., edge weights $\{W_{ij}\}_{i,j=1}^{l+u}$. However, this adjacency matrix doesn't take into consideration the information carried by labeled examples. The regularization term $I(G)$, especially for binary case classifiers, is proportional to the number of separated neighbors, that is, the number of connected pairs that are classified differently by decision boundary. Therefore for labeled examples x_i and x_j, if they are of the same label, they should not be separated by the decision boundary, so we can redefine the relationship between x_i and x_j by strengthening it. If x_i and x_j have the different labels, we can weaken it.

3.2 Definition of Label Information Graph

In the manifold learning, one of the key assumptions is that the data lie on a low dimensional manifold M and this manifold can be approximated by a weighted graph constructed with all the labeled and unlabeled examples. So the performance of the learning algorithm significantly depends on how the graph is constructed.

We consider all the sample points $\{x_1, x_2, \cdots, x_{l+u}\}$, including both the labeled and unlabeled examples. When the support of P_X is a compact submanifold M, the geometry structure can be approximated using the Graph Laplacian with both labeled and unlabeled examples. The Least Squares algorithm solves the problem with the squared loss function $\sum_{i=1}^{l} V(x_i, y_i, f) = \sum_{i=1}^{l}(y_i - f(x_i))^2$, which is based on the minimizing the error on the labeled examples. It is important to observe that

$$2(l-1)\sum_{i=1}^{l}(y_i - f(x_i))^2 \geq \sum_{i,j=1}^{l}((y_i - f(x_i))-(y_j - f(x_j)))^2 \qquad (3)$$

$$= \sum_{i,j=1}^{l}((f(x_i) - f(x_j))-(y_i - y_j))^2$$

If $\sum_{i=1}^{l}(y_i - f(x_i))^2 \to 0$, then

$$\sum_{i,j=i}^{l}((f(x_i) - f(x_j))-(y_i - y_j))^2 \to 0 \qquad (4)$$

So if $|y_i - y_j| < \delta$, then $|f(x_i) - f(x_j)| < \varepsilon$, where $\delta, \varepsilon \to 0$, and $\delta, \varepsilon > 0$.

We define $(l+u) \times (l+u)$ matrix J as follows.

$$J_{ij} = \begin{cases} 1 \text{ or } W_{ij}, & \text{if } i,j \leq l \text{ and } |y_i - y_j| < \delta \\ 0 \text{ or } -W_{ij}, & \text{if } i,j \leq l \text{ and } |y_i - y_j| \geq \delta \\ 0, & \text{otherwise} \end{cases} \qquad (5)$$

This can be seen as a label information graph $G' = (V, E')$, whose vertices are the labeled or unlabeled example points $V = \{x_1, x_2, \cdots, x_{l+u}\}$ and whose edge weights J_{ij} represent appropriate pairwise label relationship between labeled examples i and j.

According to the label information graph, the right of the equation 3 can be rewritten as follows:

$$\sum_{i,j=1}^{l}((f(x_i)-f(x_j))-(y_i-y_j))^2 = \sum_{i,j=1}^{l+u}(f(x_i)-f(x_j))^2 J_{ij} \qquad (6)$$

This term can be seen as label information carried by labeled examples and penalizes classifiers that separate the examples having the same labels.

Remark: In graph G', weight J_{ij} just represents appropriate pairwise label relationship between i and j. If labeled example i has the same label as j, they should not be separated by decision boundary. This relationship must not be represented only by element J_{ij}. For example, for large scale problems, this relationship J_{ij} can be represented by a geodesic path $J_{ik_1}, J_{k_1 k_2}, \cdots J_{k_n j}$, which can be computed by finding a shortest path $(i, k_1, k_2, \cdots k_n, j)$ from i to j in graph G'.

3.3 Classifier Based on the Modified Graph Laplacian

In this section, we consider the problem of using the manifold structure to improve the performance of the classifier f, where $f : X \in M \to Y$. In most situations, the manifold is approximated by a graph constructed with all examples and f is defined on the vertices of the graph, so a stabilizer is necessary. An important class of stabilizers is squares of norms on reproducing kernel Hilbert spaces (RKHS). The squared norm $\|f\|_K^2$ is used as stabilizer to penalize high oscillation of various types. The geometry structure of the marginal distribution P_X is incorporated as a regularization term based on the neighborhood graph [1,2]. In order to exploit the label information, equation 6 is also introduced as a penalty term based on the label information graph.

The neighborhood graph and the label information graph have the same vertices and can be incorporated together. So the optimization problem has the following objective function:

$$\min_{f \in H} H[f] = \gamma \|f\|_K^2 + \sum_{i,j=1}^{l+u}(f(x_i)-f(x_j))^2 (W_{ij}+J_{ij}) \qquad (7)$$
$$= \gamma \|f\|_K^2 + \hat{f}^T L_a \hat{f}$$

where $L_a = D - (W+J)$, D is diagonal matrix given by $D_{ii} = \sum_{j=1}^{l+u}(W_{ij}+J_{ij})$ and γ is a regularization parameter that controls the complexity of the clustering function. It has the same form as unsupervised regularization spectral clustering [1] The existence, uniqueness and an explicit formula describing the solution of this minimizing problem are given by the Representer theorem. Then the solution of the problem has the unique solution:

$$f(x) = \sum_{i=1}^{l+u} \alpha_i K(x, x_i) \qquad (8)$$

where α can be solved by an eigenvalue method and the regularization parameter γ can be selected by the approach of L-curve. For binary classification problem, classifier function f is constant within the region of input space associated with a particular class, that is $Y = \{-1, 1\}$.

3.4 Learning Algorithm

The crux of the proposed learning algorithm is to redefine the Graph Laplacian based on the clustering hypothesis and then adjust the semi-supervised classification with the labeled examples.

The complete semi-supervised learning algorithm (ReguSCoM) consists of the following five steps.

Step 1. Construct adjacency graph $G = (V, E)$ with $(l+u)$ nodes using k nearest neighbors. Choose edge weights W_{ij} with binary or heat kernel weights, construct label information graph $G' = (V, E')$, and then compute the Graph Laplacian L_a.

Step 2. Regularized semi-supervised classification. At this step, we use the objective function given by equation 7.

Step 3. Label the unlabeled examples. Firstly, we select one labeled example from $M = \{x_i, y_i\}_{i=1}^{l}$. Without loss of generality, we select $\{x_1, y_1\}$, so all the examples clustering with $\{x_1, y_1\}$ will have the same label y_1 as $\{x_1, y_1\}$, while the others will have the label different from y_1. So for every $\{x_i, y_i\} \in M$, we get a label $\hat{y}_i$.

Step 4. Compute $\sum_{i=1}^{l} |y_i - \hat{y}_i|^2$. Stop if $\sum_{i=1}^{l} |y_i - \hat{y}_i|^2 \leq threshold$, otherwise, select the i th labeled example where $i = \arg\max_i |y_i - \hat{y}_i|$.

Step 5. Adjust the weights J_{ij}. For the selected i th example, we can find the labeled examples j satisfying $\{|y_j - y_i| \leq \delta, |y_j - \hat{y}_j| \leq \varepsilon, 1 \leq j \leq l\}$, and then adjust the weight J_{ij} and re-compute the matrix L_a. Goto step 2.

4 Experimental Results

4.1 Synthetic Data

We first conducted experiments on two moons dataset. The dataset contains 200 unlabeled sample points, and all the labeled points are sampled from the unlabeled points randomly. Figure 1 (left) shows the results of unsupervised manifold regularization clustering without labeled points, where the curves represent the decision boundary. After adjusted by one labeled point for each class using Regularized Semi-supervised Classification on manifold (ReguSCoM) proposed in this paper, the decision boundary has little change as shown in Figure 1 (right). The reason lies in that this dataset has regular geometry structure and the manifold

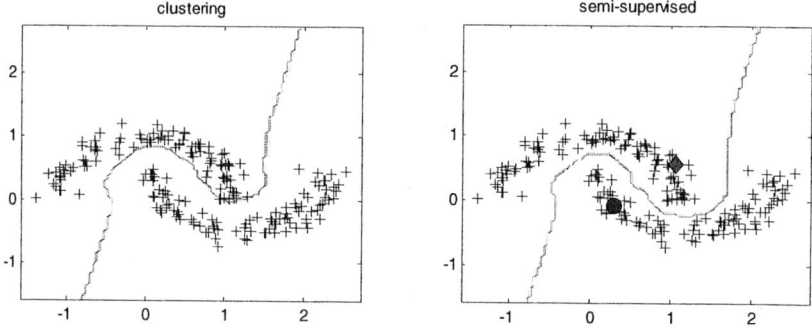

Fig. 1. The result of unsupervised regularization clustering and Regularized Semi-supervised Classification with only one labeled points for each class on two moons dataset

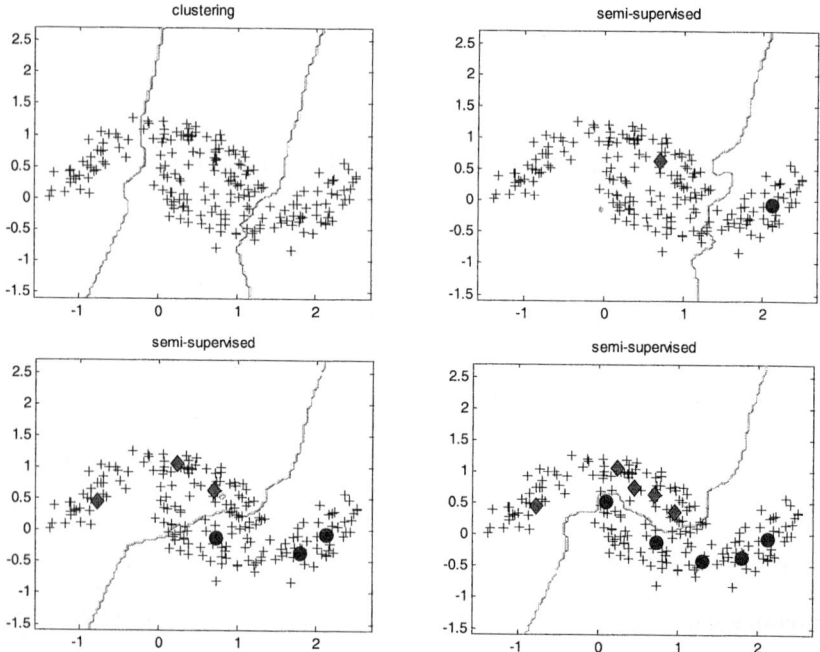

Fig. 2. Regularized Semi-supervised classification on two moons dataset added with Guassian noise and 0, 1, 3, and 5 labeled points respectively

regularization clustering can find this structure. The Graph Laplacian based algorithm can implement perfectly the cluster assumption that the decision boundary does not separate the neighbors.

Figure 2 shows the results of semi-supervised classification using ReguSCoM algorithm on two moons dataset with Guassian noise and 0, 1, 3, and 5 labeled points added respectively. With 0 labeled points it can be regarded as unsupervised manifold regularization clustering. From the figure, it is clear that unsupervised classification failed to find the optimal decision boundary. The reason is that the dataset loses the regular geometry structure when noise added. With more labeled examples added, the decision boundary can be adjusted appropriately. With only 5 labeled points for each class, the proposed algorithm can find the optimal solution shown in Figure 2.

4.2 Real World Datasets

In this section, we will show the experimental results on two real world datasets, USPS dataset and Isdolet dataset from UCI machine learning repository. We constructed the graph with 6 nearest-neighbors and used the binary weight of the edge of the neighborhood graph, that is $W_{ij} = 0 \ or \ 1$.

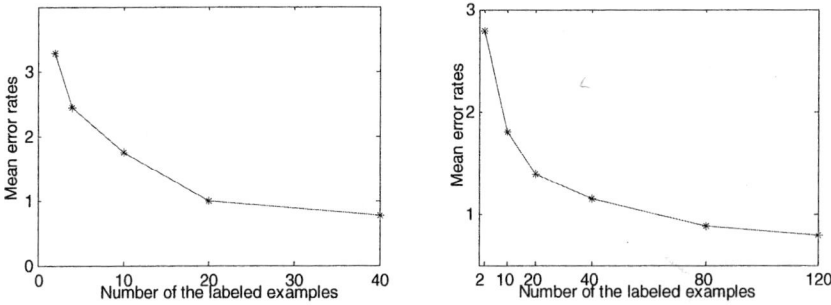

Fig. 3. Mean error rates with the number of labeled examples at the precision-recall breakeven points on Isolet (left) and USPS (right) dataset

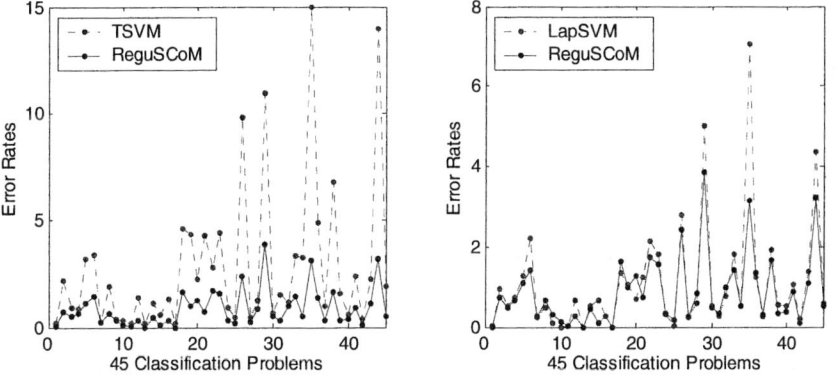

Fig. 4. Comparing the error rate of ReguSCoM, Transductive SVM, and LapSVM at the precision-recall breakeven points

We first used Isolet database of letters of the English alphabet spoken in isolation. We chose isolet1+2+3+4 dataset of 6238 examples and considered the task of binary classifying one of spoken letter from another. Figure 3 (left) shows the mean error rates with the increasing of number of labeled examples using ReguSCoM.

We also show the results of 45 binary classification problems using USPS dataset. We used the first 400 images for each handwritten digit, and processed using PCA to 100 dimensions as in [1]. Figure 3 (right) shows that the mean error rates decrease with the increase of number of labeled examples. We compare the error rate of ReguSCoM with Transductive SVM and LapSVM at the precision-recall breakeven points in the ROC curves, as shown in Figure 4. We choose Polynomial kernel of degree 3, as in [1]. Experimental results show clearly that ReguSCoM is of higher accuracy than Transductive SVM and LapSVM.

5 Conclusions

Learning from examples has been seen as an ill-posed inverse problem and semi-supervised learning is to benefit from a large number of unlabeled examples and a few labeled examples. We propose a novel regularized semi-supervised classification algorithm on manifold (ReguSCoM) in this paper. The regularization term not only represents the intrinsic geometry structure of P_X that implies the information of classification, but reflects the label information carried by labeled examples. Our method yields encouraging experimental results on both synthetic data and real world datasets and the results demonstrate effective use of both unlabeled and labeled data. In future work, we will explore the link to other semi-supervised leaning algorithms in theory and will investigate other alternative training approaches based on manifold learning to improve performance of semi-supervised learning algorithm. To attack nonlinear ill-posed inverse problem will also be part of our future work.

Acknowledgements

The research is supported by the National Natural Science Foundations of China (60373029) and the National Research Foundation for the Doctoral Program of Higher Education of China (20050004001). We would like to thank Dr. M. Belkin for useful suggestion.

References

1. Belkin M., Niyogi P., Sindhwani V. Manifold Regularization: A Geometric Framework for Learning from Examples. Department of Computer Science, University of Chicago, TR-2004-06.
2. Belkin M., Niyogi P., Sindhwani V. On Manifold Regularization. Department of Computer Science, University of Chicago, TR-2004-05.
3. Belkin M., Matveeva I., Niyogi P. Regression and Regularization on Large Graphs. In Proceedings of the Conference on Computational Learning Theory, 2004.

4. Belkin M., Niyogi P. Using Manifold Structure for Partially Labeled Classification, NIPS 2002,Vol. 15.
5. Belkin M., Niyogi P. Laplacian Eigenmaps for Dimensionality Reduction and Data Representation, Neural Computation, June 2003
6. Blum A., Mitchell T. Combining Labeled and Unlabeled Data with Co-training. In Proceedings of the Conference on Computational Learning Theory, 1998.
7. Szummer M., Jaakkola T. Partially Labeled Classification with Markov Random Walks. NIPS 2001,Vol. 14.
8. Zhu X., Ghahramani Z., Lafferty J. Semi-supervised Learning Using Gaussian Fields and Harmonic Functions. ICML 2003.
9. Blum A., Chawla S. Learning from Labeled and Unlabeled Data Using Graph Mincuts, ICML 2001.
10. Zhou D., Bousquet O, Lal TN, Weston J., Schoelkopf B., Learning with Local and Global Consistency, NIPS 2003, Vol. 16.
11. Ernesto De Vito,Lorenzo Rosasco, Andrea Caponnetto, Umberto De Giovannini, Francesca Odone.Learning from Examples as an Inverse Problem. Journal of Machine Learning Research, 6 (2005) 883–904.
12. Bousquet O., Chapelle O, Hein M. Measure Based Regularization, NIPS 2003, Vol. 16.
13. Szummer, M., Jaakkola T. Information Regularization with Partially Labeled Data. NIPS 2002,Vol. 15.
14. Krishnapuram B., Williams D., Xue Ya, Hartemink A., Carin L., Figueiredo M. A. T. On Semi-Supervised Classification. NIPS 2004,Vol. 17.
15. Kegl B., Wang Ligen. Boosting on Manifolds: Adaptive Regularization of Base Classifiers. NIPS 2004,Vol. 17.

Similarity-Based Sparse Feature Extraction Using Local Manifold Learning

Cheong Hee Park*

Dept. of Computer Science and Engineering,
Chungnam National University,
220 Gung-dong, Yuseong-gu,
Daejeon, 305-763, Korea
cheonghee@cnu.ac.kr

Abstract. Feature extraction is an important preprocessing step which is encountered in many areas such as data mining, pattern recognition and scientific visualization. In this paper, a new method for sparse feature extraction using local manifold learning is proposed. Similarities in a neighborhood are first computed to explore local geometric structures, producing sparse feature representation. Based on the constructed similarity matrix, linear dimension reduction is applied to enhance similarities among the elements in the same class and extract optimal features for classification performances. Since it only computes similarities in a neighborhood, sparsity in the similarity matrix can give computational efficiency and memory savings. Experimental results demonstrate superior performances of the proposed method.

1 Introduction

Feature extraction is an important preprocessing step which is encountered in many areas such as data mining, pattern recognition and scientific visualization [1]. Discovering intrinsic data structure embedded in high dimensional data can give a low dimensional representation preserving essential information in the original data. While Principal Component Analysis (PCA), Linear Discriminant Analysis (LDA) and Multi-dimensional Scaling (MDS) are traditional linear dimension reduction methods [1, 2, 3], recently nonlinear dimension reduction methods utilizing local geometric structures have been proposed [4, 5]. Isomap first connects paths between each data point and its neighbors and then extends them by searching for the shortest paths for each pair of data points [4]. Based on the constructed distance matrix, classical MDS finds low dimensional representation to preserve geodesic distances among data points. However, Isomap does not give optimal dimension reduction for classification, since it does not consider class information. Also some limitations in Isomap exist in its assumption that the data is connected well enough to define low dimensional geometry. But in many real situations, for example, if the data has separated classes, a small number of neighbors will not connect classes and a large number of neighbors would fail to capture nonlinear

---

* This work was supported by the Korea Research Foundation Grant funded by Korea Government (MOEHRD, Basic Research Promotion Fund) (KRF-2005-204-D00046).

structure in the data. MDS at the second stage of Isomap does not give an efficient way to compute low dimensional representation for a new data point. Moreover, MDS may not give optimal dimension reduction for classification.

In this paper, we propose a new approach which combines a linear dimension reduction and local manifold learning through the similarity-based sparse feature representation. We learn local manifolds from the neighborhood configuration. However instead of searching for shortest paths for each pair of data points, we apply a linear dimension reduction method for the similarity matrix reflecting local manifolds. Local similarity learning gives the effects of unfolding nonlinear structures in the data and a linear dimension reduction finds a optimal transformation which maximizes similarities within each classes and minimizes similarities between classes.

The rest of the paper is organized as follows. In Section 2, a new method for similarity-based feature extraction using local manifold learning is presented. In Section 3, based on the sparse similarity matrix a linear dimension reduction method, Minimum Squared Error Solution (MSE), is applied. Experimental results in Section 4 demonstrate the performance of the proposed method.

2 Similarity-Based Feature Extraction Using Local Manifold Learning

Throughout the paper, we assume that the data is given with known class labels and the problem is to assign a class label to new samples, i.e., the goal is classification. First a similarity matrix based on the local geometric structure in the data is constructed. When a natural similarity measure between data points is available, the most similar k neighbors for each data objects are kept as actual neighbors and relations with the other remaining points are disregarded, i.e. their similarities are set as zeros. Also a distance measure can be converted to a similarity measure. As in Isomap, the distance d_{ij} between two points a_i and a_j is defined as $\|a_i - a_j\|$ if one is among the k-nearest neighbors of the other or within the ϵ-radius neighborhood, otherwise $d_{ij} = \infty$. Similarity is defined from the distance by a converter function f as

$$s_{ij} = f(d_{ij}).$$

What is required for the converter function f is

$$\begin{cases} d_{ij} \leq d_{ik} \\ 0 = f(\infty). \end{cases} \Leftrightarrow f(d_{ij}) \geq f(d_{ik}), \tag{1}$$

The conditions in (1) imply that all similarities are nonnegative and the infinite distance is mapped to zero similarity. Also similarity is measured in inverse order of distances.

For a data set $A = \{a_1, \cdots, a_n\}$ and the similarity matrix

$$S = [s_{ij}]_{1 \leq i,j \leq n} = [s_1, \cdots, s_n], \tag{2}$$

each column $s_i = [s_{1i}, \cdots, s_{ni}]^T$ represents the similarities between a data point a_i and the others. Similarities among nearby points are emphasized while connections with points resided in the far distance are disregarded. Taking the column s_i as a new feature vector gives a sparse feature representation for a_i.

Nearby points a_i and a_j which belong to the same class will share a majority of neighbors and therefore s_i and s_j show similar patterns. However, nearby points can belong to different classes and some points in the same class may not be nearer than the points in the different classes as in the cases of nonlinearly structured data. Hence based on the new feature representation, we perform linear dimension reduction in order to enhance similarities among the elements in the same class and decrease them between elements belonging to different classes as discussed in Section 3.

Now we discuss several properties of the proposed method addressing detailed implementations for the optimal values for k or ϵ, and a converter function f. The requirements in (1) for a converter function f impose the inverse relationship between distance measure and similarity measure. As examples, these two functions can be used for a converter function,

$$s_{ij} = s(\{a_i, a_j\}) = \frac{1}{(1+\alpha\|a_i-a_j\|)^\beta}, \quad \alpha, \beta > 0, \tag{3}$$

$$s_{ij} = s(\{a_i, a_j\}) = \exp(-\frac{\|a_i-a_j\|^2}{2\lambda^2}), \quad \lambda \in \mathbb{R}. \tag{4}$$

The purpose of the parameter α in (3) is the normalization of distance measure. Let τ be the average of distances from each data point to the nearest neighbor. The inverse of τ was used for α in our experiments. In that case, the remaining distances are represented as a ratio of τ.

The optimal value for k should be chosen to be large enough so that the majority in the k-neighbors of data points is the members of the same class as the given point, and at the same time it should be small enough to capture nonlinear geometric structure in the data. In our implementation, k was chosen as follows. For each data point a_i, let t_i is the number of data points which have the same class labels as a_i and are nearer to a_i than any data points belonging to the different classes. Then k is determined as

$$k = \frac{1}{r}\sum_{i=1}^{r}\left(\frac{1}{n_i}\sum_{j \in N_i} t_j\right), \tag{5}$$

where N_i is the index set of data items in the class i and n_i is the number of elements in the class i. Eq. (5) computes the average number of the nearest neighbors which has same class labels as each data point. The number k chosen by Eq. (5) increases similarities among data points within each class and also decreases similarities of data points belonging to different classes. Cross-validation also can be used to determine the optimal values for any parameters.

3 Linear Dimension Reduction Based on the Similarity Feature Vectors

For the similarity matrix constructed in Section 2, any linear dimension reduction methods can be applied. In this section, we apply Minimum Squared Error Solution (MSE) [1] for the constructed sparse feature vectors.

Table 1. Class distribution in the letter image recognition data

Class	A	B	C	D	E	F	G	H	I	J	K	L	M
no. data	789	766	736	805	768	775	773	734	755	747	739	761	792
Class	N	O	P	Q	R	S	T	U	V	W	X	Y	Z
no. data	783	753	803	783	758	748	796	813	764	752	787	786	734

Let us denote a data set A as

$$A = \{a_1, \cdots, a_n\} = \{a_i^j \mid 1 \leq i \leq r, 1 \leq j \leq n_i\}, \tag{6}$$

and the similarity vectors constructed in Section 2 as $\{s_1, \cdots, a_n\}$, where each class i ($1 \leq i \leq r$) has n_i elements $\{a_i^j \mid 1 \leq j \leq n_i\}$ and the total number of data is $n = \sum_{i=1}^{r} n_i$. Minimum Squared Error Solution (MSE) finds a set of linear discriminant functions $\{g_i\}_{(1 \leq i \leq r)}$,

$$g_i(z) = w_{0i} + w_i^T z = \begin{cases} 1, & \text{if } z \in \text{ class } i \\ 0, & \text{otherwise} \end{cases}$$

which minimize the least squares error

$$\left\| \begin{bmatrix} 1 & s_1^T \\ \vdots & \vdots \\ 1 & s_n^T \end{bmatrix} \begin{bmatrix} w_{01} & \cdots & w_{0r} \\ w_1 & \cdots & w_r \end{bmatrix} - \begin{bmatrix} y_{11} & \cdots & y_{1r} \\ \vdots & & \vdots \\ y_{n1} & \cdots & y_{nr} \end{bmatrix} \right\|_F^2 \equiv \|PW - Y\|_F^2 \tag{7}$$

where $y_{ji} = 1$ if a_j belongs to the class i, and 0 otherwise [1]. The MSE solution of the problem (7) can be obtained by $W = P^+ Y$, where P^+ is the pseudo-inverse[1] of P [6]. For any new data point z and a similarity vector $u = [s(a_1, z), \cdots, s(a_n, z)]^T$, z is assigned to the class i if for all $j \neq i$

$$g_i(u) > g_j(u) \text{ where } [g_1(u), \cdots, g_r(u)] = [1, s(a_1, z), \cdots, s(a_n, z)]W.$$

We call this approach as *sparse MSE*. Since similarities are computed in a neighborhood, a similarity matrix S is very sparse. With a sparse similarity matrix S, computations utilizing sparsity can be used to save computational complexities [7, 8].

4 Experimental Results

For the experiment, letter image recognition data was downloaded from UCI Machine Leaning Repository. From the capital alphabet letters of black-and-white rectangular pixel images, 16 integer attributes were extracted [9]. The data distribution is described in Table 1. From the 26 alphabets, three data sets were composed as shown in Table 2. Each class was randomly split to the training and test sets in the ratio of 3:2 and the

---

[1] When the Singular value decomposition (SVD) of P is $P = U \Sigma V^T$, the pseudo-inverse of P is obtained as $V \Sigma^+ U^T$.

Table 2. Prediction accuracies by LDA, MSE and sparse MSE, and sparsity in a similarity matrix. The percentage of nonzero components in the similarity matrix S in sparse MSE is shown.

	LDA	MSE	Sparse MSE			
			Prediction accuracy		Sparsity	
			k-neighbors	ϵ-radius	k-neighbors	ϵ-radius
Data 1 ={A,E,I,O,U}	94.5 %	91.4 %	99.2 %	99.4 %	3.6 %	2.1 %
Data 2={A,R,U,Q,M}	94.1 %	90.4 %	97.4 %	98.7 %	3.2 %	1.8 %
Data 3={P,K,Z,Q,D}	94.4 %	91.0 %	98.3 %	99.0 %	2.9 %	3.0 %

mean prediction accuracies by 10 times random splitting to the training and test sets were computed as a performance measure.

Prediction accuracies by sparse MSE as well as LDA and MSE are shown in Table 2. In the reduced dimensional spaces by each method, the 1-NN classifier was used for classification. For sparse MSE, the converter function in (3) was used. The value k for the k-neighbors was chosen as discussed in (5) and cross-validation was used to determine the optimal values for other parameters. The percentage of nonzero components of the similarity matrix S in sparse MSE is also reported in Table 2. While the similarity matrix constructed by local manifold learning contained only nonzero components of about 4 % of the total components, sparse MSE improved classification performance greatly compared with LDA and MSE.

Note that the similarity matrix can be learned in various ways. Instead of converting Euclidean distances to similarities, similarities between data points can be defined directly without using distance measures. Hence even when the data is not represented as the vector space representation, the proposed method can be applied for any similarity measures.

References

1. R.O. Duda, P.E. Hart, and D.G. Stork. *Pattern Classification*. Wiley-interscience, New York, 2001.
2. I.T. Jolliffe. *Principal Component Analysis*. Springer-Verlag, New York, 1986.
3. T. Cox and M. Cox. *Multidimensional scaling*. Chapman & Hall, London, 1994.
4. V. d. Silva J. B. Tenenbaum and J. C. Langford. A global geometric framework for nonlinear dimensionality reduction. *Science*, 290:2319–2323, 2000.
5. S. T. Roweis and L. K. Saul. Nonlinear dimensionality reduction by locally linear embedding. *Science*, 290:2323–2326, 2000.
6. G.H. Golub and C.F. Van Loan. *Matrix Computations*. Johns Hopkins University Press, third edition, 1996.
7. G.W. Stewart. Four algorithms for the efficient computation of truncated pivoted QR approximations to a sparse matrix. *Numerische Mathematik*, 83:313–323, 1999.
8. C.C. Paige and M.A. Saunders. LSQR: An algorithm for sparse linear equations and sparse least squares. *ACM transactions on mathematical software*, 8:1:43–71, 1982.
9. P. W. Frey and D. J. Slate. Letter recognition using holland-style adaptive classifiers. *Machine learning*, 6:161–182, 1991.

Generalized Conditional Entropy and a Metric Splitting Criterion for Decision Trees

Dan A. Simovici[1] and Szymon Jaroszewicz[2]

[1] University of Massachusetts at Boston,
Dept. of Computer Science, Boston,
Massachusetts 02125,
dsim@cs.umb.edu

[2] Faculty of Computer and Information Systems,
Technical University of Szeczin, Poland
sjaroszewicz@wi.ps.pl

Abstract. We examine a new approach to building decision tree by introducing a geometric splitting criterion, based on the properties of a family of metrics on the space of partitions of a finite set. This criterion can be adapted to the characteristics of the data sets and the needs of the users and yields decision trees that have smaller sizes and fewer leaves than the trees built with standard methods and have comparable or better accuracy.

Keywords: decision tree, generalized conditional entropy, metric, metric betweenness.

1 Introduction

Decision trees constitute one of the most popular classification techniques in data mining and have been the subject of a large body of investigation. The typical construction algorithm for a decision tree starts with a training set of objects that is split recursively. The successive splits form a tree where the sets assigned to the leaves consist of objects that belong almost entirely to a single class. This allows new objects that belong to a test set to be classified into a specific class based on the path induced by the object in the decision tree which joins the root of the tree to a leaf.

Decision trees are useful classification algorithms, even though they may present problems related to overfitting and excessive data fragmentation that results in rather complex classification schemes.

A central problem in the construction of decision trees is the choice of the splitting attribute at each non-leaf node. We show that the usual splitting criterion (the information gain ratio, or the similar measure derived from the Gini index) are special cases of a more general approach. Furthermore, we propose a geometric criterion for choosing the splitting attributes that has the advantage of being adaptable to various data sets and user needs.

2 Partition Entropies

The *betweenness relation* defined by the metric space (S,d) is a ternary relation R on the set S defined by $(s,u,t) \in R$ if $d(s,u) + d(u,t) = d(s,t)$. We denote the fact that $(s,u,t) \in R$ by $[sut]$ and we say that u *is between* s *and* t.

We explore a natural link that exists between random variables and partitions of sets that allows the transfer of certain probabilistic and information-theoretical notions to partitions of sets.

Let $\mathsf{PART}(S)$ be the set of partitions of a set S. The class of all partitions of finite sets is denoted by PART. The one-block partition of S is denoted by ω_S. The partition $\{\{s\} \mid s \in S\}$ is denoted by ι_S. If $\pi, \pi' \in \mathsf{PART}(S)$, then $\pi \le \pi'$ if every block of π is included in a block of π'. Clearly, for every $\pi \in \mathsf{PART}(S)$ we have $\iota_S \le \pi \le \omega_S$.

π' covers π if $\pi \le \pi'$ and there is no partition $\theta \in \mathsf{PART}(S)$ such that $\pi < \theta < \pi'$. This fact is denoted by $\pi \prec \pi'$. It is known [1] that $\pi \prec \pi'$ if and only if π' is obtained from π by fusing two blocks of this partition into a new block.

For every two partitions π, σ both $\inf\{\pi, \sigma\}$ and $\sup\{\pi, \sigma\}$ in the partial ordered set $(\mathsf{PART}(S), \le)$ exist and are denoted by $\pi \wedge \sigma$ and $\pi \vee \sigma$, respectively. It is well known that $(\mathsf{PART}(S), \le)$ is an upper semimodular lattice.

If S, T are two disjoint and nonempty sets, $\pi \in \mathsf{PART}(S)$, $\sigma \in \mathsf{PART}(T)$, where $\pi = \{A_1, \ldots, A_m\}$, $\sigma = \{B_1, \ldots, B_n\}$, then the partition $\pi + \sigma$ is the partition of $S \cup T$ given by $\pi + \sigma = \{A_1, \ldots, A_m, B_1, \ldots, B_n\}$.

Whenever the "+" operation is defined, then it is easily seen to be associative. In other words, if S, U, V are pairwise disjoint and nonempty sets, and $\pi \in \mathsf{PART}(S)$, $\sigma \in \mathsf{PART}(U)$, $\tau \in \mathsf{PART}(V)$, then $\pi + (\sigma + \tau) = (\pi + \sigma) + \tau$. Observe that if S, U are disjoint, then $\iota_S + \iota_U = \iota_{S \cup U}$. Also, $\omega_S + \omega_U$ is the partition $\{S, U\}$ of the set $S \cup U$.

If $\pi = \{B_1, \ldots, B_m\}$, $\sigma = \{C_1, \ldots, C_n\}$ are partitions of two arbitrary sets S, U, respectively, then we denote the partition $\{B_i \times C_j \mid 1 \le i \le m, 1 \le j \le n\}$ of $S \times U$ by $\pi \times \sigma$. Note that $\iota_S \times \iota_U = \iota_{S \times U}$ and $\omega_S \times \omega_U = \omega_{S \times U}$.

Let $\pi \in \mathsf{PART}(S)$ and let $C \subseteq S$. Denote by π_C the "trace" of π on C given by $\pi_C = \{B \cap C \mid B \in \pi \text{ such that } B \cap C \ne \emptyset\}$. Clearly, $\pi_C \in \mathsf{PART}(C)$; also, if C is a block of π, then $\pi_C = \omega_C$.

A subset T of S is *pure* relative to a partition $\pi \in \mathsf{PART}(S)$ if $\pi_T = \omega_T$. In other words, T is pure relative to a partition π if T is included in some block of π.

In [2] the notion of β-entropy of a probability distribution $\mathbf{p} = (p_1, \ldots, p_n)$ was defined as:

$$\mathcal{H}_\beta(\mathbf{p}) = \frac{1}{2^{1-\beta} - 1} \left(\sum_{i=1}^{m} p_i^\beta - 1 \right),$$

where $p_1 + \cdots + p_n = 1$ and $p_i \ge 0$ for $1 \le i \le n$. In the same reference it was observed that Shannon's entropy $\mathcal{H}(\mathbf{p})$ can be obtained as $\lim_{\beta \to 1} \mathcal{H}_\beta(\pi)$.

In [3] we offered a new interpretation of the notion of entropy for finite distributions as entropies of partitions of finite sets. Our approach took advantage of the properties of the partial order of the lattice of partitions of a finite set and makes use of operations defined on partitions. We defined the $\mathcal{H}_\beta$ entropy for $\beta \in \mathbb{R}$, $\beta > 0$ as a function $\mathcal{H}_\beta : \mathsf{PART}(S) \longrightarrow \mathbb{R}_{\geq 0}$ that satisfies certain conditions. Under these conditions, we have shown in [3] that if $\pi = \{B_1, \ldots, B_n\} \in \mathsf{PART}(S)$, then

$$\mathcal{H}_\beta(\pi) = \frac{1}{2^{1-\beta} - 1} \left(\sum_{i=1}^{m} \left(\frac{|B_i|}{|S|} \right)^\beta - 1 \right).$$

In the special case, when $\beta \to 1$ we have:

$$\mathcal{H}_\beta(\pi) = - \sum_{i=1}^{m} \frac{|B_i|}{|S|} \cdot \log_2 \frac{|B_i|}{|S|}.$$

Note that if $|S| = 1$, then $\mathsf{PART}(S)$ consists of a unique partition $(\omega_S = \iota_S)$ and $\mathcal{H}_\beta(\omega_S) = 0$. Moreover, for an arbitrary finite set S we have $\mathcal{H}_\beta(\pi) = 0$ if and only if $\pi = \omega_S$.

These facts suggest that for a subset T of S the number $\mathcal{H}_\beta(\pi_T)$ can be used as a measure of the purity of the set T with respect to the partition π. If T is π-pure, then $\pi_T = \omega_T$ and, therefore, $\mathcal{H}_\beta(\pi_T) = 0$. Thus, the smaller $\mathcal{H}_\beta(\pi_T)$, the more pure the set T is.

3 Conditional β-Entropy of Partitions and Metrics on Partitions

The β-entropy defines naturally a conditional entropy of partitions. We note that the definition introduced here is an improvement over our previous definition given in [3]. Starting from conditional entropies we will be able to define a family of metrics on the set of partitions of a finite set and study the geometry of these finite metric spaces.

Definition 1. *Let $\pi, \sigma \in \mathsf{PART}(S)$ and let $\sigma = \{C_1, \ldots, C_n\}$. The β-conditional entropy is the function $\mathcal{H}_\beta : \mathsf{PART}(S)^2 \longrightarrow \mathbb{R}_{\geq 0}$ defined by:*

$$\mathcal{H}_\beta(\pi|\sigma) = \sum_{j=1}^{n} \left(\frac{|C_j|}{|S|} \right)^\beta \mathcal{H}_\beta(\pi_{C_j}),$$

for $\pi, \sigma \in \mathsf{PART}(S)$.

Observe that $\mathcal{H}_\beta(\pi|\omega_S) = \mathcal{H}_\beta(\pi)$ and that $\mathcal{H}_\beta(\omega_S|\pi) = \mathcal{H}_\beta(\pi|\iota_S) = 0$ for every partition $\pi \in \mathsf{PART}(S)$. Also, we can write:

$$\mathcal{H}_\beta(\iota_S|\sigma) = \sum_{j=1}^{n} \left(\frac{|C_j|}{|S|} \right)^\beta \mathcal{H}_\beta(\iota_{C_j}) = \frac{1}{2^{1-\beta} - 1} \left(\frac{1}{|S|^{\beta-1}} - \sum_{j=1}^{n} \left(\frac{|C_j|}{|S|} \right)^\beta \right), \quad (1)$$

where $\sigma = \{C_1, \ldots, C_n\}$. The conditional entropy can be written explicitly as:

$$\mathcal{H}_\beta(\pi|\sigma) = \frac{1}{2^{1-\beta} - 1} \sum_{i=1}^{m} \sum_{j=1}^{n} \left(\left(\frac{|B_i \cap C_j|}{|S|} \right)^\beta - \left(\frac{|C_j|}{|S|} \right)^\beta \right), \quad (2)$$

where $\pi = \{B_1, \ldots, B_m\}$.

Theorem 1. *Let π, σ be two partitions of a finite set S. We have $\mathcal{H}_\beta(\pi|\sigma) = 0$ if and only if $\sigma \leq \pi$.*

It is possible to prove that for every $\pi, \sigma \in \mathsf{PART}(S)$ we have:

$$\mathcal{H}_\beta(\pi \wedge \sigma) = \mathcal{H}_\beta(\pi|\sigma) + \mathcal{H}_\beta(\sigma) = \mathcal{H}_\beta(\sigma|\pi) + \mathcal{H}_\beta(\pi),$$

which generalizes a well-known property of Shannon's entropy.

The next result shows that the β-conditional entropy is dually monotonic with respect to its first argument and is monotonic with respect to its second argument.

Theorem 2. *Let $\pi, \sigma, \sigma' \in \mathsf{PART}(S)$, where S is a finite set. If $\sigma \leq \sigma'$, then $\mathcal{H}_\beta(\sigma|\pi) \geq \mathcal{H}_\beta(\sigma'|\pi)$ and $\mathcal{H}_\beta(\pi|\sigma) \leq \mathcal{H}_\beta(\pi|\sigma')$.*

Since $\mathcal{H}_\beta(\pi) = \mathcal{H}_\beta(\pi|\omega_S)$ it follows that if $\pi, \sigma \in \mathsf{PART}(S)$, then $\mathcal{H}_\beta(\pi) \geq \mathcal{H}_\beta(\pi|\sigma)$.

The next statement that follows from the previous theorem is useful in Section 5.

Corollary 1. *Let ξ, θ, θ' be three partitions of a finite set S. If $\theta \geq \theta'$, then*

$$\mathcal{H}_\beta(\xi \wedge \theta) - \mathcal{H}_\beta(\theta) \geq \mathcal{H}_\beta(\xi \wedge \theta') - \mathcal{H}_\beta(\theta').$$

The behavior of β-conditional entropies with respect to the "addition" of partitions is discussed in the next statement.

Theorem 3. *Let S be a finite set, π, θ be two partitions of S, where $\theta = \{D_1, \ldots, D_h\}$. If $\sigma_i \in \mathsf{PART}(D_i)$ for $1 \leq i \leq h$, then*

$$\mathcal{H}_\beta(\pi|\sigma_1 + \cdots + \sigma_h) = \sum_{i=1}^{h} \left(\frac{|D_i|}{|S|}\right)^\beta \mathcal{H}_\beta(\pi_{D_i}|\sigma_i).$$

If $\tau = \{F_1, \ldots, F_k\}$, $\sigma = \{C_1, \ldots, C_n\}$ be two partitions of S, and let $\pi_i \in \mathsf{PART}(F_i)$ for $1 \leq i \leq k$. Then,

$$\mathcal{H}_\beta(\pi_1 + \cdots + \pi_k|\sigma) = \sum_{i=1}^{k} \left(\frac{|F_i|}{|S|}\right)^\beta \mathcal{H}_\beta(\pi_i|\sigma_{F_i}) + \mathcal{H}_\beta(\tau|\sigma).$$

In [4] L. de Mántaras proved that Shannon's entropy generates a metric $d : \mathsf{PART}(S)^2 \longrightarrow \mathbb{R}^2$ given by $d(\pi, \sigma) = \mathcal{H}(\pi|\sigma) + \mathcal{H}(\sigma|\pi)$, for $\pi, \sigma \in \mathsf{PART}(S)$. We extend his result to a class of metrics that can be defined by β-entropies, thereby improving our earlier results [5].

Our central result follows.

Theorem 4. *The mapping $d_\beta : \mathsf{PART}(S)^2 \longrightarrow \mathbb{R}_{\geq 0}$ defined by: $d_\beta(\pi, \sigma) = \mathcal{H}_\beta(\pi|\sigma) + \mathcal{H}_\beta(\sigma|\pi)$ for $\pi, \sigma \in \mathsf{PART}(S)$ is a metric on $\mathsf{PART}(S)$.*

It is clear that $d_\beta(\pi, \omega_S) = \mathcal{H}_\beta(\pi)$ and $d_\beta(\pi, \iota_S) = \mathcal{H}(\iota_S|\pi)$.

The behavior of the distance d_β with respect to partition addition is discussed in the next statement.

Theorem 5. *Let S be a finite set, π, θ be two partitions of S, where $\theta = \{D_1, \ldots, D_h\}$. If $\sigma_i \in \mathsf{PART}(D_i)$ for $1 \le i \le h$, then*

$$d_\beta(\pi, \sigma_1 + \cdots + \sigma_h) = \sum_{i=1}^{h} \left(\frac{|D_i|}{|S|}\right)^\beta d_\beta(\pi_{D_i}, \sigma_i) + \mathcal{H}_\beta(\theta|\pi).$$

4 The Metric Geometry of the Partition Space

The distance between two partitions can be expressed using distances relative to the total partition or to the identity partition. Indeed, we have the following result:

Theorem 6. *Let $\pi, \sigma \in \mathsf{PART}(S)$ be two partitions. We have:*

$$\begin{aligned}d_\beta(\pi, \sigma) &= 2 \cdot d_\beta(\pi \wedge \sigma, \omega_S) - d_\beta(\pi, \omega_S) - d_\beta(\sigma, \omega_S) \\ &= d_\beta(\iota_S, \pi) + d_\beta(\iota_S, \sigma) - 2 \cdot d_\beta(\iota_S, \pi \wedge \sigma).\end{aligned}$$

From this result it follows that if $\theta \le \tau$ and we have either $d_\beta(\theta, \omega_S) = d_\beta(\tau, \omega_S)$ or $d_\beta(\iota_S, \theta) = d_\beta(\iota_S, \tau)$, then $\theta = \tau$ for every $\theta, \tau \in \mathsf{PART}(S)$.

Theorem 7. *Let $\pi, \sigma \in \mathsf{PART}(S)$. The following statements are equivalent:*

1. $\sigma \le \pi$;
2. *we have $[\sigma, \pi, \omega_S]$ in the metric space $(\mathsf{PART}(S), d_\beta)$;*
3. *we have $[\iota_S, \sigma, \pi]$ in the metric space $(\mathsf{PART}(S), d_\beta)$.*

Metrics generated by β-conditional entropies are closely related to lower valuations of the upper semi-modular lattices of partitions of finite sets. This connection was established in [6] and studied in [7, 8, 9].

A *lower valuation* on a lattice $(L, \vee, \wedge)$ is a mapping $v : L \longrightarrow \mathbb{R}$ such that $v(\pi \vee \sigma) + v(\pi \wedge \sigma) \ge v(\pi) + v(\sigma)$ for every $\pi, \sigma \in L$. If the reverse inequality is satisfied, that is, if $v(\pi \vee \sigma) + v(\pi \wedge \sigma) \le v(\pi) + v(\sigma)$ for every $\pi, \sigma \in L$, then v is referred to as an *upper valuation*.

If $v \in L$ is both a lower and upper valuation, that is, if $v(\pi \vee \sigma) + v(\pi \wedge \sigma) = v(\pi) + v(\sigma)$ for every $\pi, \sigma \in L$, then v is a valuation on L. It is known [6] that if there exists a positive valuation v on L, then L must be a modular lattice. Since the partition lattice of a set is an upper-semimodular lattice that is not modular ([6]) it is clear that positive valuations do not exist on partition lattices. However, lower and upper valuations do exist, as shown next:

Theorem 8. *Let S be a finite set. Define the mappings $v_\beta : \mathsf{PART}(S) \longrightarrow \mathbb{R}$ and let $w_\beta : \mathsf{PART}(S) \longrightarrow \mathbb{R}$ be by $v_\beta(\pi) = d_\beta(\iota_S, \pi)$ and $w_\beta(\pi) = d_\beta(\pi, \omega_S)$, respectively, for $\pi \in \mathsf{PART}(S)$. Then, v_β is a lower valuation and w_β is an upper valuation on the lattice $(\mathsf{PART}(S), \vee, \wedge)$.*

5 Metrics and Data Mining

We begin by defining the notion of *object system* as a triple $\mathcal{S} = (S, H, C)$, where S is a finite set referred to as the *training set*, $H = \{A_1, \ldots, A_n\}$ is a finite set of mappings of the form $A_i : S \longrightarrow D_i$ called the *features of* $\mathcal{S}$ for $1 \leq i \leq n$, and $C : S \longrightarrow D$ is the *classification function*. The sets $D_1, \ldots, D_n$ are supposed to contain at least two elements and they are referred as the *domains of the attributes* $A_1, \ldots, A_n$.

A set of attributes X, $X \subseteq H$ generates a mapping $\wp_X : S \longrightarrow \bigcup'\{D_i \mid A_i \in X\}$, defined by $\wp_X(t) = \{(A(t), A) \mid A \in X\}$ for every $t \in S$, where $\bigcup'$ denotes the disjoint union of a family of sets; we refer to $\wp_X$ as the *projection on* X of $\mathcal{S}$. Projections define partitions on the set of objects in a natural manner; namely if X is a set of attributes, a block B_v of the partition π^X is a non-empty set of the form $\{t \in S \mid \wp_X(t) = v\}$, where v is an element of the range of $\wp_X$.

To introduce formally the notion of decision tree we start from the notion of tree domain. A *tree domain* is a non-empty set of sequences D, over the set of natural numbers $\mathbb{N}$ such that every prefix of a sequence $s \in D$ also belongs to D, and for every $m \geq 1$, if $(p_1, \ldots, p_{m-1}, p_m) \in D$, then $(p_1, \ldots, p_{m-1}, q) \in D$ for every $q \leq p_m$. The elements of D are called the *vertices* of D. The notions of *descendant* and *ancestor* of a vertex have their usual definitions.

Let S be a finite set and let D be a tree domain. An S-tree is a function $\mathcal{T} : D \longrightarrow \mathcal{P}(S)$ such that $\mathcal{T}(\lambda) = S$, and if $u1, \ldots, um$ are the descendants of a vertex u, then the sets $\mathcal{T}(u1), \ldots, \mathcal{T}(um)$ form a partition of the set $\mathcal{T}(u)$.

A *decision tree* for an object system $\mathcal{S} = (S, H, C)$ is an S-tree $\mathcal{T}$, such that if the vertex v has the descendants $v0, \ldots, vm$, then there exists an attribute $A \in H$ (called the *splitting attribute* in v) such that $\{\mathcal{T}(vi) \mid 1 \leq i \leq m\}$ is the partition $\pi^A_{\mathcal{T}(v)}$.

Thus, each descendant vi of a vertex v corresponds to a value a of the attribute A that was used as a splitting attribute in v. If $\lambda = v_1, v_2, \ldots, v_k = u$ is the path in $\mathcal{T}$ that was used to reach the vertex u, $A_{i_1}, A_{i_2}, \ldots, A_{i_{k-1}}$ are the splitting attributes in $v_0, v_1, \ldots, v_{k-1}$ and $a_1, a_2, \ldots, a_{k-1}$ are the values that correspond to $v_2, \ldots, v_k$, respectively, then we say that u is reached by the selection:

$$A_{i_1} = a_1 \wedge \cdots \wedge A_{i_{k-1}} = a_{k-1}.$$

It is desirable that the leaves of a decision tree contain C-pure or almost C-pure sets of objects. In other words, the objects assigned to a leaf of the tree should, with few exceptions, have the the same value for the class attribute C. This amounts to asking that for each leaf w of $\mathcal{T}$ we must have $\mathcal{H}_\beta(\pi^C_{S_w})$ as close to 0 as possible. To take into account the size of the leaves note that the collection of sets of objects assigned to the leafs is a partition κ of S and that we need to minimize:

$$\sum_w \left(\frac{|S_w|}{|S|}\right)^\beta \mathcal{H}_\beta(\pi^C_{S_w}),$$

which is the conditional entropy $\mathcal{H}(\pi^C|\kappa)$. By Theorem 1 we have $\mathcal{H}(\pi^C|\kappa) = 0$ if and only if $\kappa \leq \pi^C$, which happens when the sets of objects assigned to the leafs are C-pure.

The construction of a decision tree $\mathcal{T}_\beta(\mathcal{S})$ for an object system $\mathcal{S} = (S, H, C)$ evolves in a top-down manner according to the following high-level description of a general algorithm [10]. The algorithm starts with an object system $\mathcal{S} = (S, H, C)$, a value of β and with an impurity threshold ϵ and it consists of the following steps:

1. If $\mathcal{H}_\beta(\pi_S^C) \leq \epsilon$, then return $\mathcal{T}$ as an one-vertex tree; otherwise go to 2.
2. Assign the set S to a vertex v, choose an attribute A as a splitting attribute of S (using a splitting attribute criterion to be discussed in the sequel) and apply the algorithm to the object systems $(S_{a_1}, H, C), \ldots, (S_{a_p}, H, C)$, where $S_{a_i} = \{t \in S \mid A(t) = a_i\} \neq \emptyset$. Let $\mathcal{T}_1, \ldots, \mathcal{T}_p$ the decision trees returned for the systems $\mathcal{S}_1, \ldots, \mathcal{S}_p$, respectively. Connect the roots of these trees to v.

Note that if ϵ is sufficiently small and if $\mathcal{H}_\beta(\pi_S^C) \leq \epsilon$, where $S = \mathcal{T}(u)$ is the set of objects at a node u, then there is a block Q_k of the partition π_S^C that is dominant in the set S. We refer to Q_k as the dominant class of u.

Once a decision tree $\mathcal{T}$ is built it can be used to determine the class of a new object $t \notin S$ such that the attributes of the set H are applicable. If $A_{i_1}(t) = a_1, \ldots, A_{i_{k-1}}(t) = a_{k-1}$, a leaf u was reached through the path $v_1, \ldots, v_k = u$, and $a_1, a_2, \ldots, a_{k-1}$ are the values that correspond to $v_2, \ldots, v_k$, respectively, then t is classified in the class Q_k, where Q_k is the dominant class at leaf u.

The description of the algorithm shows that the construction of a decision tree depends essentially on the method for choosing the splitting attribute. We focus next on this issue.

Classical decision tree algorithms make use of the information gain criterion or the gain ratio to choose splitting attribute. These criteria are formulated using Shannon's entropy, as their designations indicate.

In our terms, the analogue of the information gain for a vertex w and an attribute A is: $\mathcal{H}_\beta(\pi_{S_w}^C) - \mathcal{H}_\beta(\pi_{S_w}^C|\pi_{S_w}^A)$. The selected attribute is the one that realizes the highest value of this quantity. When $\beta \to 1$ we obtain the information gain linked to Shannon entropy. When $\beta = 2$ one obtains the selection criteria for the Gini index using the CART algorithm [11].

The monotonicity property of conditional entropy shows that if A, B are two attributes such that $\pi^A \leq \pi^B$ (which indicates that the domain of A has more values than the domain of B), then $\mathcal{H}_\beta(\pi_{S_w}^C|\pi_{S_w}^A) \leq \mathcal{H}_\beta(\pi_{S_w}^C|\pi_{S_w}^B)$, so the gain for A is larger than the gain for B. This highlights a well-known problem of choosing attributes based on information gain and related criteria: these criteria favor attributes with large domains, which in turn, generate bushy trees. To alleviate this problem information gain was replaced with the information gain ratio defined as:

$$\frac{\mathcal{H}_\beta(\pi_{S_w}^C) - \mathcal{H}_\beta(\pi_{S_w}^C|\pi_{S_w}^A)}{\mathcal{H}_\beta(\pi_{S_w}^A)},$$

which introduces the compensating divisor $\mathcal{H}_\beta(\pi_{S_w}^A)$.

We propose replacing the information gain and the gain ratio criteria by choosing as splitting attribute for a node w an attribute that minimizes the distance $d_\beta(\pi^C_{S_w}, \pi^A_{S_w}) = \mathcal{H}_\beta(\pi^C_{S_w}|\pi^A_{S_w}) + \mathcal{H}_\beta(\pi^A_{S_w}|\pi^C_{S_w})$. This idea has been developed by L. de Mántaras in [4] for the metric d_1 induced by Shannon's entropy. Since one could obtain better classifiers for various data sets and user needs using values of β that are different from one, our approach is an improvement of previous results.

Besides being geometrically intuitive, the minimal distance criterion has the advantage of limiting both conditional entropies $\mathcal{H}_\beta(\pi^C_{S_w}|\pi^A_{S_w})$ and $\mathcal{H}_\beta(\pi^A_{S_w}|\pi^C_{S_w})$. The first limitation insures that the choice of the splitting attribute will provide a high information gain; the second limitation insures that attributes with large domains are not favored over attributes with smaller domains.

Suppose that in the process of building a decision tree for an object system $\mathcal{S} = (S, H, C)$ we constructed a stump of the tree $\mathcal{T}$ that has n leaves and that the sets of objects that correspond to these leaves are $S_1, \ldots, S_n$. This means that we created the partition $\kappa = \{S_1, \ldots, S_n\} \in \mathsf{PART}(S)$, so $\kappa = \omega_{S_1} + \cdots + \omega_{S_n}$. We choose to split the node v_i using as splitting attribute the attribute A that minimizes the distance $d_\beta(\pi^C_{S_i}, \pi^A_{S_i})$. The new partition κ' that replaces κ is

$$\kappa' = \omega_{S_1} + \cdots + \omega_{S_{i-1}} + \pi^A_{S_i} + \omega_{S_{i+1}} + \cdots + \omega_{S_n}.$$

Note that $\kappa \geq \kappa'$. Therefore, we have:

$$d_\beta(\pi^C \wedge \kappa, \kappa) \geq d_\beta(\pi^C \wedge \kappa', \kappa').$$

This shows that as the construction of the tree advances the current partition κ gets closer to the partition $\pi^C \wedge \kappa$. More significantly, as the stump of the tree grows, κ gets closer to the class partition π^C. Indeed, by Theorem 5 we can write:

$$d_\beta(\pi^C, \kappa) = d_\beta(\pi^C, \omega_{S_1} + \cdots + \omega_{S_n})$$
$$= \sum_{j=1}^n \left(\frac{|S_j|}{|S|}\right)^\beta d_\beta(\pi^C_{S_j}, \omega_{S_j}) + \mathcal{H}_\beta(\theta|\pi^C),$$

where $\theta = \{S_1, \ldots, S_n\}$. Similarly, we can write:

$$d_\beta(\pi^C, \kappa') = d_\beta(\pi^C, \omega_{S_1} + \cdots + \omega_{S_{i-1}} + \pi^A_{S_i} + \omega_{S_{i+1}} + \cdots + \omega_{S_n})$$
$$= \sum_{j=1, j\neq i}^n \left(\frac{|S_j|}{|S|}\right)^\beta d_\beta(\pi^C_{S_j}, \omega_{S_j}) + \left(\frac{|S_i|}{|S|}\right)^\beta d_\beta(\pi^C_{S_i}, \pi^A_{S_i}) + \mathcal{H}_\beta(\theta|\pi^C).$$

These equalities imply:

$$d_\beta(\pi^C, \kappa) - d_\beta(\pi^C, \kappa') = \left(\frac{|S_i|}{|S|}\right)^\beta \left(d_\beta(\pi^C_{S_i}, \omega_{S_i}) - d_\beta(\pi^C_{S_i}, \pi^A_{S_i})\right)$$
$$= \left(\frac{|S_i|}{|S|}\right)^\beta \left(\mathcal{H}_\beta(\pi^C_{S_i}) - d_\beta(\pi^C_{S_i}, \pi^A_{S_i})\right).$$

If the choices of the node and the splitting attribute are made such that $\mathcal{H}_\beta(\pi^C_{S_i}) > d_\beta(\pi^C_{S_i}, \pi^A_{S_i})$, then the distance between π^C and the current partition κ of the tree stump will decrease. Since the distance between $\pi^C \wedge \kappa$ and κ decreases in any case when the tree is expanded it follows that the "triangle" determined by π^C, $\pi^C \wedge \kappa$, and κ will shrink during the construction of the decision tree.

6 Experimental Results

We tested our approach on a number of data sets from [12]. Due to space limitations we included only the results shown in Figure 1 which are fairly typical. Decision trees were constructed using metrics d_β, where β varied between 0.25 and 2.50. Note that for $\beta = 1$ the metric algorithm coincides with the approach of de Mántaras. We also built standard decision trees using the J48 technique of the well-known WEKA package [13].In all cases, accurracy was assessed through 10-fold cross-validation. The experimental evidence shows that β can be adapted such that accuracy is comparable, or better than the standard algorithm. The size of the trees and the number of leaves show that the proposed approach to decision trees results consistently in smaller trees with fewer leaves.

Audiology

β	accuracy	size	leaves
2.50	53.54	53	36
2.25	54.42	53	36
2.00	54.87	54	37
1.75	53.10	47	32
1.50	76.99	29	19
1.25	78.32	29	19
1.00	76.99	29	19
0.75	76.99	29	19
0.50	76.99	29	19
0.25	78.76	33	21

Hepatitis

β	accuracy	size	leaves
2.50	81.94	15	8
2.25	81.94	9	5
2.00	81.94	9	5
1.75	83.23	9	5
1.50	84.52	9	5
1.25	84.52	11	6
1.00	85.16	11	6
0.75	85.81	9	5
0.50	83.23	5	3
0.25	82.58	5	3

Primary-tumor

β	accuracy	size	leaves
2.50	34.81	50	28
2.25	35.99	31	17
2.00	37.76	33	18
1.75	36.28	29	16
1.50	41.89	40	22
1.25	42.18	38	21
1.00	42.48	81	45
0.75	41.30	48	27
0.50	43.36	62	35
0.25	44.25	56	32

Standard J4.8

Data Set	accuracy	size	leaves
Audiology	77.88	54	32
Hepatitis	83.87	21	11
Primary-tumor	39.82	88	47

Fig. 1. Experimental Results

7 Conclusion and Future Work

We introduced a family of metrics on the set of partitions of a finite set that can be used for a new splitting criterion for building decision trees. In addition to being more intuitive than the classic approach, this criterion results in decision trees that have smaller sizes and fewer leaves than the trees built with standard methods, and have comparable or better accuracy.

The value of β that results in the smallest trees seems to depend on the relative distribution of the class attribute and the values of the feature attributes of the objects. We believe that further investigations should develop numerical characteristics of data sets that allow predicting "optimal" values for β, that is, values that result in the smallest decision trees for data sets.

Another future direction is related to clustering algorithms. Since clusterings of objects can be regarded as partitions, metrics developed for partitions present an interest for the study of the dynamics of clusters, as clusters are formed during incremental algorithms [14], or as data sets evolve.

References

1. Lerman, I.C.: Classification et analyse ordinale des données. Dunod, Paris (1981)
2. Daróczy, Z.: Generalized information functions. Information and Control **16** (1970) 36–51
3. Simovici, D.A., Jaroszewicz, S.: An axiomatization of partition entropy. IEEE Transactions on Information Theory **48** (2002) 2138–2142
4. de Mántaras, R.L.: A distance-based attribute selection measure for decision tree induction. Machine Learning **6** (1991) 81–92
5. Simovici, D.A., Jaroszewicz, S.: Generalized entropy and decision trees. In: EGC 2003 - Journees francophones d'Extraction et de Gestion de Connaissances, Lyon, France (2003) 369–380
6. Birkhoff, G.: Lattice Theory. American Mathematical Society, Providence (1973)
7. Barthélemy, J., Leclerc, B.: The median procedure for partitions. In: Partitioning Data Sets, Providence, American Mathematical Society (1995) 3–34
8. Barthélemy, J.: Remarques sur les propriétés metriques des ensembles ordonnés. Math. Sci. hum. **61** (1978) 39–60
9. Monjardet, B.: Metrics on partially ordered sets – a survey. Discrete Mathematics **35** (1981) 173–184
10. Tan, P.N., Steinbach, M., Kumar, V.: Introduction to Data Mining. Pearson Addison-Wesley, Boston (2005)
11. Breiman, L., Friedman, J.H., Olshen, R.A., Stone, C.J.: Classification and Regression Trees. Chapman and Hall, Boca Raton (1998)
12. Blake, C.L., Merz, C.J.: UCI Repository of machine learning databases. University of California, Irvine, Dept. of Information and Computer Sciences, http://www.ics.uci.edu/~mlearn/MLRepository.html (1998)
13. Witten, I.H., Frank, E.: Data Mining - Practical Machine Learning Tools and Techniques. second edn. Morgan Kaufmann, San Francisco (2005)
14. Simovici, D.A., Singla, N., Kuperberg, M.: Metric incremental clustering of nominal data. In: Proceedings of ICDM 2004, Brighton, UK (2004) 523–527

RNBL-MN: A Recursive Naive Bayes Learner for Sequence Classification*

Dae-Ki Kang, Adrian Silvescu, and Vasant Honavar

Artificial Intelligence Research Laboratory,
Department of Computer Science,
Iowa State University,
Ames, IA 50011, USA
{dkkang, silvescu, honavar}@cs.iastate.edu

Abstract. Naive Bayes (NB) classifier relies on the assumption that the instances in each class can be described by a *single* generative model. This assumption can be restrictive in many real world classification tasks. We describe RNBL-MN, which relaxes this assumption by constructing a tree of Naive Bayes classifiers for sequence classification, where each individual NB classifier in the tree is based on a multinomial event model (one for each class at each node in the tree). In our experiments on protein sequence and text classification tasks, we observe that RNBL-MN substantially outperforms NB classifier. Furthermore, our experiments show that RNBL-MN outperforms C4.5 decision tree learner (using tests on sequence composition statistics as the splitting criterion) and yields accuracies that are comparable to those of support vector machines (SVM) using similar information.

1 Introduction

Naive Bayes (NB) classifiers, due to their simplicity and modest computational and training data requirements, are among the most widely used classifiers on many classification tasks, including text classification tasks [1] and macromolecular sequence classification tasks that arise in bio-informatics applications [2]. NB classifiers belong to the family of generative models (a model for generating data given a class) for classification. Instances of a class are assumed to be generated by a random process which is modeled by a generative model. The parameters of the generative model are estimated (in the case of NB) assuming independence among the attributes given the class. New instances to be classified are assigned to the class that is the most probable for the instance.

NB classifier relies on the assumption that the instances in each class can be described by a *single* generative model (i.e., probability distribution). According to Langley [3], this assumption can be restrictive in many real world classification tasks. One way to overcome this limitation while maintaining some of the

---

* Supported in part by grants from the National Science Foundation (IIS 0219699) and the National Institutes of Health (GM 066387).

computational advantages of NB classifiers is to construct a tree of NB classifiers. Each node in the tree (a NB classifier) corresponds to one set of generative models (one generative model per class), with different nodes in the tree corresponding to different generative models for a given class. Langley described a recursive NB classifier (RBC) for classifying instances that are represented by ordered tuples of nominal attribute values. RBC works analogous to a decision tree learner [4], recursively partitioning the training set at each node in the tree until the NB classifier of the node simply cannot partition the corresponding data set. Unlike in the case of the standard decision tree, the branches out of each node correspond to the most likely class lebels assigned by the NB classifier at that node. In cases where each class cannot be accurately modeled by a single Naive Bayes generative model, the subset of instances routed to one or more branches belong to more than one class. RBC models the distribution of instances in a class at each node using a Naive Bayes generative model. However, according to Langley's reports of experiments on most of the UC-Irvine benchmark data sets, the recursive NB classifier did not yield significant improvements over standard NB classifier [3].

In this paper, we revisit the idea of recursive NB classifier in the context of sequence classification tasks. We describe RNBL-MN, an algorithm for constructing a tree of Naive Bayes classifiers for sequence classification. Each NB classifier in the tree is based on a multinomial event model [1] (one for each class at each node in the tree). Our choice of the multinomial event model is influenced by its reported advantages over the multivariate event model of sequences [1] in text classification tasks. RNBL-MN works in a manner similar to Langley's RBC, recursively partitioning the training set of labeled sequences at each node in the tree until a stopping criterion is satisfied. The branches out of each node correspond to the most likely class assigned by the NB classifier at that node. As for the stopping criterion, RNBL-MN uses a conditional minimum description length (CMDL) score for the classifier [5], specifically adapted to the case of RNBL-MN based on the CMDL score for the NB classifier using the multinomial event model for sequences [6]. Previous reports by Langley [3] in the case of a recursive NB classifier (RBC) for data sets whose the instances are represented as tuples of nominal attribute values (such as the UC-Irvine benchmark data), suggested that the tree of NB classifiers offered little improvement in accuracy over the standard NB classifier. In our experiments on protein sequence and text classification tasks, we observe that RNBL-MN substantially outperforms NB classifier. Furthermore, our experiments show that RNBL-MN outperforms C4.5 decision tree learner (using tests on sequence composition statistics as the splitting criterion) and yields accuracies that are comparable to those of SVM using similar information.

The rest of the paper is organized as follows: Section 2 briefly introduces the multinomial event model for sequences; Section 3 presents RNBL-MN (recursive Naive Bayes learner based on the multinomial event model for sequences); Section 4 presents our experimental results; Section 5 concludes with summary and discussion.

2 Multinomial Event Model for Naive Bayes Sequence Classification

Consider sequences defined over a finite alphabet $\Sigma = \{w_1 \cdots w_d\}$ where $d = |\Sigma|$. For example, in the case of protein sequences, Σ can be the 20-letter amino acid alphabet ($\Sigma = \{A_1, A_2, \ldots, A_{20}\}$). In the case of text, Σ corresponds to the finite vocabulary of words. Typically, a sequence $S_j \in \Sigma^\star$ is mapped into a finite dimensional feature space D through a mapping $\Phi : \Sigma^\star \to D$.

In a multinomial event model, a sequence S_j is represented by a *bag* of elements from Σ. That is, S_j is represented by a vector D_j of frequencies of occurrences in S_j of each element of Σ. Thus, $D_j = <f_{1j}, f_{2j}, \ldots, f_{dj}, c_j>$, where $f_{ij} \in \mathbb{Z}^\star$ denotes the number of occurrences of w_i (the ith element of the alphabet Σ) in the sequence S_j. Thus, we can model the sequence S_j as a sequence of random draws from a multinomial distribution over the alphabet Σ. If we denote the probability of picking an element w_i given the class c_j by $P(w_i|c_j)$, the probability of sequence S_j given its class c_j under the multinomial event model is defined as follows:

$$P(X_1 = f_{1j}, \ldots, X_d = f_{dj}|c_j) = \left\{ \frac{(\sum_i^d f_{ij})!}{\prod_i^d (f_{ij})!} \right\} \prod_{i=1}^{d} P(w_i|c_j)^{f_{ij}}$$

(Note: To be fully correct, we would need to multiply the right hand side of the above equation by $P(N|c_j)$, the probability of drawing a sequence of a specific length $N = (\sum_i^d f_{ij})$ given the class c_j, but this is hard to do in practice.)

Given a training set of sequences, it is straightforward to estimate the probabilities $P(w_i|c_j)$ using the Laplace estimator as $\hat{P}(w_i|c_j) = p_{ij} = \frac{Count_{ij}+1}{Count_j+d}$, where $Count_{ij}$ is the number of occurrences of w_i in sequences belonging to class c_j and $Count_j$ is the total number of words in training set sequences belonging to class c_j.

3 Recursive Naive Bayes Learner Based on the Multinomial Event Model for Sequences (RNBL-MN)

3.1 RNBL-MN Algorithm

As noted above, RNBL-MN, analogous to the decision tree learner, recursively partitions the training data set using Naive Bayes classifiers at each node of the tree. The root of the tree is a Naive Bayes classifier constructed from the entire data set. The outgoing branches correspond to the different class labels, assigned by the Naive Bayes classifier.

For a given input training data set $D_0 (= D_{current})$, we create a Naive Bayes classifier n_0. We compute the CMDL score $Score_{current}$ for the classifier n_0 (See section 3.2 for details of the calculation of CMDL score for recursive Naive Bayes classifier based on the multinomial event model). The classifier n_0 partitions the data set D_0 into $|C|$ subsets based on the class labels assigned to the sequences by

RNBL-MN($D_{current}$) :
begin

1. **Input** : data set $D_0 = D_{current}$ // data set
2. Estimate probabilities given D_0 that specify the Naive Bayes classifier n_0
3. Add n_0 to the current classifier $h_{current}$ if $n_0 \notin h_{current}$
4. $Score_{current} \leftarrow CMDL(h_{current}|D_0)$ // CMDL score of the current classifier
5. Partition $D_{current}$ into $\mathbb{D} = \{D_1, D_2, \ldots, D_{|C|} | \forall_{S \in D_i} \forall_{j \neq i}, P(c_i|S) > P(c_j|S)\}$
6. For each $D_i \in \mathbb{D}$, estimate probabilities given D_i that specify the corresponding Naive Bayes classifiers n_i
7. $h_{potential} \leftarrow$ refinement of $h_{current}$ with the classifiers corresponding to each n_i based on the corresponding D_i in the previous step // see Fig. 2 for details
8. $Score_{potential} \leftarrow CMDL(h_{potential}|\sum_{i=0}^{|C|} D_i)$ // CMDL score resulting from the refined classifier
9. If $Score_{potential} > Score_{current}$ then // accept the refinement
10.   Add each n_i to $h_{current}$
11.   For each child node n_i
12.   **RNBL-MN**(D_i) // recursion
13.   End For
14. End If
15. **Output** : $h_{current}$

end.

Fig. 1. Recursive Naive Bayes Learner of Multinomial Event Model

the classifier n_0. Each such subset is in turn used to train additional Naive Bayes classifiers. At each step, the CMDL score for the resulting tree of Naive Bayes classifiers is computed and compared with the CMDL score of the classifier from the previous step. This recursive process terminates when additional refinements of the classifier yield no significant improvement in CMDL score. Fig. 1 shows the pseudo-code of RNBL-MN algorithm.

Analogous to a decision tree, the resulting classifier predicts a class label for a new sequence as follows: starting at the root of the tree, the sequence is routed along the outgoing branches of successive Naive Bayes classifiers, at each node following the branch corresponding to the most likely class label for the sequence, until a leaf node is reached. The sequence is assigned the label corresponding to the leaf node.

3.2 Conditional Minimum Description Length (CMDL) Score for Naive Bayes Classifier Based on the Multinomial Event Model

RNBL-MN employs the conditional minimum description length (CMDL) score [5], specifically adapted to the case of RNBL-MN, based on the CMDL score for NB classifier using the multinomial event model for sequences [6] as the stopping criterion.

Recall the definition of a conditional minimum description length (CMDL) score of a classifier h given a data set D [5]:

$$CMDL(h|D) = CLL(h|D) - \left\{\frac{\log|D|}{2}\right\} size(h),$$

where $size(h)$ is the size of the hypothesis h (the complexity of the model), which corresponds to the number of entries in the conditional probability tables (CPTs) of h. $CLL(h|D)$ is the conditional log likelihood of the hypothesis h given the data D, where each instance of the data has a class label $c \in C$.

When h is a Naive Bayes classifier based on a multinomial event model, the conditional log likelihood of the classifier h given data D can be estimated as follows [6]:

$$CLL(h|D) = |D| \sum_j^{|D|} \log \left\{ \frac{P(c_j) \left\{ \frac{(\sum_i^d f_{ij})!}{\prod_i^d (f_{ij})!} \right\} \prod_i^d \{p_{i,j}^{f_{ij}}\}}{\sum_k^{|C|} \left\{ P(c_k) \left\{ \frac{(\sum_i^d f_{ik})!}{\prod_i^d (f_{ik})!} \right\} \prod_i^d \{p_{i,k}^{f_{ik}}\} \right\}} \right\},$$

where $d = |\Sigma|$ is the cardinality of the vocabulary Σ, $|D|$ is the number of sequences in the data set D, $c_j \in C$ is the class label associated with the instance $S_j \in D$, f_{ij} is the integer frequency of element $w_i \in \Sigma$ in instance S_j, and $p_{i,j}$ is the estimated probability of the element w_i occurring in an instance belonging to class c_j.

The $size(h)$ for the multinomial event model is given by $size(h) = |C| + |C|d$, where $|C|$ is the number of class labels, and d is the cardinality of the vocabulary Σ.

3.3 CMDL for a Recursive Naive Bayes Classifier

We observe that in the case of a recursive Naive Bayes classifier, $CLL(h|D)$ can be decomposed in terms of the CLL scores of the individual Naive Bayes classifiers at the leaves of the tree of classifiers. Consequently, the CMDL score for the composite tree-structured classifier can be written as follows:

$$CMDL(h|D) = \sum_{node \in Leaves(h)} CLL(h_{node}|D_{node}) - \left\{\frac{\log|D|}{2}\right\} size(h),$$

where $size(h) = (|C| + |C|d)|h|$, denoting $|h|$ the number of nodes in h.

For example, Fig. 2 shows a Recursive Naive Bayes classifier consisting of 5 individual Naive Bayes classifiers. $\hat{c}_+$ and $\hat{c}_-$ are the predicted outputs of each hypothesis.

In the figure,

$$CLL(h_{current}|D) = CLL(n_{00}|D_{00}) + CLL(n_{01}|D_{01})$$

and

$$CLL(h_{potential}|D) = CLL(n_{000}|D_{000}) + CLL(n_{001}|D_{001}) + CLL(n_{01}|D_{01}),$$

where $|C|=2$, $|h_{current}| = 3$, and $|h_{potential}| = 5$.

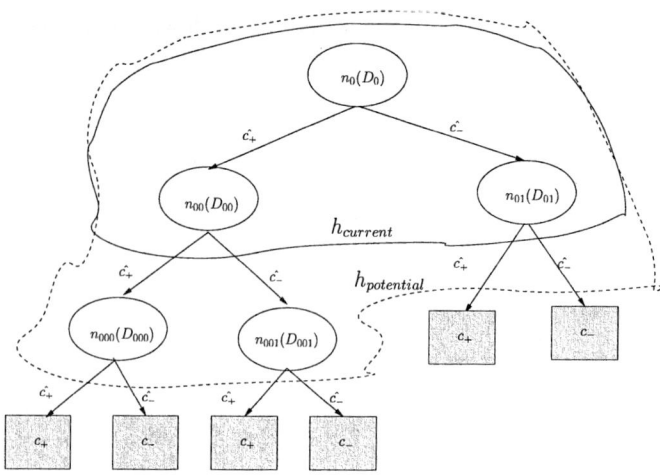

Fig. 2. Recursion tree of classifiers. Note that $h_{potential}$ is the refinement of $h_{current}$ by adding nodes $n_{000}(D_{000})$ and $n_{001}(D_{001})$ as children of $n_{00}(D_{00})$.

Using the CMDL score, we can choose the hypothesis h that effectively trades off the complexity, measured by the number of parameters, against the accuracy of classification. As is described in Fig. 1, the algorithm terminates when none of the refinements of the classifier (splits of the tree nodes) yields statistically significant improvement in the overall CMDL score.

4 Experiments

To evaluate RNBL-MN, recursive Naive Bayes learner of multinomial event model, we conducted experiments using two classification tasks: (a) assigning Reuters newswire articles to categories, (b) and classifying protein sequences in terms of their cellular localization. The results of the experiments described in this section show that the classifiers generated by RNBL-MN are typically more accurate than Naive Bayes classifiers using the multinomial model, and that RNBL-MN yields more accurate classifiers than C4.5 decision tree learner (using tests on sequence composition statistics as the splitting criterion). RNBL-MN yields accuracies that are comparable to those of linear kernel based SVM trained with the SMO algorithm [7] on a bag of letters (words) representation of sequences (text).

4.1 Reuters 21587 Text Categorization Test Collection

Reuters 21587 distribution 1.0 data set[1] consists of 12902 newswire articles in 135 overlapping topic categories. We followed the ModApte split [8] in which

---

[1] This collection is publicly available at
http://www.daviddlewis.com/resources/testcollections/reuters21578/.

9603 stories are used to train the classifier and 3299 stories to test the accuracy of the resulting classifier. We eliminated the stories that do not have any topic associated with them (i.e., no class label). As a result, 7775 stories were used for training and 3019 stories for testing the classifier.

Because each story has multiple topics (class labels), we built binary classifiers for the top ten most populous categories following the setup used in previous studies by other authors [9, 1]. In our experiments, stop words were not eliminated, and title words were not distinguished from body words. Following the widely used procedure for text classification tasks with large vocabularies, we selected top 300 features based on mutual information with class labels.

For evaluation of the classifiers, following the standard practice in text classification literature, we report the break-even points, which is the average of precision and recall when the difference between the two is minimum.

Table 1 shows the break-even points of precision and recall as a performance measure for the ten most frequent categories. The results in the table show that, RNBL-MN outperforms the other algorithms, except SVM, in terms of classification accuracy for Reuters 21587 text data set.

Table 1. Break-even point of precision and recall (a standard accuracy measure for ModApte split of Reuters 21587 data set) on the 10 largest categories of Reuters 21587 data set

Data			NBL-MN	RNBL-MN	C4.5	SVM
name	# train (+/−)	# test (+/−)	accuracy	accuracy	accuracy	accuracy
earn	2877 / 4898	1087 / 1932	94.94	96.50	95.58	**97.24**
acq	1650 / 6125	719 / 2300	89.43	**93.32**	89.29	92.91
money-fx	538 / 7237	179 / 2840	64.80	69.83	69.27	**72.07**
grain	433 / 7342	149 / 2870	74.50	**89.26**	85.23	**89.26**
crude	389 / 7386	189 / 2830	79.89	77.78	76.19	**86.77**
trade	369 / 7406	117 / 2902	59.83	70.09	61.54	**71.79**
interest	347 / 7428	131 / 2888	61.07	70.99	64.89	**73.28**
ship	197 / 7578	89 / 2930	**82.02**	**82.02**	65.17	80.90
wheat	212 / 7563	71 / 2948	57.75	73.24	**87.32**	80.28
corn	181 / 7594	56 / 2963	57.14	67.85	**92.86**	76.79

4.2 Protein Subcellular Localization Prediction

We applied RNBL-MN to two protein sequence data sets, where the goal is to predict the subcellular localization of the proteins [10, 2].

The first data set consists of 997 prokaryotic protein sequences derived from SWISS-PROT database (release 33.0) [11]. This data set includes proteins from three different subcellular locations: cytoplasmic (688 proteins), periplasmic (202 proteins), and extracellular (107 proteins).

Table 2. Localization prediction results on Prokaryotic and Eukaryotic protein sequences, calculated by 10-fold cross validation with 95% confidence interval

(a) Prokaryotic protein sequences

Algorithm	Measure	Cytoplasmic	Extracellular	Peripalsmic
NBL-MN	accuracy	88.26±2.00	93.58±1.52	81.85±2.39
	specificity	89.60±1.89	65.93±2.94	53.85±3.09
	sensitivity	93.90±1.49	**83.18±2.32**	**72.77±2.76**
RNBL-MN	accuracy	**90.67±1.81**	**94.58±1.41**	**87.76±2.03**
	specificity	**91.61±1.72**	75.73±2.66	**73.53±2.74**
	sensitivity	95.20±1.33	72.90±2.76	61.88±3.01
C4.5	accuracy	84.15±2.27	91.98±1.69	84.65±2.24
	specificity	88.58±1.97	63.37±2.99	64.00±2.98
	sensitivity	88.32±1.99	59.81±3.04	55.45±3.09
SVM	accuracy	87.26±2.07	93.78±1.50	79.74±2.49
	specificity	84.67±2.24	**89.47±1.91**	50.00±3.10
	sensitivity	**99.56±0.41**	47.66±3.1	0.50±0.44

(b) Eukaryotic protein sequences

Algorithm	Measure	Cytoplasmic	Extracellular	Mitochondrial	Nuclear
NBL-MN	accuracy	71.41±1.80	83.11±1.49	71.69±1.79	80.72±1.57
	specificity	49.55±1.99	40.23±1.95	25.86±1.74	82.06±1.53
	sensitivity	**81.29±1.55**	53.85±1.98	**61.06±1.94**	73.38±1.76
RNBL-MN	accuracy	78.12±1.64	**92.13±1.07**	**87.72±1.31**	**83.48±1.48**
	specificity	60.24±1.95	75.97±1.70	**54.44±1.98**	84.30±1.45
	sensitivity	65.79±1.89	**60.31±1.95**	43.93±1.97	**78.09±1.65**
C4.5	accuracy	**78.99±1.62**	91.18±1.13	86.57±1.36	79.85±1.60
	specificity	63.51±1.92	69.89±1.83	49.03±1.99	77.94±1.65
	sensitivity	59.80±1.95	60.00±1.95	39.25±1.94	77.30±1.67
SVM	accuracy	71.98±1.79	86.69±1.35	86.77±1.35	79.36±1.61
	specificity	**83.33±1.48**	**100.00±0.00**	N/A	**87.53±1.31**
	sensitivity	0.73±0.34	0.62±0.31	0.00±0.00	63.35±1.92

The second data set contains 2427 eukaryotic protein sequences derived from SWISS-PROT database (release 33.0) [11]. This data set includes proteins from the following four different subcellular locations: nuclear (1097 proteins), cytoplasmic (684 proteins), mitochondrial (321 proteins), extracellular (325 proteins).

The accuracy, sensitivity, and specificity of the classifiers (estimated using 10-fold cross-validation) on the two data sets [2] are shown in Table 2. The results show that RNBL-MN generally outperforms C4.5, and compares favorably with SVM. Specificity of SVM for 'Mitochondrial' is "N/A", because the SVM classifier always outputs negative when most of the instances in the data set have negative class label (imbalanced), which leads its specificity to be undefined.

---

[2] These two datasets are available to download at
http://www.doe-mbi.ucla.edu/~astrid/astrid.html.

5 Related Work and Conclusion

5.1 Related Work

As noted earlier, Langley [3] investigated recursive Bayesian classifiers for the instances described by tuples of nominal attribute values. RNBL-MN reported in this paper works with a multinomial event model for sequence classification.

Kohavi [12] introduced NBTree algorithm, a hybrid of a decision tree and Naive Bayes classifiers for instances represented using tuples of nominal attributes. NBTree evaluates the attributes available at each node to decide whether to continue building a decision tree or to terminate with a Naive Bayes classifier. In contrast, RNBL-MN algorithm, like Langley's RBC, builds a decision tree, whose nodes are all Naive Bayes Classifiers.

Gama and Brazdil [13] proposed an algorithm that generates a cascade of classifiers. Their algorithm combines Naive Bayes, C4.5 decision tree and linear discriminants, and introduces a new attribute at each stage of the cascade. They performed experiments on several UCI data sets [14] for classifying instances represented as tuples of nominal attribute values. In contrast, RNBL-MN recursively applies the Naive Bayes classifier based on the multinomial event model for sequences.

5.2 Summary and Conclusion

RNBL-MN algorithm described in this paper relaxes the *single generative model per class* assumption of NB classifiers, while maintaining some of their computational advantages. RNBL-MN constructs a tree of Naive Bayes classifiers for sequence classification. It works in a manner similar to Langley's RBC [3], recursively partitioning the training set of labeled sequences at each node in the tree until a stopping criterion is satisfied. RNBL-MN employs the conditional minimum description length (CMDL) score for the classifier [5], specifically adapted to the case of RNBL-MN classifier based on the CMDL score for the Naive Bayes classifier using the multinomial event model [6] as the stopping criterion. Previous reports by Langley [3] in the case of a recursive NB classifier (RBC) on data sets whose instances were represented by tuples of nominal attribute values (such as the UC-Irvine benchmark data) had suggested that the tree of NB classifiers offered little improvement in accuracy over the standard NB classifier. In contrast, we observe that on protein sequence and text classification tasks, RNBL-MN substantially outperforms the NB classifier. Furthermore, our experiments show that RNBL-MN outperforms C4.5 decision tree learner (using tests on sequence composition statistics as the splitting criterion) and yields accuracies that are comparable to those of SVM using similar information.

Given the relatively modest computational requirements of RNBL-MN relative to SVM, RNBL-MN is an attractive alternative to SVM in training classifiers on extremely large data sets of sequences or documents. Our results raise the possibility that Langley's RBC might outperform NB on more complex data sets in which the *one generative model per class* assumption is violated, especially if RBC is modified to use an appropriate CMDL criterion.

References

1. McCallum, A., Nigam, K.: A comparison of event models for naive bayes text classification. In: AAAI-98 Workshop on Learning for Text Categorization. (1998)
2. Andorf, C., Silvescu, A., Dobbs, D., Honavar, V.: Learning classifiers for assigning protein sequences to gene ontology functional families. In: 5th International Conference on Knowledge Based Computer Systems. (2004) 256–265
3. Langley, P.: Induction of recursive bayesian classifiers. In: Proc. of the European Conf. on Machine Learning, London, UK, Springer-Verlag (1993) 153–164
4. Quinlan, J.R.: C4.5: Programs for machine learning. Morgan Kaufmann Publishers Inc., San Francisco, CA, USA (1993)
5. Friedman, N., Geiger, D., Goldszmidt, M.: Bayesian network classifiers. Machine Learning **29** (1997) 131–163
6. Kang, D.K., Zhang, J., Silvescu, A., Honavar, V.: Multinomial event model based abstraction for sequence and text classification. In: 6th International Symposium on Abstraction, Reformulation and Approximation. (2005) 134–148
7. Platt, J.C.: Fast training of support vector machines using sequential minimal optimization. Advances in kernel methods: support vector learning (1999) 185–208
8. Apté, C., Damerau, F., Weiss, S.M.: Towards language independent automated learning of text categorization models. In: 17th annual international ACM SIGIR conference on Research and development in information retrieval. (1994) 23–30
9. Dumais, S., Platt, J., Heckerman, D., Sahami, M.: Inductive learning algorithms and representations for text categorization. In: Proceedings of the 7th international conference on Information and knowledge management, ACM Press (1998) 148–155
10. Reinhardt, A., Hubbard, T.: Using neural networks for prediction of the subcellular location of proteins. Nucleic Acids Research **26** (1998) 2230–2236
11. Bairoch, A., Apweiler, R.: The SWISS-PROT protein sequence database and its supplement TrEMBL in 2000. Nucleic Acids Research **28** (2000) 45–48
12. Kohavi, R.: Scaling up the accuracy of Naive Bayes classifiers: a decision-tree hybrid. In: Proc. of the 2nd International Conference on Knowledge Discovery and Data Mining. (1996) 202–207
13. Gama, J., Brazdil, P.: Cascade generalization. Machine Learning **41** (2000) 315–343
14. Blake, C., Merz, C.: UCI repository of machine learning databases (1998)

TRIPPER: Rule Learning Using Taxonomies

Flavian Vasile, Adrian Silvescu, Dae-Ki Kang, and Vasant Honavar

Artificial Intelligence Research Laboratory,
Department of Computer Science, Iowa State University,
Ames, IA 50011, USA
{flavian, silvescu, dkkang, honavar}@cs.iastate.edu

Abstract. In many application domains, there is a need for learning algorithms that generate accurate as well as comprehensible classifiers. In this paper, we present TRIPPER - a rule induction algorithm that extends RIPPER, a widely used rule-learning algorithm. TRIPPER exploits knowledge in the form of taxonomies over the values of features used to describe data. We compare the performance of TRIPPER with that of RIPPER on benchmark datasets from the Reuters 21578 corpus using WordNet (a human-generated taxonomy) to guide rule induction by TRIPPER. Our experiments show that the rules generated by TRIPPER are generally more comprehensible and compact and in the large majority of cases at least as accurate as those generated by RIPPER.

1 Introduction

Knowledge discovery aims at constructing predictive models from data that are both accurate and comprehensible. Use of prior knowledge in the form of taxonomies over attribute values offers an attractive approach to this problem.

Several authors have explored the use of taxonomies defined over attribute values to guide learning. Zhang and Honavar developed a Decision Tree [8] and a Naive Bayes [9] learning algorithm that exploit user-supplied feature value taxonomies. Kang et al [2] introduced WTL, Word Taxonomy Learner for automatically deriving taxonomies from data and a Word Taxonomy-guided Naive Bayes (WTNBL-MN) algorithm for document classification. Michalski [7] has proposed a general framework of attributional calculus that can be seen as an alternative way of representing rules containing abstractions. Additional references to related work can be found in [9,11]. Against this background, we present a rule induction method that exploits user-supplied knowledge in the form of attribute value taxonomies to generate rules at higher levels of abstraction, named TRIPPER (Taxonomical RIPPER). We report results of experiments that demonstrate the promise of the proposed approach on a widely used benchmark data set (the Reuters text classification data set [10]).

2 Method

RIPPER (*Repeated Incremental Pruning to Produce Error Reduction*), was proposed by Cohen [1]. It consists of two main stages: the first stage constructs an initial ruleset using a rule induction algorithm called IREP* [4]; the second stage further optimizes the ruleset initially obtained. These stages are repeated for k times. IREP*[1] is called inside RIPPER-k for k times, and at each iteration, the current dataset is randomly partitioned in two subsets: a growing set, that usually consists of 2/3 of the examples and a pruning set, consisting in the remaining 1/3. These subsets are used for two different purposes: the growing set is used for the initial rule construction (the rule growth phase) and the pruning set is used for the pruning (the rule pruning phase). IREP* uses MDL[5] as a criterion for stopping the process.

The rule growth phase: The initial form of a rule is just a head (the class value) and an empty antecedent. At each step, the best condition based on its information gain is added to the antecedent. The stopping criterion for adding conditions is either obtaining an empty set of positive instances that are not covered or not being able to improve the information gain score.

The rule pruning phase: Pruning is an attempt to prevent the rules from being too specific. Pruning is done accordingly to a scoring metric denoted by v^*.

IREP* chooses the candidate literals for pruning based on a score v^* which is applied to all the prefixes of the antecedent of the rule on the pruning data:

$$v^*(rule, prunepos, prunenef) = \frac{p - n}{p + n} \quad (1)$$

where p / n denote the total number of positive / negative instances covered by the rule. The prefix with the highest v^* score becomes the antecedent of the final rule.

Before introducing **TRIPPER**, it is helpful to formally define a taxonomy:

Taxonomy: *Let $S = \{v1, v2, ... vn\}$ be a set of feature values. Let T be a directed tree where children(i) denotes the set of nodes that have incoming arrows to the node i. A node i is called leaf if it has no children. A taxonomy Tax(T,S) is a mapping which assigns to a node i of the tree T a subset S' of S with the following properties:*

$$Tax(T, S)(i) = \bigcup_{j \in children(i)} Tax(T, S)(j) \quad (2)$$

$$Leaves(T) = S \quad (3)$$

1. TRIPPER(G) - improvement at rule growth phase: Introducing the taxonomical knowledge at the rule-growth phase is a straightforward process we call **feature space augmentation.** The augmentation process takes all the interior nodes of the attribute value taxonomy and adds them to the set of candidate literals used for the growth phase.

2. TRIPPER(G+P) - improvement at rule pruning phase: A more general version of feature selection than pruning is abstraction: in the case of abstraction, instead of casting the problem as a matter of preserving or discarding a feature, we are able to choose from a whole range of levels of specificity for the feature under consideration.

The effect on the resulting rule can be observed in the following example:
[original rule]   - *(rate = t) and (bank = t) and **(dollar = t)** => is_interest*
[pruned rule]   - *(rate = t) and (bank =t) and **(any_concept = t)** => is_interest*
[abstracted rule] - *(rate = t) and (bank = t) and **(monetary_unit= t)** => is_interest*

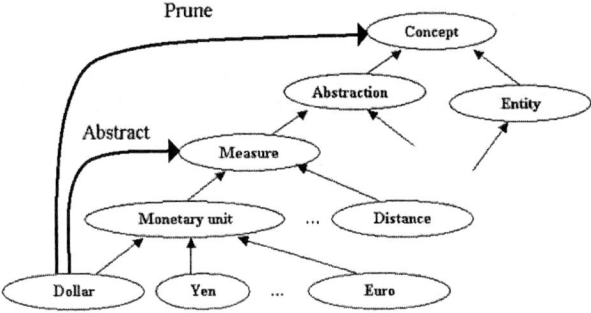

Fig. 1. Taxonomy over a set of nouns. Pruning and abstraction on a taxonomy.

Example 1: Variants of a classification rule for the class "interest"

The algorithm **Prune_by_abstraction** (fig.2.) uses exactly this idea to incrementally search for useful abstractions for the literals in the suffix to be pruned according to the v^* score of the rule prefixes.

```
Prune-by-abstraction(Rule,PruneData)
```
```
PrunedRule=PruneRule(Rule,PruneData)
Score=v*(PrunedRule,PruneData)
PrunePos=GePrunePos(PrunedRule), Level=0
While(improvement)
        Improvement=false, Increase(Level)
        For j:=PrunePos to size(Rule)
          AbstrRule=PrunedRule
          For i:=j to size(Rule)
             Literal=Rule(i)
             AbstrRule:=AbstrRule^Abstract(Literal,
             Level)
          If(v*(AbstrRule, PruneData)>Score)
             Update(Score)
             WinRule=AbstrRule, Improvement=true
Return WinRule
```

Fig. 2. Prune by Abstraction pseudocode

3 Experiments

Experimental setup: Experiments were performed on the benchmark dataset Reuters 21578 using the *ModApte* split [10] of training and testing data. Following the experimental setup used in [6], only the ten biggest classes in the dataset were used. As in [6], only the 300 best features were used as inputs to the classifier. The experiments compare RIPPER with TRIPPER (G+P) . The text-specific taxonomies used for our experiments on the Reuters dataset comes from WordNet[3], using only the hypernimy relation that stands for "isa" relation between concepts.

Results: Our experiments show that: (a) TRIPPER (G+P) outperforms, or matches RIPPER in terms of *break-even point* on the Reuters dataset (Table 3-1) in a majority (8 out of 10) of classes; (b) TRIPPER generates more abstract (and often more comprehensible) rules than RIPPER: Table 3-2 shows some of the abstract literals discovered to be important for 3 of the 10 classes. Furthermore, the rules generated by TRIPPER(G+P) are often more concise than those generated by RIPPER (results not shown) [11].

Table 3-1. Comparison of performance (break even point) of TRIPPER and RIPPER using WN

Class	Acq	Corn	Crud	Earn	Grn.	Inter	Mon	Ship	Trd.	Wht.
Trip.	**86.3**	**85.7**	**82.5**	**95.1**	87.9	**71.5**	**70.4**	**80.9**	58.9	**84.5**
Ripp.	85.3	83.9	79.3	94	**90.6**	58.7	65.3	73	**68.3**	83

Table 3-2. Abstract literals from WordNet

Class subject	Abstract literals
Crude Oil	assets, chemical_phenomenon, chemical_element, financial_gain, macromolecule, magnitude_relation, process, worker
Money, Foreign Exchange	artifact, assets, businessperson, document, institution, location, medium_of_exchange, measure, organization, signal, social_ event, solid
Trade	assembly, assets, calendar_month, change_of_magnitude, mass_unit, outgo, signal

The usefulness of abstraction is confirmed by the prevalence of abstract literals in almost all the rules of every ruleset. Both of the phases (growth and pruning) generated improvements (results not shown) [11], lending empirical support for the idea that both of the extensions are useful.

4 Conclusions

TRIPPER is a taxonomy-based extension of the popular rule-induction algorithm RIPPER [1]. The key ingredients of TRIPPER are: the use of an augmented set of features based on taxonomies defined over values of the original features (WordNet in the case of text classification) in the growth phase and the replacement of pruning, as an overfitting avoidance method, with the more general method of abstraction guided by a taxonomy over the features. The experiments briefly summarized in this paper show that TRIPPER generally outperforms RIPPER on the Reuters text classification task in terms of break-even points, while generating potentially more comprehensible rule sets than RIPPER. It is worth noting that on the Reuters dataset, TRIPPER slightly outperforms WTNBL [2] in terms of break-even points on 7 out of 10 classes.

The additional computation cost of TRIPPER is small when compared with RIPPER, consisting in an additional multiplicative factor that represents the height of the largest taxonomy, which in the average case scales logarithmically with the number of feature values.

References

1. Cohen, W. W.: Fast effective rule induction. *Proceedings of International Conference on Machine Learning,* Lake Tahoe, CA. (1995)
2. Kang, D.-K., Silvescu, A., Zhang, J., Honavar, V.: Generation of Attribute Value Taxonomies from Data for Data-Driven Construction of Accurate and Compact Classifiers, *Proceedings of the 4th IEEE International Conference on Data Mining,* Brighton, UK. (2004)
3. Fellbaum, C: WordNet, An Electronic Lexical Database. *The MIT Press.* (1998)
4. Fürnkranz, J., Widmer, G: Incremental reduced error pruning. *Proceedings of International Conference on Machine Learning.* New Brunswick, NJ. (1994)
5. Quinlan, J. R.: MDL and categorical theories. *Proceedings of International Conference on Machine Learning,* Lake Tahoe, CA. (1995)
6. McCallum, A., Nigam, K.: A comparison of event models for naive bayes text classification. In: *AAAI-98 Workshop on Learning for Text Categorization.* (1998) 3-5.
7. Michalski, R. S.: Attributional Calculus: A Logic and Representation Language for Natural Induction, *Reports of the Machine Learning and Inference Laboratory*, MLI 04-2, George Mason University, Fairfax, VA. (2004)
8. Zhang, J., Honavar, V.: Learning decision tree classifiers from attribute value taxonomies and partially specified data. *Proceedings of International Conference on Machine Learning*, Washington, DC. (2003)
9. Zhang, J., Honavar, V.: AVT-NBL 2004: An algorithm for learning compact and accurate naive bayes classifiers from feature value taxonomies and data, *Proceedings of the Fourth IEEE International Conference on Data Mining,* Brighton, UK. (2004)
10. Apte, C., Damerau, F., Weiss Sholom, .M.: Towards language independent automated learning of text categorization models. *SIGIR '94*, Springer-Verlag New York, Inc. (1994) 23-30.
11. Vasile, F, Silvescu, A, Kang, D.-K., Honavar V.: TRIPPER: Rule learning using taxonomies, Tehnical Report ISU-CS-TR, Department of Computer Science, Iowa State University, Jan.2006. (Publicly available at http://www.cs.iastate.edu/~flavian/tripper_long.pdf)

Using Weighted Nearest Neighbor to Benefit from Unlabeled Data

Kurt Driessens[1,2], Peter Reutemann[2],
Bernhard Pfahringer[2], and Claire Leschi[3]

[1] Department of Computer Science,
K.U. Leuven, Belgium
[2] Department of Computer Science, University of Waikato,
Hamilton, New Zealand
[3] Institut National des Sciences Appliquees, Lyon, France

Abstract. The development of data-mining applications such as text-classification and molecular profiling has shown the need for machine learning algorithms that can benefit from both labeled and unlabeled data, where often the unlabeled examples greatly outnumber the labeled examples. In this paper we present a two-stage classifier that improves its predictive accuracy by making use of the available unlabeled data. It uses a weighted nearest neighbor classification algorithm using the combined example-sets as a knowledge base. The examples from the unlabeled set are "pre-labeled" by an initial classifier that is build using the limited available training data. By choosing appropriate weights for this pre-labeled data, the nearest neighbor classifier consistently improves on the original classifier.

1 Introduction

The combination of supervised and unsupervised learning [1] is a growing subfield of Machine Learning. Applications such as text- or image-mining and molecular profiling have revealed application areas that yield very little (and often expensive) labeled data but often plenty of unlabeled data. As traditional machine learning algorithms are not able to use and benefit from the information available in the unlabeled data, custom built algorithms should be able to outperform them. Current research in semi-supervised learning using algorithms such as Co-Training [2] or more recent approaches based on graph representations [3] confirms that this is indeed possible.

Most of the semi-supervised learning approaches use the labeled and unlabeled data simultaneously or at least in close collaboration. Roughly speaking, the unlabeled data provides information about the structure of the domain, i.e. it helps to capture the underlying distribution of the data. The challenge for the algorithms can be viewed as realizing a kind of trade-off between robustness and information gain [1]. To make use of unlabeled data, one must make assumptions, either implicitly or explicitly. As reported in [3], the key to semi-supervised learning is the prior assumption of consistency, that allows for exploiting the geometric structure of the data distribution. Close data points should belong to the same class and decision boundaries should lie in regions of low data density; this is also called the "cluster assumption".

In this paper, we introduce a very simple two-stage approach that uses the available unlabeled data to improve on the predictions made when learning only from the labeled examples. In a first stage, it uses an off-the-shelf classifier to build a model based on the small amount of available training data, and in the second stage it uses that model to transform the available unlabeled data into a weighted "pre-labeled" data-set that together with the original data is used in a nearest neighbor classifier. We will show that the proposed algorithm improves on the classifier built in stage 1, especially in cases where much more unlabeled data is available compared to the labeled data.

The rest of the paper is structured as follows: in section 2 we describe a few related semi-supervised learning techniques. Section 3 introduces the proposed algorithm in detail. In section 4 we show experimental results using an array of different classifiers used in the first stage. Section 5 concludes and presents some directions for future work.

2 Learning from Labeled and Unlabeled Data

Early methods in semi-supervised learning were using mixture models (in which each mixture component represents exactly one class) and extensions of the EM algorithm [4]. More recent approaches belong to one of the following categories: self-training, co-training, transductive SVMs, split learning, and graph-based methods. In the self-training approach, a classifier is trained on the labeled data and then used to classify the unlabeled ones. The most confident (now labeled) unlabeled points are added to the training set, together with their predictive labels, and the process is repeated until convergence [5]. Approaches based on co-training [2] assume that the features describing the objects can be divided in two subsets such that each of them is sufficient to train a good classifier, and that the two sets are conditionally independent given the class attribute. Two classifiers are iteratively trained, each on one set, and they teach each other with a respective subset of unlabeled data and their highest confidence predictions. The transductive SVMs [6] are a "natural" extension of SVMs to the semi-supervised learning scheme. They aim at finding a labeling of the unlabeled data so that the decision boundary has a maximum margin on the original labeled data and on the (newly labeled) unlabeled data.

Graph-based methods attempt to capture the underlying structure of the data with a graph whose vertices are the available data (both labeled and unlabeled) and whose (possibly weighted) edges encode the pairwise relationships among this data. Examples of recent work in that direction include Markov random walks [7], cluster kernels [8], and regularization on graphs [3]. The learning problem on graphs can generally be viewed as an estimation problem of a classifying function f which should be close to a given function y on the labeled data and smooth on the whole graph. Different graph-based methods mainly vary by their choice of the loss function and the regularizer [9]. For example, the work on graph cuts [10] minimizes the cost of a cut in the graph for a two-class problem, while [11] minimizes the normalized cut cost and [12, 3] minimize a quadratic cost. As noticed in [9], these differences are not actually crucial. What is far more important is the construction and quality of the graph, which should reflect domain knowledge through the similarity function used to assign edges and their weights.

Collective classification [13] is an ILP approach that uses the relational structure of the combined labeled and unlabeled data-set to enhance classification accuracy. With relational approaches, the predicted label of an example will often be influenced by the labels of related examples. The idea behind collective classification is that the predicted labels of a test-example should also be influenced by the predictions made for related test-examples. The algorithm presented in this paper is closely related to this, but works on non-relational data by using a distance and the nearest neighbor relation that results from it.

Also related to our approach, although originally not used in a transductive setting, is the work by [14]. Also using two stages, in the first stage an ensemble of neural networks is trained on the available data and the resulting model is used to generate random, extra training examples for a decision tree algorithm in the second stage. This approach could be easily adapted to the transductive setting by using the test set instead of randomly generated examples.

3 YATSI

The YATSI algorithm[1] that we present in this paper will incorporate ideas from different algorithms that were discussed in the previous section. Since we really like the idea of giving the user the option to choose from a number of machine learning algorithms (like it is possible in co-training), we will develop a technique that builds on top of any standard machine learning algorithm. To incorporate the general idea behind collective classification, we use a nearest neighbor approach and the distance between as a way of relating them to each other.

The YATSI classifier (See Algorithm 1 for high-level pseudo-code) uses both labeled and unlabeled data in a two-stage set-up[2]. In the first stage a standard, off-the-self, classifier (or regression-algorithm) is trained on the available training data. Since this kind of data is limited in the specific application areas we are looking at, it is best to choose an algorithm that can learn a model well using only a small amount of learning data.

In the second stage, the model generated from the learning data is used to "pre-label" all the examples in the test set. These pre-labeled examples are then

---

[1] YATSI was developed during a time when we were experimenting with a number of multi-stage classifiers. At the time, we referred to the presented algorithm as: "Yet Another Two-Stage Idea", hence the name YATSI.
[2] We will use the terms *labeled, unlabeled* and *pre-labeled* examples for the examples in the training set, the test set and the test set after it has been temporarily labeled in stage 1, respectively.

Algorithm 1 High level pseudo code for the two-stage YATSI algorithm.

Input: a set of labeled data D_l and a set of unlabeled data D_u, an off-the-shelf classifier C and a nearest neighbor number K; let $N = |D_l|$ and $M = |D_u|$
Step 1:
   Train the classifier C using D_l to produce the model M_l
   Use the model M_l to "pre-label" all the examples from D_u
   Assign weights of 1.0 to every example in D_l
         and of $F \times \frac{N}{M}$ to all the examples in D_u
   Merge the two sets D_l and D_u into D
Step 2:
   For every example that needs a prediction:
     Find the K-nearest neighbors to the example from D to produce set NN
     For each class:
        Sum the weights of the examples from NN that belong to that class
     Predict the class with the largest sum of weights.

---

used together with the original training data in a weighted nearest neighbor algorithm. The weights used by the nearest neighbor classifier are meant to limit the amount of trust the algorithm puts into the labels generated by the model from the first step. As a default value, we set the weights of the training data to 1.0 and the weights of the pre-labeled test-data to N/M with N the number of training examples and M the number of test-examples. Conceptually, this gives equal weights to the whole train- and the whole test-set. By adding a parameter F to the algorithm that will cause the weight of the test-examples to be set to $F * (N/M)$, it becomes possible to vary the influence one wants to give to the unlabeled data and the classifier built in step 1. Values of F between 0.0 and 1.0 will lower the influence on the test-data and the learned model from the first step, values larger than 1.0 will increase their influence. In the experiments, we will test values ranging from 0.01 to 10. An F-value of 10.0 will adjust the weights of the individual examples such as to give the total test-set 10 times the weight of the total training set.

3.1 Weighted Nearest Neighbor

In the previous section we stated the way we add a label and a weight to every example in the dataset that will be used for nearest neighbor classification. There are different ways in which to use weights for nearest neighbor classification. One way is to make the distance dependent on the weight of the examples. An obvious way would be to divide the standard distance by the weight of the example [15]. This would make it harder for examples with a small weight to influence the prediction. However, when using k-nearest-neighbor prediction, this approach will change the identity of the k selected examples and in a set-up like the one provided by YATSI , where only 2 different weights are available, it could prevent the examples with the lower weight to ever be part of the k closest examples.

    Another way of incorporating weights in nearest neighbor predictions is that once the k nearest neighbors are selected, we choose to use the weights of the examples as a measure for their influence on the total vote. Instead of counting the number of neighbors that belong to each class, we sum their weight and predict the class with the largest weight. By normalizing the sums of the weights, so that they all add up to 1, we get an indication of the probability for each of the available classes. Note though, that the distance to an example does not influence its contribution in the vote. Once an example makes it into the set of the k closest examples, its contribution is only influenced by its weight.

For continuous class-values, where predictions are made using the sum

$$\frac{\sum_j \frac{t_j}{dist_{ij}}}{\sum_j \frac{1}{dist_{ij}}}$$

over all examples in the dataset with t_j being the target value of example j and $dist_{ij}$ being the distance between examples i and j, both ways of incorporating the weights of examples are equivalent. As such, although we have not yet implemented this and do not have any experimental results, YATSI can be used for predicting continuous target values as well without major changes.

3.2 Other Nearest Neighbor Issues

For our experiments, we fixed the number of nearest neighbor to 10. This is not a requirement for the YATSI algorithm. Cross-validation on the labeled training examples could be used to adapt the number of nearest neighbors. However, the resulting values of k might be misleading because of the large amount of extra examples that will be available in the second step of the YATSI algorithm.

Since the algorithm is designed to work in applications where the amount of labeled training data is limited, one can get away with less efficient algorithms in the first step. As we expect the amount of test data to greatly exceed that of the training data, most of the computational complexity will lie in the search for nearest neighbors, as this search spans the combined sets of examples.

YATSI will therefore greatly benefit from using efficient nearest neighbor search algorithms. Currently, we use KD-trees [16] to speed up the nearest neighbor search. However, recently a lot of research effort has gone into the development of more efficient search strategies for nearest neighbors, which can be directly applied to the YATSI algorithm. Examples of such search strategies are cover trees [17] and ball trees [18].

4 Experimental Results

We evaluated YATSI using a number of datasets from the UCI-repository. We created labeled and unlabeled sets by splitting the available data into randomly chosen subsets. We ran experiments with 1%, 5%, 10% and 20% of the available data labeled (the training set) and the rest available as the test-set. In general, we collected results from 29 different data set, except for the 1%-99% case split, where the 8 smallest data-set were removed because a 1% sub-set was not large enough to train a classifier on.

The design of YATSI does not specify any specific algorithm to be used in the first step. We ran experiments with an array of algorithms that are all available in WEKA consisting of:

AdaBoostM1: This is a straightforward implementation of the AdaBoostM1 algorithm. In the experiments reported we used J48 both with default parameter settings and without pruning as a base learner, and performed 10 iterations.

J48: This is Weka's reimplementation of the original C4.5 algorithm. Default parameter settings were used except for the confidence, which was set to the values 0.25, 0.50 and 0.75. We also ran experiments without pruning the trees.

Logistic: A straightforward implementation of logistic regression run with Weka's default parameter settings.

RandomForest: An implementation of Breiman's RandomForest algorithm, but based on randomized REPTrees (instead of CART). At each split the best of $log(n_{attrs})$ randomly chosen attributes is selected. The ensemble size was set to 10 and 100.

SMO: Weka's implementation of the SMO algorithm for training support vector machines. Linear, quadratic and cubic kernels and a cost value of 1.0 were used.

IB1: A standard nearest-neighbor algorithm using Euclidean distance with all attributes being normalized into a $[0, 1]$ range.

We also collected results for different values of the weighting parameter F ranging from 0.1, i.e., giving 10 times as much weight to the training set as to the test-set, to 10.0 which does the exact opposite. We also ran some experiments that used no weights at all. These values used for the weighting parameter are a bit extreme but will give a good illustration of the behavior of the YATSI algorithm. These experiments treat all the "pre-labeled" test-set examples exactly like training examples. Therefore, in the 1%-99% split case, the total weight of the test-set would be almost 100 times as big as that of the training-set.

We expect the performance of YATSI to go down with the performance of the classifier trained on the labeled data in stage 1 as the amount of available training data decreases, but we expect (and will show) that the performance degrades slower, i.e., that YATSI is able to improve on the results obtained by only learning from the labeled data. To get statistically sound results, we repeated every experiment 20 times.

Table 1 shows the number of statistically significant wins, draws and losses of YATSI versus the classifier trained on the training-set in stage 1. For J48, we show the results for the experiment with the confidence set to 0.75. This is higher than normal so this setting generates slightly larger trees, which seems to be appropriate for the very small training sets that we use. Higher levels of pruning could even lead to empty trees in extreme cases. Overall, all the J48 experiments showed the same trend. The results shown for the RandomForest experiments are those with an ensemble size of 100. The ones with ensemble size 10 were similar with a slightly bigger advantage for YATSI . On the SMO

Table 1. Number of statistically significant wins, draws and losses (in that order) in predictive accuracy of YATSI vs. the classifier trained in stage 1, for different values of the weighting parameter.(Tested with a paired t-test, confidence level 0.05, two tailed).

Base Classifier	% labeled data	$F = 0.1$	$F = 1.0$	$F = 10.0$	No Weights
J48	1%	14/7/0	14/7/0	13/8/0	6/15/0
	5%	15/13/1	16/12/1	15/9/5	14/9/6
	10%	16/8/5	16/7/6	15/7/7	16/7/6
	20%	18/4/7	18/4/7	13/6/10	15/6/8
RandomForest	1%	10/10/1	10/10/1	9/11/1	7/12/2
	5%	9/11/9	9/11/9	10/10/9	10/10/9
	10%	6/10/13	10/7/12	9/6/14	10/6/13
	20%	5/9/15	9/8/12	7/5/17	10/13/16
Logistic	1%	13/7/1	13/7/1	13/7/1	11/8/2
	5%	17/9/3	15/11/3	15/12/2	15/11/3
	10%	17/8/4	18/7/4	12/13/4	14/11/4
	20%	13/8/8	15/9/5	12/6/11	14/7/8
SMO	1%	11/8/2	11/8/2	11/8/2	10/9/2
	5%	8/19/2	7/20/2	9/15/5	9/12/8
	10%	5/17/7	8/17/4	9/12/8	10/11/8
	20%	6/14/9	9/12/8	8/5/16	7/11/11
AdaBoost (J48)	1%	13/8/0	13/8/0	13/8/0	6/15/0
	5%	15/13/1	15/13/1	13/12/4	12/13/4
	10%	12/10/7	14/7/8	15/7/7	12/10/7
	20%	11/10/8	13/8/8	12/7/10	12/8/9
IB1	1%	6/12/3	6/12/3	7/11/3	7/11/3
	5%	12/12/5	12/12/5	12/9/8	13/9/7
	10%	13/13/3	14/11/4	11/7/11	15/4/10
	20%	12/10/7	13/9/7	12/6/11	13/7/9

experiments, we show the results for the linear kernel experiments. For quadratic and cubic kernels, YATSI produces less of an advantage, mostly due to the fact that the SMO predictions get better and YATSI is not able to improve on them, but performs equal to the SMO algorithm more often. For AdaBoost, the shown results are obtained with the standard settings for J48; a range of different parameter values for AdaBoost produced almost identical results.

Overall, the results show that YATSI often improves on the results of the base classifier. Especially when very little of the data is labeled, YATSI gains a lot from having the unlabeled data available. When the percentage of labeled data increases, YATSI loses some of its advantage, but for the most part performs comparable if not better than the base classifier. The exception seems to be when one uses Random Forests. The weighted nearest neighbor approach of YATSI loses some of the accuracy obtained by voting over the ensemble of trees.

To give more of an indication of the actual improvements reached by YATSI in terms of predictive accuracy, Table 2 shows the actual predictive accuracies from the experiments with 5%-95% splits when one uses J48 as the classifier in stage 1. To gain additional insights into the results, we compared error rates for

Table 2. Predictive accuracies of J48 and YATSI using J48 as the stage 1 classifier averaged over 20 runs of the experiments. The data-sets were split into training- and test-set with a 5%-95% ratio. Significant improvements or degradations were tested with a two-tailed 5% confidence interval.

Dataset	J48	$F = 0.1$	$F = 1.0$	$F = 10.0$	No Weights
iris	75.73	87.18 o	87.15 o	84.52 o	83.79 o
ionosphere	76.63	74.60	74.60	72.21 •	72.23 •
lymphography	62.73	63.37	63.41	60.99	60.77 •
labor	60.27	65.88 o	66.25 o	60.27	60.27
hungarian-14-heart-disease	76.73	75.74	75.74	77.00	76.69
cleveland-14-heart-disease	68.72	73.83 o	73.79 o	73.37 o	72.56 o
hepatitis	73.77	78.16	78.03	77.93	77.89
heart-statlog	68.50	71.28 o	71.34 o	70.91 o	70.66 o
vote	93.80	91.60 •	91.58 •	91.86 •	91.83 •
vehicle	52.41	55.07 o	55.08 o	53.94 o	53.24 o
zoo	57.27	72.31 o	72.37 o	59.84 o	59.79 o
vowel	33.69	33.55	33.55	29.65 •	28.65 •
sonar	60.83	62.43	62.48	60.76	60.33
primary-tumor	19.72	24.55 o	23.75 o	20.31	20.06
soybean	47.49	65.56 o	65.59 o	53.25 o	52.77 o
balance-scale	68.74	74.08 o	74.07 o	69.72 o	69.53 o
autos	38.51	40.26	40.31	38.72	38.46
wisconsin-breast-cancer	90.52	94.65 o	94.62 o	94.56 o	94.43 o
breast-cancer	64.69	66.52	67.11 o	67.69 o	67.69 o
anneal.ORIG	76.38	77.22	76.88	74.80	74.67
anneal	87.69	87.70	87.70	86.81 •	86.81 •
audiology	43.63	43.50	43.69	40.50 •	40.36 •
pima-diabetes	66.86	68.18 o	68.53 o	68.60 o	68.14 o
german-credit	65.18	67.55 o	67.53 o	68.27 o	68.46 o
Glass	42.94	48.74 o	48.74 o	44.74 o	44.32
ecoli	65.49	73.31 o	73.34 o	71.70 o	71.65 o
horse-colic.ORIG	62.79	63.19	63.26	64.08	64.20
horse-colic	75.57	76.12	76.22	78.13 o	78.30 o
credit-rating	79.69	81.53 o	81.51 o	82.72 o	82.52 o

o, • statistically significant improvement or degradation

J48 and YATSI(J48) using different values for the weighting parameter F and with the percentage of labeled examples varying between 1% and 20%[3]. General trends are obvious, like the fact that more labels usually lead to globally better results, or that with a very small number of labels J48 usually performs worse than YATSI but that J48 can outperform YATSI when given more labeled data. With regard to the weighting parameter F we see that values of 0.1 and 1.0 consistently perform better than a value of 10 or without using weights, which indicates the advantage of taking a cautious approach that puts more trust into the originally supplied labels over the labels generated by the first stage classifier.

---

[3] For plots we refer to an extended version of this paper available online: http://www.cs.kuleuven.be/~kurtd/papers/2005_pakdd_driessens_extented.pdf

As already stated, all previous experiments were run with the number of nearest neighbors for the second stage fixed to 10. Because of the use of weights and the large difference in weights between training and test examples, we thought it might make sense to use a larger number of nearest neighbors, so we also performed experiments with 20 and 50 nearest neighbors in the 1% labeled training data case. Overall, these experiments showed very little difference with the 10 nearest neighbor ones. When there was a difference, there was a little improvement for low values of F (0.1 or 1.0) and a small loss for the cases where a high weight was given to the test-examples ($F = 10.0$ or no weights used at all).

5 Conclusions and Further Work

We have presented a simple two-stage idea that benefits from the availability of unlabeled data to improve on predictive accuracies of standard classifiers. YATSI uses an off-the-shelf classification or regression algorithm in a first step and uses weighted nearest neighbor on the combined set of training data and "pre-labeled" test data for actual predictions. Experimental results obtained from both a large array of different classifiers used in the first step, different amounts of available unlabeled data and a relatively large selection of data-sets show that YATSI will usually improve on or match the predictive performance of the base classifier used generated in the first stage. These improvements are largest in cases where there is a lot more unlabeled data available than there is labeled data.

The YATSI algorithm in its current form is quite simple and therefore a number of further improvements are possible. Some ideas have already been presented in section 3 such as the inclusion of a more efficient nearest neighbor search algorithm or the use of cross validation to determine the best number of nearest neighbors to use. Also, the current weighting scheme does not allow the user to stress the relative importance of different classes. Appropriate weighting schemes for cost-sensitive settings could be easily integrated into the YATSI algorithm. More elaborate extensions could include some sort of EM-algorithm that tries to match the "pre-labels" of test-examples with the eventually predicted values. Distance functions different to simple Euclidean distance could encode specialized domain knowledge and thus help improving classification performance. These directions would relate YATSI more closely to both graph-based and kernel-based methods of semi-supervised learning.

References

1. Seeger, M.: Learning with labeled and unlabeled data. Technical report, Edinburgh University (2001)
2. Blum, A., Mitchell, T.: Combining labeled and unlabeled data with co-training. In: COLT: Proceedings of the Workshop on Computational Learning Theory, Morgan Kaufmann (1998) 92–100
3. Zhou, D., Bousquet, O., Lal, T., Weston, J., Schölkopf, B.: Learning with local and global consistency. In: Proceedings of the Annual Conf. on Neural Information Processing Systems, NIPS. (2004)

4. Nigam, K., McCallum, A., Thrun, S., Mitchell, T.: Text classification from labeled and unlabeled documents using em. Machine Learning **39** (2000) 103–134
5. Rosenberg, C., Hebert, M., Schneiderman, H.: Semi-supervised self-training of object detection models. In: 7th IEEE Workshop on Applications of Computer Vision / IEEE Workshop on Motion and Video Computing, 5-7 January 2005, Breckenridge, CO, USA, IEEE Computer Society (2005) 29–36
6. Joachims, T.: Transductive inference for text classification using support vector machines. In Bratko, I., Džeroski, S., eds.: Proceedings of ICML99, 16th International Conference on Machine Learning, Morgan Kaufmann (1999) 200–209
7. Szummer, M., Jaakkola, T.: Partially labeled classification with markov random walks. In Dietterich, T., Becker, S., Ghahramani, Z., eds.: Advances in Neural Information Processing Systems 14 [Neural Information Processing Systems, NIPS 2001, December 3-8, 2001, Vancouver and Whistler, British Columbia, Canada], Cambridge, MA, MIT Press (2001) 945–952
8. Chapelle, O., Weston, J., Schölkopf, B.: Cluster kernels for semi-supervised learning. In Becker, S., Thrun, S., Obermayer, K., eds.: Advances in Neural Information Processing Systems 15 [Neural Information Processing Systems, NIPS 2002, December 9-14, 2002, Vancouver, British Columbia, Canada], Cambridge, MA, MIT Press (2002) 585–592
9. Zhu, X.: Semi-supervised learning with graphs. PhD thesis, Carnegie Mellon University, School of Computer Science, Pittsburgh, Pennsylvania (PA), USA (2005)
10. Blum, A., Chawla, S.: Learning from labeled and unlabeled data using graph mincuts. In Brodley, C., Pohoreckyj Danyluk, A., eds.: Proceedings of the Eighteenth International Conference on Machine Learning (ICML 2001), Williams College, Williamstown, MA, USA, June 28 - July 1, 2001, Morgan Kaufmann (2001) 19–26
11. Joachims, T.: Transductive learning via spectral graph partitioning. In Fawcett, T., Mishra, N., eds.: Machine Learning, Proceedings of the Twentieth International Conference (ICML 2003), August 21-24, 2003, Washington, DC, USA, AAAI Press (2003) 290–297
12. Zhu, X., Ghahramani, Z., Lafferty, J.: Semi-supervised searning using gaussian fields and harmonic functions. In Fawcett, T., Mishra, N., eds.: Machine Learning, Proceedings of the Twentieth International Conference (ICML 2003), August 21-24, 2003, Washington, DC, USA, AAAI Press (2003) 912–919
13. Neville, J., Jensen, D.: Collective classification with relational dependency networks. In: Proceedings of the Second International Workshop on Multi-Relational Data-Mining. (2003)
14. Zhou, Z.H., Jiang, Y.: Nec4.5: neural ensemble based c4.5. IEEE Transactions on Knowledge and Data Engineering **16** (2004) 770–773
15. Blum, A., Chawla, S.: Learning from labeled and unlabeled data using graph mincuts. In: Proceedings of the Eighteenth International Conference on Machine Learning, Morgan Kaufmann (2001)
16. Friedman, J., Bentley, J., Finkel, R.: An algorithm for finding best matches in logarithmic expected time. ACM Transactions on Mathematical Software **3** (1977) 209–226
17. Beygelzimer, A., Kakade, S., Langford, J.: Cover trees for nearest neighbor. In pre-print, available from www.cs.rochester.edu/u/beygel/publications.html (2005)
18. Omohundro, S.: Efficient algorithms with nearal network behavior. Journal of Complex Systems **1** (1987) 273–347

Constructive Meta-level Feature Selection Method Based on Method Repositories

Hidenao Abe[1] and Takahira Yamaguchi[2]

[1] Department of Medical Informatics, Shimane University,
89-1 Enya-cho Izumo Shimane, 693-8501, Japan
abe@med.shimane-u.ac.jp
[2] Faculty of Science and Technology, Keio University,
3-14-1 Hiyoshi Kohoku Yokohama, 223-8522, Japan
yamaguti@ae.keio.ac.jp

Abstract. Feature selection is one of key issues related with data pre-processing of classification task in a data mining process. Although many efforts have been done to improve typical feature selection algorithms (FSAs), such as filter methods and wrapper methods, it is hard for just one FSA to manage its performances to various datasets. To above problems, we propose another way to support feature selection procedure, constructing proper FSAs to each given dataset. Here is discussed constructive meta-level feature selection that re-constructs proper FSAs with a method repository every given datasets, de-composing representative FSAs into methods. After implementing the constructive meta-level feature selection system, we show how constructive meta-level feature selection goes well with 32 UCI common data sets, comparing with typical FSAs on their accuracies. As the result, our system shows the highest performance on accuracies and the availability to construct a proper FSA to each given data set automatically.

1 Introduction

Feature selection is one of the key procedures to get a better result from the data mining process. However, it is difficult to determine the relevant feature subset before the mining procedure. At practical data mining situations, data miners often face a problem to choose the best feature subset for a given data set. If it contains irrelevant or/and redundant features, a data miner can't get any satisfactory results from mining/machine learning scheme. Irrelevant features not only lead to lower performance of the results, but also preclude finding potentially existing useful knowledge. Besides, redundant features not affect the performance of classification task, but influence the readability of the mining result. To choose a relevant feature subset, data miners have to take trial-and-error testing, expertise for the given feature set, or/and heavy domain knowledge for the given data set.

Feature selection algorithms (FSAs) have been developed to select a relevant feature subset automatically as a data pre-processing in a data mining process.

The performance of FSA is always affected by a given data set. To keep their performance higher, a user often tries to execute prepared FSAs to his/her dataset exhaustively. Thus a proper FSA selection is still costly work in a data mining process, and this is one of the bottle necks of data mining processes.

To above problems, we have developed a novel feature selection scheme based on constructive meta-level processing. We have developed a system to construct proper FSAs to each given data set with this scheme, which consists of decomposition of FSAs and re-construction of them. To de-compose current FSAs into functional parts called 'methods', we have analyzed currently representative FSAs. Then we have constructed the feature selection method repository, to re-construct a proper FSA to a given data set.

After constructing the feature selection method repository, we have implemented a system to choose a proper FSA to each given data set, searching possible FSAs obtained by the method repository for the best one. Taking this system, we have done a case study to evaluate the performance of FSAs on 32 UCI common data sets. As the result, the performance of FSAs has achieved the best performance, comparing with representative higher performed FSAs.

2 Related Work

After constructing a feature set to describe each instance more correctly, we take a FSA to select an adequate feature subset for a prepared learning algorithm.

To improve classification tasks at data mining, many FSAs have been developed [2, 3, 4]. As shown in the survey done by Hall [5], wrapper methods [6] such as forward selection and backward elimination have high performance with high computational costs. Besides, filter methods such as Relief [7, 8], Information Gain and FOCUS [9] can be executed more quickly with lower performance than that of wrapper methods. Some advanced wrapper methods such as CFS [10], which executes a substitute evaluator instead of a learned evaluator, have lower computational costs than wrapper methods. However, these performances are still non-practical, comparing with wrapper methods.

We also developed a novel FSA called 'Seed Method' [1]. Seed Method has achieved both of practical computational cost and practical performance, because it improves wrapper forward selection method, determining a proper staring feature subset for given feature set. With an adequate starting subset, this method can reduce the search space of 2^n feature subsets obtained by n features. To determine an adequate starting subset, the method extracts a feature subset with Relief.F and C4.5 decision tree [11] from given feature set.

Although studies done by [6, 12, 13] have shown each way to characterize FSAs, they have never discussed any way to construct a proper FSA to a given data set. So, a data miner still selects FSA with exhaustive executions of prepared FSAs, depending on his/her expertise. Weka [14] and Yale [15] provide many feature selection components and frameworks to users. We can construct several hundred FSAs with these materials. However, they never support to choose a proper one.

3 Constructive Meta-level Processing Scheme Based on Method Repositories

At the field of meta-learning, there are many studies about selective meta-learning scheme. There are two approaches as selective meta-learning. One includes bagging [16] and boosting [17], combining base-level classifiers from multiple training data with different distributions. In these meta-learning schemes, we should select just one learning algorithm to learn base-level classifiers. The other approach includes voting, stacking [18] and cascading [19], which combines base-level classifiers from different learning algorithms. METAL [20] and IDA [21] are also selective meta-learning approach, selecting a proper learning algorithm to the given data set with a heuristic score, which is called meta-knowledge.

Constructive meta-level processing scheme [22] takes meta-learning approach, which controls objective process with meta-knowledge as shown in Fig.1. In this scheme, we construct a meta-knowledge, representing with method repositories. The meta-knowledge consists of information of functional parts, restrictions of combinations of each functional part, and the ways to re-construct object algorithms with the functional parts.

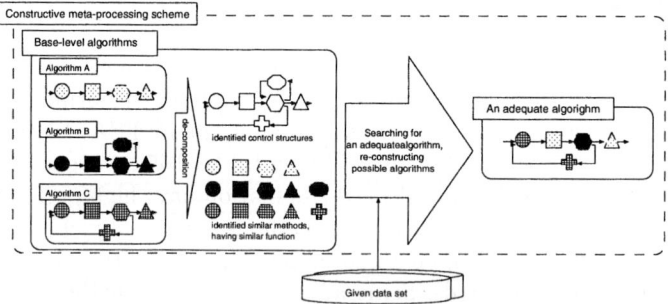

Fig. 1. An overview of constructive meta-level processing scheme

3.1 Issues to Implement a Method Repository

To build up a method repository, we should consider the following three major issues: how to de-compose prepared algorithms into functional parts, how to restrict the combinations of the functional parts, and how to re-construct a proper algorithm to a given data set.

To implement a feature selection method repository, we have considered above issues to identify feature selection methods(FSMs) in typical FSAs. Fortunately, FSAs have a nature as a search problem on possible combinations of features, which is pointed out in some papers [6, 12, 13]. With this nature, we have been able to identify generic methods in FSAs. Then we have also identified specific FSMs, which get into each implemented functional parts [1]. At the same time, we have also defined data types which are input/output/referenced for these

---

[1] For example, these functions are corresponded to Java classes in Weka.

methods. Thus we have organized these methods into a hierarchy of FSMs and a data type hierarchy. With these hierarchies, the system constructs FSAs to a given data set, searching possible FSAs obtained by the method repository for a proper one.

4 Implementation of the Constructive Meta-level Feature Selection Scheme

To implement constructive meta-level feature selection scheme, we have to build a feature selection method repository and the system to construct proper FSAs to given data sets with the feature selection method repository.

4.1 Constructing a Feature Selection Method Repository

Firstly, we have identified the following four generic methods: determining initial set, evaluating attribute subset, testing a search termination of attribute subsets and attribute subset search operation. This identification is based on what FSAs can be assumed one kind of search problems. Considering the four generic methods, we have analyzed representative FSAs implemented in Weka[14] attribute selection package[2]. Then we have build up a feature selection method repository.

After identifying 26 specific methods from Weka, we have described restrictions to re-construct FSAs. The restriction has defined with input data type, output data type, reference data type, pre-method and post-method for each method. With this description, we have defined control structures with these generic four methods as shown in Fig.2.

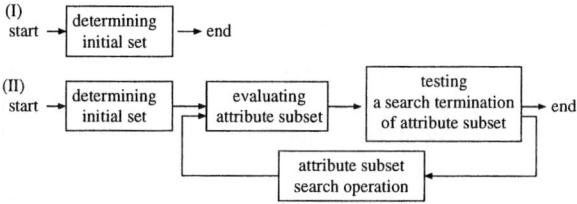

Fig. 2. Identified control structures on the four generic methods

The control structure (I) corresponds ordinary that of filter approach FSAs. Besides, with the control structure (II), we can construct hybrid FSAs, which is combined wrapper and filter FSAs. Of course, we can also construct analyzed filter and wrapper FSAs with these control structure.

At the same time, we have also defined method hierarchy, articulating each method. Fig.3 shows us the method hierarchy of feature selection. Each method has been articulated with the following roles: input data type, output data type, reference data type, pre-method, and post-method. With these roles, we have also defined combinations of FSMs.

---

[2] We have taken weka-3-4-5 in this time.

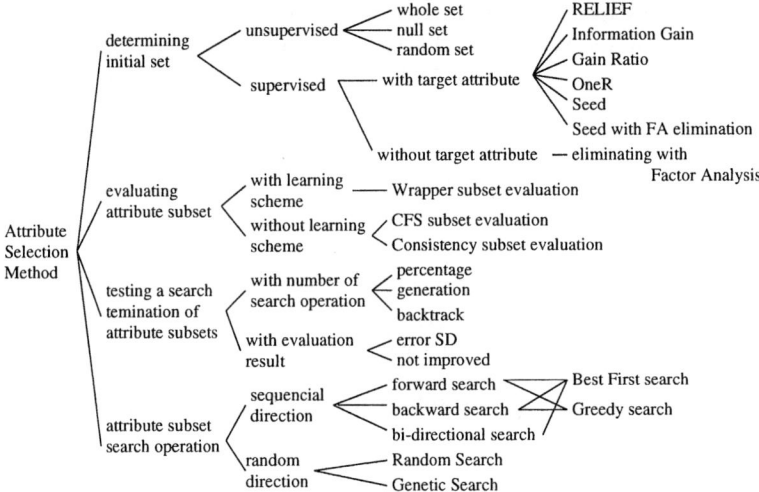

Fig. 3. The feature selection method hierarchy

To articulate data types for input, output and reference of methods, we have also defined data type hierarchy as shown in Fig.4.

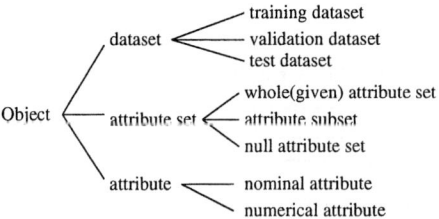

Fig. 4. The hierarchy of data types for the feature selection methods

4.2 The System to Construct a Proper FSA with a Feature Selection Method Repository

To re-construct a proper FSA to given data set, the system have to search possible FSAs obtained by the FSM repository for the most proper one. This process is also one of the search problems. Then we have designed the system with the following procedures: construction, instantiation, compilation, test, and refinement. The system chooses a proper FSA with these procedures as shown in Fig.5.

Each function of procedures is described in detail as follows: **Construction** procedure constructs a specification of the initial feature selection algorithm, selecting each specific method at random. **Instantiation** procedure transforms constructed or refined specifications to the intermediate codes. **Compilation** procedure compiles the intermediate codes to executable codes such as commands for Weka. **Go & Test** procedure executes the executable codes to the

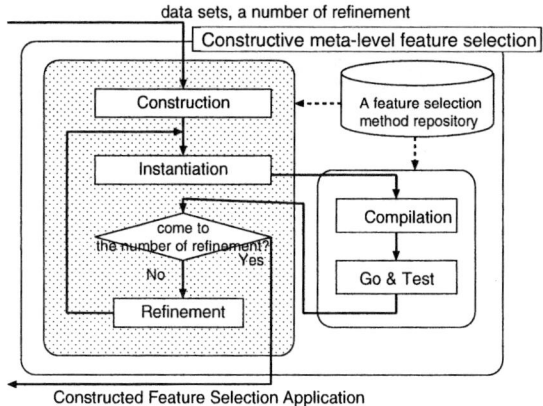

Fig. 5. An overview of constructive meta-level feature selection system

given data set to estimate the performance of FSAs. If the number of refinement doesn't come to the given limitation number **Refinement** procedure refines specifications of executed FSAs with some search operations.

5 Evaluation on UCI Common Data Sets

After implementing the feature selection method repository and the system to construct proper FSAs to given data sets, we have done a case study to evaluate an availability of our constructive meta-level feature selection scheme.

In this case study, we have taken 32 common data sets from UCI ML repository [23], which are distributed with Weka. With the implemented feature selection method repository, the system has been able to construct 292 FSAs. The system has searched specification space of possible FSAs for the best FSA to each data set with the following configuration of GA operation at 'Refinement' procedure:

Population size. Each generation has τ individuals.
Selection. We take roulette selection to select 60% individuals for parents.
Crossover. Each pair of parents is crossed over single point, which is selected at random.
Mutation. Just one gene of selected child is mutated, selecting just one child with the probability 2%.
Elite Preservation. The best individual is preserved on each generation.

5.1 The Process to Select a FSA

Firstly, the system selects proper FSAs to each data set, estimating the actual performance with the performance of n-fold cross validation. The selection phase has done at 'Go & Test' procedure in Fig.5. This selection phase has been repeated multiple times in each construction of FSA with our system. Finally, the system output just one FSA, which has the highest 'evaluation score' as shown in Fig.6.

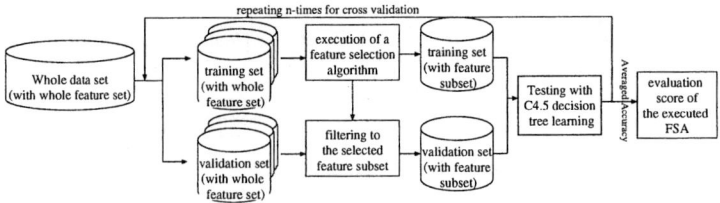

Fig. 6. Computing evaluation scores of each spec for GA in 'Refinement' procedure

We have taken averaged predictive accuracy $EstAcc(D)$ of n-fold cross validation from predictive accuracies $acc(evd_i)$ for each validation data set evd_i as the following formulations:

$$EstAcc(D) = \frac{\sum_{i=1}^{n} acc(evd_i)}{n} \; zwacc(evd_i) = \frac{crr(evd_i)}{size(evd_i)} \times 100$$

$acc(evd_i)$ is a percentage score from the number of correctly predicted instances $crr(evd_i)$ and size of each validation set $size(evd_i)$.

According to this evaluation scores, the GA refinement searched for proper FSAs to each given data set. We have set up population size $\tau = 10$ and maximum generation $N = 10$ in this case study. So this set of GA operations has repeated maximum 10 times to each data set. Finally, the best FSA included in a final generation has been selected as output of our constructive meta-level feature selection system.

5.2 The Process of the Evaluation

We have designed the process of this evaluation for representative FSAs and constructed FSAs to each data set as shown in Fig.7.

In this evaluation, we have applied each FSA to each whole data set. Then n-fold cross validation have been performed on each data set with selected feature subset. The performances of each data set $Acc(D)$ have been averaged predictive accuracies $acc(vd_i)$ from each fold as the following formulations:

$$Acc(D) = \frac{\sum_{i=1}^{n} acc(vd_i)}{n} \; zwacc(vd_i) = \frac{crr(vd_i)}{size(vd_i)} \times 100$$

Where vd_i means i-th validation set of the n-fold cross validation.

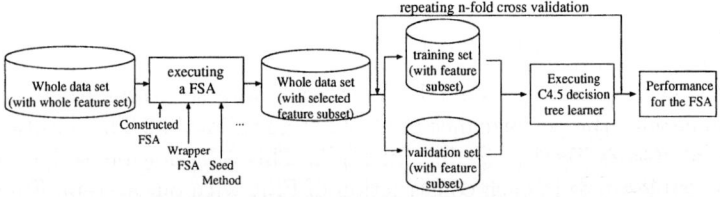

Fig. 7. Evaluation framework for the accuracy comparison

We have compared the performance of our constructive meta-level feature selection system with the following FSAs: **Whole feature set, Seed method,** and **Genetic Search**[24]. All of them have been evaluated with the same way as shown in the evaluation phase of Fig.7. We had done wrapper forward selection, Relief.F, Seed method and 'Genetic Search' to the data sets previously. Then the two methods were selected because of their higher performance.

5.3 Results and Discussions of the Evaluation

Table1 shows us the accuracies from whole feature set, subset selected by seed method, subset selected by 'Genetic Search' and subset selected by FSAs which constructed with our constructive meta-level feature selection system. Each score is the averaged accuracy calculated from 10-fold cross validation. The significance of the average for all of the data sets has tested with t-test. The comparison between the averages of our system and the other FSAs shows the statistically significant difference, where $p < 0.05$ for the other FSAs.

Table1 also shows us the result of the best performances, comparing among performances of the FSAs. To the 17 data sets, FSAs composed by our system have achieved the best performance. To breast-cancer, colic, hepatitis, ionosphere, iris

Table 1. The performances of the feature selection algorithms on the UCI common data sets. Each score means averaged accuracies(%) with 10-fold cross validation. '*' means the best accuracy within this evaluation.

datasets	whole feature set	seed method	genetic search with wrapper	FSAs composed by our syste
anneal	98.44	98.33	*98.78	98.55
audiology	77.87	*80.14	77.51	77.51
autos	81.95	81.48	80.00	*84.90
balance-scale	76.64	*78.72	*78.72	*78.72
breast-cancer	*75.52	*75.52	72.71	74.14
breast-w	94.56	*94.85	94.57	94.71
colic	*85.33	85.08	84.54	84.81
credit-a	86.09	84.93	84.49	*86.67
credit-g	70.50	71.90	70.60	*74.60
diabetes	73.83	*76.56	75.00	75.25
heart-c	77.56	81.82	76.95	*83.19
heart-h	80.95	*82.66	80.6	81.98
heart-statlog	76.67	82.22	82.22	*85.56
hepatitis	*83.87	79.33	80.67	83.23
hypothyroid	99.58	99.60	97.88	*99.63
ionosphere	*91.45	89.17	88.32	91.44
iris	*96.00	92.67	92.00	94.00
kr-vs-kp	*99.44	98.81	98.78	99.37
letter	87.98	*88.4	83.56	87.62
lymph	77.03	75.67	75.67	*81.10
mushroom	*100.0	*100.0	*100.0	*100.0
primary-tumor	39.82	41.00	*43.94	43.06
segment	96.93	*96.97	96.80	*96.97
sick	98.81	98.75	98.86	*98.94
soner	71.15	75.45	72.12	*75.48
soybean	91.51	92.24	91.51	*92.39
splice	94.08	94.29	*94.42	94.14
vehicle	72.46	70.92	71.99	*73.30
vote	96.32	95.85	95.62	*96.53
vowel	81.51	83.03	*83.64	*83.64
waveform-5000	75.08	*77.16	75.98	76.74
zoo	92.08	92.00	92.09	*96.00
Average	84.41	84.86	84.08	85.76

```
Input: Whole feature set  F, training data set  Tr
Output: Feature subset for the training data set  Fsub
Prameters: number of backtracks=5

begin:
  Feature set f;
  f = determining_initial_set_with_FA+Seed(F);
  int i=0;
  double[] evaluations;
  while(1){
    evaluations[] = feature_subset_evaluation_with_CFS(f);
    (f,i) = backward_elimination(evaluations,f);
    if(number_of_backtracks(i,5)==true){ break; }
  }
  return f;
end:
```

Fig. 8. Pseudo-code of the feature selection algorithm for heart-statlog

and kr-vs-kp, whole feature set wins selected feature subsets, because all of the evaluated FSAs have not been able to select whole feature sets. They tend to output smaller feature subset, because they believe in that there are some irrelevant features in the given feature set. If we had defined the control structure for filter method Fig.2, the system would have selected whole feature subset with 'whole set' method in Fig.3.

To anneal, audiology, breast-w, diabetes, heart-h, letter, primary-tumor, splice and waveform-5000, FSAs composed by our system have not achieved the best performance, comparing with the other FSAs. The evaluation scores to estimate actual performances have not worked correctly on these cases. However, these disadvantages are not significant differences statistically.

Fig.8 shows us the FSA composed by our system to heart-statlog data set. This algorithm consists of initial set determination with 'seed method' & elimination unique features using Factor Analysis result, feature subset evaluation with CFS method, backward elimination, and stopping with the number of backtracks[3]. Although this algorithm bases on backward elimination method, the combination of methods has been never seen in any study of FSAs. As this example, our system has been also able to construct a novel FSA automatically, reconstructing feature selection methods on the repository.

6 Conclusion

We present a novel meta-level feature selection approach based on constructive meta-level processing with method repositories. This scheme chooses a proper FSA to the given data set, re-constructing the FSA with a FSMs repository.

To evaluate the availability of our approach, we have done an empirical experiment with 32 UCI common data sets. Our constructive meta-level feature selection system has significantly outperformed than representative FSAs, which have higher performance compared with the other FSAs. The result also shows that our constructive meta-level feature selection system have been able to construct a proper algorithm to given feature set automatically.

---

[3] The number has been set up five.

As feature work, we will improve criterion to choose a proper FSA, considering search time to select a proper one, execution time of selected FSA and its performance.

References

[1] Komori, M., Abe, H., Yamaguchi, T.: A new feature selection method based on dynamic inclemental extension of seed features. In: Proceedings of Knowledge-Based Software Engineering. (2002) 291–296
[2] John, G.H., Kohavi, R., Pfleger, K.: Irrelevant features and the subset selection problem. In: International Conference on Machine Learning. (1994) 121–129
[3] John, G.H.: Enhancements to the data mining process. PhD thesis, Computer Science Department, Stanford University (1997)
[4] Liu, H., Motoda, H.: Feature Selection for Knowledge Discovery and Data Mining. Kluwer Academic Publishers (1998)
[5] Hall, M.A.: Benchmarking attribute selection techniques for data mining. Technical Report Working Paper 00/10, Department of Computer Science, University of Waikato (2000)
[6] Kohavi, R., John, G.H.: Wrappers for feature subset selection. Artificial Intelligence **97** (1997) 273–324
[7] Kira, K., Rendell, L.: A practical approach to feature selection. In Sleeman, D., Edwards, P., eds.: Proceedings of the Ninth International Conference on Machine Learning. (1992) 249–256
[8] Kononenko, I.: Estimating attributes: Analysis and extensions of relief. In: Proceedings of the 1994 European Conference on Machine Learning. (1994) 171–182
[9] Alumualim, H., Dietterich, T.G.: Learning boolean concepts in the presence of many irrelevant features. Artificial Intelligence **69** (1994) 279–305
[10] Hall, M.: Correlation-based Feature Selection for Machine Learning. PhD thesis, Department of Computer Science, University of Waikato (1998)
[11] Quinlan, J.R.: Programs for Machine Learning. Morgan Kaufmann (1992)
[12] Langley, P.: Selection of relevant features in machine learning. In: Proceedings of the AAAI Fall Symposium on Relevance. (1994)
[13] Molina, L.C., Beranche, L., Nebot, A.: Feature selection algorithms: A survey and experimental evaluation. In: Proceedings of the 2002 Internatiolan Conference on Data Mining. (2002) 306–313
[14] Witten, I., Frank, E.: Data Mining: Practical machine learning tools and techniques with Java implementations. Morgan Kaufmann (2000)
[15] Mierswa, I., Klinkenberg, R., Fischer, S., Ritthoff, O.: A Flexible Platform for Knowledge Discovery Experiments: YALE – Yet Another Learning Environment. In: LLWA 03 - Tagungsband der GI-Workshop-Woche Lernen - Lehren - Wissen - Adaptivität. (2003)
[16] Breiman, L.: Bagging predictors. Machine Learning **24** (1996) 123–140
[17] Freund, Y., Schapire, R.E.: A decision-theoretic generalization of on-line learning and an application to boosting. In: Proceedings the Second European Conference on Computational Learning Theory. (1995)
[18] Wolpert, D.: Stacked generalization. Neural Network **5** (1992) 241–260
[19] Gama, J., Brazdil, P.: Cascade generalization. Machine Learning **41** (2000) 315–343
[20] METAL: http://www.metal-kdd.org/. (2002)

21. Bernstein, A., Provost, F.: An intelligent assistant for knowledge discovery process. In: IJCAI 2001 Workshop on Wrappers for Performance Enhancement in KDD. (2001)
22. Abe, H., Yamaguchi, T.: Constructive meta-learning with machine learning method repositories. In: Proceedings of the seventeenth International Conference on Industrial and Engineering Applications of Artificial Intelligence and Expert Systems. (2004) 502–511
23. Blake, C.L., Merz, C.J.: UCI Repository of machine learning databases. http://www.ics.uci.edu/~mlearn/MLRepository.html (1998)
24. Vafaie, H., Jong, K.D.: Genetic algorithms as a tool for feature selection in machine learning. In: Proceedings of the fourth International Conference on Tools with Artificial Intelligence. (1992) 200–204

Variable Randomness in Decision Tree Ensembles

Fei Tony Liu and Kai Ming Ting

Gippsland School of Information Technology,
Monash University,
Churchill, 3842, Australia
{Tony.Liu, KaiMing.Ting}@infotech.monash.edu.au

Abstract. In this paper, we propose Max-diverse.α, which has a mechanism to control the degrees of randomness in decision tree ensembles. This control gives an ensemble the means to balance the two conflicting functions of a random random ensemble, i.e., the abilities to model non-axis-parallel boundary and eliminate irrelevant features. We find that this control is more sensitive to the one provided by Random Forests. Using progressive training errors, we are able to estimate an appropriate randomness for any given data prior to any predictive tasks. Experiment results show that Max-diverse.α is significantly better than Random Forests and Max-diverse Ensemble, and it is comparable to the state-of-the-art C5 boosting.

1 Introduction

Random tree ensembles utilize randomization techniques such as data perturbation, random sampling and random feature selection to create diverse individual trees. Examples of such are Bagging [1], Randomized Trees [2], Random Subspace [3], Decision Tree randomization [4], Random Forests [5], Random Decision Tree [6] and Max-diverse Ensemble [7]. Based on Breiman's analysis, randomization increases ensemble diversity and reduces the strength of individual learners [5]. Many studies suggested that choosing a proper degree of diversity or strength would greatly enhance the ensemble accuracy [8, 9]. Among the above-mentioned implementations, Random Forests provides a parameter to vary the degree of randomness. To some extents the parameter affects the diversity and the average strength of individual trees. However, Breiman concluded that ensemble accuracy is insensitive to the different values of the parameter [5]. This leaves several open questions to be addressed: (1) What are the effects of different degrees of randomness? (2) Is there a better way to control the amount of randomness used? (3) What is the appropriate level of randomness for a given problem?

The spectrum of randomness in the above-mentioned implementations can be conceptualized as a continuum ranging from highly deterministic to completely random. Max-diverse Ensemble represents the extreme of complete-randomness. It completely randomizes the feature selection process which is geared toward maximizing tree diversity. Max-diverse Ensemble has been shown to be comparable to Random Forests in terms of accuracy [7]. Analytically, it has a lower time-complexity since the feature selection is completely random and does not require bootstrap sampling. In order to answer the questions listed, a study into the effect of variable randomness is needed. Due to Max-diverse Ensemble's complete-randomness and outstanding performance, we are motivated to use it as the upper limit to generate variable randomness. The lower limit is simply set by a conventional deterministic decision tree. In this case, C4.5 [10] is selected. In this paper, these two limits set the range of the variable randomness and provide the necessary platform for our study.

The rest of this paper is organized as follows. Section 2 gives a brief account of contemporary random decision tree ensembles, including the conventional random tree ensembles and complete-random tree ensemble. In section 3, we investigate the strengths and weaknesses of Max-diverse Ensemble [7] which serve as a primer to better understand the effects of variable randomness. Section 4 introduces Max-diverse.α, a novel variable-random approach which is capable of estimating an appropriate randomness for any given data set. In section 5, we empirically compare the proposed Max-diverse.α with Max-diverse Ensemble, Random Forests and C5 boosting. This is then followed by discussion and conclusions in the last two sections.

2 Contemporary Random Decision Tree Ensembles

One of the attractive characteristics of the random decision tree ensembles is the anti-overfitting property[5]. For decision tree ensembles, the posterior probability is estimated either by *voting* or *probability averaging*. By the Law of Large Numbers, *voting* and *probability averaging* approach the true posterior probability, when the number of trees becomes large.

Furthermore, Breiman's analysis on strength and diversity [5] provides a way to analyse the performance of different decision tree ensembles. Breiman gives

$$\text{PE} \geq \bar{\rho}(1 - s^2)/s^2 \tag{1}$$

where PE is the generalization error for an ensemble, $\bar{\rho}$ is the mean correlation among trees, and s is the average strength of individual trees in an ensemble. Strength s corresponds to the accuracy of individual trees and correlation $\bar{\rho}$ corresponds to similarity of tree structures in an ensemble. Correlation is the opposite of diversity. In essence, equation (1) suggests that diversified and accurate individual trees create accurate ensemble. Also, Buttrey and Kobayashi conjectured that strength s and correlation $\bar{\rho}$ form a non-linear relationship in which diversity increases in the expense of s and vice versa [11]. In a nutshell, algorithms that build single decision tree strive to build the most accurate one. To achieve diversity, single decision trees in an ensemble sacrifice some of their strength to allow for variation. As it stands, finding the right balance of strength and diversity is the key challenge in developing a good performing random tree ensemble.

Random Forests [5] is a popular implementation of random tree ensemble. Building on Bagging [1], Random Forests uses an F parameter to further randomize the feature selection process. In the tree construction phase, prior to selecting a feature test for a tree node, Random Forests randomly pre-selects F number of features. The F number of features are then fed to a deterministic feature selection to select the best feature test. F is recommended to be the first integer less than $log_2 m + 1$, where m is the total number of features in a training set [5]. In summary, Bagging makes use of bootstrap sampling to enhance the accuracy of a single decision tree. Random Forests injects the randomized feature selection process into Bagging to achieve higher degree of randomness to further improve ensemble accuracy.

In contrast, algorithms that generate complete-random trees such as **Max-diverse Ensemble** [7] do not use any deterministic feature selection at all. It achieves the highest degree of diversity, as it can generate any possible trees that have no empty leaves. [1] With complete-random feature selection, the test feature for each node is randomly selected from available features. Max-diverse Ensemble grows unpruned trees and combines their predictions by probability averaging. Explaining using Breiman's equation in (1), Max-diverse Ensemble lowers the generalization error PE by lowering correlation $\bar{\rho}$ through increased diversity, and compensating poorer performing individual trees.

3 Strengths and Weaknesses of Complete-Random Tree Ensemble

As complete-random trees are the upper limit of variable randomness, In this section, we discuss the strengths and weaknesses of complete-random tree ensemble which will allow us to see the changing effects of variable randomness. In the first subsection, we reveal Max-diverse's strengths in modeling non-axis-parallel boundaries and capturing small details. The problem of small disjuncts is also covered in this subsection as a side effect of capturing small details. In the second subsection, we show Max-diverse's weakness in its inability to eliminate irrelevant features.

3.1 Modeling Non-axis-parallel Boundary

This section focuses on the relationship between randomness and the ability to model non-axis-parallel boundaries. For single decision trees, the feature test separates feature values using a simple logical test, so the decision boundary must be aligned to one of the feature axes. When constructing a decision tree, it first looks for the most significant structure in the instance space. With further division in the instance space, its accumulated information gain does not allow it to form certain partitions with less information. Hence, small details are neglected.

---

[1] Bagging and Random Forests consider a smaller set of trees because the number of possible trees are further constrained by the feature selection criterion.

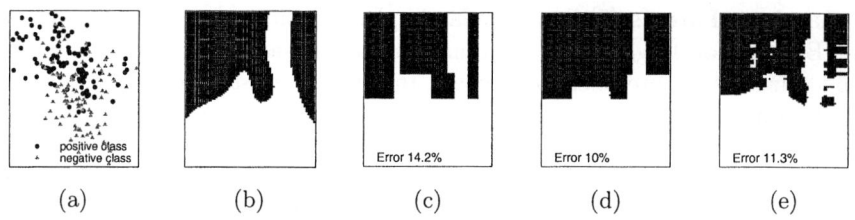

Fig. 1. Gaussian mixture (a) training data and (b) the optimal boundary. Shaded area denotes positive class; cleared area denotes negative class. (c) When classifying using a single unpruned decision tree, C4.5, the decision boundary is rectangular in shape. (d) The decision boundary of Random Forests ($F = 1$) retains the rectangularity from the single decision tree. (e) Max-diverse Ensemble fits a non-axis-parallel decision boundary to the training data, forming non-rectangular boundaries.

To visualize these limitations, we employ a Gaussian mixture data set from [12]. The training data [2] and optimal boundary are given in Figure 1a and Figure 1b. Figure 1c shows the classification of a single unpruned decision tree (C4.5). As expected, the decision boundaries appear to be axis-parallel.

Contemporary random decision tree ensembles overcome this limitation by overlapping the decision boundaries to model a non-axis-parallel boundary. To be effective, trees have to be different from each other or to be diverse to model any non-axis-parallel boundary. This effect is similar to *analog-to-digital conversion*. As the quality of the analog-to-digital conversion is dependent on the sampling rate, the quality of the decision boundary is dependent on the diversity of the ensemble.

Figure 1 also shows the effects of increasing randomness in feature selection. The decision boundary of Random Forests in Figure 1d is still far from the optimal boundary and still exhibits rectangular shapes. On the other hand, Max-diverse Ensemble's decision boundary in Figure 1e appears to resemble the optimal boundary more closely. It captures small details presented in the training data, though it also suffers from small disjuncts [13], preventing it from reaching higher accuracy.

3.2 The Effect of Irrelevant Features

One of the consequences of complete-randomness in feature selection is that there is no means to avoid irrelevant features. In ordinary decision trees, avoiding irrelevant features is usually done through feature selection. For example, Max-diverse Ensemble performs poorly on the *dna* data set with an error rate of 28.8% with an average tree size[3] of 2555. It is noteworthy that *dna* data has sixty features. On average, an ordinary decision tree only uses four features to classify a test case. Therefore, an ordinary single decision tree regards the

---

[2] Gaussian mixture data set has 200 training samples and 6831 testing samples.
[3] Tree size is the average number of nodes (including internal nodes and leaf nodes) of single trees in an ensemble.

other fifty-six features as irrelevant or unnecessary when making a classification. Random Forests performs well with *dna* data as it employs an feature selection which filters out irrelevant features. This results in an error rate of 12.9% and an average tree size of 421, indicating Random Forests' ability to filter out irrelevant features. In contrast, Max-diverse Ensemble has a much larger tree size as a consequence of its inability to identify irrelevant features.

The question then presents itself as to how best to harness the advantage of Max-diverse Ensemble which is modeling non-axis-parallel boundaries and yet manage the effects of small disjuncts and its inability to eliminate irrelevant features. One possible approach is to essentially adjust the randomness in such a way that balances these conflicting requirements. Randomness can be conceptualized as a factor ranging from 0 to 1, where 0 is completely random and 1 is the most deterministic. Therefore, an ensemble with a more deterministic feature selection will be characterized by rectangular decision boundaries and the ability to eliminate irrelevant features. An ensemble with a more stochastic approach to feature selection will be characterized by non-axis-parallel boundaries with a weaker ability to eliminate irrelevant features. In the following section, we will introduce a novel mechanism to adjust Max-diverse Ensemble's randomness in order to optimize the predictive accuracy; this mechanism is more sensitive to the existing method used in Random Forests.

4   Variable Randomness

Variable randomness provides the flexibility to produce different kinds of tree ensembles to suit the different characteristics of individual data sets. It serves as a mechanism for adjusting the balance of strength and diversity of individual trees in decision tree construction. To fine tune the randomness in decision tree ensembles, we introduce a parameter α into Max-diverse Ensemble. This results in an algorithm called Max-diverse.α, which employs both complete-random and deterministic feature selections. It splits the feature selection process into two stages at every node. The first stage of the algorithm decides which method to use, *complete-random* or *deterministic*; the second stage proceeds with the selected criterion to perform subsequent feature selection. α is the probability of choosing the deterministic feature selection, used in the first stage,

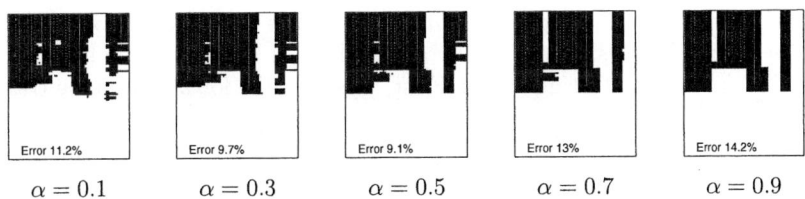

Fig. 2. Classify Gaussian mixture using Max-diverse.α with different α values. Notes the changes in decision boundary rectangularity as α changes.

Algorithm 1. The feature selection process in Max-diverse.α

INPUT S: Training set, α: probability for using deterministic feature selection
OUTPUT T: tree node
let r be a random value, $0 < r \leq 1$;
if $r \leq \alpha$ **then**
　　$T = \text{deterministic\_feature\_selection}(S)$;
else
　　$T = \text{complete\_random\_selection}(S)$;
end if
return T;

where $0 \leq \alpha \leq 1$. Algorithm 1 illustrates the proposed two-stage process. α also approximates the percentage of deterministic feature selection used in a tree.

To illustrate the effect of α, Figure 2 demonstrates changes in decision boundary. Note that the boundary rectangularity and the effect of small disjuncts (i.e., small pockets) change when α changes. When α ranges between 0.3 and 0.5 in this example, error rates drop below 10%.

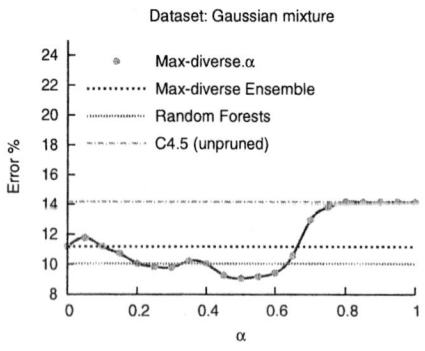

Fig. 3. Test errors versus α using Max-diverse.α in comparison with C4.5, Random Forests and Max-diverse Ensemble

Figure 3 shows the testing error rates when plotting against different α values for Max-diverse.α and three other methods. It shows that there is a relatively wide range of α in which Max-diverse.α can achieve lower error rates than C4.5, Random Forests and Max-diverse Ensemble. More importantly, Figure 3 demonstrates the sensitivity of ensembles' accuracy toward different degrees of randomness.

Therefore, picking an effective α value prior to building an ensemble is essential for practical applications as the effective range of α values is data dependent. We observe from our initial investigation that most of the optimal α values were found in $0 \leq \alpha \leq 0.5$. At this stage, there is no optimal way to estimate this α value using any data characteristics from training data. In this paper, a simple estimation procedure is proposed, based on the average training errors. There are two difficulties in using training errors to estimate or predict the testing errors. First, it is possible for all ensemble training errors of different α to reach zero, making them very hard to compare. Second, when $\alpha > 0.5$, the deterministic test selection fits tree structures to the training samples, creating exceptionally low training errors which bias the selection. To overcome these difficulties, we estimate an effective α based on the average of progressive training errors. When constructing an ensemble, progressive training errors can be obtained by evaluating training data after adding each decision tree into the ensemble. The average progressive training error reflects the rate of training error convergence from first tree to the last tree. So, the lower the average progressive training error is the better the performance of an ensemble. An estimated $\hat{\alpha}$ for each data set is generated as follows:

$$\hat{\alpha} = \arg\min_{0 \leq \alpha \leq 0.5}(\frac{1}{t} \sum_{i=1}^{t} err(\alpha, i, S)) \tag{2}$$

where t is the total number of trees in an ensemble, $err()$ returns the training error rate of an ensemble of size i, set at α and the training samples S. After obtaining $\hat{\alpha}$, Max-diverse.α employs the model with $\hat{\alpha}$ for actual predictive tasks.

5 Experiment

The experiment compares four different ensemble methods of unpruned trees: Max-diverse.α, Max-diverse Ensemble, Random Forests and C5 boosting [10], where the last three are used as benchmark classifiers. One hundred trees are used in each ensemble for each data set. A ten-fold cross-validation is conducted for each data set and the average error rate is reported. Note that all ensembles are given exactly the same folds for training and evaluation. In estimating $\hat{\alpha}$, we sample eleven α values from 0 to 0.5 in steps of 0.05, that is $\alpha = \{0, 0.05, ..., 0.5\}$.

Forty-five data sets from UCI repository [14] are used in this experiment. Table 1 presents the data properties and the results from the experiment. Table 2 provides pair-wise comparisons among the four methods in terms of the number of data sets in which one ensemble wins, loses and draws over the other ensemble. We summarize the result as follows:

- Compared to Max-diverse Ensemble, Max-diverse.α wins in *thirty two* data sets, loses in *twelve* data sets and draws in *one* data set. This is significant in a sign test at 95% confidence level.
- Compared to Random Forests, Max-diverse.α wins in *twenty six* data sets, loses in *fourteen* data sets and draws in *five* data sets. This is also significant in a sign test at 95% confidence level.
- Compared to C5 Boosting, Max-diverse.α wins in *twenty one* data sets, loses in *twenty four* data sets and draws in none. C5 Boosting and Max-diverse.α are comparable to each other.

6 Discussion

In section 4, our analysis shows clearly that varying the degree of randomness (using α) has a significant impact on the performance of the ensemble. To understand the insensitivity of F parameter in Random Forests, it is thus important to identify the differences between Max-diverse.α and Random Forests that result in their different behaviours.

Table 1. Data sets properties and experimental results reported in average error rate (%) of ten-fold cross-validation. In each data set, the best error rate among the four methods is bold faced.

datasets	size	#att.	#class labels	Max-diverse.α	C5 Boosting	Max-diverse Ensemble	Random Forests
abalone	4177	8	2	30.5	31.1	30.2	**29.5**
anneal	898	38	6	**1.1**	5.0	1.4	23.8
audiology	226	69	23	15.8	**15.0**	17.7	33.7
auto	205	25	7	15.7	**15.6**	22.5	19.0
balance	625	4	3	15.7	18.9	**12.3**	19.7
breast-w	699	10	2	3.7	3.1	**2.4**	3.4
breast-y	286	9	2	**25.5**	26.9	25.9	28.6
chess	3196	35	2	0.5	**0.3**	1.6	0.9
cleveland	303	13	5	42.9	41.6	41.6	**39.6**
coding	20000	15	2	16.5	**15.4**	16.8	17.7
credit-a	690	13	2	**12.6**	14.3	13.0	14.5
credit-g	1000	24	2	23.5	**22.4**	25.7	24.3
dna	3186	60	3	5.1	4.8	26.5	**3.7**
echo	133	7	2	34.9	37.4	**34.2**	34.3
flare	1066	10	2	18.9	**17.5**	19.2	18.3
glass	214	9	7	22.8	**21.4**	22.9	25.3
hayes	160	4	3	18.1	16.9	21.9	**14.4**
hepatitis	155	19	2	20.0	**14.1**	15.5	16.7
horse	368	22	2	**13.6**	22.5	17.9	15.2
hypo	3163	25	2	1.1	**0.8**	1.7	**0.8**
ionosphere	351	34	2	5.7	**5.4**	8.5	6.3
iris	150	4	3	5.3	**4.0**	4.7	5.3
labor	57	16	2	5.0	15.7	**3.3**	14.0
led24	3200	24	10	28.3	**27.8**	30.3	28.3
led7	3200	7	10	**26.6**	28.1	26.9	26.7
liver	345	6	2	**25.8**	29.6	27.9	32.5
lymph	148	18	4	15.0	19.1	**14.3**	18.2
nursery	12960	8	5	**0.7**	0.9	2.2	1.4
pima	768	8	2	24.3	25.0	24.6	**23.4**
post	90	8	3	37.8	**30.0**	36.7	43.3
primary	339	17	22	56.3	56.9	57.2	**55.2**
satimage	6435	36	7	8.5	**8.1**	10.4	8.3
segment	2310	19	7	**1.6**	1.8	3.1	2.3
sick	3163	25	2	2.3	**2.2**	5.7	2.3
solar	323	12	6	30.0	**25.7**	30.3	28.8
sonar	208	60	2	**15.4**	15.9	15.9	18.7
soybean	683	35	19	**5.4**	6.2	6.0	11.7
threeOf9	512	9	2	0.2	**0.0**	0.6	1.2
tic-tac-toe	958	9	2	2.1	**1.2**	9.7	6.3
vehicle	846	18	4	24.2	**23.3**	27.1	24.2
vote	435	16	2	**4.4**	4.8	5.3	**4.4**
waveform21	5000	21	3	15.2	15.6	**14.7**	15.7
waveform40	5000	40	3	15.8	**15.1**	17.0	16.0
wine	178	13	3	4.0	5.6	**1.1**	1.7
zoo	101	16	7	**2.0**	3.0	**2.0**	2.9
			mean	**15.6**	15.9	16.8	17.4

Random Forests and Max-diverse.α differ in how the feature selections are applied in each decision node. Random Forests applies both random and deterministic feature selections in each node; but Max-diverse.α only applies one of

Table 2. A pair-wise comparison of four ensemble methods in terms of the number of wins, losses and draws. Scores are read from top to left. Significant scores using a sign test at 95% confidence are bold faced.

wins, losses, draws	Max-diverse.α	C5 Boosting	Random Forests
Max-diverse Ensemble	**32**,12,1	26,17,2	24,21,0
Random Forests	**26**,14,5	**30**,14,1	
C5 Boosting	21,24,0		

the two methods in each node. α controls the probability in which the deterministic (or random) feature selection is applied in each node; whereas the mixed application of the two selection processes in each node constrains the 'amount' of randomness in Random Forests. To explain this, F only controls the number of features to be randomly selected. Once the best feature is selected in the first place, no matter what F is, the deterministic feature selection would choose the best feature. In effect, the randomness only applies to which of the best features are selected in F features. This explains the insensitivity that Breiman has observed.

When F is set to 1, Random Forests appears to be identical to Max-diverse Ensemble. However, the deterministic feature selection used in Random Forests has a second function to stop splitting insensible nodes. It means that trees grown with Random Forests are restricted by the second function of the deterministic selection criterion. Complete-random selection, however, ignores any selection criterion: it keeps on splitting until further split is impossible. Together with the use of bootstrap samples, these are the fundamental differences between Random Forests ($F = 1$) and Max-diverse.α ($\alpha = 0$).

7 Conclusions and Future Works

We contribute to identify the strengths and weaknesses of complete-random ensemble. The ability to model non-axis parallel boundary is the key distinctive strength of complete-random ensemble; but it lacks the ability to eliminate irrelevant attributes. Motivated to balance these abilities, we propose Max-diverse.α, a variable-random model capable of estimating an appropriate randomness for any given data. This gives the ensemble the ability to overcome the weaknesses of complete-randomness, i.e., it helps to eliminate irrelevant features and reduce the effect of small disjuncts.

Our answers to the three questions posted in the introduction are as follows:

1. Using Max-diverse.α, we are able to (i) explore the changes in forming non-axis-parallel boundaries as α varies and (ii) understand that the ability to eliminate irrelevant features changes with the degrees of randomness.
2. Max-diverse.α uses the α factor to control the amount of randomness used in the ensemble generation process. The α factor is a better alternative to Random Forests' F parameter in which α covers the full spectrum of variable randomness from completely random to pure deterministic, which

gives a fine granularity representing any level of randomness whereas F only accepts integers limited by the number of features.
3. To choose an appropriate α value for a given task, we introduce an estimation procedure based on progressive training errors. Using progressive training errors, Max-diverse.α is able to select an $\hat{\alpha}$ prior to its predictive tasks. Our experiment shows that Max-diverse.α is significantly better than Max-diverse Ensemble and Random Forests. It is also comparable to C5 boosting.

In the near future, we will explore ways to improve the efficiency of $\hat{\alpha}$ estimation, avoid selecting irrelevant feature in such a way that diversity is preserved, and reduce further the negative impact of small disjuncts.

Acknowledgement. Special thanks to Julie Murray who helps to make this paper more readable.

References

1. Breiman, L.: Bagging predictors. Machine Learning **24** (1996) 123–140
2. Amit, Y., Geman, D.: Shape quantization and recognition with randomized trees. Neural Computation **9** (1997) 1545–1588
3. Ho, T.K.: The random subspace method for constructing decision forests. IEEE Transactions on Pattern Analysis and Machine Intelligence **20** (1998) 832–844
4. Dietterich, T.G.: An experimental comparison of three methods for constructing ensembles of decision trees: Bagging, boosting, and randomization. Machine Learning **40** (2000) 139–157
5. Breiman, L.: Random forests. Machine Learning **45** (2001) 5–32
6. Fan, W., Wang, H., Yu, P.S., Ma, S.: Is random model better? on its accuracy and efficiency. Third IEEE International Conference on Data Mining (2003) 51–58
7. Liu, F.T., Ting, K.M., Fan, W.: Maximizing tree diversity by building complete-random decision trees. Advances in Knowledge Discovery and Data Mining, 9th Pacific-Asia Conference, PAKDD 2005 (2005) 605–610
8. Kuncheva, L.I., Whitaker, C.J.: Measures of diversity in classifier ensembles and their relationship with the ensemble accuracy. Machine Learning **51** (2003) 181–207
9. Ji, C., Ma, S.: Combinations of weak classifiers. In: IEEE Transactions on Neural Networks. Volume 8. (1997) 494–500
10. Quinlan, J.R.: C4.5 : programs for machine learning. Morgan Kaufmann, San Mateo, Calif. (1993) The latest version of C5 is available from http://www.rulequest.com.
11. Buttrey, S., Kobayashi, I.: On strength and correlation in random forests. In: Proceedings of the 2003 Joint Statistical Meetings. (2003)
12. Hastie, T., Tibshirani, R., Friedman, J.: The elements of statistical learning : Data mining, Inference, and Prediction. Springer-Verlag (2001)
13. Holte, R.C., Acker, L., Porter, B.W.: Concept learning and the problem of small disjuncts. IJCAI (1989) 813–818
14. Blake, C., Merz, C.: UCI repository of machine learning databases (1998)

Further Improving Emerging Pattern Based Classifiers Via Bagging

Hongjian Fan[1], Ming Fan[2], Kotagiri Ramamohanarao[1], and Mengxu Liu[2]

[1] Department of CSSE, The University of Melbourne,
Parkville, Vic 3052, Australia
{hfan, rao}@csse.unimelb.edu.au
[2] Department of Computer Science, Zhengzhou University,
Zhengzhou, China
{mfan, mxliu}@zzu.edu.cn

Abstract. Emerging Patterns (EPs) are those itemsets whose supports in one class are significantly higher than their supports in the other class. In this paper we investigate how to "bag" EP-based classifiers to build effective ensembles. We design a new scoring function based on growth rates to increase the diversity of individual classifiers and an effective scheme to combine the power of ensemble members. The experimental results confirm that our method of "bagging" EP-based classifiers can produce a more accurate and noise tolerant classifier ensemble.

Keywords: emerging patterns, classification, bagging, ensemble learning.

1 Introduction

Classification is one of the fundamental tasks in machine learning that has been studied substantially over decades. Recent studies [1, 8, 9] show that classification ensemble learning techniques such as Bagging [2] and Boosting [6] are very powerful for increasing accuracy by generating and aggregating multiple classifiers.

Classification based on patterns is a relatively new methodology. Patterns are conjunctions of simple conditions, where each conjunct is a test of the value of one of the attributes. Emerging Patterns (EPs) [4] are defined as multivariate features (i.e., patterns or itemsets) whose supports (or frequencies) change *significantly* from one class to another. As a relatively new family of classifiers, EP-based classifiers such as the CAEP classifier [5] and the JEP-classifier [7] are not only highly accurate but also easy to understand. It is an interesting question how to combine multiple EP-based classifiers to further improve the classification accuracy.

Bagging of previous EP-based classifiers (such as the CAEP classifier and the JEP-Classifier) does not work because of the following reasons: (1) these classifiers - using a scoring function that aggregates supports - heavily biased toward the support of EPs; (2) the supports remain relatively stable with respect to different samples. These properties are very similar to the Naive Bayes (NB) classifier, as it is remarked in [1] that NB is "very stable". It is well recognized that an important pre-requisite for classification ensemble learning to reduce test error is to generate a diversity of ensemble members. Therefore, our aim is to produce multiple diverse EP-based classifiers with respect to different bootstrap samples. Our solution is a new scoring function for EPs-based classifiers. The key idea is to abandon the use of support in the scoring function, while making good use of the discriminating information (i.e., growth rates) contained in EPs. Our scoring function not only maintains the high accuracy, but also makes the classifiers diverse with respect to different bootstrap samples. We also develop a new method for combining the knowledge learned in each individual classifier. Instead of simply using majority voting, we only consider the votes of member classifiers that have good knowledge about a specific test - if a classifier does not have enough knowledge about the test, its right of voting is deprived. We carried out experiments on a number of benchmark datasets to study the performance of our new scoring function and voting scheme. The results show that our method of creating ensembles often improve classifier performance vs. learning a classifier over the entire dataset directly.

We highlight the following contributions. First, we studied bagging of the EP-based classifiers for the first time. Our analysis shows that CAEP classifier and JEP-Classifier are stable inducers due to their scoring function favoring EPs' support rather than EPs' discriminating power (growth rates). Second, we proposed a new scoring function for EP-based classifiers, which maintains the excellent accuracy while increasing the diversity of ensemble members. Both t-tests and wilcoxon rank sum tests show that the bagged ensemble of the new-scoring-function based classifiers often significantly improves classification performance over an individual classifier. What is more, our ensemble classifiers are superior to other ensemble methods such as bagged C4.5, boosted C4.5 and RandomForest [3]. Lastly, we designed a new scheme to combine the outputs of ensemble members. Different from the static weighting of bagging and boosting, we assign weights to member classifiers dynamically – **instance-based**, based on whether they have specific knowledge to classify the test. Our scheme can also be applied to combine the outputs of other rule based classifiers.

2 A New Scoring Function for EP-Based Classifiers

We assume any instance is represented by an itemset. We say an instance S contains another X, if $X \subseteq S$. The support of X in a dataset D, $supp_D(X)$, is $count_D(X)/|D|$, where $count_D(X)$ is the number of instances in D containing X.

We first use a two-class problem to illustrate the main idea of our scoring function and then discuss how to generalize it in the case of more than two classes. Let the training dataset D contain two classes: $D = D_i \cup D_{\bar{i}}$. Suppose X is an EP of class C_i and S is a test to classify. We define $GrowthRate_i(X) = supp_i(X)/supp_{\bar{i}}(X)$. If $X \not\subseteq S$, we can not use X to determine whether S belongs to class C_i. However, if $X \subseteq S$, we can use it effectively: we predict that S belongs to class C_i with confidence of $GrowthRate_i(X)/(GrowthRate_i(X) + 1)$. This is because

$$\frac{GrowthRate_i(X)}{GrowthRate_i(X) + 1} = \frac{supp_i(X)/supp_{\bar{i}}(X)}{supp_i(X)/supp_{\bar{i}}(X) + 1} = \frac{supp_i(X)}{supp_i(X) + supp_{\bar{i}}(X)}.$$

Similarly, we predict that S does not belong to class C_i (belonging to $C_{\bar{i}}$ instead) with confidence of $1/(GrowthRate_i(X) + 1) = supp_{\bar{i}}(X)/(supp_i(X) + supp_{\bar{i}}(X))$.

Note that if X is a JEP (where $GrowthRate_i(X) = \infty$), we let $GrowthRate_i(X)/(GrowthRate_i(X)+1) = 1$ and $1/(GrowthRate_i(X)+1) = 0$.

To determine whether S belongs to class C_i, we may also consider EPs of class $C_{\bar{i}}$. Let Y be an EP of $C_{\bar{i}}$ and $Y \subseteq S$. Note that $GrowthRate_{\bar{i}}(Y) = supp_{\bar{i}}(Y)/supp_i(Y)$. Using Y, we predict that S belongs to C_i with confidence of $1/(GrowthRate_{\bar{i}}(Y)+1) = supp_i(X)/(supp_i(X)+supp_{\bar{i}}(X))$. When Y has large growth rate, the impact of Y on the final decision is very small and hence negligible. However, when its growth rate is relatively small (e.g., $GrowthRate_{\bar{i}}(Y) < 5$), its impact should be considered.

For a k-class ($k \geq 2$) problem, where $D = D_1 \cup D_2 \cup \cdots \cup D_k$, we use the one-against-all class binarization technique to handle it. For each class D_i, we discover a set $E(C_i)$ of EPs from $(D - D_i)$ to D_i, and a set $E(\bar{C}_i)$ of EPs from D_i to $(D - D_i)$, where $\bar{C}_i$ refers to the non-C_i class $(D - D_i)$. We then use the following scoring function.

Definition 1. *Given a test instance T, a set $E(C_i)$ of EPs of data class C_i and a set $E(\bar{C}_i)$ of EPs of data class non-C_i, the **score** of T for the class C_i is defined as*

$$score(T, C_i) = \sum_{X \subseteq T, X \in E(C_i)} \frac{GrowthRate_i(X)}{GrowthRate_i(X)+1} + \sum_{Y \subseteq T, Y \in E(\bar{C}_i)} \frac{1}{GrowthRate_{\bar{i}}(Y)+1}.$$

Note $GrowthRate(X) = supp_{C_i}(X)/supp_{\bar{C}_i}(X)$ since $X \in E(C_i)$; $GrowthRate(Y) = supp_{\bar{C}_i}(Y)/supp_{C_i}(Y)$ since $Y \in E(\bar{C}_i)$.

Then we have the following: $score(T, C_i)$

$$= \sum_{X \subseteq T, X \in E(C_i)} \frac{supp_{C_i}(X)/supp_{\bar{C}_i}(X)}{supp_{C_i}(X)/supp_{\bar{C}_i}(X)+1} + \sum_{Y \subseteq T, Y \in E(\bar{C}_i)} \frac{1}{1+supp_{\bar{C}_i}(Y)/supp_{C_i}(Y)}$$

$$= \sum_{X \subseteq T, X \in E(C_i)} \frac{supp_{C_i}(X)}{supp_{C_i}(X)+supp_{\bar{C}_i}(X)} + \sum_{Y \subseteq T, Y \in E(\bar{C}_i)} \frac{supp_{C_i}(Y)}{supp_{C_i}(Y)+supp_{\bar{C}_i}(Y)}$$

$$\therefore score(T, C_i) = \sum_{X \subseteq T,\ X \in E(C_i) \cup E(\bar{C}_i)} \frac{supp_{C_i}(X)}{supp_{C_i}(X)+supp_{\bar{C}_i}(X)}.$$

Let the impact of an EP be its support in class C_i divided by the support across all classes. The impact measures how much more frequently an EP appear in its home class than in the whole dataset. The above formula effectively means summing up the contributions of all EPs that are contained in the test.

3 An Improved Voting Scheme for Classifier Combination

Given a number of independently learned EP-based classifiers, we must combine their knowledge effectively. A reasonable combining scheme is to simply let all the classifiers vote equally for the class to be predicted. However, some member classifiers may have no EPs to use to classify a test instance (where the scores for all classes will be zero). These classifiers should be deprived of their rights to vote. The ensemble scheme is formally shown in Definition 2.

Definition 2. *Given the ensemble classifier C^* (the combination of N classifiers built from N bagged training datasets $C_1, C_2, \cdots, C_N$) and a test instance $T = (x_t, y_t)$*

Table 1. Win/Draw/Loss record

EP_{base} vs Alternatives using direct accuracy comparison			
EP_{base} Vs	C4.5	SVM	JEP-C
Win/Draw/Loss	18, 0, 9	14, 0, 13	18, 1, 8
using t-tests for significance			
EP_{base} Vs	C4.5	SVM	JEP-C
Win/Draw/Loss	12, 9, 6	7, 11, 9	7, 16, 4
using Wilcoxon signed rank test			
EP_{base} Vs	C4.5	SVM	JEP-C
Win/Draw/Loss	13, 8, 6	10, 7, 10	11, 9, 7

EP_{bag} vs Alternatives using direct accuracy comparison			
EP_{bag} Vs	$C4.5_{bag}$	$C4.5_{boost}$	Forest
Win/Draw/Loss	18, 2, 7	20, 1, 6	17, 2, 8
using t-tests for significance			
EP_{bag} Vs	$C4.5_{bag}$	$C4.5_{boost}$	Forest
Win/Draw/Loss	12, 11, 4	15, 9, 3	11, 12, 4
using Wilcoxon signed rank test			
EP_{bag} Vs	$C4.5_{bag}$	$C4.5_{boost}$	Forest
Win/Draw/Loss	17, 6, 4	18, 6, 3	14, 9, 4

with labels $y_t \in Y = \{1, 2, \cdots, k\}$, the final classification of the ensemble is $C^*(T) = \arg\max_{y \in Y} \sum_{i=1}^{TN} \delta(C_i(T) = y)$, where $C_i(T)$ represents the output of classifier C_i for the test T, $\delta(\text{true}) = 1$ and $\delta(\text{false}) = 0$. Note that $C_i(T) = -1$ when C_i fails to classify T; otherwise, $C_i(T) = j, j \in Y = \{1, 2, \cdots, k\}$.

Our voting scheme is different from the static weighting of bagging and boosting. It assigns weights to member classifiers dynamically – **instance-based**, based on whether they have specific knowledge to classify the test. Our scheme can also be applied to combine the outputs of other rule based classifiers.

4 Experimental Evaluation

We evaluate the proposed approaches to learning by experiments on 27 well-known datasets from the UCI Machine Learning Repository. We use WEKA [10]'s Java implementation of C4.5, SVM, RandomForest, bagging and boosting. The accuracy was obtained by using the methodology of *stratified* ten-fold cross-validation (CV-10).

Since we will use the newly proposed scoring function (definition 1) as the base classifier (denoted as EP_{base}) to create classifier ensembles, we investigate its performance first. We do not provide detailed classifier accuracy due to the space constraint. Instead we present a win/draw/loss summary in Table 1 (left part) to compare overall performance of EP_{base} against each other classifier (C4.5, SVM, JEP-Classifier). We find that EP_{base} achieves an average accuracy similar to other classifiers (SVM and JEP-Classifier) and higher than C4.5.

Then we investigate the performance of bagging our new EP-based classifier. We choose 51 bags, generating 51 diverse ensemble members. The ensemble classifier is denoted as EP_{bag}. The results clearly show that EP_{bag} is superior to single EP-based classifier: t-tests show that EP_{bag} is significantly better than EP_{base} on 14 datasets and never significantly worse on the remaining 13 datasets. The improvement is due to the diversity of ensemble members. EP_{bag} is also superior to bagged C4.5, boosted C4.5 and RandomForest, as validated by t-tests and Wilcoxon signed rank test for significance (Table 1 right part).

The number of trials TN is equal to the number of classifiers built. We plot the effect of TN on accuracy in Figure 1. Not surprisingly, as TN increases, the performance of

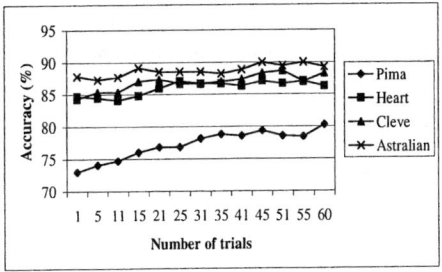 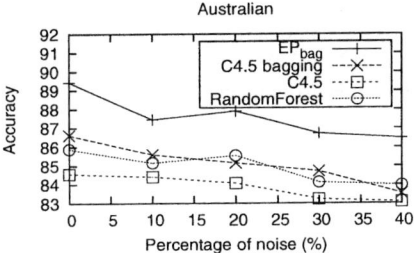

Fig. 1. Accuracy with respect to number of trials

Fig. 2. The effect of increasing noise on classification accuracy

the ensemble classifier usually improves, although there are fluctuations. We expect the ensemble of EP-based classifiers maintains the ability of noise tolerance. From Figure 2, we see clearly that the EP-ensemble classifier has good noise tolerance and consistently achieves higher accuracy than C4.5 and RandomForest across all noise levels.

5 Conclusions

In this paper, we discussed why the "bagging" of CAEP and JEP-Classifier produces no gain. Based on the analysis, we propose a new scoring function to use EPs in classification. This new EP classifier is not only highly correct, but also give diversified outputs on different bootstrap samples. The two characteristics of our new EP classifier are important for the success of creating ensembles of them. We also develop a new, dynamic (instance based) voting scheme to combine the output. This voting scheme can be applied to combine the results of other rule-based classifiers. The experiments show that our method is able to create very effective ensembles of EP-based classifiers.

References

1. Eric Bauer and Ron Kohavi. An empirical comparison of voting classification algorithms: Bagging, boosting, and variants. *Machine Learning*, 36(1-2):105–142, 1999.
2. Leo Breiman. Bagging predictors. *Machine Learning*, 24(2):123–140, 1996.
3. Leo Breiman. Random forests. *Machine Learning*, 45(1):5–32, 2001.
4. Guozhu Dong and Jinyan Li. Efficient mining of emerging patterns: Discovering trends and differences. In *Proc. 5th ACM SIGKDD (KDD'99)*, pages 43–52, San Diego, CA, Aug 1999.
5. Guozhu Dong, Xiuzhen Zhang, Limsoon Wong, and Jinyan Li. Caep: Classification by aggregating emerging patterns. In *Proc. 2nd Int'l Conf. on Discovery Science (DS'99)*, pages 30–42, Tokyo, Japan, Dec 1999.
6. Yoav Freund and Robert E. Schapire. Experiments with a new boosting algorithm. In *Proc. Thirteenth Int'l Conf. on Machine Learning*, pages 148–156. Morgan Kaufmann, 1996.
7. Jinyan Li, Guozhu Dong, and Kotagiri Ramamohanarao. Making use of the most expressive jumping emerging patterns for classification. *Knowl. Inf. Syst.*, 3(2):131–145, 2001.
8. G. I Webb. Multiboosting: A technique for combining boosting and wagging. *Machine Learning*, 40(2):159–196, 2000.

9. G.I. Webb and Z. Zheng. Multistrategy ensemble learning: Reducing error by combining ensemble learning techniques. *IEEE Transactions on Knowledge and Data Engineering*, 16(8):980–991, 2004.
10. Ian H. Witten and Eibe Frank. *Data Mining: Practical Machine Learning Tools and Techniques with Java Implementations.* Morgan Kaufmann, San Francisco, CA, 1999.

Improving on Bagging with Input Smearing

Eibe Frank and Bernhard Pfahringer

Department of Computer Science,
University of Waikato,
Hamilton, New Zealand
{eibe, bernhard}@cs.waikato.ac.nz

Abstract. Bagging is an ensemble learning method that has proved to be a useful tool in the arsenal of machine learning practitioners. Commonly applied in conjunction with decision tree learners to build an ensemble of decision trees, it often leads to reduced errors in the predictions when compared to using a single tree. A single tree is built from a training set of size N. Bagging is based on the idea that, ideally, we would like to eliminate the variance due to a particular training set by combining trees built from all training sets of size N. However, in practice, only one training set is available, and bagging simulates this platonic method by sampling with replacement from the original training data to form new training sets. In this paper we pursue the idea of sampling from a kernel density estimator of the underlying distribution to form new training sets, in addition to sampling from the data itself. This can be viewed as "smearing out" the resampled training data to generate new datasets, and the amount of "smear" is controlled by a parameter. We show that the resulting method, called "input smearing", can lead to improved results when compared to bagging. We present results for both classification and regression problems.

1 Introduction

Ensembles of multiple prediction models, generated by repeatedly applying a base learning algorithm, have been shown to often improve predictive performance when compared to applying the base learning algorithm by itself. Ensemble generation methods differ in the processes used for generating multiple different base models from the same set of data. One possibility is to modify the input to the base learner in different ways so that different models are generated. This can be done by resampling or reweighting instances [1, 2], by sampling from the set of attributes [3], by generating artificial data [4], or by flipping the class labels [5]. A different possibility is to modify the base learner so that different models can be generated from the same data. This is typically done by turning the base learner into a randomized version of itself, e.g. by choosing randomly among the best splits at each node of a decision tree [6]. This paper investigates an ensemble learning method that belongs to the former category. We call it "input smearing" because we randomly modify the attribute values of an instance, thus smearing it out in instance space. We show that, when combined with bagging, this method can improve on using bagging alone, if the amount of smearing is chosen appropriately for each dataset. We show that this can be reliably achieved using internal cross-validation, and present results for classification and regression problems.

The motivation for using input smearing is that it may be possible to increase the diversity of the ensemble by modifying the input even more than bagging does. The aim of ensemble generation is a set of classifiers such that they are simultaneously as different to each other as possible while remaining as accurate as possible when viewed individually. Independence—or "diversity"—is important because ensemble learning can only improve on individual classifiers when their errors are not correlated. Obviously these two aims—maximum accuracy of the individual predictors and minimum correlation of erroneous predictions—conflict with each other, as two perfect classifiers would be rather similar, and two maximally different classifiers could not at the same time both be very accurate. This necessary balance between diversity and accuracy has been investigated in various papers including [7], which among other findings reported that bagged trees are usually much more uniform than boosted trees. But it was also found that increasing levels of noise lead to much more diverse bagged trees, and that bagging starts to outperform boosted trees for high noise levels.

Commonly the attribute values of the examples are not modified in any way in the ensemble generation process. One exception to this "rule" is called "output smearing" [5], which modifies the class labels of examples by adding a controlled amount of noise. In this paper we investigate the complimentary process of applying "smearing" not to the output variable, but to the input variables. Initial experiments showed that smearing alone could not consistently improve on bagging. This lead us to the idea of combining smearing and bagging, by smearing the subsamples involved in the bagging process. The amount of smearing enables us to control the diversity in the ensemble, and more smearing increases the diversity compared to bagging alone. However, more smearing also means that the individual ensemble members become less accurate. Our results show that cross-validation can be used to reliably determine an appropriate amount of smearing.

This paper is structured as follows. In Section 2 we discuss previous work on using artificial data in machine learning and explain the process of "input smearing" in detail. Section 3 presents our empirical results on classification and regression datasets, and Section 4 discussed related work. Section 5 summarizes our findings and points out directions for future work.

2 Using Artificial Training Data

One way of viewing input smearing is that artificial examples are generated to aid the learning process. Generating meaningful artificial examples may seem straightforward, but it is actually not that simple. The main issue is the problem of generating meaningful class values or labels for fully artificially generated examples. Theoretically, if the full joint distribution of all attributes including the class attribute were known, examples could simply be drawn according to this full joint distribution, and their class labels would automatically be meaningful. Unfortunately this distribution is not available for practical learning problems.

This "labelling" problem is the most likely explanation as to why artificially generated training examples are rarely used. One exception is the approach reported in [8]. This work is actually not concerned with improving the predictive accuracy of an ensemble, but instead tries to generate a single tree with similar performance to an ensemble generated by an ensemble learning method. The aim is to have a comprehensible model with a similar predictive performance as the original ensemble. The method generates artificial examples and uses the induced ensemble to label the new examples. It has been shown that large sets of artificial examples can lead to a large single tree capable of approximating the predictive behaviour of the original ensemble.

Another exception is the work presented in [9], which investigates the problem of very skewed class distributions in inductive learning. One common idea is oversampling of the minority class to even out the class distribution, and [9] takes that one step further by generating new artificial examples for the minority class. This is done by randomly selecting a pair of examples from the minority class, and then choosing an arbitrary point along the line connecting the original pair. Furthermore the method makes sure that there is no example from the majority class closer to the new point than any of the minority examples. The main drawback of this method is that it is very conservative, and that it relies on nearest neighbour computation, which is of questionable value in higher-dimensional settings. In the case of highly skewed class distributions such conservativeness might be appropriate, but in more general settings it is rather limiting.

Finally, the Decorate algorithm [4] creates artificial examples adaptively as an ensemble of classifiers is being built. It assigns labels to these examples by choosing those labels that the existing ensemble is least likely to predict. It is currently unclear why this method works well in practice [4].

We have chosen a very simple method for generating artificial data to improve ensemble learning. Our method addresses the labelling problem in a similar fashion as what has been done for skewed class distributions, taking the original data as the starting point. However, we then simply modify the attribute values of a chosen instance by adding random attribute noise. The method we present here combines bagging with this modification for generating artificial data. More specifically, as in bagging, training examples are drawn with replacement from the original training set until we have a new training set that has the same size as the original data. The next step is new: instead of using this new dataset as the input for the base learning algorithm, we modify it further by perturbing the attribute values of all instances by a small amount (excluding the class attribute). This perturbed data is then fed into the base learning algorithm to generate one ensemble member. The same process is repeated with different random number seeds to generate different datasets, and thus different ensemble members.

This method is very simple and applicable to both classification and regression problems (because the dependent variable is not modified), but we have not yet specified how exactly the modification of the original instances is performed. In this paper we make one simplification: we restrict our attention to datasets with numeric attributes. Although the process of input smearing can be applied to nominal data as well (by changing a given attribute value with a certain probability to a different value) it can be more naturally applied with numeric attributes because they imply a notion of distance. To modify the numeric attribute values of an instance we simply add Gaussian noise to them. We take the variance of an attribute into account by scaling the amount of noise based on this variance (using Gaussian noise with the same variance for every attribute would obviously not work, given that attributes in practical datasets are often on different scales). More specifically, we transform an attribute value $a_{original}$ into a smeared value $a_{smeared}$ based on

$$a_{smeared} = a_{original} + p * N(0, \sigma_a),$$

where σ_a is the estimated global standard deviation for attribute $a_{original}$, and p is a user-specifiable parameter that determines the amount of noise to add. The original class value is left intact.

Usually the value of the smearing parameter is greater than zero but the optimum value depends on the data. Cross-validation is an obvious method for finding an appropriate value in a purely data-dependent fashion, and as we will see in the next section, it chooses quite different values depending on the dataset. In the experiments reported below we employed internal cross-validation in conjunction with a simple grid search, evaluating different values for p in a range of values that is explored in equal-size steps. As it turns out, there are datasets where no smearing ($p = 0$) is required to achieve maximum accuracy.

Another view of input smearing is that we employ a kernel density estimate of the data, placing a Gaussian kernel on every training instance, and then sample from this estimate of the joint distribution of the attribute values. We choose an appropriate kernel width by evaluating the cross-validated accuracy of the resulting ensemble (and combine the smearing process with bagging) but an alternative approach would be to first fit a kernel density estimate to the data by some regularized likelihood method, and then use the resulting kernel widths to generate a smeared ensemble. A potential drawback of our method is that the amount of noise is fixed for every attribute (although it is adjusted based on the attributes' scales). It may be that performance can be improved further by introducing a smearing parameter for every attribute and tuning those smearing parameters individually. Using an approach based on kernel density estimation may make this computationally feasible.

Note that, compared to using bagging alone, the computational complexity remains unchanged. Modifying the attribute values can be done in time linear in the number of attributes and instances. The cross-validation-based grid search for the optimal smearing parameter increases the runtime by a large constant factor but it may be possible to improve on this using a more sophisticated search strategy in place of grid search.

Figure 1 shows the pseudo code for building an ensemble using input smearing. The process for making a prediction (as well as the type of base learner

```
method inputSmearing(Dataset D, Ensemble size n, Smearing parameter p)

  compute standard deviation σ_a for each attribute a in the data
  repeat n times
    sample dataset R of size |D| from D using sampling with replacement
    S = ∅
    for each instance x in R
      for each attribute a in R
        x'_a = x_a + p * N(0, σ_a)
      add x' to S
    apply based learner to S and add resulting model to committee
```

Fig. 1. Algorithm for generating an ensemble using input smearing

employed) depends on whether we want to tackle a regression problem or a classification problem. In the case of regression we simply average the predicted numeric values from the base models to derive an ensemble prediction. In the case of classification, we average the class probability estimates obtained from the base models, and predict the class for which the average probability is maximum. (In the experiments reported in the next section we use exactly the same method for bagging.)

3 Experimental Results

In this section we conduct experiments on both classification and regression problems to compare input smearing to bagging. As a baseline we also present results for the underlying base learning algorithm when used to produce a single model. The main parameter needed for input smearing, the noise threshold p, is set automatically using cross-validation, as explained above. We will see that this automated process reliably chooses appropriate values. Consequently input smearing competes well with bagging.

3.1 Classification

Our comparison is based on 22 classification problems from the UCI repository [10]. We selected those problems that exhibit only numeric attributes. Missing values (present in one attribute of one of the 22 datasets, the breast-w data) are not modified by our implementation of smearing.

Input smearing was applied in conjunction with unpruned decision trees built using the fast REPTree decision tree learner in Weka. REPTree is a simple tree learner that uses the information gain heuristic to choose an attribute and a binary split on numeric attributes. It avoids repeated re-sorting at the nodes of the tree, and is thus faster than C4.5. We performed ten iterations to build ten ensemble members. Internal 5-fold cross-validation was used to choose an appropriate parameter value for the smearing parameter p for each training set. To identify a good parameter value we used a simple grid search that evaluated values 0, 0.05, 0.1, 0.15, 0.2, 0.25, and 0.3. This automated parameter estimation

Table 1. Input smearing applied to classification problems

Dataset	Input smearing	Bagging	Unpruned tree	C4.5	Parameter value
balance-scale	85.8±3.6	81.2±3.8•	78.5±4.4•	78.1±4.1•	0.27±0.05
breast-w	96.0±2.1	95.5±2.0	93.7±2.3•	94.9±2.3	0.19±0.10
ecoli	84.7±5.6	83.1±5.4	82.0±5.5	82.8±5.3	0.22±0.08
glass	74.9±9.3	76.5±9.1	69.7±8.9	68.1±8.2•	0.06±0.07
hayes-roth	81.1±9.3	80.7±9.6	84.1±9.5	79.0±8.4	0.10±0.12
heart-statlog	80.8±6.5	78.8±6.6	74.9±7.3•	78.6±7.1	0.19±0.10
ionosphere	91.6±5.4	91.0±4.6	89.6±5.0	90.0±5.0	0.15±0.09
iris	96.1±5.0	95.3±5.5	94.3±5.6	95.4±5.4	0.17±0.10
letter	92.1±0.7	91.9±0.7	87.9±0.7•	88.1±0.8•	0.14±0.04
liver-disorders	69.0±7.0	69.8±7.7	64.5±8.1	66.2±7.8	0.08±0.08
mfeat	77.6±2.6	73.5±2.6•	68.5±3.1•	71.4±2.7•	0.28±0.03
optdigits	95.9±0.9	94.9±1.1•	90.8±1.2•	90.6±1.1•	0.29±0.02
page-blocks	97.2±0.6	97.3±0.6	96.8±0.6•	97.0±0.7	0.02±0.03
pendigits	98.4±0.4	98.1±0.5•	96.4±0.5•	96.5±0.6•	0.16±0.04
pima-diabetes	75.3±4.4	75.0±4.8	71.1±4.6•	73.8±5.3	0.18±0.09
segment	97.4±1.0	97.5±1.1	96.6±1.3	96.8±1.2	0.01±0.02
sonar	81.5±8.5	81.3±8.2	77.5±9.0	74.3±9.5	0.14±0.09
spambase	94.6±1.0	94.6±1.0	92.8±1.3•	92.7±1.2•	0.00±0.00
spectf	88.5±5.1	89.3±4.7	86.0±5.3	84.8±5.8	0.03±0.05
vehicle	74.9±4.1	75.0±4.5	72.4±4.5	73.4±4.2	0.15±0.09
waveform	82.6±1.8	81.8±1.9	75.3±2.0•	75.3±1.9•	0.25±0.06
wine	95.5±4.5	95.4±4.7	93.9±6.0	92.7±6.6	0.17±0.11

• denotes a statistically significant degradation compared to input smearing

adds a large computational overhead, but prevents the user from bad choices, and might also provide valuable insights into both the data as well as the example generation process.

Table 1 lists the estimated classification accuracy in percent correct, obtained as averages over 100 runs of the stratified hold-out method. In each run 90% of the data was used for training and 10% was used for testing. The corrected resampled t-test [11] was used to perform pairwise comparison between algorithms.

Apart from the results for input smearing, the table also lists results for bagging, unpruned decision trees generated using REPTree, and pruned C4.5 trees. It also shows the average parameter value chosen by the internal cross-validation, and the standard deviation for each of the statistics across the 100 runs. Bagging was applied in conjunction with the same base learner and the same number of iterations as input smearing.

Analyzing the results of Table 1, we see that "input smearing" can improve the predictive accuracy of single trees for about half of all the datasets, and also significantly outperforms bagging four times. More importantly, it never performs significantly worse than any of the other algorithms. The average values chosen for p vary from 0 up to 0.29. Given that the latter value is quite close to the upper boundary of the range that we searched in our experiments, it may be possible that larger values would result in further improvements for the datasets where such a large value was chosen. For all datasets except one a non-zero parameter value is chosen, with *spambase* being the sole exception. We can only speculate why smearing does not work for this dataset. Most likely the noise generation process is not appropriate for this dataset, which consists solely of counts of word occurrences. These are non-negative and generally follow a power law [12]. A more specialized distribution like the Poisson distribution may be more appropriate for smearing in this case. Alternatively, the input variables could also be preprocessed by a logarithmic transformation, which is common practice in statistics for dealing with counts.

One method for analysing the behaviour of a modelling technique is the so-called bias-variance decomposition (see e.g. [13]), which tries to explain the total prediction error as the sum of three different sources of error: bias (i.e. how close is the average model to the actual function?), variance (i.e. how much do the models' guesses "bounce around"?), and intrinsic noise (the Bayes error).

Using the specific approach described in [13], a bias-variance decomposition was computed for all the classification datasets used above for both input smearing and bagging. We would expect that input smearing exhibits a higher bias than bagging on average, as it modifies the input distribution of all attributes. To verify this hypothesis, the relative contribution of bias compared to variance was computed for both methods on each dataset. More specifically, we computed

$$relativeBias = bias/(bias + variance).$$

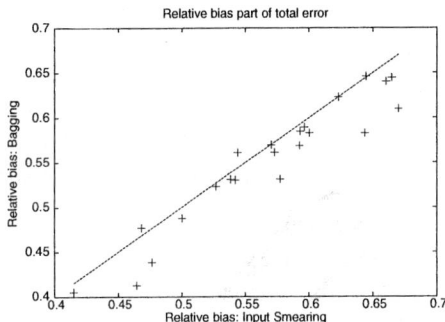

Fig. 2. Relative bias: smearing vs. bagging

In Figure 2 we plot the relative bias of bagging over the relative bias of input smearing. Points below the diagonal indicate cases where smearing exhibits a higher relative bias than bagging. This is the case for most datasets. Some points are very close to the diagonal or exactly on the diagonal. One of these points represents the spambase dataset, where the threshold value of 0.0 effectively turns input smearing into bagging.

3.2 Regression

Classification is not the only application of input smearing. In the following we investigate its performance when applied in conjunction with a state-of-the-art tree learner for regression problems. This comparison is based on a collection of 23 regression problems [14] that are routinely used as benchmarks for evaluating regression algorithms.

We employed the same evaluation framework as in the classification case: ensembles are of size ten and random train/test splits of 90%/10% are repeated

Table 2. Input smearing applied to regression problems

Dataset	Input smearing	Bagging	Pruned model trees	Unpruned model trees	Parameter value
2dplanes	22.9±0.3	23.2±0.3 •	22.7±0.3 ○	23.3±0.3 •	0.30±0.00
ailerons	39.2±1.3	39.2±1.3	39.9±1.2 •	41.1±1.3 •	0.00±0.00
bank32nh	68.5±2.3	69.0±2.4 •	67.0±2.5 ○	74.5±2.8 •	0.19±0.07
bank8FM	19.4±0.7	19.5±0.7 •	20.0±0.7 •	20.4±0.7 •	0.06±0.02
cal-housing	44.0±1.6	44.0±1.6	48.5±2.1 •	46.4±1.8 •	0.00±0.00
cpu-act	13.2±1.0	13.8±1.2 •	14.7±1.7 •	15.3±2.5 •	0.14±0.04
cpu-small	16.1±1.3	16.2±1.4	17.4±2.0 •	17.7±2.3 •	0.07±0.03
delta-ailerons	53.2±2.1	53.2±2.2	54.4±2.1 •	54.5±2.2 •	0.06±0.03
delta-elevators	59.8±1.4	60.0±1.5 •	60.1±1.4 •	61.0±1.6 •	0.18±0.05
diabetes-numeric	94.4±39.4	94.9±42.0	98.5±49.5•	96.8±44.7	0.13±0.10
elevators	34.1±6.1	33.4±1.2	32.1±1.2 ○	35.5±1.3 •	0.01±0.02
fried	25.9±0.4	26.1±0.4 •	27.8±0.5 •	28.1±0.5 •	0.05±0.01
house-16H	62.6±4.6	62.0±4.5 ○	68.0±3.2 •	66.7±3.6 •	0.01±0.02
house-8L	57.7±7.0	57.7±7.0	59.7±3.5 •	59.7±3.6 •	0.00±0.01
kin8nm	53.7±1.6	54.4±1.8 •	60.9±2.1 •	59.9±2.1 •	0.10±0.03
machine-cpu	36.0±12.3	35.7±11.8	40.5±18.5•	36.0±14.1	0.14±0.12
pol	13.6±1.0	13.5±1.0	15.2±1.2 •	14.8±1.1 •	0.02±0.02
puma32H	26.0±0.8	26.1±0.8 •	27.1±0.8 •	27.5±0.9 •	0.05±0.01
puma8NH	56.9±1.5	57.7±3.9 •	57.0±1.6	59.1±1.8 •	0.12±0.03
pyrim	58.5±21.3	57.3±21.7	64.9±26.2•	58.8±25.2	0.09±0.11
stock	13.9±1.9	14.2±2.5	14.4±2.5 •	14.3±2.6 •	0.07±0.03
triazines	79.8±13.9	79.6±13.9	84.0±17.4•	81.4±17.8	0.00±0.03
wisconsin	94.4±11.0	95.1±10.4	98.1±12.4•	98.7±12.8•	0.19±0.10

•/○ denote a statistically significant degradation/improvement wrt input smearing.

100 times (in this case without applying stratification, of course). Performance is measured based on the root relative squared error. A value of zero would indicate perfect prediction, and values larger than 100 indicate performance worse than simply predicting the global mean of the class-values obtained from the training data. Unpruned M5 model trees [15], generated using the M5' model tree learner in Weka [16], were used as the base learner for input smearing and bagging, and we compare to single unpruned and pruned M5 model trees. Again, the noise parameter p was determined automatically by internal five-fold cross-validation using a grid search on the values 0, 0.05, 0.1, 0.15, 0.2, 0.25, and 0.3.

Again, analyzing the results of Table 2, we see that input smearing almost always improves prediction over single model trees. However, it is significantly worse than a single pruned tree on three datasets. Compared to bagging, significant improvements are achieved 39% of the time, with only one significant loss. As with classification, the average smearing parameter values chosen by cross-validation are well below 0.3 in most cases, except for one dataset (2dplanes), where an even larger parameter value may have been chosen if it had been available. Again there is one dataset where zero is chosen consistently. As we are not familiar with the actual meaning of the attributes in this dataset (ailerons), we cannot make such strong claims as for the spambase dataset, but at least one third of all attributes in this dataset again appear to be based on counts, and another third of all attributes is almost constant, i.e. clearly not normally distributed either. Inspecting the attribute distributions for the only other two datasets with smearing parameter values close to 0 (house-8L and triazines) reveals that in both datasets a majority of attributes again is not normally distributed.

4 Related Work

In this section we discuss related work but restrict our attention to ensemble generation methods. We do not repeat the discussion of methods that have already been discussed in Section 2. In terms of ensemble generating methods we only list and discuss methods that modify the data in some way.

- Bagging [1] has its origin in bootstrap sampling in statistics, which produces robust estimates of population statistics by trying to simulate averaging over all possible datasets of a given size. Sets are generated by sampling with replacement. Bagging can reduce the variance of a learner, but it cannot reduce its bias.
- Dagging [17] is an alternative to bagging that combines classifiers induced on disjoint subsets of the data. It is especially appropriate when either the data originally comes from disjoint sources, or when data is plentiful, i.e. when the learning algorithm has reached the plateau on the learning curve. Like bagging, dagging could potentially be combined with input smearing to increase diversity.
- Output smearing [5] adds a controlled amount of noise to the output or dependent attribute only. The empirical results in [5] show that is works surprisingly well as an ensemble generator. An interesting question for future work is whether input and output smearing can be combined successfully.
- Random feature subsets [3, 18] work particularly well for so-called stable algorithms like the nearest neighbour classifier, where bagging does not achieve much improvement. Random feature projections [19] may have some potential in this setting as well.

5 Conclusions

We have described a new method for ensemble generation, called input smearing, that works by sampling from a kernel density estimator of the underlying distribution to form new training sets, in addition to resampling from the data itself like in bagging. Our experimental results show that it is possible to obtain significant improvements in predictive accuracy when applying input smearing instead of bagging (which can be viewed as a special case of input smearing in our implementation). Our results also show that it is possible to use cross-validation to determine an appropriate amount of smearing on a per-dataset basis.

Input smearing using Gaussian noise is not necessarily the best choice. An avenue for future work is to investigate the effect of other distributions in input smearing, and to choose an appropriate distribution based on the data. Such a more sophisticated approach should also make it possible to generalize input smearing to other attribute types and structured input.

References

1. Breiman, L.: Bagging predictors. Machine Learning **24** (1996) 123–140
2. Freund, Y., Schapire, R.E.: Experiments with a new boosting algorithm. In: Thirteenth Int Conf on Machine Learning. (1996) 148–156
3. Bay, S.D.: Nearest neighbor classification from multiple feature subsets. Intelligent Data Analysis **3** (1999) 191–209
4. Melville, P., Mooney, R.J.: Creating diversity in ensembles using artificial data. Journal of Information Fusion (Special Issue on Diversity in Multiple Classifier Systems) **6/1** (2004) 99–111
5. Breiman, L.: Randomizing outputs to increase prediction accuracy. Machine Learning **40** (2000) 229–242
6. Breiman, L.: Random forests. Machine Learning **45** (2001) 5–32
7. T.Dietterich: An experimental comparison of three methods for constructing ensembles of decision trees: Bagging, boosting, and randomization. Machine Learning **40** (2000) 139–157
8. Domingos, P.: Knowledge acquisition from examples via multiple models. In: Proc. 14th Int Conf on Machine Learning. (1997) 98–106
9. N.V. Chawla, K.W.Bowyer, L., W.P.Kegelmeyer: Smote: Synthetic minority oversampling technique. Journal of Artificial Intelligence Research **16** (2002) 321–357
10. D.J. Newman, S. Hettich, C.B., Merz, C.: UCI repository of machine learning databases (1998)
11. C.Nadeau, Y.Bengio: Inference for the generalization error. Machine Learning **52** (2003) 239–281
12. Rennie, J.D.M., Shih, L., Teevan, J., Karger, D.R.: Tackling the poor assumptions of naive Bayes text classifiers. In: Proc Twentieth Int Conf on Machine Learning, AAAI Press (2003) 616–623
13. Kohavi, R., Wolpert, D.H.: Bias plus variance decomposition for zero-one loss functions. In: Proc Thirteenth Int Conf on Machine Learning. (1996) 275–283
14. Torgo, L.: Regression datasets (2005) [www.liacc.up.pt/~ltorgo/Regression].
15. Quinlan, J.R.: Learning with Continuous Classes. In: Proc 5th Australian Joint Conf on Artificial Intelligence, World Scientific (1992) 343–348
16. Wang, Y., Witten, I.: Inducing model trees for continuous classes. In: Proc of Poster Papers, European Conf on Machine Learning. (1997)
17. Ting, K., Witten., I.: Stacking bagged and dagged models. In: Fourteenth Int Conf on Machine Learning (ICML07). (1997) 367–375
18. Ho, T.K.: The random subspace method for constructing decision forests. IEEE Transactions on Pattern Analysis and Machine Intelligence **20** (1998) 832–844
19. Achlioptas, D.: Database-friendly random projections. In: Twentieth ACM Symposium on Principles of Database Systems. (2001) 274–281

Boosting Prediction Accuracy on Imbalanced Datasets with SVM Ensembles

Yang Liu, Aijun An, and Xiangji Huang

Department of Computer Science and Engineering,
York University,
Toronto, Ontario, M3J 1P3, Canada
{yliu, aan, jhuang}@cs.yorku.ca

Abstract. Learning from imbalanced datasets is inherently difficult due to lack of information about the minority class. In this paper, we study the performance of SVMs, which have gained great success in many real applications, in the imbalanced data context. Through empirical analysis, we show that SVMs suffer from biased decision boundaries, and that their prediction performance drops dramatically when the data is highly skewed. We propose to combine an integrated sampling technique with an ensemble of SVMs to improve the prediction performance. The integrated sampling technique combines both over-sampling and under-sampling techniques. Through empirical study, we show that our method outperforms individual SVMs as well as several other state-of-the-art classifiers.

1 Introduction

Many real-world datasets are imbalanced, in which most of the cases belong to a larger class and far fewer cases belong to a smaller, yet usually more interesting class. Examples of applications with such datasets include searching for oil spills in radar images [1], telephone fraudulent detection [2], credit card fraudulent detection diagnosis of rare diseases, and network intrusion detection. In such applications, the cost is high when a classifier misclassifies the small (positive) class instances.

Despite the importance of handling imbalanced datasets, most current classification systems tend to optimize the overall accuracy without considering the relative distribution of each class. As a result, these systems tend to misclassify minority class examples when the data is highly skewed. Techniques have been proposed to handle the problem. Approaches for addressing the problem can be divided into two main directions: sampling approaches and algorithm-based approaches. Generally, sampling approaches include methods that over-sample the minority class to match the size of the majority class [3,4], and methods that under-sample the majority class to match the size of the minority class [1,5,6,7]. Algorithmic-based approaches are designed to improve a classifier's performance based on their inherent characteristics.

This paper is concerned with improving the performance of the Support Vector Machines (SVMs) on imbalanced data sets. SVMs have gained success in many applications, such as text mining and hand-writing recognition. However, when the data is highly imbalanced, the decision boundary obtained from the training data is biased toward the minority class. Most approaches proposed to address this problem have been algorithm-based [8, 9, 10], which attempt to adjust the decision boundary through modifying the decision function.

We take a complementary approach and study the use of sampling as well as ensemble techniques to improve SVM's performance. First, our observation indicates that using over-sampling alone as proposed in previous work (e.g. SMOTE [10]) can introduce excessive noise and lead to ambiguity along decision boundaries. We propose to integrate the two types of sampling strategies by starting with over-sampling the minority class to a moderate extent, followed by under-sampling the majority class to the similar size. This is to provide the learner with more robust training data. We show by empirical results that the proposed sampling approach outperforms over-sampling alone irrespective of the parameter selection. We further consider using an ensemble of SVMs to boost the performance. A collection of SVMs are trained individually on the processed data, and the final prediction is obtained by combining the results from those individual SVMs. In this way, more robust results can be obtained by reducing the randomness induced by a single classifier, as well as by alleviating the information loss due to sampling.

2 Related Work

Sampling is a popular strategy to handle the class imbalance problem since it straightforwardly re-balances the data at the data processing stage, and therefore can be employed with any classification algorithm [1, 3, 4, 5, 6, 7]. As one of the successful oversampling methods, the SMOTE algorithm [11] over-samples the minority class by generating interpolated data. It first searches for the K-nearest-neighbors for each minority instance, and for each neighbor, randomly selects a point from the line connecting the neighbor and the instance itself, which will serve as a new minority instance. By adding the "new" minority instances into training data, it is expected that the over-fitting problem can be alleviated. SMOTE has been reported to achieve favorable results in many classification algorithms [11, 12]. Algorithm-based approaches include methods in which existing learning algorithms are tailored to improve the performance for imbalanced datasets. For example, some algorithms consider class distributions or use cost functions for decision tree inductions [6, 13, 14].

SVMs have established themselves as a successful approach for various machine learning tasks. The class imbalance issue has also been addressed in the literature. Through empirical study, Wu et al. [9] report that when the data is highly imbalanced, the decision boundary determined by the training data is largely biased toward the minority class. As a result, the false negative rate that associates with the minority class might be high. To compensate for the skewness, they propose to enlarge the resolution around the decision boundary by revising kernel functions. Furthermore, Veropoulos et al. [8] use pre-specified penalty constants on Lagrange multipliers for different classes; Akbani et al. [10] combine SVMs with SMOTE over-sampling and cost sensitive learning. In contrast, Japkowicz et al. [15] argue that SVMs are immune to the skewness of the data, because the classification decision boundary is determined only by a small quantity of support vectors. Consequently, the large volume of instances belonging to the majority class might be considered redundant. In this paper, we will demonstrate that the decision boundary changes as imbalance ratios vary, and discuss its implications.

Using an ensemble of classifiers to boost classification performance has also been reported to be effective in the context of imbalanced data. This strategy usually makes use of a collection of individually trained classifiers whose prediction results are integrated to make the final decision. The work in this direction includes that Chen et al. [6] use random forest to unite the results of decision trees induced from bootstrapping the training data, and that Guo et al [4] apply data boosting to improve the performance on hard examples that are difficult to classify. However, most current studies are confined to decision tree inductions instead of other classifiers, e.g, SVM. Moreover, decision-tree-based algorithms might be ill-suited for the class imbalance problem as they favor short trees.

3 Background

3.1 Support Vector Machines

In this section we briefly describe the basic concepts in two-class SVM classification. Assume that there is a collection of n training instances $Tr = \{x_i, y_i\}$, where $\mathbf{x_i} \in \mathcal{R}^N$ and $y_i \in \{-1, 1\}$ for $i = 1, \ldots, n$. Suppose that we can find some hyperplane which linearly separates the positive from negative examples in a feature space. The points $\mathbf{x}$ belonging to the hyperplane must satisfy $\mathbf{w} \cdot \mathbf{x} + b = 0$, where $\mathbf{w}$ is normal to the hyperplane and b is the intercept. To achieve this, given a kernel function K, a linear SVM searches for Lagrange multiplier α_i ($i = 1, ..., n$) in Lagrangian

$$L_p \equiv \frac{1}{2}||w||^2 - \sum_{i=1}^{n} \alpha_i y_i(x_i \cdot w + b) + \sum_{i=1}^{n} \alpha_i \quad (1)$$

such that the margin between two classes $\frac{2}{||w||}$ is maximized in the feature space [16]. In addition, in the α_i optimizing process, Karush Kuhn Tucker (KKT) conditions which require $\sum_{i=1}^{n} \alpha_i y_i = 0$, must be satisfied.[1] To predict the class label for a new case x, we need to compute the sign of $f(x) = \sum_{i=1}^{n} y_i \alpha_i K(x, x_i) + b$. If the sign function is greater than zero, x belongs to the positive class, and the negative otherwise.

In SVMs, support vectors (SVs) are of crutial importance to the training set. They lie closest to the decision boundary; thus form the margin between

---

[1] In the case of non-separable data, 1-norm soft-margin SVMs minimize the Lagrangian $L_p = \frac{1}{2}||w||^2 + C\sum_i \xi_i - \sum_i \alpha_i\{y_i(x_i \cdot w + b) - 1 + \xi_i\} - \sum_i \mu_i \xi_i$, where ξ_i, $i \in [1, n]$ are positive slack variables, C is selected by users with a larger C indicating a higher penalty to errors, and μ_i are Lagrange multipliers to enforce ξ_i being positive. Similarly, corresponding KKT conditions have to be met for the purpose of optimization.

two sides. If all other training data were removed, and training was repeated, the same separating hyperplane would still be constructed. Note that there is a Lagrange multiplier α_i for each training instance. In this context, SVs correspond to those points for which $\alpha_i > 0$; other training instances have $\alpha_i = 0$. This fact gives us the advantage of classifying by learning with only a small number of SVs, as all we need to know is the position of the decision boundary which lies right in the middle of the margin; other training points can be considered redundant. Further, it is of prime interest in the class imbalance problem because SVMs could be less affected by the negative instances that lie far away from the decision boundary even if there are many of them.

3.2 Effects of Class Imbalance on SVMs

We conducted a series of experiments to investigate how the decision boundaries are affected by the imbalance ratio, i.e., the ratio between the number of negative examples and positive examples. We start with classifying a balanced training dataset, and detect that the real decision boundary is close to the "ideal boundary", as it is almost of equal length to both sides. We then reform successive new datasets with different degrees of data skewness by removing instances from the positive and add instances to the negative. Figure 1 reflects the data distribution when imbalance ratios vary from 10:1 to 300:1, where crosses and circles represent the instances from positive and negative classes respectively. From Figure 1 (a), we find that if the imbalance ratio is moderate, the boundary will still be close to the "ideal boundary". This observation demonstrates SVMs could be robust and self-adjusting; and is thus able to alleviate the problem arising from moderate imbalance. Nonetheless, as the imbalance ratio becomes larger and larger, as illustrated in Figure 1 (b) and (c), the boundaries get evidently biased toward the minority class. As a consequence, making predictions with such a system may lead to a high false negative rate.

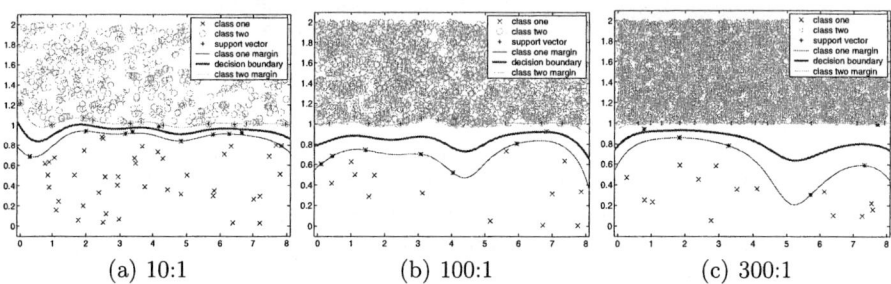

Fig. 1. Boundary changes with different imbalance ratios

4 Re-balancing the Data

We have shown that SVMs may perform well while the imbalance ratio is moderate. Nonetheless, their performance could still suffer from the extreme data

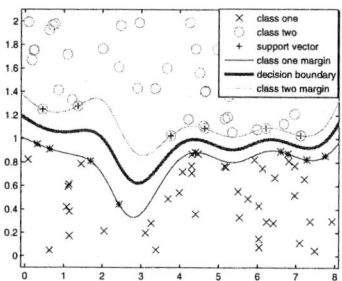

Fig. 2. Under-sampling majority instances

skewness. To cope with this problem, in this section, we study the use of sampling techniques to balance the data.

4.1 Undersampling

Under-sampling approaches have been reported to outperform over-sampling approaches in previous literatures. However, under-sampling throws away potentially useful information in the majority class; it thus could make the decision boundary trembling dramatically. For example, given the imbalance ratio as 100:1, in order to get a close match for the minority, it might be undesirable to throw away 99% of majority instances. Figure 2 illustrates such a scenario, where the majority class is undersampled to keep the same size as the minority, but a considerable amount of SVs lie far away from the ideal boundary $y = 1$. Accordingly, predicting with such SVMs may lead to low accuracies.

4.2 Oversampling

Considering that simply replicating the minority instances tends to induce overfitting, using interpolated data is often preferred in the hope of supplying additional and meaningful information on the positive class. SMOTE is the method that has been mostly cited along this line.

However, the improvement of integrating SVMs with the SMOTE algorithm can be limited due to its dependence on the proper selection of the number of nearest neighbors K as well as imbalance ratios. Basically, the value of K determines how many new data points will be added into the interpolated dataset. Figure 3 shows how the decision boundary will change with different K values. Figure 3 (a) shows the original class distribution while the imbalance ratio is 100:1. Figure 3 (b) demonstrates that the classification boundary is relatively smoothed when K has a small value; nonetheless, it is still biased toward the minority class. This is due to SMOTE actually providing little information of the minority; hence the oversampling in this case should be considered as a type of "phantom-transduction". When the interpolated dataset is considerably enlarged as K increases, as shown in Figure 3 (c), ambiguities could arise along the current boundary, because SMOTE makes the assumption that the instance between a positive class instance and its nearest neighbors is also positive. However

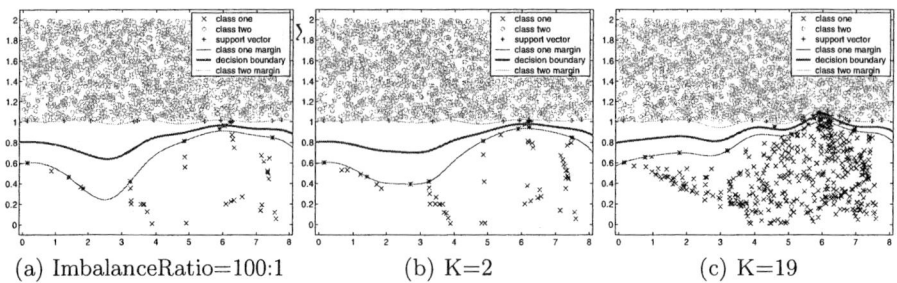

Fig. 3. Using SMOTE with different K values

it may not be always true in practice. As a positive instance is very close to the boundary, its nearest neighbor is likely to be negative, and this possibility may increase as K and imbalance ratio become larger. Consequently, the new data instance, which actually belongs to the negative class, is mis-labeled as positive, and the induced decision boundary, as shown in Figure 3 (c), could be inversely distorted to the majority class.

4.3 Combination of Two Types of Samplings

To address the problems arising from using each of the two types of sampling approaches alone, we integrate them together. Given an imbalance ratio, we first over-sample the minority instances with SMOTE to some extent, and then under-sample the majority class so that both sides have the same or similar amount of instances. To under-sample the majority class, we use the bootstrap sampling approach with all available majority instances, provided that the size of the new majority class is the same as that of the minority class after running SMOTE. The benefit of doing so is that this approach inherits the strength of both strategies, and alleviates the over-fitting and information loss problems.

In addition, to avoid taking risks of inducing ambuities along the decision boundary, we choose to filter out the "impure" data firstly before sampling. In this context, an instance is defined to be "impure", if and only if two of its three nearest neighbors provide different class labels other than that of itself. This idea is motivated by the *Edited Nearest Neighbor Rule* [7], which was originally used to remove unwanted instances from the majority. In our work, however, to further reduce the uncertainty from both classes, such a filtering process is taken on each side.

5 Ensemble of SVMs

In this section, we present a method that uses an ensemble of SVM classifiers integrated with a re-balancing technique that combines both over-sampling and under-sampling. Re-balancing is still necessary in this context since in learning from extremely imbalanced data, it is very likely that a bootstrap sample used to train an SVM in the ensemble is composed of few or even none of the minority instances. Hence, each component learner of the ensemble would suffer from severe skewness, and the improvement of using an ensemble would be confined. Our proposed method, called *EnSVM*, is illustrated in Figure 4. As described in

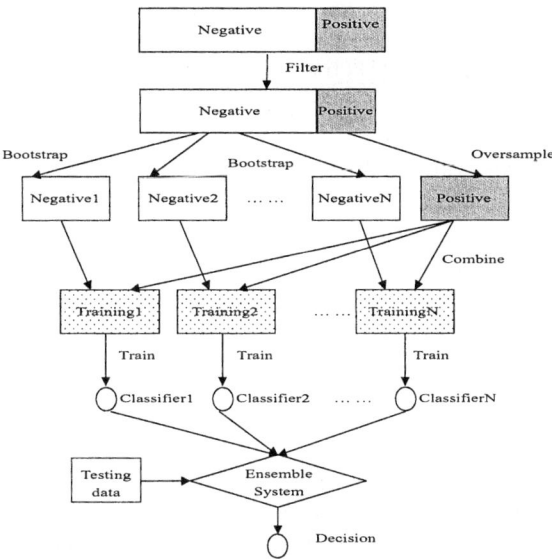

Fig. 4. EnSVM algorithm

Section 4.3, we start re-balancing the data by filtering out impurities which may induce ambiguities. Then, the minority class is over-sampled with the SMOTE method to smooth the decision boundary. That is, for each positive instance, it finds the K nearest neighbors, draws a line between the instance and each of its K nearest neighbors, and then randomly selects a point on each line to use as a new positive instance. In this way, $K \times n$ new positive instances are added to the training data, where n is the number of positive instances in the original training data. After that, we under-sample the majority class instances N times to generate N bootstrap samples so that each bootstrap sample has the same or similar size with the over-sampled positive instances. Then, each bootstrap sample (of the majority class) is combined with the over-sampled positive instances to form a training set to train an SVM. Therefore, N SVMs can be obtained from N different training sets. Finally, the N SVMs are combined to make a prediction on a test example by casting a *majority vote* from the ensemble of SVMs. In our experiments reported below, we set N to be 10.

6 Empirical Evaluation

In this section, we first introduce the evaluation measures used in our study, and then describe the datasets. After that, we report the experimental results that compare our proposed approach with other methods.

6.1 Evaluation Measures

The evaluation measures used in our experiments are based on the *Confusion Matrix*. Table 1 illustrates a confusion matrix for a two class problem with *pos-*

Table 1. Two-class confusion matrix

	Predicted Positive	Predicted Negative
Actual Positive	TP(True Positive)	FN(False Negative)
Actual Negative	FP(False Positive)	TN(True Negative)

itive and *negative* class values. With this matrix, our performance measures are expressed as follows:

- $g\text{-}mean = \sqrt{a^- \times a^+}$, where $a^- = \frac{TN}{TN+FP}$ and $a^+ = \frac{TP}{TP+FN}$;
- $F\text{-}measure = \frac{2 \times Precision \times Recall}{Precision + Recall}$, where $precision = \frac{TP}{TP+FP}$ and $recall = \frac{TP}{TP+FN}$.

G-mean is based on the recalls on both classes. The benefit of selecting this metric is that it can measure how balanced the combination scheme is. If a classifier is highly biased toward one class (such as the majority class), the *g-mean* value is low. For example, if $a^+ = 0$ and $a^- = 1$, which means none of the positive examples is identified, *g-mean*=0. In addition, *F-measure* combines the recall and precision on the positive class. It measures the overall performance on the minority class. Besides, we utilize the ROC analysis [17] to assist the evaluation. A ROC curve demonstrates a trade off between true positive and false positive rates provided with different classification parameters. Informally, one point in ROC space is superior to another if it is closer to the northwest corner (TP is higher, but FP is lower). Thus, ROC curves allow for a visual comparison of classifiers: the larger the area below the ROC curve, the higher classification potential of the classifier.

6.2 Benchmark Data

We use five datasets as our testbeds. Four of the datasets are from the UCI Machine Learning Repository and another dataset is a medical compound dataset (mcd) collected by National Cancer Institute (NCI) for discovering new compounds capable of inhibiting the HIV virus. The four UCI datasets are *spambase, letter-recognition, pima-indians-diabetes* and *abalone*. Each dataset in this study is randomly split into training and test subsets of the same size, where a stratified manner is employed to ensure that the training and test sets have the same imbalance ratio. Table 2 shows the characteristics of the five datasets. The first

Table 2. Benchmark datasets

Dataset	Datapoints	Attributes	ImbalanceRatio
letter	20000	16	2:1
pima	768	9	2:1
spambase	3068	57	10:1
abalone	4280	8	40:1
mcd	29508	6	100:1

three datasets (letter, pima, and spambase) are mildly imbalanced, while the next two (abalone and mcd) are very imbalanced. These datasets were carefully selected to (1) fulfill the requirements that they are obtained in real applications, (2) distinct from feature characteristics, and vary in size and imbalance ratio, and (3) maintain sufficient amount of instances in each individual class to keep the classification performance.

6.3 Experimental Results

In this section, we compare the performance of our proposed *EnSVM* method with those of five other methods: 1) single SVM without re-sampling the data, 2) single SVM with over-sampling using SMOTE [10] (without applying cost functions), 3) random forest with balanced training data from under-sampling [6], 4) random forest with our combined sampling method, and 5) single SVM with our combined sampling method. In our experiments, for all the SVMs, we employed Gaussian RBF kernels of the form $K(x_i, x_j) = \exp(-\gamma |x_i - x_j|^2)$ of C-SVMs. For each method we repeated our experiments ten times, computed average g-mean values and F-measures.

Table 3. Performance in terms of g-mean

Dataset	SVM	SMOTE K=1	SMOTE K=highest	RandForest[1]	RandForest[2]	AvgSVM	EnSVM K=1	EnSVM K=highest
letter	0.9551	0.9552	0.9552	0.9121	0.9281	0.9563	**0.9566**	**0.9566**
pima	0.6119	0.7320	0.7320	0.7358	0.7002	0.7419	**0.7503**	**0.7503**
spam	0.8303	0.8364	0.8580	0.8593	**0.9050**	0.8592	0.8616	0.8988
abalone	0.6423	0.6280	0.8094	0.7358	0.7678	0.8041	**0.8958**	0.8311
mcd	0.4500	0.4496	0.5952	0.5896	0.5968	0.5931	0.5951	**0.6039**

Results in terms of g-mean are shown in Table 3, where *SVM* denotes the single SVM method with the original training data, *SMOTE* represents oversampling the minority class and then training a system with single SVMs, *RandForest*[1] denotes undersampling the majority class and then making an ensemble with C4.5 decision trees, *RandForest*[2] denotes sampling data with our combined method, followed by forming an ensemble with C4.5, *AvgSVM* denotes the average performance of 10 single SVMs with our sampling method, and *EnSVM* is our ensemble method with the combined sampling method. For the first two datasets, the K values for *SMOTE* and *EnSVM* can only be set to be 1 since their imbalance ratio is 2:1. For each of other datasets, we test two K values: the smallest value, which always equals to 1, and the highest value. The latter will depend on the imbalance ratios of three datasets, which are 9, 39, and 99 respectively. From the results we can see that *EnSVM* achieves the best results on all the datasets except on the spam dataset for which *RandForest*[2] is the best. [2]

Table 4 shows the performance for each method in terms of F-measure. We find that *EnSVM* deserves the highest value on all five datasets. In particular, a big improvement is made on the datasets where the imbalance ratios are large. By comparing the results from the four SVM methods, we can see that (1) using SMOTE to over-sample the data is better than SVM without sampling; (2) using our combined sampling method with single SVMs is better than using only over-sampling with SMOTE; and (3) using the ensemble method together with the combined sampling method achieve the best results. By comparing the two Random Forest methods, using the combined sampling method is better than

---

[2] In Table 3, from top to bottom, the optimal γ obtained empirically in using SVMs is 1.0×10^{-2}, 5.0×10^{-5}, 7.0×10^{2}, and 10^2 respectively. In addition, C is set to be 1000 for each case.

Table 4. Performance in F-measure

Dataset	SVM	SMOTE K=1	SMOTE K=highest	RandForest[1]	RandForest[2]	AvgSVM	EnSVM K=1	EnSVM K=highest
letter	0.9548	0.9549	0.9549	0.9111	0.9268	0.9406	**0.9563**	**0.9563**
pima	0.5664	0.7135	0.7135	0.7098	0.6165	0.7259	**0.7357**	**0.7357**
spam	0.8164	0.8238	0.8492	0.8512	0.8751	0.7498	0.8553	**0.8950**
abalone	0.5843	0.5659	0.7938	0.7938	0.7426	0.7875	**0.8940**	0.8190
mcd	0.3367	0.3364	0.5285	0.5285	0.5286	0.5274	0.5272	**0.5415**

using only the under-sampling method on most datasets. Moreover, between the Random Forest method and the ensemble of SVMs method, the latter performs better.

In addition to the imbalance ratio, the selection of K may also impact on the prediction accuracy of *SMOTE* and *EnSVM*. To make a better understanding, we present a *ROC* analysis result with the *spambase* dataset. This dataset is considered since it has a moderate imbalance ratio and instance volume. The original *spambase* has an imbalance ratio of 10; therefore, in this experiment, we test K from 1 to 9, and depict the ROC curves of the two approaches in Figure 5. Clearly, compared to simply over-sampling the minority instances, *EnSVM* generates a better result. We also test how the g-mean value may change with different Ks in *SMOTE* and *EnSVM*. The *abalone* and *mcd* datasets are used in this case as they hold large imbalance ratios and allow K to vary in relatively large ranges. We set parameter K to vary from 1 to 39 for the *abalone* dataset and from 1 to 99 for the *mcd* dataset. As shown in Figures 6.3 (a) and (b), the prediction performance of *EnSVM* is superior to simply applying the *SMOTE* algorithm with respect to each K value. Moreover, we can see that the optimal K value can be difficult to determine in both *SMOTE* and *EnSVM*. For *EnSVM*, when K is small, we get *better* neighbors for the oversampling process, so the prediction performance can be dramatically improved. Further, when K is big, more noise is likely to be introduced, but a larger training data set is generated using *EnSVM* and less information is lost. Consequently, it becomes a trade off between inducing more noise and losing less information. Nonetheless, our method is better than *SMOTE* with all K values.

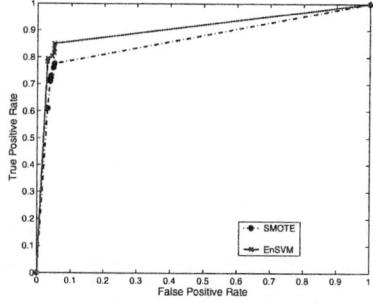

Fig. 5. ROC curve of spambase dataset

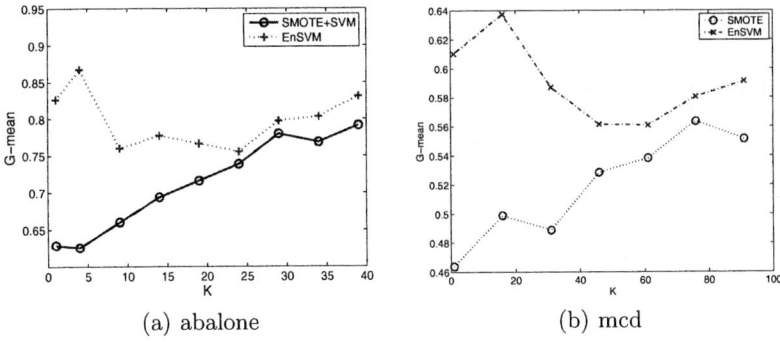

(a) abalone (b) mcd

Fig. 6. G-mean wrt. different K values

7 Conclusions

This paper introduces a new approach to learning from imbalanced datasets through making an ensemble of SVM classifiers and combining both oversampling and under-sampling techniques. We first show in this study that using SVMs for class prediction can be influenced by the data imbalance, although SVMs can adjust itself well to some degree of data imbalance. To cope with the problem, re-balancing the data is a promising direction, but both undersampling and oversampling have limitations. In our approach, we integrate the two types of sampling strategies together. Over-sampling the minority class provides complementary knowledge for the training data, and under-sampling alleviates over-fitting problem. In addition, we make an ensemble of SVMs to enhance the prediction performance by casting a majority vote. Through extensive experiments with real application data, our proposed method is shown to be effective and better than several other methods with different data sampling methods or different ensemble methods. We are now working on a method for automatically determining the value of K based on the data set characteristics in order to optimize the performance of *EnSVM*.

References

1. Kubat, M., Holte, R.C., Matwin, S.: Machine learning for the detection of oil spills in satellite radar images. Mach. Learn. **30** (1998) 195–215
2. Fawcett, T., Provost, F.J.: Adaptive fraud detection. Data Mining and Knowledge Discovery **1** (1997) 291–316
3. Ling, C.X., Li, C.: Data mining for direct marketing: Problems and solutions. In: KDD. (1998) 73–79
4. Guo, H., Viktor, H.L.: Learning from imbalanced data sets with boosting and data generation: the databoost-im approach. SIGKDD Explorations **6** (2004) 30–39
5. Kubat, M., Matwin, S.: Addressing the curse of imbalanced training sets: one-sided selection. In: Proc. 14th International Conference on Machine Learning. (1997) 179–186

6. Chen, C., Liaw, A., Breiman, L.: Using random forest to learn imbalanced data. Technical Report 666, Statistics Department, University of California at Berkeley (2004)
7. Wilson, D.R., Martinez, T.R.: Reduction techniques for instance-basedlearning algorithms. Mach. Learn. **38** (2000) 257–286
8. Veropoulos, K., Cristianini, N., Campbell, C.: Controlling the sensitivity of support vector machines. In: International Joint Conference on Artificial Intelligence(IJCAI99). (1999)
9. Wu, G., Chang, E.Y.: Aligning boundary in kernel space for learning imbalanced dataset. In: ICDM. (2004) 265–272
10. Akbani, R., Kwek, S., Japkowicz, N.: Applying support vector machines to imbalanced datasets. In: ECML. (2004) 39–50
11. Chawla, N.V., Bowyer, K.W., Hall, L.O., Kegelmeyer, W.P.: Smote: Synthetic minority over-sampling technique. J. Artif. Intell. Res. (JAIR) **16** (2002) 321–357
12. Chawla, N.V., Lazarevic, A., Hall, L.O., Bowyer, K.W.: Smoteboost: Improving prediction of the minority class in boosting. In: PKDD. (2003) 107–119
13. Weiss, G.M., Provost, F.J.: Learning when training data are costly: The effect of class distribution on tree induction. J. Artif. Intell. Res. (JAIR) **19** (2003) 315–354
14. Drummond, C., Holte, R.C.: C4.5, class imbalance, and cost sensitivity: Why under-sampling beats over-sampling. In: Workshop on Learning from Imbalanced Datasets II held in conjunction with ICML'2003. (2003)
15. Japkowicz, N., Stephen, S.: The class imbalance problem: A systematic study. Intell. Data Anal. **6** (2002) 429–449
16. Burges, C.J.C.: A tutorial on support vector machines for pattern recognition. Data Mining and Knowledge Discovery **2** (1998) 121–167
17. Swets, J.: Measuring the accuracy of diagnostic systems. Science **240** (1988) 1285–1293

DeLiClu: Boosting Robustness, Completeness, Usability, and Efficiency of Hierarchical Clustering by a Closest Pair Ranking

Elke Achtert, Christian Böhm, and Peer Kröger

Institute for Computer Science, University of Munich, Germany
{achtert, boehm, kroegerp}@dbs.ifi.lmu.de

Abstract. Hierarchical clustering algorithms, e.g. Single-Link or OPTICS compute the hierarchical clustering structure of data sets and visualize those structures by means of dendrograms and reachability plots. Both types of algorithms have their own drawbacks. Single-Link suffers from the well-known single-link effect and is not robust against noise objects. Furthermore, the interpretability of the resulting dendrogram deteriorates heavily with increasing database size. OPTICS overcomes these limitations by using a density estimator for data grouping and computing a reachability diagram which provides a clear presentation of the hierarchical clustering structure even for large data sets. However, it requires a non-intuitive parameter ε that has significant impact on the performance of the algorithm and the accuracy of the results. In this paper, we propose a novel and efficient k-nearest neighbor join closest-pair ranking algorithm to overcome the problems of both worlds. Our density-link clustering algorithm uses a similar density estimator for data grouping, but does not require the ε parameter of OPTICS and thus produces the optimal result w.r.t. accuracy. In addition, it provides a significant performance boosting over Single-Link and OPTICS. Our experiments show both, the improvement of accuracy as well as the efficiency acceleration of our method compared to Single-Link and OPTICS.

1 Introduction

Hierarchical clustering methods determine a complex, nested cluster structure which can be examined at different levels of generality or detail. The complex cluster structure can be visualized by concepts like dendrograms or reachability diagrams. The most well-known hierarchical clustering method is Single-Link [1] and its variants like Complete-Link and Average-Link [2]. Single-Link suffers from the so-called single-link effect which means that a single noise object bridging the gap between two actual clusters can hamper the algorithm in detecting the correct cluster structure. The time complexity of Single-Link and its variants is at least quadratic in the number of objects.

Another hierarchical clustering algorithm is OPTICS [3], which follows the idea of density-based clustering [4], i.e. clusters are regions of high data density separated by regions of lower density. OPTICS solves some of the problems of Single-Link but only to the expense of introducing new parameters *minPts* and ε. The latter is not very intuitive and critical for both, performance of the algorithm and accuracy of the result.

If ε is chosen too low, fundamental information about the cluster structure is lost, if it is chosen too high the performance of the algorithm decreases dramatically.

In this paper, we introduce a novel hierarchical clustering algorithm *DeLiClu* (Density Linked Clustering) that combines the advantages of OPTICS and Single-Link by fading out their drawbacks. Our algorithm is based on a closest pair ranking (CPR). The objective of a CPR algorithm is: given two sets R and S of feature vectors, determine in a first step that pair of objects $(r, s) \in (R \times S)$ having minimum distance, in the next step the second pair, and so on. Well-known CPR algorithms like [5] operate on static data sets which are not subject to insertions or deletions after initialization of the ranking. Our new DeLiClu algorithm, however, needs a ranking algorithm where after each fetch operation for a new pair (r, s) the object s is deleted from S and inserted into R. We show how the ranking algorithm can be modified to allow the required update operations without much additional overhead and how Single-Link can be implemented on top of a CPR. This allows the use of an index structure which makes the algorithm more efficient without introducing the parameter ε like OPTICS does. Finally, we describe how the density-estimator of OPTICS can be integrated into our solution.

The rest of this paper is organized as follows: Sec. 2 discusses related work. In Sect 3 our novel algorithm is described. Sec. 4 presents an experimental evaluation. Sec. 5 concludes the paper.

2 Related Work

Hierarchical Clustering. Hierarchical clustering algorithms produce a nested sequence of clusters, resulting in a binary tree-like representation, a so-called dendrogram. The root of the dendrogram represents one single cluster, containing the n data points of the entire data set. Each of the n leaves of the dendrogram corresponds to one single cluster which contains only one data point. Hierarchical clustering algorithms primarily differ in the way they determine the similarity between clusters. The most common method is the Single-Link method [1] which measures the similarity between two clusters by the similarity of the closest pair of data points belonging to different clusters. This approach suffers from the so-called single-link effect, i.e. if there is a chain of points between two clusters then the two clusters may not be separated. In the Complete-Link method the distance between two clusters is the maximum of all pairwise distances between the data points in the two clusters. Average-Link clustering merges in each step the pair of clusters having the smallest average pairwise distance of data points in the two clusters. A major drawback of the traditional hierarchical clustering methods is that dendrograms are not really suitable to display the full hierarchy for data sets of more than a few hundred compounds. Even for a small amount of data, a reasonable interpretation of the dendrogram is almost impossible due to its complexity. The single-link effect can also be seen in the figure: as an impact of the connection line between the two clusters Single-Link computes no clearly separated clusters.

OPTICS [3] is another hierarchical clustering algorithm, but uses the concept of density based clustering and thus reduces significantly the single-link effect. Additionally, OPTICS is specifically designed to be based on range queries which can be efficiently supported by index-based access structures. The density estimator used by OPTICS

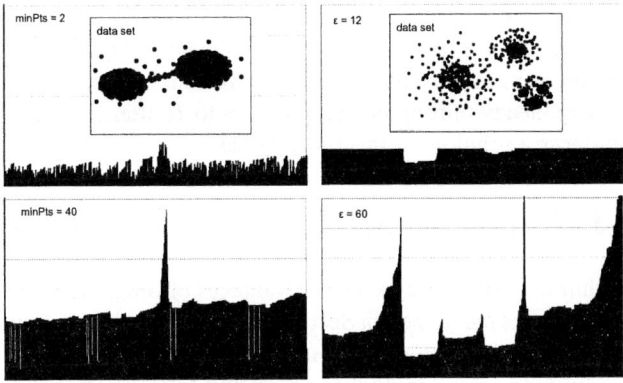

Fig. 1. Impact of parameters *minPts* and ε

consists of two values for each object, the core distance and the reachability distance w.r.t. parameters *minPts* $\in \mathbb{N}$ and $\varepsilon \in \mathbb{R}$. The clustering result can be displayed in a so-called reachability plot that is more appropriate for very large data sets than a dendrogram. A reachability plot consists of the reachability values on the y-axis of all objects plotted according to the cluster order on the x-axis. The "valleys" in the plot represent the clusters, since objects within a cluster have lower reachability distances than objects outside a cluster. Figure 1 shows examples of reachability plots with different parameter settings for ε and *minPts*. The effect of *minPts* to the resulting cluster structure is depicted in the left part of Figure 1. The upper part shows a reachability plot resulting from an OPTICS run with *minPts* $= 2$ where no meaningful cluster structure has been detected. If the value of *minPts* is increased as in the lower part of the figure, the two clusters in the data set can be seen as valleys in the reachability plot. The second parameter ε is much more difficult to determine but has a considerable impact on the efficiency and the accuracy of OPTICS. If ε is chosen too small, fundamental information about the cluster structure will be lost. The right part of figure 1 shows this effect in the upper diagram where the information about clusters consisting of data points with reachability values greater than $\varepsilon = 12$ is no longer existent.

Closest Pair Ranking. The closest pair problem is a classical problem of computational geometry [6]. The intention is to find those two points from given data sets R and S whose mutual distance is the smallest. The CPR determines in the first step that pair of objects in $R \times S$ having the smallest distance, in the next step the second pair, etc. The number of pairs to be reported is *a priori* unknown. In the database context the CPR problem was introduced first in [5], calling it distance join. An incremental algorithm based on the R-Tree family is proposed. For each data set R and S a spatial index is constructed as input. The basic algorithm traverses the two index structures, starting at the root of the two trees. The visited pairs of nodes are kept in a priority queue sorted by their distances. If the first entry of the priority queue exists of a pair of data points, then the pair is reported as the next closest pair. Otherwise, the pair is expanded and all possible pairs formed by inspecting the children of the two nodes are inserted into the priority queue. The algorithm terminates if all closest pairs are reported or the query is stopped by the user. CPR algorithms operate on static data sets, i.e. they do not support insertions or deletions of objects after initializing the ranking query. Our new DeLiClu algorithm, however, needs shifting object s from S to R after reporting pair (r, s). In Section 3 we propose a solution for this special case.

3 Density-Linked Clustering

Our new algorithm DeLiClu combines the advantages of Single-Link and OPTICS by fading out the drawbacks mentioned in Section 2. To achieve these requirements we introduce a density-smoothing factor *minPts* into hierarchical clustering and use as representation of the clustering result reachability plots like OPTICS. In contrast to OPTICS we avoid the introduction of the non-intuitive parameter ε which is critical for both, performance of the algorithm and completeness of the result. In addition, we improve the performance over both algorithms by applying powerful database primitives such as the similarity join and a CPR, and by applying index structures for feature spaces.

3.1 General Idea of DeLiClu

Typical hierarchical clustering algorithms work as follows: They keep two separate sets of points, those points which have already been placed in the cluster structure and those which have not. In each step, one point of the latter set is selected and placed in the first set. The algorithm always selects that point which minimizes the distance to any of the points in the first set. Assume the algorithm has already done part of its work, and some of the points have already been placed in the cluster structure. What actually happens then is that the closest pair is selected between the set R of those points which are already assigned to clusters and the set S of the points which are not yet processed. This means, we can also reformulate the main loop of the general algorithm into:

determine the closest pair $(r, s) \in (R \times S)$;
migrate s from S into R;
append s to cluster structure / reachability plot;

Note that we still have to render more precisely what exactly we mean by the notion *closest pair* because we have to integrate the density-based smoothing factor *minPts* into this notion. Additionally, since the reachability plot shows for each object its reachability distance we have to define a proper density distance for our DeLiClu algorithm. However, this will be done in Section 3.3 and until then, we simply mean the closest pair according to the Euclidean distance and assign each object with its closest pair or nearest neighbor distance to the reachability plot.

If the closest pair from $(R \times S)$ would be determined in each step from scratch, we would do a lot of unnecessary work. Instead, we like to save the status of processing from one call of the closest pair determination to the next one. But since we migrate object s from S to R after the closest pair (r, s) has been reported, we need a ranking algorithm which supports insertions or deletions after initialization. We show in the next section how the standard algorithm [5] can be extended to allow the required object migration during the ranking. The core of our DeLiClu clustering algorithm now is:

1. Let R contain an arbitrary start object from data set $\mathcal{D}$;
2. Let S be $\mathcal{D} \setminus R$;
3. Initialize the CPR over $(R \times S)$;
4. Take the next pair (r, s) from the ranking;
5. Migrate s from S into R;
6. Append s to the reachability plot;
7. Continue with step (4) until all points are handled;

The critical remaining aspects are the migration of point s from S into R (step 5) and the introduction of the density-based smoothing factor *minPts* and a proper density distance definition.

3.2 Closest Pair Ranking with Object Migration

The original algorithm for CPR without object migration requires the two data sets to be stored in hierarchical index structures such as R-trees [7]. The algorithm uses a priority queue into which pairs of nodes and pairs of data objects can be inserted. The entries in the priority queue are ordered by ascending distances between the pair of objects (nodes, respectively) in the data space. Upon each request, the algorithm dequeues the top pair. If it is a pair of data objects, it is reported as the result of the request. Otherwise, the pair is expanded, i.e. for all pairs of child nodes the distances are determined and the pairs are inserted into the queue. Several strategies exist to decide which of the elements of a pair is expanded (left, right, or both). We assume here a symmetric expansion of both elements of the pair. Further, we assume that both indexes have exactly the same structure. Although the tree for R initially contains only the arbitrarily chosen start element, we use a full copy of the directory of S for convenience, because this method facilitates insertion of any element of S into R. We simply use the same path as in the tree storing S. No complex insert and split algorithm has to be applied for insertion.

Whenever a new element s is inserted into the index storing the data set R, we have to determine a suitable path $P = (root, node_1, ..., node_h, s)$ from the root to a leaf node for this element (including the element itself). Comparing the nodes of this path with the nodes of the index for S, we observe that some node pairs might already have been inserted into the priority queue, others may not. Some of the pairs (e.g. $(root_R, root_S)$) might have even already been removed from the priority queue. We call such removed pairs *processed*. Processed pairs are a little bit problematic because they require catch-up work for migrated objects. Processed pairs can be easily found by traversing the tree S top-down. A pair should be in the queue if the parent pair has already been processed (i.e. has a distance smaller than the current top element of the priority queue), but the pair itself has a distance higher than the top element.

After a pair of objects (r, o) has been processed, the formerly not handled object o is migrated from S to the set of already processed objects R. The catch-up work which now has to be done consists of the insertion of all pairs of objects (nodes, respectively) $(o, s) \in R \times S$ into the priority queue for which the parent pair of nodes $(o.parent, s.parent)$ has already been processed. The complete recursive method is called reInsertExpanded and is shown in Figure 2. Initially, reInsertExpanded is called with the complete path of the migrated object o in R and the root node of S.

---
reInsertExpanded(Object[] $path$**, Object** o**)**

if $(path[0], o)$ is a pair of objects **then**
  insert the pair $(path[0], o)$ into priority queue;
if $(path[0], o)$ is a pair of nodes and has not yet been expanded **then**
  insert the pair $(path[0], o)$ into priority queue;
if $(path[0], o)$ is a pair of nodes and has already been expanded **then**
  determine all child nodes o_{child} of o;
  reInsertExpanded(tail($path$), o_{child});

Fig. 2. Algorithm reInsertExpanded

3.3 The Density Estimator MinPts

Until now, we have re-engineered the Single-Link method without applying any density estimator for enhancing the robustness. Our re-engineering has great impact on the performance of the algorithm because now a powerful database primitive is applied to accelerate the algorithm. We will show in Section 4 that the performance is significantly improved. But our new implementation also offers an easy way to integrate the idea of the density estimator *minPts* into the algorithm without using the difficult parameter ε of OPTICS. To determine the reachability distance of an object shown in the reachability plot we consider additionally the k-nearest neighbor distance of the point where $k = minPts$. We call this distance density distance and it is formally defined as follows:

Definition 1 (density distance). *Let $\mathcal{D}$ be a set of objects, $q \in \mathcal{D}$ and* DIST *be a distance function on objects in $\mathcal{D}$. For minPts $\in \mathbb{N}$, minPts $\leq |\mathcal{D}|$ let r be the minPts-nearest neighbor of q w.r.t.* DIST*. The* density distance *of an object $p \in \mathcal{D}$ relative from object q w.r.t. minPts is defined as*

$$\text{DENDIST}_{minPts}(p, q) = \max\{\text{DIST}(q, r), \text{DIST}(q, p)\}.$$

The density distance of of an object p relative from object q is an asymmetric distance measure that takes the density around p into account and is defined as the maximum value of the *minPts*-nearest neighbor distance of p and the distance between p and q. Obviously, the density distance of DeLiClu is equivalent to the reachability distance of OPTICS w.r.t. the same parameter *minPts* and parameter $\varepsilon = \infty$. Our algorithm DeLiClu can adopt the density-based smoothing factor *minPts* by ordering the priority queue using the density distance rather than the Euclidean distance. The rest of the algorithm remains unchanged. Obviously, this modification can be done without introducing the parameter ε. The cluster hierarchy is always determined completely, unlike in OPTICS. And in contrast to OPTICS a guaranteed complete cluster result is not payed with performance deterioration.

The k-nearest neighbor distance where $k = minPts$ can be determined for all points in a preprocessing step which applies a k-nearest neighbor join of the data set. Some methods have been proposed for this purpose [8,9] but unfortunately none for the simple R-tree and its variants. Therefore, we apply a new algorithm which is described in the next section.

3.4 The k-NN Join on the R-Tree

The k-nn join combines each of the points of R with its k nearest neighbors in S. Algorithms for the k-nn join have been reported in [8] and in [9]. The first algorithm is based on the MuX-index structure [10], the latter is on top of a grid order. Unfortunately, there is no k-nn join algorithm for the R-tree family. Thus, in the following we present a k-nn join algorithm based on the R-tree [7] and its variants, e.g. R*-tree[11].

Formally we define the k-nn join as follows:

Definition 2 (k-**nn join** $R \ltimes S$). *Let R and S be sets of objects, and* DIST *be a distance function between objects in R and S. $R \ltimes S$ is the smallest subset of $R \times S$ that contains for each point of R at least k points of S and for which the following condition holds:*

$$\forall (r, s) \in R \ltimes S, \forall (r, s') \in R \times S \setminus R \ltimes S : \text{DIST}(r, s) < \text{DIST}(r, s')$$

Essentially, the k-nn join combines each point of the data set R with its k-nearest neighbors in the data set R. Each point of R appears in the result set exactly k times. Points of S may appear once, more than once (if a point is among the k-nearest neighbors of several points in R) or not at all (if a point does not belong to the k-nearest neighbors of any point in R).

For the k-nn join $R \ltimes S$ based on the R-tree it is assumed that each data set R and S is stored in an index structure belonging to the R-tree family. The data set R of which the nearest neighbors are searched for each point is denoted as the *outer point set*. Consequently, S is the *inner point set*. The data pages of R and S are processed in two nested loops whereas each data page of the outer set R is accessed exactly once. The outer loop iterates over all data pages pr of the outer point set R which are accessed in an arbitrary order. For each data page pr, the data pages ps of the inner point set S are sorted in ascending order to their distance to pr. For each point r stored in the data page pr, a data structure for the k- nearest neighbor distances, short a k-nn distance list, is allocated. The distances of candidate points are maintained in these k-nn distance lists until they are either discarded and replaced by smaller distances of better candidate points or until they are confirmed to be the actual nearest neighbor distances of the corresponding point. A distance is confirmed if it is guaranteed that the database cannot contain any points being closer to the given object than this distance. The last distance value in the k-nn distance list belonging to a point r is the (actual) k-nn distance of r: points and data pages beyond that distance need not to be considered. The pruning distance of a data page is the maximum (actual) k-nn distance of all points stored in this page. All data pages $ps \in S$ having a distance from a given data page $pr \in R$ that exceeds the pruning distance of the data page pr can be safely neglected as join-partners of that data page pr. Thus, in the inner loop only those data pages ps have to be considered having a distance to the current data page pr less or equal than the pruning distance of pr. Analogous, all points s of a data page ps having a distance to a current point r greater than the current k-nn distance of r can be safely pruned and do not have to be taken into consideration as candidate points.

3.5 Algorithm DeLiClu

The algorithm DeLiClu is given in Figure 3. In a preprocessing step, the k-nearest neighbor distance for all points is determined as described in Section 3.4. In the follow-

---
DeLiClu(SetOfObjects S)

kNNJoin(S,S);
copy the index storing S to the index storing R;
s := start object $\in S$;
write (s, ∞) to output;
migrate s from S to R;
add pair $(S.root, R.root)$ to priority queue;
while $S \neq \emptyset$ **do**
   p:= minimum pair in priority queue;
   if $p = (n_S, n_R)$ is a pair of nodes **then**
      insert all combinations of $(n_S.children, n_r.children)$ into priority queue;
   else $p = (s,r)$ is a pair of objects
      write $(s, denDist(s,r))$ to output;
      reInsertExpanded($path(s)$, *root*);

Fig. 3. Algorithm DeLiClu

ing R, denotes the set of objects already processed and S indicates the set of objects which are still not yet handled. The algorithm starts with an arbitrary chosen start object $s \in S$, migrates s from S to R and writes s with a density distance of infinity to output. Note that migration of s from S to R means, that s is stored in the index structure of R in the same path as in S. Thus, we do not need any complex insert or split algorithm upon object migration. The two index structures of R and S only need to have the same structure, i.e. the same directory and data nodes although the tree for R initially contains no point.

The algorithm uses a priority queue into which pairs of nodes and pairs of data objects from $S \times R$ can be inserted. The entries in the priority queue are sorted in ascending order by the distance between the nodes of the pair or the density distance between the objects of the pair. The first pair inserted into the queue is the pair of nodes existing of the root of the index of S and the root of the index of R. In each step, the top pair having minimum distance is dequeued from the priority queue. If it is a pair (n_s, n_r) of nodes, the pair will be expanded, i.e. all combinations of the children of n_s with the children of n_r are inserted into the priority queue. Otherwise, if the top pair of the priority queue consists of a pair (s, r) of data objects from $S \times R$, the not yet processed object $s \in S$ is written to output with the density distance $\text{DENDIST}_{minPts}(s, r)$. Afterwards, s is migrated from S to R. As described in Section 3.2, objects belonging to already expanded nodes of the path of s have to be reinserted into the priority queue by invoking the algorithm *reinsertExpanded* (see Figure 2). The algorithm terminates if all objects are moved from S to R.

4 Experimental Evaluation

All experiments have been performed on Linux workstations with two 64-bit 1.8 GHz CPU and 8 GB main memory. We used a disk with a transfer rate of 45 MB/s, a seek time of 4 ms and a latency delay of 2 ms. For either technique a LRU cache of about 50% of the data set size was allocated. The OPTICS algorithm was supported by an

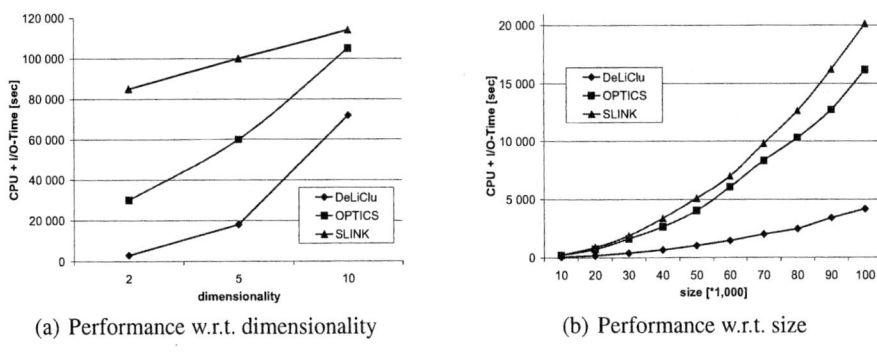

(a) Performance w.r.t. dimensionality   (b) Performance w.r.t. size

Fig. 4. Performance analysis

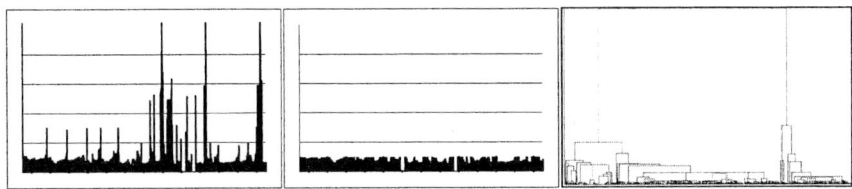

Fig. 5. Comparison of accuracy on real-world data set (El Nino data set)

R-tree index structure. Unless otherwise specified, the *minPts* parameter of DeLiClu and OPTICS was set to 5. The ε-parameter of OPTICS was set to the optimal value w.r.t. accuracy. Performance is presented in terms of the elapsed time including I/O and CPU-time. Beside synthetic data sets, we used a data set containing 500,000 5D featurevectors generated from the SEQUOIA benchmark and the El Nino data set from the UCI KDD data repository, containing about 800 9D data objects.

Performance speed-up. We first compared the performance of the methods. As it can be seen in Figure 4(a) DeLiClu significantly outperforms OPTICS and SLINK w.r.t. the dimensionality of the database. In Figure 4(b), we can observe that DeLiClu also outperforms SLINK and OPTICS w.r.t. the number of data objects is. Obviously, the speed-up of DeLiClu grows significantly with increasing database size. Similar results can be made on the SEQUOIA benchmark (results are not shown due to space limitations). DeLiClu achieved a speed-up factor of more than 20 over OPTICS and a speed-up factor of more than 50 over SLINK.

Improvement of accuracy. The significant effect of parameter ε on the results of the OPTICS algorithm is shown in Figure 5 (El Nino data). The left part of the figure shows a reachability plot resulting from the new algorithm DeLiClu, the middle part of the figure shows a reachability plot resulting from an OPTICS run with parameter ε chosen too small. For this experiment, ε was set to a value for which the runtime of OPTICS was approximately the same as for DeLiClu. Apparently, OPTICS lost a significant part of the whole cluster information due to the wrongly chosen ε. The interpretability of the dendrogram depicted in the right part of the figure is very weak in comparison with the reachability plot resulting from the DeLiClu algorithm. DeLiClu generates strongly separated clusters which cannot be seen in the dendrogram. Similar results have been achieved on the SEQUOIA benchmark.

5 Conclusions

We proposed the new algorithm DeLiClu based on a novel closest pair ranking algorithm that efficiently computes the hierarchical cluster structure. DeLiClu shows improved robustness over Single-Link w.r.t. noise and avoids the single-link effect by using a density estimator. In contrast to OPTICS it guarantees the complete determination of the cluster structure. It has an improved usability over OPTICS by avoiding the non-intuitive parameter ε. Our experimental evaluation showes that DeLiClu significantly outperforms Single-Link and OPTICS in terms of robustness, completeness, usability and efficiency.

References

1. Sibson, R.: SLINK: An optimally efficient algorithm for the single-link cluster method. The Computer Journal **16** (1973)
2. Jain, A.K., Dubes, R.C.: Algorithms for Clustering Data. Prentice Hall (1988)
3. Ankerst, M., Breunig, M.M., Kriegel, H.P., Sander, J.: OPTICS: Ordering points to identify the clustering structure. In: Proc. SIGMOD. (1999)
4. Ester, M., Kriegel, H.P., Sander, J., Xu, X.: A density-based algorithm for discovering clusters in large spatial databases with noise. In: Proc. KDD. (1996)
5. Hjaltason, G.R., Samet, H.: Incremental distance join algorithms for spatial databases. In: Proc. SIGMOD. (1998)
6. Preparata, F.P., Shamos, M.I.: Computational Geometry: An Introduction. Springer Verlag (1985)
7. Guttman, A.: R-Trees: A dynamic index structure for spatial searching. In: Proc. SIGMOD. (1984)
8. Böhm, C., Krebs, F.: The k-nearest neighbor join: Turbo charging the KDD process. KAIS **6** (2004)
9. Xia, C., Lu, H., Ooi, B.C., Hu, J.: GORDER: An efficient method for KNN join processing. In: Proc. VLDB. (2004)
10. Böhm, C., Kriegel, H.P.: A cost model and index archtecture for the similarity join. In: Proc. ICDE. (2001)
11. Beckmann, N., Kriegel, H.P., Schneider, R., Seeger, B.: The R*-Tree: An efficient and robust access method for points and rectangles. In: Proc. SIGMOD. (1990)

Iterative Clustering Analysis for Grouping Missing Data in Gene Expression Profiles

Dae-Won Kim[1,*] and Bo-Yeong Kang[2]

[1] School of Computer Science and Engineering, Chung-Ang University,
Heukseok-dong, Dongjak-gu, 155-756, Seoul, Korea
dwkim@cau.ac.kr
[2] Center of Healthcare Ontology R&D, Seoul National University,
Yeongeon-dong, Jongro-gu, Seoul, Korea

Abstract. Clustering has been used as a popular technique for finding groups of genes that show similar expression patterns under multiple experimental conditions. Because a clustering method requires a complete data matrix as an input, we must estimate the missing values using an imputation method in the preprocessing step of clustering. However, a common limitation of these conventional approach is that once the estimates of missing values are fixed in the preprocessing step, they are not changed during subsequent process of clustering. Badly estimated missing values obtained in data preprocessing are likely to deteriorate the quality and reliability of clustering results. Thus, a new clustering method is required for improving missing values during iterative clustering process.

1 Introduction

Since Eisen et al. first used the hierarchical clustering method to find groups of coexpressed genes [16], numerous methods have been studied for clustering gene expression data: self-organizing map [23], k-means clustering [24], graph-theoretic approach [25], mutual information approach [22], fuzzy c-means clustering [14], diametrical clustering [15], quantum clustering with singular value decomposition [8], bagged clustering [9], CLICK [21], and GK [20]. However, the analysis results obtained by clustering methods will be influenced by missing values in microarray experiments, and thus it is not always possible to correctly analyze the clustering results due to the incompleteness of data sets. The problem of missing values have various causes, including dust or scratches on the slide, image corruption, spotting problems [2, 5]. Ouyang et al. [3] pointed out that most of the microarray experiments contain some missing entries and more than 90 % of rows (genes) are affected.

To convert incomplete microarray experiments to a complete data matrix that is required as an input for a clustering method, we must handle the missing values before calculating clustering. To this end, typically we have either removed the genes with missing values or estimated the missing values using an imputation

---

[*] Corresponding author.

prior to cluster analysis. Of the methods proposed, several imputation methods have been demonstrating their effectiveness in building the complete matrix of clustering: missing values are replaced by zeros [4] or by the average expression value over the row (gene). Troyanskaya et al. [2] presented two correlation-based imputation methods: a singular value decomposition based method (SVDimpute) and weighted K-nearest neighbors (KNNimpute). Besides, a classical Expectation Maximization approach (EMimpute) exploits the maximum likelihood of the convariance of the data for estimating the missing values [5, 3]. However, a common limitation of existing approaches for clustering incomplete microarray data is that the estimation of missing values must be calculated in the preprocessing step of clustering. Once the estimates are found, they are not changed during the subsequent steps of clustering. Thus badly estimated missing values during data preprocessing can deteriorate the quality and reliability of clustering results, and therefore drive the clustering method to fall into a local minimum; it prevents missing values from being imputed by better estimates during the iterative clustering process.

To minimize the influence of bad imputation, in the present study we developed a method for clustering incomplete microarray data, which iteratively finds better estimates of missing values during clustering process. Incomplete gene expression data is used as an input without any prior imputation. This method preserves the uncertainty inherent in the missing values for longer before final decisions are made, and is therefore less prone to falling into local optima in comparison to conventional imputation-based clustering methods. To achieve this, a method for measuring the distance between a cluster centroid and a row (a gene with missing values) is proposed, along with a method for estimating the missing attributes using all available information in each iteration.

2 The Proposed Method

The objective of the proposed method is to classify a data set $X=\{x_1, x_2, \ldots, x_n\}$ in p-dimensional space into k disjoint and homogeneous clusters represented as $C = \{C_1, C_2, \ldots, C_k\}$. Here each data point $x_j = [x_{j1}, x_{j2}, \ldots, x_{jp}]$ $(1 \leq j \leq n)$ is the expression vector of the j-th gene over p-different environmental conditions or samples. A data point with some missing conditions or samples is referred to as an incomplete gene; a gene x_j is incomplete if x_{jl} is missing for $\exists 1 \leq l \leq p$, i.e., an incomplete gene $x_1 = [0.75, 0.73, ?, 0.21]$ where x_{13} is missing. A gene expression data set X is referred to as an incomplete data set if X contains at least one incomplete gene expression vector.

To find better estimates of missing values and improve the clustering result during iterative clustering process, in each iteration we exploit the information of current clusters such as cluster centroids and all available non-missing values. For example, a missing value x_{jl} is estimated using the corresponding l-th attribute value of the cluster centroid to which x_j is closest in each iteration. To improve the estimates during each iteration, the proposed method attempts to optimize the objective function with respect to the missing values, which is often referred to as the alternating optimization (AO) scheme. The objective of the proposed method is obtained by minimizing the function J_m:

$$\min \left\{ J_m(U, V) = \sum_{i=1}^{k} \sum_{j=1}^{n} (\mu_{ij})^m D_{ij} \right\} \quad (1)$$

where

$$D_{ij} = \|x_j - v_i\|^2 \quad (2)$$

is the distance between x_j and v_i,

$$V = [v_1, v_2, \ldots, v_k] \quad (3)$$

is a vector of the centroids of the clusters $C_1, C_2, \ldots, C_k$,

$$U = [\mu_{ij}] = \begin{bmatrix} \mu_{11} & \mu_{12} & \cdots & \mu_{1n} \\ \mu_{21} & \mu_{22} & \cdots & \mu_{2n} \\ \vdots & \vdots & \vdots & \vdots \\ \mu_{k1} & \mu_{k2} & \cdots & \mu_{kn} \end{bmatrix} \quad (4)$$

is a fuzzy partition matrix of X satisfying the following constraints,

$$\mu_{ij} \in [0,1],\ 1 \leq i \leq k,\ 1 \leq j \leq n,$$

$$\sum_{i=1}^{k} \mu_{ij} = 1,\ 1 \leq j \leq n, \quad (5)$$

$$0 < \sum_{j=1}^{n} \mu_{ij} < n,\ 1 \leq i \leq k.$$

and

$$m \in [1, \infty) \quad (6)$$

is a weighting exponent that controls the membership degree μ_{ij} of each data point x_j to the cluster C_i. As $m \to 1$, J_1 produces a hard partition where $\mu_{ij} \in \{0,1\}$. As m approaches infinity, J_∞ produces a maximum fuzzy partition where $\mu_{ij} = 1/k$. This fuzzy k-means-type approach has advantages of differentiating how closely a gene belongs to each cluster [14] and being robust to the noise in microarray data [7] because it makes soft decisions in each iteration through the use of membership functions.

Under this formulation, missing values are regarded as optimization parameters over which the functional J_m is minimized. To obtain a feasible solution by minimizing Eq. 1, the distance D_{ij} between an incomplete gene x_j and a cluster centroid v_i must be calculated as:

$$D_{ij} = \frac{p}{\sum_{l=1}^{p} w_{jl}} \sum_{l=1}^{p} (x_{jl} - v_{il})^2 w_{jl} \quad (7)$$

where

$$w_{jl} = \begin{cases} 1 & \text{if } x_{jl} \text{ is non-missing} \\ 1 - \exp(-t/\tau) & \text{if } x_{jl} \text{ is missing} \end{cases} \quad (8)$$

We differentiate the missing attribute values from the non-missing values in calculating D_{ij}. The fraction part in Eq. 7 indicates that D_{ij} is inversely proportional to the number of non-missing attributes used where p is the number of attributes. w_{jl} indicates the confidence degree with which l-th attribute of x_j contributes to D_{ij}; specifically, $w_{jl} = 1$ if x_{jl} is non-missing and $0 \leq w_{jl} < 1$ otherwise. The exponential decay, $\exp(-t/\tau)$, represents the reciprocal of the influence of the missing attribute x_{jl} on discrete time t where τ is a time constant. At the initial iteration ($t = 0$), w_{jl} has a value of 0. As time t (i.e., the number of iterations) increases, the exponent part decreases fast, and thus w_{jl} approaches 1. Let us consider an incomplete data point $x_1 = [0.75, 0.73, ?, 0.21]$ where initially x_{13} is missing. Suppose that x_{13} is estimated as a value of 0.52 after two iterations; then x_1 has a vector of $[0.75, 0.73, 0.52, 0.21]$. From this vector, we see that x_{13} participates in calculating the distance to cluster centroids less than the other three values because it is now being estimated. Besides, the influence of x_{13} to D_{i1} is increased as the iteration continues because its estimate is improved by an iterative optimization.

Using D_{ij} in Eq. 7, the saddle point of J_m is obtained by considering the constraint Eq. 5 as the Lagrange multipliers:

$$\nabla J_m(U, V, \lambda)$$
$$= \sum_{i=1}^{k} \sum_{j=1}^{n} (\mu_{ij})^m D_{ij} + \sum_{j=1}^{n} \lambda_j \left[\sum_{i=1}^{k} \mu_{ij} - 1 \right] \quad (9)$$

and by setting $\nabla J_m = 0$. If $D_{ij} > 0$ for all i, j and $m > 1$, then (U, V) may minimize J_m only if,

$$\mu_{ij} = \left[\sum_{z=1}^{k} \left(\frac{D_{ij}}{D_{iz}} \right)^{2/(m-1)} \right]^{-1}, \quad (10)$$
$$1 \leq i \leq k; 1 \leq j \leq n,$$

and

$$v_i = \frac{\sum_{j=1}^{n} (\mu_{ij})^m x_j}{\sum_{j=1}^{n} (\mu_{ij})^m}, \quad 1 \leq i \leq k. \quad (11)$$

This solution also satisfies the remaining constraints of Eq. 5. Along with the optimization of the cluster centroids and membership degrees in Eqs. 10 and 11, missing values are optimized during each iteration to minimize the functional J_m. In this study, we optimize the missing values by minimizing the function $J(x_j)$ presented by [1]:

$$J(x_j) = \sum_{i=1}^{k} (\mu_{ij})^m \|x_j - v_i\|_A^2 \quad (12)$$

Table 1. Comparison of the clustering performance of the KNNimpute, EMimpute-based clustering methods and proposed method for the yeast cell-cycle data set of [12]. For the data sets with different percentages of missing values, the z-scores [19] of all methods are specified. The number of clusters is $k = 5$, and the k-means, SOM, bclust methods were tested based on the data obtained by KNNimpute using $K = 10, 15, 20$.

Method \ %missing	Cell-cycle data				
	5%	10%	15%	20%	25%
KNNimpute(K=10)+k-means	23.0	19.7	21.9	26.6	24.8
KNNimpute(K=15)+k-means	23.7	21.6	21.1	24.8	21.2
KNNimpute(K=20)+k-means	26.5	21.3	22.6	24.6	22.2
KNNimpute(K=10)+SOM	15.8	17.9	14.4	21.9	14.8
KNNimpute(K=15)+SOM	27.0	17.9	14.4	21.9	14.8
KNNimpute(K=20)+SOM	20.0	22.2	16.5	15.8	22.6
KNNimpute(K=10)+BagClust	27.7	40.7	24.2	24.2	20.8
KNNimpute(K=15)+BagClust	31.2	20.8	27.0	22.8	30.6
KNNimpute(K=20)+BagClust	26.7	28.1	24.8	21.4	23.4
EMimpute+k-means	23.2	20.3	22.7	20.0	21.1
EMimpute+SOM	18.1	17.3	17.3	18.8	16.0
Proposed	35.5	32.2	27.3	19.5	14.5

By setting $\nabla J = 0$ with respect to the missing attributes of x_j, a missing value x_{jl} is calculated as:

$$x_{jl} = \frac{\sum_{i=1}^{k}(\mu_{ij})^m v_{il}}{\sum_{i=1}^{k}(\mu_{ij})^m}, \quad 1 \leq i \leq k. \tag{13}$$

By Eq. 13, x_{jl} is estimated by the weighted mean of all cluster centroids in each iteration. At the initial iteration, x_{jl} is initialized with the corresponding attribute of the cluster centroid to which x_j has the highest membership degree.

This method iteratively improves a sequence of sets of clusters until no further improvement in $J_m(U, V)$ is possible. It loops through the estimates for $V_t \rightarrow U_{t+1} \rightarrow V_{t+1}$ and terminates on $\|V_{t+1} - V_t\| \leq \epsilon$. Equivalently, the initialization of the algorithm can be done on U_0, and the iterates become $U_t \rightarrow V_{t+1} \rightarrow U_{t+1}$, with the termination criterion $\|U_{t+1} - U_t\| \leq \epsilon$. This way of alternating optimization using membership computation makes the present method be less prone to falling into local minima than conventional clustering methods.

3 Experimental Results

3.1 Data Sets and Implementation Parameters

To test the effectiveness with which the proposed method clusters incomplete microarray data, we applied the proposed method and conventional imputation-based clustering methods to three published yeast data sets and compared the performance of each method.

Table 2. Comparison of the clustering performance of the KNNimpute, EMimpute-based clustering methods and proposed method for the yeast sporulation data set of [13]. For the data sets with different percentages of missing values, the z-scores [19] of all methods are specified. The number of clusters is $k = 5$, and the k-means, SOM, bclust methods were tested based on the data obtained by KNNimpute using $K = 10, 15, 20$.

Method \ %missing	Sporulation data				
	5%	10%	15%	20%	25%
KNNimpute(K=10)+k-means	2.1	0.8	2.9	1.7	0.1
KNNimpute(K=15)+k-means	2.0	1.2	2.5	3.3	0.5
KNNimpute(K=20)+k-means	2.2	0.9	2.4	2.8	0.4
KNNimpute(K=10)+SOM	1.2	1.7	1.8	1.5	3.0
KNNimpute(K=15)+SOM	0.9	1.9	1.5	2.0	2.4
KNNimpute(K=20)+SOM	1.7	1.7	1.7	1.6	2.7
KNNimpute(K=10)+BagClust	1.2	1.2	0.5	1.1	0.1
KNNimpute(K=15)+BagClust	0.9	1.2	1.3	1.5	0.1
KNNimpute(K=20)+BagClust	0.6	0.7	0.9	1.2	0.8
EMimpute+k-means	1.8	2.1	1.7	2.3	1.3
EMimpute+SOM	0.5	0.9	1.4	2.7	1.9
Proposed	51.1	46.7	49.0	34.3	46.5

The data sets employed were the yeast cell-cycle data set of Cho et al. [12], the yeast sporulation data set of Chu et al. [13], and the yeast Calcineurin-regulation data set of Yoshimoto et al. [11]. The Cho data set contains the expression profiles of 6,200 yeast genes measured at 17 time points over two complete cell cycles. We used the same selection of 2,945 genes made by Tavazoie et al. [24] in which the data for two time points (90 and 100 min) were removed. The Chu data set consists of the expression levels of the yeast genes measured at seven time points during sporulation. Of the 6,116 gene expressions analyzed by Eisen et al. [16], 3,020 significant genes obtained through two-fold change were used. The Yoshimoto's Calcineurin data set contains the expression profiles of 6,102 yeast genes at 24 experiments by the presence and absence of Ca^{2+}, Na^+, CRZ1, and FK506. These three data sets were preprocessed for the test by randomly removing 5–25% (5, 10, 15, 20 and 25) of the data in order to create incomplete matrices.

To cluster these incomplete data sets with conventional methods, we first estimated the missing values using the widely used KNNimpute [2] and EMimpute [5, 3]. For the estimated matrices yielded by each imputation method, we used EXPANDER [21] software that implements many clustering methods, of which we investigated the results of the k-means and SOM methods, along with the results of the bagged clustering (BagClust) [9]. In these experiments, the parameters used in the proposed method were $\epsilon = 0.001, m = 2.5$, and $\tau = 100$. The KNNimpute was tested with $K = 10, 15, 20$; these values were chosen because they have been overwhelmingly favored in previous studies [2]. In the tests reported here, we analyzed the performance of each approach at the number of clusters of $k = 5$.

Table 3. Comparison of the clustering performance of the KNNimpute, EMimpute-based clustering methods and proposed method for the yeast Calcineurin data set of [11]. For the data sets with different percentages of missing values, the z-scores [19] of all methods are specified. The number of clusters is $k = 5$, and the k-means, SOM, bclust methods were tested based on the data obtained by KNNimpute using $K = 10, 15, 20$.

	Calcineurin data				
Method \ %missing	5%	10%	15%	20%	25%
KNNimpute(K=10)+k-means	30.0	32.8	32.2	25.6	28.2
KNNimpute(K=15)+k-means	30.1	32.3	23.0	26.4	27.9
KNNimpute(K=20)+k-means	32.1	32.2	24.6	29.5	29.7
KNNimpute(K=10)+SOM	50.9	49.0	55.6	49.0	49.7
KNNimpute(K=15)+SOM	44.6	50.5	48.9	49.6	49.0
KNNimpute(K=20)+SOM	48.3	53.3	59.4	56.9	46.2
KNNimpute(K=10)+BagClust	4.6	44.3	47.4	48.3	34.3
KNNimpute(K=15)+BagClust	40.5	22.4	37.4	53.1	38.4
KNNimpute(K=20)+BagClust	44.6	37.4	38.5	38.2	10.1
EMimpute+k-means	31.9	31.5	27.6	23.5	23.4
EMimpute+SOM	41.5	49.7	49.1	51.3	57.9
Proposed	79.0	77.5	71.0	70.7	66.4

3.2 Comparison of Clustering Performance

To show the performance of imputation, most of imputation methods proposed to date, including KNNimpute and EMimpute, have examined the the root mean squared error (RMSE) between the true values and the imputed values. However, as Bo et al. pointed out [5], the RMSE is limited to study the impact of missing value imputation on cluster analysis. To make this study more informative regarding how large an impact the imputation method has on cluster analysis, in the present work the clustering results obtained using the alternative imputations were evaluated by comparing gene annotations using the z-score [19, 5]. Besides, we analyzed the cluster qualities using the figure of merits (FOMs) for an internal validation [26]. Firstly, the z-score [19] is calculated by investigating the relation between a clustering result and the functional annotation of the genes in the cluster. To achieve this, this score uses the *Saccharomyces* Genome Database (SGD) annotation of the yeast genes, along with the gene ontology developed by the Gene Ontology Consortium [17, 18]. A higher score of z indicates that genes are better clustered by function, indicating a more biologically significant clustering result.

Table 1 shows the clustering results of the KNNimpute/EMimpute-based clustering methods and proposed method for the yeast cell-cycle data set. The z-score of each method is listed with respect to the percentages of missing values (5-25%). The number of neighbors in the KNNimpute was $K = 10, 15, 20$. The k-means method using KNNimpute gave z-scores from 19.7% to 26.6%. The z-scores of the SOM using KNNimpute were ranged from 14.4 to 27.0. The BagClust using KNNimpute outperformed the other methods at 10% missing values. Compared

Table 4. Comparison of clustering performance of the KNNimpute, EMimpute-based methods and proposed method for the yeast cell-cycle data set. The number of clusters is $k = 5$. The figure of merits (FOMs) of each method at 5-25% missing data are specified. The KNNimpute are tested with $K = 10$.

Method \ %missing	5%	10%	15%	20%	25%
KNNimpute(K=10)+k-means	6.95	6.87	6.60	6.64	5.95
KNNimpute(K=10)+SOM	6.80	6.84	6.67	6.75	6.60
KNNimpute(K=10)+BagClust	7.15	7.13	6.89	6.86	6.93
EMimpute+k-means	7.10	6.69	6.77	6.56	6.86
EMimpute+SOM	6.73	6.75	6.75	6.72	6.43
Proposed	3.69	3.90	3.74	3.98	3.32

to these methods, the proposed method provided better clustering performance at low missing values; the z-scores were varied from 14.5 to 35.5. At 5% missing value, it is observed that the proposed method showed its best z-score of 35.5. Of the other methods, the EMimpute-based SOM method provided the best $z = 18.1$, whereas the BagClust method using KNNimpute yielded the best $z = 31.2$ at $K = 15$.

Table 2 shows the clustering performance of the KNNimpute/EMimpute-based clustering methods and proposed method for the yeast sporulation data set. On the whole, the three KNNimpute-based clustering methods showed similar tendency for all missing values. In comparison to these methods, it is evident that the proposed clustering method shows markedly better performance, giving z-scores of more than 34.0 for all missing values; it provided significantly better clustering performance than other methods, giving $z = 51.1$ at 5% and $z = 46.7$ at 10%. The best z-values of the KNNimpute-based and EMimpute-based methods were $z = 2.2$ and $z = 1.8$ at 5% missing value respectively.

Table 3 shows the clustering results of the KNNimpute/EMimpte-based clustering methods and proposed method for the yeast Calcineurin data set. The proposed method also gave improved and more stable performance compared to the imputation-based clustering methods, with z-scores of more than 70 for all missing values. Of the conventional methods, the KNNimpute-based SOM method using $K = 10$ achieved its best z-scores of $z = 50.9$ and $z = 49.0$ at 5% and 10% missing values respectively. From the three tests, we see that the proposed method is the most effective of the methods considered; it provides the highest z-value for most cases. The KNNimpute-based clustering methods achieved better z-scores than the EMimpute-based methods; the KNNimpute-based BagClust showed better z-scores for the cell-cycle data set and the KNNimpute-based SOM for the Calcineurin data set.

Besides the assessment using the z-score, we quantified the clustering result of each method using the figure of merit (FOM) that is an estimate of the predictive power of a clustering method [26]. A lower value of FOM represents a well clustered result, indicating that a clustering method has high predictive power. Table 4 lists the results of FOMs of six clustering methods for the yeast cell-cycle data set. Of the methods considered, the proposed method provides the lowest FOMs for 5-25% missing of data. The KNNimputed-based k-means method showed better FOMs than other methods for 15-20% of missing data, whereas the EMimpute-based SOM gave lower scores at 5% and 10% missing. The KNNimputed-based BagClust proved the most ineffective of the methods considered. The results of the comparison tests indicate that the proposed method gave markedly better clustering performance than the other imputation-based methods considered, highlighting the effectiveness and potential of the proposed method.

4 Conclusion

Clustering has been used as a popular technique for analysis of large amounts of microarray gene expression data, and many clustering methods have been developed in biological research. However, conventional clustering methods have required a complete data matrix as input even if many microarray data sets are incomplete due to the problem of missing values. In such cases, typically either genes with missing values have been removed or the missing values have been estimated using imputation methods prior to the cluster analysis. In the present study, we focused on the bad influence of the earlier imputation on the subsequent cluster analysis. To address this problem, we have presented the proposed method of clustering incomplete gene expression data. By taking the alternative optimization approach, the missing values are considered as additional parameters for optimization. The evaluation results based on gene annotations have shown that the proposed method is the superior and effective method for clustering incomplete gene expression data. Besides the issues mentioned in present work, we initialized missing values with the corresponding attributes of the cluster centroid to which the incomplete data point is closest. Although this way of initialization is considered appropriate, further work examining the impact of different initializations on clustering performance is needed.

References

1. Hathaway,R.J., Bezdek,J.C.: Fuzzy c-means clustering of incomplete data. IEEE Transactions on Systems, Man, and Cybernetics–Part B: Cybernetics 31 (2001) 735–744
2. Troyanskaya,O., Cantor,M., Sherlock,G. et al.: Missing value estimation methods for DNA microarrays. Bioinformatics 17 (2001) 520–525
3. Ouyang,M., Welsh,W.J., Georgopoulos,P.: Guassian mixture clustering and imputation of microarray data. Bioinformatics 20 (2004) 917–923
4. Alizadeh,A.A., Eisen,M.B., David,R.E. et al.: Distinct types of diffuse large B-cell lymphoma identified by gene expression profiling. Nature 403 (2000) 503–511
5. Bo,T.H., Dysvik,B., Jonassen,I.: LSimpute: accurate estimation of missing values in microarray data with least square methods. Nucleic Acids Research 32 (2004) e34
6. Dumitrescu,D., Lazzerini,B., Jain,L.C.: Fuzzy Sets and Their Applications to Clustering and Traning. CRC Press, Florida (2000)

7. Fuschik,M.E.: Methods for Knowledge Discovery in Microarray Data. Ph.D. Thesis, University of Otago (2003)
8. Horn,D., Axel,I.: Novel clustering algorithm for microarray expression data in a truncated SVD space. Bioinformatics 19 (2003) 1110–1115
9. Dudoit,S., Fridlyand,J.: Bagging to improve the accuracy of a clustering procedure. Bioinformatics 19 (2003) 1090–1099
10. Mizuguchi,G., Shen,X., Landry,J. et al.: ATP-driven exchange of histone H2AZ variant catalyzed by SWR1 chromatin remodeling complex. Science 303 (2004) 343–348
11. Yoshimoto,H., Saltsman,K., Gasch,A.P. et al.: Genome-wide analysis of gene expression regulated by the Calcineurin/Crz1p signaling pathway in Saccharomyces cerevisiae. The Journal of Biological Chemistry 277 (2002) 31079–31088
12. Cho,R.J., Campbell,M.J., Winzeler,E.A. et al.: A genome-wide transcriptional analysis of the mitotic cell cycle. Mol. Cell 2 (1998) 65–73
13. Chu,S., DeRish,J., Eisen,M. et al.: The transcriptional program of sporulation in budding yeast. Science 282 (1998) 699–705
14. Dembele,D., Kastner,P.: Fuzzy c-means method for clustering microarray data. Bioinformatics 19 (2003) 973–980
15. Dhilon,I.S., Marcotte,E.M., Roshan,U.: Diametrical clustering for identifying anticorrelated gene clusters. Bioinformatics 19 (2003) 1612–1619
16. Eisen,M., Spellman,P.T., Brown,P.O. et al.: Cluster analysis and display of genome-wide expression patterns. Proc. Natl. Acad. Sci. USA 95 (1998) 14863–14868
17. Ashburner,M., Ball,C.A., Blake,J.A. et al.: Gene Ontology: tool for the unification of biology. Nat. Genet. 25 (2000) 25–29
18. Issel-Tarver,L., Christie,K.R., Dolinski,K. et al.: *Saccharomyces* genome database. Methods Enzymol 350 (2002) 329–346
19. Gibbons,F.D., Roth,F.P.: Judging the quality of gene expression-based clustering methods using gene annotation. Genome Res. 12 (2002) 1574–1581
20. Kim,D.W., Lee,K.H., Lee,D.: Detecting clusters of different geometrical shapes in microarray gene expression data. Bioinformatics 21 (2005) 1927–1934
21. Sharan,R., Maron-Katz,A., Shamir,R.: CLICK and EXPANDER: a system for clustering and visualizing gene expression data. Bioinformatics 19 (2003) 1787–1799
22. Steuer,R., Kurths,J., Daub,C.O. et al.: The mutual information: Detecting and evaluating dependencies between variables. Bioinformatics 18 (2002) S231-S240
23. Tamayo,P., Slonim,D., Mesirov,J. et al.: Interpreting patters of gene expression with self-organizing maps - methods and application to hematopoietic differentiation. Proc. Natl. Acad. Sci. USA 96 (1999) 2907–2912
24. Tavazoie,S., Hughes,J.D., Campbell,M.J. et al.: Systematic determination of genetic network architecture. Nat. Genet. 22 (1999) 281–285
25. Xu,Y., Olman,V., Xu,D.: Clustering gene expression data using a graph-theoretic approach - an application of minimum spanning trees. Bioinformatics 17 (2001) 309–318
26. Yeung,K., Haynor,D.R., Ruzzo,W.L.: Validating clustering for gene expression data. Bioinformatics 17 (2001) 309–318

An EM-Approach for Clustering Multi-Instance Objects

Hans-Peter Kriegel, Alexey Pryakhin, and Matthias Schubert

Institute for Informatics,
University of Munich,
D-80538 Munich, Germany
{kriegel, pryakhin, schubert}@dbs.ifi.lmu.de

Abstract. In many data mining applications the data objects are modeled as sets of feature vectors or multi-instance objects. In this paper, we present an expectation maximization approach for clustering multi-instance objects. We therefore present a statistical process that models multi-instance objects. Furthermore, we present M-steps and E-steps for EM clustering and a method for finding a good initial model. In our experimental evaluation, we demonstrate that the new EM algorithm is capable to increase the cluster quality for three real world data sets compared to a k-medoid clustering.

1 Introduction

In modern data mining applications, the complexity of analyzed data objects is increasing rapidly. Molecules are analyzed more precisely and with respect to all of their possible spatial conformations [1]. Earth observation satellites are able to take images with higher resolutions and in a variety of spectra which was not possible some years before. Data mining started to analyze complete websites instead of single documents [2]. All of these application domains are examples for which the complexity demands a richer object representation than single feature vectors. Thus, for these application domains, an object is often described as a set of feature vectors or a multi-instance (MI) object. For example, a molecule can be represented by a set of feature vectors where each vector describes one spatial conformation or a website can be analyzed as a set of word vectors corresponding to its HTML documents.

As a result the research community started to develop techniques for multi-instance learning that where capable to analyze multi-instance objects. One of the first publications in this area [1,3] was focussed to a special task called multi-instance learning. In this task the appearance of one positive instance within a multi-instance object is sufficient to indicate that the object belongs to the positive class. Besides classical multi-instance learning, some approaches like [4,5] aim at more general problems. However, all of the mentioned approaches are based on a setting having a set of labeled bags to train a learning algorithms.

In this paper, we focus on clustering unlabeled sets of feature vectors. To cluster those objects, the common approach so far is to select some distance measures for point sets like [6, 7] and then apply a distance-based clustering algorithm e.g. k-medoid methods like CLARANS [8] or a density-based algorithm like DBSCAN[9]. However, this approach does not yield expressive cluster models. Depending on the used algorithm, we might have some representative for some cluster, but we do not have a good model for describing the mechanism behind this clustering. To overcome this problem, we will refer to the model of multi-instance objects that was introduced in [5] stating that a multi-instance object of a particular class (or in our problem each cluster) needs to provide instances belonging to a certain concept or several concepts. We will adapt this view of multi-instance objects to clustering. Therefore, we propose a statistical model that is based on 2 steps. In the first step, we use a standard EM Clustering algorithm on the union set of all multi-instance objects. Thus, we determine a mixture model describing the instances of all multi-instance objects. Assuming that each of the found clusters within each mixture model corresponds to some valid concept, we now can derive distributions for the clustering of multi-instance objects. For this second step, we assume that a multi-instance object containing k instances can be modeled as k draws from the mixture model over the instances. Thus, each cluster of multi-instance objects is described by a distribution over the instance clusters derived in the first step and some prior probability. For example, for the classical multi-instance learning task, it can be expected that there is at least one instance cluster that is very unlikely to appear in the multi-instance clusters corresponding to the negative bags.

The rest of the paper is organized as following: In section 2, we will survey previous work in data mining with multi-instance objects and give a brief introduction to EM clustering. Section 3 will describe our statistical model for multi-instance data. In section 4, this model is employed for EM clustering. To demonstrate the usefulness of our approach, section 5 contains the results on several real world data sets. Section 6 concludes the paper with a summary and directions for future work.

2 Related Work

Data Mining in multi-instance objects has so far been predominantly examined in the classification section. In [1] Dietterich et al. defined the problem of multi-instance learning for drug prediction and provided a specialized algorithm to solve this particular task by learning axis parallel rectangles. In the following years, new algorithms increasing the performance for this special task were introduces [3]. In [5] a more general method for handling multi-instance objects was introduced that is applicable for a wider variety of multi-instance problems. This model considers several concepts for each class and requires certain cardinalities for the instances belonging to the concepts in order to specify a class of MI objects. Additionally, to this model [10] proposes more general kernel functions for MI comparing MI objects.

For clustering multi-instance objects, it is possible to use distance functions for sets of objects like [6, 7]. Having such a distance measure, it is possible to cluster multi-instance objects with k-medoid methods like PAM and CLARANS [11] or employ density-based clustering approaches like DBSCAN [9]. Though this method yields the possibility to partition multi-instance objects into clusters, the clustering model consists of representative objects in the best case. Another problem of this approach is that the selection of a meaningful distance measure has an important impact of the resulting clustering. For example, netflow-distance [7] demands that all instances within two compared objects are somehow similar, whereas for the minimal Hausdorff [12] distance the indication of similarity is only dependent on the closest pair.

In this paper, we introduce an algorithm for clustering multi-instance objects that optimizes probability distributions to describe the data set. Part of this work is based on expectation maximization (EM) clustering for ordinary feature vectors using Gaussians. Details about this algorithm can be found in [13]. In [14], a method for producing a good initial mixture is presented which is based on multiple sampling. It is empirically shown that using this method, the EM algorithm achieves accurate clustering results.

3 A Statistical Model for Multi-Instance Objects

In this section, we will introduce our model for multi-instance clustering. Therefore, we will first of all define the terms instance and multi-instance (MI) object.

Definition 1 (instance and MI object). *Let F be a feature space. Then, $i \in F$ is called an instance in F. A multi-instance (MI) object o in F is given by an arbitrary sized set of instances $o = i_1, .., i_k$ with $i_j \in F$. To denote the unique MI object an instance i belongs to, we will write $MiObj(i)$.*

To cluster multi-instance objects using an EM approach, we first of all need a statistical process that models sets of multi-instance objects. Since multi-instance objects consist of single instances in some feature space, we begin with modeling the data distribution in the feature space of instances. Therefore, we first of all define the instance set of a set of multi-instance objects:

Definition 2 (Instance Set). *Given a database DB of multi-instance Objects $o = i_1, \ldots, i_k$, the corresponding instance set $I_{DB} = \bigcup_{DB} o$ is the union of all multi-instance objects.*

To model the data distribution in the instance space, we assume a mixture model of k independent statistical processes. For example, an instance set consisting of feature vectors could be described by a mixture of Gaussians.

Definition 3 (Instance Model). *Let DB be a data set consisting of multi-instance objects o and let I_{DB} be its instance set. Then, an instance model IM for DB is given by a mixture model of k statistical processes that can be described by a prior probability $Pr[k_j]$ for each component k_j and the necessary parameters for the process corresponding to k_j, e.g. a mean vector μ_j and co-variance matrix M_j for Gaussian processes.*

After describing the instance set, we can now turn to the description of multi-instance objects. Our solution is based on the idea of modeling a cluster of multi-instance objects as a multinomial distribution over the components of the mixture model of instances. For each instance and each concept, the probability that the instance belongs to this concept is considered as result of one draw. If the number n of instances within an object o is considered to be important as well, we can integrate this into our model as well by considering some distribution over the number of draws, e.g. a binomial distribution. To conclude, a mixture model of multi-instance clusters can be described by a set of multinomial distributions over the components of a mixture model of instances. A multi-instance object is thus derived in the following way:

1. Select a multi-instance cluster c_i w.r.t. some prior distribution over the set of all clusters C.
2. Derive the number of instances n within the multi-instance object w.r.t some distribution depending on the chosen cluster c_i.
3. Repeat n-times:
   (a) Select some model component k_j within the mixture model of instances w.r.t. the multi-instance cluster specific distribution.
   (b) Generate an instance, w.r.t. to the distribution corresponding to component k_j.

Formally, the underlying model for multi-instance data sets can be defined as follows:

Definition 4 (Multi-Instance Model). *A multi-instance model M over the instance model IM is defined by a set C of l processes over I_{DB}. Each of these processes c_i is described by a prior probability $Pr[c_i]$, a distribution over the number of instances in the bag $Pr[Card(o) | c_i]$ and an conditional probability describing the likelihood that a multi-instance object o belonging to process c_i contains an instance belonging to the component $k_l \in IM$. The probability of an object o in the model M is calculated as following:*

$$Pr[o] = \sum_{c_i \in C} Pr[c_i] \cdot Pr[Card(o)|c_i] \cdot \prod_{i \in o} \prod_{k \in MI} Pr[k|c_i]^{Pr[k|i]}$$

The conditional probability of process c_i under the condition of a given multi-instance object o can be calculated by:

$$Pr[c_i|o] = \frac{1}{Pr[o]} \cdot Pr[c_i] \cdot Pr[Card(o)|c_i] \cdot \prod_{i \in o} \prod_{k \in MI} Pr[k|c_i]^{Pr[k|i]}$$

Let us note that the occurrence of an instance within the data object is only dependent on the cluster of instances it is derived from. Thus, we do not assume any dependencies between the instances of the same objects. Another important characteristic of the model is that we assume the same set of instance clusters for all multi-instance clusters. Figure 3 displays an example of a two dimensional multi-instance data set corresponding to this model. This assumption leads to the following 3 step approach for multi-instance EM clustering.

4 EM-Clustering for Multi-Instance Objects

After introducing a general statistical process for multi-instance objects, we will now introduce an EM algorithm that fits the distribution parameters to a given set of multi-instance objects. Our method works in 3 steps:

1. Derive a Mixture Model for the Instance Set.
2. Calculate a start partitioning.
3. Use the new EM algorithm to optimize the start partitioning.

4.1 Generating a Mixture Model for the Instance Set

To find a mixture of the instance space, we can employ a standard EM approach as proposed in section 2. For general feature vectors, we can describe the instance set as a mixture of Gaussians. If the feature space is sparse using a mixture of multinomial processes usually provides better results. If the number of clusters in the instance is already known, we can simply employ EM clustering. However, if we do not know how many clusters are hidden within the instance set, we need to employ a method for determining a suitable number of processes like [15].

4.2 Finding a Start Partitioning of Multi-Instance Objects

After deriving a description of the instance space, we now determine a good start partitioning for the final clustering step. A good start partitioning is very important for finding a good cluster model. Since EM algorithms usually do not achieve a global maximum likelihood, a suitable start partitioning has an important impact on both, the likelihood of the cluster and the runtime of the algorithm. The versions for EM in ordinary feature spaces often use k-means clustering for finding a suitable start partitioning. However, since we cluster sets of instances instead of single instances, we cannot use this approach directly.

To overcome this problem, we proceed as follows. For each multi-instance object we determine a so-called confidence summary vector in the following way.

Definition 5 (Confidence Summary Vector). *Let IM be an instance model over database DB containing k processes and let o be a multi-instance object. Then the confidence summary vector $\overrightarrow{csv}(o)$ of o is a k dimensional vector that is calculated as follows:*

$$csv_j(o) = \sum_{i \in o} Pr[k_j] \cdot Pr[i|k_j]$$

After building the confidence summary vector for each object, we can now employ k-means to cluster the multi-instance objects. Though the resulting clustering might not be optimal, the objects within one cluster should yield similar distributions over the components of the underlying instance model.

4.3 EM for Clustering Multi-Instance Objects

In this final step, the start partitioning for the data set is optimized using the EM algorithm. We therefore describe a suitable expectation and maximization step and then employ an iterative method. The likelihood of the complete model M can be calculated by adding up the log-likelihoods of the occurrence of each data object in each clusters. Thus, our model is (locally) optimal if we obtain a maximum for the the following log-likelihood term.

Definition 6 (Log-Likelihood for M).

$$E(M) = \sum_{o \in DB} \log \sum_{c_i \in M} Pr[c_i|o]$$

To determine $Pr[c_i|o]$, we proceed as mentioned in definition 4. Thus, we can easily calculate $E(M)$ in the expectation step for a given set of distribution parameters and an instance model. To improve the distribution parameters, we employ the following updates to the distribution parameters in the maximization step:

$$W_{c_i} = Pr[c_i] = \frac{1}{Card(DB)} \sum_{o \in DB} Pr[c_i|o]$$

where W_{c_i} denotes the prior probability of a cluster of multi-instance objects.

To estimate the number of instances contained in an MI object belonging to cluster c_i, we can employ a binomial distribution determined by the parameter l_{c_i}. The parameters are updated as follows:

$$l_{c_i} = \frac{\sum_{o \in DB} Pr[c_i|o] \cdot Card(o)}{Card(DB)} \cdot \frac{1}{MAXLENGTH}$$

where $MAXLENGTH$ is the maximum number of instances for any MI object in the database.

Finally, to estimate the relative number of instances drawn from concept k_j for MI objects belonging to cluster c_i, we derive the parameter updates in the following way:

$$P_{k_j,c_i} = Pr[k_j|c_i] = \frac{\sum_{o \in DB} (Pr[c_i|o] \cdot \sum_{u \in o} Pr[u|k_j])}{\sum_{o \in DB} \sum_{u \in o} Pr[u|k_j]}$$

Using these update steps, the algorithm is terminated after the improvement of $E(M)$ is less than a given value σ. Since the last step of our algorithm is a modification of EM clustering based on multinomial processes, our algorithm always converges against a local maximum value for $E(M)$.

5 Evaluation

All algorithms are implemented in Java 1.5. The experiments described below are carried out on a work station that is equipped with two 1.8 GHz Opteron processors and 8 GB main memory.

Table 1. Details of the test environments

	Data Set 1 (DS1)	Data Set 2 (DS2)	Data Set 3 (DS3)
Name	Brenda	MUSK 1	MUSK 2
Number of MI-Objects	6082	92	102
Average Number of Instances per MI-Object	1.977	5.2	64.7
Number of MI-Object classes	6	2	2

Our experiments were performed on 3 different real world data sets. The properties of each test bed are illustrated in Table 1. The Brenda data set contains of enzymes taken from the protein data bank (PDB) [1]. Each enzyme comprises several chains given by amino acid sequences. In order to derive feature vectors from the amino acid sequences, we employed the approach described in [16]. The basic idea is to use local (20 amino acids) and global (6 exchange groups) characterization of amino acid sequences. In order to construct a meaningful feature space, we formed all possible 1-grams for each kind of characteristic. This approach provided us with 26 dimensional histograms for each chain. To obtain the class labels for each enzyme we used a mapping from PDB to the enzyme class numbers from the comprehensive enzyme information system BRENDA [2].

MUSK 1 and MUSK 2 data sets come from UCI repository [17] and describe a set of molecules. The MI-objects in MUSK 1 and MUSK 2 data sets are judged by human experts to be in musks or non-musks class. The feature vectors of MUSK data sets have 166 numerical attributes that describe these molecules depending on the exact shape or conformation of the molecule.

To measure the effectiveness, we considered the agreement of the calculated clusterings to the given class systems. To do so, we calculated three quality measures namely precision, F-measure and average entropy. In order to calculate the precision and F-Measure, we proceeded as follows. For each cluster c_i found by a clustering algorithm, its class assignment $Class(c_i)$ is determined by the class label of objects belonging to c_i that are in the majority. Then, we calculated the Precision within all clusters w.r.t. the determined class assignments by using the following formulas.

$$Precision = \frac{\sum_{c_i \in C} Card(\{o | (c_i = \arg\max_{c_j \in C} Pr[c_j|o]) \wedge Class(o) = Class(c_i)\})}{Card(DB)}$$

$$Avg.Entropy = \sum_{c_i \in C}(Card(c_i) * (-\sum_{Class_j} p_{j,i} log(p_{j,i})))/Card(DB)$$

In addition, we measured the average entropy over all clusters. This quality measure is based on the impurity of a cluster c_i w.r.t. the class labels of objects

---
[1] http://www.rcsb.org/pdb/
[2] http://www.brenda.uni-koeln.de/

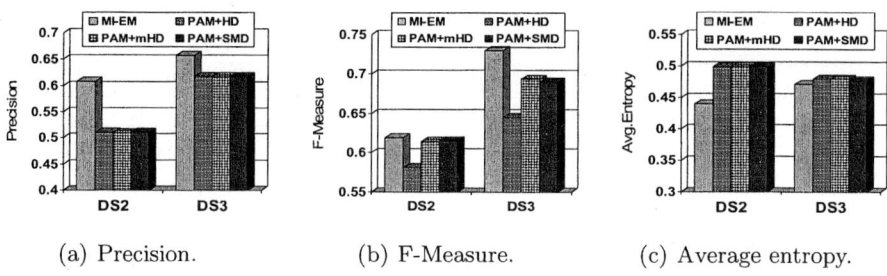

Fig. 1. Effectiveness evaluation on DS2 and DS3 where no. of clusters is 2

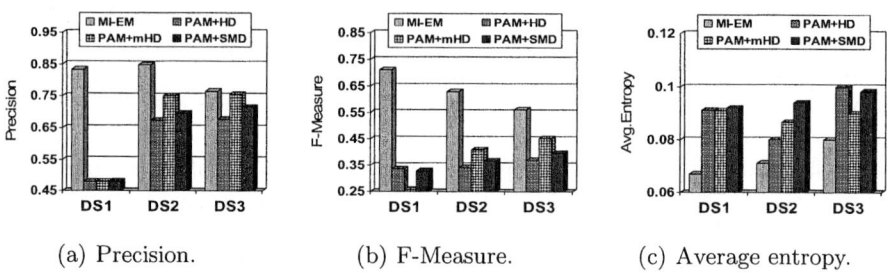

Fig. 2. Effectiveness evaluation on DS1, DS2 and DS3 where no. of clusters is 8

belonging to c_i. Let $p_{j,i}$ be the relative frequency of the class label $Class_j$ in the cluster c_i. We calculate average entropy as following.

In order to demonstrate that the proposed clustering approach for multi-instance objects outperforms standard clustering algorithms working on a suitable distance functions, we compared precision, F-Measure and average entropy of the MI-EM with that of k-medoid clustering algorithm (PAM). To enable cluster analysis of multi-instance objects by PAM, we used the Hausdorff distance (HD)[6], the minimum Hausdorff distance (mHD)[12] and the Sum of Minimum Distances (SMD)[6]. Due to the fact that the data set DS1 has 6 classes and the data sets DS2 and DS3 have 2 classes, we investigated the effectiveness of the cluster analysis where the number of clusters is equal to or slightly than the number of the desired classes. Thus, we set in our experiments the number of clusters equal to 6 and 8 for DS1, and equal to 2, 6 and 8 for the data sets DS2 and DS3. The results of our comparison are illustrated in Figures 1,3 and 2.

In all our experiments, PAM working on distance functions suitable for multi-instance objects achieved a significantly lower precision than MI-EM. For example, the MI-EM algorithm reached a precision of 0.833 on DS1 and the number of clusters equal to 8 (cf. Figure 2(a)). In contrast to the result of MI-EM, the precision calculated for clusterings found by all competitors lies between 0.478 and 0.48. Furthermore, MI-EM obtained in all experiments higher or comparable values of F-Measures. This fact indicates that the cluster structure found by applying of the proposed EM-based approach is more exact w.r.t. precision and

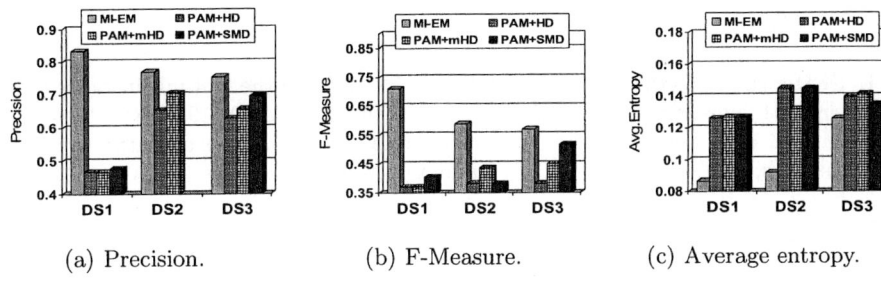

(a) Precision. (b) F-Measure. (c) Average entropy.

Fig. 3. Effectiveness evaluation on DS1, DS2 and DS3 where no. of clusters is 6

recall than that found by PAM with 3 different MI distance functions. For example, the F-Measure calculated for MI-EM clustering of DS2 with 8 clusters is 0.63 whereas PAM clustering with different MI distance functions shows values between 0.341 and 0.41 (cf. Figure 2(b)). Finally, the values of average entropy observed by the MI-EM results are considerably lower than those of PAM on HD, mHD and SMD. The lower values of average entropy imply a lower level of impurity in the cluster structures detected by applying MI-EM.

To summarize, the values of the different quality measures observed on real world data sets when varying the number of clusters show that the proposed EM-based approach for cluster analysis of MI-objects outperforms the considered competitors w.r.t. effectiveness.

6 Conclusions

In this paper, we described an approach for statistical clustering of MI objects. Our approach models instances as members of concepts in some underlying feature space. Each concept is modeled by a statistical process in this feature space, e.g. a Gaussian. A multi-instance object can now be considered as the result of selecting several times a concept and generating an instance with the corresponding process. Clusters of multi-instance objects can now be described as multinomial distributions over the concepts. In other words, different clusters are described by having different probabilities for the underlying concepts. An additional aspect is the length of the MI object. To derive MI clusters corresponding to this model, we introduce a three step approach. In the first step we derive a mixture model describing concepts in the instance space. The second step finds a good initialization for the target distribution by subsuming each MI object by a so-called confidence summary vector (csv) and afterwards clustering these csvs using the k-means method. In the final, step we employ a final EM clustering step optimizing the distribution for each cluster of MI objects. To evaluate our method, we compared our clustering approach to clustering MI objects with the k-medoid clustering algorithm PAM for 3 different similarity measures. The results demonstrate that the found clustering model offers better cluster qualities w.r.t. to the provided reference clusterings.

References

1. Dietterich, T., Lathrop, R., Lozano-Perez, T.: "Solving the multiple instance problem with axis-parallel rectangles". Artificial Intelligence **89** (1997) 31–71
2. Kriegel, H.P., Schubert, M.: "Classification of websites as sets of feature vectors". In: Proc. IASTED Int. Conf. on Databases and Applications (DBA 2004), Innsbruck, Austria. (2004)
3. Zhou, Z.H.: "Multi-Instance Learning: A Survey". Technical Report, AI Lab, Computer Science a. Technology Department, Nanjing University, Nanjing, China (2004)
4. Ruffo, G.: Learning single and multiple instance decision tree for computer security applications. PhD thesis, Department of Computer Science, University of Turin, Torino,Italy (2000)
5. Weidmann, N., Frank, E., Pfahringer, B.: "A Two-Level Learning Method for Generalized Multi-instance Problems". In: Proc. ECML 2003, Cavtat-Dubrovnik,Cr. (2003) 468–479
6. Eiter, T., Mannila, H.: "Distance Measures for Point Sets and Their Computation". Acta Informatica **34** (1997) 103–133
7. Ramon, J., Bruynooghe, M.: "A polynomial time computable metric between points sets". Acta Informatica **37** (2001) 765–780
8. Han, J., Kamber, M.: "Data Mining Concepts and Techniques". Morgan Kaufmann Publishers (2001)
9. Ester, M., Kriegel, H.P., Sander, J., Xu, X.: "A Density-Based Algorithm for Discovering Clusters in Large Spatial Databases with Noise". In: Proc. Int. Conf. on Knowledge Discovery and Data Mining (KDD). (1996) 291–316
10. Gärtner, T., Flach, P., Kowalczyk, A., Smola, A.: "Multi-Instance Kernels". (2002) 179–186
11. Ng, R., Han, J.: "Efficient and Effective Clustering Methods for Spatial Data Mining". In: Proc. Int. Conf. on Very Large Databases (VLDB). (1994) 144–155
12. Wang, J., Zucker, J.: "Solving Multiple-Instance Problem: A Lazy Learning Approach". (2000) 1119–1125
13. Han, J., Kamber, M.: Data Mining: Concepts and Techniques. Academic Press (2001)
14. Fayyad, U., Reina, C., Bradley, P.: "Initialization of Iterative Refinement Clustering Algorithms". In: Proc. Int. Conf. on Knowledge Discovery in Databases (KDD). (1998)
15. Smyth, P.: Clustering using monte carlo cross-validation. In: KDD. (1996) 126–133
16. Wang, J.T.L., Ma, Q., Shasha, D., Wu, C.H.: New techniques for extracting features from protein sequences. IBM Syst. J. **40** (2001) 426–441
17. D.J. Newman, S. Hettich, C.B., Merz, C.: UCI repository of machine learning databases (1998)

Mining Maximal Correlated Member Clusters in High Dimensional Database*

Lizheng Jiang[1], Dongqing Yang[1], Shiwei Tang[1,2], Xiuli Ma[2], and Dehui Zhang[2]

[1] School of Electronics Engineering and Computer Science, Peking University,
Beijing 100871, China
jianglz@cis.pku.edu.cn
[2] National Laboratory on Machine Perception, School of Electronics Engineering and Computer Science, Peking University, Beijing 100871, China
{dqyang, tsw}@pku.edu.cn, {maxl, dhzhang}@cis.pku.edu.cn

Abstract. Mining high dimensional data is an urgent problem of great practical importance. Although some data mining models such as frequent patterns and clusters have been proven to be very successful for analyzing very large data sets, they have some limitations. Frequent patterns are inadequate to describe the quantitative correlations among nominal members. Traditional cluster models ignore distances of some pairs of members, so a pair of members in one big cluster may be far away. As a combination and complementary of both techniques, we propose the Maximal-Correlated-Member-Cluster (MCMC) model in this paper. The MCMC model is based on a statistical measure reflecting the relationship of nominal variables, and every pair of members in one cluster satisfy unified constraints. Moreover, in order to improve algorithm's efficiency, we introduce pruning techniques to reduce the search space. In the first phase, a Tri-correlation inequation is used to eliminate unrelated member pairs, and in the second phase, an Inverse-Order-Enumeration-Tree (IOET) method is designed to share common computations. Experiments over both synthetic datasets and real life datasets are performed to examine our algorithm's performance. The results show that our algorithm has much higher efficiency than the naïve algorithm, and this model can discover meaningful correlated patterns in high dimensional database.

1 Introduction

Information system generates a lot of data in different industries, such as manufacturing, retail, financial services, transportation, telecommunication, utilities, and healthcare. Many of these historical data are high dimensional data, which have a large number of dimensions. There are needs to analyze and mine these high dimensional data to find patterns, general trends and anomalies for many applications. But the curse of dimensionality makes many existing data mining algorithms become

---

* This work is Supported by the National Natural Science Foundation of China under Grant No.60473072 and Grant No. 60473051.

computationally intractable and therefore inapplicable in many real applications. In this paper, we try to design a novel model to mine correlated member clusters in the high dimensional database environment.

We use an example to explain some concepts.

Example: Analysis of economic data.

We will analyze Chinese industrial production statistical data. The data is organized in a multidimensional database that has 3 dimensions: *Product*, *City*, and *Month*, and 1 measure: *Production*. The measure is the total production amount for (*Product*, *City*, *Month*).

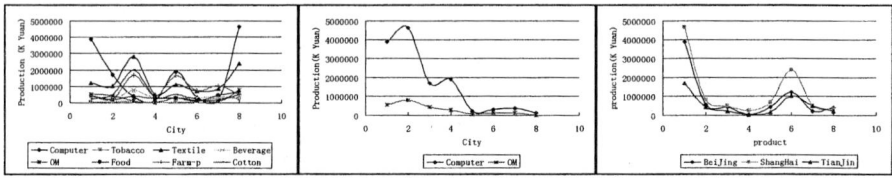

Fig. 1. Original production series

Fig. 2. Correlated product series

Fig. 3. Correlated city series

Fig1 plots the production series of 8 products (*computer*, *tobacco*, *office-machine*, etc.) in different cities (*BeiJing*, *TianJin*, *ShangHai*, etc.). Among these series, some of them are correlated. Two correlated product series (*Computer* and *Office Machine*) are shown in Fig2, and correlated city series (*BeiJing*, *ShangHai*, and *TianJin*) are shown in Fig3. These correlated series exhibit similar patterns. The curve goes up as its correlated series increases, and it goes down as its correlated series decreases.

As we have seen, some products or cities may have correlated patterns. Their production levels rise and fall coherently under a subset of conditions, that is, they exhibit fluctuation of a similar shape when conditions change. Discovering these correlated products or cities is helpful for us to perform more intensive research work. The question is how to find such correlated patterns among a great number of series.

Pearson's coefficient is a statistical measure that represents the degree of linear correlation between variable vectors, which is used in many kinds of applications. In this example, we use Pearson's coefficient to define the similarity of datasets corresponding to members .

Our work is first related to correlation mining. Correlation analysis and mining has played an important role in data mining applications. A common data-mining task is the search for associations of item sets in a database of transactions. There have been many works about association relations, since Agrawal et al. proposed association rule mining [1]. The works in first category are about fast algorithms for association rule mining, such as Apriori algorithm to generate frequent item sets [2] and frequent-pattern tree approach to mine frequent patterns without generating candidate item sets [4]. The works in the second category adopt other interesting measure to mine association rules of specific interest, or employ artful methods to reduce the number of rules. These works include constraint-based association pattern mining [8] [10], frequent closed patterns [11], maximal frequent patterns [3], and condensed frequent pattern base [7]. In nature, association rules generated by frequent item sets represents the relationships of concurrence in historical transactions. It reflects the relationships of binary variables, but doesn't describe relationships among nominal variables.

In the second place, our work is related to pair-wise clustering models based on pattern similarity. In paper [9], Wang et al. defined the similarity of objects by *pscore*, and proposed the *p-cluster* model to discover clusters that exhibited similarity of patterns. Because *pscore* only considered strict shifting patterns or strict scaling patterns, Liu et al. [6] designed a more flexible *op-cluster* model to find patterns that preserved orders in attributes. Paper [5] chose *Pearson's correlation coefficient* as a coherence measure to mine coherent patterns in the GST(Gene-Sample-Time) microarray data. All these models calculate the similarity of every pair of members. The weak is that these algorithms have low efficiency for high dimensional data.

Given *m objects*, pair-wise clustering models calculate *Pearson's coefficients* for all $C_m^2 = \frac{m*(m-1)}{2}$ pairs of objects. In this paper, as an optimizing technique, we prove a Tri-correlation inequation and design a heuristic approach to prune the unrelated member pairs.

Generating Maximal-Correlated-Member-Cluster (MCMC) algorithm is similar to Max-miner [3] that uses a set-enumeration tree algorithm to mine maximal frequent item sets. In Max-miner, the maximal height of set-enumeration tree is m, and the maximal number of tree branches is $O(2^m)$. Ordinary set-enumeration tree is infeasible for high dimensional data. Instead, we design an Inverse-Order-Enumeration-Tree (IOET) algorithm, in which the tails of MCMC are generated first, and then the heads are added to them. Two advantages will benefit the IOET algorithm. The first one is that we start from a small member set and then expand it. The second one is that redundant sub branches can be detected and pruned as soon as possible.

In summary, our work has the following contributions.

1. This paper proposes a MCMC model to mine correlated member sets from high dimensional database. MCMC model borrows *Pearson's Correlation coefficient* as the similarity measure of members, which is applicable to not only binary variables, but also nominal variables.

2. In order to compute correlated member pairs efficiently, we prove a tri-correlation inequation (Lemma2) in theory, which can be used to prune a lot of unrelated member pairs without calculating their coefficients.

3. We design an IOET algorithm to generate complete MCMCs from correlated member pairs. Compared to the set-enumeration tree algorithm in Max-Miner, IOET algorithm will reduce the search space dramatically.

The rest of the paper is organized as the following. Section 2 describes our model and gives some relative definitions. Algorithms are explained in section 3. Section 4 presents our experiment results. Section 5 summarizes our work.

2 Problem Description and Formulation

In statistics, a measure of correlation is a numerical grade, which describes the degree of a relationship among variables. Support of frequent item sets and *Jaccard coefficient* are measures for binary variables. *Kendall's Tau* and *Spearman's Rank Correlation Coefficient* represents relationships among ordinal variables. *Pearson's Correlation Coefficient* measures relationships among nominal variables.

2.1 Pearson's Correlation Coefficient and Its Property

Pearson's coefficient describes the linear relationship between two variables. Given two vectors $X=(x_1, x_2, \ldots, x_n)$, $Y=(y_1, y_2, \ldots, y_n)$, their *Pearson's coefficient* is:

$$r(X,Y) = \frac{\sum_{i=1}^{n}(x_i-\bar{x})(y_i-\bar{y})}{\sqrt{\sum_{i=1}^{n}(x_i-\bar{x})^2} \cdot \sqrt{\sum_{i=1}^{n}(y_i-\bar{y})^2}}, \bar{x}=\frac{\sum_{i=1}^{n}x_i}{n}, \bar{y}=\frac{\sum_{i=1}^{n}y_i}{n} \tag{1}$$

Lemma 1 (Linear invariability): given two variable series $X=(x_1, x_2, \ldots, x_n)$, $Y=(y_1, y_2, \ldots, y_n)$ and any nonzero constants k_1, k_2, $k_1X=(k_1x_1, k_1x_2, \ldots, k_1x_n)$, $k_2Y=(k_2y_1, k_2y_2, \ldots, k_2y_n)$, we have $r(k_1X, k_2Y) = r(X,Y)$.

Proof: obvious.

Lemma 2 (Tri-correlation inequation): given variable series $X=(x_1, x_2, \ldots, x_n)$, $Y=(y_1, y_2, \ldots, y_n)$, and $Z=(z_1, z_2, \ldots, z_n)$, for any $0<\sigma\leq 1$, if $r(x,y)\geq\sigma$, $r(x,z)\geq\sigma$, we have $r(y,z)\geq 2\sigma^2-1$.

Proof: We transform variable series $X=(x_1, x_2, \ldots, x_n)$, $Y=(y_1, y_2, \ldots, y_n)$, and $Z=(z_1, z_2, \ldots, z_n)$ to new variable series $X'=(x_1-\bar{x}, x_2-\bar{x}, \ldots, x_n-\bar{x})$, $Y'=(y_1-\bar{y}, y_2-\bar{y}, \ldots, y_n-\bar{y})$, and $Z'=(z_1-\bar{z}, z_2-\bar{z}, \ldots, z_n-\bar{z})$. For the purpose of simplicity, we won't distinguish X, Y, and Z from X', Y' and Z'.

For any number x, y, z, w_1, w_2, and w_3, we have
$(w_1x-w_2y-w_3z)^2 = (w_1x)^2 - 2w_1w_2xy - 2w_1w_3xz + 2w_2w_3yz + (w_2y)^2 + (w_3z)^2 \geq 0$.
$2w_2w_3yz \geq 2w_1w_2xy + 2w_1w_3xz - (w_1x)^2 - (w_2y)^2 - (w_3z)^2$.

$$\frac{2\sum_{i=1}^{n}w_2w_3y_iz_i}{\sqrt{\sum_{i=1}^{n}(w_2y_i)^2}\sqrt{\sum_{i=1}^{n}(w_3z_i)^2}} \geq \frac{2\sum_{i=1}^{n}w_1w_2x_iy_i + 2\sum_{i=1}^{n}w_1w_3x_iz_i - \sum_{i=1}^{n}(w_1x_i)^2 - \sum_{i=1}^{n}(w_2y_i)^2 - \sum_{i=1}^{n}(w_3z_i)^2}{\sqrt{\sum_{i=1}^{n}(w_2y_i)^2}\sqrt{\sum_{i=1}^{n}(w_3z_i)^2}}$$

$$2r(y,z) \geq \frac{2r(x,y)\sqrt{\sum_{i=1}^{n}(w_1x_i)^2}\sqrt{\sum_{i=1}^{n}(w_2y_i)^2} + 2r(x,z)\sqrt{\sum_{i=1}^{n}(w_1x_i)^2}\sqrt{\sum_{i=1}^{n}(w_3z_i)^2} - \sum_{i=1}^{n}(w_1x_i)^2 - \sum_{i=1}^{n}(w_2y_i)^2 - \sum_{i=1}^{n}(w_3z_i)^2}{\sqrt{\sum_{i=1}^{n}(w_2y_i)^2}\sqrt{\sum_{i=1}^{n}(w_3z_i)^2}}$$

Let $w_1 = \dfrac{k}{\sqrt{\sum_{i=1}^{n} x_i^2}}, w_2 = \dfrac{1}{\sqrt{\sum_{i=1}^{n} y_i^2}}, w_3 = \dfrac{1}{\sqrt{\sum_{i=1}^{n} z_i^2}}$, because $r(x, y) \geq \sigma$, $r(x, z) \geq \sigma$, we have

$r(y,z) \geq 2k\sigma - \dfrac{1}{2}k^2 - 1 = -\dfrac{1}{2}(k - 2\sigma)^2 + 2\sigma^2 - 1$.

Let $k=2\sigma$, $r(y, z) \geq 2\sigma^2 - 1$.
So the lemma is proven.

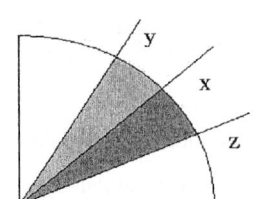

2.2 Correlated Member Clusters

Let $A = \{A_1, A_2, \ldots, A_m\}$ be a set of attributes, and a database DB is a $m*n$ relational table DB= $\{ R_1, R_2, \ldots, R_n \}$, where $R_i(i \in [1 .. n])$ is a record (row), which contains a set of values in attributes, $R_i=(a_{i1}, a_{i2}, \ldots, a_{im})$, $a_{ij} \in$ Domain(A_j). Database DB can also be viewed as a set of columns DB=$\{ A_1, A_2, \ldots, A_m \}$, where $A_j(j \in [1 .. m])$ is a value series for attribute A_j in records (column), $A_j=(a_{1j}, a_{2j}, \ldots, a_{nj})$. Members may be rows or columns depending on the analyzer's view. In this paper, we choose columns as members, and rows as features of members.

Given a user-specified minimum correlation threshold σ and a database with m members and n features, DB=$\{A_1, A_2, \ldots, A_m\}$, the member set is A=$\{A_1, A_2, \ldots, A_m\}$, we define the following terminologies.

Definition 1: Correlated-Member-Pair (CMP)

A member pair P=$\{A_s, A_t\}$ $P \subseteq A$ is a Correlated-Member-Pair, if their *Pearson's coefficient* is above the threshold σ, $r(A_s, A_t) \geq \sigma$.

Definition 2: Correlated-Member-Cluster (CMC)

A member set K=$\{A_p, \ldots, A_q\}$ $K \subseteq A$ is a Correlated-Member-Cluster, if $\forall A_s \in K, A_t \in K$, their *Pearson's coefficient* is above the threshold σ, $r(A_s, A_t) \geq \sigma$.

Definition 3: Maximal-Correlated-Member-Cluster (MCMC)

If a member set J=$\{A_u, \ldots, A_v\}$ $J \subseteq A$ is a Correlated-Member-Cluster, and for its any super set S $J \subset S \subseteq A$, S is not a Correlated-Member-Cluster, J is a Maximal-Correlated-Member-Cluster.

From the definitions, we have the following facts:

(1) A member pair $\{A_s, A_s\}$ that is represented by $\{A_s\}$ is a trivial CMP. In this paper we won't consider these trivial CMP.

(2) A CMP is also a CMC.

(3) The MCMC is a concise representation of many CMC. For a t-member MCMC, it contains $2^t - 1$ different sub CMC.

Problem definition:
Given a user-specified minimum correlation threshold σ, the problem is to mine all MCMCs with correlations above predefined threshold σ from database DB.

3 MCMC Algorithms

In this section, we present the algorithm to mine the complete MCMCs from databases. It's a two-phase method. In the first phase, procedure Calculate-CMP()

calculates Pearson's coefficient for member pairs one by one and gets all CMP. The members are sorted by the number of their correlated members in descending order. In the second phase, according to the result members' order, procedure Construct-MCMC-Tree() constructs a MCMC-Tree in an inverse order, and then travel the MCMC-Tree to generate the complete MCMCs.

3.1 Calculate Correlated Member Pairs

Procedure Calculate-CMP() scans database and calculates Pearson's coefficient for member pairs. The processing order of members in the member set $A=\{A_1, A_2, ..., A_m\}$ is originally defined to be A_1-A_2- ... - A_m. Member A_1 is first processed, and the *Pearson's coefficients* of pairs, (A_1, A_2), ..., (A_1, A_m) are calculated. Given a user defined thresholdσ, we partition the set $\{A_2, A_3, ..., A_m\}$ into 3 groups G_1, G_2, and G_3 by rule (2).

For any member A_p being processed currently,

If $r(A_p, A_i) \geq \sigma$, $A_i \in G_1$;
if $\sigma > r(A_p, A_i) \geq 2\sigma^2 - 1$, or (A_p, A_i) is marked unrelated, $A_i \in G_2$;
if $2\sigma^2 - 1 > r(A_p, A_i)$, $A_i \in G_3$. (2)

Member A_p's correlated members are in group G_1. According to Lemma2, for any member in G_1, members in group G_3 are its unrelated members. We mark member pairs in $G_1 \times G_3$ as unrelated pairs to indicate that it's unnecessary to calculate their coefficients in the following steps. This technique will prune some member pairs.

We process members A_1, A_2, ..., A_m one by one.

For each member $A_i \in A$, its possible correlated member set is $S(A_i)$. Card($S(A_i)$) is the number of members in set $S(A_i)$. We sort members of set $A=\{A_1, A_2, ..., A_m\}$ in descending order according to Card($S(A_i)$). A^* is the sorted member set, which defines the member's order that we will follow in the procedure of Construct-MCMC-Tree(). Members in $S(A_i)$ are sorted in the same order, and $S(A_i)$ contains members correlated to the member A_i. Only members after A_i are included in $S(A_i)$, while others before A_i are eliminated.

```
      Procedure Calculate-CMP()
      Input: database DB, threshold ;
      Output: sorted member set A*, S(A_1), ..., S(A_m);
 1    {A={A_1,A_2,...,A_m};
 2    For i=1 to m do
 3    {find possible corr member pairs (A[i],A_t) and put it to CS;
 4      Scan database(DB);
 5      Calculate r(A[i],A_t) for pairs in CS using formula (1);
 6      Partition set {A[i+1],...,A[m]} to G_1,G_2,and G_3 by rule (2);
 7      Mark member pair (A_j,A_k)∈G_1XG_3 as unrelated;
 8      For each member A_s in G_1 {S(A_i)=S(A_i)∪{A_s}; S(A_s)=S(A_s)∪{A_i};}
 9    }
10    A*=sort(A); /*in descending order according to card(S(A_i))*/
11    For each member A_i in A* do
12    { sort S(A_i) according to the order of A*;
13      delete members before A_i from S(A_i);
14    }
15    }
```

3.2 IOET Algorithm

A CMC is a member set, and in the given members' order, it can be represented by a sequence. For an example, the member set {b,c,g,h} is a CMC, and its sequence is α={bcgh}. The sequence can be divided to two parts: *head* and *tail*. Its head part *head*(α)={b}, and its tail part *tail*(α)={cgf}. We enumerate all MCMCs by their heads in a Maximal-Correlated-Member-Cluster-Tree (MCMC-Tree).

Definition 4: Maximal-Correlated-Member-Cluster-Tree (MCMC-Tree)

Given a member set A={A_1, A_2, . . . , A_m}, a MCMC-Tree is a 3 levels tree. Nodes in different levels are defined as the following:

(1) In level-1, there's only one {root} node that has pointers to level-2 nodes.
(2) In level-2, nodes are indexed by members in set A. Each member A_i refers to a node Node(A_i). Node(A_i) contains all possible members correlated with A_i. Node(A_i).content=S(A_i). Node(A_i).index=A_i. There are m nodes in level-2.
(3) In level-3, nodes are indexed by members as the same as their fathers' indexies. Node(A_t) contains Local-MCMC sequences { β_1,... β_k}, head(β_i)= A_t, and β_i is a Local-MCMC in sub set {A_t, …, A_m}.

Constructing MCMC-Tree

In procedure Calculate-CMP(), we get the possible correlated members set S(A_i) for each member A_i. S(A_i) contains the possible members that immediately follows A_i in the MCMC sequence. We can use S(A_i) to expand MCMC sequence headed by A_i. Similar works in paper [3] proposed a set-enumeration tree technique to expand sets over an ordered and finite member domain. It appends possible suffix one by one from a head to get the maximal sets. Noted that the max number of combinations is 2^m-1 in worst, it will be expensive for a big m. This technique isn't applicable to high dimensional data. Here, we design an Inverse-Order-Enumeration-Tree (IOET) algorithm to construct the MCMC-Tree and get the complete MCMC.

The output of procedure Calculate-CMP() A* defines the members sequential order, which is the order of MCMC sequences, and set S(A_i) is also sorted by this order. For the expression simplicity, the members order in A* is assumed to be A_1-A_2-. . . - A_m. In the MCMC-Tree, nodes in level-2 are constructed in the same order. When we construct nodes in level-3, we won't follow the order A_1-A_2- . . . - A_m, but in an inverse order A_m-A_{m-1}- . . . – A_1. We generate Local-MCMCs headed by A_m, A_{m-1}, . . ., A_1 one by one. The advantage of inverse order algorithm is that when we generate a Local-MCMC α headed by A_i, *tail*(α) are already calculated.

We use an example to illustrate the MCMC-Tree construction procedure.

A sorted member set is A={a,b,c,d,e,f,g,h}. Inverse order set A*={h,g,f,e,d,c,b,a}.

Table 1.

Member (Head)	Correlated members	Local-MCMC
{h}	Null	{h}
{g}	Null	{g}
{f}	{g,h}	{fg, fh}
{e}	{f,g,h}	{efg, efh, eg, eh}
{d}	{e,f,g}	{defg, dfg, dg}
{c}	{e,f}	{cef, ef}
{b}	{c,d,f,g}	{bcf, bdfg, bfg, bg}
{a}	{b,d,f,h}	{abf, abdf, adf, afh, ah}

Considering member b as an example, S(b)={c,d,f,g}, when we construct node(b) in level-3, all Local-MCMC headed by {c}, {d}, {f}, {g} are already generated. In order to generate MCMC sequences α headed by {b}, tails of α are: S(b)∩{c}.content={cf}, S(b)∩{d}.content={dfg}, S(b)∩{f}.content={fg}, and S(b)∩{g}.content={g}. Finally, we get {bcf,bdfg,bfg,bg} after we add {b} as the head. Sequences {~~bfg~~, ~~bg~~} are eliminated, because they are subsequence of {bdfg}.

After we generate all nodes in level-3, we travel these nodes in ordinary order, and output MCMC. Local-MCMC that is a subset of MCMC is eliminated, such as {~~h~~}, {~~g~~}, {~~fg~~, ~~fh~~} etc.

```
   Procedure Construct-MCMC-Tree()
   Input: ordered member set A*, S(A₁), ..., S(Aₘ);
   Output: all MCMC;
1  { MCMC-Tree={root};
2    For i=1 to m do /*generate level-2 nodes*/
3    { K.index=A*[i]; K.content=S(A*[i]);Insert(K, CMC-Tree);}
4    For each Node K in level-2 do /*by inverse order*/
5    { create empty node T; T.index=K.index;
6      For each member B in K.content do
7      { find node L in level-3 that L.index=B;
8        For each seq in L.content do
9        { seq1=Sequence(S(T.index)•Set(seq));
10         seq2=Catenate(T.index,seq1);
11         Append(T.content,seq2);
12       }
13     }
14     For each seq in T.content do
15       If (seq isn't Local-MCMC) {elminate seq};
16     Insert(T,K);
17   }
18   For each Node K in level-3 do /*K.index in Ordinary order*/
19     For each seq in K.content do
20       If (seq is a MCMC) {output seq};
21 }
```

4 Experiments and Analysis

We implement the algorithm in Microsoft visual c++ 6.0 on the windows2000 platform with a 1.7 GHz CPU and 512 MB main memory. First, we generate the synthetic data sets in tabular forms. A data set is a relational table that has m columns (members) and n rows (records). In order to evaluate the performance of the algorithm, we test the algorithm on these synthetic data sets as we change numbers m, n, and user predefined threshold σ.

As the main algorithm contains 2 major subroutines Calculate-CMP() and Construct-MCMC-Tree(), we will examine their performance separately.

Performance of procedure Calculate-CMP()

For procedure Calculate-CMP() and the original algorithm without pruning, Fig4a illustrates the CPU time cost when the number of columns increases from 100 to 10k, and Fig4b shows the CPU time cost when the number of rows increases from 200 to 10k. Form experiment results, we can see that procedure Calculate-CMP() outperforms the original algorithm, and has a good scalability with the number of columns and rows.

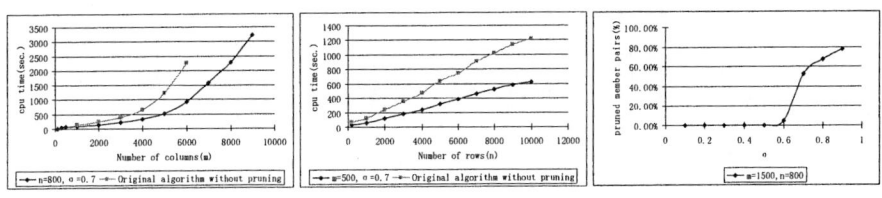

Fig. 4a.   Fig. 4b.   Fig. 4c.

Fig4c shows the percentage of unrelated member pairs being pruned for different user predefined threshold σ. We can see that about 80% member pairs are pruned when σ=0.9. The percentage is defined to be: $\dfrac{number\ of\ pruned\ member\ pairs}{C_m^2}$.

Performance of procedure Construct-MCMC-Tree()

Fig5a compares the CPU time cost of IOET algorithm with the original enumeration-tree algorithm. Using the original algorithm, the CPU time cost rises rapidly when the number of columns is above 100. Because the time cost of enumeration-tree algorithm is $O(2^m)$ in nature, m is quite a bottleneck for such algorithms. However, the IOET algorithm shows extraordinary scalability when the number members over 1k. Fig5b displays the number of MCMCs generated when the number of columns increases. We notice that the number of result MCMCs is in a reasonable range (about 1400 when m=9k and σ=0.7).

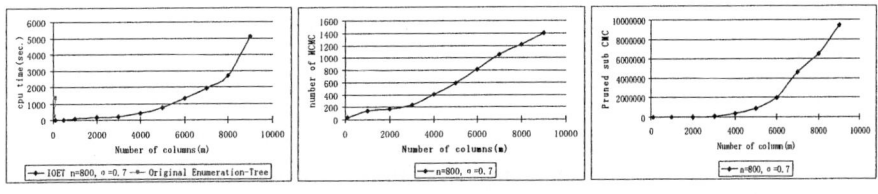

Fig. 5a.   Fig. 5b.   Fig. 5c.

Although there're not other works exactly as the same as ours, we compare the IOET algorithm with the traditional enumeration-tree algorithm. Works in [5][3][9] use the similar enumeration-tree algorithms, and we notice that the column number of data set in their experiments is about 100. Our experiment tests IOET algorithm on data sets of 1k-9k columns. In our experiments, there are a quite large number of eliminated sub CMCs (more than 8M in Fig5c), and IOET cut these branches in search space as soon as possible in order to prevent them from growing exponentially. It will help to explain the reason that IOET has an excellent performance.

Experiment on real life data set

Back to the example, we experiment on economic data set that has 99 columns and 29 rows. The results are shown below. Tab2 illustrates the effectiveness of our pruning techniques.

Table 2.

σ	Pruned Member pairs	Number of Pruned CMC	Number of MCMC	Max length of MCMC
0.7	1877	1343	28	26
0.8	2914	569	10	18
0.9	3665	562	16	18

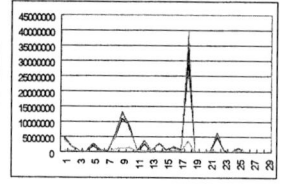
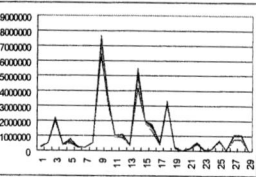

Fig. 6a. Fig. 6b.

Fig6a plots a result MCMC series that include 18 members, and Fig6b plots another result MCMC series that include 8 members. It is clear that correlated members in a MCMC exhibit similar trend patterns, while different MCMCs show different patterns.

5 Conclusion

Correlation mining has been studied widely and intensively since association rule mining was first proposed in 1993, and now it attracts more attentions than ever time before. Another useful tool for similarity search, pattern recognition, and trend analysis is clustering model, which defines closeness of nominal variables by distance (Minkowski, Manhattan or Euclidean) or similarity (cosine) measures. In this paper, we propose the MCMC model for the first time to find correlated member clusters based on a statistical measure. This extended model will discover patterns of rise and fall among data series, which will benefit a lot of applications.

Mining MCMCs from high dimensional database is an interesting and challenging problem. Just as the Frequent Item Set model and the pair-wise cluster model based on pattern similarity, the MCMC model considers all combinations of members. Its computational complexity problem is getting worse than traditional clustering models. For this reason, we design optimizing algorithms to make the MCMC model be applicable to high dimensional data (more than 1k members).

Discovering MCMCs is the first stage of data analyzing. From MCMCs, we will deduce the hierarchy of members naturally. Then we will employ other methods to inspect those correlated data intensively to find trends or anomalies. Generally, the MCMC model is a very useful tool in correlation mining, and can be used in a wide range of applications.

References

[1] R. Agrawal, T. Imielinski, and A. Swami. Mining Association Rules between Sets of Items in Large Databases. In Proc. of 1993 Int. Conf. on Management of Data (SIGMOD'93), pp. 207-216, 1993.

[2] R. Agrawal, and R. Srikant. Fast Algorithms for Mining Association Rules in Large Databases. In Proc. of 1994 Int. Conf. Very Large Data Bases (VLDB'94), pp. 487-499, 1994.

[3] R. J. Bayardo. Efficiently mining long patterns from databases. In Proc. of 1998 Int. Conf. on Management of Data (SIGMOD'98), pp.85-93, 1998.

[4] J. Han, J. Pei, and Y. Yin. Mining Frequent Patterns Without Candidate Generation. In Proc. of 2000 Int. Conf. on Management of Data (SIGMOD'00), pp. 1-12, 2000.

[5] D. Jiang, J. Pei, and A. Zhang. Mining Coherent Gene Clusters from Gene-Sample-Time Microarray Data. In Proc. of Int. Conf. Knowledge Discovery and Data Mining (KDD '04), pp. 430-439, 2004.

[6] J. Liu, and W. Wang. OP-Cluster:Clustering by Tendency in High Dimensional Space. In Proc. of the 3rd IEEE Int. Conf. on Data Mining (ICDM'03)), pp. 187-194, 2003.

[7] J. Pei, G. Dong, W. Zou, and J. Han. Mining Condensed Frequent Pattern Bases. Knowledge and Information Systems, Vol. 6 No. 5, pp. 570-594, Springer-Verlag, 2004.

[8] J. Pei, J. Han, and W. Wang. Mining sequential patterns with constraints in large databases. In Proc. of ACM Conf. on Information and Knowledge Management (CIKM'02), pp. 18-25, 2002.

[9] H. Wang, W. Wang, J. Yang, and P. S. Yu. Clustering by Pattern Similarity in Large. Data Sets. In Proc. of 2002 Int. Conf. on Management of Data (SIGMOD'02), pp. 418-427, 2002.

[10] H. Xiong, S. Shekhar, P. N. Tan, and V. Kumar. Exploiting a Support-based Upper Bound of Pearson's Correlation Coefficient for Efficiently Identifying Strongly Correlated Pairs. In Proc. of 2004 Int. Conf. Knowledge Discovery and Data Mining (KDD'04), pp. 334-343, 2004.

[11] M. J. Zaki. Generating non-redundant association rules. In Proc. of 2000 Int. Conf. Knowledge Discovery and Data Mining (KDD'00), pp. 34-43, 2000.

Hierarchical Clustering Based on Mathematical Optimization

Le Hoai Minh[1], Le Thi Hoai An[1], and Pham Dinh Tao[2]

[1] Laboratory of Theoretical and Applied Computer Science - LITA EA 3097,
UFR MIM, University of Paul Verlaine - Metz, Ile de Saulcy, 57045 Metz, France
lehoai@univ-metz.fr, lethi@univ-metz.fr
http://lita.sciences.univ-metz.fr/~lethi/

[2] Laboratory of Modelling, Optimization & Operations Research,
National Institute for Applied Sciences - Rouen,
BP 08, Place Emile Blondel F 76131 Mont Saint Aignan Cedex, France
pham@insa-rouen.fr

Abstract. In this paper a novel optimization model for bilevel hierarchical clustering has been proposed. This is a hard nonconvex, nonsmooth optimization problem for which we investigate an efficient technique based on DC (Difference of Convex functions) programming and DCA (DC optimization Algorithm). Preliminary numerical results on some artificial and real-world databases show the efficiency and the superiority of this approach with respect to related existing methods.

Keywords: nonconvexe optimization, nonsmooth optimization, DC programming, DCA, Bilevel hierarchical clustering, K-means.

1 Introduction

Multilevel hierarchical clustering consists of grouping data objects into a hierarchy of clusters. It has a long history (see e.g. [2], [5], [15]) and has many important applications in various domains, since many kinds of data, including observational data collected in the human and biological sciences, have a hierarchical, nested, or clustered structure. Hierarchical clustering algorithms are useful to determine hierarchical multicast trees in the network topology identification, Grid computing using in e-Science, e-Medicine or e-Commerce, Multimedia conferencing, Large-scale dissemination of timely information, ...

A hierarchical clustering of a set of objects can be described as a tree, in which the leaves are precisely the objects to be clustered. A hierarchical clustering scheme produces a sequence of clusterings in which each clustering is nested into the next clustering in the sequence. Standard existing methods for Multilevel hierarchical clustering are often based upon nonhierarchical clustering algorithms coupled with several iterative control strategies to repeatedly modify an initial clustering (reordering, and reclustering) in search of a better one.

To our knowledge, while mathematical programming is widely used for nonhierarchical clustering problems there exist a few optimization models and techniques for multilevel hierarchical clustering ones. Except the work in [14] we have not found other approaches using mathematical programming model for multilevel hierarchical clustering.

In this paper we investigate an efficient optimization approach for a model of this class, that is bilevel hierarchical clustering. The problem can be stated as follows. Given a set $\mathcal{A}$ of p objects $\mathcal{A} := \{a_j \in \mathbb{R}^n : j = 1, ..., p\}$, a measured distance, and an integer k. We are to choose $k + 1$ members in $\mathcal{A}$, one as the total centre (the root of the tree) and others as centres of k disjoint clusters, and assign other members of $\mathcal{A}$ to their closest centre. The total centre is defined as the closest object to all centres (in the sense that the sum of distances between it and all centres is the smallest).

Our approach is based on mathematical optimization via DC (Difference of Convex functions) programming - which deals with DC programs, i.e., the minimization of a DC function over a convex set - and DC optimization Algorithm called DCA. They were introduced by Pham Dinh Tao in their preliminary form in 1986 and have been extensively developed since 1994 by Le Thi Hoai An and Pham Dinh Tao to become now classic and more and more popular (see e.g. [7], [8] - [12], [16], [17] and references therein). DCA has been successfully applied to many large-scale (smooth or nonsmooth) nonconvex programs in various domains of applied sciences, in particular in data analysis and data mining ([1], [6], [11], [19], [20]), for which it provides very often a global solution and proves to be more robust and efficient than standard methods.

We propose in this work a new optimization formulation that seems to be appropriate for hierarchical clustering. This is a nonsmooth, nonconvex problem and can be reformulated as a DC program which we then suggested using DC programming approach and DCA to solve. Preliminary numerical results on some artificial and real-world databases demonstrate that the proposed algorithm is very promising and more efficient than some existing optimization based clustering algorithms.

The paper is organized as follows. Section 2 introduces a novel optimization model for the bilevel hierarchical clustering problem. Section 3 deals with DC programming and DCA for solving the underlying bilevel hierarchical clustering problem. For the reader's convenience, at the beginning of this section we provide a brief introduction to DC programming and DCA. Computational results are reported in the last section.

2 Optimization Formulation

In [14] the authors have proposed two nonsmooth, nonconvex optimization models for the bilevel hierarchical clustering problem in the context of determining a multicast group. They considered the set $\mathcal{A}$ as the set of p nodes in the plane, and the measured distance is the Euclidean distance. The disadvantages of their models are the following:

- first, the total centre is determined according to other centres - this is not natural for bilevel clustering;

- second, in their approach using the artificial centres the constraints do not ensure that the total centre is in the set $\mathcal{A}$;
- third, these problems can be formulated as DC programs, but *it is not suitable for the search of resulting DCA in explicit form.*

In this work we introduce a novel model that seems to be more appropriate: we search simultaneously the total centre and other centres. Moreover, by considering the squared Euclidean distance as the measured distance we get a DC program for which DCA is explicitly determined and very inexpensive.

Denoting by x_i, $i = 1, \ldots, k$ the centre of clusters in the second level and x_{k+1} the total centre we can formulate the problem in the form

$$\min\left\{\sum_{j=1}^{p} \min_{i=1\ldots k} \|x_i - a_j\|^2 + \sum_{i=1}^{k} \|x_{k+1} - x_i\|^2 \text{ s.t. } \sum_{i=1}^{k+1} \min_{j=1\ldots p} \|x_i - a_j\|^2 = 0\right\}. \quad (1)$$

The objective function containing the two terms is nonsmooth and nonconvex. The first term is a cluster function while the second term presents the distance between the total centre and the other centres. The constraint ensures that all centres are in the set $\mathcal{A}$. The advantage of this formulation is that all centres are found in the same time.

This is a hard constrained global optimization problem. Using penalty technique in DC programming ([8], [12]) leads us to the more tractable unconstrained nonsmooth nonconvex optimization problem ($\tau > 0$ is the penalty parameter):

$$\min\left\{\frac{1}{2}\sum_{j=1}^{p} \min_{i=1..k} \|x_i - a_j\|^2 + \frac{1}{2}\sum_{i=1}^{k} \|x_{k+1} - x_i\|^2 \right.$$
$$\left. + \frac{\tau}{2}\sum_{i=1}^{k+1} \min_{j=1\ldots p} \|x_i - a_j\|^2 \text{ s.t. } x_i \in \mathbb{R}^n\right\}. \quad (2)$$

We will prove in Section 3 that this problem can be reformulated as a DC program and show how to use DCA for solving it.

3 DC Programming and DCA for Bilevel Hierarchical Clustering

3.1 A Brief Presentation of DC Programming and DCA

To give the reader an easy understanding of the theory of DC programming & DCA and our motivation to use them for solving Problem (2), we briefly outline these tools in this section. Let $\Gamma_0(\mathbb{R}^n)$ denote the convex cone of all lower semicontinuous proper convex functions on $\mathbb{R}^n$. The vector space of DC functions, $DC(\mathbb{R}^n) = \Gamma_0(\mathbb{R}^n) - \Gamma_0(\mathbb{R}^n)$, is quite large to contain almost real life objective functions and is closed under all the operations usually considered in optimization.

Consider the general DC program

$$\alpha = \inf\{f(x) := g(x) - h(x) \ : \ x \in \mathbb{R}^n\} \quad (P_{dc})$$

with $g, h \in \Gamma_0(\mathbb{R}^n)$. Such a function f is called DC function, and $g - h$, DC decomposition of f while the convex functions g and h are DC components of f.

If g or h are polyhedral convex functions then (P_{dc}) is called a polyhedral DC program.

It should be noted that a constrained DC program whose feasible set C is convex can always be transformed into an unconstrained DC program by adding the indicator function χ_C of C ($\chi_C(x) = 0$ if $x \in C, +\infty$ otherwise) to the first DC component g.

Let

$$g^*(y) := \sup\{\langle x, y \rangle - g(x) \ : \ x \in \mathbb{R}^n\}$$

be the conjugate function of g. By using the fact that every function $h \in \Gamma_0(\mathbb{R}^n)$ is characterized as a pointwise supremum of a collection of affine functions, say

$$h(x) := \sup\{\langle x, y \rangle - h^*(y) : y \in \mathbb{R}^n\},$$

we have

$$\alpha = \inf_{x \in \mathbb{R}^n}\{g(x) - \sup_{y \in \mathbb{R}^n}\{\langle x, y \rangle - h^*(y)\}\} = \inf\{\alpha(y) : y \in \mathbb{R}^n\}$$

with $\alpha(y) := \inf_{x \in \mathbb{R}^n}\{g(x) - [\langle x, y \rangle - h^*(y)]\}$ (P_y).

It is clear that (P_y) is a convex program and

$$\alpha(y) = h^*(y) - g^*(y) \text{ if } y \in \text{dom } h^*, +\infty \text{ otherwise.}$$

Finally we state the dual program of (P_{dc})

$$\alpha = \inf\{h^*(y) - g^*(y) : y \in \text{dom } h^*\}$$

that is written, in virtue of the natural convention in DC programming, say $+\infty = +\infty - (+\infty)$:

$$\alpha = \inf\{h^*(y) - g^*(y) : y \in Y\}. \quad (D_{dc})$$

We observe the perfect symmetry between primal and dual DC programs: the dual to (D_{dc}) is exactly (P_{dc}).

DC programming investigates the structure of the vector space $DC(\mathbb{R}^n)$, DC duality and optimality conditions for DC programs. The complexity of DC programs resides, of course, in the lack of practical optimal globality conditions. We developed instead the following necessary local optimality conditions for DC programs in their primal part, by symmetry their dual part is trivial (see [8] - [12], [16], [17] and references therein):

$$\partial h(x^*) \cap \partial g(x^*) \neq \emptyset \qquad (3)$$

(such a point x^* is called *critical point* of $g - h$ or for (P_{dc})), and

$$\emptyset \neq \partial h(x^*) \subset \partial g(x^*). \qquad (4)$$

The condition (4) is also sufficient for many classes of DC programs. In particular it is sufficient for the next cases quite often encountered in practice:

- In polyhedral DC programs with h being a polyhedral convex function (see [8] - [12], [16], [17] and references therein). In this case, if h is differentiable at a critical point x^*, then x^* is actually a local minimizer for (P_{dc}). Since a convex function is differentiable everywhere except for a set of measure zero, one can say that a critical point x^* is almost always a local minimizer for (P_{dc}).
- In case the function f is locally convex at x^* ([10], [12]).

Based on local optimality conditions and duality in DC programming, the DCA consists in the construction of two sequences $\{x^k\}$ and $\{y^k\}$, candidates to be optimal solutions of primal and dual programs respectively, such that the sequences $\{g(x^k) - h(x^k)\}$ and $\{h^*(y^k) - g^*(y^k)\}$ are decreasing, and $\{x^k\}$ (resp. $\{y^k\}$) converges to a primal feasible solution $\widetilde{x}$ (resp. a dual feasible solution $\widetilde{y}$) verifying local optimality conditions and

$$\widetilde{x} \in \partial g^*(\widetilde{y}), \quad \widetilde{y} \in \partial h(\widetilde{x}). \qquad (5)$$

These two sequences $\{x^k\}$ and $\{y^k\}$ are determined in the way that x^{k+1} (resp. y^k) is a solution to the convex program (P_k) (resp. (D_k)) defined by

$$\inf_{x \in \mathbb{R}^n} \{g(x) - h(x^k) - \langle x - x^k, y^k \rangle\} \ (P_k)$$

$$\inf_{y \in \mathbb{R}^n} \{h^*(y) - g^*(y^{k-1}) - \langle y - y^{k-1}, x^k \rangle\} \ (D_k).$$

The *interpretation* of DCA is simple: at each iteration one replaces in the primal DC program (P_{dc}) the second component h by its affine minorization $h_k(x) := h(x^k) + \langle x - x^k, y^k \rangle$ at a neighbourhood of x^k to give birth to the convex program (P_k) whose the solution set is nothing but $\partial g^*(y^k)$. Likewise, the second DC component g^* of the dual DC program (D_{dc}) is replaced by its affine minorization $(g^*)_k(y) := g^*(y^k) + \langle y - y^k, x^{k+1} \rangle$ at a neighbourhood of y^k to obtain the convex program (D_k) whose $\partial h(x^{k+1})$ is the solution set. DCA performs so a double linearization with the help of the subgradients of h and g^* and the DCA then yields the next scheme:

$$y^k \in \partial h(x^k); \quad x^{k+1} \in \partial g^*(y^k). \qquad (6)$$

First of all, it is worth noting that our works involve the convex DC components g and h but not the DC function f itself. Moreover, a DC function f has *infinitely many DC decompositions which have crucial impacts on the qualities* (speed of convergence, robustness, efficiency, globality of computed solutions,...) of DCA. For a given DC program, the choice of *optimal* DC decompositions is still open. Of course, this depends strongly on the very specific structure of the problem being considered. In order to tackle the large scale setting, one tries in practice to choose g and h such that sequences $\{x^k\}$ and $\{y^k\}$ can be easily calculated, i.e. either they are in explicit form or their computations are inexpensive.

It is proved in [8] - [12], [16], [17]) that DCA is a descent method without linesearch which enjoys the following properties:

i) The sequences $\{g(x^k) - h(x^k)\}$ and $\{h^*(y^k) - g^*(y^k)\}$ are decreasing.
ii) If the optimal value α of problem (P_{dc}) is finite and the infinite sequences $\{x^k\}$ and $\{y^k\}$ are bounded, then every limit point $\tilde{x}$ (resp. $\tilde{y}$) of the sequence $\{x^k\}$ (resp. $\{y^k\}$) is a critical point of $g - h$ (resp. $h^* - g^*$).
iii) DCA has a linear convergence for general DC programs.
iv) DCA has a finite convergence for polyhedral DC programs.

For a complete study of DC programming and DCA the redear is referred to [7], [8] - [12], [16], [17] and references therein. The solution of a nonconvex program by DCA must be composed of two stages: the search of an *appropriate* DC decomposition and that of a *good* initial point. We shall apply *all these DC enhancement features* to solve problem (2) in its equivalent DC program given in the next.

3.2 Solving Problem (2) by DCA

To simplify related computations in DCA for solving problem (2) we will work on the vector space $\mathbb{R}^{(k+1)\times n}$ of $((k+1) \times n)$ real matrices. The variables are then $X \in \mathbb{R}^{(k+1)\times n}$ whose i^{th} row X_i is equal to x_i for $i = 1, ..., k+1$:

$$\mathbb{R}^{(k+1)\times n} \ni X \leftrightarrow (X_1, X_2, \ldots, X_{k+1}) \in (\mathbb{R}^n)^{k+1}, \quad X_i \in \mathbb{R}^n, (i = 1, .., k+1).$$

The Euclidean structure of $\mathbb{R}^{(k+1)\times n}$ is defined with the help of the usual scalar product and its Euclidean norm:

$$\langle X, Y \rangle := Tr(X^T Y) = \sum_{i=1}^{k} \langle X_i, Y_i \rangle, \quad \|X\|^2 := \sum_{i=1}^{k} \langle X_i, X_i \rangle = \sum_{i=1}^{k+1} \|X_i\|^2$$

(Tr denotes the trace of a square matrix). We will reformulate problem (2) as a DC program in the matrix space $\mathbb{R}^{(k+1)\times n}$ and then describe DCA for solving it.

DC Formulation of (2). According to the property

$$\min_{i=1,\ldots,k} \|x_i - a_j\|^2 = \sum_{i=1}^{k} \|x_i - a_j\|^2 - \max_{r=1,\ldots,k} \sum_{\substack{i=1 \\ i \neq r}}^{i \neq r} \|x_i - a_j\|^2$$

we can write the objective function of (2), denoted F, as

$$F(X) = \frac{1}{2}\sum_{j=1}^{p}\sum_{i=1}^{k}\|X_i-a_j\|^2 - \frac{1}{2}\sum_{j=1}^{p}\max_{i=1..k}\sum_{r=1,r\neq i}^{k}\|X_r-a_j\|^2 + \frac{\tau}{2}\sum_{i=1}^{k+1}\sum_{j=1}^{p}\|X_i-a_j\|^2$$

$$-\frac{\tau}{2}\sum_{i=1}^{k+1}\max_{j=1..p}\sum_{s=1,s\neq j}^{p}\|X_i-a_s\|^2 + \frac{1}{2}\sum_{i=1}^{k}\|X_{k+1}-X_i\|^2$$

$$= \frac{\tau+1}{2}\sum_{j=1}^{p}\sum_{i=1}^{k+1}\|X_i-a_j\|^2 + \frac{1}{2}\sum_{i=1}^{k}\|X_{k+1}-X_i\|^2$$

$$-\frac{1}{2}\sum_{j=1}^{p}\max_{i=1..k}\sum_{r=1,r\neq i}^{k}\|X_r-a_j\|^2 - \frac{\tau}{2}\sum_{i=1}^{k+1}\max_{j=1..p}\sum_{s=1,s\neq j}^{p}\|X_i-a_s\|^2$$

$$-\frac{1}{2}\sum_{j=1}^{p}\|X_{k+1}-a_j\|^2 = G(X) - H(X),$$

where

$$G(X) = \frac{\tau+1}{2}\sum_{j=1}^{p}\sum_{i=1}^{k+1}\|X_i-a_j\|^2 + \frac{1}{2}\sum_{i=1}^{k}\|X_{k+1}-X_i\|^2,$$

$$H(X) = \frac{1}{2}\sum_{j=1}^{p}\max_{i=1..k}\sum_{r=1,r\neq i}^{k}\|X_r-a_j\|^2 + \frac{\tau}{2}\sum_{i=1}^{k+1}\max_{j=1..p}\sum_{s=1,s\neq j}^{p}\|X_i-a_s\|^2$$

$$+\frac{1}{2}\sum_{j=1}^{p}\|X_{k+1}-a_j\|^2. \qquad (7)$$

It is easy to see that G and H are convex functions and then (2) is DC program in the form

$$\min\left\{G(X) - H(X) : X \in \mathbb{R}^{(k+1)\times n}\right\}. \qquad (8)$$

According to Section 3.1, determining the DCA scheme applied to (8) amounts to computing the two sequences $\{X^{(l)}\}$ and $\{Y^{(l)}\}$ in $\mathbb{R}^{(k+1)\times n}$ such that

$$Y^{(l)} \in \partial H(X^{(l)}), \quad X^{(l+1)} \in \partial G^*(Y^{(l)}).$$

We shall present below the computation of $\partial H(X)$ and $\partial G^*(Y)$.

Computing of $\partial H(X)$. We have

$$\partial H(X) = \partial H_1(X) + \partial H_2(X) + \partial H_3(X) \qquad (9)$$

where

$$H_1 := \sum_{i=1}^{p} h_j^1, \quad h_j^1 := \max_{i=1,\ldots,k} h_{j,i}^1, \quad h_{j,i}^1(X) := \frac{1}{2}\sum_{r=1,r\neq i}^{k}\|X_r-a_j\|^2 \qquad (10)$$

$$H_2 := \sum_{i=1}^{k+1} h_i^2, \quad h_i^2 := \max_{j=1,\ldots,p} h_{i,j}^2, \quad h_{i,j}^2(X) := \frac{\tau}{2} \sum_{s=1, s\neq j}^{p} \|X_i - a_s\|^2,$$

$$H_3(X) := \frac{1}{2} \sum_{j=1}^{p} \|X_{k+1} - a_j\|^2. \tag{11}$$

The functions $h_{j,i}^1$ are differentiable and

$$[\nabla h_{j,i}^1(X)]_l = 0 \text{ if } l \in \{i, k+1\}, \quad X_l - a_j \text{ otherwise.} \tag{12}$$

Hence the subdifferential of H_1 can be explicitly determined as follows: (co denotes the convex hull)

$$\partial H_1(X) = \sum_{i=1}^{p} \partial h_j^1(X), \quad \partial h_j^1(X) = co\{\partial h_{j,i}^1(X) : h_{j,i}^1(X) = h_j^1(X)\}. \tag{13}$$

Likewise we have

$$\partial H_2(X) = \sum_{j=1}^{k+1} \partial h_i^2(X), \quad \partial h_i^2(X) = co\{\partial h_{i,j}^2(X) : h_{i,j}^2(X) = h_i^2(X)\}, \tag{14}$$

and the functions $h_{i,j}^2$ are differentiable of which the derivative is computed as

$$[\nabla h_{i,j}^2(X)]_l = (p-1)X_l - \sum_{s=1, s\neq j}^{p} a_s \text{ if } l = i, \quad 0 \text{ otherwise.} \tag{15}$$

The subdifferential of $H_2(X)$ is therefore also explicitly determined. Finally for H_3 we get

$$[\nabla H_3(X)]_l = pX_{k+1} - \sum_{j=1}^{p} a_j \text{ if } l = k+1, \quad 0 \text{ otherwise.} \tag{16}$$

Computing of $\partial G^*(X)$. Let G_1 and G_2 be the functions defined by

$$G_1(X) := \frac{\tau+1}{2} \sum_{j=1}^{p} \sum_{i=1}^{k+1} \|X_i - a_j\|^2, \quad G_2(X) := \frac{1}{2} \sum_{i=1}^{k} \|X_{k+1} - X_i\|^2. \tag{17}$$

Then, according to (7):

$$G(X) = (\tau+1)G_1(X) + G_2(X). \tag{18}$$

Let $A^{(j)} \in \mathbb{R}^{(k+1)\times n}$ be the matrix whose all rows are equal to a_j. We can write G_1 in the form

$$G_1(X) = \frac{1}{2} \sum_{j=1}^{p} \sum_{i=1}^{k+1} \|X_i - a_j\|^2 = \frac{1}{2} \sum_{j=1}^{p} \|X - A^j\|^2. \tag{19}$$

On the other hand we can express G_2 as

$$G_2(X) = \tfrac{1}{2}\sum_{i=1}^{k}\|X_{k+1} - X_i\|^2 = \tfrac{1}{2}\sum_{i=1}^{k+1}\|X_{k+1} - X_i\|^2 = \tfrac{1}{2}\|WX\|^2, \qquad (20)$$

where $W = (w_{ij}) \in \mathbb{R}^{(k+1)\times(k+1)}$ is the matrix defined by

$$w_{ij} = -1 \text{ if } i = j, \text{ for } j = 1,\ldots,k, \quad 1 \text{ if } j = k+1, \text{ for } i = 1,\ldots,k,$$
$$0 \text{ otherwise}. \qquad (21)$$

The convex function G is then a positive definite quadratic form on $\mathbb{R}^{(k+1)\times n}$ and its gradient is given by

$$\nabla G(X) = (\tau+1)\sum_{j=1}^{p}(X - A^j) + W^TWX = [(\tau+1)pI + W^TW]X - (\tau+1)A \qquad (22)$$

with $A := \sum_{j=1}^{p} A^{(j)}$, i.e., $A_i = \sum_{j=1}^{p} a_j$, $i = 1,...,k+1$. Since $X = \nabla G^*(Y)$ iff $Y = \nabla G(X)$, we get

$$Y = [(\tau+1)pI + W^TW]X - (\tau+1)A \quad \text{or} \quad [(\tau+1)pI + W^TW]X = Y + (\tau+1)A.$$

This permits us to compute explicitly X as follows:

$$X_i = \frac{B_i + X_{k+1}}{1+c} \text{ for } i = 1...k, \quad X_{k+1} = \frac{(1+c)B_{k+1} + \sum_{l=1}^{k} B_l}{(1+c)(k+c) - k}, \qquad (23)$$

with $B = Y + (\tau+1)A$ and $c = (\tau+1)p$.

In the matrix space $\mathbb{R}^{(k+1)\times n}$, according to (7), (18), (19) and (20) the DC program (8) then is minimizing the difference of the simple convex quadratic function and the nonsmooth convex function. This nice feature is very convenient for applying DCA, which consists in solving a sequence of approximate convex quadratic programs whose solutions are explicit.

We can now describe our DCA scheme for solving (2).

Algorithm DCA Initialization.
Let $X^{(0)} \in \mathbb{R}^{(k+1)\times n}$ and $\epsilon > 0$ be small enough. Set $l = 0$.
Repeat

- Compute $Y^{(l)} \in \partial H(X(^{(l)}))$ with the help of the formulations ((9) - (16));
- Compute $X^{(l+1)} \in \partial G^*(Y^{(l)})$ via (23) ;
- Set $l = l + 1$

Until $\|X^{(l)} - X^{(l-1)}\| \leq \epsilon(\|X^{(l)}\| + 1)$ or $|F(X^{(l)}) - F(X^{(l-1)})| \leq \epsilon(|F(X^{(l)})| + 1|)$.

Find again the real centres. Let X^* be the solution obtained by DCA and let $x_i^* = (X^*)_i$, $i = 1, ..., k+1$. Then the *real* centres $\bar{x}_i$ for $i = 1, ..., k+1$ (corresponding to a solution of problem (1) are determined by

$$\bar{x}_i = \operatorname{argmin} \{\|x_i^* - a_j\|^2 : j = 1, ..., p\}. \tag{24}$$

How to find a good initial point for DCA. Finding a good starting point is important for DCA to reach global solutions. For this, we combine alternatively the two procedures by exploiting simultaneously the efficiency of DCA and the K-means algorithm. More precisely, starting with a point $X^{(0)}$ with $X_i^{(0)}$ randomly chosen among the points in $\mathcal{A}$ we perform one iteration of DCA, namely set $Y^{(0)} \in \partial H(X^{(0)})$ and $Z^{(1)} \in \partial G^*(Y^{(0)})$, and then improve $Z^{(1)}$ by one iteration of K-means to obtain $X^{(1)}$. We note that at each iteration DCA returns $k+1$ "centres" while K-means return k "centres" of clusters from which the "total centre" is determined via the formula (25) below. This procedure can then be repeated some times to provide a good initial point for the main DCA as will be shown in numerical simulations.

The combined DCA - K-means procedure, denoted **IP**, to find a good initial point for the main DCA is described as follows:

Procedure IP: let q be a positive integer.

Let $X^{(0)} \in \mathbb{R}^{k \times n}$ such that $X_i^{(0)}$ is randomly chosen among the points of $\mathcal{A}$.
For **t = 0, 1, ..., q** do
   t1. Compute $Y^{(t)}$ by the formulations ((9) - (16)) and $X^{(t+1)}$ by (23);
   t2. Assign each point $a_j \in \mathcal{A}$ into the cluster that has the closest centre $X_1^{(t+1)}, ..., X_k^{(t+1)}$. Let π_i the cluster of the centre $X_i^{(t+1)}$, $i = 1, ...k$.
   t3. For each $i \in \{1, ..., k\}$ recompute Z_i as the centres of the cluster π_i:

$$Z_i := \arg\min \left\{ \sum_{a_j \in \pi_i} \|y - a_j\|^2 : y \in \mathbb{R}^n \right\},$$

and set $Z_{k+1} = \arg\min {}_{a_j \in \mathcal{A}} \sum_{i=1}^{k} \|Z_i - a_j\|^2.$

Update $X_i^{(t+1)} := Z_i$ for $i = 1, ...k+1$.
enddo
Ouput: set $X^{(0)} := X^{(q)}$.

We note (from several numerical tests) that the alternative DCA - K-means procedure is better than the combination of the complete K-means (until the convergence) and DCA.

4 Numerical Experiments

Our experiments are composed of two sets of data. The first data set is the geographical locations of 51 North American cities studied in [3], [4], [14] with

$k = 6$. Those works consist in investigating the hierarchical clustering algorithms for multicast group hierarchies. We got this data from the picture included in [3].

In the first numerical experiment we compare our algorithm **DCAIP** (DCA with the procedure IP for finding the initial point) with an optimization method based on K-means algorithm denoted **OKM**. We take $q = 5$ in the procedure IP, $\epsilon = 10^{-6}$, and $\tau = 2$ (the penalty parameter in (2)).

In **OKM** we used the code of K-means algorithm which is available on the web site: "http://www.fas.umonteral.ca/biol/legendre/" for finding the centres of clusters at the second level. The nearest city to this "centre" is then taken as the *real centre* ($\bar{x}_i$ for $i = 1, ..., k$) that serves the other cities in the cluster. The total centre $\bar{x}_{k+1}$ is determined by the next way:

$$\bar{x}_{k+1} = \arg \min_{j=1,...,p} \sum_{i=1}^{k} \|\bar{x}_i - a_j\|^2. \qquad (25)$$

Since the K-means clustering algorithm is a heuristic technique and is influenced by the choice of initial centres, we have run **DCAIP** and **OKM** ten times from the same initial centres that are randomly chosen from the set . The total costs given by the algorithms are reported in Table 1 (left). The total cost of the tree is computed as

$$\sum_{i=1}^{k} \sum_{j \in \mathcal{A}_i} \|\bar{x}_i - a_j\| + \sum_{i=1}^{k} \|\bar{x}_{k+1} - \bar{x}_i\|, \qquad (26)$$

where $\mathcal{A}_i$ is the cluster with the centre $\bar{x}_i$ for $i = 1, \ldots, k$.

In Table 1 (right) we present the best results given by the algorithms proposed in [3] (**KMC**) and in [14] (**1-km**) and **DCAIP** for this dataset. In [3] the algorithm **KMC** has been proposed for multilevel hierarchical clustering where

Table 1. Results for geographical locations, **DCAIP** and **OKM** (left), the best result of **DCAIP**, **OKM**, **1-km** and **KMC** (right)

Initial point	DCAIP	OKM
1	298	318
2	317	320
3	314	357
4	312	318
5	310	368
6	317	320
7	314	320
8	314	318
9	305	334
10	303	318

DCAIP	OKM	1-km([14])	KMC ([3])
298	318	308	345

the hierarchical trees are formed by repeated application of K-means algorithm at each hierarchical level. The procedure is beginning at the top layer where all members are partitioned into k clusters. From each of the k clusters found, a representative member is chosen to act as a server. The top-level servers become children of the source (the root of the tree). Each cluster is then again decomposed using clustering algorithms to form a new layer of sub-clusters, whose servers become children of the server in the cluster just partitioned. And so on, until a suitable terminating condition is reached. Some variants of **KMC** have been proposed in [4]. We note that **KMC** is a variant of **OKM** in which the K-means algorithm returns the "Euclidean centre" and the root of the tree is the Euclidean centre of the six servers.

In [14] the authors have proposed four variants of their optimization algorithms, based on the derivative-free discrete gradient method, for two nonsmooth nonconvex problems. They have compared their algorithms and the optimization algorithm based on K-means with the same initial points. The **1-km** algorithm (a version of their optimization algorithm with the initial point given by K-means) provides the best results among their four variants algorithms.

The best total cost given by **DCAIP**, **OKM** and **1-km** among ten tests with different initial points and the one of **KMC** among two tests (6 distributed throughout the data set, and 6 in South West) is reported in Table 1 (right).

In the second numerical experiment we use a randomly generated database with up to 50000 objects in higher dimensional spaces. We first generate k centres of clusters. The points of each cluster are randomly generated in a circle whose centre is the centre of this cluster. The numbers of points in clusters are randomly chosen. In Table 2 (left) we present the total cost given by **DCAIP** and **OKM** with the same initial point.

For testing the efficiency of procedure **IP** we perform two versions of **DCA** with and without procedure **IP**. We run **DCAIP** and **DCA** on ten test prob-

Table 2. Numerical results on the random data, comparison between **DCAIP** and **OKM** (left), **DCAIP** and **DCA** (right)

Data (p,n,k)	DCAIP	OKM
(100,2,5)	322	330
(500,2,8)	333	348
(1000,8,10)	183	228
(2000, 3, 20)	1965	2391
(5000,5, 10)	3851	4428
(5000, 20, 6)	18244	21612
(10000, 20, 7)	43699	45239
(20000, 30, 12)	107987	119829
(50000, 20, 20)	282099	345553

Data (p, n, k)	DCAIP			DCA		
	Cost	iter	CPU	Cost	iter	CPU
(51,2,6)	298	80	0.010	320	75	0.010
(100,2, 5)	322	10	0.010	323	10	0.010
(500, 2, 8)	333	80	0.053	333	82	0.060
(1000, 8, 10)	183	10	0.086	196	12	0.092
(2000, 3, 20)	1965	10	0.1	2024	15	0.16
(5000, 5, 10)	3851	72	0.74	4108	87	0.99
(5000, 20, 6)	18244	46	2,4	19342	50	2.6
(10000, 20, 7)	43699	66	7,80	43879	63	7.8
(20000, 30, 12)	107987	74	37	108124	74	37
(50000, 20, 20)	282099	182	351	289987	189	371

lems. The results are reported in Table 2 (right). Here "iter" denotes the number of iterations of the algorithm and all CPU are computed in seconds.

From numerical experiments we see that DCA is always the best for both dataset, and it is very inexpensive: it solves problems with large dimension in a short time. On the other hand, Procedure **IP** is efficient for finding a good starting point for DCA.

5 Conclusion

We have proposed, for solving a bilevel clustering problem with the squared Euclidean distance, a new and efficient approach based on DC programming and DCA. The considered hierarchical clustering problem has been formulated as a DC program in the suitable matrix space and with a natural choice of DC decomposition in order to make simpler and so much less expensive the computations in the resulting DCA. It fortunately turns out that our algorithm **DCA** is explicit, and very inexpensive. An interesting procedure that combines DCA and K-means is introduced for initializing DCA. Preliminary numerical simulations show the robustness, the efficiency and the superiority of our algorithm with respect to other optimization based clustering algorithms. The efficiency of our approach comes from two facts:

- The optimization model is appropriate for multilevel clustering : it requires the search for all centres in the same time;
- The optimization algorithm DCA is very suitable to this model.

The efficiency of DCA suggests to us investigating it in the solution of other models of bilevel clustering problems as well as the higher level for hierarchical clustering. Works in these directions are in progress.

References

1. M. Tayeb Belghiti, Le Thi Hoai An and Pham Dinh Tao, *Clustering via DC programming and DCA*. Modelling, Computation and Optimization in Information Systems and Management Sciences Hermes Science Publishing, pp. 499-507 (2004).
2. D. Fisher, *Iterative optimization and simplification of hierarchical clusterings*, Journal of Artificial Intelligence Research, vol. 4, pp. 147-180, 1996.
3. Gill Waters and Sei Guan Lim, *Applying clustering algorithms to multicast group hierarchies*, Technical Report No. 4-03 August 2003.
4. Gill Waters, John Crawford, and Sei Guan Lim, *Optimising multicast structures for grid computing*, Computer Communications 27 (2004) 1389?1400.
5. A. K. Jain, M. N. Murty, and P. J. Flynn, *Data clustering: a review*, ACM Computing Surveys, vol. 31, no. 3, pp. 264-323, 1999.
6. Julia Neumann, Christoph Schnörr, Gabriele Steidl, *SVM-based Feature Selection by Direct Objective Minimisation, Pattern Recognition*, Proc. of 26th DAGM Symposium, LNCS, Springer, August 2004.
7. LE Thi Hoai An, Contribution à l'optimisation non convexe et l'optimisation globale: Théorie, Algorithmes et Applications, Habilitation, July 1997, Université de Rouen.

8. Le Thi Hoai An and Pham Dinh Tao, *Solving a class of linearly constrained indefinite quadratic problems by DC algorithms,* Journal of Global Optimization, Vol 11, No 3, pp 253-285, 1997.
9. Le Thi Hoai An, Pham Dinh Tao, Le Dung Muu, *Exact penalty in DC programming.* Vietnam Journal of Mathematics, 27:2 (1999), pp. 169-178.
10. Le Thi Hoai An and Pham Dinh Tao, *DC Programming: Theory, Algorithms and Applications. The State of the Art.* Proceedings of The First International Workshop on Global Constrained Optimization and Constraint Satisfaction (Cocos' 02), 28 pages, Valbonne-Sophia Antipolis, France, October 2-4, 2002.
11. Le Thi Hoai An and Pham Dinh Tao, *Large Scale Molecular Optimization from distances matrices by a DC optimization approach,* SIAM Journal of Optimization,Volume 14, Number 1, 2003, pp.77-116.
12. Le Thi Hoai An and Pham Dinh Tao, *The DC (difference of convex functions) Programming and DCA revisited with DC models of real world nonconvex optimization problems,* Annals of Operations Research 2005, Vol 133, pp. 23-46.
13. Le Thi Hoai An, Pham Dinh Tao, Huynh Van Ngai, *Exact penalty techniques in DC programming.* Submitted
14. Long Jia, A. Bagirov, I. Ouveysi, A.M. Rubinov, *Optimization based clustering algorithms in Multicast group hierarchies,* Proceedings of the Australian Telecommunications, Networks and Applications Conference (ATNAC), 2003, Melbourne Australia, (published on CD, ISNB 0-646-42229-4).
15. F. Murtagh, *A survey of recent advances in hierarchical clustering algorithms,* The Computer Journal, vol. 26, no. 4, 1983.
16. Pham Dinh Tao and Le Thi Hoai An, *Convex analysis approach to d.c. programming: Theory, Algorithms and Applications,* Acta Mathematica Vietnamica, dedicated to Professor Hoang Tuy on the occasion of his 70th birthday, Vol.22, Number 1 (1997), pp. 289-355.
17. Pham Dinh Tao and Le Thi Hoai An, *DC optimization algorithms for solving the trust region subproblem,* SIAM J. Optimization, Vol. 8, pp. 476-505 (1998).
18. Tina Wong, Randy Katz, Steven McCanne, *A Preference Clustering Protocol for Large-Scale Multicast Applications,* Proceedings of the First International COST264 Workshop on Networked Group Communication, 1999, pp 1-18.
19. Stefan Weber, Thomas Schüle, Christoph Schnörr, *Prior Learning and Convex-Concave Regularization of Binary Tomography Electr.* Notes in Discr. Math., 20:313-327, 2005. December 2003.
20. Stefan Weber, Christoph Schnörr, Thomas Schüle, Joachim Hornegger, *Binary Tomography by Iterating Linear Programs,* R. Klette, R. Kozera, L. Noakes and J. Weickert (Eds.), *Computational Imaging and Vision - Geometric Properties from Incomplete Data,* Kluwer Academic Press 2005.

Clustering Multi-represented Objects Using Combination Trees

Elke Achtert, Hans-Peter Kriegel, Alexey Pryakhin,
and Matthias Schubert

Institute for Computer Science,
University of Munich, Germany
{achtert, kriegel, pryakhin, schubert}@dbs.ifi.lmu.de

Abstract. When clustering complex objects, there often exist various feature transformations and thus multiple object representations. To cluster multi-represented objects, dedicated data mining algorithms have been shown to achieve improved results. In this paper, we will introduce combination trees for describing arbitrary semantic relationships which can be used to extend the hierarchical clustering algorithm OPTICS to handle multi-represented data objects. To back up the usability of our proposed method, we present encouraging results on real world data sets.

1 Introduction

In modern data mining applications, there often exists no universal feature representation that can be used to express similarity between all possible objects in a meaningful way. Thus, recent data mining approaches employ multiple representations to achieve more general results that are based on a variety of aspects. In this paper, we distinguish two types of representations and show how to combine sets of representations containing both types using so-called combination trees. The combination trees are build with respect to domain knowledge and describe multiple semantics. To employ combination trees for clustering, we introduce a multi-represented version of the hierarchical density-based clustering algorithm OPTICS. OPTICS derives so-called cluster orderings and is quite insensitive to the parameter selection. The introduced version of OPTICS is capable to derive meaningful cluster hierarchies with respect to an arbitrary combination tree. The rest of this paper is organized as follows. Section 2 surveys related work. In Section 3, we define combination trees. Section 4 describes a multi-represented version of OPTICS which is based on combination trees. In Section 5, we provide encouraging experimental results.

2 Related Work

In [1] an algorithm for spectral clustering of multi-represented objects is proposed. [2] introduces Expectation Maximization (EM) clustering and agglomerative clustering for multi-represented data. Finally, [3] introduces the framework of reinforcement clustering, which is applicable to multi-represented objects.

However, these three approaches do not consider any semantic aspects of the underlying data spaces. In [4], DBSCAN [5] has been adapted to multi-represented objects distinguishing two possible semantics. However, DBSCAN has several drawbacks leading to the development of OPTICS[6] which is the algorithm the method proposed in this paper is based on.

3 Handling Semantics

In [4], there were two general methods to combine multiple representation for density based clustering, called union and intersection method. The union method states that an object is an union core-object if there are at least k data objects in the union of the local ε-neighborhoods. The intersection method was defined analogously. However, it is not clear which method is better suited to compare an arbitrary set of representations. In [7], the suitability of representations for one or the other combination method is discussed. As a result, two aspects of a data space can be distinguished, the precision space and recall space property. An examples for a good precision space are word vectors because documents containing the same set of words usually describe the same content. An example for a recall space are color histograms because two images having a similar content usually have similar color distributions. Furthermore, we can state that precision spaces should be combined using the union method and recall spaces should be combined using the intersection method. The result of combining recall spaces improves the precision and the result of combining precision spaces improves the recall. Thus, we can successively group representation of both types and construct a so-called combination tree according to the following formalization:

Definition 1 (Combination Tree). *Let $R = \{R_1, \ldots, R_m\}$. A combination tree CT for R is a tree of arbitrary degree fulfilling the following conditions:*

- *CT.root denotes the root of the combination tree CT.*
- *Let n be a node of CT, then n.label denotes the label of n and n.children denotes the children of n.*
- *The leaves are labeled with representations, i.e. for each leaf $n \in CT$: $n.label \in \{R_1, \ldots, R_m\}$.*
- *The inner nodes are labeled with either the union or the intersection operator, i.e. for each inner node $n \in CT$: $n.label \in \{\cup, \cap\}$.*

4 Hierarchical Clustering of Multi-represented Objects

In order to obtain the comparability of distances, we normalize the distance in representation R_i with respect to the mean value μ_i^{orig} of the original distance d_i^{orig}. The algorithm OPTICS [6] works like an extended DBSCAN algorithm, computing the density-connected clusters w.r.t. all parameters ε_i that are smaller than a generic value of ε. OPTICS does not assign cluster memberships, but stores the order in which the objects have been processed and the information can be used to assign cluster memberships. This information consists of two values for each object, its core distance and its reachability distance. To compute these information during a run of OPTICS on multi-represented objects, we must adapt the core distance and reachability distance predicates of OPTICS to our multi-represented approach. In the following, we will show how we can use a combination tree CT for a given set of representations R to cluster multi-represented objects. The (global) distance between two objects $o, p \in \mathcal{D}$ w.r.t. a combination tree CT is defined as the combination of the distances of the nodes of CT.

Definition 2 (distance w.r.t. CT). Let $o, p \in \mathcal{D}$, $R = \{R_1, \ldots, R_m\}$, d_i be the distance function of R_i, CT be a combination tree for R, and let n be a node in CT, i.e. $n.label \in \{\cup, \cap, R_1, \ldots, R_m\}$.

The distance between o and p w.r.t. node $n \in CT$, denoted by $d^n(o, p)$, is recursively defined by

$$d^n(o,p) = \begin{cases} \min_{c \in n.children} \{d^c(o,p)\} & \text{if } n.label = \cup \\ \max_{c \in n.children} \{d^c(o,p)\} & \text{if } n.label = \cap \\ d_i(o,p) & \text{if } n.label = R_i \end{cases}$$

The distance between o and p w.r.t. CT, denoted by $d_{CT}(o, p)$, is defined by

$$d_{CT}(o, p) = d^{CT.root}(o, p)$$

The (global) ε-neighborhood of an object $o \in \mathcal{D}$ w.r.t. a combination tree CT is defined as the combination of the ε-neighborhoods of the nodes of CT.

Definition 3 (ε-neighborhood w.r.t. CT). Let $o \in \mathcal{D}$, $\varepsilon \in \mathbb{R}^+$, $R = \{R_1, \ldots, R_m\}$, CT be a combination tree for R, and let n be a node in CT, i.e. $n.label \in \{\cup, \cap, R_1, \ldots, R_m\}$.

The ε-neighborhood of o w.r.t. node $n \in CT$, denoted by $\mathcal{N}_\varepsilon^n(o)$, is recursively defined by

$$\mathcal{N}_\varepsilon^n(o) = \begin{cases} \bigcup_{c \in n.children} \mathcal{N}_\varepsilon^c(o) & \text{if } n.label = \cup \\ \bigcap_{c \in n.children} \mathcal{N}_\varepsilon^c(o) & \text{if } n.label = \cap \\ \mathcal{N}_\varepsilon^{R_i}(o) & \text{if } n.label = R_i \end{cases}$$

The ε-neighborhood of o w.r.t. CT, denoted by $\mathcal{N}_{CT,\varepsilon}(o)$, is defined by

$$\mathcal{N}_{CT,\varepsilon}(o) = \mathcal{N}_\varepsilon^{CT.root}(o)$$

Since the core distance predicate of OPTICS is based on the concept of k-nearest neighbor (k-NN) distances, we have to redefine the k-nearest neighbor distance of an object o w.r.t. a combination tree CT.

Definition 4 (k-NN distance w.r.t. CT). Let $o \in \mathcal{D}$, $k \in \mathbb{N}$, $|\mathcal{D}| \geq k$, $R = \{R_1, \ldots, R_m\}$, CT be a combination tree for R, and let n be a node in CT, i.e. $n.label \in \{\cup, \cap, R_1, \ldots, R_m\}$.

The k-nearest neighbors of o w.r.t. CT is the smallest set $NN_{CT,k}(o) \subseteq \mathcal{D}$ that contains (at least) k objects and for which the following condition holds:

$$\forall p \in NN_{CT,k}(o), \forall q \in \mathcal{D} - NN_{CT,k}(o) : d_{CT}(o,p) < d_{CT}(o,q).$$

The k-nearest neighbor distance of o w.r.t. CT, denoted by $\text{NN-DIST}_{CT,k}(o)$, is defined as follows:

$$\text{NN-DIST}_{CT,k}(o) = \max\{d_{CT}(o,q)\} \mid q \in NN_{CT,k}(o)\}.$$

Now, we can adopt the core distance definition from OPTICS to our combination approach: If the ε-neighborhood w.r.t. CT of an object o contains at least k objects, the core distance of o is defined as the k-nearest neighbor distance of o. Otherwise, the core distance is infinity.

Definition 5 (core distance w.r.t. CT). *Let $o \in \mathcal{D}$, $k \in \mathbb{N}$, $|\mathcal{D}| \geq k$, $R = \{R_1, \ldots, R_m\}$, CT be a combination tree for R, and let n be a node in CT, i.e. $n.label \in \{\cup, \cap, R_1, \ldots, R_m\}$.*
The core distance of o w.r.t. CT, ε and k, denoted by $\text{CORE}_{CT,\varepsilon,k}(o)$, is defined by

$$\text{CORE}_{CT,\varepsilon,k}(o) = \begin{cases} \text{NN-DIST}_{CT,k}(o) & \text{if } |\mathcal{N}_{CT,\varepsilon}(o)| \geq k \\ \infty & \text{otherwise.} \end{cases}$$

The reachability distance of an object $p \in \mathcal{D}$ from $o \in \mathcal{D}$ w.r.t. CT is an asymmetric distance measure that is defined as the maximum value of the core distance of o and the distance between p and o.

Definition 6 (reachability distance w.r.t. CT). *Let $o, p \in \mathcal{D}$, $k \in \mathbb{N}$, $|\mathcal{D}| \geq k$, $R = \{R_1, \ldots, R_m\}$, CT be a combination tree for R, and let n be a node in CT, i.e. $n.label \in \{\cup, \cap, R_1, \ldots, R_m\}$.*
The reachability distance of o to p w.r.t. CT, ε, and k, denoted by $\text{REACH}_{CT,\varepsilon,k}(p,o)$, is defined by

$$\text{REACH}_{CT,\varepsilon,k}(p,o) = \max\{\text{CORE}_{CT,\varepsilon,k}(p), d_{CT}(o,p)\}$$

5 Performance Evaluation

We implemented the proposed clustering algorithm in Java 1.5 and ran several experiments on a work station with two 1.8 GHz Opteron processors and 8 GB main memory. The experiments were performed on protein data that is described by text descriptions (R_1) and amino-acid sequences (R_2). We employed entries of the Swissprot protein database [1] belonging to 5 functional groups (cf. Table 1). As reference clustering, we employed the classes of Gene Ontology [2]. To evaluate the derived cluster structure C, we extracted flat clusters from OPTICS plots

---
[1] http://us.expasy.org/sprot/sprot-top.html
[2] www.geneontology.org

Table 1. Description of the protein data sets and results

	Set 1	Set 2	Set 3	Set 4	Set 5
Name	Isomerase	Lyase	Signal Transducer	Oxidoreductase	Transferase
No. of Classes	16	35	39	49	62
No. of Objects	501	1640	2208	3399	4086
$R_1 \cup R_2$	**0.66**	**0.56**	**0.43**	**0.50**	**0.38**
R_1	0.61	0.54	0.32	0.46	0.35
R_2	0.31	0.25	0.36	0.39	0.24
CFS	0.62	0.46	0.28	0.41	0.29
RCL	0.55	0.43	0.25	0.33	0.19

and applied the following quality measure for comparing different clusterings w.r.t. the reference clustering K: $Q_K(C) = \sum_{C_i \in C} \frac{|C_i|}{|DB|} \cdot (1 - entropy_K(C_i))$. We employed an combination tree describing the union of both representations. As first comparison partners, we clustered text and sequences separately using only one of the representations. A second approach combines the features of both representations into a common feature space (CFS) and employs the cosine distance to relate the resulting feature vectors. Additionally, we compared reinforcement clustering (RCL) using DBSCAN as underlying cluster algorithm. For reinforcement clustering, we ran 10 iterations and tried several values of the weighting parameter α. The ε-parameters were set sufficiently large and we chose $k = 2$. Table 1 displays the derived quality for our method and the four competitive methods mentioned above. As it can be seen, our method clearly outperforms any of the other algorithms.

Another, set of experiments were performed on a data set of images being described by 4 representations. The OPTICS clustering based on a 2 level combination trees achieved encouraging results as well. More information about these experiments can be found in [7].

References

1. De Sa, V.R.: Spectral Clustering with two Views. In: Proc. ICML Workshop. (2005)
2. Bickel, S., Scheffer, T.: Multi-View Clustering. In: Proc. ICDM. (2004)
3. Wang, J., Zeng, H., Chen, Z., Lu, H., Tao, L., Ma, W.: ReCoM: Reinforcement clustering of multi-type interrelated data objects. In: Proc. SIGIR. (2003)
4. Kailing, K., Kriegel, H.P., Pryakhin, A., Schubert, M.: Clustering Multi-represented Objects with Noise. In: Proc. PAKDD. (2004)
5. Ester, M., Kriegel, H.P., Sander, J., Xu, X.: "A Density-Based Algorithm for Discovering Clusters in Large Spatial Databases with Noise". In: Proc. KDD. (1996)
6. Ankerst, M., Breunig, M.M., Kriegel, H.P., Sander, J.: OPTICS: Ordering Points to Identify the Clustering Structure. In: Proc. SIGMOD. (1999)
7. Achtert, E., Kriegel, H.P., Pryakhin, A., Schubert, M.: Hierarchical Density-Based Clustering for Multi-Represented Objects. In: Workshop on Mining Complex Data (MCD 2005)at ICDM05, Houston, TX, USA. (2005)

Parallel Density-Based Clustering of Complex Objects

Stefan Brecheisen, Hans-Peter Kriegel, and Martin Pfeifle

Institute for Informatics, University of Munich
{brecheis, kriegel, pfeifle}@dbs.ifi.lmu.de

Abstract. In many scientific, engineering or multimedia applications, complex distance functions are used to measure similarity accurately. Furthermore, there often exist simpler lower-bounding distance functions, which can be computed much more efficiently. In this paper, we will show how these simple distance functions can be used to parallelize the density-based clustering algorithm DBSCAN. First, the data is partitioned based on an enumeration calculated by the hierarchical clustering algorithm OPTICS, so that similar objects have adjacent enumeration values. We use the fact that clustering based on lower-bounding distance values conservatively approximates the exact clustering. By integrating the multi-step query processing paradigm directly into the clustering algorithms, the clustering on the slaves can be carried out very efficiently. Finally, we show that the different result sets computed by the various slaves can effectively and efficiently be merged to a global result by means of cluster connectivity graphs. In an experimental evaluation based on real-world test data sets, we demonstrate the benefits of our approach.

1 Introduction

Density-based clustering algorithms like DBSCAN [1] are based on ε-range queries for each database object. Thereby, each range query requires a lot of distance calculations. When working with complex objects, e.g. trees, point sets, and graphs, often complex time-consuming distance functions are used to measure similarity accurately. As these distance calculations are the time-limiting factor of the clustering algorithm, the ultimate goal is to save as many as possible of these complex distance calculations.

Recently an approach was presented for the efficient density-based clustering of complex objects [2]. The core idea of this approach is to integrate the multi-step query processing paradigm directly into the clustering algorithm rather than using it "only" for accelerating range queries. In this paper, we present a sophisticated parallelization of this approach. Similar to the area of join processing where there is an increasing interest in algorithms which do not assume the existence of any index structure, we propose an approach for parallel DBSCAN which does not rely on the pre-clustering of index structures.

First, the data is partitioned according to the clustering result carried out on cheaply computable distance functions. The resulting approximated clustering

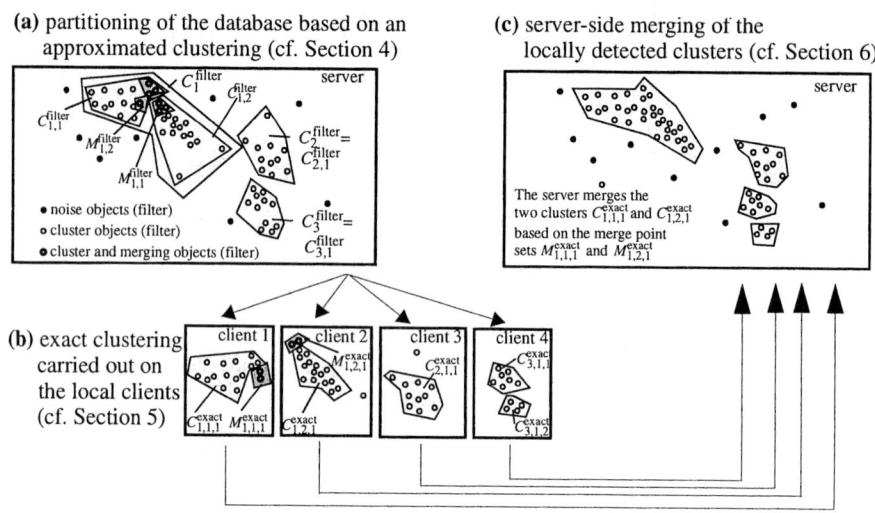

Fig. 1. Basic idea of parallel density-based clustering

conservatively approximates the exact clustering. The objects of the conservative cluster approximations are then distributed onto the available slaves in such a way that each slave has to cluster the same amount of objects, and that the objects to be clustered are close to each other. Note that already at this early stage, we can detect some noise objects which do not have to be transmitted to the local clients. In addition to the objects to be clustered by a client, we send some filter merge points to this client. These filter merge points are also determined based on approximated distance functions. (cf. Figure 1a).

Second, each client carries out the clustering independently of all the other clients. No further communication is necessary throughout this second step. The presented local clustering approach also takes advantage of the approximating lower-bounding distance functions. The detected clusters and the detected exact merge point sets are then transmitted to the server (cf. Figure 1b).

Finally, the server determines the correct clustering result by merging the locally detected clusters. This final merging step is based on the exact merge points detected by the clients. Based on these merge points, cluster connectivity graphs are created. In these graphs, the nodes represent the locally detected clusters. Two local clusters are connected by an edge if a merge point of one cluster is a core object in the other cluster (cf. Figure 1c).

The remainder of this paper is organized as follows. In Section 2, we shortly sketch the work from the literature related to our approach. In Sections 3, 4 and 5, we explain the server-side partitioning algorithm, the client-side clustering algorithm, and the server-side merging of the results from the clients, respectively. In Section 6, we present a detailed experimental evaluation based on real world test data sets. We close the paper in Section 7 with a short summary and a note on future work.

2 Related Work

Complex Object Representations. Complex object representations, like high-dimensional feature vectors [3], vector sets [4], trees or graphs [5], are helpful to model real world objects accurately. The similarity between these complex object representations is often measured by means of expensive distance function, e.g. the edit distance. For a more detailed survey on this topic, we refer the interested reader to [6].

Clustering. Given a set of objects with a distance function on them, an interesting data mining question is, whether these objects naturally form groups (called clusters) and what these groups look like. Data mining algorithms that try to answer this question are called clustering algorithms. For a detailed overview on clustering, we refer the interested reader to [7].

Density-Based Clustering. Density based clustering algorithms apply a local cluster criterion to detect clusters. Clusters are regarded as regions in the data space in which the objects are dense, and which are separated by regions of low object density (noise). One of the most prominent representatives of this clustering paradigm is DBSCAN [1].

Density-Based Clustering of Complex Objects. In [2] a detailed overview can be found describing several approaches for the efficient density-based clustering of complex object. Furthermore, in [2] a new approach was introduced which performs expensive exact distance computations only when the information provided by simple distance computations is not enough to compute the exact clustering. In Section 4, we will use an adaption of this approach for the efficient clustering on the various slaves.

Parallel Density-Based Clustering of Complex Objects. To the best of our knowledge there does not exist any work in this area.

3 Server-Side Data Partitioning

The key idea of density-based clustering is that for each object of a cluster the neighborhood of a given radius ε has to contain at least a minimum number of *MinPts* objects, i.e. the cardinality of the neighborhood has to exceed a given threshold. An object p is called *directly density-reachable* from object q w.r.t. ε and *MinPts* in a set of objects D, if $p \in \mathcal{N}_\varepsilon(q))$ and $|\mathcal{N}_\varepsilon(q)| \geq$ *MinPts*, where $\mathcal{N}_\varepsilon(q)$ denotes the subset of D contained in the ε-neighborhood of q. The condition $|\mathcal{N}_\varepsilon(q)| \geq$ *MinPts* is called the *core object condition*. If this condition holds for an object q, then we call q a *core object*. Other objects can be directly density-reachable only from core objects. An object p is called *density-reachable* from an object q w.r.t. ε and *MinPts* in the set of objects D, if there is a chain of objects $p_1, \ldots, p_n$, $p_1 = q$, $p_n = p$, such that $p_i \in D$ and p_{i+1} is directly density-reachable from p_i w.r.t. ε and *MinPts*. Object p is *density-connected* to object q w.r.t. ε and *MinPts* in the set of objects D, if there is an object $o \in D$ such that both p and q are density-reachable from o. Density-reachability is the transitive closure of direct density-reachability and is not necessarily symmetric. On the other hand, density-connectivity is a symmetric relation.

DBSCAN. A flat density-based *cluster* is defined as a set of density-connected objects which is maximal w.r.t. density-reachability. Thus a cluster contains not only core objects but also border objects that do not satisfy the core object condition. The *noise* is the set of objects not contained in any cluster.

OPTICS. While the partitioning density-based clustering algorithm DBSCAN can only identify a flat clustering, the newer algorithm OPTICS [8] computes an ordering of the points augmented by the so-called *reachability-distance*. The reachability-distance basically denotes the smallest distance of the current object q to any core object which belongs to the current cluster and which has already been processed. The clusters detected by DBSCAN can also be found in the OPTICS ordering when using the same parametrization, i.e. the same ε and *MinPts* values. For an initial clustering with OPTICS based on the lower-bounding filter distances the following two lemmas hold.

Lemma 1. *Let $C_1^{exact}, \ldots, C_n^{exact}$ be the clusters detected by OPTICS based on the exact distances, and let $C_1^{filter}, \ldots, C_m^{filter}$ be the clusters detected by OPTICS based on the lower-bounding filter distances. Then the following statement holds:*

$$\forall i \in \{1, \ldots, n\} \exists j \in \{1, \ldots, m\} : C_i^{exact} \subseteq C_j^{filter}.$$

Proof. Let $N_\varepsilon^{filter}(o)$ denote the ε-neighborhood of o according to the filter distances, and let $N_\varepsilon^{exact}(o)$ denote the ε-neighborhood according to the exact distances. Due to the lower-bounding filter property $N_\varepsilon^{exact}(o) \subseteq N_\varepsilon^{filter}(o)$ holds. Therefore, each object o which is a core object based on the exact distances is also a core object based on the lower-bounding filter distances. Furthermore, each object p which is directly density-reachable from o according to the exact distances is also directly density-reachable according to the filter functions. Induction on this property shows that if p is density-reachable from o based on the exact distances, it also holds for the filter distances. Therefore, all objects which are in one cluster according to the exact distances are also in one cluster according to the approximated distances.

Lemma 2. *Let $noise^{exact}$ denote the noise objects detected by OPTICS based on the exact distances and let $noise^{filter}$ denote the noise objects detected by OPTICS based on the lower-bounding filter distances. Then the following statement holds:*

$$noise^{filter} \subseteq noise^{exact}.$$

Proof. An object p is a noise object if it is not included in the ε-neighborhood of any core object. Again, let $N_\varepsilon^{filter}(o)$ and $N_\varepsilon^{exact}(o)$ denote the ε-neighborhood of o according to the filter distances and the exact distances, respectively. Due to the lower-bounding filter property $N_\varepsilon^{exact}(o) \subseteq N_\varepsilon^{filter}(o)$ holds. Therefore, if $p \notin N_\varepsilon^{filter}(o)$, it cannot be included in $N_\varepsilon^{exact}(o)$, proving the lemma.

Both Lemma 1 and Lemma 2 are helpful to partition the data onto the different slaves. Lemma 1 shows that exact clusters are conservatively approximated by the clusters resulting from a clustering on the lower-bounding distance functions.

On the other hand, Lemma 2 shows that exact noise is progressively approximated by the set of noise objects resulting from an approximated clustering. For this reason, noise objects according to the filter distances do not have to be transmitted to the slaves, as we already know that they are also noise objects according to the exact distances. All other N objects have to be refined by the P available slave processors. Let $C_1^{filter}, \ldots, C_m^{filter}$ be the approximated clusters resulting from an initial clustering with OPTICS. In this approach, we assign $P_{slave} = \sum_{i=1}^{m} |C_i^{filter}|/P$ objects to each of the P slaves. We do this partitioning online while carrying out the OPTICS algorithm. At each time during the clustering algorithm, OPTICS knows the slave j having received the smallest number L_j of objects up to now, i.e. the client j has the highest free capacity $C_j = P_{slave} - L_j$. OPTICS stops the current clustering at two different event points: In the first case, a cluster C_i^{filter} of cardinality $|C_i^{filter}| \leq C_j$ was completely determined. This cluster is sent to the slave j. In the second case, OPTICS determined C_j more points belonging to the current cluster C_i^{filter}. These points are grouped together to a filter cluster $C_{i,j}^{filter}$. Then, we transmit the cluster $C_{i,j}^{filter}$ along with the filter merge points $M_{i,j}^{filter}$ to the slave j. The set $M_{i,j}^{filter}$ can be determined throughout the clustering of the set $C_{i,j}^{filter}$ and can be defined as follows.

Definition 1 (filter merge points). *Let C_i^{filter} be a cluster which is split during an OPTICS run into n clusters $C_{i,1}^{filter}, \ldots, C_{i,n}^{filter}$. Then, the filter merge points $M_{i,j}^{filter}$ for a partial filter cluster $C_{i,j}^{filter}$ are defined as follows: $M_{i,j}^{filter} = \{q \in C_i^{filter} - C_{i,j}^{filter} \mid \exists p \in C_{i,j}^{filter} : q$ is directly density-reachable from $p\}$.*

The filter merge points $M_{i,j}^{filter}$ are necessary in order to decide whether objects $o \in C_{i,j}^{filter}$ are core objects. Furthermore, a subset $M_{i,j}^{exact} \subseteq M_{i,j}^{filter}$ is used to merge exact clusters in the final merge step (cf. Section 5).

4 Client-Side Clustering

Each of the filter clusters $C_{i,j}^{filter}$ is clustered independently on the exact distances by the assigned slave j. For clustering these filter clusters, we adapt the approach presented in [2], so that it can also handle the additional merge points $M_{i,j}^{filter}$. The main idea of the client-side clustering approach is to carry out the range queries based on the lower-bounding filter distances instead of using the expensive exact distances. Thereto, we do not use the simple seedlist of the original DBSCAN algorithm, but we use a list of lists, called *Xseedlist*. The *Xseedlist* consists of an ordered *object list OL*. Each entry $(o, T, PL) \in OL$ contains a flag T indicating whether $o \in C_{i,j}^{filter}$ ($T = $ C) or $o \in M_{i,j}^{filter}$ ($T = $ M). Each entry of the *predecessor list PL* consists of the following information: a *predecessor* o_p of o, which is a core object already added to the current cluster, and the *predecessor distance*, which is equal to the filter distance $d_f(o, o_p)$ between the two objects.

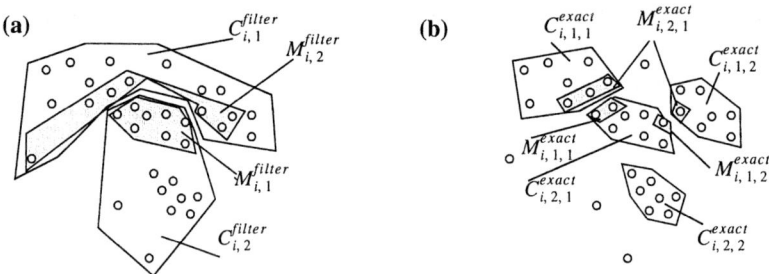

During the server-side *partitioning step*, the cluster C_i^{filter} is split into two clusters $C_{i,1}^{filter}$ and $C_{i,2}^{filter}$ with their corresponding merge point sets.

During the server-side *merge step*, the cluster $C_{i,1,1}^{exact}$, $C_{i,1,2}^{exact}$, and $C_{i,2,1}^{exact}$ are merged based on their exact merge point sets to a cluster $C_{i,1}^{exact} \subseteq C_i^{filter}$. Furthermore, there exists a cluster $C_{i,2}^{exact} = C_{i,2,2}^{exact} \subseteq C_i^{filter}$.

Fig. 2. Server-side partitioning step (a) and merge step (b)

The result of the extended DBSCAN algorithm is a set of exact clusters $C_{i,j,l}^{exact} \subseteq C_{i,j}^{filter}$ along with their additional exact merge points $M_{i,j,l}^{exact} \subseteq M_{i,j}^{filter}$. To expand a cluster $C_{i,j,l}^{exact}$ we take the first element (o, T, PL) from OL and set o_p to the nearest predecessor object in PL.

Let us first assume that $T = C$ holds. If $PL = NIL$ holds, we add o to $C_{i,j,l}^{exact}$, delete o from OL, carry out a range query around o, and try to expand the cluster $C_{i,j,l}^{exact}$. If $PL \neq NIL$ holds, we compute $d_o(o, o_p)$. If $d_o(o, o_p) \leq \varepsilon$, we proceed as in the case where $PL = NIL$ holds. If $d_o(o, o_p) > \varepsilon$ and length of $PL > 1$ hold, we delete the first entry from PL. If $d_o(o, o_p) > \varepsilon$ and length of $PL = 1$ hold, we delete o from OL. Iteratively, we try to expand the current cluster $C_{i,j,l}^{exact}$ by examining the first entry of OL until OL is empty.

Let us now assume that $T = M$ holds. If $PL = NIL$ holds, we add o to $M_{i,j,l}^{exact}$, delete o from OL, and try to expand the exact merge point set $M_{i,j,l}^{exact}$. If $PL \neq NIL$ holds, we compute $d_o(o, o_p)$. If $d_o(o, o_p) \leq \varepsilon$, we proceed as in the case where $PL = NIL$ holds. If $d_o(o, o_p) > \varepsilon$ and length of $PL > 1$ hold, we delete the first entry from PL. If $d_o(o, o_p) > \varepsilon$ and length of $PL = 1$ hold, we delete o from OL. Iteratively, we try to expand the current exact merge point set $M_{i,j,l}^{exact}$ by examining the first entry of OL until OL is empty.

5 Server-Side Merging

Obviously, we only have to carry out the merge process for those clusters C_i^{filter} which were split in several clusters $C_{i,j}^{filter}$. The client detects that each of these clusters $C_{i,j}^{filter}$ contains t clusters $C_{i,j,1}^{exact}, \ldots, C_{i,j,t}^{exact}$. Note that t can also be equal to 0, i.e. no exact cluster is contained in the cluster $C_{i,j}^{filter}$. For each of the t exact clusters $C_{i,j,l}^{exact}$ there also exists a corresponding set of exact merge points $M_{i,j,l}^{exact} \subseteq M_{i,j}^{filter}$ (cf. Figure 2) defined as follows.

Definition 2 (exact merge points). Let $C_{i,j}^{filter}$ be a cluster to be refined on the slave with the corresponding merge point set $M_{i,j}^{filter}$. Let $C_{i,j,l}^{exact} \subseteq C_{i,j}^{filter}$ be an exact cluster determined during the client-side refinement clustering. Then, we determine the set $M_{i,j,l}^{exact} \subseteq M_{i,j}^{filter}$ of exact merge points where $M_{i,j,l}^{exact} = \{q \in M_{i,j}^{filter} \mid \exists p \in C_{i,j,l}^{exact} : q \text{ is directly density-reachable from } p\}$.

Based on these exact merge point sets and the exact clusters, we can define a "cluster connectivity graph".

Definition 3 (cluster connectivity graph). Let C_i^{filter} be a cluster which was refined on one of the s different slaves. Let $C_{i,j,l}^{exact} \subseteq C_{i,j}^{filter} \subseteq C_i^{filter}$ be an exact cluster determined by slave j along with the corresponding merge point sets $M_{i,j,l}^{exact} \subseteq M_{i,j}^{filter}$. Then a graph $G_i = (V_i, E_i)$ is called a cluster connectivity graph for C_i^{filter} iff the following statements hold:

- $V_i = \{C_{i,1,1}^{exact}, \ldots, C_{i,1,n_1}^{exact}, \ldots, C_{i,s,1}^{exact}, \ldots, C_{i,s,n_s}^{exact}\}$.
- $E_i = \{(C_{i,j,l}^{exact}, C_{i,j',l'}^{exact}) \mid \exists p \in M_{i,j,l}^{exact} : p \in C_{i,j',l'}^{exact} \wedge p \text{ is a core point}\}$.

Note that two clusters $C_{i,j,l}^{exact}$ and $C_{i,j',l'}^{exact}$ from the same slave $j = j'$ are never connected by an edge. Such a connection of the two clusters would already have taken place throughout the refinement clustering on the slave j. Based on the connectivity graphs G_i for the approximated clusterings C_i^{filter}, we can determine the *database connectivity graph*.

Definition 4 (database connectivity graph). Let C_i^{filter} be one of n approximated clusters along with the corresponding cluster connectivity graph $G_i = (V_i, E_i)$. Then we call $G = (\bigcup_{i=1}^n V_i, \bigcup_{i=1}^n E_i)$ the database connectivity graph.

The database connectivity graph is nothing else but the union of the connectivity graphs of the approximated clusters. Based on the above definition, we state the central lemma of this paper.

Lemma 3. *Let G be the database connectivity graph. Then the determination of all maximal connected subgraphs of G is equivalent to a DBSCAN clustering carried out on the exact distances.*

Proof. For each object o the client-side clustering determines correctly, whether it is a core object, a border object, or a noise object. Note, that we assign a border object which is directly density-reachable from core objects of different clusters redundantly to all of these clusters. Therefore, the only remaining issue is to show that two core objects which are directly density-reachable to each other are in the same maximal connected subgraph. By induction, according to the definition of density-reachability, two clusters then contain the same core objects. Obviously, two core objects o_1 and o_2 are directly density-reachable if they are either in the same exact cluster $C_{i,j,l}^{exact}$ or if $o_1 \in C_{i,j,l}^{exact}$ and $o_2 \in M_{i,j,l} exact$ resulting in an edge of the database connectivity graph. Therefore, depth-first traversals through all of the connectivity graphs G_i corresponding to a filter cluster C_i^{filter} create the correct clustering result where each subgraph corresponds to one cluster.

6 Experimental Evaluation

In this section, we present a detailed experimental evaluation based on real-world data sets. We used CAD data represented by 81-dimensional feature vectors [3] and vector sets where each element consists of 7 6D vectors [4]. Furthermore, we used graphs [5] to represent image data. The used distance functions can be characterized as follows: (i) The exact distance computations on the graphs are very expensive. On the other hand, the filter is rather selective and can efficiently be computed. (ii) The exact distance computations on the feature vectors and vector sets are also very expensive as normalization aspects for the CAD objects are taken into account [4,3]. As a filter for the feature vectors we use their Euclidean norms [9] which is not very selective, but can be computed very efficiently. The filter used for the vector sets is more selective than the filter for the feature vectors, but also computationally more expensive. If not otherwise stated, we used 3,000 complex objects from each data set.

The original OPTICS and DBSCAN algorithms, their extensions introduced in this paper, and the used filter and exact distances functions were implemented in Java 1.4. The experiments were run on a workstation with a Xeon 2.4 GHz processor and 2 GB main memory. All experiments were run sequentially on one computer. Thereby, the overall time for the client-side clustering is determined by the slowest slave. If not otherwise stated, we chose an ε-parameter yielding as many flat clusters as possible, and the *MinPts*-parameter was set to 5.

Characteristics of the partitioning step. Figure 3 compares the number of merge points for different split techniques applied to filter clusters. As explained in Section 3, we split a filter cluster during the partitioning step along the ordering produced by OPTICS. Note that OPTICS always walks through a cluster by visiting the densest areas first. Figure 3 shows that this kind of split strategy yields considerably less merge points than a split strategy which arbitrarily groups objects from a filter cluster together. Thus, the figure proves the good clustering properties of our metric space filling curve OPTICS.

Dependency on the Number of Slaves. Figure 4 shows the absolute runtimes of our parallel DBSCAN approach dependent on the number of available slaves for the vector sets and for the graph dataset. The figure shows the accumulated times after the partitioning, client-side clustering, and the merge step. The partitioning

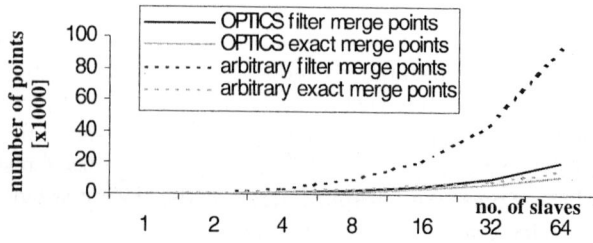

Fig. 3. Number of merge points w.r.t. a varying number of slaves for the graph dataset

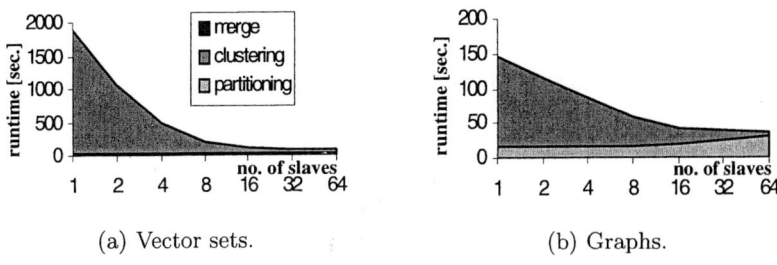

Fig. 4. Absolute runtimes w.r.t. a varying number of slaves

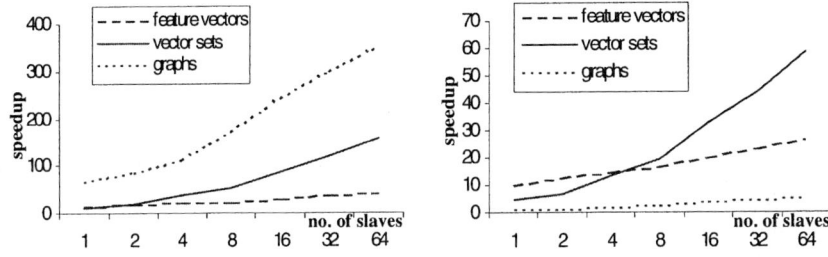

Fig. 5. Overall speedup w.r.t. a varying number of slaves

times also include simulated communication times for the transfer of the objects to the slaves in a 100 Mbit LAN. No communication costs arise from the client-side clustering step, as each client already received all needed filter merge points. A growing number of slaves leads to a significant speedup of the client-side clustering. A lower bound of the achievable total runtime is given by the time needed for the initial partitioning step. It is worth to note the time needed for the final merging step is negligible even for a high number of slaves. Although the number of exact merge points grows with an increasing number of slaves (cf. Figure 3), the merge step remains cheap.

Speedup. Finally, Figure 5 depicts the speedup achieved by our new parallel DB-SCAN approach based on a server-side partitioning with OPTICS. We compared this approach to a DBSCAN approach based on a full table scan and compared to a DBSCAN approach based on the traditional multi-step query processing paradigm. The figure shows that for the feature vectors we achieve a speedup of one order of magnitude already when only one slave is available. In the case of the graph dataset we have a speedup of 67 compared to DBSCAN based on a full table scan. These results demonstrate the suitability of the client-side clustering approach. For the vector sets the benefits of using several slaves can clearly be seen. For instance, our approach achieves a speedup of 4 for one slave and a speedup of 20 for eight slaves compared to DBSCAN based on traditional multi-step range queries.

7 Conclusions

In this paper, we applied the novel concept of using efficiently computable lower-bounding distance functions for the parallelization of data mining algorithms to the density-based clustering algorithm DBSCAN. For partitioning the data, we used the hierarchical clustering algorithm OPTICS as a kind of space filling curve for general metric objects, which provides the foundation for a fair and suitable partitioning strategy. We showed how the local clients can carry out their clustering efficiently by integrating the multi-step query processing paradigm directly into the clustering algorithm. Based on the concept of merge points, we constructed a global cluster connectivity graph from which the final clustering result can easily be derived. In the experimental evaluation, we demonstrated that our new approach is able to efficiently cluster metric objects. We showed that if several slaves are available, the benefits achieved by the full computational power of the slaves easily outweigh the additional costs of partitioning and merging by the master. In our future work, we will demonstrate that also other data mining algorithms can beneficially be parallelized based on lower-bounding distance functions.

References

1. Ester, M., Kriegel, H.P., Sander, J., Xu, X.: "A Density-Based Algorithm for Discovering Clusters in Large Spatial Databases with Noise". In: Proc. 2nd Int. Conf. on Knowledge Discovery and Data Mining (KDD'96), Portland, OR. (1996) 291–316
2. Brecheisen, S., Kriegel, H.P., Pfeifle, M.: "Efficient Density-Based Clustering of Complex Objects". In: Proc. 4th IEEE Int. Conf. on Data Mining (ICDM'04), Brighton, UK. (2004) 43–50
3. Kriegel, H.P., Kröger, P., Mashael, Z., Pfeifle, M., Pötke, M., Seidl, T.: "Effective Similarity Search on Voxelized CAD Objects". In: Proc. 8th Int. Conf. on Database Systems for Advanced Applications (DASFAA'03), Kyoto, Japan. (2003) 27–36
4. Kriegel, H.P., Brecheisen, S., Kröger, P., Pfeifle, M., Schubert, M.: "Using Sets of Feature Vectors for Similarity Search on Voxelized CAD Objects". In: Proc. ACM SIGMOD Int. Conf. on Management of Data (SIGMOD'03), San Diego, CA. (2003) 587–598
5. Kriegel, H.P., Schönauer, S.: "Similarity Search in Structured Data". In: Proc. 5th Int. Conf. on Data Warehousing and Knowledge Discovery (DaWaK'03), Prague, Czech Republic. (2003) 309–319
6. Kailing, K.: New Techniques for Clustering Complex Objects. PhD thesis, Institute for Computer Science, University of Munich (2004)
7. Jain, A.K., Murty, M.N., Flynn, P.J.: "Data Clustering: A Review". ACM Computing Surveys **31(3)** (1999) 265–323
8. Ankerst, M., Breunig, M.M., Kriegel, H.P., Sander, J.: "OPTICS: Ordering Points to Identify the Clustering Structure". In: Proc. ACM SIGMOD Int. Conf. on Management of Data (SIGMOD'99), Philadelphia, PA. (1999) 49–60
9. Fonseca, M.J., Jorge, J.A.: "Indexing High-Dimensional Data for Content-Based Retrieval in Large Databases". In: Proc. 8th Int. Conf. on Database Systems for Advanced Applications (DASFAA'03), Kyoto, Japan. (2003) 267–274

Neighborhood Density Method for Selecting Initial Cluster Centers in K-Means Clustering

Yunming Ye[1], Joshua Zhexue Huang[2], Xiaojun Chen[1], Shuigeng Zhou[3,*], Graham Williams[4], and Xiaofei Xu[1]

[1] Shenzhen Graduate School,
Harbin Institute of Technology, Shenzhen 518055, China
yym_sjtu@yahoo.com.cn
[2] E-Business Technology Institute,
University of Hong Kong, Pokfulam Road, Hong Kong
jhuang@eti.hku.hk
[3] Department of Computer Science and Engineering,
Fudan University, Shanghai 200433, China
sgzhou@fudan.edu.cn
[4] Australian Taxation Office, Australia
graham.williams@ato.gov.au

Abstract. This paper presents a new method for effectively selecting initial cluster centers in k-means clustering. This method identifies the high density neighborhoods from the data first and then selects the central points of the neighborhoods as initial centers. The recently published Neighborhood-Based Clustering (*NBC*) algorithm is used to search for high density neighborhoods. The new clustering algorithm *NK-means* integrates *NBC* into the k-means clustering process to improve the performance of the k-means algorithm while preserving the k-means efficiency. *NBC* is enhanced with a new cell-based neighborhood search method to accelerate the search for initial cluster centers. A merging method is employed to filter out insignificant initial centers to avoid too many clusters being generated. Experimental results on synthetic data sets have shown significant improvements in clustering accuracy in comparison with the random k-means and the refinement k-means algorithms.

Keywords: Clustering, k-means, Neighborhood-Based Clustering, Initial Cluster Center Selection.

1 Introduction

k-means clustering [1, 2] is one of the most widely used clustering methods in data mining, due to its efficiency and scalability in clustering large datasets. One well known problem of using k-means is selecting initial cluster centers for the iterative clustering process. Given a proper k, the clustering result of k-means

---

* Shuigeng Zhou was supported by the National Natural Science Foundation of China (NSFC) under grants No.60373019 and No.60573183.

is very sensitive to the selection of initial cluster centers because different initial centers often result in very different clusterings. In k-means clustering and other clustering methods, it is assumed that clusters distribute with certain high density in the data. Therefore, the k-means clustering process would produce a better clustering result if the initial cluster centers were taken from each high density area in the data. However, the currently used initial cluster center selection methods can hardly achieve this. Better selection of initial cluster centers for k-means clustering is still an interesting research problem because of the importance of k-means clustering in real word applications [3, 4, 5, 6, 7, 8, 9]

In this paper, we propose a neighborhood density method for effectively selecting initial cluster centers in k-means clustering. The method is to use the recently published Neighborhood-Based Clustering (*NBC*) algorithm [10] to search for high density neighborhoods from the data. *NBC* not only identifies all high density neighborhoods but also gives the central points of each neighborhood. Therefore, the neighborhood central points are used as the initial cluster centers. Since *NBC* determines neighborhoods based on local density, clusters of different densities are taken into account. A new clustering algorithm called *NK-means* is developed to integrate *NBC* into the k-means clustering process to improve the performance of the k-means algorithm while preserving the k-means efficiency. To enhance *NBC*'s search for dense neighborhoods, we have developed a new cell-based neighborhood search method to accelerate the search for initial cluster centers. A merging method is also employed to filter out insignificant initial centers to avoid too many clusters to be generated. Because the initial cluster centers are taken from the dense areas of the data, *NBC* enables the k-means clustering process to take less iterations to arrive at a near optimal solution, therefore, improving k-means clustering accuracy and efficiency.

We experimented *NK-means* with synthetic data. In comparison with the simple *k-means* and the refinement *k-means* algorithms [6], *NK-means* produced more accurate clustering results. It also showed a linear scalability in clustering data with varying sizes and dimensions. These results demonstrated that using the neighborhood density method to select initial cluster centers can significantly improve the performance of the k-means clustering process.

The rest of this paper is organized as follows. Section 2 describes *NBC* for initial cluster center selection, and the enhancement of *NBC* on search method. The merging process of insignificant initial clusters is also discussed. Section 3 defines the *NK-means* algorithm. Experimental results and analysis are presented in Section 4. In Section 5, we summarize this work and point out the future work.

2 Neighborhood Density Based Selection for Initial Cluster Centers

2.1 Search for Initial Cluster Centers with *NBC*

The Neighborhood Based Clustering algorithm, or *NBC*, is a density based clustering method [10]. Unlike other density based methods such as *DBSCAN* [11],

NBC finds clusters from data with respect to the local density instead of the global density. As such, it is able to discover clusters in different densities.

The locally dense neighborhood of a given point p is identified by the Neighborhood Density Factor (NDF), defined as:

$$NDF(p) = \frac{|R - kNB(p)|}{|kNB(p)|} \quad (1)$$

where $kNB(p)$ is the set of p's k-nearest neighbor points, and $R - kNB(p)$ is the set of the reverse k-nearest neighbor points of p. $R - kNB(p)$ is defined as the set of points whose k-nearest neighborhoods contain p. The value of $NDF(p)$ measures the local density of the object p. Intuitively, the larger $|R - kNB(p)|$ is, the more neighborhoods that contain p in their k-nearest neighbors, the denser p's neighborhood is. Generally speaking, $NDF(p) > 1$ indicates that p is located in a dense area. $NDF(p) < 1$ indicates that p is in a sparse area. If $NDF(p) = 1$, then p is located in an area where points are evenly distributed in space. The details of the (NBC) algorithm is given in [10].

Given a data set $\mathbf{X}$, we can use NBC to find all locally dense areas. In each dense area, we select its center as the candidate initial cluster center in k-means clustering.

2.2 Merging Candidate Clusters

In the original NBC algorithm, the size of a neighborhood is specified by an input parameter. We use k^{nbc} for this parameter here to distinguish the k parameter of the k-means algorithm. k^{nbc} specifies the minimal number of points in a neighborhood and controls the granularity of the final clusters by NBC. If k^{nbc} is set large, a few large clusters are found. If k^{nbc} is set small, many small clusters will be generated.

Let $\{C'_1, C'_2, \ldots, C'_i, \ldots, C'_{k'}\}$ be k' candidate clusters generated from a sample data by NBC. Assume k' is greater than the expected cluster number k. Each cluster C_i contains a set of points $\{x_1, x_2, \ldots, x_j, \ldots, x_{n_i}\}$. The radius of cluster C_i is defined as:

$$r_i = max_{j=1}^{n_i} \|x_j - z_i\|_2, \quad x_j \in C_i \quad (2)$$

where z_i is the center of cluster C_i and $\|x_j - z_i\|_2$ represents the distance between the object x_j and z_i. The similarity between two clusters c_i and c_j is calculated as

$$d(c_i, c_j) = \frac{\|z_j - z_i\|_2}{r_i + r_j} \quad (3)$$

To reduce the number of candidate clusters k' to the expected number k, we can iteratively merge the two most similar clusters according to Formula (3). One merging procedure of the entire merging process is given in Table 1.

Each run of the merging procedure merges two most similar clusters. To get the final k clusters the merging procedure is repeated $k' - k$ times in the $NK-means$ algorithm (see Table 2). Steps 1-6 of this procedure allocate the data points of the entire data set into k' initial clusters. Steps 7-10 recompute k' new

Table 1. The pseudo-code of the cluster merging procedure

Input: X -original data set, Z'- Centroids before merging,
Output: C'-the resulted clusters after merging, where $|C| - |C'| = 1$
1. for each object $x_j \in X$ do {
2.   for each $z_i \in Z'$ do {
3. $d_{j,i}$=calculateDistance(x_j, Z_i);
4.   }
5.   assign x_j to the cluster C_i with minimal $d_{j,i}$;
6. }   //The step 1-6 will build a set of new clusters C
7. for each cluster $C_i \in C$ do {
8.   recomputed the centroid z_i for the cluster C_i;
9.   calculate the cluster radius r_i according to formula (2);
10. }
11. for i=1 to $|C|$+1
12.   for j=1 to $|C|$+1
13.   compute the cluster dissimilarity $d(c_i, c_j)$ based on formula (3);
14. merge the two clusters with lowest $d(c_i, c_j)$ to build C';
15. end;

---

cluster centers and radius. Steps 11-15 merges the most similar clusters according to Formula (3).

2.3 Enhancement of *NBC* for Neighborhood Search

To identify the dense neighborhoods in data requires to calculate the *NDF* value for every point. This is a very time-consuming process. In *NBC*, a cell-based approach is adopted to facilitate the calculation of *NDF* and the *k*-nearest neighborhood search [10]. In this approach, the data space is divided into hypercube cells of equal sides in each dimension. Search for dense neighborhoods is conducted in the cells instead of the entire space, so that the search time is reduced.

Let n be the number of points in the data set and m the number of dimensions of the data space. Given k^{nbc} as the number of points in a neighborhood, the ideal way is to divide the data space into n/k^{nbc} cells and each cell contains only one dense neighborhood with k^{nbc} points. To obtain the same number of divisions in each dimension, the number of intervals in each dimension is calculated as:

$$\gamma = \lfloor \sqrt[m]{n/k^{nbc}} \rfloor + 1 \qquad (4)$$

Each dimension can be divided into γ equal intervals, and the n points will be divided by each dimension into γ subsets, $\{p_{i1}, p_{i2}, \ldots, p_{ij}, \ldots, p_{i\gamma}\}$ where

$\gamma \geq 2$. Because each dimension is equally divided, the data density in each subset p_{ij} will be very different, depending on the distribution of the data. The problems of this approach are that it results in more cells to search because $\gamma^m >> n/k^{nbc}$ in high dimensional data and that it is still time consuming in searching dense neighborhoods in high density cells. To solve these two problems, we use a density-aware approach to divide the dense areas into more cells and the sparse areas into few cells. In this way, we can obtain a division with the number of cells close to n/k^{nbc}. The search efficiency is improved significantly.

From the initial equal division of γ^m cells, we define a distribution balance factor for each dimension as:

$$\xi_i = \sqrt[\gamma]{\prod_{j=1}^{\gamma} \frac{s(p_{ij}) + \sigma}{n}} \quad (5)$$

where ξ_i denotes the distribution balance factor for dimension i, $s(p_{ij})$ is the number of points in cell(i,j), n is the number of points in the data set, σ is a normalization factor to avoid the zero value of $s(p_{ij})$.

After sorting the distribution balance factors for all dimensions as $\xi'_1 \geq \xi'_2 \geq \cdots \geq \xi'_m$, we calculate the relative division rate for each dimension as:

$$\mu_i = \begin{cases} \theta & i=m \\ \frac{\xi'_i}{\xi'_m} & 1 \leq i < m \end{cases} \quad (6)$$

where θ is the base number, defined as:

$$\theta = max\{1, \sqrt[m]{\frac{n/k^{nbc}}{\prod_{i=1}^{m-1} \mu_i}}\} \quad (7)$$

According to the relative division rate μ, each dimension will be divided into $\mu_i \cdot \theta$ intervals, and the total number of cells will be:

$$\omega = \prod_{i=1}^{L} \mu_i \cdot \theta \quad (8)$$

We only select the first L dimensions to divide, which makes $\prod_{i=1}^{L} \mu_i \cdot \theta \geq n/k^{nbc}$ and $\prod_{i=1}^{L-1} \mu_i \cdot \theta < n/k^{nbc}$. Therefore, the total number of cells will be close to n/k^{nbc}.

Based on the above calculations, we can obtain a set of new cells based on merging the sparse cells and redividing the dense cells of the initial division. We scan all the cells to calculate the number of points in each cell. The cells with few points will be merged with adjacent cells, while the cells with too many points (more than k^{nbc}) will be recursively re-divided using a cell adjustment method.

Fig. 1 illustrates the process of this approach. The left figure is a data set with 1600 points. The middle figure is the initial division (first step). As the

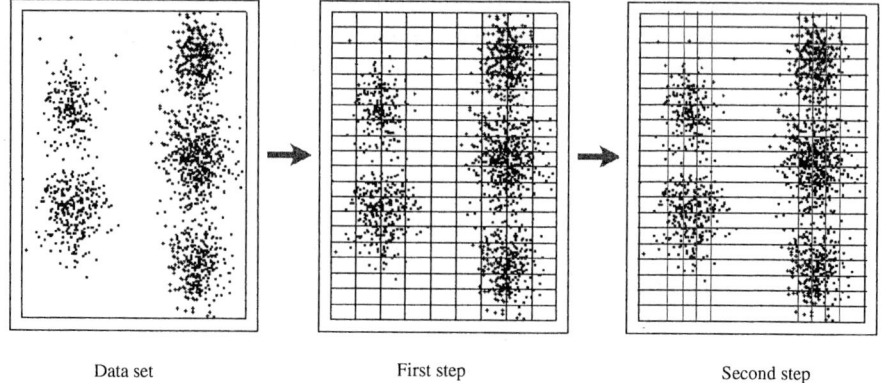

Fig. 1. Illustration of the enhancement of the data space division

data set distributes more evenly on the vertical dimension than on the horizontal dimension, the vertical dimension is divided into 20 intervals ($k^{nbc} = 10$) while the horizontal dimension into 9 intervals. In the readjustment (second step), the dense areas are re-divided into more cells while sparse areas are merged into fewer cells. Because the dense cells contain less points, the search for dense neighborhoods can be improved. Searching the kNB in sparse and wide cells will consider the neighbor cells, which is also very efficient.

3 The *NK-Means* Clustering Algorithm

The *NK-means* clustering algorithm combines the enhanced *NBC* algorithm with the k-means algorithm to cluster larger high dimensional data. The enhanced *NK-means* is used to select initial cluster centers from the sample data for the k-means algorithm to cluster the large data. Since the enhanced *NK-means* produces better initial cluster centers that are taken from the high density areas of the data, starting from these initial clusters, the k-means algorithm will produce better clustering results. Because sample data is used and the search method is enhanced, the initial cluster selection does not add too much computation burden to the entire clustering process but significantly improve the k-means clustering results. This has been demonstrated by our experiments.

Table 2 gives the pseudo-code of the *NK-means* algorithm. Given an input data set X, a sample rate and an expected cluster number k, the algorithm first takes a sample from X. Then, the sample is fed to the enhanced *NBC* algorithm to produce a set of initial candidate clusters. The third step is to calculate the centers of the candidate clusters. After the centers are calculated, they are readjusted with the entire data set X and the cluster merging process starts in Steps 4-7 (note that the merging procedure in Table 1 is executed for $k' - k$ times). After the merging process, a new set of initial cluster centers are obtained and used in the k-means algorithm in Step 9 as the initial cluster

Table 2. The pseudo-code of *NK-means*

Input: X -data set, r_s-sample rate, k-expected cluster number
Output: C-the resulted clusters
1. X' = sample(X, r_s);
2. $C' = NBC(X', k)$;
3. $Z' = calculateCentroid(X', C')$;
4. while ($|C'| > k$){ //the merging procedure runs ($k' - k$) times
5. C'=mergeClusters(X, Z'); //the merging procedure, see Table 1
6. Z'= calculateCentroid(X, C');
7. }
8. $Z = Z'$;
9. C =k-means(k, Z, X);
10. end;

centers to cluster the entire data set X. In the next section, we will show the experiment results of the *NK-means* algorithm.

4 Experiments

We have implemented the *NK-means* clustering algorithm in Java and conducted experiments with synthetic data. In these experiments, we compared *NK-means* with two other *k-means* algorithms with different initial cluster center selection methods: *random k-means* using the simple initial cluster center selection method [12], and Bradly's *refinement k-means* algorithm [6]. We also conducted scalability tests of *NK-means* against different data sizes and dimensions.

We used Matlab to generate synthetic data sets with mixture Gaussian distributions. We first carried out experiments on a two-dimension data set that contained 8,000 objects in eight inherent clusters. To test the robustness of the algorithm, we also added some noise into the data set. Fig. 2(a) shows the distribution of this data set with a noise rate of 10%. The solid cycles are the real centers of the clusters.

Fig. 2(b), 2(c) and 2(d) show the clustering results from the random *k-means*, the refinement *k-means* and the *NK-means* respectively. The solid cycles represent the inherent cluster centers while the star symbols in these figures give the initial cluster centers selected by the three clustering algorithms. We can observe from the figures that the initial cluster centers selected by the *NK-means* are very close to the inherent cluster centers in the data. Some of the initial cluster centers selected by the random *k-means*, the refinement *k-means* were located outside of some inherent clusters. For example, no initial cluster centers were selected from the two middle inherent clusters in Fig. 2(b). Because of this, the two inherent clusters were clustered into one cluster by the random *k-means*.

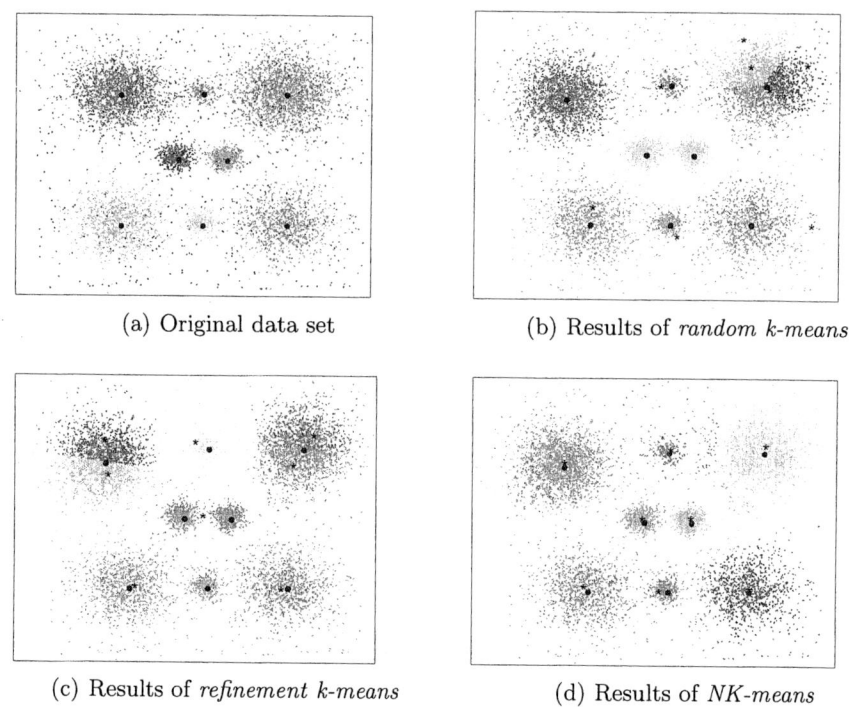

Fig. 2. Comparisons of three clustering results from a 2D data set

Four initial cluster centers were selected from the large inherent cluster on the upright corner of Fig. 2(b). This cluster was clustered into 3 clusters. Therefore, the random *k-means* could not recover the eight inherent clusters because of the bad selection of the initial cluster centers.

Fig. 2(c) shows that the refinement *k-means* could not recover the inherent clusters neither, because of the improper selection of the initial cluster centers. In this case, the two large clusters on the top were clustered into four small clusters, while the two middle small inherent clusters were clustered as one cluster. From Fig. 2(d), we can see that all eight inherent clusters were completely recovered by *NK-means*, due to the good selection of the initial cluster centers.

Table 3 lists the locations of real cluster centers and the final centers found by the three clustering algorithms. The final cluster centers by the three algorithms were calculated as the average values of 100 runs on the same data set. The final cluster centers by the *NK-means* were clearly very close to the real cluster centers, while the final cluster centers by other two algorithms were different.

To test the scalability of *NK-means*, we generated one data set with 600,000 normally distributed points in six clusters, and with additional 60,000 noise points. Each point is described in 10 dimensions.

Fig. 3 shows the scalability test results against the number of points and the number of dimensions in data. Fig. 3(a) plots the running time against

Table 3. The final cluster centers found by the three algorithms

Cluster Center	Real Center	Center by NK-means	Center by Random K-means	Center by Refinement K-means
Center 1	(10.129,-9.950)	(10.121,-9.988)	(8.010,-6.184)	(10.141,-9.279)
Center 2	(2.796,-9.995)	(-0.238,-9.680)	(-0.954,-5.468)	(-0.820,-4.784)
Center 3	(-9.945, -10.046)	(-9.938,-10.201)	(-9.267,-7.514)	(-9.829,-9.690)
Center 4	(2.937, -0.022)	(2.945,-0.234)	(3.302,2.515)	(2.029,2.670)
Center 5	(-2.974, 0.065)	(-2.691,-0.068)	(-3.240,1.521)	(-2.320,3.069)
Center 6	(12.003, 10.062)	(11.778,9.954)	(11.653,9.106)	(12.264,10.179)
Center 7	(0.069, 10.079)	(0.011,10.053)	(0.854,9.243)	(0.511,10.409)
Center 8	(-13.027, 7.991)	(-12.799,8.071)	(-11.140,6.751)	(-13.095,7.977)
Deviation	-	1.984	4.352	3.781

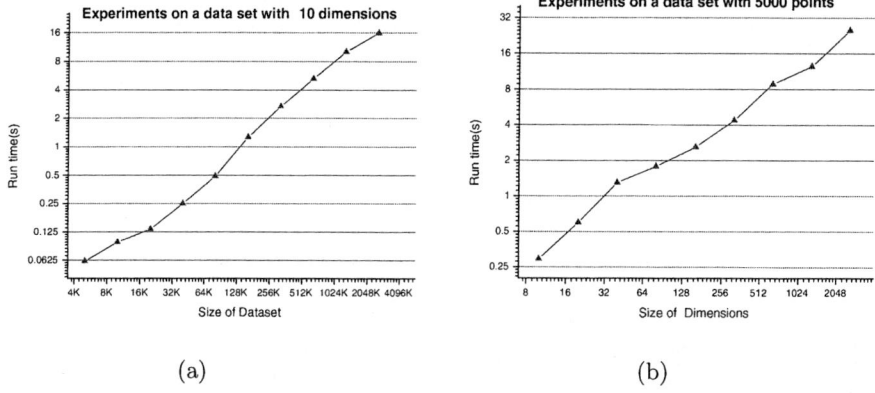

Fig. 3. Scalability against the data size and dimensions

different numbers of points, while Fig. 3(b) is the running time against different dimensions. These results show that the running time of *NK-means* linearly increased with the number of points and the number of dimensions. This property indicates that *NK-means* is scalable to large high-dimensional data.

5 Conclusions

In this paper, we have proposed a new neighborhood density method for selecting initial cluster centers for *k-means* clustering. We have presented the *NK-means* algorithm that makes use of the neighborhood-based clustering algorithm to select initial cluster centers and use the centers as input to the *k-means* clustering algorithm to improve the clustering performance of *k*-means. We have shown the experiments on both synthetic and real data to demonstrate that *NK-means* was superior to the other two algorithms: the random *k-means* and the refinement *k-means*.

We have also discussed the enhancement of *NBC*'s neighborhood search method and the merging process to generate the initial cluster centers. This enhancement enables *NBC* to take a larger sample which can result in better initial cluster centers. The next stage is to develop a termination method in the merging process to automatically generate the expected number of clusters k which has been a long standing problem in k-means clustering.

References

1. Jain, A., Murty, M., P.J., F.: Data clustering: A review. ACM Computing Surveys **31** (1999) 264–323
2. P., B.: Survey of clustering data mining techniques. In: Technical Report, Accrue Software, Inc. (2002)
3. Katsavounidis, I., Kuo, C., Zhang, Z.: A new initialization technique for generalized lloyd iteration. IEEE Signal Processing Letters **1** (1994) 144–146
4. Pena, J., Lozano, J., Larranaga, P.: An empirical comparison of four initialization methods for the k-means algorithm. Pattern Recognition Letters **20** (1999) 1027–104
5. Tou, J., Gonzalez, R.: Pattern recognition principles. In: Addison- Wesley, Massachusetts. (1974)
6. Bradley, P., Fayyad, U.: Refining initial points for kmeans clustering. In: Proceedings of 15th International Conference on Machine Learning. (1998)
7. Meila, M., Heckerman, D.: An experimental comparison of several clustering and initialization methods. In: Proceedings of the Fourteenth Conference on Uncertainty in Artificial Intelligence. (1998)
8. He, J., Lan, M., Tan, C., Sung, S., Low, H.: Initialization of cluster refinement algorithms: A review and comparative study. In: Proceedings of International Joint Conference on Neural Networks. (2004)
9. Kaufman, L.: Finding groups in data: an introduction to cluster analysis. In: Wiley, New York. (1990)
10. Zhou, S., Zhao, Y., Guan, J., Huang, J.: Nbc: A neighborhood based clustering algorithm. In: Proceedings of PAKDD'05. (2005)
11. Ester, M., Kriegel, H., Sander, J., Xu, X.: A density-based algorithm for discovering clusters in large spatial databases with noise. In: Proceedings of 1998 Int. Conf. Knowledge Discovery and Data Mining (KDD'96). (1996)
12. MacQueen, J.: Some methods for classification and analysis of multivariate observations. In: Proceedings of 5-th Berkeley Symposium on Mathematical Statistics and Probability. (1967)

Uncertain Data Mining: An Example in Clustering Location Data

Michael Chau[1], Reynold Cheng[2], Ben Kao[3], and Jackey Ng[1]

[1] School of Business, The University of Hong Kong, Pokfulam, Hong Kong
mchau@business.hku.hk, jackeyng@hkusua.hku.hk
[2] Department of Computing, Hong Kong Polytechnic University, Kowloon, Hong Kong
csckcheng@comp.polyu.edu.hk
[3] Department of Computer Science, The University of Hong Kong, Pokfulam, Hong Kong
kao@cs.hku.hk

Abstract. Data uncertainty is an inherent property in various applications due to reasons such as outdated sources or imprecise measurement. When data mining techniques are applied to these data, their uncertainty has to be considered to obtain high quality results. We present UK-means clustering, an algorithm that enhances the K-means algorithm to handle data uncertainty. We apply UK-means to the particular pattern of moving-object uncertainty. Experimental results show that by considering uncertainty, a clustering algorithm can produce more accurate results.

1 Introduction

In applications that require interaction with the physical world, such as location-based services [6] and sensor monitoring [3], data uncertainty is an inherent property due to measurement inaccuracy, sampling discrepancy, outdated data sources, or other errors. Although much research effort has been directed towards the management of uncertain data in databases, few researchers have addressed the issue of mining uncertain data. We note that with uncertainty, data values are no longer atomic. To apply traditional data mining techniques, uncertain data has to be *summarized* into atomic values. Unfortunately, discrepancy in the summarized recorded values and the actual values could seriously affect the quality of the mining results. Figure 1 illustrates this problem when a clustering algorithm is applied to moving objects with location uncertainty. If we solely rely on the recorded values, many objects could possibly be put into wrong clusters. Even worse, each member of a cluster would change the cluster centroids, thus resulting in more errors.

We suggest incorporating uncertainty information, such as the probability density functions (pdf) of uncertain data, into existing data mining methods so that the mining results could resemble closer to the results obtained as if actual data were used in the mining process [2]. In this paper we study how uncertainty can be incorporated in data mining by using data clustering as a motivating example. In particular, we study one of the most popular clustering methods – K-means clustering.

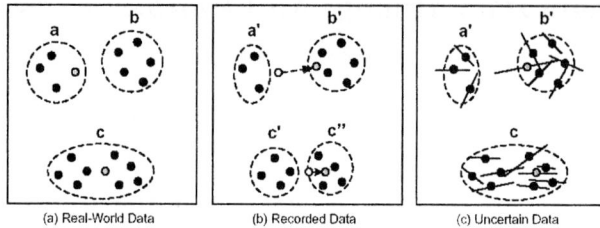

Fig. 1. (a) The real-world data are partitioned into three clusters (a, b, c). (b) The recorded locations of some objects (shaded) are not the same as their true location, thus creating clusters a', b', c' and c''. (c) When line uncertainty is considered, clusters a', b' and c are produced. The clustering result is closer to that of (a) than (b) is.

2 Related Work

There is significant research interest in data uncertainty management in recent years. Most work has been devoted to "imprecise queries", which provide probabilistic guarantees over correctness of answers. For example, in [4], indexing solutions for range queries over uncertain data have been proposed. The same authors also proposed solutions for aggregate queries such as nearest-neighbor queries in [3]. Notice that all these works have applied the study of uncertain data management to simple database queries, instead of to the more complicated data analysis and mining problems.

Clusterization has been well studied in data mining research. However, only a few studies on data mining or data clustering for uncertain data have been reported. Hamdan and Govaert have addressed the problem of fitting mixture densities to uncertain data for clustering using the EM algorithm [5]. However, the model cannot be readily applied to other clustering algorithms and is rather customized for EM. Clustering on interval data also has been studied. However, the pdf of the interval is not taken into account in most of the metrics used. Another related area of research is fuzzy clustering. In fuzzy clustering, a cluster is represented by a fuzzy subset of a set of objects. Each object has a "degree of belongingness" for each cluster. In other words, an object can belong to more than one cluster, each with a different degree. The fuzzy c-means algorithm was one of the most widely used fuzzy clustering method [1].

3 Clustering on Data with Uncertainty

Problem Definition: Let S be a set of V-dimensional vectors x_i, where $i = 1$ to n, representing the attribute values of all the records in the clustering application. Each record o_i is associated with a probability density function (pdf), $f_i(\mathbf{x})$, which is the pdf of o_i's attribute values $\mathbf{x}$ at time t. The clustering problem is to find a set C of clusters C_j, where $j = 1$ to K, with cluster means $\mathbf{c}_j$ based on similarity. Different clustering algorithms have different objective functions, but the general idea is to minimize the distance between objects in the same cluster while maximizing the distance between objects in different clusters. Minimization of intra-cluster distance can also be viewed as the minimization of the distance between each data $\mathbf{x}_i$ and the cluster means $\mathbf{c}_j$ of the cluster C_j that $\mathbf{x}_i$ is assigned to.

To consider data uncertainty in the clustering process, we propose a clustering algorithm with the goal of minimizing the *expected* sum of squared errors E(SSE). Note that a data object $\mathbf{x}_i$ is specified by an uncertainty region with an uncertainty pdf $f(\mathbf{x}_i)$. Given a set of clusters, C_j's the expected SSE can be calculated as follow:

$$E\left(\sum_{j=1}^{k}\sum_{i\in C_j}\|\mathbf{c}_j - \mathbf{x}_i\|^2\right) = \sum_{j=1}^{k}\sum_{i\in C_j}\int\|\mathbf{c}_j - \mathbf{x}_i\|^2 f(\mathbf{x}_i)d\mathbf{x}_i \qquad (1)$$

where $\|\cdot\|$ is a distance metric between a data point $\mathbf{x}_i$ and a cluster mean $\mathbf{c}_j$.

Cluster means are given by:

$$\mathbf{c}_j = E\left(\frac{1}{|C_j|}\sum_{i\in C_j}\mathbf{x}_i\right) = \frac{1}{|C_j|}\sum_{i\in C_j}\int\mathbf{x}_i f(\mathbf{x}_i)d\mathbf{x}_i \qquad (2)$$

We propose a new K-means algorithm, called UK-means, for clustering uncertain data:

1. Assign initial values for cluster means $\mathbf{c}_1$ to $\mathbf{c}_K$
2. **repeat**
3.   **for** $i = 1$ to n **do**
4.   Assign each data $\mathbf{x}_i$ to cluster C_j where $E(\|\mathbf{c}_j - \mathbf{x}_i\|)$ is the minimum
5.   **end for**
6.   **for** $j = 1$ to K **do**
7.   Recalculate cluster mean $\mathbf{c}_j$ of cluster C_j
8.   **end for**
9. **until** convergence
10. **return** C

The main difference between UK-mean clustering and the traditional K-means clustering lies in the computation of distance and clusters. In particular, UK-means compute the *expected* distance and cluster centroids based on the data uncertainty model. Convergence can be defined based on different criteria.

In Step 4, it is often difficult to determine $E(\|\mathbf{c}_j - \mathbf{x}_i\|)$ algebraically. In particular, the variety of geometric shapes of uncertainty regions (e.g., line, circle) and different uncertainty pdf imply that numerical integration methods are necessary. We propose to use the squared expected distance $E(\|\mathbf{c}_j - \mathbf{x}_i\|^2)$, which is much easier to obtain.

4 UK-Means Clustering for Moving Objects with Uncertainty

The UK-means algorithm presented in the last section is applicable to any uncertainty region and pdf. In this section, we describe how the proposed algorithm can be applied to uncertainty models specific to moving objects that are moving in a two-dimensional space. According to [4] and [6], there are two types of moving-object uncertainty, namely line-moving uncertainty and free-moving uncertainty. In line-moving uncertainty, an object moves at a velocity vector, which is smaller than V_{max}, along a fixed direction. Line-moving uncertainty can be unidirectional or bidirectional. The free-moving uncertainty model assumes that an object cannot move beyond a certain speed, V_{max}. Given that the current position of the object is (h,k) at time t_0, the object's location is uniformly distributed within a circle of radius $V_{max} \times (t-t_0)$.

Suppose we have a centroid $\mathbf{c} = (p, q)$ and a data object $\mathbf{x}$ specified by a line uncertainty region with a uniform distribution. Let the end points of the line segment uncertainty be (a,b) and (c,d). The line equation can be parametrized by $(a + t(c - a), b + t(d - b))$, where t is between $[0,1]$. Let the uncertainty pdf be $f(t)$. Also, let the distance of the line segment uncertainty be $D = \sqrt{(c-a)^2 + (d-b)^2}$. We have:

$$E(\|\mathbf{c} - \mathbf{x}\|^2) = \int_0^1 f(t)(D^2 t^2 + Bt + C)dt \tag{3}$$

where $B = 2[(c - a)(a - p) + (d - b)(b - q)]$, $C = (p - a)^2 + (q - b)^2$
If $f(t)$ is uniform, then $f(t) = 1$, and the above becomes:

$$E(\text{distance of line uncertainty from centroid}^2) = \frac{D^2}{3} + \frac{B}{2} + C \tag{4}$$

For free-moving uncertainty, suppose we have a centroid $\mathbf{c} = (p, q)$ and a data object $\mathbf{x}$ specified by a circle uncertainty region with a uniform distribution. Suppose the circle uncertainty has center (h, k) and radius R. Let the uncertainty pdf of the circle be $f(r,\theta)$. Then we have:

$$E(\|\mathbf{c} - \mathbf{x}\|^2) = \int_0^R \int_0^{2\pi} f(r,\theta)(A\cos\theta + B\sin\theta + C) r \, d\theta \, dr \tag{5}$$

where $A = 2r(h - p)$, $B = 2r(k - q)$, $C = r^2 + (h - p)^2 + (k - q)^2$

We are thus able to compute the expected squared distance easily for line-moving and free-moving object uncertainty. The use of uniform distribution is only a specific example here. When the pdf's are not uniform (e.g., Gaussian), sampling techniques can be used to estimate $E(\|\mathbf{c}_j - \mathbf{x}_i\|)$.

5 Experiments

In our experiments, we simulate a scenario in which a system that tracks the locations of a set of moving objects has taken a snapshot of these locations [2]. This location data is stored in a set called **recorded**. Each object assumes an uncertainty model captured in **uncertainty**. We compare two clustering approaches: (1) apply K-means to **recorded** and (2) apply UK-means to **recorded** + **uncertainty**. We first generated a set of random data points in a 100 x 100 2D space as **recorded**. For each data point, we then randomly generated its uncertainty according to a chosen uncertainty model. We also generated **actual** — the *actual locations* of the objects based on **recorded** and **uncertainty**, simulating the scenario that the objects have moved away from their original locations as registered in **recorded**. We remark that *ideally*, a system should know **actual** and apply K-means on the actual locations. Hence, we compute and compare the cluster outputs of the following data sets:

(1) **recorded** (using classicial K-means)
(2) **recorded + uncertainty** (using UK-means)
(3) **actual** (using classical K-means)

We use the Adjusted Rand Index (ARI) to measure the similarity between the clustering results [7]. A higher ARI value indicates a higher degree of similarity between two sets of clusters. We compare the ARI between the sets of clusters created in (2) and (3) and the ARI between those created in (1) and (3). Due to limited space, only the results of unidirectional line uncertainty are reported here.

The number of objects (n), number of clusters (K), and the maximum distance an object can move (d) were varied during the experiment. Table 1 shows the different experiment results by varying d while keeping $n = 1000$ and $K = 20$. Under each set of different parameter settings, 500 rounds were run and the results were averaged. In each round, the sets of **recorded**, **uncertainty**, and **actual** were first generated and the same set of data was used for the three clustering processes. The same set of initial centroids were also used in each of the three processes in order to avoid any bias.

The UK-means algorithm consistently showed a higher ARI than the traditional K-means algorithm applied on the recorded data. Pairwise t-tests were conducted and the results showed that the difference in the ARI values of the two methods was significant ($p < 0.000001$ for all cases). The results demonstrated that the UK-means algorithm can give a set of clusters that could be a better prediction of the clusters that would be produced if the real-world data were available.

Table 1. Experiment results

d	1.5	2.5	5	7.5	10	20	50
ARI (UK-means)	0.740	0.733	0.689	0.652	0.632	0.506	0.311
ARI (K-means)	0.715	0.700	0.626	0.573	0.523	0.351	0.121
% of improvement	3.58%	4.77%	10.03%	13.84%	20.82%	44.34%	155.75%

6 Conclusions and Future Work

In this paper we present the UK-means algorithm, which aims at improving the accuracy of clustering by considering the uncertainty associated with data. Although in this paper we only present clustering algorithms for uncertain data with uniform distribution, the model can be generalized to other distribution (e.g., by using sampling techniques). We also suggest that our concept of using expected distance could be applied to other clustering approaches (such as nearest neighbor clustering and self-organizing maps) and other data mining techniques (such as data classification).

Acknowledgement

We thank David Cheung (University of Hong Kong), Edward Hung (Hong Kong Polytechnic University) and Kevin Yip (Yale University) for their helpful comments.

References

1. Bezdek, J. C.: Pattern Recognition with Fuzzy Objective Function Algorithms. Plenum Press, New York (1981).
2. Chau, M., Cheng, R., and Kao, B.: Uncertain Data Mining: A New Research Direction. In Proc. Workshop on the Sciences of the Artificial, Hualien, Taiwan (2005).
3. Cheng, R., Kalashnikov, D., and Prabhakar, S.: Querying Imprecise Data in Moving Object Environments. IEEE TKDE, 16(9) (2004) 1112-1127.
4. Cheng, R., Xia, X., Prabhakar, S., Shah, R. and Vitter, J.: Efficient Indexing Methods for Probabilistic Threshold Queries over Uncertain Data. In Proc. VLDB, 2004.
5. Hamdan, H. and Govaert, G.: Mixture Model Clustering of Uncertain Data. IEEE International Conference on Fuzzy Systems (2005) 879-884.
6. Wolfson, O., Sistla, P., Chamberlain, S. and Yesha, Y.: Updating and Querying Databases that Track Mobile Units. Distributed and Parallel Databases, 7(3), 1999.
7. Yeung, K. and Ruzzo, W.: An Empirical Study on Principal Component Analysis for Clustering Gene Expression Data. Bioinformatics 17(9) (2001) 763-774.

Parallel Randomized Support Vector Machine

Yumao Lu and Vwani Roychowdhury

University of California, Los Angeles, CA 90095, USA

Abstract. A parallel support vector machine based on randomized sampling technique is proposed in this paper. We modeled a new LP-type problem so that it works for general linear-nonseparable SVM training problems unlike the previous work [2]. A unique priority based sampling mechanism is used so that we can prove an average convergence rate that is so far the fastest bounded convergence rate to the best of our knowledge. The numerical results on synthesized data and a real geometric database show that our algorithm has good scalability.

1 Introduction

Sampling theory has a long successful history in optimization [6, 1]. The application to the SVM training problem is first proposed by Balcazar et al. in 2001 [2]. However, Balcazar assumed that the SVM training problem is a separable problem or a problem that can be transformed to an equivalent separable problem by assuming an arbitrary small regularization factor γ (D and $1/k$ in [2] and [3]). They also stated that there were number of implementation difficulties so that no relevant results could be provided [3].

We model a LP-type problem such that the general linear nonseparable problem can be covered by our randomized support vector machine (RSVM). In order to take advantage of distributed computing facilities, we proposed a novel parallel randomized SVM (PRSVM) in which multiple working sets can be worked on simultaneously. The basic idea of the PRSVM is to randomly shuffle the training vectors among a network based on a carefully designed priority and weighting mechanism and to solve the multiple local problems simultaneously. Unlike the previous works on parallel SVM [7, 10] that lacks of a convergence bound, our algorithm, the PRSVM, on average, converges to the global optimum classifier/regressor in less than $(6\delta \ln(N + 6r(C-1)\delta)/C$ iterations, where δ denotes the underlying combinatorial dimension, N denotes the total number of training vector, C denotes the number of working sites, and r denotes the size for a working set. Since the RSVM is a special case of PRSVM, our proof naturally works for the RSVM. Note that, when $C = 1$, our result reduces to Balcazar's bound [3].

This paper is organized as follows. The support vector machine is introduced and formulated in the next section. Then, we present the parallel randomized support vector machine algorithm. The theoretical global convergence is given in the fourth section followed by a presentation of a successful application. We conclude our result in Section 6.

2 Support Vector Machine and Randomized Sampling

We prepare fundamentals and basic notations on SVM and randomized sampling technique in this section.

2.1 Support Vector Machine

Let us first consider a simple linear separation problem. We are seeking a hyperplane to separate of a set of positively and negatively labeled training data. The hyperplane is defined by $w^T x_i - b = 0$ with parameter $w \in \mathbf{R}^m$ and $b \in \mathbf{R}$ such that $y_i(w^T x_i - b) > 1$ for $i = 1, ..., N$ where $x_i \in \mathbf{R}^m$ is a training data point and $y_i \in \{+1, -1\}$ denotes the class of the vector x_i. The margin is defined by the distance of the two parallel hyperplanes $w^T x - b = 1$ and $w^T x - b = -1$, i.e. $2/||w||_2$. The margin is related to the generalization of the classifier [12]. The support vector machine (SVM) is in fact a quadratic programming problem, which maximizes the margin over the parameters of the linear classifier. For general nonseparable problems, a set of slack variables $\mu_i, i = 1, ..., N$ are introduced. The SVM problem is defined as follows:

$$\begin{aligned} &\text{minimize } (1/2)w^T w + \gamma \mathbf{1}^T \mu \\ &\text{subject to } y_i(w^T x_i - b) \leq 1 - \mu_i, \ i = 1, ..., N \\ &\mu \geq 0 \end{aligned} \quad (1)$$

where the scalar γ is usually empirically selected to reduce the testing error rate. To simplify notations, we define $v_i = (x_i, -1)$, $\theta = (w, b)$, and a matrix X as

$$Z = [(y_1 v_1) \ (y_2 v_2) \ ... \ (y_N v_N)]^T.$$

The dual of problem (1) is shown as follows:

$$\begin{aligned} &\text{maximize } -(1/2)\alpha^T Z Z^T \alpha + \mathbf{1}^T \alpha \\ &\text{subject to } 0 \leq \alpha \leq \gamma \mathbf{1}. \end{aligned} \quad (2)$$

A nonlinear kernel function can be used for nonlinear separation of the training data. In that case, the gram matrix ZZ^T is replaced by a kernel matrix $k(x, \tilde{x}) \in \mathbf{R}^{N \times N}$. Our PRSVM that is described in the following section can be kernelized and therefore is able to keep the full advantages of the SVM.

2.2 The Sampling Lemma, LP-Type Problem and KKT Condition

An abstract problem is denoted by $(\mathcal{S}, \phi)$. Let $\mathcal{X}$ be the set of training vector. That is, each element of $\mathcal{X}$ is a row vector of the matrix X. Throughout this paper, we use $\mathcal{CALLIGRAPHIC}$ style letters to denote sets of the row vectors of a matrix denoted by the same letter with *italian* style. Here, ϕ is a mapping from a given subset $\mathcal{X}_R$ of $\mathcal{X}$ to the local solution of problem (1) with constraints corresponding to X_R and $\mathcal{S}$ is of size N. Define

$$\begin{aligned} \mathcal{V}(\mathcal{R}) &:= \{s \in \mathcal{S} \setminus \mathcal{R} | \phi(\mathcal{R} \cup \{s\}) \neq \phi(\mathcal{R})\}, \\ \mathcal{E}(\mathcal{R}) &:= \{s \in \mathcal{R} | \phi(\mathcal{R} \setminus \{s\}) \neq \phi(\mathcal{R})\}. \end{aligned}$$

The elements of $\mathcal{V}(\mathcal{R})$ are called violators of $\mathcal{R}$ and the elements of $\mathcal{E}(\mathcal{R})$ are called extremes in $\mathcal{R}$. By definition, we have

$$s \text{ violates } \mathcal{R} \Leftrightarrow s \text{ is extreme in } \mathcal{R} \cup \{s\}.$$

For a random sample $\mathcal{R}$ of size r, we consider the expected values

$$v_r := E_{|\mathcal{R}|=r}(|\mathcal{V}_\mathcal{R}|)$$
$$e_r := E_{|\mathcal{R}|=r}(|\mathcal{E}_\mathcal{R}|)$$

Gartner proved the following sampling lemma [9]:

Lemma 1. *(Sampling Lemma). For $0 \leq r < N$,*

$$\frac{v_r}{N-r} = \frac{e_{r+1}}{r+1}.$$

Proof. By definitions, we have

$$\binom{N}{r} v_r = \sum_\mathcal{R} \sum_{s \in \mathcal{S} \setminus \mathcal{R}} [s \text{ violates } \mathcal{R}]$$
$$= \sum_\mathcal{R} \sum_{s \in \mathcal{S} \setminus \mathcal{R}} [s \text{ is extreme in } \mathcal{R} \cup \{s\}]$$
$$= \sum_\mathcal{Q} \sum_{s \in \mathcal{Q}} [s \text{ is extreme in } \mathcal{Q}]$$
$$= \binom{N}{r+1} e_{r+1},$$

where [.] is the indicator variable for the event in brackets and the last row follows the fact that the set $\mathcal{Q}$ has $r+1$ elements. The Lemma immediately follows. □

The problem $(\mathcal{S}, \phi)$ is said to be a LP-type problem if ϕ is monotone and local (see Definition 3.1 in [9]). Balcazar proved that the problem (1) is a LP-type problem [2]. So is the problem (2). We use the same definitions given by [9] to define the *basis* and *combinatorial dimension* as follows. For any $\mathcal{R} \subseteq \mathcal{S}$, a *basis* of $\mathcal{R}$ is a inclusion-minimal subset $\mathcal{B} \subseteq \mathcal{R}$ with $\phi(\mathcal{B}) = \phi(\mathcal{R})$. The *combinatorial dimension* of $(\mathcal{S},\phi)$, denoted by δ, is the size of a largest basis of $\mathcal{S}$. For a LP-type problem $(\mathcal{S},\phi)$ with combinatorial dimension δ, the sampling lemma yields

$$v_r \leq \delta \frac{N-r}{r+1}. \tag{3}$$

This follows that $|\mathcal{E}(\mathcal{R})| \leq \delta$.

Then, we are able to relate the definitions of the extremes, violators and the basis to our general SVM training problem (1) or (2). For any local solution θ^p or α^p of problem $(\mathcal{X}_p, \phi)$, the basis is the support vector set, $\mathcal{SV}_p$. The violators of the local solutions will be the vectors that violate the Karush-Kuhn-Tucker (KKT) necessary and sufficient optimality conditions. The KKT conditions for the problem (1) and (2) are listed as follows:

$$Z\theta^* \geq 1 - \mu^*,\ \mu^* \geq 0,\ 0 \leq \alpha^* \leq \gamma\mathbf{1},$$

$$\theta^* = Z^T \alpha^*, \ (\gamma - \alpha_i^*)\mu_i^* = 0, \ i = 1, \ldots, N.$$

Since the μ_i and α_i for the training vector x_i is always 0 for $x_i \in \mathcal{X} \setminus \mathcal{X}_p$, the only condition needed to be tested is

$$\theta^{p^T} z_i \geq 1$$

or

$$\alpha^{p^T} Z_p z_i \geq 1.$$

Any training vector that violates the above condition is called a violator to $(\mathcal{X}_p, \phi)$. The size of the largest basis, δ is naturally the largest number of support vectors for all subproblems $(\mathcal{X}_p, \phi)$, $\mathcal{X}_p \subseteq \mathcal{X}$. For separable problems, δ is bounded by one plus the lifted dimension, i.e., $\delta \leq n+1$. For general nonseparable problems, we do not know the bound for δ before we actually solve the problem. What we can do is to set a sufficiently large number to bound δ from above.

3 Algorithm

We consider the following problem: the training data are distributed in $C+1$ sites, where there are C working sets and 1 nonworking set. Each working site is assigned a priority number $p = 1, 2, \ldots, C$. We also assume that each working site contains r training vectors, where $r \geq 6\delta^2$ and δ denotes the combinatorial dimension of the SVM problem.

Define a function $u(.)$ to record the number of copies of elements of a training set. For training set $\mathcal{X}$, we define a set $\mathcal{W}$ such that $\mathcal{W}$ contains the virtually duplicated copies of the training vectors. We have $|\mathcal{W}| = u(\mathcal{X})$. We also define the virtual set $\mathcal{W}_p$ corresponding to training set $\mathcal{X}_p$ at site p.

Our parallel randomized support vector machine (PRSVM) works as follows.

Initialization Training vectors $\mathcal{X}$ are randomly distributed to $C+1$ sites. Assign priorities to all sites such that each site gets a unique priority number. Set $u(\{x_i\}) = 1$, $\forall i$. Hence, $u(\mathcal{X}) = N$. We have $|\mathcal{X}_p| = |\mathcal{W}_p|$ for all p. Set $t = 0$.

Iteration Each iteration consists of the following steps.
   Repeat for $t = 1, 2, \ldots$

1. Randomly distribute the training vectors over the working sites according to $u(\mathcal{X})$ as follows. Let $\mathcal{S}^1 = \mathcal{W}$.
   For $p = 1 : C$
   Choose r training vectors, $\mathcal{W}_p$ from $\mathcal{S}^p$ uniformly (and make sure $r \geq 6\delta^2$); $\mathcal{S}^{p+1} := \mathcal{S}^p \setminus \mathcal{W}_p$;
   End For
2. Each site with priority p, $p \leq C$ solves the local partial problem and record the solution θ^p. Send this solution to all other sites q, $q \neq p$.

3. Each site with priority q, $q = 1, ..., C + 1$, checks the solution θ^p from site with higher priority $p, p < q$. Define $\mathcal{V}_{q,p}$ to be the training vectors in the site with priority q that violate the KKT condition corresponding to solution (w^p, b^p), $q \neq p$. That is,

$$\mathcal{V}_{q,p} := \{x_i | \theta^{p^T}([x_i; 1])y_i < 1, x_i \in \mathcal{X}_q, x_i \notin \mathcal{X}_p\}$$

4. If $\sum_{q=p+1}^{C+1} u(\mathcal{V}_{q,p}) \leq |\mathcal{S}^p|/(3\delta)$ then $u(\{x_i\}) = 2u(\{x_i\})$, for all $x_i \in \mathcal{V}_{q,p}$, $\forall q \neq p, \forall p$;

until $\cup_{q \neq p} \mathcal{V}_{q,p} = \emptyset$ for some p.
Return the solution θ^p.

The priority setting of working sets actually defines the order of sampling. The highest priority server gets the first sampled batch of data, lower one gets the second batch and so on. This kind of sequential behavior is designed to help define violators and extremes clearly under a multiple working site configuration.

Step 2 involves a merging procedure. If $u(\{x_i\})$ copies of vector x_i are sampled to a working set $\mathcal{W}_p$, only one copy of x_i is included in the optimization problem $(\mathcal{X}_p, \phi)$ that we are solving, while we record this number of copies as a weight of this training vector.

The merging procedure has two properties:

Property 1. A training vector that is not in working set $\mathcal{X}_p$ must not be a violator of the problem $(\mathcal{X}_p, \phi)$ if one or more copies of this vector are included in the working set $\mathcal{X}_p$. That is, $x_i \notin \mathcal{V}(\mathcal{X}_p)$, if $x_i \in \mathcal{X}_p$.

Property 2. If multiple copies of a vector x_i are sampled to a working set $\mathcal{X}_p$, none of those of vectors can be the extreme of the problem $(\mathcal{X}_p, \phi)$. That is, $x_i \notin \mathcal{E}(\mathcal{X}_p)$ if $u(\{x_i\}) > 1$ at site p.

The above two properties follow immediately by definitions of violators and extremes.

One may note that the merging procedure actually constructs an abstract problem $(\mathcal{W}_p, \phi')$ such that $\phi'(\mathcal{W}_p) = \phi(\mathcal{X}_p)$. By definition, $(\mathcal{W}_p, \phi')$ is a LP-type problem and has the same combinatorial dimension, δ, as the problem $(\mathcal{X}_p, \phi)$. If the set of violators of $(\mathcal{X}_p, \phi)$ is $\mathcal{V}_p$, the number of violators of $(\mathcal{W}_p, \phi')$ is $u(\mathcal{V}_p)$.

Step 4 plays the key role in this algorithm. It says that if the number of violators of the LP-type problem $(\mathcal{W}_p, \phi')$ is not too large, we double the weights of the violators of $(\mathcal{W}_p, \phi')$ in all sites. Otherwise, we keep the weights untouched since the violators already have enough weights to be sampled to a working site.

One may note when $C = 1$, the PRSVM is reduced to the RSVM. However, our RSVM is different from the randomized support vector machine training algorithm in [2] in several ways. First, our RSVM is capable of solving general nonseparable problems, while Balcazar's method has to transfer nonseparable problems to an equivalent separable problems by assuming an arbitrarily small γ. Second, our RSVM merges examples after sampling them. Duplicated examples are not allowed in the optimization steps. Third, we test the KKT conditions to identify a violator instead of identifying a misclassified point. In our RSVM, a correctly classified example may also be a violator if this example violates the KKT condition.

4 Proof of the Average Convergence Rate

We prove the average number of iterations executed in our algorithm, PRSVM, is bounded by $(6\delta/C)\ln(N+6r(C-1)\delta)$ in this section. This proof is a generalization of the one given in [2]. The result of the tradition RSVM becomes a special case of our PRSVM.

Theorem 1. *For general SVM training problem the average number of iterations executed in the PRSVM algorithm is bounded by $(6\delta/C)\ln(N+6r(C-1)\delta)$.*

Proof. We consider an update to be successful if the if-condition in the step 4 holds in an iteration. One iteration has C updates, successful or not.

We first show the bound of the number of successful updates. Let $\mathcal{V}_p$ denote the set of violators from site with priority $q \geq p$ for the solution θ^p. By this definition, we have

$$u(\mathcal{V}_p) = \sum_{q=p+1}^{C+1} u(\mathcal{V}_{q,p})$$

Since the if-condition holds, we have

$$\sum_{q=p+1}^{C+1} u(\mathcal{V}_{q,p}) \leq u(S^p)/(3\delta) \leq u(\mathcal{X})/(3\delta).$$

By noting that the total number of training vectors including duplicated ones in each working sites is always r for any iterations, we have

$$\sum_{q=1}^{p-1} u(\mathcal{V}_{q,p}) \leq r(p-1) \leq r(C-1)$$

and

$$\sum_{q \neq p} u(\mathcal{V}_{q,p}) = \sum_{q=p+1}^{C+1} u(\mathcal{V}_{q,p}) + \sum_{q=1}^{p-1} u(\mathcal{V}_{q,p})$$
$$= u(\mathcal{V}_p) + \sum_{q=1}^{p-1} u(\mathcal{V}_{q,p})$$

Therefore, at each successful update, we have

$$u_k(\mathcal{X}) \leq u_{k-1}(\mathcal{X})(1+\frac{1}{3\delta}) + 2r(C-1).$$

where k denotes the number of successful updates. Since $u_0(\mathcal{X}) = N$, after k successful updates, we have

$$u_k(\mathcal{X}) \leq N(1+\tfrac{1}{3\delta})^k + 2r(C-1)3\delta[(1+\tfrac{1}{3\delta})^k - 1]$$
$$< (N+6r(C-1)\delta)(1+\tfrac{1}{3\delta})^k$$

Let $\mathcal{X}_0$ be the set of support vectors of the original problem (1) or (2). At each successful iterations, some x_i of $\mathcal{X}_0$ must not be in $\mathcal{X}_p$. Hence, $u(\{x_i\})$ gets doubled. Since, $|\mathcal{X}_0| \leq \delta$, there is some x_i in $\mathcal{X}_0$ that gets doubled at least once every δ successful updates. That is, after k successful updates, $u(\{x_i\}) \geq 2^{k/\delta}$.

Therefore, we have

$$2^{\frac{k}{\delta}} \leq u(\mathcal{X}) \leq (N + 6r(C-1)\delta)(1 + \frac{1}{3\delta})^k.$$

By simple algebra, we have

$$k \leq 3\delta \ln(N + 6r(C-1)\delta).$$

That is, the algorithm terminates within less than $3\delta \ln(N+6r(C-1)\delta)$ successful updates.

The rest is to prove that the probability of a successful update is higher than one half. By sampling lemma, the bound (3), we have

$$\mathrm{Exp}(u(\mathcal{V}_p)) \leq \frac{(u(\mathcal{S}^p)-r)\delta}{r+1}$$
$$< \frac{u(\mathcal{S}^p)}{6\delta}$$

By Markov equality, we have

$$\mathrm{Pro}\{u(\mathcal{V}_p) \leq \frac{u(\mathcal{S}^p)}{3\delta}\}$$
$$\geq \mathrm{Pro}\{u(\mathcal{V}_p) \leq 2\mathrm{Exp}(u(\mathcal{V}_p))\}$$
$$\geq \frac{1}{2}.$$

This implies that the expected number of updates is at most twice as large as the number of successful updates, i.e., $K \leq 6\delta \ln(N + 6r(C-1)\delta)$, where K denotes the total number of updates. Note that, at the end of each iteration, we have

$$K = Ct.$$

Therefore, the PRSVM algorithm guarantees, on average, within $(6\delta/C)\ln(N + 6r(C-1)\delta)$ steps, that all the support vectors are contained by one of the C working sites. For separable problems, we have $\delta \leq n+1$. For general nonseparable problems, we have δ is bounded by the number of support vectors. □

The bound of average convergence rate $(6\delta/C)\ln(N+6r(C-1)\delta)$ clearly shows the linear scalability if $N >> \delta$. This can be true if the number of support vector is very limited.

5 Simulations and Applications

We analysis our PRSVM by using synthesized data and a real-world geographic information system (GIS) database.

Through out this section, the machine we used has a Pentium IV 2.26G CPU and 512M RAM. The operation system is Windows XP. The SVM[light] [11] version 6.01 was used as the local SVM solver. Parallel computing is virtually simulated in a single machine. Therefore, we ignore any communication overhead.

5.1 Synthesized Demonstration

We demonstrate our RSVM (reduced PRSVM when $C = 1$) training procedure by using a synthesized two-dimensional training data set. This data set consists of 1000 data points: 500 positive and 500 negative. Each class is generated from an independent Gaussian distribution. Random noise is added.

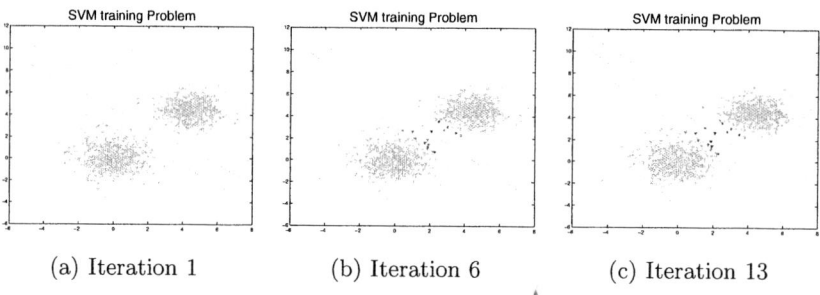

(a) Iteration 1   (b) Iteration 6   (c) Iteration 13

Fig. 1. Weights of training vectors in iterations. Darker points denote higher weights.

We set the sample size r to be 100 and the regularization factor γ to be 0.2. The RSVM converges in 13 iteration. In order to demonstrate the weighting procedure, we choose three iterations (iteration 1, iteration 6 and iteration 13) and plot the weights of the training vectors in Fig. 1. The darker a point appears, the higher weight the training sample has. Fig. 1 shows that how those "important" points stand out and get higher and higher probability to be sampled.

5.2 Application in a Geographic Information System Database

We select covtype, a geographic information system database, from the UCI Repository of machine learning databases as our PRSVM applications [5]. The covtype database consists of 581,012 instances. There are 12 measures but 54 columns of data: 10 quantitative variables, 4 binary wilderness areas and 40 binary soil type variables [4]. There are totally 7 classes. We scale all quantitative variables to [0,1] and keep binary variable unchanged. We select 287831 training vectors and use our PRSVM to classify class 4 against the rest. This is a very suitable database for testing PRSVM since the database has huge number of training data and the number of SVs is limited.

We set the size of working size r to be 60000, the regularization factor γ to be 0.2. We try three cases with $C = 1$, $C = 2$ and $C = 4$ and compare the learning time with the SVM^{light} in Table 1. The results show that our implementation of RSVM and PRSVM achieves comparable result with the reported fastest algorithm SVM^{light}, though they cannot beat SVM^{light} in terms of computing speed for now. However, the lack of a theoretical convergence bound makes SVM^{light} not always preferable.

Table 1. Algorithm performance comparison of SVM^{light}, RSVM and PRSVM

Algorithm	C	Number of Iterations	Learning Time (CPU Seconds)
SVM^{light}	1	-	11.7
RSVM	1	27	47.32
PRSVM	2	10	20.81
	4	7	15.52

We plot the number of violators and support vectors (extremes) in each iterations in Fig. 2 to compare the performance of different number of working sites. The results show the scalability of our method. The numerical results match the theoretical result very well.

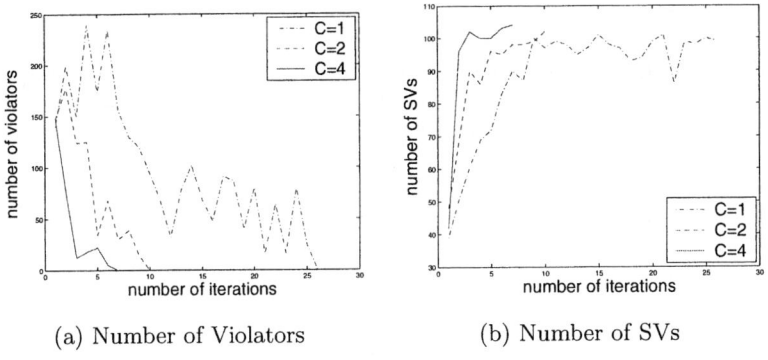

(a) Number of Violators   (b) Number of SVs

Fig. 2. Number of violators and SVs found in each iterations of PRSVM

This figure shows the effect of adding more servers. The system with more servers will find the support vectors much faster than that with less servers.

6 Conclusions

The proposed PRSVM has the following advantages over previous works. It is able to solve general nonseparable SVM training problems. This is achieved by using KKT condition as the criterion of identifying violators and extremes. Second, our algorithm supports multiple working sets that may work parallel. Multiple working sets have more freedom than normal gradient based parallel algorithms since no synchronization and no special solver is required. Our PRSVM also has a provable and fast average convergence bound. Last, our numerical results show that multiple working sets have scalable computing advantage. The provable convergence bound and scalable results make our algorithm more preferable in some applications.

Further research is going to be conducted to accelerate the performance of the PRSVM. Intuitively, the weighting mechanism may be able to be improved so that the initial iterations play a more determinant role.

References

1. Ilan Adler and Ron Shamir. A randomized scheme for speeding up algorithms for linear and convex programming with high constraints-to-variable ratio. *Mathematical Programming*, 1993.
2. Jose Balcazar, Yang Dai, Junichi Tanaka, and Osamu Watanabe. Provably fast training algorithm for support vector machines. *Proceedings of First IEEE International Conference on Data Mining (ICDM01)*, 2001.
3. Jose Balcazar, Yang Dai, and Osamu Watanabe. Provably fast support vector regression using random sampling. *Proceedings of SIAM Workshop on Discrete Mathematics and Data Mining*, April 2001.
4. Jock A. Blackard and Denis J. Dean. Comparative accuracies of artificial neural networks and discriminant analysis in predicting forest cover type from cartographic variables. *Computer and Electronics in Agriculture*, 24, 1999.
5. C.L. Blake and C.J. Merz. UCI repository of machine learning databases, 1998.
6. Kenneth L. Clarkson. Las vegas algorithms for linear and integer programming when the dimension is small. *Proceeding of 29th IEEE Symposium on Foundations of Computer Science (FOCS'88)*, 1988.
7. Ronan Collobert, Samy Bengio, and Yoshua Bengio. A parallel mixture of svms for very large scale problems. In *Neural Information Processing Systems*, pages 633–640, 2001.
8. Tatjana Eitrich and Bruno Lang. Shared memory parallel support vector machine learning. Technical report, ZAM Publications on Parallel Applications, 2005.
9. Bernd Gartner and Emo Welzl. A simple sampling lemma: Analysis and applications in geometric optimization. *Proceeding of the 16th Annual ACM Symposium on Computational Geometry (SCG)*, 2000.
10. Hans Peter Graf, Eric Cosatto, Leon Bottou, Igor Dourdanovic, and Vladimir Vapnik. Parallel support vector machine: The cascade svm. In *Advances in Neural Information Processing Systems*, 2005.
11. Thorsten Joachims. Making large-scale svm learning practical. *Advances in Kernel Methods - Support Vector Learning*, pages 169–184, 1998.
12. Vladimir Vapnik. *The Nature of Statistical Learning Theory*. Springer Verlag, New York, 1995.

ε-Tube Based Pattern Selection for Support Vector Machines

Dongil Kim and Sungzoon Cho*

Department of Industrial Engineering, College of Engineering, Seoul National University,
San 56-1, Shillim Dong, Kwanak-Gu, Seoul 151-744, South Korea
{dikim01, zoon}@snu.ac.kr

Abstract. The training time complexity of Support Vector Regression (SVR) is $O(N^3)$. Hence, it takes long time to train a large dataset. In this paper, we propose a pattern selection method to reduce the training time of SVR. With multiple bootstrap samples, we estimate ε-tube. Probabilities are computed for each pattern to fall inside ε-tube. Those patterns with higher probabilities are selected stochastically. To evaluate the new method, the experiments for 4 datasets have been done. The proposed method resulted in the best performance among all methods, and even its performance was found stable.

1 Introduction

Support Vector Machine (SVM), developed by Vapnik based on the Structural Risk Minimization (SRM) principle [1], has performed with a great generalization accuracy [2]. SVR, a modified version of SVM, was developed to estimate regression functions [3]. Both SVM and SVR are capable of solving non-linear problems.

For a brief review of SVR, consider a regression function $f(x)$ to be estimated with training patterns $\{(x_i, y_i)\}$

$$f(x) = w \cdot x + b \quad \text{with } w, x \in R^N, b \in R \tag{1}$$

$$\text{where } \{(x_1, y_1), \cdots, (x_n, y_n)\} \subset R^N \times R. \tag{2}$$

SVR is moved around to include training patterns inside ε-*insensitive tube* (ε-tube). By the SRM principle, the generalization accuracy is optimized by the flatness of the regression function. Since the flatness is guaranteed on small w, SVR is moved to minimize the norm, $\|w\|^2$. An optimization problem could be formulated with constraints where C, ε, and ξ, ξ^* are trade-off cost between empirical error and the flatness, size of ε-tube and slack variables, respectively, for the following soft margin problem.

$$\text{Minimize } \frac{1}{2}\|w\|^2 + C\sum_{i=1}^{n}(\xi_i + \xi_i^*) \tag{3}$$

---

* Corresponding author.

$$\text{Subject to} \quad y_i - w \cdot x_i - b \leq \varepsilon + \xi_i$$
$$w \cdot x_i + b - y_i \leq \varepsilon + \xi_i^*$$
$$\xi_i, \xi_i^* \geq 0$$

Hence, SVR is trained by minimizing $\|w\|^2$ with including training patterns inside the ε-tube.

It takes $O(N^3)$ to solve the optimization problem of Eq. (3), thus the training time complexity of SVR is also $O(N^3)$. If the number of training patterns increases, the training time increases more radically, i.e. in a cubic proportion.

So far, many algorithms such as Chunking, SMO, SVM[light] and SOR have been proposed to reduce the training time with time complexity $T \cdot O(Nq+q)$ where T is the number of iterations and q is the size of working set. However, their training time complexity is still strongly related to the number of training patterns [4].

Another direction of research efforts focuses on reducing the number of patterns. But reducing the number of training patterns is likely to result in information loss, i.e. the generalization performance of SVM deteriorates. What is desired is to reduce the number of training patterns without accuracy loss. Instead of training SVM with all the patterns, only those with "more information" can be selected and used for training. Such methods include NPPS $(O(N^2))$ [5] and Fast NPPS $(O(vN))$ [6]. However, NPPS approaches were developed for classification problem, not regression problem.

In 2004, a pattern reduction method for regression tasks was proposed, which is called HSVM [7]. The training patterns are split into k groups. Then the similarity is calculated between every pattern with the center pattern of each group. The pattern is selected if similarity (i.e. reverse of their euclidean distances) of the pattern is larger than a pre-fixed threshold. Finally, patterns that are far away from each group's center are rejected from a training pattern set. However, too much accuracy loss occurred.

The k-NN based pattern selection method was also proposed that employed entropy and variability. It has reduced the number of patterns while keeping accuracy more or less same [8].

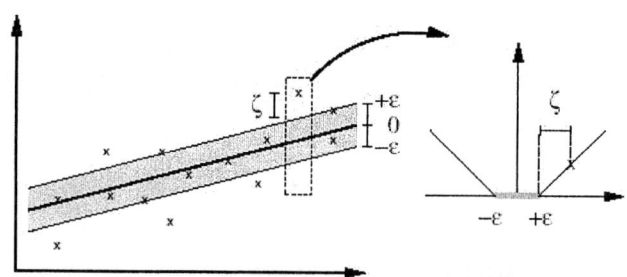

Fig. 1. ε-tube and ε-loss Foundation of SVR [9]

In this paper, we propose an ε-tube based pattern selection method for SVR with a goal of minimum or no loss of accuracy. SVR makes ε-tube on training patterns, and the center-line of ε-tube is estimated on the regression function (see Fig. 1). Thus, by employing those patterns inside ε-tube, which preserve the shape of ε-tube, we can get the same regression function with a significantly smaller number of patterns (see Fig. 2). Of course, before training, we do not know the exact ε-tube. Thus ε-tube is estimated from multiple bootstrap samples.

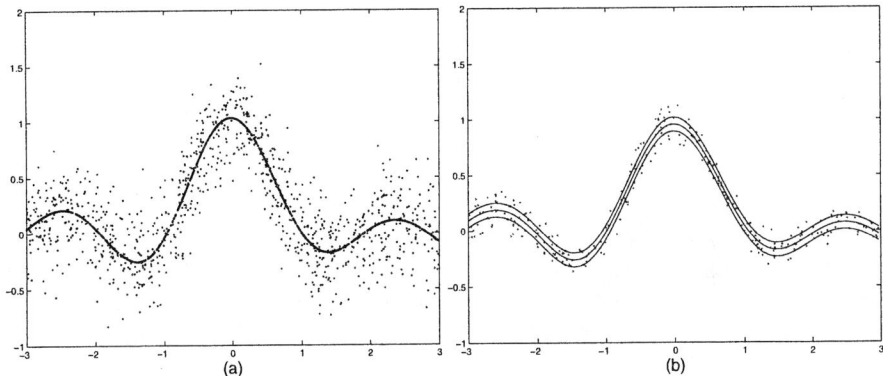

Fig. 2. (a) The regression function after training original pattern set, and (b) The regression function after training ONLY patterns inside estimated ε-tube

Two artificial datasets and two real-world datasets were used for experiments. HSVM and random sampling method were used on benchmark methods. We compared the respective results in terms of the training time and mean squared regression error.

The remaining of this paper is organized as follows. In Section 2, we provide the main idea of the proposed method and state the algorithm. In Section 3, we present details of datasets and parameters for experiment as well as the result. In Section 4, we summarize the result and conclude the paper with a remark on limitations and future research directions.

2 Stochastic Pattern Selection Method

SVR trains patterns based on ε–loss function foundation. SVR makes ε–tube on the training patterns. The patterns in ε–tube are not counted as error, and patterns out of ε–tube, i.e. Support Vectors (SVs), are used for training. In addition, SVR estimates the regression function as the center-line of ε–tube. Hence, if ε–tube can be estimated before training, we can find the regression function with only those patterns inside ε–tube. However, removing all patterns outside ε–tube could lead to reduction of ε–tube itself, thus it is desirable to keep some of "outside" patterns for training. Hence, we

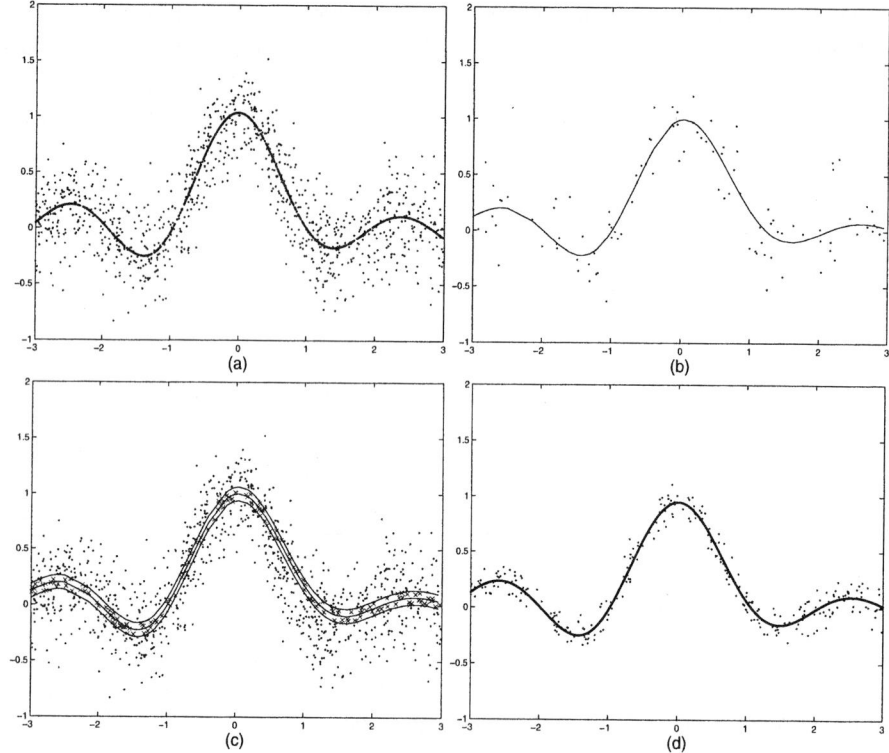

Fig. 3. (a) Original dataset and an SVR trained on it, (b) A bootstrap sample and an SVR trained on it, (c) Original dataset and ε–tube of (b)'s SVR, and (d) Selected patterns and an SVR trained on them

defined a "fitness" probability for each pattern based on its location with respect to ε–tube and then selected patterns stochastically.

We made k bootstrap samples of size l ($l<n$) from original training pattern set (D). We trained an SVR with each bootstrap sample and obtained k SVR regression functions. Each regression function was used to see if a training pattern is located inside ε–tube. Each training pattern in D is located inside a minimum of zero ε–tubes to a maximum of k ε–tubes. Let m_j denote the number of times that pattern j is found inside an ε–tube. We use m_j as the likelihood that pattern j is actually located inside the real ε–tube. Each m_j is converted to a probability, p_j as in Eq. (4). Since we want to select patterns inside ε–tube, pattern j is selected with a probability of p_j (see Fig. 3)

$$p_j = \frac{m_j}{\sum_{i=1}^{n} m_i}. \tag{4}$$

The algorithm is presented in Fig. 4.

1. Initialize the number of bootstrap samples, k
   Initialize the number of patterns in each bootstrap sample, l
   Initialize the number of patterns to be selected, s
2. Make k bootstrap samples, D_i $(i=1...k)$, from
   the original dataset D by random sampling without replacement
3. Train SVR f_i with D_i, $\forall i$
4. Count the number of times m_j that pattern j is found
   inside ε –tube of f_i
5. Convert m_j to p_j according to Eq. (4)
6. Select s patterns stochastically from D without replacement based
   on p_j
7. Train final SVR with s selected patterns

Fig. 4. ε–tube based pattern selection algorithm

3 Experiment Results

We used two artificial datasets and two real-world datasets to show the performance of the proposed method. Artificial dataset 1 (Fig. 5 (a)) was sampled from a math function given in Eq. (5) used in [10]. The input variable x was drawn uniformly from the interval [-3, 3] and additive noise ξ was sampled from $N(0, 0.5^2)$. The pattern set consists of 1,000 training patterns and 1,000 testing patterns.

$$y = (\sin \pi x) / \pi x + \xi \tag{5}$$

A more realistic artificial dataset 2 (Fig. 5 (b)) was sampled from a math function given in Eq. (6). The input variable x was drawn from the interval [0, 10] under

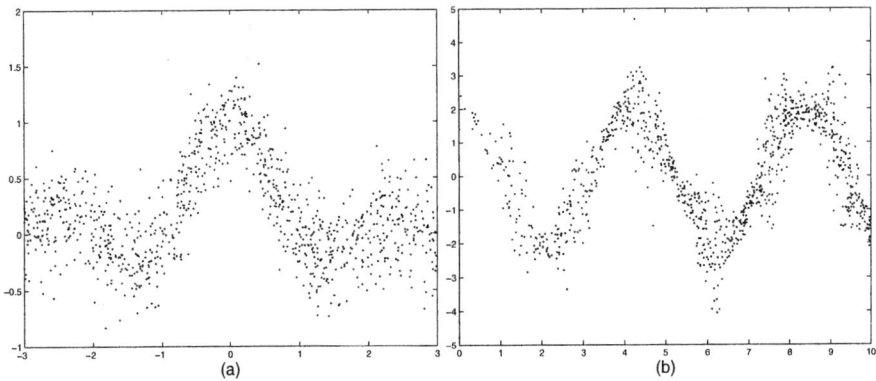

Fig. 5. (a) Artificial dataset1, and (b) Artificial dataset2

Beta(1.5, 1). A additive noise ξ_1 was sampled from $N(0,0.5^2)$, and the other additive noise ξ_2 was sampled from $N(0,sin2(x+1)^2)$. In short, the variance of noise varied with *x*. Artificial dataset 2 consists of 1,000 training patterns and 1,000 testing patterns.

$$y = 2\cos(15x) + (\xi_1 + \xi_2) \tag{6}$$

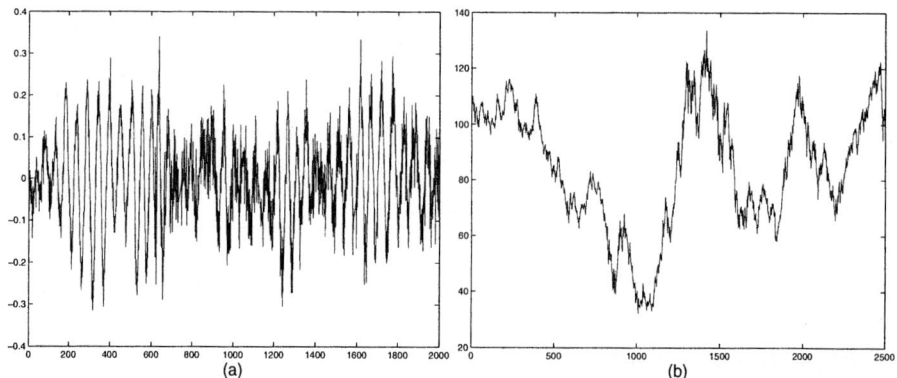

Fig. 6. (a) The SantaFe E dataset, and (b) The KOSPI200 dataset

One of real-world datasets came from Santa Fe competition (Fig. 6 (a)) [11]. We used 1,500 patterns for training and 500 patterns for testing. We used another real-world dataset, KOSPI200 dataset (Fig. 6 (b)) [12]. KOSPI200 is a weighted average of 200 stock prices of Korean stock market. We gathered 2,500 daily patterns between 1995~2004. The first 2,000 patterns were used for training, while the last 500 patterns were used for testing. Both real-world datasets are time series datasets. Hence, we reformulated the problem as a regression problem by using 10 previous values to estimate the following one value, which is a typical way to solve time series problems. Real-world datasets were normalized.

We set hyper-parameters of SVR, the trade-off cost *C* and the size of ε–tube ε based on [13]. The functions of the parameter setting are given in Eq. (7) and Eq. (8). RBF kernel was used as a kernel function and kernel parameter σ was fixed to 1.0 for all experiments.

$$c = \max[(\overline{y} + 3\sigma_y), (\overline{y} - 3\sigma_y)] \tag{7}$$

$$\varepsilon = 3\sigma_{noise}\sqrt{\frac{\ln(n)}{n}} \tag{8}$$

The parameters of the proposed method were set as follows. The number of bootstrap samples *k* was set to two values : 10 and 100. The other parameter that controls the number of patterns in a bootstrap sample *l* was set to 10% of the number of patterns in dataset, *n*. The number of selected patterns *s* was set to 10%, 30%, 50%, and 70% of *n*.

HSVM and random sampling were also implemented to be compared with the proposed method. HSVM has a threshold parameter to set. So, we tried various threshold values and found the appropriate thresholds that resulted in a similar number of patterns as s mentioned above for comparison. We tried to keep the same interval of thresholds for each experiment. Mean Squared Error (MSE) was used as a measure of accuracy. Each setting was repeated 10 times and the result is an average of these.

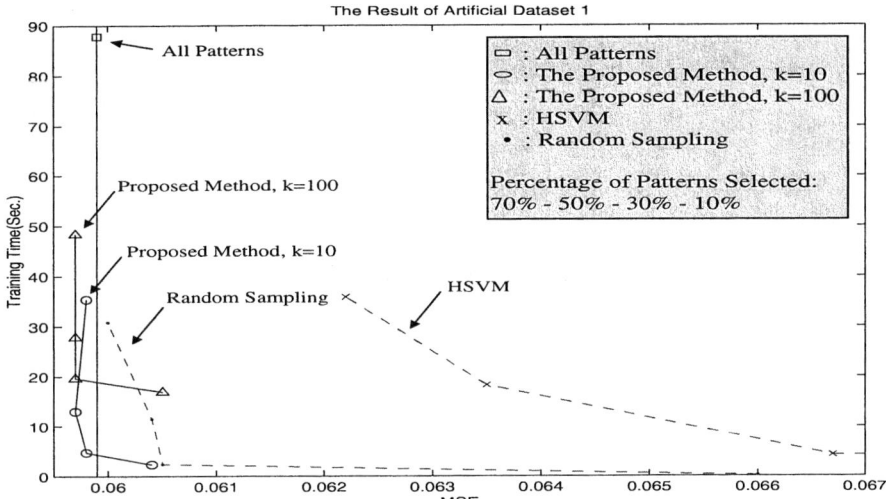

Fig. 7. The Result of Artificial Dataset 1 ($c=1.5058$, $\varepsilon=0.1247$)

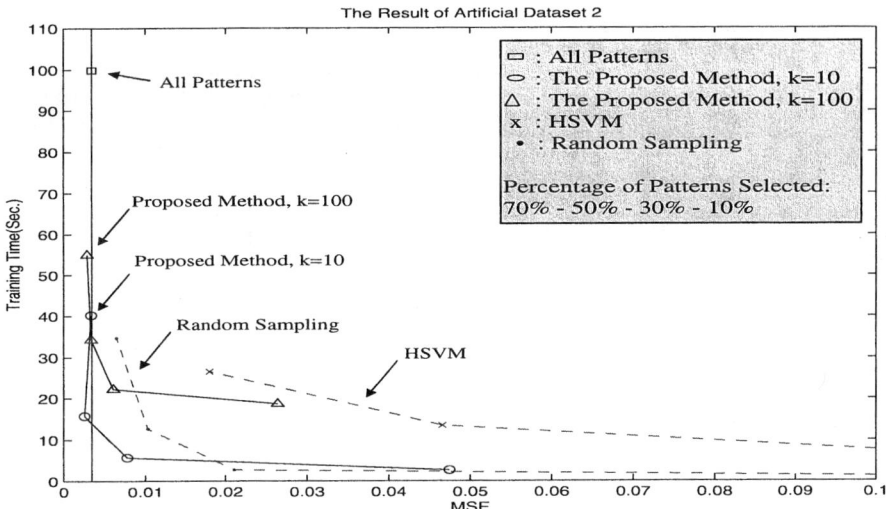

Fig. 8. The Result of Artificial Dataset 2 ($c=4.6723$, $\varepsilon=0.1643$)

Fig. 7 shows the experimental result of artificial dataset 1. The MSE and training time in seconds pairs are plotted that correspond to 70, 50, 30 and 10 percents of patterns selected, respectively. Results from different methods are shown by different shapes of dots. The proposed method was more accurate than HSVM and random sampling given a same amount of training time. The SVRs trained with as low as 30% of the patterns selected by the proposed method resulted in a smaller MSE than SVR

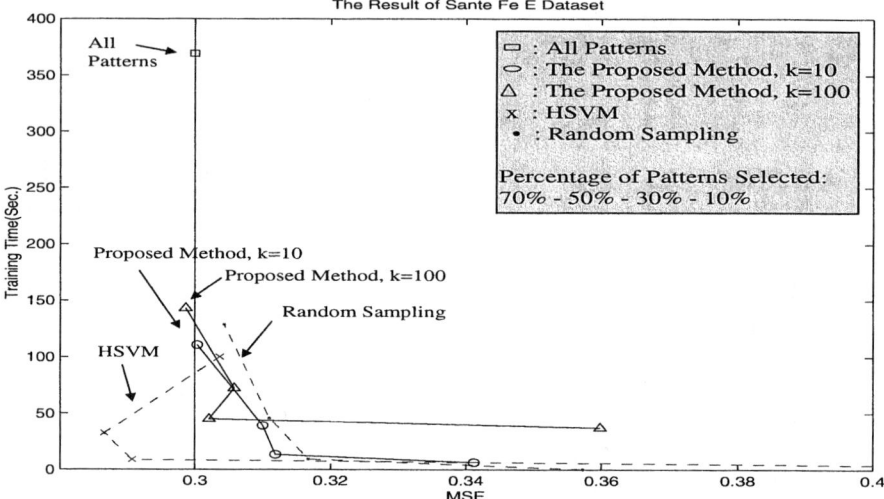

Fig. 9. The Result of Santa Fe E Dataset ($c=2.6553$, $\varepsilon=0.2974$)

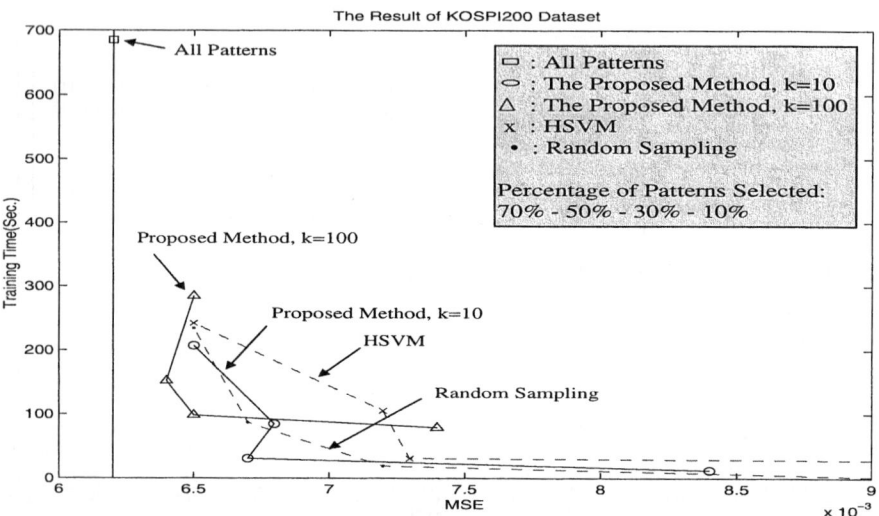

Fig. 10. The Result of KOSPI200 Dataset ($c=3.3112$, $\varepsilon=0.0631$)

trained with 100% of patterns. Many of noisy patterns seemed to be removed. Fig. 8 shows the experimental result of artificial dataset 2. The proposed method clearly outperformed HSVM and random sampling.

The experimental result of Santa Fe E dataset is shown in Fig. 9. The SVRs trained by the proposed sampling method did better than those by random sampling for all percentages and than HSVM for 70% and 10%. The SVR trained by HSVM did very well for some percentages but not for others. Its results seem rather unstable. The experimental result of KOSPI200 dataset is shown in Fig. 10. Similar results were obtained.

4 Conclusion

This paper provides a new pattern selection method to reduce training time of SVR. We selected a subset of patterns that are important for training and accomplished reducing the training time without accuracy loss. Two artificial datasets and two real-world datasets, Santa Fe E dataset and KOSPI200 dataset were employed for comparison. The results showed that the generalization performance of the proposed method was better than HSVM and random sampling. In addition, the proposed method was found quite stable.

There are some limitations of the current work. First, there is no guideline in determining parameters k and l. "Reasonable" numbers were set in the experiments. But, some guideline should be obtained from more experiments. Second, SVR's hyper-parameters could be set differently after pattern selection. A better generalization performance could have been obtained. Finally, a more extensive experiment involving large scale datasets is due. Pattern selection is most useful when a huge dataset is available which simply prevents powerful SVR from being used due to time complexity.

References

1. Vapnik, V., The Nature of Statistical Learning Theory, Springer, New York (1995)
2. Cristianini, N., Shawe-Taylor, J., An Introduction to Support Vector Machines, Cambridge University Press, Cambridge, UK (2000)
3. Drucker, H., Burges, C. J. C., Kaufman, L., Smola, A., Vapnik, V., Support Vector Regression Machines, In: Mozer, M. C., Jordan, M. I., Petsche, T.(eds.): Advances in Neural Information Processing System 9, MIT Press, Cambridge, MA (1997) 155-161
4. Platt, J. C., Fast Training of Support Vector Machines Using Sequential Minimal Optimization, Advanced in Kernel Methods; Support Vector Machines, MIT Press, Cambridge, MA (1999) 185-208
5. Shin, H., Cho, S., Pattern Selection for Support Vector Classifiers, Lecture Notes in Computer Science 2412 (2002) 469-474
6. Shin, H., Cho, S., Fast Pattern Selection Algorithm for Support Vector Classifiers: Time Complexity Analysis, Lecture Notes in Computer Science 2690 (2003) 1008-1015
7. Wang, W., Xu, Zongben., A Heuristic Training for Support Vector Regression, Neurocomputing 61 (2004) 259-275

8. Sun, J., Cho, S., Pattern Selection for Support Vector Regression based on Sparseness and Variability, Submitted (2005)
9. Smola, A., Schölkopf, B., A Tutorial on Support Vector Regression, NeuroCOLT Technical Report NC-TR-98-030, Royal Holloway College, University of London, UK (2002)
10. Chalimourda, A., Schölkopf, B., Smola, A., Experimentally Optimal ν in Support Vector Regression for Different Noise Models and Parameter Settings, Neural Networks 17 (2004) 127-141
11. Santa Fe Dataset : http://www-psych.stanford.edu/~andreas/Time-Series/SantaFe.html
12. KOSPI200 Dataset from Korea Stock Market : http://sm.krx.co.kr
13. Cherkassky, V., Ma, Y., Practical Selection of SVM Parameters and Noise Estimation for SVM Regression, Neural Networks 17 (2004) 113-126

Self-adaptive Two-Phase Support Vector Clustering for Multi-Relational Data Mining

Ping Ling[1,2], Yan Wang[1], and Chun-Guang Zhou [1,*]

[1] College of Computer Science, Jilin University, Key Laboratory of Symbol Computation and Knowledge Engineering of the Ministry of Education, Changchun 130012, China
cgzhou@jlu.edu.cn
[2] School of Computer Science, Xuzhou Normal University, Xuzhou, 221116, China
lingicehan@yahoo.com.cn

Abstract. This paper proposes a novel Self-Adaptive Two-Phase Support Vector Clustering algorithm (STPSVC) to cluster multi-relational data. The algorithm produces an appreciate description of cluster contours and then extracts cluster centers information by iteratively performing classification procedure. An adaptive Kernel function is designed to find a desired width parameter for diverse dispersions. Experimental results indicate that the designed Kernel can capture multi-relational features well and STPSVC is of fine performance.

1 Introduction

Multi-Relational Data Mining (MRDM) [1] looks for patterns in Multi-Relational (MR) environment, namely multi connected table scenario. It requires mining from multi tables directly. Kernel function is often used in MRDM due to its fine quality of defining a nonlinear map from original space to feature space. For example, Support Vector Clustering (SVC) [2] uses non-bounded Support Vector (nbSV) to describe cluster contours. Support Vector Machine (SVM) [3] finds the optimal decision interface for highly structured data.

This paper presents an algorithm STPSVC that is equipped with a multi-relational version Kernel to solve MRDM clustering task. STPSVC produces cluster contours firstly. Then a classification procedure is executed to find cluster centers information. The final cluster assignment is determined according to the affinity between data and cluster centers without suffering the expensive operations used in traditional SVC.

2 STPSVC Algorithm

STPSVC performs SVC procedure firstly to produce nbSVs and bSVs that are used to describe cluster contour bands. Then an iterative classification procedure to separate set of SVs from set of inner-clusters points is appended so as to extract cluster center information. The final cluster assignment is finished by computing similarity between point and each cluster center. Algorithm pseudocodes are as following:

---

* Corresponding author.

1) **SVC** produces $\{nbSV\}$ and $\{bSV\}$;
2) $A = \{data\} - \{nbSV\} - \{bSV\}$; $\quad\quad B = \{nbSV\} + \{bSV\}$;
3) While (*iteration _ condition*;
4) **SVM** (*A*, *B*);
5) $A = A + B - new\{nbSV\} - new\{bSV\}$; $\quad B = new\{nbSV\} + new\{bSV\}$;
6) End
7) Aggregate centers according to $Affinity(Cen\_i, Cen\_j)$;
8) Clusters' assignment according to $Affinity(x, Cen\_i)$.

Line 1 is Phase 1. It executes SVC process. Both nbSVs and bSVs are united together to give a broad description for decision bands. Phase 2 is from Line 2 to Line 6. It iteratively performs SVM classification between the set of points located inside clusters, *A*, and the set of SVs, *B*. Update *A* as the new cluster contents, and *B* as the new decision bands. This makes dividing interfaces generated in each run move closer and closer to central zones. We give the heuristic of the upper limit of runs of SVM as $i \leq \frac{\sqrt{MinSize}}{coef}$, where *MinSize* is the size estimate of the minimum cluster. *MinSize* is evaluated by following steps: a) Sort rows of Kernel matrix in a descending order. b) Find $gap(i) = max_j\{k(i,j)-k(i,j-1)\}$. c) $MinSize = \min\{gap(i)\}$. Clearly, $gap(i)$ is the natural size estimate of the inherent cluster that contains point x_i, and *MinSize* is the rough approximation of the smallest cluster size. Parameter *coef* indicates the width of decisive bands. It is set as 2 in this paper. For two cluster centers: $Cen\_i = \{si_1, si_2, \ldots si_n\}$ and $Cen\_i = \{sj_1, sj_2, \ldots sj_m\}$, where si_t and sj_t, are SVs, their affinity is: $Affinity(Cen\_i, Cen\_j) = \sum_{u=1}^{n}\sum_{v=1}^{m}\frac{k(si_u, sj_v)}{m \cdot n}$. Cluster assignment is decided according to: $Affinity(x, Cen\_i) = \sum_{u=1}^{n}\frac{k(x, si_u)}{n}$.

3 Kernel Definition

In MR environment a **main table** is mainly investigated. Its relational information is implicated by **association keys (AK)**. AK includes **foreign key (FK)** and **referenced key (RK)**. We think each key stands one expanding direction of data description, and affinity produced by each key can be considered as a local affinity. All local affinities are collected in a desirable way that:

$$K(R_{i\_}, R_{j\_}) = k(R_{ci\_}, R_{cj\_}) \cdot \frac{k_{FK}(R_{i\_}, R_{j\_}) + k_{RK}(R_{i\_}, R_{j\_})}{|FK| + |RK|}. \tag{1}$$

In (1), $k_{FK}(R_{i\_}, R_{j\_}) = \sum_{u=1}^{|FK|} k(R_{ui\_}^+, R_{uj\_}^+)$. $k_{RK}(R_{i\_}, R_{j\_}) = \sum_{v=1}^{|RK|} k(R_{vi\_}^+, R_{vj\_}^+)$. (2)

In (2), where $\quad k(R_{vi\_}^+, R_{vj\_}^+) = \prod_{m=1}^{|H_v|} k_{set}(R_{vi\_m}^+, R_{vj\_m}^+)$. (3)

In (3), where
$$k_{set}(S_1, S_2) = \frac{\sum_{x \in S_1, y \in S_2} k(x, y)}{|S_1 \times S_2|}. \quad (4)$$

The global Kernel affinity is expressed by the product of affinity coming from Main table and the affinities from expanding information descriptions. More in details, for the map from *FK* to *RK*, an object corresponds to a single expanding instance. To measure local affinity between two objects in this expanding direction, their respective expanding instances can be introduced into the elementary Kernel directly. As to the map from *RK* to *FK*, there might be multi expanding tables because a *RK* might correspond to multi *FKs* of tables. If we denote by H_v the set of tables that a *RK* v corresponds to, then the affinity produced by v is the product of local affinities generated by each table in H_v. This point is shown in (3), where $k_{set}(R^+_{vi\_m}, R^+_{vj\_m})$ is to compute the local affinity from table m. Moreover, under the map from a *RK* to a *FK*, one object could correspond to multi instances. So the local affinity of two objects reduces to affinity of their expanding instances sets, which is shown in (4). Gaussian Kernel serves as the elementary Kernel. According to the theorem of Kernel construction, formula (1) is of positive semi-definite property.

Here, for each expanding table, we find q that satisfies the following constrain:

$$\max(\exp(-q \| x_p - x_{in} \|^2) - \exp(-q \| x_{out} - x_p \|^2)). \quad (5)$$

In objection function (5), x_p is found according to below steps: a) Sort rows of distance matrix D in an ascending order. b) Find the maximum gap between adjacent entries: $D(i,j)$ and $D(i,j+1)$ for each row, and denote the column index j as $gap(i)=\max_j\{D(i,j+1)-D(i,j)\}$. c) Let $p=\min_i\{gap(i)\}$, then x_p is the point that produces max gap feature. Find x_{out} that produces $D(p,gap(p))$ with x_p, namely $\| x_p - x_{out} \| = D(p,gap(p))$. x_{out} is considered as the nearest point outside its group. Find x_{in} that fulfills $\| x_p - x_{in} \| = D(p,gap(p)-1)$. x_{in} is considered as the furthest point within its same group. Formula (5) tries to find the top gap between inner-cluster affinity and inter-cluster affinity. If this gap arrives at maximum, there would produce high intra-cluster affinity and low inter-cluster affinity, and cluster contours can be revealed more apparently. And width parameter q under this setting is expected to be suitable.

4 Experimental Results

Firstly, STPSVC is applied on some real datasets: IRIS [4], WINE [4], and Breast Cancer (BC) [4]. The performance of STPSVC is compared in Table 1 with some other clustering algorithms: classical K-means; traditional SVC, Girolami method [5] and NJW [6]. For each algorithm, the minimum number of incorrectly clustered points is documented. Note that the errors of SVC in WINE and BC datasets are offered by us under the same experiment conditions. There, pure TPSVC just performs the two phases without the tuning approach. Its scale q in Gaussian Kernel is set by searching in some space. (With 2G P4 CPU PC, 256M memory, WinXP, MatLab7.0)

For two TPSVC algorithms, TPSVC is a little better than STPSVC. And STPSVC shows finer quality in IRIS and BC dataset. But as far as WINE dataset is concerned.

Table 1. Empirical error comparison on real datasets (N is dataset size. D is the dimensionality. To STPSVC and TPSVC, we set C_{svc}= 0.8, C_{svm}= 0.8, and 2 runs of SVM.)

Dataset	Properties	K-means	Girolami	SVC	NJW	STPSVC	TPSVC
IRIS	N = 150 D = 4	16	8	14	14	6 (q=0.2603)	4 (q=0.512)
WINE	N = 178 D = 13	5	3	12	3	9 (q=0.0573)	5 (q=0.317)
BC	N = 638 D = 9	26	20	32	22	24 (q=0.01)	21 (q=0.405)

Table 2. Classification Accuracy Comparison on MUSK1 (%)

Dataset	TILDE	SVM-MM	SVM-MI	SVM
MUSK1	87.00	91.60	86.40	92.3 (C_{svm} = 2.3)

Table 3. Clustering Accuracy on MUSK1 (%) (C_{svc}= 0.8, C_{svm}= 0.8, and 2 runs of SVM)

STPSVC	TPSVC
90.87 (q = 4.5455)	93.12 (q = 5.146)

Table 4. Clustering on Student database(C_{svc} = 0.8, C_{svm} = 5.8 and 3 runs SVM)

	Settings	Number of clusters	Number of SVs	Cluster size
STPSVC	q = 0.425	4	6,12,11,8	36,130,189,32
TPSVC	q = 0.6311	4	8,13,11,7	43,140,176,28

The fact that 178 points cover 13 dimensions leads to the weak connection information in neighboring context. So of STPSVC has a higher error than others.

Now STPSVC is applied on a relational problem: MUSK [7]. We use MUSK1 version. We fix the data and develop a two-level relation frame for it. To test quality of the designed Kernel, we employ it into classical SVM and compare it with other classifiers: TILDE [8], SVM-MM and SVM-MI [9]. Their accuracy ratios with ten fold cross-validation are in Table 2. Our SVM procedure achieves better result, which shows the designed Kernel can grasp relational features effectively. Then, STPSVC and TPSVC are performed, with comparison in Table 3. It is easy to see the tuning approach is competitive with the searching one but with ease of parameterization.

Finally, STPSVC is conducted on the document data of 387 students coming from some grade of Software College, Jilin University. This database contains six tables, Student, Rank, Classtype, Agegroup, Work and Activities, where Student Main table. The relationship among tables is shown in Fig 1. Table 4 gives clustering results produced by STPSVC and TPSVC. To examine the effect of algorithm, in Fig 2, the statistics information of Main table, score data, is demonstrated after being processed by a weighted averaging method. Based on it, all students can be divided into four groups, which can be referred as 4 groups. And this intuitive analysis coincides with the results of STPSVC by and large, which forms 4 clusters. But when we investigate the content of corresponding clusters and group, we find that their content details differ and STPSVC provides result that is more agreed with the true comments.

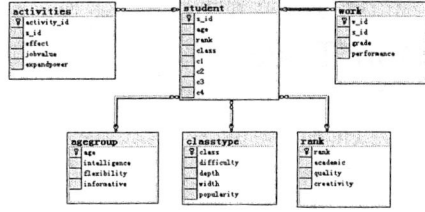 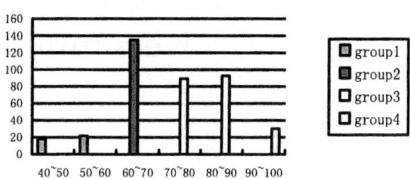

Fig. 1. Relation Schema of Student Database   **Fig. 2.** Statistics on Weighted Score

5 Conclusion and Prospect

A novel STPSVC algorithm is presented in this paper. It obtains contour descriptions of clusters in Phase 1, and then performs SVM iterations between set of SVs and set of points inside clusters to find SVs that are located closer to cluster core zones. Future working directions is to utilize Kernel further to design suitable expression of relational schema and to develop potent algorithms.

Acknowledgement

This work is supported by the National Natural Science Foundation of China under Grant No. 60433020; 985 Project: Technological Creation Support of Computation and Software Science; and the Key Laboratory for Symbol Computation and Knowledge Engineering of the National Education Ministry of China.

References

1. D˘zeroski, S.: Multi-Relational Data Mining: An Introduction. ACM SIGKDD Explorations Newsletter Vol. 5, Issue 1, (2003)
2. Tax,R., D.M.J., Duin, P.W.: Data Domain Description using Support Vectors. Proceedings of European Symposium on Artificial Neural Networks, Bruges, Belgium, (1999) 251-256
3. Cristianini, N, Shawe-Taylor, J.: An introduction to Support Vector Machines. Cambridge University Press, London (2000)
4. http://www.ics.uci.edu/~mlearn/MLSummary.html
5. Girolami, M.: Mercer Kernel-Based Clustering in Feature Space. IEEE Trans. on Neural Networks, Vol. 13(3) (2002) 780-784
6. Ng, A., Jordan, M., Weiss, Y.: On Spectral Clustering: Analysis and An Algorithm. Advances in Neural Information Processing Systems, Cambridge, MA: MIT Press (2002)
7. Dietterich, T.G., Lathrop, R.H., Lozano-Perez, T.: Solving the Multiple Instance Problem with Axis-Parallel rectangles. Artificial Intelligence, Vol. 89(1-2) (1997) 31-71
8. Bloedorn, E., Michalski, R.: Data Driven Constructive Induction. IEEE Intelligent Systems, Vol. 13(2) (1998) 30-37
9. Gaertner, T., Flach, P., Kowalczyk, A., Smola, A.: Multi-instance Kernels. Proceedings of the 19th International Conference on Machine, (2002) 179-186

One-Class Support Vector Machines for Recommendation Tasks

Yasutoshi Yajima

Tokyo Institute of Technology,
Department of Industrial Engineering and Management,
Ookayama, Meguro-ku, Tokyo 152-8552, Japan
yasutosi@me.titech.ac.jp

Abstract. The present paper proposes new approaches for recommendation tasks based on one-class support vector machines (1-SVMs) with graph kernels generated from a Laplacian matrix. We introduce new formulations for the 1-SVM that can manipulate graph kernels quite efficiently. We demonstrate that the proposed formulations fully utilize the sparse structure of the Laplacian matrix, which enables the proposed approaches to be applied to recommendation tasks having a large number of customers and products in practical computational times. Results of various numerical experiments demonstrating the high performance of the proposed approaches are presented.

1 Introduction

Recently, the importance of recommender systems has increased rapidly with the growing availability of online information on the Web. Customers visiting the largest e-commerce sites often have difficulty in finding a particular item among the enormous number of products for sale. Many recommender systems [5, 8] have been installed to filter out irrelevant products and locate products that might be of interest to individual customers.

Collaborative filtering is one of the most successful technologies for recommendation tasks, in which customer ratings on products or historical records of purchased products are exploited to extract the preferences of individuals. Collaborative filtering calculates similarities between customers based on the customer rating, or the purchased products patterns of each individual. Collaborative filtering then finds a set of the most similar patterns, and recommends products for a particular individual. In the present paper, we provide new approaches for recommendation tasks using *kernels* defined on a graph that represents the relationships between the products.

Very recently, Fouss et al. [3] introduced a graph kernel, referred to as the commute time kernel and directly applied the kernel-based dissimilarities to the recommendation task. More precisely, they defined the kernel over a bipartite graph with two sets of nodes corresponding to a set of customers and products. They placed edges between the customer nodes and the product nodes when the customer has purchased the product. They defined a random walk model over this graph by assigning the transition probabilities over the edges. They showed that the average commute time between the two nodes is given by the kernel and that it can be used as a distance measure between the corresponding customer and product.

In the present paper, we use the 1-SVM with graph-based kernels to select relevant products for each customer. We introduce new formulations for the 1-SVM that can efficiently manipulate several recently developed graph kernels, such as [11, 10, 1, 4]. In addition, we show that a special case of our formulation does not require any optimization calculations. More importantly, the new kernel matrix is significantly smaller than that of the method reported in [3], which enables us to apply the present approach to large e-commerce sites with a practical amount of computation.

In Sect. 2, we briefly review the standard formulation of the 1-SVM and its basic settings for recommendation tasks. In Sect. 3, we describe various graph kernels, and in Sect. 4, we introduce new formulations for the 1-SVM. Experiments using a movie dataset are presented in Sect. 5, and conclusions are presented in Sect. 6.

2 1-SVM for Recommendation

The SVM was originally designed as a method for two-class classification problems. In this section, we will describe a variant of the SVM, called the one-class SVM (1-SVM) [7], which can handle problems that consider a single class of data points.

Suppose that we have a set of N-dimensional data points $x_j \in \mathbf{R}^N$ ($j = 1, 2, \ldots, l$). Also, assume that we have a function $\phi(\cdot) : \mathbf{R}^N \mapsto \mathcal{F}$ that maps the data points into a higher-dimensional *feature space*, denoted by $\mathcal{F}$. Hereinafter, for simplicity, we denote the mapped image $\phi(x_j)$ as ϕ_j. Let $w \in \mathcal{F}$ and $\rho \in \mathbf{R}$. Also, the inner product in $\mathcal{F}$ is denoted as $\langle \cdot, \cdot \rangle$. The purpose of the 1-SVM is to calculate a hyperplane that holds most of the data points in its positive side, i.e., $\langle w, \phi_j \rangle - \rho > 0$.

Introducing additional variables $\boldsymbol{\xi} = (\xi_1, \xi_2, \ldots, \xi_l)^T$, w and ρ are obtained by solving the following quadratic programming problem:

$$\left| \begin{array}{l} \text{Min.} \ \frac{1}{2} \langle w, w \rangle + \frac{1}{\nu l} \sum_{j=1}^{l} \xi_j - \rho \\ \text{s.t.} \ \ \langle w, \phi_j \rangle + \xi_j \geq \rho, \ \ \xi_j \geq 0, \ j = 1, \ldots, l, \end{array} \right. \quad (1)$$

where $\nu \in (0, 1]$ is a predefined positive parameter. Let (w^*, ρ^*) denote an optimal solution of the problem (1). When a data point, the mapped image of which is denoted by ϕ, belongs to the negative side of the hyperplane, i.e., $\langle w^*, \phi \rangle + \rho^* < 0$, the pattern can be considered to be different from the given single class of data points.

The objective of the recommendation task is to find products that have not yet been purchased but that would likely be purchased by a specific customer, hereinafter referred to as an *active* customer. Suppose that we are given a set of products $P = \{1, 2, \ldots, M\}$ and that, for each product $j \in P$, the associated feature vectors $\phi_j \in \mathcal{F}$ are obtained. In addition, let $P(a) \subseteq P$ be a subset of indices that are rated as preferable products, or that have actually been purchased by the active customer a. For simplicity, let us assume that $P(a)$ consists of l products and is denoted as $P(a) = \{1, 2, \ldots, l\}$, which is treated as a set of the single class of data points in the problem (1). Let $(\boldsymbol{w}^*, \rho^*)$ denote an optimal solution of (1). Then, for each product i that has not been purchased, i.e., $i \in P \setminus P(a)$, the distance from the hyperplane calculated as $(\langle \boldsymbol{w}^*, \phi_i \rangle + \rho^*)/\langle \boldsymbol{w}^*, \boldsymbol{w}^* \rangle$ can be used as a preference score of the product i. Ignoring the constants, one can use the inner product $\langle \boldsymbol{w}^*, \phi_i \rangle$ as a *score* to rank the product i for the specific active customer a.

Generating a nonlinear map $\phi(\cdot)$ is quite important in SVM. Usually, this is done implicitly by *kernels* that are naturally introduced by the following dual formulation of the problem (1).

$$\begin{vmatrix} \text{Max.} & -\frac{1}{2} \sum_{i=1}^{l} \sum_{j=1}^{l} \langle \phi_i, \phi_j \rangle \alpha_i \alpha_j \\ \text{s.t.} & \sum_{j=1}^{l} \alpha_j = 1, \quad 0 \leq \alpha_j \leq \frac{1}{\nu l}, \; j = 1, 2, \ldots, l, \end{vmatrix} \quad (2)$$

where $\alpha_1, \alpha_2, \cdots, \alpha_l$ are dual variables. Note that the dual formulation can be defined using only the values of the inner products, without knowing the mapped image ϕ_i, explicitly. In addition, let $(\alpha_1^*, \alpha_2^*, \cdots, \alpha_l^*)$ be the solution to the dual problem. Then, the associated optimal primal solution is given as $\boldsymbol{w}^* = \sum_{j=1}^{l} \alpha_j^* \phi_j$, which immediately implies that the score of the product i is given by $\langle \boldsymbol{w}^*, \phi_i \rangle = \sum_{j=1}^{l} \alpha_j^* \langle \phi_i, \phi_j \rangle$.

Let $K = \{K_{ij}\}$ be a symmetric matrix called a kernel matrix, which consists of the inner products $\langle \phi_i, \phi_j \rangle$ as the $i - j$ element. Any positive semidefinite matrices K can be used as kernel matrices. It has been shown that positive semidefiniteness ensures the existence of the mapped points, ϕ_is (see, for example, [9]).

3 Laplacian of a Graph and Associated Kernel

Recently, several studies [11, 10, 1, 4] have reported the development of kernels using weighted graphs. In this section, we will review such kernels.

First, let us introduce a weighted graph $G(V, E)$ having a set of nodes V and a set of undirected edges E. The set of nodes V corresponds to a set of data items such as products in a recommendation task. For each edge $(i, j) \in E$, a positive weight $b_{ij} > 0$ representing the similarity between the two nodes $i, j \in V$ is assigned. We assume that the larger the weight b_{ij}, the greater the similarity between the two nodes. Let M be the number of nodes in V, and let B be an $M \times M$ symmetric matrix with elements b_{ij} for $(i, j) \in E$. Note that if there exists no edge between i and j, then we set $b_{ij} = 0$.

Next, let us introduce the Laplacian matrix L of the graph $G(V, E)$ as $L = D - B$, where D is a diagonal matrix, the diagonal elements d_{ii} of which are the sum of the ith row of B, i.e., $d_{ii} = \sum_j b_{ij}$. Throughout this paper, we assume that the graph $G(V, E)$ is connected.

There are several methods for generating kernel matrices based on L. Fouss et al. [3] considered a random walk model on the graph G, in which, for each edge (i,j), the transition probability p_{ij} is defined as $p_{ij} = b_{ij}/\sum_{k=1}^{M} b_{ik}$. They considered the average commute time $n(i,j)$, which represents the average number of steps that a random walker, starting from node i, will take to enter node j for the first time and then return to node i. They indicated that the average commute time $n(i,j)$ can be used as a dissimilarity measure between any two data points corresponding to the nodes of the graph, and that $n(i,j)$ is given as $n(i,j) = V_G \left(l_{ii}^+ + l_{jj}^+ - 2l_{ij}^+ \right)$, where $V_G = \sum_{i,j} b_{ij}$ and l_{ij}^+ is the $i-j$ element of the Moore-Penrose pseudoinverse of L, which is denoted by L^+. Fouss et al. [3] also showed that as long as the graph is connected, the pseudoinverse L^+ is explicitly given as follows:

$$L^+ = \left(L - ee^T/M \right)^{-1} + ee^T/M, \tag{3}$$

where e is a vector of all ones. Since L is positive semidefinite [2], so is its pseudoinverse L^+, which implies that L^+ can act as a kernel matrix [3].

Here, L and L^+ share the common eigenvectors. Let $v_1, v_2, \ldots, v_M$ and $\lambda_1, \lambda_2, \ldots, \lambda_M$ be the eigenvectors and the corresponding eigenvalues of L, respectively. It is well-known that L is decomposed into $L = \sum_{i=1}^{M} \lambda_i (v_i^T v_i)$, and that the pseudoinverse is also given as

$$L^+ = \sum_{i=1}^{M} \lambda_i^+ (v_i^T v_i), \quad \text{where } \lambda^+ = \begin{cases} \lambda^{-1} & \text{if } \lambda \neq 0 \\ 0 & \text{if } \lambda = 0. \end{cases} \tag{4}$$

Several variants of the above equation have been proposed. Smola & Kondor [10] introduced the following regularized Laplacian kernel matrix

$$K_1 = \sum_{i=1}^{M} (1 + t\lambda_i)^{-1} v_i^T v_i = \sum_{k=0}^{\infty} t^k (-L)^k = (I + tL)^{-1}. \tag{5}$$

Moreover, by introducing the modified Laplacian $L_\gamma = \gamma D - B$ with a parameter $0 \leq \gamma \leq 1$, Ito et al. [4] defined the modified Laplacian regularized kernel matrix as

$$K_2 = (I + tL_\gamma)^{-1}. \tag{6}$$

In particular, when $\gamma = 0$ this kernel matrix is the von Neumann diffusion kernel, which is defined as

$$K_3 = \sum_{k=0}^{\infty} t^k B^k = (I - tB)^{-1}. \tag{7}$$

4 Learning 1-SVMs with Graph Kernels

Next, we will describe recommendation methods based on the 1-SVM using the kernel matrices K described in the previous section. Recall that we are given a set of M products $P = \{1, 2, \ldots, M\}$ and a subset $P(a) \subseteq P$, which have been purchased by the active customer a. We assume that $P(a) = \{1, 2, \ldots, l\}$. In addition, the elements of the kernel matrix K represent the inner products of the feature vectors corresponding to the products.

Let us first rewrite the primal formulation. To this end, introducing M variables $\boldsymbol{\alpha} = (\alpha_1, \cdots, \alpha_M)^T$, let us assume that $w \in \mathcal{F}$ is given as a linear combination of M points as $w = \sum_{j=1}^{M} \alpha_j \phi_j$ satisfying $\sum_{j=1}^{M} \alpha_j = 1$. Substituting these equations into the primal problem (1), the following is obtained:

$$\begin{vmatrix} \text{Min. } \frac{1}{2}\boldsymbol{\alpha}^T K \boldsymbol{\alpha} + \frac{1}{\nu l}\sum_{i=1}^{l} \xi_i - \rho \\ \text{s.t. } \left\langle \sum_{i=1}^{M} \alpha_i \phi_i, \phi_j \right\rangle + \xi_j \geq \rho, \quad \xi_j \geq 0, \quad j = 1, 2, \ldots, l, \\ e^T \boldsymbol{\alpha} = 1. \end{vmatrix} \quad (8)$$

Let $\boldsymbol{\alpha}^*$ be an optimal solution of this problem, the preference score of the product i is given as the ith element of the vector $K\boldsymbol{\alpha}^*$, i.e., $\sum_{j=1}^{M} \alpha_j^* \langle \phi_i, \phi_j \rangle = (K\boldsymbol{\alpha}^*)_i$.

Here, generating the kernel matrices given in Sect. 3 requires calculation of the inverse of the matrices as described in (3) and (5) through (7). The inverse operations require a significant computational effort, which prevents us from using these kernel matrices for the recommendation tasks when the number of products is large. Moreover, in general, these kernel matrices become fully dense, which causes difficulty in holding the kernel matrices in memory during the time required for solving the problem (8). In the subsequent subsections, however, we will propose new formulations of 1-SVMs which can handle the kernel matrices defined by (3) and (6) efficiently.

4.1 Modified Laplacian Regularized Kernel

Suppose that the kernel matrix K is the modified Laplacian regularized kernel matrix given by (6), which includes the regularized Laplacian kernel matrix (5) and the von Neumann diffusion kernel matrix (7) as the special cases.

Let us first introduce a new vector of variables $\boldsymbol{\beta} = (\beta_1, \beta_2, \ldots, \beta_M)^T \in \mathbf{R}^M$, and define $\boldsymbol{\beta} \equiv K\boldsymbol{\alpha}$. Note that $\beta_j = (K\boldsymbol{\alpha})_j = \left\langle \sum_{i=1}^{M} \alpha_i \phi_i, \phi_j \right\rangle$ holds for each j. It follows that $\boldsymbol{\alpha} = K^{-1}\boldsymbol{\beta} = (I + tL_\gamma)\boldsymbol{\beta}$ holds. The equality constraint $e^T \boldsymbol{\alpha} = 1$ in (8) can then be verified to be $(e - t(\gamma - 1)d)^T \boldsymbol{\beta} = 1$ where $d = De = Be$. Furthermore, a straightforward calculation reveals that $\boldsymbol{\alpha}^T K \boldsymbol{\alpha} = \boldsymbol{\beta}^T (I + tL_\gamma) \boldsymbol{\beta}$.

Therefore, the problem (8) can be equivalently formulated with respect to the new variable $\boldsymbol{\beta}$ as follows:

$$\begin{vmatrix} \text{Min. } \frac{1}{2}\boldsymbol{\beta}^T (I + tL_\gamma) \boldsymbol{\beta} - \rho + \frac{1}{\nu l}\sum_{i=1}^{l} \xi_i \\ \text{s.t. } \beta_j + \xi_j \geq \rho, \quad \xi_j \geq 0, \quad j = 1, 2, \ldots, l, \\ (e - t(\gamma - 1)d)^T \boldsymbol{\beta} = 1. \end{vmatrix} \quad (9)$$

Here, it should be emphasized that we can formulate the 1-SVM without the inversion calculations.

4.2 Commute Time Kernel

When we use the commute time kernel matrix L^+ as K in (8), a simpler formulation can also be derived. First, as in the previous section, let us introduce a vector of variables $\boldsymbol{\beta} = (\beta_1, \beta_2, \ldots, \beta_M)^T$, and let us define

$$\boldsymbol{\beta} \equiv \left(L^+ - ee^T/M\right) \boldsymbol{\alpha} + e/M. \tag{10}$$

For each j, if $\boldsymbol{\alpha}$ satisfies the constraint $e^T \boldsymbol{\alpha} = 1$ of the problem (8), then $\beta_j = \left\langle \phi_j, \sum_{i=1}^M \alpha_i \phi_i \right\rangle$ holds. Therefore, it follows from (3) and (10) that $\boldsymbol{\alpha} = \left(L - \frac{ee^T}{M}\right)\left(\boldsymbol{\beta} - \frac{e}{M}\right)$ holds. In addition, we can easily verify that the constraint $e^T\boldsymbol{\alpha} = 1$ in (8) is written as $e^T\boldsymbol{\beta} = 0$. Furthermore, $\boldsymbol{\alpha}^T L^+ \boldsymbol{\alpha} = \boldsymbol{\beta}^T L \boldsymbol{\beta}$ holds if $\boldsymbol{\beta}$ satisfies $e^T \boldsymbol{\beta} = 0$. Therefore, the primal problem (8) can be equivalently formulated as follows:

$$\left| \begin{array}{l} \text{Min. } \frac{1}{2} \boldsymbol{\beta}^T L \boldsymbol{\beta} - \rho + \frac{1}{\nu l} \sum_{i=1}^l \xi_i \\ \text{s.t. } \beta_j + \xi_j \geq \rho, \ \xi_j \geq 0, \ j = 1, 2, \ldots, l, \\ e^T \boldsymbol{\beta} = 0. \end{array} \right. \tag{11}$$

Let $(\boldsymbol{\beta}^*, \boldsymbol{\xi}^*, \rho^*)$ be an optimal solution of the problem (11). We have the following lemma.

Lemma 1. *The optimal solution $(\boldsymbol{\beta}^*, \boldsymbol{\xi}^*, \rho^*)$ satisfies $\beta_j^* \leq \rho^*$ for all $j = 1, \ldots, M$.*

Proof. Let $\bar{\beta} \equiv \max\{\beta_j^* \mid j = 1, 2, \ldots, M\}$. For the purpose of contradiction, let us assume that $\bar{\beta} > \rho^*$. We will show that a better solution can be constructed. Let $I \equiv \{i | \beta_i^* = \bar{\beta}\}$. Note that $\xi_i^* = 0$ for any $i \in I$. In addition, for a sufficiently small $\epsilon > 0$, let us define a new solution $\hat{\boldsymbol{\beta}} = (\hat{\beta}_1, \hat{\beta}_2, \cdots, \hat{\beta}_M)$, where $\hat{\beta}_i \equiv \bar{\beta} - \epsilon$ if $i \in I$, and $\hat{\beta}_i \equiv \beta_i^* + \frac{|I|}{M-|I|}\epsilon$ if $i \notin I$. Here, $\hat{\boldsymbol{\beta}}$ satisfies $e^T \hat{\boldsymbol{\beta}} = 0$ and, for all $i = 1, 2, \ldots, l$, when ϵ is sufficiently small, $\hat{\beta}_i + \xi_i^* \geq \rho^*$ holds true. Therefore, $(\hat{\boldsymbol{\beta}}, \boldsymbol{\xi}^*, \rho^*)$ is a feasible solution of the problem (11). It is easy to verify that the objective value of the solution $(\hat{\boldsymbol{\beta}}, \boldsymbol{\xi}^*, \rho^*)$ is better than that of $(\boldsymbol{\beta}^*, \boldsymbol{\xi}^*, \rho^*)$, which is a contradiction. □

From Lemma 1, the following corollary can be obtained.

Corollary 1. *The optimal solution $(\boldsymbol{\beta}^*, \boldsymbol{\xi}^*, \rho^*)$ of the problem (11) satisfies $\beta_j^* + \xi_j^* = \rho^*$ for all $j = 1, 2, \ldots, l$.*

Consequently, by substituting $\xi_j = \rho - \beta_j$, the problem (11) can be simplified as follows:

$$\left| \begin{array}{l} \text{Min. } \frac{1}{2} \boldsymbol{\beta}^T L \boldsymbol{\beta} + \frac{1-\nu}{\nu} \rho - \frac{1}{\nu l} \sum_{i=1}^l \beta_i \\ \text{s.t. } \beta_j \leq \rho, \ j = 1, 2, \ldots, l, \\ e^T \boldsymbol{\beta} = 0. \end{array} \right. \tag{12}$$

4.3 Some Special Cases

It has been shown that the 1-SVM formulation given in (1) can be solved analytically when $\nu = 1.0$. This is also true for our formulation given in (8). We have the following lemma:

Lemma 2. *Let $(\alpha^*, \xi^*, \rho^*)$ be an optimal solution of (8) with $\nu = 1.0$, i.e.,*

$$\left| \begin{array}{l} \text{Min. } \frac{1}{2}\alpha^T K \alpha + \frac{1}{l}\sum_{j=1}^{l} \xi_j - \rho \\ \text{s.t. } \left\langle \sum_{i=1}^{M} \alpha_i \phi_i, \phi_j \right\rangle + \xi_j \geq \rho, \quad \xi_j \geq 0, \; j = 1, \ldots, l, \\ e^T \alpha = 1. \end{array} \right. \qquad (13)$$

Then, for all $j = 1, 2, \ldots, l$, the inequalities $\left\langle \sum_{i=1}^{M} \alpha_i^* \phi_i, \phi_j \right\rangle \leq \rho^*$ hold true.

Proof. Let us assume, to the contrary, that there exists an index k such that $\left\langle \sum_{i=1}^{M} \alpha_i^* \phi_i, \phi_k \right\rangle > \rho^*$. It should be noted that $\xi_k^* = 0$.

Next, let $\Delta \equiv \left\langle \sum_{i=1}^{M} \alpha_i^* \phi_i, \phi_k \right\rangle - \rho^* > 0$. Then, we can define a new solution $\hat{\xi} = (\hat{\xi}_1, \ldots, \hat{\xi}_l)$ and $\hat{\rho}$ as follows:

$$\hat{\xi}_j = \begin{cases} \xi_j^* + \Delta & \text{if } j \neq k, \\ \xi_k^* & \text{if } j = k, \end{cases} \quad \text{and} \quad \hat{\rho} = \rho^* + \Delta.$$

The solution $(\alpha^*, \hat{\xi}, \hat{\rho})$ also satisfies the constraints of the problem (13). In particular, we note that the equality $\left\langle \sum_{i=1}^{M} \alpha_i^* \phi_i, \phi_k \right\rangle + \hat{\xi}_k = \hat{\rho}$ holds true because $\hat{\xi}_k = 0$. Straightforward calculations show that the objective value of $(\alpha^*, \hat{\xi}, \hat{\rho})$ is better than that of $(\alpha^*, \xi^*, \rho^*)$, which is a contradiction. This completes the proof. □

This lemma also ensures that $\xi_j^* = \rho^* - \left\langle \sum_{i=1}^{M} \alpha_i^* \phi_i, \phi_j \right\rangle$ holds for each $j = 1, 2, \ldots, l$. Then, substituting these equations into the objective function of the problem (13), the following formulation is obtained:

$$\left| \begin{array}{l} \text{Min. } W(\alpha) = \frac{1}{2}\alpha^T K \alpha - \frac{1}{l} y^T K \alpha \\ \text{s.t. } e^T \alpha = 1, \end{array} \right. \qquad (14)$$

where $y = (y_1, y_2, \ldots, y_M)^T$ is an M-dimensional vector such that $y_1 = y_2 = \cdots = y_l = 1$ and $y_{l+1} = y_{l+2} = \cdots = y_M = 0$. Note that y is a binary vector representing the purchased products by the active customer.

The problem (14) can be solved analytically. Since the gradient of the objective function $W(\alpha)$ is described as $\nabla W(\alpha) = K\alpha - \frac{1}{l}Ky$, a stationary point of $W(\alpha)$ is given as $\alpha = \frac{1}{l}y$, which happens to satisfy the constraint $e^T \alpha = 1$. Therefore, the problem (14) is solved.

5 Computational Experiments

To evaluate the performances of the proposed approaches, numerical experiments are conducted using a real-world dataset. We use the MovieLens dataset developed at the University of Minnesota. This dataset contains 1,000,209 ratings of approximately 3,900 movies made by 6,040 customers. We use 100,000 randomly selected ratings [6] containing 943 customers and 1682 movies. This set of ratings is divided into five subsets to perform five-fold cross-validation. The divided dataset can be retrieved from http://www.grouplens.org/data/. Moreover, in order to demonstrate the scalability of the proposed approach, we use the original full dataset, which is also randomly divided into five subsets to perform the cross-validation.

In these experiments, all of the rating values are converted into binary values, indicating whether a customer has rated a movie. This conversion has been used in several papers, including [6, 3]. Let M and N be the number of products and customers, respectively. Then the dataset is represented as an $N \times N$ binary matrix A, where the $i-j$ element $A_{ij} = 1$ if customer i has watched movie j.

In order to generate the graph-based kernels, we first construct a k-nearest neighbor graph $G(V, E)$ where the set of nodes V corresponds to that of the movies. For each node $j \in V$, let $\boldsymbol{A}_j$ denote the jth column vector of matrix A. Based on the cosine similarities $\frac{\boldsymbol{A}_i^T \boldsymbol{A}_j}{\|\boldsymbol{A}_i\|\|\boldsymbol{A}_j\|}$ between movie i and movie j, when movie i is among the k nearest neighbors of movie j, or when movie j is among those of movie i, we place an edge $(i,j) \in E$ and assign a unit weight $b_{ij} = 1$. We report the results obtained by the kernel matrices given in (3) and (5).

For each kernel matrix, we solve the 1-SVM with the parameter $\nu = 1$ for generating the preference scores, which can be achieved by solving a system of linear equations as described in Sect. 4.3. More precisely, for each active customer a, let $\boldsymbol{y}_a \in \mathbf{R}^M$ be an M-dimensional binary vector representing the purchased products by active customer a. Then, the preference score of each product i is given as the ith element of the vector $\left(L - \boldsymbol{e}\boldsymbol{e}^T/M\right)^{-1} \boldsymbol{y}_a$ when we use the kernel matrix (3), or of the vector $(I + tL)^{-1} \boldsymbol{y}_a$ when we use (5).

The cross-validation is conducted using the training and test set splits described above. We first calculate the score using the training set. Note that, for each active customer, the movies contained in the corresponding test set are not contained in the training set. Then, if the score is ideally correct, these movies have to be ranked higher than any other movies not watched in the training set. For comparison, the performance of the proposed method is evaluated in the manner described in [3] using the degree of agreement, which is the proportion of pairs ranked in the correct order with respect to the total number of pairs. Therefore, a degree of agreement of 0.5 will be generated by the random ranking, whereas a degree of agreement of 1.0 is the correct ranking.

The average degrees of agreement of the five-fold cross validation are given in Figs. 1 through 3. Figures 1 and 2 show the results for the 100,000 selected ratings, and Fig. 3 shows the results for the full MovieLens dataset with more than one million ratings. Figure 1 shows the results obtained by the kernel matrix

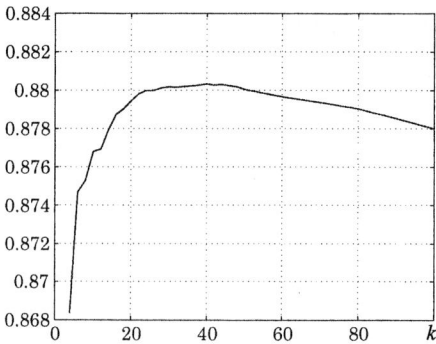

Fig. 1. Results obtained by kernel (3)

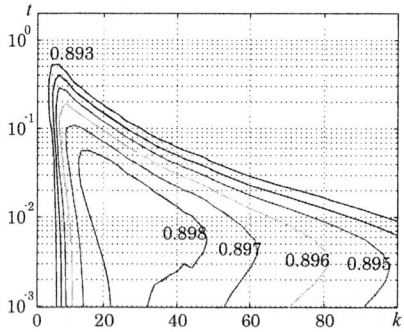

Fig. 2. Selected dataset with kernel (5)   **Fig. 3.** Full dataset with kernel (5)

(3), and Figs. 2 and 3 show the results by (5). The kernel matrices are constructed by changing the number of neighbors ranging from $k = 4$ to $k = 100$, as well as the parameter t in (5), which ranges from $t = 2^{-10}$ to $t = 2^{10}$. Note that contour lines that are less than 0.893 are omitted from Fig. 2, and those that are less than 0.911 are omitted from Fig. 3.

For comparison, we also perform the same five-fold cross-validation using a previously proposed scoring method [3]. In this case, the average degree of agreement is 0.8780, which is approximately the same as the results of the kernel matrix (3), but is significantly less than that obtained by (5). It should be emphasized that the proposed method offers better performance in a wide range of parameter settings (See Fig. 2). Furthermore, the kernel matrix used in [3] is generated from a large graph, the nodes of which corresponds to all of the product and customers. When the full movie dataset is considered, the size of the kernel matrix is approximately $10,000 \times 10,000$, which can not be handled due to memory constraints. The present kernel matrix, however, is defined by a graph with nodes corresponding only to the products and does not depend on the number of the customers, which is another advantage of the proposed method.

6 Conclusion

We have introduced a new method for recommendation tasks based on the 1-SVM. Using special structures of graph kernels, we show that the 1-SVM can be formulated as rather simple quadratic programming problems. In addition, the formulations can take advantage of the sparsity of the Laplacian matrix. Numerical experiments indicate that the quality, of our recommendations is high, as is the scalability of the method, which can handle tasks with over one million ratings.

Acknowledgments

This study was supported in part by Grants-in-Aid for Scientific Research (16201032 and 16510106) from JSPS.

References

[1] M. Belkin and P. Niyogi. Semi-supervised learning on Riemannian manifolds. *Machine Learning*, 56:209–239, 2004.
[2] F. R. Chung. *Spectral Graph Theory*. American Mathematical Society, 1997.
[3] F. Fouss, A. Pirotte, and M. Saerens. A novel way of computing dissimilarities between nodes of a graph, with application to collaborative filtering. In *ECML/SAWM*, pages 26–37, 2004.
[4] T. Ito, M. Shimbo, T. Kudo, and Y. Matsumoto. Application of kernels to link analysis. In *KDD '05*, pages 586–592.
[5] P. Resnick, N. Iacovou, M. Suchak, P. Bergstorm, and J. Riedl. GroupLens: An Open Architecture for Collaborative Filtering of Netnews. In *Proceedings of ACM 1994 Conference on Computer Supported Cooperative Work*, pages 175–186, 1994.
[6] B. Sarwar, G. Karypis, J. Konstan, and J. Riedl. Analysis of recommendation algorithms for e-commerce. In *EC '00: Proceedings of the 2nd ACM Conference on Electronic Commerce*, pages 158–167, 2000.
[7] B. Schölkopf, J. C. Platt, J. Shawe-Taylor, A. J. Smola, and R. C. Williamson. Estimating the support of a high-dimensional distribution. *Neural Computation*, 13:1443–1471, 2001.
[8] U. Shardanand and P. Maes. Social information filtering: Algorithms for automating "word of mouth". In *ACM CHI'95*, pages 210–217.
[9] J. Shawe-Taylor and N. Cristianini. *Kernel Methods for Pattern Analysis*. Cambridge University Press, Cambridge, 2004.
[10] A. Smola and I. Kondor. Kernels and regularization on graphs. In *COLT*, 2003.
[11] M. Szummer and T. Jaakkola. Partially labeled classification with Markov random walks. In *Advances in Neural Information Processing Systems*, volume 14, pages 945–952, 2002.

Heterogeneous Information Integration in Hierarchical Text Classification

Huai-Yuan Yang[1,2,*], Tie-Yan Liu[1], Li Gao[2], and Wei-Ying Ma[1]

[1] Microsoft Research Asia, 5F Sigma Center, No. 49 Zhichun Road, Haidian District,
Beijing, 100080, P.R. China
{tyliu, wyma}@microsoft.com
http://research.microsoft.com/users/tyliu/
[2] Department of Scientific & Engineering Computing, School of Mathematical Sciences,
Peking University, Beijing, 100871, P.R. China
{goat, gaol}@pku.edu.cn

Abstract. Previous work has shown that considering the category distance in the taxonomy tree can improve the performance of text classifiers. In this paper, we propose a new approach to further integrate more categorical information in the text corpus using the principle of multi-objective programming (MOP). That is, we not only consider the distance between categories defined by the branching of the taxonomy tree, but also consider the similarity between categories defined by the document/term distributions in the feature space. Consequently, we get a refined category distance by using MOP to leverage these two kinds of information. Experiments on both synthetic and real-world datasets demonstrated the effectiveness of the proposed algorithm in hierarchical text classification.

1 Introduction

Text Classification (TC) is a process of assigning text documents into one or more topical categories. It is an important research problem in information retrieval and machine learning. In the past two decades, TC has attracted a lot of research efforts from different research communities. As a result, many TC algorithms have been proposed, such as Naïve Bayes, Support Vector Machines (SVM) and their variations [6][11].

More recently, with the explosive growth of the World Wide Web, hierarchical classification [3][7][10] has been widely used to facilitate the browsing and maintaining of large-scale Web page corpora such as the Yahoo! Directory and the Open Directory Project (ODP). Other than simply using the hierarchical taxonomy to organize classifiers, empirical studies also showed that by exploiting the distance between categories (i.e. the path length between categories) in the taxonomy tree, the

---

[*] This work was performed at Microsoft Research Asia.

classification performance can be improved [2][4][5]. For example, [2] showed that bounding the margin between two classifiers as a function of the corresponding category distance can achieve obvious performance increase. Inspired by this result, we propose to use more categorical information in the data corpus to further improve the classification performance.

Actually, as we know, the path length in the taxonomy tree is totally based on the prior knowledge of the human editors. Therefore it is not necessarily consistent with the real data distribution. This phenomenon is especially serious for those multi-label datasets. So, to further improve the performance, we should also leverage the document distribution of a category in the feature space. For this purpose, we propose an algorithm to integrate these two types of category relations, by using the principle of multi-objective programming (MOP). In particular, we embed the categories into a new Euclidean space in order that their distance in this new space can preserve the similarities defined both in the taxonomy tree and by the document/term distributions as much as possible. To get this embedding, we construct a two-objective optimization problem: one objective is to minimize the difference between the category distance in the new space and the corresponding path length in the taxonomy tree, while the other is to minimize the difference between the category similarity in the new space and in the original feature space. By solving this MOP problem, we eventually get a *refined category distance* (**RCD**) to improve existing hierarchical classifiers such as Hieron [2]. In addition, if the dimension of this embedded Euclidean space is equal to the dimension of term space, we can regard it as a translation of the categories in the original term space. Thus we actually derive a new vector space model (called *refined text vectors* (**RTV**)), which can also help improve the hierarchical classifiers. Experiments on both synthetic and real-world datasets showed the effectiveness of the *refined category distance* and *refined text vectors*.

The rest of this paper is organized as follows: In Section 2, the basic idea of our algorithm is proposed. Then a trick for complexity reduction is discussed in Section 3. In Section 4, experimental results are presented to evaluate our algorithm. Concluding remarks and future work are discussed in the last section.

2 Heterogeneous Information Integration

2.1 General Approach

As mentioned in the introduction, there are two types of relations between categories in hierarchical classification, which can be illustrated as in Fig.1. The first type of relation is the hierarchy of categories, while the second is category-document and document-term relations. Our basic idea is to integrate these two types of category relations by using multi-objective optimization. In particular, we propose to embed the categories into a new space in which they preserve the similarities defined both by the path length in the taxonomy tree and by the document/term distributions in the feature space as much as possible.

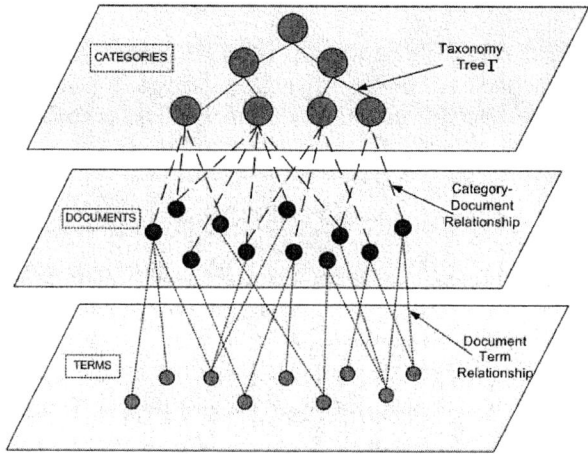

Fig. 1. Heterogeneous relations in a text corpus. The relationships between categories and terms are indirect and can be obtained by combining category-document and document-term relationships.

Mathematically, for any pair of nodes (representing categories) i and j in the taxonomy tree, let $d(i, j)$ denote the path length (the number of edges in the path) from i to j in Γ. Denote $A = [d(i, j)]$ the corresponding path length matrix. Let n be the number of categories and m be the number of terms. Let B denote the relationship ($n*m$) matrix between categories and terms and let $b(i)$ be the i-th row of B[1]. Suppose there is a (k-dimension) Euclidean space, in which the representations of the categories are $X = [x_1, x_2, ..., x_n]^T$, where n is the number of categories and each x_i is a k-dimension vector. Then our proposed algorithm can be written as in (1), where the first objective is to minimize the difference between the distance of categories calculated in the new Euclidean space and the path length in the taxonomy tree, while the second objective is to minimize the difference between the distance of categories calculated in the original feature space and the new Euclidean space.

$$\min_X \| A - D_X \|_F^2 \quad (1)$$
$$\min_X \| BB^T - XX^T \|_F^2$$

where $D_X = [\| x_i - x_j \|_F]_{n \times n}$, and $\| \cdot \|_F$ is the F-norm

It is clear that this is a multi-objective programming (MOP) problem. Without loss of generality and for simplicity, we convert this MOP problem to a single-objective one by means of linear combination as follows.

$$\min_X \alpha \| A - D_X \|_F^2 + (1-\alpha) \| BB^T - XX^T \|_F^2 \quad (2)$$

---

[1] In our definition, each row $b(i)$ of B is simply calculated as the mean vector of all the documents in category i.

Note that the optimization problem (2) is of very large scale because there are nk variables to tune and n may be as large as hundreds of thousands[2]. Many existing optimization algorithms [1] can hardly handle such kind of large-scale problems because they need second-order information in the optimization process, which corresponds to space complexity of $O(n^2k^2)$. To tackle this problem, we use a recently-proposed method, named Global Barzilai and Borwein (GBB) algorithm [9] in our approach, which has been proven to require space complexity of only $O(nk)$. By solving (2) in this way, actually we get a new distance matrix (D_X), called the *refined category distance* (**RCD**) matrix. This matrix can be used directly as the category distances in hierarchical classifier such as Hieron [2].

2.2 Further Discussion

As discussed in the above subsection, by solving (2), we embed the categories into a new (k-dimension) Euclidean space and k is usually smaller than the dimension of the original term space. However, it may be interesting to discuss what will happen if k is equal to the dimension of the original term space (m). Actually, in such a special case, we can regard the new embedding space as just a translation of the categories in the original term space. And accordingly, we can come out another approach to improve hierarchical classifiers as follows.

Denote $x(i)$ the embedding of category i in the new space, denote $v_j(i)$ the j-th document belonging to category i in the original term space, and denote $b(i)$ the row vector in B corresponding to category i. Then, we can refine the document vectors as follows,

$$v_j^*(i) = v_j(i) + (b(i) - x(i)) \qquad (3)$$

Actually (3) can be explained as that we shift the mean of all the documents in a category by considering the information contained in the hierarchical taxonomy. After this shift, the resultant *refined text vectors* (**RTV**) can be used as the new feature representations fed to the hierarchical classifiers for training. Then for testing, the instance will be shifted by $b(i)-x(i)$ before being tested by the classifier for category i.

To sum up, we take **RCD** and **RTV** as two manners of information integration in hierarchical text classification. Both their effectiveness was tested in our experiments.

3 Complexity Reduction

Considering that many real hierarchical text corpora have tens of thousands of categories, the complexity is still very high even if we use the GBB algorithm. To further reduce the complexity, in this section, we will propose some tricks based on matrix decomposition. Note that the following discussions are meaningful only if $k \leq n$. Otherwise, we assume that the complexity has not been high enough and the corresponding optimization problem can be solved efficiently already.

---

[2] There are about 300,000 categories in the Yahoo! Directory.

First of all, we will conduct eigenvalue decomposition (EVD) for BB^T. Actually if we only calculate k eigenvalues and their associated eigenvectors of BB^T, we will have the following approximation of BB^T

$$BB^T \approx U_k \Sigma_k^2 U_k^T \qquad (4)$$

Where U_k is an $n \times k$ matrix, Σ_k is a $k \times k$ square matrix and $U_k \Sigma_k$ is also an $n \times k$ matrix which has the same dimension with X.

Then if we can make $\| U_k \Sigma_k - X \|_F^2$ sufficiently small,[3] we are able to guarantee that $\| BB^T - XX^T \|_F^2$ is also very small due to the characteristics of eigenvalue decomposition. With this fact, we can simplify our second objective function from $\min_X \| BB^T - XX^T \|_F^2$ to $\min_X \| U_k \Sigma_k - X \|_F^2$, and the overall objective turns to be

$$\min_X \alpha \| A - D_X \|_F^2 + (1-\alpha) \| U_k \Sigma_k - X \|_F^2 \qquad (5)$$

In practical, U_k and Σ_k can be calculated through the singular value decomposition (SVD) of B as well, which can be much more efficiently computed than the EVD of BB^T.

$$B \approx U_k \Sigma_k V_k^T \qquad (6)$$

One may find that the above singular value decomposition can actually be regarded as the spectral embedding of the category-term bipartite graph (See Fig.2). This implies some problem of our aforementioned method for computation reduction because the graph shown in Fig.2 sometimes is too sparse and even unconnected. In such a case, the SVD will not be as robust as we expect.

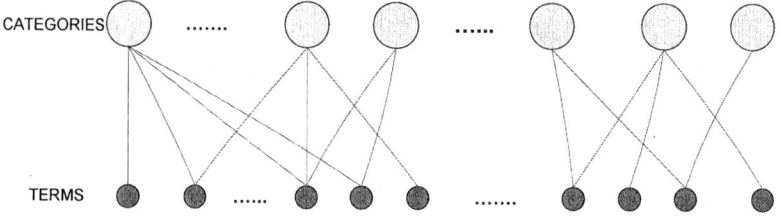

Fig. 2. Bipartite graph of category-term relationship

To tackle this problem, we add a smoothing item to matrix B before conducting SVD, so as to improve the connectivity of its corresponding bipartite graph:

$$B^{new} = \beta B + (1-\beta) \frac{1}{n} ee^T \qquad (7)$$

---

[3] Where k is the dimension of our embeddings of X in the new space.

where $e = [1,1,..,1]^T$. Actually the same trick as above has been widely used in many other works such as PageRank [8] and so on.

4 Experiments

4.1 Experiment Setting

In this section, we present our experimental evaluation of the proposed algorithms. First of all, we will introduce the experimental settings.

In our experiments, Hicron was used as the baseline for testing the effectiveness of *refined category distance* (**RCD**) and *refined text vectors* (**RTV**). Hieron is a large margin hierarchical classifier, which enforces a margin among multiple categories. The basic optimization formulation of Hieron is as given in (8).

$$\min \frac{1}{2} \sum_{v \in Y} \| w^v - w_i^v \|^2 \qquad (8)$$

$$s.t. \sum_{v \in P(y_i)} w^v \cdot x_i - \sum_{u \in P(\hat{y}_i)} w^u \cdot x_i \geq dist(y_i, \hat{y}_i)$$

where $\hat{y}_i$ is the predicted category label of training example x_i, y_i is the real category label, $dist(y_i, \hat{y}_i)$ is a distance measure between these two categories. In [2], two versions of Hieron were proposed. The first one simplified the distance between any two categories to 1 (denoted by Flat Hieron) and the second used the path length between two categories in the taxonomy tree (or the tree distance) as the category distance (denoted by Tree-Hieron).

To evaluate the performance of our first method, we replaced the tree distance in the Tree-Hieron by the *refined category distance* and other elements remained the same as the standard Tree-Hieron classifier. Note that in this evaluation, we set $k=1000$ for the **RCD** method. And to evaluave our second method, we used the *refined text vectors* as the training input, and other elements remained the same as the standard Tree-Hieron classifier. Note that in this evaluation, we set k equal to the dimension of the original term space. For the evaluation, we used both Micro-averaged F1 and Macro-averaged F1[4] (denoted by MicroF1 and MacroF1 in brief) as the metrics.

In our experiments, both synthetic and real-world data sets were used. The synthetic datasets are very similar to that used in [2], which were generated as follows. First, a symmetric ternary tree of depth 4 was constructed as the taxonomy hierarchy. This hierarchy contains 121 vertices, each of which was assigned a base vector w^u (where u represents a vertex). Then each example was generated by

---

[4] MicroF1 and MacroF1 are two popular evaluation criteria for multi-class text classification, which definitions are $MicroF1 = \frac{\sum_{i=1}^{n} TP_i}{\sum_{i=1}^{n}(TP_i + FP_i)}, MacroF1 = (\sum_{i=1}^{n} \frac{TP_i}{TP_i + FP_i})/n$, where TP_i is the number of documents correctly classified into category i; FP_i is the number of documents wrongly classified into category i. [10] [12]

setting $(x, y) = (\sum_{u \in P(y)} w^u + \eta, y)$, where $P(y)$ represents the path from the root to a leaf node y, and η is a random vector sampled from the distribution $N(0, 0.16)$. Furthermore, we "disturbed" the above synthetic dataset by randomly selecting 20 pairs of category centers, and pulling them closer to each other by 30 percent. With this strategy, we generated two synthetic datasets of different sizes. Each category in the first dataset (denoted by DS1) contains 10 training documents and 5 test documents, while that in the second dataset (denoted by DS2) contains 20 training documents and 10 test documents. For the real-world dataset, the 20NG [13] dataset was used. We randomly divided the documents in each category of the 20NG dataset into a training set and a test set with a ratio of 6:4. To avoid the bias of one single training/test set partitioning, we partitioned the dataset for 10 times and reported the average performance accordingly. As can be seen, for either the synthetic or the real-world data set, the number of categories is only tens or hundreds. Since this number is smaller than k, we actually did not apply the tricks described in Section 3. However, those deductions are surely meaningful for those who want to conduct experiments with much larger scales.

4.2 Experimental Results on the Synthetic Datasets

In this subsection, we report the performance of our methods on the synthetic datasets. As can be seen in Fig.3, the curve for the **RCD** method is very smooth, indicating that the classification performance does not depend heavily on the parameter α. Without loss of generality, we set $\alpha = 0.5$ in our further experiments. And comparatively speaking, the curve for **RTV** drops significantly when α is very close to 1. This is because it is not reasonable to modify the original document vectors too much with the human-defined taxonomy tree which is very subjective and not data dependent.

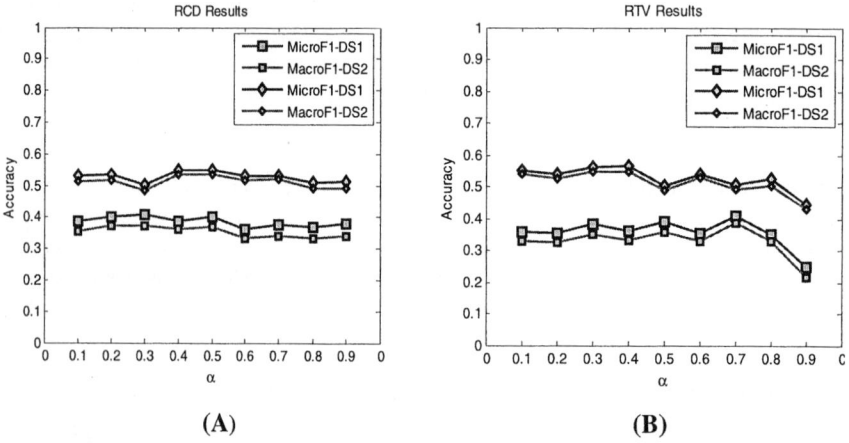

Fig. 3. Classification performance of **RCD** and **RTV** with respect to different α

Further comparisons with the Hieron baselines are shown in Table 1. From this table, we can see that by utilizing the tree distance, Tree-Hieron outperformed Flat Hieron. And both **RCD** and **RTV** led to much higher classification performances. This improvement is consistent regardless of the size of the data set.

Table 1. Comparison of different methods on the sythetic data sets

Method	DS1		DS2	
	MicroF1	**MacroF1**	**MicroF1**	**MacroF1**
Flat Hieron	0.34321	0.31023	0.51234	0.48993
Tree-Hieron	0.35802	0.33310	0.52839	0.51420
RCD	**0.40246**	**0.37016**	0.54691	0.53398
RTV	0.36049	0.33291	**0.56543**	**0.54857**

4.3 Experimental Results on the 20NG Dataset

In this subsection, we report the experimental results on the 20NG dataset.

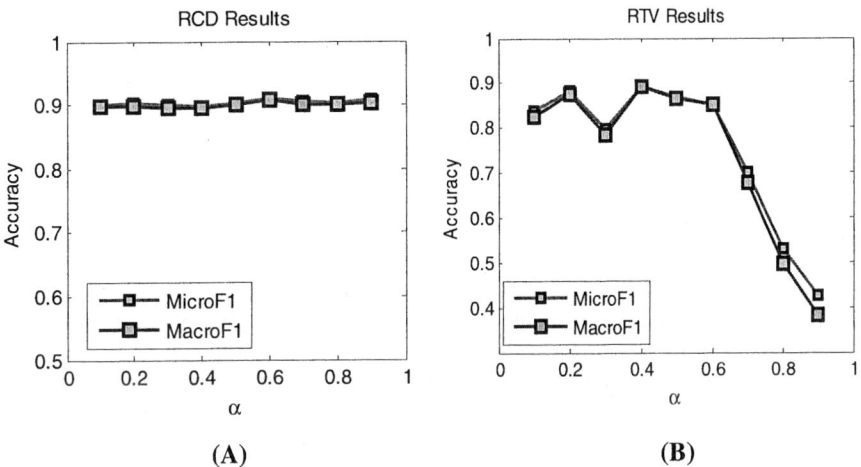

Fig. 4. The classification performance of **RCD** and **RTV** with respect to different α

From Fig.4 we can draw very similar conclusion to what we have got in Section 4.2. That is, the performance of the **RCD** method does not depend heavily on the value of α, while **RTV** prefers a smaller α to guarantee its high classification accuracy. Furthermore, from the comparison listed in Table 2 we can see that the improvement of classification accuracy is even more significant as compared to that on the synthetic dataset. For example, the MicroF1 of flat Hieron and Tree-Hieron are only 0.78 and 0.83 respectively, while the MicroF1 of **RTV** is about 0.89 and the MicroF1 of **RCD** is even more than 0.91.

Table 2. The comparison of different methods on the 20NG dataset with different training/test set partitions

Method	Mean of MicroF1	Variance of MicroF1	Mean of MacroF1	Variance of MacroF1
Flat Hieron	0.78130	0.00288	0.76593	0.00277
Tree-Hieron	0.83402	0.00192	0.81796	0.00248
RCD	**0.91091**	0.00038	**0.90793**	0.00053
RTV	0.89197	5.62E-05	0.88976	8.64E-05

Besides, we have another interesting observation from Table 2: when we conducted our experiments for 10 times, the variances of the classification performance for different classifiers are quite different. As can be seen, our **RCD** and **RTV** methods performed stable with very small variances, while the variances of Flat Hieron and Tree-Hieron are much larger. Our explanation to this is as follows. Since we randomly sampled the training and test set, in some cases the tree distance used in Tree-Hieron (or the identical distance in Flat Hieron) may be consistent with the training data while in other cases it may be rather inconsistent. Comparatively speaking, by introducing our MOP formulation, we can better adapt to the real data distribution thus the corresponding classification becomes much more robust.

To sum up, our experiments show that it is very benefitcial to leverage the infotmation contained in both the taxonomy tree and the data distriution, either in terms of classification performance, or in terms of the robustness of the classifiers.

5 Conclusion and Future Work

In this paper, we proposed an algorithm for the integration of heterogeneous information in the application of hierarchical text classification, which is based on multi-objective optimization. Experiments on both synthetic and real-world datasets showed that the proposed approach can improve both the classification performance and the robustness of the classifiers. For the future work, we plan to investigate whether the same idea can be used in other applications, such as the mining of click-through data, and the analysis of scientific citation graph.

References

1. Boyd, S., and Vandenberghe, L.: Convex Optimization. Cambridge University Press, 2004.
2. Dekel, O., Keshet, J., Singer, Y.: Large Margin Hierarchical Classification, In Proceedings of the 21st International Conference on Machine Learning (2004)
3. Dumais, S., Chen, H. Hierarchical Classification of Web Content, In Proc. SIGIR, 256-263,2000
4. Huang, K., Yang, H., King, I., Lyu, M.R.: Learning Large Margin Classifiers Locally and Globally, In Proceedings of the 21st International Conference on Machine Learning (2004)

5. Hofmann, T., Cai, L., Ciaramita, M: Learning with Taxonomies: Classifying Documents and Words, In Conference on Neural Information Processing Systems (NIPS).
6. Lewis, D.D.: Naïve (Bayes) at Forty: the Independence Assumption in Information Retrieval. In ECML, 1998
7. Liu, TY., Yang, Y., Wan, H., Zeng, HJ., Chen, Z., Ma, WY.: Support Vector Machines Classification with Very Large Scale Taxonomy, SIGKDD Explorations, Special Issue on Text Mining and Natural Language Processing, vol.7, issue.1, pp36~43, 2005.
8. Page, L., Brin, S., Motwani, R., Winograd, T.: The PageRank Citation Ranking: Bring Order to the Web. Technical Report, Stanford University, CA, 1998.
9. Raydan, M.: The BarziLai and Borwein Gradient Method for Large Scale Unconstrained Minimization Problem. SIAM J.OPIM, 1997.
10. Sun, A., Lim, E.P.: Hierarchical Text Classification and Evaluation. In Proceedings of the 2001 IEEE International Conference on Data Mining.
11. Vapnik, V.: Statistical Learning Theory. Wiley, New York, 1998.
12. Y. Yang. An evaluation of statistical approaches to text categorization. Information Retrieval, 1(1-2):69–90, 1999.
13. http://people.csail.mit.edu/~jrenie/20Newsgroups

FISA: Feature-Based Instance Selection for Imbalanced Text Classification

Aixin Sun[1], Ee-Peng Lim[1], Boualem Benatallah[2], and Mahbub Hassan[2]

[1] School of Computer Engineering, Nanyang Technological University, Singapore
{axsun, aseplim}@ntu.edu.sg
[2] School of Computer Science and Engineering,
University of New South Wales, NSW 2052, Australia
{boualem, mahbub}@cse.unsw.edu.au

Abstract. Support Vector Machines (SVM) classifiers are widely used in text classification tasks and these tasks often involve imbalanced training. In this paper, we specifically address the cases where negative training documents significantly outnumber the positive ones. A generic algorithm known as FISA (Feature-based Instance Selection Algorithm), is proposed to select only a subset of negative training documents for training a SVM classifier. With a smaller carefully selected training set, a SVM classifier can be more efficiently trained while delivering comparable or better classification accuracy. In our experiments on the 20-Newsgroups dataset, using only 35% negative training examples and 60% learning time, methods based on FISA delivered much better classification accuracy than those methods using all negative training documents.

1 Introduction

Studies have shown that imbalanced training data can adversely affect classification accuracy of a classifier [7]. In particular, SVM classifiers are known to favor negative decisions when trained with significantly larger proportion of negative examples [1, 11]. In multi-label classification problem using SVM classifiers, imbalanced training data can often be caused by the *one-against-all* learning strategy. That is, with positive training examples given for each category, the one-against-all strategy trains SVM classifier of the category using the training examples belonging to the category as positive examples, and all training examples not belonging to the category as negative examples. In our study, we address the problem of imbalanced text classification using SVM classifiers with one-against-all strategy.

We focus on the under-sampling approach and propose a generic algorithm known as FISA (Feature-based Instance Selection Algorithm), to select only a subset of negative training documents for training SVM classifier. FISA operates in two steps: *feature discriminative power computation* and *instance selection*. In the first step, the discriminative power of each feature is computed using some feature selection technique. In the second step, for each negative training document, a representativeness score is computed based on both the number of discriminative features appearing in the document and their discriminative powers. The higher the score, the more significant the document in representing the negative training examples, and hence more useful in learning SVM classifiers. Given a smaller training set consisting of only negative training documents with high representativeness scores, a SVM classifier will take a much shorter time to learn while delivering comparable or even better classification accuracy.

We evaluated FISA on the 20-Newsgroups dataset. Two FISA methods using feature selection techniques Odds Ratio and Information Gain have been evaluated, known as FOR and FIG respectively. FOR and FIG were compared with baseline SVM, Different Error Cost (DEC) method and Stratified Random Instance Selection (SRIS) method. Both FOR and FIG delivered significantly better classification accuracies than DEC using only 35% negative training examples and 60% learning time required by DEC. Our experiments also showed that random selection of negative training examples compromised the classification accuracy.

The rest of the paper is organized as follows. We survey related work in Section 2 and discuss FISA in Section 3, followed by experiments and results in Section 4. We finally conclude this paper in Section 5.

2 Related Work

The two main approaches to address the imbalanced classification problems are the data-level approach and the algorithmic-level approach. Data-level approach includes under-sampling methods that select only a subset of negative instances for training [3, 5, 6], and over-sampling methods that synthetically generate positive training instances [2]. Nevertheless, studies have shown that over-sampling with replacement does not significantly improve the classification accuracy. For methods using the algorithmic-level approach, one can assign different classification-error costs on positive/negative training instances, or modify the classifier-specific parameters [1, 11].

One extreme case in imbalanced text classification is to use one-class SVM classifiers [8, 10]. One-class SVM learns from positive training documents only and totally ignore the negative training documents. However, Manevitz and Yousef [8] demonstrated that one-class SVM is very sensitive to the choice of feature representation (e.g., binary or *tfidf*) and SVM kernels.

3 Feature-Based Instance Selection

Given a target category c_i, a set of positive training documents Tr_i^+ and a much larger set of negative training documents Tr_i^-, say $|Tr_i^-| \geq 10 \times |Tr_i^+|$, the problem is to select a subset of negative training documents from Tr_i^-, denoted by Ts_i^-, such that the classification accuracy of a SVM classifier learned using Tr_i^+ and Ts_i^- is comparable with (or hopefully better than) the one learned using Tr_i^+ and Tr_i^- while reducing the learning time. Note that, in this paper, $|S|$ denotes the number of elements in the set S.

The training of a SVM classifier involves finding a hyperplane that separates positive training examples from the negative ones with the widest margin. As the hyperplane is defined by both the positive and negative training examples, intuitively, the hyperplane lies in the boundaries between the positive and negative training examples; most importantly, the negative training examples used to define the hyperplane (i.e., the support vectors) are the ones that are close to the positive examples. Given the large set of negative training documents, many of them are expected to be far away from the positive ones and are less useful in SVM classifier traning. These negative training documents are known as *less representative* examples with respect to the target category. We therefore try to remove these less representative examples to obtain more balanced positive/negative training examples and to achieve comparable or better classification accuracy using shorter learning time.

The proposed FISA algorithm includes a feature discriminative power computation step and an instance selection step. In the first step, a feature selection technique is applied to compute the discriminative power of each term feature. Most feature selection techniques rooted in information theory can be used. For each category c_i, a feature selection technique computes the discriminative power of term t_k, denoted by $\delta(t_k|c_i)$. Note that $\delta(t_k|c_i)$ needs to be computed *only* if t_k appears in at least one positive training document in c_i. In the second step, the representativeness of each negative training document is computed. Those with representativeness scores larger than a threshold r_θ will be selected to learn a SVM classifier. The **representativeness** of a document d_j with respect to a category c_i, denoted by $r(c_i|d_j)$, is defined as the average discriminative powers of the features found in d_j (see Equation 1 where w_{jk} is the weight of term feature t_k in document d_j).

$$r(c_i|d_j) = \frac{\sum_{t_k \in d_j, t_k \in F_i} w_{jk} \times \delta(t_k|c_i)}{\sum_{t_k \in d_j} w_{jk}} \quad (1)$$

To determine the document representativeness threshold, we adopt the concept of quality control from statistics [9].

$$r_\theta = \frac{1}{|Tr_i^-|} \sum_{d_j \in Tr_i^-} r(c_i|d_j) + z \times \frac{\sigma}{\sqrt{|Tr_i^-|}} \quad (2)$$

In Equation 2, σ is the standard deviation of representativeness scores of all negative training documents. Given a huge number of negative training documents, we can assume that their representativeness scores follow a normal distribution and the z parameter determines the proportion of documents to be selected.

Note that, feature selection technique is applied in FISA for feature discriminative power computation only; the final training of SVM classifiers actually involves all the features of positive training and the selected negative training instances. This is because SVM is known to perform well without feature selection [4].

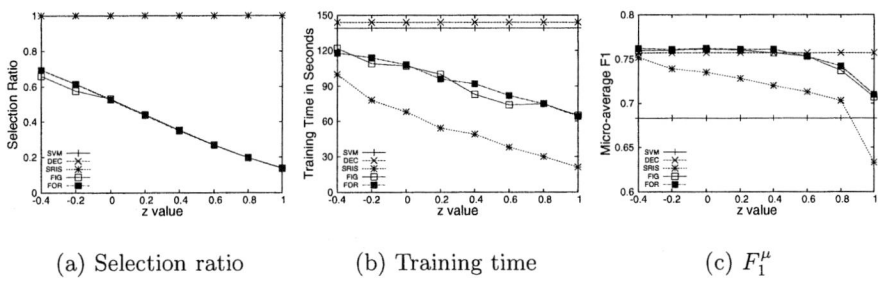

(a) Selection ratio   (b) Training time   (c) F_1^μ

Fig. 1. SVM, DEC, SRIS, FIG and FOR against different z values

4 Experiments

We evaluated FISA with two well-studied feature selection techniques, namely, Odds Ratio (OR) and Information Gain (IG). Those two FISA methods are therefore known as FOR and FIG respectively. FOR and FIG were compared with baseline *SVM, Different Error Cost* (DEC), and *Stratified Random Instance Selection* (SRIS) methods. In our experiments, SVM^{light} was used as the baseline classifier for those five methods. DEC method was implemented by adjusting the *cost-factor* (parameter j) in SVM^{light} to be the ratio of the number of negative training examples over positive ones. The same cost-factor setting was also applied to FOR, FIG, and SRIS after instance selection in these methods. For a fair comparison, the number of instances selected by SRIS was the larger one selected by FOR and FIG.

The experiments were conducted on 20-Newsgroups[1] dataset with different z values from -0.4 to 1.0. Binary document representation was used after stopword removal and term stemming. The percentage of the selected negative training documents (e.g., selection ratio), training time[2] and micro-averaged F_1 (denoted by F_1^μ) of these five methods are shown in Figures 1(a), 1(b), and 1(c) respectively.

The larger the z the fewer negative examples were selected in training as expected (see Figure 1(a)). Particularly, when $z = 0.4$, only about 35% of negative training examples were used for FOR, FIG and SRIS. In terms of training time, SRIS was clearly the winner as no document representativeness computation was required. Figure 1(b) also shows that smaller number of training documents led to less training time. When $z = 0.4$, FIG and FOR used about only 60% of training time required by DEC or baseline SVM. In terms of classificatoin accuracy, baseline SVM was clearly the worst. The F_1^μ of SRIS decreases as z increases. An incease of z, on the other hand, had little effect on FOR and FIG when z was not greater than 0.4. When $z = 0.4$, the two FISA methods delievered better F_1^μ than DEC using 35% of the latter's negative training documents

---

[1] http://www.gia.ist.utl.pt/~acardoso/datasets/

[2] Training time includes I/O time, CPU time for instance selection, and SVM training time. PC configuration: CPU 3GHz, RAM 1GB, OS Windows 2000 SP4.

and 60% of its training time. This experiment shows that with carefully selected less number of training instances, faster and better classification results can be achieved.

5 Conclusion and Future Work

In this paper, we studied imbalanced text classification using SVM classifiers with one-against-all learning strategy. We proposed a generic algorithm known as FISA to select instances based on well-studied feature selection methods. Our experiment results on the 20-Newsgroups dataset confirmed that instance selection was useful for efficient and effective text classification using SVM classifiers. The major limitation of the proposed FISA algorithm is that duplicates or nearly duplicated documents receive similar representativeness scores and therefore could all be selected. However, the training of a SVM classifier does not benefit much from duplicated documents. Addressing this limitation will be part of our future research.

References

1. J. Brank, M. Grobelnik, N. Milic-Frayling, and D. Mladenic. Training text classifiers with SVM on very few positive examples. Technical Report MSR-TR-2003-34, Microsoft Research, April 2003.
2. N. V. Chawla, K. W. Bowyer, L. O. Hall, and W. P. Kegelmeyer. SMOTE: Synthetic minority over-sampling technique. *J. of Artificial Intelligence Research*, 16:321–357, 2002.
3. C.-M. Chen, H.-M. Lee, and M.-T. Kao. Multi-class svm with negative data selection for web page classification. In *Proc. of IEEE Joint Conf. on Neural Networks*, pages 2047 – 2052, Budapest, Hungary, 2004.
4. G. Forman. An extensive empirical study of feature selection metrics for text classification. *J. of Machine Learning Research*, 3:1289–1305, 2003.
5. D. Fragoudis, D. Meretakis, and S. Likothanassis. Integrating feature and instance selection for text classification. In *Proc. of ACM SIGKDD'02*, pages 501–506, Canada, 2002.
6. M. Kubat and S. Matwin. Addressing the curse of imbalanced training sets: One-sided selection. In *Proc. of ICML'97*, pages 179–186, 1997.
7. H. Liu and H. Motoda. On issues of instance selection. *Data Mining and Knowledge Discovery*, 6:115–130, 2002.
8. L. M. Manevitz and M. Yousef. One-class svms for document classification. *J. of Machine Learning Research*, 2:139–154, 2002.
9. D. C. Montgomery. *Introduction to Statistical Quality Control*. Wiley, 4th edition, 2000.
10. B. Raskutti and A. Kowalczyk. Extreme re-balancing for svms: a case study. *SIGKDD Explorations Newsletter*, 6(1):60–69, 2004.
11. G. Wu and E. Y. Chang. Kba: Kernel boundary alignment considering imbalanced data distribution. *IEEE TKDE*, 17(6):786–795, June 2005.

Dynamic Category Profiling for Text Filtering and Classification

Rey-Long Liu

Department of Medical Informatics, Tzu Chi University,
Hualien, Taiwan, R.O.C.
rlliutcu@mail.tcu.edu.tw

Abstract. Information is often represented in text form and classified into categories for efficient browsing, retrieval, and dissemination. Unfortunately, automatic classifiers may conduct many misclassifications. One of the reasons is that the documents for training the classifiers are mainly from the categories, leading the classifiers to derive category profiles for distinguishing each category from others, rather than measuring the extent to which a document's content overlaps that of a category. To tackle the problem, we present a technique DP4FC to help various classifiers to improve the mining of category profiles. Upon receiving a document, DP4FC helps to create dynamic category profiles with respect to the document, and accordingly helps to make proper filtering and classification decisions. Theoretical analysis and empirical results show that DP4FC may make a classifier's performance both better and more stable.

1 Introduction

Information is often represented in text form and classified into multiple categories for efficient browsing, retrieval, and dissemination. In such an information space, each category often contains several documents about a specific topic, and hence lots of documents may be entered at any time, but only a small portion of the documents may be classified into some categories. Therefore, text filtering (TF) and text classification (TC) should be integrated together to autonomously classify suitable documents into suitable categories.

One of the popular ways to achieve integrated TF and TC was to delegate a classifier to each category. The classifier was associated with a threshold, and upon receiving a document, it could autonomously make a yes-no decision for the corresponding category. Conceptually, a document was "accepted" by the classifier if its degree of acceptance (DOA) with respect to the category (e.g. similarity with the category or probability of belonging to the category) was higher than or equal to the corresponding threshold; otherwise it was "rejected." With the help of the thresholds, TF was actually achieved in the course of TC. Each document could be classified into zero, one, or several categories.

Unfortunately, perfect estimation of DOA values could not be expected [1] [7] [15], since no classifiers may be perfectly tuned. Therefore, a document that is believed to be similar to (different from) a category could not always get a higher (lower) DOA value with respect to the category. Obviously, improper DOA estimations may heavily deteriorate the performance of both TF and TC. Traditionally, DOA values were often estimated in the space whose dimensions were specified by a set of features (keywords). Therefore, a document that gets a higher DOA value with respect to a category under a feature set may get a very low DOA value with respect to the category under another feature set. Feature selection is thus one of the most important issues related to the tackling of improper DOA estimations.

In this paper, we explore how various classifiers' performances may be improved by employing more suitable features to distinguish relevant documents from non-relevant documents for each category. This goal differs from many previous related attempts, which aimed at improving the thresholding process (e.g. [7]) and the document selection process such as boosting [10], adaptive resampling ([4]), and query zoning [11]. The research result of the paper may be used to complement the previous techniques for integrated TF and TC.

In the next section, we present an observation that provides significant hints to tackle the problem. Accordingly, we develop a novel approach DP4FC (Dynamic Profiling for Filtering Classification) that helps to dynamically create the profile of each category so that the performance of TC and TF may be improved (ref. Section 3). Empirical evaluation was conducted to evaluate DP4FC under different circumstances (ref. Section 4). DP4FC was shown to be competent in helping the underlying classifier to achieve both better and stable performances in TF and TC.

2 Misclassifications of Documents: An Observation

Feature selection, which is an important issues related to DOA estimations, was often an experimental issue in previous studies [8] [9] [14]. There was no standard guideline to construct a perfect feature set. Some studies maintained an evolvable feature set covering *all* features currently seen (e.g. [2]). However, inappropriate features may introduce inefficiency [14] and poor performance [9] in TC.

Moreover, even a feature set may be perfectly tuned to distinguish among the categories, it is not necessarily suitable to filter out those documents not belonging to *all* the categories. This is due to the goal of feature selection: selecting those features that may be used to distinguish a category from others. Under such a goal, whether a feature may be selected mainly depends on the content relatedness among the categories, without paying much attention to how the contents of a category c and a document d overlap with each other. If d (c) talks too much information not in c (d), d should not be classified into c, even though d mentions some content of c. To tackle the problem, features should be *dynamically* selected in response to *each individual* input document (rather than training documents in the categories). This task motivates the research in the paper.

More specially, the observation suggests a *dynamic profiling* strategy to avoid misclassifying a document d into a category c: (1) selecting those terms that have occurrences in c but not in the document, and conversely (2) selecting those terms that have occurrences in the document but not in c. Therefore, each category should have a feature set, which is *dynamic* in the sense that it is reconstructed once a test document is entered.

Dynamic profiling may complement the functionality of those classifiers that aim to distinguish c from other categories by building a *static* profile for each category. The profile is static in the sense that it is often composed of those terms that are discriminative for the categories, and hence does not vary for each input document. Dynamic profiling complements the classifiers by considering another issue: how d (c) talks about those contents not in c (d). If d lacks important contents of c or talks much information not about c, it could not be classified into c, even though it mentions some discriminative contents of c.

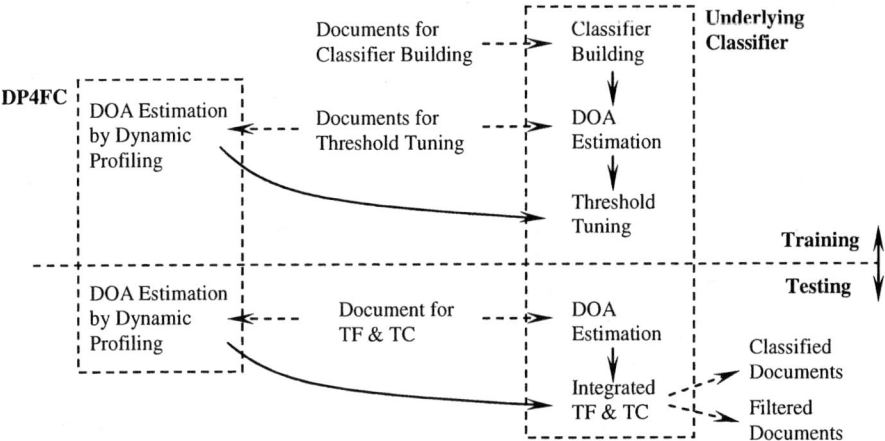

Fig. 1. Associating various classifiers with DP4FC

3 Dynamic Profiling for Filtering and Classification

Based on the above analysis, we develop a dynamic profiling technique DP4FC (Dynamic Profiling for Filtering Classification) to complement various classifiers to improve the performances of integrated TF and TC. Figure 1 illustrates the introduction of DP4FC to a classifier. In training, DP4FC joins the thresholding process, while in testing, DP4FC joins the process of making TF and TC decisions.

Both the underlying classifier and DP4FC estimate each document's DOA with respect to each category. The key point is that DOA values estimated by DP4FC are based on dynamic profiling, which aims to measure the extent to which a document's content overlaps that of a category. The algorithm is depicted in Table 1. Given a category c and a document d, it considers tow kinds of terms: those terms that occurs in c but not in d (ref. Step 2), and those terms that occur in d but not in c (ref. Step 3). Once a term t of the two kinds is found, the DOA value is reduced by its strength, which is estimated by a modified *tf×idf* (term frequency × inverse document frequency) technique (ref. Steps 2.1 and 3.1). The term frequency is replaced by the support of t in c (i.e. P(t|c) if t only occurs in c, ref. Step 2.1) or d (i.e. P(t|d) if t only occurs in d, ref. Step 3.1). P(t|c) is computed by [times t appears in c / total number of terms in c], and P(t|d) is computed by [times t appears in d / total number of terms in *d*]. On the other hand, the inverse document frequency (IDF) of *t* is modified to consider *d* as an additional training document, or more specially, IDF of *t* is computed by [(total number of training documents + 1) / number of documents (including *d* and training documents) in which *t* appears]. Therefore, a smaller DOA value indicates that *d* (*c*) talks more important information not in *c* (*d*), and hence indicates that we have a lower confidence to classify *d* into *c*, no matter whether *c* is the most suitable category for *d* or not.

Table 1. DOA estimation by dynamic profiling

Procedure *DOAEstimationByDP(c, d)*, where
   (1) *c* is a category,
   (2) *d* is a document for thresholding or testing
Return: DOA value of *d* with respect to *c*
Begin
   (**1**) DOAbyDP = 0;
   (**2**) For each term *t* in *c* but not in *d*, do
     (**2.1**) DOAReduction = Support(*t*, *c*) × $\log_2$(IDF of *t* in training data and *d*);
     (**2.2**) DOAbyDP = DOAbyDP - DOAReduction;
   (**3**) For each term *t* in *d* but not in *c*, do
     (**3.1**) DOAReduction = Support(*d*, *c*) × $\log_2$(IDF of *t* in training data and *d*);
     (**3.2**) DOAbyDP = DOAbyDP - DOAReduction;
   (**4**) Return DOAbyDP;
End.

With the DOA estimation, DP4FC may join the thresholding process to help the underlying classifier to derive proper thresholds for each individual category. The basic idea is that, each category has two thresholds: one for thresholding the DOA values produced by DP4FC, while the other is for thresholding the original DOA values produced by the underlying classifier. The former helps to filter out those irrelevant documents that would otherwise be noises for the setting the latter. The two thresholds work together in the hope to optimize the category's performance in a predefined criterion (e.g. F_1 = [2PR] / [P+R]).

Upon receiving a document to be filtered or classified, its two DOA values (i.e. by DP4FC and the underlying classifier) are produced, and the corresponding thresholds are consulted. The document may be classified into a category only if both DOA values are higher than or equal to their corresponding thresholds. That is, DP4FC and the underlying classifier actually work together to complement each other to make proper TF and TC decisions.

Time-complexity of dynamic profiling deserves analysis. The realization of DP4FC requires two main components: thresholding (conducted in training only) and DOA estimation (conducted in both training and testing, ref. Figure 1). As noted above, in thresholding, each document receives two DOA values, which are produced by DP4FC and the underlying classifier, respectively. Therefore, suppose a category has *n* documents used for thresholding, DP4FC needs to compute *n*×*n* combinations of DOA values. On the other hand, in DOA estimation, DP4FC needs to check those terms in category *c* but not in document *d* (ref. Step 2 in Table 1), and vice versa (ref.

Step 3 in Table 1). Therefore, there are at most $x + y$ computations, where x is the number of terms in c and y is the number of terms in d. DP4FC is thus efficient enough to realize the idea of dynamic profiling.

4 Experiments

Experiments were designed to investigate the contributions of DP4FC. To conduct objective and thorough investigation, DP4FC was evaluated under different circumstances, including (1) different sources of experimental data, (2) different kinds of test data, (3) different settings of training data, and (4) different settings for the classifier. Table 2 summarizes the different circumstances, which are to be explained in the following subsections.

Table 2. Experimental designs for thorough investigation

Aspects	Settings
(1) Source of experimental data	(A) Reuter-21578
	(B) Yahoo text hierarchy
(2) Split of test data	(A) In-space test data (for evaluating TC)
	(B) Out-space test data (for evaluating TF)
(3) Split of the training data for classifier building (CB) and threshold tuning (TT)	(A) 50% for CB; 50% for TT (with 2-fold cross validation)
	(B) 80% for CB; 20% for TT (with 5-fold cross validation)
(4) Parameter settings for the classifier	Different sizes of feature sets on which the classification methodologies were built

4.1 Experimental Data

Experimental data came from Reuter-21578, which was a public collection for related studies (http://www.daviddlewis.com/resources/testcollections/reuters21578). There were 135 categories (topics) in the collection. We employed the ModLewis split, which skipped unused documents and separated the documents into two parts based on their time of being written: (1) the *test* set, which consisted of the documents after April 8, 1987 (inclusive), and (2) the *training* set, which consisted of the documents before April 7, 1987 (inclusive). The test set was further split into two subsets: (1) the *in-space* subset, which consisted of 3022 test documents that belong to some of the categories (i.e. fall into the category space), and (2) the *out-space* subset, which consisted of 3168 documents that belong to none of the categories. They helped to investigate the systems' performances in TC and TF, respectively. An integrated TF and TC system should (1) properly classify in-space documents, and (2) properly filter out out-space documents.

As suggested by previous studies (e.g. [13]), the training set was randomly split into two subsets as well: the *classifier building* subset and the *threshold tuning* (or validation) subset. The former was used to build the classifier (to be described later), while the latter was used to tune a threshold for each category. Therefore, to guarantee that each category had at least one document for classifier building and one document for threshold tuning, we removed those categories that had fewer than 2 training documents, and hence 95 categories remained. Among the 95 categories, 12 categories had no test documents. From both theoretical and practical standpoints, these categories deserve investigation [5], although they were excluded by several previous studies (e.g. [13]). After removing those documents to which no categories were assigned (i.e. not belonging to any of the 95 categories), the training set contained 7780 documents. Moreover, since previous studies did not suggest the way of setting the documents for classifier building and threshold tuning, we will try different settings to conduct more thorough investigation: 50%-50% and 80%-20%, in which 2-fold and 5-fold cross validation were conducted, respectively. That is, 50% (80%) of the data was used for classifier building, and the remaining 50% (20%) of the data was used for threshold tuning, and the process repeated 2 (5) times so that each training document was used for threshold tuning exactly one time.

Moreover, to test those out-space documents that are less related to the categories, we randomly sample 370 documents from a text hierarchy extracted from http://www.yahoo.com [6]. The documents were randomly extracted from the categories of *science, computers and Internet,* and *society and culture,* and hence were less related to the content of the Reuters categories. With the help of the Yahoo out-space documents, we may measure the system's TF performance in processing those out-space documents with different degrees of relatedness to the Reuters categories.

4.2 Evaluation Criteria

The classification of in-space test documents and the filtering of out-space test documents require different evaluation criteria. For the former, we employed precision (P) and recall (R). Both P and R were common evaluation criteria in previous studies. P was estimated by [total number of correct classifications / total number of classifications made], while R was estimated by [total number of correct classifications / total number of correct classifications that should be made]. To integrate P and R into a single measure, the well-known F-measure was employed: $F_\beta = [(\beta^2+1)PR] / [\beta^2 P+R]$, where β is a parameter governing the relative importance of P and R. As in many studies, we set β to 1 (i.e. the F_1 measure), placing the same emphasis on P and R.

Note that P, R, and F_1 were "micro-averaged" rather than "macro-averaged". Macro-averaged F_β was the average of the F_β values for *individual* categories, where the F_β value for a category c was computed based on precision and recall for c [13]. It was not employed in the experiment, since we included those categories that had no test documents (for the reasons noted above, ref. Section 4.1), making precision and recall values for these categories incomputable (since the denominators for computing the values could be zero).

On the other hand, to evaluate the filtering of out-space test documents, we employed two criteria: filtering ratio (FR) and average number of misclassifications for misclassified out-space documents (AM). FR was estimated by [number of out-space documents filtered out / number of out-space documents], while AM was estimated by [total number of misclassifications / number of out-space documents misclassified into the category space]. An integrated TF and TC system should reject more out-space documents (i.e. higher FR) and avoid misclassifying out-space documents into many categories (i.e. lower AM). As P and R, FR and AM complemented each other by focusing on different aspects. For example, suppose there are M out-space documents, and system A misclassifies 1 out-space document into 2 categories, and system B misclassifies 2 out-space documents into 2 categories. Although both systems make 2 misclassifications, system A is better in FR ([M-1]/M vs. [M-2]/M), while system B is better in AM (2/1 vs. 2/2). FR and AM may thus support more in-depth comparison of system performances.

4.3 The Underlying Classifier

Each category c was associated with a classifier, which was based on the Rocchio method (RO). Upon receiving a document d, the classifier estimated the similarity between d and c (i.e. DOA of d with respect to c) in order to make a binary decision for d: accepting d or rejecting d. The system that applied DP4FC to RO was named RO+DP4FC. By comparing the performances of RO and RO+DP4FC, we may identify the contributions of DP4FC.

RO was commonly employed in TC (e.g. [12]), TF (e.g. [10] [11]), and retrieval (e.g. [3]). Some studies even showed that its performances were more promising in several ways (e.g. [6] [7]). RO constructed a vector for each category, and the similarity between a document d and a category c was estimated using the cosine similarity between the vector of d and the vector of c. More specially, the vector for a category c was constructed by considering both relevant documents and non-relevant documents of c: $\eta_1 * \sum_{Doc \in P} Doc/|P| - \eta_2 * \sum_{Doc \in N} Doc/|N|$, where P was the set of vectors for relevant documents (i.e. the documents in c), while N was the set of vectors for non-relevant documents (i.e. the documents not in c). We set $\eta_1=16$ and $\eta_2=4$, since the setting was shown to be promising in previous studies (e.g. [12]).

RO required a fixed (predefined) feature set, which was built using the documents for classifier building. The features were selected according to their weights, which were estimated by the χ^2 (chi-square) weighting technique. The technique has been shown to be more promising than others [14]. As noted above, there is no perfect way to determine the size of the feature set. Therefore, to conduct more thorough investigation, we tried 5 feature set sizes, including 1000, 5000, 10000, 15000, and 20000 (there were about 20000 different features in the 2-fold training data).

To make TF and TC decisions, RO also required a thresholding strategy to set a threshold for each category. As in many previous studies (e.g. [10] [13] [15]), RO tuned a relative threshold for each category by analyzing document-category similarities. The threshold tuning documents were used to tune each relative threshold. As suggested by many studies (e.g. [13]), the thresholds were tuned in the hope to optimize the system's performance with respect to F_1.

4.4 Result and Discussion

Figure 2 illustrates the performance (in F_1) for in-space documents, while Figure 3 and Figure 4 illustrates the performance for out-space documents (FR and AM, respectively). The results indicate the following contributions provided by DP4FC:

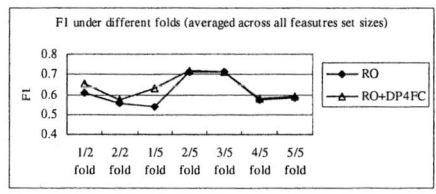 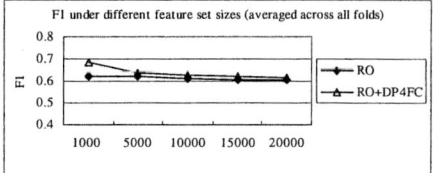

Fig. 2. Performance (in F_1) in processing in-space documents

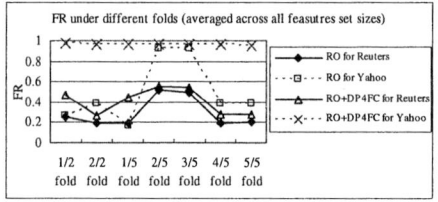 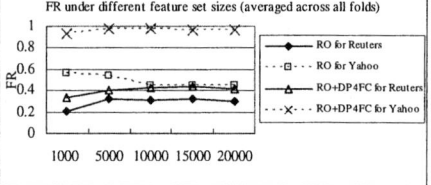

Fig. 3. Performance (in FR) in processing out-space documents

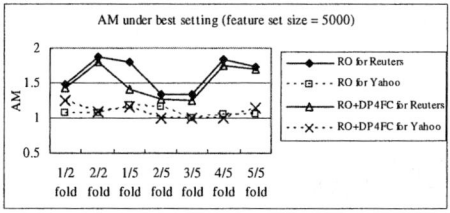

Fig. 4. Performance (in AM) in processing out-space documents

(1) For in-space documents, DP4FC helped RO to achieve better performances. As shown in Figure 2, RO+DP4FC outperformed RO under all different circumstances (i.e. different folds and feature set sizes). When comparing the average performances under all the circumstances, it provided 4.2% improvement in F_1 (0.6383 vs. 0.6127).

(2) For out-space documents from Reuters, DP4FC helped RO to achieve both better and stable performances. As shown in Figure 3, under all different circumstances, DP4FC+RO filtered out more Reuters out-space documents than RO. When comparing the average performances under all the circumstances, DP4FC provided 38.1% improvement in FR (0.4039 vs. 0.2924). Moreover, as shown in Figure 4, DP4FC also provided contributions in reducing AM (9.3% improvement, 1.5080 vs. 1.6617).

(3) For out-space documents from Yahoo, DP4FC helped RO to achieve both better and stable performances as well. As shown in Figure 3, under all different circumstances, DP4FC successfully filtered out almost all out-space documents from Yahoo (recall that the documents from Yahoo were less related to the categories). On the other hand, the performance of RO dramatically oscillated under different circumstances. When comparing the average performances under all the circumstances, DP4FC provided 95.9% improvement in FR (0.9637 vs. 0.4920). As

shown in Figure 4, both systems achieved a similar performance in AM (1.1504 vs. 1.1364).
(4) In the 2-fold experiment, even under the setting that leads RO to achieve the best performance in processing in-space documents, DP4FC provided significant contributions. RO achieved its best performance in F_1 when the feature set size was 5000 in the 1st fold. Under such a setting, DP4FC provided (A) 5.5% improvement in F_1 (0.6609 vs. 0.6267), (B) 46.4% improvement in Reuters FR (0.5073 vs. 0.3466), and (C) 204.2% improvement in Yahoo FR (0.9784 vs. 0.3216).
(5) In the 5-fold experiment, even under the setting that leads RO to achieve the best performance in processing in-space documents, DP4FC provided significant contributions as well. RO achieved its best performance in F_1 when the feature set size was 5000 in the 3rd fold. In this case, DP4FC provided (A) similar performance in F_1 (0.7222 vs. 0.7245), (B) 8.3% improvement in Reuters FR (0.5290 vs. 0.4886), and (C) 2.5% improvement in Yahoo FR (0.9784 vs. 0.9541).

5 Conclusion

Given an information space spanned by a set of categories, lots of documents may be entered at any time, but only a small portion of them may be classified into the information space. Misclassification of documents into the information space may deteriorate the management, dissemination, and retrieval of information. We thus present a technique DP4FC to complement and enhance a classifier's capability in mining category profiles. Instead of distinguishing a category from other categories, DP4FC measures whether a document d (a category c) talks too much information not in a category c (d), since in that case d could not be classified into c, even though d mentions some discriminative content of c. To achieve that, DP4FC helps the underlying classifier to create dynamic category profiles with respect to each individual document. It then works with the classifier to set proper thresholds, and accordingly make proper TF and TC decisions. Empirical results show that DP4FC may help the underlying classifier to achieve both better and more stable performances. The contributions are of both theoretical and practical significance to the classification of suitable information into suitable categories.

Acknowledgement

This research was supported by the National Science Council of the Republic of China under the grants NSC 94-2213-E-320 -001.

References

1. Arampatzis, A., Beney, J., Koster, C. H. A., and Weide, T. P. van der: Incrementality, Half-life, and Threshold Optimization for Adaptive Document Filtering. In Proceedings of the 9th Text Retrieval Conference (2000), pp. 589-600, Gaithersburg, Maryland
2. Cohen W. W. and Singer Y.: Context-Sensitive Mining Methods for Text Categorization. In Proceedings of the 19th annual international ACM SIGIR conference on research and development in information retrieval (1996), Zurich, Switzerland

3. Iwayama, M.: Relevance Feedback with a Small Number of Relevance Judgments: Incremental Relevance Feedback vs. Document Clustering. In Proceedings of the 23rd annual international ACM SIGIR conference on research and development in information retrieval (2000), pp. 10-16. Athens, Greece
4. Iyengar, V. S., Apte, C., and Zhang, T.: Active Learning using Adaptive Resampling. In Proceedings of the 6th ACM SIGKDD International Conference on Knowledge Discovery and Data Mining (2000), pp. 91-98. Boston, Massachusetts
5. Lewis D. D.: Reuters-21578 text categorization test collection Distribution 1.0 README file (v 1.2). http://www.daviddlewis.com/resources/testcollections/reuters21578_(1997)
6. Liu R.-L. and Lin W.-J.: Incremental Mining of Information Interest for Personalized Web Scanning, Information Systems (2005), Vol. 30, Issue 8, 630-648
7. Liu R.-L. and Lin W.-J.: Adaptive Sampling for Thresholding in Document Filtering and Classification, Information Processing and Management (2004), Vol. 41, Issue 4, 745-758
8. Mladenić D., Brank J., Grobelnik M., and Milic-Frayling N.: Feature Selection using Linear Classifier Weights: Interaction with Classification Models, In Proceedings of the 27th annual international ACM SIGIR conference on research and development in information retrieval (2004), pp. 234-241, Sheffield, South Yorkshire, UK
9. Mladenić D. and Grobelnik M.: Feature Selection for Classification based on Text Hierarchy, Proc. of the Conference on Automated Learning and Discovery (1998)
10. Schapire, R. E., Singer, Y., and Singhal, A.: Boosting and Rocchio Applied to Text Filtering. In Proceedings of the 21st annual international ACM SIGIR conference on research and development in information retrieval (1998), pp. 215-223. Melbourne, Australia
11. Singhal, A., Mitra, M., and Buckley, C.: Learning Routing Queries in a Query Zone. In Proceedings of the 20th annual international ACM SIGIR conference on research and development in information retrieval (1997), pp. 25-32. Philadelphia, Pennsylvania
12. Wu, H., Phang, T. H., Liu, B., and Li, X.: A Refinement Approach to Handling Model Misfit in Text Categorization. In Proceedings of the 8th ACM SIGKDD International Conference on Knowledge Discovery and Data Mining (2002), pp. 207-216. Edmonton, Alberta, Canada
13. Yang, Y.: A Study of Thresholding Strategies for Text Categorization. In Proceedings of the 24th annual international ACM SIGIR conference on research and development in information retrieval (2001), pp. 137-145. New Orleans, Louisiana
14. Yang, Y. and Pedersen, J. O.: A Comparative Study on Feature Selection in Text Categorization. In Proceedings of the 14th International Conference on Machine Learning (1997), pp. 412-420. Nashville, Tennessee
15. Zhang, Y. and Callan, J.: Maximum Likelihood Estimation for Filtering Thresholds. In Proceedings of the 24th annual international ACM SIGIR conference on research and development in information retrieval (2001), pp. 294-302. New Orleans, Louisiana

Detecting Citation Types Using Finite-State Machines

Minh-Hoang Le, Tu-Bao Ho, and Yoshiteru Nakamori

School of Knowledge Science,
Japan Advanced Institute of Science and Technology,
1-1, Asahidai, Nomi, Ishikawa 923-1292, Japan
{hoangle, bao, nakamori}@jaist.ac.jp

Abstract. This paper presents a method to extract citation types from scientific articles, viewed as an intrinsic part of emerging trend detection (ETD) in scientific literature. There are two main contributions in this work: (1) Definition of six categories (types) of citations in the literature that are extractable, human-understandable, and appropriate for building the interest and utility functions in emerging trend detection models, and (2) A method to classify citation types using finite-state machines which does not require user-interactions or explicit knowledge. The experimental comparative evaluations show the high performance of the method and the proposed ETD model shows the crucial role of classified citation types in the detection of emerging trends in scientific literature.

1 Introduction

Emerging trend detection (ETD) is a new and challenging problem in text mining. ETD is commonly defined as "detecting topic areas which are growing in interest and utility over time" [1]. Recently, several ETD models have been proposed [2, 3] in which the ETD process can be viewed in three phases: topic representation, identification, and verification. Each topic — the ETD central notion — is usually represented by a set of temporal features in the phase of *topic representation*. These features are then extracted from document databases using text-processing methods in the *topic identification* phase. After that, the *topic verification* phase plays the role of monitoring these features over time and classifying the topic by using interest and utility functions [1].

One very significant task for ETD is to find emerging research trends from a collection of scientific articles. This can help researchers quickly understand the occurrence and the tendency of a scientific topic, and thus they can, for example, find the most recent, related topics in the research domain. However, existing ETD models are still poor in representing research topics and inappropriate for determining and ranking interest and utility. Motivated by the need of a more appropriate model for emerging trend detection from scientific corpora, our ultimate target is to build an ETD model which has a richer representation scheme for topics, and to use citation information as one of the characteristics of the ETD model.

Citations appear very frequently in scientific articles and most of digital libraries now organize their papers in the structure of citation indexes [4]. By examining the citations inside an article, we can reveal relationships between articles, draw attention to important corrections of published work and identify significant improvements or criticisms of earlier work [5, 6]. However, this is still very difficult for researchers because the large and increasing number of articles prevents them from reading everything in the published literature. There is a clear need for new tools to identify the types of citation relationships that indicate the reasons for citation in a human-understandable way [7].

The purpose of identifying the reasons for citations (citation type detection - CTD) varies according to the main objective of each research. The method of Nanba and Okumura [8] uses an heuristic sentence selection and pre-defined cue phrases to classify citations into three categories for supporting a system of automatic review articles. To extend the usage of linguistic patterns, Teufel [9] uses formulaic expressions, agent patterns and semantic verb classes instead of cue phrases to determine the corresponding class for a sentence. Although both these works show the usefulness of linguistic patterns in citation type detection, the manual construction of linguistic patterns is obviously a rather time-consuming task. It also involves some conflicts that are difficult to be resolved. For example, the method of Pham and Hoffmann [10] has to eliminate such conflicts and send to human experts for providing rules that resolved them.

The available methods do not appear to be integrated into an ETD process because of two main limitations: the first is their definitions of citation types are not appropriate for evaluating the interest and utility of topics; the second reason is the manual construction of linguistic patterns must depend on the corpus. This makes the detection process become inflexible when applying to other corpora. The work presented in this paper is an intrinsic part of the construction of an emerging trend detection model for scientific corpora, for that we propose an automatic method for detecting citation types. The significant differences of our method compared to other works are: (1) the defined six categories of the reasons for citations which support the detection of emerging trends by tracing the development of a topic and clarify the relationship between articles; (2) our method using finite-state machines can detect citation types without any need for user-interactions or explicit knowledge about linguistic patterns as were required in [8, 9, 10].

In the following section, we first define the six citation types and then propose a method for detecting citation types using two kinds of finite-state machines: HMMs and MEMMs. Section 3 describes the experimental comparative evaluations. In Section 4, we briefly introduce our proposed ETD model and the integration of citation types into the interest and utility functions. Conclusions and future works are given in the last section.

2 A Method for Citation Type Detection

2.1 Definition of Citation Types

Given a paragraph containing citations (we call this paragraph the citing area), we want to detect why the cited paper is mentioned in the purpose of the authors written in this paragraph. It is well known that there are many reasons for citations (citation types). To classify citing areas using citation relationships, we also have to consider the citation types. For example, in [11], Weinstock proposed 15 categories for the common reasons of citations, to build a system for the automatic generation of review articles, Nanba and Okumura [8] classified the reasons for citations into three categories while Pham and Hoffmann [10] used four types of citations for building a citation map between articles.

In order to support researchers in tracing the development of a topic over time as well as clarify the relationship between articles, we classified citation types into the following six main categories (or classes), which are important for emerging trend detection:

Type 1: The paper is based on the cited work; it means that the citation shows other researchers' theories or methods as the theoretical basis for the current work. (corresponding to Nanba's type B)
Type 2: The paper is a part of the cited work
Type 3: The cited work supports this work
Type 4: The paper points out problems or gaps in the cited work (corresponding to Nanba's type C, Pham's type Limitation)
Type 5: The cited work is compared with the current work
Type 6: Other citations

Note that these classes are overlapping, meaning that a citation area may belong to two or more classes. We will choose the most suitable class label for a citation area and also measure the likelihood of each citing area on a class. Details of technique are discussed in the following sections.

2.2 Citation Type Detection Using Finite-State Machines

In this section, we describe the method for detecting citation types. The detection process can be described as follows: Given a citing area consisting of several sentences, we apply finite-state machines to compute the likelihood of each sentence on each class. After that, we evaluate the importance of each sentence and combine these values to identify the corresponding class for this citing area. We present here two methods to evaluate the above likelihood using hidden Markov models and maximum-entropy Markov models, after that we will introduce the sentence-weighting strategy to identify class label for a given citing area.

Sentence Evaluation Using Hidden Markov Models. A hidden Markov model (HMM) is a finite-state automaton with stochastic state transitions and observations whereby a sequence of observations is emitted along the transitions of states over time [12]. A HMM $\lambda = (A, B, \Pi)$ is defined on a set of states S, a set of possible observations O and three probability distributions: a state transition probability to $s_j \in S$ from $s_i \in S$: $a_{ij} = P(s_j|s_i)$; an observation probability distribution $b_j(o) = P(o|s_j)$ for $o \in O$, $s_j \in S$; and an initial state distribution for each state $s_i \in S$: $\pi_i = P(q_1 = s_i)$.

In most text-processing tasks using HMMs, people often use word-based models, i.e, each word (or n-gram) is one observation. The main drawback of these methods is the machine cannot accept unknown observation symbols or accepts them with a very low probability of emission functions. For example, if we consider each English word as an observation, the model trained by the sentence "*The man walks so fast*" may produce 0 or a small value depending on the training algorithm when computing the likelihood of the sentence "*The man goes so fast*" even though the meaning of the second sentence can be implied from the semantics of the trained sentence. This problem occurs not only with finite-state machines, but also with all word-based methods.

One solution to this problem is enlarging the training set so as to cover all possible cases of synonymy and hyponymy. However, it is difficult to build a large training set and it also increases the complexity of training phase. For example, the method using cue phrases [8] has to construct a very long list of cue phrases; the rule-based method [10] has to add many rules to the rule set in order to achieve high accuracy.

To overcome the drawback of the aforementioned solution, we still use word-based models, but after the training phase, we re-adjust the emission functions of the HMMs so as to deal with the synonym and hyponym of words by:

$$\bar{b}_j(o) = \max_{o' \subseteq o} b_j(o') \qquad (1)$$

where $o' \subseteq o$ means the word o' is a hyponym or synonym of the word o.

For detecting citation types, we used six HMMs, each HMM consisting of n states: $S = \{s_1, s_2, ..., s_n\}$ and accepting the set of English words including "\cite" as the set of observations O. In the following explanation, we denoted q_t and o_t as the state of the model and the observation at time t, respectively.

In the training phase, we have a number of training sentences for each class. These sentences are used as the input of the training algorithm for estimating model parameters. The standard method to train HMMs is the EM algorithm, also known in HMM context as the Baum-Welch algorithm [12]. However, we use the Viterbi training (VT) algorithm instead of EM to avoid expensive computation in practice. The VT algorithm just takes the single most likely path and maximize the probability of emitting the observation sequence along its corresponding path. The details of the Viterbi Training Algorithm is described in [12].

Given an unknown sentence O and six trained HMMs corresponding to six classes, we compute how well the sentence O matches these HMMs by calculating the probability of generating sentence O along its best path on each HMM:

$$P^*(O|\lambda) = \arg\max_Q P(O,Q|\lambda) = P\left(O, Q^{(O)}|\lambda\right) \quad (2)$$

where $Q^{(O)}$ is the state sequence found by Viterbi algorithms.

Sentence Evaluation Using Maximum-Entropy Markov Models. The structure of maximum-entropy Markov models (MEMMs) is similar to that of hidden Markov models, but instead of transition and observation probabilities, we have only one single function $P(s|s',o)$ which provides the probability of the current state s given the previous state s' and the current observation o. This complex function is often separated in to $\|S\|$ transition functions $P_{s'}(s|o)$. In contrast to HMMs, in which the current observation only depends on the current state, in MEMMs, the current observation may also depend on the previous state. It means the observations is associated with state transition rather than with states [13].

In MEMMs, each transition function $P_{s'}(s,o)$ is often represented in exponential form:

$$P_{s'}(s,o) = \frac{1}{Z(o,s')} \exp\left(\sum_a \gamma_a f_a(o,s)\right) \quad (3)$$

where f_a is a feature, γ_a is a parameter to be learned and $Z(o,s')$ is the normalizing factor that makes the distribution sum to one across all next state s.

To find the corresponding state sequence to an observation sequence, we can still use an efficient dynamic programming algorithm by modifying some equations of the Viterbi algorithm for HMMs [13]. To train a MEMM, we first split the training data into (state-observation) pairs relevant to the transitions from each state s', then apply the Generalized Iterative Scaling method (GIS) [14] to estimate the transition function for state s' (f'_s).

To measure how well a sentence matches a MEMM, we first organize all word concepts in a concept hierarchy, in which each node in the hierarchy consists of a word and its synonyms and a sub-concept is represented by a descendant of its parent concepts. The synonymy and hyponymy relationships between words are represented by feature functions of MEMMs:

$$f_{(c,q)}(w,s) = \begin{cases} 1, & if\ (s=q) \wedge (w \in c) \\ 0, & otherwise \end{cases} \quad (4)$$

where c represented for a node in the concept hierarchy, w is a word and $w \in c$ means the concept c accepts the word w as its synonym or hyponym.

Similar to HMMs we can find the best path for a given sentence O and use $P^*(O|\lambda)$ to measure how well the sentence O matches the MEMM λ

Weighting Sentences and Classification of Citing Areas. Consider a kind of finite-state machine, HMM or MEMM. We have a total of six machines $\{\lambda_i\}_{i=1}^6$ corresponding to six classes. Given an unknown sentence O, we find the best state sequence Q_i^O corresponding to O in each machine λ_i and compute the likelihood $P^*(O|\lambda_i) = P(O, Q_i^O|\lambda_i)$ to measure how closely the sentence O matches the machine λ_i.

A citing area might consist of many sentences; each sentence can match all six machines with different levels. We need to combine these likelihoods in order to determine which class is suitable for the entire citing area. To this end, we want to determine the importance of each sentence in evaluating the citing area.

Given a sentence O, and a finite-state machine λ_i, we compute $P^*(O|\lambda_i)$ and define:

$$P^{(O)}(\lambda_i) = \frac{P^*(O|\lambda_i)}{\sum_{j=1}^{6} P^*(O|\lambda_j)} \quad (5)$$

as the probability of selecting the model λ_i given the sentence O. The entropy of this probability distribution is:

$$H^{(O)} = -\sum_{i=1}^{6} P^{(O)}(\lambda_i) \log_2 P^{(O)}(\lambda_i) \quad (6)$$

As the entropy $H^{(O)}$ becomes larger, the chance of selecting the model corresponding to sentence O becomes more uncertain, and the the role O plays in determining class label for the citing area becomes less important. Thus, we can weight each sentence O in the citing area by

$$Weight(O) = \frac{\log_2 6 - H^{(O)}}{\log_2 6}; (0 \leq Weight(O) \leq 1) \quad (7)$$

If the citing area C consists of m sentences: $O^1, O^2, \ldots O^m$. The corresponding citation type for this citing area is:

$$Type(C) = \arg \max_{1 \leq i \leq 6} \sum_{j=1}^{m} Weight(O^j).P^*(O^j|\lambda_i) \quad (8)$$

To use citation types more flexibly, instead of assigning a class label for a given citing area, we can compute how closely a given citing area matches a category i by measuring the likelihood:

$$L(C|i) = \frac{\sum_{j=1}^{m} Weight(O^j).P^*(O^j|\lambda_i)}{\sum_{i'=1}^{6} \sum_{j=1}^{m} Weight(O^j).P^*(O^j|\lambda_{i'})} \quad (9)$$

Making a model that analyzes the entire citing area requires many complicated computations and a very large training set. Like other methods, our method segments the citing area into sentences and classifies it by evaluating the sentences. However, instead of selecting only one sentence for evaluating the whole citing area, we evaluate the likelihood of each sentence on each class, and use the weight of each sentence to combine these likelihoods in a reasonable way.

From theoretical viewpoint, before doing experiments, it is worth noting that our method can be extended to deal with more citation types. It takes into account the problem of word synonymy and hyponymy, allows overlapping between classes and works without any user-interactions or pre-defined linguistic patterns. That can be viewed as a significant difference between our citation type detection method and previous works.

3 Experiments

We designed two experiments for two purposes: first, we want to evaluate if the model using FSMs is more appropriate than other methods using linguistic patterns in the task of detecting citation types; secondly we want to compare two methods using HMMs and MEMMs and discuss the advantages and drawbacks of each model in practice.

The concept hierarchy is built from WordNet [15] in which each node – a concept – consists of a word and its synonyms, a sub concept (hyponym) is placed in the hierarchy as a descendant of its parent concepts. These experiments used HMMs and MEMMs with 25 states (This is the average of number of words in each sentence). Increasing the number of states may improve the classification results, but requires longer computational time in the training and testing phases.

3.1 Experiment 1

This experiment is used to evaluate if our method achieves higher accuracy compared to Nanba and Okumura's method when running in the same conditions. The data set provided by Nanba and Okumura in [8] consists of 282 citing area for training and 100 citing area for testing. We use the same definition of citation types as they defined: B, C and O and select training sentences according to their sentence selection strategy. Table 1 shows the accuracy of Nanba and Okumura's method comparing to our methods.

Table 1. The accuracies of Nanba and Okumura's method, HMMs, and MEMMs

	Nanba				HMMs				MEMMs			
	C	B	O	(%)	C	B	O	(%)	C	B	O	(%)
16 citations type C	12	0	4	75.0	14	0	2	87.5	14	0	2	87.5
32 citations type B	2	25	5	78.1	0	25	7	78.1	0	26	6	81.3
52 citations type O	1	5	46	88.5	3	1	48	92.3	1	1	50	96.1

Running under the same conditions, our method using HMMs and MEMMs based on concept-representation achieve higher accuracy than Nanba's method. Although the set of cue phrases is well designed for this dataset, Nanba's method still has the problem of synonymy and hyponymy, that why our method using concept-representation can result in higher accuracy.

3.2 Experiment 2

This experiment is used to compare the performance of two methods using HMMs and MEMMs. To this end, we collect 9000 papers from two main sources: ACM Digital Library and Science Direct, and randomly select 811 citing areas for this experiment. For a limited number of sentences for training, we randomly selected sentences from these 811 citing area and run the experiment 10 times before taking an average of accuracy. Table 2 shows the detection accuracies of the methods using HMMs and MEMMs.

Table 2. The accuracies of two methods using HMMs and MEMMs

Number of training sentences	HMMs (%)	MEMMs (%)
100	60.1	61.4
200	67.1	67.2
300	72.6	73.8
400	79.9	79.6
500	84.9	86.6
600	90.4	91.8
700	95.2	95.9
800	99.5	99.7
811	100.0	100.0

The method using MEMMs produced slightly better result than HMMs as shown in Table 2. In addition, the method using MEMMs requires lower computation time for the training phase: it takes 7918 seconds for training MEMMs with 800 sentence compared to 20168 seconds taken by the VT algorithm. The main reason is not only the different characteristics of HMM training and MEMM training algorithms, but also because we must re-distribute the emission functions of HMMs to deal with the synonymy and hyponymy relationships between words while we can model these relations by feature functions of MEMMs.

4 Integration of CTD into the ETD Model

Because the details of our ETD model is out of the scope of this paper, we will briefly describe the structure of the ETD model and the key idea of building the interest and utility functions to detect emerging trends, including the integration of citation types into ETD process.

In our ETD model, each topic t_i in T is a node in the topic hierarchy, which is associated with a time series:

$$t_i = \left(t_i^1, t_i^2, \ldots, t_i^\Delta\right)$$

where Δ is the length of the trial period.

Given a year k^{th} in the trial period, we denoted t_i^k as the topic t_i in this year. Each t_i^k is a vector in 6-dimensional space: $t_i^k = \left(t_i^k(1), t_i^k(2), t_i^k(3), t_i^k(4), t_i^k(5), t_i^k(6)\right)$, where:

- $t_i^k(1)$: determine how often the topic t_i is mentioned in the year k^{th}
- $t_i^k(2)$: the weight of citations type 1, 3, and 5 in the year k^{th} to t_i
- $t_i^k(3)$: the number of citations in the year k^{th} to t_i
- $t_i^k(4)$: the influence of t_i on other topics in the year k^{th}
- $t_i^k(5)$: the weight of author of t_i in the year k^{th}
- $t_i^k(6)$: the weight of journal/proceedings talking about t_i in the year k^{th}

The topic verification module will monitor these features along the time-series to evaluate the growth in interest and utility of the topic. In our ETD model, the growths of all six time-series $\{t_i^k(j)\}_k$ ($1 \leq j \leq 6$) are independently evaluated and integrated into interest and/or utility functions. In concrete terms, the growth in interest of each topic is evaluated using four time-series $\{t_i^k(1)\}_k$, $\{t_i^k(3)\}_k$, $\{t_i^k(5)\}_k$, and $\{t_i^k(6)\}_k$; similarly, the growth in utility of each topic is evaluated using $\{t_i^k(2)\}_k$, $\{t_i^k(4)\}_k$, $\{t_i^k(5)\}_k$, and $\{t_i^k(6)\}_k$.

The citation information is used in both the interest and utility functions. Only citation types 1, 3, and 5 are integrated into the utility function while the number of citations, regardless of citation type, is used to evaluate the interest of each topic. We then consider each pair $(time, value)$ as a data point, then use regression analysis to predict the dependence of values on the time. The simplest way is to apply linear regression on all data points and use the slope co-efficient of the regression equation to evaluate the global tendency of the time-series.

Citation types can help us understand the research context, select papers for background reading, and identify problems or gaps in related works. In addition, as the topics of recent papers are not always novel and attractive, using citation information is an appropriate way to find the most recent and important topics in a research domain. The integration of these usages of citations into the emerging trend detection process is our ongoing work.

5 Conclusion

We have proposed a method to detect the reasons for citations. By defining six classes of citation types, we developed a method using finite-state machines to evaluate how closely a citing area matches a class. Our method is robust to the problem of synonymy and hyponymy, achieved better accuracy that previous works. In addition, our method using finite-state machines requires neither user-interactions nor explicit knowledge about cue phrases, so it has more flexibility to be extended. We believe this method can be improved and applied to other text-processing tasks, such as named-entity classification, document ranking, text segmentation, emerging trend detection, etc.

References

1. April Kontostathis, Leon Galitsky, William M. Pottenger, Soma Roy, and Daniel J. Phelps. A survey of emerging trend detection in textual data mining. In Michael Berry, editor, *A Comprehensive Survey of Text Mining*, chapter 9. Springer-Verlag, 2003.
2. William M. Pottenger and Ting-Hao Yang. Detecting emerging concepts in textual data mining. *Computational information retrieval*, pages 89–105, 2001.
3. Russell Swan and James Allan. Automatic generation of overview timelines. In *SIGIR '00: Proceedings of the 23rd annual international ACM SIGIR conference on Research and development in information retrieval*, pages 49–56, New York, NY, USA, 2000. ACM Press.
4. H. Small. Co-citation in the scientific literature: A new measure of the relationship between two documents. *Journal of the American Society of Information Science*, 24:265–269, 1973.
5. Steve Lawrence, C. Lee Giles, and Kurt Bollacker. Digital libraries and autonomous citation indexing. *IEEE Computer*, 32(6):67–71, 1999.
6. Ronald N. Kostoff, J. Antonio del Rio, James A. Humenik, Esther Ofilia Garcia, and Ana Maria Ramirez. Citation mining: integrating text mining and bibliometrics for research user profiling. *Journal of the American Society for Information Science and Technology*, 52(13):1148–1156, 2001.
7. David R. Gevry. Detection of emerging trends: Automation of domain expert practices, 2002.
8. Hidetsugu Nanba and Manabu Okumura. Towards multi-paper summarization using reference information. In *Proceedings of 16th International Joint Conference on Artificial Intelligence – IJCAI'99*, pages 926–931, 1999.
9. Simone Teufel. *Argumentative Zoning: Information Extraction from Scientific Text*. PhD thesis, University of Edinburgh, 1999.
10. Son Bao Pham and Achim G. Hoffmann. A new approach for scientific citation classification using cue phrases. In *Australian Conference on Artificial Intelligence*, pages 759–771, 2003.
11. Melvin Weinstock. Citation indexes. *Encyclopedia of Library and Information Science*, 5:16–41, 1971.
12. Lawrence R. Rabiner. A tutorial on hidden markov models and selected applications in speech recognition. In *Proceedings of the IEEE*, volume 77:2, pages 257–286. IEEE, 1989.
13. Andrew McCallum, Dayne Freitag, and Fernando Pereira. Maximum entropy Markov models for information extraction and segmentation. In *Proceedings of the 17th International Conference on Machine Learning*, pages 591–598, 2000.
14. J.N. Darroch and D. Ratcliff. Generalized iterative scaling for log-linear models. *The Annals of Mathematical Statistics*, pages 1470–1480, 1972.
15. Wordnet: A lexical database for the english language. http://wordnet.princeton.edu.

A Systematic Study of Parameter Correlations in Large Scale Duplicate Document Detection

Shaozhi Ye[1],*, Ji-Rong Wen[2], and Wei-Ying Ma[2]

[1] Department of Computer Science, University of California, Davis
sye@ucdavis.edu
[2] Microsoft Research Asia
{jrwen, wyma}@microsoft.com

Abstract. Although much work has been done on duplicate document detection (DDD) and its applications, we observe the absence of a systematic study of the performance and scalability of large-scale DDD. It is still unclear how various parameters of DDD, such as similarity threshold, precision/recall requirement, sampling ratio, document size, correlate mutually. In this paper, correlations among several most important parameters of DDD are studied and the impact of sampling ratio is of most interest since it heavily affects the accuracy and scalability of DDD algorithms. An empirical analysis is conducted on a million documents from the TREC .GOV collection. Experimental results show that even using the same sampling ratio, the precision of DDD varies greatly on documents with different size. Based on this observation, an adaptive sampling strategy for DDD is proposed, which minimizes the sampling ratio within the constraint of a given precision threshold. We believe the insights from our analysis are helpful for guiding the future large scale DDD work.

1 Introduction

Duplicate pages and mirrored web sites are phenomenal on the web. For example, it was reported that more than 250 sites mirrored the documents of Linux Document Project (LDP)[1]. Broder *et al.* clustered the duplicated and nearly-duplicated documents in 30 millions documents and got 3.6 millions clusters containing 12.1 millions documents [1]. Bharat and Broder reported that about 10% of hosts were mirrored to various extents in a study involving 238,000 hosts [2].

Because of the high duplication of Web documents, it is important to detect duplicated and nearly duplicated documents in many applications, such as crawling, ranking, clustering, archiving, and caching. On the other hand, the tremendous volume of web pages challenges the performance and scalability of DDD algorithms. As far as we know, Broder *et al.* for the first time proposed a DDD algorithm for large-scale documents sets in [1]. Many applications and

---
* This work was conducted when this author visited Microsoft Research Asia.
[1] http://www.linuxdoc.org

following research, such as [2] [3] [4] [5] [6], later adopted this algorithm for its simplicity and efficiency.

While much work has been done on both DDD algorithms and their applications, little has been explored about the factors affecting their performance and scalability. Meanwhile, because of the huge volume data, all prior work makes some kinds of tradeoffs in DDD. How do these tradeoffs affect accuracy? To our best knowledge, no previous work conducts any systematic analysis on correlations among different parameters of DDD, and none of them provides a formal evaluation of their tradeoff choices.

This paper studies several of the most important parameters of DDD algorithms and their correlations. These parameters include similarity threshold, precision/recall requirement, sampling ratio, document size. Among them, sampling ratio is of most interest, for it greatly affects the accuracy and scalability of DDD algorithms.

To uncover the correlations of parameters, an empirical analysis is conducted in this paper. The TREC .GOV collection[2] are used as our testing dataset. Although the volume of this collection is much smaller than the whole Web, we believe that this collection to some extent represents the Web well for DDD algorithms [7]. Experiment results show that even using the same sampling ratio, the precision of DDD in documents of different size varies greatly. To be more specific, small sampling ratio heavily hurts the accuracy of DDD for small documents. Based on this observation, we propose an adaptive sampling method for DDD which uses dynamic sampling ratio for different document size with constraint of given precision thresholds. We believe that our analysis is helpful for guiding the future DDD work.

The remainder of this paper is organized as follows. Section 2 reviews the prior work on DDD. Section 3 describes the duplicate detection algorithm and the definition of *document similarity* used in this paper. Section 4 presents the experimental results on parameter correlations, and then proposes an adaptive sampling strategy. Finally we conclude this paper with Section 6.

2 Prior Work

The prior work of duplicate document detection can be partitioned into two categories based on the ways to calculate document similarity, shingle based and term based algorithms, both of which can be applied offline and online. We review these algorithms in this section.

2.1 Shingle Based Algorithms

The algorithms, such as [8] [9] [1] [10] [2] [3] [11] [5] [6], are based on the concept of *shingle*. A shingle is a set of contiguous terms in a document. Each document is divided into multiple shingles and a hash value is assigned to each shingle. By sorting these hash values, shingles with the same hash value are grouped

---

[2] http://es.csiro.au/TRECWeb/govinfo.html

together. Then the resemblance of two documents is calculated based on the number of shingles they share.

Because of the large size of the document collections to be examined, several sampling strategies have been proposed to reduce the number of shingles to compare. Heintze selects shingles with the smallest N hash values and removes shingles with high frequencies [9]. Broder et al. samples one of 25 shingles by selecting the shingles whose value modulo 25 is zero and choose at most 400 shingles for each document [1]. In this way they process 30 millions web pages in 10 days. Another more efficient alternative is also proposed in [1], which combines several shingles into a *supershingle* and computes the hash values of supershingles. Although the supershingle algorithm is much faster, the authors noted that it does not work well for small documents and no detailed results of this algorithm are reported. In [10][11], exact copies are removed in advance and then every two or four lines of document are made as a shingle. Fetterly et al. use five-gram as a shingle and apply a 64-bit hash to get fingerprints of shingles, then employ 84 different hash functions to construct a feature vector for each document [4][5]. More precisely, they apply 84 different(randomly selected but fixed thereafter) one-to-one functions to produce shingle fingerprints of each document. For each function, they retain the shingle with numerically smallest hash value of its fingerprints. Thus a vector of 84 shingles is constructed for each document. Then the 84 shingles are separated into six supershingles, in other words, each supershingle contains 14 adjacent shingles. The documents having two supershingles in common are clustered as nearly-duplicate documents. Fetterly et al. processed 150M web pages by using this method. We summarize some of the previous work in Table 1.

To deal with the large-scale data, almost all the previous work employs sampling strategies. However, none of them provides an analysis of how their sampling strategies affect the accuracy of DDD algorithms. On the other hand, sampling has to be adopted to scale up with the index volume of search engines. So it is important to study the impact of sampling in DDD.

Table 1. Parameters used in Prior Work

Work	Volume of Documents Set	Shingling Strategy	Hash Function	Similarity Threshold
Broder97[1]	30M	10-gram	40-bit	0.5
Shivakumar98[10], Cho00[11]	24M 25M	entire document, two or four lines	32-bit	25 or 15 shingles in common
Fetterly03[4][5]	150M	5-gram	64-bit	two supershingles in common

	Sampling Ratio/Strategy
Broder97[1]	1/25 and at most 400 shingles per document
Shivakumar98[10] and Cho00[11]	No Sampling
Fetterly03[4][5]	14 shingles per supershingle six supershingles per document

2.2 Term Based Algorithms

Term based algorithms [12] [13] [14] use individual terms/words as the basic unit, instead of continuous k-gram shingles. Cosine similarity between document vectors is usually used to calculate similarity between documents. Many IR techniques, especially feature selection, are used in these algorithms, which makes them much more complex than shingle-based algorithms. The largest set processed by term based algorithms contains only about 500K web pages [12].

Term based DDD algorithms work well for small-scale IR systems and most of them also achieve good performance when used in online DDD. But for search engines which need to answer over 100M queries everyday, online methods are not a good choice because of their prohibitive computing cost. Meanwhile, in some applications, we have to do DDD offline. In this paper, we focus on shingle based approaches and do not discuss more about term based and online methods.

3 Algorithm

Although much work has been done on DDD algorithms and many applications employ DDD techniques, there is no systematic analysis on how the parameters in DDD correlate, such as accuracy, similarity and sampling ratio. And there is also no formal study on the accuracy and scalability of DDD. This paper aims to explore these problems. We choose the method in [1] for analysis since many DDD algorithms and applications follow it, while we believe our conclusions can also guide other DDD algorithms especially in sampling strategies.

3.1 Document Similarity

Since the exactly duplicate documents, which have no differences between two documents, are easily to identify by comparing the fingerprints of the whole document, this paper focuses on nearly duplicates, which have slightly differences between two documents. We choose the *resemblance* in [1] as our document similarity metric for its widely usage in DDD. However, we believe the conclusions based on this similarity can be easily extended to other metrics of document similarity.

The *resemblance* given by [1] is defined as follows. Each document is viewed as a sequence of words and is transformed into a canonical sequence of tokens. This canonical form ignores minor details such as formatting and HTML tags. Then every document D is associated with a set of subsequences of token $S(D, w)$. A contiguous subsequence in D is called a *shingle*. Given a document D we define its w-shingling $S(D, w)$ as the union of all unique shingles with size w contained in D. Thus, for instance, the 4-shingling of (a, rose, is, a, rose, is, a, rose) is the set {(a, rose, is, a), (rose, is, a, rose), (is, a, rose, is)}.

For a given shingle size, the *resemblance* r of two documents A and B is defined as:

$$r(A, B) = \frac{|S(A) \cap S(B)|}{|S(A) \cup S(B)|}. \tag{1}$$

Where $|S|$ represents the number of elements in the set S.

In our experiments, the shingle size w is set to 10, the same as that in [1]. Different shingle size affects the performance of DDD. Generally, greater w results in higher precision and lower recall. In our own experiences, although greater w produces fewer shingles for each document, greater w also hurts the recall of DDD. So a moderate w is usually chosen to get a balance between precision and recall.

3.2 Hash Function

32-bit and 40-bit Rabin [15] hash functions are used in some of the prior work [1] [10] [11] [2] [3]. However, for large scale dataset with several millions of documents and several billions of shingles, 32-bit or 40-bit hash may produce many false positives. A 40-bit message digest has the probability $1/2$ that a collision (false positive) is found with just over 2^{20} (about a million) random hashes [16]. In this paper, we use the well known 128-bit MD5 hash for both document fingerprints and shingle fingerprints, which generates many fewer false positives for it requires 2^{64} hashes for a collision with $1/2$ probability.

4 Experiments

4.1 Data Description

There are several datasets used in prior work, most of which are not public available. [12] chooses 2GB NIST web pages and TREC disks 4&5 collections as their testing data, but these two sets contain only 240k and 530k documents respectively. In this paper we choose the TREC .GOV collection as our testing dataset since it contains about a million documents and is widely used in Web related research. Table 2 summarizes the main properties of this dataset.

Table 2. Summary of the TREC .GOV Collection

HTML Documents	1,053,034
Total Size	12.9 GB
Average Document Size	13.2 KB
Average Words per Document	699

4.2 Data Preprocessing

First we canonicalize each document by removing all HTML formatting information. Special characters such as HT (Horizontal Tab), LF (Line Feed) and CR (Carriage Return) are converted into spaces, and continuous spaces are replaced by one space. Thus each document is converted into a string of words separated by single spaces.

Then we remove the exact duplicates from the Web collection since we focus on detecting nearly-duplicate documents. By calculating MD5 hash for each document, we cluster exactly duplicate documents, then choose a document from each cluster as the representative and remove the other documents in the cluster. As a result, 94,309 documents are removed from the collection and the final set contains 958,725 documents.

The documents are divided into 11 groups based on the number of words they contain, as shown in Table 3.

Table 3. 11 Groups of Documents

Group	Words in Document	Number of Documents	Shingles in Group
0	0-500	651,983	118,247,397
1	500-1000	153,741	105,876,410
2	1000-2000	78,590	107,785,579
3	2000-3000	28,917	69,980,491
4	3000-4000	14,669	50,329,605
5	4000-5000	8,808	39,165,329
6	5000-6000	5,636	30,760,394
7	6000-7000	3,833	24,750,365
8	7000-8000	2,790	20,796,424
9	8000-9000	1,983	16,770,544
10	>9000	7,775	93,564,410

4.3 Implementation

We implement the algorithm in [1] and run DDD experiments with different similarity thresholds and sampling ratios for each group.

We use three machines with 4GB memory and 1T SCSI disks, one with Intel 2GHz Xeon CPU and the other two with 3GHz Xeon CPU. It takes us two weeks to run about 400 trials of DDD experiments with different combinations of parameters.

Broder et al. [1] processes 30 millions web pages in 10 CPU days. There are two main tradeoffs in their approach. First, they sample one out of 25 shingles and at most 400 shingles are used for each document. They also discard *common shingles* which are shared by more than 1,000 documents. Second, they divide the data into pieces to fit the main memory. However, [1] does not give the size of each piece. It just mentions that "the final file containing the list of the documents in each cluster took up less than 100Mbytes." Thus we believe that the size of each piece can not be too large, and small pieces hurt the recall of DDD since duplicates across different clusters are missed. Moreover, although the CPU speed has been greatly improved since then, the speed of ram and disk advances not so much. So our experiments are rather time consuming although we use much more powerful hardware than theirs.

4.4 Experimental Results

For evaluation we use the result without sampling as the ground truth and compare the result using sampling with this ground truth to calculate the precision. If two documents are judged as duplicates in the result using sampling while they are not judged as duplicates in the result without sampling, it is a false positive. The precision of a trial is calculated by the ratio between the number of correctly detected duplicate document pairs and the number of total detected duplicate pairs in this trial.

For sampling experiments, we make use of the module of the numerical hash value to select shingles. For example, when using 1/2 sampling ratio, we select the shingles whose hash value modulo two is zero, that is, the singles with even hash value. We also run multiple trials for each sampling ratio. For example, when the sampling ratio is 1/2, we run two trials by selecting shingles with odd and even hash value respectively and then calculate the average performance of these two trials. Thus, when the sampling ratio is $1/n$, we run n trials by selecting the singles with different remainders. In our experiments, we count the number of both selected shingles and total shingles and find that the selection ratio is consisted with the given sampling ratio. And there are only slight differences between the precision of different trials with the same sampling ratio, which verifies that MD5 is a good hash function for this sampling task.

The experimental results of 1/4 and 1/16 sampling ratio are shown in Figure 1(a) and 1(b).

As shown in Figure 1(a), precision of DDD decreases with the increasing of similarity threshold. The curve of Group 0, documents having fewer than 500 words, decreases significantly. In Figure 1(b), the highest precision on Group 0 is lower than 0.8 no matter what similarity threshold is used. Also, the precision on several groups with small documents drops dramatically when the similarity threshold is higher than 0.9. The low precision on groups with small documents proves that small documents are sensitive to sampling and it is hard for them

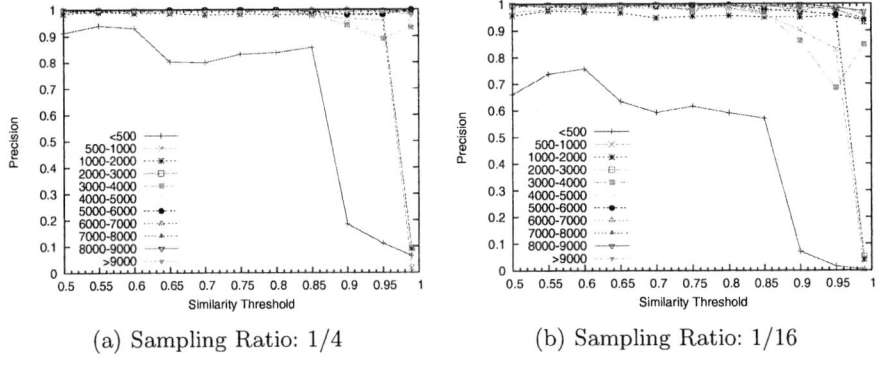

(a) Sampling Ratio: 1/4 (b) Sampling Ratio: 1/16

Fig. 1. Precision with Different Similarity Thresholds

to achieve good precision when small sampling ratio or high similarity threshold is required. On the other hand, for groups with large documents, the precision is high and stable even when the similarity threshold is high and sampling ratio is small. We also ran experiments with sampling ratio 1/2 and 1/8, which show the similar properties as 1/4 and 1/16 sampling ratios.

4.5 Adaptive Sampling Strategy

Based on above observations, we propose an adaptive sampling strategy that applies small sampling ratio on large documents and large sampling ratio on small documents. To show the power of our sampling strategy, we conduct the following experiment. We partition the TREC .GOV collection into 11 groups as previous experiments. For every group we minimize the sampling ratio out of 1/2, 1/4, 1/8, 1/16, subjected to different given precisions ranging from 0.5 to 0.99, thus we minimize the total shingles which we have to process. For example, with the precision requirement 0.8 and similarity threshold 0.6, we choose 1/8 sampling ratio for Group 0 and 1/16 sampling ratio for the other groups, so only 8% of the total shingles have to be processed. As shown in Figure 2, our algorithm greatly reduces the shingles to process and thus can deal with larger scale documents sets than the previous unified sampling strategy.

Due to the well known long tailed distribution of web document size, small documents consist of a large proportion of the whole documents collection. In our experiments, the documents having fewer than 500 words consist of 68% of the whole collection. For higher precision we can not do small sampling in these small documents, otherwise it would greatly hurt the overall precision. Fortunately these small documents consist of only 17% shingles, thus our adaptive sampling

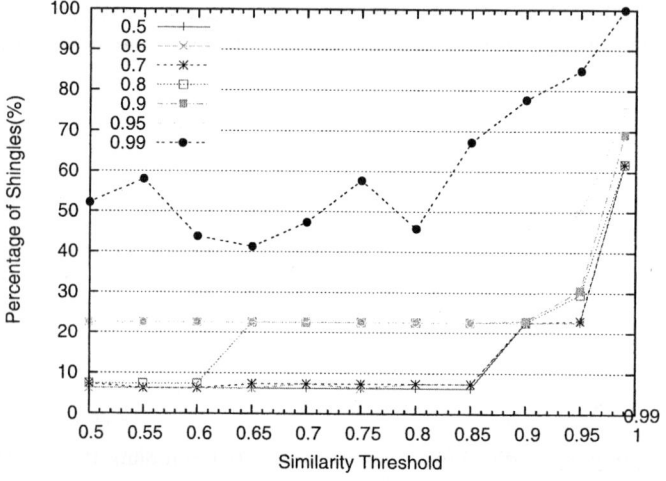

Fig. 2. Adaptive Sampling with Different Precision Thresholds

strategy greatly reduces the total shingles to process by applying small sampling ratio on large documents.

4.6 Summary of Parameter Correlations

Here we give a summary of the correlations between precision and other parameters.

- Similarity Threshold: precision drops with the increase of similarity threshold., especially when the threshold is higher than 0.9. When high similarity threshold, greater than 0.9, is required, sampling ratio should be increased to achieve a good precision.
- Sampling Ratio: precision drops with the decreasing of sampling ratio, especially for small documents containing fewer than 500 words. When dealing with small documents, either similarity threshold should be decreased or sampling ratio should be raised.
- Document Size: small documents are more sensitive to similarity threshold and sampling ratio than large documents. Sampling ratio can be decreased when dealing with large documents to reduce the shingles in computation.

Generally, sampling ratio does not hurt recall because sampling only generates false positives. While for small documents, recall may drop because some of the documents have no shingle sampled by chance.

5 Conclusion and Future Work

Although much work has been done on duplicate document detection and many applications employ this technique, little has been explored on the performance and scalability of DDD. In this paper, a systematic study on parameter correlations in DDD is conducted and several most important parameters of DDD are analyzed.

Our experiment results show that small sampling ratio hurts the precision of DDD, especially for small documents which consist of a major fraction of the whole Web. Based on this observation, an adaptive sampling strategy is proposed, which minimizes the sampling ratio of documents with constraint of given precision thresholds, making DDD feasible to deal with large scale documents collections. We believe the observations in our work are helpful in guiding the future DDD work.

References

1. Broder, A.Z., Glassman, S.C., Manasse, M.S., Zweig, G.: Syntactic clustering of the Web. In: Proceedings of the 6th International World Wide Web Conference (WWW). (1997)
2. Bharat, K., Broder, A.Z.: Mirror, mirror on the Web: A study of host pairs with replicated content. In: Proceedings of the 8th International World Wide Web Conference (WWW). (1999) 501–512

3. Bharat, K., Broder, A.Z., Dean, J., Henzinger, M.R.: A comparison of techniques to find mirrored hosts on the WWW. Journal of the American Society for Information Science (JASIS) **51**(12) (2000) 1114–1122
4. Fetterly, D., Manasse, M., Najork, M., Wiener, J.: A large-scale study of the evolution of web pages. In: Proceedings of the 12th International World Wide Web Conference (WWW). (2003) 669–678
5. Fetterly, D., Manasse, M., Najork, M.: On the evolution of clusters of near-duplicate web pages. In: Proceedings of the 1st Latin American Web Congress (LA-Web). (2003) 37–45
6. Ye, S., Song, R., Wen, J.R., Ma, W.Y.: A query-dependent duplicate detection approach for large scale search engines. In: Proceedings of the 6th Asia-Pacific Web Conference (APWeb). (2004) 48–58
7. Soboroff, I.: Do TREC Web collections look like the Web? SIGIR Forum **36**(2) (2002) 23–31
8. Brin, S., Davis, J., Garcia-Molina, H.: Copy detection mechanisms for digital documents. In: Proceedings of the 1995 ACM International Conference on Management of Data (SIGMOD). (1995) 398–409
9. Heintze, N.: Scalable document fingerprinting. In: Proceedings of the 2nd USENIX Electronic Commerce Workshop. (1996) 191–200
10. Shivakumar, N., Garcia-Molina, H.: Finding near-replicas of documents and servers on the Web. In: Proceedings of the 1st International Workshop on World Wide Web and Databases (WebDB). (1998) 204–212
11. Cho, J., Shivakumar, N., Garcia-Molina, H.: Finding replicated Web collections. In: Proceedings of the 2000 ACM International Conference on Management of Data (SIGMOD). (2000) 355–366
12. Chowdhury, A., Frieder, O., Grossman, D., McCabe, M.C.: Collection statistics for fast duplicate document detection. ACM Trans. Inf. Syst. **20**(2) (2002) 171–191
13. Cooper, J.W., Coden, A., Brown, E.W.: Detecting similar documents using salient terms. In: Proceedings of the 11th ACM International Conference on Information and Knowledge Management (CIKM). (2002) 245–251
14. Conrad, J.G., Guo, X.S., Schriber, C.P.: Online duplicate document detection: signature reliability in a dynamic retrieval environment. In: Proceedings of the 12th International Conference on Information and knowledge management (CIKM). (2003) 443–452
15. Rabin, M.: Fingerprinting by random polynomials. Technical report tr-15-81, Center for Research in Computing Technology, Harvard University (1981)
16. Feller, W. In: An Introduction to Probability Theory and Its Applications. 3rd edn. Volume 1. Wiley (1968) 31–32

Comparison of Documents Classification Techniques to Classify Medical Reports

F.H. Saad, B. de la Iglesia, and G.D. Bell

School of Computing Sciences, University of East Anglia, Norwich NR4 7TJ, UK
{fathi.saad, bli}@uea.ac.uk, gdb@cmp.uea.ac.uk

Abstract. This paper addresses a real world problem: the classification of text documents in the medical domain. There are a number of approaches to classifying text documents. Here, we use a *partially supervised classification* approach and argue that it is effective and computationally efficient for real-world problems. The approach uses a two-step strategy to cut down on the effort required to label each document for classification. Only a small set of positive documents are labeled initially, with others being labeled automatically as a result of the first step. The second step builds the actual text classifier. There are a number of methods that have been proposed for each step. A comprehensive evaluation of various combinations of methods is conducted to compare their performances using real world medical documents. The results show that using EM based methods to build the classifier yields better results than SVM. We also experimentally show that careful selection of a subset of features to represent the documents can improve the performance of the classifiers.

Keywords: Text classification, partially supervised classification, labeled and unlabeled data, medical data mining, and features reduction.

1 Introduction

Medical data is often presented, at least partially in the form of free text (e.g. medical reports attached to patients' records). Such documents contain important information about patients, disease progression and management, but are difficult to analyse with conventional data mining techniques due to their unstructured or semi-structured nature. Medical staff may have a number of interesting questions that can be asked of such data, but they certainly need automatic methods for reading, categorising and analyzing thousands of electronic patients' reports.

The Gastroenterology unit of a local hospital had just such a problem as they collected electronic reports on thousands of colonoscopy procedures, but could not give answer to simple questions, such as the percentage of successful colonoscopies undertaken. Colonoscopy refers to the passage of the colonoscope to the entire large intestine, from the lowest part (the caecum) through the colon to the small intestine. This constitutes a complete examination. The aim of colonoscopy is to check for medical problems such as bleeding, colon cancer, polyps, colitis, etc. [6]. After each colonoscopy procedure, the endoscopist writes a detailed report about the current status of the examined part of the body and the result of the procedure itself. The information contained in this report is extremely valuable for clinical purposes but difficult to handle due to the lack of structure. The procedure can be classified as successful or unsuccessful depending on what the clinicians claim they have been able to examine and the reasons for any limited examinations. Classifying colonoscopy procedure reports into categories is a text classification task.

Text classification is defined as the process of assigning pre-defined category labels to documents based on what a classifier has learned from training examples [9]. For binary classification, the classifier should identify the documents of the class of interest *(positive documents)* from a set of mixed documents. There can also be multi-class problems in which the classifier has to distinguish documents from each of several classes. To build a text classifier is may be necessary to manually label a set of documents and then use a learning algorithm to produce a classification [15]. This approach, called supervised learning [8], has the problem of the considerable effort required to manually label a large number of training examples for every class, particularly for multi-class problems. An alternative approach called *partially supervised classification* has recently been introduced [1, 2] for binary classification problems, and earlier [7] for multi-class problems. It is based on the use of a large set of unlabeled documents and a small set of labelled documents for every class so as to reduce the labelling effort. It is also possible to take this idea further and use only positive and unlabeled documents to learn a classification [10], cutting down more on the labelling effort. This approach is based on a two-step strategy. Step 1 identifies the positive documents from the unlabeled documents, and step 2 builds the final classifier. There are a number of algorithms that are applicable in step 1 and step 2. Deciding on what algorithms should be applied is not a trivial task, but is required for the effective application of the technique to real-world data.

The main purpose of this paper is to perform a practical evaluation of partially supervised classification. The methods available in each step of the process will be tested in combination. The combination that produces the best performance according to some evaluation measures will be recommended. The evaluation will be performed through a real-world medical problem: the classification of a set of colonoscopy reports. For further efficiency, we will also experiment on reducing the set of features used to represent a document.

2 Partially Supervised Classification

The partially supervised classification approach uses a reduced set of positive documents, P, and a large set of unlabeled documents, U. There is initially no labeling of negative documents. The first step of the text classification is therefore to identify a reliable set of negative documents, RN, from the unlabeled documents. This can be achieved by a number of algorithms; in this paper we used *Rocchio (ROC)* [11], *Naïve Bayesian classifier (NB)* [12] and *Spy* [2]. Step 2 consists of iteratively applying a classification algorithm to the newly labeled data. Since some documents are still in the unlabeled set, $U- RN$, the chosen classifier is applied repeatedly to the data with the intention of extracting more possible negative data at each iteration and improving the overall performance of the classifier. The procedure will stop when no further negative documents are found in the unlabeled set, $U-RN$. There are two classifiers used in this step: *Expectation-Maximization* (EM) [16, 19] and *Support Vector Machines* (SVM) [13]. The algorithms were selected based on their availability to the authors.

3 Dataset, Text Representation and Performance Measures

For these experiments we used real world medical documents collected from the Gastroenterology unit of a local hospital. These documents contain information on colonoscopy procedures including preparation of the bowel, features of the colon identified in examination, abnormalities found during examination with their description, patient's reaction to the procedure, etc. The number of documents in this collection is 4,876. 25% of these documents were selected using a *1-in-4* sampling strategy to be used as test documents. The rest (75%) were used to create training sets as follows: 120 documents from the positive class were selected as the positive set. The rest of the documents were used as the unlabeled set.

The most frequently used method to represent text is *bag-of-words* representation where all words from the set of documents are taken and no ordering of words or any structure of text is used [4]. Each distinct word corresponds to a feature of the set of documents. Each feature weighted using *term frequency-inverse document frequency (tf-idf)* [20] which is refined model of term frequency.

Four different measures were used to evaluate the performance of different classifiers: *precision, recall, F-measure* and *accuracy* [14].

4 Documents Pre-processing

Not all the words in the documents are important, so they may degrade the classifier's performance. In addition, representing small set of documents that may have hundreds of different words using *bag-of words* approach will generate a huge feature space and thus will increase the processing time. To solve these problems, approaches to reduce the feature space dimension are needed. We used three approaches:

1. As a result of consulting an expert in the domain field, we removed unhelpful sentences;
2. We have removed stop words from all data sets using stop-lists;
3. We stemmed the words using Porter's suffix-stripping algorithm [3].

The total number of words before applying any of the feature reduction approaches is 319689 word, after applying the three approaches only 154999 words left. That means the total number of words reduced to 48.5%.

5 Results and Analysis

As we mentioned earlier, the main objective of this paper is to find what is the best strategy for partially supervised classification for a real-world application. It will then be possible to test the claim that his method is effective and computationally efficient [2] using a challenging medical problem. The combination of different methods used in step 1 (spy, NB and ROC) and step 2 (SVM and EM) will produce six techniques (classifiers) when we used one method for step1 and one method for step 2. These six classifiers will be investigated and evaluated in our experiments. The results shown in Table 1 illustrate the recall, precision, F-measure and accuracy obtained by different classifiers.

Table 1. The recall, precision, F-measure and accuracy results obtained by different classifiers

	Recall %	Precision %	F-Measure %	Accuracy %
ROC-SVM	66.10	93.60	77.48	94.42
NB-SVM	33.33	98.33	49.79	90.24
SPY-SVM	57.06	95.28	71.38	93.36
ROC-EM	85.88	85.88	85.88	95.90
NB-EM	79.66	90.39	84.69	95.82
S-EM	84.18	87.65	85.88	95.98

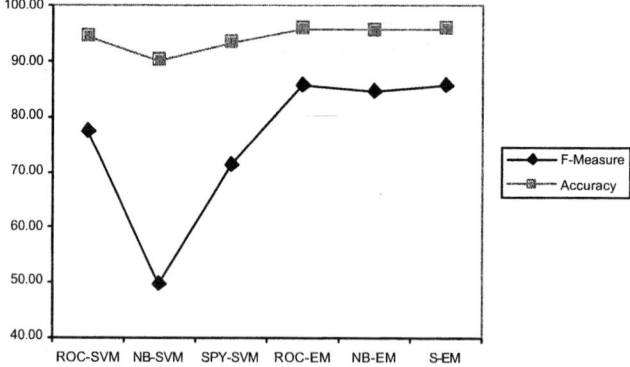

Fig. 1. The F-measure and accuracy results for six classifiers

Figure 1 illustrates graphically the F-measure and accuracy results for the six classifiers. The axes x and y represent the classification techniques and the percentage of the F-measure and the accuracy respectively. The main observation from Table 2 and Figure 1is that the best results are obtained by classifiers using EM in step 2, regardless of the technique used in step 1. In addition, if we compared the F-measure and accuracy results obtained by SVM and EM we find that EM significantly outperforms SVM. We also observe that when NB is used in step 1 to identify the RN set, it produces the worst results in term of accuracy and F-measure. Spy-SVM also underperforms. This may be due to a small positive set, resulting in a small number of spies added to U. This in turn produces a poor RN set. In the case of S-EM the problem is ameliorated since EM used in step 2 will first fill the missing data. According to both the F-measure and accuracy, the highest results are obtained by S-EM, but ROC-EM and NB-EM performed very close to it with less than 0.2% difference. It is worth noting that those classifiers represent the best balance of recall and precision but lower precision than can otherwise be obtained.

Another set of experiments was conducted to attempt to improve the performance of different classifiers by reducing the number of features used. As shown in table 1, the final total number of distinct features in the collection is 2,636. The frequencies of these features vary from the highest frequency 7,111 to the lowest frequency of 1. 1,124 of these features occurred only once. The previous set of experiments was repeated with a reduced feature set. In each case, only the γ top features according to their frequency will be selected to build the classifier. The four values of γ used are 100, 200, 300 and 500.

Table 2 shows the resulting accuracy (acc.) and F-measure (f-m) values respectively for these sets or experiments. Figure 2 and Figure 3 depict the same values graphically. The x axe in both figures represents the six classification techniques, and y axe in Figure 2 represent the percentage of the accuracy and in Figure 3 represents the F-measure values.

Table 2. Accuracy and F-measure results of the six classifiers for four values of γ: 100, 200, 300 and 500 top features.

	ROC-SVM		NB-SVM		Spy-SVM		ROC-EM		NB-EM		S-EM	
	Acc.	f-m	Acc.	f-m	Acc.	f-m	Acc.	f-m	Acc.	f-m	Acc.	f-m
All features	94.42	77.48	90.24	49.79	93.36	71.38	95.9	85.88	95.82	84.69	95.98	85.88
γ = Top 100 features	95.24	82.74	93.6	72.54	93.6	72.54	89.75	70.17	91.56	73.79	91.31	73.37
γ = Top 200 features	94.75	79.08	93.27	70.5	94.91	79.74	96.1	86.67	96	86.21	96.2	86.87
γ = Top 300 features	94.91	79.74	91.96	62.88	93.93	74.66	95.57	85.67	95.82	85.22	95.82	85.47
γ = Top 500 features	95	80	90.48	52.07	93.6	72.73	95.9	86	96.1	86.46	96.06	86.21

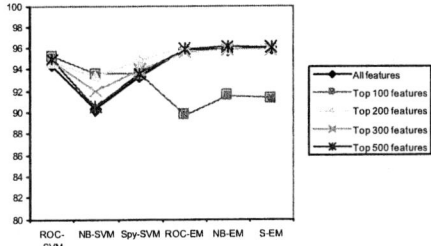
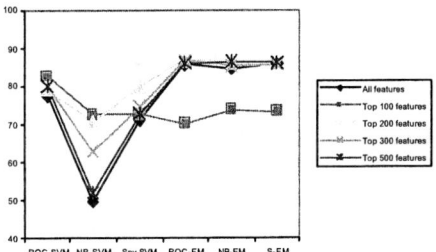

Fig. 2. Accuracy results of the six classifiers for γ= 100, 200, 300 and 500

Fig. 3. F-measure results of the six classifiers for γ= 100, 200, 300 and 500

Using the top 100 features improved the performance of the SVM based methods but significantly degraded the performance of the EM based methods. This may indicate that a set of 100 features is too small to produce and revise good probabilistic labels of the documents in *U-RN* when the EM method is used.

The results obtained using the top 200 features slightly improve the performance of a number of classifiers whilst producing no significant deterioration in others. Larger feature sets (γ=300 and 500) did not provide significantly improved results and in some cases produced slightly worse results. The main observations from the last set of experiments are: (1) Selecting a reduced set of features to represent the documents can improve the performance of all classifiers based on F-measure and accuracy; (2) A very reduced feature set may affect the performance of certain classifiers such as EM; and (3) Finding a *sufficient* set of features can improve performance while also increasing efficiency, but it may require some experimentation.

6 Conclusions

The objective of the research to test partially supervised classification on a real world problem. To this effect, a number of experiments were conducted to evaluate the performance of different methods within the two-step approach. The approach has the advantage of requiring only a small set of labeled positive documents to operate. Our experimental results showed that using EM to build the text classifier in the second step yielded the best results, regardless of the method used to identify negative documents in the first step. We also experimentally showed that the careful use of feature selection can improve the performance and should obviously improve efficiency. In our case, selecting the top 200 features to represent the documents yielded satisfactory result for all classifiers.

Our results are very competitive for this real world problem and could be used to automatically label and classify medical reports. We believe the method is widely applicable to other text classification problems in the medical domain that requires two-class or binary classification.

Acknowledgement

This work was supported by the Engineering and Physical Sciences Research Council (EPSRC) grant number GR/T04298/01.

References

1. Bing Liu, Yang Dai, Xiaoli Li, Wee Sun Lee and Philip Yu. *"Building Text Classifiers Using Positive and Unlabeled Examples"*. Proceedings of the Third IEEE International Conference on Data Mining (ICDM-03), Melbourne, Florida, 2003.
2. Bing Liu, Wee Sun Lee, Philip S Yu and Xiaoli Li. *"Partially Supervised Classification of Text Document"s*. Proceedings of the Nineteenth International Conference on Mach ine Learning (ICML-2002), Sydney, Australia. 2002.
3. Porter, M.F., *"An algorithm for suffix stripping"*, Program; automated library and information systems, 14(3), 130-137, 1980.
4. Benbrahim, H. and Barmer, M.A. *"Neighborhood Exploitation in Hypertext Categorization"*. In Research and Development in Intelligent Systems XXI. Springer-Verlag, 2005.
5. David B. Aronow, Fangfang Feng. *"Ad-Hoc Classification of Electronic Clinical Documents"*. D-Lib Magazine. ISSN 1082-9873. 1997.
6. C.J. Bowles, R Leicester, C. Romaya, E Swarbrick, C. B. Williams and O. Epstein. *"A Prospective Study of Colonoscopy Practice in the UK today: are we Adequately Prepared for national colorectal Cancer Screening Tomorrow?"* International Journal of Gastroenterology and Hepatology, 2003.

7. Nigam K., McCallum A., Thrun S., and Mitchell T. *"Learning to Classify Text from Labeled and Unlabeled documents"*. *AAAI-98*. pp 792-799. AAAI Press. Menlo Park, US. 1998.
8. Yang Y., and Liu X., *"Are-examination of Text Categorization Methods"*, Special Interest Group of Information Retrieval (SIGIR), 1999.
9. David D. Lewis, *"Representation and Learning in Information Retrieval"*, PhD Thesis, Department of Computer and Information Science, University of Massachusetts, 1992.
10. Denis F., *"PAC Learning from Positive Statistical Quires"*, ALT, pp 112-126. 1998.
11. Rocchio J., *"Relevant Feedback in Information Retrieval,* The smart retrieval system-experiments in automatic document processing". Englewood Cliffs, NJ, 1971
12. McCallum A., and Nigam K., *"A Comparison of Event Models for Naïve Bayes Text Classification"*. In AAAI-98 Workshop on Learning for Text Categorization, 1998.
13. Xiaoli Li, Bing Liu. *"Learning to classify text using positive and unlabeled data"*. Proceedings of Eighteenth International Joint Conference on Artificial Intelligence (IJCAI-03), Acapulco, Mexico. 2003.
14. David D. Lewis, *"Evaluating Text Categorization"*. Proceedings of the Speechand Natural Language Workshop Asilomar, Morgan Kaufmann, pp 312-318. 1991.
15. Gao Cong, Wee Sun Lee, Haoran Wu, Bing Liu. *"Semi-supervised Text Classification Using Partitioned EM"*. 11th International Conference on Database Systems for Advanced Applications (DASFAA), pp 482-493. 2004.
16. A. Dempster, N. M. Laird, and Rubin D., *"Maximum Likelihood from Incomplete Data via EM Algorithm"*. Journal of the Royal Statistical Society, 1997
17. Lewis, D., and Ringuette, M. *"A Comparison of Two Learning Algorithms for Text Categorization"*. 3rd annual symposium on document analysis and information retrieval, pp 81-93, 1994.
18. Joachim, T. *"Making Large Scale SVM Learning Practical"*. Advances in Kernel Methods - Support Vector Learning, 1999.
19. Nigam, K., McCallum, A., Thrun, S., Mitchell, T. *"Text Classification from Labeled and Unlabeled Documents Using EM"*. *Machine Learning*, 103-134, 2000.
20. Salton, G. and McGill, M. *"Introduction to Modern Information Retrieval"*. McGraw-Hill. 1983.

XCLS: A Fast and Effective Clustering Algorithm for Heterogenous XML Documents

Richi Nayak and Sumei Xu

School of Information Systems, Queensland University of Technology,
Brisbane, Australia
r.nayak@qut.edu.au

Abstract. We present a novel clustering algorithm to group the XML documents by similar structures. We introduce a *Level structure* format to represent the XML documents for efficient processing. We develop a *global* criterion function that do not require the pair-wise similarity to be computed between two individual documents, rather measures the similarity at clustering level utilising structural information of the XML documents. The experimental analysis shows the method to be fast and accurate.

1 Introduction

The eXtensible Markup Language (XML) has become a standard language for data representation and exchange [11]. With the continuous growth in XML data sources, the ability to manage collections of XML documents and discover knowledge from them for decision support becomes increasingly important. Several databases tools are developed to deliver, store and querying XML data [2,4,10]. However they do require efficient data management techniques such as indexing based on structural similarity to support an effective document storage and retrieval. The clustering of XML documents according to their structural similarity facilitates these applications.

Mining of XML documents significantly differs from structured data mining and text mining [9]. XML allows the representation of semi-structured and hierarchal data containing not only the values of individual items but also the relationships between data items. Element tags and their nesting therein dictate the structure of an XML document. The inherent flexibility of XML, in both structure and semantics, poses new challenges to find similarity among XML data.

Research on measuring the similarity of XML documents is gaining momentum [1,3,6,7,8]. Most of these methods rely on the notion of tree edit distance developed in combinational pattern matching – finding common structures in tree collection [14]. (A document is usually represented as a tree structure.) These methods are built on pair-wise similarity between documents/trees. The similarity is measured using the *local* functions between each pairs of objects to minimise the intra-cluster similarity and maximize the inter-cluster similarity. The similarity value between each pair of trees is mapped into a *similarity matrix*. This matrix becomes the input to the clustering process using either the hierarchical agglomerative or k-means clustering algorithms [5]. They are generally computationally expensive when the data sources are large due to the need of pair wise similarity matching among diverse documents.

Our strategy is quite different from these pair-wise clustering approaches. It is inspired by the clustering algorithms developed for transactional data, LargeItem [13] and Clope [12], that do not need to compute a pair wise similarity. These methods define the clustering criterion functions on the cluster level calling *global* similarity measures to optimize the cluster parameters. Each new object is compared against the existing clusters instead of comparing against the individual objects. Since the computations of these global metrics are much faster than that of pair-wise similarities, global approaches are efficient. However, these methods are not suitable for XML documents, as they do not consider the hierarchical structure of a document, (i.e. the level positions, context or relationships of elements).

This paper presents the **XML** documents Clustering with Level Similarity (**XCLS**) algorithm to group the heterogenous XML documents according to similar structure using *global* similarity measures. We develop a *Level structure* format that represents the documents for efficient clustering. The novel *global* criterion function, called LevelSim measures the similarity at clustering level utilising the hierarchal relationships between elements of documents. The experimental results show the XCLS to be an accurate, fast and scalable technique for grouping XML documents.

2 XML Documents Clustering with LevelSim (XCLS)

2.1 Level Structure: Inferring of XML Documents Structure

In a heterogeneous and flexible environment as the Web, it is not appropriate to assume that each XML document on the web has a schema that defines its structure definition. Additionally if they have one, many documents depart from their structure definition through multiple modifications. For XCLS to be used for general Web documents, the structural information within the document is inferred. The documents and schemas are first represented as labelled trees. We define a novel concept of the **level structure** to show the level and the elements in each level of a tree structure, preserving the hierarchy and the context of elements of the documents. The focus is on paths of elements with content values (i.e. leaves in a document tree), without considering attributes in an XML document. Figure 1 shows a XML document (X_Movie) and its corresponding structural tree (T_Movie). In order to enhance the clustering speed, the name of each element is denoted by a distinct integer. The Figure 2 shows the level structure for T_Movie.

The contents of a cluster preserving the hierarchical information of document are also represented as a level structure. Each level of a cluster contains a collection of elements of the same level for all documents within the cluster. The figure 4 shows a tree structure of a document on Actor information and its corresponding level structure. The Figure 3 shows the level structure of a cluster containing both the Movie and Actor documents. Each block in this structure contains information including element value, level in the hierarchy, its original tree identification, etc.

2.2 Clustering *Global* Criterion Function with Level Similarity (LevelSim)

Considering the level information and elements' relationships/context of XML data, a new solution for measuring structural similarity between two XML objects (cluster to tree, tree to tree, cluster to cluster) is developed which is called Level Similarity

(LevelSim). It measures the common items in each corresponding level, and allocates different weight according to the level (i.e. high level (e.g. root) has more weight than low level (e.g. leaf)). Elements are matched according to the level information of each object. The order of matching between two objects is important due to the structural information present in an XML document. The LevelSim when matching object 1 (tree) to object 2 (cluster) is defined as:

$$LevelSim_{1 \rightarrow 2} = \frac{0.5 \times ComWeight_1 + 0.5 \times ComWeight_2}{TreeWeight}$$

$$LevelSim_{1 \rightarrow 2} = \frac{0.5 \times \sum_{i=0}^{L-1} N_1^i \times (r)^{L-i-1} + 0.5 \times \sum_{j=0}^{L-1} N_2^j \times (r)^{L-j-1}}{\sum_{k=0}^{L-1} N^k \times (r)^{L-k-1}}$$

$ComWeight_1$ and $ComWeight_2$ denote the total weight of the common elements in all levels considering the level information of object 1 and object 2 respectively;

$TreeWeight$ denotes the total weight of all items in each level of the tree (object 1);

N_1^i and N_2^j denotes the number of common elements in level i of object 1 and level j of object 2 respectively; N^k denotes the number of elements in level k of the tree.

r is the increasing factor of weight, which is usually larger than 1 to indicate that the higher level elements have larger than lower level elements called as "Base Weight";

L is the number of levels in the tree.

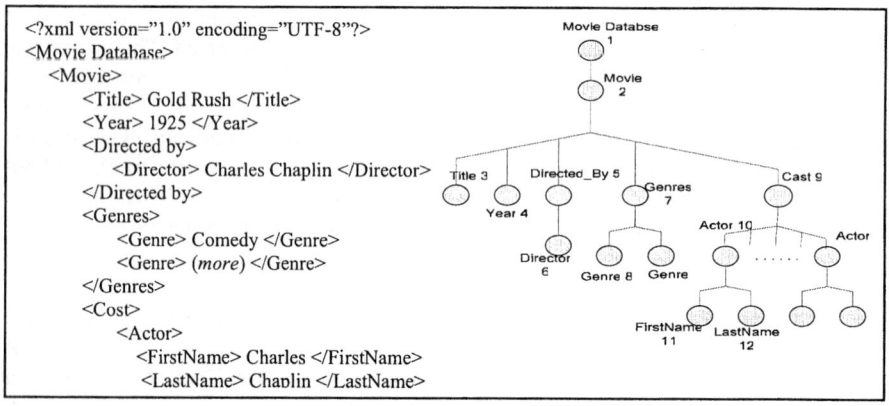

Fig. 1. An XML Document (X_Movie) & its tree representation (T_Movie)

$LevelSim$ yields the values between 0 and 1; 0 indicates completely different objects and 1 indicates homogenous objects. The operation $LevelSim$ is not transitive. There are some cases when one object may be a part of the other sharing a large similarity. In order to solve this problem, the $LevelSim_{1 \rightarrow 2}$ and $LevelSim_{2 \rightarrow 1}$ are both measured and the larger value between two is chosen:

$$LevelSim = LevelSim_{1 \rightarrow 2} > LevelSim_{2 \rightarrow 1} ? LevelSim_{1 \rightarrow 2} : LevelSim_{2 \rightarrow 1}.$$

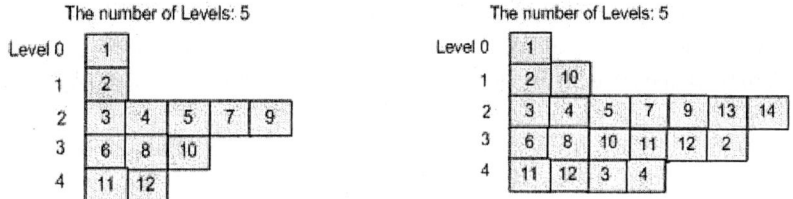

Fig. 2. Level structure for T_Movie   **Fig. 3.** Level structure of a cluster

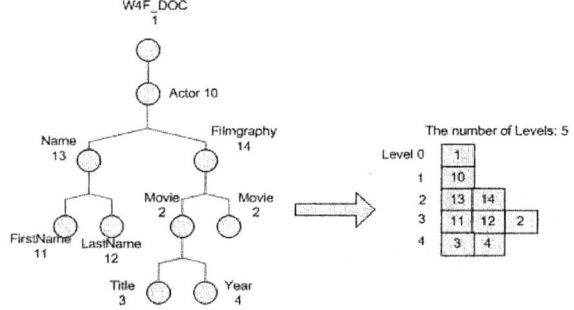

Fig. 4. T_Actor and its level structure

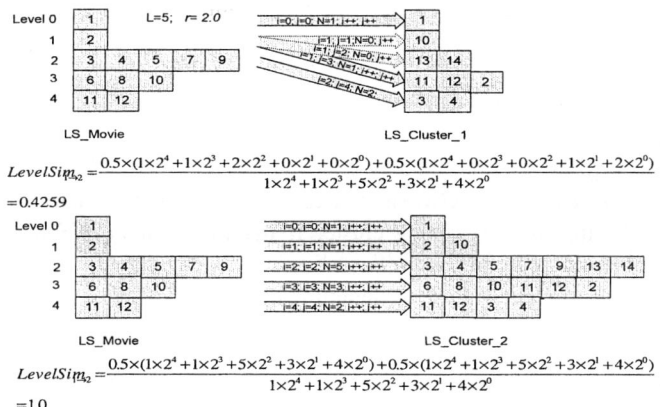

$$LevelSim_{1,2} = \frac{0.5 \times (1 \times 2^4 + 1 \times 2^3 + 2 \times 2^2 + 0 \times 2^1 + 0 \times 2^0) + 0.5 \times (1 \times 2^4 + 0 \times 2^3 + 0 \times 2^2 + 1 \times 2^1 + 2 \times 2^0)}{1 \times 2^4 + 1 \times 2^3 + 5 \times 2^2 + 3 \times 2^1 + 4 \times 2^0}$$
$$= 0.4259$$

$$LevelSim_{1,2} = \frac{0.5 \times (1 \times 2^4 + 1 \times 2^3 + 5 \times 2^2 + 3 \times 2^1 + 4 \times 2^0) + 0.5 \times (1 \times 2^4 + 1 \times 2^3 + 5 \times 2^2 + 3 \times 2^1 + 4 \times 2^0)}{1 \times 2^4 + 1 \times 2^3 + 5 \times 2^2 + 3 \times 2^1 + 4 \times 2^0}$$
$$= 1.0$$

Fig. 5. Two different cases showing the process of matching a tree to a cluster

2.3 The Process of Structure Matching Between Two Objects

The steps to match elements of a tree (object 1) to elements of a cluster (object 2) are:

1. Start with searching common elements in the 1st level of both objects. If at least one common element is found, mark the number of common elements with the level number in object 1 (N_1^0) and in object 2 (N_2^0), then go to step 2. Otherwise, go to step 3.

2. Move both objects to next level (level i++, level j++) and search common elements in these new levels; If at least one common element is found, mark the number of common elements with the level number in object 1 (N_1^i) and in object 2 (N_2^j), then go to step 2. Otherwise, go to step 3.
3. Only move object 2 to next level (level j), then search common elements in the original level (i) of object 1 and the new level (j) of object 2. If at least one common element is found, mark the number of common elements with the level number in object 1 (N_1^i) and in object 2 (N_2^j), then go to step 2. Otherwise, go to step 3.
4. Repeat the process until all levels in either object have been matched.

After completion of structure matching the Level Similarity (LevelSim) is computed.

The Figure 5 shows two cases of matching object 1 (a tree T_Movie) to object 2. In the first case, object 2 is a cluster only containing the tree T_Actor. In the second case, object 2 is a cluster containing both T_Actor and T_Movie.

2.4 Clustering with Level Similarity

This section discusses the algorithm that groups the XML structures according to *LevelSim* values. The task is to group each XML document into an existing cluster that have the maximum LevelSim or to a new cluster. The figure 6 outlines the algorithm that includes two phases of allocation and reassignment. In the allocation phase, clusters are progressively formed driven by the criterion function *LevelSim*. In the reassignment phase, only a few iterations are required to refine the clustering and

---

/*Phase 1 – Allocation*/
For all XML trees to be clustered
- read the next tree (represented as level structure);
- compute the LevelSim between the tree and each existing cluster;
- assign the tree to an existing cluster if maximum of *LevelSim(s)* is found between two objects and *LevelSim > LevelSim_Threshold*;
- otherwise, form a new cluster containing the tree.

/*Phase 2 – Reassignment** (adjustment) */
For all XML trees
- read the next tree (i.e. level structure);
- compute the LevelSim between the tree and each existing cluster;
- reassign the tree to an existing cluster if maximum of *LevelSim(s)* is found between two objects and *LevelSim > LevelSim_Threshold*;
- otherwise, form a new cluster containing the tree.

/*Stop if there is ano improvement in two iterations*/

Fig. 6. The sketch of XCLS core clustering algorithm

optimize the *LevelSim*. The XCLS algorithm uses a user-defined threshold *LevelSim_Threshold* below which the cohesion between two objects is not considered. This threshold (between 0 and 1) can be set according to the application requirement, if only highly homogenous documents are to be grouped the threshold is set higher (near 1) otherwise it is set at a lower value (near 0).

3 Experiments and Results

The data used in experiments are 460 XML documents downloaded from the Wisconisn's XML data bank (http://www.cs.wisc.edu/hiagara/data.html) and the XML data repository (http://www.cs.washington.edu/research/xmldatasets/). The data set includes various domains such as (Movie (#Documents: 74), University (22), Automobile (208), Bibliography (16), Company (38), Hospitality message (24), Travel (10), Order (10), Auction data (4), Appointment (2), Document page (15), Bookstore (2), Play (20), Club (12), Medical (2), and Nutrition (1). The number of nodes varies form 10 to 1000 in these sources. The nesting level varies from 2 to 50. Majority of these domains consists of a number of different documents that have structural and semantic differences. Hence, even though documents are from the same domain, they might not be considered similar enough to be grouped into the same clusters.

3.1 Evaluation Criteria

The two commonly used evaluation methods are utilised: (1) the intra-cluster and inter-cluster quality and (2) the *FScore* measure.

The **intra-cluster similarity** measures the cohesion within a cluster, how close the documents within a cluster are. This is computed by measuring the level similarity between a pair of trees (i.e. XML document structures) within a cluster. The intra-cluster similarity of a cluster C_i is the average of all pair-wise level similarities within the cluster:
$$IntraSim(C_i) = \frac{\sum_{i=1}^{n}\sum_{j=i+1}^{n} LevelSim_{i,j}}{0.5 \times n \times (n-1)}$$
where n is the number of trees in C_i.

The intra-cluster similarity of a clustering solution $C = \{C_1, C_2 ... C_k\}$ is the average of the intra-cluster similarities of all clusters taking into account the number of trees within each cluster:
$$IntraSim = \frac{\sum_{i=1}^{k} IntraSim(C_i) \times n_i}{N}$$
where n_i is the number of trees in C_i, N is the total number of trees and k is the number of clusters in the solution.

The **inter-cluster similarity** measures the separation among different clusters. It is computed by measuring the level similarity between two clusters. The inter-cluster similarity of the clustering solution is the average of all pair-wise level similarities of two clusters. The Level Similarity between two clusters is defined as similar to two trees, using the objects as clusters. The inter-cluster similarity for the clustering solution $C = \{C_1, C_2 ... C_k\}$ is:

$$InterSim = \frac{\sum_{i=1}^{k}\sum_{j=i+1}^{k} LevelSim_{i,j}}{0.5 \times k \times (k-1)}$$ where k is the number of clusters in the clustering.

Precision and **recall** are external cluster quality evaluation based on the comparison of clusters' classes to known external classes. Given a XML document category Z_r with the n_r number of similar XML documents, and a cluster C_i with the n_i number of similar XML documents categorised by XCLS. Let n^r_i be the number of documents in cluster C_i belonging to Z_r, then precision (correctness) is defined as: $p(Z_r, C_i) = n^r_i / n_i$ and recall (accuracy) is defined as: $r(Z_r, C_i) = n^r_i / n_r$. The **FScore** combining precision and recall with equal weights is defined as:

$$F(Z_r, C_i) = \frac{p(Z_r, C_i) \times r(Z_r, C_i)}{p(Z_r, C_i) + r(Z_r, C_i)} = \frac{2n^r_i}{n_i + n_r}$$

The FScore value of a category Z_r is the *maximum* FScore value attained in any clusters of the clustering solution. Hence the FScore of the overall clustering solution is then defined to be the sum of the individual class FScore weighted differently according to the number of documents in the class:

$$FScore = \frac{\sum_{r=1}^{q} n_r F(Z_r, C_i)}{n}$$ where q is the total number of XML document clusters.

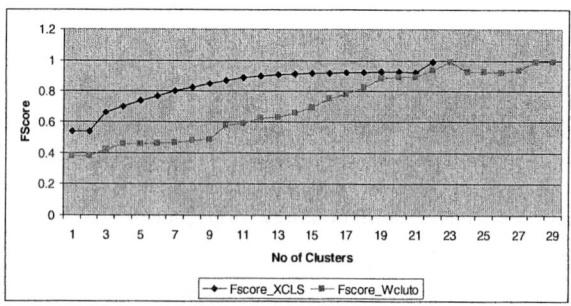

Fig. 7. The Fscore Performance of XCLS vs Wcluto

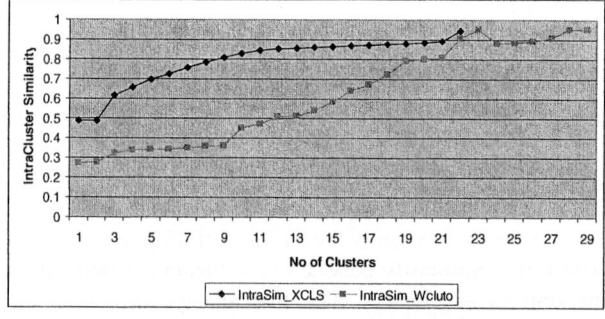

Fig. 8. The IntraSimilarity Performance of XCLS vs Wcluto

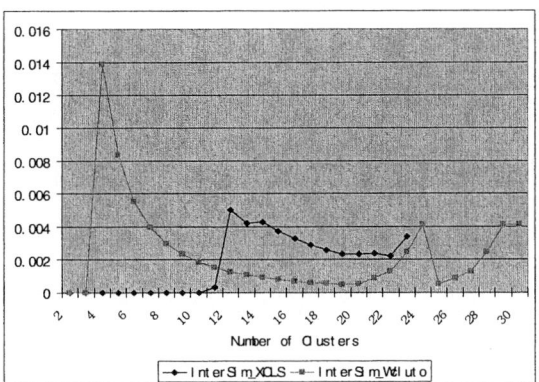

Fig. 9. The Inter Similarity Performance of XCLS vs Wcluto

3.2 Experimental Evaluation for Accuracy of Clustering

To show the comparison between pair-wise similarity algorithms and XCLS, a similarity matrix is generated by measuring the similarity between each pair of documents in the database. The constrained hierarchal agglomerative clustering algorithm Wcluto, [15] is used to group documents from this similarity matrix. The figure 7 shows the value of **FScore** near 1 as the given document set is clustered into groups according to the natural distribution of domains in the input data sources. XCLS achieves the **intra-class similarity** value to 1 as the number of clusters increases (figure 8). Due to the nature of the XCLS algorithm, documents are allocated to the same cluster only if there is any similarity exists, otherwise the new clusters are being formed. This causes the **inter-class similarity** between clusters to be near 0 from the very beginning of the process (figure 9). This also proves that XCLS does not need many iterations in the second phase, only minor adjustments are made in consecutive passes.

3.3 Scalability Evaluation

Space Complexity. The XCLS require only the information of the current document (in process) and a small amount of information of each cluster in the RAM. The tree's information, called tree features, includes the number of levels and its level structure containing all distinct elements in each level. The cluster's information, called cluster features, includes the number of trees, the level structure containing all distinct elements in each level of the cluster. Since just one tree structure is kept in RAM, only the memory consumed by level structures of clusters need to be analysed for the space complexity. Suppose the maximum number of levels is N and the average number of elements in a level of the level structure is M, the total memory required for the level structure in a cluster is approximately N*M*8 bytes using array of 2*4-byte integers (4-byte for element id, 4-byte for occurrences). Therefore, XML document sources with up to 50 levels, average of 20 elements in a level of a level structure and with a clustering of $1k$ can be fit into a 8M (50*10*8*$1k$) RAM.

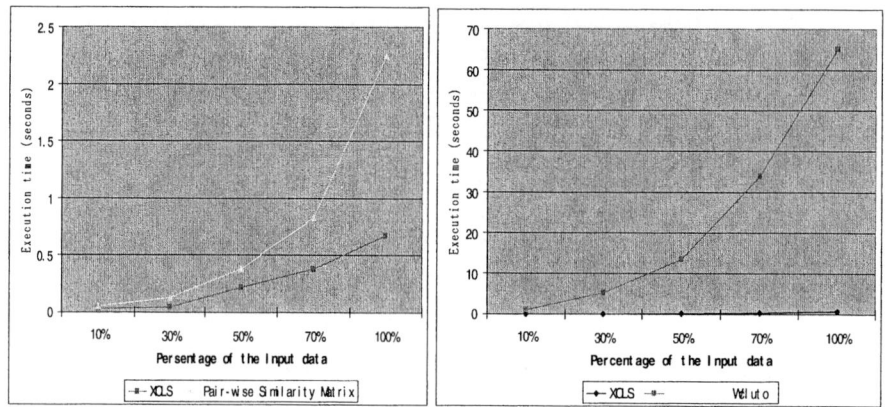

Fig. 10. The execution time of XCLS, similarity matrix generation and Wcluto

Time Complexity. The time complexity of pair-wise clustering algorithms is at least $O(m^2)$, where m is the number of elements in the documents. This is infeasible for large amount of data. XCLS computes the structure similarity between the document structure and clusters avoiding the need of pair-wise comparison. Its time complexity is $O(m \times c \times p \times n)$: m is number of elements in documents; c is number of clusters; p is number of iterations; n is number of distinct elements in clusters. The documents grouped into a cluster should have similar structures and elements. So the number of distinct elements in clusters should always be less than the distinct elements in documents. The number of iterations is usually small and its maximum can be configured. (In our experiments, we never required more than two passes. The maximum is set as 6.) Therefore, if the number of clusters is less than the number of documents (that is usually the case) the time cost is **linear** to the number of documents. The execution time of XCLS (including both pre-processing and clustering), time of generating the similarity matrix between each pairs of documents in the data set, and execution time of Wcluto (including the similarity matrix generation and clustering) in Figure 10 shows the effectiveness of XCLS.

4 Conclusions and Future Work

A novel algorithm for clustering heterogeneous XML documents by their structures called XCLS is presented based on the intuitive idea of the global criterion function *LevelSim*. XCLS does not compute pair-wise structural similarity between two XML documents to get the matrix for clustering; instead, it computes the *LevelSim* to quantify the structural similarity between a XML document and existing clusters and groups the XML document to the cluster with the maximum level similarity.

The *LevelSim* emphasizes different importance of elements in different level positions by allocating different weight to them. The hierarchical relationships of elements are also considered by only counting common elements sharing common ancestors. The derivation of level structure from a tree is straightforward; and the computation of *LevelSim* is quite effective.

This simple idea behind XCLS makes it accurate, fast and memory saving in clustering. The experiments shows that XCLS is a scalable (linear time cost), robust (independent of the data input order and less sensitive to parameters) and effective (inter-cluster similarity close to 0, intra-cluster similarity close to 1, the FScore value close to 1) clustering algorithm for diverse and heterogenous XML documents.

XCLS can be widely used in creating hierarchical index of a large number documents for browsing, discovering elements patterns when describing a specific object, retrieving relating information for a query quickly or creating learning model for documents classification. XCLS can be used to enhance the speed and accuracy of fast searching or locating of XML documents.

XCLS needs some future work to improve its effectiveness. XCLS ignored the sematic similarity among documents, which is impractical in the flexible environment on web since people may use different tags to describe the same thing. As WordNet can organize English words into synonym sets and defined different relations link the synonym sets, it can be added to the pre-processing phase to recognize the semantic similarity among elements.

References

1. Bertino, E., Guerrini, G. & Mesiti, M. (2004). A Matching Algorithm for Measuring the Structural Similarity between an XML Document and a DTD and its applications. Information Systems, 29(1): 23-46.
2. Boag S. Chamberlin D, Fernández M, Florescu D, Robie J and Siméon J. "XQuery 1.0: An XML Query Language" W3C Working Draft, September, 2005. http://www.w3.org/TR/2005/WD-xquery-20050915/
3. Flescu, S., Manco, G., Masciari, E., Pontieri, L., & Pugliese, A. (2005). *Fast Detection of XML Structural Similarities.* IEEE Transaction on Knowledge and Data Engineering, Vol 7 (2), pp 160-175.
4. Guardalben, G. (2004), *Integrating XML and Relational Database Technologies: A Position Paper*, HiT Software Inc, retrieved May 1st ,2005, from http://www.hitsw.com/products_services/whitepapers/integrating_xml_rdb/integrating_xml_white_paper.pdf.
5. Jain, A. K., Murty, M. N., & Flynn, P. J. (1999). Data Clustering: A Review. *ACM Computing Surveys (CSUR), 31*(3), 264-323.
6. Leung, H.-p., Chung, F.-l., & Chan, S. C.-f. (2005). On the use of hierarchical information in sequential mining-based XML document similarity computation. *Knowledge and Information Systems, 7*(4),pp 476-498.
7. Nayak R and Iryadi W (2006). XMine: A methodology for mining XML structure. To appear in *The Eighth Asia Pacific Web Conference.* January 2006, China.
8. Nayak R & Xia, F. B. (2004). "Automatic integration of heterogenous XML-schemas", *Proceedings of the International Conferences on Information Integration and Web-based Applications & Services.* Jakarta, Indonesia, Sec 27-29, pp. 427-437.
9. Nayak, R., Witt, R., and Tonev, A. (2002) Data Mining and XML documents, International Conference on Internet Computing, USA.
10. Xylem L. (2001). Xylem: A dynamic Warehouse for XML data of the Web," IDEAS'01, pp3-7, 2001.
11. Yergeau, F, Bray T, Paoli J, Sperberg-McQueen, C M and Maler E. (2004). Extensible Markup Language (XML) 1.0 (Third Edition) W3C Recommendation, February 2004, http://www.w3.org/TR/2004/REC-XML-20040204/

12. Ying Y, Guan X and You J. (2002), CLOPE: A Fast and effective clustering algorithm for transactional data,
13. Wang, K., Xu, C. (1999), *Clustering Transactions Using Large Items*, in the proceedings of ACM CIKM-99, Kansas, Missouri.
14. Zhang, K., & Shasha, D. (1989). *Simple Fast Algorithms for the Editing Distance Between Trees and Related Problems*. SIAM Journal Computing, 18(6), 1245-1262.
15. Zhao, Y., & Karypis, G. (2002). *Evaluation of Hierarchical Clustering Algorithms for Document Datasets*. The 2002 ACM CIKM, USA.

Clustering Large Collection of Biomedical Literature Based on Ontology-Enriched Bipartite Graph Representation and Mutual Refinement Strategy*

Illhoi Yoo and Xiaohua Hu

College of Information Science and Technology, Drexel University,
Philadelphia, PA, 19104, USA
`iy28@drexel.edu, thu@cis.drexel.edu`

Abstract. In this paper we introduce a novel document clustering approach that solves some major problems of traditional document clustering approaches. Instead of depending on traditional vector space model, this approach represents a set of documents as bipartite graphs using domain knowledge in ontology. In this representation, the concepts of the documents are classified according to their relationships with documents that are reflected on the bipartite graph. Using the concept groups, documents are clustered based on the concepts' contribution to each document. Through the mutual-refinement relationship with concept groups and document groups, the two groups are recursively refined. Our experimental results on MEDLINE articles show that our approach outperforms two leading document clustering algorithms: BiSecting K-means and CLUTO. In addition to its decent performance, our approach provides a meaningful explanation for each document cluster by identifying its most contributing concepts, thus helps users to understand and interpret documents and clustering results.

1 Introduction

Document clustering was initially investigated for improving information retrieval (IR) performance (i.e. precision and recall) because similar documents grouped by document clustering tend to be relevant to the same user queries [1] [2]. However, because document clustering was too slow or infeasible for very large document sets in early days, it was not widely used in IR systems [3]. As faster clustering algorithms have been introduced and those have been adopted in document clustering, document clustering has been recently used to facilitate nearest-neighbor search [4], to support an interactive document browsing paradigm [3] [5] [6], and to construct hierarchical topic structures [7]. Thus, as information grows exponentially, document clustering plays a more important role for IR and text mining communities.

However, traditional document clustering approaches have four main problems. First, when the approaches represent documents based on the bag of word model, they

---

* This research work is supported in part from the NSF Career grant (NSF IIS 0448023). NSF CCF 0514679 and the PA Dept of Health Tobacco Settlement Formula Grant (#240205, 240196).

use all words/terms in documents. As Wang et al pointed out [8], only a small number of words/terms in documents have distinguishable power on clustering documents. Words/terms with distinguishable power are normally the concepts in the domain related to the documents. Second, the approaches do not consider semantically related words/terms (e.g. synonyms or hyper/hyponyms). For instance, they treat {Cancer, Tumor, Neoplasm, Malignancy} as the different terms even though all these words have similar meaning. Third, the approaches cannot provide an explanation of why a document is grouped into one of document clusters [9] because they pursue similarity-based mechanism on clustering, which does not produce any models or rules for document clusters. Lastly, the approaches are based on vector space model. The use of vector space representation on document clustering causes the two main problems. The first problem is that the vector space model assumes all the dimensions on the space are considered independently. In other words, the model assumes that words/terms are mutually independent in a document. However, most words/terms in a document are related to each other. The second problem is that clustering in high dimensional space significantly hampers the similarity detection for objects (here, documents) because the distance between every pair of objects tends to the same regardless of data distributions and distance functions [10]. Thus, it dramatically decreases clustering performance.

These problems have motivated this study. In this paper, we introduce a novel document clustering approach that solves all the four problems stated above. The rest of the paper is organized as follows. Section 2 surveys the related work. In section 3, we propose a novel graph-based document clustering approach that uses domain knowledge in ontology. An extensive experimental evaluation on MEDLINE articles is conducted and the results are reported in section 4. Finally, we conclude the paper with the three main contributions and future work.

2 Related Work

Many document clustering approaches have been developed for several decades. Most of document clustering approaches are based on vector space representation and apply various clustering algorithms to the representation. To this end, the approaches can be categorized according to what kind of clustering algorithms are used. Thus, we classify the approaches into hierarchical and partitional [11].

Hierarchical agglomerative clustering algorithms were used for document clustering. The algorithms successively merge the most similar objects based on the pairwise distances between objects until a termination condition holds. Thus, the algorithms can be classified by the way they pick the pair of objects for calculating the similarity measure; for example, single-link, complete-link, and average-link. Partitional clustering algorithms (especially K-means) are the most widely-used algorithms in document clustering [12]. Most of the algorithms first randomly select k centroids and then decompose the objects into k disjoint groups through iteratively relocating objects based on the similarity between the centroids and the objects. The clusters become optimal in terms of certain criterion functions.

There are some hybrid document clustering approaches that combine hierarchical and partitional clustering algorithms. For instance, Buckshot [3] is basically K-means but Buckshot uses average-link to set cluster centroids with the assumption that hierarchical clustering algorithms provide superior clustering quality to K-means. In order to create cluster centroids, Buckshot first picks $\sqrt{kn}$ objects randomly and then uses average-link algorithm; to make the overall complexity linear, Buckshot selects $\sqrt{kn}$ objects. However, as Larsen & Aone [13] pointed out that using hierarchical algorithm for centroids does not significantly improve the overall clustering quality, compared with the random selection of centroids.

Recently, Hotho et al. introduced the semantic document clustering approach that uses background knowledge [9]. The authors apply ontology during the construction of vector space representation by mapping terms in documents to ontology concepts and then aggregating concepts based on the concept hierarchy, which is called concept selection and aggregation (COSA). As a result of COSA, they resolve a synonym problem and introduce more general concepts on vector space to easily identify related topics [9]. Because they cannot reduce the dimensionality (i.e. the document features) on vector space, it still suffers from *"Curse of Dimensionality"*. In addition, COSA cannot reflect the relationships among the concepts on vector space due to the limitation of vector space model.

3 The Proposed Approach: COBRA

We present a novel approach for *C*lustering *O*ntology-enriched *B*ipartite G*r*aph Represent*a*tion, called COBRA. The proposed approach consists of three main steps: (1) bipartite graph representation for documents through concept mapping, (2) initial clustering by combining co-occurrence concepts based on their semantic similarities on concept hierarchy and document subsets that share co-occurrence concepts, and (3) mutual refinement strategy for concept groups and document clusters. Before discussing these three main components in detail we first briefly discuss Medical Subject Headings (MeSH) as a biomedical ontology due to its importance in our approach.

Medical Subject Headings (MeSH), published by the National Library of Medicine in 1954, mainly consists of the controlled vocabulary and MeSH Tree. The controlled vocabulary contains several different types of terms. Among them Descriptor and Entry terms are used in this research because only they can be used for graph representation. Descriptor terms are main concepts or main headings. Entry terms are the synonyms or the related terms to descriptors. For example, "Neoplasms" as a descriptor has the following entry terms {"Cancer", "Cancers", "Neoplasm", "Tumors", "Tumor", "Benign Neoplasms", "Neoplasms, Benign", "Benign Neoplasm", "Neoplasm, Benign"}. MeSH descriptors are organized in MeSH Tree, which can be seen as MeSH Concept Hierarchy. In MeSH Tree there are 15 categories (e.g. category A for anatomic terms) and each category is further divided into subcategory. For each subcategory, corresponding descriptors are hierarchically arranged from most general to most specific. In fact, because descriptors normally appear in more than one place in the tree, they are represented in a graph rather than a tree. In addition to its ontology role, MeSH descriptors were originally used to index MEDLINE articles. For this purpose around 10 to 20 MeSH terms are manually assigned to each article (after reading full papers). On the assignment of MeSH terms to articles around 3 to 5 MeSH terms are set as "MajorTopic" which primarily represent an article.

3.1 Bipartite Graphical Representation for Documents Through Concept Mapping

Every document clustering method first needs to convert documents into proper format (e.g. document*term matrix). Since we recognize documents as a set of concepts that have their complex internal semantic relationships and assume that documents could be clustered based on what concepts they contain, we represent a set of documents as a bipartite graph to indicate the relationships between concepts and documents on the graph.

This procedure takes the following three steps: concept mapping in documents, detection of co-occurrence concepts, and construction of bipartite graph representations with co-occurrence concepts. Firstly, it maps terms in each document into MeSH concepts. In order to reduce unnecessary search for MeSH concepts, it removes stop words from each document and generates three gram-words as the candidates of MeSH Entry terms. After matching the candidates with Entry terms it replaces Entry terms with Descriptor terms, which is called *concept aggregation*. Then it filters out some MeSH terms that are too general (e.g. HUMAN, WOMEN or MEN) or too common over MEDLINE articles (e.g. ENGLISH ABSTRACT or DOUBLE-BLIND METHOD); see [14] for details. We assume that those terms do not have distinguishable power on clustering documents.

In the second step, it finds out co-occurrence concepts from sets of concept pairs in each document based on the number of times they appear in documents. Co-occurrence terms have long been used in document retrieval systems to identify indexing terms during query expansion [15] [16]. We use co-occurrence concepts instead of concepts because co-occurrence concepts contain some semantic associations between concepts and thus they are regarded more important than single concept.

The remaining problem for co-occurrence concepts is how to set the threshold value for co-occurrence counts; concept pairs whose co-occurrence counts equal or bigger than the value are considered as co-occurrence concepts. Because the threshold value fairly depends on documents or query to retrieve documents, we develop a simple algorithm to detect reasonable threshold value instead of just setting a fixed value. This algorithm tries to finds bisecting point in one-dimensional data. It first sorts the data, takes as centroids the two end objects, and then assigns the remaining objects to the two centroids based on the distances with dynamic centroids update; because the data (co-occurrence counts) was already sorted, it does not need any iteration like other partitional clustering algorithms. After obtaining the threshold value co-occurrence concepts are mirrored as edges on the graph and their co-occurrence counts are used as edge weights.

In the third step, it constructs a bipartite graph. Given the graph $G = (V_D + V_{CC}, E)$, V_D indicates a set of documents, V_{CC} represents a set of co-occurrence concepts in documents and E indicates the relationships between two vertices. Weights can be optionally specified on edges. In that case one should provide a sophisticated weight scheme to measure the contribution of concepts to each document. However, such a weight scheme may not be appropriate especially for small size of documents, such as Medline abstracts. In addition, the scheme requires $|V_D| * |V_C|$ complexity. Thus, we draw an unweighted bipartite graph.

3.2 Initial Clustering by Combining Co-occurrence Concepts

Here, COBRA generates initial clusters for the next step by combining co-occurrence concepts. Since similar documents share the same or semantically similar co-occurrence concepts, COBRA combines co-occurrence concepts and then cluster documents based on their similarities to k co-occurrence concept groups. On combining them there are two ways to measure the similarity between co-occurrence concepts: their semantic similarity on the concept hierarchy (sim_{cc}) and the overlap of their document sets (sim_{doc}). We integrate the two measures with weights. Given two co-occurrence concepts (CC_i & CC_j), the similarity is defined as ($\lambda=0.5$ in the experiments)

$$sim(CC_i, CC_j) = \lambda \cdot sim_{cc}(CC_i, CC_j) + (1-\lambda) \cdot sim_{doc}(CC_i, CC_j), \text{ with } \lambda \in [0,1] \text{ as weights}$$

The semantic similarity between two co-occurrence concepts (CC_i & CC_j) on concept hierarchy (sim_{cc}) is the average similarity of four concept pairs. C^p indicates the set of parent concepts of C concept on the concept hierarchy. sim_{doc} is built on the information theoretic based measure [17]. It is defined as the ratio between the amount of information needed to state the commonality of co-occurrence concepts and the information needed to fully describe what the co-occurrence concepts are in terms of the number of relevant documents.

$$sim_{cc}(CC_i, CC_j) = \frac{\sum_{C_i \in CC_i, C_j \in CC_j} \frac{C_i^p \cap C_j^p}{C_i^p \cup C_j^p}}{|CC_i| + |CC_j|}$$

$$sim_{doc}(CC_i, CC_j) = \frac{|docs_{CC_i} \cap docs_{CC_j}|}{|docs_{CC_i} \cup docs_{CC_j}|},$$

where $docs_{CC_i}$ implies a set of documents that contain CC_i co-occurrence concept.

Based on average-link clustering algorithm that uses the integrated similarity function, COBRA combines co-occurrence concepts until we get k co-occurrence concept groups. For initial document clusters COBRA links each document to k co-occurrence concept groups based on its similarity to k groups. This similarity is simply measured by the number of times co-occurrence concepts in each document appear in each of k groups. A document is assigned to the most similar co-occurrence concept group. For example, suppose there are two co-occurrence concept groups ($CCG_1=\{CC_1, CC_2, CC_3\}$, $CCG_2=\{CC_4, CC_5\}$) and a document has CC_2, CC_3, and CC_5. Then, the document is assigned to CCG_1.

3.3 Mutual Refinement Strategy for Document Clustering

Through the procedures above COBRA generates initial clusters. However, this clustering cannot correct erroneous decisions like hierarchical clustering methods. In other words, once clustering procedures are performed, the clustering results are never refined further even if the procedures are based on local optimization.

In this procedure COBRA "purifies" the initial document clusters by mutually refining k co-occurrence concept groups and k document clusters. The basic idea of the mutual refinement strategy for document clustering is the followings.

- A co-occurrence concept should be linked to the document cluster to which the co-occurrence concept makes the best contribution.
- A document cluster should be related to co-occurrence concepts that make significant contributions to the document cluster.

For this mutual refinement strategy we draw another bipartite graph. Given the graph $G = (V_{DC}+V_{CC}, E)$, V_{DC} indicates a set of (k) document clusters, V_{CC} represents a set of co-occurrence concepts in documents and E indicates the relationships between two vertices. We specify weights on edges so that we measure the contribution of co-occurrence concepts to each document cluster. This contribution is defined as the ratio between the amount of information needed to state the co-occurrence concepts in a document cluster and the total information in the document cluster in terms of the number of documents.

$$cntrb(CC_i, DC_k) = \frac{Size(docs_{DC_k}^{CC_i})}{Size(DC_k)},$$

where *Size* function returns the number of relevant documents, $docs_{DC_k}^{CC_i}$ indicates a set of documents with co-occurrence concept (CC_i) in the document cluster (DC_k).

After each refinement, using k new co-occurrence concept groups, each document is reassigned to the proper document cluster in the same way used for generating initial clusters. This mutual refinement iteration continues until no further changes occur on the document clusters.

4 Experimental Evaluation

In order to measure the performance of COBRA, we conduct experiments on public MEDLINE documents (abstracts). For the experiments first we collect several abstract sets about various diseases from PubMed. Specifically, we use "MajorTopic" tag along with the disease MeSH terms as queries to PubMed (see Section 3 for the tag in detail). Table 1 shows each document set and its size. After retrieving the data sets, we generate various document combinations whose numbers of classes are 2 to 10 using the document sets. Each document set used for the combinations is later used as an answer key on the performance measure.

There are a number of clustering evaluation methods. Among them we use misclassification index (MI) [18] as a measure of cluster quality since MI intuitively shows the overall quality of generated clusters. MI is the ratio of the number of misclassified objects to the size of the whole data set [18]; thus, 0% MI means the perfect clustering.

We evaluate our approach to see how much COBRA provides better clustering results compared with two leading document clustering approaches, and to check if the mutual refinement strategy is able to improve clustering quality.

Table 1. Document Sets

Document Sets	# of Docs	Document Sets	# of Docs
Gout	642	Otitis	5,233
Chickenpox	1,083	Osteoporosis	8,754
Raynaud Disease	1,153	Osteoarthritis	8,987
Insomnia	1,352	Parkinson Disease	9,933
Jaundice	1,486	Alzheimer Disease	18,033
Hepatitis B	1,815	Diabetes Type2	18,726
Hay Fever	2,632	AIDS	19,671
Kidney Calculi	3,071	Depressive Disorder	19,926
Impotence	3,092	Prostatic Neoplasms	23,639
AMD	3,277	Coronary Heart Disease	53,664
Migraine	4,174	Breast Neoplasms	56,075

4.1 Comparison of COBRA, BiSecting K-Means and CLUTO

We apply COBRA to MEDLINE articles to compare its performance with two leading document clustering approaches BiSecting K-means and CLUTO's *v*cluster (http://www-users.cs.umn.edu/~karypis/cluto). Two recent document clustering studies showed BiSecting K-means outperforms traditional hierarchical clustering method and K-means on various document sets from TREC, Reuters, WebACE, etc, [12] [19]. A recent comparative study showed CLUTO's *v*cluster outperforms several model-based document clustering algorithms [20]; none of studies have compared the two approaches.

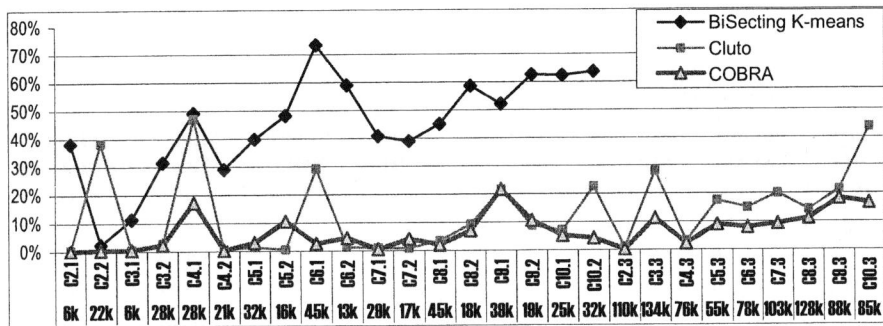

Fig. 1. Comparison of MI for BiSecting K-means, CLUTO, and COBRA (MI on X-axis and Corpus ID and Corpus Size on Y axis); Cx.y, where x indicates *k*, and y is a sequence number. BiSecting K-means failed to cluster the corpora whose size are more than 45k. Because BiSecting K-means produces different results every time due to its random initialization, BiSecting K-means is run ten times and the average values of MIs are used for the comparison.

For the experiments we generated the various document collections using document sets in Table 1. These corpora include very large corpus sets (Cx.3 as Corpus ID in Figure 1) whose size are more than 50k; most document clustering studies

[13][19][20][21] used at most 8.3k to 20k size corpora for their experiments. Figure 1 shows MI results (smaller is better) for the three approaches. Table 2 shows averages of MIs as overall clustering performance index and standard deviation of MIs as the clustering performance consistence index for the approaches. These experiment results indicate that COBRA outperforms BiSecting K-means and CLUTO. As Table 2 shows, COBRA consistently produces better clustering results for various corpus sets. CLUTO yields more or less comparable clustering results with COBRA. But sometimes (for C2.2, C4.1, C6.1, C10.2, C3.3, & C10.3) CLUTO outputs poor clusters. We believe that a prestigious document clustering should consistently produce high-quality clustering results for various document sets.

Table 2. Simple Statistical Analysis of Experiment Results

	Average of MIs	Standard Deviation of MIs
BiSecting K-means	44.77%	0.18%
CLUTO	13.30%	0.14%
COBRA	**6.78%**	**0.06%**

4.2 Evaluation of Mutual Refinement Strategy on Document Clustering

We evaluate mutual refinement strategy (MRS) to check if MRS is able to improve overall clustering quality. For this evaluation we measured MIs before and after MRS process. Table 3 shows MI improvement through mutual refinement strategy (MRS). We notice that MRS significantly improves the performance of COBRA. We also observe that, without this iterative MRS, COBRA still yields comparable performance with CLUTO.

Table 3. MI Improvements through Mutual Refinement Strategy (MRS)

Corpus ID	Before MRS	After MRS	MI Improvements	Corpus ID	Before MRS	After MRS	MI Improvements
C2.1	0.15%	0.15%	0.00%	C6.3	13.06%	7.99%	38.82%
C2.2	6.70%	0.41%	93.88%	C7.1	2.50%	0.52%	79.20%
C2.3	0.12%	0.16%	-33.33%	C7.2	5.46%	4.21%	22.89%
C3.1	0.61%	0.51%	16.39%	C7.3	7.23%	9.27%	-28.22%
C3.2	3.66%	2.36%	35.52%	C8.1	2.68%	2.00%	25.37%
C3.3	23.07%	11.24%	51.28%	C8.2	10.40%	7.04%	32.31%
C4.1	17.16%	17.18%	-0.12%	C8.3	15.59%	11.15%	28.48%
C4.2	0.95%	0.35%	63.16%	C9.1	28.15%	21.60%	23.27%
C4.3	1.93%	2.29%	-18.65%	C9.2	12.11%	10.58%	12.63%
C5.1	27.52%	3.05%	88.92%	C9.3	29.19%	18.15%	37.82%
C5.2	24.96%	10.61%	57.49%	C10.1	6.42%	5.17%	19.47%
C5.3	25.65%	8.93%	65.19%	C10.2	18.09%	4.29%	76.29%
C6.1	6.52%	2.60%	60.12%	C10.3	13.64%	16.57%	-21.48%
C6.2	13.21%	4.58%	65.33%	**AVG**	**11.73%**	**6.78%**	**33.04%**

5 Conclusions

In this paper, we mainly discussed how ontology is incorporated into document clustering procedures and how ontology-enriched bipartite graph representation and mutual refinement strategy improves the document clustering results. The main contributions of this paper are fourfold. First, COBRA becomes a new leading document clustering approach in terms of performance. Second, we introduce a new way of the use of domain knowledge in ontology on document clustering without depending on vector space model. Third, COBRA provides a meaningful explanation for each document cluster by identifying its most contributing co-occurrence concepts. Fourth, we introduce mutual refinement strategy to improve clustering quality. The strategy can be applied to virtually every document clustering approach.

References

1. van Rijsbergen, C. J. (1979). Information Retrieval, 2nd edition, London: Buttersworth. (http://www.dcs.gla.ac.uk/Keith/Preface.html)
2. Willett, P. (1988). Recent trends in hierarchical document clustering: A critical review. Information Processing & Management, Vol. 24, No. 5, pp. 577-597.
3. Cutting, D., Karger, D., Pedersen, J. and Tukey, J. (1992). Scatter/Gather: A Cluster-based Approach to Browsing Large Document Collections, SIGIR '92, pp. 318-329.
4. Buckley, C. and Lewit, A. F. 1985. Optimization of inverted vector searches. In Proceedings of SIGIR-85. pp. 97–110.
5. Hearst, M. A. and Pedersen, J. O. 1996. Reexamining the cluster hypothesis: Scatter/Gather on retrieval results. In Proceedings of SIGIR-96. pp. 76–84. Zurich, Switzerland.
6. Zamir O., Etzioni O.: Web Document Clustering: A Feasibility Demonstration, Proc. ACM SIGIR 98, 1998, pp. 46-54.
7. Koller, D. and Sahami, M. 1997. Hierarchically classifying documents using very few words. In Proceedings of ICML-97. pp. 170–176. Nashville, TN.
8. Bill B. Wang, R I. (Bob) McKay, Hussein A. Abbass, Michael Barlow. Learning Text Classifier using the Domain Concept Hierarchy. In Proceedings of International Conference on Communications, Circuits and Systems 2002, China.
9. Hotho, A., Maedche A., and Staab S. (2002). Text Clustering Based on Good Aggregations. Künstliche Intelligenz (KI), 16(4), p. 48-54
10. Beyer, K., Goldstein, J., Ramakrishnan, R., & Shaft, U. (1999). When is nearest neighbor meaningful?. Proceedings of 7th International Conference on Database Theory, pp 217-235.
11. Kaufman L., and Rousseeuw P.J. 1990. "Finding Groups in Data: an Introduction to Cluster Analysis". John Wiley & Sons.
12. Steinbach, M., Karypis, G., and Kumar, V. (2000). A Comparison of Document Clustering Techniques. Technical Report #00-034. Department of Computer Science and Engineering, University of Minnesota.
13. Bjorner Larsen and Chinatsu Aone, Fast and Effective Text Mining Using Linear-time Document Clustering, KDD-99, San Diego, California, 1999.
14. Hu X., Mining Novel Connections from Large Online Digital Library Using Biomedical Ontologies, Library Management Journal, 26(4/5), 2005, pp. 261-270.

15. Harper, D.J., and van Rijsbergen, C. J. (1978). Evaluation of feedback in document retrieval using co-occurrence data. Journal of Documentation, 34, 189-216
16. Van Rijsbergen, C.J., Harper, D.J. and Porter, M.F. (1981). The selection of good search terms. Information Processing and Management, 17, 77-91.
17. D. Lin. An information-theoretic definition of similarity. In Proceedings of the Fifteenth International Conference on Machine Learning, 1998, 296-304.
18. Zeng, Y., Tang, J., Garcia-Frias, J. and Gao, G.R. (2002): An Adaptive Meta-Clustering Approach: Combining The Information From Different Clustering Results, CSB2002 IEEE Computer Society Bioinformatics Conference Proceedings 276-287.
19. F. Beil, M. Ester and X. Xu: "Frequent Term-Based Text Clustering", 8th ACM SIGKDD International Conference on Knowledge Discovery and Data Mining, July 23-26, 2002, Edmonton, Alberta, Canada
20. Zhong, S., & Ghosh, J. (2003). A comparative study of generative models for document clustering. Proceedings of the workshop on Clustering High Dimensional Data and Its Applications in SIAM Data Mining Conference.
21. Patrick Pantel, Dekang Lin: Document clustering with committees. SIGIR 2002: 199-206
22. Jinze Liu, Wei Wang, and Jiong Yang: A framework for ontology-driven subspace clustering, Proceedings of the tenth ACM SIGKDD international conference on Knowledge discovery and data mining, 2004, pp. 623-628.

Level-Biased Statistics in the Hierarchical Structure of the Web

Guang Feng[1,*], Tie-Yan Liu[2], Xu-Dong Zhang[1], and Wei-Ying Ma[2]

[1] Microsoft Research Asia, No. 49 Zhichun Road, Haidian District,
Beijing 100080, P.R. China
{tyliu, wyma}@microsoft.com

[2] MSPLAB, Department of Electronic Engineering, Tsinghua University,
Beijing 100084, P.R. China
fengg03@mails.tsinghua.edu.cn,
zhangxd@tsinghua.edu.cn

Abstract. In the literature of web search and mining, researchers used to consider the World Wide Web as a flat network, in which each page as well as each hyperlink is treated identically. However, it is the common knowledge that the Web is organized with a natural hierarchical structure according to the URLs of pages. Exploring the hierarchical structure, we found several level-biased characteristics of the Web. First, the distribution of pages over levels has a spindle shape. Second, the average indegree in each level decreases sharply when the level goes down. Third, although the indegree distributions in deeper levels obey the same power law with the global indegree distribution, the top levels show a quite different statistical characteristic. We believe that these new discoveries might be essential to the Web, and by taking use of them, the current web search and mining technologies could be improved and thus better services to the web users could be provided.

1 Introduction

The World Wide Web has been investigated deeply in the past decade because of its explosive growth and significant power in changing the style of people's daily lives. By exploring the link structure of the Web [10], researchers found many exciting characteristics, such as small world [14], highly clustering [14][8] and scale free [1]. Small world means that there is always a relatively short path between any two web pages. And highly clustering means that a web page's neighbors are also probable to become neighbors. Scale free, which is well studied in many scientific areas, means that the probability that a page is pointed by k other pages decays as a power law, following $P(k) \sim k^{-\gamma}$, regardless of the scale of the web page collections. With the belief that these characteristics have discovered some principles of web evolution, many generative graph models [1][14] were proposed to illustrate how these characteristics could be reproduced by simple rules.

---

* This work was performed at Microsoft Research Asia.

Although most of the aforementioned works were done on analyzing the link structure of the Web, The Web structure is not only featured by hyperlinks since the URLs also contain rich structural information. By utilizing the directory depth in URLs, one can naturally reconstruct the hierarchical relationships among web pages and represent the Web as a hyperlinked forest. This concept on the Web has been widely used in many previous works in the literature of web modeling and mining. Ravasz and Barabási [12] proposed a hierarchical network model. Laura et al [9] proposed a multi-level layer model. Eiron and McCurley [5] gave a widely study on the hierarchical structure of the Web. They proved that the hierarchical structure is closely related to the link structure of the Web.

Rather than the methodologies employed in the above works on hierarchical Web modeling, in this study, we investigated the hierarchical structure of the Web in a more explicit way: we dispatched web pages into certain levels by analyzing their URLs and discussed the roles of levels in the hierarchical structure of the Web. As a result, we found several novel and interesting statistical characteristics of the Web, which have not yet been discovered. First, the distribution of pages over levels has a spindle shape. That is, most web pages locate in the middle levels of the hierarchical Web, so there are much fewer pages in the top and very deep levels. Second, the average indegree in each level decreases sharply when the level goes down. Third, although the indegree distributions in deeper levels obey the same power law with the global indegree distribution, the top levels show a quite different statistical characteristics (less skewed, which means certain fairness in attracting in-links).

We believe that the aforementioned new discoveries with respect to the levels in the hierarchical structure might also be essential to the Web. By taking use of them, the current web search and mining technologies [2][3][7] could be improved and thus better services to the web users could be provided.

The rest of the paper is organized as follows. In Section 2, we analyzed the hierarchical structure of the Web. In Section 3, we exhibited the level-dependent characteristics through the statistics on well-known webpage collections and tried to give our explanations to them. In Section 4, we gave the conclusions and future work discussions.

2 Reconstructing Hierarchical Structure of the Web

Most of the artificial complex systems are organized with the hierarchical structures [13], such as geographical districts, governmental branches and so on. It is not only for the feasibility of searching but also for the efficiency of administration. As one of the artificial complex systems, since its birthday, the Web has been constructed and organized with a hierarchical structure. In this section, we presented the exhibition of the hierarchical structure of the Web and then described how we reconstructed the hierarchical Web from the experimental datasets.

2.1 Hierarchical Structure of the Web

Firstly, the Web can be divided into a number of domains according to the services, such as *.com*, *.edu* and so on. After that, each domain can be further divided into many sub domains, such as *microsoft.com*, *ibm.com* and so on. As a result, the domain system forms a hierarchical structure. Although these divisions are engaged for the convenience of management at the very beginning, their senses have gone far beyond that.

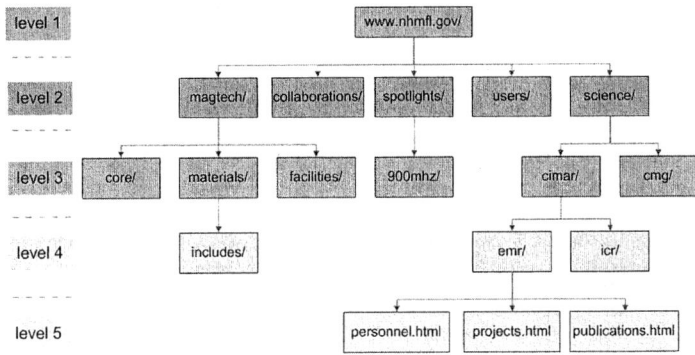

Fig. 1. Hierarchical structure of the website

Secondly, from another point of view, the Web actually consists of large numbers of websites, each of which is organized with hierarchical tree structure. For example, a piece of the sitemap of http://www.nhmfl.gov/ is shown in Fig. 1. Obviously, the site in this figure is divided into five levels, where the portal of the website corresponds to the first level (denoted by Level 1). For other pages, their levels will be determined by their relationship with the portal page. Intuitively, we can get their level properties as shown in Fig. 1.

As the Web is hierarchical and we can clearly define the levels in the hierarchical structure, we believe that there must be some novel level-dependent characteristics. To verify this, we conducted some statistics over two well-known web page collections.

2.2 Datasets and Mapping Strategy

Our statistics were conducted over two well-known benchmark datasets for TREC Web track, the .GOV corpus and the .GOV2 corpus. These two corpora were both crawled from the ".gov" domain. The first one contains about 1M web pages and the second one contains about 25M ones.

We firstly got the indegree distributions of the two corpora as shown in Fig. 2 in order to justify whether the datasets are representative. From this figure, we can clearly see that the indegree distributions indeed follow the power law with

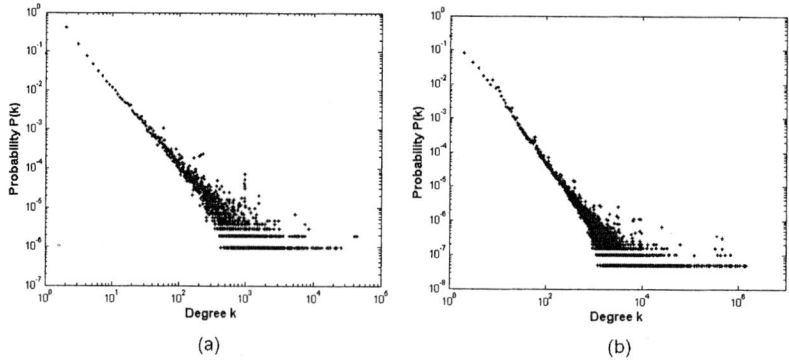

Fig. 2. Global indegree distribution. (a).GOV and (b).GOV2.

an exponent $\gamma = 2.0 \pm 0.1$. This result is quite in accordance with the previous conclusions [3], where γ is around 2.1. Thus, we are confident that the datasets are representative and the statistical results on them are convincing.

In order to mine level-dependent characteristics, we adopted the following strategies for URL analysis. Evidently, if a page's URL is formatted like http://www.aaa.com/, it will belong to Level 1. If a page's URL is formatted like http://www.aaa.com/bbb/, it should belong to Level 2. If all the URLs have such regularity, we will be able to decide the level of a page only by using the number of slashes in its URL. However, the URLs are not always as regular as we hope. To tackle this problem, we designed the following algorithm to extract the level property from the URL information.

Algorithm for Level Extraction
1. *URL regularization and Noise Reduction.*
   (a) Remove the string after '?'.
   (b) Remove "http://www." in the front of the URL.
   (c) Remove the string formatted as "<name>.<suffix>" if it appears in the end of the URL. <name> is one instance in {index, home, default, main} and <suffix> is one instance in {html, htm, asp, aspx, php, pl}.
   (d) Attach a slash to the end of the URL if there is not any.
2. *Level Decision.*
   (a) Extract the number of slashes, denoted by s.
   (b) Extract the string before the first slash. If the number of dots in this string is d, the value of the page level is determined by $s + d - 1$.

3 Level-Biased Characteristics of the Web

After dispatching the pages to a hierarchical structure by the algorithm proposed in Section 2.2, we found that there are totally 17 levels in the .GOV corpus and 21 levels in the .GOV2 corpus. In each dataset, the first ten levels contain more 90% pages so that in the latter paragraphs we will only show the features of the first ten levels in the visualization of the statistical results.

3.1 Spindle Distribution of Pages over Levels

Our first concern is surly how many pages in each level. Fig. 3 shows our corresponding statistical results: the distribution of pages with respect to levels has a spindle shape. That is, there is a dominative level in the middle of the hierarchical structure containing the most pages (the fifth level in the .GOV corpus and the seventh level in the .GOV2 corpus), and starting from this dominative level, the proportion of page numbers decreases when going to either the higher or the deeper levels. For example, in Fig. 3(a), we can see that over 85% pages reside in the middle four levels (3, 4, 5 and 6) while in Fig. 3(b) over 70% pages reside in the middle five levels (4, 5, 6, 7 and 8).

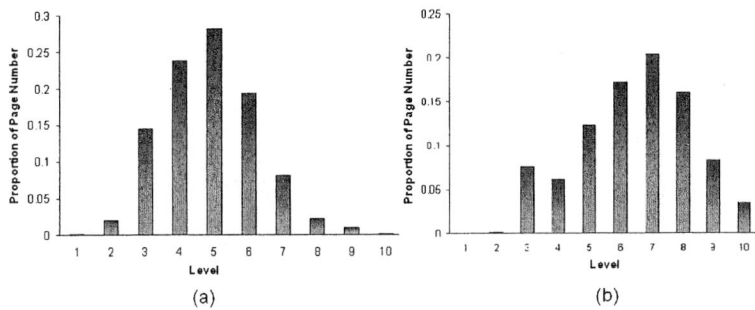

Fig. 3. Spindle distribution of page numbers over levels. (a).GOV and (b).GOV2.

As we know, a page may have several child pages in the hierarchical structure. Therefore, it seems that the total number of pages in each level should always increase when the level goes down. However, our statistical results on real datasets do not support this imagination. Actually, the Web does not look like a triangle but like a spindle.

As for the gap between imagination and real statistics, we provided our explanation to this from the viewpoint of the evolution of the Web. Suppose the Web grows in an incremental manner that new pages are added one by one. In the initial state, there is only a virtual page located in Level 0 which is above the first level. During the growth of the Web, new pages will be created in different levels with different probabilities. Since in the hierarchical structure, there is definite parent-child relationship among the pages, we had better decide who its parent is when adding a new page. It can be proved that a spindle distribution of page numbers will be generated if the parent selection of a new page is fair to all the existing pages. This could be represented by the following theorem.

Theorem 1. *If the probability that a new page is put in Level i is proportional to the number of pages in Level i-1 (this is equivalent to that the existing pages will get a new child with the same probability), the distribution of pages with respect to levels will have a spindle shape. In particular, it will obey a Poisson distribution.*

Proof. Let X_i denote the number of pages in Level i. According to the growth rule, there are totally t pages at time t. The growth rate of X_i is

$$\frac{dX_i}{dt} = \frac{X_{i-1}}{t}. \qquad (1)$$

This equation can be considered to be a generative Polya's urn model[4]. Because Level 0 always contains one virtual page, we can easily get

$$\frac{dX_0}{dt} = 0, X_0 = 1. \qquad (2)$$

With these initial conditions, we can get the general solution of X_i as follows.

$$X_i = \frac{1}{i!}(\ln t)^i. \qquad (3)$$

In the time t, the proportion of the pages in Level k is

$$P(k) = \frac{X_k}{t} = e^{-\ln t} \cdot \frac{(\ln t)^k}{k!}. \qquad (4)$$

As one can see, $P(k)$ is exactly a Poisson distribution with $\lambda = \ln t$. □

Actually, it is really an interesting conclusion that the uneven distribution is caused by a fair generative process. Based on this characteristic and the corresponding explanation, we might say that the Web is not always dominated by the law of rich-get-richer [1], sometimes there are also some fair aspects [11].

3.2 Decreasing Average Indegree Along with the Increasing Level

The spindle distribution tells us that the number of pages in the high levels is small. Then a next question is whether the number of in-links in these levels is also small. Our statistics show a negative answer to this question that the average indegree of the high-level pages is much larger than the low-level pages. That is, web authors prefer to point to the pages in the high levels when they create new web pages.

Take the .GOV corpus for example. There are 616 pages and 462,723 in-links in the first level, which is equivalent to about 751 indegrees per page. However, for the fifth level, although it contains 296,500 pages, it only has 2,230,431 in-links, or about 7.5 indegrees per page. Overall speaking, when the level goes down, the average indegree decreases sharply. The same conclusion can also be obtained from the .GOV2 corpus. If we plot the average indegree in a double logarithm coordinate, it is nearly a straight line (see Fig. 4(a)(b)) . In other words, the average indegree over levels has a smoothed power law form.

As for the power law form, although we may have thought of the advantage of the high-level pages in attracting hyperlinks, we may not think that the difference is so significant that it almost obeys a power law. Actually, the top 5 levels have attracted more than 80% in-links. In other words, a random surfer will visit the

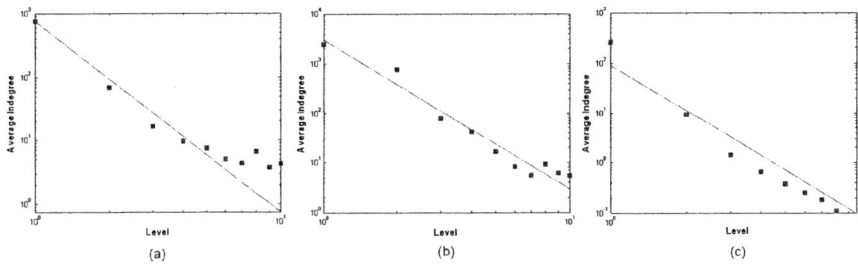

Fig. 4. Average indegree.(a).GOV, (b).GOV2 and (c).GOV after removing the navigational links.

top-5-level pages with a very high probability thus the pages in these levels will have very high popularity or importance.

As for the term "smoothed", we are meaning that the tail of the curve does not match a power law exactly. We believe that it is because of the navigational link. As we known, in the Web, a part of hyperlinks are created only for the navigation purpose, but not for endorsement. These hyperlinks are usually referred to as navigational links. Even if the number of preferentially attached links decreases sharply when the level goes down, the number of navigational links may keep stable. As a result, the proportion of navigational links becomes larger and larger when the proportion of preferential links becomes smaller and smaller. Therefore, the curves in Fig. 4(a)(b) are actually smoothed at the tails. If removing the navigational links, the absolute amount of the average indegree in each level decreases evidently. However, it still has a power law form. The experimental result on the .GOV corpus is shown in Fig. 4(c). That is, the predominance of the high-level pages in attracting links can not be affected by removing the navigational links. Furthermore, the average indegree is no long smoothed. It justifies that the in-links of the low-level pages mainly consist of the navigational links.

3.3 Distinct Indegree Distribution in the First Level

From Section 3.1 and 3.2, we have found that the top levels in the hierarchical Web have many distinct properties. In this section, we will investigate some more details about the indegree distribution in the top levels, other than only an average value, to see whether the rule of link attachment is also distinct.

In Fig. 5 and Fig. 6, we showed the indegree distribution of web pages in each level of the .GOV and the .GOV2 corpora. From these figures, we can find that although the indegree distributions in deeper levels follow the same power law with the global one, the first level is quite different because its exponent is significantly less. As we know, the exponent of the power law distribution can reflect the predominance of rich pages for attracting hyperlinks from other pages. The bigger the exponent is, the more predominant the richer are. Therefore, we can predict that the pages in the first level will take rather fairness in attracting new in-links. In other words, the gap between the poor and the rich in the first

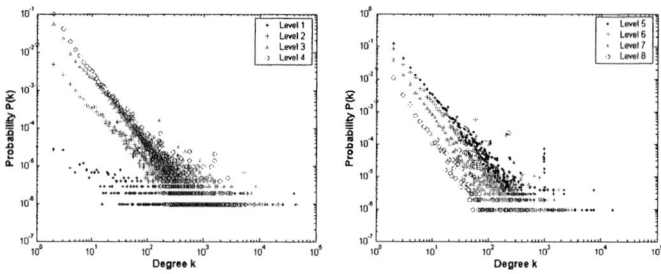

Fig. 5. Indegree distribution in each level of the .GOV corpus

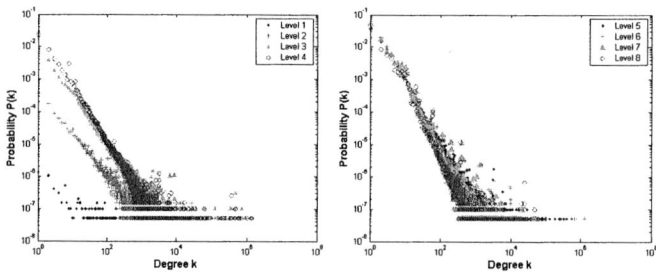

Fig. 6. Indegree distribution in each level of the .GOV2 corpus

level is much smaller than that in other levels. There could be several possible explanations to it.

First, as mentioned in Section 3.2, many pages usually have a hyperlink pointing to the homepage of its website. Therefore, although a homepage might not be very popular, the low-level pages in the same website will still have many hyperlinks pointing to it. As a result, the indegrees of the first-level pages are always very high. After the above arguments, one may have a further question whether this kind of intra-site links also causes the different exponent of the first level. To validate it, we removed the navigational links once again and recalculated the number of in-links of the web pages. The corresponding results as shown in Fig. 7 indicate that the indegree distribution of the first level is still very different. Therefore we can come to the conclusion that there must be some other reasons rather than the intra-site links that caused the differences.

Then, second, we would like to point out that the artificial factors nowadays might be one of the sounded reasons for the above phenomenon. As we know, today's Web is no longer an environment with fair competition. There are many methods that can increase the indegree of the homepage of a website, such as spam, search engine, paid advertisement, and so on. The webmasters will choose to utilize these methods to make their website browsed by more and more surfers. Note that, in such a way, a website (and its homepage) can become very popular even before it has attracted many in-links. This is surely a crack to the rich get richer concept. This partially explains why the indegree distribution of the first

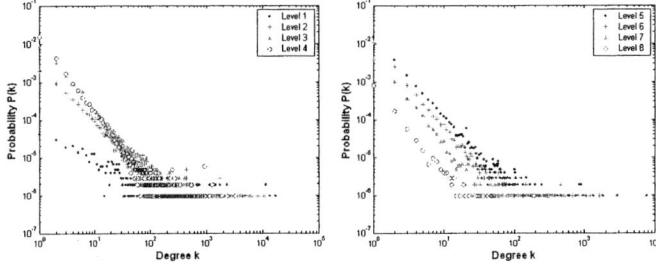

Fig. 7. Indegree distribution in each level of the .GOV corpus after removing the navigational links

level is so different. Actually this explanation also validates the usefulness of the research works on anti-spam raised in recent years.

To summarize, from the three level-dependent characteristics discovered in this paper, we can get a common knowledge that the pages in each level should not be treated equally, especially for the top levels. Although these three characteristics might not portray the hierarchical structure of the Web completely, it has provided several beneficial hints for many web applications [6]. It is definitely worth finding more such characteristics in the hierarchical structure to further give help to the applications and researches on the Web.

4 Conclusions

In this paper, we explicitly mapped the web pages to a hierarchical structure by their URLs. Exploring the hierarchical structure of the Web, we found three level-biased characteristics of the Web. These characteristics, together with our explanations to them, may provide very helpful hints for the current web search and mining technologies. There are still many substantial characteristics undiscovered. We plan to conduct further works on the hierarchical Web and give useful information to other branches of the researches on the Web.

Acknowledgements

The work of Guang Feng and Xu-Dong Zhang was funded by the Joint Key Lab on Media and Network Technology set up by Microsoft and Chinese Ministry of Education in Tsinghua University.

References

1. Barabási, A.-L., and Albert, R.: Emergence of scaling in random networks. Science, vol. 286, pp. 509-512, 1999.
2. Brin, S., Page, L., Motwami, R.and Winograd, T.: The PageRank citation ranking: bring order to the web. Technical report, Computer Science Department, Stanford University, 1998.

3. Broder, A. Z., Kumar, S. R., Maghoul, F., Raghavan, P., Rajagopalan, S., Stata, R., Tomkins, A., and Wiener, J.: Graph structure in the web: experiments and models. In Proc. of the 9th WWW Conference, pp. 309-320, 2000.
4. Chung, F., Handjani, S. and Jungreis, D.: Generalizations of Polya's urn problem. Annals of Combinatorics 7, pp. 141-153, 2003.
5. Eiron, N. and McCurley, K.: Link structure of hierarchical information networks. Proc. Third Workshop on Algorithms and Models for the Web-Graph, 2004.
6. Feng, G., Liu, T.-Y., Zhang, X.-D., Qin. T., Gao, B., Ma, W.-Y.: Level-based link analysis. In Proc. of the 7th Asia Pacific Web Conference, pp. 183-194, 2005.
7. Kleinberg, J.: Authoritative sources in a hyperlinked environment. Journal of the ACM, vol. 46, No. 5, pp. 604-622, 1999.
8. Klemm, K., and Eguiluz, V. M.: Highly clustered scale-free networks. Phys. Rev. E 65, 036123, 2002.
9. Laura, L., Leonardi, S., Caldarelli, G. and Rios, P. D. L.: A multi-layer model for the web graph. In 2nd International Workshop on Web Dynamics, Honolulu, 2002.
10. Newman, M. E. J.: The structure and function of complex networks. SIAM Review, vol. 45, pp. 167-256, 2003.
11. Pennock, D. M., Flake, G. W., Lawrence, S., Giles, C. L., and Glover, E. J.: Winners don't take all: Characterizing the competition for links on the Web. Proceedings of the National Academy of Sciences, 2002.
12. Ravasz, E., and Barabasi, A.-L.: Hierarchical organization in complex networks, Phys. Rev. E 67, 026112, 2003.
13. Simon, H. A.: The Sciences of the Artifical. MIT Press,Cambridge, MA, 3rd edition, 1981.
14. Watts, D. J., and Strogatz, S. H.: Collective dynamics of 'small world' networks. Nature, vol. 393, pp. 440-442, 1998.

CLEOPATRA: Evolutionary Pattern-Based Clustering of Web Usage Data

Qiankun Zhao[1], Sourav S Bhowmick[1], and Le Gruenwald[2,*]

[1] CAIS, Nanyang Technological University, Singapore
[2] University of Oklahoma, Norman, USA
qkzhao@pmail.ntu.edu.sg, assourav@ntu.edu.sg, ggruenwald@ou.edu

Abstract. Existing web usage mining techniques focus only on discovering knowledge based on the statistical measures obtained from the *static* characteristics of web usage data. They do not consider the dynamic nature of web usage data. In this paper, we present an algorithm called CLEOPATRA (**CL**ustering of **Ev**Olutionary **PA**tTe**R**n-based web **A**ccess sequences) to cluster web access sequences ($\mathcal{WAS}s$) based on their *evolutionary patterns*. In this approach, Web access sequences that have similar change patterns in their support counts in the history are grouped into the same cluster. The intuition is that often $\mathcal{WAS}s$ are event/task-driven. As a result, $\mathcal{WAS}s$ related to the same event/task are expected to be accessed in similar ways over time. Such clusters are useful for several applications such as intelligent web site maintenance and personalized web services.

1 Introduction

Recently, web usage mining has become an active area of research and commercialization [3,6,10]. Often, web usage mining provides insight about user behaviors that helps optimizing the website for increased customer loyalty and e-business effectiveness. Applications of web usage mining are widespread, ranging from usage characterization, web site performance improvement, personalization, adaptive site modification, to market intelligence [1].

Generally, the web usage mining process can be considered as a three-phase process, which consists of *data preparation, pattern discovery*, and *pattern analysis* [10]. In the first phase, the web log data are transformed into sequences of events (called Web Access Sequences ($\mathcal{WAS}s$)) based on the identification of users and the corresponding timestamps [1]. Figure 1(a) shows an example of such $\mathcal{WAS}$s. Here S_ID represents a sequence id and a $\mathcal{WAS}$ such as $\langle a, b, d, c, a, f, g \rangle$ denotes a visiting sequence from web page a to pages b, d, c, a, f and finally to page g. Each sub-table in Figure 1(a) records the collection of $\mathcal{WAS}$s for a particular month. In the second phase, statistical methods and/or

---

* This material is based upon work supported by (while serving at) the National Science Foundation (NSF). Any opinion, findings, and conclusions or recommendations expressed in this material are those of the authors and do not necessarily reflect the views of the NSF.

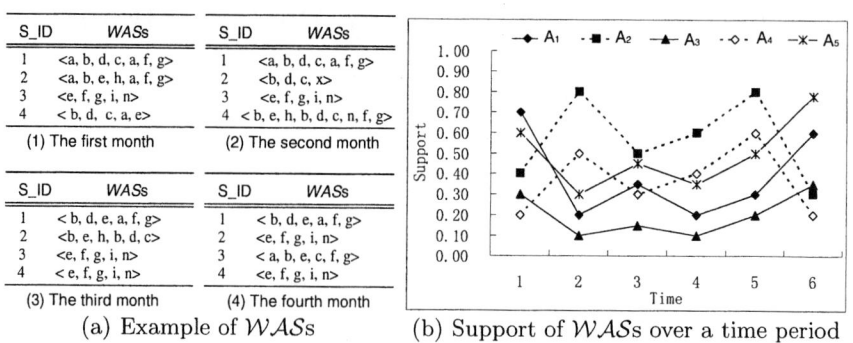

(a) Example of $\mathcal{WAS}$s   (b) Support of $\mathcal{WAS}$s over a time period

Fig. 1.

data mining techniques are applied to extract interesting patterns such as *Web Access Patterns* (WAPs)[7]. A WAP is a sequential pattern in a large set of $\mathcal{WAS}$s, which is visited frequently by users [7], that is, given a support threshold ξ and a set of $\mathcal{WAS}$s (denoted as $\mathcal{A}$), a sequence W is a WAP if W appears as a *subsequence*[1] in at least $\xi \times |\mathcal{A}|$ web access sequences of $\mathcal{A}$. Lastly, these patterns are used for further analysis in the third phase, which is application dependent.

From Figure 1(a), it is obvious that web usage data is dynamic in nature. For instance, the $\mathcal{WAS}$ ⟨ b, d, e, a, f, g ⟩ did not exist in the first and second months but appeared in the third and fourth months. The dynamic behaviors of $\mathcal{WAS}$s can be attributed to various factors, such as changes to web content and users' interest, arrival of new web visitors, and effects of real life events.

In particular, the dynamic nature of $\mathcal{WAS}$ data leads to two challenging problems in the context of web usage mining: maintenance of web usage mining results and discovering novel knowledge [11]. In this paper, we focus on discovering novel knowledge from historical $\mathcal{WAS}$s. Particularly, we focus on clustering of $\mathcal{WAS}$s based on the characteristics of their evolution over time. The intuition behind this is that $\mathcal{WAS}$s are event/task driven. Consequently, $\mathcal{WAS}$s related to the same event/tasks are expected to be accessed in a similar way over time. For example, consider Figure 1(b), which depicts the support values (y-axis) of five $\mathcal{WAS}$s (denoted as A_1, A_2, A_3, A_4, and A_5) from time period 1 to 6 (x-axis). Note that i in the x-axis represents a time period (e.g., day, week, month etc.) and not a particular time point. It can be observed that evolutionary pattern of the supports for A_1, A_3, and A_5 are very similar over time (like the letter "W"). Similarly, the evolutionary patterns of supports for A_2 and A_4 are similar (like the letter "M"). However, the "W" and "M" clusters cannot be discovered by existing web usage mining techniques due to the fact that they focus only on knowledge discovery from snapshot data and maintenance of the knowledge with the changes to the data source. To extract those clusters, in this paper, we

---

[1] If there are two $\mathcal{WAS}$s $A_1 = \langle B, E, A \rangle$ and $A_2 = \langle A, B, C, E, A \rangle$, then A_1 is a subsequence of A_2.

propose the CLEOPATRA (**CL**ustering of **Ev**Olutionary **PAT**te**R**n-based web **A**ccess sequences) algorithm.

The CLEOPATRA clustering results can be useful in many applications, two of which are given below.

Intelligent Web Site Maintenance: With the massive amount of data on the web, it is critical to maintain a well-structured web site in order to increase customer loyalty. Recently web usage mining techniques have been successfully used as a key solution to this issue [3]. However, none of these techniques exploits the dynamic nature of $\mathcal{WAS}$s to restructure web sites. The CLEOPATRA clustering results can be used by web site administrators to maintain a well-structured web site. For example, consider the "W" cluster of $\mathcal{WAS}$s in Figure 1(b), which includes A_1, A_3, and A_5. By analyzing the evolutionary patterns, the web site administrator can figure out the possible reasons (such as promotions, release of new products, and holidays) for such patterns. Accordingly, the structure of the web site can be modified.

User Segmentation: User segmentation is to cluster web users based on the corresponding $\mathcal{WAS}$s to provide personalized services [4,3]. Existing works either use sequence-based distance or probability models to measure the distance between $\mathcal{WAS}$s [4,3]. However, none of them has taken the dynamic nature of $\mathcal{WAS}$s into account. For instance, two users may have the same list of $\mathcal{WAS}$s that belong to two topics, T_1 and T_2, having the same support. Using existing segmentation techniques, the two users will be grouped into the same cluster. However, they may have different preferences. For example, the first user may be currently interested in T_2 as most of the $\mathcal{WAS}$s about T_1 were accessed long time ago, while the second user may be currently interested in T_1 as most of the $\mathcal{WAS}$s about T_2 were also accessed long time ago. By taking the temporal information into account, the user segmentation can be more accurate as users in the same group are not only expected to have similar $\mathcal{WAS}$s but also evolutionary patterns of those $\mathcal{WAS}$s are expected to be similar as well.

The contributions of this paper can be summarized as follows:

- This is the first approach to cluster $\mathcal{WAS}$s based on the evolutionary patterns of their support counts.
- We proposed an algorithm called CLEOPATRA for clustering $\mathcal{WAS}$s based on the evolutionary patterns. Also, the performance of the algorithm is evaluated with real life web usage dataset.

2 Problem Statement

In general, web log data can be considered as sequences of web pages with *session identifiers* [1]. Formally, let $P = \{p_1, p_2, \ldots, p_m\}$ be a set of web pages. A *session* S is an ordered list of pages accessed by a user, i.e., $S = \langle (p_1, t_1), (p_2, t_2), \ldots, (p_n, t_n) \rangle$, where $p_i \in P$, t_i is the time when the page p_i is accessed and $t_i \leq t_{i+1}$ $\forall\ i = 1, 2, 3, \ldots, n-1$. Each session is associated with a unique identifier, called session ID. A *web access sequence* ($\mathcal{WAS}$), denoted as

A, is a sequence of consecutive pages in a session, that is, $A = \langle p_1, p_2, p_3, \ldots, p_n \rangle$ where n is called the *length* of the $\mathcal{WAS}$.

The access sequence $W = \langle p'_1, p'_2, p'_3, \ldots, p'_m \rangle$ is called a *web access pattern* (WAP) of a $\mathcal{WAS}$ $A = \langle p_1, p_2, p_3, \ldots, p_n \rangle$, denoted as $W \subseteq A$, if and only if there exist $1 \leq i_1 \leq i_2 \leq \ldots \leq i_m \leq n$ such that $p'_j = p_{i_j}$ for $1 \leq j \leq m$.

A $\mathcal{WAS}$ *group*, denoted as G, is a bag of $\mathcal{WAS}$s that occurred during a specific time period. Let t_s and t_e be the start and end times of a period. Then, $G = [A_1, A_2, \ldots, A_k]$ where p_i is included in $\mathcal{WAS}$ A_j for $1 \leq j \leq k$ and p_i was visited between t_e and t_s. For instance, we can partition the set of $\mathcal{WAS}$s on a daily, weekly or monthly basis, where the timestamps for all the $\mathcal{WAS}$s in a specific $\mathcal{WAS}$ group are within a day, a week, or a month. Consider the $\mathcal{WAS}$s in Figure 1(a) as an example. They can be partitioned into four $\mathcal{WAS}$ groups on a monthly basis, where $\mathcal{WAS}$s, the timestamps of which are in the same month, are partitioned into the same $\mathcal{WAS}$ group. The *size* of G, denoted as $|G|$, reflects the number of $\mathcal{WAS}$s in G.

Given a $\mathcal{WAS}$ group G, the *support* of a $\mathcal{WAS}$ A in G is $\Phi_G(A) = \frac{|\{A_i | A \subseteq A_i \in G\}|}{|G|}$. When the $\mathcal{WAS}$ group G is obvious from the context, the support is denoted as $\Phi(A)$. Similarly, when the $\mathcal{WAS}$ A is obvious from the context, the support is denoted as Φ.

In our investigation, the historical web log data is divided into a sequence of $\mathcal{WAS}$ groups. Let $H_G = \langle G_1, G_2, G_3, \ldots, G_k \rangle$ be a sequence of k $\mathcal{WAS}$ groups generated from the historical web log data. Given a $\mathcal{WAS}$ A, let $H_A = \langle \Phi_1(A), \Phi_2(A), \Phi_3(A), \ldots, \Phi_k(A) \rangle$ be the sequence of support values of A in H_G. Then, the *degree of dynamic* (denoted as $\omega(A)$) and *version dynamic* (denoted as $\chi(A)$) of A are defined to summarize the changes of support values in the history (defined later in Section 3.1). Moreover, an *evolutionary pattern-based distance* (denoted as $\mathcal{D}$) is defined as the *Euclidian* distance between $\mathcal{WAS}$s based on their *version dynamic* values.

Given a collection of $\mathcal{WAS}$s, with an evolutionary pattern-based distance $\mathcal{D}$ and the degree of dynamic, the objective of the CLEOPATRA algorithm is to partition $\mathcal{WAS}$s into clusters such that $\mathcal{WAS}$s within the same cluster are more similar/closer to each other than to $\mathcal{WAS}$s in other clusters.

3 Representation of Historical $\mathcal{WAS}$s

Given a $\mathcal{WAS}$ denoted as $A = \langle p_1, p_2, p_3, \ldots, p_n \rangle$, in this paper, we use an unordered tree called $\mathcal{WAS}$ *tree* to represent the $\mathcal{WAS}$. A $\mathcal{WAS}$ tree is defined as $T_A = (r, N, E)$, where r is the root of the tree that represents web page p_1; $N = \{p_1, p_2, \cdots, p_n\}$ is the set of nodes; and E is the set of edges in the maximal forward sequences of A. An example of a $\mathcal{WAS}$ tree is shown in Figure 2(a), which corresponds to the first $\mathcal{WAS}$ shown in Figure 1(a).

As a result, a $\mathcal{WAS}$ group consists of a bag of $\mathcal{WAS}$ trees. Here, all occurrences of the same $\mathcal{WAS}$ within a $\mathcal{WAS}$ group are considered identical. Then the $\mathcal{WAS}$ group can also be represented as an unordered tree by merging the

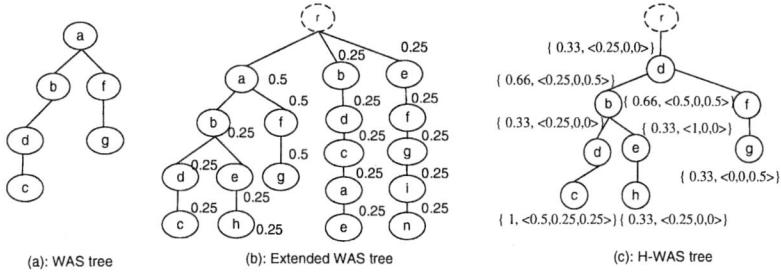

Fig. 2. Examples

$\mathcal{WAS}$ trees. We propose an *extended $\mathcal{WAS}$ tree* to record the aggregated support information about the bag of $\mathcal{WAS}$s within a $\mathcal{WAS}$ group.

Definition 1 [Extended $\mathcal{WAS}$ Tree]. Let $G = [A_1, A_2, \ldots, A_k]$ be a bag of $\mathcal{WAS}$s, where each $\mathcal{WAS}$ A_i, $1 \leq i \leq k$, is represented as a tree $T_{A_i} = (r_i, N_i, E_i)$. Then, the extended $\mathcal{WAS}$ is defined as $T_G = (r, N, E, \Theta)$, where $N = N_1 \cup N_j \cdots \cup N_k$; $E = E_1 \cup E_j \cdots \cup E_k$; r is a virtual root; and Θ is a function that maps each node in N to the support of the corresponding $\mathcal{WAS}$. □

Consider the first $\mathcal{WAS}$ group in Figure 1(a). The corresponding *extended $\mathcal{WAS}$ tree* is shown in Figure 2(b), where the value associated with each node is the Θ value. Next, we propose to merge the sequence of extended $\mathcal{WAS}$ trees into an historical $\mathcal{WAS}$ tree, called *H-$\mathcal{WAS}$ tree*.

Definition 2 [H-$\mathcal{WAS}$ Tree]. Let $H_G = \langle G_1, G_2, G_3, \ldots, G_k \rangle$ be a sequence of k $\mathcal{WAS}$ groups, where each $\mathcal{WAS}$ group G_i, $1 \leq i \leq k$, is represented as an extended $\mathcal{WAS}$ tree, $T_{G_i} = (r_i, N_i, E_i, \Theta)$. Then, the H-$\mathcal{WAS}$ tree is defined as $H_G = (r, N, E, \wp)$, where r is a virtual root; $N = N_1 \cup N_j \cdots \cup N_k$; $E = E_1 \cup E_j \cdots \cup E_k$; and $\wp$ is a function that maps each node in N to the sequence of historical support values of the corresponding $\mathcal{WAS}$. □

Note that, in the H-$\mathcal{WAS}$ tree there is a sequence of support values for each node; while there is only one support value for each node in the extended $\mathcal{WAS}$. In this paper, rather than using the entire sequence of support values, we propose two metrics called *version dynamic* and *degree of dynamic* to summarize the history of support values.

Definition 3 [Degree of Dynamic]. Given a $\mathcal{WAS}$, A, with the corresponding support count sequence $H_A = \langle \Phi_1(A), \Phi_2(A), \cdots \Phi_n(A) \rangle$, the degree of dynamic, denoted as $\omega(A)$, is defined as:

$$\omega(A) = \frac{1}{n-1} * \sum_{i=1}^{n-1} d_i \quad \text{where} \quad d_i = \begin{cases} 1, & \text{if } \Phi_i(A) \neq \Phi_{i+1}(A); \\ 0, & \text{otherwise} \end{cases}$$

□

Definition 4 [Version Dynamic]. Given a $\mathcal{WAS}$, A, with the corresponding support count sequence $H_A = \langle \Phi_1(A), \Phi_2(A), \cdots \Phi_n(A) \rangle$, the version dynamic, denoted as $\chi(A)$, is defined as a sequence $\chi(A)=\langle\chi_1(A), \chi_2(A), \cdots, \chi_{n-1}(A)\rangle$, where $\chi_i(A) = \frac{|\Phi_i(A)-\Phi_{i+1}(A)|}{max\{\Phi_i(A),\Phi_{i+1}(A)\}}$, for $1 \leq i < n\text{-}1$. □

Figure 2(c) shows a part of an H-$\mathcal{WAS}$ tree, where the associated values are the corresponding degree of dynamic value, and the sequence of version dynamic values. The degree of dynamic measures how frequently the $\mathcal{WAS}$ changed and the version dynamic measures how significant are the changes in the history. Furthermore, based on the version dynamic metric, we propose an *evolutionary pattern-based distance* to measure the relationships between $\mathcal{WAS}$s.

Definition 5 [Evolutionary Pattern-based Distance]. *Given two $\mathcal{WAS}$s (A_1 and A_2), the evolutionary pattern-based distance between A_1 and A_2, denoted as $\mathcal{D}(A_1, A_2)$, is defined as:*

$$\mathcal{D}(A_1, A_2) = \sqrt{(\chi'_1(A_1) - \chi'_1(A_2))^2 + \cdots + (\chi'_{n-k+1}(A_1) - \chi'_{n-k+1}(A_2))^2}$$

where $\chi'_i(A_j) = \frac{1}{k}\sum_i^{i+k-1}(\frac{\chi_i(A_j)-\overline{\chi(A_j)}}{\sigma(A_j)})$, k is the user defined window size, $\overline{\chi(A_j)}$ and $\sigma(A_j)$ are the average support count value and standard deviation of $\chi(A)$. □

Note that, the above evolutionary pattern-based distance measure is actually the Euclidean distance between the smoothed $\chi(A)$ sequence using the moving average. This distance measure can handle $\mathcal{WAS}$s with *different baseline, scale,* and *time offset*. Such properties are highly desired in this specific problem for the following reasons. Firstly, the average $\chi(A)$, which can be viewed as the baseline for the $\chi(A)$ sequence, for $\mathcal{WAS}$s that are related to the same event/task may vary a lot while their evolutionary patterns are similar. Secondly, the effects of event/task on different $\mathcal{WAS}$s can be different, which makes the scales of changes ($\chi(A)$) to those $\mathcal{WAS}$s different. Thirdly, there may be a different time delays for different $\mathcal{WAS}$s related to the same event/task, which may cause the time offset among $\chi(A)$ sequences.

4 Cleopatra Algorithm

The Cleopatra algorithm consists of three major phases: the *H-$\mathcal{WAS}$ tree construction* phase, the *node-based clustering* phase, and the *subtree-based clustering* phase. The objective of the *H-$\mathcal{WAS}$ tree construction* phase is to represent the $\mathcal{WAS}$s as trees and merge them into a single tree structure that records both the structural and temporal information. As the *H-$\mathcal{WAS}$ tree construction* has been discussed in [11], we focus on the clustering phases.

Node-based Clustering Phase: The objective of this phase is to categorize individual nodes with similar evolutionary patterns in the H-$\mathcal{WAS}$ tree into clusters. Note that individual nodes represent $\mathcal{WAS}$s from the root to the current nodes. *Hereafter, clustering individual nodes refer to clustering $\mathcal{WAS}$s that starts*

```
Input: H-WAS tree: H
Output: a set of clusters C
1: C'=DBSCAN(H, ω(A))
2: for all Node pairs (N_i, N_j) in cluster c'_i ∈ C' do
3:    calculate D(N_i, N_j)
4: end for
5: C = DBSCAN (c'_i, D), ∀ c'_i ∈ C'
6: for Stop = False do
7:    C'=Split(C)
8:    C'=Merge(C')
9: end for
10: Return(C)
```

Fig. 3. Node-based Clustering Algorithm

```
Input: A set of clusters C, distance threshold ε for DBSCAN
Output: Refined clusters C'
1: for cluster C_j ∈ C do
2:    calculate the centroid point C(C_j)
3: end for
4: for all C_j, C_k ∈ C & C(C_j) ≠ C(C_k) do
5:    if D(C(C_j), C(C_k)) < 2 * ε then
6:       merge them into a new cluster
7:       calculate the new centroid point
8:    end if
9: end for
10: Return clusters C'
```

Fig. 4. Merging Operation

from the root and ends at the corresponding leaf nodes. This algorithm is shown in Figure 3 and consists of two phases, a *two-level clustering* phase and an *iterative refinement* phase. In the first phase, given an H-$\mathcal{WAS}$ tree, firstly, it is *clustered* based on the *degree of dynamic* associated with the individual nodes. Then, using the *evolutionary pattern-based distance*, the degree of dynamic based clustering results are further partitioned into smaller clusters. In the second phase, the iterative refinement phase, the *merging* and *splitting* algorithms are used to refine the quality of the clustering results. The reason is that in the first phase, the two metrics *degree of dynamic* and *evolutionary pattern-based distance* are used separately, when the merging and splitting operations converge, the results will be more accurate.

Note that we use the DBSCAN algorithm [2] to cluster the individual nodes in the H-$\mathcal{WAS}$ tree in this phase for the following reasons. First, the DBSCAN algorithm needs no prior knowledge about the number of clusters in the data collection. This is an advantage of the density-based clustering algorithms. Secondly, the naive DBSCAN approach has the time complexity of $O(N \log N)$, where N is the total number of points in the database, using spatial indexing techniques. Moreover, the DBSCAN algorithm is able to discover clusters with arbitrary shapes and is efficient for very large database. Notice that here the distances between nodes in the H-$\mathcal{WAS}$ tree are the Euclidean distances calculated based on the smoothed $\chi(A)$ sequence generated using the moving average.

In the first phase, the reason for designing a two-level clustering algorithm is to avoid computational cost. In the first level, the degree of dynamic values are used for producing a preliminary results as the degree of dynamic values are easier to obtain while the cost for calculating the evolutionary pattern-based distances are relatively more expensive. By doing this, the computational cost for calculating the evolutionary pattern-based distances for nodes that are not expected to be in the same cluster can be reduced.

In the second phase, the merging and splitting operations are proposed to refine the clustering results in the first phase. The intuition behind is that it is possible that the first level of *degree of dynamic* based clustering results may not fully reflect the *evolution pattern-based distances* between the nodes. Using this iterative merging and splitting operations, which will converge to certain results, we can guarantee that node-based clustering results are accurate, which is the foundation for the sub-tree based clustering in the next phase.

Specifically, merging operation is shown in Figure 4. Firstly, for each cluster a virtual centroid is obtained. Then, the distances between those centroids are calculated using the proposed evolutionary pattern-based distance measure. For clusters whose centroids are within a distance of $2*\epsilon$ will be merged together to form a new cluster, where ϵ is the radius parameter for the DBSCAN algorithm [2]. After that, the splitting operations is then performed on the new clustering results to split them into new clusters if possible. This splitting process is based on the DBSCAN algorithm as well.

Subtree-based Clustering Phase: The output of the node-based clustering phase is a set of clusters that consist of sets of individual nodes with similar change patterns. However, given a cluster, the relations between individual nodes are not captured. In this section, the individual nodes within clusters are merged together to form subtrees, which can represent higher level concepts or objects. Note that, the subtree construction process is guided by not only the links in the H-$\mathcal{WAS}$ tree, but evolution patterns of these nodes should be similar. For a given node in the cluster, to measure the number of nodes that have similar evolution patterns with it, the *evolutionary degree* is defined as follows.

Definition 6 [Evolutionary Degree]. *Let $C = NodeClust(H)$ be a function that implements the node-based clustering phase where H is the H-$\mathcal{WAS}$ tree and C is the set of clusters returned by the function. Let $B(i,j) = Edge(n_i, n_j)$ be a function that takes in two nodes n_i and n_j and returns 1 if there exists or 0 if there does not exist an edge (n_i, n_j) in H. Let $C_x = \{n_1, n_2, \cdots, n_{|C_x|}\}$ and $C_x \in C$. Then, the evolutionary degree of $n_i \in C_x$ (denoted as $E_\bullet(n_i)$) is defined as follows: $E_\bullet(n_i) = \sum_{j=1}^{|C_x|} B(i,j)$, where $i \neq j$ and $0 < j \leq |C_x|$* □

From the above definition, it can be observed that nodes that have large evolutionary degree are expected to form large subtrees. In this section, we propose to extract the list of subtrees for each cluster. Firstly, nodes in each cluster are ranked based on the evolutionary degree in descending order. Then, to ensure that $\mathcal{WAS}$s in the same subtree have similar evolutionary patterns with each other, the *intra similarity* is defined as follows.

Definition 7 [Intra Similarity]. *Let $C = NodeClust(H)$ and $C = \{C_1, C_2, \cdots, C_n\}$. Let t_j be a subtree of H and $\mathcal{N}_t$ be the set of nodes in t_j. Let $\mathcal{K} = \{K_1, K_2, \cdots, K_i\}$, where $K_r = |\mathcal{N}_t \cap C_r| \ \forall \ 0 \leq r \leq i$ and $r \leq n$. Then, the intra similarity of t_j, denoted as $\mathcal{IS}(t_j)$, is defined as: $Max(\mathcal{K}) / |\mathcal{N}_t|$, where $Max(\mathcal{K})$ is the maximum value in $\mathcal{K}$.* □

Definition 8 [Cluster Subtree]. *Let $t_j = (N_j, A_j)$ be a subtree of H such that $N_j \subseteq C_x$ and $C_x \in C$ where $C = NodeClust(H)$. Then t_j is a cluster subtree if $\mathcal{IS}(t_j) \geq \beta$ where β is a user-defined threshold.* □

The algorithm for extracting subtree clusters is presented in Figure 5. The input of the subtree-based clustering algorithm is a set of clusters with sorted nodes.

```
Input: Clusters with sorted nodes C, IS
       threshold β
Output: Clusters of subtrees CoS
1: for all cluster C_j ∈ C do
2:    for all node n_x with the largest
      E_•(n_x) where E_•(n_x) > 0 do
3:       prune all the leaf nodes that are in
         different cluster with n_x iteratively
4:       calculate the IS of the subtree
         rooted at n_x
5:       if IS (Tree(n_x)) ≥ β then
6:          insert this subtree into the CoS list
7:          prune all the leaf nodes in this sub-
            tree from this cluster
8:       else
9:          E_•(n_x) = -1
10:      end if
11:   end for
12: end for
13: Return(CoS)
```

Dataset	ε	k	β	H_{avg}	H_{min}	S_{avg}	S_{max}	$|CoS|$
UoS	0.05	30	0.8	0.81	0.16	0.21	0.46	46
UoS	0.10	60	0.8	0.79	0.13	0.23	0.51	38
UoS	0.15	60	0.75	0.78	0.17	0.19	0.48	34
UoS	0.20	90	0.75	0.78	0.14	0.20	0.46	36
Calgary	0.05	30	0.8	0.80	0.14	0.23	0.45	71
Calgary	0.10	60	0.8	0.79	0.15	0.16	0.38	68
Calgary	0.15	60	0.75	0.71	0.13	0.17	0.38	63
Calgary	0.20	90	0.75	0.75	0.06	0.13	0.32	62

Fig. 5. Subtree-based Clustering   **Fig. 6.** Experimental Results

Firstly, the node with maximum evolutionary degree is selected and the corresponding subtree that includes all the nodes that are connected to that nodes is constructed and tested against the threshold value of IS. If this subtree is a cluster subtree, then all the nodes in this subtree are eliminated from the list of subtrees in that cluster. Otherwise, if this subtree is not a cluster subtree, then the evolutionary degree of this node is set to -1. This process iterates till all the nodes in the subtree are tested.

5 Performance Evaluation

In this section, we evaluate our proposed clustering algorithm with two real datasets, the web log *UoS* and *Calgary*, obtained from the Internet Traffic Archive [5]. The *UoS* records the historical visiting patterns for University of Saskatchewan from June 1, 1995 to December 31, 1995, a total of 214 days. In this seven month period there were 2,408,625 requests. The *Calgary* logs were collected from October 24, 1994 through October 11, 1995, a total of 353 days. There were 726,739 requests. Both of them have 1 second resolution. The web access patterns are transformed into a sequence of extended $\mathcal{WAS}$ trees with a duration of one day. All the following experiments are carried out on a PC with Intel Pentium 4, 1.7Ghz CPU, and 512MB RAM.

Our experiments focus on two aspects: the quality and novelty of the clustering results. To evaluate the quality of the our clustering results, two quality metrics, *Homogeneity* and *Separation* [9, 8], are used. Here we review the metrics: $H_{avg} = \frac{1}{M} \sum_{i<j,\ C(A_i)=C(A_j)} S(A_i, A_j)$; $H_{min} = \min_{C' \in C} \frac{2*\sum_{i<j \in C'} S(A_i,A_j)}{|C'|*(|C'|-1)}$; $S_{avg} = \frac{2}{n(n-1)-2M} \sum_{i<j,\ C(A_i) \neq C(A_j)} S(A_i, A_j)$; and $S_{max} = \max_{C,C' \in C} \sum_{A_i \in C,\ A_j \in C'} S(A_i, A_j)|C|*|C'|$, where n is the total number of $\mathcal{WAS}$ subtrees; A_i is the ith $\mathcal{WAS}$ subtree; M is the total number of node pairs that are within

the same cluster; C is the set of clusters in the result and $|C|$ is the size of the set; $C(A_i)$ is the cluster to which A_i belongs. Note that, here we transform the evolutionary pattern-based distance to the similarity measure S such that we can use the above cluster quality metrics. That is, $S(A_i, A_j) = e^{-\mathcal{D}(A_i, A_j)}$. The larger homogeneity implies a better result, while a larger separation shows a worse result.

Figure 6 shows the quality of the clustering results with different parameters for the DBSCAN algorithm, size of moving window in the moving average, and the intra similarity threshold. The reason of using the above cluster quality metrics is that, due to privacy reasons, the original URLs of web pages in the web usage dataset are not available. Therefore, the ground truth of the clusters are not available. However, from the values in Figure 6, compared with the corresponding values in other applications that using above quality metrics, the quality of our results is comparable to the results in [9, 8].

Considering the novelty of our clustering results, although there is no quantified measures, we have the following observations. First, in the CLEOPATRA clustering results, we found many $\mathcal{WAS}$ pairs that are in the same cluster are very far away in the H-$\mathcal{WAS}$-tree while the evolutionary patterns are quite similar. Such clustering results can be useful for exploring the hidden factors that lead to the evolution of the corresponding $\mathcal{WAS}$s. Second, the overall structures of the clusters are quite similar in the CLEOPATRA clustering result. This means that suppose we have two clusters C_1 and C_2, where $C_1 = \{A_1, A_2, A_3\}$ and $C_2 = \{A_4, A_5, A_6\}$, although A_1, A_2, and A_3 may not be siblings or connected but pairs such as $\{A_1, A_4\}$, $\{A_2, A_5\}$, and $\{A_3, A_6\}$ are siblings or connected.

6 Conclusions

This work is motivated by the fact that existing web usage mining techniques only focus on mining snapshot web usage data and maintaining of the mining results incrementally. They do not consider the dynamic nature of web usage data. In this paper, we proposed the first approach of clustering historical $\mathcal{WAS}$s based on the evolutionary patterns. Experiments with real life datasets show CLEOPATRA can efficiently produce high quality clusters that cannot be discovered using existing web usage mining techniques.

Acknowledgements. We thank Dr Mukesh Mohania from IBM India Research Lab for the feedbacks on the initial draft of this paper.

References

1. M.-S. Chen, J. S. Park, and P. S. Yu. Efficient data mining for path traversal patterns. *TKDE*, 10(2):209–221, 1998.
2. M. Ester, H.-P. Kriegel, J. Sander and X. Xu. A Density-Based Algorithm for Discovering Clusters in Large Spatial Databases with Noise. In *KDD*, pages 226-231, 1996.

3. S. Gunduz and M. T. Ozsu. A web page prediction model based on click-stream tree representation of user behavior. In *Proc. of SIGKDD*, pages 535–540, 2003.
4. X. Jin, Y. Zhou, and B. Mobasher. Web usage mining based on probabilistic latent semantic analysis. In *Proceedings of ACM SIGKDD*, pages 197–205, 2004.
5. L. B. N. Laboratory. Internet traffic archive, http://ita.ee.lbl.gov/, 2004.
6. T. Li, Q. Yang, and K. Wang:. Classification pruning for web-request prediction. In *Proc. of WWW*, 2001.
7. J. Pei, J. Han, B. Mortazavi-asl, and H. Zhu. Mining access patterns efficiently from web logs. In *Proc. of PAKDD*, pages 396–407, 2000.
8. R. Sharan, A. M. Katz, and R. Shamir. Click and expander: a system for clustering and visualizing gene expression data. *Bioinformatics*, 19(14):1787–1799, 2003.
9. R. Sharan and R. Shamir. Center click: A clustering algorithm with applications to gene expression analysis. In *ISMB*, pages 307–316, 2000.
10. J. Srivastava, R. Cooley, and P.-N. Tan. Web usage mining: Discovery and applications of usage patterns from web data. *KDD Exploration*, 1(2):12–23, 2000.
11. Q. Zhao and S. S. Bhowmick and L. Gruenwald. WAM-Miner: In the Search of Web Access Motifs from Historical Web Log Data. In *Proceedings of CIKM* 2005.

Extracting and Summarizing Hot Item Features Across Different Auction Web Sites*

Tak-Lam Wong, Wai Lam, and Shing-Kit Chan

Department of Systems Engineering and Engineering Management,
The Chinese University of Hong Kong, Shatin,
Hong Kong
{wongtl, wlam, skchan}@se.cuhk.edu.hk

Abstract. Online auction Web sites are fast changing and highly dynamic. It is difficult to digest the poorly organized and vast amount of information contained in the auction sites. We develop a *unified* framework aiming at automatically extracting the product features and summarizing the hot item features across different auction Web sites. One challenge of this problem is to extract useful information from the product descriptions provided by the sellers, which vary largely in the layout format. We formulate the problem as a single graph labeling problem using conditional random fields which can model the relationship among the neighbouring tokens in a Web page, the tokens from different pages, as well as various information such as the hot item features across different auction sites. We have conducted extensive experiments from several real-world auction Web sites to demonstrate the effectiveness of our framework.

1 Introduction

The easily accessible Internet creates a profit-generating market place and convenient shopping environment for many users. One example is the online auction Web sites such as *ebay.com*. Individual sellers place items for bidding in the auction Web sites. Potential buyers can then start bidding the items by setting the prices that they are willing to pay. The item is then sold to the one with the highest bid at the end of the bidding period. Online auction Web sites are becoming increasingly popular. According to the press release from ebay.com, they currently have 147 million community members and approximately 50 million items for sale at any given time[1]. Several reasons account for the popularity of

---

* The work described in this paper is substantially supported by grants from the Research Grant Council of the Hong Kong Special Administrative Region, China (Project Nos: CUHK 4179/03E and CUHK4193/04E), the Direct Grant of the Faculty of Engineering, CUHK (Project Code: 2050363), and CUHK Strategic Grant (No: 4410001). This work is also affiliated with the Microsoft-CUHK Joint Laboratory for Human-centric Computing and Interface Technologies.
[1] The article was posted on May 25th, 2005 and was accessible in http://biz.yahoo.com/bw/050525/255399.html?.v=1.

the online auction business. One reason is that sellers do not need to set up and promote for their own Web sites for selling the items and hence reduce the cost. Another reason is that potential buyers can ask the price for the items depending on their budgets and have a chance to successfully buy the items at a lower price if they can bid the right price at the right time.

Since online auction Web sites have a large number of sellers and potential buyers with tremendous number of items from different categories listed for bidding at any time, they are fast changing, highly dynamic, and complex systems. For example, a digital camera may receive a large number of bids ranging from few US dollars to few hundreds US dollars in just one or two days. The mutual influences of the items can be seen from the fact that an item being sold may be seriously affected if another similar item is placed for bidding with a lower bidding price. Therefore, acquiring the up-to-date and accurate information in the auction Web sites offers many potential benefits.

It is useful for both sellers and potential buyers to digest the huge amount of continuously changing information. For example, when a seller intends to place an item for bidding, he/she is required to set a start bidding price. Some sellers may set the start bidding price with their subjective expectations. This can easily result in either that the start bidding price is set too high and hence the chance of the item being sold may be very slim, or that the start bidding price is set too low and hence the return may decrease. Some other sellers may manually analyze the items currently listed for bidding and their prices before setting the start bidding price. However, this manual process for analyzing the vast amount of information is tedious. Besides the sellers, it is also beneficial for a potential buyer to obtain up-to-date, detailed, and accurate information to assist the decision. For example, before bidding for a particular item, the potential buyer may study the description of the item, and other similar items listed for bidding. After certain investigation, he/she can then decide on the amount of money for this bid. Due to the highly dynamic and fast changing nature of the online auction Web sites, rapid decision is essential. If the potential buyer spends too long time for analysis, he/she may either lose the opportunity for successfully buying the items, or need to pay a higher cost.

We develop a framework which can automatically extract and summarize the hot item features across different auction Web sites to assist the sellers and the buyers in decision marking. One objective of our framework is to characterize the popularity of an item listed for bidding. Intuitively, a hot item is the item which attracts many potential buyers for bidding. However, we should not measure the popularity of an item solely by its number of bids because of the following reasons. First, the number of bids on a hot item may be affected by the presence of another similar item listed with a lower price. Both of these items actually attract many buyers' interest and should be considered as hot items. Second, from Auction Software Review, we know that about one-forth of the items receive only one bid at the end of the auction period and potential buyers like to place the bid in the last minute [1]. Therefore, our approach for characterizing the popularity of the items is based on the product features of the items. For

Fig. 1. A sample of Web page about the bidding of a digital camera collected from *ebay.com*

Fig. 2. Another sample of Web page about the bidding of a digital camera collected from *ebay.com*

example, a possible product feature of a digital camera may be "4 megapixel resolution". Our approach can automatically discover the product features from the descriptions provided by the sellers. However, the diversified format of the descriptions can range from regular format such as tables to unstructured free texts, making the extraction task difficult. For example, Figures 1 and 2 depict two Web pages collected from ebay.com. These two Web pages are about the auction of digital cameras. However, the descriptions provided by the sellers are very different in layout format.

Our framework is able to collaboratively discover and summarize the hot item features across different auction Web sites. We formulate the product feature extraction task and hot item feature summarization task as a single graph labeling problem using conditional random fields (CRF) [2]. One characteristic of this graph is that it can model the relationship between the inter-dependence between the neighbouring tokens in the Web page, as well as the tokens in different Web pages. As a result, Web pages collected from different Web sites can then be considered under a coherent model improving the extraction quality. This also leads to another characteristic that various information such as the hot item feature information can be easily integrated in the graphical structure. We have conducted extensive experiments on several real-world auction Web sites to demonstrate the effectiveness of our framework.

2 Related Work

Ghani and Simmons proposed a closely related work on end-price prediction from auction Web sites [3]. They predict the price of the items at the end of the bidding period using four different kinds of features. The first kind of features is related to the sellers such as the seller rating. The second kind of feature is related to auction such as the first bid price of the item. The third kind of feature is related to the item. This kind of features consists of the indicators of the occurrence of certain phrases such as "like new" in the title. The last kind of feature is called temporal feature which is obtained from the recent history of the same item. They compared different machine learning techniques such as neural network and decision tree for the end-price prediction based on these features. Our proposed framework is different from their work in several aspects. First, the objective of their approach is to predict end-price whereas our framework is to extract and summarize the hot item features. Second, their approach assumes that each item placed for bidding is independent. However, as mentioned in Section 1, the items actually have a lot of mutual influences.

Hui and Liu [4] have investigated the task of summarizing customer reviews posted on the Web sites which is similar to sentiment classification [5]. Their objective is to classify sentences with subjective orientation. They make use of opinion terms such as "prefect", "good" as clues and extract the frequent features of the product from the reviews. Popescu and Etzioi [6] also conducted similar research. They first made use of the extraction system called KnowItAll [7] to extract the explicit features of the product. Next the extracted explicit features are utilized to identify the opinion or orientation from the reviews. Both of these two methods apply linguistic techniques and focus on the sentences which are largely grammatical. In contrast, the proposed work in this paper is to discover the product features of the hot items from the descriptions provided by the sellers in the auction Web sites. Such descriptions can be vary largely in layout format ranging from rigid table to free texts. Our work is also different from the research work on text summarization [8] whose objective is to produce text summary from text documents.

For semi-structured documents such as Web pages, different information extraction techniques have been proposed [9, 10]. Wrapper is a popular information extraction method and it usually consists of a set of extraction rules which can identify the attributes of interest from Web documents. Several machine learning methods have been developed for automatic wrapper generation by learning extraction models from training examples and achieve promising results [11, 12, 13, 14]. All these methods suffer from one common shortcoming in that the learned wrapper can only extract the attributes specified in the training examples. For example, if we just annotate the start time, end time, location, and speaker in the training examples in the seminar announcement domain, the learned wrapper can only extract these four attributes. Some other useful information such as the title of the seminar will not be extracted. In our previous work, we extended the traditional information extraction technique for discovering new attributes in Web pages [15]. It should be noted that our objective of hot item feature extraction and summarization is different from the objective of ordinary information extraction since our goal is not only to extract the product features, but also to generate the summary for the hot items across different auction Web sites. Some techniques have also been developed for fully automatic information extraction from Web pages without using any training examples such as IEPAD [16], MDR [17], Roadrunner [18]. Both IEPAD and MDR assume that the input Web pages contain multiple records and make use of the repeated patterns for extraction. However, a Web page normally consists of one item for bidding in the auction sites. Roadrunner does not require the

Web pages contain multiple records. However, the Web pages are required to have similar layout format and this is rare in auction Web sites.

Recently, various techniques have been proposed for collectively conducting information extraction and data mining [19]. For example, Wellner et al. proposed an approach for extracting different fields in citation and solving the citation matching problem using conditional random fields [20]. McCallum and Wellner also proposed an approach to extracting proper nouns and linking the extracted proper nouns using a single model [21]. Bunescu and Mooney proposed to use relational Markov networks to collectively extract information from documents [22].

3 Graphical Model for Hot Item Feature Mining

3.1 Model Formulation

In CRF, each node in the graph represents a variable and each edge represents the inter-dependence between the connected variables. Suppose we collect a set of Web pages from the auction Web sites and we wish to discover the hot item features. Figure 3 shows a simplified CRF model automatically constructed for the hot item feature mining application. The size of the graph is much larger when dealing with real data. There are two kinds of nodes. The shaded nodes represent observable variables while the unshaded nodes represent unobservable variables. Suppose we have a collection of Web pages $\mathbb{P}$. As mentioned above, a Web page, $M \in \mathbb{P}$, can be regarded as a set of text fragments denoted by $\mathbb{S}^M$ and each text fragment is considered as a sequence of tokens. For a particular

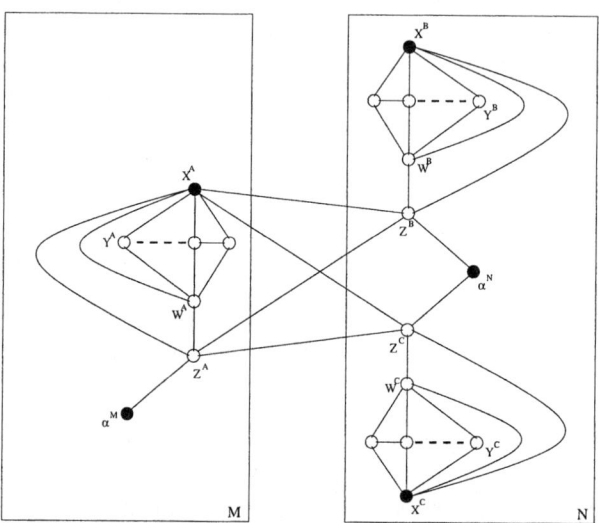

Fig. 3. Our proposed conditional random fields model for product feature extraction and hot item feature mining across different auction Web sites

sequence $A \in \mathbb{S}^M$, each token is actually composed of two kinds of information. The first kind of information is the observation of the tokens such as the content characteristics or the context characteristics. This information can be observed and is represented by the observable variable X^A. The second kind of information is the labeling information of the token. In product feature extraction, each token is labeled with either *product feature* or *normal text*. This information is hidden and is represented by the unobservable variable Y^A. Notice that X^A and Y^A actually represent a sequence of variables X_i^A and Y_i^A respectively where $0 < i < L$ and L denotes the number of the tokens in the sequence A. A node denoted by W^A represents the identified product features in the sequence A. Each Y_i^A is connected to Y_{i-1}^A, Y_{i+1}^A, X^A, and W^A as shown in Figure 3 since the tag label of each token is inter-dependent with the tag labels of the neighbouring tokens, the observation of the sequence, and the product features. There is another unobservable node called Z^A which refers to the hot item feature found in the sequence. An observable variable denoted by α^M in Figure 3 represents the number of bids of the item listed in page M. In page M, Z^A is connected with W^A and X^A because a hot item feature is related to the observation and the product feature found in the sequence. Z^A is also connected to α^M because a hot item feature is inter-dependent with the number of bids of the item listed in page M. For example, it is likely that the product feature is a hot item feature if the item receives high number of bids from the potential buyers. In Figure 3, the sequence $B, C \in \mathbb{S}^N$ are collected from the same page $N \in \mathbb{P}$ and $N \neq M$. As mentioned in Section 1, a hot item is not only related to its number of bids, but also related to other items listed for bidding. Therefore, X^B and Z^B, as well as X^C and Z^C in page N are also connected to the Z^A in page M.

Once the undirected graph is constructed, the conditional probability of a particular configuration of the hidden variables, given the values of all the observed variables can be written as follows:

$$P(y|x) = \frac{1}{Z} \prod_{C(x,y) \in \mathbb{C}(x,y)} \Phi(C(x,y)) \qquad (1)$$

where x and y are the set of observable variables and the set of unobservable variables respectively, $\mathbb{C}(x,y)$ refers to the set of cliques of the graph. A clique is defined as the maximal complete subgraph. $\Phi(C(x,y))$ refers to the clique potential for $C(x,y)$. Z is called the partition function defined as:

$$Z = \sum_y \prod_{C(x,y) \in \mathbb{C}(x,y)} \Phi(C(x,y)) \qquad (2)$$

We define the clique potential as a linear exponential function as follows:

$$\Phi(C(x,y)) = \exp \sum_i \gamma_i f_i(x,y) \qquad (3)$$

where $f_i(x,y)$ and γ_i are the i-th binary feature and the associated weight respectively. For example, $f_i(x,y)$ equals to one if the underlying token is "resolution" and the tag label is product feature and equals to zero otherwise in the digital camera domain. Hence, Equation 1 can be written as follows:

$$P(y|x) = \frac{1}{Z} \exp \sum_i \gamma_i f_i(x, y) \qquad (4)$$

Given the set of γ_i, one can find the optimal labeling of the unobserved variables of the graph via conducting inference. The graph typically consists of a large number of combination for the labels of all the unobservable variables. Hence, direct computation of the probability of a particular labeling of the unobservable variables is infeasible. The inference can be carried out by the message passing algorithm, also known as the sum-product algorithm, by transforming the graph into junction tree or factor graph [23]. By finding the configuration of the hidden variables achieving the highest conditional probability stated in Equation 1, the hot item features can then be discovered from these Web pages.

3.2 Adaptive Training of CRF

Learning in CRF refers to estimating the value of the weights γ_i associated with each f_i in Equation 4. Suppose we have a set of training examples denoted by Tra for which the actual labels of the variables are known. We define the log likelihood function as follows:

$$\mathcal{L}(\gamma_i) = \sum_{j=1}^{j<|Tra|} \{\sum_i \gamma_i f_i(x^{(j)}, y^{(j)}) - log(Z)\} \qquad (5)$$

where $|Tra|$ and $(x^{(j)}, y^{(j)})$ denotes the number of training examples and the j-th training example respectively. Maximum likelihood approach aims at finding the set of γ_i which maximize Equation 5. It can be shown that Equation 5 is convex and achieves maximum when the following condition holds:

$$\frac{\nabla \mathcal{L}(\gamma_i)}{\nabla \gamma_i} = \sum_{j=1}^{j<|Tra|} f_i(x^{(j)}, y^{(j)}) - \sum_{j=1}^{j<|Tra|} \sum_{y'} f_i(x^{(j)}, y') P(y'|x^{(j)})$$
$$= 0 \qquad (6)$$

Therefore, one can obtain the set of γ_i achieving the maximum of Equation 5 by using iterative methods such as conjugate gradient methods or voted perceptron algorithm [24]. In particular, Figure 4 shows the outline of voted perceptron algorithm for learning the parameters. In essence, the voted perceptron algorithm estimates the weight by iteratively minimizing the following expression:

$$\left| \sum_{j=1}^{j<Tra} f_i(x^{(j)}, y^{(j)}) - \sum_{j=1}^{j<Tra} f_i(x^{(j)}, \hat{y}^{(j)}) \right| \qquad (7)$$

where $\hat{y}^{(j)}$ is the predicted labeling using the current weighting.

However, recall that one objective of our framework is to extract the previously unseen product feature contained in the Web pages. To achieve this, we

```
# Original voted perceptron algorithm for learning CRF
Input: Training examples: $Tra$; Number of iteration: $K$
       Learning rate: $\rho$; Initial parameter set: $\gamma_i^0$
Output: The final parameter set: $\gamma_i^K$
Algorithm:
1. for $k = 0 \ldots K - 1$
2.    for $j = 1 \ldots |Tra|$
3.       $\hat{y}^{(j),k} = \arg\max_{y'} P(y'|x^{(j)}; \gamma_i^k)$
4.       $\gamma_i^{k+1} \leftarrow \gamma_i^k + \rho \left\{ f_i(x^{(j)}, y^{(j)}) - f_i(x^{(j)}, \hat{y}^{(j),k}) \right\}$
5.       $k \leftarrow k + 1$
6.    end for
7. end for
8. return $\gamma_i^K$
```

Fig. 4. The outline of the supervised voted perceptron learning algorithm for CRF

```
# EM based voted perceptron algorithm for learning CRF
Input: Training examples: $Tra$; Number of iteration: $K$
       Learning rate: $\rho$; Initial parameter set: $\gamma_i^0$
Output: The final parameter set: $\gamma_i^K$
Algorithm:
1.  $\gamma_i^* \leftarrow \gamma_i^0$
2.  until convergence
    E-step:
3.     for $j = 1 \ldots |Tra|$
4.        $P(y'|x^{(j)}) = \frac{1}{Z} \exp \sum_i \gamma_i^* f_i(x^{(j)}, y')$
5.     end for
    M-step:
6.     for $k = 0 \ldots K - 1$
7.        for $j = 1 \ldots |Tra|$
8.           $\hat{y}^{(j),k} = \arg\max_{y'} P(y'|x^{(j)}; \gamma_i^k)$
9.           $\gamma_i^{k+1} \leftarrow \gamma_i^k + \rho \left\{ \sum_{y'} f_i(x^{(j)}, y') P(y'|x^{(j)}) - f_i(x^{(j)}, \hat{y}^{(j),k}) \right\}$
10.          $k \leftarrow k + 1$
11.       end for
12.    end for
13.    $\gamma_i^* \leftarrow \gamma_i^K$
14. end until
15. return $\gamma_i^*$
```

Fig. 5. The outline of our EM based voted perceptron learning algorithm for CRF

exploit the clue embodied in the context characteristic such as the layout format of the extracted data. However, the extracted data cannot be directly used because they involve uncertainty. To tackle this problem, we treat the extracted data as unlabeled data and develop an expectation-maximization (EM) based voted perceptron algorithm as shown in Figure 5. In the E-step of our algorithm, we estimate the probability of the labeling of the unobservable variables. In the M-step, we employ the voted perceptron algorithm augmented with the following weight updating function:

$$\gamma_i^{k+1} \leftarrow \gamma_i^k + \rho \left\{ \sum_{y'} f_i(x^{(j)}, y') P(y'|x^{(j)}; \gamma_i^k) - f_i(x^{(j)}, \hat{y}^{(j),k}) \right\} \qquad (8)$$

Compared with the algorithm stated in Figure 4, our EM based voted perceptron algorithm estimates the weight by iteratively diminishing the following expression:

$$\left| \sum_{j=1}^{j<Tra} \sum_{y'} f_i(x^{(j)}, y') P(y'|x^{(j)}; \gamma_i^*) - \sum_{j=1}^{j<Tra} f_i(x^{(j)}, \hat{y}^{(j)}) \right| \quad (9)$$

The first term of Equation 9 (i.e., $\sum_{j=1}^{j<Tra} \sum_{y'} f_i(x^{(j)}, y') P(y'|x^{(j)}; \gamma_i^*)$) is the expectation value of $f_i(x^{(j)}, y')$ and it approaches to the first term of Equation 7 (i.e., $\sum_{j=1}^{j<Tra} f_i(x^{(j)}, y^{(j)})$) when the data set is sufficiently large.

4 Experimental Results

We conducted experiments on three real-world auction Web sites in the digital camera domain to demonstrate the effectiveness of our framework. The three auction Web sites are *www.ebay.com*, *auctions.yahoo.com*, and *www.ubid.com*. We collected 50 Web pages from each of the auction sites for the evaluation. Each Web page contains an item listed for bidding and the remaining bidding time is less than an hour. We conducted two sets of experiments to evaluate our approach to product feature extraction and hot item feature summarization.

We manually annotated the product features in the Web pages. These annotated product features were served as the gold standard in our evaluation. We randomly chose 5 pages from each of the Web sites (a total of 15 Web pages) to produce the set of training examples to train our model as described in Section 3.2. The trained model is then applied to the remaining Web pages to extract the product features of the items. *Recall(R), precision(P), and F-measure(F)* are adopted as the evaluation metrics. Recall is defined as the number of items for which the system correctly identified divided by the total number of actual items. Precision is defined as the number of items for which the system correctly identified divided by the total number of items it extracts. F-measure is defined as $2PR/(P+R)$. Table 1 depicts the extraction performance of our approach. Our approach achieves about 81% and 75% for average precision and recall respectively. This shows that our approach can effectively leverage the content and context characteristics to extract the product features.

Next, we employ our framework to generate the summary of the hot item features. To increase comprehensibility, we generate the summary by outputting the text fragments containing the hot item features instead of individual token. Table 2 shows some text fragments extracted. We manually investigate the items listed for bidding in the auction Web sites and find that over 70% of the items receiving at least one bid from the potential buyers contain at least three of the reported product features mentioned in the summary. This demonstrates that the summary generated is very helpful for the auction Web site participants.

Table 1. The experimental results of our approach to extracting product features. (P, R, and F refer to the precision, recall, and F-measure respectively. Ave. refers to the average of extraction performance.)

	P	R	F
ebay	0.77	0.62	0.69
yahoo	0.87	0.88	0.87
ubid	0.78	0.75	0.76
Ave.	0.81	0.75	0.78

Table 2. Some of the text fragments containing the hot item features in the digital camera domain

Text fragments about the hot item features
Digital Zoom 8X
For Mac or Windows
15 - 25 fps (for 640 x 480 pixels)
2 . 0 " TFT LCD Screen
About 120g (without battery and SD card)
Add SD flash memory cards up to 1GB to store over 1000 pictures
Bundled Kits : Camera Bag

5 Conclusions

We have developed a unified framework which is able to extract and summarize the hot item features across different auction Web sites. Our system can assist sellers and potential buyers in making decision. One challenge of this problem is to extract information from the product descriptions provided by different sellers, which vary largely in the layout format. We formulate the problem as a single graph labeling problem employing conditional random fields. The solution is then obtained by conducting inference in the graph. One characteristic of our framework is to extract the previously unseen product features by making use of the clue embodied in the layout format of the extracted data. We have designed an EM based voted perceptron algorithm to handle the uncertainty involved. Extensive experiments from several real-world auction Web sites have been conducted to demonstrate the effectiveness of our framework.

References

1. Auction Sotware Review. In: http://www.auctionsoftwarereview.com/article-ebay-statistics.asp. (2003)
2. Lafferty, J., McCallum, A., Pereira, F.: Conditional random fields: Probabilistic models for segmenting and labeling sequence data. In: Proceedings of Eighteenth International Conference on Machine Learning (ICML). (2001) 282–289
3. Ghani, R.: Price prediction and insurance for online auctions. In: Proceedings of the Eleventh ACM SIGKDD International Conference on Knowledge Discovery and Data Mining (SIGKDD). (2005) 411–418

4. Hu, M., Liu, B.: Mining and summarizing customer reviews. In: Proceedings of the Tenth ACM SIGKDD International Conference on Knowledge Discovery and Data Mining (SIGKDD). (2004) 168–177
5. Yi, J., Niblack, W.: Sentiment mining in WebFountain. In: Proceedings of the 21st International Conference on Data Engineering (ICDE). (2005) 1073–1083
6. Popescu, A., Etzioni, O.: Extracting product features and opinions from reviews. In: Proceedings of the Human Language Technology Conference Conference on Empirical Methods in Natural Language Processing. (2005)
7. Etzioni, O., Cafarella, M., Kok, S., Popescu, A., Shaked, T., Soderland, S., Weld, D., Yates, A.: Unsupservised named-entity extraction from the web: An experimental study. Artificial Intelligence **165(1)** (2005) 91–134
8. Mani, I., Maybury, M.: Advances in Automatic Text Summarization. MIT Press, Cambridge, MA (1999)
9. Kushmerick, N., Thomas, B.: Adaptive information extraction: Core technologies for information agents. In: Intelligents Information Agents R&D In Europe: An AgentLink Perspective. (2002) 79–103
10. Muslea, I., Minton, S., Knoblock, C.: Hierarchical wrapper induction for semistructured information sources. Journal of Autonomous Agents and Multi-Agent Systems **4(1-2)** (2001) 93–114
11. Agichtein, E., Ganti, V.: Mining reference tables for automatic text segmentation. In: Proceedings of the Tenth ACM SIGKDD International Conference on Knowledge Discovery and Data Mining (SIGKDD). (2004) 20–29
12. Crescenzi, V., Mecca, G.: Automatic information extraction from large websites. Journal of the ACM **51(5)** (2004) 731–779
13. Freitag, D., McCallum, A.: Information extraction with HMM structures learned by stochastic optimization. In: Proceedings of the Seventeenth National Conference on Artificial Intelligence (AAAI). (2000)
14. Kushmerick, N.: Wrapper induction: Efficiency and expressiveness. Artificial Intelligence **118(1-2)** (2000) 15–68
15. Wong, T.L., Lam, W.: A probabilistic approach for adapting information extraction wrappers and discovering new attributes. In: Proceedings of the 2004 IEEE International Conference on Data Mining (ICDM). (2004) 257–264
16. Chang, C., Lui, S.C.: IEPAD: information extraction based on pattern discovery. In: Proceedings of the Tenth International Conference on World Wide Web (WWW). (2001) 681–688
17. Liu, B., Grossman, R., Zhai, Y.: Mining data records in web pages. In: Proceedings of the Ninth ACM SIGKDD International Conference on Knowledge Discovery and Data Mining (SIGKDD). (2003) 601–606
18. Crescenzi, V., Mecca, G., Merialdo, P.: ROADRUNNER: Towards automatic data extraction from large web sites. In: Proceedings of the 27th Very Large Databases Conference (VLDB). (2001) 109–118
19. McCallum, A., Jensen, D.: A note on the unification of information extraction and data mining using conditional-probability, relational models. In: Proceedings of the IJCAI Workshop on Learning Statistical Models from Relational Data. (2003)
20. Wellner, B., McCallum, A., Peng, F., Hay, M.: An integrated, conditional model of information extraction and coreference with application to citation matching. In: Proceedings of the 20th Conference on Uncertainty in Artificial Intelligence (UAI). (2004) 593–601
21. McCallum, A., Wellner, B.: Toward conditional models of identity uncertainty with application to proper noun coreference. In: Proceedings of the IJCAI Workshop on Information Integration on the Web. (2003)

22. Bunescu, R., Mooney, R.: Collective information extraction with relational markov networkds. In: Proceedings of the 42nd Annual Meeting of the Association for Computational Linguistics (ACL). (2004) 439–446
23. Kschischang, F., Frey, B., Loeliger, H.: Factor graphs and the sum-product algorithm. IEEE Transaction on Information Theory **47(2)** (2001) 498–519
24. Collins, M.: Ranking algorithms for named-entity extraction: Boosting and the voted perceptron. In: Proceedings of the Annual Meeting of the Association for Computational Linguistics (ACL). (2002) 489–496

Clustering Web Sessions by Levels of Page Similarity

Caren Moraes Nichele and Karin Becker

Programa de Pós Graduação em Ciências da Computação (PPGCC),
Pontifícia Universidade Católica do Rio Grande do Sul (PUCRS), Porto Alegre, RS, Brazil

Abstract. Session similarity is a key issue in web session clustering. Existing approaches vary on session representation and similarity computation. However, they do not consider the similarity between pages, which is crucial due to the semantic gap between URLs and corresponding application events. This paper presents a domain taxonomy-based clustering approach, which extends the WLCS technique by integrating page similarity to compute session similarity. The approach can be applied to both usage and navigation clustering purposes.

1 Introduction

Web Usage Mining (WUM) applies data mining techniques to discover patterns from web server logs. WUM process is divided into three phases [2]: pre-processing, pattern discovery, and pattern analysis. The categorization of visitors' behavior based on their interaction in a website is a key issue in WUM. Several works [1, 3, 5, 7] leverage clustering techniques with the purpose of characterizing user behavior during navigation. The goals of session clustering can be roughly classified as *usage* and *navigation*. However, most clustering techniques do not consider the meaning in the application domain of accessed pages, in order to measure the similarity between web sessions. Page semantics is frequently considered in the pre-processing phase, in data enrichment tasks, in which URLs are mapped into domain concepts [1, 3]. This approach is static in the sense that a new perspective of a URL (e.g. more generalized concept), to obtain better clustering results, often implies re-processing data.

This paper presents Generalized Conceptual Session Clustering (GCSC) technique, which extends WLCS [1] in two ways: 1) it considers page similarity for session similarity computation, and 2) it deals with both usage and navigation clustering purposes. The goal is to improve the quality of resulting clusters and to ease their interpretation.

The remainder of this paper is structured as follows: Section 2 presents related work and Section 3 presents a review of WLCS algorithm [1]. Section 4 describes the GCSC approach. Preliminary experimental results are discussed in Section 5. Conclusions and future work are addressed in Section 6.

2 Related Work

Most works are concerned on how user sessions are represented according to the clustering objectives, and how session similarity is computed. Works such as [3, 7] are concerned with *usage clustering*, and hence represent web sessions as a set of (weighted) pages, disregarding their order of access. *Navigation clustering* is addressed by [1, 5, 9], which explicitly consider the user trajectory. Hence, session is represented in terms of visited pages, weight, order of access, and revisits.

A critical issue in WUM is the semantic gap between events in the site, and their related URL at server log. Semantic approaches conceptualize the application domain through a semantic model (i.e. taxonomy or ontology) which represents pages in terms of concepts and their relationships [8]. Works such as [1, 3] adopt a domain taxonomy in the pre-processing phase to transform URLs into concepts. However, page similarity is not explicitly used during clustering, for session similarity computation.

3 Weighted Longest Common Subsequences

WLCS [1] establishes the similarity of sessions considering both the similarity of their overlap region weighted by the time spent, and the importance of this region within each session. User trajectory is defined as a n-length sequence $(j=\{1,...,n\})$ of ordered pairs, given by $s_i = \{(c_1^i, \tau(c_1^i)),...,(c_n^i, \tau(c_n^i))\}$, where c_j^i is the j-th visited page in the session s_i and $\tau(c_j^i)$ is the associated time spent. WLCS discovers the path intersection by applying a Longest Common Subsequence (LCS) algorithm to find the longest common path between any two sessions. Then, given a specific overlap for sessions s_1 and s_2, WLCS obtains two functions, $l^{s_1}(i)$ and $l^{s_2}(i)$, where $i=(1,....L)$ and L is the length of overlap region, in order to retrieve the related page indices in each session. Then, considering the time spent on those pages, WLCS combines two measurement components: *similarity* and *importance*. The former computes how similar two sessions are in the overlap region, and the later, how important the overlap region is to each session. The total session similarity is the product of *similarity* and *importance*. WLCS assumes total equality between pages in order to find the overlap region.

4 Generalized Conceptual Session Clustering

Generalized Conceptual Session Clustering (GCSC) is a clustering approach that extends WLCS in two aspects: a) it considers page similarity during the computation of session similarity, and b) it deals homogeneously with any clustering goals, as well as time and binary weights, by applying the proper post-processing operations over the input sessions. GCSC is based on a taxonomy representing domain events for addressing the semantic gap between URLs and application events. GCSC assumes the following pre-processing tasks over input log data:

– Log file is submitted to typical pre-processing tasks [2], and clustering-specific post-processing tasks [1, 3, 7].
– A domain taxonomy exists, which reflects expert knowledge about the domain. Data enrichment activities map URLs into concepts of the domain hierarchy.

4.1 Session Representation

As in [1], the user trajectory is represented as the sequence of ordered pair of pages and weight. We assume both time and binary weights. This session representation

(Figure 1 A) obviously suits the navigation clustering objective. Hence, each index j of any session s_i represents the order in which a given concept was accessed in the session, including its revisits. The generalization proposed by GCSC with regard to session clustering purposes uses this same session representation for usage clustering, in which neither the order of visited concepts nor their revisits are relevant. Using appropriated pre-processing tasks, concepts are re-arranged in the session according to their proximity in the domain taxonomy to emulate a common usage sequence, as shown in Figure 1 (B). If time weights are used, they are added in case of revisits.

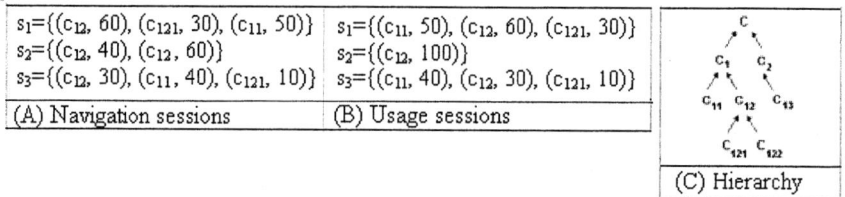

Fig. 1. Session representation and domain hierarchy example

4.2 Page Similarity Measure

This work leverages domain semantics, represented by the taxonomy, to compute the similarity between the concepts in terms of their location in the hierarchy. The adopted similarity function (Formula 1) is an adaptation of the GVSM (Generalized Vector-Space Model) [4] element similarity, where c_1 and c_2 are concepts in the hierarchy, *LCA* is the lowest common ancestor for c_1 and c_2, and *depth* is the number of edges from the concept to the top of hierarchy.

$$sim(c_1,c_2) = \frac{2 \times depth(LCA(c_1,c_2))}{depth(c_1) + depth(c_2)} \qquad (1)$$

4.3 Similarity-Based WLCS

Another component of GCSC is SWLCS (Similarity-based WLCS), which extends WLCS by considering the similarity between concepts when computing the LCS between sessions. The equality comparison of WLCS was replaced by a similarity comparison (Formula 1) considering an additional input, namely *similarity threshold*. The similarity threshold allows defining the minimum similarity required in order to include two different, but similar concepts into the LCS. For example, consider the conceptual hierarchy of Figure 1 (C), sessions s_1 and s_2 of Figure 1 (B) and a similarity threshold of 0.5. The LCS between s_1 and s_2 is defined by the session indices $l^{s_1}(1)=1$, $l^{s_1}(2)=3$ and $l^{s_2}(1)=1$, $l^{s_2}(2)=2$, because c_{11} and c_{12} similarity meets the threshold value. It should be noticed that the definition of a threshold value is not a simple task, since it is strongly dependent on depth of the hierarchy. This discussion is out of the scope of this paper scope. Given that any two sessions may have more than one LCS [1] and time spent in each page of the LCS is used to compute session similarity, SWLCS computes session similarity based on two LCS, yielding the highest similarity value.

GSCS was implemented as operators in LogPrep [6], a customizable WUM pre-processing tool, which is organized in terms of pre-processing tasks. For each type of task, it provides one or more operators that implement some pre-processing technique. We extended LogPrep with a new task (clustering transformations) and two new operators that define respectively usage and navigation clustering goals, as well as weight types, together with some clustering-specific post-processing tasks of [6].

5 Case Study

We leverage the distance education department domain of our university (PUCRS-Virtual) as case study. For a given web course, we developed a domain taxonomy that organizes 173 concepts through generalization relationships, which was validated by a domain expert of PUCRS-Virtual. Due to space limitation, the experiment discussed here compares WLCS and SWLCS algorithms for *navigation clustering* goal, considering time weighted sessions. For this purpose, we created 11 hypothetical sessions (Table 2) out of real log data, where each session s_i is composed of pairs of concept and weight, and concepts correspond to URLs mapped into the taxonomy. The mean time technique [7] was used to infer the time of the last access in each session.

We run SWLCS over this data twice, with a similarity threshold of 0.7 and 1.0, the later to produce WLCS results. Then, resulting similarity matrixes were submitted to a graph-based clustering algorithm (http://www.users.cs.umn.edu/~karypis/cluto). Comparing the clusters produced (Table 3), along with the intuitive meaning of domain concepts, it is possible to state that SWLCS appropriately grouped s_1 and s_2 because they have a common chat ancestor in their overlap region. Likewise, s_7 is grouped with s_8, s_9, s_{10} and s_{11} since they have common forum and email services.

Table 1. Hypotetical web sessions

Sessions		Legend	
$s_1=\{(c_1,2593),(c_2,718),(c_3,0)\}$	$s_2=\{(c_4,1177),(c_5,102),(e_1,0)\}$	c_j:	Chat
$s_3=\{(t_1,45),(t_2,186),(m_1,0)\}$	$s_4=\{(t_1,98),(t_3,57),(m_2,24),(m_3,0\}$	e_j:	Email
$s_5=\{(m_1,48),(q_1,0)\}$	$s_6=\{(m_1,101),(q_1,0)\}$	f_j:	Forum
$s_7=\{(f_1,80),(e_5,72)\}$	$s_8=\{(f_2,43),(e_3,118),(e_4,0)\}$	q_j:	Quiz
$s_9=\{(e_5,102),(e_3,178),(e_4,0)\}$	$s_{10}=\{(e_6,58),(e_3,70),(e_7,72),(e_8,0)\}$	m_j:	Material
$s_{11}=\{(e_5,42),(e_3,58),(e_4,0)\}$		t_j:	Tasks
		m_j:	Site map

Table 2. Experiment results

WLCS	SWLCS
Cluster1: s_3, s_4, s_5, s_6	Cluster1: s_5, s_6
Homework tasks, material, site map and quiz	Site map and quiz
Cluster2: s_2	Cluster2: s_3, s_4
Chat and email services	Homework tasks and materials
Cluster3: s_7	Cluster3: s_1, s_2
Forum and email services	Chat and email services
Cluster4: $s_1, s_8, s_9, s_{10}, s_{11}$	Cluster4: $s_7, s_8, s_9, s_{10}, s_{11}$
Email, forum and chat services	Email and forum services

Additionally, SWLCS enables to distinguish between visitors who browse general content (s_5 and s_6) from the ones who access didactical content and perform homework tasks (s_3 and s_4). Hence, it is possible to state that SWLCS revealed in this experiment more meaningful clusters with regard to the domain. Other experiments, not addressed in this paper due to space limitations, have displayed similar improvements. Notice that the same results could have be obtained by WLCS, if data enrichment tasks statically generalize concepts that are generalized dynamically by SWLCS.

6 Conclusions

GCSC is a clustering approach that extends WLCS to consider page similarity, and for both navigation and usage clustering objectives. The main contribution is to leverage the advantages of the domain taxonomy for computing session similarity. Concept generalization is dynamically considered during session similarity computation, as opposed to static, data enrichment-based approaches. Preliminary results have shown encouraging results, but several issues need further experimentation: the effects of concepts ordering in the usage sessions when computing LCS, bigger data samples, threshold definition, etc. This work is part of a research that aims at using WUM techniques to provide a learning monitoring behavior for distance education, targeted at instructors. Future work includes cluster interpretation using the conceptual representation of sessions, for both navigation and usage clusters.

References

1. Banerjee, A., Ghosh, J.: Clickstream Clustering Using Weighted Longest Common Subsequences. In Proceedings of the Web Mining Workshop at the 1st SIAM Conference on Data Mining, Chicago (2001).
2. Cooley, R.; Mobasher, B., Srivastava, J.: Data Preparation for Mining Word Wide Web Browsing Patterns. Knowledge and Information Systems (1999).
3. Fu, Y.; Sandhu, K.; Shih, M.: A Generalization-Based Approach to Clustering of Web Usage Sessions. In Masand and Spiliopoulou (Eds), Web Usage Analysis and User Profiling, Lecture Notes in Artificial Intelligence (2000), v.1836, 21-38.
4. Ganesan, P.; Garcia-Molina, H.; Widom, J.: Exploiting Hierarchical Domain Structure to Compute Similarity. ACM TOIS (2003), v.21, n.1, 64-93.
5. Gündüz, S.; Özsu, M. T.: A Web page prediction model based on click-stream tree representation of user behavior. In Proceedings of the 9th ACM (SIGKDD) (2003) 535-540.
6. Marquardt, C., Becker, K.; Ruiz, D.: A Pre-processing Tool for Web Usage Mining in the Distance Education Domain. In Proceedings of the 8th IDEAS, Coimbra, (2004) 78-87.
7. Mobasher, B.: Web Usage Mining and Personalization. Draft Chapter in Practical Handbook of Internet Computing, Munindar P. Singh (ed.), CRC Press (2004).
8. Stume, G.; Berendt, B.; Hotho, A.: Usage Mining for and on the Semantic Web. In Proceedings of NSF Workshop, Baltimore (2002) 77-86.
9. Wang, W.; Zaiane, O. Z.: Clustering Web Sessions by Sequence Alignment. International Workshop on DEXA (2002) 394-398.

iWED: An Integrated Multigraph Cut-Based Approach for Detecting Events from a Website

Qiankun Zhao, Sourav S Bhowmick, and Aixin Sun

CAIS, Nanyang Technological University, Singapore
qkzhao@pmail.ntu.edu.sg, assourav@ntu.edu.sg, axsun@ntu.edu.sg

Abstract. The web is a sensor of the real world. Often, content of web pages correspond to real world objects or events whereas the web usage data reflect users' opinions and actions to the corresponding events. Moreover, the *evolution patterns* of the web usage data may reflect the evolution of the corresponding events over time. In this paper, we present two variants of iWED(**I**ntegrated **W**eb **E**vent **D**etector) algorithm to extract events from website data by integrating *author-centric data* and *visitor-centric data*. We model the website related data as a *multigraph*, where each vertex represents a web page and each edge represents the *relationship* between the connected web pages in terms of *structure, semantic,* and/or *usage pattern*. Then, the problem of event detection is to extract *strongly connected subgraphs* from the multigraph to represent real world events. We solve this problem by adopting the normalized graph cut algorithm. Experiments show that the *usage patterns* play an important role in iWED algorithms and can produce high quality results.

1 Introduction

The web has invaded our lives. In some sense, the web is a sensor of the real world. Specifically, it has been observed that events and objects are often represented by sets of web pages but not individual web pages [4, 8]. Consequently, a large body of literature has focused on extracting real world events or objects from web data [4, 5, 8, 11, 12]. These approaches can be classified into two groups: *structure-based* extraction and *content-based* extraction. In the structure-based approaches, the website structures, hyperlink structures, and URLs are used to extract sets of web pages corresponding to events and objects [4, 8]. In the content-based extraction, content of web pages are segmented and categorized into subgroups that correspond to different topics, events, and stories using techniques such as natural language processing and probability models [1, 11, 12]. At the same time, such extraction results have been proved useful in many applications such as organizing the website structure [8], restructuring the web search results [4], terrorism event detection [9], and *Photo Story* and *Chronicle* [5].

Data associated with a set of web pages in a web site can be classified into two types: *author-centric* and *visitor-centric*. Author-centric data refers to a set of hyperlinked web pages that describes certain object or event, while *visitor-centric* data refers to web access sequences of these pages and describes how the

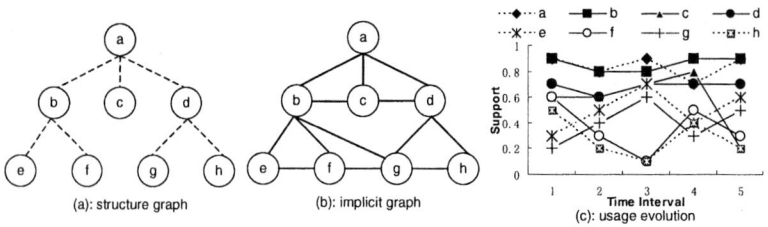

Fig. 1. Web data representation

web pages are accessed in the history. Observe that author-centric data describes authors' point of view while visitor-centric data reflects the web visitors' point of view.

We observed that existing event and object extraction approaches only analyzed the author-centric data. *These techniques ignore visitor-centric data.* However, often it may not be possible to distinguish different events related to the same topic by using the author-centric data alone. This is because events belonging to the same topic often share a set of keywords and the pages containing these different events are often connected by hyperlinks. For example, web pages talking about different *car accidents* tend to share keywords like *car, accidents,* and *crash*. Also, these pages may be connected as they belong to the same topic (car accident). Hence, it is difficult to distinguish one car accident from another based on only keywords and hyperlink structure.

In this paper, we consider visitor-centric data along with author-centric data to detect real-world events. In other words, we integrate visitor-centric and author-centric data to distinguish different events under the same topic. The major differences between our event detection approach and the related research [1, 11, 12, 5, 4, 2] are twofold. First, all the above works focus on either the author-centric or the visitor-centric data, while our approach incorporates the visitor-centric data along with the author-centric data. Second, the temporal property of the visitor-centric data is utilized in our approach to improve the event detection accuracy.

For example, suppose Figure 1(a) shows a subset of hyperlinked web pages; Figure 1(b) shows the *implicit links* extracted from the corresponding usage data; and Figure 1(c) shows the *evolution pattern* of web usage data (the y-axis shows the frequency of a web page being accessed over the time intervals shown in the x-axis). Here, there is an *implicit link* between two web pages if and only if they were accessed consecutively in the web access sequences [10]. The *evolution pattern* of web usage data refers to how the web pages changed in the history in terms of their supports [13].

It can be observed that from only Figure 1(a), it is difficult to distinguish sibling pages such as e and f even if they correspond to different events. However, with the evolution of web usage data as shown in Figure 1(c), connected web pages with similar content but corresponding to different events can be distinguished. For example, in Figure 1(c), pages e and g have similar evolution pattern while pages e and f have different evolution pattern. At the same, web pages that are not connected by hyperlinks but corresponding to the same event can be identified using implicit links in Figure 1(b), since they are expected to be accessed together. As shown in Figure 1(b), the implicit link between web pages b and g, which are not connected by hyperlink in Figure 1(a), implies that b and g have a possibility to represent the same event. In this paper, we focus on detecting events in a specific website as it is extremely difficult to gather web usage data of the entire web. The contributions of this paper are as follows.

- To the best of our knowledge, this is the first approach that detects website level events by integrating web structure, web content, and web usage data and its evolution patterns.
- A *multigraph* is proposed to model website related data in terms of structure, semantics, and usage patterns by integrating the *author-centric* and *visitor-centric* data.
- We present two variants of iWED algorithm, called *fusion-based graph cut* and *level-wise graph cut*, to detect events from the multigraph. These algorithms are inspired by the normalized graph cut algorithm widely used in image and video object extraction [6]. Experiment results show that the iWED event detection algorithms can produce high quality results.

2 Website Data Representation and Problem Statement

In this section, we first discuss how to represent web structure, web content, and web usage data of a web site using *structure graph*, *content graph*, and *usage graph*, respectively. Then, we present how these three types of graphs are integrated using a *multigraph*, followed by the problem statement of website-based event detection.

2.1 Structure Graph

The web structure data here refers to the set of web pages and hyperlinks between them. It can be modelled as a *structure graph*, $G_s = \langle V_s, E_s \rangle$, where each vertex in V_s is a web page and each edge in E_s represents the *structure similarity* (will be defined later) between the two pages that are connected by this edge. Note that the *structure similarity* is defined to reflect the similarity between web pages in terms of structure. The intuition is "two web pages are structurally *similar* if they are linked with *similar* web pages" [3]. As the base case, we consider a web page maximally similar to itself, to which we can assign a structure similarity score of *1*. With this intuition, given two web pages i and j in V_s, the *structure similarity* is defined as:

$$S_s(i,j) = \frac{C}{|D(i)| * |D(j)|} \sum_{m=1}^{|D(i)|} \sum_{n=1}^{|D(j)|} S_s(D_m(i), D_n(j))$$

Here C is a constant between 0 and 1, $|D(i)|$ is the degree of vertex i in the graph and $D_m(i)$ is the mth neighbor of vertex i. It is obvious that this similarity is an iterative function where similarities between web pages are propagated through recursions. That is, the value of $S_s(i,j)$ in the tth iteration, denoted by S_{s_t}, is based on the values of the t-1th iteration. More over it has been proved that this recursive function is nondecreasing and it will converge eventually [3]. We initialize the recursions with S_{s_0}: if $i=j$, then $S_{s_0}(i,j)=1$; otherwise $S_{s_0}(i,j)=0$.

2.2 Content Graph

The web content data refers to the content of each web page. The web content data is modelled as a *content graph*, $G_c = \langle V_c, E_c \rangle$, where each vertex in V_c is a web page and each edge in E_c represents the *semantic similarity* between two pages. It has been experimentally proven that *cosine measure* is one of the best measures for web content clustering [7]. Hence, we use the cosine measure to quantify *semantic similarity* between two pages. Given a web page i, using some stemming algorithm, it will be represented as a vector, $\vec{X_i}$, which corresponds to the *TF.IDF* of the terms after stemming [7]. Then, the *semantic similarity* between two web pages i and j, denoted as $S_c(i,j)$, is defined as:

$$S_c(i,j) = \frac{(\vec{X_i} \bullet \vec{X_j})}{||\vec{X_i}|| \cdot ||\vec{X_j}||}$$

where $(\vec{X_i} \bullet \vec{X_j})$ is the dot product of the two vectors and $||\vec{X_j}||$ denote the length of vector $\vec{X_j}$.

2.3 Usage Graph

The usage data refers to the access log of the web pages. It also can be modelled as a graph, called *usage graph*, $G_u = \langle V_u, E_u \rangle$, where each vertex in V_u is a web page and each edge in E_u represents the *usage pattern-based similarity* between two pages. Firstly, we review some of the literature in web usage mining.

In general, web usage data records the interactions between web users and the web server. A web access sequence ($\mathcal{WAS}$) is an ordered list of pages accessed by a user, denoted by $A = \langle (p_1,t_1), (p_2,t_2), \ldots, (p_n,t_n) \rangle$, where p_i is a web page, t_i is the time when p_i was accessed and $t_i \leq t_{i+1} \; \forall \; i = 1,2,3,\ldots,n-1$. Similar to [13], the $\mathcal{WAS}$s can be represented as a sequences of $\mathcal{WAS}$ *group* based on the user-defined time interval. A $\mathcal{WAS}$ *group* (denoted by G) is a bag of $\mathcal{WAS}$s that occurred during a specific time period. Let t_s and t_e be the start and end times of a period. Then, $G = [A_1, A_2, \ldots, A_k]$ such that $\forall \; p_i \in A_j, 1 < j \leq k$, p_i was visited between t_e and t_s. As a result, the historical web log data is divided into a sequence of $\mathcal{WAS}$ groups. Let $H_G = \langle \; G_1, G_2, G_3, \ldots, G_k \; \rangle$ be a sequence of k $\mathcal{WAS}$ groups generated from the historical web log data. Given a web page i, let $H_i = \langle \; \Phi_1(i), \Phi_2(i), \Phi_3(i), \ldots, \Phi_k(i) \; \rangle$ be the sequence of support values of i in H_G. Note that, for $1 \leq t \leq k$, $\Phi_t(i) = \frac{\mathcal{N}}{|G_t|}$, where $\mathcal{N}$ is the number of $\mathcal{WAS}$s that contain i.

Given two web pages, i and j, with the corresponding web usage data, the *usage pattern-based similarity*, denoted by $S_u(i,j)$, is defined as:

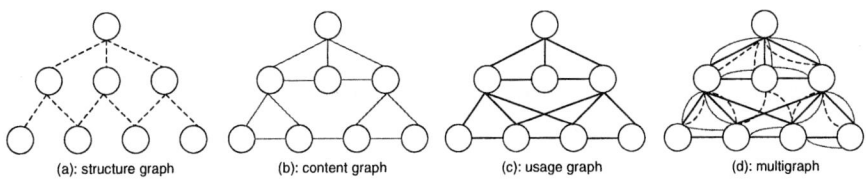

(a): structure graph   (b): content graph   (c): usage graph   (d): multigraph

Fig. 2. Web data representation

$$S_u(i,j) = \lambda \times e^{-D} + (1-\lambda) \times \frac{\sum_{t=1}^{k}(\Phi_t(\langle i,j \rangle) + \Phi_t(\langle j,i \rangle))}{\sum_{t=1}^{k}(\Phi_t(i) \cup \Phi_t(j))},$$

where $D = \sqrt{\sum_{t=1}^{k} |\Phi_t(i) - \Phi_t(j)|^2}$. Note that, the usage pattern-based similarity is a linear combination of the *evolution pattern-based similarity* and the *implicit link-based similarity*. The evolution pattern-based similarity is denoted by e^{-D}, where D is the *Euclidian distance* between the support sequences $H(i)$ and $H(j)$. The implicit link-based similarity is represented as the percentage of $\mathcal{WAS}$s that contain i and j consecutively against the total number of $\mathcal{WAS}$s that contain at least one of i and j. Here, λ and $1-\lambda$ are the weights of evolution pattern-based similarity and implicit link-based similarity, respectively. It is obvious that both the evolution pattern-based similarity and implicit link-based similarity are within the range of 0 to 1. Similarly, the usage pattern-based similarity is between 0 and 1.

2.4 Multigraph

We merge the above three graphs into a *multigraph*, which includes web structure, web content, and web usage data in a website. A *multigraph* is a graph whose edges are unordered pairs of vertices, and the same pair of vertices can be connected by multiple edges. In this case, there are three edges for each pair of vertices. These three edges represent the edges of structure graph, content graph, and usage graph, respectively.

Definition 1 [Multigraph]. *A multigraph is represented as a 3-tuple $M = \langle V, E, f \rangle$, where V is a set vertices, E a set of edges, and f is a function $f(e_i) = \{\{u,v\}|u,v \in V; u \neq v\}$ that takes an edge $e_i \in E$ and returns the set of web pages u and v that are connected by e_i. Two edges e_i and e_j are called parallel or multiple edges iff $f(e_i) = f(e_j)$.*

An example of the multigraph representation of website data is shown in Figure 2 with the corresponding structure graph, content graph, and usage graph. Note that, the similarities between disconnected web pages are *0* and the weights of the edges represent the corresponding similarity values.

Website-based Event Detection Problem: Based on the multigraph representation of the website related data, each real world event corresponds to a strongly connected subgraph in the multigraph. That is, a real world event can be represented as a set of structurally and semantically strongly connected web pages with similar usage patterns in the multigraph. The website based event detection problem is to extract such subgraphs from the multigraph representation.

3 iWED Algorithms

In this section, we present the iWED event detection algorithms based on the multigraph representation of the website data. To extract the strongly connected subgraphs from a graph, different graph cut algorithms have been proposed. In this paper, we adopt the normalized graph cut algorithm, which is widely used in object extraction from image data and frame segmentation of video data [6].

The three similarity measures, S_s, S_c, and S_u, introduced in Section 2 can be classified into two categories: *topic similarity* and *evolution similarity*. Topic similarity is the combination of the structure similarity (S_s) and the semantic similarity (S_c), while evolution similarity is the usage pattern-based similarity (S_u). Based on those two categories, we propose two variants of iWED algorithm for cutting the multigraph. The first approach, called the *fusion approach*, fuses the two types of similarity measures together and cuts the graph by treating the multiedges between two vertices as a single edge. The second approach, called the *level-wise approach*, cuts the graph with the two similarity measures separately. We now elaborate on these two approaches.

Fusion Approach: The fusion approach, denoted by FUS, integrates the three similarity measures together using linear combination with different weights. Such kind of fusion has been extensively used in combining different types of similarity measures in web content analysis [3]. In the fusion approach, a new similarity S is proposed as: $S = \alpha S_s + \beta S_c + \gamma S_u$, where α, β, γ are the weights for the corresponding similarity measures, and $\alpha+\beta+\gamma=1$. Then, the multigraph is transformed to a normal graph, where the weight of each edge is represented by S. The graph is then cut using the normalized graph cut algorithm.

Level-wise Approach: In the level-wise approach, the topic similarity and the evolution similarity are used to cut the multigraph separately. Note that, the topic similarity, denoted as S^T, defined as the fusion of structure similarity and semantic similarity. There are two alternative level-wise approaches. In the first approach, denoted by LTF (Level-wise Topic First), the multigraph is cut based on the topic similarity, which corresponds to only two types of edges in the multigraph, and the result, C_T, is returned. Then, each subgraph in C_T is cut again based on the evolution similarity and the final result, C_F, is returned. In the second approach, denoted by LEF (Level-wise Evolution First), the multigraph is first cut based on the evolution similarity and the result, C_E, is returned. Then, each subgraph in C_E is cut again using the topic similarity and the result C_F is returned. The underlying intuition is that, in the first approach, web pages are clustered into semantic topics before they are clustered into events as each event is expected to be a set of semantically similar web pages that have similar usage patterns. In the second approach, firstly web pages that correspond to similar types of events are gathered together and then clustered based on their semantic relationships.

For both the fusion approach and the level-wise approach, we present the clustering results with a hierarchical structure. That is, at the first recursion of the 2-way graph cut algorithm, there are two partitions. After that each partition is further cut into two child partitions and so on. However, not all the subgraphs correspond to real world events. To identify real world events and exclude outliers, we propose an *intra-cluster similarity measure*, $\mathcal{S}_{intra}(G')$, for any subgraph G':

$$\mathcal{S}_{intra}(G') = \frac{2\sum_{i}^{|G'|} \mathcal{S}(i,j)}{|G'| \times (|G'|-1)},$$

where $i \neq j$ and $i, j \in G'$. Based on this similarity measure, a threshold τ in the range of $[0, 1]$, is proposed to distinguish the event-based subgraph and the non-event-based subgraph. A subgraph, G' in the cut results corresponds to a real world event if and only if $\mathcal{S}_{intra}(G') \geq \tau$.

4 Performance Evaluation

In this section, the experimental results are presented to show the performance of our proposed event detection approaches. The three approaches, FUS, LTF, and LEF, are implemented and compared to the *baseline* approach, B_L, which only takes the structure and content of web pages using the corresponding similarity measures proposed in Section 2.

In our experiments, a synthetic e-commerce website dataset is used. Even though there are some real web usage datasets available, but due to privacy issues the original URLs and web pages are not available and they cannot be used in our experiments. The synthetic dataset we generated consists of 300 products and 2000 unique web pages. The 300 products belong to 5 categories, where the content of the web pages are generated according the attributes of products in different categories (we use the schema extracted from http://www.bargaincity.com.sg, which is one of the biggest e-commerce websites in Singapore). The usage data are generated in three steps. Firstly, the web access sequences are generated using uniform random generation. Then, we synthesize a list of 100 events (20 burst events such as one day only promotion and release of new products, 40 periodic events such as weekend promotion and new semester promotion, 20 increasing events such as price of a popular product keeps decreasing, 20 decreasing events such as some products are fading out of the market). Lastly, some noise access sequences are randomly inserted into the web usage data to mimic the real life usage data. In total, there are 10,000,000 unique page requests in the synthetic web usage data, which are partitioned into 100 access groups.

4.1 Evaluation Measures

As the event detection results are set of events, which consist of sets of web pages, it is different from existing classification algorithms. Although, we have the set of labelled events with corresponding web pages, the precision and recall measures in our event detection approach are different from the ones commonly used in classification tasks for the following reasons. Since an event consists of many web pages, the event may be detected but the corresponding web pages may not be accurate. That is, some pages may be missed and/or some non-related pages may be included. For example, given a real world event $E = \{P_1, P_2, P_3, P_4, P_5\}$, there may be one corresponding event $E' = \{P_1, P_3, P_4, P_7, P_8\}$ in the detection results. Moreover, for one real world event, there may be more than two corresponding events in the results. For example, given a real world event $E = \{P_1, P_2, P_3, P_4, P_5\}$, there may be two corresponding events $E' = \{P_1, P_3, P_4, P_7, P_8\}$ and $E'' = \{P_2, P_5, P_9\}$ in the detection results. We propose extended precision/recall measure for event detection based on the commonly-used precision/recall from IR.

Let $\mathcal{E} = \{E_1, E_2, \cdots, E_n\}$ be the set of detected events based on our proposed approach and $\mathcal{E}' = \{E'_1, E'_2, \cdots, E'_m\}$ be the set of labelled events in the dataset, where each event E_i consists of a set of web pages $\{P_{i1}, P_{i2}, \cdots, P_{ik}\}$. For each E_i, the corresponding real event E'_j with the largest value of $|E_i \cap E'_j|$ is selected, $|E_i|$ is the number of pages included in that event while $|E_i \cap E'_j|$ is the number of common pages included in both E_i and E'_j. Also, for each real world event E'_j, the corresponding event E_i with the largest value of $|E_i \cap E'_j|$ is selected from the results. Moreover, for different events in the real world, their corresponding events in the results should be different and vise versa. Then, the precision and recall are defined as:

$$Pr = \frac{\sum_i^{|\mathcal{E}|} \frac{|E_i \cap E'_j|}{|E_i|}}{|\mathcal{E}|} \quad Re = \frac{\sum_j^{|\mathcal{E}'|} \frac{|E_i \cap E'_j|}{|E'_j|}}{|\mathcal{E}'|}$$

4.2 Experimental Results

Two sets of experiments have been conducted to evaluate our proposed event detection approaches. Firstly, comparison of our proposed event detection approaches with the baseline approach is presented. Secondly, we show the effects of intra-similarity threshold τ on the quality of the detected events. Within each set of results, both the overall performance and the performance for each type of events are presented. Lastly, we discuss about how to set the fusion parameters in the *FUS* approach. Note that, the λ value in the usage pattern-based similarity is set to 0.5 for the following experiments.

Table 1(a) shows the performance of the four approaches with the precision, recall, and F_1 measure[1]. It can be observed that the *LEF*, *FUS*, and *LTF* approaches outperform the baseline approach, B_L, which shows the improvement of integrating the usage data and their evolution patterns. Among our proposed

---

[1] The F_1 measure is computed as $F_1 = \frac{2*Pr*Re}{Pr+Re}$.

Table 1. Event Detection Results

(a) All events

Alg	Pr	Re	F_1
B_L	0.376	0.108	0.168
FUS	**0.729**	**0.696**	**0.712**
LTF	0.591	0.412	0.486
LEF	0.684	0.625	0.653

(b) Burst events

Alg	Pr	Re	F_1
B_L	0.531	0.192	0.282
FUS	**0.892**	**0.751**	**0.815**
LTF	0.674	0.582	0.625
LEF	0.873	0.749	0.806

(c) Periodic events

Alg	Pr	Re	F_1
B_L	0.227	0.098	0.137
FUS	**0.678**	**0.622**	**0.649**
LTF	0.535	0.491	0.512
LEF	0.647	0.562	0.602

(d) In/Decreasing events

Alg	Pr	Re	F_1
B_L	0.483	0.298	0.364
FUS	**0.912**	**0.895**	**0.904**
LTF	0.692	0.769	0.728
LEF	0.875	0.864	0.869

(e) FUS

τ	Pr	Re	F_1
0.1	0.314	0.452	0.371
0.3	0.729	0.696	0.712
0.5	0.758	**0.712**	0.734
0.7	**0.841**	0.709	**0.769**
0.9	0.413	0.422	0.417

(f) LEF

τ	Pr	Re	F_1
0.1	0.279	0.354	0.312
0.3	0.591	0.412	0.486
0.5	0.681	0.527	0.594
0.7	**0.748**	**0.699**	**0.723**
0.9	0.324	0.435	0.371

approaches, the LEF and FUS archive better performances than the LTF approach. This is because some of the synthetic events in our dataset usually cover more than one semantic topic. Tables 1(b), (c), and (d) show the performance of our approaches with respect to different types of events.

In the above experiments, weights of the three similarity measures are set to 0.31, 0.20, and 0.49, which are experimentally proved to be the optimal values for our dataset. The threshold for intra-cluster similarity is set to 0.6. Tables 1(e) and (f) show the quality of the event detection results of the FUS and LEF approaches by varying the corresponding τ values. The results are for all types of events. Observe that the effects of threshold τ are similar for the three types of events. When the value of τ increases from 0.3 to 0.7, the quality of the event detection results becomes better; when the value of τ increases from 0.7 to 0.9, the quality of the event detection results becomes worse. This is because when the threshold for intra-cluster similarity is too small/large, the number of events detected may be too many/few. While the number of real world event is fixed, the performance of the approaches decreases when the threshold is close to the two extremes.

From the results shown in Table 1, it is evident that the FUS approach performs relatively better than other approaches in most cases. This is because, in the FUS approach, the weights of different types of similarities can be tuned. In our experiments, we show the average results of the FUS approach. It can be observed that the usage pattern-based similarity significantly improves the clustering results. Moreover, we observed that the structure similarity is less important than the usage pattern-based similarity but more important than the content similarity.

5 Conclusions

This work is motivated by the fact that existing event and object detection approaches only analyze the content and structure data of a website. In this paper, we integrate the author-centric and visitor-centric data to detect real-world events. Experimental results show that our proposed approaches can produce promising results.

References

1. J. Allan, C. Wade, and A. Bolivar. Retrieval and novelty detection at the sentence level. In *SIGIR*, 314–321, 2003.
2. S. Gunduz and M. T. Ozsu. A web page prediction model based on click-stream tree representation of user behavior. In *SIGKDD*, 535–540, 2003.
3. G. Jeh and J. Widom. Simrank: a measure of structural-context similarity. In *SIGKDD*, 538–543, 2002.
4. W.-S. Li, K. S. Candan, Q. Vu, and D. Agrawal. Retrieving and organizing web pages by "information unit". In *WWW*, 230–244, 2001.
5. Z. Li, M. Li, and B. Wang. Probabilistic model of retrospective news event detection. In *SIGIR*, 106–113, 2005.
6. J. Shi and J. Malik. Normalized cuts and image segmentation. *IEEE Trans PAMI*, 22(8):888–905, 2000.
7. A. Strehl, J. Ghosh, and R. Mooney. Impact of similarity measures on web-page clustering. In *AAAI Workshop of AI for Web Search*, pages 58–64, 2000.
8. A. Sun and E.-P. Lim. Web unit mining: finding and classifying subgraphs of web pages. In *CIKM*, 108–115, 2003.
9. Z. Sun, E.-P Lim, K. Chang, T.-K. Ong and R. K. Gunaratna Event-Driven Document Selection for Terrorism Information Extraction. In *IEEE ISI*, 37–48, 2005.
10. G.-R. Xue, H.-J. Zeng, Z. Chen, W.-Y. Ma, H.-J. Zhang and C.-J. Lu. Implicit link analysis for small web search. In *SIGIR*, 56–63, 2003.
11. Y. Yang, J. Zhang, J. Carbonell, and C. Jin. Topic-conditioned novelty detection. In *SIGKDD*, 688–693, 2002.
12. J. Zhang, Z. Ghahramani, and Y. Yang. A probabilistic model for online document clustering with application to novelty detection. In *NIPS*, 1617–1624. 2005.
13. Q. Zhao and S. S. Bhowmick and L. Gruenwald. WAM-Miner: In the Search of Web Access Motifs from Historical Web Log Data. In *Proceedings of CIKM*, 421–428, 2005.

Enhancing Duplicate Collection Detection Through Replica Boundary Discovery

Zhigang Zhang[1], Weijia Jia[1], and Xiaoming Li[2]

[1] Department of Computer Science,
City University of Hong Kong, 83 Tat Chee Avenue, Kowloon, Hong Kong
zhigang_chn@hotmail.com, itjia@cityu.edu.hk
[2] Institute of Network Computing and Information Systems,
School of Electronics Engineering and Computer Science, Peking University, Beijing, China
lxm@pku.edu.cn

Abstract. Web documents are widely replicated on the Internet. These replicated documents bring potential problems to Web based information systems. So replica detection on the Web is an indispensable task. The challenge is to find these duplicated collections from a very large data set with limited hardware resources in acceptable time. In this paper, we first introduce the notion of *replica boundary* to roughly reflect the situation of the replicas; then we propose an effective and efficient approach to discover the boundary of the replicas. The advantages of the proposed approach include: first, it dramatically reduces pair-wise document similarity computation, making it much faster than traditional replicated document detection approaches; second, it can identify the boundary of the replicated collections accurately, demonstrating to what extent two collections are replicated. On two web page sets containing 24 million and 30 million Web pages respectively, we evaluated the accuracy of the approach.

1 Introduction

The information explosion on the Web has led to a proliferation of documents that are identical or almost identical. Studies [3, 14] revealed that 30% to 45% of the Web consisted of replicated pages. The replica proliferation causes many problems to Web based information systems, including consuming excess bandwidth and crawling time, wasting disk to store redundant data, slowing down the indexing and retrieval time, and impairing the quality of retrieval results. Large concentrations of replicated documents may also skew the content distribution statistics with potentially harmful consequences to machine learning applications [6] and Web mining applications[16].

Most previous studies [2,3,4,7,15,16] investigated this problem at two different granularities: duplicate hosts and duplicate documents. However, through analyzing the duplicated data recorded in the replica reduction process of our search engine [17], we obtained an important observation that *resembling page collections often exists in two hosts that are obviously not mirror sites, even more serious than duplicate hosts*. A typical example is the how-to documents of Linux. By submitting queries such as 'matlab plot' or 'linux nfs howto' to the large search engines such as Google, Alta Vista and MSN, many duplicated pages can be found even in the top ranked results. Most previous research has ignored the existence of replicated sub-collections in two distinct sites. Duplicate host detection methods [7, 16] obviously skip over these replicated collections. Although the sub-collections can be removed by pair-wise similarity computation at the document granularity, considering the scale of the Web, performing such a massive amount of computation is highly time-consuming. A better way to efficiently remove such replicated sub-collections is to consider them as a whole, introducing an intermediate granularity, the directory granularity.

In order to clarify the complex situation of replication on the Web, we start with some definitions.

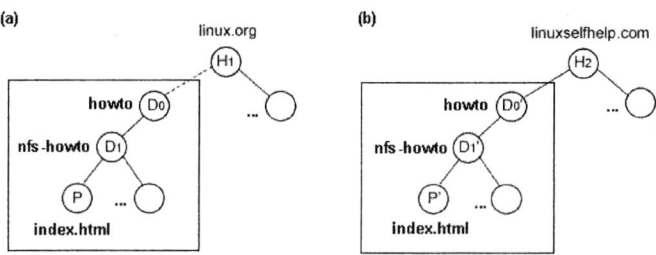

Fig. 1. Example of replicated directories and replica boundaries for (a) http://www.linux.org/docs/ldp/howto/NFS-HOWTO/ (b) http://www.linuxselfhelp.com/HOWTO/NFS- HOWTO/

Directories. In most hosts, Web pages are organized into nested *directories*, and the path indicates the location of the directory in the tree structure of the entire site. The website managers tend to design one or more *index page*s to organize the pages within the directory. Index pages serve as access portals to the related Web page collection. So, the index page can be regarded as the representative of the directory.

Replicated Directories. According to our observation, the directory trees corresponding to replicated sub-collections are often identical as well. We define *replicated directories* as follows: given two directories D and D' belonging to two distinct hosts H and H', D and D' are replicated directories if every page P nested in D can find an identical counterpoint P' in D'. Given the example shown in Figure 1, "www.linux.org" (H_1) and "www.linuxselfhelp.com" (H_2) are not duplicate hosts. But all pages nested in directories "www.linux.org/docs/ldp/howto/" (D_0) and "www.linuxselfhelp.com/ howto" (D'_0) are identical, and thus replicated directories.

Replica Boundary. Now we can define replica boundary as follows: given a directory tree of a Web page collection, a replica boundary of the collection is a node nearest to the root, from which all nested Web pages and sub-directories are replicated with another collection. In the previous example, directory pair (D_0, D'_0) are replica boundaries of hosts H_1 and H_2 respectively. By this definition, duplicate hosts can be regarded as a special case of replicated directories. Replica boundary serves as a brief description of the replica situation between two hosts.

In this paper, we proposed an effective and efficient approach to discover the replicated directories and the replica boundaries. The advantages of the proposed approach include: first, it dramatically reduces pair-wise document similarity computation, making it much faster than traditional replicated document detection approaches; second, it can identify the boundary of the replicated collections accurately, demonstrating to what extent two collections are replicated.

2 Related Works

A number of approaches for near replica detection have been proposed. They can be partitioned into two categories according to the granularities at which the replica is discovered: near replicated collection detection [4,7,16] and near replicated document detection [1, 2, 3, 8, 9, 10, 11, 12, 14].

Two typical approaches were proposed for replicated collection detection in [4, 7]. In [4], the authors proposed a clustering based approach to find replicated Webpage collections. This approach has the following drawbacks: first, the generation of trivial clusters is a bottom-up clustering process, which is very time consuming; second, the approach merges the clusters according to their link relations, the merge condition is too strict, because generally not all the Web pages within a Website can be fetched back in the crawling process, resulting in a difference in the link situations for two identical collections. In [7], the authors proposed an efficient approach to discover mirror hosts on the Web. This approach depends mostly on the syntactic analysis of the URL strings, and requires retrieval and content analysis for only a small number of pages. The approach suffers from the following drawbacks: the approach can only identify whether two hosts are totally mirroring or partially mirroring, but for the partially mirrors, it can not identify which part are replicated. In [16], the authors proposed and compared several algorithms that identify mirrored hosts on the Web based only on URL strings and the hyperlink structure. These algorithms can discovery the duplicated host very efficiently, but they can not find the directory level replicas. And the duplicated host can not be discovered when the path is renamed.

If not for the size of the Web, detecting near replica documents would be a traditional information retrieval problem. All the traditional text clustering approaches can be used to find near replicated web pages. Some more efficient approaches [1, 2, 3, 8, 9, 10, 11, 12, 14] are proposed to find replicated documents (web pages). These approaches can be roughly classified as [6]: similarity-based and signature-based. Despite the efforts to accelerate the computational time, all the proposed methods need to analyze the content of every Web page and depend on large scale pair-wise similarity computation, which makes the approaches very time-consuming.

3 Approaches for Replica Detection

We first verify an assumption by an experiment. Based on the assumption, we propose the replica directory detection approach. In our approach, we identify all the index pages in the collection, and discover candidate replicated directories by finding replicated index pages. Then, we use coordinate sampling to check the replicate status of the candidate directories. After obtaining the replicated directories, the replica boundaries are discovered.

3.1 Assumption Verification

Nearly all the efficient duplicated host detection approaches [7, 16] are based on one assumption: if one host is replicated, the relative path and filename of the pages within the host will not be altered. However, many replicas are not at host granularity, but at directory granularity. In this section, we design an experiment to verify whether the assumption still works when directories are replicated.

In the experiment, we first used a bottom-up algorithm to discover the replicated directories in the data set. After getting the replicated top directories in the data set, we can check whether the relative paths and filenames are changed or not. Based on one hundred of replicated directory pair samples, we got the following statistical result. All characters in the URLs were first converted to lower case before comparison. The result shows that 95% of the replicated directories kept the identical path names. All of the rest 5% were replica boundaries and made only one variation, the renaming of the replica boundary. Though the replica boundaries were renamed, the names all sub-directories and nested files remained unaltered.

3.2 Index Page Identification

Index pages are the representatives of directories. Instead of analyzing all URLs at hand, we concentrate on index pages to facilitate our algorithm. In [13], much information, including URL information, link information and text information, is considered to assist in finding the entry pages of websites. By constructing the decision tree on the training data, the author found that only the length of URL (the number of slashes in the URL), the slash information, and whether URL ends with some special keywords were necessarily retained. The URLs identified as entry pages by the obtained decision tree are regarded as index pages in our approach.

3.3 Finding Replicated Directories

As mentioned above, index pages can be regarded as the representatives of the corresponding directories. Generally, if two directories are replicas, their index pages are likely to be replicas as well, but this is not necessarily true vice versa. So when two index pages are identified as replicas, we need to check the pages in the two corresponding directories to justify the decision. If the two directories are replicated to each other, we call the two index pages *collective replicas*, because they reflect not only the replica relation between themselves but also the replica relation between the two Web page collections in the two directories. Otherwise, we call two index pages *individual replicas*. Our approach for finding replicated directories consists of the following three steps:

Step 1, finding the replicated index pages. The method introduced in [8] is used here. It is trivially true that, the replica relation obtained by such a replica identification method is an equivalent relation (reflexive, transitive and symmetrical). Thus, grouping the pages with identical fingerprints into one cluster constructs a partition of the whole page collection.

Step 2, coordinate sampling from candidate replicated directories. Given two replicated index pages, P_1 (http://x1/x2/index.html) and P_2 (http://y1/y2/index.html), the corresponding directories are D_1 (http://x1/x2/) and D_2 (http://y1/y2/), we select a certain number of pages (α) at random from D_1. By generating one fingerprint per page, a fingerprint set composed of the α fingerprints, $F_1=\{fp_1,...,fp_\alpha\}$, is created to represent D_1. Then the corresponding pages from D_2 with the same relative path and filenames are selected, and the set $F_2=\{fp_1,...,fp_m\}$ (m<=α) is generated to represent D_2. Again, the case of the letters in the URLs is ignored. The intuition here is that if a directory is replicated, the paths and filenames of the pages in the directory will not be renamed (verified in Section 3.1).

Searching records in a large database repeatedly is time consuming. By using a two-level hash method, we can accomplish the sampling for all candidate index page pairs in one scan of the database. As generating fingerprints requires analyzing each sample page, we can further improve the efficiency by representing the directory with the fingerprints of sample URL strings rather than the content of pages. This method is also evaluated in the experiments.

Step 3, evaluating the replica status. As the index pages identified as replicas by the previous step may not be collective replicas, we need to evaluate the replica status according the sampling results in Step 2. It leads to a split of the original partitioned groups into even smaller clusters.

Since not all the pages on the Web can be fetched, the sample fingerprint sets may not be the same (m<=α) even though two directories are replicated to each other. The replica status can be measured by computing the *resemblance* of the two index pages via the following formula (1):

$$r(F_1, F_2) = \frac{|F_1 \cap F_2|}{|F_1 \cup F_2|} \quad (1)$$

If r (F_1,F_2) is larger than a threshold β, P_1 and P_2 are collective replicas and the corresponding directories are replicated to each other; otherwise, they are individual replicas. For the clusters generated in Step 1, we further partition them into smaller clusters where the index pages in same cluster are collective replicas to each other.

3.4 Replica Boundary Detection

According to the definition of replicated directory and replica boundary, the replicated directory closest to the root is the replica boundary in all the replicated directories discovered. However, the replicated directories are identified based on sampling in the previous stage. Sampling leads to a measure of similarity between two candidate directories. And due to the sampling method, the closer the candidate directories are to the root, the less proportion of the pages are examined for the directory. Thus, we cannot jump to the conclusion of replicated directory, especially when large amount of sampling is impractical.

Next, we proposed the replica boundary detection algorithm that considers the replica information in all the directories. The replica boundary detection task can be formulated as follows: given groups of replicated directories, $G=\{g_1:\{d_{11},d_{12},...,d_{1n}\},..., g_k:\{d_{k1},d_{k2},...,d_{km}\}\}$, by parsing the paths of the index pages we construct another group of directory trees, $T=\{T_1,..., T_l\}$, indicating the structural relationship between the corresponding directories. Thus, the goal is to find pairs of nodes (index pages) <d_i, d_i'>

in T, where all offspring nodes of d_i' are collective replicas to the corresponding nodes of d_i and their parent nodes are not.

To accomplish this task, we propose a top down boundary identification algorithm. In this algorithm we use *containment* $C(T_1, T_2)$ to estimate how much directory tree T_1 is contained in T_2. First the directory tree is represented by the set of all nodes in the tree.

$$C(T_1,T_2) = \frac{|\{node_i \mid node_i \in T_1 \wedge (\exists node_j \mid node_j \in T_2 \wedge CR(node_i, node_j))\}|}{|T_1|} \quad (2)$$

where $CR(node_i, node_j)$ means that $node_i$ and $node_j$ are two replicated directories, $|\cdot|$ is the cardinality of a set. The value of containment naturally ranges from 0 to 1. If the containment $C(T_1, T_2)$ is equal to 1, T_1 is fully contained in T_2. It is evidence that "fully contained" is a partial order relation, maintaining the transitive property. As the crawler cannot fetch the entire Web, some deep sub-directories may not be fetched by the crawler, which affects the containment of two replicated directories. To handle this situation, we define a threshold γ ($0<\gamma<1$), if $C(T_1, T_2)$ is larger than γ, we say that T_1 is "fully contained" by T_2 with the confidence degree of γ. Since the loosed "fully contained" is also a partial order relation, we can achieve fast redundant directory removal by deleting the "contained" directories in a linear scan rather than requiring square time.

The matrix C that records the containment of directory pairs is called the containment matrix and the entry c_{ij} of matrix C is the value of containment (T_i, T_j). Given the containment matrix, the replica boundary can be found by the following algorithm:

Algorithm 1: Replica boundary identification algorithm

```
Input:    containment matrix C and collective replicated index page clusters G
Output: replica boundary list List
Procedure
0 Putting the root nodes of all directory trees into queue Q.
1 List={ }
2 While Q is not empty
3 {      take out one node n_i from Q  //we use t_i to denote the directory tree with root n_i
4        If (there exist a node n_x in the same cluster with n_i satisfying C(n_i,n_x)> γ)
5        {      //tree t_i is a replica of t_x, so t_i can be eliminated.
6               List=List ∪ {n_i}
7               Eliminating all nodes in t_i (including t_i) from all clusters in G }
8        Else
9        {      //tree t_i is not a replica of other directory trees
10              For all nodes n_j in the same cluster g with n_i
11              {      If (C(n_j, n_i)> γ)
12                     {      //n_j is a replica of n_i
13                            List=List ∪ {n_j}
14                            Eliminating all nodes in t_j (including t_j) from all groups in G }}
15              If (g still contains more than one node)
16              {      Inserting all the children of n_i into Q
17                     delete n_i from Q}}}
18 Return List
```

4 Experiments

We evaluated the accuracy of our approach on two Web page sets containing 24 million and 30 million web pages respectively. The two Web page sets were crawled by a breadth-first strategy from distinct sets of seeds.

4.1 Accuracy Evaluation of Replicated Directory Detection

We adopted *precision* and *recall*, which are widely used in the information retrieval area to evaluate our replicated directory detection approach.

$$precision = \frac{\#of\ directory\ pairs\ identified\ as\ replicated\ correctly}{\#of\ all\ directory\ pairs\ identified\ as\ replicated}$$

$$recall = \frac{\#of\ directory\ pairs\ identified\ as\ replicated\ correctly}{\#of\ all\ the\ replicated\ directory\ pairs\ in\ the\ dataset}$$

We verify whether two directories are really replicated through checking if a set of paths sampled from one directory are all valid on the other directory and yield highly similar documents, and vice versa. In the approach, two parameters are used: the sampling sum (α) and the resemblance threshold (β). The experimental results shown in Table 1 illustrated how the variability of the sampling sum (α) and resemblance threshold (β) affect the performance.

Table 1. Replicated directory detection based on web page content sampling

Data set 1				Data set 2			
α	β	precision	relative recall	α	β	precision	relative recall
0	0	90%	100%	0	0	87%	100%
1	1	100%	61%	1	1	100%	70%
2	0.5	100%	85%	2	0.5	100%	88%
2	1	100%	34%	2	1	100%	52%
3	0.33	100%	92%	3	0.33	100%	94%
3	0.67	100%	62%	3	0.67	100%	74%
3	1	100%	24%	3	1	100%	38%
4	0.25	100%	96%	4	0.25	100%	95%
4	0.5	100%	80%	4	0.5	100%	87%
4	0.75	100%	57%	4	0.75	100%	73%
4	1	100%	11%	4	1	100%	28%

From the data shown in Table 1, we can see that the changes of the precision on the two data sets are similar. The precision reaches 90% and 87% when no sampling is used in the approach ($\alpha = 0$). We checked the false cases when $\alpha=0$ and $\beta=0$, and found that the error results from the wrong decision on whether two index pages were replicas, due to the near replicated document detection method. We can also see that when sampling is used ($\alpha > 0$), the precision can reach 100%. And the variations of α and β can only affect the recall value. That means if two index pages are replicated to each other and there exists at least one extra replicated page pairs on the two directories, the two directories can be regarded as replicated safely. As for the recall, we can see that the value of the recall varied greatly with the variation of the parameters and data sets. The reason is that the crawler has not fetched all the pages in a host, especially when the data set is not large enough. It can not be guaranteed that the two replicated pages in the two replicated directories are all fetched in the data set, which affects the result of sampling.

To improve the efficiency of the algorithm, we can represent the directory using the fingerprints of sample URL strings rather than the content of pages, which can save the time on fingerprint computation. From the data shown in Table 2, we can see that such an alternative only affects the accuracy performance slightly. *In conclusion, if the index pages of two directories are replicated, we can identify the replica status of the two directories by sampling and considering only the URL strings; as for the number of samples, 4 samples can achieve satisfactory performance.*

Table 2. Replicated directory detection based on pure URL string sampling

\multicolumn{4}{c}{Data set 1}		\multicolumn{4}{c}{Data set 2}	
α	β	precision	relative recall	α	β	precision	relative recall
0	0	90%	100%	0	0	87%	100%
1	1	100%	64%	1	1	98%	72%
2	0.5	100%	87%	2	0.5	98%	88%
2	1	100%	39%	2	1	98%	54%
3	0.33	100%	93%	3	0.33	98%	94%
3	0.67	100%	64%	3	0.67	98%	75%
3	1	100%	29%	3	1	98%	43%
4	0.25	100%	96%	4	0.25	98%	95%
4	0.5	100%	81%	4	0.5	98%	88%
4	0.75	100%	63%	4	0.75	98%	77%
4	1	100%	20%	4	1	98%	35%

4.2 Accuracy Evaluation of Replica Boundary Detection

The accuracy is defined as following:

$$precision = \frac{\# \text{ of correctly identified replica boundary pairs}}{\# \text{ of all identified replica boundary pairs}}$$

If two directories are identified as a boundary pair, we need check the following two conditions to verify whether the two directories are really a replica boundary pair:

1. whether the two directories are replicated to each other;
2. whether the parent directories of these two directories are **not** replicated.

If the two conditions are both satisfied, the two directories are a boundary pairs.

The containment threshold γ affects the accuracy of replica boundary detection in two directions. On one hand, with the rising of γ's value, the accuracy is increasing as the containment relation is more and more sufficient; On the other hand, due to the incompleteness of the data set, with the rising of γ's value, more and more top level replicated directories are excluded and their child directories are regarded as replica boundary, which makes the accuracy decrease.

As shown in Figure 2, the effect of the latter one is the dominant factor. Even though the accuracies of the algorithm on the two data sets differ larger and larger when the value of γ is rising, they are very similar when γ is small. Especially, the accuracy of the algorithm on the two data sets all reached 97% when γ is equal to 0. According to the boundary detection algorithm, all the top replicated directories are regarded as replica boundaries when γ is equal to 0. That means if two index pages are identified as replicated to each other, they are likely to be totally replicated rather than partially replicated. So, the top-down algorithm is suitable.

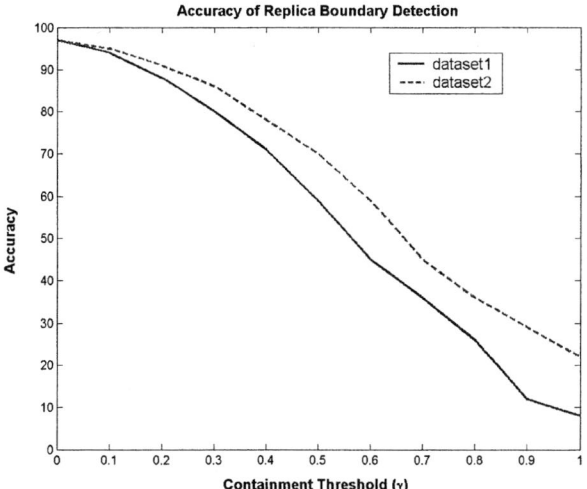

Fig. 2. Accuracy of replica boundary detection

5 Conclusions and Future Work

In this paper, we propose an intermediate granularity of replication on the Web at the directory level, and present an effective approach to discover the replicated directories and the replica boundaries. Experimental results show the effectiveness of the approach. The conclusions obtained in the experiment can also be viewed as some of the conventions on the Web that can be used in the related applications. For future work, we plan to extend the current approach so that it can handle the data in a distributed environment. We would also like to explore the online replica detection approach, with which we can avoid fetching the duplicate collections.

Acknowledgements

This work is sponsored in part by City University of Hong Kong strategic grants (7001709) and CityU FSE Funding for Research (9610027), in part by 973 National Basic Research Program, Minister of Science and Technology of China under Grant No. 2003CB317003, and in part by an NSFC grant (60573166).

References

[1] A.Z.Broder. On the resemblance and containment of documents. In Proceedings of Compression and Complexity of Sequences 1997, pages 21-29. IEEE Computer Society
[2] A. Z. Broder. Identifying and Filtering Near-Duplicate Documents. In Combinatorial Pattern Matching, 11th Annual Symposium, pages 1-10, June 2000
[3] Z. Broder, S. C. Glassman, M. S. Manasse, and G. eig. Syntactic clustering of the Web. In Proceedings of the sixth International World Wide Web Conference, pages 391-404, 1997
[4] Junghoo Cho, Narayanan Shivakumar, Hector Garcia-Molina: Finding Replicated Web Collections. SIGMOD Conference 2000: 355-366
[5] N. Heintze. Scalable Document Fingerprinting. Proceedings of the Second USENIX Workshop on Electronic Commerce, pages 191-200 1996
[6] Aleksander Ko?cz, Abdur Chowdhury, Joshua Alspector. Improved robustness of signature-based near-replica detection via lexicon randomization. Proceedings of the 2004 ACM SIGKDD Conference, Pages 605-610, 2004.
[7] Krishna Bharat and Andrei Z. Broder. Mirror, Mirror, on the Web: A study of host pairs with replicated content. In Proceedings of 8th International Conference on World Wide Web (WWW'99), May 1999
[8] Zhigang Zhang, Jing Chen and Xiaoming Li, "A Preprocessing Framework and Approach for Web Applications", Journal of Web Engineering, Vol.2 NO.3 pp175-191□2004.
[9] A. Chowdhury, O. Frieder, D. A. Grossman, and M. C. McCabe. Collection statistics for fast duplicated document detection. ACM Transactions on Information Systems, 20(2): 171-191, 2002.
[10] S. Brin, J. Davis, and Garcia-Molina. Copy detection mechanisms for digital documents. In Proceedings of the ACM SIGMOD Annual Conference, San Francisco, CA, May 1995
[11] N. Shivakumar, H. Garcia-Molina. SCAM: A Copy Detection Mechanism for Digital Documents. Proceedings of the 2nd International Conference on Theory and Practice of Digital Libraries, 1995.
[12] N. Shivakumar, H. Garcia-Molina. Building a Scalable and Accurate Copy Detection Mechanism. Proceedings of the 3nd International Conference on Theory and Practice of Digital Libraries, 1996,
[13] Wensi Xi, Edward A. Fox, Roy P. Tan, Jiang Shu: Machine Learning Approach for Homepage Finding Task. In Proceedings of the 9th International Symposium on String Processing and Information Retrieval. Lisbon Portugal, 11-15, Sep 2002 pp 145-159.
[14] N. Shivakumar, H. Garcia-Molina. Finding near-replicas of documents on the Web. In Proceedings of Workshop on Web Databases (WebDB'98), March 1998.
[15] M. Henzinger, R. Motwani, Silverstein. Challenges in Web Search Engines. In Proceedings of the 18th International Joint Conference on Artificial Intelligene, 2003.
[16] K. Bharat, A. Broder, J. Dean, M. R. Henzinger. A Comparison of Techniques to find mirrored hosts on the WWW. Journal of the American Society for Information Science Vol. 51, No. 12: 1114-1122.
[17] Tianwang Web search engine, http://e.pku.edu.cn.

Summarization and Visualization of Communication Patterns in a Large-Scale Social Network

Preetha Appan[1], Hari Sundaram[1], and Belle Tseng[2]

[1] Arts Media and Engineering Program, Arizona State University, Tempe AZ 85281
{Preetha.Appan, Hari.Sundaram}@asu.edu
[2] NEC Research, Cupertino CA
belle@sv.nec-labs.com

Abstract. This paper deals with the problem of summarization and visualization of communication patterns in a large scale corporate social network. The solution to the problem can have significant impact in understanding large scale social network dynamics. There are three key aspects to our approach. First we propose a ring based network representation scheme – the insight is that visual displays of *temporal dynamics* of large scale social networks can be accomplished *without using graph based layout mechanisms*. Second, we detect three specific network activity patterns – *periodicity, isolated* and *widespread* patterns at multiple time scales. For each pattern we develop specific visualizations within the overall ring based framework. Finally we develop an activity pattern ranking scheme and a visualization that enables us to summarize key social network activities in a single snapshot. We have validated our approach by using the large Enron corpus – we have excellent activity detection results, and very good preliminary user study results for the visualization.

1 Introduction

This paper deals with the problem of summarization and visualization of large scale social network communication patterns. Understanding large scale social networks is an emerging area of research [9]. The problem is made difficult due to the large size of the network and the long term duration of these networks. Hence visualization and summarization tools that enable users to gain insight into the dynamic behavior of these networks are extremely important.

There has been extensive work in visualization of graph data. Various graph layout algorithms have been developed to enable exploration of large graphs [6]. However these visualizations are for a single large scale graphs. Tools developed to visualize graph data that change over time, show only one graph at a single time instance with a slider to move the graph forward / backward in time. However understanding the temporal dynamics in the network is difficult. Prior work in analysis of communication has focused on issues such as the information propagation in blogs [5] and community structure detection. However prior work does not explore email communication patterns that are influenced by both time and people. There also has been little focus on summarizing key social network activity patterns though visual means. There has been prior work in innovative visualizations for data analysis [4,7]. We focus on two aspects not addressed before – (a) closely coupling the results of the visualization to the specifics of the social network activity patterns, (b) providing a systematic framework for summarizing the entire social network communication predicated on a topic.

We address the summarization and visualization problems by solving three sub-problems. (a) defining a ring visualization framework for social network activity representation. (b) detecting and visualizing three specific activity patterns and (c) providing a single snapshot summary of the entire network activity. Our visualization framework is inspired by the observation that natural phenomena (ref. Fig. 1) can compactly summarize long term activity. The key insight is that compact representation of large scale networks, need not require graph based visualizations. We develop a ring based visualization and summarization framework, that displays relationships between people, time and topic.

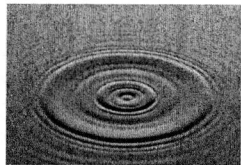

Fig. 1. Summarization: Ripples in water provide a compact snapshot view of temporal activity

We detect three specific activity patterns – *periodicity*, *isolated events* and *widespread growth* at multiple time scales and develop specific ring based visualizations for each activity. The summarization framework allows us to represent key activity patterns over the entire duration of the network in a compact manner. Periodic patterns in time are detected using local maxima of message activity. Regularity in people refers to people who appear frequently in the conversation – this is detected using set intersection techniques. Isolated patterns are detected using constrained global maxima detection, while distributed growth can be detected using a multi-scale message activity analysis (more details in [2]). We have conducted experiments over the large Enron corpus, and preliminary user studies on the visualization, with excellent results.

The rest of this paper is organized as follows. In the next Section we present our approach to visualization of large scale network activity. In Section 3, we discuss our activity pattern detection algorithms. In Section 4 we present out summarization algorithm. In Section 5 we discuss our experimental results and then present our conclusions.

2 The Visualization Problem

In this Section we will present our visualization framework. The central innovation in our approach is that visual displays of temporal dynamics of large scale social networks can be accomplished without using graph based layout mechanisms.

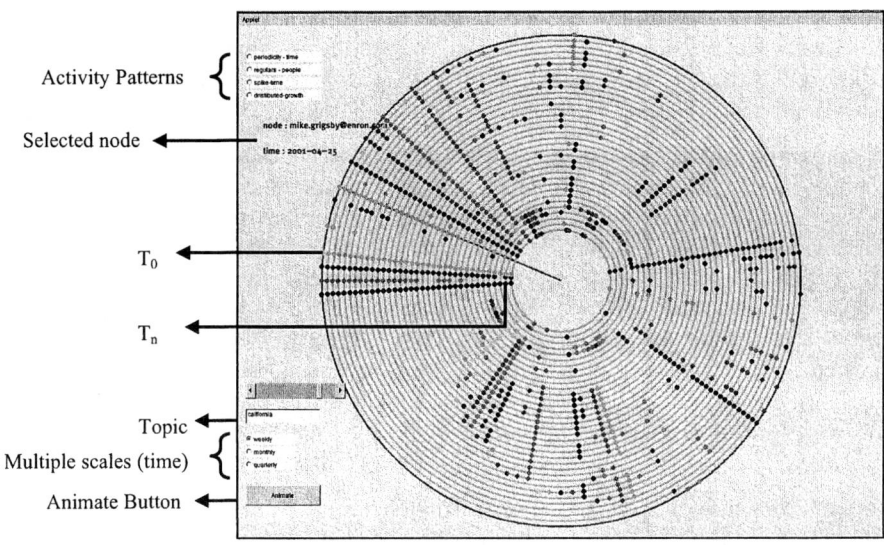

Fig. 2. Ring Visualization – Time is represented as concentric circles with the innermost circle representing most recent time slot and outermost ring the oldest time, similar to natural phenomenon. People are indicated with colored dots whose radial location is consistent over time, thus making it very easy to understand how people communicating about a topic vary over time.

2.1 Graphs Do Not Reveal Network Temporal Dynamics

We are addressing the problem of visualizing email communication amongst members of in a large scale social network over an extended period of time. In our system, we are using the Enron Data set. This dataset was collected and prepared by the CALO Project [3]. It contains email data from about 150 users, from senior management of Enron, organized into folders. The corpus contains a total of about 0.5M messages. In our system, we create an edge between two nodes (people) if there is evidence of communication between them. Graphs can be very useful to understand the structural properties of any social network – i.e. who talks to whom.. However, graphs do not reveal the temporal dynamics of the communication in the social network. This is because a graph represents the state of the network at a *single* time instant. This can be a significant problem in large networks such as the Enron corpus that has large user set (150) and significant communication activity over a long duration. Simple techniques such as animation, graph aggregation will not work well.

2.2 Our Approach: Rings

The intuition behind our visualization comes from observing phenomena in the physical world. For example, as seen in Figure 1 we observe ripples in water start from the center and radiate outward. We observe that the growth or spread of energy in these phenomena happens in a radial direction starting from the innermost towards the outermost ring. This creates two constraints – (a) temporal: the outermost ring /

ripple represents the earliest temporal event. (b) rotational: the relative orientation of each ring is not arbitrary – a line from the center to the outermost ring corresponds to a direction of energy flow. Before we describe the visualization, we briefly discuss message preprocessing.

Message preprocessing: We assume that the user provides a topic, i.e. a keyword. We then find all messages in the corpus relevant to the given topic. Since our focus is on the visualization rather than information retrieval, we are using a simple keyword match algorithm on the subject line to find all messages pertaining to a given topic. The messages obtained are then ordered in time as well as associated with a set of people - the sender and the set of recipients. In our system, users can browse through three scales of time – 'weekly', 'monthly' and "quarterly". We divide the users into three categories from their email ids – (a) network members – employees amongst the 150 users whose emails contributed to the data set (b) other employees – other employees of Enron not part of the initial 150 people and (c) external – people outside of Enron.

Design Elements: We designed a visualization that indicates multiple graphs that vary over time, in a single snapshot (Figure 2). We now explain the design elements of our visualization.

- *Time*: In our visualization, time is represented as concentric circles, with the innermost circle indicating the latest time slot. Additionally, we can also show activity over multiple time scales.
- *People*: Each person is represented as a distinctly colored dot whose radial location is maintained over different circles. Since people in general form an unordered set, we assign a default ordering along the clockwise direction in the order they *first* appeared in messages sorted in time.
- *Activity:* The message density per time slot is mapped to the color intensity of the circles representing them. Higher the message density, darker the color of the circle that represents the corresponding time slot.
- *Animation:* The visualization can be animated to show the evolution of people talking to each other over time. Time slots indicating more recent activity about the topic are added from the innermost ring and move outward, reminiscent of ripples in water.

The graph structure: The graph structure is not obvious when using rings. We have dealt with this issue by indicating the actual communication graph structure when the user clicks on a particular node in a certain ring. To bring the graph into focus, the rest of the nodes in other time rings are dimmed out by changing their color saturation.

3 Activity Patterns

We now discuss the detection and visualization of three specific temporal communication patterns (periodic, isolated and widespread) in a social network to help summarize the activity with respect to a certain topic. The activity patterns we describe are an extension of the chatter and spiky communication patterns in blogspace that are described in [5]. We add two novel patterns – *distributed growth* and *regulars in people*, to the spiky patterns described there. Also while [5] looked at variations in communication over time, the activity patterns we describe depend on both time and people. We assume that we are given the topic and the corresponding set of relevant messages (ref. Section 2.2).

3.1 Periodic Activity Patterns

We shall detect periodic patterns that are regular over time, as well as regular over people. Periodic patterns over time refer to high message activity in the network relevant to a particular topic that appears in regular time intervals.

Detecting periodicity in time: Periodic patterns in time are revealed by detecting the local maxima in message activity and then imposing simple temporal constraints on the maxima. The periodicity detection algorithm proceeds as follows. First all messages are ordered in time, and then grouped according to any chosen scale (weekly, monthly, quarterly). Then each time slot is given an activity score using the following equation:

$$S(t_i) = \sum_{j=1}^{N} P_j(t_i), \qquad (1)$$

Where $S(t_i)$ is the score given to the ith time slot, N is the total number of people involved in messages about the topic, $P_j(t_i)$ is one if the jth node is present in message communication at time instant t_i, zero otherwise. This score is high on time slots involving large number of messages *and* recipients. All the local maxima from the time series scores obtained using eq. (1) are marked as "peaks." Temporal distances are computed from each peak to every other peak and stored in a table per peak. The local distance tables are then combined to construct a global histogram of distances. We consider a period to be valid only if there exists at least one set of three peaks at the same temporal distance from each other. For example if $d(p_1,p_2) = d(p_2,p_3) = d(p_3,p_4) = d_1$, then d_1 is considered a valid period. This removes spurious maxima. The algorithm gives a list of periods and the number of peaks that are participants with that period.

Temporal periodic patterns are easily understood using rings. Every time period that corresponds to a peak and part of the top three detected period sets is colored with a distinct color.

Detecting regulars in people: Regularity in people refers to the set of people who occur together, frequently, over the duration of the topic. This can be detected using a set intersection algorithm. Consider N to be the total number of people exchanging emails about the topic. We iteratively find all subsets S_k from these N people, that occurred together more than q_l times over all time slots. The threshold q_l is fixed according to the time scale. These subsets of people form groups that are the 'regulars' to the topic. To visualize the set people who appear together, all nodes (representing people) in the visualization that are part of the same set are colored with the same color. This is done only after the user explicitly selects this pattern to be revealed. Their radius is also increased and the background is dimmed out in order to prominently display the regulars in the topic.

3.2 Isolated Patterns

We now show how we can detect and visualize isolated patterns (also referred to as spikes) over time and people. Isolated patterns over time refer to significant message activity over a short time window. Isolated patterns over people refer to information generators – a small set of people, who contribute to most of the messages.

Detecting spikes in time: A spike in time is characterized by three conditions: (a) there exists local maxima in activity, (b) the message activity exceeds a certain threshold and (c) the activity exhibits a sharp rise and fall in small time duration.

In order to find such spikes in time, we first begin with the ordered set of all messages relevant to the particular topic. We use equation (1) to calculate the score of each time slot, which depends on both the number of messages and the number of recipients per message. The global maxima verified with the above constraints are then visualized. In order to indicate spikes in time for the given topic to the user, we highlight the time slot in which the maxima occurred. We additionally increase the radius of all nodes representing people communicating in that time slot.

Spikes in people: Spikes in people refer to the information generators – a small set of people who send a large percentage of messages relevant to the topic. We now define two measures α (sender coverage) and β (message coverage), to be as follows.

$$\alpha(N_s) = \frac{N_s}{N_0}; \beta(N_s) = \frac{1}{M}\sum_{j=1}^{N_s} m_j, \qquad (2)$$

where N_s is the number of unique senders, and m_j is the number of messages contributed by the jth sender. Given a certain threshold for β, we can find a corresponding N_s – the minimum number of senders required to generate those messages. The information generator set is determined by determining N_s using equation (2) such that these values of $\beta \geq \beta_0$ and $\alpha \leq \alpha_0$ are satisfied. These N_s senders are then the *information generators* for the given topic. We determined the thresholds ($\beta_0 = 0.65$ and $\alpha_0 = 0.15$) using a training set [2]. Spikes in people in are indicated our visualization, by increasing the size of nodes that are spikes in people in *all* times that they occur. We also place them along equidistant radial lines. Details of our algorithm to detect and visualize distributed growth can be found in [2].

4 Summarization

In this Section we discuss the problem of summarizing the key activity patterns in a single snapshot. The solution involves two steps – (a) the detection and ranking of activity patterns and (b) developing a single representative snapshot. The visualization problem is difficult since the activity patterns need not co-occur within the same time window.

4.1 Ranking the Activity Patterns

Each activity detector (ref. Section 3) returns a set of detected activities. We thus need to develop measures to order the activities within each set. We now discuss a systematic ranking measure for each activity pattern.

- *Periodic in time*: each period is associated with a frequency (the number of message activity peaks that are in that period), which is used to pick the top three periods.
- *Regulars in people*: Each set of people has a corresponding frequency which is the number of times they occurred together. We use the *average* closeness centrality [8], as a measure to rank the different 'regular sets' of people.
- *Isolated in time*: The message activity score from equation (1) implicitly ranks the sets of the spike in time patterns.
- *Isolated in people*: The *number* of information generators is used to rank the sets. Smaller the set of information generators, higher the rank.
- *Distributed growth*: The size of the time window of growth is used as a ranking mechanism. Larger the time window, higher the rank.

We will now discuss how we construct a summary snapshot to indicate all the key activity patterns, given a certain topic.

4.2 Constructing the Summary Snapshot

Each of the activity patterns detected could occur at different times as well as involve different people. Hence indicating all of them within the same screen is difficult, especially if the time range is bigger than the maximum that can be shown in the

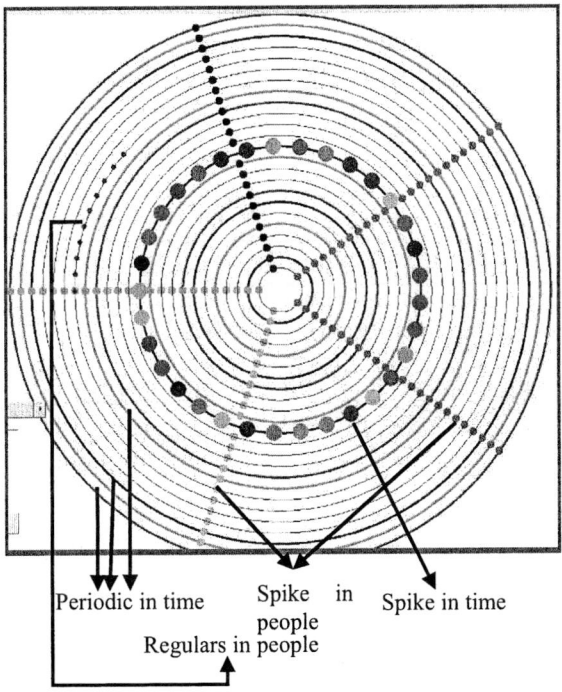

Fig. 3. Summary Snapshot – Indicating four different activity patterns in a representative snapshot for the query 'California' grouped monthly

available display area. Instead, we have constructed a *representative* summary snapshot that only visually indicates the key patterns for messages of the given topic, but does not correspond to the actual time of when the pattern occurred. This is an interactive summary, where the user can then click on the pattern of interest to go to the ring visualization and see the actual time period corresponding to the activity pattern of interest. The Figure 3 shows an example summary snapshot of the query 'California' in the monthly scale.

5 Experiments

The visualization and summarization framework was implemented in Java and Processing [1] with the Mysql database as the backend. In order to evaluate the system, we conducted a preliminary user study with five graduate students. Users were asked to interact with the system by executing several text queries (e.g. 'power crisis', 'California', 'trading' etc.). They were then asked to answer questions on various aspects of the system on a scale of one to seven. The results are summarized in Table 1 and indicate that users find the activity patterns as well as the visualization to be very useful in understanding email communication. Users also suggested various improvements such as (a) visualizing relationships between a single person, time and the topic, rather than the entire social network, (b) comparing communication activity for multiple topics in the same visualization.

Table 1. Preliminary user studies

Interface Aspects	Score
User Friendliness	5.0 / 7
Activity Patterns	6.25 / 7
Multi-scale analysis	5.75 / 7
Helps understand relationships between topics, people and time	6.0 / 7
Summary snapshot	6.25 / 7

We acknowledge that this is only a preliminary evaluation – the actual study would involve applying our visualization technique to emails from an organization and allowing *members of the same organization* to evaluate whether the visualization was able to communicate temporal patterns well. We also ran activity pattern detection algorithms on 100 queries on the Enron data. The detailed results can be found in [2].

6 Conclusion

In this paper, we proposed a framework for visualization and summarization of email communication activity in large social networks. The framework addressed three challenges (a) visualization (b) activity pattern detection and (c) summarization. The novel ring visualization scheme depicts multiple graphs in the same snapshot and enables users to understand communication activity that varies over multiple scales in time. We also defined and detected three classes of communication activity patterns that depend on people and time – (a) periodic (b) isolated and (c) distributed. We discussed visualization of these patterns using the ring visualization. The detected activity patterns are then summarized by ranking activity patterns and constructing a single snapshot that communicates all key activity patterns to the user. Preliminary experiments and user study results are promising and we plan to conduct further extensive evaluation.

References

[1] *Processing* http://proce55ing.net.
[2] P. APPAN, H. SUNDARAM and B. TSENG (2005). *Summarization and Visualization of Communication Patterns in a Large Social network*. Arts Media and Engineering Program, ASU, AME-TR-2005-12, May 2005.
[3] CALO *http://www.ai.sri.com/project/CALO*
[4] J. V. CARLIS and J. A. KONSTAN (1998). *Interactive visualization of serial periodic data*, Proc. of the 11th annual ACM symposium on UIST, 29-38, San Francisco.
[5] D. GRUHL, R. GUHA, D. LIBEN-NOWELL and A. TOMKINS (2004). *Information Diffusion through Blogspace*, Proceedings of the 13th international conference on World Wide Web,
[6] I. HERMAN, M. DELEST and G. MELANCON (2000). *Graph visualization and navigation in information visualization: A survey*. IEEE Transactions on Visualization and Computer Graphics **6(1)**.
[7] D. A. KEIM, J. SCHNEIDEWIND and M. SIPS (2004). *CircleView: a new approach for visualizing time-related multidimensional data sets*, Proc. Advanced visual interfaces, 179-182, Gallipoli, Italy.
[8] M. E. J. NEWMAN (2003). *A measure of betweenness centrality based on random walks*. Social Networks.
[9] J. R. TYLER, D. M. WILKINSON and B. A. HUBERMAN (2003). *Email as spectroscopy: automated discovery of community structure within organizations*. Communities and technologies: 81 - 96.

Patterns of Influence in a Recommendation Network*

Jure Leskovec[1], Ajit Singh[1], and Jon Kleinberg[2,**]

[1] School of Computer Science, Carnegie Mellon University
{jure, ajit}@cs.cmu.edu
[2] Department of Computer Science, Cornell University
kleinber@cs.cornell.edu

Abstract. Information cascades are phenomena in which individuals adopt a new action or idea due to influence by others. As such a process spreads through an underlying social network, it can result in widespread adoption overall. We consider information cascades in the context of recommendations, and in particular study the patterns of cascading recommendations that arise in large social networks. We investigate a large person-to-person recommendation network, consisting of four million people who made sixteen million recommendations on half a million products. Such a dataset allows us to pose a number of fundamental questions: What kinds of cascades arise frequently in real life? What features distinguish them? We enumerate and count cascade subgraphs on large directed graphs; as one component of this, we develop a novel efficient heuristic based on graph isomorphism testing that scales to large datasets. We discover novel patterns: the distribution of cascade sizes is approximately heavy-tailed; cascades tend to be shallow, but occasional large bursts of propagation can occur. The relative abundance of different cascade subgraphs suggests subtle properties of the underlying social network and recommendation process.

1 Introduction

The social network of interactions among a group of individuals plays a fundamental role in the spread of information, ideas, and influence. Such effects have been observed in many cases, when an idea or action gains sudden widespread popularity through word-of-mouth or "viral marketing" effects. To take a recent example from the technology domain, free e-mail services such as Microsoft's Hotmail and later Google's Gmail achieved wide usage largely through referrals, rather than direct advertising. (Gmail achieved wide usage at a time when the

---

* Work partially supported by the National Science Foundation under Grants No. IIS-0209107 IIS-0205224 INT-0318547 SENSOR-0329549 EF-0331657IIS-0326322 CCF-0325453, IIS-0329064, CNS-0403340, CCR-0122581, a David and Lucile Packard Foundation Fellowship, and also by the Pennsylvania Infrastructure Technology Alliance (PITA). This publication only reflects the authors' views.
** The work of the third author was performed in part while on sabbatical leave at Carnegie Mellon University.

only way to obtain an account was through a referral.) One also finds many examples in weblogs (blogs), where a piece of information spreads rapidly from one blogger to another before eventually being picked up by the mass media.

Information cascades are phenomena in which an action or idea becomes widely adopted due to influence by others [3, 5, 6]. Cascades are also known as "fads" or "resonance." Cascades have been studied for many years by sociologists concerned with the *diffusion of innovation* [15]; more recently, researchers in several fields have investigated cascades for the purpose of selecting trendsetters for viral marketing [9, 14], finding inoculation targets in epidemiology [13], and explaining trends in blogspace [2, 7, 10]. Despite much empirical work in the social sciences on datasets of moderate size, the difficulty in obtaining data has limited the extent of analysis on very large-scale, complete datasets representing cascades. Here we look at the patterns of influence in a large-scale, real recommendation network and examine the topological structure of cascades.

We address a set of related questions: What kinds of cascades arise frequently in real life? Are they like trees, stars, or something else? And how do they reflect properties of their underlying network environment? We describe (in Section 3) a large person-to-person recommendation network, consisting of 4 million people who made 16 million recommendations on half a million products. To analyze the data, we first create graphs where incoming edges influence the creation of outgoing edges. Then, we enumerate and count all possible cascade subgraphs, using an algorithm developed in Section 4. There, we propose an approximate heuristic for graph isomorphism involving the degree distribution and the eigenvalues of the adjacency matrix that scales to large datasets. We apply the algorithm to the recommendation dataset, and analyze it in Section 5.

We find novel patterns, and the analysis of the results gives us insight into the cascade formation process. We find that the distribution of cascade sizes can be approximated by a heavy-tailed distribution. Generally cascades are shallow but occasional large bursts also occur. The cascade sub-patterns reveal mostly small tree-like subgraphs; however we observe differences in connectivity, density, and the shape of cascades across product types. Indeed, the frequency of different cascade subgraphs is not a simple consequence of differences in size or density; rather, we find instances where denser subgraphs are more frequent than sparser ones, in a manner suggestive of properties in the underlying social network and recomendation process.

2 Related Work

To our knowledge, this is the first large-scale study of cascades in a real recommendation network. We believe the lack of prior studies is due in large part to the difficulty in acquiring large recommendation network datasets without link ambiguity from a real-world setting.

Most work on extracting cascades from large-scale on-line data has been done in the blog domain [1, 7, 10]. The authors in this domain note that, while information propagates between blogs, examples of genuine cascading behavior appeared relatively rare. This may, however, be due in part to the Web-crawling and text analysis techniques used to infer relationships among pages. In our dataset, all the recommendations are stored as database transactions, and we know that no records are missing. Associated with each recommendation is the product involved, and the time the recommendation was made. Studies of blogspace either spend a lot of effort mining topics from posts [2, 7] or consider only the properties of blogspace as a graph of unlabeled URLs [1, 10]. Temporally evolving graphs are explored in [4]. Theoretical analysis of cascades on random graphs is provided in [16], using a threshold model. Analysis based on thresholding as well as alternative probabilistic models of node activation is considered in [9, 14]. Note that this analytical work posits a known network; in the present paper, we are able to observe the cascades but not the underlying social network.

In our work we need to efficiently enumerate and count cascade subgraphs. This is an instance of the general issue of *frequent subgraph mining* [8, 11, 17]; however, most of the prior work in this area is focused on graphs that are richly labeled and undirected, often motivated by applications to chemical compound and bioinformatics datasets. While our data has labels as well, we are specifically interested in enumerating subgraphs based purely on their structures, so heuristics for pruning the search space using node and edge labels cannot be applied. Another crucial difference is that we have additional temporal constraints on cascades. We take advantage of the specific problem domain and develop efficient algorithms for extracting "temporally consistent" subgraphs, as well as heuristics for approximate graph isomorphism testing.

3 The Recommendation Network

We study a recommendation network dataset from a large on-line retailer. During the period covered by the dataset, each time a person purchased a book, DVD, video, or music product, he or she was given the option of sending an e-mail message recommending the item to friends. The first recipient to purchase the item received a discount, and the sender received a referral credit with monetary value. A person could make recommendations on a product only after purchasing it. Since each sender had an incentive for making effective referrals, it is natural to hypothesize that this dataset is a good source of cascades.

Each recommendation is annotated with the product recommended, the time the recommendation was sent, whether it resulted in a purchase, and the date of purchase (if applicable). Customer information is anonymized; no demographic or uniquely identifying information is available.

We represent this relational dataset as a labeled directed multigraph: nodes represent customers, and a directed edge (v, w) with label (p, t) means that node v recommended product p to customer w at time t. (For convenience, we will sometimes denote this edge by (v, w, p, t).) The typical edge generation process is as follows: a node (person) v buys product p at time t, and then recommends it to nodes $\{w_1, \ldots, w_n\}$. These nodes w_i can then buy the product (with the option to recommend it to others). Note that even if all nodes w_i buy the product,

Table 1. Product group recommendation network statistics: p: number of products, n: number of nodes, e: total number of edges (recommendations), e_u: number of unique edges, b_t: total number of purchases, b_r: purchases made through recommendations

Group	p	n	e	e_u	b_t	b_r
Book	103,161	2,863,977	5,741,611	2,097,809	2,859,096	83,113
DVD	19,829	805,285	8,180,393	962,341	837,300	75,421
Music	393,598	794,148	1,443,847	585,738	712,673	10,576
Video	26,131	239,583	280,270	160,683	165,109	1,376
Full network	542,719	3,943,084	15,646,121	3,153,676	4,574,178	170,486

only the first purchaser will get the discount, which is marked by a purchase flag (*buy-bit*). However, we can infer purchases by others of the w_i by seeing if they generated subsequent recommendations for it. (Recall that one had to buy the product in order to be allowed to recommend it).

The recommendation network consists of 15,646,121 recommendations made among 3,943,084 distinct users from June 2001 to May 2003 (711 days). A total of 542,719 different products belonging to four product categories (Books, DVDs, Music, and Videos) were recommended. For a detailed analysis of the customer recommendation behavior in this dataset, see [12].

We extract per-group recommendation networks by taking the edge-induced subgraph formed by all the products of a given category. Table 1 describes the four networks. The DVD network contains the most recommendations; but the book network involves more customers. On average a node in the DVD network made more than 10 recommendations; on average a book or music node made about two recommendations.

There can be multiple recommendations between the nodes, and by counting only unique edges (e_u), we find that only DVDs have more edges than nodes. In summary, all networks are very sparsely linked, but those pairs of users who exchanged recommendations often did so several times. Moreover, exploration of the social network was rather poor. At the end of the two-year period, the largest connected component contained fewer than 2.5% of the nodes.

The last two columns of Table 1 show the total number of purchases (b_t) and the purchases that occurred in response to a recommendation (b_r). Observe that for DVDs 9% of purchases are associated with a recommendation, for books 3%, music 1.5% and video less than 1%.

Overall, then, while book recommendations appear quite influential, most readers do not appear to make many of them. The DVD network, while smaller, is significantly denser and can be viewed as having a qualitatively richer structure.

4 Proposed Method

In this section we present the algorithms and techniques developed to efficiently enumerate and count frequent recommendation patterns in a large graph, including an approximate heuristic for subgraph isomorphism.

One might imagine cascades to be trees or near-trees. In fact, we find that recommendations create essentially arbitrary graphs: there can be multiple recommendations on the same product or multiple product recommendations between the same pair of nodes; there are multiple purchases of the same product by the same individual (this is natural given that many items are purchased as gifts); and one also finds many cycles.

To find cascades we need to identify cases when incoming recommendations cause purchases and further outgoing recommendations. Recommendations into a node u that precede a purchase can be posited to have potentially influenced the purchase. There are two ways to establish this. If an edge is marked by a purchase flag, we assume the recommendation influenced the purchase. Alternately, the existence of two directed edges (u, v, p, t) and (v, w, p, t') for $t' > t$ suggests cascade behavior. That is, node v receives a recommendation for product p at time t and then makes recommendation for the same product at a *later* time t'. (Recall that a node makes recommendations at the time of purchase.)

First we create a separate graph of recommendations for each product. To find cascades we propose the following two-step procedure:

Delete late recommendations: given a single product recommendation network, for every node we delete all incoming recommendations (edges) that happen after the first purchase of a product. This removes all recommendations of the product a person received after the first purchase, i.e. keeps only recommendations that potentially influenced the purchase. Now for every node the time of all incoming edges is strictly smaller than the time of all outgoing edges.

Delete no-purchase nodes: Preliminary analysis showed that the majority of recommendations do not produce cascades. We observed many star-like patterns where the center node recommends to a large number of people, none of whom purchase the product. To prevent this type of large but shallow pattern, we delete all nodes that did not purchase the product.

After deleting late recommendations each connected component in the undirected version of the graph can be viewed as a cascade, since all directed paths in the component are time-increasing (*i.e.*, a cascade subgraph contains only directed paths with strictly increasing edge times). Deleting no-purchase nodes ensures that we include only nodes whose behavior was potentially affected by the cascade (as evidenced by the fact that they made a purchase).

Cascade enumeration: Next we enumerate all possible cascades. By the discussion in the previous paragraph, the undirected components correspond to maximal cascades, but simply enumerating components makes it difficult to reason about the smaller building blocks of the cascades. Thus, we instead focus on enumerating all *local cascades*: For every node we explore the cascade in the (undirected) neighborhood of the node. Thus, for every node v, we create the subgraphs induced on nodes reachable by up to h steps forward or backward from v (stopping at h that includes all reachable nodes). One can think of this as capturing the local structure of the cascade at increasing distances around v.

Approximate graph isomorphism: An essential step in counting cascades is determining whether a new cascade is isomorphic to a previously discovered one. No polynomial-time algorithm is known for the graph isomorphism problem, and so we resort to an approximate, heuristic solution. For each graph we create a *signature*. A good signature is one where isomorphic graphs have the same signature, but where few non-isomorphic graphs share the same signature.

We propose a multi-level approach where the computational complexity (and accuracy) of the graph isomorphism resolution depends on the size of the graph. For smaller graphs we perform an exact isomorphism test; as the size of the graph increases this becomes prohibitively expensive so we use gradually simpler but faster techniques which give only approximate solutions. Another technique employed is to create an efficiently computable signature for each graph, use hashing on this signature value, and then use more expensive isomorphism tests only on graphs with the same signature.

For every graph we create a signature which is composed of the number of nodes, the number of edges, and the sorted in- and out-degree sequences. For graphs with fewer than 500 nodes, we also include the singular values of the adjacency matrix (via singular value decomposition). We then hash the graphs using the signatures. Additionally, for graphs with fewer than 9 nodes we perform exact isomorphism checking. When the exact isomorphism check is used, we keep a list of all variants of graphs that collided (have the same signature). Since we first hash, we perform the isomorphism check only on graphs with the same signatures, and so the number of true isomorphism checks is small.

Note that a small minority of cascades are larger than 9 nodes, so for most of the subgraphs we get the exact solution; as the cascade size increases the number of occurrences decreases, and this is where we use an approximate solution.

We performed a small set of experiments to evaluate the proposed approximate graph isomorphism algorithm. Given a graph with 8 nodes and 12 edges 100,000 brute-force evaluations of graph-isomorphism took under 40 seconds on a standard desktop machine. In the second experiment we generated 100,000 random graphs (using the Erdős-Rényi model), each of them with a randomly chosen number of nodes between 4 to 20 and twice as many edges. The counting took 50 seconds. In this experiment we observed at most 53 non-isomorphic graphs (5 nodes, 10 edges) with the same signature.

5 Patterns of Recommendation

5.1 Size Distribution of Cascades

First, we discuss results on the size of the cascades, measured by the number of nodes. As in all experiments we create per-product recommendation networks, delete late recommendations and no-purchase nodes, and then perform the analysis. Figure 1 shows the distribution of cascade sizes for the four product types.

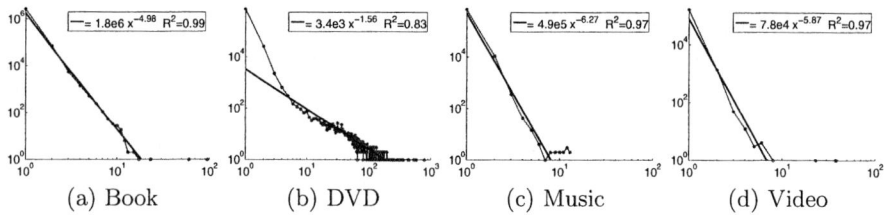

(a) Book (b) DVD (c) Music (d) Video

Fig. 1. Size distribution of the cascades for the four product types (log size of cascade vs. log count). Superimposed line presents a power-fit. R^2 is the coefficient of determination.

The size of cascades follows a heavy-tailed distribution. For books the largest cascade has 95 nodes and 231 edges. For DVDs the largest cascade is eight times larger ($n = 791, e = 5544$). The cascades involving music or videos are much smaller; the largest cascades are $n = 13, e = 56$ and $n = 37, e = 169$ respectively.

DVDs had the highest proportion of large cascades, and the plot for DVDs in Figure 1(b) has an interesting transition in its behavior. For smaller cascade sizes, in the size range consistent with most of the book, music, and video cascades, the DVD distribution has a power-law fit with slope -4.5, comparable to the other three product types. For larger cascades, which are observed in abundance only for DVDs, the distribution flattens to a slope of -1.5.

The cascade size distributions suggest that the simplest branching process models will not suffice to explain the underlying cascade process; a family of richer models is proposed in [12], in which the success probability increases when collisions occur among cascades, and cascade sizes follow a power law with exponent -1. We have also found that the cascade size distribution follows a heavy-tailed distribution in sales frequencies [12], with the number of purchases decaying as a function of rank faster than the number of recommendations does.

5.2 Frequent Cascade Subgraphs

What kinds of cascades arise frequently in real life? Are they like trees, stars, long chains, or something else? We now explore the building blocks of the cascades, by performing the described procedure: for each product recommendation graph, we first identify cascades by deleting late recommendations and no-purchase nodes. Then, for each node we create a subgraph on nodes at distance up to h hops, where h varies from 1 up to the value where all nodes are reached. We count the graphs using the approximate graph isomorphism technique from Section 4.

General observations: For books we identified 122,657 cascades, of which 959 are topologically different. There are 213 cascades that occur at least ten times. For DVDs we identified 289,055 cascades, 87,614 are topologically different, and 3,015 cascades occur at least ten times. For music we identified 13,330 cascades, 158 were topologically different, and only 23 cascades occurred at least ten times. Videos were the least rich, with 1,928 subgraphs containing 109 unique patterns, and only 12 subgraphs occurring at least ten times.

The frequency of different cascades concurs with observations made in connection with Figure 1 and Table 1, where DVDs had the largest and richest set of cascades. Also, even though the music network is three times larger than the video network, it does not exhibit much larger topological variety.

Analysis of frequent cascade patterns: Table 2 shows ranks R and frequencies F of 22 cascades for the 4 product types, including all subgraphs with at most four nodes and four edges. Cascades are ordered by size. 14 cascade patterns can be observed in all four product types; Table 2 includes 10 of them.

Table 2. Frequent cascades for the 4 product types. We show all graphs up to 4 nodes and 4 edges. Ordered by size. For each graph we show rank (R) and frequency (F).

Id	Graph	Nodes	Edges	Book R	Book F	DVD R	DVD F	Music R	Music F	Video R	Video F
G_1		2	1	1	86,430	1	36,863	1	11,518	1	1,425
G_2		3	2	2	10,573	4	3,238	2	492	5	33
G_3		3	2	3	5,089	2	5,147	3	389	3	61
G_4		3	2	6	1,593	5	2,419	5	115	22	4
G_5		3	3	4	3,115	3	4,746	4	201	2	63
G_6		4	3	5	2,769	15	505	6	55	20	5
G_7		4	3	8	726	25	416	7	30	27	4
G_8		4	3	10	598	7	909	8	25	0	0
G_9		4	3	12	398	33	312	13	12	0	0
G_{10}		4	3	13	362	22	424	9	18	26	4
G_{11}		4	3	18	156	37	276	53	4	0	0
G_{12}		4	3	29	82	24	418	28	8	0	0
G_{13}		4	3	92	21	12	549	54	4	0	0
G_{14}		4	4	9	625	11	552	31	7	13	8
G_{15}		4	4	22	112	16	495	10	15	0	0
G_{16}		4	4	23	111	20	435	57	3	0	0
G_{17}		4	4	26	85	17	485	83	2	0	0
G_{18}		4	4	30	79	9	706	32	7	29	3
G_{19}		4	4	37	64	38	273	24	9	0	0
G_{20}		4	4	47	51	955	28	0	0	0	0
G_{21}		4	4	90	21	857	31	0	0	0	0
G_{22}		4	4	91	21	1368	20	0	0	0	0

The most common cascade, G_1, represents a single recommendation. This pattern accounts for 70% of all book cascades, 86.4% of all music cascades, 74% of all video cascades, but just 12.8% of DVD cascades. The chain of three nodes (G_3) accounts for 4.1% of book cascades, about 3% of video and music cascades, but only 1.8% of DVD cascades. DVD cascades tend to be most densely linked.

Comparing G_2 and G_4 shows that simple splits are more frequent than collisions. For books there are 6.6 times more splits than collisions; for DVDs this factor drops to 1.3; and it is 4.2 and 8.25 for music and videos respectively. Very similar observations hold for splits and collisions on 4 nodes (G_6 and G_{13}); however notice that for DVDs the collision of 3 nodes (G_{13}) is slightly more frequent than the split (G_6). Another such example of reversed graphs are G_7, G_{11} and G_8, G_{12}. Again, the split pattern is more frequent than the collision. The ratio is more unbalanced for books (1 collision per 7 splits) than for DVDs (1 to 2).

Graphs from G_{14} to G_{19} all have a triangle, with one additional node attached. Again, except for DVDs, splits of recommendations (G_{14} and G_{15}) are more frequent than collisions (G_{18}, G_{19}). For DVDs the most frequent sub-graph of the set is G_{18} (involving a collision), followed by G_{14} and G_{15}.

Finally, Figure 2 shows typical classes of cascades. Graphs G_{23} and G_{27} show the case when two people have the same set of friends but do not recommend to each other. Similarly, in graphs G_{24} and G_{26}, the top node recommends to a set of people, and then one purchases and recommends to the same set. Flat cascades are also found (G_{25}, G_{28}, G_{29}) – a person recommends, a number of people respond (and purchase a product), but the cascade does not propagate. Graph G_{30} shows an illustrative example of a cascade that is quite intricate.

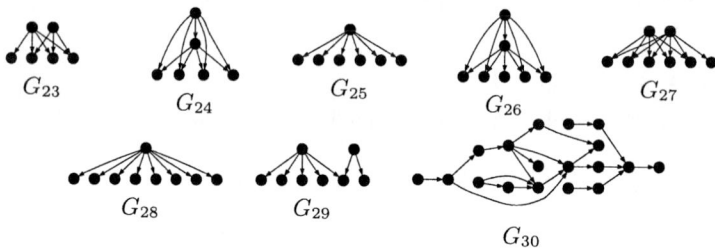

Fig. 2. Typical classes of cascades. G_{23}, G_{27}: nodes recommending to the same set of people, but not each other. G_{24}, G_{26}: one node recommends to another, and both recommend to the same community. G_{25}, G_{28}, G_{29}: a flat cascade. G_{30} is an example of a large cascade.

A concluding, general observation is that the frequency of cascade subgraphs does not simply decrease monotonically in the number of nodes and edges; for example, G_5 is more frequent than either of its subgraphs G_2 and G_4 in DVDs and videos (and more frequent than G_4 in books and music). Thus, frequency appears to reflect properties of the underlying social network (the clustering of people who know each other), as well as properties of the ways in which recommendations typically get made (e.g. splits are more common than collisions).

6 Conclusion

A basic premise behind the study of social networks is that interaction leads to complex collective behavior. Cascades are a form of collective behavior that has been analyzed both empirically and theoretically, but for which the study of complete, large-scale datasets has been limited. We have shown that cascades exist in a large real-world recommendation dataset, and have investigated some of their structural features.

We developed a scalable algorithm and set of techniques to illustrate the existence of cascades, and to measure their frequencies. From our experiments, we found that most cascades are small, but large bursts can occur; that cascade sizes approximately follow a heavy-tailed distribution; that the frequency of different cascade subgraphs depends on the product type; and that these frequencies do not simply decrease monotonically for denser subgraphs, but rather reflect more subtle features of the domain in which the recommendations are operating.

References

1. L. Adamic and N. Glance. The political blogosphere and the 2004 US election: Divided they blog. Report, March 2005.
2. E. Adar and L. A. Adamic. Tracking information epidemics in blogspace. 2005.
3. S. Bikhchandani, D. Hirshleifer, and I. Welch. A theory of fads, fashion, custom, and cultural change as informational cascades. *J. of Political Economy*, (5), 1992.
4. P. Desikan and J. Srivastava. Mining temporally evolving graphs. In *WebKDD*, 2004.
5. J. Goldenberg, B. Libai, and E. Muller. Talk of the network: A complex systems look at the underlying process of word-of-mouth. *Marketing Letters*, 12, 2001.
6. M. Granovetter. Threshold models of collective behavior. *American Journal of Sociology*, 83(6):1420–1443, 1978.
7. D. Gruhl, R. Guha, D. Liben-Nowell, and A. Tomkins. Information diffusion through blogspace. *SIGKDD Explorations*, 6(2):43–52, Dec 2004.
8. A. Inokuchi, T. Washio, and H. Motoda. An apriori-based algorithm for mining frequent substructures from graph data. In *PKDD '00*, pages 13–23, 2000.
9. D. Kempe, J. Kleinberg, and E. Tardos. Maximizing the spread of influence through a social network. In *KDD '03*, 2003.
10. R. Kumar, J. Novak, P. Raghavan, and A. Tomkins. On the bursty evolution of blogspace. In *WWW '03*, pages 568–576. ACM Press, 2003.
11. M. Kuramochi and G. Karypis. An efficient algorithm for discovering frequent subgraphs. *IEEE Trans. on Knowledge and Data Engineering*, 16(9), 2004.
12. J. Leskovec, L. Adamic, and B. Huberman. The dynamics of viral marketing. 2005.
13. M. Newman. The spread of epidemic disease on networks. *Phys. Rev. E*, 66, 2002.
14. M. Richardson and P. Domingos. Mining knowledge-sharing sites for viral marketing. In *KDD '02*, 2002.
15. E. Rogers. Diffusion of innovations (4th ed.). Free Press, 1995.
16. D. Watts. A simple model of global cascades on random networks. *PNAS*, 2002.
17. X. Yan and J. Han. gspan: Graph-based substructure pattern mining. In *ICDM '02*, pages 721–724, 2002.

Constructing Decision Trees for Graph-Structured Data by Chunkingless Graph-Based Induction

Phu Chien Nguyen, Kouzou Ohara, Akira Mogi,
Hiroshi Motoda, and Takashi Washio

Institute of Scientific and Industrial Research, Osaka University,
8-1 Mihogaoka, Ibaraki, Osaka, 567-0047, Japan
{chien, ohara, mogi, motoda, washio}@ar.sanken.osaka-u.ac.jp

Abstract. Chunkingless Graph-Based Induction (Cl-GBI) is a machine learning technique proposed for the purpose of extracting typical patterns from graph-structured data. This method is regarded as an improved version of Graph-Based Induction (GBI) which employs stepwise pair expansion (pairwise chunking) to extract typical patterns from graph-structured data, and can find overlapping patterns that cannot not be found by GBI. In this paper, we propose an algorithm for constructing decision trees for graph-structured data using Cl-GBI. This decision tree construction algorithm, called Decision Tree Chunkingless Graph-Based Induction (DT-ClGBI), can construct decision trees from graph-structured datasets while simultaneously constructing attributes useful for classification using Cl-GBI internally. Since patterns extracted by Cl-GBI are considered as attributes of a graph, and their existence/non-existence are used as attribute values, DT-ClGBI can be conceived as a tree generator equipped with feature construction capability. Experiments were conducted on synthetic and real-world graph-structured datasets showing the effectiveness of the algorithm.

1 Introduction

In recent years, there has been much research work on data mining in seeking for better performance. Better performance includes mining from structured data, which is a new challenge. Since structure is represented by proper relations and a graph can easily represent relations, knowledge discovery from graph-structured data poses a general problem for mining from structured data.

Chunkingless Graph-Based Induction (Cl-GBI) [4] is a machine learning technique which was devised for the purpose of extracting typical patterns (subgraphs) from graph-structured data. Cl-GBI is regarded as an improved version of Graph-Based Induction (GBI) [8] which extracts typical patterns from graph-structured data by recursively chunking two adjoining nodes. However, Cl-GBI does not employ this pairwise chunking strategy. Instead, the most frequent pairs are regarded as new nodes and given new node labels in the subsequent steps but none of them is chunked. In other words, they are used as pseudo nodes, thus allowing extraction of overlapping subgraphs. It was shown in [4] that Cl-GBI can extract more typical substructures than Beam-wise Graph-Based Induction (B-GBI) [4] which is an enhanced version of GBI adopting the beam search.

On the other hand, a majority of methods widely used for data mining are for data that do not have structure and that are represented by attribute-value pairs. Decision trees [5, 6], and induction rules [1] relate attribute values to target classes. Association rules often used in data mining also use this attribute-value pair representation. These methods can induce rules such that they are easy to understand. However, the attribute-value pair representation is not suitable to represent a more general data structure such as graph-structured data, which means that most of useful methods in data mining are not directly applicable to graph-structured data.

In this paper, we propose an algorithm to construct decision trees for graph structured data using Cl-GBI. This decision tree construction algorithm, called Decision Tree Chunkingless Graph-Based Induction (DT-ClGBI), is a revised version of our previous algorithm called Decision Tree Graph-Based Induction (DT-GBI) [2], and can construct decision trees for graph-structured datasets while simultaneously constructing substructures used as attributes for the classification task by means of Cl-GBI instead of B-GBI adopted in DT-GBI. In this context, substructures mean subgraphs or patterns that appear in a given graph database. Patterns extracted by Cl-GBI are regarded as attributes of graphs and their existence/non-existence are used as attribute values. Namely, DT-ClGBI does not require the user to define available substructures in advance. Since attributes (features) are constructed while a classifier is being constructed, DT-ClGBI can be conceived as a method for feature construction. Using both synthetic and real-world graph-structured datasets, we experimentally show DT-ClGBI can construct decision trees from graph-structured data that achieve reasonably good predictive accuracy.

This paper is organized as follows: Section 2 briefly describes the framework of Cl-GBI. Section 3 proposes DT-ClGBI and explains its working mechanism of how a decision tree is constructed and used for classification. The performance of DT-ClGBI is experimentally evaluated and reported in Section 4. Finally, Section 5 concludes the paper.

2 Graph-Based Induction Revisited

2.1 Graph-Based Induction (GBI)

GBI [8] employs stepwise pair expansion (pairwise chunking) to extract typical patterns from graph-structured data. Later an enhanced version of GBI, named Beam-wise GBI (B-GBI) [3], adopting the beam search was proposed to increase the search space, thus extracting more discriminative patterns while keeping the computational complexity within a tolerant level. Since the search in GBI is greedy and no backtracking is made, which patterns are extracted by GBI depends on which pairs are selected for chunking. This means that patterns that partially overlap can no longer be extracted, and thus there can be many patterns which are not extracted by GBI. B-GBI can help alleviate this problem, but cannot solve it completely because the chunking process is still involved.

2.2 Chunkingless Graph-Based Induction (Cl-GBI)

Cl-GBI [4] was developed to cope with the problem of overlapping subgraphs incurred by both GBI and B-GBI. Cl-GBI employs a "chunkingless chunking" strategy, where frequent pairs are never chunked but used as pseudo nodes in the subsequent steps, thus allowing extraction of overlapping subgraphs. As in B-GBI, the Cl-GBI approach can handle both directed and undirected graphs as well as both general and induced subgraphs. It can also extract typical patterns in either a single large graph or a graph database. The algorithm of Cl-GBI is briefly described as follows. For a detailed mathematical treatment of Cl-GBI the reader is referred to [4].

Given a graph database, two natural numbers b (beam width) and N_e, and a frequency threshold θ, the "chunkingless chunking" strategy repeats the following three steps N_e times, each of which is referred to as a level (N_e is thus the number of levels).

Step 1. Extract all the pairs consisting of two connected nodes in the graphs, register their positions using node id (identifier) sets, and count their frequencies. From the 2nd level on, extract all the pairs consisting of two connected nodes with at least one node being a new pseudo node.

Step 2. Select the b most frequent pairs from among the pairs extracted at Step 1 (from the 2nd level on, from among the unselected pairs in the previous levels and the newly extracted pairs). Each of the b selected pairs is registered as a new node. If either or both nodes of the selected pair are not original but pseudo nodes, they are restored to the original patterns before registration.

Step 3. Assign a new label to each pair selected at Step 2 but do not rewrite the graphs. Go back to Step 1.

All the pairs extracted at Step 1 in all the levels (i.e. level 1 to level N_e), including those that are not used as pseudo nodes, are ranked based on a typicality criterion using a discriminative function such as information gain [5] or gain ratio [6]. Those pairs that have frequency count below θ are eliminated, which means that there are three parameters b, N_e, θ to control the search.

The output of Cl-GBI algorithm is a set of ranked typical patterns, each of which comes together with the positions of all its occurrences in each transaction of the graph database as well as the numbers of occurrences.

3 Decision Tree Cl-GBI (DT-ClGBI)

3.1 Decision Tree for Graph-Structured Data

As mentioned in Section 1, the attribute-value pair representation is not suitable for graph-structured data, although both attributes and their values are essential

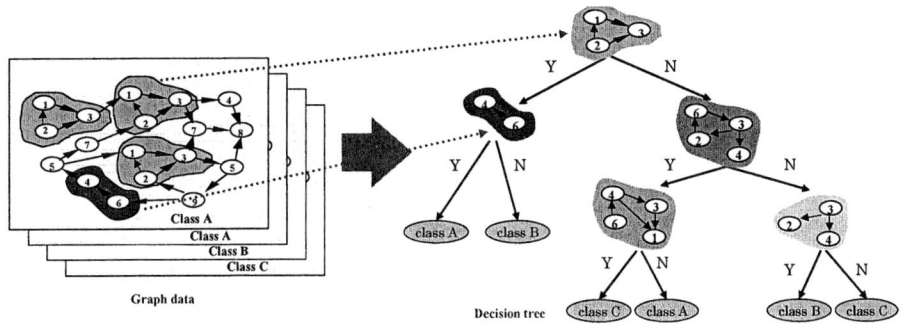

Fig. 1. Decision tree for classifying graph-structured data

for a classification or prediction task because a class is related to some attribute values in most cases. In a decision tree, each node and a branch connecting the node to its child node correspond to an attribute and one of its attribute values, respectively. Thus, to formulate the construction of a decision tree for a graph-structured dataset, we define attributes and their values as follows:

- attribute: a pattern/subgraph in graph-structured data,
- value of an attribute: existence/non-existence of the pattern in each graph.

Since the value of an attribute is either yes (the pattern corresponding to the attribute exists in the graph) or no (the pattern does not exist), the resulting decision tree is represented as a binary tree. Namely, data (graphs) are divided into two groups: one consists of graphs with the pattern, and the other consists of graphs without it. Figure 1 illustrates the decision tree constructed based on this approach. One remaining question is how to determine patterns which are used as attributes for graph-structured data. Our approach to this question is described in the next subsection.

3.2 Feature Construction by Cl-GBI

The algorithm we propose here, called Decision Tree Chunkingless Graph-Based Induction (DT-ClGBI), utilizes Cl-GBI to extract patterns from graph-structured data and use them as attributes for a classification task, whereas our previous algorithm, Decision Tree Graph-Based Induction (DT-GBI), adopted B-GBI to extract patterns. Namely, DT-ClGBI invokes Cl-GBI at each node of a decision tree, and selects the most discriminative pattern from those which were extracted by Cl-GBI. Then the data (graphs) are divided into two groups, i.e., one with the pattern and the other without the pattern as described above. For each group, the same process is recursively applied until the group contains graphs of a single class like the ordinary decision tree construction method such as C4.5 [6]. The algorithm of DT-ClGBI is summarized in Fig. 2.

In DT-ClGBI, each of the parameters of Cl-GBI, b, N_e, and θ, can be set to different values at different nodes in a decision tree. All patterns extracted at a

```
DT-ClGBI(D)
   INPUT
   D: a graph database
   begin
      Create a node DT for D
      if termination condition reached
         return DT
      else
         P := Cl-GBI(D) (with b, N_e, and θ specified)
         Select the most discriminative pattern p from P
         Divide D into D_y (with p) and D_n (without p)
         for D_i := D_y, D_n
            DT_i := DT-ClGBI(D_i)
            Augment DT by attaching DT_i as its child
            along yes/no branch
         return DT
   end
```

Fig. 2. Algorithm of DT-ClGBI

node are inherited to its descendant nodes to prevent a pattern that has already been extracted in the node from being extracted again in its descendants. This means that, as the construction of a decision tree progresses, the number of patterns to be considered at a node progressively increases, and the size of a pattern newly extracted can be larger than existing patterns. Thus, although initial patterns at the start of search consist of two nodes and the link between them, attributes useful for the classification task can be gradually grown up into larger patterns (subgraphs) by applying Cl-GBI recursively. In this sense, DT-ClGBI can be conceived as a method for feature construction, since features, i.e., attributes (patterns) useful for the classification task, are constructed during the application of DT-ClGBI.

However, recursive partitioning of data until each subset in the partition contains data of a single class often results in overfitting to the training data and thus degrades the predictive accuracy of resulting decision trees. To avoid overfitting, and improve predictive accuracy, DT-ClGBI incorporates "pessimistic pruning" used in C4.5 [6] that prunes an overfitted tree based on the confidence interval for binomial distribution. This pruning is a postprocess that follows the algorithm in Fig. 2.

Note that the criterion for selecting a pair that becomes a pseudo node in Cl-GBI and the criterion for selecting a discriminative pattern in DT-ClGBI can be different. In the following experiments, frequency of a pair is used as the former criterion, and information gain of a pattern is used as the latter criterion[1].

3.3 Classification Using the Constructed Decision Tree

Unseen new graph data must be classified once the decision tree has been constructed. Here the problem of subgraph isomorphism arises to test if the input graph contains the pattern (subgraph) specified in the test node of the tree. To alleviate this problem, we utilize Cl-GBI again. Theoretically, if the test pattern actually exists in the input graph, Cl-GBI can find it by setting the beam width b

---

[1] We did not use information gain ratio because DT-ClGBI constructs a binary tree.

and the number of levels N_e large enough and by setting the frequency threshold to 0. However, note that nodes and links that never appear in the test pattern are never used to form the test pattern in Cl-GBI. Therefore, we can remove such nodes and links from the input graph before applying Cl-GBI to reduce its running time. This approach is summarized as follows:

Step 1. Remove nodes and links that never appear in the test pattern from the input graph.
Step 2. Apply Cl-GBI to the resulting input graph setting the parameters b and N_e large enough, while setting the parameter θ to 0.
Step 3. Test if one of the canonical labels of extracted patterns with the same size as the test pattern is equal to the canonical label of the test pattern.

In general, Step 1 results in a small graph and Cl-GBI can run very quickly without any constraints on N_e and b. However, if we need to set these constraints, we may not be able to obtain the correct answer because we don't know how large these parameters should be. In that sense, this procedure can be regarded as an approximate solution to the subgraph isomorphism problem.

4 Experimental Evaluation of DT-ClGBI

To evaluate the performance of DT-ClGBI, we conducted some experiments on both synthetic and real-world datasets consisting of directed graphs.

4.1 Synthetic Datasets

Data Preparation. Synthetic datasets were artificially generated in a random manner. The number of nodes in a graph is determined by the gaussian distribution having the average of T and the standard deviation of 1. The links are attached randomly with the probability of p. The node labels and link labels are randomly determined with equal probability. The number of node labels and the number of link labels are denoted as L_V and L_E, respectively. The total number of transactions is kept fixed as GD.

Two datasets of directed graphs having the average size of 30 and 40, each of which has $GD = 300$, $L_V = 5$, $L_E = 10$, $p = 20\%$, were generated and are represented as $T30$ and $T40$, respectively. Each dataset was equally divided into two classes, namely "active" and "inactive". Similarly, L basic patterns of connected subgraphs having the average size of I, where $L = 4$ and $I = 4$, were generated. The number of basic patterns to be embedded in a transaction G_t of the class "active", N_t, was randomly selected in the range between 1 and L. Each of these N_t basic patterns was in turn chosen from the set of L basic patterns by equal probability, i.e. $1/L$, and overlaid on that transaction. This means that each transaction of the class "active" includes from 1 to L basic subgraphs, some of them may happen to be the same. We also check if there is any basic subgraph included in a transaction of the class "inactive" by Cl-GBI as described in Section 3.3. If there is, the involved node and link labels

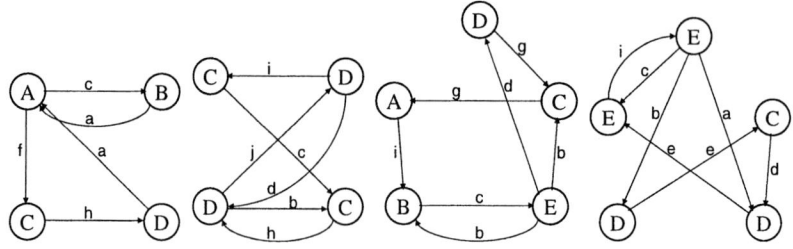

Fig. 3. Example of 4 basic subgraphs

are changed in a way that the basic pattern no longer exists in the transaction. In other words, basic subgraphs are those which discriminate the two classes. This does not necessarily mean that any subgraph of the basic patterns is not discriminative enough. We have not checked that all the subgraphs of the basic patterns appear in both active and inactive data. Figure 3 shows the 4 basic subgraphs which were embedded in the transactions of the class "active".

Experiments. For these two synthetic datasets, the classification task is to classify two classes "active" and "inactive" using DT-ClGBI by a single run of 10-fold cross validation (CV). The final prediction error rate was evaluated by the average of 10 estimates of the prediction error (a total of 10 decision trees).

The first experiment was conducted to confirm that the most discriminative patterns with respect to the used index can be extracted by Cl-GBI not only at the root node, but also at each internal node itself. To this end, we compared the predictive accuracy and the tree size obtained by two different settings for DT-ClGBI described as follows. In the first setting, i.e. setting 1, a decision tree is constructed by applying Cl-GBI at the root node only, with $N_e = 2$. At the other nodes, we simply recalculate information gain for those patterns that have already been discovered at the root node. In the other case, i.e., setting 2, Cl-GBI is invoked at the root node with $N_e = 2$ and at other nodes with $N_e = 1$. In addition, the total number of levels of Cl-GBI in the second setting is limited to 6 to keep the computation time at a tolerant level. Whenever the total number of levels reaches this limitation, Cl-GBI is no longer used for extracting patterns. Instead, only the existing patterns are employed for constructing the decision tree thereafter. Note that beam width is set to 5 in both settings.

Results of the first experiment are summarized in Table 1, and it is shown that the second setting obtains higher predictive accuracy. Moreover, we observe

Table 1. Comparisons of different settings for DT-ClGBI

Dataset	Setting 1			Setting 2		
	Training error	Test error	Average of tree sizes	Training error	Test error	Average of tree sizes
T30	0.22%	1.33%	17.2	0%	0%	9.2
T40	0.37%	5%	18	0.15%	3.33%	12.8

Table 2. Comparisons of DT-ClGBI and DT-GBI

Dataset	DT-GBI			DT-ClGBI		
	Training error	Test error	Average of tree sizes	Training error	Test error	Average of tree sizes
T30	1.41%	7.67%	24	0%	0%	9
T40	3.15%	7.67%	18.2	0%	0.67%	9

that the decision trees constructed by the second setting have smaller sizes in most cycles of the 10-fold CV for both datasets. The result reveals that invoking of Cl-GBI at internal nodes is needed to improve the predictive accuracy of DT-ClGBI, as well as to reduce the tree size. Intuitively, the search space increases by applying Cl-GBI at the internal nodes in addition to the root node. As a result, more discriminative patterns which have not been discovered in the previous steps are discovered at these nodes. In other words, applying Cl-GBI at only the root node cannot help enumerate all the necessary patterns unless N_e and b are set large enough. For example, in the decision tree constructed by the first run of 10-fold CV on the dataset $T30$ using the second setting, the classifying pattern for a node at the third level was not found at the root node but its parent node. If N_e is set large enough, the necessary pattern should be able to be found at the root node. This pattern, if found at the root node, should give smaller information gain at the root node but Cl-GBI retains this and passes down to the lower node. The question is how to find this pattern where it is needed without running Cl-GBI using all the dataset.

The second experiment focused on the comparisons between DT-ClGBI and DT-GBI [2], also in terms of the predictive accuracy and the tree size. Here beam width is also set to 5 in both cases. For DT-GBI, the number of levels of B-GBI at any node of a decision tree is kept fixed as 4. It should be noted that, whenever being invoked for constructing a decision tree by DT-GBI, B-GBI starts extracting typical patterns from the beginning, i.e. no inheritance is employed, because the graphs that pass down to the yes branch have been chunked by the test pattern. On the other hand, the number of levels of Cl-GBI is 4 at the root node and 1 at the other nodes of a decision tree in the case of DT-ClGBI. In addition, the total number of levels of Cl-GBI is limited to 8, which means that the number of levels performed by the feature construction tool in DT-ClGBI is much less than that in DT-GBI.

Table 2 shows the results of the second experiment. It can be seen that DT-ClGBI achieves lower prediction error for both datasets. We also observe that, for each dataset, the decision trees constructed by DT-ClGBI have smaller sizes in most cycles of the 10-fold CV. The higher predictive accuracy of DT-ClGBI and the simpler decision trees obtained by this method can be explained by the improvement of Cl-GBI over B-GBI, and the inheritance of previously extracted patterns at an internal node (in a decision tree) from its predecessors. It is known that Cl-GBI resolves the problem of overlapping patterns incurred by B-GBI, thus resulting in more typical patterns extracted by Cl-GBI. In addition, the computation time of DT-ClGBI was found to be less than half of that required by DT-GBI with the above settings, which is mainly due to the fact that the feature construction tool for DT-ClGBI was not run from the scratch at any internal node as in DT-GBI. This implies that DT-ClGBI performs better than DT-GBI while requiring much less computation resource.

It should also be noted that the size of the embedded graphs in these two datasets is 4 or 5. Setting $N_e = 2$ as in the first experiment means that the maximum size of patterns we can get at the root node is 4. Considering the beam width, it is unlikely that the embedded patterns are found at the root node. Even $N_e = 4$ as in the second experiment, these basic patterns cannot be found. However, the substructures of the embedded graphs are discriminative enough as shown in the two experiments. Due to the downward closure property, these substructures were embedded in the transactions of the class "active".

4.2 Real-World Datasets

Finally, we verified if DT-ClGBI can construct decision trees that also achieve reasonably good predictive accuracy on a real-world dataset. For that purpose, we used the hepatitis dataset as in [2]. The classification task here is to classify patients into two classes, "LC" (Liver Cirrhosis) and "nonLC" (non Liver Cirrhosis) based on their fibrosis stages, which are categorized into five stages in the dataset: F0 (normal), F1, F2, F3, and F4 (severe = Liver Cirrhosis). All 43 patients at F4 stage were used as the class "LC", while all 4 patients at F0 stage and 61 patients at F1 stage were used as the class "nonLC". This ratio of "LC" to "nonLC" was determined based on [7]. The records for each patient were converted into a directed graph as described in [2]. The resulting graph database has 108 graph transactions, and the average size of a graph transaction is 316.2 and 386.4 in terms of the number of nodes and of links, respectively.

Through some preliminary experiments on this database using DT-ClGBI, we found that existence of some graphs often makes the resulting decision tree too complicated and worsen the predictive accuracy. This has led us to adopt a two step approach, first to divide the patients into "typical" and "non-typical", and second to construct a decision tree for each group of the patients. To divide the patients in the first step, we ran 10-fold cross validation of DT-ClGBI on this database, varying its parameters b and N_e in the ranges of $\{5, 6, 8, 10\}$ and $\{6, 8, 10, 12\}$, respectively. Note that these values are only for the root node and we did not run Cl-GBI at the succeeding nodes. The frequency threshold θ was fixed to 10%. Namely, we conducted 10-fold cross validation 16 times with different combinations of these parameters, and obtained totally 160 decision trees in this step. Then we classified graphs whose average error rate is 0% into "typical", and the others into "non-typical". As a result, for the class "LC", 28 graphs are classified into the subset "typical" and the other 15 graphs into "non-typical", while for the class "nonLC", 48 graphs are classified into "typical" and 17 graphs into "non-typical".

In the second step, we applied DT-ClGBI to each subset again adopting the best parameter setting in the first step with respect to the predictive accuracy, where $b = 8$ and $N_e = 10$. The predictive accuracy (average of 10-CV) for the subset "typical" is 97.4%, and that for "non-typical" is 78.1%. The overall accuracy is 91.7%, which is much better than the accuracy obtained by applying DT-ClGBI to the original dataset with $b = 8$ and $N_e = 10$, i.e., 83.4%. We can find typical features for a patient with Liver Cirrhosis in the extracted patterns such as "GOT is High" or "PLT is Low". From these results, we can say that DT-ClGBI can achieve reasonably good predictive accuracy on a real-world dataset and extract discriminative features embedded in the dataset as subpatterns.

5 Conclusions

We have proposed an algorithm called DT-ClGBI, which can construct decision trees for graph-structured data using Cl-GBI. In DT-ClGBI, substructures, or patterns useful for a classification task are constructed on the fly by means of Cl-GBI during the construction process of a decision tree. The experimental results using synthetic and real-world datasets showed that decision trees constructed by DT-ClGBI achieve good predictive accuracy for graph-structured data. The good predictive accuracy of DT-ClGBI is mainly attributed to the fact that Cl-GBI can give the correct number of occurrences of a pattern in each transaction of the graph database due to its capability of extracting overlapping patterns, which is very useful for algorithms such as DT-ClGBI that need correct counting. Also, the inheritance of previously extracted patterns at an internal node from its predecessors is shown helpful.

References

1. Clark, P. and Niblett, T. 1989. The CN2 Induction Algorithm, *Machine Learning*, **3**(4): 261–283.
2. Geamsakul, W., Yoshida, T., Ohara, K., Motoda, H., Washio, T., Takabayashi, K., Yokoi, H. 2005. Constructing a Decision Tree for Graph-Structured Data and Its Applications. *Fundamenta Informaticae*, **66**(1-2): 131–160.
3. Matsuda, T., Motoda, H., Yoshida, T., Washio, T. 2002. Mining Patterns from Structured Data by Beam-wise Graph-Based Induction, In *Proc. DS 2002*, pp. 422–429.
4. Nguyen, P.C., Ohara, K., Motoda, H., and Washio, T. 2005. Cl-GBI: A Novel Approach for Extracting Typical Patterns from Graph-Structured Data, In *Proc. PAKDD 2005*, pp. 639–649.
5. Quinlan, J.R. 1986. Induction of Decision Trees, *Machine Learning*, **1**: 81–106.
6. Quinlan, J.R. 1993. *C4.5: Programs for Machine Learning*, Morgan Kaufmann.
7. Yamada, Y., Suzuki, E., Yokoi, H., and Takabayashi, K. 2003. Decision-tree Induction from Time-series Data Based on a Standard-example Split Test, In *Proc. ICML 2003*, pp. 840–847.
8. Yoshida, K. and Motoda, M. 1995. CLIP: Concept Learning from Inference Patterns, *Artificial Intelligence*, **75**(1): 63–92.

Combining Smooth Graphs with Semi-supervised Classification

Xueyuan Zhou and Chunping Li

School of Software, Tsinghua University,
Beijing 100084, P.R. China
zhou-xy03@mails.tsinghua.edu.cn
cli@tsinghua.edu.cn

Abstract. In semi-supervised classification, many methods use the graph representation of data. Based on the graph, different methods, e.g. random walk model, spectral cluster, Markov chain, and regularization theory etc., are employed to design classification algorithms. However, all these methods use the form of graphs constructed directly from data, e.g. kNN graph. In reality, data is only the observation with noise of hidden variables. Classification results using data directly from the observation may be biased by noise. Therefore, filtering the noise before using any classification methods can give a better classification. We propose a novel method to filter the noise in high dimension data by smoothing the graph. The analysis is given from the aspects of spectral theory, Markov chain, and regularization. We show that our method can reduce the high frequency components of the graph, and also has an explanation from regularization view. A graph volume based parameter learning method can be efficiently applied to classification. Experiments on artificial and real world data set indicate that our method has a superior classification accuracy.

1 Introduction

In semi-supervised classification, graph-based methods have drawn great attention recently. This kind of methods maps the data points into vertices of a graph, and a weighted graph is formed after defining a similarity function between points. Graph methods are nonparametric, discriminative, and transductive in nature. Many methods employ this graph representation of data [2, 6, 7, 10, 8, 9].

Although these works employ many variants of graph formation, e.g. kNN and eNN graphs, exponential and cosine weights, they are all formed directly from the data points. We know that all the data points are only the observation of some hidden variables, and the noise can not be avoided in observed values. To build a graph which reflects the data distribution better, we should filter the noise first, and then classify the unlabeled data.

In many practical applications of data classification and data mining, we usually make necessary assumptions to facilitate our analysis and help us to find more applicable methods. Labels smoothness [11] and cluster assumption

[2,3] are of this kind. As we know, nearby data points in the same cluster tend to share the same label, and the density of one cluster changes slowly inside one cluster. So we assume that the real distribution of data points in the same cluster consists of lower frequency components, and higher frequency ones of the distribution are more likely to be the noise.

We utilize the Markov random walk model to construct such a graph. Based on our assumption, theoretical analysis shows that our graph has less high frequency noise, and a smoother transition matrix. After the filtering step, we use a graph-based classification method to classify data points.

In this paper, we introduce a new form of smooth graph in Sect. 2. In Sect. 3, we give the analysis of our graph from spectral theory, Markov chain, and regularization theory. The classification and parameter learning method are given in Sect. 4. With the experiment analysis of the artificial and real world data sets, the results and evaluations are shown in Sect. 5. In Sect. 6, we give the concluding remark.

2 Smooth Graph

2.1 Data Representation

In this paper, we use the Markov random walk on graph model to build our graph. Let the graph $G = (V, E)$ be a pair consisting of a set V of vertices, and a set E of edges joining some pairs of vertices. For each $x \in V$, we may consider the set N_x of neighbors of x, formed by vertices y with an edge joining x to y. The random walk is based on this graph, where the step from x to y ($y \in N_x$) has probability p_{xy}. Under the assumption of Markovian property, such a random walk on graph can be viewed as a Markov chain.

In the Markov random walk for classification, data points are mapped into the vertices in a graph or states in a Markov chain. The transition probability can be seen as the similarity between data points. Given that a dataset consists of data pairs $\{(x_1, y_1), \cdots, (x_n, y_n)\}$, after the definition of the weights of edges, we can construct a graph from input data x_i. For instance, a typical weight of edge is defined as

$$w_{ij} = exp(-\frac{d(x_i, x_j)}{\sigma^2}) \qquad (1)$$

where $d(\cdot)$ can be Euclidean distance or other distance measure, and σ is the parameter for exponential weight. Then let the one-step transition probability be

$$p_{ij} = \frac{w_{ij}}{\sum_k w_{ik}} \qquad (2)$$

We can get the weight matrix $W = [w_{ij}]$ and transition matrix $P = [p_{ij}]$. In the matrix form, we have $P = D^{-1}W$, where D is a diagonal matrix with $D_{ii} = \sum_j w_{ij}$. Some semi-supervised classification methods [7,9,10] are based on such a basic representation. In this paper, we assume $w_{ii} = 0$, which forms a non-lazy random walk.

2.2 Smooth Graph

From the Markov random walk, we treat that the walker starts walking from some point according to the transition matrix as a diffusion process. And this process can smooth the graph as its step increases. We know that the step of the walk corresponds to the power of transition matrix P. If the matrix P is from graph G, then we propose the following transition matrix of graph $G^{(m)}$ for semi-supervised classification.

$$P^{(m)} = P^m \qquad (3)$$

Therefore, $P \times P$ is the transition matrix of another graph $G^{(2)}$, and $P \times P \times P$ is the transition matrix of graph $G^{(3)}$, and so on.

3 Analysis for Smoothness

All the following discussions are based on the assumption that real world data has a smooth distribution, which can be also viewed as data density changes slowly in a connected region.

3.1 Spectral Analysis

Spectral theory and harmonic analysis have been used to analyze high dimension data [1,4]. Dimension reduction, visualization, and many machine learning methods for high dimension data can be derived from this framework. Our analysis is given from the frequency point of view.

The smoothness of distribution for some data sets can be captured by the transition matrix P, sparse region with lower transition probability and dense region with higher one. For a smooth distribution, the transition probabilities inside a cluster should be relative high, and should not change greatly from point to point. On the other hand, the probabilities between clusters should be relatively small. However, in real world data, for the reason of noise and high dimensionality of data point, the transition probabilities inside a cluster vary greatly. In semi-supervised classification, there are fewer labeled data points, and the noise can easily result in wrong classification.

If the matrix P is composed from different frequencies, based on our assumption, we can filter the high frequency components to smooth data distributions. From spectral theory we know that matrix P can be decomposed as $P = \Phi \Lambda \Phi^{-1}$, where Λ is a diagonal matrix with eigenvalue λ_i of P, Φ is composed of the eigenvectors corresponding to each eigenvalue.

From spectral theory and harmonic analysis we know that the eigenfunctions can be interpreted as a generalization of the Fourier harmonics on the manifold defined by the data points [1]. In our problem setting, smaller eigenvalues correspond to higher frequency eigenfunctions, and larger eigenvalues correspond to lower ones. By eigen decomposition, we can rewrite $P^{(m)}$ as:

$$P^{(m)} = P^m = \Phi \Lambda \Phi^{-1} \cdots \Phi \Lambda \Phi^{-1} = \Phi \Lambda^m \Phi^{-1} \qquad (4)$$

It is easy to see:
$$p_{ij}^{(m)} = \sum_k \lambda_k^m \phi_{ik} \psi_{kj} \qquad (5)$$

where ϕ_{ik} and ψ_{kj} are the elements in Φ and Φ^{-1}. For a probability matrix, we have $\lambda_0 = 1 > \lambda_1 > \lambda_2 \cdots > \lambda_n \geq 0$. From equation (5) we know that as the power increases, smaller eigenvalues decay greatly, and larger ones stay relatively larger. From the frequency view, that is to say, the power of P acts as a low-pass filter, reducing the higher frequency components while retaining lower ones. Furthermore, the filtering process can be controlled with the parameter m.

3.2 Markov Chain

In this subsection, we show that the transition matrix will become smoother and smoother as m increases. We firstly map the graph representation of data into a Markov chain. The vertices set V of the graph is mapped into the state set I of the chain. The weight of the graph G defined in Sect. 2 is mapped into transition probability between states in Markov chain.

From the view of Markov chain, if P^m exists when $m \to \infty$, there is an uniform distribution π_j ($1 \leq j \leq n$) over all the data, where $n = |I|$, total number of states in I or vertices in the graph G. That means the probabilities from any data point to one fixed point are the same. In this case, we say the graph is "flat". We can define the smoothness of a graph according to its "flat" state if it exists. The smoothness function is defined as:

$$Q^{(m)} = \sum_{i,j=1}^{n} (p_{ij}^{(m)} - \pi_j)^2 \qquad (6)$$

$Q^{(m)}$ reflects the smoothness of the graph. The smaller value of $Q^{(m)}$ means a smoother graph. We can predict that graph $G^{(2)}$ is smoother than $G^{(1)}$. This can be explained from the view of Markov chain analysis. We give a brief proof here.

By treating a connected graph G as the Markov chain, it is easy to satisfy the following conditions: it is a finite-state Markov chain with no two disjoint closed sets, and it is aperiodic. After mapping the graph into a Markov chain, we have the following result [5]: there exist a probability distribution $\{\pi_j, j \in I\}$ and numbers $\alpha > 0$ and $0 < \beta < 1$ such that, for all $i, j \in I$,

$$|p_{ij}^{(n)} - \pi_j| \leq \alpha \beta^n, \quad n = 1, 2, \cdots \qquad (7)$$

In particular,

$$\lim_{n \to \infty} p_{ij}^{(n)} = \pi_j \quad for\ all\ \ i,j \in I \qquad (8)$$

In (7), $p_{ij}^{(n)}$ is an element in the matrix P^n. From (7), we know that as n gets larger, $p_{ij}^{(n)}$ gets closer to the fixed value π_j.

From (6) and (7), we have

$$Q^{(m)} = \sum_{i,j=1}^{n} (p_{ij}^{(m)} - \pi_j)^2 \leq \sum_{i,j=1}^{n} \alpha^2 \beta^{2m} = n^2 \alpha^2 \beta^{2m}$$

For a fixed graph, n is a constant. Therefore, $Q^{(m)}$ gets smaller and smaller as m grows. Then we can say that the graph gets smoother and smoother.

Different m values will result in different graphs. In special case when $m = 1$, it is the original graph G. However, m should not get too large. When $m \to \infty$, P^m will become an uniform distribution, which provides no information about classification.

The following example can illustrate this point. Assume that a is the transition matrix of some graph G.

$$\mathbf{a} = \begin{pmatrix} 0.40 & 0.60 & 0.00 & 0.00 \\ 0.20 & 0.70 & 0.10 & 0.00 \\ 0.00 & 0.10 & 0.60 & 0.30 \\ 0.00 & 0.00 & 0.80 & 0.20 \end{pmatrix} \quad \mathbf{a}^4 = \begin{pmatrix} 0.22 & 0.60 & 0.14 & 0.04 \\ 0.20 & 0.55 & 0.19 & 0.06 \\ 0.05 & 0.19 & 0.54 & 0.22 \\ 0.03 & 0.16 & 0.58 & 0.23 \end{pmatrix}$$

$$\mathbf{a}^{16} = \begin{pmatrix} 0.14 & 0.41 & 0.33 & 0.12 \\ 0.14 & 0.40 & 0.34 & 0.12 \\ 0.11 & 0.34 & 0.40 & 0.15 \\ 0.11 & 0.33 & 0.41 & 0.15 \end{pmatrix} \quad \mathbf{a}^{64} = \begin{pmatrix} 0.12 & 0.37 & 0.37 & 0.14 \\ 0.12 & 0.37 & 0.37 & 0.14 \\ 0.12 & 0.37 & 0.37 & 0.14 \\ 0.12 & 0.37 & 0.37 & 0.14 \end{pmatrix}$$

We can see that $Q^{(1)} = 1.00$, $Q^{(4)} = 0.39$, $Q^{(16)} = 0.01$, $Q^{(64)} = 0.00$. As m increases, the graph gets flatter and flatter and at last becomes an uniform distribution for each column.

3.3 Regularization

In graph-based semi-supervised classification, the key problem is to estimate the probability for each class. In order to solve the "ill-posed problem" in estimating the probability, regularization is proposed as a solution. Under this framework, many methods can be viewed as to estimate a smooth function f on the graph. This function should be close to the given values on the labeled data points, and at the same time it should be smooth on the whole graph.

One typical method minimizes the following smoothness function:

$$S(f) = \sum_{ij} w_{ij} (f_i - f_j)^2 \tag{9}$$

The solution for the graph G is vector $\mathbf{f}^{(0)}$, whose weight matrix is $W^{(0)}$. For the graph $G^{(m)}$, solution is $\mathbf{f}^{(m)}$ with weight matrix $W^{(m)}$. The minimization forces the f_i and f_j to be close with biger w_{ij}. From (2) we know that w_{ij} is associated with p_{ij}. As m goes up, from the analysis above, some data points far away but in the same cluster will have a higher transition probabilities $p_{ij}^{(m)}$. This means that their corresponding weight $w_{ij}^{(m)}$ increases. With Euclidean distance, the path between i and j is shortened. From the smoothness function $S(f)$, f_i and f_j are forced to be closer than before. Although we notice that as m increases, bigger weights will be reduced gradually, as long as we pick up a proper value for m, it is feasible to both keep the local structure and introduce global information.

4 Classification

Many graph based methods for semi-supervised classification can be viewed as Markov random walks [7, 9, 10]. These methods have clear explanations and have done well in semi-supervised classification. In order to take advantage of our graph representation, we employ a random walk related method, i.e. harmonic function in [10] to label those unlabeled data points.

4.1 Parameter Learning

The parameters in our model are m, and σ in equation (1) if selected. Zhu [10] has proposed a method for learning σ. We will focus on how to learn parameter m in our model.

There are many methods that can be used for parameter learning in the graph-based semi-supervised classification. We propose a graph volume based method. In semi-supervised classification, labeled data usually has a relatively small size, e.g. in case of only two labeled data points for binary classification, one for each class. In this case, we can not fully trust these two points, because they may be noises or biased by noises greatly. Therefore, many parameter learning methods, which rely on only labeled data, can not be used here. Based on the cluster assumption, we know that when two clusters are separated well, no lower density region exists in any cluster. The connection inside a cluster is stronger than that between clusters. Based on this intuition, we propose the following method to select a proper m: let C_i be a subset of the vertices of the graph G. We define

$$vol(C_k) = \sum_{i,j=1}^{|C_k|} p(x_i, x_j) \quad i,j \in C_k \tag{10}$$

The following function can be used to measure the connection inside a class.

$$g(C_k) = \frac{vol(C_k)}{|C_k|^2} \tag{11}$$

We call $g(C_k)$ the cohesion factor of the class C_k, and it can be viewed as the density of C_k. Since our classification method can be viewed as a random walk, labels are propagated from the labeled data points to unlabeled ones. As long as the classification is wrong, there will be low density region inside one class, and $vol(C_k)$ will be small. When $g(C_k)$ becomes relative large and stable, that is $\Delta g(C_k) < \epsilon$, we can stop the walk and pick up the value of m at this time.

Average distance and its variants are frequently used to find a proper m [7]. This kind of methods computes the average distance between each pair of data points in different classes. When the classes are well separated, the average distance should be larger than other cases. However, this method is influenced by the shapes of clusters more than our method. For instance, average distance method might not be suitable to be used in the two moon data set in Fig.1. Because of the interwoven shape, when two classes are well separated, the average distance may not be the largest one. But our method is independent of the shapes of classes.

5 Experimental Results

5.1 Artificial Data

We test our method on the switch or two-moon data [7] with two labeled points, one for each class. The weight is formed using equation (1), and $d(\cdot)$ is Euclidean distance. From Fig.1 we can see that, as the value of m gets larger and larger, the graph becomes smoother. Furthermore, when $m = 1$ and $\sigma = 0.04 \sim 0.06$, classification can be totally correct using the method in [10]. However, keeping the accuracy at 100%, the smooth graph enlarges the range of parameter σ to $\sigma = 0.04 \sim 0.45$.

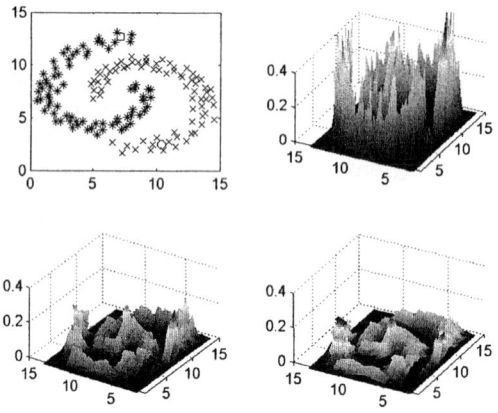

Fig. 1. Smooth graph on artificial data. $\sigma = 0.45$. Right up: $m = 1$; Left bottom: $m = 5$; Right bottom: $m = 9$

The smoothness here is different from ones in [9, 10]. In this paper, we use smoothness to describe the transition probabilities between points. A rough transition probabilities between points might easily spread the errors. If the bridge noise has a high transition probability, it will bring more error. However, a smoother transition probability can reduce this error.

5.2 Text Classification

We also apply our method to text classification, with few labeled documents but many unlabeled ones. Text documents are represented with high dimension vectors, which are usually quite sparse. We expect to construct a smooth graph for the classification.

We test our method with real-world data set 20 Newsgroups. In order to make a comparison, the data and setting for Windows vs Mac are the same as [2, 7]. From 2 to 128 labeled data points are randomly selected to form X_L. We test on 100 randomly splits balanced for class labels. The value of m ranges from 1 to 16. We show experimental results under different m values. From the experimental results, we have the following observations:

1.The smooth graph has a high classification accuracy.

In Fig.2, Smooth is the accuracy of our method with learned parameter m. MAM is the result of [7] and CK is the result of [2]. We can see a clear advantage of our smooth graph methods, especially when the number of labeled examples is relatively small. Table 1 shows the classification results of our methods and original harmonic function [10], where $m = 1$ is the result of harmonic function and learned m is the result of our method.

2. Different m affects results greatly.

From Fig.3, we can see that different steps affect the classification results greatly. When labeled examples are fewer, more steps give a better classification. But when labeled examples get more, accuracy after more steps falls slightly. Because when there are more labeled examples, fewer steps are needed to reach labeled examples, and more steps result in a flatter graph. If the graph is too "flat", it may not be good for classification. When steps become infinite, then

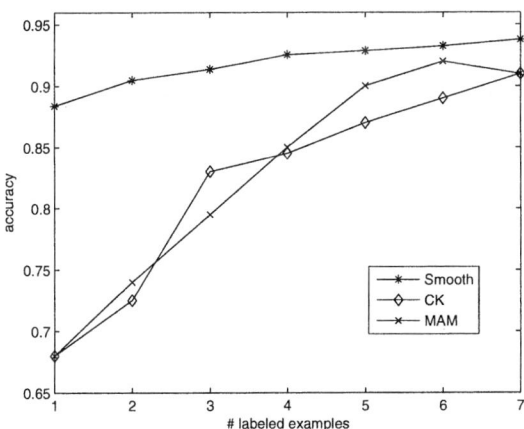

Fig. 2. Classification results of smooth graph on text data set Windows vs Mac with learned m.

Table 1. Classification accuracy. Left two columns: Electronics vs Space; Right two columns: Baseball vs Hocky.

Labeled examples	m=1	learned m	m=1	learned m
2	0.640	0.913	0.648	0.848
4	0.687	0.935	0.700	0.901
8	0.750	0.948	0.763	0.927
16	0.807	0.952	0.824	0.933
32	0.872	0.948	0.880	0.943
64	0.915	0.950	0.921	0.944
128	0.945	0.955	0.946	0.946

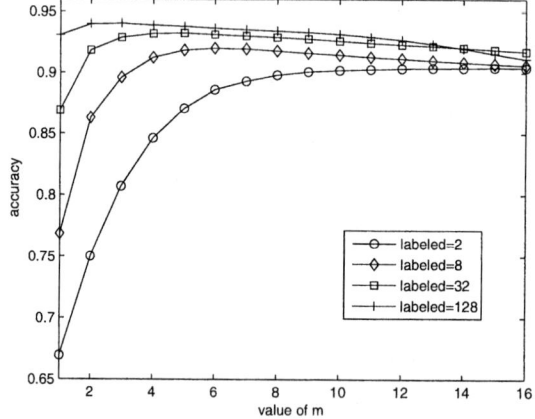

Fig. 3. Classification accuracy of full connected graph under different m. From bottom to up, $2, 8, 32$ and 128 labeled examples.

the graph becomes an uniform distribution. In the case we can not make a classification.

6 Conclusions

In this paper, we propose a new form of graph, i.e., smooth graph, which aims to filter the high frequency noise with smooth transition matrix. This graph is constructed using Markov random walk model. By the power of transition matrix P, we obtain a smooth graph. From spectral theory, Markov chain, and regularization theory, we show that the high frequency components of this graph are reduced.

Based on the smooth graph, a semi-supervised classification method that combines the smooth graph and the graph-based method has been developed and applied to text classification. Results from the artificial and real data are convincing and illustrate that smooth graph fits graph-based semi-supervised classification better, which has a clear advantage over several other methods.

Acknowledgements

We are grateful to Stéphane Lafon for the helpful conversation on harmonic analysis. And we are also grateful to Olivier Chapelle for the data set Windows vs Mac. This work was supported by Chinese 973 Research Project under grant No. 2004CB719401.

References

1. Belkin, M., Niyogi, P.: Semi-supervised Learning on Riemannian Manifolds. Machine Learning, Volume 56, Special Issue on Clustering, (2004) 209-239
2. Chapelle, O., Weston, J., Schölkopf, B.: Cluster Kernels for Semi-Supervised Learning. Advances in Neural Information Processing Systems 15. MIT Press, Cambridge, MA, USA (2003) 585-592
3. Chapelle, O. and A. Zien: Semi-Supervised Classification by Low Density Separation. Proceedings of the Tenth International Workshop on Artificial Intelligence and Statistics (2005) 57-64
4. Coifman, R.R., Lafon, S., Lee, A.B., Maggioni, M., Nadler, B., Warner, F., and Zucker, S.W.: Geometric Diffusions as a Tool for Harmonic Analysis and Structure Definition of Data, Proceedings of the National Academy of Sciences (2005)
5. Henk C. T.:Stochastic Models: An Algorithmic Approach, John Wiley & Sons (1994)
6. Meila, M., Shi, J.: Learning Segmentation by Random Walks. In Neural Information Processing Systems 13 (2000) 873-879
7. Szummer, M., Jaakkola, T.: Partially labeled Classification with Markov Random Walks. Neural Information Processing Systems (NIPS), Vol 14 (2001)
8. Zhou, D., Schölkopf, B.: Learning from Labeled and Unlabeled Data Using Random Walks. Proceedings of the 26th DAGM Symposium. Springer, Berlin, Germany (2004) 237-244
9. Zhou, D. et al.: Learning with Local and Global Consistency. In Advances in Neural Information Processing System 16. MIT Press, Cambridge, MA, USA (2004) 321-328
10. Zhu, X., Lafferty, J., Ghahramani, Z.: Semi-Supervised Learning Using Gaussian Fields and Harmonic Function. In The Twentieth International Conference on Machine Learning (2003)
11. Zhu, X.: Semi-Supervised Learning with Graphs. Doctoral Thesis. CMU-LTI-05-192, Carnegie Mellon University (2005)

Network Data Mining: Discovering Patterns of Interaction Between Attributes

John Galloway[1,2] and Simeon J. Simoff[3,4]

[1] Complex Systems Research Centre, University of Technology Sydney,
PO Box 123, Broadway NSW 2007, Australia
john.galloway@uts.edu.au
[2] Chief Scientist, NetMap Analytics Pty Ltd,
52 Atchison Street, St Leonards NSW 2065, Australia
[3] Faculty of Information Technology, University of Technology Sydney,
PO Box 123, Broadway NSW 2007, Australia
simeon@it.uts.edu.au
[4] Electronic Markets Group, Institute for Information and Communication Technologies
http://research.it.uts.edu.au/emarkets

Abstract. Network Data Mining identifies emergent networks between myriads of individual data items and utilises special statistical algorithms that aid visualisation of 'emergent' patterns and trends in the linkage. It complements predictive data mining methods and methods for outlier detection, which assume the independence between the attributes and the independence between the values of these attributes. Many problems, however, especially phenomena of a more complex nature, are not well suited for these methods. For example, in the analysis of transaction data there are no known suspicious transactions. This paper presents a human-centred methodology and supporting techniques that address the issues of depicting implicit relationships between data attributes and/or specific values of these attributes. The methodology and corresponding techniques are illustrated on a case study from the area of security.

1 Introduction

A large volume of data with different storage systems, multiple formats and all manner of internal complexity can *often hide more than it reveals* to the data mining techniques, focused on building descriptive or/and predictive computational models [1]. There are several measures of model quality [2], with the accuracy of predictions remaining as a key measure of model quality, rather than *the theory that may explain the phenomena*. However, many areas require deeper understanding of the phenomena. In addition to the explicitly coded relationships, there often are *implicit* relationships between the entities described by the data set, especially in the realm of transactions data. The structure of such relationships between the individual entities can be revealed by *network models*. During recent years there has been an increasing input from physicists [3], mathematicians [4] and organisational scientists [5] to network research, with the focus shifting to large scale networks (with millions of links and nodes), their statistical properties and explanatory power, the discovery of such models and their use in explaining different phenomena. *This change of scale of the network models indicates a need for change in the analytics approach* [4].

Network Data Mining (NDM) addresses this challenge. We define *network data mining as the process of discovering emergent network patterns and models in large and complex data sets*. NDM addresses the "loss of detail" problem and the "independency of attributes" assumption in predictive modelling. In many areas (e.g. science, fraud, intelligence, to name the few) the *loss of detail* can defeat the whole purpose of the analysis as it is often in the detail where the most valuable information is hidden [6]. The "independency of attributes" assumption is accepted in many classifier building algorithms, for example, Naïve Bayes techniques [7]. The logic is clear: by missing detail or making the wrong assumptions or simply by being unable to define what is normal, an organisation that relies solely upon predictive data mining may fail to discover critical information buried in its data. Further we present a human-centred knowledge discovery methodology, that addresses these issues, and a case study that follows this methodology and illustrates the solutions that the network data mining approach and technology offers.

2 Network Data Mining – The Methodology

The overall NDM process is illustrated in Fig. 1. The techniques supporting this methodology are implemented in the NetMap visual analytics engine[1].The main NDM methodological steps and accompanying assumptions approach include:

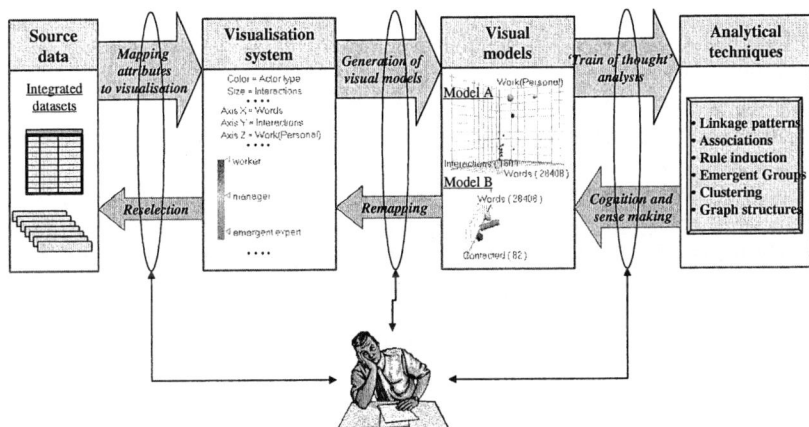

Fig. 1. Network data mining as a human-centered knowledge discovery process

Specify sources of data and modelling: NDM integrates data in order to obtain a single address space, a common view, for disparately sourced data. Having decided the sources, multiple data models are created depending on the input fields.
Visualisation and Generation of Visual Models: Visualisation of entities and links consists of a set of various consistent visual models that facilitate the discovery capabilities of the analyst – essential when dealing with millions data points.

---

[1] This premier technology is developed by NetMap Analytics (http://www.netmap.com).

'Train of thought' analysis: The analyst acts similarly to Donald Schön's "reflective practitioner" [8]. Data miners put together visual pieces of information and create new chain of inquiries interacting with the network slices of the data set.

Cognition and sense-making: Intuition and cognition are integral, and are harnessed in the analytical process (an argument well supported in [9]).

Discovery and remapping: An emergent process, not prescriptive one. To realise its full value the discovery phase needs to be repeated at regular intervals so that new irregularities that arise and variations on old patterns can be identified and fed into exception detection processes.

Reselection and converting patterns to knowledge: Any linkage *pattern* observed on screen is simply an observation of potential interest. For example, in retail, the perpetrators of a scam had taken it in turns to report levels of refunds always just under the limits no matter what the limits were varied to over an extensive period. NDM depicted collusive and periodic reporting linkages to supervisors. Such patterns are termed *scenarios* and characterised as definable and re-usable patterns. Their value is that they are patterns that have now become 'known'. Hence they can be defined, stored in a knowledge base, and applied during other data mining processes.

3 Example of Network Data Mining Approach in Fraud Detection

The NDM approach is illustrated with an application to a real world case, presented as a reflective analysis of the analyst's steps, following the main methodological steps.

Specify sources of data and modelling: The case involved analysis of approximately twelve months of motor vehicle insurance claims from one company. Initially there were no known persons or transactions of interest.

Visualisation and Generation of Visual Models: The analyst first built a set of linkages between persons, addresses, claim numbers, telephone numbers, and bank accounts into which claims monies had been paid, as shown in Fig. 2 and Fig. 3.

Cognition and sense-making (and 'Train of thought' analysis): The analyst was able to quickly focus *where* to look in the myriad of linkages by eliminating the 'regular' small triangles of data comprised of a person, a claim number and an address, all fully inter-linked (most people just had one claim and one address). The 'bumps' looked as though they were 'irregularities'. The 'bumps' comprised persons linked to multiple claims and/or addresses (see Fig. 4).

Discovery and remapping: Four emergent groups were identified (see Fig. 5). An emergent group comprises closely inter-related data items; they have more links within the group than outside to any other group.

'Train of thought' analysis: The emergent group on the right in Fig. 5 comprised five people called Simons, three claims and four addresses, all closely inter-related. They were linked across to another group at 11 o'clock comprised of more people called Simons and somebody called Wesson. That Wesson (initial A) was linked down to the group at 5 o'clock to E Wesson via a common claim. That in turn took the analyst over to the address at 4 o'clock and then to an 'A Verman' at 9 o'clock. This 'train of thought' analysis led the analyst to Verman. Following her intuition she wanted to look at Verman more closely (although she could not explain why).

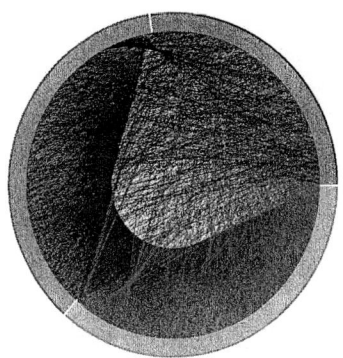

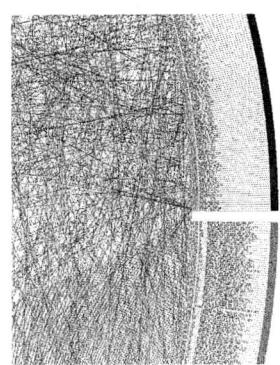

Fig. 2. An initial macro view of links between persons, addresses and claim numbers

Fig. 3. Close up at approximately 3 o'clock of the linked data shown in Fig. 2

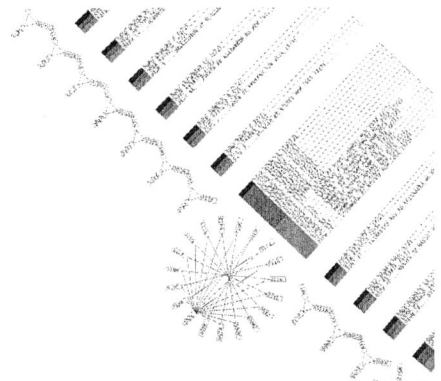

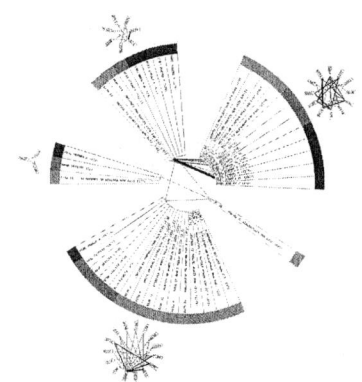

Fig. 4. 'Regular triangles' and 'irregular bumps'

Fig. 5. Emergent groups

Fig. 6. A potential suspect on the left (Verman) with all his indirect linkage to other data items

Reselection and converting patterns to knowledge: The analyst hypothesised that Verman was a few steps removed from the activity of the Simons. She firstly took Verman and stepped out to obtain all his indirect linkage (see Fig. 6). The analyst then added extra linkage (see Fig. 7, in this case, only two extra fields: bank account information and telephone numbers). The analyst quickly discovered one extra and crucial link that helped her to qualify Verman – one of his two telephone numbers was also linked to A Wesson (see Fig. 8). That additional link provided the 'tipping point', the extra knowledge that gave her sufficient confidence to recommend that Verman be investigated. This subsequently led to his conviction on fraud charges.

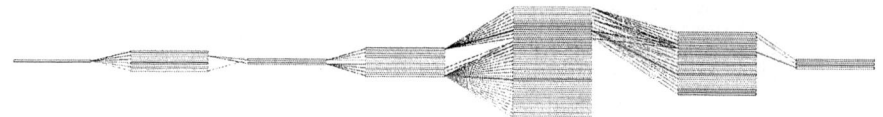

Fig. 7. Enrichment of the linkage in Fig. 6 by adding in extra fields of data

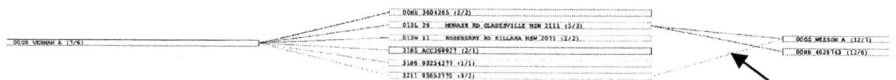

Fig. 8. Close up of portion of Fig. 7, showing the extra 'tell tale' link (arrowed: a telephone number in common)

4 Conclusion

This paper has described the concept of network data mining and presented a case study that illustrates its real-world implementation, its distinction from other analytical technics, and also its distinction from social network analysis. Network data mining involves a human-centred process which harnesses the intuitive powers of the human intellect in conjunction with unique algorithms to facilitate the intuition. The extra information in Fig. 8 led to an investigation and then the arrest and successful prosecution of Verman, who would have escaped detection with traditional exception detection methods. He had only had one claim, no 'red flag' information was involved, and nothing particularly anomalous occurred with respect to him. A complementary usage of a frequent pattern mining algorithm could have revealed a rule based on all names occurring more often than expected in the telephone population. The result being that the name Simons would be flagged. Future work aims at deeper integration of both approaches.

References

1. Nong, Y., ed. *The Handbook of Data Mining*. 2003, Lawrence Erlbaum Associates: Mahwah, New Jersey. 689.
2. Weiss, S.M. and T. Zhang, *Performance analysis and evaluation*, in *The Handbook of Data Mining*, Y. Nong, Editor. 2003, Lawrence Erlbaum Associates: Mahwah, New Jersey.
3. Albert, R. and A.-L. Barabási, *Statistical mechanics of complex networks*. Reviews of Modern Physics, 2002. **74**(January 2002): p. 47-97.
4. Newman, M.E.J., *The structure and function of complex networks*. SIAM Review, 2003. **45**: p. 167-256.
5. Borgatti, S.P., *The network paradigm in organizational research: A review and typology*. Journal of Management, 2003. **29**(6): p. 991-1013.
6. Fayyad, U.M., *Editorial*. ACM SIGKDD Explorations, 2003. **5**(2): p. 1-3.
7. Ramoni, M.F. and P. Sebastiani, *Bayesian methods for intelligent data analysis*, in *Intelligent Data Analysis: An Introduction*, M. Berthold and D.J. Hand, Editors. 2003, Springer: New York, NY. p. 131-168.
8. Schön, D., *Educating The Reflective Practitioner*. 1991, San Francisco: Jossey Bass.
9. Martin, J., *After the Internet: Alien Intelligence*. 2000, Washington, DC: Capital Press. 480.

SGPM: Static Group Pattern Mining Using Apriori-Like Sliding Window

John Goh[1], David Taniar[1], and Ee-Peng Lim[2]

[1] Monash University, Clayton, Vic 3800, Australia
{Jen.Goh, David.Taniar}@infotech.monash.edu.au
[2] Nanyang Technological University, Singapore 639798, Singapore
aseplim@ntu.edu.sg

Abstract. Mobile user data mining is a field that focuses on extracting interesting pattern and knowledge out from data generated by mobile users. Group pattern is a type of mobile user data mining method. In group pattern mining, group patterns from a given user movement database is found based on spatio-temporal distances. In this paper, we propose an improvement of efficiency using area method for locating mobile users and using *sliding window* for *static group pattern mining*. This reduces the complexity of valid group pattern mining problem. We support the use of static method, which uses areas and *sliding windows* instead to find group patterns thus reducing the complexity of the mining problem.

1 Introduction

Modern society is increasingly adopting mobile phones [15]. Mobile phone is increasing complex, and providing more user oriented services to mobile users and thus is becoming more and more beneficial to have a mobile phone [16]. Mobile phones are usually carried by a single user, and are personalized to that particular mobile user. As mobile phones now can be personalized and tracked [3, 14], it opens up a new dimension of data mining, called mobile user data mining [5, 17, 18], in which interesting knowledge can be mined from the record of the mobile user's background, places visited, and details of the places visited.

Data mining focuses on methods and algorithms in order to extract interesting patterns and knowledge from mobile users. Data mining have since been applied into different areas such as temporal domain [4, 7, 12, 13], spatial temporal domain [10, 11], and market basket analysis domain such as association rules [1, 8, 9] and sequential patterns [2].

Group pattern [17, 18] developed by Wang et al. is useful in determining grouping information over a large geographical location, a large number of mobile users and over a large duration of time series through data mining. However, one major limitation of group pattern is that it uses Euclidean distance to determine the relative proximity among mobile users. This is a method which becomes a limitation when the size of total number of mobile users through the time horizon becomes large, leading to complex dataset and reduced efficiency. The rationale behind group pattern is such that human beings physically close together over a certain time occurring frequently can be deemed as close socially [6].

In real life mobile environment there are obstacles, which will be termed as static objects for the rest of this paper. These static objects are such as things that do not move in the mobile environment. For example, walls, doors, phone booths, floors are all static objects. As group pattern uses Euclidean distance, or direct distance between two mobile users in order to determine their social proximity, the weakness is that if two mobile users is separated by a wall (i.e. between two classroom), they will be deemed to be as a close group. The result of this is that there will be more group pattern generated in the end of the process than the true number of group pattern there really is. This is because people separated by a wall are principally not close together.

2 Background

Data source for group pattern [17, 18] mining is a user movement database defined by $D = (D_1, D_2, ..., D_M)$, where D_i is a time series containing tuples $(t, (x, y, z))$ denoting the (x, y, z) values respectively of user u_i at time point t. For conformance to previous definition, we denote the location of a user u_i at time t by $u_i[t].p$ and his/her (x, y, z) values at time t by $u_i[t].x$, $u_i[t].y$, $u_i[t].z$ respectively. It is also assumed that all user locations are known at every time point and the interval between t and $t+1$ is fixed.

Definition 1. *Given a set of users G, a maximum distance threshold max_dis, and a minimum time duration threshold min_dur, a set of consecutive time points $[t_a, t_b]$ is called a **valid segment** of G, if*

1. $\forall u_i, u_j \ni G, \forall t, t_a \leq t \leq t_b, d(u_i[t].p, u_j[t].p) \leq max\_dis$;
2. $t_a = 0$ or $\exists u_i, u_j \ni G, d(u_i[t_a-1].p, u_j[t_a-1].p) > max\_dis$;
3. $t_b = N - 1$ or $\exists u_i, u_j \ni G, d(u_i[t_b+1].p, u_j[t_b+1].p) > max\_dis$;
4. $(t_b - t_a + 1) \geq min\_dur$;

The distance fuction, $d()$, is defined to return the Euclidean distance between two points, i.e., $d(u_i[t].p, u_j[t].p) =$

$$\sqrt{(u_i[t].x - u_j[t].x)^2 + (u_i[t].y - u_j[t].y)^2 + (u_i[t].z - u_j[t].z)^2}$$

Consider the user movement database in Table 1. For min_dur = 3 and max_dis = 10, [5,8] is a valid segment of the set of users, {u2, u4}.

Definition 2. *Given a database D, a group of users G, thresholds max_dis and min_dur, we say that G, max_dis and min_dur form a **group pattern**, denoted by P = < G, max_dis, min_dur >, if G has a valid segment.*

In the interest of space, algorithm AGP [17, 18] is not shown. Valid segments of the group pattern P are therefore the valid segments of its G component. Group pattern with k users is also known as **k-group pattern**. In a user movement database, a group pattern [17, 18] may have multiple valid-segments. The combined length of these valid segments is called the weight count of the pattern. We quantify the significance of the pattern by comparing its *weight count* with the overall time duration.

Definition 3. *Let P be a group pattern with valid segments $s_1, ..., s_n$, and N denotes the number of time points in the database, the **weight** of P is defined as*:

$$weight(P) = \frac{\sum_{i=1}^{n} |s_i|}{N} \quad (1)$$

If the weight of a group pattern [17, 18] exceeds a threshold *min_wei*, we call it a **valid group pattern**, and the corresponding group of users a **valid group**. For example, considering the user movement database D in Table 1, if *min_wei* = 50%, the group pattern $P = <\{u_2, u_3, u_4\}, 10, 3>$ is a valid group pattern, since it has valid segments {[1,3], [6,8]} and its weight is $6/10 \geq 0.5$.

Definition 4. *Given a database D, thresholds max_dis, min_dur, and min_wei, the problem of finding all the valid group patterns (or simply valid groups) is known as **valid group (pattern) mining**.*

3 Proposed Method: Static Group Pattern Mining (SGPM)

Group pattern [17, 18] mining is defined in Section 2. This proposal proposes a way of mining without using Euclidean distance. Euclidean distance is a formula to calculate the distance in a two dimensional space. The use of Eucilidean distance means more calculation, and also Eucilidean distance is prone to problems where two mobile users are separated by an obstacle, such as a wall. In this paper, we focuses on the issue of redefining group pattern mining, while the issue of obstacles has been proposed and addressed in another contribution.

First, we re-define how the data in mobile devices are collected. For each mobile device, it is assumed that the mobile device have some form of memory and global positioning system function, and internal system clock to determine the current time and location. In the previous proposed group pattern, data is collected as a stream for each and every second throughout the time. This automatically translates to a huge and immense amount of source data to be mined. Consider each mobile user generates a piece of coordinate (x, y) in the set of integer, the data keeps incrementing at all times. Data source for group pattern mining is a user movement database defined by $D = (D_1, D_2, ..., D_M)$, where D_i is a time series containing tuples $(t, (x, y))$ denoting the (x, y) values respectively of user u_i at time point t. For conformance to previous definition, we denote the location of a user u_i at time t by $u_i[t].p$ and his/her (x, y) values at time t by $u_i[t].x$, $u_i[t].y$ respectively. It is also assumed that all user locations are known at every time point and the interval between t and $t+1$ is fixed.

Assumption 1. *Given a mobile device $\Re$, it is assumed that $\Re$ is equipped with a location identification system, such as global positioning system where it could* determine its position in earth, or otherwise determine which room the mobile device is located in a shopping mall.

Assumption 2. *Given a mobile device $\Re$, it is assumed that $\Re$ is equipped with brief processing capability, and data recording facility. $\Re$ will roam around the mobile environment, and subsequently records down the user movement activity accordance to definition 1, and subsequently uploaded to the mobile user data mining centre when the recording facility is full, for mobile user data mining.*

Definition 5 (Location of Interest). *Given a mobile device $\Re$, duration threshold $\wp$ is defined. $\wp$ is an integer value that represents time unit, which can be second, minute or hour. It is set to a value that if a mobile user stops in a location for $\wp$ duration of time, then the mobile user has shown some interest in this particular location. If a mobile user spent more than $\wp$ in a location, that location is also known as location of interest (**LOI**).*

Definition 6 (Data Recording Conditions). *Given a mobile device $\Re$, variables t_{start}, t_{stop}, $t_{current}$, $t_{threshold}$, $v_{threshold}$, $v_{current}$ are defined. For $\Re$, in order to save processing time and storage space, user movement data is not recorded if the mobile user moving at a velocity $v_{current}$ where $v_{current} > v_{threshold}$.*

Explanation: This is because if the mobile user is travelling fast, it is unlikely that the mobile user have interest in the location, but merely travelling from one point to another. If $v_{current} < v_{threshold}$ which means that mobile user slows down or stationery for $t_{threshold}$ duration of time, then user movement recording starts, such that $t_{start} = t_{current} - t_{threshold}$. Once $\Re$ moves, where $v_{current} > v_{threshold}$ recording stops, such that $t_{stop} = t_{current}$.

Definition 7 (Movement Data Format). *Recording of user movement database will be represented in the format of: u_x : [$a_i(t_{start} : t_{stop}, ..., t_{start} : t_{stop}), ..., a_j(t_{start} : t_{stop}, ..., t_{start} : t_{stop})$] where user x ($u_x$) visited area $a_i ... a_j$ where in a_i, user u_x is present for a set of ($t_{start} : t_{stop}$) duration of time, where each $t_{start} : t_{stop}$ is such that $t_{stop} - t_{start} > t_{threshold}$.*

Example: For example: $u_1 : a_1(0 : 5, 21 : 30), a_2(6 : 10), a_3(11 : 20)$, which represent user u_1 have visited area a_1 from time $0 : 5$ and $21 : 30$, and area a_2 from time $6 : 10$ and area a_3 from time $11 : 20$. This means user u_1 have visited a_1, a_2, a_3, and back to $a1$ in sequence. We define each of this record r_n.

Definition 8 (Valid Segment). *Given a set of mobile devices $\Re$, area A, t_{start}, t_{stop}, $t_{current}$, $t_{threshold}$, $v_{threshold}$, $v_{current}$, each record r_n is called a **valid segment** of G. We wish to remove the definition of weight in previous group pattern proposal, as weight is no longer required. Given database D, threshold $t_{threshold}$, $v_{threshold}$, area $a_{current}$, time $t_{current}$, t_{start}, t_{stop}, the problem of finding all the valid group patterns (or simply valid groups) is known as valid group (pattern) mining.*

4 SGPM Mining: Algorithm ASGP

We propose Apriori-like Static Group Pattern (*ASGP*) mining algorithm for the purpose of finding all valid group patterns. *ASGP* is an algorithm for the mining problem of Static Group Pattern Mining (*SGPM*). *ASGP* utilizes sliding window concept and also *Apriori* combination generation concept in order to mine all valid group patterns. Sliding window is a window defined by the size of $t_{duration}$. Let total time in the time series be t_{total}. Sliding window will starts from $t = 0$, until $t = t_{total} - t_{threshold}$. Each slide will involve the sliding window reference time $t_{ref} = t_{ref} + 1$.

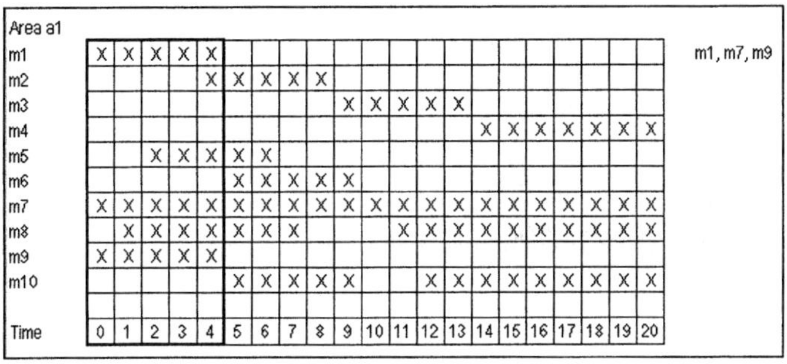

Fig. 1. Demonstration of *Sliding-Window*

Figure 1 illustrates the sliding window and the dataset in order to find all valid groups patterns. Dataset is grouped by area, which the illustration shows all mobile users ($m_1, ..., m_{10}$) who have visited area a_1 from time (0, ..., 20). For each area, (i.e. area a_1), only mobile users who have stayed in this area longer than $t_{threshold}$, is recorded through the definition in mobile devices. Sliding window is shown in $t = 0$ ... $t = 4$, where it is illustrated as a highlighted border. There are altogether 10 mobile users, ($m_1, ..., m_{10}$), and the total time ranges from $t = 0 ... t = 20$. There are 17 passes altogether. For each pass, the sliding window will examine the mobile users in the sliding on whether they have stayed in this sliding window for the total duration of time (i.e. mobile user must stay from $t = 0$ to $t = 4$ in this window to be recorded). Illustration above shows that only mobile user m_1, m_7 and m_9 satisfied this requirement, and subsequently registered. These will be recorded as a transaction t_n in each pass.

Next pass for the sliding window is to slide the window one step forward, and now the sliding window have a coverage from $t = 1$ to $t = 5$. This process is repeated until the sliding window covers from $t = 16$ to $t = 20$. For each pass, a set of mobile users who satisfied to be close at the same time for the time_threshold duration is registered. A list of them will be displayed here. We call them valid groups, as defined in the group pattern definition paper. Figure 2 illustrates.

Figure 3 shows the *support* counting for mobile users and its subsequent vertical representation of *support*. Support threshold *support*$_{threshold}$ is defined. Mobile user m

Pass 1 ($t = 1 \ldots 4$)	m_1, m_7, m_9
Pass 2 ($t = 2 \ldots 5$)	m_7, m_8
Pass 3 ($t = 3 \ldots 6$)	m_5, m_7, m_8
Pass 4 ($t = 4 \ldots 7$)	m_7, m_8
Pass 5 ($t = 5 \ldots 8$)	m_2, m_7
Pass 6 ($t = 6 \ldots 9$)	m_6, m_7, m_{10}
Pass 7 ($t = 7 \ldots 10$)	m_7
Pass 8 ($t = 8 \ldots 11$)	m_7
Pass 9 ($t = 9 \ldots 12$)	m_7
Pass 10 ($t = 10 \ldots 13$)	m_3, m_7
Pass 11 ($t = 11 \ldots 14$)	m_7
Pass 12 ($t = 12 \ldots 15$)	m_7, m_8
Pass 13 ($t = 13 \ldots 16$)	m_7, m_8, m_{10}
Pass 14 ($t = 14 \ldots 17$)	m_7, m_8, m_{10}
Pass 15 ($t = 15 \ldots 18$)	m_4, m_7, m_8, m_{10}
Pass 16 ($t = 16 \ldots 19$)	m_4, m_7, m_8, m_{10}
Pass 17 ($t = 17 \ldots 20$)	m_4, m_7, m_8, m_{10}

Fig. 2. Records of transaction for all sliding window passes

Support for Mobile Users	Vertical Representation of Support
m_1: 1	m_1: 1
m_2: 1	m_2: 5
m_3: 1	m_3: 10
m_4: 3	m_4: 15, 16, 17
m_5: 1	m_5: 3
m_6: 1	m_6: 6
m_7: 17	m_7: 1, 2, 3, 4, 5, 6, 7, 8, 9, 10, 11, 12, 13, 14, 15, 16, 17
m_8: 9	m_8: 2, 3, 4, 12, 13, 14, 15, 16, 17
m_9: 1	m_9: 1
m_{10}: 6	m_{10}: 6, 13, 14, 15, 16, 17

Fig. 3. Calculating support for mobile users and vertical representation of support

is not considered if their *support* $m_{\text{support}} <$ *support*$_{\text{threshold}}$. Let *support*$_{\text{threshold}}$ be 3, only m_4, m_7, m_8, and m_{10} will be considered. Algorithm now will proceed taking the supported mobile users to generate *k-2* itemset from (m_4, m_7, m_8, and m_{10}). The subsequent combination for *k-2* itemset are [(m_4, m_7), (m_4, m_8), (m_4, m_{10}), (m_7, m_8), (m_7, m_{10}) and (m_8, m_{10})].

Figure 4 illustrates the valid static group pattern mining (*SGPM*) process. The defined *support* = 3. It is now time to test the confidence of valid groups for high degree of *confidence*. Confidence is defined as *confidence*$_{\text{threshold}}$, and *confidence* for a particular combination of mobile user itemset such as (m_7, m_8, m_{10}) is defined as:

$$\frac{\sup port(m_7 \cap m_8 \cap m_{10})}{\sup port(m_7 \cup m_8 \cup m_{10})}$$

k-2 itemsets	k-3 itemsets	k=4 itemsets
$(m_4, m_7) : 3$	$(m_4, m_7, m_8) : 3$	$(m_4, m_7, m_8, m_{10}) : 3$
$(m_4, m_8) : 3$	$(m_7, m_8, m_{10}) : 5$	
$(m_4, m_{10}) : 3$	$(m_4, m_8, m_{10}) : 3$	
$(m_7, m_8) : 9$	$(m_4, m_7, m_{10}) : 3$	
$(m_7, m_{10}) : 4$		
$(m_8, m_{10}) : 5$		
Support = 3	Support = 3	Support = 3
∴ Select All	∴ Select All	∴ Select (m_4, m_7, m_8, m_{10})

Fig. 4. Valid static group pattern mining process demonstration

Support for maximal itemset $(m_4 \cap m_7 \cap m_8 \cap m_{10})$ is 3. Support for $(m_4 \cup m_7 \cup m_8 \cup m_{10})$ is 17. The confidence of valid group pattern (m_4, m_7, m_8, m_{10}) is *17%*. Confidence is used to confirm that within the whole time horizon for that *area* a_1, from $t = 0$ to $t = 20$, altogether 17 records generated from the sliding window, the ratio of (m_4, m_7, m_8, m_{10}) is present within the same transaction, compared to transactions containing either one of m_4, m_7, m_8 or m_{10}. Confidence is therefore, subject to the size of *time horizon*, and the *frequency* of occurrence of individual item. In the interest of space, we do not show how this problem is dealt in this paper.

Algorithm *Sliding-Window*
Input: User movement database grouped by area a_n, variable $t_{threshold}$
Output: m_{record} of mobile users who is present in the whole sliding window
```
01      result = ∅;
02      S_width = t_threshold; // defining width of sliding window
03      for (S_ref = 0; (S_ref + t_threshold != t_horizon); t ++) do begin
04           for (m_i = 1; m_i ≤ m_j; m ++) do begin
05                for (m_i.start; m_i.start < m_i.finish; m_i ++) do begin
06                     if (m_i.t_ref == ∅) skip;
07                     append(result, m_i);
08                end for
09           end for
10      end for
11      return result;
```
Fig. 5. Algorithm *Sliding-Window*

Figure 5 represents algorithm *Sliding-Window* where the sliding window is defined by $t_{threshold}$, and the program code for how the sliding window slides through the database. In order for a mobile user to be recorded, a mobile user must be within the sliding window, be present at all times from sliding window t_{ref} to $t_{ref} + t_{threshold}$. If the mobile user is not present, it will not be recorded. If the mobile user is present at all times, it will be recorded for *AGSP* algorithm.

Algorithm AGSP
Input: *result* from algorithm *Sliding-Window*, $support_{threshold}$
Output: List supported itemsets
01   R_1 = {large *r*-itemsets} // R gathered from result
02   **for** (k=2; $R_{k-1} \neq \emptyset$; k ++) do begin
03   R_k = apriori-gen(R_{k-1});
04   **for** all transactions $t \: \varepsilon \: R$ do begin
05   R_t = subset (R_k, t)
06   **for** all candidates $r \: \varepsilon \: R_t$ do begin
07   r.count++;
08   R_k = {$r \: \varepsilon \: R_k$ | $r.count \geq support_{threshold}$}
09   **return** R_k;

Fig. 6. Algorithm *AGSP*

Figure 6 shows the algorithm *AGSP* where the result from *Sliding-Window* algorithm is given in order to generate a list of frequent combinations of itemsets similar to *Apriori* algorithm. For instance, only mobile users who have $support \geq support_{threshold}$ is considered for combination generation. The process is repeated until no further combinations can be generated, and the resulting output is a combination of mobile users (m_i, ..., m_j) where they are highly supported from the result generated from *Sliding-Window* algorithm.

Resulting output is a combination of valid group pattern, where (m_i, ..., m_j) is located within the same area, near to each other, for a good duration $t_{threshold}$. This shows evidence of them being close together frequently enough within the same area and time for at least $t_{threshold}$. In order to find out the ratio of time that this combination (m_i, ..., m_j) over the total duration of records R from *sliding-window*, apply the formula of *confidence* = ($m_i, \cap ... \cap, m_j$) / ($m_i, \cup ... \cup, m_j$).

5 Evaluation

In this section, we evaluate and compare the performance between *ASGP* and *AGP* algorithms. The experiments has been conducted using synthetically generated user movement database on a Pentium IV machine with a CPU clock rate of 2.8 Ghz, and 504 MB of main memory. Note that both dataset and program are executed in main memory so that it represents execution time without bottlenecks from disk access. We compare the time it requires *AGP* algorithm and *ASGP* algorithm to access from user movement database, perform mining and generating the result of all the valid group patterns.

5.1 Dataset

Since real dataset are not available, we have implemented a synthetic user movement database generator for our experiment. Figure 7 shows the parameters used in performance evaluation for dataset *T5.I2.D1000*, *T10.I2.D1000*, *T5.I4.D1000*, *T10.I4.D1000*. Fig 8 represents the input parameters, where *D* represents the number of records, *T* represents the average size of record, and *I* represent the average size of maximal potentially large item sets.

Dataset	D	T	I	Size (MB)
T5.I2.D1000	1000	5	2	9.76
T10.I2.D1000	1000	10	2	19.53
T5.I4.D1000	1000	5	4	18.25
T10.I4.D1000	1000	10	4	39.06

Fig. 7. Dataset parameters for performance evaluation

5.2 Results

Figure 8 illustrates the evaluation results for *T5.I2.D1000, T10.I2.D1000, T5.I4.D1000, and T10.I4.D1000* respectively. It can be observed that on all occasions, algorithm *ASGP* takes shorter time to generate valid group patterns. When the *support* threshold is set to very high (i.e. 1.0) both algorithm takes roughly the same time to generate result, because there are very limited amount of candidates in the dataset for traversal. As the *support* is reduced from 1.0 to 0.1 through each decrement of 0.1, the number of potential candidates becomes larger and larger. *ASGP* takes a shorter time than *AGP* generally from *support* = 0.9 to *support* = 0.3, and after this both algorithms takes roughly the same time to generate valid group patterns. This is because *support* is low, and there are many potential candidates, and more processing time required. Nevertheless, algorithm *ASGP* still outperforms algorithm *AGP* at a varying degree, from slightly quicker for very large dataset to much quicker for a moderate sized dataset.

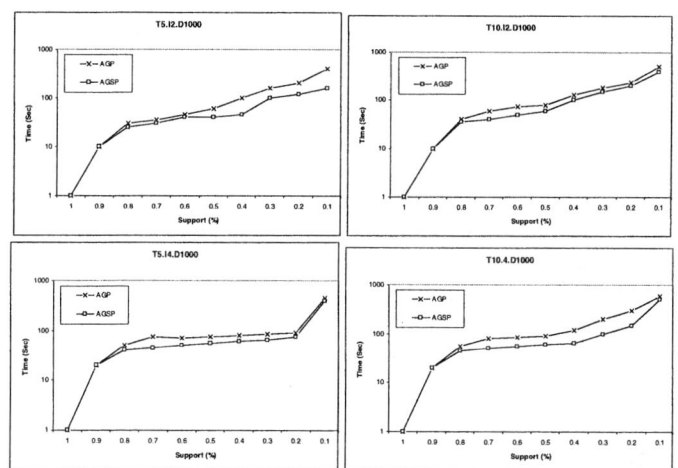

Fig. 8. Execution time required between ASGP and AGP algorithms

6 Conclusion

This paper reports an innovative redefinition of group pattern mining, called Static Group Pattern Mining (*SGPM*). The objective of this research work is to address the bottlenecks of *AGP* algorithm. Instead of using Eucilidean distance and calculate the distance for each and every pair of mobile users over the time horizon, *SGPM* uses the concept of area, sliding window and Apriori-like algorithm to find all valid groups, and valid group patterns. Performance evaluations have shown that *SGPM* have quicker execution time than *AGP* algorithm in 4 out of 4 cases. Future work from here is to further improve the execution time of valid group pattern mining problem and addressing obstacle issues in the mobile environment.

References

1. R. Agrawal and R. Srikat. Fast Algorithms for Mining Association Rules. In Proc. of the 20th VLDB, 1994.
2. R. Agrawal and R. Srikat. Mining Sequential Patterns. In Proc. of 11th ICDE, 1995.
3. B. Hofmann-Wellenhof, H. Lichtenegger, and J. Collins. Global Positioning System: Theory and Practice. Springer-Verlag Wien New York, third revised edition, 1994.
4. S. Chakrabarti, S. Sarawagi, and B. Dom. Mining Surprising Patterns using Temporal Description Length. In Proc. of 24th VLDB, 1998.
5. L. Forlizzi, R. H. Guting, E. Nardelli, and M. Schneider. A Data Model and Data Structures for Moving Objects Databases. ACM SIGMOD Record, 2000.
6. D. R. Forsyth. Group Dynamics. Wadsworth, Belmont, CA, 1999.
7. J. Han, G. Dong, and Y. Yin. Efficient Mining of Partial Periodic Patterns in Time Series Database. In Proc. of 15th ICDE, 1999.
8. J. Han, J. Pei, and Y. Yin. Mining Frequent Patterns Without Candidate Generation. In Proc. of ACM SIGMOD, 2000.
9. J. Han and A. W. Plank. Background for Association Rules and Cost Estimate of Selected Mining Algorithms, In Proc. of the 5th CIKM, 1996.
10. K. Koperski and J. Han. Discovery of Spatial Association Rules in Geographical Information Databases. In Proc of 4th Int Symp. on Advances in Spatial Databases, 1995.
11. J. F. Roddick and B. G. Lees. Paradigms for Spatial and Spatio-Temporal Data Mining. Geographic Data Mining and Knowledge Discovery, 2001.
12. J. F. Roddick and M. Spiliopoulou. A Survey of Temporal Knowledge Discovery Paradigms and Methods. IEEE Trans. on Knowledge and Data Engineering, 2002.
13. Wei Wang, Jiong Yang, and P. S. Yu. InfoMiner+: Mining Partial Periodic Patterns in Time Series Data. IEEE Transaction on Knowledge and Data Engineering, 2002.
14. Paul Zarchan. Global Positioning System: Theory and Applications, vol I. American Institute of Aeronautics and Astronautics, 1996.
15. Reed Electronics Research RER – The mobile phone industry – a strategic overview, October 2002.
16. U. Varshney, R. Vetter, and R. Kalakota. Mobile commerce: A new frontier. IEEE Computer: Special Issue on E-commerce, pages 32-38, October 2000.
17. Yida Wang, Ee-Peng Lim, and San-Yih Hwang. On Mining Group Patterns from Mobile Users. In Proc. of the 14th International Conference on Database and Expert Systems Applications – DEXA 2003, Prague, Czech Republic, 1-5 Sep, Lecture Notes in Computer Science vol. 2736, pp. 287-296, 2003.
18. Yida Wang, Ee-Peng Lim, and San-Yih Hwang. Efficient Group Pattern Mining Using Data Summarization. In Proc. of the 15th International Conference on Database and Expert Systems Applications – DEXA 2004, Beijing, China, Lecture Notes in Computer Science vol. 2973, pp. 895-907, 2004.

Mining Temporal Indirect Associations

Ling Chen[1,2], Sourav S. Bhowmick[1], and Jinyan Li[2]

[1] School of Computer Engineering,
Nanyang Technological University, 639798, Singapore
[2] Institute for Infocomm Research, 119613, Singapore

Abstract. This paper presents a novel pattern called *temporal indirect association*. An indirect association pattern refers to a pair of items that rarely occur together but highly depend on the presence of a mediator itemset. The existing model of indirect association does not consider the *lifespan* of items. Consequently, some discovered patterns may be invalid while some useful patterns may not be covered. To overcome this drawback, in this paper, we take into account the lifespan of items to extend the current model to be temporal. An algorithm, *MG-Growth*, that finds the set of mediators in *pattern-growth* manner is developed. Then, we extend the framework of the algorithm to discover temporal indirect associations. Our experimental results demonstrated the efficiency and effectiveness of the proposed algorithms.

1 Introduction

Association rule mining was initially introduced by Agrawal et al. [1]. Traditional association rules discover knowledge from frequent itemsets, i.e., a set of items frequently occur together. However, it has been noted that some of the infrequent itemsets may provide useful insight about the data as well. In [6], a particular type of patterns called *indirect associations* was proposed. A pair of items, x and y, is said to be indirectly associated via an itemset M if they rarely occur together while their respective occurrence highly depends on the presence of the itemset M.

As observed in [3], a notable feature of transaction data is that they are temporal, e.g. transaction products have *lifespan*. The current model of indirect associations does not take into account the lifetime of items, which might lead to some unfair measurement. We explain the incurred problems with the following illustrative examples.

Example 1. *Without considering the lifespan of transaction items, some discovered indirect associations may not be valid.* Figure 1 (a) shows the publication date of a set of web pages. Figure 1 (b) is an example database where each record is a set of pages visited in a web user session. Let the support threshold be 0.4. Based on traditional indirect associations, a pair of two pages is an infrequent itempair if the absolute support of the pair is less than $\lceil 12 \times 0.4 \rceil = 5$, where 12 is the size of the complete database. Since the absolute support of $\{A, E\}$ is 3 (< 5), traditional indirect association will discover indirect associations for this pair of items, via some mediators, such as $\{C\}$. However, since page E is published in Aug 05, it is unfair to compute support of itempairs containing page E with respect to the complete database, which contains records starting from Jul 05. Actually, $\{A, E\}$ is frequent with respect to the set of records from

Page	Publication Date
A	May-05
B	May-05
C	Jun-05
D	Jul-05
E	Aug-05
F	Sep-05

(a)

Session Date	Session ID	Pages
Jul-05	1	B C
	2	A B D
	3	A B C
	4	B C D
Aug-05	5	B D
	6	B C E
	7	A C D E
	8	A B C E
Sep-05	9	A C E
	10	A F
	11	B C E F
	12	A B C F

(b)

p_1, p_2, p_3

Fig. 1. Motivating example

Aug 05, e.g. absolute support of $\{A, E\}$ is $3 \geq \lceil 7 \times 0.4 \rceil = 3$. Thus, traditional indirect associations discovered for $\{A, E\}$ are not valid.

Example 2. *Without considering the lifespan of transaction items, some valid indirect associations may not be covered.* The traditional indirect association model discovers a pair of items as an indirect association pattern only if there exists an itemset M that occurs frequently together with the two items respectively. Since the pair of itemset $\{E, F\}$ in Figure 1 is infrequent, we need to search whether there exists a mediator itemset M such that E and F are indirectly associated via M. Consider the itemset $\{B, C\}$. Since the absolute support values of $\{E, B, C\}$ and $\{F, B, C\}$ are 3 and 2 respectively, both of which are less than $\lceil 12 \times 0.4 \rceil = 5$, $\{B, C\}$ will not be considered as a candidate mediator of $\{E, F\}$. However, since page E was published in Aug 05, itemset $\{E, B, C\}$ is frequent w.r.t. the set of records from Aug 05, so does itemset $\{F, B, C\}$. Thus, $\{B, C\}$ should be considered as a candidate mediator while traditional indirect association misses it.

Therefore, considering the lifespan of items, the current indirect association model is not able to discover the complete set of valid indirect association patterns. In this paper, we incorporate time in the current model of indirect associations to discover Informally, we discover a pair of items, x and y, as an indirect association pattern via a mediator M only if 1) x and y are infrequent in their *maximal common existing period*; 2) the occurrence of x (resp. y) depends on M in their maximal common existing period as well. Particularly, we call such type of patterns as *temporal indirect associations*. Temporal indirect associations are useful in the applications of traditional indirect associations, such as competitive product analysis [6] and Web usage mining [5], when the lifespan of items are taken into account.

The main contributions of this paper are summarized as follows.

- We proposed the notion of temporal indirect association considering lifespan of items.
- We designed a novel algorithm to discover indirect association patterns and extended the framework of the algorithm to discover temporal indirect association patterns.
- We implemented the developed algorithms and conducted extensive experiments to evaluate the performance of the algorithms.

2 Problem Statement

Considering the time factor, each transaction item is associated with a lifetime. Similar to the definition in [4], we associate each item with a starting time but no ending time as most applications are interested in existing items. Thus, we define a temporal transaction database as follows.

Definition 1 (Temporal Transaction Database). *Let $P = <p_1, \cdots, p_n>$ be a sequence of continuous time periods such that each period is a particular time granularity, e.g. month, quarter, year etc. $\forall\ 1 \leq i \leq j \leq n$, p_i occurs before p_j, denoted as $p_i \leq p_j$. Given a temporal item x, its starting period is denoted as $S(x)$. Given a temporal itemset X, $S(X) = \max(\{S(x)\})$, where $x \in X$. Let I be a set of temporal items s.t. $\forall x \in I, S(x) \leq p_n$. Let T be a temporal transaction, $T \subseteq I$. The occurring period of T is denoted as $O(T)$. Then, $D = \{T|p_1 \leq O(T) \leq p_n\}$ is temporal transaction database on I over P.*

For example, Figure 1 (b) is a temporal transaction database D over three periods, $P = <p_1, p_2, p_3>$, in accordance with the "month" granularity. $I = \{A, B, C, D, E, F\}$, where each item is associated with a starting period. For example, $S(F) = p_3$. Each transaction in D is also associated with an occurring period. For example, for the 8th transaction $T = \{A, B, C, E\}$, $O(T) = p_2$.

For the purpose of incorporating lifespan of items, the measures involved in traditional indirect association, *support* and *dependence* [6], need to be extended to be temporal. We now define the temporal measures as follows.

Definition 2 (Temporal Support). *Let D be a temporal transaction database on I over $P = <p_1, \cdots, p_n>$. Let X be a set of temporal items, $X \subseteq I$. The temporal support of X with respect to the subset of D from the period p_i, denoted as $TSup(X, p_i)$, is defined as:*

$$TSup(X, p_i) = \frac{|\{T|X \subseteq T, O(T) \geq p_i, T \in D\}|}{|\{T|O(T) \geq p_i, T \in D\}|}$$

Then the temporal support of X, denoted as $TSup(X)$, can be computed as $TSup(X, S(X))$.

That is, the temporal support of an itemset X is the ratio of the number of transactions that support X to the number of transactions that occur from the starting period of X. For example, consider the temporal transaction database in Figure 1. Let $X = \{B, C, E\}$. Then, $S(X) = p_2$ (because of E). $TSup(X) = 3/7$ since it is supported by three transactions while there are seven transactions starting from p_2.

Definition 3 (Temporal Dependence). *Let D be a temporal transaction database on I over $P = <p_1, \cdots, p_n>$. Let X, Y be two temporal itemsets, $X \subseteq I, Y \subseteq I$. The temporal dependence between X and Y, denoted as $TDep(X,Y)$, is defined as:*

$$TDep(X,Y) = \frac{TSup(X \cup Y)}{\sqrt{TSup(X, S(X \cup Y))TSup(Y, S(X \cup Y))}}$$

Since the correlation between two attributes makes sense only when both attributes exist, we calculate the probability of X and Y (in the denominator) with respect to the subset of D from the period where $X \cup Y$ starts. Similar to the traditional definition of dependence in [6], the value of temporal dependence ranges from 0 to 1. The higher the value of temporal dependence, the more positive correlation between the two itemsets. For example, consider the two temporal itemsets $X = \{B,C\}$ and $Y = \{E\}$ in Figure 1. As computed above, $S(X \cup Y) = p_2$, $TSup(X \cup Y) = 3/7$. Since $TSup(X,p_2)$ is 4/7 and $TSup(Y,p_2)$ is 5/7, the $TDep(X,Y) = \frac{3/7}{\sqrt{4/7 \times 5/7}} = 0.67$.

Based on the temporal support and temporal dependence extended above, the temporal indirect association can be defined as follows.

Definition 4 (Temporal Indirect Association). *A temporal itempair $\{x,y\}$ is a temporal indirect association pattern via a temporal mediator M, denoted as $< x,y|M >$, if the following conditions are satisfied:*

1. $TSup(\{x,y\}) < t_s$ *(Itempair Support Condition).*
2. $TSup(\{x\} \cup M) \geq t_f$, $TSup(\{y\} \cup M) \geq t_f$ *(Mediator Support Condition).*
3. $TDep(\{x\},M) \geq t_d$, $TDep(\{y\},M) \geq t_d$ *(Mediator Dependence Condition).*

where t_s, t_f, t_d are user defined itempair support threshold, mediator support threshold and mediator dependence threshold respectively.

For example, consider the pair of temporal items $\{E,F\}$ in Figure 1. Let user defined thresholds t_s, t_f, t_d be 0.4, 0.4 and 0.6 respectively. Since $TSup(\{E,F\}) = 1/3 < 0.4$, $\{E,F\}$ is an infrequent itempair. Consider $\{B,C\}$ as a candidate mediator. $TSup(\{E,B,C\}) = 3/7 \geq 0.4$, $TSup(\{F,B,C\}) = 2/3 \geq 0.4$. Meanwhile, $TDep(\{E\},\{B,C\}) = 0.67 \geq 0.6$ and $TDep(\{F\},\{B,C\}) = 0.82 \geq 0.6$. Thus, $< E,F|\{B,C\} >$ is a temporal indirect association pattern.

Problem Statement. Let D be a temporal transaction database over a sequence of time periods $P = <p_1,\cdots,p_n>$. Given user defined thresholds t_s, t_f and t_d, the problem of **temporal indirect association mining** is to discover the complete set of patterns s.t. each pattern $< x,y|M >$ satisfies the conditions: 1) $TSup(\{x,y\}) < t_s$; 2) $TSup(\{x\} \cup M) \geq t_f$, $TSup(\{y\} \cup M) \geq t_f$; 3) $TDep(\{x\},M) \geq t_d$, $TDep(\{y\},M) \geq t_d$.

3 Algorithm

In this section, we discuss the algorithm for temporal indirect association mining. We first present a novel algorithm for indirect association mining and then extend it to support temporal transaction database.

3.1 Indirect Association Mining

An algorithm called *HI-Mine* was proposed in [7] to use the *divide-and-conquer* strategy to discover mediators. However, *HI-Mine* generates a complete set of mediators for each item x although some of the mediators are useless, e.g. there exists no item y such

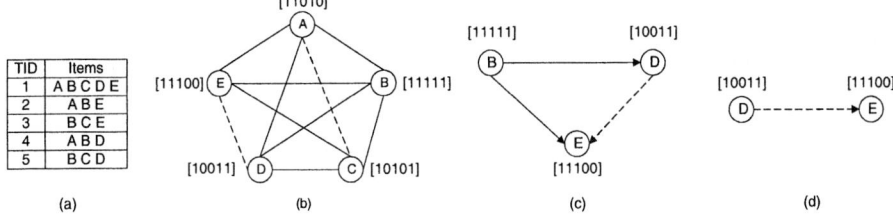

Fig. 2. Indirect association mining

that $\{x, y\}$ is infrequent and y depends on these mediators as well. Our algorithm addresses this problem by generating a mediator only if there exists an infrequent itempair such that both items depend on it.

Basically, we first construct a *frequency graph* which is used to find *infrequent itempairs* and items that are *possible mediators* of each infrequent itempair. For each infrequent itempair, we then construct a *mediator graph* with these possible mediator items. Then, the complete set of mediators for the infrequent itempair will be generated from the mediator graph.

We use a vertical bitmap representation for the database. For example, consider the transaction database in Figure 2 (a). The bitmap for item A is $[11010]$. Then a *frequency graph* can be defined as follows (For the clarity of exposition, we assume $t_s = t_f$ in the following. The algorithm in Figure 3 explains the situation when $t_s \neq t_f$. Let t_s and t_f be absolute support threshold).

Definition 5 (Frequency Graph). *Given a database D on itemset I, and the user defined mediator (itempair) support threshold t_f, a frequency graph, denoted as $FG = (N, E)$, can be constructed such that N is a set of nodes representing frequent items $\{x | b(x) \geq t_f, x \in I\}$ and E is a set of edges representing itempairs. Each node x is associated with the bitmap $b(x)$. Each edge (x, y) is frequent if $b(x) \cap b(y) \geq t_f$. Otherwise, it is infrequent.*

For example, let the threshold t_f be 2. All individual items in the database in Figure 2 (a) are frequent and the constructed *frequency graph* is shown in Figure 2 (b) where infrequent edges are drawn in dashed lines.

Traverse edges in a *frequency graph*. For each infrequent edge, which corresponds to an infrequent itempair, we collect a set of *candidate mediator nodes*.

Definition 6 (Candidate Mediator Node). *Given a frequency graph $FG = (N, E)$, for an infrequent edge $(x, y) \in E$, its candidate mediator nodes, denoted as $MN(x, y)$, is a set of nodes: $\{n | b(n) \cap b(x) \geq t_f, b(n) \cap b(y) \geq t_f, n \in N\}$.*

For example, for the infrequent edge (A, C) in Figure 2 (b), $MN(A, C) = \{B, D, E\}$. Then, a *mediator graph* for an infrequent edge can be constructed with the set of candidate mediator nodes.

Definition 7 (Mediator Graph). *Given a frequency graph FG and an infrequent edge (x, y), a mediator graph created for (x, y) is a directed graph, denoted as $MG(x, y) = (N, E)$, where N is a set of nodes such that $N = MN(x, y)$ and E is a set of directed*

(a) MG-Growth

Input: Database D, t_s, t_f and t_d
Output: The complete set of indirect associations S
Description:
1: Scan D to find $F_1 = \{x|Sup(x) \geq t_f\}$.
2: Construct the frequency graph FG with F_1.
3: **for** each edge (x, y) in FG **do**
4:    **if** $Sup(x, y) < t_s$ **then**
5:      Construct mediator graph $MG(x, y)$
6:      **if** $MG(x, y) \neq \emptyset$ **then**
7:        MGrowth($MG(x, y), M, 0, C$)
8:        $S = S \cup C$
9:      **end if**
10:   **end if**
11:   **return** S
12: **end for**
13: **function** MGrowth($MG(x, y), M, dep, C$)
14:   **for** each node n in $MG(x, y)$ **do**
15:     $M[dep] = n; dep ++$
16:     **if** $Sup(n, x) \geq t_f$ && $Dep(n, x) \geq t_d$
       && $Sup(n, y) \geq t_f$ && $Dep(n, y) \geq t_d$
       then
17:        $C = C \cup \{<x, y|M>\}$
18:     **end if**
19:     Construct conditional mediator graph $MG_n(x, y)$
20:     **if** $MG_n(x, y) \neq \emptyset$ **then**
21:        MGrowth($MG_n(x, y), M, dep, C$)
22:     **end if**
23:     $dep --$
24:   **end for**
25: **end function**

(b) TMG-Growth

Input: Temporal transaction database D, t_s, t_f and t_d
Output: The complete set of indirect associations S
Description:
1: Scan D to find $F_1 = \{x|TSup(x) \geq t_f\}$.
2: Construct the frequency graph FG with F_1.
3: **for** each edge (x, y) $s.t. S(x) = p_i, S(y) = p_j$
   in FG **do**
4:    **if** $TSup(x, y) < t_s$ **then**
5:      Construct mediator graphs
      $\{MG^{p_i}(x, y), \cdots, MG^{p_n}(x, y)\}$
6:      **for** each graph $MG^{p_k}(x, y) \neq \emptyset$ **do**
7:        TMGrowth($MG^{p_k}(x, y), M, 0, C$)
8:        $S = S \cup C$
9:      **end for**
10:   **end if**
11:   **return** S
12: **end for**
13: **function** TMGrowth($MG^{p_k}(x, y), M, dep, C$)
14:   **for** each node n in $MG^{p_k}(x, y)$ **do**
15:     **if** $dep == 0$ && n is non-extendable **then**
16:        return;
17:     **end if**
18:     $M[dep] = n; dep ++$
19:     **if** $TSup(n, x) \geq t_f$ && $TDep(n, y) \geq t_d$
       && $TSup(n, y) \geq t_f$ && $TDep(n, y) \geq t_d$
       then
20:        $C = C \cup \{<x, y|M>\}$
21:     **end if**
22:     Construct $MG_n^{p_k}(x, y)$
23:     **if** $MG_n^{p_k}(x, y) \neq \emptyset$ **then**
24:        TMGrowth($MG_n^{p_k}(x, y), M, dep, C$)
25:     **end if**
26:     $dep --$
27:   **end for**
28: **end function**

Fig. 3. Algorithms of *MG-Growth* and *TMG-Growth*

edges. Each node n is associated with a bitmap $b(n)$ as in FG. Each edge $(m \rightarrow n)$, originating from m if m precedes n according to lexicographical order, is frequent if $b(m) \cap b(n) \geq t_f$.

For example, the mediator graph constructed for infrequent edge (A, C) is shown in Figure 2 (c). Likewise, infrequent edges are shown in dashed lines.

From the mediator graph $MG(A, C)$, we now present how to compute the set of mediators for infrequent itempair $\{A, C\}$. Let the threshold of support be 0.4 and threshold of dependence 0.6. We first consider the candidate mediator node B. $support(\{A, B\}) = 3/5$ because $b(A) \cap b(B) = 3$. $dependence(A, B) \frac{support(\{A,B\})}{\sqrt{support(A) \times support(B)}} = \frac{3}{\sqrt{3 \times 5}} = 0.77$. The support and the dependence between C and B can be calculated similarly and we discover an indirect association pattern $<A, C|\{B\}>$.

The remaining nodes in the mediator graph that have frequent edges originating from node B consist of $B's$ *conditional mediator base*, from which we construct $B's$ *conditional mediator graph*. For each node n in the conditional mediator graph of node B, its bitmap is updated by joining with the bitmap of node B. After that, each edge $(m \rightarrow n)$ is frequent if $b(m) \cap b(n) \geq t_f$. For example, Figure 2 (d) shows the

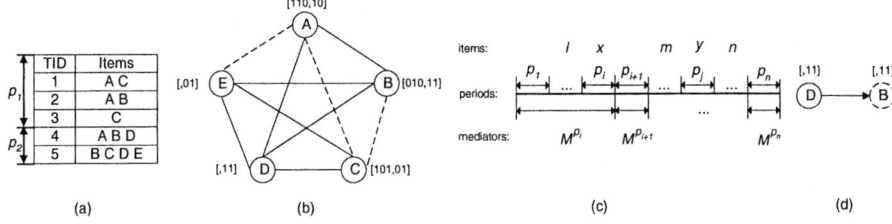

Fig. 4. Temporal transaction database

conditional mediator graph of node B. Then, we compute the mediators involving B, such as $\{BD\}$ and $\{BE\}$, for itempair $\{A, C\}$. Similarly, the support and dependence between A and $\{BD\}$ can be calculated by joining $b(A)$ with $b(D)$ (Note that $b(D)$ represents the support of $\{BD\}$ now) while the support and dependence between C and $\{B, D\}$ can be computed with $b(C) \cap b(D)$. The complete algorithm, *MG-Growth*, is given in Figure 3 (a).

3.2 Temporal Indirect Association Mining

Based on the measure of *temporal support*, a frequency graph consisting of frequent items can be constructed similarly. For example, let the threshold of temporal support be 0.4. The constructed frequency graph is shown in Figure 4 (b).

Before discussing how to construct a mediator graph for an infrequent itempair, we highlight that the downward closure property does not hold for mediator discovery in temporal indirect association mining, e.g even if B is not a mediator of the infrequent itempair $\{A, C\}$, it is possible that $\{BD\}$ is a mediator of $\{A, C\}$. Hence, in order to discover the complete set of mediators for each infrequent itempair, we divide the set of mediators according to their lifespan. Given a sequence of periods $P = <p_1, \cdots, p_n>$ as shown in Figure 4 (c), the complete set of mediators M of an infrequent itempair $\{x, y\}$, where $S(x) = p_i$ and $S(y) = p_j$ $(p_i \leq p_j)$, can be divided into $n - i + 1$ subsets as shown in the figure: $M = M^{p_i} \cup M^{p_{i+1}} \cup \cdots \cup M^{p_n}$, where $M^{p_i} = \{X | X \in M, S(X) \leq p_i\}$ and $\forall p_{i+1} \leq p_k \leq p_n, M^{p_k} = \{X | X \in M, S(X) = p_k\}$. When discovering mediators of M^{p_i}, we use the two corresponding subsets of database as counting bases (for computing temporal support and temporal dependence of x and mediators, y and mediators respectively). We create different temporal mediator graphs for discovering different subsets of mediators.

Consider the frequency graph in Figure 4 (b). We now explain how to discover mediators for the infrequent edge (A, C) where $S(A) = S(C) = p_1$. First, we construct the mediator graph for mining M^{p_1}, which involves item B only. Since edge (B, C) is infrequent, there is no candidate mediator nodes and the graph is empty. Then, we construct the mediator graph for mining M^{p_2}, which involves items D and B because the edge (B, C) turns to be frequent with respect to the subset of database from p_2. Note that, D is an extendable mediator node while B is non-extendable[1]. The constructed

---
[1] See the definitions of extendable and non-extendable mediator nodes in our online version [2].

mediator graph is shown in Figure 4 (d), where non-extendable nodes are depicted in dashed lines. From this graph, we recursively examine whether $\{D\}$ and $\{D, B\}$ are mediators of $\{A, C\}$. The algorithm for mining temporal indirect associations is shown in Figure 3 (b).

4 Performance Evaluation

In this section, we evaluate the performance of developed algorithms. All experiments are conducted on a 2GHz P4 machine with 512M main memory, which runs Microsoft Windows XP. All the algorithms are implemented in C++. In order to obtain comparable experimental results, the method we employed to generate synthetic datasets is similar to the one used in prior works [7]. Without loss of generality, we use the notation $Tx.Iy.Dz$ to represent a data set where the number of transactions is z, the average size of transaction is x and the average size of potentially large itemsets is y. Additionally, we use the notation $Tx.Iy.Dz.Pn$ to represent a temporal transaction database which is over a sequence of n periods.

Comparison of *MG-Growth* and *HI-Mine*. we compare the performance of *MG-Growth* with *HI-Mine*, which is the clear winner of the other existing algorithms [7]. We ran experiments on two datasets: $T10.I5.D10K$ and $T10.I5.D20K$. The threshold of t_s and t_f are set as the same. The threshold of t_d is set as 0.1. The results are shown in Figure 5. *MG-Growth* is more efficient than *HI-Mine*, especially when t_f (t_s) is small. This is because when the threshold is small, there are more frequent individual items. Consequently, *HI-Mine* needs to discover all the set the mediators for more items no matter whether these mediators are useful or not. On the contrary, *MG-Growth* discovers a mediator only if it is depended on by an infrequent itempair. Thus, the performance of *MG-Growth* will not deteriorate significantly with the decrease of mediator (itempair) support threshold.

We further examine the scale-up feature of *MG-Growth*. Figure 5 (c) shows the results with the variation of data size from $200K$ to $1M$. The scale-up performance under two different thresholds of t_f are studied. The execution times are normalized with respect to the execution time for the data set of $200K$. We observed that the run time of *MG-Growth* increases slightly with the growth of data size, which demonstrated the good scalability of *MG-Growth*.

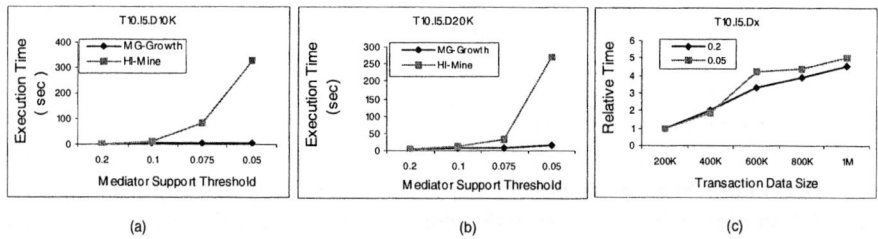

Fig. 5. Experimental Results I

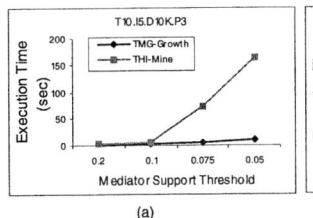 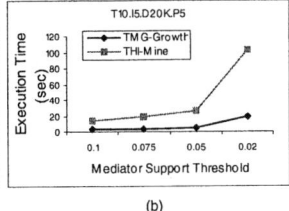 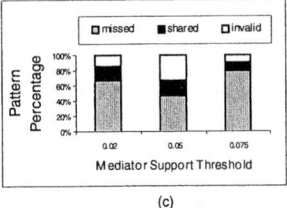

Fig. 6. Experimental Results II

Comparison of *TMG-Growth* and *THI-Mine*. In order to evaluate the performance of the temporal version of *MG-Growth*, *TMG-Growth*, we also extend the *HI-Mine* to support temporal transaction database [2]. Correspondingly, we denote the temporal version of *HI-Mine* as *THI-Mine*. We compare the performance of *TMG-Growth* and *THI-Mine* with respect to two datasets: $T10.I5.D10K.P3$ and $T10.I5.D20K.P5$. Figures 6 (a) and (b) present the results respectively. Obviously, the temporal version of *MG-Growth* outperforms the temporal version of *HI-Mine* as well. When the number of periods increases, the gap between the two algorithms is apparent even if the mediator support threshold is large.

We evaluate the quality of temporal indirect association patterns by comparing the results of the traditional model and the temporal model on the same temporal transaction database. Figure 6 (c) shows the results with respect to the variation of t_f threshold, where black blocks depict the percentage of patterns shared by two models, white blocks depict the percentage of patterns missed by the traditional model and the gray blocks depict the percentage of invalid patterns. It can be observed that the set of temporal indirect association patterns is significantly different from the results of the traditional model.

5 Conclusions

In this paper, we take into account the lifespan of items to explore a new model of temporal indirect association. We first develop an algorithm *MG-Growth* for indirect association mining. Under *MG-Growth*, a set of mediators are generated only if both items in an infrequent itempair depend on them. Then, we extend the framework of *MG-Growth* so that mediators starting from different periods are discovered separately. Our experimental results showed that *MG-Growth* outperforms the existing algorithm significantly and its extended version discovers the temporal indirect association pattern efficiently.

References

1. R. Agrawal, T. Imielinski, and A. Swami. Mining association rules between sets of items in large databases. In *Proc. of ACM SIGMOD*, 1993.
2. L. Chen, S. Bhowmick, and J. Li. Temporal indirect association mining. In *http://www.cais.ntu.edu.sg/~sourav/papers/TIA_PAKDD-05.pdf*.

3. X. Chen, I. Petrounias, and H. Heathfield. Discovery of association rules in temporal databases. In *Proc. of IADT*, 1998.
4. C. Lee, M. Chen, and C. Lin. Progressive partition miner: an efficient algorithm for mining general temporal association rules. In *IEEE TKDE vol.15, no. 4*, 2003.
5. P. Tan and V. Kumar. Mining indirect associations in web data. In *Proc. of WebKDD*, 2001.
6. P. Tan, V. Kumar, and J. Srivastava. Indirect association: mining higher order dependencies in data. In *Proc. of PKDD*, 2000.
7. Q. Wan and A. An. Efficient mining of indirect associations using hi-mine. In *Proc. of Canadian Conference on AI*, 2003.

Mining Top-K Frequent Closed Itemsets Is Not in APX

Chienwen Wu

National Taipei University of Technology, Taipei 10643, Taiwan, R.O.C

Abstract. Mining top-k frequent closed itemsets was initially proposed and exactly solved by Wang et al. [IEEE Transactions on Knowledge and Data Engineering 17 (2005) 652-664]. However, in the literature, no research has ever considered the complexity of this problem. In this paper, we present a set of proofs showing that, in the general case, the problem of mining top-k frequent closed itemsets is not in **APX**. This indicates that heuristic algorithms rather than exact algorithms are preferred to solve the problem.

1 Introduction

In recent years, frequent itemset mining has been studied intensively. Conventional frequent itemset mining requires the user to specify a *min_support* threshold and aims at discovering subsets of items that occur together at least *min_support* times in a database. In practical applications [1], setting an appropriate *min_support* threshold is no easy task. If *min_support* is set to be too large, no itemsets will be generated. If *min_support* is set to be too small, an overwhelming number of itemsets may be generated. Most of the time, it needs repeated trials and errors to come up with a proper *min_support* threshold.

In order to remove this restriction, Wang et al. [1] proposed the problem of mining top-k frequent closed itemsets. As opposed to specifying a *min_support* threshold, Wang et al. [1] allows the miner to specify the desired number of interesting itemsets, which is much easier for the miner to specify.

We follow the notations in Wang et al. [1] for the problem description. Let $I = \{i_1, i_2, \ldots, i_n\}$ be a set of items. An itemset X is defined to be a nonempty subset of I. The length of an itemset X is the number of items contained in X. X is called an l-itemset if its length is l. A transaction is a tuple $<tid, X>$, where tid is a transaction identifier and X is an itemset. A transaction database TDB is a set of transactions. We say that an itemset X is contained in transaction $<tid, X>$ if $X \subseteq Y$. Given a transaction database TDB, the support of an itemset X, denoted as $sup(X)$, is the number of transactions in TDB which contain X.

An itemset X is called a closed itemset if there exists no proper superset Y of X with $sup(X)=sup(Y)$. A closed itemset X is a top-k frequent closed itemset of minimal length min_l if there exists no more than $(k$ -1$)$ closed itemsets of length at least min_l whose support is higher than that of X.

The problem is to mine top-k frequent closed itemsets of minimal length min_l in a transaction database TDB. For clarity of presentation, the problem is called the **TFCI** problem in this paper. Each instance of the **TFCI** problem is represented by $<I, TDB, min\_l, k>$.

Wang et al. [1] proposed an exact algorithm based on FP-tree [2] to solve the **TFCI** problem. An extensive performance studies had been performed by Wang et al. [1]. The results show that their algorithm offers very high performance.

In this paper, we show that, in the general case, the **TFCI** problem is not in **APX**. This implies that heuristic algorithms rather than exact algorithms are preferable to solve the **TFCI** problem in reasonable run time.

No existing work has ever considered the inapproximability issue of the **TFCI** problem. Most existing works consider the NP-completeness of mining specific types of frequent itemsets and association rules. Gunopulos et. al [3] first proved that the problem of deciding whether there is a maximal itemset of length at least t and support at least σ is NP-complete. Zaki et. al [4] further considered the complexity of several variants of the same problem. Angiulli et. al [5] considered the complexity of mining association rules with categorical and quantitative attributes. However, these works never address the inapproximability issues.

Jermaine [6] first considered the inapproximability issue of itemset mining. He showed that mining the itemset that maximizes some correlation function is not approximable. Our work is distinct from Jermaine [6] in the following two aspects: (1) we consider the top-k closed itemsets that have the best support, intead of the itemset that maximizes some correlation function (2) we not only show that the **TFCI** problem is not approximable but also strengthen the inapproximability result by showing that the **TFCI** problem is not in **APX**.

2 The Non-APX Result

In this section, we will focus on proving that mining top-1 frequent closed itemset is not in **APX**, which immediately implies that mining top-k frequent closed itemsets is not in **APX**.

We will show that if, for some constant $r \geq 1$, there is a polynomial-time r-approximation algorithm for mining top-1 frequent closed itemset, we can solve the **CLIQUE** problem in polynomial-time, a contradiction to the fact that the **CLIQUE** problem is NP-Complete [7]. The **CLIQUE** problem we consider is as follows.

CLIQUE

Instance: An undirected graph $G = (V, E)$ and an integer k.

Question: Does there exist a k-clique in G, i.e., a subset of vertices $C \subseteq V$ such that $|C| = k$ and, for any $u, v \in C$, $\{u, v\} \in E$?

For any instance y and any algorithm A, we use $OPT(y)$ to represent an optimal solution of y and use $A(y)$ to represent the solution obtained by applying A on y. We say that A is a polynomial-time r-approximation algorithm for mining top-1 frequent closed itemset if, for any instance y and a constant $r \geq$

1, A when applied to input (y, r) returns an approximate solution $A(y)$ of y in time polynomial in $|y|$ such that $\frac{sup(OPT(y))}{sup(A(y))} \leq r$.

We now suppose that we are given a **CLIQUE** problem instance $x = <G, k>$ where $G = (V, E)$, $V = \{v_1, v_2, \ldots, v_n\}$, and $E = \{e_i \mid e_i \subseteq V \text{ and } |e_i| = 2, 1 \leq i \leq m\}$. Also we suppose that we are given a polynomial-time r-approximation algorithm A for mining top-1 frequent closed itemset for some constant $r \geq 1$. We construct from x a **TFCI** problem instance $y = <V, TDB, n-k, 1>$, where the transaction database $TDB = \{<(i_1, i_2, \ldots, i_a), (V - \bigcup_{j=1}^{a} e_{i_j})> \mid 1 \leq i_j \leq m$ for all $1 \leq j \leq a\}$ and $a = \left\lfloor \frac{\log r}{\log(\frac{k(k-1)}{2}) - \log(\frac{k(k-1)}{2} - 1)} \right\rfloor + 1$.

In the next, we show that we can determine whether there is a k-clique in G in polynomial time by checking if $sup(A(y)) \geq (\frac{k(k-1)}{2})^a \times \frac{1}{r}$.

We begin by introducing some useful definitions and lemmas. Let S be an arbitrary subset of V. We define $e(S)$ as the number of edges in E that connect vertices in S. Formally, $e(S) = |\{i \mid e_i \subseteq S\}|$.

Consider the following lemma that relates $sup(V - S)$ and $e(S)$.

Lemma 1. $sup(V - S) = (e(S))^a$.

Proof. $sup(V - S) = |\{<(i_1, \ldots, i_a), (V - \bigcup_{j=1}^{a} e_{i_j})> \mid 1 \leq i_j \leq m$ for $1 \leq j \leq a$ and $(V-S) \subseteq (V - \bigcup_{j=1}^{a} e_{i_j})\}|$

$= |\{<(i_1, \ldots, i_a), (V - \bigcup_{j=1}^{a} e_{i_j})> \mid 1 \leq i_j \leq m$ for $1 \leq j \leq a$ and $(\bigcup_{j=1}^{a} e_{i_j}) \subseteq S\}|$

$= |\{<(i_1, \ldots, i_a), (V - \bigcup_{j=1}^{a} e_{i_j})> \mid e_{i_j} \subseteq S$ for $1 \leq i_j \leq m, 1 \leq j \leq a\}|$

$= |\{i \mid e_i \subseteq S\}|^a$

$= e(S)^a$

Lemma 2 presents some useful property of closed itemset.

Lemma 2. *Let X be an arbitrary itemset. Then, there exists an itemset Y such that Y is a closed itemset, $X \subseteq Y$ and $sup(X) = sup(Y)$.*

Proof. Without loss of generality, let $U = \{Y \mid X \subseteq Y$ and $sup(X) = sup(Y)\}$. We will prove the lemma by showing that some itemset in U is closed. The proof is by contradiction.

We note that $U \neq \emptyset$ because $X \in U$. We assume that every itemset in U is not closed. Let Y be an arbitrary itemset in U. Since Y is not closed, by the definition of closed itemset, there exists an itemset Z such that $Y \subset Z$ and $sup(Y) = sup(Z)$. Hence, $X \subseteq Y \subset Z$ and $sup(X) = sup(Y) = sup(Z)$, which mean $Z \in U$.

This implies that every itemset in U has a proper superset in U, a falsity. This completes the proof.

Lemma 3. $(\frac{k(k-1)}{2} - 1)^a < (\frac{k(k-1)}{2})^a \times \frac{1}{r}$.

Proof. The lemma is a direct result of $a > \frac{\log r}{\log(\frac{k(k-1)}{2}) - \log(\frac{k(k-1)}{2} - 1)}$, which is derived from $a = \left\lfloor \frac{\log r}{\log(\frac{k(k-1)}{2}) - \log(\frac{k(k-1)}{2} - 1)} \right\rfloor + 1$.

Based on the above three lemmas, we show that we can determine whether there is a k-clique in G in polynomial time by checking if $sup(A(y)) \geq (\frac{k(k-1)}{2})^a \times \frac{1}{r}$. Theorem 1 and Theorem 2 are provided for this fact.

Theorem 1. *If the graph G has a clique of size k, then $sup(A(y)) \geq (\frac{k(k-1)}{2})^a \times \frac{1}{r}$.*

Proof. Let S be an arbitrary clique of size k in G. Obviously, $e(S) = \frac{k(k-1)}{2}$. By Lemma 1, $sup(V-S) = (\frac{k(k-1)}{2})^a$.

By Lemma 2, there exists an itemset Y such that Y is a closed itemset, $(V-S) \subseteq Y$ and $sup(Y) = sup(V-S)$. Since $|Y| \geq |V-S| = n-k$, Y is a feasible solution of y. Therefore, $sup(OPT(y)) \geq sup(Y) = sup(V-S) = (\frac{k(k-1)}{2})^a$.

Since A is a polynomial-time r-approximation algorithm for the instance y, we have $\frac{sup(OPT(y))}{sup(A(y))} \leq r$. Hence, $sup(A(y)) \geq sup(OPT(y)) \times \frac{1}{r} \geq (\frac{k(k-1)}{2})^a \times \frac{1}{r}$. This concludes the proof.

Theorem 2. *If the graph G has no clique of size k, then $sup(A(y)) < (\frac{k(k-1)}{2})^a \times \frac{1}{r}$.*

Proof. Since $A(y)$ is a solution of y, we have $|A(y)| \geq n-k$ and $|V - A(y)| \leq k$. Since there is no clique of size k in G, $e(V - A(y)) \leq (\frac{k(k-1)}{2} - 1)$. By Lemma 1 and Lemma 3, we have $sup(A(y)) = (e(V - A(y)))^a \leq (\frac{k(k-1)}{2} - 1)^a < (\frac{k(k-1)}{2})^a \times \frac{1}{r}$. This completes the proof.

Based on Theorem 1 and Theorem 2, if we had a polynomial-time r-approximation algorithm A for top-1 frequent closed itemset mining for some constant r, we could use it to decide whether G has a k-clique as follows: we apply the polynomial-time r-approximation algorithm A on the constructed instance y of top-1 frequent closed itemset mining corresponding to G and we answer yes if and only if $sup(A(y)) \geq (\frac{k(k-1)}{2})^a \times \frac{1}{r}$. However, this contradicts to the fact that the **CLQIUE** problem is NP-Complete [7]. The next two theorems immediately follow.

Theorem 3. *The problem of mining top-1 closed itemset is not in* **APX** *unless* **P = NP**.

Theorem 4. *The* **TFCI** *problem is not in* **APX** *unless* **P = NP**.

3 Conclusions

We have provided in this paper a set of proofs showing that there is no **APX** for mining top-k frequent closed itemsets. The result indicates that heuristic algorithms, instead of exact algorithms, are preferred to solve the problem.

Acknowledgements

The author would like to thank the anonymous referees for their helpful comments. This research was supported by the National Science Council of the Republic of China under the grant NSC 94-2213-E-027-030.

References

1. Wang, J., Han, J., Lu, Y., Tzvetkov P.: TFP: An Algorithm for Mining Top-K Frequent Closed Itemsets. IEEE T. KNOWL. DATA EN. **17** (2005) 652–664
2. Han, J., Pei, J., Yin, Y., Mao, R.: Mining Frequent Patterns without Candidate Generation: A Frequent-pattern Tree Approach. DATA MIN. KNOWL. DISC. **8** (2004) 53–87
3. Gunopulos, D., Khardon, R., Mannila, H., Saluja, S.: Discovering All Most Specific Sentences. ACM T. on DATABASE SYST. **28** (2003) 140-174
4. Zaki, M. and Ogihara, M.: Theoretical foundations of association rules. in: Proceedings of Third SIGMOD'98 Workshop on Research Issues in Data Mining and Knowledge Discovery, Seattle, USA. (1998) 71-78
5. Angiulli, F., Ianni, G., Palopoli L.: On the Complexity of Inducing Categorical and Quantitative Association Rules. THEOR. COMPUT. SCI. **314** (2004) 217-249
6. Jermaine, C.: Finding the Most Interesting Correlations in a Database: How Hard Can it Be?. INFORM. SYST. **30** (2005) 21–46
7. Ausiello, G., Crescenzi, P., Gambosi, G., Kann, V., Marchetti-Spaccamela, A. and Protasi, M.: Complexity and Approximation: Combinatorial Optimization Problems and Their Approximability Properties, Springer Verlag. (1999)

Quality-Aware Association Rule Mining

Laure Berti-Équille

IRISA, Campus Universitaire de Beaulieu,
Rennes 35042, France
berti@irisa.fr

Abstract. The quality of discovered association rules is commonly evaluated by interestingness measures (commonly support and confidence) with the purpose of supplying subsidies to the user in the understanding and use of the new discovered knowledge. Low-quality datasets have a very bad impact over the quality of the discovered association rules, and one might legitimately wonder whether a so-called "interesting" rule noted LHS -> RHS is meaningful when 30 % of LHS data are not up-to-date anymore, 20% of RHS data are not accurate, and 15% of LHS data come from a data source that is well-known for its bad credibility. In this paper we propose to integrate data quality measures for effective and quality-aware association rule mining and we propose a cost-based probabilistic model for selecting legitimately interesting rules. Experiments on the challenging KDD-CUP-98 datasets show for different variations of data quality indicators the corresponding cost and quality of discovered association rules that can be legitimately (or not) selected.

1 Introduction

Quality in data mining results critically depends on the preparation and on the quality of analyzed datasets [10]. Indeed data mining processes and applications require various forms of data preparation, correction and consolidation combining complex data transformation operations and cleaning techniques [11], because the data input to the mining algorithms is assumed to conform to "nice" data distributions, containing no missing, inconsistent or incorrect values [15]. This leaves a large gap between the available "dirty" data and the available machinery to process and analyze the data for discovering added-value knowledge and decision making [9]. Data quality is a multidimensional, complex and morphing concept [4]. Since a decade, there has been a significant amount of work in the area of information and data quality management initiated by several research communities (database, statistics, workflow management, knowledge management), ranging from techniques in assessing information quality [13] to building large-scale data integration systems over heterogeneous data sources with different degrees of quality and trust. In error-free data warehouses or database-backed information systems with perfectly clean data, knowledge discovery techniques (such as clustering, mining association rules or visualization) can be relevantly used as decision making processes to automatically derive new knowledge patterns and new concepts from data. Unfortunately, most of the time, these data are neither rigorously chosen from the various heterogeneous sources with different degrees of quality and trust, nor carefully controlled for quality [9]. Deficiencies in data quality still are a burning issue in many application areas, and become acute for practical applications of knowledge discovery and data mining techniques [5]. We illustrate this idea with the following example in the context of association rule mining. Among traditional descriptive data mining techniques, association rule mining identifies intra-transaction patterns in a database and describes how much the presence of a set of attributes in a database's record (*i.e.*, a transaction) implicates the presence of other distinct set of attributes in the same record (respectively the same transaction). The quality of discovered association rules is commonly evaluated by interestingness measures (namely support and confidence). The support of a rule measures the occurrence frequency of the pattern in the rule while the confidence is the measure of the strength of implication. The problem of mining association rules is to generate all association rules that have support and confidence greater than the user-specified minimum support and confidence thresholds. Besides support and confidence, other interestingness measures have been proposed in the literature for knowledge quality evaluation with the purpose of supplying subsidies to the user in the understanding and use of the new discovered knowledge [12], [7]. But, to illustrate the impact of low-quality data over discovered association rule quality, one might legitimately wonder whether a so-called "interesting" rule noted $LHS \rightarrow RHS$ is meaningful when 30 % of *LHS* data are not up-to-date anymore, 20% of *RHS* data are not accurate, and 15% of *LHS* data come from a data source that is well-known for its bad credibility. Our assumption is that interestingness measures are not self-sufficient for representing association rule quality. Association rule quality should also integrate the measures of the quality of data the rule is computed from with considering the probability that the deficiencies in data quality may be adequately detected. The twofold contribution of this paper is to propose a method for scoring association rule quality and a probabilistic cost model that predicts the cost of low-quality data over the quality of discovered association rules. This model is used to select so-called "legitimately interesting" rules. We evaluate our approach using the KDD-Cup-98 dataset.

The rest of the paper is organized as follows. Section 2 gives a brief overview on data quality characterization and management. In Section 3, we present our decision model for estimating the cost of low-quality data on association rule mining. In Section 4, we evaluate our approach using the KDD-Cup-98 dataset. Section 5 provides concluding remarks and guidelines for future extensions of this work.

2 An Overview of Data Quality Characterization and Management

Maintaining a certain level of quality of data is challenging and can not be limited to one-shot approaches addressing simpler abstract versions the real problems of dirty or low-quality data [4]. Solving them requires highly domain- and context-dependent information and also human expertise. Classically, the database literature refers to data quality management as ensuring: *i)* syntactic correctness (e.g., constraints enforcement, that prevent "garbage data" from being entered into the database) and *ii)* semantic correctness (*i.e.*, data in the database truthfully reflect the real world situation). This traditional approach of data quality management has lead to techniques such as integrity constraints, concurrency control and schema integration for distributed and heterogeneous information systems. But since a decade, literature on data and information quality across different research communities (including databases, statistics, workflow management and knowledge engineering) proposed a plethora of:

- **Data quality dimensions** and **classifications** with various definitions depending on authors and application contexts [1], [13], on the audience type or on the architecture of systems (e.g. for data warehouses [6])
- **Data quality metrics** [4],
- **Conceptual data quality models** [6], [1],
- **Frameworks** and **methodologies** for cleaning data [11], for improving or assessing data quality in databases [6] or using data mining techniques to detect anomalies [3], [5], [10], [8].

The most frequently mentioned data quality dimensions in the literature are accuracy, completeness, timeliness and consistency [1].

3 Probabilistic Cost Model for Quality-Driven Selection of Interesting Association Rules

Our initial assumption is that the quality of an association rule depends on the quality of the data which the rule is computed from. This section will present the formal definitions of our model that introduces data quality indicators and combines them for determining the quality of association rules.

3.1 Preliminary Definitions for Association Rule Quality

Let I be a superset of items. An association rule R is an implication of the form: $LHS \rightarrow RHS$ where $LHS \subseteq I$, $RHS \subseteq I$ and $LHS \cap RHS = \emptyset$. LHS and RHS are conjunctions of variables such as the extension of LHS is: $g(LHS) = x_1 \wedge x_2 \wedge ... \wedge x_n$ and the extension of Y is $g(RHS) = y_1 \wedge y_2 \wedge ... \wedge y_{n'}$.

Let j ($j=1, 2, ..., k$) be the dimensions of data quality (e.g., data completeness, freshness, accuracy, consistency, completeness, credibility, etc.). Let $q_j(I_i) \in [min_{ij}, max_{ij}]$ be a scoring value for the dataset I_i on the quality dimension j ($I_i \subseteq I$). The vector, that keeps the values of all quality dimensions for each dataset I_i (normalized in [0,1]) is called quality vector and noted $q(I_i)$. The set of all possible quality vectors is called quality space Q.

Definition 1. Association Rule Quality
The quality of the association rule R is defined by a fusion function denoted "o_j" specific for each quality dimension j that merges the components of the quality vectors of the datasets constituting the extension of the right-hand and left-hand sides of the rule. The quality of the rule R is k-dimensional vector such as:

$$Quality\ (R) = \begin{pmatrix} q_1(R) \\ q_2(R) \\ \vdots \\ q_k(R) \end{pmatrix} = \begin{pmatrix} q_1(LHS)\circ_1 q_1(RHS) \\ q_2(LHS)\circ_2 q_2(RHS) \\ \vdots \\ q_k(LHS)\circ_k q_k(RHS) \end{pmatrix} \quad (1)$$

$$= \begin{pmatrix} q_1(x_1)\circ_1 q_1(x_2)\circ_1 \cdots \circ_1 q_1(x_n)\circ_1 q_1(y_1)\circ_1 q_1(y_2)\circ_1 \cdots \circ_1 q_1(y_n) \\ q_2(x_1)\circ_2 q_2(x_2)\circ_2 \cdots \circ_2 q_2(x_n)\circ_2 q_2(y_1)\circ_2 q_2(y_2)\circ_2 \cdots \circ_2 q_2(y_n) \\ \vdots \\ q_k(x_1)\circ_k q_k(x_2)\circ_k \cdots \circ_k q_k(x_n)\circ_k q_k(y_1)\circ_k q_k(y_2)\circ_k \cdots \circ_k q_k(y_n) \end{pmatrix}$$

The average quality of the association rule R denoted $\bar{q}(R)$ can be computed by the weighted sum of the quality dimensions of the quality vector components of the rule:

$$\bar{q}(R) = \sum_{j=1}^{k} w_j . q_j(R) \quad (2)$$

with w_j the weight of the quality dimension j. We assume the weights are normalized:

$$\sum_{j=1}^{k} w_j = 1 \ \forall j = 1, 2, \ldots k \quad (3)$$

Definition 2. Fusion Function per Quality Dimension
Let T be the domain of values of the quality score $\bar{q}(I_i)$ for the dataset I_i on the quality dimension j. The fusion function denoted "o_j" is commutative and associative such as $o_j: T \times T \rightarrow T$. The fusion function may have different definitions depending on the considered quality dimension j in order to suit the properties of each quality criterion. Table 1 presents several examples of definition for the fusion function allowing the combination of quality scores per quality dimension for two datasets noted x and y over the four quality dimensions; freshness, accuracy, completeness, consistency.

Table 1. Different fusion functions for merging quality scores per dimension

j	DATA QUALITY DIMENSION	FUSION FUNCTION "o_j"	QUALITY DIMENSION OF THE RULE $x \rightarrow y$
1	Freshness	$\min[q_1(x), q_1(y)]$	The freshness of the association rule $x \rightarrow y$ is estimated pessimistically as the lower score of freshness of the 2 data sets composing the rule.
2	Accuracy	$q_2(x) \cdot q_2(y)$	The accuracy of the association rule $x \rightarrow y$ is estimated as the probability of accuracy of the two data sets x and y of the rule.
3	Completeness	$q_3(x) + q_3(y) - q_3(x) \cdot q_3(y)$	The completeness of the association rule $x \rightarrow y$ is estimated as the probability that one of the two data sets of the rule is complete.
4	Consistency	$\max[q_4(x), q_4(y)]$	The consistency of the association rule $x \rightarrow y$ is estimated optimistically as the higher score of consistency of the 2 data sets composing the rule.

We consider that selecting an association rule is a decision that designates the rule as legitimately interesting (noted D_1), potentially interesting (D_2), or not interesting (D_3) based both on good interestingness measures and on the actual quality of the datasets composing the left-hand and right-hand sides of the rule. Consider the item $x \in LHS \cup RHS$ of a given association rule, we use $P_{CE}(x)$ to denote the probability that the item x will be classified as "erroneous" (or "polluted" and "with low-quality"), e.g., freshness, accuracy, etc. and $P_{CC}(x)$ denotes the probability that the item x will be classified as "correct" (i.e., "with correct quality" in the range of acceptable values for

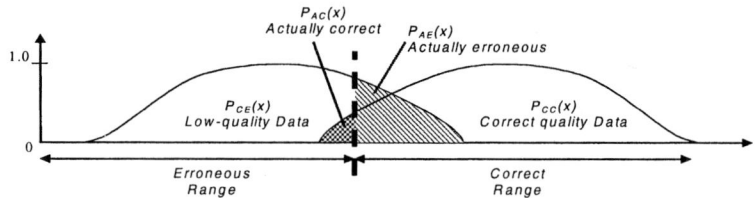

Fig. 1. Probabilities of detection of correct and low-quality data

each pre-selected quality dimension). Also, $P_{AE}(x)$ represents the probability that the item x is "actually erroneous" (AE) but detected correct, and $P_{AC}(x)$ represents the probability that it is "actually correct" (AC) but detected erroneous (see Figure 1).

For an arbitrary average quality vector $\bar{q} \in Q$ on the datasets in $LHS \cup RHS$ of the rule, we denote by $P(\bar{q} \in Q \mid CC)$ or $f_{CC}(\bar{q})$ the conditional probability that the average quality vector $\bar{q}$ corresponds to the datasets that are classified as correct (CC). Similarly, we denote by $P(\bar{q} \in Q \mid CE)$ or $f_{CE}(\bar{q})$ the conditional probability that the average quality vector $\bar{q}$ corresponds to the datasets that are classified erroneous (CE). We denote by d the decision of the predicted class of the rule (i.e., legitimately interesting D_1, potentially interesting D_2, or not interesting D_3), and by s the actual status of quality of the datasets upon which the rule has been computed. Let us also denote by $P(d=D_i, s=j)$ and $P(d=D_i \mid s=j)$ correspondingly, the joint and the conditional probability that the decision D_i is taken, when the actual status of data quality (i.e., CC, CE, AE, AC) is j. We also denote by c_{ij} the cost of making a decision D_i for classifying an association rule with the actual data quality status j of the datasets composing the two parts of the rule. Based on the example presented in Table 3 where we can see how the cost of decisions could affect the result of selection among interesting association rules, we need to minimize the mean cost $\bar{c}$ that results from making such a decision. In Table 3, c_{10} is the cost of a confident decision (D_1) for the selection of a rule based on correct-quality data (CC). c_{21} is the cost of a neutral decision (D_2) for the selection of a rule based on low-quality data (CE). c_{33} is the cost of a suspicious decision (D_3) of selecting a rule based on low-quality data but actually detected as correct (AC). The corresponding mean cost $\bar{c}$ is written as follows:

$$\bar{c} = c_{10}.P(d=D_1, s=CC) + c_{20}.P(d=D_2, s=CC) + c_{30}.P(d=D_3, s=CC) \\ + c_{11}.P(d=D_1, s=CE) + c_{21}.P(d=D_2, s=CE) + c_{31}.P(d=D_3, s=CE) \\ + c_{12}.P(d=D_1, s=AE) + c_{22}.P(d=D_2, s=AE) + c_{32}.P(d=D_3, s=AE) \\ + c_{13}.P(d=D_1, s=AC) + c_{23}.P(d=D_2, s=AC) + c_{33}.P(d=D_3, s=AC) \quad (4)$$

From the Bayes theorem, the following is true:

$$P(d=D_i, s=j) = P(d=D_i \mid s=j).P(s=j) \quad (5)$$

where $i=1,2,3$ and $j=CC,CE,AE,AC$. Let us also assume that $\bar{q}$ is the average quality vector drawn randomly from the space of all quality vectors of the datasets of the rule. The following equality holds for the conditional probability $P(d=D_i \mid s=j)$:

$$P(d=D_i \mid s=j) = \sum_{\bar{q} \in Q_i} f_j(\bar{q}). \quad (6)$$

where $i=1,2,3$ and $j=CC,CE,AE,AC$. f_j is the probability density of the quality vectors when the actual data quality status is j. We also denote the a priori probability of CC or else $P(s=CC)$ as π^0, the a priori probability of $P(s=AC)=\pi^0_{AC}$, the a priori probability of $P(s=AE)=\pi^0_{AE}$ and the a priori probability of $P(s=CE)=1-(\pi^0+\pi^0_{AE}+\pi^0_{AC})$. The mean cost $\overline{c}$ in Eq. (4) based on Eq. (5) is written as follows:

$$\begin{aligned}\overline{c} = & \, c_{10}.P(d=D_1|s=CC).P(s=CC) + c_{20}.P(d=D_2|s=CC).P(s=CC) \\ & + c_{30}.P(d=D_3|s=CC).P(s=CC) + c_{11}.P(d=D_1|s=CE).P(s=CE) \\ & + c_{21}.P(d=D_2|s=CE).P(s=CE) + c_{31}.P(d=D_3|s=CE).P(s=CE) \\ & + c_{12}.P(d=D_1|s=AE).P(s=AE) + c_{22}.P(d=D_2|s=AE).P(s=AE) \\ & + c_{32}.P(d=D_3|s=AE).P(s=AE) + c_{13}.P(d=D_1|s=AC).P(s=AC) \\ & + c_{23}.P(d=D_2|s=AC).P(s=AC) + c_{33}.P(d=D_3|s=AC).P(s=AC)\end{aligned} \quad (7)$$

and by using Eq. (6) and dropping the dependent vector variable $\overline{q}$, Eq. (7) becomes:

$$\begin{aligned}\overline{c} = & \sum_{q \in Q_1} \left[f_{CC}.c_{10}.\pi^0 + f_{CE}.c_{11}.(1-(\pi^0+\pi^0_{AC}+\pi^0_{AE})) + f_{AE}.c_{12}.\pi^0_{AE} + f_{AC}.c_{13}.\pi^0_{AC} \right] \\ & + \sum_{q \in Q_2} \left[f_{CC}.c_{20}.\pi^0 + f_{CE}.c_{21}.(1-(\pi^0+\pi^0_{AC}+\pi^0_{AE})) + f_{AE}.c_{22}.\pi^0_{AE} + f_{AC}.c_{23}.\pi^0_{AC} \right] \\ & + \sum_{q \in Q_3} \left[f_{CC}.c_{30}.\pi^0 + f_{CE}.c_{31}.(1-(\pi^0+\pi^0_{AC}+\pi^0_{AE})) + f_{AE}.c_{32}.\pi^0_{AE} + f_{AC}.c_{33}.\pi^0_{AC} \right]\end{aligned} \quad (8)$$

For the sake of simplicity for the following of the paper, let's now consider the case of the absence of the misclassification region (*i.e.*, f_{AC}, f_{AE} are null and $\pi^0_{AE}=\pi^0_{AC}=0$). Without misclassification region $P(s=CE)$ could be simplified as $1-\pi^0$. Every point $\overline{q}$ in the quality space Q belongs to the partitions of quality Q_1 or Q_2 or Q_3 that correspond respectively to partitions of the decision space: D_1 or D_2 or D_3 in such a way that its contribution to the mean cost is minimum. This will lead to the optimal selection for the three sets of rules which we denote by D^0_1, D^0_2 and D^0_3. Based on this observation, a point $\overline{q}$ that represents the quality of a rule defined in Eq. (2) is assigned to one of the three optimal areas as follows:

$$D^0_1 = \left\{ \overline{q} : \frac{f_{CE}}{f_{CC}} \leq \frac{\pi^0}{1-\pi^0}.\frac{c_{30}-c_{10}}{c_{11}-c_{31}} \text{ and, } \frac{f_{CE}}{f_{CC}} \leq \frac{\pi^0}{1-\pi^0}.\frac{c_{20}-c_{10}}{c_{11}-c_{21}} \right\} \quad (9)$$

$$D^0_2 = \left\{ \overline{q} : \frac{f_{CE}}{f_{CC}} \geq \frac{\pi^0}{1-\pi^0}.\frac{c_{20}-c_{10}}{c_{11}-c_{21}} \text{ and, } \frac{f_{CE}}{f_{CC}} \leq \frac{\pi^0}{1-\pi^0}.\frac{c_{30}-c_{20}}{c_{21}-c_{31}} \right\}$$

$$D^0_3 = \left\{ \overline{q} : \frac{f_{CE}}{f_{CC}} \geq \frac{\pi^0}{1-\pi^0}.\frac{c_{30}-c_{10}}{c_{11}-c_{31}} \text{ and, } \frac{f_{CE}}{f_{CC}} \geq \frac{\pi^0}{1-\pi^0}.\frac{c_{30}-c_{20}}{c_{21}-c_{31}} \right\}$$

The inequalities of Eq. (9) give rise to three different threshold values L, P and N (respectively for legitimately, potentially and not interesting rules) in the decision space as defined in Eq. (10):

$$L = \frac{\pi^0}{1-\pi^0}.\frac{c_{30}-c_{10}}{c_{11}-c_{31}}, \quad P = \frac{\pi^0}{1-\pi^0}.\frac{c_{20}-c_{10}}{c_{11}-c_{21}}, \text{ and } N = \frac{\pi^0}{1-\pi^0}.\frac{c_{30}-c_{20}}{c_{21}-c_{31}} \quad (10)$$

4 Experiments and Results

In order to validate and evaluate our decision model, we built an experimental system. The system relies on a data generator that automatically generates data quality metadata with a priori known characteristics. This system also allows us to perform controlled studies so as to establish data quality indicators and quality variations on datasets and on discovered association rules which are assigned to the decision areas D_1, D_2 or D_3. In the set of experiments that we present, we make use the KDD-CUP-98[1] dataset from the UCI repository. The KDD-Cup-98 dataset contains 191,779 records about individuals contacted in the 1997 mailing campaign. Each record is described by 479 non-target variables and two target variables indicating the "respond"/"not respond" classes and the actual donation in dollars. About 5% of records are "respond" records and the rest are "not respond" records. The KDD-Cup-98 competition task was to build a prediction model of the donation amount. The participants were contested on the sum of actual profit $\Sigma(\textit{actual donation} - \$0.68)$ over the validation records with predicted donation greater than the mailing cost $0.68 (see [14] for details). Because we ignored the quality of the data collected during this campaign, we generated synthetic data quality indicators with different distributions representative of common data pollutions. In this experiment, our goal is to demonstrate that data quality variations may have a great impact on the significance of KDD-Cup-98 results (*i.e.*, the top ten discovered "respond" rules) and we use different assumptions on data quality indicators that do not affect the top ten list of discovered association rules but that significantly change the reliability (and quality) of this mining result and also the cost of the decisions relying on these rules. The variable names, definitions, estimated probabilities and average quality score per attribute are given in Table 2. For the sake of simplicity, we suppose that the quality dimension scores are uniformly representative of the quality of the attribute value domain. The average quality per attribute in Table 2 is computed from the equi-weighted function given in Eq. (2). $f_{CC}(\overline{q}(I_i))$ (also noted f_{CC} in Table 2) is the probability density that the dataset I_i is "correct" when the average quality score of I_i is $\overline{q}(I_i)$. $f_{CE}(\overline{q}(I_i))$ is the probability density that the dataset I_i is "erroneous" when the average quality score of I_i is $\overline{q}(I_i)$. Table 3 shows tentative unit costs developed by the staff of the direct marketing department on the basis of consideration of the consequences of the decisions on selecting and using the discovered association rules. Without misclassification problem, the costs c_{12}, c_{13}, c_{22}, c_{23}, c_{32}, and c_{33} are null; the cost c_{30} of a suspicious decision for rule selection based on correct data is $500. Based on the values assigned to the various costs in Table 2, we also assume that the a priori probability that a certain quality vector belongs to CC equals the a priori probability that the same vector belongs to CE. For this reason, the ratio $\frac{\pi^0}{1-\pi^0}$ in Eq. (9) and (10) equals 1. By using Eq. (10) and Table 3, we compute the values of the three decision thresholds for rule selection for the a priori probability $\pi^0 = 0.200$ without misclassification and we obtain: L=0.125, P = 0.0131579 and N = 2.25. In order to be consistent with the conditional independency of the quality vector components we also need to take the logarithms of the thresholds values. By doing this we obtain: $log(L)$=-0.9031; $log(P)$ = -1.8808 and $log(N)$ = 0.3522. Based on the values for these thresholds, we can assign the rules to one of the three decision areas. The top 10 a priori association rules discovered by Wang *et al.* [14] are given in Table 4 with the

---

[1] *http://kdd.ics.uci.edu/databases/kddcup98/kddcup98.html* for the dataset and *http://www.kdnuggets.com/meetings/kdd98/kdd-cup-98.html* for the results.

confidence, the support (in number of records), and the profit. Table 4 also shows the score per quality dimension, the average quality and the cost of selecting the association rule. The scores are computed from the definitions of the quality dimensions given in Table 1. The costs are computed from Eq. (8). It's very interesting to notice that the predicted profit per rule may be considerably affected by the cost of the rule computed from low-quality data (e.g., the second best rule R2 whose predicted profit is $61.73 has a cost of $109.5 and thus is classified as "not interesting" due to the bad quality of its datasets). Let us now introduce different variations on the average quality of the datasets composing the rules. Based on the cost Table 3, Figure 2 shows the behavior of the decision cost of rule selection when data quality varies from the initial average quality down to -10%, -30%, and -50% and up to +10%, +30% and +50% for a priori probability $\pi^0 = 0.200$ and without misclassification. In Figure 2 we observe that the quality degradation of the datasets composing the rules increases the cost of these rules with variable amplitudes.

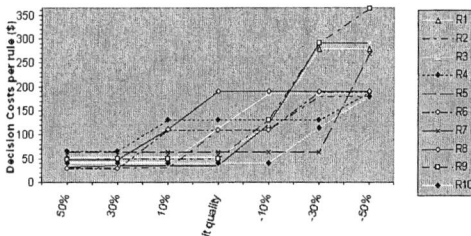 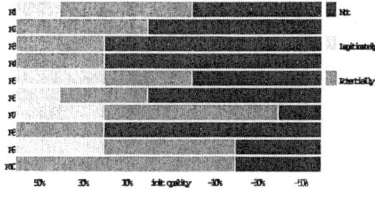

Fig. 2. Decision cost for rule selection with different data quality variations without misclassification for $\pi^0 = 0.200$

Fig. 3. Decision status on rule selection for data quality variations for $\pi^0 = 0.200$

Data quality amelioration implies a stabilization trend of the decision cost for legitimately interesting rule selection. Another interesting result is shown in Figure 3 where the decisions for rule selection change simultaneously with the data quality variations. Among the top 10 interesting rule discovered by Wang et al. [14] with the initial data quality (noted Init Qual), 5 rules (R1, R5, R7, R9 and R10) are potentially worth being selected based on their average data quality. Increasing data quality up to +30%, 3 rules were legitimately interesting (R5, R7 and R9). This observation offers two (among others) interesting research perspectives for both association rule mining and data quality management: first, for proposing a post-filtering rule process based on data quality indicators and decision costs for rule selection and secondly, for the optimal scheduling of data quality improvement activities (e.g., cleaning) driven and tuned by the rule pruning step. Additionally to the interestingness measures the three thresholds can be used as a predictive technique for quality awareness in association rule mining for the appropriate selection of legitimately interesting rules based on the data quality indicators.

Table 2. KDD-Cup-98 dataset with quality measures and estimatedprobabilities

Attribute	Definition	Quality					f_{CC}	f_{CE}
		Fresh.	Accur.	Compl.	Consi.	Average		
AGE904	Average Age of Population	0,50	0,21	0,39	0,73	0,46	0,9	0,05
CHIL2	Percent Children Age 7 - 13	0,16	0,99	0,75	0,71	0,65	0,95	0,1
DMA	DMA Code	0,49	0,58	0,16	0,95	0,55	0,95	0,01
EIC16	Percent Employed in Public Administration	0,03	0,56	0,33	0,61	0,38	0,98	0,01
EIC4	Percent Employed in Manufacturing	0,17	0,37	0,87	0,15	0,39	0,9	0,2
ETH1	Percent White	0,21	0,76	0,50	0,53	0,50	0,55	0,15
ETH13	Percent Mexican	0,52	0,77	0,87	0,79	0,74	0,9	0,6
ETHC4	Percent Black < Age 15	0,84	0,52	0,32	0,35	0,51	0,95	0,45
HC6	Percent Owner Occupied Structures Built Since 1970	0,47	0,96	0,74	0,11	0,57	0,98	0,03
HHD1	Percent Households w/ Related Children	0,61	0,95	0,27	0,08	0,48	0,96	0,41
HU3	Percent Occupied Housing Units	0,07	0,40	0,18	0,57	0,30	0,94	0,53
HUPA1	Percent Housing Units w/ 2 thru 9 Units at the Address	0,76	0,85	0,96	0,93	0,88	0,95	0,52
HVP5	Percent Home Value >= $50,000	0,99	0,88	0,38	0,95	0,80	0,94	0,05
NUMCHLD	NUMBER OF CHILDREN	0,44	0,23	0,53	0,50	0,42	0,96	0,17
POP903	Number of Households	0,77	0,52	0,74	0,61	0,66	0,87	0,15
RAMNT_22	Dollar amount of the gift for 95XK	0,37	0,95	0,95	0,75	0,76	0,84	0,25
RFA_11	Donor's RFA status as of 96X1 promotion date	0,59	0,34	0,34	0,76	0,51	0,95	0,12
RFA_14	Donor's RFA status as of 95NK promotion date	0,60	0,69	0,24	0,10	0,41	0,95	0,13
RFA_23	Donor's RFA status as of 94FS promotion date	0,34	0,01	0,23	0,63	0,30	0,97	0,55
RHP2	Average Number of Rooms per Housing Unit	0,66	0,72	0,08	0,26	0,43	0,98	0,2
TPE11	Mean Travel Time to Work in minutes	0,20	0,26	0,78	0,32	0,39	0,85	0,05
WEALTH2	Wealth Rating	0,24	0,82	0,41	0,58	0,51	0,87	0,05

Table 3. Costs of various decisions for classifying association rules

Decision for Rule Selection	Cost#	Data Quality Status	Cost without misclassification
D_1	c_{10}	CC	$0.00
	c_{11}	CE	$1 000.00
	c_{12}	AE	$0.00
	c_{13}	AC	$0.00
D_2	c_{20}	CC	$50.00
	c_{21}	CE	$50.00
	c_{22}	AE	$0.00
	c_{23}	AC	$0.00
D_3	c_{30}	CC	$500.00
	c_{31}	CE	$0.00
	c_{32}	AE	$0.00
	c_{33}	AC	$0.00

Table 4. The top 10 "respond" rules by Wang et al. [14] with quality, cost, and decision area

#	Association Rule	(Conf. ; Supp.)	Profit (Wang et al. 2005)	Quality					Cost	Decision Area
				Fresh.	Accur.	Compl.	Consi.	Average		
1	ETHC4=[2.5,4.5], ETH1=[22.84,29.76], HC6=[60.91,68.53]	(0.11; 13)	$81.11	0,21	0,38	0,79	0,53	0,48	$ 53	potentially
2	RFA_14=f1d, ETH1=[29.76,36.69]	(0.17; 8)	$61.73	0,21	0,52	0,62	0,53	0,47	$109.5	not
3	HHD1=[24.33,28.91], EIC4=[33.72,37.36]	(0.12;12)	$47.07	0,17	0,35	0,90	0,15	0,39	$113	not
4	RFA_23=s2g, ETH13=[27.34,31.23]	(0.12;16)	$40.82	0,34	0,01	0,90	0,79	0,51	$130	not
5	EIC16=[11.25,13.12], CHIL2=[33,35.33], HC6=[45.69,53.30]	(0.16;11)	$35.17	0,03	0,53	0,77	0,71	0,51	$ 34.7	potentially
6	RHP2=[36.72,40.45], AGE904=[42.2,44.9]	(0.16;7)	$28.71	0,50	0,15	0,44	0,73	0,46	$109	not
7	HVP5=[56.07,63.23], ETH13=[31.23,35.61], RAMNT_22=[7.90,10.36]	(0.14;10)	$24.32	0,37	0,65	0,68	0,95	0,66	$ 62.8	potentially
8	NUMCHLD=[2.5,3.25], HU3=[66.27,70.36]	(0.08;31)	$19.32	0,07	0,09	0,61	0,57	0,34	$190	not
9	RFA_11=f1g, DMA=[743,766.8], POP903=[4088.208,4391.917], WEALTH2=[6.428571,7.714286]	(0.25;8)	$17.59	0,24	0,08	0,72	0,95	0,50	$ 49.6	potentially
10	HUPA1=[41.81+,], TPE11=[27,64,31.58]	(0.23;9)	$9.46	0,20	0,22	0,99	0,93	0,59	$ 40.8	potentially

5 Conclusion

The original contribution of this paper is twofold: first, we propose a method for scoring the quality of association rules that combines and integrates measures of data quality; secondly, we propose a probabilistic cost model for estimating the cost of selecting "legitimately (or not) interesting" association rules based on correct- or low-quality data. The model defines the thresholds of three decision areas for the predicted class of the discovered rules (*i.e.*, legitimately interesting, potentially interesting, or not interesting). To validate our approach, our experiments on the KDD-Cup-98 dataset consisted of: *i)* generating synthetic data quality indicators, *ii)* computing the average quality of the top ten association rules discovered by Wang *et al.* [14], *iii)* computing the cost of selecting low-quality rules and the decision areas they belong to, *iv)* examining the cost and the decision status for rule selection when the quality of underlying data varies. Our experiments confirm our original assumption that is: interestingness measures are not self-sufficient and the quality of association rules depends on the quality of the data which the rules are computed from. Data quality includes various dimensions (such as data freshness, accuracy, completeness, etc.) which should be also considered for effective and quality-aware mining. Our future plans regarding this work, are to study the optimality of our decision model, to propose error estimation and to validate the model with experiments on large biomedical datasets (see [2]) with on-line collecting and computing operational data quality indicators with the aim to select high-quality and interesting association rules.

References

1. Batini C., Catarci T. and Scannapiceco M., A Survey of Data Quality Issues in Cooperative Information Systems, *Tutorial, Intl. Conf. on Conceptual Modeling (ER)*, 2004.
2. Berti-Equille L., Moussouni F., Quality-Aware Integration and Warehousing of Genomic Data., *Proc. of the Intl. Conf. on Information Quality*, M.I.T., Cambridge, U.S.A., 2005.
3. Dasu T. and Johnson T., Hunting of the Snark: Finding Data Glitches with Data Mining Methods, *Intl. Conf. on Information Quality*, M.I.T., Cambridge, M.A., U.S.A., 1999.
4. Dasu T., Johnson T., *Exploratory Data Mining and Data Cleaning*, Wiley, 2003.
5. Hipp J., Guntzer U., and Grimmer U., Data Quality Mining - Making a Virtue of Necessity. *Proc. of the Workshop on Research Issues in Data Mining and Knowledge Discovery (DMKD2001)*, Santa Barbara, CA, U.S.A, May 20th, 2001.
6. Jeusfeld, M. A., Quix C., Jarke M., Design and Analysis of Quality Information for Data Warehouses, *17th Intl. Conf. on Conceptual Modeling (ER'98)*, Singapore, 1998.
7. Lavrač N., Flach P.A., Zupan B., Rule Evaluation Measures: A Unifying View, *ILP*, p. 174-185, 1999.
8. Lübbers D., Grimmer U. and Jarke M., Systematic Development of Data Mining-Based Data Quality Tools, *Proc. of the Intl. VLDB Conf.*, p. 548-559, 2003.
9. Pearson R.K., Data Mining in Face of Contaminated and Incomplete Records, *Proc. of SIAM Intl. Conf. Data Mining*, 2002.
10. Pyle D., Data Preparation for Data Mining, Morgan Kaufmann, 1999.
11. Rahm E., Do H., Data Cleaning: Problems and Current Approaches, *IEEE Data Eng. Bull.* 23(4): 3-13, 2000.
12. Tan P-N., Kumar V. and Srivastava J., Selecting the Right Interestingness Measure for Association Patterns, *Proc. of Intl. KDD Conf.*, p. 32-41, 2002.
13. Wang R., Storey V., Firth C., A Framework for Analysis of Data Quality Research, *IEEE TKDE*, 7(4): 670-677, 1995.
14. Wang K., Zhou S., Yang Q. and, Yeung J.M.S., Mining Customer Value: from Association Rules to Direct Marketing, *J. of Data Mining and Knowledge Discovery*, 2005.
15. Zhang C., Yang Q. and Liu B. (Eds). Introduction: Special Section on Intelligent Data Preparation, *IEEE Transactions on Knowledge and Data Engineering,* 17(9), 2005.

IMB3-Miner: Mining Induced/Embedded Subtrees by Constraining the Level of Embedding

Henry Tan[1], Tharam S. Dillon[1], Fedja Hadzic[1], Elizabeth Chang[2], and Ling Feng[3]

[1] University of Technology Sydney, Faculty of Information Technology, Sydney, Australia
{henryws, tharam, fhadzic}@it.uts.edu.au
[2] Curtin University of Technology, School of Information System, Perth, Australia
Elizabeth.Chang@cbs.curtin.edu.au
[3] University of Twente, Department of Computer Science, Enschede, Netherlands
ling@cs.utwente.nl

Abstract. Tree mining has recently attracted a lot of interest in areas such as Bioinformatics, XML mining, Web mining, etc. We are mainly concerned with mining frequent induced and embedded subtrees. While more interesting patterns can be obtained when mining embedded subtrees, unfortunately mining such embedding relationships can be very costly. In this paper, we propose an efficient approach to tackle the complexity of mining embedded subtrees by utilizing a novel *Embedding List* representation, *Tree Model Guided* enumeration, and introducing the *Level of Embedding* constraint. Thus, when it is too costly to mine all frequent embedded subtrees, one can decrease the level of embedding constraint gradually up to 1, from which all the obtained frequent subtrees are induced subtrees. Our experiments with both synthetic and real datasets against two known algorithms for mining induced and embedded subtrees, FREQT and TreeMiner, demonstrate the effectiveness and the efficiency of the technique.

1 Introduction

Research in both theory and applications of data mining is expanding driven by a need to consider more complex structures, relationships and semantics expressed in the data [2,3,4,6,8,9,12,15,17]. As the complexity of the structures to be discovered increases, more informative patterns could be extracted [15]. A tree is a special type of graph that has attracted a considerable amount of interest [3,8,9,11,12,17]. Tree mining has gained interest in areas such as Bioinformatics, XML mining, Web mining, etc. In general, most of the formally represented information in these domains is of a tree structured form and XML is commonly used. Tan et. al. [8] suggested that XML association rule mining can be recast as mining frequent subtrees in a database of XML documents. Wang and Liu [13] developed an algorithm to mine frequently occurring induced subtrees in XML documents. Feng et. al. [4] extend the notion of associated items to XML fragments to present associations among trees.

The two known types of subtrees are *induced* and *embedded* [3,8,9,17]. An Induced subtree preserves the parent-child relationships of each node in the original tree whereas an embedded subtree preserves not only the parent-child relationships but also the ancestor-descendant relationships over several levels. Induced subtrees are a subset of embedded subtrees and the complexity of mining embedded subtrees is higher than mining induced subtrees [3,9,17].

In this study, we are mainly concerned with mining frequent embedded subtrees from a database of rooted ordered labeled subtrees. Our primary objectives are as follows: (1) to develop an efficient and scalable technique (2) to provide a method to control and limit the inherent complexity present in mining frequent embedded subtrees. To achieve the first objective, we utilize a novel tree representation called *Embedding List (EL)*, and employ an optimal enumeration strategy called *Tree Model Guided (TMG)*. The second objective can be attained by restricting the maximum level of embedding that can occur in each embedded subtree. The level of embedding is defined as the length of a path between two nodes that form an ancestor-descendant relationship. Intuitively, when the level of embedding inherent in the database of trees is high, numerous numbers of embedded subtrees exist. Thus, when it is too costly to mine all frequent embedded subtrees, one can restrict the level of embedding gradually up to 1, from which all the obtained frequent subtrees are induced subtrees.

The two known enumeration strategies are enumeration by extension and join [3]. Recently, Zaki [17] adapted the join enumeration strategy for mining frequent embedded rooted ordered subtrees. An idea of utilizing a tree model for efficient enumeration appeared in [14]. The approach uses the XML schema to guide the candidate generation so that all candidates generated are valid because they conform to the schema. The concept of schema guided candidate generation is generalized into tree model guided (TMG) candidate generation for mining embedded rooted ordered labeled subtrees [8,10]. TMG can be applied to any data with clearly defined semantics that have tree like structures. It ensures that only valid candidates which conform to the actual tree structure of the data are generated. The enumeration strategy used by TMG is a specialization of the right most path extension approach [2,8,9,10]. It is different from the one that is proposed in FREQT [2] as TMG enumerates embedded subtrees and FREQT enumerates only induced subtrees. The right most path extension method is reported to be complete and all valid candidates are enumerated at most once (non-redundant) [2,8,9]. This is in contrast to the incomplete method TreeFinder [11] that uses an Inductive Logic Programming approach to mine unordered, embedded subtrees. The extension approach utilized in the TMG generates fewer candidates as opposed to the join approach [8,9].

In section 2 the problem decomposition is given. Section 3 describes the details of the algorithm. We empirically evaluate the performance of the algorithms and study their properties in section 4, and the paper is concluded in section 5.

2 Problem Definitions

A tree can be denoted as $T(r,V,L,E)$, where (1) $r \in V$ is the root node; (2) V is the set of vertices or nodes; (3) L is the set of labels of vertices, for any vertex $v \in V$, $L(v)$ is the label of v; and (4) E is the set of edges in the tree. *Parent* of node v, *parent(v)*, is defined as the predecessor of node v. There is only one parent for each v in the tree. A node v can have one or more *children, children(v)*, which are defined as its successors. If a path exists from node p to node q, then p is an *ancestor* of q and q is a descendant of p. The number of children of a node is commonly termed as *fan-out/degree* of the node, *degree(v)*. A node without any child is a *leaf* node; otherwise, it is an *internal* node. If for each internal node, all the children are ordered, then the tree is an *ordered tree*. The *height* of a node is the length of the path from a node to its furthest leaf. The *rightmost path* of T is defined as the path connecting the *rightmost leaf* with the root node. The *size* of a tree is determined by the number of nodes in the tree. *Uniform tree T(n,r)* is a tree with height equal to n and all of its internal nodes have degree r. All trees considered in this paper are rooted ordered labeled.

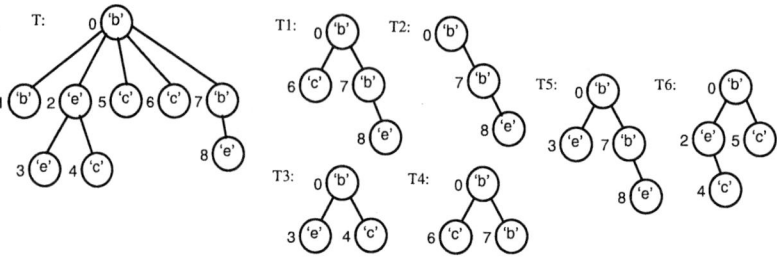

Fig. 1. Example of induced subtrees (*T1, T2, T4, T6*) and embedded subtrees (*T3, T5*) of tree T

Induced Subtree. A tree T'(r', V', L', E') is an ordered induced subtree of a tree T (r, V, L, E) iff (1) V'⊆V, (2) E'⊆E, (3) L'⊆L and L'(v)=L(v), (4) ∀v'∈V', ∀v∈ V and v' is not the root node parent(v')=parent(v), (5) the left-to-right ordering among the siblings in T' should be preserved. Induced subtree T' of T can be obtained by repeatedly removing leaf nodes or the root node if its removal doesn't create a forest in T.

Embedded Subtree. A tree T'(r', V', L', E') is an ordered embedded subtree of a tree T(r, V, L, E) if and only if it satisfies property 1, 2, 3, 5 of induced subtree and it generalizes property (4) such that ∀v'∈V', ∀v∈ V and v' is not the root node ancestor(v') = ancestor (v).

Level of Embedding (Φ). If T'(r', V', L', E') is an embedded subtree of T, the *level of embedding (Φ)* is defined as the length of a path between two nodes p and q, where p∈ V' and q∈V', and p and q form an ancestor-descendant relationship from p to q. We could define induced subtree T as an embedded subtree with maximum Φ that can occur in T equals to 1, since the level of embedding of two nodes that form parent-child relationship equals to 1.

For instance in fig 2 the level of embedding, Φ, between node at position 0 and node at position 5 in tree T is 3, whereas between node 0 and node 2, 3, and 4 is equal to 2. According to our definition of induced and embedded subtree previously, S1 is an example of an induced subtree and S2, S3, and S4 are examples of embedded subtrees.

Transaction based vs occurrence match support. We say that an embedded subtree t is supported by transaction $k \subseteq K$ in database of tree T_{db} as $t \prec k$. If there are L occur-

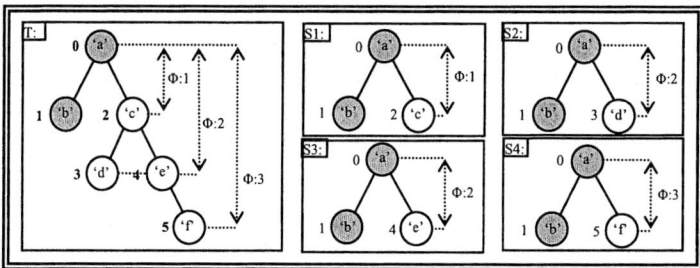

Fig. 2. Illustration of restricting the level of embedding when generating *S1-4* subtrees from subtree '*a b*' with OC 0:[0,1] of tree *T*

rences of t in k, a function $g(t,k)$ denotes the number of occurrences of t in transaction k. For *transaction based support*, $t \prec k = 1$ when there exists at least one occurrence of t in k, i.e. $g(t,k) \geq 1$. In other words, it only checks for existence of an item in a transaction. For *occurrence match support*, $t \prec k$ corresponds to the number of all occurrences of t in k, $t \prec k = g(t,k)$. Suppose that there are N transactions, k_1 to k_N, of trees in T_{db}, the support of embedded subtree t in T_{db} is defined as:

$$\sum_{i=1}^{N} t \prec k_i \tag{1}$$

Transaction based support has been used in [3,12,17]. However occurrence match support has been less utilized and discussed. In this study we are in particular interested in exploring the application and the challenge of using occurrence match support. Occurrence match support takes repetition of items in a transaction into account whilst transaction based support only checks for existence of items in a transaction. There has not been any general consensus which support definition is used for which application. However, it is intuitive to say that whenever repetition of items in each transaction is to be accounted for and order is important, occurrence match support would be more applicable. Generally, transaction based support is very applicable for relational data.

String encoding (φ). We utilize the pre-ordering string encoding (φ) as utilized in [8,9,17]. We denote encoding of subtree T as $\varphi(T)$. For each node in T (fig. 1), its label is shown as a single-quoted symbol inside the circle whereas its pre-order position is shown as indexes at the left/right side of the circle. From fig. 1, $\varphi(T1)$:'b c / b e / /'; $\varphi(T3)$:'b e / c /', etc. We could omit backtrack symbols after the last node, i.e. $\varphi(T1)$:'b c / b e'. We refer to a group of subtrees with the same encoding L as *candidate subtree* C_L. A subtree with k number of nodes is denoted as *k-subtree*. Throughout the paper, the '+' operator is used to conceptualize an operation of appending two or more tree encodings. However, this operator should be contrasted with the conventional string append operator, as in tree string encoding the backtrack symbols needs to be computed accordingly.

Mining (induced/embedded) frequent subtrees. Let T_{db} be a tree database consisting of N transactions of trees, K_N. The task of frequent (induced/embedded) subtree mining from T_{db} with given minimum support (σ), is to find all the candidate (induced/embedded) subtrees that occur at least σ times in T_{db}. Based on the downward-closure lemma [1], every sub-pattern of a frequent pattern is also frequent. In relational data, given a frequent itemset all its subsets are also frequent. A question however arises of whether the same principle applies to tree structured data when the occurrence match support definition is used. To show that the same principle doesn't apply, we need to find a counter-example.

Lemma 1. Given a tree database T_{db}, if there exist candidate subtrees C_L and $C_{L'}$, where $C_L \subseteq C_{L'}$, such that $C_{L'}$ is frequent and C_L is infrequent, we say that $C_{L'}$ is a *pseudo-frequent candidate subtree*. In the light of the downward closure lemma these candidate subtrees are infrequent because one or more of its subtrees are infrequent.

Lemma 2. The antimonotone property of frequent patterns suggests that the frequency of a superpattern is less than or equal to the frequency of a subpattern. If pseudo-frequent candidate subtrees exist then the antimonotone property does not hold for frequent subtree mining.

From fig. 1, suppose that the minimum support σ is set to 2. Consider a candidate subtree C_L where L:'b c / b'. When an embedded subtree is considered, there are 3 occurrences of C_L that occur at position {(0, 4, 7), (0, 5, 7), (0, 6, 7)}. On the other hand, when an induced subtree is considered, there are only 2 occurrences of C_L that occur at position {(0, 5, 7), (0, 6, 7)}. With σ equal to 2, C_L is frequent for both induced and embedded types. By extending C_L with node 8 we obtain $C_{L'}$ where L':$L+$'e' = 'b c / b e'. In the light of lemma 1, $C_{L'}$ is a pseudo-frequent candidate subtree because we can find a subtree of $C_{L'}$ whose encoding 'b b e' at position (0, 7, 8) is infrequent. This holds for both induced and embedded subtrees. In other words, lemma 1 holds whenever occurrence match support is used. Subsequently, since pseudo-frequent candidate subtrees exist, according to lemma 2, the antimonotone property does not hold for frequent subtree mining when occurrence match support is used. Hence, when mining induced and embedded subtrees, there can be frequent subtrees with one or more of its subsets infrequent. This is different to flat relational data where there are only 1-to-1 relationships between items in each transaction. Tree structured data has a hierarchical structure where 1-to-many relationships can occur. This multiplication between one node to its many children/descendants makes the antimonotone property not hold for tree structured data. This makes full (k-1) pruning should be performed at each iteration when generating k-subtrees from a (k-1)-subtree when occurrence match support is used to avoid generating pseudo-frequent subtrees.

3 IMB3-Miner Algorithms

Database scanning. The process of frequent subtree mining is initiated by scanning a tree database, T_{db}, and generating a global pre-order sequence D in memory (*dictionary*). The dictionary consists of each node in T_{db} following the pre-order traversal indexing. For each node its position, label, right-most leaf position (scope), and parent position are stored. An item in the dictionary D at position i is referred to as $D[i]$. The notion of position of an item refers to its index position in the dictionary. When generating the dictionary, we compute all the frequent 1-subtrees, F_1. After the dictionary is constructed our approach does not require further database scanning.

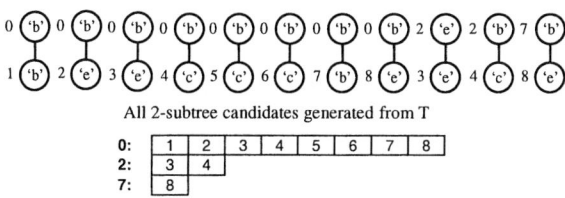

All 2-subtree candidates generated from T

0:	1	2	3	4	5	6	7	8
2:	3	4						
7:	8							

Fig. 3. The EL representation of T in fig 1

Constructing Embedding List (EL). For each frequent internal node in F_1, a list is generated which stores its descendant nodes' hyperlinks [12] in pre-order traversal ordering such that the embedding relationships between nodes are preserved. The notion of hyperlinks of nodes refers here to the positions of nodes in the dictionary. For a given internal node at position i, such ordering reflects the enumeration sequence of generating 2-subtree candidates rooted at i (fig 3). Hereafter, we call this list as *embedded list (EL)*. We use notation *i-EL* to refer to an embedded list of node at position i. The position of an item in EL is referred to as *slot*. Thus, *i-EL[n]* refers to the item in the list at slot n. Whereas |*i-EL*| refers to the size of the embedded list of node at position i. Fig 3 illustrates an example of the EL representation of tree T (fig. 1). In fig 3, *0-EL* for example refers to the list: $0:[1,2,3,4,5,6,7,8]$, where *0-EL[0]*=1 and *0-EL[6]*=7.

Occurrence Coordinate (OC). When generating k-subtree candidates from (k-1)-subtree, we consider only frequent (k-1)-subtrees for extension. Each occurrence of k-subtree in T_{db} is encoded as *occurrence coordinate* $r:[e_1,...e_{k-1}]$; r refers to k-subtree root position and $e_1,...,e_{k-1}$ refer to slots in *r-EL*. Each e_i corresponds to node $(i+1)$ in k-subtree and $e_1 < e_{k-1}$. We refer to e_{k-1} as *tail* slot. From fig. 1 & 3, the OC of 3-subtree (T2) with encoding 'b b e' is encoded as $0:[6,7]$; 4-subtrees T1 with encoding 'b c / b e' are encoded as $0:[5,6,7]$, and so on. Each OC of a subtree describes an instance of each occurrence of the subtree in T_{db}. Hence, each *candidate instance* has an OC associated with it.

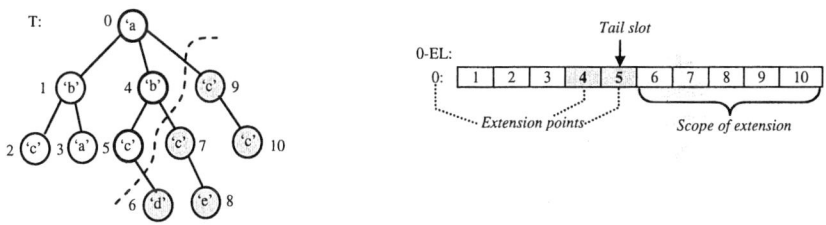

Fig. 4. TMG enumeration: extending (k-1)-subtree t_{k-1} where $\varphi(t_{k-1})$: 'a b / b c' ($0:[1,4,5]$) with nodes at position 6, 7, 8, 9, and 10

TMG enumeration formulation. TMG is a specialization of right most path extension method which has been reported to be complete and all valid candidates are enumerated at most once (non-redundant) [2,8,9,10]. To enumerate all embedded k-subtrees from a (k-1)-subtree, TMG enumeration approach extends one node at the time to the right most path of (k-1)-subtree. We refer to each node in the right most path as an extension point. One important property of EL is that the positions of nodes are stored in pre-order manner. Hence, given a (k-1)-subtree with known tail slot, the subsequent slots in EL will form the *scope of extension* from i to j. All embedded k-subtree are generated by attaching a node at position i to j to the (k-1)-subtree. Suppose $l(i)$ denotes a labeling function of node at position i. Given frequent (k-1)-subtree t_{k-1} with $\varphi(t_{k-1}):L$, the root position r, tail position t, and occurrence coordinate $r:[m,...,n]$, k-subtrees are generated by extending t_{k-1} with $j \in r\text{-}EL$ such that $t<j\leq |r\text{-}EL|\text{-}1$. Thus its occurrence coordinate becomes $r:[m,...,n,j]$ and its encoding becomes $L':L+l(i)$ where $i=r\text{-}EL[j]$ and $m<n<j$. To restrict the level of embedding of each node, at each extension a check is performed if the level of embedding is less or equal to the specified Φ. Only when the level of embedding of a node at position j to its extension point is less than Φ, the extension is performed. From fig 4, suppose that Φ is set to 1, when we extend a subtree with OC *0:[0,3,4]* with node at position 6, 7, and 9 *(0:[5], 0:[6], 0:[8])*, the level of embedding between nodes at position 6, 7, and 9 to their extension point equals to 1 ($\leq \Phi$), and thus should not be pruned. However when it is extended with node at position 8 and 10 *(0:[7], 0:[9])* the level of embedding between node at position 8 and 10 to their extension points is>2 ($\geq \Phi$), and thus should be pruned.

Pruning. When using occurrence match support there can be pseudo-frequent candidate subtrees generated when generating k-subtrees from (k-1)-subtrees. To make sure that all generated subtrees do not contain infrequent subtrees, full (k-1) pruning must be performed. The rationale of this has been discussed in [9,17]. From this point onward we refer to full (k-1) pruning as *full pruning*. This implies that at most (k-1) numbers of (k-1)-subtrees need to be generated from the currently expanding k-subtrees. An exception is made whenever the Φ constraint is set to 1, i.e. mining induced subtree, we only need to generate l numbers of (k-1)-subtrees where $l<(k-1)$ and l equal to the number of leaf nodes in k-subtrees. When the removal of root node of k-subtree doesn't generate a *forest* [8,9,17] then an additional (k-1)-subtree is generated by taking the root node off from the expanding k-subtree. The expanding k-subtree is pruned when at least one (k-1)-subtree is infrequent, otherwise it is added to the frequent k-subtree set. This ensures that the method generates no pseudo-frequent subtrees. While full pruning is easily done in a breadth first search (BFS) based method, it is a challenge for a depth first search (DFS) based approach such as VTreeMiner (VTM). As a consequence, VTM performs opportunistic pruning [17]. Doing full pruning is quite time consuming and expensive. Further, to accelerate full pruning, a caching technique is used by checking whether a candidate is already in the frequent k-subtree. If a (k-1)-subtree candidate is already in the frequent k-subtree set, it is known that all its (k-1)-subtrees are frequent, and hence only one comparison is made.

Vertical Occurrence List. To determine if a subtree is frequent, we count the occurrences of that subtree and check if it is greater or equal to the specified minimum support σ. We say that a candidate subtree with encoding L has a frequency n if there

```
Inputs    : T_db(Tree database),•(min.support),Φ(max. level of embedding)
Outputs   : F_k(Frequent subtrees), D(dictionary)
{D, F_1}  : DatabaseScanning (T_db)
{EL, F_2} : ConstructEmbeddedList (F_1,D,Φ)
k=3
while( |F_k| ≥ 0 )
   F_k = GenerateCandidateSubtrees(F_{k-1,}Φ)
   k = k+1

GenerateCandidateSubtrees(F_{k-1},Φ):
for each frequent k-subtree t_{k-1} ∈ F_{k-1}
   L_{k-1} = GetEncoding (t_{k-1})
   VOL-t_{k-1} = GetVOL(t_{k-1})
   for each occurrence coordinate oc_{k-1} (r:[m,...n]) ∈ VOL-t_{k-1}
         for (j = n+1 to |r-EL|-1 )
            if( EmbeddingLevel(j) ≤ Φ) then
                  {oc_k, L_k} = TMG-extend( oc_{k-1},L_{k-1}, j )
               if( Contains(L_k, F_k) )
                  Insert( hashkey(L_k), oc_k, F_k )
               else
                  If( k-1Pruning (L_k) == false)
                     Insert( hashkey(L_k), oc_k, F_k )
return F_k
```

Fig. 5. Pseudo-code of IMB3-Miner algorithm

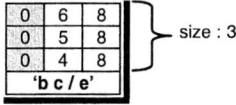

Fig. 6. VOL example of a subtree ('b c / e') of T in fig 1

are n instances of subtrees in the database with the same encoding L. Each occurrence of a subtree is stored as an occurrence coordinate, as previously described in [9]. Computing the frequency of a subtree can be easily determined from the size of the list that stores each occurrence of a subtree. We call such a list as *vertical occurrence list* (VOL). *VOL(L)* denotes the vertical occurrence list of a subtree with encoding L. The frequency of a subtree with encoding L is denoted as |*VOL(L)*|. When transaction based support is used the occurrence of each subtree is grouped by its transaction IDs and the support count corresponds to the number of unique transactions in the VOL.

4 Results and Discussions

We compare IMB3-Miner (IMB3), FREQT (FT) for mining induced subtrees and MB3-Miner (MB3), X3-Miner (X3), VTreeMiner (VTM) and PatternMatcher (PM) for mining embedded subtrees. We created a synthetic database of trees with varying: max. size (s), max. height (h), max. fan-out (f), and number of transactions (|T_r|). Notation XXX–T, XXX-C, and XXX–F are used to denote execution time (including data preprocessing, variables declaration, etc), number of candidate subtrees |C|, and the number of frequent candidate subtrees |F| obtained from the XXX approach respectively. Additionally, IMB3-(NP)-dx notation is used where x refers to the level of embedding $Φ$ and (NP) is optionally used to indicate that full pruning is not performed. The minimum support σ is denoted as (sxx), where xx is the minimum frequency. Occurrence match support was used for all algorithms; Experiments were run on 3Ghz (Intel-CPU), 2Gb RAM, Mandrake 10.2 Linux machine and used GNU g++ (3.4.3) for compilation.

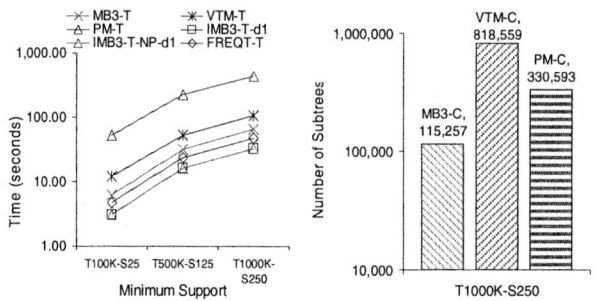

Fig. 7. Scalability test: (a) time performance (b) number of subtrees |C|

Scalability (s:10,h:3,f:3). $|T_r|$ was varied to 100K, 500K & 1000K, with σ set to 25, 125 and 250, respectively. We can see that all algorithms are well scalable (fig 6a). MB3 outperforms VTM & PM for mining embedded subtrees and IMB3 outperforms FT for mining induced subtrees. For $|T_r|$:1000K at σ:250, it can be seen that VTM and PM generate more candidates (VTM-C & PM-C) by using the join approach (fig 6b). Those extra candidates are invalid, i.e. they do not conform to the tree model.

Pseudo-frequent (s:9,h:2,f:5,$|T_r|$:1). We created a dataset that corresponds to the tree T in fig 1 to illustrate the importance of full pruning when occurrence match support is used. We set σ to 2 and compare the number of frequent subtrees generated by various algorithms. From fig 8 we can see that the number of frequent subtrees detected by VTM (DFS) is larger in comparison to PM, MB3 and X3 (BFS). The difference comes from the fact that the three BFS based algorithms perform full pruning whereas DFS based approach such as VTM relies on opportunistic pruning which does not prune pseudo-frequent candidate subtrees. Fig 7 shows that FT & IMB3-NP generate more frequent induced subtrees in comparison to IMB3. This is because they don't perform full pruning, and as such generate extra pseudo-frequent subtrees.

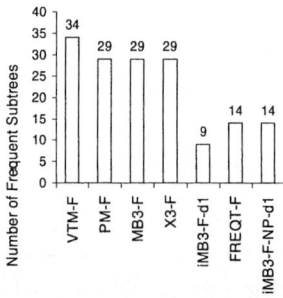

Fig. 8. Pseudo-frequent test: number of frequent subtrees |F|

Deep Tree (s:28,h:17,f:3,|T_r|:10,000) & Wide Tree (s:428,h:3,f:50,|T_r|:6,000). For deep trees (273,090 nodes), when comparing the algorithms for mining frequent embedded subtrees, MB3 has the best performance (fig 9a). VTM aborts when σ<150 where the number of frequent subtrees increases significantly when σ is decreased (fig 9b). At σ:150, VTM generates a superfluous 688x more frequent subtrees compared to MB3 and PM. In regards to mining frequent induced subtrees, fig 8a shows that IMB3 has a slight better time performance than FT. At s80, FT starts to generate pseudo-frequent candidates. For wide tree (1,303,424 nodes), the DFS based approach like VTM outperforms MB3 as expected (fig 9c). However, VTM fails to finish the task when σ<7. We omit IMB3 & FT because the support threshold at which they produce interesting results is too low for embedded subtrees algorithms. In general, the DFS and BFS based approaches suffer from, deep and wide trees respectively.

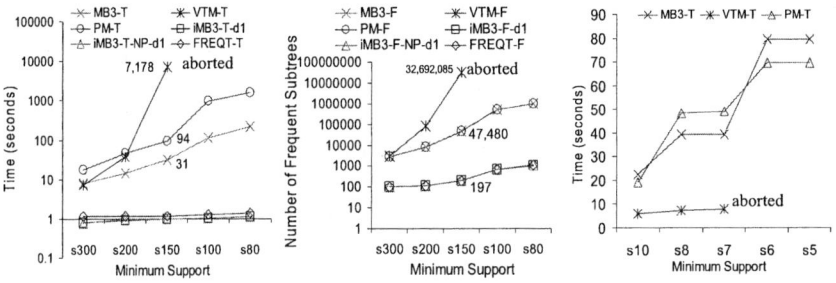

Fig. 9. (a) deep tree time performance (b) deep tree number of frequent subtrees, (c) wide tree time performance

CSLogs (s:214,h:28,f:21). The dataset was used by Zaki in [17]. When used for occurrence match support, the tested algorithms had problems in returning frequent subtrees. Hence, the dataset was trimmed. At |T_r|:32,241, interesting results started to appear. VTM aborts when σ<200 due to numerous numbers of candidates generated. The usefulness of constraining the level of embedding is demonstrated in fig. 10b & 10c. From fig 10b, we can see that the number of frequent subtrees generated by FT & IMB3-NP is identical. Both FT & IMB3-NP generate pseudo-frequent subtrees as they do not perform full pruning. Because of this, the number of frequent induced subtrees detected by FT & IMB3-NP can unexpectedly exceed the number of frequent embedded subtrees found by MB3 & PM (fig 10b, s80).

Fig 10a shows that both IMB3-NP & IMB3 outperform FT. A large time increase for FT and IMB3-NP is observed at σ:200 as a large number of pseudo-frequent subtrees are generated (fig 10b). We also compare the results from VTM, PM & MB3 to the result obtained when the level of embedding is restricted to 6 (IMB3-d6) (fig 10c). By restricting the embedding level, we expect to decrease the execution time without missing many frequent subtrees. The complete set of frequent subtrees was detected at σ≥200, while only less than 2% were missed with σ<200. Overall, MB3 and its variants have the best performance.

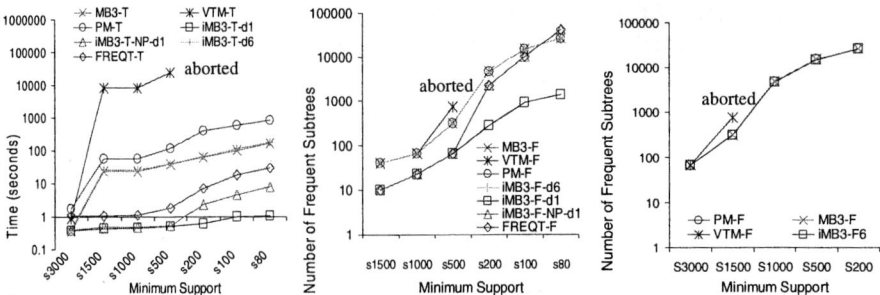

Fig. 10. 54% transactions of original CSLogs data [17] (a) time performance (b) number of frequent subtrees (c) number of frequent subtrees for unconstrained vs constrained approach

Overall Discussion. All MB3 and its variants demonstrate high performance and scalability which comes from the efficient use of the EL representation and the optimal TMG approach that ensures only valid candidates are generated. The join approach utilized in VTM & PM could generate many invalid subtrees which degrades the performance. MB3 performs expensive full pruning, whereas VTM utilizes less expensive opportunistic pruning but suffers from the trade-off that it generates many pseudo-frequent candidate subtrees. This can cause memory blow up and serious performance problem (fig 8a & 10). In the context of association mining, regardless of which approach is used, for a given dataset with minimum support σ, the discovered frequent patterns should be identical and consistent. Assuming pseudo-frequent subtrees are infrequent, techniques that don't perform full pruning would have limited applicability to association rule mining. When representing subtrees, FT [5] uses string labels. VTM, PM, and MB3 (and its variants) use integer labels. When a hashtable is used for candidate frequency counting, hashing integer labels is reported to be faster than hashing string labels especially for long patterns [10]. As we can see, IMB3 & IMB3-NP always outperform FT. When experimenting with the level of embedding constraint (fig 10c), we have found that restricting the level of embedding at a particular level leads to speed increases at the low cost of missing a very small percentage of frequent subtrees while providing a good estimate could be found by restricting the level of embedding.

5 Conclusions

In this study we have provided some detailed discussions about various theoretical and performance issues of the different approaches. We proposed an efficient approach to tackle the complexity of mining embedded subtrees by utilizing a novel *Embedding List* representation, *Tree Model Guided* enumeration, and introducing *Level of Embedding* constraint. High performance and scalability of the proposed approach was demonstrated in our experiments by contrasting it with the state of the art algorithms Tree-Miner and FREQT. Specifically, we studied the problem of embedded subtrees rather than just induced subtrees. Further, we studied the notion of using occurrence match support instead of the simpler transaction based support. We use both synthetic and real datasets in the experimental studies.

Acknowledgement

A special thanks to Prof. M. J. Zaki [17] for providing us the TreeMiner source code and discussing the results obtained from it with us.

References

1. Agrawal, R., Srikant, R.: Fast Algo. for Mining Assoc. Rules. In Proc. the 20th VLDB (1994) 487–499.
2. Abe, K., Kawasoe, S., Asai, T., Arimura, H., Arikawa, S.: Optimized Substructure Discovery for Semistructured Data. In Proc. PKDD'02 (2002) 1–14
3. Chi, Y., Nijssen, S., Muntz, R.R., Kok. J.N.: Frequent Subtree Mining An Overview. Fundamenta Informaticae, Special Issue on Graph and Tree Mining (2005)
4. Feng, L., Dillon, T.S., Weigand, H., Chang, E.: An XML-Enabled Assoc. Rule Framework. In Proc. of DEXA'03 (2003) 88-97
5. Kudo, T.: FREQT Implementation, http://www.chasen.org/~taku/software/freqt/ (2003)
6. Kuramochi, M., Karypis, G.: An Efficient Algo. for Discovering Freq. Subgraphs. IEEE Transactions Knowledge and Data Engineering (2004), Vol. 16, No. 9, 1038-1051
7. Sidhu, A. S., Dillon T. S., et al.: Protein Ontology: Vocabulary for Protein Data. 3rd IEEE ICITA'05. Sydney, (2005) Vol. 1, 465-469.
8. Tan, H., Dillon, T.S., Feng, L., Chang, E., Hadzic, F.: X3-Miner: Mining Patterns from XML Database. In Proc. Data Mining '05. Skiathos, Greece (2005)
9. Tan, H., Dillon, T.S., Hadzic, F., Chang, E., Feng, L.: MB3-Miner: mining eMBedded subTREEs using Tree Model Guided candidate generation. MCD'05. Houston, USA (2005)
10. Tan, H., Dillon, T.S., Hadzic, F., Feng, L., Chang, E.: TMG: Tree Model Guided Candidate Generation. Data Mining'06, Prague, Czech Republic (2006) (Submitted)
11. Termier, A., Rousset, M-C., Sebag, M.: Treefinder: A First Step Towards XML Data Mining. In Proc. IEEE ICDM'02 (2002)
12. Wang, C., Hong, M., Pei, J., Zhou, H., Wang, W., Shi, B.: Efficient Pattern-Growth Methods for Frequent Tree Pattern Mining. In Proc. of PAKDD'04 (2004)
13. Wang, K., Liu, H.: Discovering Typical Structures of Documents: A Road Map Approach. In Proc. ACM SIGIR Conf. Information Retrieval (1998)
14. Yang, L.H., Lee, M.L., Hsu, W.: Efficient Mining of XML Query Patterns for Caching. In Proc. the 29th VLDB Conf. (2003)
15. Zhang, J., Ling, T. W., Bruckner, R.M., Tjoa, A.M., Liu, H.: On Efficient and Effective Association Rule Mining from XML Data. In Proc. of DEXA'04 (2004) 497 - 507
16. Zaki, M.J.: Fast Vertical Mining Using Diffsets. In. Proc. of SIGKDD'03 (2003)
17. Zaki, M.J.: Efficiently Mining Frequent Trees in a Forest: Algorithms and Applications. In IEEE Transaction on Knowledge and Data Engineering (2005), Vol. 17, No. 8, 1021-1035

Maintaining Frequent Itemsets over High-Speed Data Streams*

James Cheng, Yiping Ke, and Wilfred Ng

Department of Computer Science,
Hong Kong University of Science and Technology,
Clear Water Bay, Kowloon, Hong Kong, China
{csjames, keyiping, wilfred}@cs.ust.hk

Abstract. We propose a false-negative approach to approximate the set of *frequent itemsets* (*FIs*) over a sliding window. Existing approximate algorithms use an error parameter, ϵ, to control the accuracy of the mining result. However, the use of ϵ leads to a dilemma. A smaller ϵ gives a more accurate mining result but higher computational complexity, while increasing ϵ degrades the mining accuracy. We address this dilemma by introducing a progressively increasing minimum support function. When an itemset is retained in the window longer, we require its minimum support to approach the minimum support of an FI. Thus, the number of potential FIs to be maintained is greatly reduced. Our experiments show that our algorithm not only attains highly accurate mining results, but also runs significantly faster and consumes less memory than do existing algorithms for mining FIs over a sliding window.

1 Introduction

Frequent itemset (*FI*) mining is fundamental to many important data mining tasks. Recently, the increasing prominence of data streams has led to the study of online mining of FIs [5]. Due to the constraints on both memory consumption and processing efficiency of stream processing, together with the exploratory nature of FI mining, research studies have sought to approximate FIs over streams.

Existing approximation techniques for mining FIs are mainly *false-positive* [5,4,1,2]. These approaches use an *error parameter*, ϵ, to control the quality of the approximation. However, the use of ϵ leads to a dilemma. A smaller ϵ gives a more accurate mining result. Unfortunately, a smaller ϵ also results in an enormously larger number of itemsets to be maintained, thereby drastically increasing the memory consumption and lowering processing efficiency. A *false-negative* approach [6] is proposed recently to address this dilemma. However, the method focuses on the entire history of a stream and does not distinguish recent itemsets from old ones.

---
* This work is partially supported by RGC CERG under grant number HKUST6185/02E and HKUST6185/03E.

We propose a false-negative approach to mine FIs over high-speed data streams. Our method places greater importance on recent data by adopting a sliding window model. To tackle the problem introduced by the use of ϵ, we consider ϵ as a *relaxed minimum support threshold* and propose to progressively increase the value of ϵ for an itemset as it is kept longer in a window. In this way, the number of itemsets to be maintained is greatly reduced, thereby saving both memory and processing power. We design a progressively increasing minimum support function and devise an algorithm to mine FIs over a sliding window. Our experiments show that our approach obtains highly accurate mining results even with a large ϵ, so that the mining efficiency is significantly improved. In most cases, our algorithm runs significantly faster and consumes less memory than do the state-of-the-art algorithms [5, 2], while attains the same level of accuracy.

2 Preliminaries

Let $\mathcal{I} = \{x_1, x_2, \ldots, x_m\}$ be a set of items. An *itemset* is a subset of $\mathcal{I}$. A *transaction*, X, is an itemset and X *supports* an itemset, Y, if $X \supseteq Y$. A *transaction data stream* is a continuous sequence of transactions. We denote a *time unit* in the stream as t_i, within which a variable number of transactions may arrive. A *window* or a *time interval* in the stream is a set of successive time units, denoted as $T = \langle t_i, \ldots, t_j \rangle$, where $i \leq j$, or simply $T = t_i$ if $i = j$. A *sliding window* in the stream is a window that slides forward for every time unit. The window at each slide has a fixed number, w, of time units and w is called the *size* of the window. In this paper, we use t_τ to denote the *current time unit*. Thus, the *current window* is $W = \langle t_{\tau-w+1}, \ldots, t_\tau \rangle$.

We define $trans(T)$ as the set of transactions that arrive on the stream in a time interval T and $|trans(T)|$ as the number of transactions in $trans(T)$. The *support* of an itemset X over T, denoted as $sup(X, T)$, is the number of transactions in $trans(T)$ that support X. Given a predefined *Minimum Support Threshold (MST)*, σ ($0 \leq \sigma \leq 1$), we say that X is a *frequent itemset (FI)* over T if $sup(X, T) \geq \sigma |trans(T)|$.

Given a transaction data stream and an MST σ, the problem of *FI mining over a sliding window* is to find the set of all FIs over the window at each slide.

3 A Progressively Increasing MST Function

Existing approaches [5, 4, 2] use an error parameter, ϵ, to control the mining accuracy, which leads to a dilemma. We tackle this problem by considering $\epsilon = r\sigma$ as a relaxed MST, where r ($0 \leq r \leq 1$) is the *relaxation rate*, to mine the set of FIs over each time unit t in the sliding window. Since all itemsets whose support is less than $r\sigma|trans(t)|$ are discarded, we define the *computed support* as follows.

Definition 1 (Computed Support). The *computed support* of an itemset X over a time unit t is defined as follows:

$$\widetilde{sup}(X, t) = \begin{cases} 0 & \text{if } sup(X, t) < r\sigma|trans(t)| \\ sup(X, t) & \text{otherwise.} \end{cases}$$

The *computed support* of X over *a time interval* $T = \langle t_j, \ldots, t_l \rangle$ is defined as

$$\widetilde{sup}(X,T) = \sum_{i=j}^{l} \widetilde{sup}(X,t_i).$$

□

Based on the computed support of an itemset, we apply *a progressively increasing MST function* to define *a semi-frequent itemset*.

Definition 2 (Semi-Frequent Itemset). Let $W = \langle t_{\tau-w+1}, \ldots, t_\tau \rangle$ be a window of size w and $T^k = \langle t_{\tau-k+1}, \ldots, t_\tau \rangle$, where $1 \le k \le w$, be the most recent k time units in W. We define a *progressively increasing* function

$$minsup(k) = \lceil m_k \times r_k \rceil,$$

where $m_k = \sigma|trans(T^k)|$ and $r_k = (\frac{1-r}{w})(k-1) + r$.

An itemset X is a *semi-frequent itemset* (*semi-FI*) over W if $\widetilde{sup}(X,T^k) \ge minsup(k)$, where $k = \tau - o + 1$ and t_o is the oldest time unit such that $\widetilde{sup}(X,t_o) > 0$. □

The first term m_k in the *minsup* function in Definition 2 is the minimum support required for an FI over T^k, while the second term r_k progressively increases the relaxed MST $r\sigma$ at the rate of $((1-r)/w)$ for each older time unit in the window. We keep X in the window only if its computed support over T^k is no less than $minsup(k)$, where T^k is the time interval starting from the time unit t_o, in which the support of X is computed, up to the current time unit t_τ.

4 Mining FIs over a Sliding Window

We use a prefix tree to keep the semi-FIs. A node in the prefix tree represents an itemset, X, and has three fields: (1) *item* which is the last item of X; (2) $uid(X)$ which is the ID of the time unit, $t_{uid(X)}$, in which X is inserted into the prefix tree; (3) $\widetilde{sup}(X)$ which is the computed support of X since $t_{uid(X)}$.

The algorithm for mining FIs over a sliding window, *MineSW*, is given in Algorithm 1, which is self-explanatory.

Algorithm 1 (MineSW)
Input: *(1) An empty prefix tree. (2) σ, r and w. (3) A transaction data stream.*
Output: *An approximate set of FIs of the window at each slide.*

1. Mine all FIs over each time unit using a relaxed MST $r\sigma$.
2. **Initialization:** For each of the first w time units, t_i ($1 \le i \le w$), mine all FIs from $trans(t_i)$. For each mined itemset, X, check if X is in the prefix tree.
   (a) If X is in the prefix tree, perform the following operations: (i) Add $\widetilde{sup}(X,t_i)$ to $\widetilde{sup}(X)$; (ii) If $\widetilde{sup}(X) < minsup(i - uid(X) + 1)$, remove X from the prefix tree and stop mining the supersets of X from $trans(t_i)$.
   (b) If X is not in the prefix tree, create a new node for X in the prefix tree with $uid(X) = i$ and $\widetilde{sup}(X) = \widetilde{sup}(X,t_i)$.
3. **Incremental Update:**
   – For each expiring time unit, $t_{\tau-w+1}$, mine all FIs from $trans(t_{\tau-w+1})$. For each mined itemset, X:

- If X is in the prefix tree and $\tau - uid(X) + 1 \geq w$, subtract $\widetilde{sup}(X, t_{\tau-w+1})$ from $\widetilde{sup}(X)$. Otherwise, stop mining the supersets of X from $trans(t_{\tau-w+1})$.
- If $\widetilde{sup}(X)$ becomes 0, remove X from the prefix tree. Otherwise, set $uid(X) = \tau - w + 2$.
  - For each incoming time unit, t_τ, mine all FIs from $trans(t_\tau)$. For each mined itemset, X, check if X is in the prefix tree.
    (a) If X is in the prefix tree, perform the following operations: (i) Add $\widetilde{sup}(X, t_\tau)$ to $\widetilde{sup}(X)$; (ii) If either $\tau - uid(X) + 1 \leq w$ and $\widetilde{sup}(X) < minsup(\tau - uid(X) + 1)$, or $\tau - uid(X) + 1 > w$ and $\widetilde{sup}(X) < minsup(w)$, remove X from the prefix tree and stop mining the supersets of X from $trans(t_\tau)$.
    (b) If X is not in the prefix tree, create a new node for X in the prefix tree with $uid(X) = \tau$ and $\widetilde{sup}(X) = \widetilde{sup}(X, t_\tau)$.
4. **Pruning and Outputting:** Scan the prefix tree once. For each itemset X visited:
   - Remove X and its descendants from the prefix tree if (1) $\tau - uid(X) + 1 \leq w$ and $\widetilde{sup}(X) < minsup(\tau - uid(X) + 1)$, or (2) $\tau - uid(X) + 1 > w$ and $\widetilde{sup}(X) < minsup(w)$.
   - Output X if $\widetilde{sup}(X) \geq \sigma|trans(W)|$ (we can thus set $minsup(w) = \sigma|trans(W)|$ to prune more itemsets).

5 Experimental Evaluation

We run our experiments on a Sun Ultra-SPARC III with 900 MHz CPU and 4GB RAM. We compare our algorithm *MineSW* with a variant of the *Lossy Counting* algorithm [5] applied in the sliding window model, denoted as *LCSW*. We remark that LCSW, which updates a batch of incoming/expiring transactions at each window slide, is different from the algorithm proposed by Chang and Lee [2], which updates on each incoming/expiring transaction. We implement both algorithms and find that the algorithm by Chang and Lee is much slower than LCSW and runs out of our 4GB memory. We generate two types of data streams, t10i4 and t15i6, using a generator [3] that modifies the IBM data generator.

We first find (see details in [3]) that when r increases from 0.1 to 1, the precision of LCSW ($\epsilon = r\sigma$ in LCSW) drops from 98% to around 10%, while the recall of MineSW only drops from 99% to around 90%. This result reveals that the estimation mechanism of the Lossy Counting algorithm relies on ϵ to control the mining accuracy, while our progressively increasing *minsup* function maintains a high accuracy which is only slightly affected by the change in r. Since increasing r means faster mining process and less memory consumption, we can use a larger r to obtain highly accurate mining results at much faster speed and less memory consumption.

We test $r = 0.1$ and $r = 0.5$ for MineSW. According to Lossy Counting [5], a good choice of ϵ is 0.1σ and hence we set $r = 0.1$ for LCSW. Fig. 1 (a) and (b) show that for all σ, the precision of LCSW is over 94% and the recall of MineSW is over 96% (mostly over 99%). The recall of MineSW ($r = 0.5$) is only slightly lower than that of MineSW ($r = 0.1$). However, Fig. 2 (a) and (b) show that MineSW ($r = 0.5$) is significantly faster than MineSW ($r = 0.1$) and LCSW, especially when σ is small. Fig. 3 (a) and (b) show the memory consumption of

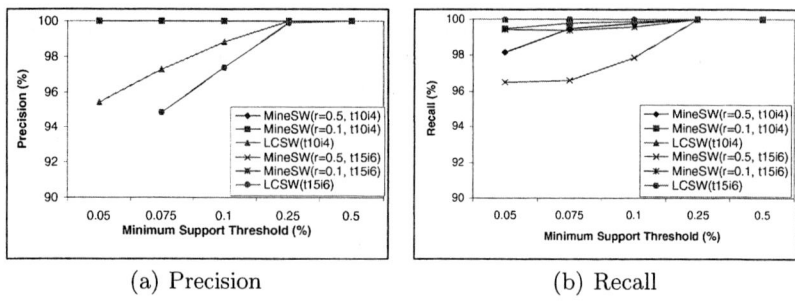

Fig. 1. Precision and Recall with Varying Minimum Support Threshold

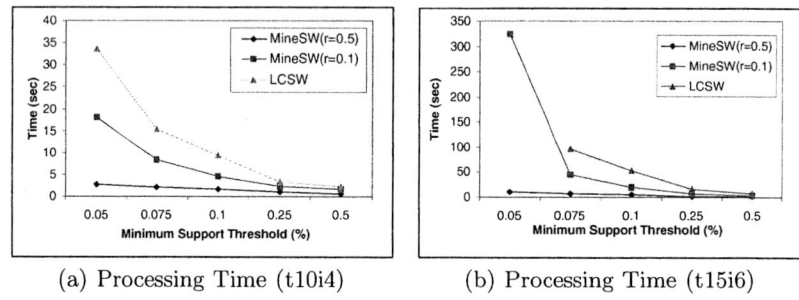

Fig. 2. Processing Time with Varying Minimum Support Threshold

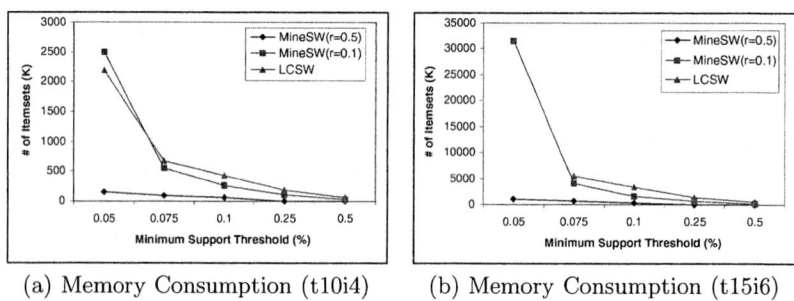

Fig. 3. Memory Consumption with Varying Minimum Support Threshold

the algorithms in terms of the number of itemsets maintained at the end of each slide. The number of itemsets kept by MineSW ($r = 0.1$) is about 1.5 times less than that of LCSW, while that kept by MineSW ($r = 0.5$) is less than that of LCSW by up to several orders of magnitude.

6 Conclusions

We propose a progressively increasing minimum support function, which allows us to increase ϵ at the expense of only slightly degraded accuracy, but significantly improves the mining efficiency and saves memory usage. We verify, by extensive experiments, that our algorithm is significantly faster and consumes less memory than existing algorithms, while attains the same level of accuracy. When applications require highly accurate mining results, our experiments show that by setting $\epsilon = 0.1\sigma$ (a rule-of-thumb choice of ϵ in Lossy Counting [5]), our algorithm attains 100% precision and over 99.99% recall.

References

1. J. H. Chang and W. S. Lee. estWin: Adaptively Monitoring the Recent Change of Frequent Itemsets over Online Data Streams. In *Proc. of CIKM*, 2003.
2. J. H. Chang and W. S. Lee. A Sliding Window method for Finding Recently Frequent Itemsets over Online Data Streams. In *Journal of Information Science and Engineering*, Vol. 20, No. 4, July, 2004.
3. J. Cheng, Y. Ke, and W. Ng. Maintaining Frequent Itemsets over High-Speed Data Streams. *Technical Report*, http://www.cs.ust.hk/~csjames/pakdd06tr.pdf.
4. H. Li, S. Lee, and M. Shan. An Efficient Algorithm for Mining Frequent Itemsets over the Entire History of Data Streams. In *Proc. of First International Workshop on Knowledge Discovery in Data Streams*, 2004.
5. G. S. Manku and R. Motwani. Approximate Frequency Counts over Data Streams. In *Proc. of VLDB*, 2002.
6. J. Yu, Z. Chong, H. Lu, and A. Zhou. False Positive or False Negative: Mining Frequent Itemsets from High Speed Transactional Data Streams. In *VLDB*, 2004.

Generalized Disjunction-Free Representation of Frequents Patterns with at Most k Negations*

Marzena Kryszkiewicz

Institute of Computer Science, Warsaw University of Technology,
Nowowiejska 15/19, Warsaw 00-665, Poland
mkr@ii.pw.edu.pl

Abstract. The discovery of frequent patterns and their representations has attracted a lot of attention in the data mining community. An extensive research has been carried out mainly in discovering positive patterns. Recently, the generalized disjunction–free representation GDFLR of all frequent patterns both with and without negation has been proposed. There are cases, however, when a user is interested in patterns with a restricted number of negated items. In this paper, we offer the k-GDFLR representation as an adaptation of GDFLR, which represents all frequent patterns with at most k negated items. Algorithms discovering this representation are discussed as well. The experimental results show that k-GDFLR is more concise than GDFLR.

1 Introduction

Discovering of frequent patterns in large databases is an important data mining problem. The problem was introduced in [1] for a sales transaction database. Frequent patterns were defined there as sets of items that are purchased together frequently. Frequent patterns are commonly used for building association rules. Patterns and association rules can be generalized by admitting negation. A sample rule with negation could state that 75% of customers who buy coke also buy chips and neither beer nor milk. Admitting negation usually results in abundance of mined patterns. It is thus preferable to discover and store a possibly small fraction of patterns from which one can derive all other significant patterns. In [2-3], a generalized disjunction-free literal set representation (GDFLR) was offered as a lossless representation of all frequent patterns, both with and without negation. GDFLR is by orders of magnitude more concise than all frequent patterns [2]. Nevertheless, GDFLR can still be numerous. In addition, the patterns with large number of negated items might not be of interest to a user at all. In this paper, we propose the k-GDFLR representation as an adaptation of GDFLR to represent all frequent patterns with at most k negated items. We introduce necessary modifications in algorithms discovering the GDFLR representation to adapt them to discover frequent patterns with a restricted number of negated items. Finally, we present the results of the experiments testing the conciseness of the new representation w.r.t. the number of allowed negated items.

---

* Research has been supported by grant No 3 T11C 002 29 received from Polish Ministry of Education and Science.

2 Basic Notions

Let us analyze sample transactional database D presented in Table 1. Each row in D reports items that were bought by a customer during a single visit to a supermarket.

Table 1. Sample database D

Id	Transaction
T_1	{abce}
T_2	{abcef}
T_3	{abch}
T_4	{abe}
T_5	{acfh}
T_6	{bef}
T_7	{h}
T_8	{af}

As follows from Table 1, items a and b were purchased together in four transactions. The number of transactions in which set of items $\{x_1, ..., x_n\}$ occurs is called its *support* and is denoted by $sup(\{x_1, ..., x_n\})$. A set of items is called a *frequent pattern* if its support exceeds a user-specified threshold (*minSup*). Otherwise, it is called *infrequent*. Clearly, the support of a pattern never exceeds the supports of its subsets. Thus, subsets of a frequent pattern are also frequent, and supersets of an infrequent pattern are infrequent.

A pattern consisting of items $x_1, ..., x_m$ and negations of items $x_{m+1}, ..., x_n$ is denoted by $\{x_1, ..., x_m, -x_{m+1}, ..., -x_n\}$. The *support of pattern* $\{x_1, ..., x_m, -x_{m+1}, ..., -x_n\}$ is defined as the number of transactions in which all items in set $\{x_1, ..., x_m\}$ occur and no item in set $\{x_{m+1}, ..., x_n\}$ occurs. A pattern X is called *positive*, if it does not contain any negated item. Otherwise, X is called a *pattern with negation(s)*. A pattern obtained from X by negating any number of items in X is called a *variation of X*.

One can easily note that $sup(X \cup \{(-x)\}) = sup(X) - sup(X \cup \{x\})$ [6]. Multiple usage of this property enables calculation of the supports of patterns with any number of negated items from the supports of their positive variations and their subsets [6]:

$$sup(\{x_1,...,x_m\} \cup \{-x_{m+1},...,-x_n\}) = \Sigma_{Z \subseteq \{x_{m+1}, ..., x_n\}} (-1)^{|Z|} \times sup(\{x_1,...,x_m\} \cup Z) \quad (1)$$

Nevertheless, the knowledge of the supports of only frequent positive patterns may be insufficient to derive the supports of all frequent patterns with negation [6].

A *generalized disjunctive rule* based on a positive pattern $X = \{x_1, ..., x_n\}$ is defined as an expression of the form $x_1 ... x_m \rightarrow x_{m+1} \vee ... \vee x_n$, where $\{x_1, ..., x_m\} \cap \{x_{m+1}, ..., x_n\} = \emptyset$ and $\{x_{m+1}, ..., x_n\} \neq \emptyset$. We say that a transaction *supports* rule $r: x_1 ... x_m \rightarrow x_{m+1} \vee ... \vee x_n$ if it contains all items in $\{x_1, ..., x_m\}$ and at least one item in $\{x_{m+1}, ..., x_n\}$. We say that a transaction *violates* rule r if it contains all items in $\{x_1, ..., x_m\}$ and no item in $\{x_{m+1}, ..., x_n\}$. The number of transactions violating rule r is called its *error* and is denoted by $err(r)$. It was shown in [2] that $err(r)$ is determinable from the supports of subsets of $\{x_1, ..., x_m, x_{m+1}, ..., x_n\}$:

$$err(x_1 ... x_m \rightarrow x_{m+1} \vee ... \vee x_n) = \Sigma_{Z \subseteq \{x_{m+1}, ..., x_n\}} (-1)^{|Z|} \times sup(\{x_1, ..., x_m\} \cup Z) \quad (2)$$

The following equation follows immediately from Eq. 1 and Eq. 2:

$$err(x_1 ... x_m \rightarrow x_{m+1} \vee ... \vee x_n) = sup(\{x_1, ..., x_m\} \cup \{-x_{m+1}, ..., -x_n\}) \quad (3)$$

Rule $x_1 ... x_m \rightarrow x_{m+1} \vee ... \vee x_n$ is an implication ($x_1 ... x_m \Rightarrow x_{m+1} \vee ... \vee x_n$) if $err(x_1 ... x_m \rightarrow x_{m+1} \vee ... \vee x_n) = 0$. Clearly, if $x_1 ... x_m \rightarrow x_{m+1} \vee ... \vee x_n$ is an implication, then $x_1 ... x_m z \rightarrow x_{m+1} \vee ... \vee x_n$ and $x_1 ... x_m \rightarrow x_{m+1} \vee ... \vee x_n \vee z$, which are based on a superset of $\{x_1, ..., x_n\}$, are also implications. Such implications can be used for calculating the supports of patterns on which they are based. E.g., $ac \Rightarrow b \vee f$ implies that $sup(\{ac\}) = sup(\{abc\}) + sup(\{acf\}) - sup(\{abcf\})$. Hence, the support of pattern $\{abcf\}$ is determinable from the supports of its proper subsets. In general, if there is an implication based on a positive pattern, then the support of this pattern is derivable from the supports of its proper subsets [5]. Each such pattern is called a *generalized disjunctive set*. Otherwise, it is called a *generalized disjunction-free set*. It can be noted that supersets of generalized disjunctive sets are generalized disjunctive and subsets of generalized disjunction-free sets are generalized disjunction-free [5].

3 Representing Frequent Patterns with at Most k Negations

In this section, we offer a *generalized disjunction-free literal set representation of frequent patterns with at most k negations* (k-GDFLR) as consisting of the following components:

1. the *main component* (k-Main) containing each positive pattern (stored with its support) that has at least one frequent variation with at most k negations and is neither generalized disjunctive nor has support equal 0;
2. the *infrequent border* (k-IBd^-) containing each positive pattern all variations of which that have at most k negations are infrequent and all proper subsets of which belong to k-Main;
3. the *generalized disjunctive border* (k-DBd^-) containing each positive pattern (stored with its support and/or implication) that is either generalized disjunctive or has support equal 0, has at least one frequent variation with at most k negations and has all its proper subsets in k-Main.

It can be observed that k-GDFLR equals GDFLR (please, see [2-3] for the definition of GDFLR) for $k = \infty$ (unrestricted number of negated items allowed).

Lemma 3.1.
a) If P belongs to k-Main, then all subsets of P belong to k-Main.
b) If P belongs to k-IBd^-, then for each superset of P all its variations with at most k negations are infrequent.
c) If P belongs to k-DBd^-, then each superset of P is either generalized disjunctive or has support equal 0.

Lemma 3.2.
a) k-Main $\cup$ k-DBd^- is the set containing each positive pattern that has at least one frequent variation with at most k negations and all proper subsets in k-Main.
b) k-Main $\cup$ k-DBd^- $\cup$ k-IBd^- is the set containing each positive pattern all proper subsets of which belong to k-Main.
c) k-DBd^- $\cup$ k-IBd^- is the set containing each positive pattern not contained in k-Main, but having all proper subsets in k-Main.
d) $2^I \setminus (k\text{-Main} \cup k\text{-}DBd^- \cup k\text{-}IBd^-) = \{X \subseteq I | \exists Y \subset X \ Y \in k\text{-}DBd^- \cup k\text{-}IBd^-\}$.

As follows from Lemma 3.2d, positive patterns that do not belong to k-GDFLR have at least one proper subset in the border k-DBd^- $\cup$ k-IBd^-.

Lemma 3.3 (calculating supports of positive patterns based on k-GDFLR). The k-GDFLR representation is sufficient to determine for any positive pattern X, if it is frequent, and if so, enables determining its support. In addition, k-GDFLR determines the support of each positive pattern X that does not have subsets in k-IBd^-.

Proof: Analogous to the proof of Lemma 3.3.2 in [2]. □

Eventually, we conclude that k-GDFLR is a lossless representation of all frequent patterns with at most k negations.

Theorem 3.1 (calculating supports of patterns with at most k negations based on k-GDFLR). The k-GDFLR representation determines for any pattern X with at most k negations whether it is frequent, and if so, enables determining its support. In addition, the k-GDFLR representation determines the support of each pattern X with at most k negations that does not have subsets in k-IBd^-.

Proof (constructive): Let X be a pattern with at most k negations and $Pos(X)$ be a positive variation of X. If $\exists Y \in k$-IBd^- $Y \subseteq Pos(X)$, then X is not frequent. Otherwise, $Pos(X)$ and all its subsets do not have subsets in k-IBd^-. Thus, by Lemma 3.3, k-GDFLR determines the supports of $Pos(X)$ and all its proper subsets. In accordance with Eq. 1, this suffices to calculate $sup(X)$. □

Below we determine upper bounds on the length of elements in k-GDFLR.

Theorem 3.2.
a) $\forall Z \in k$-$Main$, $|Z| \leq \lfloor \log_2(|D| - k\text{-}minSup) \rfloor$.
b) $\forall Z \in k$-$DBd^- \cup k$-IBd^-, $|Z| \leq \lfloor \log_2(|D| - minSup) \rfloor + 1$.

Proof: Analogous to the proof of Theorem 3.3.3 in [2]. □

4 Discovering the k-GDFLR Representation

```
Algorithm k-GDFLR-Apriori(support threshold minSup; maximal allowed number of negations k);
k-Main = { }; k-DBd⁻ = { }; k-IBd⁻ = {∅};                                    // initialize GDFLR
if |D| > minSup then begin
    ∅.sup = |D|; move ∅ from k-IBd⁻ to k-Main₀;  X₁ = {1 item patterns};
    for (i = 1; Xᵢ ≠ ∅; i++) do begin
        calculate the supports of i item patterns in Xᵢ within one scan of database D;
        forall candidates X∈ Xᵢ do begin
            /* calculate the errors of all generalized dis. rules based on X with at most k disjuncts (by Eq. 2)*/
            Errs[1-k] = Errors-of-rules(X, k-Main, [1 .. k]);
            if max({X.sup}∪Errs[1-k]) ≤ minSup then
                /* all variations of X with at most k negations are infrequent (by Eq. 3) */
                add X to k-IBd⁻ᵢ
            else begin
                /* calculate the errors of all generalized disjunctive rules based on X (by Eq. 2) */
                Errs = Errs[1-k] ∪ Errors-of-rules(X, k-Main, [k+1 .. |X|]);
                if min({X.sup}∪Errs) = 0 then add X to k-DBd⁻ᵢ    // there is a generalized dis. variation of X
                else add X to k-Mainᵢ endIf
            endif
        endfor;
        /* create new i+1 item candidates by merging i item patterns in k-Mainᵢ */
        Xᵢ₊₁ = {X⊆I| ∃Y,Z ∈ k-Mainᵢ (|Y∩Z| = i−1 ∧ X = Y∪Z)};
        /* remain only those candidates that have all i item subsets in k-Mainᵢ */
        Xᵢ₊₁ = Xᵢ₊₁ \ {X∈Xᵢ₊₁| ∃Y⊆X (|Y| = k ∧ Y∉ k-Mainᵢ}
    endfor
endif;
return <∪ᵢ k-Mainᵢ, ∪ᵢ k-DBd⁻ᵢ, ∪ᵢ k-IBd⁻ᵢ>;
```

Here, we offer the *k-GDFLR-Apriori* algorithm to discover *k*-GDFLR as a modification of *GDFLR-Apriori* [2]. The differences between *k-GDFLR-Apriori* and *GDFLR-Apriori* are highlighted in the code. Similar modification of the *GDFLR-SO-Apriori* algorithm [4], which is an alternative to *GDFLR-Apriori*, would result in obtaining the *k-GDFLR-SO-Apriori* algorithm adapted to discovering *k*-GDFLR.

5 Experimental Results

Let us present the experimental results obtained for the benchmark data sets: *mushroom* (8124 transactions of length 23 items; 119 distinct items) and *connect-4* (67557 transactions of length 43 items; 129 distinct items). Fig. 1 shows the cardinality of *k*-GDFLR for $k = 0$ (no negation allowed), 2, 3, and ∞.

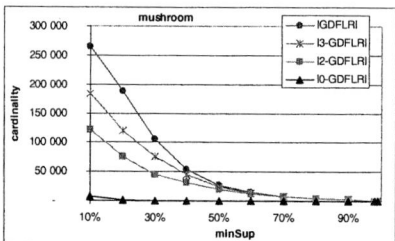 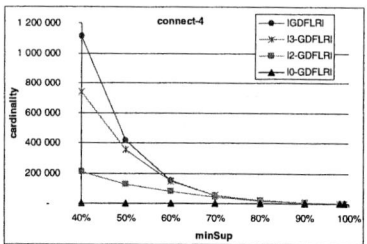

Fig. 1. Cardinalities of the representations (linear scale)

6 Conclusions

We have offered the lossless *k*-GDFLR representation of all frequent patterns with a restricted number of negated items. We have shown how to adapt the algorithms discovering GDFLR to discover *k*-GDFLR. The conducted experiments prove that *k*-GDFLR is more concise than GDFLR especially for low support values and low *k*.

References

[1] Agrawal, R., Imielinski, T., Swami, A.: Mining Associations Rules between Sets of Items in Large Databases. In: Proc. of the ACM SIGMOD, Washington, USA (1993) 207–216
[2] Kryszkiewicz, M.: Generalized Disjunction-Free Representation of Frequent Patterns with Negation. JETAI, Taylor & Francis Group, UK (2005) 63–82
[3] Kryszkiewicz, M.: Reasoning about Frequent Patterns with Negation. Encyclopedia of Data Warehousing and Mining. Idea Group Reference (2005) 941–946
[4] Kryszkiewicz, M., Cichon, K.: Support Oriented Discovery of Generalized Disjunction-Free Representation of Frequent Patterns with Negation. In: Proc. of PAKDD'05 (2005) 672-682
[5] Kryszkiewicz, M., Gajek, M.: Concise Representation of Frequent Patterns based on Generalized Disjunction-Free Generators. In: Proc. of PAKDD'02 (2002) 159–171
[6] Toivonen, H.: Discovery of Frequent Patterns in Large Data Collections. Ph.D. Thesis, Report A-1996-5, University of Helsinki (1996)

Mining Interesting Imperfectly Sporadic Rules

Yun Sing Koh, Nathan Rountree, and Richard O'Keefe

Department of Computer Science, University of Otago, New Zealand
{ykoh, rountree, ok}@cs.otago.ac.nz

Abstract. Detecting association rules with low support but high confidence is a difficult data mining problem. To find such rules using approaches like the Apriori algorithm, *minimum* support must be set very low, which results in a large amount of redundant rules. We are interested in *sporadic* rules; i.e. those that fall below a *maximum* support level but above the level of support expected from random coincidence. In this paper we introduce an algorithm called MIISR to find a particular type of sporadic rule efficiently: where the support of the antecedent as a whole falls below maximum support, but where items may have quite high support individually. Our proposed method uses item constraints and coincidence pruning to discover these rules in reasonable time.

1 Introduction

There are many association mining algorithms dedicated to *frequent* itemset mining [1, 2, 3, 4, 5]. These algorithms are defined in such a way that they only find rules with high support and high confidence. A much less explored area in association mining is *infrequent* itemset mining. Recently, Koh and Rountree [6] proposed the Apriori-Inverse algorithm to mine infrequent itemsets without generating any frequent rules. It captures so-called sporadic rules using *maximum* support (maxsup) and minimum confidence (minconf) thresholds. The support of itemsets forming each rule is below the maxsup threshold but above a user-defined minimum absolute support value. They define the notion of *perfectly sporadic* rules, where the itemset forming each rule consists only of items that are below the maxsup threshold. In contrast, *imperfectly sporadic* rules consist of individual items with high support but the support of the intersection of the items is low.

Apriori-Inverse is not able to find imperfectly sporadic rules because it never considers itemsets that have support above maxsup; therefore no subset of any itemset that it generates can have support above maxsup. Apriori will miss these rules, because the support for the itemsets forming the rules is too low. Apriori-Inverse will miss them as well, because the support for the individual items is too high. Therefore, both algorithms will miss rules of the form $AB \rightarrow C$, where A and B are individually common, but AB is rare and C is rare. This, for example, is the situation where two symptoms—both of which commonly occur alone—occur together only rarely; but when they do, the combination indicates a rare and serious disease with high confidence. We consider this to be a very interesting type of rule to be able to find.

The aim of our research is to develop a technique to mine imperfectly sporadic rules efficiently. To force any variant of the Apriori algorithm [2] to find imperfectly sporadic rules, the minimum support threshold must be set very low. This in turn drastically increases the running time of the algorithm, due to a combinatorial explosion in the number of frequent itemsets. Apriori-Inverse suffers the same problem in reverse: maximum support has to be set so high that too many itemsets qualify as sporadic.

In this paper, we propose an algorithm called MIISR (Mining Interesting Imperfectly Sporadic Rules) to find imperfectly sporadic rules using item constraints: we capture rules with a single-item consequent below the maxsup threshold. The maxsup threshold is used to identify all items that are considered rare. These items are then considered to be the only possible consequents for all rules that will be generated. Items in the transactions containing the consequent are then detected. The items found are used to form antecedents that have strong associations with the consequents.

Inherently we are looking for rules with low support that could make them indistinguishable from coincidences (that is, situations where items fall together no more often than would be allowed by chance). Hence, we use coincidence pruning to remove the occurrences of coincidental itemsets. For an itemset to be considered non-coincidental it must have support above a minimum absolute support (minabssup) value which is generated using a variant of Fisher's exact test. The rest of this paper is organised as follows. Definitions pertinent to infrequent itemset mining and a review of related work are given in Section 2. The MIISR algorithm and an explanation of coincidence pruning is presented in Section 3. In Section 4, we evaluate MIISR on synthetic and real datasets, and in Section 5 we conclude the paper.

2 Related Work

The following is a formal statement of association rule mining for transaction databases. Let $I = \{i_1, i_2, \ldots, i_m\}$ be the universe of items and D be a set of transactions, where each transaction T is a set of items such that $T \subseteq I$. An association rule is an implication of the form $X \rightarrow Y$, where $X \subset I, Y \subset I$, and $X \cap Y = \emptyset$. X is referred to as the *antecedent* of the rule, and Y as the *consequent*. The rule $X \rightarrow Y$ holds in the transaction set D with *confidence* $c\%$ if $c\%$ of transactions in D that contain X also contain Y. The rule $X \rightarrow Y$ has *support* $s\%$ in the transaction set D, if $s\%$ of transactions in D contain XY [2]. Throughout this paper we shall use XY to denote an itemset that contains both X and Y.

One way of forcing low-support items to take part in mined rules is by imposing *item constraints*; i.e., providing a list of those items that may or may not take part in a rule and then modifying the mining process to take advantage of that information [7, 8, 9, 12, 11]. One of the restrictions that may be imposed is called *consequent constraint-based rule mining*. Among these we shall discuss Dense-Miner, EP (Emerging Pattern), and Fixed-Consequent ARM (Association Rule Mining).

Bayardo et al. [9] proposed a consequent constraint-based rule mining approach called Dense-Miner. They require mined rules to have a given consequent C specified by the user. This approach introduces an additional metric called *improvement*. The *improvement* of a rule is defined as the minimum difference between its confidence and the confidence of any proper sub-rule with the same consequent.

$$\text{imp}(A \rightarrow C) = \min(\forall A' \subset A, \text{conf}(A \rightarrow C) - \text{conf}(A' \rightarrow C))$$

If the *imp* of a rule is greater than 0, then removing any non-empty combination of items from the antecedent will lower the confidence by at least the improvement.

Emerging pattern (EP) was proposed by Li et al. [12]. Given a known consequent T, they use a dataset partitioning approach to find "top", "zero-confidence", and "μ-level confidence" rules. The dataset D is divided into sub-datasets D_1 and D_2; where D_1 consists of the transactions containing T and D_2 consists of transactions which do not contain T. All items in T are then removed from D_1 and D_2. Using the transformed dataset, EP then finds all itemsets X which occur in D_1 but not in D_2. For each X, the rule $X \to T$ is a "top rule" in D with confidence of 100%. On the other hand, for all itemsets Z that only occur in D_2, all transactions in D which contain Z must not contain T. Therefore $Z \to T$ has a negative association and is a "zero-confidence" rule. For "μ-level confidence" rules $Y \to T$ the confidences are greater than or equal to $1 - \mu$.

Rahal et al. [11] propose a slightly different approach. Fixed-Consequent ARM generates minimal confidence rules using SE trees and P-trees. Given two rules R_1 and R_2, with confidence values higher than the confidence threshold, where R_1 is $A \to C$ and R_2 is $AB \to C$, R_1 is preferred, because the antecedent of R_2 is a superset of the antecedent of R_1. The support of R_1 is greater than or equal to R_2. R_1 is considered a minimal rule and R_2 is considered a non-minimal rule. The algorithm was devised to generate the *highest* support rules that match the user specified minimum confidence threshold without having the user specify any support threshold.

The drawback to all three approaches is that they are only useful when we have prior knowledge that a particular consequent is of interest. For our application, we are interested in searching for imperfectly sporadic rules, without having to wade through a lot of rules that have high support, without generating a large number of trivial rules, and without needing prior knowledge of which consequents ought to be interesting.

3 Proposal of MIISR

In the previous section, the techniques discussed might generate some imperfectly sporadic rules, if there is prior knowledge of a rare and interesting consequent, and if minsup is set low enough. So, rather than address the problem in the context of frequent itemset mining, we suggest explicitly treating it as a problem of *infrequent* itemset mining. Hence we propose the MIISR algorithm to mine interesting imperfectly sporadic rules. This algorithm uses the same definition of maxsup as in Apriori-Inverse [6]. Since itemsets lose support as they grow larger, and our guiding constraint is maxsup rather than minsup, we can no longer rely on a downward-closure principle.

We begin by searching for any individual items below maxsup and using these as candidate consequents. For each candidate consequent, we then generate candidate antecedents from the items within the same transactions. Because we are dealing with candidate itemsets with low support, it is possible that we will see items occurring together in transactions about as many times as chance would allow—we refer to this situation as a *coincidence*. Itemsets that occur within the database due to coincidence do not add meaningful information and should be ignored. Hence we identify a minabssup value to filter out these itemsets. The imperfectly sporadic rules are then generated in a similar fashion to Apriori. As we are storing the transactional dataset in an inverted index, we note that our method does not require dataset partitioning.

3.1 Imperfectly Sporadic Rules

A rule is considered imperfectly sporadic if it meets the requirements of maxsup and minconf but contains any items that have support above maxsup. For instance, suppose we had an itemset AB with support(A)=12%, support(B)=10%, and support(AB)= 10%, with maxsup = 11% and minconf = 75%. Both $A \rightarrow B$ (confidence = 92%) and $B \rightarrow A$ (confidence = 100%) are sporadic in that they have low support and high confidence. *Imperfectly* sporadic rules are defined as in [6]:

Definition:

$A \rightarrow B$ is *imperfectly sporadic* for maxsup s and minconf c iff

$$\text{confidence}(A \rightarrow B) \geq c, \quad \text{and}$$
$$\text{support}(AB) < s, \quad \text{and}$$
$$\exists x : x \in (AB), \quad \text{support}(x) \geq s$$

Some imperfectly sporadic rules could be completely trivial or uninteresting: for instance, when the antecedent is rare but the consequent has support of 100%. We can characterise four different types of imperfectly sporadic rule:

Type 1 rules have both frequent *and* infrequent itemsets in antecedent and consequent. They may end up sporadic due to two or more frequent items occurring *together* infrequently.
Type 2 rules have *only* frequent itemsets in both antecedent and consequent. They too may be sporadic due to two or more items occurring together infrequently.
Type 3 rules have consequents that contain *only* infrequent itemsets; they will only be imperfectly sporadic if there are frequent items in the antecedent.
Type 4 rules have antecedents that contain *only* infrequent itemsets; they will only be imperfectly sporadic if there are frequent items in the consequent.

We would prefer a technique that finds imperfectly sporadic rules that are *interesting*: for instance when the items in the antecedent are above maxsup but the intersection of these items is below maxsup and the consequent has a support below maxsup. Clearly Type 3 rules are interesting under this definition, and the rest of this paper describes our attempt to generate them in a reasonably efficient manner. An example of a Type 3 rule is *fever, stiff neck, rash* $\rightarrow$ *meningitis*, where *fever*, *stiff neck*, and *rash* are common separately, but just occasionally occur together. When they do, one can diagnose *meningitis* with some confidence, even though meningitis is quite rare.

3.2 MIISR Overview

Broadly speaking, the MIISR algorithm performs the following steps. On the first pass through the database an inverted index is built using items as keys and the transaction IDs as the data. At this point, the support of each unique item (the 1-itemsets) in the database is available as the length of each data chain. Items that fall under maxsup are identified and recorded as candidate consequents. For each candidate consequent found, we use the items that reside in the same transactions as the candidate consequent to extend the $(k-1)$-itemsets in precisely the same manner as Apriori to generate candidate k-itemsets. These extensions are considered to be candidate antecedents. We then check the candidate antecedent itemsets against the inverted index to ensure they meet the minimum absolute support requirement and prune them out if they do not. This candidate generation process is repeated until no further candidate antecedents are produced.

3.3 Minimum Absolute Support Value

When searching for rare itemsets, we consider two circumstances: occurrences of itemsets due to some non-random process that is generating them, or occurrences of itemsets by random collision (coincidence). It is important to distinguish between them, as itemsets that have a low support but high confidence that seem interesting may be occurring due to chance and should be considered "noise". Clearly it makes sense only to consider candidate itemsets that appear together more often than coincidence. We define coincidence in the following way: for N transactions in which the antecedent A occurs in a transactions and consequent B occurs in b transactions, we can calculate the probability that A and B will occur together exactly c times by chance. We refer to this as "probability of chance collision" [10]. We can calculate this probability using Pcc in (1). The probability that A and B will occur together exactly c times is:

$$\text{Pcc}\left(c|N,a,b\right) = \frac{\binom{a}{c}\binom{N-a}{b-c}}{\binom{N}{b}} \quad (1)$$

For example, given $N = 1000$, $A = B = 500$, and $AB = 250$, we are able to determine that the probability that A and B will occur exactly 250 times is 0.05. This equation is the usual calculation for exact probability of a 2×2 contingency table [13]. Now, we want the least number of collisions above which Pcc is smaller than some small value p (say, 0.0001). This is:

$$\text{minabssup}(N,a,b,p) = \min\left\{m \Big| \sum_{i=0}^{i=m} \text{Pcc}(i|N,a,b) \geq 1.0 - p\right\} \quad (2)$$

This formula amounts to inverting the usual sense of Fisher's exact test [13]. Usually a 2×2 contingency table is provided and a p-value calculated. However here we are providing two of the four values and a p-value, and calculating the minimum value to complete the table. By selecting the minabssup value for each itemset we are able to prune out associations that appear in the dataset by chance. We calculate the cumulative Pcc of AB together m times (beginning from 0 and incremented by 1). We stop the incrementation when the cumulative value of $Pcc \geq 1.0 - p$ and m is set as the minabssup value. For example given that we set $N = 1000$, $A = B = 500$, and $p = 0.0001$, minabsup value is 274. Candidate itemsets that appear above the minabssup requirement are considered somewhat interesting, and worth retaining to evaluate their confidence.

3.4 Exclusory Constraint

Even after pruning out candidate antecedents that are indistinguishable from coincidence, a considerable number of itemsets can still be produced for a modest dataset. One solution to this problem is to prune out a larger number of itemsets before generating new candidates. Another solution, and the one we adopt, is to prohibit current candidates from being extended if their extensions are not likely to produce interesting rules. Here we introduce an *exclusory* constraint for this purpose. Given an imperfectly sporadic rule $A \to C$ which has high confidence, it becomes less meaningful if $C \to A$ has confidence that is too low. For a candidate antecedent A to be considered worth expanding with respect to consequent C, it must therefore meet this requirement:

$$\text{conf}(C \to A) > \max(\text{minconf}, 1 - \sup(C)) \tag{3}$$

Once the confidence of $C \to A$ falls below minconf or $1 - \sup(C)$, A may still produce an interesting rule with C, but AZ is unlikely to do so no matter what Z is. Thus, we wish to keep A in the pool of candidate itemsets, but we do not wish to extend it—we *exclude* it from the next round of candidate generation.

3.5 The MIISR Algorithm

Having defined how to calculate the minimum absolute support (minabssup) necessary to consider a rule to be non-coincidental, and a procedure to prevent candidate antecedents from being extended if they are unlikely to produce interesting rules, we are now able to define an algorithm for Mining Interesting Imperfectly Sporadic Rules: MIISR.

```
Algorithm for Mining Interesting Imperfectly Sporadic Rules (MIISR)
Input: Transaction Database D, maxsup value, minconf value
Output: Antecedent Itemsets for Imperfectly Sporadic Rules,
        indexed by Consequent

(1) Generate inverted index Idx of (item, [TID-list]) from D.
(2) Generate candidate consequent items:
        C ← ∅
        for each item i ∈ Idx do begin
            if count(Idx,i)/|D| < maxsup
                then C ← C ∪ i
        end
(3) Generate candidate antecedent itemsets:
        A ← ∅
        for each item i ∈ C do begin
            A_{i,1} ← ∅
            (3.1) Generate candidate antecedent itemsets of size 1:
            for each item j ∈ {items in the same trans as i} do begin
                if j ∉ A_{i,1} and count(Idx,j,i) > minabssup(N, i, j, 0.0001)
                    then A_{i,1} ← A_{i,1} ∪ j
            end
            (3.2) Find A_{i,k}, the set of k-antecedent-itemsets where k ≥ 2:
            for (k ← 2; A_{i,k-1} ≠ ∅; k ← k+1) do begin
                A_{i,k} ← ∅
                for each j ∈ {itemsets that extend A_{i,k-1}} do begin
                    if all subsets of j of size k − 1 ∈ A_{i,k-1}
                       and count(Idx,j,i) > minabssup(N, i, j, 0.0001)
                        then A_{i,k} ← A_{i,k} ∪ j
                end
            end
        end
        return A
```

In Section 3.2 of MIISR, "itemsets that extend $A_{i,k-1}$" refers to the same process that Apriori uses to turn candidate itemsets of size $k-1$ into itemsets of size k. That is, itemsets that share all but their last item are used to form a new itemset with the same prefix, the suffix of the first itemset, and the suffix of the second itemset. The only difference is that candidate antecedents that do not meet the exclusory requirement outlined in Section 3.4 are never considered for extension. The "count" function, when given one itemset argument in addition to the inverted index, returns the number of transactions in which the itemset occurs. When given two itemset arguments, it returns the number of transactions in which *both* of the itemsets may be found.

The result of MIISR is a data structure indexed by all 1-itemsets that fall under maximum support. We need not return C, the list of candidate consequents, since if A_i is non-empty, $i \in C$. For each of these items i, A contains a list of antecedents such that $A_{i,j} \rightarrow i$ should be an imperfectly sporadic rule. Since MIISR does not restrict the antecedents to containing only frequent itemsets, some perfectly sporadic rules may also be produced.

4 Experiments

To assess the performance of MIISR in discovering Type 3 imperfectly sporadic rules, we developed a synthetic data generator which deliberately injects imperfectly sporadic itemsets. We then tested the MIISR algorithm on six different datasets from the UCI Machine Learning Repository [14].

Our synthetic data generator is a modified version of the data generator proposed by Agrawal and Srikant [2]. In real databases there may be both frequent and infrequent itemsets and rules, but we are only interested in the imperfectly sporadic itemsets. Table 1 summarizes the characteristics of several of the datasets generated during our tests. Since we are deliberately ignoring large itemsets, we left $|\overline{I}|$ set to 2 for all experiments.

Table 1. Parameter settings

Name	$|T|$	$|L|$	$|S|$	$|D|$	Size in Megabytes
T20.L100.S30.D10K	20	100	30	10K	1.4
T60.L60.S20.D10K	60	60	20	10K	2.9
T10.L30.S10.D1000K	10	30	10	1000K	48.0

To create a dataset D, our synthetic data generation program takes the following parameters: number of transactions $|D|$, average size of transactions $|\overline{T}|$, average size of large itemsets $|\overline{I}|$, number of large itemsets $|L|$, number of imperfectly sporadic itemsets $|S|$, and number of items N. We first determine the size of the next transaction which is generated using a Poisson distribution with mean as the average size of the transaction. We then fill the transactions with items. Each transaction is assigned a series of potential large itemsets and/or an imperfectly sporadic itemset. Each itemset in T has a weight associated with it, which corresponds to the probability that this itemset will be picked.

4.1 Results

Three different experiments varying either the number of transactions, average size of transactions or number of imperfectly sporadic itemsets injected were conducted to assess the efficiency and scalability of MIISR. The datasets used were generated using the synthetic data generator described in the previous section. Maxsup was set to 0.10 and minconf to 0.90.

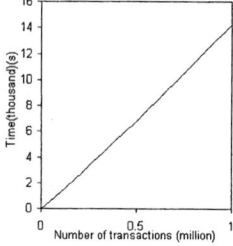

Fig. 1. Scale-up based on the number of transactions

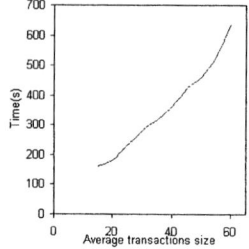

Fig. 2. Scale-up based on the average size of transactions

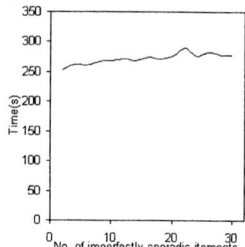

Fig. 3. Scale-up based number of imperfectly sporadic itemsets

In the first experiment, we varied the number of transactions from 10^3 to 10^6 over the T10.L30.S10 dataset. In Figure 1 we see that the time taken to process the data seems to increase linearly with the number of transactions. Figure 2 shows the execution time taken to process a dataset while varying the average size of transactions, from 15 to 60 with an increment of 5, for the L60.S20.D10K dataset. Note that as the average size of transactions increased, the execution time increased somewhat worse than linearly. However, the curvature is gentle over a reasonably practical range of values. In the third experiment, we investigate the scale-up of the number of deliberately injected imperfectly sporadic itemsets which ranged from 2 to 30 with an interval of 2 for the T20.L100.D10K datasets. Figure 3 shows the results of the execution time to process a dataset with a varying number of injected imperfectly sporadic itemsets. Although the number of rules found for each dataset increases as the number of imperfectly sporadic itemsets increases, notice that the fluctuation of the runtime taken is quite a small percentage of the total runtime. The difference between the maximum and minimum time taken is 38 seconds.

Testing of the MIISR algorithm was also carried out using six different datasets from the UCI Machine Learning Repository [14]. Table 2 displays results from using MIISR with and without the exclusory constraint. Each row of the table represents an attempt to find the number of imperfectly sporadic rules (with minconf 0.95, and lift greater than 1.0) from the database named in the left-most column. In the table, *accept* represents the number of itemsets below maxsup but above the minabssup value and *reject* represents the number of itemsets below the minabssup value. The number of accepted and rejected itemsets depends on the the amount of noise within a certain dataset.

Table 2. Comparison of MIISR with and without exclusory constraint

Dataset	Maxsup	MIISR						MIISR with Exclusory Constraint					
		Rules	Passes	Avg Sporadic Sets	Itemset Analysed		Time (sec)	Rules	Passes	Avg Sporadic Sets	Itemset Analysed		Time (sec)
					Accept	Reject					Accept	Reject	
TeachingEval.	0.100	5	16	1	15	421	0.48	5	7	1	15	421	0.52
Bridges	0.100	3	5	1	5	2261	1.04	2	3	1	4	2261	1.64
Zoo	0.100	8	11	3	29	1926	0.98	1	9	2	19	1926	1.52
Flag	0.100	215	47	8	356	10964	5.32	3	14	2	48	10962	6.93
Soybean-Large	0.100	66922	247	339	83585	42090	150.14	4631	129	50	6585	1946	9.77
Mushroom	0.005	3981966	207	20236	4188691	11716	8887.15	488274	678	1694	1149372	891	908.86

Using MIISR, we were able to find imperfectly sporadic rules below maxsup of 0.10 for the datasets Teaching Assistant Evaluation, Bridges, Zoo, Flag, and Soybean-Large within reasonable time (below 200 seconds). However for the Mushroom dataset, maxsup was lowered to 0.005. Due to the nature of the Mushroom dataset, where strong association holds among most of the items, coincidence pruning is not able to prune out many itemsets. Consequently, we end up with a very large amount of imperfectly sporadic rules, some of which might be less interesting than others.

Finally, we note that both the proportion of itemsets accepted by the minabssup constraint and the number of rules accepted by the exclusory constraint seem to be entirely dataset-dependent.

5 Conclusion and Future Work

Very few existing algorithms try to find infrequent itemsets, despite the fact that the most potentially interesting things that happen in a database are likely to happen infrequently. In this paper, we present a new algorithm called MIISR for discovering imperfectly sporadic rules. We are particularly interested in infrequently occurring associations of frequent itemsets giving rise to infrequent consequents. The supports of imperfectly sporadic rules are by definition low, and we therefore run the risk of accepting as interesting rules things that have only fallen together by chance. For this reason, the minimum absolute support value proposed in the paper plays an important role because it does not allow rules that have only chance association to be generated. We acknowledge that this approach, and that based on the exclusory constraint, are heuristic, but we believe that to be unavoidable when generating low support rules. Since the number of low support rules can be very large, but those that are likely to be interesting not as common, exhaustive techniques tend rapidly to fall into pathological cases.

Currently this approach performs fairly efficiently on synthetic datasets and medium sized real datasets. Our future work will deal with examining the problem of transaction length, to which MIISR seems most sensitive in the synthetic datasets. More importantly, we need to try to characterise the UCI Mushroom dataset in terms of imperfectly sporadic rules, and determine an approach that would narrow down the candidate antecedent itemsets even further. If it turns out that there simply *are* a lot of high confidence, low support rules in that database, then we need to investigate interestingness metrics that could be used to generate a better subset of them. Hence we are interested in determining the actual interestingness of imperfectly sporadic rules in real domains.

References

1. Agrawal, R., Imielinski, T., Swami, A.: Mining association rules between sets of items in large databases. In Buneman, P., Jajodia, S., eds.: Proceedings of the 1993 ACM SIGMOD International Conference on Management of Data. (1993) 207–216
2. Agrawal, R., Srikant, R.: Fast algorithms for mining association rules. In Bocca, J.B., Jarke, M., Zaniolo, C., eds.: Proceedings of the 20th International Conference on Very Large Data Bases, (VLDB'94). (1994) 487–499
3. Srikant, R., Agrawal, R.: Mining Generalized Association Rules. In Dayal, U., Gray, P.M.D., Nishio, S., eds.: Proceedings of the 21st International Conference on Very Large Data Bases, (VLDB95). (1995) 407–419
4. Uno, T., Kiyomi, M., Arimura, H.: LCM ver. 2: Efficient Mining Algorithms for Frequent/Closed/Maximal Itemsets. In: Proceedings of the IEEE ICDM Workshop on Frequent Itemset Mining Implementations (FIMI'04). (2004)
5. Flouvat, F., Marchi, F.D., Petit, J.M.: ABS: Adaptive borders search of frequent itemsets. In: Proceedings of the IEEE ICDM Workshop on Frequent Itemset Mining Implementations (FIMI'04). (2004)
6. Koh, Y.S., Rountree, N.: Finding sporadic rules using Apriori-Inverse. In: Advances in Knowledge Discovery and Data Mining, 9th Pacific-Asia Conference (PAKDD 2005). (2005) 97–106
7. Srikant, R., Vu, Q., Agrawal, R.: Mining association rules with item constraints. In: Proceedings of the Third International Conference on Knowledge Discovery and Data Mining KDD-97. (1997) 67–73
8. Ng, R.T., Lakshmanan, L.V.S., Han, J., Pang, A.: Exploratory mining and pruning optimizations of constrained associations rules. In: Proceedings of the 1998 ACM SIGMOD international conference on Management of data, New York, NY, USA, ACM Press (1998) 13–24
9. Bayardo(Jr.), R.J., Agrawal, R., Gunopulos, D.: Constraint-based rule mining in large, dense databases. Data Mining and Knowledge Discovery **4** (2000) 217–240
10. Koh, Y.S., Rountree, N., O'Keefe, R.: Finding Non-Coincidental Sporadic Rules Using Apriori-Inverse. To appear in International Journal of Data Warehousing and Mining **2** (2006) 38–54
11. Rahal, I., Ren, D., Wu, W., Perrizo, W.: Mining confident minimal rules with fixed consequents. In: Proceedings of the 16th IEEE International Conference on Tools with Artifical Intelligence (ICTAI 2004). (2004)
12. Li, J., Zhang, X., Dong, G., Ramamohanarao, K., Sun, Q.: Efficient mining of high confidence association rules without support threshold. In: Proceedings of the 3rd European Conference on Principle and Practice of Knowledge Discovery in Databases (PKDD). (1999) 406–411
13. Weisstein, E.: Fishers exact test. MathWorldA Wolfram Web Resource. http://mathworld.wolfram.com/FishersExactTest.html (2005)
14. Newman, D.J., Hettich, S., Blake, C.L., Merz, C.J.: UCI repository of machine learning databases. http://www.ics.uci.edu/~mlearn/MLRepository.html, University of California, Irvine, Department of Information and Computer Sciences (1998)

Improved Negative-Border Online Mining Approaches

Ching-Yao Wang[1], Shian-Shyong Tseng[2], and Tzung-Pei Hong[3]

[1] Information & Communications Research Lab, Industrial Technology Research Institute,
Hsinchu, Taiwan, 31040, R.O.C.
simon.cis89g@nctu.edu.tw
[2] Department of Computer Science, National Chiao-Tung University,
Hsinchu, Taiwan, 30010, R.O.C.
sstseng@cis.nctu.edu.tw
[3] Department of Electrical Engineering, National University of Kaohsiung,
Kaohsiung, Taiwan, 811, R.O.C.
tphong@nuk.edu.tw

Abstract. In the past, we proposed an *extended multidimensional pattern relation* (EMPR) to structurally and systematically store previously mining information for each inserted block of data, and designed a *negative-border online mining* (NOM) approach to provide ad-hoc, query-driven and online mining supports. In this paper, we try to use appropriate data structures and design efficient algorithms to improve the performance of the NOM approach. The *lattice* data structure is utilized to organize and maintain all candidate itemsets such that the candidate itemsets with the same proper subsets can be considered at the same time. The derived *lattice-based* NOM (LNOM) approach will require only one scan of the itemsets stored in EMPR, thus saving much computation time. In addition, a hashing technique is used to further improve the performance of the NOM approach since many itemsets stored in EMPR may be useless for calculating the counts of candidates. At last, experimental results show the effect of the improved NOM approaches.

1 Introduction

Some researchers have recently developed incremental mining and online mining approaches to maintain association rules without re-processing an entire database whenever the database is updated [3][4][6][10] or user-specified parameters are changed [1]. In general, data under decision-support consideration usually evolve in a systematic way. For example, the data in a data warehouse may be inserted or deleted in a block during an interval of a month. In the past, we proposed the *multidimensional pattern relation* (MPR) [11] to structurally and systematically store additional context information and mining information for each inserted block of data. MPR is conceptually similar to the construction of a data warehouse for OLAP [7][13], except it is not used to store data but mined patterns. We also extended the mining information in MPR by including *negative pattern sets* (candidate sets which are not large) and developed a *negative-border online mining* (NOM) approach based on the *extended multidimensional pattern relation* (EMPR) especially for blocks of data with different item sets [12].

The NOM approach needs to calculate the appearing counts and the non-appearing upper-bound counts of the candidate itemsets derived from matched tuples. A straightforward way for finding these values is to process matched tuples one after one for each candidate itemset. The computation cost will, however, become large along with the increase of the itemsets kept in EMPR and the candidate itemsets to be considered. In fact, in the NOM approach, many candidate itemsets with the same subsets can be processed at the same time. On the other hand, many itemsets kept in the matched tuples are useless for calculating the counts of candidates since they are not the subsets of candidates and can be omitted. In this paper, we thus try to use appropriate data structures and design efficient algorithms to improve the performance of the NOM approach.

At first, the problem of calculating the appearing and upper-bound counts of candidate itemsets in a matched tuple is conceptually modeled by a graph and converted into a *directed-minimum-spanning-tree* problem. The *spanning-tree-count-calculating* (STCC) algorithm is then proposed to find the *directed minimum spanning tree*. The *lattice* data structure [1] is utilized to organize and maintain all candidate itemsets such that the candidate itemsets with the same proper subsets can be considered at the same time. Consequently, by the STCC algorithm, the proposed *lattice-based* NOM (LNOM) approach requires only one scan of the itemsets for each matched tuple in Phase 1. In addition, the hashing technique is used to filter out a part of itemsets in the matched tuples which are useless for calculating the counts of candidates. The computational time can thus be further reduced.

Table 1. An EMPR example based on the minimum support of 5%

ID	Region	Branch	Time	No_Trans	Frequent_Pattern_Set (Itemset, Support)	Frequent_Pattern_Set (Itemset, Support)
1	CA	San Francisco	2003/10	10000	(A,10%),(B,11%),(C,9%), (AB,8%),(AC,7%),(BC,6%) ,(ABC,6%)	(D,2%)
2	CA	San Francisco	2003/11	15000	(A,5%),(B,7%),(C,5%)	(D,1%),(AB,2%),(AC,2%) ,(BC,1%)
3	CA	Los Angeles	2003/10	20000	(A,8%),(B,6%),(F,5%)	(C,2%),(D,3%),(AB,3%), (AF,4%),(BF,3%)
4	CA	Los Angeles	2003/11	25000	(A,5%),(C,6%),(F,7%), (AF,6%),(CF,5%)	(B,3%),(D,4%),(F,2%), (AC,3%)
5	NY	New York	2003/10	18000	(B,8%),(C,7%),(BC,6%)	(A,2%),(D,2%)
6	NY	New York	2003/11	18500	(B,8%),(C,6%)	(A,4%),(D,2%),(BC,3%)

2 Review of the NOM Approach

2.1 The Extended Multidimensional Pattern Relation (EMPR)

Each tuple in EMPR comes from a block of data in the database to be processed. EMPR consists of two major types of information. One is the *context information* used to represent the contexts of each individual block of data. The other is the *mining information* used to record the available information mined from each individual block of data by a batch mining algorithm. Given an initial minimum support s, the set of previously mined large itemsets with supports for a block of data D is called a *frequent pattern set (fps)* for D; the set of previously mined *negative itemsets* with supports is called a *negative pattern set (nps)* for D. The latter consists of the itemsets which are candidates but do not have enough supports [8].

Example 1: Table 1 shows an EMPR based on an initial minimum support set at 5%. The tuple with $ID = 1$ shows that seven large itemsets $(A, 10\%)$, $(B, 11\%)$, $(C, 9\%)$, $(AB, 8\%)$, $(AC, 7\%)$, $(BC, 6\%)$, $(ABC, 6\%)$ and one negative itemset $(D, 2\%)$ are discovered from 10000 transactions under the contexts of *Region* = **CA**, *Branch* = **San Francisco** and *Time* = **2003/10**.

2.2 The Negative-Border Online Mining (NOM) Approach

Assume an EMPR based on an initial minimum support s includes m tuples $\{t_1, t_2, \ldots, t_m\}$. Given a mining request q which consists of a set of contexts cx_q, a new minimum support s_q ($s_q \geq s$) and a new minimum confidence $conf_q$, the NOM approach can effectively and efficiently derive the association rules simultaneously satisfying s_q, $conf_q$ and cx_q by three consecutive phases, *generation of candidate itemsets*, *reduction of candidate itemsets*, and *generation of association rules*.

The phase for generation of candidate itemsets first selects the tuples in EMPR satisfying cx_q (called matched tuples), collects the itemsets kept in these matched tuples whose supports are larger than or equal to s_q as the set of candidate itemsets, and calculates the *appearing count*, $Count_x^{appearing}$, and the *non-appearing upper-bound count*, $Count_{\bar{x}}^{UB}$, for each candidate. The appearing count and the non-appearing upper-bound count of a candidate itemset x can be represented as follows:

$$Count_x^{appearing} = \sum_{t_i \in mt \ \& \ x \in t_i.fps \cup t_i.nps} t_i.trans * t_i.s_x, \text{ and} \tag{1}$$

$$Count_{\bar{x}}^{UB} = \sum_{t_i \in mt \ \& \ x \notin t_i.fps \cup t_i.nps} min(t_i.trans * s - 1, t_i.trans * \min_{\forall x' \subset x}(t_i.s_{x'})), \tag{2}$$

where mt denotes the set of matched tuples, $t_i.fps$ denotes the *frequent pattern set* in t_i, $t_i.nps$ denotes the *negative pattern set* in t_i, $t_i.trans$ denotes the number of transactions kept in t_i, and $t_i.s_x$ denotes the actual support of x in t_i. After that, the phase for reduction of candidate itemsets calculates the *upper-bound supports* of candidate itemsets and adopts two pruning strategies to reduce the candidate number. The *upper-bound support* s_x^{UB} of a candidate itemset x is defined as follows:

$$s_x^{UB} = \frac{Count_x^{appearing} + Count_{\bar{x}}^{UB}}{Match\_Trans}, \tag{3}$$

where *Match_Trans* denotes the number of transactions in the matched tuples. The first pruning strategy will remove the candidate itemsets whose upper-bound supports are less than s_q. The second pruning strategy will put the ones which appear in all the matched tuples and have upper-bound supports larger than or equal to s_q into the set of final large itemsets. Finally, the phase for generation of association rules reprocesses, if necessary, the remaining candidate itemsets against the underlying database, and derives the association rules satisfying $conf_q$ from all the final large itemsets found.

Example 2: For the EMPR given in Table 1, assume a mining request q wants to get the patterns with the contexts cx_q of *Region* = **CA** and *Time* = **2003/10** and satisfying the minimum support $s_q = 5.5\%$. Phase 1 finds the set of candidate itemsets $\{\{A\}, \{B\}, \{C\}, \{AB\}, \{AC\}, \{BC\}, \{ABC\}\}$, which is the union of the itemsets appearing in the frequent pattern sets and with their supports larger than 5.5%. In Phase 2, the upper-bound supports of these candidate itemsets are calculated. According to the calculation results, the itemsets $\{C\}$, $\{AB\}$, $\{AC\}$, $\{BC\}$ and $\{ABC\}$ will be pruned, and the itemsets $\{A\}$ and $\{B\}$ will be put into the set of final large itemsets. No remaining candidate itemsets need to be further processed in Phase 3 for this example.

3 Improving the Performance of the NOM Approach

The NOM approach needs to calculate the appearing counts and the non-appearing upper-bound counts of the candidate itemsets derived from matched tuples. Assume k is the number of matched tuples, m is the average number of itemsets in the k matched tuples, and n is the number of candidate itemsets generated from the k matched tuples. The computation cost will be $O(knm)$ when the candidate itemsets are processed one by one. It will become large along with the increase of the itemsets kept in EMPR and the candidate itemsets to be considered.

In fact, many candidate itemsets with the same subsets can be calculated at the same time. For example, in Tuple 3 of Example 2, the appearing count of the candidate itemset $\{C\}$ and the upper-bound counts of the candidate itemsets $\{AC\}$, $\{BC\}$ and $\{ABC\}$ can be calculated at the same time because they have the same subset $\{C\}$. On the other hand, many itemsets kept in matched tuples are useless for calculating the counts of candidates. For example, in Example 2, the itemsets $\{D\}$, $\{F\}$, $\{AF\}$ and $\{BF\}$ are not the subsets of the candidate itemsets and can be omitted.

4 The Proposed Lattice-Based NOM (LNOM) Approach

The problem of calculating the appearing and upper-bound counts of candidate itemsets in a matched tuple t can be conceptually modeled by a graph. Let $G = (V, E)$ be a directed graph, where V is the set of vertices representing all candidate itemsets and E is the set of directed edges representing *a-proper-subset-of* relationships between pairs of candidate itemsets. For each edge $(u, v) \in E$, a weight $w(u, v)$ specifies the possible upper-bound count of the candidate itemset v estimated from the candidate itemset u. Given a new vertex r representing the pseudo starting vertex, we make a new graph $G' = (V', E')$, where $V' = V \cup \{r\}$, $E' = E \cup \{(r, u): u \in V\}$. For each edge (r, u), if u appears in t, the appearing count of u is assigned as the weight $w(r, u)$. For the case that u does not appear in t, meaning it is collected from the other matched tuple(s), then $w(r, u) = 0$ if there exists one item contained in u but not contained in t and $w(r, u) = t.trans*s-1$ otherwise, where s is the initial minimum support for deriving EMPR.

For each vertex other than r in G', the smallest weight on all its incoming edges is its tight upper-bound count. The count-calculation problem can thus be easily thought of as the *directed-minimum-spanning-tree* problem [5], which wishes to find a rooted directed spanning tree $T = (V', S')$ from G', such that S' is a subset of E' and $\sum_{(u,v) \in S} w(u,v)$ is a minimum. In this section, the *spanning-tree-count-calculating* (STCC) algorithm is thus proposed based on the above concept for efficiently finding the counts of all candidate itemsets in a tuple. The STCC algorithm first selects an itemset appearing in t and with the smallest support. It then estimates the upper-bound count of each itemset reachable from the selected one in the graph, and thus avoids recalculating the counts of these traversed vertices in the future. This requires only one scan of the itemsets in t if they have been sorted according to their supports.

The STCC algorithm can be efficiently implemented by the *lattice* data structure [1], which organizes all candidate itemsets in a systematic way. For each candidate itemset x, a corresponding vertex u_x associated with a pair of values ($Count_x^{appearing}$, $Count_x^{UB}$) is built in the lattice. A directed edge is generated from u_x to u_y if y can be derived by adding an item to x. By the connected edges, the proposed *lattice-based NOM* approach (called LNOM) can not only restrict the number of candidate itemsets to be examined, but also easily consider candidate itemsets with the same proper subsets at the same time.

5 Using the Hashing Technique to Reduce Computation Cost Further

Many itemsets kept in matched tuples, especially negative itemsets, may be useless for calculating the counts of candidate itemsets. Negative itemsets are formed by excluding large itemsets from the candidates which are generated in a level-wise way [10]. In general, the set of candidate itemsets generated level-wisely is usually much larger than the set of large itemsets found, especially in the early stage of candidate generation [2][9]. In this section, we shall utilize the hashing technique to filter out a part of useless itemsets to be considered in Phase 1.

Take the *direct hashing function* as an example. Each bucket of the hash table consists of only an integer to represent how many candidate itemsets have been hashed into this bucket. 0 denotes that no candidate itemsets have been hashed into this bucket. After a hash table is constructed from all the candidate itemsets, it can then be used to filter out a part of useless itemsets in a tuple. When an itemset of a matched tuple is selected, the NOM approach calculates its hash value and finds its corresponding bucket. If the value stored in the target bucket is equal to 0, the itemset must be useless since it is not a candidate itemset. It can thus be directly omitted. Otherwise, rescanning the candidate itemsets is necessary to determine whether it is a candidate. Furthermore, the corresponding value in the bucket of the itemset which is assured to be a candidate will be decreased by one. The next itemset of the same tuple is then checked according to the modified hash table, which can thus raise the probability for a useless itemset to be filtered out. After a tuple is processed, the hash table is restored to its original state, which is then used for another tuple.

6 The LNOM Algorithm with a Direct Hashing Function

The algorithm of LNOM approach with a direct hashing function is stated below.
The LNOM approach with a direct hashing function:
INPUT: An EMPR based on an initial minimum support s, and a mining request q with a set of contexts cx_q, a minimum support s_q ($s_q \geq s$) and a minimum confidence $conf_q$.
OUTPUT: A set of association rules satisfying the mining request q.
Phase 1: Generation of candidate itemsets:
STEP 1: Set $C = \phi$ and $Match\_Trans = 0$, where C is a lattice used to maintain the set of candidate itemsets and $Match\_Trans$ is a variable used to keep the total number of transactions in the matched tuples which have been processed.
STEP 2: Initialize two equal-sized hash tables HT_1 and HT_2 with all the bucket values being zero.
STEP 3: For each tuple t in EMPR, do the following substeps:
    STEP 3-1: If t satisfies cx_q, put it into the matched set and do STEP 3-2; otherwise, repeat STEP 3 to process the next tuple.
    STEP 3-2: For each itemset $x \in t.fps$, if $x \notin C$ and $t.s_x \geq s_q$, set $HT_1[h(x)] = HT_1[h(x)] + 1$, insert x into C with $Count_x^{appearing} = 0$ and $Count_{\bar{x}}^{UB} = 0$, and add edges to its parents and children, where $HT_1[h(x)]$ denotes the value stored in the bucket corresponding to the hash value $h(x)$ of x in HT_1.
STEP 4: For each tuple t in the matched set, do the following substeps:
    STEP 4-1: Set $ProcessedSet = \phi$, where $ProcessedSet$ is a set used to keep the itemsets in C which have been processed.
    STEP 4-2: Restore the bucket values in HT_2 to those in HT_1 and set $Match\_Trans = Match\_Trans + t.trans$.
    STEP 4-3: Select an itemset x with the smallest support $t.s_x$ from t.
    STEP 4-4: If $HT_2[h(x)] \neq 0$ and $x \in C$, set $Count_x^{appearing} = Count_x^{appearing} + t.trans * t.s_x$, $HT_2[h(x)] = HT_2[h(x)] - 1$, $ProcessedSet = ProcessedSet \cup \{x\}$, and do STEP 4-5; otherwise, do nothing and go to STEP 4-6.
    STEP 4-5: For each itemset y in the proper superset of x in C and $y \notin ProcessedSet$, set
$$Count_{\bar{y}}^{UB} = Count_{\bar{y}}^{UB} + \min(t.trans * s - 1, t.trans * t.s_x), HT_2[h(y)] = HT_2[h(y)] - 1,$$
and $ProcessedSet = ProcessedSet \cup \{y\}$.
    STEP 4-6: Repeat STEPs 4-3 and 4-4 until all itemsets in t are processed.
Phase 2: Reduction of candidate itemsets:
STEP 5: Set $k = 1$, where k is used to keep the number of items in a candidate itemset currently being processed.
STEP 6: For each itemset $x \in C_k$, do the following substeps:
    STEP 6-1: Calculate the upper-bound support s_x^{UB} by the formula:
$$s_x^{UB} = \frac{Count_x^{appearing} + Count_{\bar{x}}^{UB}}{Match\_Trans}.$$
    STEP 6-2: If $s_x^{UB} < s_q$, set $C = C - \{y \mid y \in C \text{ and } x \subseteq y\}$.
    STEP 6-3: If $s_x^{UB} = \frac{Count_x^{appearing}}{Match\_Trans}$ and $s_x^{UB} \geq s_q$, then set $L = L \cup \{x\}$ and $C = C - \{x\}$.
STEP 7: Set $k = k + 1$.
STEP 8: Repeat STEPs 6 and 7 until all candidate itemsets are processed.
Phase 3: Generation of association rules:
STEP 9: For each $x \in C$, re-process each underlying block of data D_i for tuple t_i in which x does not appear to get $Count_{\bar{x}}^{appearing}$, and then calculate the actual support of x by the following formula:
$$s_x = \frac{Count_x^{appearing} + Count_{\bar{x}}^{appearing}}{Match\_Trans}.$$

STEP 10: If $s_x < s_q$, then set $C = C - \{x\}$; otherwise, set $L = L \cup \{x\}$ and $C = C - \{x\}$.
STEP 11: Derive the association rules satisfying $conf_q$ from the set of large itemsets L.

7 Experiments

The experiments were conducted in Java on a workstation with dual XEON 2.8GHz processors and 2048MB main memory, running the RedHat 9.0 operating system. Several synthetic datasets were used. The synthetic datasets were generated by a generator similar to that used in [2]. Table 2 listed the four groups of synthetic datasets generated and used in our experiments. Each dataset was treated as a block of data in the database. For example, Group 1 contained ten blocks of data, from $T20I8D100KL^1$ to $T20I8D100KL^{10}$, each consisting of 100000 transactions averaging 20 items and generated according to 400 to 490 maximal potentially large itemsets with an average size of 8 from a total of 200 items.

Table 2. The four groups of synthetic dataset

Group	Size	Datasets	D	T	I	L	N
1	10	$T20I8D100KL^1$ to $T20I8D100KL^{10}$	100000	20	8	400 to 490	200
2	10	$T20I8D100KN^1$ to $T20I8D100KN^{10}$	100000	20	8	400	200 to 290
3	5	$T10I8D500KL^1$ to $T10I8D500KL^5$	500000	10	8	400 to 560	200
4	5	$T10I8D500KN^1$ to $T10I8D500KN^5$	500000	10	8	400	200 to 360

An EMPR was first derived from each group of synthetic datasets. These are summarized in Table 3.

Table 3. Mining information for the four groups

Group	Initial minimum support	Average length of maximal large itemsets	Average size of large itemsets	Average size of negative itemsets
1	2%	9	12127	55625
2	2%	11	18534	49318
3	2%	5	799	11899
4	2%	8	869	14488

For showing the influence of the number of negative itemsets on execution time, the NOM algorithms using no negative itemsets ($NOM(0)$) and all negative itemsets ($NOM(A)$) from the stored negative pattern sets in the EMPR were run. Fig. 1(a) to Fig. 1(d) shows the execution times for the two NOM algorithms on Groups 1 to 4, where the query support is set at 2.4%. For Group 1, most candidate itemsets appeared in nearly all tuples in EMPR such that the negative itemsets provided little help in calculating counts of candidates. This can be easily seen from Fig. 1(a) that the execution time by $NOM(0)$ was less than that by $NOM(A)$. For Group 2, most candidate itemsets appeared in only one or few tuples in EMPR. The effect of negative itemsets on finding tight upper-bound supports thus become apparent. However, since the computation cost in Phase 1 was much larger than that in Phase 3, the execution time by *NOM*(0) was still less than that by *NOM*(A) as shown in Fig. 1(b). Even so, it can be observed from Fig. 1(c) and Fig. 1(d) that *NOM*(0) did not always outperform *NOM*(A) for Groups 3 and 4, This phenomena is especially when the size of candidate itemsets is small and the size of underlying data is large.

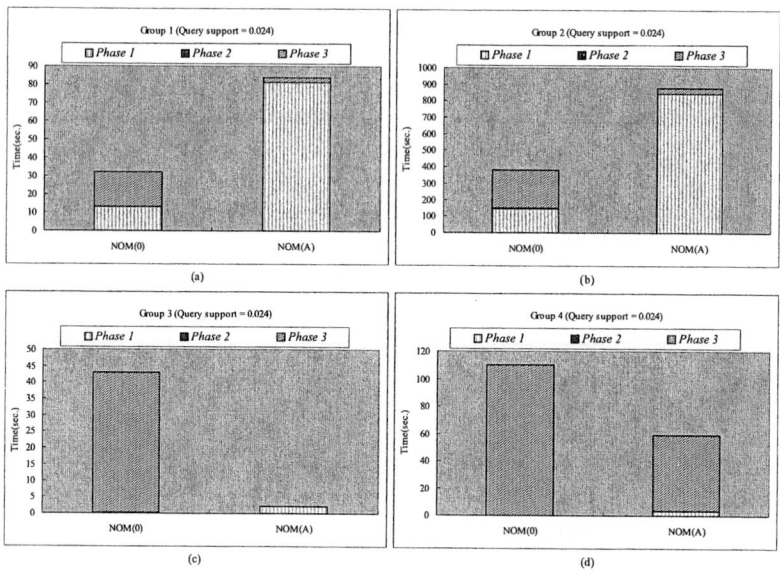

Fig. 1. The influence of the number of negative itemsets on execution time of the NOM algorithm for Groups 1 to 4

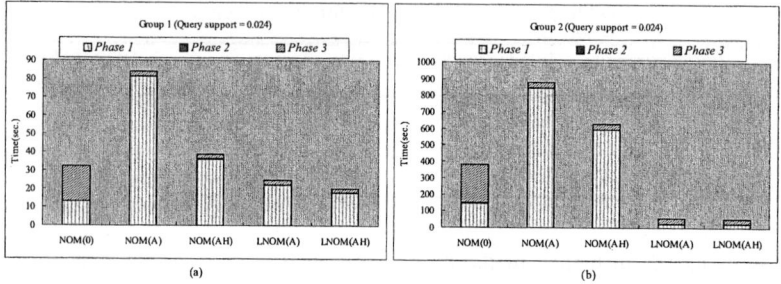

Fig. 2. Execution times spent by the NOM and LNOM algorithms on Groups 1 to 2

The performance of the NOM algorithm with a direct hashing function was then evaluated. Let *NOM*(AH) denote running *NOM*(A) with a direct hashing function. The execution times on Groups 1 to 2 are shown in Fig. 2(a) and Fig. 2(b), where the query support is set at 2.4% and the size of the hash table is about 10K. It can be easily seen that the computation time in Phase 1 of the NOM algorithm can be efficiently reduced by the hashing technique. Next, experiments were made to show the effect of using the lattice data structure on the NOM algorithm. The execution time of the NOM algorithm was compared with that of the LNOM algorithm with and without a direct hashing function. The query support is set at 2.4% and the size of the hash table is about 10K. The results for Groups 1 to 2 are also shown in Fig. 2(a) and Fig. 2(b), where *LNOM*(A) and *LNOM*(AH) respectively denote running LNOM algorithm with and without a direct hashing function. It is easily seen that the execution time by the LNOM algorithm was always much less than that by the NOM algorithm.

8 Conclusion

For providing ad-hoc, query-driven and online mining supports, the NOM approach utilized three phases to acquire interesting association rules by aggregating related mining information from EMPR. In Phase 1, for each candidate itemset, the NOM approach needs to find its smallest-support subset from the information of itemsets kept in each related mining information to calculate its appearing count or its upper-bound count. When the candidate itemsets are processed one by one, this cost is considerably tremendous. For overcoming this problem, in this paper, we have developed a *lattice-based* NOM (LNOM) approach to consider candidate itemsets with the same proper subset at the same time, and utilized the hashing technique to reduce the number of itemsets kept in the matched tuples to be considered.

Acknowledgements

This research was partially supported by the National Science Council of the Republic of China under Grand No. NSC93-2752-E-009-006-PAE.

References

1. Aggarwal, C.C., Yu, P.S.: A New Approach to Online Generation of Association Rules, IEEE Transactions on Knowledge and Data Engineering, Vol. 13, No. 4, pp. 527-540, 2001.
2. Agrawal, R., Srikant, R.: Fast Algorithm for Mining Association Rules, The ACM International Conference on Very Large Data Bases, pp. 487-499, 1994.
3. Aref, W.G., Elfeky, M.G., Elmagarmid, A.K.: Incremental, Online, and Merge Mining of Partial Periodic Patterns in Time-series Databases, IEEE Transactions on Knowledge and Data Engineering, Vol. 16, No. 3, pp. 332-342, 2004.
4. Cheung, D.W., Han, J., Ng, V.T., Wong, C.Y.: Maintenance of Discovered Association Rules in Large Databases: An Incremental Updating Approach, The IEEE International Conference on Data Engineering, pp. 106-114, 1996.
5. Gabow, H.N., Galil, Z., Spencer, T., Tarjan, R.E.: Efficient Algorithms for Finding Minimum Spanning Trees in Undirected and Directed Graphs, Combinatorica, Vol. 6, No. 2, pp. 109-122, 1986.
6. Hong, T.P., Wang, C.Y., Tao, Y.H.: A New Incremental Data Mining Algorithm Using Pre-large Itemsets, An International Journal: Intelligent Data Analysis, pp. 111-129, 2001.
7. Immon, W.H.: Building the Data Warehouse, Wiley Computer Publishing, 1996.

8. Mannila, H., Toivonen, H.: On an Algorithm for Finding All Interesting Sentences, The European Meeting on Cybernetics and Systems Research, pp. 973-978, 1996.
9. Park, J.S., Chen, M.S., Yu, P.S.: Using a Hash-based Method with Transaction Trimming for Mining Association Rules, IEEE Transactions on Knowledge and Data Engineering, Vol. 9, No. 5, pp. 812-825, 1997.
10. Thomas, S., Bodagala, S., Alsabti, K., Ranka, S.: An Efficient Algorithm for the Incremental Update of Association Rules in Large Databases, The International Conference on Knowledge Discovery and Data Mining, pp. 263-266, 1997.
11. Wang, C.Y., Tseng, S.S., Hong, T.P.: Flexible Online Association Rule Mining Based on Multidimensional Pattern Relations, to appear in An International Journal: Information Sciences, 2005.
12. Wang, C.Y., Tseng, S.S., Hong, T.P., Chu, Y.S.: Online Generation of Association Rules under Multidimensional Consideration Based on Negative-border, to appear in Journal of Information Science and Engineering, 2005.
13. Widom, J.: Research Problems in Data Warehousing, The ACM International Conference on Information and Knowledge Management, pp. 25-30, 1995.

Association-Based Dissimilarity Measures for Categorical Data: Limitation and Improvement

Si Quang Le[1,2], Tu Bao Ho[1], and Le Sy Vinh[3,4]

[1] Japan Advanced Institute of Science and Technology,
Tatsunokuchi, Ishikawa 923-1292, Japan
[2] LIRMM, Montpellier Cedex 5, France
[3] John von Neumann Institute for Computing, Juelich, Germany
[4] American Museum of Natural History, New York, USA
{quang, bao}@jaist.ac.jp, vinh@cs.uni-duesseldorf.de

Abstract. Measuring the similarity for categorical data is a challenging task in data mining due to the poor structure of categorical data. This paper presents a dissimilarity measure for categorical data based on the relations among attributes. This measure not only has the advantage of value variance but also overcomes the limitations of condition the probability-based measure when applied to databases whose attributes are independent. Experiments with 30 databases also showed that the proposed measure boosted the accuracy of Nearest Neighbor classification in comparison with other tested measures.

1 Introduction

The most common similarity measures for categorical data are binary vector-based methods ([1] and references therein). These methods transform each data object into a binary vector, at which each bit indicates the presence or absence of a possible attribute value. Then the similarity between two objects is estimated by the similarity between two corresponding binary vectors. These methods contain two main drawbacks: (1) the transformation of data objects into binary vectors may leave out many subtleties of the data; (2) they do not consider the correlations between attributes that typically exist in real-life data and are potentially concerned with the difference among attribute values.

Recently, Le and Ho presented the condition probability-based measure [2] based on relations among condition probability distributions of attributes. Their experiments showed that the condition probability-based measure gave better results than other tested methods. However, the method did not work properly with databases whose attributes are likely independent.

In this paper, we propose a dissimilarity measure for categorical data based on three main points. First, we employ the idea from text mining [3] and taxonomy [4] that is to weight attribute values by its frequency in databases and in its conditional distributions. Second, the dissimilarity among weighted condition distributions of attributes are taken into account to estimate the dissimilarity between categorical values. This point is inherited from [2] but we now consider dissimilarity between weighted condition distributions instead of condition probability distributions, as was used in [2]. Finally, we include the dissimilarity between weighted condition distributions and weighted distributions of attributes. The intuition behind the idea is that the dissimilarity between the weighted condition distributions conditioned on a value, and the weighted distributions of attributes shows reliability when considering the weighted condition distributions as the preventative of this value. The proposed measure overcomes the limitations of the condition probability-based measure [2] when applied to databases whose attributes are independent. Experiments with 30 databases also showed that the proposed measure boosted the accuracy of Nearest Neighbor classification [5] in comparison with the condition probability-based measure, and the binary vector-based measures.

2 Association-Based Dissimilarity Measure

Let $A_1, \ldots, A_m$ be m attributes and $D \subseteq dom(A_1) \times \ldots \times dom(A_m)$ be a database, $N = |D|$. Denote $fr(A_i = x_i)$ the frequency of value x_i of attribute A_i, $x_i \in dom(A_i)$. Also denote $fr(A_j = x_j | A_i = x_i)$ the condition frequency of value x_j of attribute A_j given that attribute A_i holds value x_i. Consider value x_j of an attribute A_j, its weight $w(x_j)$ can simply be defined as

$$w(x_j) = (1 + fr(A_j = x_j)/N) * log_2 N / fr(A_j = x_j)$$

We restrict the term weight $w(x_j)$ of value x_j of attribute A_j to conditional weight $w(x_j | A_i = x_i)$ of value x_j given that attribute A_i holds value x_i.

$$w(x_j | A_i = x_i) = \begin{cases} 0 & \text{if } fr(A_j = x_j | A_i = x_i) = 0 \\ \left(1 + \frac{fr(A_j = x_j | A_i = x_i)}{fr A_i = x_i}\right) * log_2 \frac{N}{fr(A_j = x_j)} & \text{otherwise} \end{cases}$$

The intuition is that the less frequency $fr(A_j = x_j)$ in database D but more conditional frequency $fr(A_j = x_j | A_i = x_i)$, the greater the conditional weight $w(x_j | A_i = x_i)$. This idea is commonly used in text mining([3] and references therein) and taxonomy [4].

Denote weight vector $W_j = (w(x_j) : x_j \in dom(A_j))$ and $W_j^{x_i} = (w(x_j | A_i = x_i) : x_j \in dom(A_j))$. We can consider these vectors as approximations of weight distribution and conditional weight distribution of values of A_j, respectively, and called them hereafter weight distribution and conditional weight distribution of A_j. Denote $Eucl(W_j^{x_i}, W_j^{y_i})$ the Euclidean distance between these two distributions.

Definition 1. *The dissimilarity of two values x_i and y_i of an attribute A_i, denoted $\delta(x_i, y_i)$, is defined as*

$$\delta(x_i, y_i) = \begin{cases} 0 & \text{if } x_i = y_i \\ \frac{1}{m-1} \sum_{A_j, A_j \neq A_i} \frac{e^{Eucl(W_j^{x_i}, W_j^{y_i})}}{e^{Eucl(W_j, W_j^{x_i})} \times e^{Eucl(W_j, W_j^{y_i})}} & \text{otherwise} \end{cases}$$

It can be seen from Definition 1 that the dissimilarity between two different values x_i and y_i is based on three factors. The first one is the dissimilarity between the conditional weight distribution $W_j^{x_i}$ and $W_j^{y_i}$ with the assumption that the less dissimilar between the conditional weight distribution $W_j^{x_i}$ of A_j when $A_i = x_i$, and $W_j^{y_i}$ when $A_i = y_i$, the less dissimilar between x_i and y_i. The second (third) factor is the dissimilarity between the weight distribution W_j of A_j, and $W_j^{x_i}$ when $A_i = x_i$, ($W_j^{x_i}$ when $A_i = y_i$) with the assumption that the more dissimilar between W_j and $W_j^{x_i}$ the more reliable when considering the dissimilarity between $W_j^{x_i}$ and $W_j^{y_i}$ as the dissimilarity between x_i and y_i. The exponential function is used with the purpose to be applicable when dissimilarities between weight distributions are 0.

Definition 2. *The dissimilarity of two objects* **x** *and* **y** *is defined as the average dissimilarity of their attribute value pairs.*

$$\phi(\mathbf{x}, \mathbf{y}) = \frac{1}{m} \sum_{i=1}^{m} \delta(x_i, y_i)$$

3 Characteristics and Complexity

3.1 Characteristics

Proposition 1. *Given any objects* **x** *and* **y**, *it holds true for*

1. $\phi(\mathbf{x}, \mathbf{y}) \geq 0$
2. $\phi(\mathbf{x}, \mathbf{y}) = 0$ *if and only if* $\mathbf{x} = \mathbf{y}$
3. $\phi(\mathbf{x}, \mathbf{y}) = \phi(\mathbf{y}, \mathbf{x})$

Let J be Jaccard similarity measure $J(\mathbf{x}, \mathbf{y}) = \frac{a}{m}$ where a is the number of identical value pairs of **x** and **y**.

Proposition 2. *If m attributes $A_1, \ldots, A_m$ are all independent of each other, then $\phi = 1 - J$.*

Hence, $\phi(.)$ can be applied to databases whose attributes are absolutely independent. This overcomes the limitations of measures based on relations among attributes, the condition probability-based measures [2].

3.2 Algorithm and Complexity

Now we present a three-step algorithm to measure the dissimilarities of all pairs of data objects of a data set D. At the first step, the weighted contributions of attributes and condition weighted contributions are estimated. Then, the dissimilarities between value pairs are computed based on the weighted condition contributions and the weighted contributions. Finally, dissimilarities between data objects are determined by Definition 2. Obviously, the complexity of the algorithm is $O(nm^2) + O(m_v^3) + O(n^2m) = O(n^2m)$, the same as the complexity of the condition-based dissimilarity measure [2].

Table 1. Databases and Nearest Neighbor results

	Name	Proposed μ_1	S_θ or T_θ μ_0	z	Cond. Prob. Based μ_0	z
1	audiology	75.10	74.90	0.17	64.45	7.44
2	balance-scale	64.88	64.75	0.17	46.82	22.14
3	breastda	96.55	95.89	2.24	95.97	2.00
4	cmc	45.92	43.73	3.73	46.05	-0.21
5	crxda	80.17	78.61	2.42	83.71	-5.71
6	diabetesda	73.38	73.43	-0.07	72.01	1.92
7	germanda	70.40	70.17	0.38	67.70	4.52
8	glassda	74.60	73.92	0.51	71.39	2.57
9	heartda	80.67	77.52	3.03	79.56	1.07
10	horseda	77.07	74.63	2.54	73.30	4.12
11	ionoda	93.93	93.22	1.28	91.99	3.31
12	irisda	95.10	92.77	2.91	95.10	0.01
13	isonod	93.65	93.71	-0.10	92.08	2.66
14	krkopt	41.87	37.67	33.24	46.90	-5.00
15	krvskp	93.41	90.92	13.24	89.62	17.50
16	monks	64.50	77.48	-13.92	50.39	14.16
17	page-blocksd	96.66	96.50	1.58	96.80	-1.41
18	pimada	71.90	66.34	7.72	70.83	1.51
19	primary-tumor	35.58	33.48	1.99	34.78	0.72
20	promoters	82.94	81.14	1.21	85.69	-32.72
21	sickda	96.92	96.12	5.51	97.15	-1.79
22	sonarda	80.01	79.96	0.04	78.12	1.55
23	soybeanl	92.02	91.89	0.21	91.08	1.39
24	splice	77.99	75.93	6.78	87.32	-11.03
25	ttt	100.00	81.02	58.95	98.22	16.17
26	vehicleda	71.79	69.33	3.74	69.76	2.79
27	vote	93.31	92.02	2.54	93.96	-1.38
28	waveformda	74.06	71.65	8.55	77.12	-5.71
29	yeastd	49.93	44.46	9.62	44.50	8.89
30	zoo	97.22	96.21	1.35	97.25	-0.04
	Aver. acc.	78.05%	76.31%		76.32	

4 Evaluations

In this section we show the merit of our approach when it is applied to real data. To this end, we compared the proposed measure with the binary vector-based measures of two families S_θ and T_θ [1], and the condition probability-based measure [2]. S_θ and T_θ include most of the popular binary-based similarity measures (see [6]). To compare similarity measures, we combined these measures with the popular distance-based data mining method, nearest neighbor classifier (NN), and analyze the accuracies of NN.

4.1 Methodology and Databases

We compared the accuracy of NN in combination with the proposed measure (denoted μ_1) with the accuracy of NN when combined with a similarity measure of family T_θ or S_θ, or the condition probability-based measure (denoted by μ_0) [2], using the 10-time, 10-fold cross-validation strategy (see [6] for more detail).

To avoid bias on data selection, we used 30 data sets from UCI [7] (see Table 1), for which numerical attributes are automatically discretized using the data mining system CBA [8].

4.2 Results and Discussion

Table 1 shows experimental results of 30 databases including accuracies of NN with the proposed measure (the third column), accuracies of NN with binary based-measures and Z values when comparing with those of NN with the proposed measures (the fourth and fifth columns), and accuracies of NN with conditional probability-based measure and Z values when comparing with those of NN with the proposed measure (the last two columns).

Consider 95% significant level ($Z = 1.64$). As can be seen from Table 1 that, the accuracy of NN with the proposed measure δ is higher than the accuracy of NN with a measure of S_θ or T_θ in 17 databases and only lower than in one. Similarly, 14 over 30 databases NN with the proposed measure δ is better than with the conditional probability-based measure. However, we observed 6 cases in which NN with the conditional probability-based measure outperforms NN with the proposed measure. The number of databases for which NN with the proposed dissimilarity measures is the most accurate is 17, while that number is 3 for a measure of S_θ or T_θ, and 9 for the condition probability-based measure.

The average accuracy of NN with the proposed measure δ over 30 databases is 1.74% (1.73%) higher than NN with a measure of S_θ or T_θ (the condition probability-based measure).

In a nutshell, the proposed measure boosts the accuracy of NN in comparison with the condition probability-based measure and binary vector-based measures. However, none of the measures outperforms completely the others.

5 Conclusion

In this paper, we presented a similarity measure for categorical data based on relations between attributes. Experiments with a large amount of data showed that the nearest neighbor classification using this proposed measure achieved higher accuracy than using the condition probability-based measure and binary vector-based methods. More importantly, the proposed method overcomes the limitations of the condition probability-based measure when applied to databases whose attributes are independent.

Acknowledgments

We appreciate professor Judith Steeh at Japan Advanced Institute of Science and Technology, Japan for helpful comments on the manuscript.

References

1. J.C. Gower and P. Logondro. Metric and euclidean properties of dissimilarity coefficients. *Journal of classification*, 3:5–48, 1986.
2. S.Q. Le and T.B Ho. Conditional probability distribution-based dissimilarity measure for categorical data. In *8th Pacific-Asia Conference on Knowledge Discovery and Data Mining PAKDD*, pages 580–589. LNAI 3056, Springer, 2004.
3. M. Aono M. Kobayashi. Vector space models for search and cluster mining. *Survey of Text Mining: clustering, classification and retrieval, Springer NY*, pages 103–122, 2004.
4. D.W. Goodall. A new similarity index based on probability. *Biometrics*, 22:882–907, 1966.

5. T.M. Cover and P.E Hart. Nearest neighbor pattern classification. *IEEE Transactions on Information Theory*, 13:21–27, 1967.
6. S.Q Le and T.B Ho. An association-based dissimilarity measure for categorical data. *Pattern Recognition Letters*, 26(16):2549–2557, 2005.
7. C.L. Blake and C.J. Merz. (uci) repository of machine learning databases, 1998.
8. B. Liu, W. Hsu, and Y. Ma. Integrating classification and association rule mining. In *Knowledge Discovery and Data Mining*, pages 80–86, 1998.

Is Frequency Enough for Decision Makers to Make Decisions?

Shichao Zhang[1], Jeffrey Xu Yu[2], Jingli Lu[3], and Chengqi Zhang[1]

[1] University of Technology Sydney, Australia
{zhangsc, chengqi}@it.uts.edu.au
[2] Chinese University of Hong Kong, China
yu@se.cuhk.edu.hk
[3] Monash University, Australia
Jingli.Lu@infotech.monash.edu.au

Abstract. There are many advanced techniques that can efficiently mine frequent itemsets using a minimum support. However, the question that remains unanswered is whether the minimum support can really help decision makers to make decisions. In this paper, we study four summary queries for frequent itemsets mining, namely, 1) finding a support-average of itemsets, 2) finding a support-quantile of itemsets, 3) finding the number of itemsets that greater/less than the support-average, i.e., an approximated distribution of itemsets, and 4) finding the relative frequency of an itemset. With these queries, a decision maker will know whether an itemset in question is greater/less than the support-quantile; the distribution of itemsets; and the frequentness of an itemset. Processing these summary queries is challenging, because the minimum-support constraint cannot be used to prune infrequent itemsets.

1 Introduction

Frequent itemsets mining is one of the fundamental research topics in data mining and is rooted in market basket analysis [Roddick&Rice 2001][Wang et al 2001]. A frequent itemset is actionable if its support is greater than or equal to a user-specified threshold, called a minimum support, denoted τ. For example, ({beer, diaper}, 0.15%) is a frequent itemset in a transactional dataset when $\tau = 0.1\%$. Data marketers and decision-makers are interested in knowing how to evaluate the frequency of an itemset, for example {beer, diaper}.

In order to provide more information sources for decision makers to effectively make decisions, in this paper, we propose novel approaches to report *statistical itemset parameters*. We specify summary queries (statistical parameters), namely, 1) finding a support-average of itemsets, 2) finding a support-quantile of itemsets, 3) finding the number of itemsets that greater/less than the support-quantile, i.e., an approximated distribution of itemsets, and 4) finding the frequentness of an itemset. Taken the itemset, {beer, diaper}, as an example. We

---

* This research is partially supported by Australian large ARC grants (DP0449535, DP0559536 and DP0667060), a China NSFC major research program (60496327), and a China NSFC grant (60463003).

can provide a decision maker the following information. The frequency of {beer, diaper} is greater than the support-quantile. The number of itemsets that have a support less than the support-average is more than the number of itemsets that have a support greater than the support-average. In addition, we can also provide a frequentness of the itemset if all frequent itemsets, their supports are greater than τ, are normalized into a range of [0,1].

The rest of this paper is organized as follows. In Section 2, the problem statement is given with an example. Section 3 reports the results of our experimental studies. We conclude this paper in Section 4.

2 Problem Description

Let $I = \{x_1, x_2, \cdots, x_N\}$ be a set of items, and $N = |I|$. An itemset X is a subset of items I, $X \subseteq I$. Let $\mathcal{I}$ be the set of all possible non-empty itemsets of I. The size of $\mathcal{I}$ is then $2^N - 1$. A transactional dataset D is a set of transactions where a transaction is a set of items. Let the number of transactions in D that contain an itemset X be $s(X)$. The support of an itemset, X, is defined as $sup(X) = s(X)/|D|$. An itemset X is a frequent itemset, if and only if $sup(X) \geq \tau$, where τ is a threshold called a minimum support threshold.

We call an itemset, X, *effective*, if $sup(X) > 0$. Let s_{min} and s_{max} be the minimum and maximum supports of all possible effective itemsets in D, respectively, then $s_{min} \leq sup(X) \leq s_{max}$. It is important to note that s_{min} is the minimum value of effective itemsets in D and is different from the minimum support threshold, τ, which is a user-given parameter.

In the literature, the problem of frequent itemsets mining is to find the complete set of frequent itemsets in a given transactional dataset with respect to a given support threshold, τ. In this paper, we do not discuss how to efficiently mine frequent itemsets. Instead, we focus ourselves on how to efficiently answer the summary queries. Let $\mathcal{N}$ be the number of effective itemsets existing in a transactional dataset D. Four summary queries are given below.

Support-average query: We consider how to find the average of supports for all effective itemsets, X, in D, which is defined below.

$$s_{avg} = \frac{\sum_{X \in \mathcal{I}} sup(X)}{\mathcal{N}} \quad (1)$$

Note: the minimum support threshold τ is not used.

Support-quantile query: Let IS be the set of itemsets in dataset D. For any itemset A in IS, assume

$$IS_A^+ = \{x | x \in IS \wedge supp(x) \geq supp(A)\}$$

and

$$IS_A^- = \{x | x \in IS \wedge supp(x) < supp(A)\}$$

It is commonsense that an itemset B is frequent (or large) in D if $\frac{|IS_B^+|}{|IS|} < \frac{1}{2}$. This means that if B is frequent in D, then the support of most itemsets in IS is lesser than that of B. We take this as the **first support constraint** of frequent itemsets.

From this first constraint, there is a support quantile, written to s_{quan}, such that $IS_{squan}^- \approx IS_{squan}^+$, where

$$IS_{quan}^+ = \{x | x \in IS \land supp(x) > s_{quan}\}$$

and

$$IS_{quan}^- = \{x | x \in IS \land supp(x) \leq s_{quan}\}$$

That is,

$$s_{quan} = sup(A), \text{ when } \frac{|IS_{sup(A)}^+|}{|IS_{sup(A)}^-|} \approx 1 \qquad (2)$$

Support-lean query: We consider a distribution of all effective itemsets in D, called an *itemset distribution*. In an itemset distribution for a transactional dataset D, the x-axis represents supports in the range of s_{min} and s_{max}, and the y-axis shows the number of effective itemsets that have the corresponding support given in the x-axis. The median of supports in the x-axis is $(s_{min} + s_{max})/2$. We want to know the position of s_{avg} in the itemset distribution, and introduce a notion called *lean*, denoted L.

$$L = \frac{L_- - L_+}{\mathcal{N}} \qquad (3)$$

where, L_- is the number of effective itemsets, $X \in \mathcal{I}$, whose support is less than s_{avg} in the transactional dataset D, and L_+ is the number of effective itemsets, $X \in \mathcal{D}$, whose support is greater than s_{avg} in the transactional dataset D.

In other words, when $L_- = L_+$, $L = 0$ and $s_{avg} = s_{quan}$; when $L_- > L_+$, $L > 0$ and $s_{avg} > s_{quan}$; and when $L_- < L_+$, $L < 0$ and $s_{avg} < s_{quan}$. We call the itemsets distribution of D *left gradient* when $L > 0$, and *right gradient* when $L < 0$, respectively. Like the support-average and support-quantile queries, support-lean query is irrelevant to τ.

Relative-support query: We consider a relative frequency query with respect to a minimum support threshold. In detail, given a minimum threshold, τ, we can find all frequent itemsets, X, $\tau \leq sup(X) \leq s_{max}$. If we take $sup(X) = 0$ for those itemsets, X, such as $sup(X) < \tau$, and take $sup(X) = 1$ for those itemsets, X, such as $sup(X) = s_{max}$, what is the relative minimum support $sup(X)$, for any itemsets X found in between. In other words, it normalizes the supports of itemsets in $[\tau, s_{max}]$ onto $[0, 1]$. We model it as $\delta(X, \tau)$. Here, $\delta(X, \tau) = 0$ if $sup(X) < \tau$, $\delta(X, \tau) = 1$ if $sup(X) = s_{max}$, and any $\delta(X, \tau)$ is between 0 and 1.

Example 1. Let $I = \{a, b, c, d, e\}$. Suppose there is a transactional dataset with 5 transactions, $D = \{(a, b, d), (b, c), (b, d), (b, c, d, e), (a, d)\}$. Among all

$2^5 - 1$ itemsets, there are 19 effective itemsets, namely, {a, b, c, d, e, ab, ad, bc, bd, be, cd, ce, de, abd, bcd, bce, bde, cde, bcde}. The minimum and maximum support, s_{min} and s_{max}, are 0.2 (the support of bcde), and 0.8 (the support of d), respectively. Note: the set of non-effective itemsets, for example, includes ae, because sup(ae) = 0. Here, $s_{avg} = (31/5)/19 = 0.33$ and $s_{quan} = sup(e) = 0.2$. Because $L = (12-7)/19$, the transactional dataset is left gradient. It implies that more itemsets have a support less than the support-average value $s_{avg} = 0.33$.

The problems of processing the above four summary queries are important, because they provide global information for users to understand the transactional dataset to be mined and to select a proper minimum support threshold to mine. The problems are also challenging, because the computation costs are high. For example, the first three, support-average, support-quantile and support-lean queries, do not use the minimum-support threshold, τ. Therefore, the pruning techniques used in frequent itemsets mining can not be effectively used. The last requests high computational overhead after frequent itemsets mining. In this paper, we emphasize on approximations for the three summary queries (see next section for the efficiency).

3 Experimental Studies

We evaluate our testing results using a relative error rate, ϵ, such as $\epsilon = \hat{E} - E)/E$, where E is the true value and $\hat{E}$ is the approximated value.

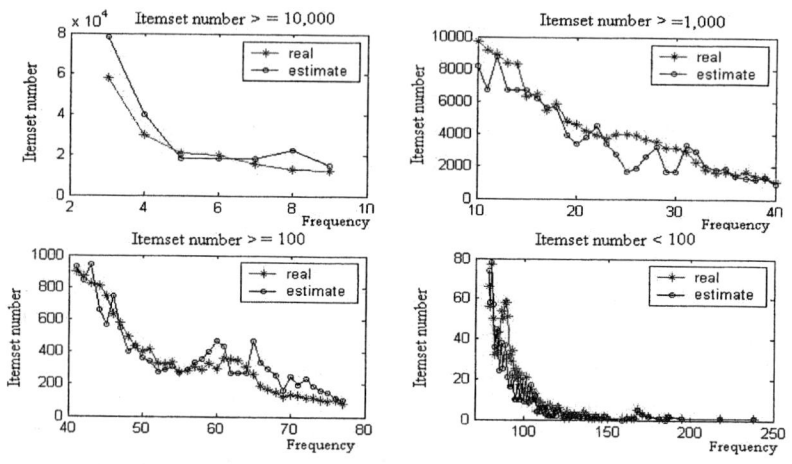

Fig. 1. Itemset distribution in TDB_1 and its sample D_1

We use a real dataset taken from UCI/CENSUS-INCOME, denoted TDB_1. The number of transactions in TDB_1 is 32,562, and the number of items is 104. We selected 9 attributes (2, 4, 6-10, 14, 15) from TDB_1. The average size of frequent itemsets is 6. We take a simple, D_1, where $|D_1| = 1,612$. The sample size is 5% of the real dataset.

We use the sample D_1 to approximate the itemset distribution in TDB_1.

We show the itemset distribution in Fig. 1 using four subfigures where the itemsets with support less than 2 are omitted. Here, for simplicity, we use the actual count, $s(X)$, instead of the support $sup(X)$ for an itemset X. Fig. 1 (a), (b), (c) and (c) show the count ranges in 2-9, 10-40, 40-79, and 80-250, respectively.

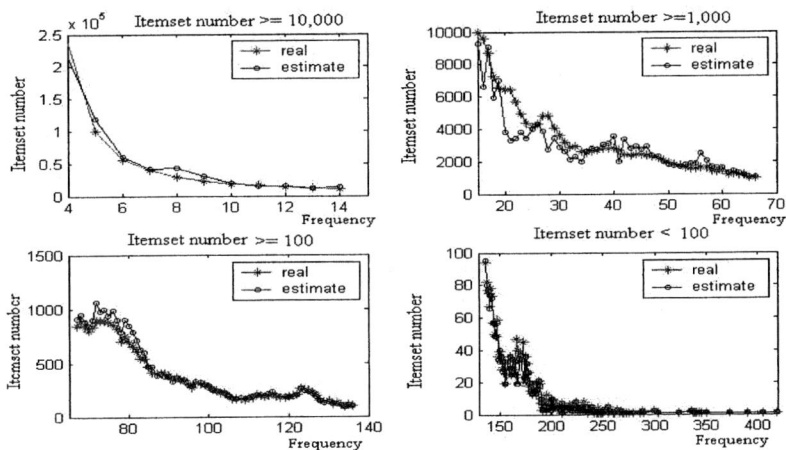

Fig. 2. Itemset distribution in TDB_2 and its sample D_2

4 Conclusion

We have considered four summary queries, and we provide approximation solutions. The new techniques proposed unleash the power of frequent itemsets by providing useful statistical parameters. As shown in our experimental studies, our approximation solutions can achieve high satisfactory level for answering the four summary queries efficiently.

References

[Roddick&Rice 2001] J. F. Roddick and S. Rice (2001), What's Interesting About Cricket? – On Thresholds and Anticipation in Discovered Rules. *SIGKDD Explorations*, 3(1): 1-5.

[Wang et al 2001] K. Wang, Y. He, D. Cheung, F. Chin (2001), Mining Confident Rules without Support Requirement. In *Proceedings of CIKM*, Atlanta.

Ramp: High Performance Frequent Itemset Mining with Efficient Bit-Vector Projection Technique

Shariq Bashir and Abdul Rauf Baig

FAST-National University of Computer and Emerging Sciences, Islamabad, Pakistan
shariqadel@yahoo.com, rauf.baig@nu.edu.pk

Abstract. Mining frequent itemset using bit-vector representation approach is very efficient for small dense datasets, but highly inefficient for sparse datasets due to lack of any efficient bit-vector projection technique. In this paper we present a novel efficient bit-vector projection technique, for sparse and dense datasets. We also present a new frequent itemset mining algorithm *Ramp* (Real Algorithm for Mining Patterns) using bit-vector representation approach and our bit-vector projection technique. The performance of the *Ramp* is compared with the current best frequent itemset mining algorithms. Different experimental results on sparse datasets show that mining frequent itemset using *Ramp* is faster than the current best algorithms.

1 Introduction

Association rules mining introduced by Agrawal [1], has now become one of main pillar of data mining and knowledge discovery tasks, and it is successfully applied in sequential pattern mining, emerging pattern mining, multidimensional pattern mining, classification, maximal and closed pattern mining. Using the support-confidence framework, the problem of mining the complete association rules from transactional dataset is divided into two parts – (a) finding frequent itemsets, and (b) generating association rules from frequent itemsets. Among them part (a) is considered to be the most time consuming process, requires heaviest frequency counting operation for each itemset.

As indicated in [8], MAFIA a maximal itemset mining algorithm [3] (using bit-vector representation approach) is considered to be most efficient algorithm for small dense dataset mining. The main components of MAFIA are its traversal of search space by depth first search, filtering infrequent items from node's tail by dynamic reordering [3] and frequent items representation using vertical bit-vectors. To check the frequency (support) of itemsets it performs a bitwise-AND (bitwise-∧) operation on head and tail item bit-vectors regions. Since 32-bit CPU supports 32-bit ∧ per operation, hence each region of item bit-vector is composed of 32-bits (*represents 32 transactions*). Calculating frequency using bit-vectors is efficient when the dataset is dense, but highly inefficient when the items bit-vectors contain more zeros than ones, resulting in many useless counting operations, which usually happens in the case of sparse datasets. To handle the bit-vectors sparseness problem, MAFIA proposed a bit-vector projection technique known as projected bitmap. The main deficiency of projection using projected bitmap technique is that, it requires a high processing cost (time) for its creation. Due to this reason, MAFIA used adaptive compression [4], since projection is done only when saving from the compressed bitmaps outweigh the cost of projection. However, with adaptive compression, projection cannot be possible on all nodes of search space.

In this paper we present a novel bit-vector projection technique, we call it *Projected-Bit-Regions* (PBR) bit-vector projection technique and its implementation *Ramp (itemset mining algorithm)*. The main advantages of projection using PBR are that – (a) it consumes a very small processing cost and memory space for projection, and (b) it can be easily applied on all nodes of search space without requiring any adaptive approach. In section 2.2 to 2.4 we present some efficient implementation techniques for *Ramp*, which we experienced in our implementation. Our different experimental results on sparse dataset suggest that mining frequent itemset using *Ramp* is faster than the current best algorithms which marked good scores on FIMI03 and FIMI04 [7]: fpgrowth-zhu [6], AFOPT [8], PatriciaMine [10], AIM [5], MAFIA [3], fpgrowth-borgelt, Eclat-borgelt [2], Apriori-borgelt [2]. This shows the effectiveness of our PBR bit-vector projection technique.

2 Bit-Vector Projection with PBR (Projected-Bit-Regions)

For efficient projection of bit-vectors, the goal of projection should be such as, to bitwise-∧ only those regions of head bit-vector ⟨bitmap(head)⟩ with tail item X bit-vector ⟨bitmap(X)⟩ which contains a value greater than zero and skip all others. Obviously for doing this, our counting procedure must be so powerful and have some information which guides it, that which regions are important and which ones it can skip. To achieve this goal, we propose a novel bit-vector projection technique PBR (Projected-Bit-Regions). With projection using PBR, each node Y of search space contains an array of valid region indexes PBR⟨Y⟩ which guide the frequency counting procedure to traverse only those regions which contain an index in array and skip all other. Figure 1 show the code of itemset frequency calculation using PBR technique. As clear from Figure 1, line 2 first retrieves a valid region index ℓ in ⟨bitmap (head)⟩ and line 3 apply a bitwise-∧ on ⟨bitmap (head)⟩ with ⟨bitmap (X)⟩ on region ℓ.

One main advantage of bit-vector projection using PBR is that, it consumes a very small processing cost for its creation, thereby can be easily applied on all nodes of search space. At any node, projection of child nodes can be created either at the time of frequency calculation if pure depth first search is used, or at the time of creating head bit-vector if dynamic reordering is used. The strategy of creating $PBR_{(X)}$ at node n for each tail item X is that, when the PBR of ⟨bitmap(n)⟩ are bitwise-∧ with ⟨bitmap(X)⟩ a simple check is perform on each bitwise-∧ result. If the value of result is greater than zero, then an index is allocated in $PBR_{(n.head\ \cup\ X)}$. The set of all indexes which contain a value greater than zero makes the projection of *{head* $\cup$ *X}* node.

2.1 Ramp: Itemset Mining Algorithm

The basic strategy of *Ramp* for mining frequent itemset is that, it traverses search space in depth first order. At any node *n*, infrequent items from tail are removed by dynamic reordering and new node m for every tail item X in tail n, is generated such as $m.head = n \cup X$ and $m.tail = n.tail - X$. Items in $m.tail$ are reordered by increasing support which keeps the search space as small as possible. For frequency counting, item X bit-vector is bitwise-∧ with $n.head$ bit-vector on $PBR_{\langle n \rangle}$. The pseudo code of *Ramp* is described in Figure 1.

Ramp (Node n)
(1)   for each item X in $n.tail$
(2)   for each region index ℓ in $PBR_{\langle n \rangle}$
(3)   AND-result = bit-vector[ℓ] ∧ head-bit-vector of n [ℓ]
(4)   support[X] = support[X] + number of ones(AND-result)
(5)   remove infrequent items from $n.tail$, reorder them by increasing support
(6)   for each item X in $n.tail$
(7) $m.head = n.head \cup X$
(8) $m.tail = n.tail - X$
(9)   for each region index ℓ in $PBR_{\langle n \rangle}$
(10)   AND-result = bit-vector[ℓ] ∧ head-bit-vector[ℓ]
(11)   if AND-result > 0
(12)   insert ℓ in $PBR_{\langle m \rangle}$
(13)   head bit-vector of m [ℓ] = AND-result
(14)   Ramp (m)

Fig. 1. Pseudo code of *Ramp* for mining all frequent itemset

2.2 Increasing Projected Bit-Regions Density

The bit-vector projection technique described in section 2 does not provide any compaction or compression mechanism to increase the density in bit-vector regions. As a result, on the sparse dataset only one or two bits are set in each bit-vector region, which not only increase the projection length but also it is not possible to achieve true 32bit CPU performance. To increase the density in bit-vector regions the *Ramp* starts with an array-list [9]. Next at root node, a bit-vector representation for each frequent item is created which provide a sufficient compression and compaction in bit-vectors regions. Sufficient improvements are obtained in *Ramp* by using this approach.

2.3 2-Itemset Pair

There are two methods to check whether current itemset is frequent or infrequent – (a) to directly compute its frequency from its PBR (b) by 2-Itemset pair. If any 2-Itemset pair of any itemset is found infrequent, then by following Apriori [1] property itemset is consider to be as infrequent. In AIM [5] almost the same approach was used with the name *efficient initialization*. However AIM used this approach only for those itemsets which contain a length equal to two. In *Ramp* we extend the basic approach and apply 2-Itemset pair approach also on those itemsets which contain a length more than two. We know any itemset which contains a length more than two, is the superset of its entire 2-Itemset pairs. Before counting its frequency from *TDB*, *Ramp* checks its

2-Itemset pairs. If any pair is found infrequent then that itemset is automatically considered to be infrequent

2.4 Writing Frequent Itemsets to Output File

When the dataset is dense and contains millions of frequent itemsets on low support threshold, almost 90% of overall mining time is spent on writing frequent itemsets to output file. We have noted hat some of previous implementations e.g. AFOPT [8], PatriciaMine [10], fpgrowth-zhu [6] write output itemsets one by one, which increases the context switch and disk rotation times and degrades their algorithm performance. A better approach which we use in *Ramp* is to write itemsets to output file only when a sufficient number of itemsets are mined in memory. In *Ramp* we find that, writing itemsets using this approach sufficiently decreases the processing time of algorithm.

3 Computation Experiments

The implementation of Ramp-all is coded in C language, and the experiments are done on Pentium4 3.2 GHz CPU with 512MB memory. The performance measure is the execution time of the algorithms datasets with different support thresholds on two benchmark datasets (available at http://fimi.cs.helsinki.fi/data/). Figures 2 and 3 show the performance curves of all algorithms. As we can see from Figures, the *Ramp-all* outperforms the other algorithms on almost all support level thresholds, and gives global best performance. The performance improvements of *Ramp-all* over other algorithms are significant at low support thresholds.

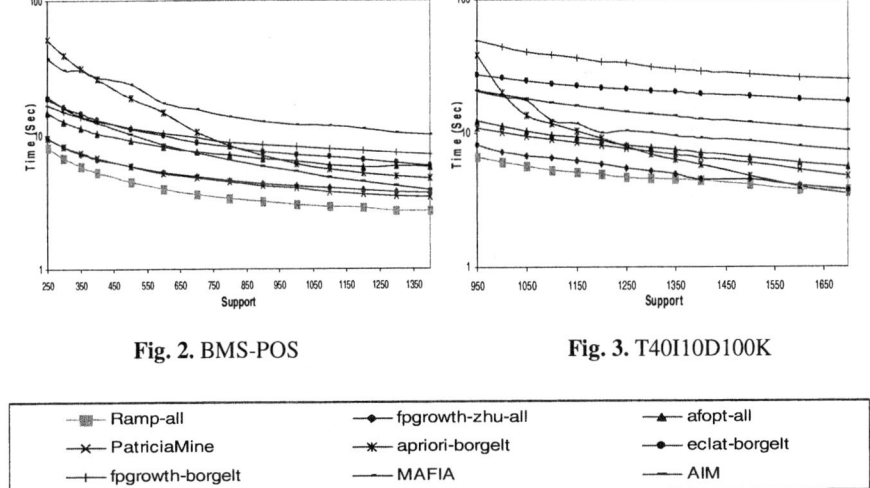

Fig. 2. BMS-POS   Fig. 3. T40I10D100K

4 Conclusion

Mining frequent itemset using bit-vector representation approach is very efficient for dense datasets, but highly inefficient for sparse datasets due to lack of any efficient bit-vector projection technique. In this paper we present a novel efficient bit-vector projection technique, which is better than the previous projected bitmap projection technique. The main advantages of our bit-vector projection technique are that, it does not require any rebuilding threshold or does not depend on any adaptive approach for projection, and can be easily applicable on all nodes of search space. We also present a new frequent itemset mining algorithm *Ramp* using our bit-vector projection technique. Different experiments on benchmark datasets show that *Ramp* is faster than the current best frequent itemset algorithms, which show the effectiveness of our bit-vector projection technique.

References

1. R. Agrawal, R. Srikant. Fast algorithms for mining association rules. In *VLDB'94*, Santiago, Chile, 1994.
2. C. Borgelt. Efficient Implementation of Eclat and Apriori. In *IEEE ICDM'03 Workshop FIMI'03*, Melbourne, Florida, USA, 2003.
3. D. Burdick, M. Calimlim, J. Gehrke. MAFIA: A Maximal Frequent Itemset Algorithm for Transactional Databases. In *ICDE'01*, Heidelberg, Germany, 2001.
4. D. Burdick, M, Calimlim, J. Flannick, J. Gehrke, T. Yiu. MAFIA: A Performance Study of Mining Maximal Frequent Itemsets. In *IEEE ICDM'03 Workshop FIMI'03*, Melbourne, Florida, USA, 2003.
5. A. Fiat, S. Shporer. AIM: Another Itemset Miner. In *IEEE ICDM'03 Workshop FIMI'03*, Melbourne, Florida, USA, 2003.
6. G. Grahne, J. Zhu. Efficiently Using Prefix-trees in Mining Frequent Itemsets. In *IEEE ICDM'03 Workshop FIMI'03*, Melbourne, Florida, USA, 2003.
7. Proc. IEEE ICDM Workshop Frequent Itemset Mining Implementations, B. Goethals, M.J. Zaki, eds. CEUR Workshop Proc., vol. 80, Nov. 2003, http://CEUR-WS.org/Vol-90.
8. G. Liu, H. Lu, J.X. Yu, W. Wei, X. Xiao. AFOPT: An Efficient Implementation of Pattern Growth Approach. In *IEEE ICDM '03 Workshop FIMI'03*, Melbourne, Florida, USA, 2003.
9. J. Pei, J. Han, H. Lu, S. Nishio, S. Tang, D. Yang. H-Mine: Hyper-structure mining of frequent patterns in large databases. In *ICDM'01*, San Jose, California, USA, 2001.
10. A. Pietracaprina, D. Zandolin. Mining Frequent Itemsets using Patricia Tries. In *IEEE ICDM'03 Workshop FIMI'03*, Melbourne, Florida, USA, 2003.

Evaluating a Rule Evaluation Support Method Based on Objective Rule Evaluation Indices

Hidenao Abe[1], Shusaku Tsumoto[1], Miho Ohsaki[2], and Takahira Yamaguchi[3]

[1] Department of Medical Informatics, Shimane University, School of Medicine,
89-1 Enya-cho, Izumo, Shimane 693-8501, Japan
abe@med.shimane-u.ac.jp, tsumoto@computer.org
[2] Faculty of Engineering, Doshisha University,
1-3 Tataramiyakodani, Kyo-Tanabe, Kyoto 610-0321, Japan
mohsaki@mail.doshisha.ac.jp
[3] Faculty of Science and Technology, Keio University,
3-14-1 Hiyoshi, Kohoku Yokohama, Kanagawa 223-8522, Japan
yamaguti@ae.keio.ac.jp

Abstract. In this paper, we present an evaluation of novel rule evaluation support method for post-processing of mined results with rule evaluation models based on objective indices. Post-processing of mined results is one of the key issues in a data mining process. However, it is difficult for human experts to evaluate many thousands of rules from a large dataset with noises completely. To reduce the costs of rule evaluation task, we have developed the rule evaluation support method with rule evaluation models, which are obtained with objective indices of mined classification rules and evaluations of a human expert for each rule. To evaluate performances of learning algorithms for constructing rule evaluation models, we have done a case study on the meningitis data mining as an actual problem. Furthermore, we have also evaluated our method on four rulesets from the four kinds of UCI datasets.

1 Introduction

In recent years, huge data are easily stored on information systems in natural science, social science and business domains, developing information technologies. With these huge data, people hope to utilize them for their purposes. Besides, data mining techniques have been widely known as a process for utilizing stored data on database systems, combining different kinds of technologies such as database technologies, statistical methods and machine learning methods. Especially, IF-THEN rules, which are produced by rule induction algorithms, are discussed as one of highly usable and readable output of data mining. However, to large dataset with hundreds attributes including noises, the process often obtains many thousands of rules. From such huge rule set, it is difficult for human experts to find out valuable knowledge which are rarely included in the rule set.

To support such a rule selection, many efforts have done using objective rule evaluation indices such as recall, precision, and other interestingness measurements (we call them 'objective indices' later). However, it is also difficult to estimate a criterion of a human expert with single objective rule evaluation index, because his/her subjective criterion such as interestingness and importance for his/her purpose is influenced by the amount of his/her knowledge and/or a passage of time.

To above issues, we have been developed an adaptive rule evaluation support method for human experts with rule evaluation models, which predict experts' criteria based on objective indices, re-using results of evaluations of human experts. In Section 3, we describe the rule evaluation model construction method based on objective indices. Then we present a performance comparison of learning algorithms for constructing rule evaluation models in Section 4. With the results of the comparison, we discuss about the availability of our rule evaluation model construction approach.

2 Related Work

Many efforts have done to select valuable rules from mined large rule set based on objective rule evaluation indexes. Some of these works suggest the indexes to discover interesting rules from such a large amount of rules.

Focusing on interesting rule selection with objective indexes, researchers have developed more than forty objective indexes based on number of instances, probability, statistics values, information quantity, distance of rules or their attributes, and complexity of a rule[11, 21, 23]. Most of these indexes are used to remove meaningless rules rather than to discover really interesting ones for a human expert, because they can not include domain knowledge. In contrast, a dozen of subjective indexes estimate how a rule fits with a belief, a bias or a rule template formulated beforehand by a human expert. Although these subjective indexes are useful to discover really interesting rules to some extent due to their built-in domain knowledge, they depend on the precondition that a human expert is able to clearly formulate his/her interest. Although interestingness indexes were verified their availabilities on each suggested domain, nobody has validated their availabilities on the other domains or/and characteristics related to the background of a given dataset.

Ohsaki et. al[15] investigated the relation between objective indexes and real human interests, taking real data mining results and their human evaluations. In this work, the comparison shows that it is difficult to predict real human interests with a single objective index. Based on the result, they indicated the possibility of logical combination of the objective indexes to predict real human interests more exactly.

3 Rule Evaluation Support with Rule Evaluation Model Based on Objective Indices

We considered the process of modeling rule evaluations of human experts as the process to clear up relationships between the human evaluations and features

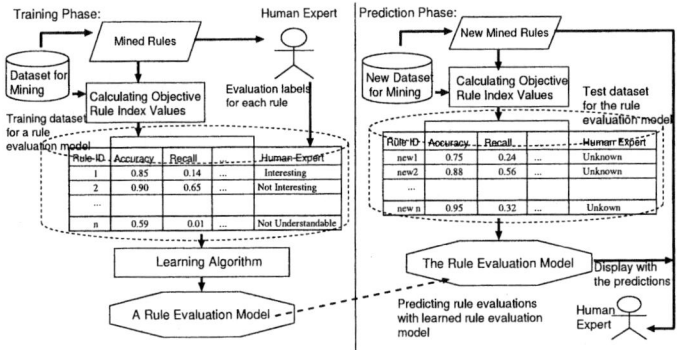

Fig. 1. Overview of the construction method of rule evaluation models

of input if-then rules. With this consideration, we decided that the process of rule evaluation model construction can be implemented as a learning task. Fig.1 shows the process of rule evaluation model construction based on re-use of human evaluations and objective indices for each mined rule.

At the training phase, attributes of a meta-level training data set is obtained by objective indices such as recall, precision and other rule evaluation values. The human evaluations for each rule are joined as class of each instance. To obtain this data set, a human expert has to evaluate the whole or part of input rules at least once. After obtaining the training data set, its rule evaluation model is constructed by a learning algorithm. At the prediction phase, a human expert receives predictions for new rules based on their values of the objective indices. Since the task of rule evaluation models is a prediction, we need to choose a learning algorithm with higher accuracy as same as current classification problems.

4 Performance Comparisons of Learning Algorithms for Rule Model Construction

To predict human evaluation labels of a new rule based on objective indices more exactly, we have to construct a rule evaluation model, which has higher predictive accuracy.

In this section, we firstly present the result of an empirical evaluation with the dataset from the result of a meningitis data mining[9]. Then to confirm the performance of our approach on the other datasets, we evaluated the five algorithms on four rule sets from four kinds of UCI benchmark datasets [10]. With the experimental results, we discuss about the following three view points: accuracies of rule evaluation models, analyzing learning curves of the learning algorithms, and contents of learned rule evaluation models.

As an evaluation of accuracies of rule evaluation models, we have compared predictive accuracies on the whole dataset and Leave-One-Out. The accuracy of a

Table 1. The objective rule evaluation indices for classification rules used in this research. **P:** Probability of the antecedent and/or consequent of a rule. **S:** Statistical variable based on P. **I:** Information of the antecedent and/or consequent of a rule. **N:** Number of instances included in the antecedent and/or consequent of a rule. **D:** Distance of a rule from the others based on rule attributes.

Theory	Index Name (**Abbreviation**) [Reference Number of Literature]
P	Coverage(**Coverage**), Prevalence(**Prevalence**)
	Precision(**Precision**), Recall(**Recall**)
	Support(**Support**), Specificity(**Specificity**)
	Accuracy(**Accuracy**), Lift(**Lift**)
	Leverage(**Leverage**), Added Value(**Added Value**)[21]
	Klösgen's Interestingness(**KI**)[14], Relative Risk(**RR**)[1]
	Brin's Interest(**BI**)[2], Brin's Conviction(**BC**)[2]
	Certainty Factor(**CF**)[21], Jaccard Coefficient(**Jaccard**)[21]
	F-Measure(**F-M**)[19], Odds Ratio(**OR**)[21]
	Yule's Q(**YuleQ**)[21], Yule's Y(**YuleY**)[21]
	Kappa(**Kappa**)[21], Collective Strength(**CST**)[21]
	Gray andOrlowska's Interestingness weighting Dependency(**GOI**)[7]
	Gini Gain(**Gini**)[21], Credibility(**Credibility**)[8]
S	χ^2 Measure for One Quadrant(χ^2-**M1**)[6]
	χ^2 Measure for Four Quadrant(χ^2-**M4**)[6]
I	J-Measure(**J-M**)[20], K-Measure(**K-M**)[15]
	Mutual Information(**MI**)[21]
	Yao and Liu's Interestingness 1 based on one-way support(**YLI1**)[23]
	Yao and Liu's Interestingness 2 based on two-way support(**YLI2**)[23]
	Yao and Zhong's Interestingness(**YZI**)[23]
N	Cosine Similarity(**CSI**)[21], Laplace Correction(**LC**)[21]
	ϕ Coefficient(ϕ)[21], Piatetsky-Shapiro's Interestingness(**PSI**)[16]
D	Gago and Bento's Interestingness(**GBI**)[5]
	Peculiarity(**Peculiarity**)[24]

validation dataset D is calculated with correctly predicted instances $Correct(D)$ as $Acc(D) = (Correct(D)/|D|) \times 100$, where $|D|$ means the size of the dataset. Recalls of class i on a validation dataset is calculated with correctly predicted instances about the class $Correct(D_i)$ as $Recall(D_i) = (Correct(D_i)/|D_i|) \times 100$, where $|D_i|$ means the size of instances with class i. Also the precision of class i is calculated with the size of instances predicted i as $Precision(D_i) = (Correct(D_i)/Predicted(D_i)) \times 100$.

As for learning curves, we obtained learning curves about accuracies to the whole training dataset to evaluate whether each learning algorithm can perform in early stage of a process of rule evaluations. Accuracies from randomly subsampled training datasets are averaged with 10 times trials on each percentage of subset.

Looking at elements of the rule evaluation models on the meningitis data mining result, we consider the characteristics of objective indices, which are used in these rule evaluation models.

To construct a dataset to learn a rule evaluation model, values of objective indices have been calculated for each rule, taking 39 objective indices as shown in Table1. Thus each dataset for each rule set has the same number of instances as the rule set. Each instance consists of 40 attributes including the class attribute.

To these dataset, we applied five learning algorithms to compare their performance as a rule evaluation model construction method. We used the following learning algorithms from Weka[22]: C4.5 decision tree learner[18] called J4.8, neural network learner with back propagation (BPNN)[12], support vector machines (SVM)[1][17], classification via linear regressions (CLR)[2][3], and OneR[13].

4.1 Constructing Rule Evaluation Models on an Actual Datamining Result

In this case study, we have taken 244 rules, which are mined from six dataset about six kinds of diagnostic problems as shown in Table2. These datasets are consisted of appearances of meningitis patients as attributes and diagnoses for each patient as class. Each rule set was mined with each proper rule induction algorithm composed by a constructive meta-learning system called CAMLET[9]. For each rule, we labeled three evaluations (I:Interesting, NI:Not-Interesting, NU:Not-Understandable), according to evaluation comments from a medical expert.

Table 2. Description of the meningitis datasets and their datamining results

Dataset	#Attributes	#Class	#Mined rules	#'I' rules	#'NI' rules	#'NU' rules
Diag	29	6	53	15	38	0
C_Cource	40	12	22	3	18	1
Culture+diag	31	12	57	7	48	2
Diag2	29	2	35	8	27	0
Course	40	2	53	12	38	3
Cult_find	29	2	24	3	18	3
TOTAL	—	—	244	48	187	9

Comparison on Classification Performances. In this section, we show the result of the comparisons of accuracies on the whole dataset, recall of each class label, and precisions of each class label. Since Leave-One-Out holds just one test instance and remains as the training dataset repeatedly for each instance of a given dataset, we can evaluate the performance of a learning algorithm to a new dataset without any ambiguity.

The results of the performances of the five learning algorithms to the whole training dataset and the results of Leave-One-Out are also shown in Table3. All of the accuracies, Recalls of I and NI, and Precisions of I and NI are higher than predicting default labels.

Comparing with the accuracy of OneR, the other learning algorithms achieve equal or higher performance with combination of multiple objective indices than sorting with single objective index. Looking at Recall values on class I, BPNN have achieved the highest performance. As for the other algorithms, they show lower performance than OneR, because they have tended to be learned classification patterns for the major class NI.

The accuracies of Leave-One-Out shows robustness of each learning algorithm. These learning algorithms have achieved from 75.8% to 81.9%. However, these

---

[1] The kernel function was set up polynomial kernel.
[2] We set up the elimination of collinear attributes and the model selection with greedy search based on Akaike Information Metric.

Table 3. Accuracies(%), Recalls(%) and Precisions(%) of the five learning algorithms

	On the whole training dataset						Leave-One-Out						
		Recall of		Precision of				Recall of		Precision of		
	Acc.	I	NI	NU	I	NI	NU	Acc.	I	NI	NU	I	NI	NU
J4.8	85.7	41.7	97.9	66.7	80.0	86.3	85.7	79.1	29.2	95.7	0.0	63.6	82.5	0.0
BPNN	86.9	81.3	89.8	55.6	65.0	94.9	71.4	77.5	39.6	90.9	0.0	50.0	85.9	0.0
SVM	81.6	35.4	97.3	0.0	68.0	83.5	0.0	81.6	35.4	97.3	0.0	68.0	83.5	0.0
CLR	82.8	41.7	97.3	0.0	71.4	84.3	0.0	80.3	35.4	95.7	0.0	60.7	82.9	0.0
OneR	82.0	56.3	92.5	0.0	57.4	87.8	0.0	75.8	27.1	92.0	0.0	37.1	82.3	0.0

learning algorithms have not been able to classify the instances with class NU, because it is difficult to predict a minor class label in this dataset.

Learning Curves of the Learning Algorithms. Since the rule evaluation model construction method needs evaluations of mined rules by a human expert, we have investigated learning curves of each learning algorithm to estimate minimum training subset to construct a valid rule evaluation model. The upper table in Fig.2 shows accuracies to the whole training dataset with each subset of training dataset. The percentages of achievements for each learning algorithm, comparing with the accuracy with the whole dataset, are shown in the lower chart of Fig.2.

As shown in these results, SVM and CLR, which learn hype-planes, achieves grater than 95% with only less than 10% of training subset. Although decision tree learner and BPNN could learn better classifier to the whole dataset than

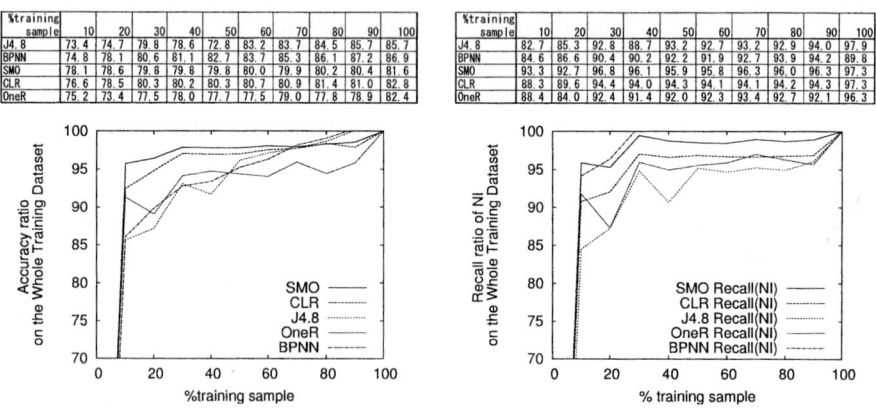

Fig. 2. Learning Curves of accuracies(%) on the learning algorithms with sub-sampled training dataset: The left table shows accuracies(%) on each training dataset to the whole dataset. The left graph shows their achievement ratio(%). Also the right table shows recalls(%), and the graph shows their achievement ratio(%).

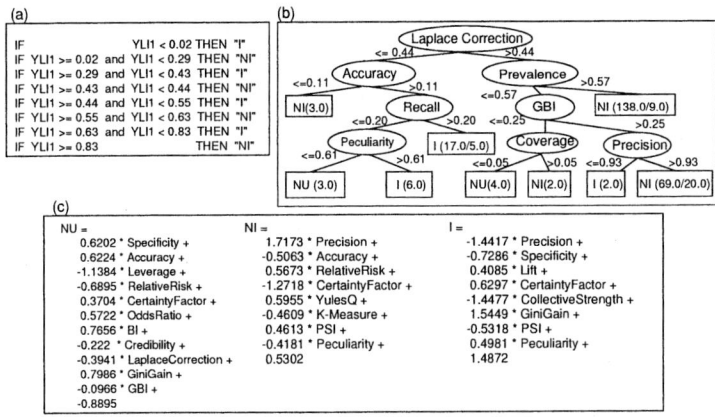

Fig. 3. Learned models to the meningitis data mining result dataset

these hyper-plane learners, they need more training instances to learn accurate classifiers.

To eliminate known ordinary knowledge from large rule set, it is needed to classify non-interesting rules correctly. The right upper table in Fig.2 shows percentages of recalls on NI. The right lower chart in Fig.2 also shows the percentages of achievements on recall of NI, comparing with the recall of NI on the whole training dataset. Looking at this result, we can eliminate NI rules with rule evaluation models from SVM and BPNN even if there is only 10% of rule evaluations by a human expert. This is guaranteed with no less than 80% precisions of all learning algorithms.

Rule Evaluation Models on the Actual Datamining Result Dataset.
In this section, we present rule evaluation models to the whole dataset learned with OneR, J4.8 and CLR, because they are represented as explicit models such as a rule set, a decision tree, and a set of linear models.

Fig.3 shows rule evaluation models on the actual data mining result: The rule set of OneR is shown in Fig.3(a), Fig.3(b) shows the decision tree learned with J4.8, and Fig.3(c) shows linear models to classify each class.

Looking at indices used in learned rule evaluation models, they are not only the group of indices increasing with a correctness of a rule, but also they are used some different groups of indices on different models. Almost indices such as YLI1, Laplace Correction, Accuracy, Precision, Recall, Coverage, PSI and Gini Gain are the former type of indices on the models. The later indices are GBI and Peculiality, which sums up difference of antecedents between one rule and the other rules in the same ruleset. This corresponds to the comment from the human expert. He said that he evaluated these rules not only correctness but also his interest based on his expertise.

4.2 Constructing Rule Evaluation Models on Artificial Evaluation Labels

We have also evaluated our rule evaluation model construction method with rule sets from four datasets of UCI Machine Learning Repository to confirm the lower limit performances on probabilistic class distributions.

We selected the following four datasets: Mushroom, Heart, Internet Advertisement Identification (called InternetAd later) and Letter. With these datasets, we obtained rule sets with bagged PART, which repeatedly executes PART[4] to bootstrapped training sub-sample datasets.

To these rule sets, we calculated the 39 objective indices as attributes of each rule. As for the class of these datasets, we set up three class distributions with multinomial distribution. Table4 shows us the datasets with three different class distributions. The class distribution for 'Distribution I' is $P = (0.35, 0.3, 0.3)$ where p_i is the probability for class i. Thus the number of class i in each instance D_j become $p_i D_j$. As the same way, the probability vector of 'Distribution II' is $P = (0.3, 0.5, 0.2)$, and 'Distribution III' is $P = (0.3, 0.65, 0.05)$.

Table 4. Datasets of the rule sets learned from the UCI benchmark datasets

	#Mined Rules	#Class labels L1	L2	L3	%Def. class
Distribution I		(0.30)	(0.35)	(0.35)	
Mushroom	30	8	14	8	46.7
InternetAd	107	26	39	42	39.3
Heart	318	97	128	93	40.3
Letter	6340	1908	2163	2269	35.8
Distribution II		(0.30)	(0.50)	(0.20)	
Mushroom	30	11	16	3	53.3
InternetAd	107	30	53	24	49.5
Heart	318	99	140	79	44.0
Letter	6340	1890	3198	1252	50.4
Distribution III		(0.30)	(0.65)	(0.05)	
Mushroom	30	7	21	2	70.0
InternetAd	107	24	79	9	73.8
Heart	318	98	205	15	64.5
Letter	6340	1947	4062	331	64.1

Accuracy Comparison on Classification Performances. To above datasets, we have attempted the five learning algorithms to estimate whether their classification results can go to or beyond the accuracies of just predicting each default class. The left table of Table5 shows the accuracies of the five learning algorithms to each class distribution of the three datasets. As shown in Table5, J48 and BPNN always work better than just predicting a default class. However, their performances are suffered from probabilistic class distributions to larger datasets such as Heart and Letter.

Evaluation on Learning Curves. As same as evaluations of learning curves on the meningitis rule set, we have estimated the minimum training subsets for a valid model, which works better than just predicting a default class.

The right table in Table5 shows sizes of minimum training subsets, which can be constructed more accurate rule evaluation models than percentages of a default class by each learning algorithm. To smaller dataset, such as Mushroom

Table 5. Accuracies(%) on whole training datasets labeled with three different distributions(The left table). Number of minimum training sub-samples to outperform %Def. class(The right table).

	J48	BPNN	SVM	CLR	OneR		J48	BPNN	SVM	CLR	OneR
Distribution I						Distribution I					
Mushroom	80.0	93.3	56.7	66.7	53.3	Mushroom	8	8	12	18	14
InternetAd	84.1	82.2	29.9	53.3	60.7	InternetAd	14	14	-	30	14
Heart	78.0	75.8	40.3	42.5	54.7	Heart	42	31	66	114	98
Letter	36.8	36.4	30.1	36.6	52.1	Letter	189	217	-	955	305
Distribution II						Distribution II					
Mushroom	93.3	93.3	80.0	80.0	76.7	Mushroom	6	4	4	6	12
InternetAd	73.8	79.4	49.5	59.8	60.7	InternetAd	24	24	52	42	70
Heart	72.3	69.2	35.9	47.8	55.7	Heart	52	40	-	104	92
Letter	51.0	51.0	50.4	50.4	57.0	Letter	897	>1000	451	-	>1000
Distribution III						Distribution III					
Mushroom	93.3	96.7	70.0	70.0	76.7	Mushroom	22	14	22	28	22
InternetAd	86.0	90.7	70.1	69.2	72.0	InternetAd	80	66	-	-	-
Heart	78.0	77.7	64.5	65.7	71.4	Heart	114	94	142	318	182
Letter	64.1	64.3	64.1	64.1	68.3	Letter	>1000	>1000	998	>1000	>1000

and InternetAd, they can construct valid models with less than 20% of given training datasets. However, to larger dataset, they need more training subsets to construct valid models, because their performances with whole training dataset fall to the percentages of default class of each dataset as shown in the left table in Table5.

5 Conclusion

In this paper, we have described rule evaluation support method with rule evaluation models to predict evaluations for an IF-THEN rule based on objective indices, re-using evaluations of a human expert.

As the result of the performance comparison with the five learning algorithms, rule evaluation models have achieved higher accuracies than just predicting each default class. Considering the difference between the actual evaluation labeling and the artificial evaluation labeling, it is shown that the medical expert evaluated with noticing particular relations between an antecedent and a class/another antecedent in each rule. In the estimation of robustness to a new rule with Leave-One-Out, we have achieved more than 75.8% with these learning algorithms. On the evaluation with learning curves to the dataset of the actual datamining result, SVM and CLR have achieved more than 95% of achievement ratio compared to the accuracy of the whole training dataset with less than 10% of subset of the training dataset with certain human evaluations. These results indicate the availability of this rule evaluation support method for a human expert.

As future work, we will introduce a selection method of learning algorithms to construct a proper rule evaluation model according to each situation. We also apply this rule evaluation support method to estimate other data mining result such as decision tree, rule set, and committee of them with objective indices, which evaluate whole mining results.

References

1. Ali, K., Manganaris,S., Srikant, R.: Partial Classification Using Association Rules. Proc. of Int. Conf. on Knowledge Discovery and Data Mining KDD-1997 (1997) 115–118
2. Brin, S., Motwani, R., Ullman, J., Tsur, S.: Dynamic itemset counting and implication rules for market basket data. Proc. of ACM SIGMOD Int. Conf. on Management of Data (1997) 255–264
3. Frank, E., Wang, Y., Inglis, S., Holmes, G., and Witten, I. H.: Using model trees for classification, Machine Learning, Vol.32, No.1 (1998) 63–76
4. Frank, E, Witten, I. H., Generating accurate rule sets without global optimization, in Proc. of the Fifteenth International Conference on Machine Learning, (1998) 144–151
5. Gago, P., Bento, C.: A Metric for Selection of the Most Promising Rules. Proc. of Euro. Conf. on the Principles of Data Mining and Knowledge Discovery PKDD-1998 (1998) 19–27
6. Goodman, L. A., Kruskal, W. H.: Measures of association for cross classifications. Springer Series in Statistics, 1, Springer-Verlag (1979)
7. Gray, B., Orlowska, M. E.: CCAIIA: Clustering Categorical Attributes into Interesting Association Rules. Proc. of Pacific-Asia Conf. on Knowledge Discovery and Data Mining PAKDD-1998 (1998) 132–143
8. Hamilton, H. J., Shan, N., Ziarko, W.: Machine Learning of Credible Classifications. Proc. of Australian Conf. on Artificial Intelligence AI-1997 (1997) 330–339
9. Hatazawa, H., Negishi, N., Suyama, A., Tsumoto, S., and Yamaguchi, T.: Knowledge Discovery Support from a Meningoencephalitis Database Using an Automatic Composition Tool for Inductive Applications, in Proc. of KDD Challenge 2000 in conjunction with PAKDD2000 (2000) 28–33
10. Hettich, S., Blake, C. L., and Merz, C. J.: UCI Repository of machine learning databases [http://www.ics.uci.edu/~mlearn/MLRepository.html], Irvine, CA: University of California, Department of Information and Computer Science, (1998).
11. Hilderman, R. J. and Hamilton, H. J.: Knowledge Discovery and Measure of Interest, Kluwe Academic Publishers (2001)
12. Hinton, G. E.: "Learning distributed representations of concepts", *Proceedings of 8th Annual Conference of the Cognitive Science Society*, Amherest, MA. REprinted in R.G.M.Morris (ed.) (1986)
13. Holte, R. C.: Very simple classification rules perform well on most commonly used datasets, Machine Learning, Vol. 11 (1993) 63–91
14. Klösgen, W.: Explora: A Multipattern and Multistrategy Discovery Assistant. in Fayyad, U. M., Piatetsky-Shapiro, G., Smyth, P., Uthurusamy R. (Eds.): Advances in Knowledge Discovery and Data Mining. AAAI/MIT Press, California (1996) 249–271
15. Ohsaki, M., Kitaguchi, S., Kume, S., Yokoi, H., and Yamaguchi, T.: Evaluation of Rule Interestingness Measures with a Clinical Dataset on Hepatitis, in Proc. of ECML/PKDD 2004, LNAI3202 (2004) 362–373
16. Piatetsky-Shapiro, G.: Discovery, Analysis and Presentation of Strong Rules. in Piatetsky-Shapiro, G., Frawley, W. J. (eds.): Knowledge Discovery in Databases. AAAI/MIT Press (1991) 229–248
17. Platt, J.: Fast Training of Support Vector Machines using Sequential Minimal Optimization, Advances in Kernel Methods - Support Vector Learning, B. Schölkopf, C. Burges, and A. Smola, eds., MIT Press (1999) 185–208

18. Quinlan, R.: C4.5: Programs for Machine Learning, Morgan Kaufmann Publishers, (1993)
19. Rijsbergen, C.: Information Retrieval, Chapter 7, Butterworths, London, (1979) http://www.dcs.gla.ac.uk/Keith/Chapter.7/Ch.7.html
20. Smyth, P., Goodman, R. M.: Rule Induction using Information Theory. in Piatetsky-Shapiro, G., Frawley, W. J. (eds.): Knowledge Discovery in Databases. AAAI/MIT Press (1991) 159–176
21. Tan, P. N., Kumar V., Srivastava, J.: Selecting the Right Interestingness Measure for Association Patterns. Proc. of Int. Conf. on Knowledge Discovery and Data Mining KDD-2002 (2002) 32–41
22. Witten, I. H and Frank, E.: DataMining: Practical Machine Learning Tools and Techniques with Java Implementations, Morgan Kaufmann, (2000)
23. Yao, Y. Y. Zhong, N.: An Analysis of Quantitative Measures Associated with Rules. Proc. of Pacific-Asia Conf. on Knowledge Discovery and Data Mining PAKDD-1999 (1999) 479–488
24. Zhong, N., Yao, Y. Y., Ohshima, M.: Peculiarity Oriented Multi-Database Mining. IEEE Trans. on Knowledge and Data Engineering, 15, 4, (2003) 952–960

Scoring Method for Tumor Prediction from Microarray Data Using an Evolutionary Fuzzy Classifier

Shinn-Ying Ho[1], Chih-Hung Hsieh[1], Kuan-Wei Chen[1], Hui-Ling Huang[2], Hung-Ming Chen[3], and Shinn-Jang Ho[4]

[1] Institute of Bioinformatics, National Chiao Tung University, Hsinchu, Taiwan
syho@mail.nctu.edu.tw
[2] Department of Information Management, Jin Wen Institute of Technology, Hsin-Tien, Taipei, Taiwan
[3] Institute of Information Engineering and Computer Science, Feng Chia University, Taichung, Taiwan
[4] Department of Automation Engineering, National Formosa University, Huwei, Yunlin, Taiwan

Abstract. In this paper, we propose a novel scoring method for tumor prediction using an evolutionary fuzzy classifier which can provide accurate and interpretable information. The merits of the proposed method are threefold. 1) The score ranged in [0, 100] can further illustrate the degree of tumor status in contrast to the conventional tumor classifier. 2) The derived score system can be used as a tumor classifier using a system-suggested or human-specified threshold value. 3) The derived classifier with a compact fuzzy rule base can generate an interpretable and accurate prediction result. The effectiveness of the proposed method is evaluated and compared using two well-known datasets from microarray data and an existing tumor classifier. It is shown by computer simulation that the proposed scoring method is effective using ROC curves of classification.

1 Introduction

Microarray gene expression profiling technology is one of the most important research topics in clinical diagnosis of disease [1]-[4]. There are a lot of machine learning algorithms, such as support vector machines, neural networks, and logistic regression, which have been used in the tumor classification from gene expression data. Soinov *et al.* [5] and Li *et al.* [6] used tree structures to classify microarray samples. Hvidsten *et al.* [7] proposed learning rule based models of biological process from gene expression time profiles. Vinterbo *et al.* [8] presented a rule-induction and filtering strategy to obtain an accurate, small and interpretable fuzzy classifier using a grid partition of feature space, compared with logistic regression models.

In pattern recognition problems, the scoring ability is important not only to quantify the certainty grades of samples belonging to each class, but also to help researchers to finding out the true active samples and filtering out the background noise [9]. Liu *et al.* [10] proposed a scoring algorithm based on negative entropy to position specific frequency matrix (PSFM) and Markov model to predict protein-DNA binding site.

Murvai *et al.* [11] used a probabilistic scoring method for protein domain identification. Jensen and Liu [12] proposed a bayesian scoring function approach to motif discovery.

In this study, a completely new and effective scoring method for tumor prediction from microarray data is investigated. It is necessary to cope with the following difficulties in designing the scoring system, described below. 1) It is desirable to select a *minimal* number of relevant genes while maintaining the highest accuracy for designing tumor classifiers, which is essential for developing inexpensive diagnostic tests. 2) The desrived scores can faithfully respond to accurate tumor classification with an interpretable manner. To achieve the above-mentioned goals, we propose a scoring method based on an interpretable fuzzy classifier (named iSFC, interpretable scoring fuzzy classifier).

The design of iSFC has three classification and one scoring function objectives to be simultaneously optimized: maximal classification accuracy, minimal number of rules, minimal number of used features, and maximal area under a ROC curve. High performance of iSFC arises from that the flexible membership function, simplified fuzzy rule, and rule/gene selection are simultaneously optimized [13]. An intelligent genetic algorithm (IGA) is used to efficiently solve the design problem with a large number of tuning parameters [14].

The performance of iSFC is evaluated using two benchmark datasets. It is shown that iSFC has concisely interpretable rules and better performance than the existing Vinterbo's classifier [8]. iSFC is also comparable to some non-rule-based methods using a large number of genes in terms of accuracy performance. Furthermore, the efficient scoring ability of iSFC is evaluated using the mean areas under ROC curves having 0.984 and 0.930 for training and test data, respectively.

2 The Scoring Method Using iSFC

2.1 Membership Function and Fuzzy Partition

The classifier design of iSFC uses flexible generic parameterized fuzzy regions which can be determined by flexible generic parameterized membership functions (FGPMFs) and a hyperbox-type fuzzy partition of feature space. Each fuzzy region corresponds to a parameterized fuzzy rule. In this study, each value of gene expression is normalized into a real number in the unit interval [0, 1]. An FGPMF with a single fuzzy set is defined as

$$\mu(x) = \begin{cases} 0 & \text{if } x \leq a \text{ or } x \geq d \\ (x-a)/(b-a) & \text{if } a < x < b \\ (d-x)/(d-c) & \text{if } c < x < d \\ 1 & \text{if } b \leq x \leq c \end{cases} \quad (1)$$

where $x \in [0, 1]$ and $a \leq b \leq c \leq d$. The variables a, b, c, and d determining the shape of a trapezoidal fuzzy set are the parameters to be optimized. Five parameters V^1, V^2, ..., $V^5 \in [0, 1]$ without constraints instead of a, b, c, and d are encoded into a chromosome for facilitating IGA. Let an additional variable $L=V^1$ which determines the location of the fuzzy set characterizing the occurrence of training patterns. When V^i are obtained, variables a, b, c, and d can be derived as follows: $a = L - (V^2 + V^3)$, $b = L - V^3$, $c = L + V^4$, and $d = L + (V^4 + V^5)$. This transformation can always make the derived values of a, b, c, and d feasible and reduce interactions among encoded parameters of chromosomes.

2.2 Fuzzy Rule and Fuzzy Reasoning Method

The following fuzzy if-then rules for n-dimensional pattern classification problems are used in the design of iSFC:

R_j: If x_1 is A_{j1} and ... and x_n is A_{jn} then Class CL_j with CF_j, $j = 1,...,N$

where R_j is a rule label, x_i denotes a gene variable, A_{ji} is an antecedent fuzzy set, C is a number of classes, $CL_j \in \{1, ..., C\}$ denotes a consequent class label, CF_j is a certainty grade of this rule in the unit interval [0, 1], and N is a number of initial fuzzy rules in the training phase.

To enhance interpretability of fuzzy rules, linguistic variables in fuzzy rules can be used. Each variable x_i has a linguistic set $U = \{L, ML, M, MH, H\}$. Each linguistic value of x_i equally represents 1/5 of the domain [0, 1]. Following the quantization criterion, we can consider genes to be regulated according to a qualitative level. For example, x_i is Low for down-regulated genes; x_i is Medium for neutral genes; and x_i is High for up-regulated genes. An antecedent fuzzy set $A_{ji} \in A_u$ where A_u denotes a set of the subsets of U. Examples of linguistic antecedent fuzzy sets are shown in Fig. 1.

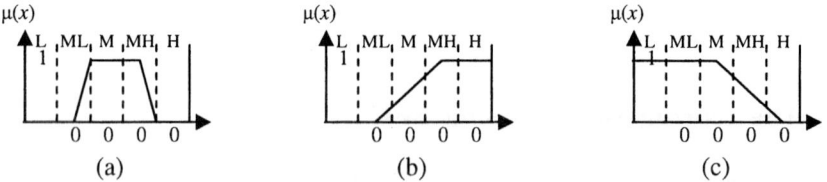

Fig. 1. Examples of an antecedent fuzzy set A_{ji} with linguistic values (L: low, ML: medium low, M: medium, MH: medium high, H: high). (a) A_{ji} represents {ML, M, MH}. (b) A_{ji} represents {ML, M, MH, H}, i.e., not Low. (c) A_{ji} represents {L, ML, M, MH, H } or ALL.

In the training phase, all the variables CL_j and CF_j are treated as parametric genes encoded in chromosomes and their near-optimal values are obtained using IGA. The following fuzzy reasoning method is adopted to determine the score of an input pattern $x_p = (x_{p1}, x_{p2}, ..., x_{pn})$ based on voting using multiple fuzzy if-then rules:

Step 1) Calculate the difference of certainty grades, DCG_p, between Class 1 and Class 2 for the input pattern x_p as follows:

$$DCG_p = \sum_{\substack{R_j \in FC \\ CL_j = \text{Class } 1}} \mu_j(x_p) \cdot CF_j - \sum_{\substack{R_j \in FC \\ CL_j = \text{Class } 2}} \mu_j(x_p) \cdot CF_j \quad (2)$$

where FC denotes the fuzzy classifier, the scalar value $\mu_j(x_p) = \mu_{j1}(x_{p1}) \cdot ... \cdot \mu_{jn}(x_{pn})$, and $\mu_{ji}(\cdot)$ represents the membership function of the antecedent fuzzy set A_{ji}.

Step 2) Normalize all *DCG* of training samples to [0, 100] and the normalized value is defined as the classification score, $Score_p$, of the corresponding sample, x_p.

Step 3) If $Score_p$ is greater or equal to a boundary threshold, δ_{score}, then classify x_p as Class 1, otherwise classify x_p as Class 2. In this study, the boundary threshold, δ_{score}, is set to 50.

2.3 Fitness Function and Chromosome Representation

In this study, we define the fitness function (or objective function) of IGA as

$$\max \; Fit(FC) = NP - W_r \cdot N_r - W_f \cdot N_f \tag{3}$$

where W_r and W_f are positive weights. In this study, we use $W_r = 0.1$ and $W_f = 0.001$ [13].

A chromosome consists of control genes for selecting useful features and significant fuzzy rules, and parametric genes for encoding the membership functions and fuzzy rules. The control genes comprise two types of parameters. One is parameter r_j, $j = 1,\ldots, N$, represented by one bit for eliminating unnecessary fuzzy rules. If $r_j = 0$, the fuzzy rule R_j is excluded from the rule base. Otherwise, R_j is included. The other is parameter f_i, $i = 1,\ldots, n$, represented by one bit for eliminating useless features. If $f_i = 0$, the feature x_i is excluded from the classifier. Otherwise, x_i is included. The parametric genes consist of three types:

1) $V_{ji}^k \in [0,1]$, $k = 1,\ldots, 5$, for determining the antecedent fuzzy set A_{ji} for each feature variable x_i in rule R_j;
2) $C_j \in \{1,\ldots, C\}$ for determining the consequent class of rule R_j;
3) $CF_j \in [0,1]$ for determining the certainty grade of rule R_j;

where $j = 1,\ldots, N$ and $i = 1,\ldots, n$. A rule base with N fuzzy rules is represented as an individual, as shown in Fig. 2. The number of encoding parameters to be optimized is equal to $N_p = n + 3N + 5Nn$. A chromosome representation uses a binary string for encoding control and parametric genes. There are 8 bits for encoding one of parameters V_{ji}^k and CF_j. Since each fuzzy region defines a fuzzy rule, the setting of number N is independent of value n but dependent on the number of fuzzy regions. Generally, N is set to the maximal number of possible fuzzy regions. In this study, N is set to $3C$. The design of an efficient fuzzy classifier is formulated as a large-scale parameters optimization problem (LPOP). If the optimal or near-optimal solution to the LPOP can be found, an efficient fuzzy classifier can be obtained.

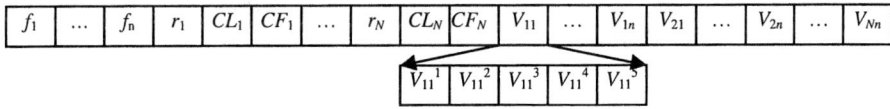

Fig. 2. Chromosome representation

2.4 IGA for Designing iSFC

The used intelligent genetic algorithm (IGA) is a specific variant of the intelligent evolutionary algorithm [14] to solve the design problem of iSFC. The main difference between IGA and the traditional GA is an efficient intelligent crossover operation. The intelligent crossover is based on orthogonal experimental design to solve intractable optimization problems comprising lots of design parameters.

Orthogonal Experimental Design. The two-level orthogonal arrays (OAs) used in IGA are described below. Let there be α factors, with two levels each. The total number of level combinations is 2^α for a complete factorial experiment. To use an OA of α factors, we obtain an integer $M = 2^{\lceil \log_2(\alpha+1) \rceil}$ where the bracket represents an upper ceiling operation, build an OA $L_M(2^{M-1})$ with M rows and $M-1$ columns, use the first α columns, and ignore the other $M-\alpha-1$ columns. OA can reduce the number of level combinations for factor analysis. The number of OA combinations required to analyze all individual factors is only $M = O(\alpha)$, where $\alpha+1 \leq M \leq 2\alpha$.

After proper tabulation of experimental results, the summarized data are analyzed using factor analysis to determine the relative effects of levels of various factors as follows. Let y_t denote a objective function value of the combination t, where $t = 1, \ldots, M$. Define the main effect of factor i with level k as S_{ik} where $i = 1, \ldots, \alpha$:

$$S_{ik} = \sum_{t=1}^{M} y_t W_t \qquad (4)$$

where $W_t = 1$ if the level of factor i of combination t is k; otherwise, $W_t = 0$. Consider that the objective function is to be maximized. For the two-level OA, level 1 of factor i makes a better contribution to the objective function than level 2 of factor i does when $S_{i1} > S_{i2}$. If $S_{i1} < S_{i2}$, level 2 is better. If $S_{i1} = S_{i2}$, levels 1 and 2 have the same contribution. The main effect reveals the individual effect of a factor. The most effective factor i has the largest main effect difference $MED_i = |S_{i1} - S_{i2}|$. After the better one of two levels of each factor is determined, an efficient combination consisting of all factors with the better levels can be easily derived.

Intelligent Crossover. All parameters are encoded into a chromosome using binary codes. Like traditional GAs, two parents P_1 and P_2 produce two children C_1 and C_2 in one crossover operation. Let all encoded parameters be randomly assigned into α groups where each group is treated as a factor. The following steps describe the intelligent crossover operation.

Step 1: Use the first α columns of an OA $L_M(2^{M-1})$
Step 2: Let levels 1 and 2 of factor i represent the i_{th} groups of parameters coming from parents P_1 and P_2, respectively.
Step 3: Evaluate the fitness values y_t for experiment t where $t = 2, \ldots, M$. The value y_1 is the fitness value of P_1.
Step 4: Compute the main effect S_{ik} where $i = 1, \ldots, \alpha$ and $k = 1, 2$.
Step 5: Determine the better one of two levels of each factor.

Step 6: The chromosome of C_1 is formed using the combination of the better genes from the derived corresponding parents.
Step 7: The chromosome of C_2 is formed similarly as C_1, except that the factor with the smallest main effect difference adopts the other level.
Step 8: The best two individuals among P_1, P_2, C_1, C_2, and M-1 combinations of OA are used as the final children C_1 and C_2 for elitist strategy.

One intelligent crossover operation takes M+1 fitness evaluations, where $a+1 \leq M \leq 2a$, to explore the search space of 2^a combinations.

Intelligent Genetic Algorithm. The used IGA is given as follows:

Step 1: Randomly generate an initial population with N_{pop} individuals.
Step 2: Evaluate fitness values of all individuals. Let I_{best} be the best individual in the population.
Step 3: Use the simple ranking selection that replaces the worst $P_s \cdot N_{pop}$ individuals with the best $P_s \cdot N_{pop}$ individuals to form a new population, where P_s is a selection probability.
Step 4: Randomly select $P_c \cdot N_{pop}$ individuals including I_{best}, where P_c is a crossover probability. Perform intelligent crossover operations for all selected pairs of parents.
Step 5: Apply a conventional bit-inverse mutation operator to the population using a mutation probability P_m. To prevent the best fitness value from deteriorating, mutation is not applied to the best individual.
Step 6: Termination test: If a pre-specified termination condition is satisfied, stop the algorithm. Otherwise, go to step 2.

3 Experimental Results

The parameter settings of IGA from [13] were used: N_{pop}=20, P_c=0.7, P_s=1-P_c, P_m=0.01. The stopping condition is to use 100×N_p fitness evaluations. All the experimental results are the averaged values of 30 independent runs. In each run, a ten-ford cross validation test (10-CV) is used.

Two benchmark data sets from [15] were used. For comparison, we adopted the same Wilcoxon rank sum test with [8] as a non-parametric feature pre-selection method for the unbalanced data set, such as microarray gene expression data. In this study, we have pre-selected n=10, 15 and 20 features (genes) to evaluate the performance of our method on various values of n. After computer simulation, the results of iSFC using the two data sets for n=10, 15 and 20 revealed no significant difference. In the following experiments, we used the data sets with n=15. Table 1 shows the used two data sets and the number N_p of parameters to be optimized using IGA.

Table 1. The two data sets with numerical feature values proposed in [16]

Dataset	Type	C	# of samples	# of genes	N_p
1	DLBCL	2	77	5469	483
2	prostate tumor	2	102	10509	483

Two experiments are conducted to evaluate iSFC. Experiment 1 is to compare the performances of iSFC with the Vinterbo's fuzzy classifier [8]. Experiment 2 is to compare the performances of iSFC with the non-rule-based classifiers in [13]. Due to the high classification performance of iSFC, the scoring ability is also enhanced in terms of ROC curves.

3.1 Experiment 1

For easy comparisons, we conducted two evaluations on the Vinterbo's method using different numbers of pre-selected features. One is to use 200 features (V200), which is the same with that in [8]. The other is to use 15 features (V15), which is the same with that of the proposed method.

Table 2 shows the statistical results (average and standard deviation) of iSFC, V200, and V15 in terms of training accuracy ($TrCR$), test accuracy ($TeCR$), number of rules (N_r), number of features (N_f), and rule number per class (N_r/C). The results of V200 and V15 were obtained by using the execution file provided by S. A. Vinterbo et al. [8].

From these results, we can obviously observe that iSFC is more compact and accurate using 15 candidate features than the Vinterbo's classifier using 200 candidate features in terms of $TrCR$ (97.93% vs. 83.72%), $TeCR$ (89.67% vs. 83.50%), N_r (1.92 vs. 2.80), and N_r/C (0.96 vs. 1.40). On the other hand, V200 is better than V15 in test accuracy but worse in training accuracy and using more candidate features and computation time. Moreover, the classifiers V200 compare favorably to those of logistic regression models, one of the standard classification methodologies applied in the biomedical domain [8].

Table 2. The statistical results of iSFC and the Vinterbo's classifier in terms of training accuracy, test accuracy, number of rules, number of features rule number per class

Data set	Method	$TrCR$ (%)	$TeCR$ (%)	N_r	N_f	N_r/C
DLBCL	iSFC	97.73±0.51	89.83±3.23	1.61±0.20	3.37±0.17	**0.81**
	V200	85.91	85.00	2.60	3.80	1.30
	V15	84.65	78.33	7.00	6.90	3.50
Prostate tumor	iSFC	98.12±0.19	89.50±2.86	2.22±0.19	4.36±0.26	**1.11**
	V200	81.52	82.00	3.00	**3.30**	1.50
	V15	84.46	84.00	2.90	5.10	1.45
Average	iSFC	97.93	89.67	1.92	3.87	**0.96**
	V200	83.72	83.50	2.80	**3.55**	1.40
	V15	84.56	81.17	4.95	6.00	2.48

3.2 Experiment 2

We compare the proposed method with some non-rule-based methods to evaluate the accuracy and interpretability of iSFC. Since Statnikov et al. [16] has investigated various efficient classifiers which can handle multiple classes using a very large number of genes, the reported accuracy without using gene selection can be used as an upper bound for comparisons. Table 3 shows the test accuracy comparison using 10-CV on the two data sets between iSFC and the following methods: multicategory support vector machine (SVM), k-nearest neighbors (k-NN), backpropagation neural networks (NN), and probabilistic neural networks (PNN) which are the most common methods for gene expression data analysis. The results are obtained from [13].

Table 3 indicates that the multicategory SVM with 94.75% is the most accurate classifier for tumor classification. However, it cost too much to take thousands of genes to make the classification decision such that it is not practical to implement the chips of medical test containing such lots of genes in real environment. On the other hand, iSFC needs just a few genes. It means that our method takes much less cost to make a biological test and is better in another economical view. The proposed fuzzy classifier iSFC with 89.67% using 3.87 genes on an average is worse than SVM but superior to k-NN (86.03%), NN (84.41%), and PNN (80.04%) using thousands of genes in terms of accuracy only. Because the sample sizes of microarray data are extremely small, it results in the high training accuracy (97.93%) and relatively low test accuracy (89.67%). From the viewpoint of analysis and practical applications, iSFC can serve as one of efficient tools for analysis of gene expression profiles.

Furthermore, iSFC performs well in terms of ROC curves, and results in the large areas under the ROC curve in training and test phases, *TrAUC* and *TeAUC* (0.984 and 0.930), near to 1. It reveals that iSFC has the scoring ability to efficiently differentiate each sample between two classes and effectively quantify the likelihood or certainty grades of classification for each sample. Moreover, with the ability of quantifying the certainty grades of samples belonging to each class, researcher can easily find out the real active samples and filter out the background noise. And this can lead to a more accurate experimental result. Figs. 3(a) and (b) show the score distribution histogram of test samples using the dataset prostate tumor and the corresponding ROC curve from one of the 30 10-CV runs. In this case, the test accuracy and AUC are 93.00% and 0.963, respectively.

Table 3. The test accuracies, number of used features of 10-CV, and the area under ROC curve over 2 data sets between iSFC and other non-rule based methods

Data set	Accuracy (%)					# of Used Features		Area under ROC curve (AUC) of iSFC	
	iSFC	SVM	k-NN	NN	PNN	iSFC	Non-fuzzy classifiers	TrAUC	TeAUC
DLBCL	89.83	97.50	86.96	89.64	80.89	3.37±0.17	5469	0.986±0.005	0.925±0.058
Prostate tumor	89.50	92.00	85.09	79.18	79.18	4.36±0.26	10509	0.982±0.003	0.934±0.021
Average	89.67	94.75	86.03	84.41	80.04	3.87	7989	0.984	0.930

Another advantage of iSFC is the interpretability of learning result. Fig. 4 shows an example of iSFC for the data set DLBCL using 90% samples for training and the rest for test. The classifier has 2 fuzzy rules using 5 features, genes 2, 4, 8, 11 and 14,

Where $TrCR=100\%$ and $TeCR=100\%$. The fuzzy rule R_1 tells us that when the expression of gene 8 is not greater than medium-small, the impact of gene 8 to classifying samples to Class 1 is proportional to its expression, otherwise gene 8 does not affect the classification, with a certainty grade, 0.824. On the other hand, the linguistically interpretable meaning of fuzzy rule R_2 is:

R_2: If x_2 is ALL, x_4 is not greater than medium-small, x_8 is ALL, x_{11} is not small, and x_{14} is small, then Class 2 with $CF=0.427$.

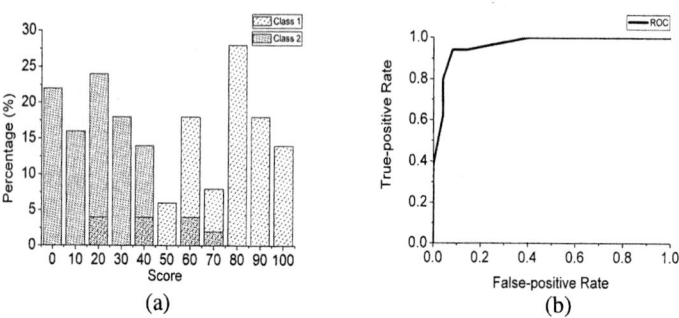

Fig. 3. The (a) score distribution histogram and (b) ROC curve of data set prostate in one run of 10-CV. The test accuracy and AUC are 93.00% and 0.963.

	Gene 2	Gene 4	Gene 8	Gene 11	Gene 14	C	CF
R_1						1	0.824
R_2						2	0.427

Fig. 4. Fuzzy rules for the classification problem over data set DLBCL using 90% samples for training and the remainder for testing. The accuracies for training and test data are both 100%

4 Conclusions

This paper has proposed an interpretable scoring fuzzy classifier, named iSFC, for microarray data analysis. The superiority of the proposed iSFC has been evaluated by computer simulation on two benchmark datasets of gene expression. The experimental results reveal that: 1) the proposed method can obtain interpretable classifiers with an accurate and compact fuzzy rule base, compared with the existing fuzzy classifier [8]; 2) iSFC using few genes is worse than SVM but superior to k-NN, NN, and PNN using thousands of genes in terms of accuracy; and 3) iSFC has an efficient scoring ability to quantify certainty grades of samples belonging to each class with the average areas under ROC curve in training and test phases (0.984, 0.930) nearly to 1.

References

[1] H. Ressom, R. Reynolds, and R. S. Varghese. Increasing the efficiency of fuzzy logic-based gene expression data analysis. Physiol Genomics, 13, (2003) 107–117.
[2] P. J. Woolf and Y. A. Wang. Fuzzy logic approach to analyzing gene expression data. Physiol Genomics, 3, (2000) 9–15.
[3] S. Kauffman, C. Peterson, B. Samuelsson, and C. Troein. Random boolean network models and the yeast transcriptional network. PNAS, 100 (25), (2003) 14796–14799.
[4] C. Creighton and S. Hanash. Mining gene expression databases for association rules. Bioinformatics, 19 (1), (2003) 79–86.
[5] L. A. Soinov, M. A. Krestyaninova, and A. Brazma. Towards reconstruction of gene networks from expression data by supervised learning. Genome Biology, 4 (R6), (2003).
[6] J. Li, H. Liu, J. R. Downing, A. E.-J. Yeoh, and L. Wong. Simple rules underlying gene expression profiles of more than six subtypes of acute lymphoblastic leukemia (all) patients. Bioinformatics, 19 (1), (2003) 71–78.
[7] T. R. Hvidsten, A. Lgreid, and J. Komorowski. Learning rulebased models of biological process from gene expression time profiles using gene ontology. Bioinformatics, 19 (9), (2003) 1116–1123.
[8] S. A. Vinterbo, E.-Y. Kim, and L. Ohno-Machado. Small, fuzzy and interpretable gene expression based classifiers. Bioinformatics, 21, (2005) 1964–1970.
[9] M. Friberg, P. Rohr, and G. Gonnet, "Scoring functions for transcription factor binding site prediction," BMC Bioinformactics, 6(84), (2005).
[10] XS. Liu, DL. Brutlag, JS. Liu. An algorithm for finding protein-DNA binding sites with applications to chromatin-immunoprecipitation microarray experiments. Nat Biotechnol, 20, (2002) 835-839.
[11] J. Murvai, K. Vlahovicek, and S. Pongor. A simple probabilistic scoring method for protein domain identification. Bioinformatics, 16(12), (2000) 1155-1156.
[12] S. T. Jensen and J.S. Liu. BioOptimizer: a Bayesian scoring function approach to motif discovery. Bioinformatics, 20(10), (2004) 1557-1564.
[13] S.-Y. Ho, H.-M. Chen, S.-J. Ho, and T.-K. Chen. Design of Accurate Classifiers with a Compact Fuzzy-Rule Base Using an Evolutionary Scatter Partition of Feature Space. IEEE Trans. Systems, Man, and Cybernetics—Part B, vol. 34, no. 2, (2004) 1031-1044.
[14] S.-Y. Ho, L.-S. Shu, and J.-H. Chen. Intelligent Evolutionary Algorithms for Large Parameter Optimization Problems. IEEE Trans. Evolutionary Computation, 8(6), (2004) 522-541.
[15] A. Statnikov, C. F. Aliferis, I. Tsamardinos, D. Hardin, and S. Levy. A comprehensive evaluation of multicategory classification methods for microarray gene expression cancer diagnosis. Bioinformatics, 21, (2005) 631–643.

Efficient Discovery of Structural Motifs from Protein Sequences with Combination of Flexible Intra- and Inter-block Gap Constraints

Chen-Ming Hsu[1], Chien-Yu Chen[2], Ching-Chi Hsu[3], and Baw-Jhiune Liu[1]

[1] Yuan Ze University, Department of Computer Science and Engineering,
Chung-Li, Taiwan, 320, R.O.C.
{cmhsu, bjliu}@saturn.yzu.edu.tw
[2] National Taiwan University, Department of Bio-Industrial Mechatronics Engineering,
Taipei, Taiwan, 106, R.O.C.
cychen@mars.csie.ntu.edu.tw
[3] Institute for Information Industry, Taipei, Taiwan, 106, R.O.C.
cchsu@iii.org.tw

Abstract. Discovering protein structural signatures directly from their primary information is a challenging task, because the residues associated with a functional motif are not necessarily clustered in one region of the sequence. This work proposes an algorithm that aims to discover conserved sequential blocks interleaved by large irregular gaps from a set of unaligned biological sequences. Different from the previous works that employ only one type of constraint on gap flexibility, we propose using combination of intra- and inter-block gap constraints to discover longer patterns with larger irregular gaps. The smaller flexible intra-block gap constraint is used to relax the restriction in local motif blocks but still keep them compact, and the larger flexible inter-block gap constraint is proposed to allow longer irregular gaps between compact motif blocks. Using two types of gap constraints for different purposes improves the efficiency of mining process while keeping high accuracy of mining results. The efficiency of the algorithm also helps to identify functional motifs that are conserved in only a small subset of the input sequences.

1 Introduction

Automatic discovery of patterns in unaligned biological sequences is an important problem in molecular biology. For a set of proteins that share a common function or structure, it is often that only a few of common residues are conserved among them [4]. In biology, a motif is a pattern that has a specific structure and is functionally significant [4]. Functional motifs are not necessarily found in only one region of the protein sequence. Instead, the conserved residues usually appear as clusters (it is called a motif block in this paper), and multiple clusters may simultaneously contribute to an important substructure [13]. Limited insertions and deletions are admitted within a motif block, and large insertions and deletions may happen between motif blocks during evolution.

Protein families can often be characterized by one or more such patterns, which each consist of one or more motif blocks [12, 18, 21, 22]. Many computational approaches have been introduced for the problem of motif identification [1, 2, 3, 6, 8, 11, 16, 17]. These approaches can be categorized based on the type of the motifs they discover, statistical or deterministic. In this paper, we focus on the problem of discovering deterministic patterns like some other web services, Pratt [6] and Teiresias [17]. A deterministic pattern can be matched or not matched by a sequence. In the mining process, a pattern is found if it matches more than a user-specified percentage of the input sequence set. This is the so-called minimum support constraint.

A sequential pattern is called sparse if a large number of wildcards exist between pattern components, and is treated as flexible, contrary to fixed, if different sequences match the same pattern with different sizes of gaps, where a gap is defined as a set of one or more successive wildcards. Discovering sparse and flexible patterns is a time-consuming task due to the large search space of solutions. So many related studies employ constraints to expedite the mining process [6, 14, 17, 20], among which the gap constraint is widely used to restrict the length of a fixed gap within some maximum and minimum values specified by the users. Jonassen *et al.* in 1995 first introduced the constraint of gap flexibility in Pratt program that allows limited variable spacing between pattern components [7]. Gaps of irregular lengths are important in biological patterns because variable sizes of loops can occur even in well-conserved regions. Setting flexibility as 2 satisfies most short patterns existing in protein sequences [7].

However, longer patterns consisting of several sequential blocks can be discovered only when a larger flexibility is allowed. According to our performance analysis on Pratt program, version 2.1 [6], we observed that the program consumes unreasonably much time when flexibility is set to a value larger than 4, as shown in Table 1. This result is because the branching factor of Pratt used in constructing the pattern tree is exponentially in proportion to the flexibility constraint. Pratt uses some other constraints, such as flexibility product, to narrow down the search space and to decrease the number of potential patterns generated. However, this largely reduces the solution space, and consequently longer patterns cannot be discovered.

Another common problem of current mining algorithms is a huge amount of memory is required for constructing a pattern tree and the associated data structures during mining process. Table 1 also shows the memory usage of Pratt v2.1 versus the flexibility constraint. This situation is getting even worse when lower support constraint is requested. Nevertheless, low supports are desired during mining process because some highly specific signatures are usually conserved in few members of a protein family.

Table 1. Performance analysis of Pratt v2.1 on a data set containing about one hundred of sequences with the support constraint set as 70%

Flexibility	Flexibility product	Memory used	Execution time
FL=2	FP=16	0.182 Gigabytes	2895 seconds
FL=3	FP=81	1.016 Gigabytes	36963 seconds
FL=4	FP=256	1.5 Gigabytes	207236 seconds
FL=5	FP=625	3.9 Gigabytes	The system crashed

This work presents the algorithm MAGIIC that aims to discover flexible long patterns from a set of unaligned biological sequences. We propose using combination of intra- and inter-block gap constraints to find patterns with large irregular gaps, provided that the derived patterns are still compact in local regions. The idea is motivated by the observation that highly conserved regions of biological sequences are usually separated by a set of large gaps with irregular lengths. The smaller flexible intra-block gap constraint is used to relax the local motif blocks but still keep them compact, and the larger inter-block gap constraint is proposed to allow longer gaps to exist between compact motif blocks. Using two types of gap constraints for different purposes largely improves the efficiency of the mining process.

2 Problem Definition

In this section, the problem of mining conserved sequential blocks with flexible intra- and inter-block gaps is defined. We first give the definition of a sequence.

Definition 1 (Sequence). A sequence over an alphabet Σ is a finite sequence of components belonging to Σ. For any sequence $\beta=\langle\beta_1...\beta_m\rangle$, a sequence α is called a subsequence of β, denoted as $\alpha <_s \beta$, if α can be obtained by deleting zero or more components from sequence β. We use $\beta[i..j]$ to denote the substring (contiguous subsequence) of β, which starts at position i and ends at position j of β, for $1 \le i \le j \le m$. In particular, $\beta[1..i]$ is the prefix of sequence β that ends at position i, and $\beta[i..m]$ is the suffix of sequence β that begins at position i. The number of components in β is denoted as $|\beta|$. □

If we segment a sequence into one or more blocks, it can be expressed as a blocked sequence. Blocks belonging to the same sequence are called sequential blocks.

Definition 2 (Blocked sequence). A sequence $\alpha = \langle\alpha_1...\alpha_m\rangle$ can be segmented into disjoint r blocks, $r \le m$, and be written as $\langle B_1...B_r\rangle$, where $B_k = \alpha[e_{k-1}+1..e_k]$, $e_0 = 0$, $e_r = m$, and $e_k > e_{k-1}+1$, for $1 \le k \le r$. □

We next define what intra- and inter-block gaps are.

Definition 3 (Intra- and inter-block gaps). Let $\beta = \langle\beta_1...\beta_m\rangle$ be a sequence, and $\alpha=\langle B_1...B_r\rangle$ be a blocked sequence provided that $\alpha <_s \beta$. If we consider the blocked sequence α as a pattern, then β serves as an instance of α. The interval between any two adjacent blocks B_i and B_{i+1} on the sequence β is called an *inter-block gap*. The interval between any two adjacent components within a block of α on the instance β is called an *intra-block gap*. □

MAGIIC employs different constraints for intra- and inter-block gaps respectively.

Definition 4 (Gap constraints). Let $\omega = (\gamma_{min}, \gamma_{max}, \tau_{min}, \tau_{max})$ be a set of constraints called gap constraints, which stand for the low and up bounds of an intra-block gap and the low and up bounds of an inter-block gap, respectively. Given a blocked sequence $\alpha=\langle B_1...B_r\rangle$, we say that α satisfies the user-defined gap constraints ω if there exists a sequence $\beta=\langle\beta_1...\beta_m\rangle$ such that α holds as a subsequence of β and the

Table 2. Parameters of MAGIIC

Parameter set θ	Description
λ	Minimum occurrences of a pattern
κ_{min}	Minimum size of a motif block
κ_{max}	Maximum size of a motif block
ω $\quad \gamma_{min}$	Low bound of an intra-block gap
γ_{max}	Up bound of an intra-block gap
τ_{min}	Low bound of an inter-block gap
τ_{max}	Up bound of an inter-block gap
n_{min}	Minimum size of a pattern
n_{max}	Maximum size of a pattern

blocks in α satisfy the constraint ω, denoted as $\alpha <_\omega \beta$. The set $\beta/(\alpha)_\omega$ stands for all the substrings of β that match pattern α under the gap constraints ω. □

MAGIIC also employs some other basic constraints associated with pattern mining, including the support and size constraints. The parameter names are given in Table 2.

Definition 5 (Support and size constraints). The support of a blocked sequence α in a sequence database D under the gap constraints ω and the size constraints (κ_{min}, κ_{max}, n_{min}, n_{max}) is defined as the number of distinct input sequences $\beta \in D$ such that $\alpha <_\omega \beta$ and the blocked sequence $\alpha = \langle B_1 ... B_r \rangle$ satisfies $n_{min} \leq |\alpha| \leq n_{max}$, and $\kappa_{min} \leq |B_i| \leq \kappa_{max}$ for $1 \leq i \leq r$. The pattern α is frequent (conserved) in sequence database D if its support is grater than λ, where λ is the minimum support constraint. □

Finally we give the problem statement as follows.

Problem Statement. Given a sequence database D and the parameter set θ listed in Table 2, the algorithm will find the complete set of conserved blocked sequences (patterns) existing in the sequence database D under the constraints in θ. □

The derived patterns are expressed in the PROSITE language [5]. The notation x(a,b), $a < b$, is used for a size-bounded gap with minimum length of a and maximum length of b, and x(a) stands for a gap with a fixed length of a. The wildcard x(a) is omitted if $a = 0$, and is written as x if $a = 1$, i.e. x = x(1).

3 Method

This paper proposes a novel algorithm called MAGIIC, which is designed based on the PrefixSpan algorithm proposed by Pei *et. al.* in 2004 [15]. The contribution of MAGIIC comes from two parts. First, MAGIIC develops a new procedure called *bounded-prefix-growth* based on the *prefix growth* procedure of the PrefixSpan algorithm. In order to identify patterns with large flexible gaps in biological data, the *bounded-prefix-growth* procedure incorporates intra- and inter-block gap constraints to speed up the mining process. Second, MAGIIC employs a newly designed projected database, called complete projected database, to guarantee that all the patterns satisfying the user-specified gap constraints will be found. In this section, we first briefly describe the PrefixSpan algorithm. After that, the concept of a complete projected database will be defined and how the intra- and inter-block gap constraints affect the scanning process of *bounded-prefix-growth* will be introduced.

PrefixSpan algorithm presents as a promising and efficient approach for many applications of sequential pattern discovery by avoiding generating a large amount of pattern candidates, and it consumes a relatively stable and small amount of memory space by using the pseudo projecting technique that records only the sequence identifiers and the associated event identifiers instead of constructing a physical projected database. The *prefix-growth* procedure of PrefixSpan employs a divided-and-conquer mechanism for pattern growing, which recursively reduces the size of the sequence database by generating the projected database of the current sequential pattern and then grows the sequential pattern in one particular projected database by exploring the local frequent components.

Fig.1 provides an example of a projected database with respect to a pattern ⟨CG⟩. Fig.1(a) shows the original database D. The projected database addressed by the PrefixSpan algorithm does not record complete information regarding gaps between sequence components, because PrefixSpan does not consider gap constraints in its mining process. As the example shown in Fig.1(b), the ⟨CG⟩'s projected database only keeps the longest substring of each sequence in D whose prefix matches the pattern ⟨CG⟩. This information is not sufficient when gaps are considered in the pattern mining process. Thus as shown in Fig.1(c), a complete projected database collects all the substrings in database D with a prefix of pattern ⟨CG⟩ that satisfies the gap constraints. We next give the definition of a complete projected database.

Seq id	1	2	3	4	5	6	7	8	9	10	11	12	13	14	15	16
S_1	C	T	G	E	Y	T	J	E	A	S	N	C	A	G	E	G
S_2	P	E	C	P	G	K	I	I	C	H	P	G	Q	G	R	K
S_3	S	C	W	V	S	Q	W	V	V	C	Q	G	W	G		

(a) The original database D

Seq id	1	2	3	4	5	6	7	8	9	10	11	12	13	14	15	16
S_1	C	T	G	E	Y	T	J	E	A	S	N	C	A	G	E	G
S_2	C	P	G	K	I	I	C	H	P	G	Q	G	R	K		
S_3	C	Q	G	W	G											

(b) The projected database of the pattern ⟨CG⟩, according to the definition of PrefixSpan

Seq id	1	2	3	4	5	6	7	8	9	10	11	12	13	14	15	16
$S_{1,1}$	C	T	G	E	Y	T	J	E	A	S	N	C	A	G	E	G
$S_{1,2}$	C	A	G	E	G											
$S_{2,1}$	C	P	G	K	I	I	C	H	P	G	Q	G	R	K		
$S_{2,2}$	C	H	P	G	Q	G	R	K								
S_3	C	Q	G	W	G											

(c) The complete projected database of the pattern ⟨CG⟩

Fig. 1. Illustration of the complete projected database

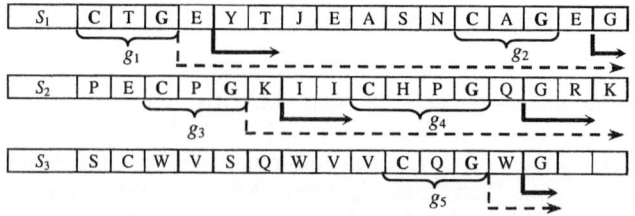

Note: $S_1/⟨CG⟩_\omega=\{g_1, g_2\}$, $S_2/⟨CG⟩_\omega=\{g_3, g_4\}$, and $S_3/⟨CG⟩_\omega=\{g_5\}$

Fig. 2. The scenario of scanning a projected database

Definition 6 (Complete projected database). Let α be a blocked sequence, and ω be a set of gap constraints. The α's complete projected database constructed from projecting the database D under gap constraints ω, denoted as $D|(\alpha)_\omega$, is a complete collection of sequences, each of which is the suffix of a sequence $\beta \in D$ and has a prefix of s, provided that $s \in \beta/(\alpha)_\omega$. □

We next describe how the *bounded-prefix-growth* procedure scans a complete projected database. The proposed procedure is called *bounded-prefix-growth* because its scanning range in the projected database is restricted by the gap constraints. In Fig.2, the dotted arrows represent the original scanning range of a projected database in PrefixSpan algorithm and the solid arrows show the scanning range of the *bounded-prefix-growth* procedure under the intra-block gap constraints ($\gamma_{min} = 1$, $\gamma_{max} = 2$). The scanning range of a complete projected database under gap constraints ω is much smaller. Most of the times, only the ranges under the intra-block gap constraints are considered when looking for the next frequent component. Only when the size of the currently growing block (the right most block) of the pattern satisfies the minimum block size constraint, larger scanning range with respect to the inter-block constraints will also be considered during the mining process.

The arguments of *bounded-prefix-growth* include a pattern as a blocked sequence α and its complete projected database $D|(\alpha)_\omega$. This procedure takes the blocked pattern α as input and tries to extend it under the user-specified constraints ω. In each call of *bounded-prefix-growth*, the search space of finding the next frequent component is bounded by the intra- and inter-block gap constraints. A component is conserved (frequent) if its occurrences in the projected database $D|(\alpha)_\omega$ satisfy the minimum support threshold. Each frequent component is appended to the current blocked sequence one at a time, and the resulted new blocked sequence (α') is used as the argument for the next call of *bounded-prefix-growth*, accompanied with a smaller projected database $D|(\alpha')_\omega$. Adding one more component to the current blocked pattern thus reduces the size of the complete projected database.

4 Results and Discussions

The performance of MAGIIC is compared with two well known packages on this problem, Teiresias [17] and Pratt v2.1 [6]. All the experiments provided here were conducted on a machine with a 3GHz Intel Pentium CPU and memory of 2GBs, running Linux Server. Regarding the parameter setting of MAGIIC, the users can set the following three constraints as a large number as long as the consuming time is acceptable on their machines. In this paper, we set both the maximum size of a motif block (κ_{max}) and the maximum size of a pattern (n_{max}) as 100, and change the up bound of an inter-block gap (τ_{max}) incrementally during mining process because this parameter affects the performance of MAGIIC significantly. Furthermore, in order to reduce the confusion of setting the other parameters, we set the low bound of an inter-block gap (τ_{min}) just one larger than the up bound of an intra-block gap (γ_{max}). Since only limited insertions and deletions within motif blocks are allowed during evolution, we set the low/up bound of an intra-block gap ($\gamma_{min}/\gamma_{max}$) as 0/2.

Table 3. Study on the first data set. (Arsenate reductase and related)

(a) Patterns discovered by different algorithms

ID	Patterns discovered by different algorithms	Support
	Pattern discovered by MAGIIC	
(1)	P-x-C-x(0,2)-S-x(0,2)-R-x(72,75)-P-x(1,2)-L-x(1,2)-R-P-I	38
	Pattern discovered by Teiresias (K=70, L=6, W=100, other parameters as default,)	
(i)	L-x(19)-P-x(4)-RPI-x(19)-L	38
	Pattern discovered by Pratt v2.1 (C%=70, PX=100, FN=4, FL=5, FP=12, other parameters as default)	
(ii)	R-x(18,20)-L-x(7)-P-x-L-x(2)-R-P-I	35
(iii)	G-x-[DEST]-x(2)-[AI]-x(2)-R-x(0,1)-K-x(4,7)-L-[ADGN]-[ILMV]-[ADEN]-x-[DEGN]-x-[FILM]-[PST]-x(3)-[FL]-x(2)-[FILM]-[IMV]-x(3)-P-x-[ILM]-[IL]-x-[RS]-P-I-[ILMV]-x-[DT]	35

(b) Executing time and usage of memory space for each algorithm with support = 70%

Method	MAGIIC	Teiresias	Pratt v2.1
Runtime in seconds	2	15	588
Memory used in Megabytes	3	15	150

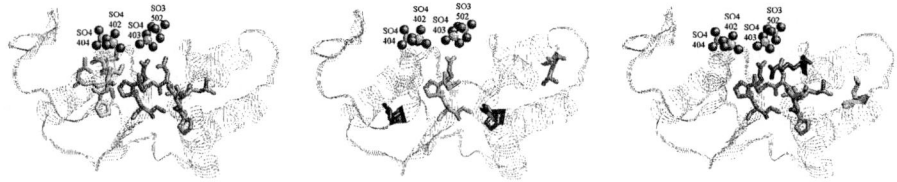

(a) Pattern (1) found by MAGIIC (b) Pattern (i) found by Teiresias (c) Pattern (ii) found by Pratt v2.1

Fig. 4. Viewing the derived patterns of the first data set in a three-dimensional structure (1I9D.pdb). The patterns are plotted in *sticks*, and blocks are shown with different colors.

This paper employs two well annotated data sets to demonstrate the capability of MAGIIC algorithm in identifying conserved structural motifs. The first input set collects 50 proteins of the InterPro family IPR006660, Arsenate reductase and related, (one fragment has been removed) from Swiss-Prot (http://www.expasy.org/sprot/) (version 48.1). With the support constraint set as 70%, the minimum size of a motif block as 4, and the up bound of an inter-block gap constraint as 100, MAGIIC found one pattern with 38 supporting sequences, as shown in Table 3(a), denoted as pattern (1). In fact, this pattern can be found as long as the user-specified constraints are more relaxed than the above settings. It can be observed in Fig. 4(a) that the two motif blocks of pattern (1) are clustered together when the protein is folded, and they are really closed to the ligands bound together with this protein. This shows that the derived motif reveals its structural and functional meanings. It has been reported in [9] that the cysteine in the first block is important in binding one of the sulfate anion (SO4). We will show in the following that it is not found by other mining programs.

Table 3(a) also provides the best pattern found by Teiresias and Pratt v2.1, labeled as pattern (i)-(iii), which are derived by using reasonable parameter settings regarding this data set. It can be observed in Fig. 4(b) and (c) that patterns (i) and (ii) do not capture the signature of the substructure with respect to the ligands bound with this protein. The main problem of Teiresias is only fixed gaps are considered. Even Pratt considers the flexibility of gaps, allowing large flexibility on every gap enlarges the search space and increases the executing time rapidly, which forces the users to give up searching for a pattern with a large gap like x(72,75). On the other hand, both Teiresias and Pratt v2.1 can handle equivalence to extend the length of the patterns, such as the pattern (iii) identified by Pratt v2.1. However, it does not really help to identify the structural motif associated with the functional site. Table 3(b) provides the executing time and memory usage of each algorithm. It is always the case that lowering support constraint enlarges the search space and thus consumes more computing time and space. Incorporating two gap constraints also makes MAGIIC more efficient than the other two packages in finding some highly specific signatures which are conserved in few members of a protein family.

The second training data was retrieved from Swiss-Prot database (release 48.1) by querying the keywords IPR001305, PF00684, and PS00637, which are associated with the CXXCXGXG domain signature of DnaJ proteins. The keywords are the entry IDs of InterPro (release 11.0), Pfam (version 18.0) and PROSITE (release 19.11) databases, respectively. We randomly select 100 proteins from the retrieved 272 sequences (totally we have 275 sequences, but three short fragments have been excluded.) And later we will show that the other 172 proteins can be completely found by using the pattern derived by MAGIIC.

With the support constraint set as 70%, the minimum size of a motif block as 4, and the up bound of an inter-block gap constraint as 20, MAGIIC found a pattern with 72 supporting sequences, denoted as pattern (1) in Table 4(a). This pattern contains four repeats of the motif block {C-x(2)-C-x-G-x-G}, which is recognized as the DnaJ central cysteine-rich (CR) domain. The DnaJ CR domain consists of two zinc centers, each of which is composed of four conserved cysteines [10]. We can see in Fig. 5(a) that each pair of the blocks in pattern (1) forms a zinc binding site, where the first and fourth blocks contribute to the first one and the second and third blocks contribute to the second one. It has been studied in [19] that the second binding site is more important than the first one. By increasing the support constraint as 100% and relaxing the minimum size of a motif block as 3, MAGIIC found another pattern, listed in Table 4(a) as pattern (2), which contains only three motif blocks. We observed that some DnaJ proteins lost the first block of the CR domain during evolution and some other lost the last block. This is consistent with the observation in [19] that the second and third blocks are more important to the function of DnaJ proteins.

The pattern of the entry PS00637 in PROSITE is provided in Table 4(a), denoted as pattern (i). As shown in Fig. 5(b), this pattern does not capture the feature of the isolated cysteine-rich domain, same as the best patterns found by Pratt v2.1 and Teiresias, also provided in Table 4(a). The selectivity of the derived patterns is evaluated by employing the ScanProsite (http://www.expasy.org/tools/scanprosite/) web service to scan protein sequences in Swiss-Prot database (release 48.1). The results are shown in Table 4(b). The precision rate is defined as TP / (TP + FP) and the recall rate as TP/ (TP + FN), where TP is short for true positives, FP for false positives, and FN for false negatives. It can be observed that the patterns found by Teiresias are not as good as MAGIIC or Pratt, and the pattern (ii) found by Pratt is specific enough to recognize the DnaJ proteins. However, pattern (2) still does not correctly capture the structural motif of the cysteine-rich domain. Also, Pratt v2.1 consumes much more resources than MAGIIC and Teiresias, as shown in Table 4(c).

Table 4. Study on the second data set (the CXXCXGXG domain signature of DnaJ proteins)

(a) Patterns discovered by different algorithms

ID	Pattern discovered by MAGIIC	Support
(1)	C-x(2)-C-x-G-x-G-x(8,14)-C-x(2)-C-x-G-x-G-x(12,19)- C-x(2)-C-x-G-x-G-x(5,12)-C-x(2)-C-x-G-x-G	72
(2)	C-x(2)-C-x-G-x-G-x(8,19)-C-x(2)-C-x-G-x(7,20)-C-x(2)-C-x-G	100

ID	Pattern of PS00637 in PROSITE database	
(i)	C-[DEGSTHKR]-x-C-x-G-x-[GK]-[AGSDM]-x(2)-[GSNKR]-x(4,6)-C-x(2,3)-C-x-G-x-G	-

Pattern discovered by Pratt (C%:70,PX:20,FL:8, FN:3, FP:256 and other parameters as default)

(ii)	G-x(7,12)-C-x(2)-C-x-G-x-G-x(6,14)-C-x(2)-C-x-G	100

Pattern discovered by Teiresias (K=70, L=4, W=30 and other parameters as default)

(iii)	C-x(2)-C-x-G-x-G-x(6)-C-x(2)-C-x-G-x-G-x(12)-P-x(14)-G	70
(iv)	C-x(2)-C-x-G-x-G	100

(b) Analysis on the selectivity and sensitivity of patterns

Method	ID	TP	FN	FP	Precision %	Recall %	Expected random matches
MAGIIC	(1)	219	53	0	100	80.50	4.342022e-14
MAGIIC	(2)	272	0	1	99.63	100	2.950594e-06
PROSITE	(i)	188	76	0	100	71.21	2.857160e-05
Pratt	(ii)	271	1	5	98.18	99.63	3.565674e-03
Teiresias	(iii)	223	49	0	100	81.98	2.180195e-07
Teiresias	(iv)	272	0	366	42.63	100	55

(c) Performance and usage of memory space with support = 70%

Method	MAGIIC	Teiresias	Pratt v2.1
Runtime in seconds	76	217	858760
Memory used in Megabytes	3	18	4000

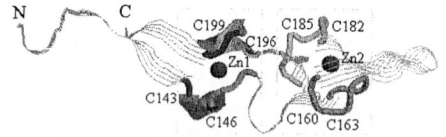
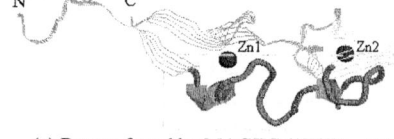

(a) Pattern found by MAGIIC (1EXK.pdb)   (a) Pattern found by MAGIIC (1EXK.pdb)

Fig. 5. Structure of the patterns for the DnaJ proteins. The zinc atoms are plotted as red spheres and the patterns in colored cartoon display.

5 Conclusion

Functional motifs composed of several sequential blocks are difficult to find. Current mining technologies might individually find each motif block but fail to connect them with large irregular gaps. On the other hand, allowing large flexible gaps might derive patterns with the conserved residues largely scattered. MAGIIC employs intra- and inter-block gap constraints to discover clusters of conserved residues present in protein sequences. The efficiency of MAGIIC remains even when the constraints are relaxed. This is important because setting lower support constrains or larger gap flexibilities helps to identify the signature of protein functional sites. The spatial information of the sequential motifs also helps to detect critical substructures of proteins that share similar functions. Thus, how to incorporate MAGIIC in the study of protein binding or protein-protein interaction deserves further study.

References

1. Blanchette, M., Schwikowski, B., Tompa, M.: An exact algorithm to identify motifs in orthologous sequences from multiple species. Proc. Int. Conf. Intell. Syst. Mol. Biol. 8 (2000) 37-45
2. Blekas, K., Fotiadis, D.I., Likas, A.: Greedy mixture learning for multiple motif discovery in biological sequences. Bioinformatics. 19 (2003) 607-617
3. Brazma, A., Jonassen, I., Eidhammer, I., Gilbert, D.: Approaches to the automatic discovery of patterns in biosequences. J. Comput. Biol. 5 (1998) 277-305
4. Eidhammer, I., Jonassen, I. Taylor, W.R.: Protein Bioinformatics: An Algorithmic Approach to Sequence and Structure Analysis. John Wiley & Sons. (2004)
5. Falquet, L., et al.: The PROSITE database, its status in 2002. Nucl. Acids Res. 30 (2002) 235-238
6. Jonassen, I.: Efficient discovery of conserved patterns using a pattern graph. Comput. Appl. Biosci. 13 (1997) 509-522
7. Jonassen, I., Collins, J.F., Higgins, D.: Finding flexible patterns in unaligned protein sequences. Protein Science. 4(8) (1995) 1587-1595
8. Liu, X., Brutlag, D.L. Liu, J.S.: BioProspector: discovering conserved DNA motifs in upstream regulatory regions of co-expressed genes. Pac. Symp. Biocomput. (2001) 127-138
9. Martin, P., et al.: Insights into the Structure, Solvation, and Mechanism of ArsC Arsenate Reductase, a Novel Arsenic Detoxification Enzyme. Structure. 9 (2001) 1071-1081, 2001.
10. Martinez-Yamout, M., Legge, G.B., Zhang, O., Wright, P.E., Dyson, H.J.: Solution structure of the cysteine-rich domain of the Escherichia coli chaperone protein DnaJ. J. Mol. Biol. 300(4) (2000) 805-818
11. Narasimhan, G., Bu, C., Gao, Y., Wang, X., Xu, N., Mathee, K.: Mining protein sequences for motifs. J. Comput. Biol. 9 (2002) 707-720
12. Neuwald, A.F., Green, P.: Detecting patterns in protein sequences. J. Mol. Biol. 239 (1994) 698-712
13. Ogiwara, A., Uchiyama, I., Yasuhiko, S., Kanehisa, M.: Construction of a dictionary of sequence motifs that characterize groups of related proteins. Protein Eng. 5 (1992) 479-488
14. Pei, J., Han, J.: Constrained frequent pattern mining : a pattern-growth view. ACM SIGKDD Explorations (Special Issue on Constraints in Data Mining). 4(1) (2002) 31-39
15. Pei, J., Han, J., Mortazavi-Asl, B., Wang, J., Pinto, H., Chen, Q., Dayal, U, Hsu, M.-C.: Mining Sequential Patterns by Pattern-Growth: The PrefixSpan Approach. IEEE Transactions on Knowledge and Data Engineering. 16 (2004) 1424-1440
16. Pevzner, P.A, Sze, S.H.: Combinatorial approaches to finding subtle signals in DNA sequences. Proc. Int. Conf. Intell. Syst. Mol. Biol. 8 (2000) 269-278
17. Rigoutsos, I, Floratos, A.: Combinatorial pattern discovery in biological sequences: The Teiresias algorithm. Bioinformatics. 14 (1998) 55-67
18. Saqi, M.A.S, Sternberg, M.J.E.: Identification of sequence motifs from a set of proteins with related function. Protein Eng. 7 (1994) 165-171
19. Shi, Y.Y., Tang, W., Hao, S.F., Wang, C.C.: Constristions of cysteine residues in Zn2 to zinc figers and thioldisulfide oxidoreductase activities of chaperone DnaJ. Biochemistry. 44 (2005) 1683-1689
20. Silvestri, C., Orlando, S., Perego, R.: A new algorithm for gap constrained sequence mining. Proceedings of the 2004 ACM Symposium on Applied Computing, special track on Data Mining. (2004) 540-547
21. Su, Q.J., Lu, L., Saxonov, S., Brutlag, D.L.: eBLOCKs: enumerating conserved protein blocks to achieve maximal sensitivity and specificity. Nucl. Acids Res. 33 (2005) D178-182
22. Wang, J.T.L., et al.: Discovering active motifs in sets of related protein sequences and using them for classification. Nucl. Acids Res. 22 (1994) 2769-2775

Finding Consensus Patterns in Very Scarce Biosequence Samples from Their Minimal Multiple Generalizations

Yen Kaow Ng[1] and Takeshi Shinohara[2]

[1] Kyushu Institute of Technology,
Graduate School of Computer Science and Systems, Iizuka, 820, Japan
kalngyk@ai.kyutech.ac.jp
[2] Kyushu Institute of Technology,
Department of Artificial Intelligence, Iizuka, 820, Japan
shino@ai.kyutech.ac.jp

Abstract. In this paper we examine the issues involved in finding consensus patterns from biosequence data of very small sample sizes, by searching for so-called *minimal multiple generalization (mmg)*, that is, a set of *syntactically minimal* patterns that accounts for all the samples. The data we use are the *sigma regulons* with more conserved consensus patterns for the bacteria *B. subtilis*. By comparing between the mmgs found over different search spaces, we found that it is possible to derive patterns close to the known consensus patterns by simply making some reasonable requirements on the kinds of patterns to obtain. We also propose some simple measures to evaluate the patterns in an mmg.

1 Introduction

Finding frequently occurring patterns (called *consensus patterns*) within a set of biosequences is a common task in molecular biology. The kinds of patterns used in the various methods may differ very significantly [3], for example compare the patterns used in the PROSITE [9] database and the TEIRESIAS algorithm [7]. The patterns we consider are strings over a finite alphabet Σ and a variable symbol "$*$", where its language is the strings obtainable by (independently) replacing all variables in the pattern with strings over Σ [8].

Our approach to consensus pattern is based on finding so-called *minimal multiple generalizations (mmg)* [2] from subclasses of the pattern languages. Given any sample set S of strings over Σ, the mmgs are the most (syntactically) specific set of up to k patterns over the given class where their languages together contain S. Arimura *et. al.* [2] gave a polynomial time algorithm **MMG** for finding such mmgs. Others [1, 11, 10, 6] have studied how well the **MMG** discovered mmgs perform in predicting gene functions. Here, we consider instead the resemblance between the discovered mmgs and well established consensus patterns.

As samples, we use the sigma factor dependent promoter sequences in *Bacillus Subtilis*, which has well established consensus patterns [4]. The difficult part in using these samples is that they are not very abundantly available, and hence it is difficult for candidate patterns to achieve sufficient levels of statistical significance. We show that by constraining our search to only patterns of slightly more specific forms, some patterns in the discovered mmgs match the known consensus patterns fairly accurately. We also note some qualities that can be found in the better matching patterns.

2 Preliminaries

Let Σ be a finite set of alphabets. A *regular pattern* is a string over $\Sigma \cup \{*\}$, where the '*' symbol is called a *variable*. A *substitution* θ for a pattern p is a set of replacements for variables in p with regular patterns. $p \preceq q$ iff p is obtainable from q via some substitution θ; $p \prec q$ iff $p \preceq q$ but not $q \preceq p$. Given two sets of patterns P and Q, $P \sqsubseteq Q$ iff for each $p \in P$, $p \preceq q$ for some $q \in Q$, and $P \sqsubset Q$ iff $P \sqsubseteq Q$ but not $Q \sqsubseteq P$. The language of a pattern p, $L(p) = \{w \in \Sigma^* \mid w \preceq p\}$. For a set of patterns P, $L(P) = \bigcup_{p \in P} L(p)$. Note that $P \sqsubseteq Q \Rightarrow L(P) \subseteq L(Q)$ [8].

Given $k \in N$, a finite set $S \subseteq \Sigma^*$ and a class of patterns $\mathcal{P} \subseteq (\Sigma \cup \{*\})^*$, a set of patterns $P \subseteq \mathcal{P}$ is a *k-minimal multiple generalization* (or *k-mmg*) for S over $\mathcal{P}$ if $\sharp P \leq k$, $S \subseteq L(P)$, and there are no other set of up to k patterns Q in $\mathcal{P}$ where $Q \sqsubset P$ and $S \subseteq L(Q)$. We use the efficient algorithm **MMG** in [2] for finding such mmgs.

Using different classes of patterns in **MMG** may change its output very significantly. Originally the regular patterns were studied in [2]. Then, the regular patterns with restricted number of variable occurrences were studied in [1]. In this work we study those classes, with the addition of so-called *range specifiers* (as in the PROSITE patterns) where a variable is written "$*(x,y)$" for some x, $y \in N$. A variable written this way, when replaced with constant strings, can only be replaced with strings of length at least x and at most y.

3 Experimental Setup

We run our tests using a straight-forward implementation of the **MMG** algorithm (downloadable from www.daisy.ai.kyutech.ac.jp/~kalngyk). The program is given as input a set of biosequences S, a maximum number k of patterns allowed in a k-mmg, and finally, a class of patterns $\mathcal{P}$. The output k-mmg is then compared with known consensus patterns. This is repeated for $k = 5$ and $k = 10$ over all combinations of our candidates for S and $\mathcal{P}$, as follows.

(For S) We use the sigma factor dependent promoter sequences in *B. subtilis* (refer [4]) for samples. These sequences (and consensus patterns) are obtained from the on-line database DBTBS [5]. We use the regulons sigD (33/51/59), sigE (62/44/51), sigH (48/37/50), sigW (32/42/45) (numbers are for sample size, average sequence length and maximum sequence length), because their sample sizes are relatively large and there are relatively long contiguous conserved regions within their consensus patterns.

(For $\mathcal{P}$) We use subclasses of the regular patterns with restricted use of variables. We also consider patterns of the following form, the rational of which will be made clear in its description: Assume that each of the target patterns consists of exactly two segments. That is, they are of the form "$*_1 v *_2 w *_3$", where v, w are non-empty strings over Σ and the $*_i$s are variables. We further assume that v is not too close to the end of the sequences, while w is not too close to the beginning of the sequences. For this reason, we place upper bounds on the maximum length of the words that can be used to replace $*_1$ and $*_3$. In this study we use the same length x as the upper bound for both $*_1$ and $*_2$, and we set x to 0.5 times the length of the longest biosequence in the input sample. This gives us the pattern "$*(0,x)\ v\ *(0,\infty)\ w\ *(0,x)$". For comparison, we also test with other classes to observe the effects of the fixed pattern form and the use of the range specifiers. In total we have the following for candidates of $\mathcal{P}$:

1. patterns of fixed form "$*(0,x)\ v\ *(0,\infty)\ w\ *(0,x)$" where $v, w \in (\Sigma \cup \{*\})^*$
2. patterns of fixed form "$*\ v\ *\ w\ *$" where $v, w \in (\Sigma \cup \{*\})^*$
3. regular patterns with at most 3 variable occurrences

To evaluate how a pattern p in a k-mmg fits a sample S, the most intuitive measure is the number of sequences in S which matches the pattern p, i.e. $\sharp(S \cap L(p))$. Another useful measure is that of *coverage*, introduced in [6], which is the total number of sequences in $L(p)$ of length up to the longest sequence in S, i.e. $\sharp(\{w \in L(p) \mid w$ is not longer than the longest sequence in $S\})$ which we shall simply refer to as the *coverage of p*. Intuitively, the coverage of p should be related to the amount of generalization committed by $L(p)$ based on the sample S, and this has been observed in [6].

4 Results

Table 1 show the mmgs obtained from the different pattern classes $\mathcal{P}$. We report only patterns in the k-mmg with either very high number of matching sequences or with very low coverage, with emphasis on the former. Each pattern listed is followed by two values (x, y), where x is the number of sequences in S which matches the pattern, and y is the coverage of the pattern (for simplicity, range specifiers have been ignored in the computation of these coverages).

Here we see that the use of range specifiers enables **MMG** to not mix up between the two segments in the pattern, which is most evident in the sigD patterns. Without the use of range specifiers, the patterns obtained from fixed form patterns (pattern class 2) are often not much better than those obtained from regular patterns (pattern class 3). We also note that good correspondence with the known consensus pattern almost always come from patterns with "high" number of matching sequences and "low" coverage, even though we currently lack a method to decide the ideal range of these values.

Table 1. MMG results from different hypothesis space

	regular pattern of fixed form with range specifiers		regular pattern of fixed form		regular pattern with at most 3 variable occurrences	
	$k = 5$	$k = 10$	$k = 5$	$k = 10$	$k = 5$	$k = 10$
sigD	*AGC*CG* (14, 2.1×10³⁵) *CAC*CG* (11, 2.0×10³⁵) *AAC*CCGATA* (10, 1.8×10³³) *TAC*CCGA* (10, 2.9×10³⁴)	*AAC*CCGATAT* (8, 4.3×10³²) *AC*ACGA* (7, 6.4×10³⁴) *TTA*CCGA* (5, 8.2×10³³) *ATAC*CCGATA* (5, 5.0×10³²)	*CCGA*C* (21, 8.2×10³⁴) *CG*CGATA* (17, 1.7×10³⁴)	*CCGGATA*GC* (9, 4.1×10³³) *ACG*CCGATAT* (8, 1.8×10³³) *GCCGATA*AC* (7, 1.0×10³³)	*TAC*CG* (18, 2.1×10³⁵) *AAAC*CGATA* (11, 2.1×10³³) *AGC*CGATA* (9, 7.2×10³³)	*AAAC*CGATAT* (8, 5.0×10³²) *ATACA*CG* (7, 1.7×10³⁴) *TTAC*CCG* (6, 2.9×10³³)
sigE	*CA*CATA* (42, 8.0×10²⁹) *TAAA*ATA* (20, 3.2×10²⁹) *CA*AATA* (18, 7.9×10²⁹)	*TAAA*ATA* (20, 3.2×10²⁹) *CA*AATA* (18, 7.9×10²⁹) *CA*TCATA* (17, 2.0×10²⁹) *CAT*CATAT* (16, 8.3×10²⁸) *TCATA*CATA* (12, 2.3×10²⁸)	*CATA*TA* (48, 8.0×10²⁹) *AATA*AAG* (18, 3.3×10²⁹) *TATA*TAG* (8, 3.1×10²⁹)	*TCATA*ATA* (23, 7.9×10²⁸) *AATA*AAG* (18, 3.3×10²⁹) *GCATA*ATA* (13, 7.9×10²⁸) *CATA*TGTA* (13, 9.2×10²⁸)	*CATA*TA* (48, 8.0×10²⁹) *AATA*TAG* (18, 3.3×10²⁹)	*AATA*AG* (22, 7.9×10²⁹) *CATA*TGTA* (13, 9.2×10²⁸) *GCATA*TATA* (13, 7.9×10²⁸)
sigH	*AGGA*GAAT* (23, 2.2×10²⁸) *GA*TAA* (22, 6.7×10²⁹) *AGGA*AAAT* (16, 2.2×10²⁸) *GGGAT*GAAT* (7, 5.4×10²⁷)	*AGGA*TAA* (16, 7.9×10²⁸) *AAGGA*AAAT* (12, 5.4×10²⁷) *AAGGA*CGAAT* (7, 1.3×10²⁷) *AGGA*TGAAT* (7, 5.4×10²⁷) *GGGAT*GAAT* (7, 5.4×10²⁷)	*AA*AAA* (32, 4.2×10²⁹) *GAA*GAAT* (23, 8.0×10²⁸) *AAAGG*GAAT* (11, 5.4×10²⁷) *TAAT*GAAT* (8, 2.2×10²⁸)	*AAAA*AAA* (12, 4.9×10²⁸) *AAAGG*GAAT* (11, 5.4×10²⁷) *GAAG*GAAT* (11, 5.4×10²⁷) *TAAT*GAAT* (10, 2.2×10²⁸) *GAATT*AAAT* (8, 2.2×10²⁸) *GAATT*AAAT* (7, 5.4×10²⁷)	*AA*AAA* (32, 4.9×10²⁹) *GAA*GAAT* (23, 8.0×10²⁸) AAAGG*AT*GAAT* (8, 1.6×10²⁶) *GAAGG*GAAT* (8, 5.4×10²⁷)	*GAA*AAAT* (16, 8.0×10²⁸) *GGAA*GAAT* (13, 2.2×10²⁸) AAAGG*AT*GAAT* (8, 1.6×10²⁶) *GAAGG*GAAT* (8, 5.4×10²⁷)
sigW	*TGAAAC*CGTA* (21, 9.9×10²³) *TGAAAC*TAG* (11, 3.7×10²⁴) *GAAACTTTT*ATAT* (8, 5.3×10²²)	*TGAAACC*ACGTAT* (8, 1.4×10²²) *TTGAAAC*TAG* (6, 8.9×10²³) *TGAAAC*TACGTATA* (5, 3.2×10²¹) *GAAACTTTTT*ATAT* (5, 2.9×10²¹)	Same as patterns obtained from regular patterns of fixed form with range specifiers		*TGAAAC*GTA* (23, 3.7×10²⁴) *TGAAAC*TA*G (12, 2.4×10²⁴) *GAAACTTT*ATAT* (8, 5.3×10²²)	A*TGAAAC*TA*G (8, 5.7×10²³) *TGAAACC*CCGTAT* (8, 1.4×10²²) *TGAAAC*CTA*G (5, 8.9×10²³) *GAAACTTTTT*ATAT* (5, 2.9×10²¹)

Consensus patterns   sigL: *TAAA*GCCGATAT*;   sigE: *CATAT*CATACA*, *ATATT*CATACA*;
     sigE: *AGGTATT*GAATT*;   sigW: *TGAAACN*CGTA*;

5 Discussions

In summary, we found that using more restrictive classes of languages in the **MMG** algorithm gave us consensus patterns that matches the known consensus patterns fairly reasonably. (Though not included in this shorter version of the paper, we also found that patterns using so-called *composite symbols*, as in the PROSITE patterns, tend to yield less resembling patterns.) Of course, this does not mean that more restrictive classes necessarily produce better patterns. For example the class generated by "$*(0,x) \; v \; w \; *(0,x)$" where $v, w \in \Sigma^*$, though more restrictive than the class "$*(0,x) \; v \; *(0,\infty) \; w \; *(0,x)$", would not produce the consensus patterns we want in our test cases.

On the other hand, the results do show some promise for using the **MMG** algorithm in finding consensus patterns, provided that a suitable pattern class is given as input to the algorithm. Since each computation of the **MMG** algorithm runs very quickly (typically in a few seconds), it is conceivable for the algorithm to be employed interactively — where a user would iteratively specify a class of patterns to use, and decide when the output is suitable for consensus patterns.

It would be ideal if we can furthermore automate this process by running the algorithm with successively more restrictive classes of languages, and stop when successive iterations provide no further improvements in the patterns obtained. To do this, we need only (1) a criteria to evaluate if a set of patterns is better than another (for example the measures we have discussed herein may be adapted for this purpose), and (2) a measure to classify how restrictive is a class of languages. Problem (1) seems to be the more difficult problem, but we note that even the classification in (2) may not be simple. For the classes of languages used in this paper, proper inclusions between the classes allow us to estimate how restrictive the classes are. However, this will not work for classes that are incomparable with respect to inclusion. For example, we could not say if the class generated by patterns of the form "$*w$" is less, more, or just as restrictive as that generated by "$w*$", since they do not include each other — though assuming homogeneity in word generation, they should intuitively be considered equally restrictive. For this reason, more theoretical work on the complexity of subclasses of the pattern languages is perhaps needed to improve the use of the **MMG** algorithm.

Acknowledgements. We thank Prof. Satoru Kuhara (Kyushu University) for introducing us to the topic.

References

1. H. Arimura, R. Fujino, T. Shinohara, and S. Arikawa. Protein motif discovery from positive examples by Minimal Multiple Generalization over regular patterns. In *Proceedings of the Genome Informatics Workshop*, pages 39–48, 1994.
2. H. Arimura, T. Shinohara, and S. Otsuki. Finding minimal generalizations for unions of pattern languages and its application to inductive inference from positive data. In *Proc. of the 11th Ann. Symp. on Theoretical Aspects of Comp. Sci. (STACS'94)*, volume 775, pages 649–660. Springer-Verlag, 1994.

3. A. Brāzma, I. Jonassen, I. Eidhammer, and D. Gilbert. Approaches to the automatic discovery of patterns in biosequences. *J. Comp. Biol.*, 5(2):277–304, 1998.
4. J. D. Helmann and C. P. Moran. *RNA Polymerase and Sigma Factors*, chapter 21, pages 289–312. American Society Microbiology, 2001.
5. Y. Makita, M. Nakao, N. Ogasawara, and K. Nakai. DBTBS: Database of transcriptional regulation in Bacillus Subtilis and its contribution to comparative genomics. *Nucl. Acids Res.*, 32:75–77, 2004.
6. Y. K. Ng, H. Ono, and T. Shinohara. Measuring over-generalization in the minimal multiple generalizations of biosequences. In *Proceedings of The Eighth Conference on Discovery Science (DS'05)*, volume 3735, pages 176–188. Springer-Verlag, 2005.
7. I. Rigoutsos and A. Floratos. Combinatorial pattern discovery in biological sequences: the TEIRESIAS algorithm. *Bioinformatics*, 14(1):55–67, 1998.
8. T. Shinohara. Polynomial time inference of extended regular pattern languages. In *RIMS Symposia on Software Science and Engineering, Kyoto, Japan*, volume 147, pages 115–127. Springer-Verlag, 1982.
9. C.J. Sigrist, L. Cerutti, N. Hulo, A. Gattiker, L. Falquet, M. Pagni, A. Bairoch, and P. Bucher. PROSITE: A documented database using patterns and profiles as motif descriptors. *Brief Bioinform.*, 3:265–274, 2002.
10. T. Takae, T. Kasai, H. Arimura, and T. Shinohara. Knowledge discovery in biosequences using sort regular patterns. Workshop on Applied Learning Theory, 1998.
11. M. Yamaguchi, S. Shimozono, and T. Shinohara. Finding minimal multiple generalization over regular patterns with alphabet indexing. In *Proceedings of the Seventh Workshop on Genome Informatics*, volume 7, pages 51–60. Universal Academy Press, 1996.

Kernels on Lists and Sets over Relational Algebra: An Application to Classification of Protein Fingerprints

Adam Woźnica, Alexandros Kalousis, and Melanie Hilario

University of Geneva, Computer Science Department,
Rue General Dufour 24, 1211 Geneva 4, Switzerland
{woznica, kalousis, hilario}@cui.unige.ch

Abstract. In this paper we propose a new class of kernels defined over extended relational algebra structures. The "extension" was recently proposed in [1] and it overcomes one of the main limitation of the standard relational algebra, i.e. difficulties in modeling lists. These new kernels belong to the class of $\Re$-Convolution kernels in the sense that the computation of the similarity between two complex objects is based on the similarities of objects' parts computed by means of sub-kernels. The complex objects (relational instances in our case) are tuples and sets and/or lists of relational instances for which elementary kernels and kernels on sets and lists are applied. The performance of this class of kernels together with the Support Vector Machines (SVM) algorithm is evaluated on the problem of classification of protein fingerprints and by combining different data representations we were able to improve the best accuracy reported so far in the literature.

1 Introduction

Recently it has been realized that one strength of the kernel-based learning paradigm is its ability to support non-vectorial input spaces, [2, 3, 4, 5]. This is mainly due to the fact that the proper definition of a kernel function enables the structured data to be embedded in some linear feature space without the explicit computation of the feature map. As a result any propositional algorithm which is based on inner products can be applied on the structured data.

In [6] we made one step in the direction of bringing kernel methods and learning from structured data together and we proposed a novel and general framework based on concepts of relational algebra for kernel-based learning over relational schema. We defined kernel functions over relational schema which are instances of $\Re$-Convolution kernels and use them as a basis for a relational instance-based learning algorithm. One of the main limitations of relational algebra representation as described in [6], is that, although it is ideal for modeling sets, it can not naturally model lists. To tackle this problem we recently proposed in [1] an extension to this representation language in such a way that it allows for modeling of lists of complex objects (relational instances). This new representation was used within the framework of distance-based learning and for the task of classification of protein fingerprints promising results were reported.

In this paper we propose new kernels over extended relational algebra language which operate directly on the structures defined in [1]. This amounts to defining new kernels on lists of relational instances. We report experimental results on a problem of classification of protein fingerprints for which we were able to improve the best accuracy reported in the literature.

2 Description of the Extended Relational Instance

Consider a general relational schema that consists of a set of relations $\mathcal{R}=\{R_1,\ldots,R_n\}$. The schema of a relation R_i is the set of attributes of R_i and we denote it as $R_i(A_1,\ldots,A_{z_i})$. A tuple (instance), R_{i_j}, of a relation R_i is a particular row in R_i with $R_{i_j} = (v_{j1}, v_{j2}, \ldots, v_{jz_i})$ and v_{jl} the value of the A_l attribute in the R_{i_j} tuple. An attribute A_k is called a *potential key* of relation R_i if it assumes a unique value for each instance of the relation. An attribute A_l of relation R_j is a *foreign key* if it references a potential key A_k of relation R_i and takes values in the domain of attribute A_k in which case we will also call the A_k a *referenced key*. A *set link* is a quadruple of the form $sl(R_i, A_k, R_j, A_l)$ where either A_l is a foreign key of R_j referencing a potential key A_k of R_i or vice versa. To be able to represent lists we define a *list link* as a quintuple $ll(R_i, A_k, R_j, A_l, LIST(A_l))$ where R_i, A_k, R_j, A_l are defined as before and $LIST(A_l)$ is a list of values from $D(A_l)$ defining the order of the elements of the list. The association between A_k and A_l encoded in sl and ll models one-to-many relations: for sl one element of R_i can be associated with a set of elements of R_j whereas for ll one element of R_i is connected with a list of elements from R_j.

We will call the set of attributes of a relation R_i that are not keys (i.e. referenced keys, foreign keys or attributes defined as keys but not referenced) *standard attributes* and denote it with $\mathcal{I}_{A,R_i}$. The notion of links is critical for our relational learner since it will provide the basis for the new types of attributes, i.e. set and list.

For a given referenced key A_k of relation R_i we denote by $SL(R_i, A_k)$ the set of links $sl(R_i, A_k, R_j, A_l)$ in which A_k is referenced by foreign key A_l of R_j. By analogy we define the set of links $ll(R_i, A_k, R_j, A_l, LIST(A_l))$ as $LL(R_i, A_k)$. By $SL(R_i) = \cup_k SL(R_i, A_k)$ we denote the set of all set links in which one of the potential keys of R_i is referenced as a foreign key by an attribute of another relation. By $LL(R_i) = \cup_k LL(R_i, A_k)$ we denote the set of all list links in which one of the potential keys of R_i is referenced as a foreign key by an attribute of another relation.

Similarly for a given foreign key A_l of R_j, $SL^{-1}(R_j, A_l)$ will return the standard link $sl(R_i, A_k, R_j, A_l)$ where A_k is a potential key of R_i referenced by the foreign key A_l of R_j. By analogy we define $LL^{-1}(R_j, A_l)$. If R_j has more than one foreign keys then by $SL^{-1}(R_j) = \cup_l SL^{-1}(R_j, A_l)$ we denote the set of all set links of R_j defined by the foreign keys of R_j. By $LL^{-1}(R_j) = \cup_l LL^{-1}(R_j, A_l)$ we define the set of all list links of R_j defined by the foreign keys of R_j.

To define a classification problem one of the relations in $\mathcal{R}$ should be defined as the *main relation*, $M.$, i.e. the relation on which the classification problem will be defined. Each instance, M_{i_j}, of the M relation will give rise to one *relational instance*, $M_{i_j}^+$, i.e. an instance that spans the different relations in $\mathcal{R}$. More precisely $M_{i_j}^+$ will have the same set of standard attributes $\mathcal{I}_{A,M}$ and the same values for these attributes as M_{i_j} has and each link $sl \in SL(M) \cup SL^{-1}(M)$ ($ll \in LL(M) \cup LL^{-1}(M)$) adds in $M_{i_j}^+$ one attribute of type set (list). The value of an attribute of type set (list) is defined based on the link sl (or ll) and it will be the set (list) of instances with which M_{i_j} is associated in some relation R_j when we follow the link sl (ll). We will denote $\mathcal{I}_{sl,R_i}$ and $\mathcal{I}_{ll,R_i}$ the set of attributes of type set and list, respectively. By recursive application of this procedure we obtain the complete description of the relational instance M_{ij}^+. More details on relational algebra representation can be found in [6, 1].

3 Kernels on Extended Relational Instances

In this Section we will introduce a new class of kernels over extended relational algebra structures which are instances of the $\Re$-Convolution kernels, [2], and are based on kernels introduced in [6].

In order to define these kernels we recall from Section 2 that a given relation R_i is divided into three parts: $\mathcal{I}_{A,R_i}, \mathcal{I}_{sl,R_i}, \mathcal{I}_{ll,R_i}$, which denote set of standard attributes, set of attributes of type set and set of attributes of type list, respectively. A relational instance R_{i_a} is $R_{i_a} = (v_{a\mathcal{I}_{A,R_i}}, v_{a\mathcal{I}_{sl,R_i}}, v_{a\mathcal{I}_{ll,R_i}})$, where $v_{a\mathcal{I}_{A,R_i}} = (v_{a1}, \ldots, v_{a|\mathcal{I}_{A,R_i}|})$ is the vector of standard attributes and $v_{a\mathcal{I}_{sl,R_i}} = (v_{s1}, \ldots, v_{s|\mathcal{I}_{sl,R_i}|})$ and $v_{a\mathcal{I}_{ll,R_i}} = (v_{l1}, \ldots, v_{l|\mathcal{I}_{ll,R_i}|})$ are vectors of attributes of type set and list.

Given this formalism we defined the *Direct Sum Kernel* on the set X (if $|\mathcal{I}_{A,R_i}| \neq 0$) as $k_\Sigma(R_{i_a}, R_{i_b}) = k_s(v_{a\mathcal{I}_{A,R_i}}, v_{b\mathcal{I}_{A,R_i}}) + \sum_{l \in \mathcal{I}_{sl,R_i}} K_{set}(v_{al}, v_{bl})$ $+ \sum_{l \in \mathcal{I}_{ll,R_i}} K_{list}(v_{al}, v_{bl})$ where $k_s(.,.)$ is an elementary kernel defined on the set $\mathcal{I}_{A,R_i}$ of the standard attributes of R_i and $K_{set}(.,.)$ and $K_{list}(.,.)$ are kernels between sets or lists, respectively. Here we use a normalized version of k_Σ (if $|\mathcal{I}_{A,R_i}| \neq 0$): $K_\Sigma(R_{i_a}, R_{i_b}) := \frac{k_\Sigma(R_{i_a}, R_{i_b})}{1+|\mathcal{I}_{sl,R_i}|+|\mathcal{I}_{ll,R_i}|}$. More details on the Direct Sum Kernels applied to relational instances can be found in [6].

In the next Section we will define kernels over sets ($K_{set}(.,.)$) and kernel over lists ($K_{list}(.,.)$) of relational instances. The computation of the final kernel is based on recursive alternating applications of $K_\Sigma(.,.)$ and kernels $K_{set}(.,.)$ and $K_{list}(.,.)$.

Kernels over Lists and Sets. Before defining kernels over lists we first introduce some helpful notation. Lets denote by **i** a sequence $i_1 \leq i_2 \leq \cdots \leq i_n$ of indices; we say that $i \in \mathbf{i}$ if i is one of the sequence indices. We denote with $l(\mathbf{i})$ the length of a sequence **i**. For a given attribute, v_{al} of type list $v_{al}[i_k]$ is its k element.

The first kernel is the *Contiguous Sublist Kernel* $K_{list1}(v_{al}, v_{bl}) = \sum_{\mathbf{i},\mathbf{j},l(\mathbf{i})=l(\mathbf{j})} \lambda^{l(\mathbf{i})}$ $\sum_{s=1,\ldots,l(\mathbf{i})} K_\Sigma(v_{al}[i_s], v_{bl}[j_s])$ which is a modified version of the *Contiguous Subtree Kernel*, [5], and where v_{al} and v_{bl} are attributes of type list, the subsequences **i** and **j** are assumed to be *contiguous* and $0 < \lambda < 1$ is a parameter penalizing longer subsequences. A slightly more general kernel is proved in [5] to be a valid kernel, which is computable in $O(mn)$ where m and n are the lengths of the lists v_{al} and v_{bl} respectively.

The other kernel on lists we experimented with is a specialized version of the kernel over basic terms from [3] which we will call the *Longest Common Sublist Kernel*. This kernel can be written as $K_{list2}(v_{al}, v_{bl}) = m + \sum_{s=1}^{m} K_\Sigma(v_{al}[i_s], v_{bl}[i_s]) + n$ where $m = min(l(v_{al}), l(v_{bl}))$, i.e. the length of the shortest list, and $n = 1$ if the lists are of the same length and 0 otherwise. This kernel in more general settings is proved to be a valid kernel in [3].

The above kernels are normalized such that the examples in the corresponding feature space have a unit norm, i.e. $K_{list}(x,y) := \frac{K_{list}(x,y)}{\sqrt{K_{list}(x,x)K_{list}(y,y)}}$.

The *Contiguous Subtree Kernel* and *Longest Common Sublist Kernel* are related to each other in the sense that they apply to the same kind of data. However the underlying notion of similarity of the *Contiguous List Kernel* and the kernel from [3] is different. In the former the overall similarity is measured by sum of the mutual similarities of all the (consecutive) sublists of the same length. The similarity between the sublists are computed by means of other kernels. On the other hand the kernel from [3] only takes the longest common contiguous sublist at the start of the two sublists into account. In that sense the former takes a more "global" view.

For attributes of type set we use the *Averaged Cross Product Kernel*, [6].

4 Experiments

We checked the performance of the SVM algorithm, [7], on a protein fingerprint classification problem. Protein fingerprints are groups of conserved motifs (regions) drawn from multiple sequences alignment that can be used as diagnostic signatures to identify and characterize collections of protein sequences, [8]. Broadly speaking, fingerprints may be diagnostic for a gene family or superfamily (united by a common function), or a domain family (united by a common structural motif). Fingerprints can be described by its component motifs and protein sequences, we are therefore confronted with a multirelational learning problem. Our approach will be different from the one presented in [8] since there the task representation is propositionalized by aggregating protein and motif characteristic over a fingerprint.

We modeled this data in the following way: the "fingerprints" (main) relation with global characteristics of the instances is associated through an one-to-many relation with the "motifs" relation. Additionally there is a number of relations with aggregated information about proteins (actually proteins IDs) associated with the main relation using one-to-one relations. Fingerprints are globally characterized by (among others) number of proteins and proportion of protein sequences that match all or only a part of the motifs in a fingerprint. Individual motifs are characterized by a number of amino acids and protein sequences, coverage (the fraction of protein sequences in the fingerprint that match the motif) and a number of features measuring motif's conservation. The last source of information are protein sequences and more precisely their *SWISS-PROT/TrEMBL* labels. Features computed on the basis of these labels can be considered as statistics computed on the set of proteins and hence they can be stored in the "fingerprints" relation. However, keeping them in separate relations provides us a way to treat missing values which is the case here since not all proteins have a SWISS-PROT entry. More information about different attributes in different relations can be found in [8].

We also defined three different data representation based on combinations of the two different types of link associations defined in Section 2. In the first two approaches we assumed that each instance from the "fingerprints" relation is associated with a *set* and *list* of instances from the "motifs" table, respectively. The latter approach assumes that the order in which motifs appear along the sequence of amino acids is important. This is justified by the fact that a motif is basically a multiple sequence alignment

Table 1. Accuracy and rank results on the protein fingerprints dataset

Elementary kernel	10-fold CV	test set	10-fold CV	test set	10-fold CV	test set
	SETS		LISTS		SETS AND LIST	
	Contiguous Sublist Kernel					
$k_{P_{p=2,a=1}}$	85.08 (4.5)	84.22	85.88 (5.5)	86.08	87.63 (8.5)	87.35
$k_{G_{\gamma=0.1}}$	83.66 (1)	81.97	83.59 (1)	83.86	86.01 (5.5)	83.99
	Longest Common Sublist Kernel					
$k_{P_{p=2,a=1}}$			85.61 (5)	86.01	86.75 (7.5)	86.35
$k_{G_{\gamma=0.1}}$			83.86 (1)	84.13	85.81 (5.5)	73.3
Def. Accuracy	54.4					

with a number of conserved regions so there exists an intrinsic notion of order (along protein sequences). In the third approach we combine the two previous representations: an instance from the "fingerprints" relation is associated both with the set and list of instances from the "motifs" table.

We follow the experimental procedure reported in [8] where 1842 fingerprints records from version 37 of the PRINTS database were split into the design and testing set. In all experiments we limited ourselves to normalized polynomial $k_{P_{p,a}}(.,.)$ (where $p = 2, a = 1$) and Gaussian RBF $k_{G_\gamma}(.,.)$ (where $\gamma = 0.1$) elementary kernels, [7]. For attributes of type lists we use the *Contiguous Sublist Kernel* ($\lambda = 0.5$) and *Longest Common Sublist Kernel* defined in Section 3. We also explored the behavior the SVM with the $C = 1$ complexity parameter. We estimate accuracy using ten-fold cross-validation and control for the statistical significance of observed differences using McNemar's test (sig. level=0.05). We also establish a ranking schema of different relational kernels as follows: in a given dataset if kernel a is significantly better than b then a is credited with one point and b with zero points; if there is no significant difference then both are credited with half point. Results are presented in table 1.

To compare the different data representations we fix a submodel and average the ranks of K_{list} (or K_{set}), k_P and k_G, ignoring their parameter settings. The average ranks for the models where motifs are represented as sets or lists are 2.75 and 3.125, respectively, whereas for the third model (instances from the "fingerprints" relation are associated both with the set and list of motifs) the average rank is 6.75. We can see that there is a clear advantage in terms of predictive performance of the third model over the others.

Here the best results are obtained for the k_P elementary kernel ($p = 2, a = 1$) together with the *Contiguous Sublist Kernel* and where the third representation of the data is used. The estimated cross-validation accuracy is 87.63 % whereas the holdout accuracy is 87.35 %. This represents a statistically significant improvement over the best accuracy previously reported in [8] by 1.72 % for cross-validation and 1.43 for the holdout test set.

5 Discussion and Future Work

In this paper we proposed a new class of kernels which extends our previous work presented in [6, 1] in the sense that these kernels are defined over a richer representation language. Our kernels can be considered as instances of the $\Re$-Convolution kernel where the subkernels are elementary kernels and kernels over sets and lists.

Although many kernels have been recently proposed for sequences over a finite alphabet (e.g. [4]), not much work has been done in defining kernels over lists of complex objects. The exceptions are kernels described in [5, 3], variants of which are proposed in Section 3. The main difference between the *Contiguous Sublist Kernel* and the *Contiguous Subtree Kernel* from [5] is that the latter uses an additional user defined matching function whereas for the objects we consider, the relation determines whether two relational instances are matchable or not. The other difference is that kernels from [5] are highly specialized whereas our kernels can be used for any classification problem. The *Longest Common Sublist Kernel* is a direct application of a kernel over basic terms defined in [3]. The main difference is that the kernel proposed in [3] was applied only for sequences over a finite alphabet (with matching kernel for sequences' elements) whereas we extended it to more complex structures where only a kernel over elements of the lists is needed. Comparison to other kernels over (general) complex structures can be found in [6].

Experiments on the protein fingerprints show that in terms of accuracy there is an advantage of relational SVM over propositional one. On the other hand using the relational approach makes it easier to preprocess the data since it has a clear relational representation. In the future work we will concentrate on designing more refined kernels for sets and lists. The other remaining challenge here is that of bringing more discriminatory information (e.g. biological literature) to bear on the classification of protein fingerprints.

References

1. Woźnica, A., Kalousis, A., Hilario, M.: Distance-based learning over extended relational algebra structures. In: In Proceedings of the 15th International Conference on Inductive Logic Programming (late breaking papers), Bonn, Germany (2005)
2. Haussler, D.: Convolution kernels on discrete structures. Technical report, UC Santa Cruz (1999)
3. Gaertner, T., Lloyd, J., Flach, P.: Kernels and distances for structured data. Machine Learning (2004)
4. Schölkopf, B., Tsuda, K., Vert, J.: Kernel Methods in Computational Biology. MIT Press series on Computational Molecular Biology. MIT Press (2003)
5. Zelenko, D., Aone, C., Richardella, A.: Kernel methods for relation extraction. Journal of Machine Learning Research 3 (2003) 1083–1106
6. Woźnica, A., Kalousis, A., Hilario, M.: Kernels over relational algebra structures. In: The Ninth Pacific-Asia Conference on Knowledge Discovery and Data, Hanoi, Vietnam (2005)
7. Schölkopf, B., Smola, A.J.: Learning with Kernels: Support Vector Machines, Regularization, Optimization, and Beyond. MIT Press, Cambridge, MA (2002)
8. Hilario, M., Mitchell, A., Kim, J.H., Bradley, P., Attwood, T.: Classifying protein fingerprints. In: Proc. 8th Conference on Principles and Practice of Knowledge Discovery in Databases, Pisa, Italy, Springer-Verlag (2004)

Mining Quantitative Maximal Hyperclique Patterns: A Summary of Results

Yaochun Huang[1], Hui Xiong[2], Weili Wu[1], and Sam Y. Sung[3]

[1] Computer Science Department, University of Texas - Dallas, USA
{yxh038100, wxw020100}@utdallas.edu
[2] MSIS Department, Rutgers University, USA
hui@rbs.rutgers.edu
[3] Dept. of Computer Science, South Texas College, USA
sysung@southtexascollege.edu

Abstract. Hyperclique patterns are groups of objects which are strongly related to each other. Indeed, the objects in a hyperclique pattern have a guaranteed level of global pairwise similarity to one another as measured by uncentered Pearson's correlation coefficient. Recent literature has provided the approach to discovering hyperclique patterns over data sets with binary attributes. In this paper, we introduce algorithms for mining maximal hyperclique patterns in large data sets containing quantitative attributes. An intuitive and simple solution is to partition quantitative attributes into binary attributes. However, there is potential information loss due to partitioning. Instead, our approach is based on a normalization scheme and can directly work on quantitative attributes. In addition, we adopt the algorithm structures of three popular association pattern mining algorithms and add a critical clique pruning technique. Finally, we compare the performance of these algorithms for finding quantitative maximal hyperclique patterns using some real-world data sets.

1 Introduction

A hyperclique pattern [9, 4] is a new type of association pattern that contains items which are highly affiliated with each other. More specifically, the presence of an item in one transaction strongly implies the presence of every other item that belongs to the same hyperclique pattern. Conceptually, the problem of mining hyperclique pattern in transaction data sets can be viewed as finding approximately all-one sub-matrix in a 0-1 matrix where each column may correspond to an item and each row may correspond to a transaction. For the rest of this paper, we refer to this problem as the binary hyperclique mining problem.

However, in many business and scientific domains, there are data sets which contain quantitative attributes (e.g. income, gene expression level). How to define and efficiently identify hyperclique patterns in data sets with quantitative attributes remains a big challenge in the literature. To this end, the focus of this paper is to address the quantitative hyperclique pattern mining problem.

To the best of our knowledge, there is no previous work on developing algorithms for finding quantitative maximal hyperclique patterns. Our approach for mining quantitative hyperclique patterns is built on top of the normalization scheme [7]. A side effect of the normalization scheme is that there is no support pruning for single items. To meet with this computational challenge, we design a **clique pruning** method to dramatically remove a large number of items which are weakly related to each other, and thus effectively improving the overall computational performance for finding quantitative hyperclique patterns. We adopt structures of three popular association pattern mining algorithms including FP-tree [5], diffEclat [10], and Mafia [3] as the bases of our algorithms. The purpose of these algorithms is to find quantitative maximal hyperclique pattern, which is a more compact representation of quantitative hyperclique patterns and is desirable for many applications, such as pattern preserving clustering [8]. A hyperclique pattern is a maximal hyperclique pattern if no superset of this pattern is a hyperclique pattern. Finally, we briefly introduce the results of using our approach on some real-world data sets.

2 Normalization and Quantitative Hyperclique Patterns

Normalization. In this paper, we adopt the normalization method proposed in [7]. For a vector $x = < x_1, x_2, \ldots, x_n >$, our normalization will turn the vector as $x' = < x'_1, x'_2, \ldots, x'_3 > = < \frac{x_1}{|x|}, \frac{x_2}{|x|}, \ldots, \frac{x_n}{|x|} >$, where $|x| = \sqrt{\sum_{k=1}^{n} x_k^2}$. After data normalization, we define the support of every individual item i, $\sigma_{L_2^2}(i) = {x'_1}^2 + {x'_2}^2 + \ldots + {x'_n}^2 = 1$ and the support of an itemset X is defined as $\sigma_{\min, L_2^2}(X) = \sum_{i \in T}(min\{T(i,j)|j \in I\})^2$, where T(i, j) means the normalized value of item j in the transaction i.

One advantage of this normalization is that the resulting support is a number between 0 and 1. Such normalization is natural in many domains, e.g., text documents. However, a side-effect of this is that individual items can no longer be pruned using a support threshold since all single items have a support of 1.

Quantitative Hyperclique Patterns. A traditional binary hyperclique pattern [9] is a frequent itemset with the additional constraint that every item in the itemset implies the presence of the remaining items with a minimum level of confidence known as the h-confidence. Specifically, we have the following:

Definition 1. *A set of attributes, X, forms a hyperclique pattern with a particular level of h-confidence, where h-confidence is defined as*

$$\mathrm{hconf}(X) = \min_{i \in X}\{\mathrm{conf}(\{i\} \to \{X - \{i\}\})\} - \sigma(X)/\max_{i \in X}\{\sigma(i)\} \quad (1)$$

Where σ is the standard support function [1].

H-confidence, just like standard support, is in the interval [0, 1] and it has the anti-monotone property; that is, the h-confidence of an itemset is greater than or equal to that of its any superset. Also, hyperclique patterns have the high affinity property, i.e., items in a pattern with a high h-confidence are guaranteed to have a high pairwise similarity as measured by the cosine metric. Additionally, there is an important relationship between h-confidence of binary hyperclique patterns and the support function $\sigma_{\min,\ L_2^2}(X)$. In particular, since $\sigma_{\min,\ L_2^2}(X)$ is equivalent to standard support for binary data, we can substitute $\sigma_{\min,\ L_2^2}(X)$ for the standard support function $\sigma(X)$ in Equation 1. It is then interesting to note that if we normalize all attributes to have an L_2 norm of 1, i.e., $\sigma_{\min,\ L_2^2}(i) = 1$ for all items i, then, by Equation 1, $\mathrm{hconf}(X) = \sigma_{\min,\ L_2^2}(X)$, since the normalization sets the support of all the item to 1, we get $\max_{i\in X}\{\sigma(i)\} =$ support (i) $=1$.

In a nutshell, finding continuous hyperclique patterns first proceeds by normalizing the attributes to have an L2 norm of 1. Then, for each row, we take the minimum of the specified attributes. Finally, we square each of these values and add them up. The resulting value is the h-confidence and is a lower bound on the pairwise cosine similarity.

3 Algorithm Descriptions

Here, we present the algorithms for mining quantitative maximal hyperclique patterns. Our algorithms are built on top of three state-of-the-art association pattern mining algorithms including FPTree [5], diffEclat [10], and Mafia [3].

Clique Pruning. We design a clique pruning method for eliminating weakly related single items. Specifically, we first compute h-confidence of all item pairs on the normalized data. For each item, we then identify the maximum h-confidence value among all pairs including this item. Finally, for a user-specified threshold, we prune all items whose maximum h-confidence is less than this threshold.

Algorithm based on FP-Tree. FP-Tree [5] is a compact tree structure which allows to identify frequent patterns without generating the candidate patterns. Here, we adopt the FP-tree algorithm for finding quantitative maximal hyperclique patterns. First, we store float values instead of integer values, since the support of the normalized data are continuous. Second, the support values should be squared before added to the FP-Tree since they have an L_2 Norm. Finally, we need to split squared transactions and make the support of preceding item not less than the successor item, before adding them into the FP-Tree.

Algorithm based on MAFIA. MAFIA [3] is a depth-first searching algorithm for mining maximal frequent patterns. For the data set with continuous attributes, we change the algorithm to store not only the tidset, but also the support (normalized data) for each transaction. For this purpose, the algorithm needs a float vector to store the support information. Each element in the vector presents the support for each transaction in order.

Algorithm based on DiffEclat. DiffEclat uses a vertical data representation, called **diffset**, for efficiently mining maximal frequent patterns[10]. The diffset only store the different set of transaction ID between the pattern and its parant pattern. The key modification that we made is to store both transaction IDs and the support information. However, for diffset, we store the support different between a pattern and its parent pattern instead of the support itself.

4 Experimental Evaluation

Experimental Setup. Our experiments were performed on two real-life gene expression data sets, Colon Cancer and NCI [2, 6]. Table 1 shows some characteristics of these gene expression data sets.

Table 1. The Characteristics of Gene Expression Data Sets

DATASET	Colon Cancer		NCI
# Gene	2000		9905
# Sample	62		68
# Class	2		9
CLASS	NAME	# SAMPLE	
C1	Tumor	40	
C2	Normal	22	

A Performance Comparison. Figure 1 (a) shows the running time of three algorithms on the Colon Cancer data set. As can be seen, when the h-confidence threshold is less than 0.35, the FP-Tree can be an order of magnitude faster than Mafia and DiffEclat is not very efficient and become unscalable when the h-confidence threshold is low. Also, Figure 1 (b) shows the performance of the proposed algorithms for mining sample patterns on the NCI data set. Similar to the observation from the Colon Cancer data set, we can also observe that when the h-confidence threshold is less than 0.5, the FPTree can be an order of magnitude faster than Mafia. However, MAFIA has a better performance when the h-confidence threshold is high. Another observation is that the performance DiffEclat is not scalable when the h-confidence threshold is low.

The Effect of Clique Pruning. Figure 2 demonstrate the effect of clique pruning on Colon and NCI data sets using the algorithm based on Mafia. As can be seen from both figures, with the increase of the clique pruning ratio, the running time is reduced significantly. The running time can be orders of magnitude faster if we target on hyperclique patterns with high affinity. Another benefit is that, the proposed algorithm can even identify patterns at a very low level support when the clique pruning ratio is at a certain level.

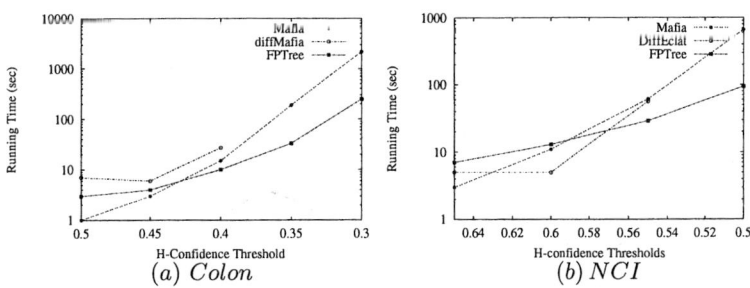

Fig. 1. The Performance Comparison on Colon and NCI data sets

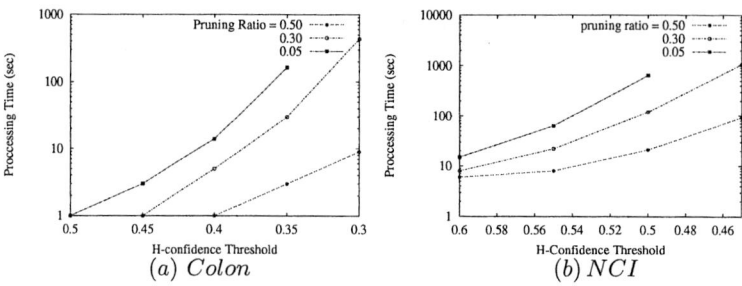

Fig. 2. The Effect of Clique Pruning on Colon Cancer and NCI Data Sets

5 Conclusions

In this paper, we addressed the problem of mining quantitative maximal hyperclique patterns in the data sets with continuous attributes. Instead of mapping continuous attributes into binary attributes, we applied a data normalization method. Also, we provided algorithms for finding quantitative maximal hyperclique patterns. These algorithms are built on top of three state-of-the-art association pattern mining algorithms and have included a clique pruning method to perform pruning for individual items. Finally, the performance of the algorithms have been demonstrated using real-world data sets.

Acknowledgement. This paper was partially supported by NSF grant #ACI-0305567 and NSF grant #CCF-0514796.

References

1. Rakesh Agrawal, Tomasz Imielinski, and Arun N. Swami. Mining association rules between sets of items in large databases. In *SIGMOD 93*, May 1993.
2. U. Alon, N. Barkai, D.A. Notterman, and et al. Broad patterns of gene expression revealed by clustering analysis of tumor and normal colon tissuese probed by oligonucleotide arrays. *Proc. Natl. Acad. Sci*, 96:6745–6750, June 1999.
3. D Burdick, M Calimlim, and J Gehrke. Mafia: A maximal frequent itemset algorithm for transactional databases. In *ICDE*, 2001.
4. Y. Huang, H. Xiong, W. Wu, and Z. Zhang. A hybrid approach for mining maximal hyperclique patterns. In *ICTAI*, 2004.
5. J.Han, J.Pei, and Y. Yin. Mining frequent patterns without candidate generation. In *ACM SIGMOD*, 2000.
6. D.T. Ross, U. Scherf, and et al. Systematic variation in gene expression patterns in human cancer cell lines. *Nature Genetics*, 24(3):227–234, 2000.
7. Michael Steinbach, Pang-Ning Tan, Hui Xiong, and Vipin Kumar. Extending the Notion of Support. In *ACM SIGKDD*, 2004.
8. H. Xiong, M. Steinbach, P. Tan, and V. Kumpar. HICAP: Hierarchial Clustering with Pattern Preservation. In *Proc. of SIAM Int'l Conf. on Data Mining*, 2004.
9. H. Xiong, P. Tan, and V. Kumar. Mining strong affinity association patterns in data sets with skewed support distribution. In *ICDM 2003, USA*, 2003.
10. Mahommed Zaki and Karam Gouda. Fast vertical mining using diffsets. In *ACM SIGKDD*, 2003.

A Nonparametric Outlier Detection for Effectively Discovering Top-N Outliers from Engineering Data

Hongqin Fan[1], Osmar R. Zaïane[2], Andrew Foss[2], and Junfeng Wu[2]

[1] Department of Civil Engineering, University of Alberta, Canada
[2] Department of Computing Science, University of Alberta, Canada

Abstract. We present a novel resolution-based outlier notion and a nonparametric outlier-mining algorithm, which can efficiently identify top listed outliers from a wide variety of datasets. The algorithm generates reasonable outlier results by taking both local and global features of a dataset into consideration. Experiments are conducted using both synthetic datasets and a real life construction equipment dataset from a large building contractor. Comparison with the current outlier mining algorithms indicates that the proposed algorithm is more effective.

1 Introduction

The term "outlier" can refer to any single data point of dubious origin or disproportionate influence. *Given a set of observations X, an outlier is an observation that is an element of this set but which is inconsistent with the majority of the data or inconsistent with a sub-group of X to which the element is meant to be similar.* The above definition has two implications: outlier vis-à-vis the majority; and outlier vis-à-vis a group of neighbours. Whether it is an interesting contaminant or dubious data entry, an outlier is often considered noise, which can have a harmful effect on statistical analysis.

Attempts have been made to remove the noisy data using various outlier mining approaches; in one example, Raz et al. [1] designed an expert system to automatically detect unlikely vehicles and erroneously classified ones from weigh-in-motion data. In contrast to noisy data, some relevant outliers contain important information on system malfunction, mismanagement, or even unpredictable phenomena (environmental or geological disaster), which should be detected for further investigation rather than discarded. In either of these two cases, the inconsistent records should first be identified as much as possible from the dataset. Data validation (range validation, single variate pattern validation etc.) can only filter out a small portion of outliers. Traditional statistical approaches including multivariate outlier detection are not applicable due to their pre-assumption of certain statistical distributions, which may not exist for datasets containing multiple clusters.

Outlier mining techniques in data mining seem to be viable solutions for outlier detection in engineering applications. Some of the popular algorithms include among others a distance-based outlier mining algorithm by Knorr and Ng [2]; a local outlier mining algorithm by Breunig et al. [3] and a connectivity-based mining algorithm by

Tang et al. [4]. However these algorithms are not widely accepted in civil engineering disciplines because they do not cater to the special features of engineering datasets, including:

- Current outlier mining algorithms need domain-dependent parameters, but these parameters are not known a priori;
- Current outlier mining algorithms need some parameters, which can only be obtained and tuned through tremendous trial-and-error effort. This is not practical for frequent time-changing applications and thus cannot be an integrated part of a real-time decision support system.
- Current outlier mining algorithms are capable of mining either global or local outliers while the engineering dataset usually contains loosely bounded clusters. It is difficult in this case to differentiate local from global outliers.
- In engineering applications, there exists a need for ranking the top-listed outliers. This is where our major focus is.

In this paper, we present a Resolution-Based outlier (RB-outlier) notion and an associated outlier detection algorithm efficient for engineering applications. The RB-outlier notion is proposed based on a nonparametric clustering algorithm called TURN* by Foss and Zaïane with the same idea of resolution change [5]. The proposed algorithm can detect and rank top-N outliers from any kind of dataset without the need for input parameters.

We also compare RB-outlier with DB-outlier and Local density based outlier (LOF-outlier) mining algorithms using both synthetic datasets and a construction equipment dataset from a large building contractor. Our experimental results show that the RB-outlier mining algorithm generates equivalent or better results than the other two competitive algorithms on all the datasets while benefiting from the absence of input parameters; the RB-outlier results seem to combine the results from both DB-outlier which looks for global outliers and LOF-outlier which looks for local outliers. Analysis on the detected outliers from the equipment datasets shows that these combined results make more sense for engineering datasets.

2 Related Work

Hawkins defines an outlier as *"an observation which deviates so much from other observations as to arouse suspicions that it was generated by a different mechanism"* [6]. Traditionally outlier detection in engineering disciplines depends on statistical approaches. After fitting a data series into a bell-shaped statistical distribution, those data points located far away from the mean (e.g. 3 standard deviations) are deemed outliers; multivariate outlier detection techniques can help identify outliers within a multivariate dataset. Other commonly used techniques are quartile methods, and visualization methods using scatter plot, etc. With regard to the data complexity and sheer data volume in engineering systems, outlier detection using statistical approaches is very inefficient and even impractical due to their limitations such as the difficulties with handling higher dimensional data, and the necessary assumption of distributions.

A distance-based definition of outliers was first proposed by Knorr and Ng. They introduced DB-outlier to identify outliers from a large database (i.e. with high dimensions and high data volume) [2]. A DB-outlier is defined as follows: "*An object O in a dataset T is considered a DB(p,D)-outlier if at least a fraction p of the objects in T lies greater than distance D from O*". The authors claim that this definition generalizes the notion of outlier defined in statistical tests for standard distributions. DB-outlier tends to find outliers in global context, which means some important outliers deviated from their local clusters are probably missed if a large number of isolated points or loosely packed clusters appear. While Knorr and Ng's definition is distribution free, it lacks a mechanism to rank outliers. Without violating the original notion of DB(p,D) outlier, Ramaswamy et al. further propose to rank each point based on its distance to its kth nearest neighbour, and use a partition-based algorithm to efficiently mine top-N outliers from a large database [7]. For them, *the outliers are the top n data elements whose distance to the kth nearest neighbour is greatest*. This definition also eliminates the need to estimate an appropriate distance D.

Another popular algorithm is a local density based outlier-mining algorithm proposed by Breunig et al. [3]. A Local Outlier Factor (LOF) is assigned for each object with respect to its surrounding neighbourhood. The LOF value depends on how the data points are closely packed in its local reachable neighbourhood. These points deep inside a dense cluster have a LOF value of approximately 1 while the isolated points have a much higher value. The authors claim that this definition also catches the spirit of the outlier definition given by Hawkins [6]. The local outlier notion seems more reasonable than DB-outlier because each data point can be measured with a numerical factor based on how the data is deviated from its genuine cluster. Therefore the outliers can be ranked as per their LOF values.

Tang et al. improved to some extent the LOF definition by using their Connectivity-based Outlier Factor (COF) for a dataset containing low density patterns. In such a case, LOF would not be effective to measure the density of an outlier with respect to its sparse neighbourhood [4].

The biggest hurdle of effectively applying these afore-mentioned outlier-mining algorithms in the engineering domain is the determination of their input parameters. All the parameters should be either known *a priori* or estimated and optimized by trial-and-error with the help of expert opinions. In particular, LOF is very sensitive to its parameter MinPts and DB-outlier results vary greatly when D and p change. Subjective results make it difficult to implement these outlier-mining algorithms for outlier detection in engineering applications.

3 Resolution-Based Outlier

TURN* is a nonparametric clustering algorithm [5]. The optimum clustering of a dataset can be obtained automatically based on resolution change: when the resolution changes on a dataset, the clusters in the dataset redistribute. All the objects are in the same cluster when the resolution is very low, meanwhile every object is a single cluster when the resolution is very high, and therefore the optimum clustering can be achieved at a point between these two extreme scenarios. The "TURN-CUT" technique was introduced to detect this critical point during the resolution change by looking for a plateau in the curve of the differential of some collected cluster statistics [5].

The same observation holds when viewing all the objects in the dataset from the perspective of outliers. All the objects are outliers when the resolution is high enough to warrant no neighbours (by distance measure) for any objects in the dataset; meanwhile all the objects are "inliers" when the resolution is low enough to have all the objects close-packed in a single cluster. If the resolution of a dataset changes, different outliers demonstrate different clustering-related behaviors during resolution change; those objects more isolated, with less neighbours, and far away from large data communities are more liable to be outliers. On the other hand, the top outliers will be merged into a cluster later when the resolution is decreased. As a result, the accumulated cluster-related properties collected on one object can be used to measure its degree of outlyingness relative to its close neighbourhood and community (reachable neighbourhoods). We first define the neighbourhood of an object:

Definition 1. Neighbourhood of Object O:

> *If an Object O has a nearest neighbouring points P along each dimension in k-dimensional dataset D and the distance between P and O is less or equal to 1, then P is defined as the close neighbour of O, all the close neighbours of P are also classified as the close neighbours of O, and so on. All these connected objects are classified as the same neighbourhood.*

The threshold value is taken as 1 to measure whether two points are close enough to become neighbours. The absolute value of this threshold is not important because the pair-wise distances between points are relative measurements during resolution change. The algorithm finds the maximum resolution S_{max} at which all the points are far enough from each other to be non-neighbours, and the minimum resolution S_{min} at which all the points are close enough to be neighbours.

Secondly we define the resolution-based outlier factor for each object:

Definition 2. Resolution-Based Outlier Factor (ROF):

> *If the resolution of a dataset changes consecutively between maximum resolution where all the points are non-neighbours, and minimum resolution where all the points are neighbours, the resolution-based outlier factor of an object is defined as the accumulated ratios of sizes of clusters containing this object in two consecutive resolutions.*

$$ROF(O) = \sum_{i=1}^{R} \frac{ClusterSize(O, r_{i-1}) - 1}{ClusterSize(O, r_i)}$$

Where $r_0, r_2...r_i...r_R$ are the resolutions at each step, R is the total number of resolution change steps from S_{max} to S_{min}, $CluserSize(O,r)$ is the number of objects in the cluster containing object O at a resolution r.

At each resolution, we cluster the points based on the distance between every two objects and the neighbourhood definition. The time complexity of clustering at each resolution is O(NLogN) as demonstrated in [5]. The cluster size of each object is set to 1 at the beginning (i.e. at Max. resolution), then the cluster size increases for the object whenever the object gets merged, or the cluster containing the object merges other objects at the next lower resolution.

The cluster size at the previous resolution is reduced by 1 for ease of comparison, which will set the ROF of an object to 0 before the object gets merged. Top N outliers are the N elements with the lowest ROF value.

Universally defining an outlier is somewhat controversial and is often subjective, relative to the application at hand. While considering locality and globality, our definition still embodies the essence of the accepted definition given by Hawkins [6] with regard to deviation from other observations.

4 Resolution-Based Outlier Mining Algorithm

Using the definitions of close neighbourhood and ROF in the previous section, a resolution-based outlier detection and ranking algorithm (RB-MINE) is proposed for mining top-N outliers in a dataset with multiple numerical attributes.

The resolution change ratio r is a percentile value used for computing the resolution change step size. The reason why this parameter is not considered an input parameter for the nonparametric clustering algorithm TURN* is explained in [5]. The same argument can be applied here to the RB-outlier mining algorithm. Indeed our experimental tests show the resolution change ratio, so long as it is kept in a moderate range, has a minor effect on the outlier results. Thus the user does not need to spend much effort to fine-tune this parameter, unlike the current popular outlier mining algorithms. Moreover, this ratio need not be static. It can decrease dynamically starting with a large step declining to a smaller step progressively. We found in our extensive experiments that fixing this resolution change to a static 10% gave good results for a wide range of data.

RB-CLUSTER
Given a resolution r and a dataset D:
1. Scale the coordinates using current resolution r.
   Current Coordinates = Original Coordinates * r
2. For each object O
   For the objects within a threshold distance of 1 from O, find the closest neighbours in each direction (+,-) along every dimension. (this can be done with a sort on each dimension)
3. Pick an unlabeled object and give it a new cluster label C.
   Initialize its neighbourhood chain n*Chain* and set the cluster size of C to 1.
4. Scan this object's close neighbours. For each neighbour:
   If the neighbour is unlabeled, give the neighbour the same cluster label C, add the neighbour to n*Chain*, and increase the cluster size of C by 1.
   If the neighbour has already been labeled as C' (C' ≠ C), change the label of all points in cluster C' to C. Increase the cluster size of C by the number of objects contained in cluster C', then delete records on cluster C'.
5. Move to the next object on *nChain*. Repeat 4 until all objects on *nChain* are checked.
6. Record the size of cluster C.
7. Repeat (3)~(6) until all objects are labeled.
8. For each object p, update the ROF value.

Fig. 1. RB-CLUSTER component algorithm

> **RB-MINE**
> Given a dataset D and the specified number N of top outliers
> 1. Find the maximum resolution, S_{max}, at which no close neighbours can be found for each object, and the minimum resolution, S_{min}, at which all the points are close neighbours in the same neighbourhood.
> 2. Starting at S_{max}, initialize the ROF value as 0 for each object.
> 3. Update $r_i = r_{i-1} + (S_{max} - S_{min})*\Delta r$, ($\Delta r$ is the resolution changing ratio).
> 4. Run RB-CLUSTER to cluster the objects at resolution r_i.
> 5. Update the ROF value for each object.
> 6. Rank objects in an increasing order of ROF, obtain top N outliers.

Fig. 2. RB-MINE algorithm

The finalized RB-outlier algorithm is comprised of a component algorithm RB-CLUSTER in Figure 1, and an overall algorithm RB-MINE in Figure 2. RB-CLUSTER clusters objects in the dataset at each resolution step while RB-MINE steers the mining process through the change of resolution and collection of ROF for each object.

5 Comparison of RB-Outlier with DB-Outlier and LOF-Outlier

One of the primary factors determining the outlier mining results is the outlier notion, which describes what an outlier is and assigns it a measuring factor, if it is possible.

For examples, DB(D,p) outlier evaluates the inconsistency of an object by judging if there are sufficient number of objects within D (or the distance to its kth nearest neighbour is closer than D, where k is the (1-p) percent of the total number of objects), therefore the algorithm searches for the specified number of nearest points in a global context; LOF-outlier measures the degree of outlyingness by taking only a restricted neighbourhood into account, LOF varies depending on how an object is deviated from its "best-guess" cluster in a local context.

The RB-outlier notion measures how an object deviates from its close neighbourhood. The definition of neighbourhood implies that this neighbourhood actually includes a series of chained neighbourhoods (we call it a "community"), as such, RB-outlier measures an object against its degree of outlyingness by taking both "global" and "local" features into account.

6 Experimental Results

To validate the RB-outlier mining results and compare with those detected by DB-outlier and LOF-outlier mining algorithms, we implemented all three algorithms in the same C++ development environment, and conducted experiments on a number of synthetic datasets as well as a real world construction equipment dataset obtained from a large building contractor. This section summarizes our experimental results and comparative analysis on a token 200-tuple 2D synthetic dataset, a 10,000-tuple 2D synthetic dataset, and a 1033-tuple 3D construction equipment dataset.

A Synthetic 2D dataset of four clusters with a total number of 200 objects: The token dataset shown in Figure 3.A, contains four distinct clusters and 20 outliers. This dataset, used in many clustering research projects for validating clustering results can be equally applied for visual validation of outlier detection. We run the three algorithms separately on this dataset, each identifying the top-10 and top-20 outliers and marking up these points in the density plot. DB-outlier was optimal with p=97/200 and LOF-outlier with MinPts=3. Figure 3.A summarizes the top outliers detected by RB-outliers.

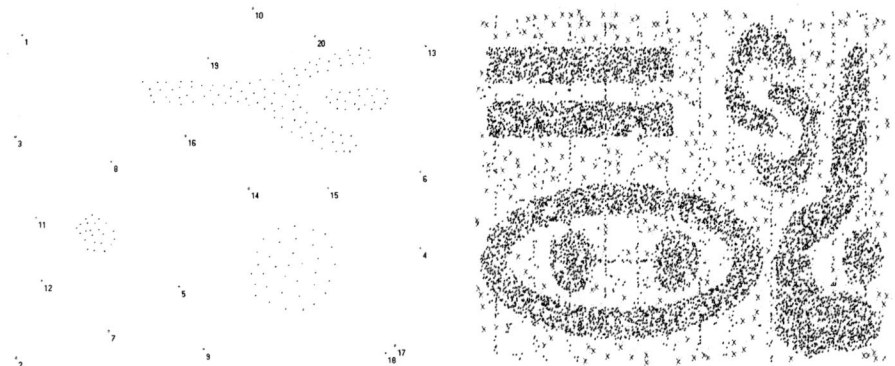

A: Top-20 outliers identified by RB-outlier in a 200-tuple synthetic dataset

B: Top-200 outliers identified by RB-outlier in a 10,000-tuple synthetic dataset

Fig. 3. Top outliers identified by RB-outlier in two synthetic datasets

Among the top-10 outliers, eight out of the top-10 outliers are the same with the three algorithms; visual judgment on the density plot confirms the "outlyingness" of the identified objects and the eight objects voted unanimously as outliers are truly isolated as compared with others. This finding indicates that all the three algorithms can effectively find outliers from this dataset. However there are some differences between the three sets of results:

1. Though top-3 outliers are exactly the same in the three sets of results, the other five unanimously identified outliers have different rankings in their perspective results. The five outliers have higher rankings in RB-outlier list.
2. The outliers No. 4 and 5 in LOF-outlier results are interpreted differently by DB-outlier and RB-outlier. In the DB-outlier set, No. 4 is ranked as No. 10 and No.5 is not included in the top-10; nevertheless neither of them is included in RB-outlier results at the top-10 level. This observation indicates that LOF-outlier tends to take objects in a small isolated cluster as outliers, the same tendency can be observed with DB-outlier. This finding can be explained by the definition for the minimum number of neighbouring points in LOF-outlier and DB-outlier: a small cluster containing number of objects less than this specified parameter tends to be classified as "a cluster of outliers."

The afore-mentioned conclusions are further enhanced by increasing the number of outliers in the top list. For example, if we look at the top-20 outliers in Figure 3.A, they are indeed detected by all the algorithms. However, RB-outlier shifts the unanimously identified outliers to higher rankings in its top-20 list.

Precision and recall are two measures in information retrieval that could be used as criteria for evaluating the outlier results. Precision is defined as the percentage of correct outliers in the set of detected outliers (i.e. the total number of correctly detected outliers divided by the number of labeled outliers); recall is the percentage of all known outliers correctly detected (i.e. total number of correctly detected outliers divided by the total number of existing outliers). For top-20, Precision and Recall are the same and all three algorithms achieve 100%. However, scrutinizing at a lower n for top-outliers, RB-outlier always has better or equivalent precision and recall.

A synthetic 10,000 object dataset containing nine clusters of different shapes: This dataset is also widely used for visually validating clustering results. The dataset contains 9 clusters of difference shapes and significant noise. This experiment aims to verify the capability of RB-outlier to distinguish outliers from "inliers" in a relatively large set and in the presence of clusters of arbitrary shapes.

In this experiment, we compared the top-200 outlier results identified by LOF-outlier and RB-outlier respectively, as marked up in Figure 3.B. Though both identify similar objects as outliers in this dataset, a detailed observation reveals that LOF-outlier biases objects in locally sparse areas, and RB-outlier takes local sparsity and global sparsity into consideration when picking up outliers. The LOF algorithm assigns a higher outlier factor for the object whose local density is lower than its neighbours. Therefore, when the number of outliers to detect is increased, LOF-outlier marks the objects at the edge of dense clusters as outliers, while RB-outlier tends to pick up outliers from isolated objects. RB-outlier's performance is indeed the sought for behaviour in construction engineering data.

A 3D construction equipment management dataset: To test-drive the RB-outlier mining algorithm and validate its usefulness in engineering datasets, we conducted experiments on a three-dimensional construction equipment dataset obtained from a large building contractor. The dataset includes characteristic attributes for 1033 pieces of equipment in the contractor's equipment fleet. Three numerical attributes are defined for each unit in addition to the identify attribute: the yearly repair/maintenance cost (yearly cost), the rate of charge, and the age. The objective of the experiments is to identify top listed inconsistent units from the dataset. The inconsistency of these units indicates the abnormal combination of the three attributes for a unit with respect to its similar equipment sub-group;

Using DB-outlier, LOF-outlier and RB-outlier algorithms, we identify the top-20 outliers from the selected equipment fleet. A comparison of the three sets of outliers determines that 11 out of 20 are unanimously identified as outliers by the three algorithms in their top-20 lists. LOF-outlier generates very similar results to DB-outlier with 16 being the same out of the top-20. This is not surprising because a large number of isolated objects appear in this dataset. LOF definition becomes similar to the DB-outlier notion for a sparsely distributed dataset: both start by looking for minimum number of neighbouring objects. Two additional units are identified as outliers by both LOF-outlier and RB-outlier; but no addition outlying units are identified by both DB-outlier and RB-outlier besides the 11 common units. RB-outlier moves all the 11 common units to higher rankings and adds 6 new units in the top-20 outlier list. Analysis of the 6 added units against their individual equipment sub-group and the entire fleet confirmed their interestingness.

The inability of identifying some outstanding outliers by DB-outlier and LOF-outlier mining algorithms can be illustrated from their outlier notions: these two algorithms evaluate the degree of outlying by drawing a hyper-sphere around each object; and the number of objects inside the hyper-sphere influences the outlier measurement of the object. The outliers harbored inside the concave of a cluster can not be identified efficiently based on this rational, subsequently some outliers become missing in the results if the two algorithms are applied on a real life dataset which may contain clusters of any arbitrary shape.

Among the top-20 outliers, Unit# 505-401, a soil cement plant (300 to 600 TPH), is detected as No.1 outlier by DB-outlier and No. 2 outlier by LOF-outlier, but is not identified as an outlier by RB-outlier. Detailed analysis of this unit finds out it is the only unit in this equipment class, therefore it is not a "true" outlier in the application. Other outliers in top-20 of DB-outliers (no. 16 and 17) are also the only units in their respective equipment class; No. 18 of LOF-outliers is identified as outlier because it is in a small 2-object cluster. These outlying units in DB-outlier and LOF-outlier results are of no particular interests; their appearance somehow degrades the general quality of the top-N outlier mining results. On the contrary, all the 11 common outlying units, the three found by RB-outlier and by another method, and the 6 additional units unique to the RB-outlier results are indeed inconsistent units relative to their individual equipment sub-group. If we consider the 11 common units as "true" outliers, and look at the top-20 or less in the outlier results, the recall is generally better for RB-outlier when we compare the three algorithms.

Execution time is not reported or discuss because all the three algorithms are comparable. While LOF uses an index structure like R* trees to find nearest neighbours, this index structure collapses with high dimensional spaces and the time complexity of all three methods becomes in the order $O(N^2)$ when the dimensionality is above 15 or 20 dimensions. In those cases, outlier analysis can benefit from linear approaches approximating k-nearest neighbours such as [8, 9, 10].

7 Summary and Conclusion

Outlier mining provides unprecedented advantages for detecting inconsistent records from a large database, which could not be possibly accomplished with traditional statistical techniques. Nevertheless the current popular outlier mining algorithms, when applied for engineering applications are neither efficient nor effective because of their domain-dependent parameters. The outlier notions in the current algorithms are targeted at either global or local outliers, but in engineering data, the data is typically noisy and clusters in the dataset are not well bounded as in the synthetic datasets used by some authors. At the same time, the current outlier mining algorithms cannot efficiently detect outliers from a dataset containing clusters of arbitrary shapes, such as in the engineering data.

The resolution-based outlier definition and the nonparametric RB-outlier mining algorithm RB-MINE introduced in this paper are more suited to engineering data when compared with the popular DB-outlier and LOF-outlier schemes. The algorithm overcomes the problems stated above and can be used for robust outlier detection in a wide variety of multi-dimensional datasets in engineering data.

The ability of mining top-N outliers based on ROF provides a mechanism of ranking by the degree of "interestingness". Problems in the records can be identified by looking at the top listed outliers for further investigation. The outlier mining technique we propose is expected to eliminate the burden of domain experts spent on estimating and tuning unknown parameters.

References

[1] Raz O., Buchheit R., Shaw M., Koopman P. and Faloutsos C. (2004). "Detecting Semantic Anomalies in Truck Weigh-in-Motion Traffic Data Using Data Mining." *Journal of Computing in Civil Engineering*, ASCE. Vol. 18, No. 4. pp.291~300.

[2] Knorr E., and Ng R. (1998). "Algorithms for Mining Distance-based Outliers in Large Datasets." *Proc. of 24th International Conference on Very Large Databases*.

[3] Breunig M., Kriegel H., Ng R. and Sander J. (2000). "LOF: Identifying Density-Based Local Outliers." *Proc. of ACM SIGMOD 2000 International Conference on Management of Data*, Dallas, TX.

[4] Tang J., Chen Z., Fu A. and Cheung D. (2002). "Enhancing Effectiveness of outlier Detections for Low Density Patterns." *Proc. of the 6th Pacific-Asia Conference on Advances in Knowledge Discovery and Data Mining*, Taipei, Taiwan. pp. 535 - 548

[5] Foss A. and Zaïane O. (2002). "A Parameterless Method for Efficiently Discovering Clusters of arbitrary Shape in Large Datasets." *Proc. of 2002 IEEE International Conference on Data Mining (ICDM'02)*, Maebashi City, Japan

[6] Hawkins D. (1980). "Identification of Outliers". Chapman and Hall, London. pp.1.

[7] Ramaswamy S., Rastogi R. and Shim K. (2000) "Efficient Algorithms for Mining Outliers from Large Data Sets." In Proc. Of the ACM SIGMOD International Conference on Management of Data, Dallas, TX.

[8] Goldstein J. and Ramakrishnan R. (2000) "Constrast Polots and P-Sphere Trees: Space vs. Time in Nearest Neighbor Searches." In Proc. 26th VLDB conference.

[9] Kushilevitz E., Ostrovsky R. and Rabani Y., (1998) "Efficient Search for Approximate Nearest Neighbor in High Dimensional Spaces", STOC'98.

[10] Liu T., Moore A.W., Gray A., and Wang K. (2004) "An Investigation of Practical Approximate Nearest Neighbor Algorithms", NIPS, December.

A Fast Greedy Algorithm for Outlier Mining

Zengyou He[1], Shengchun Deng[1], Xiaofei Xu[1], and Joshua Zhexue Huang[2]

[1] Department of Computer Science and Engineering, Harbin Institute of Technology, China
zengyouhe@yahoo.com, dsc@hit.edu.cn, xiaofei@hit.edu.cn
[2] E-Business Technology Institute, The University of Hong Kong, Hong Kong
jhuang@eti.hku.hk

Abstract. The task of outlier detection is to find small groups of data objects that are exceptional when compared with rest large amount of data. Recently, the problem of outlier detection in categorical data is defined as an optimization problem and a local-search heuristic based algorithm (LSA) is presented. However, as is the case with most iterative type algorithms, the LSA algorithm is still very time-consuming on very large datasets. In this paper, we present a very fast greedy algorithm for mining outliers under the same optimization model. Experimental results on real datasets and large synthetic datasets show that: (1) Our new algorithm has comparable performance with respect to those state-of-the-art outlier detection algorithms on identifying true outliers and (2) Our algorithm can be an order of magnitude faster than LSA algorithm.

1 Introduction

In contrast to traditional data mining task that aims to find the general pattern applicable to the majority of data, outlier detection targets the finding of the rare data whose behavior is very exceptional when compared with rest large amount of data. Studying the extraordinary behavior of outliers can uncover valuable knowledge hidden behind them and aid the decision makers to make profit or improve the service quality. Thus, mining for outliers is an important data mining research with numerous applications, including credit card fraud detection, discovery of criminal activities in electronic commerce, weather prediction, and marketing.

A well-quoted definition of outliers is firstly given by Hawkins [1]. This definition states: an outlier is an observation that deviates so much from other observations as to arouse suspicion that it was generated by a different mechanism. With increasing awareness on outlier detection in data mining literature, more concrete meanings of outliers are defined for solving problems in specific domains [3-22].

However, conventional approaches do not handle categorical data in a satisfactory manner, and most existing techniques lack for a solid theoretical foundation or assume underlying distributions that are not well suited for exploratory data mining applications. To fulfill this void, the problem of outlier detection in categorical data is defined as an optimization problem as follows [22]: finding a subset of k objects such that the expected entropy of the resultant dataset after the removal of this subset is minimized.

In the above optimization problem, an exhaustive search through all possible solutions with k outliers for the one with the minimum objective value is costly since for n objects and k outliers there are (n,k) possible solutions. To get a feel for the quality-time tradeoffs involved, a local search heuristic based algorithm (LSA) is presented in [22]. However, as is the case with most iterative type algorithms, the LSA algorithm is still very time-consuming on very large datasets.

In this paper, we present a very fast greedy algorithm for mining outliers under the same optimization model. Experimental results on real datasets and large synthetic datasets show that: (1) Our algorithm has comparable performance with respect to those state-of-the-art outlier detection algorithms on identifying true outliers and (2) Our algorithm can be an order of magnitude faster than LSA algorithm.

The organization of this paper is as follows. First, we present related work in Section 2. Problem formulation is provided in Section 3 and the greedy algorithm is introduced in Section 4. The empirical studies are provided in Section 5 and a section of concluding remarks follows.

2 Related Work

Statistical model-based methods, such as *distribution-based* methods [1,5] and *depth-based* methods [6], are rooted from the statistics community. In general, underlying distributions of data are assumed known a priori in these methods. However, such assumption is not appropriate in real data mining applications. *Distance based* methods [7-9] and *density based* methods [10,11] are recently proposed methods for mining outliers in large databases. However, they primarily focused on databases containing real-valued attributes. *Clustering-based* outlier detection techniques regarded *small* clusters as outliers [12, 14] or identified outliers by removing clusters from the original dataset [13]. *Sub-Space based* methods aim to find outliers effectively from high dimensional datasets [3,4]. *Support vector* based methods [15,16] and *neural network based* methods [17,18] are also widely used in outlier detection. *Outlier ensemble* based methods are investigated recently in [24,25].

The preceding methods may be considered as traditional in the sense that they define an outlier without regard to class membership. However, in the context of supervised learning (where data have class labels attached to them) it makes sense to define outliers by taking such information into account. The problem of class outlier detection is considered in [19-21].

3 Problem Formulation

Entropy is the measure of information and uncertainty of a random variable [2]. If X is a random variable, and $S(X)$ the set of values that X can take, and $p(x)$ the probability function of X, the entropy E (X) is defined as shown in Equation (1).

$$E(X) = -\sum_{x \in S(X)} p(x) \log(p(x)) \cdot \qquad (1)$$

The entropy of a multivariable vector $\hat{x} = \{X_1, ..., X_m\}$ can be computed as shown in Equation (2).

$$E(\hat{x}) = -\sum_{x_1 \in S(X_1)} \cdots \sum_{x_m \in S(X_m)} p(x_1, ..., x_m) \log(p(x_1, ..., x_m)) \cdot \qquad (2)$$

The problem we are trying to solve can be formulated as follows [22]. Given a dataset D of n points $\hat{p}_1, ..., \hat{p}_n$, where each point is a multidimensional vector of m categorical attributes, i.e., $\hat{p}_i = (p_i^1, ..., p_i^m)$, and given an integer k, we would like to find a subset $O \subseteq D$ with size k, in such a way that we minimize the entropy of $D - O$. That is,

$$\min_{O \subseteq D} E(D - O) \quad \text{Subject to } |O| = k. \qquad (3)$$

In this problem, we need to compute the entropy of a set of records using Equation (2). To make computation more efficient, we make a simplification in the computation of entropy of a set of records. We assume the independences among the attributes, transforming Equation (2) into Equation (4). That is, the joint probability of combined attribute values becomes the product of the probabilities of each attribute, and hence the entropy can be computed as the sum of entropies of the attributes.

$$E(\hat{x}) = -\sum_{x_1 \in S(X_1)} \cdots \sum_{x_m \in S(X_m)} p(x_1, ..., x_m) \log(p(x_1, ..., x_m)) = E(X_1) + E(X_2) + ... + E(X_n) \cdot \qquad (4)$$

4 The Greedy Algorithm

In this section, we present a greedy algorithm, denoted by greedyAlg1, which is effective and efficient on identifying outliers.

4.1 Overview

Our greedyAlg1 algorithm takes the number of desired outliers (supposed to be k) as input and selects points as outliers in a greedy manner. Initially, the set of outliers (denoted by OS) is specified to be empty and all points are marked as non-outlier. Then, we need k scans over the dataset to select k points as outliers. In each scan, for each point labeled as non-outlier, it is temporally removed from the dataset as outlier and the entropy objective is re-evaluated. A point that achieves *maximal entropy impact*, i.e., the maximal decrease in entropy experienced by removing this point, is selected as outlier in current scan and added to OS. The algorithm terminates when the size of OS reaches k.

4.2 Data Structure

Given a dataset D of n points $\hat{p}_1, \ldots, \hat{p}_n$, where each point is a multidimensional vector of m categorical attributes, we need m corresponding hash tables as our basic data structure. Each hash table has attribute values as keys and the frequencies of attribute values as referred values. Thus, in $O(1)$ expected time, we can determine the frequency of an attribute value in corresponding hash table.

4.3 The Algorithm

Fig.1 shows the greedyAlg1 algorithm. The collection of records is stored in a file on the disk and we read each record t in sequence.

In the initialization phase of the greedyAlg1 algorithm, each record is labeled as non-outlier and hash tables for attributes are also constructed and updated (Step 01-04).

In the greedy procedure, we need to scan the dataset for k times to find exact k outliers, i.e., one outlier is identified in each pass. In each scan over dataset, we read each record t that is labeled as non-outlier, its label is changed to outlier and the changed entropy value is computed. A record that achieves *maximal entropy impact* is selected as outlier in current scan and added to the set of outliers (Step 05-13).

In this algorithm, the key step is computing the changed value of entropy. In the following Theorem, we show that the decreased entropy value is only dependent on the attribute values of the record to be temporally removed.

Theorem 1: Suppose the number of records remained in D is n_l, the record $\hat{p}_i = (p_i^1, \ldots, p_i^m)$ is to be temporally removed, and the current frequency count of each attribute value p_i^j is denoted by $f(p_i^j)$. The decreased entropy value is determined by
$$\sum_{w=1}^{m} (\frac{f(p_i^w)-1}{n_l-1} \log \frac{f(p_i^w)-1}{n_l-1} - \frac{f(p_i^w)}{n_l-1} \log \frac{f(p_i^w)}{n_l-1})$$

Proof: The entropy value produced by attribute X_j before removing the record is: $\sum_{t \in S(X_j)} (f(t)/n_l) \log(f(t)/n_l)$. After removing it, the entropy value becomes
$$\sum_{t \in S(X_j), t \neq p_i^j} (-f(t)/(n_l-1)) \log(f(t)/(n_l-1)) - (f(p_i^j)-1)/(n_l-1) \log(f(p_i^j)-1)/(n_l-1)$$. Then the decreased entropy value is: $Ed(j) + \frac{f(p_i^j)-1}{n_l-1} \log \frac{f(p_i^j)-1}{n_l-1} - \frac{f(p_i^j)}{n_l-1} \log \frac{f(p_i^j)}{n_l-1}$, where $Ed(j) = \sum_{t \in S(X_j)} ((f(t)/(n_l-1)) \log(f(t)/(n_l-1)) - f(t)/n_l \log f(t)/n_l)$ is a constant in current iteration. Hence, Theorem results by considering all attributes.

With the use of hashing technique, in $O(1)$ expected time, we can determine the frequency of an attribute value in corresponding hash table. Hence, we can determine the decreased entropy value in $O(m)$ expected time since the changed value is only dependent on the attribute values of the record to be temporally removed.

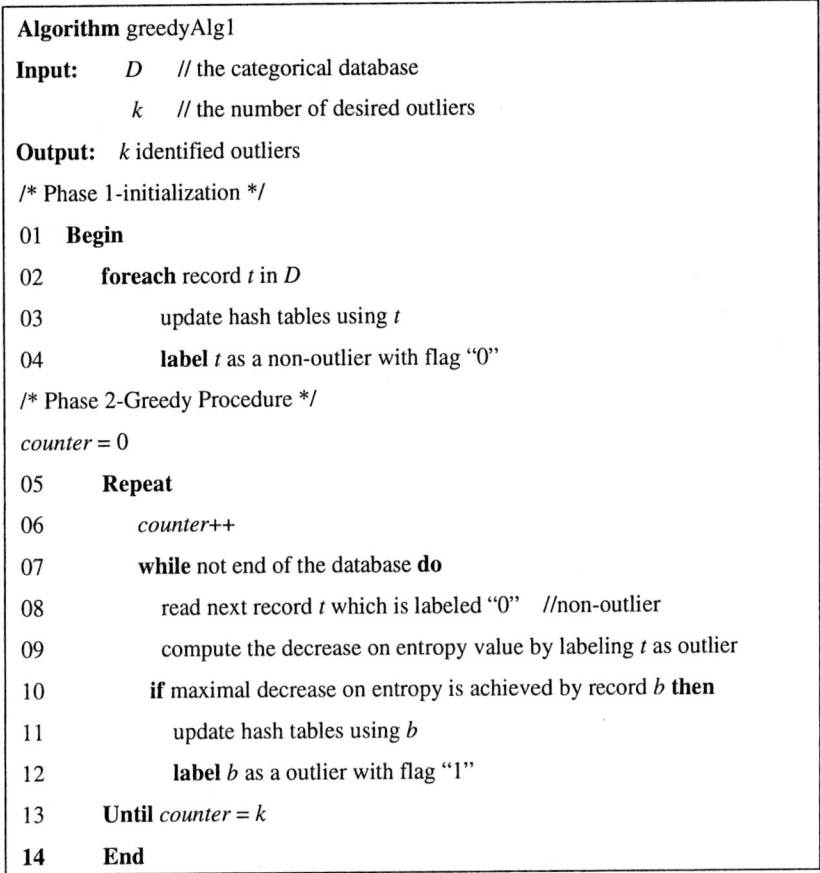

Fig. 1. The greedyAlg1 algorithm

4.4 Time and Space Complexities

Worst-case analysis: The time and space complexities of the greedyAlg1 algorithm depend on the size of dataset (n), the number of attributes (m), the size of every hash table and the number of outliers (k).

To simplify the analysis, we will assume that every attribute has the same number of distinct attributes values, p, Then, in the worst case, in the initialization phase, the time complexity is $O(nmp)$. In the greedy procedure, since the computation of value change on entropy requires at most $O(mp)$ and hence this phase has time complexity $O(nkmp)$. Totally, the algorithm has time complexity $O(nkmp)$ in worst case.

The algorithm only needs to store m hash tables and the dataset in main memory, so the space complexity of our algorithm is $O((p + n)m)$.

Practical analysis: Categorical attributes usually have *small* domains. An important of implication of the compactness of categorical domains is that the parameter, p, can be regarded to be very small. And the use of hashing technique also reduces the impact of p, as discussed previously, we can determine the frequency of an attribute value in $O(1)$ expected time, So, in practice, the time complexity of greedyAlg1 can be expected to be $O(nkm)$.

The above analysis shows that the time complexity of greedyAlg1 is linear to the size of dataset, the number of attributes and the number of outliers, which make this algorithm scalable. Previous LSA algorithm presented in [22] has the time complexity $O(nkmI)$, which is much slower than our algorithm since I (the number of iterations in LSA) is usually larger than 10.

5 Experimental Results

A comprehensive performance study has been conducted to evaluate our greedyAlg1 algorithm. In this section, we describe those experiments and their results. We ran our algorithm on real-life datasets obtained from the UCI Machine Learning Repository [23] to test its performance against other algorithms on identifying true outliers. In addition, some large synthetic datasets are used to demonstrate the scalability of our algorithm.

5.1 Experiment Design and Evaluation Method

Following the experimental setup in [22], we also used two real life datasets (*lymphography* and *cancer*) to demonstrate the effectiveness of our algorithm against *FindFPOF* algorithm [4], *FindCBLOF* algorithm [14], *KNN* algorithm [8] and LSA algorithm [22]. In addition, on the *cancer* dataset, we add the results of *RNN* based outlier detection algorithm that are reported in [17] for comparison, although we didn't implement the *RNN* based outlier detection algorithm.

For all the experiments, the two parameters needed by *FindCBLOF* algorithm are set to 90% and 5 separately as done in [14]. For the *KNN* algorithm [8], the results were obtained using the *5-nearest-neighbour*; For *FindFPOF* algorithm [4], the parameter *mini-support* for mining frequent patterns is fixed to 10%, and the maximal number of items in an itemset is set to 5. Since the LSA algorithm and greedyAlg1 are parameter-free (besides the number of desired outliers), we don't need to set any parameters.

As pointed out by Aggarwal and Yu [3], one way to test how well the outlier detection algorithm worked is to run the method on the dataset and test the percentage of points which belong to the rare classes. If outlier detection works well, it is expected that the rare classes would be over-represented in the set of points found. These kinds of classes are also interesting from a practical perspective.

Since we know the true class of each object in the test dataset, we define objects in small classes as rare cases. The number of rare cases identified is utilized as the assessment basis for comparing our algorithm with other algorithms.

5.2 Results on Lymphography Data

The first dataset used is the Lymphography data set, which has 148 instances with 18 attributes. The data set contains a total of 4 classes. Classes 2 and 3 have the largest number of instances. The remained classes are regarded as rare class labels for they are small in size. The corresponding class distribution is illustrated in Table 1.

Table 1. Class distribution of lymphography data set

Case	Class codes	Percentage of instances
Commonly Occurring Classes	2, 3	95.9%
Rare Classes	1, 4	4.1%

Table 2 shows the results produced by different algorithms. Here, the *top ratio* is ratio of the number of records specified as *top-k* outliers to that of the records in the dataset. The *coverage* is ratio of the number of detected rare classes to that of the rare classes in the dataset. For example, we let LSA algorithm find the *top 7* outliers with the top ratio of 5%. By examining these 7 points, we found that 6 of them belonged to the rare classes.

In this experiment, both the greedyAlg1 algorithm and LSA algorithm performed the best for all cases and can find all the records in rare classes when the *top ratio* reached 5%. In contrast, the *KNN* algorithm achieved this goal with the *top ratio* at 10%, which is almost the twice for that of our algorithm.

From the above results, we can see that greedyAlg1 algorithm achieves at least the same level performance as that of LSA algorithm on Lymphography data set.

Table 2. Detected rare classes in lymphography data set

Top Ratio (Number of Records)	Number of Rare Classes Included (Coverage)				
	GreedyAlg1	LSA	FindFPOF	FindCBLOF	KNN
5% (7)	6(100%)	6(100%)	5(83%)	4 (67%)	4 (67%)
10%(15)	6(100%)	6(100%)	5(83%)	4 (67%)	6(100%)
11%(16)	6(100%)	6(100%)	6(100%)	4 (67%)	6(100%)
15%(22)	6(100%)	6(100%)	6 (100%)	4 (67%)	6(100%)
20%(30)	6(100%)	6(100%)	6 (100%)	6 (100%)	6(100%)

5.3 Results on Wisconsin Breast Cancer Data

The second dataset used is the Wisconsin breast cancer data set, which has 699 instances with 9 attributes. In this experiment, all attributes are considered as categorical. Each record is labeled as *benign* (458 or 65.5%) or *malignant* (241 or 34.5%). We follow the experimental technique of Harkins, et al. [17,18] by removing some of the *malignant* records to form a very unbalanced distribution; the resultant dataset had 39 (8%) *malignant* records and 444 (92%) *benign* records (the resultant dataset is available at: http://research.cmis.csiro.au/rohanb/outliers/breast-cancer/). The corresponding class distribution is illustrated in Table 3. We also consider the *RNN* based outlier detection algorithm on this dataset, whose results are reproduced from [17,18].

Table 4 shows the results produced by the different algorithms. Clearly, among all of these algorithms, *RNN* performed the worst in most cases. In comparison to other algorithms, greedyAlg1 preformed very well in average. Hence, this experiment also demonstrates the effectiveness of greedyAlg1 algorithm.

Table 3. Class distribution of wisconsin breast cancer data set

Case	Class codes	Percentage of instances
Commonly Occurring Classes	1	92%
Rare Classes	2	8%

Table 4. Detected malignant records in wisconsin breast cancer dataset

Top Ratio (Number of Records)	Number of Rare Classes Included (Coverage)					
	GreedyAlg1	LSA	FindFPOF	FindCBLOF	RNN	KNN
1%(4)	4 (10.26%)	4 (10.26%)	3(7.69%)	4 (10.26%)	3 (7.69%)	4 (10.26%)
2%(8)	7 (17.95%)	8 (20.52%)	7 (17.95%)	7 (17.95%)	6 (15.38%)	8 (20.52%)
4%(16)	15(38.46%)	15(38.46%)	14 (35.90%)	14 (35.90%)	11 (28.21%)	16(41%)
6%(24)	22(56.41%)	22(56.41%)	21 (53.85%)	21 (53.85%)	18 (46.15%)	20(51.28%)
8%(32)	27 (69.23%)	29(74.36%)	28(71.79%)	27 (69.23%)	25 (64.10%)	27(69.23%)
10%(40)	33(84.62%)	33(84.62%)	31(79.49%)	32 (82.05%)	30 (76.92%)	32(82.05%)
12%(48)	36(92.31%)	38 (97.44%)	35 (89.74%)	35 (89.74%)	35 (89.74%)	37(94.87%)
14%(56)	39 (100%)	39 (100%)	39 (100%)	38 (97.44%)	36 (92.31%)	39 (100%)
16%(64)	39 (100%)	39 (100%)	39 (100%)	39 (100%)	36 (92.31%)	39 (100%)
18%(72)	39 (100%)	39 (100%)	39 (100%)	39 (100%)	38 (97.44%)	39 (100%)
20%(80)	39 (100%)	39 (100%)	39 (100%)	39 (100%)	38 (97.44%)	39 (100%)
25%(100)	39 (100%)	39 (100%)	39 (100%)	39 (100%)	38 (97.44%)	39 (100%)
28%(112)	39 (100%)	39 (100%)	39 (100%)	39 (100%)	39 (100%)	39 (100%)

Although the performance of greedyAlg1 algorithm on identifying true outliers on this dataset is not so good as that of the LSA algorithm in two cases, but their performance are almost identical. And as we will show in next Section, our algorithm is very fast for larger dataset, which is more important in data mining applications.

5.4 Scalability Tests

The purpose of this experiment was to test the scalability of the greedyAlg1 algorithm against LSA algorithm when handling very large datasets. A synthesized categorical dataset created with the software developed by Dana Cristofor (The source codes are public available at: http://www.cs.umb.edu/~dana/GAClust/index.html) is used. The data size (i.e., number of rows), the number of attributes and the number of classes are the major parameters in the synthesized categorical data generation, which were set to be 100,000, 10 and 10 separately. Moreover, we set the random generator seed to 5. We will refer to this synthesized dataset with name of DS1.

We tested two types of scalability of the greedyAlg1 algorithm and LSA algorithm on DS1 dataset. The first one is the scalability against the number of objects for a given number of outliers and the second is the scalability against the number of outliers for a given number of objects. Both algorithms were implemented in Java. All experiments were conducted on a Pentium4-2.4G machine with 512 M of RAM and running Windows 2000. Fig. 2 shows the results of using greedyAlg1 and LSA to find 30 outliers with different number of objects. Fig. 3 shows the results of using two algorithms to find different number of outliers on DS1 dataset.

One important observation from these figures was that the run time of greedyAlg1 algorithm tends to increase linearly as both the number of records and the number of

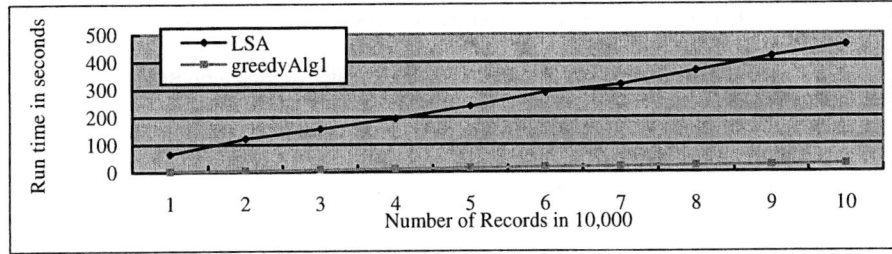

Fig. 2. Scalability to the number of objects when mining 30 outliers from DS1 dataset

outliers are increased, which verified our claim in Section 4.4. In addition, greedyAlg1 algorithm is always faster than LSA algorithm and can be at least an order of magnitude faster than LSA in most cases.

Hence, we are confident to claim that greedyAlg1 algorithm is suitable for mining very large dataset, which is very important in real data mining applications.

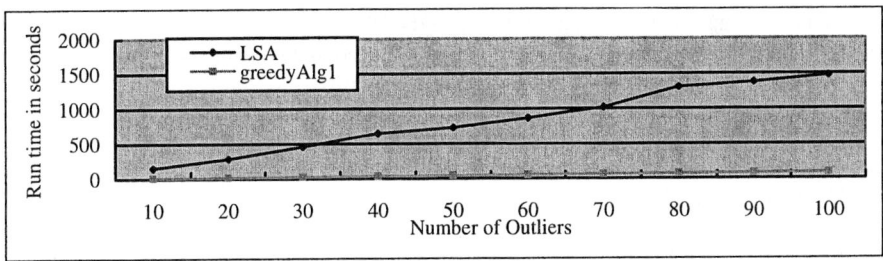

Fig. 3. Scalability to the number of outliers when mining outliers from DS1 dataset

6 Conclusions

Conventional outlier mining algorithms do not handle categorical data in a satisfactory manner. To fulfill this void, this paper presents a very fast greedy algorithm for mining outliers. Experimental results on real datasets and large synthetic datasets demonstrate the superiority of our new algorithm.

Acknowledgements

This work was supported by the High Technology Research and Development Program of China (No. 2004AA413010, No. 2004AA413030) and the IBM SUR Research Fund.

References

1. Hawkins, D.: Identification of Outliers. Chapman and Hall, Reading, London, 1980
2. Shannon, C.E.: A Mathematical Theory of Communication. Bell System Technical Journal (1948) 379-423

3. Aggarwal, C., Yu, P.: Outlier Detection for High Dimensional Data. In: Proc. of SIGMOD'01, pp. 37-46, 2001
4. He,Z., Xu, X., Huang, J., Deng, S.: A Frequent Pattern Discovery Based Method for Outlier Detection. In: Proc. of WAIM'04, LNCS 3129, pp. 726-732, 2004
5. Barnett, V., Lewis, T.: Outliers in Statistical Data. John Wiley and Sons, New York, 1994
6. Johnson, T., Kwok, I., Ng, R.; Fast Computation of 2-Dimensional Depth Contours. In: Proc. of KDD'98, pp.224-228, 1998
7. Knorr, E., Ng R., Tucakov, T.: Distance-Based Outliers: Algorithms and Applications. VLDB Journal **8(3-4)** (2000) 237-253
8. Ramaswamy, S., Rastogi, R., Kyuseok, S.: Efficient Algorithms for Mining Outliers from Large Data Sets. In: Proc. of SIGMOD'00, pp. 93-104,2000
9. Bay, S. D., Schwabacher, M.: Mining Distance Based Outliers in Near Linear Time with Randomization and a Simple Pruning Rule. In: Proc of KDD'03, pp.29-38, 2003
10. Breunig, M. M., Kriegel, H. P., Ng, R. T., Sander, J.: LOF: Identifying Density-Based Local Outliers. In: Proc. of SIGMOD'00, pp. 93-104, 2000
11. Papadimitriou, S., Kitagawa, H., Gibbons, P. B., Faloutsos, C.: Fast Outlier Detection Using the Local Correlation Integral. In: Proc of ICDE'03, 2003
12. Jiang, M. F., Tseng, S. S., Su, C. M.: Two-phase Clustering Process for Outliers Detection. Pattern Recognition Letters **22(6-7)** (2001) 691-700
13. Yu, D., Sheikholeslami, G., Zhang, A.: FindOut: Finding Out Outliers in Large Datasets. Knowledge and Information Systems **4(4)** (2002) 387-412
14. He, Z., Xu, X., Huang, J., Deng, S.: Discovering Cluster-based Local Outliers. Pattern Recognition Letters **24(9-10)** (2003) 1641-1650
15. Tax, D.M.J., Duin, R.P.W.: Support Vector Data Description. Pattern Recognition Letters **20(11-13)** (1999) 1191-1199
16. Schölkopf, B., Platt, J., Shawe-Taylor, J., Smola, A. J., Williamson, R.C.: Estimating the Support of a High Dimensional Distribution. Neural Computation **13(7)** (2001) 1443-1472
17. Harkins, S., He, H., Willams, G. J., Baster, R. A.: Outlier Detection Using Replicator Neural Networks. In: Proc. of DaWaK'02, pp. 170-180, 2002
18. Willams, G. J., Baster, R. A., He, H., Harkins, S., Gu, L.: A Comparative Study of RNN for Outlier Detection in Data Mining. In: Proc of ICDM'02, pp. 709-712, 2002
19. He, Z., Deng, S., Xu, X.: Outlier Detection Integrating Semantic Knowledge. In: Proc. of WAIM'02, LNCS 2419, pp.126-131, 2002
20. Papadimitriou, S., Faloutsos, C.: Cross-Outlier Detection. In: Proc of SSTD'03, pp.199-213, 2003
21. He, Z., Xu, X., Huang, J., Deng, S.: Mining Class Outliers: Concepts, Algorithms and Applications in CRM. Expert Systems with Applications **27(4)** (2004) 681-697
22. He, Z., Deng, S., Xu, X.: An Optimization Model for Outlier Detection in Categorical Data. In: Proc. of 2005 International Conference on Intelligent Computing, Lecture Notes in Computer Science 3644, pp.400-409, 2005
23. Merz, G., Murphy, P.: Uci Repository of Machine Learning Databases. http://www.ics.uci.edu/mlearn/MLRepository.html, 1996
24. Lazarevic, A., Kumar, V.: Feature Bagging for Outlier Detection. In: Proc. of KDD'05, pp. 157-166, 2005
25. He, Z., Deng, S., Xu, X.: A Unified Subspace Outlier Ensemble Framework for Outlier Detection. In: Proc. of WAIM'05, LNCS 3739, pp. 632-637, 2005

Ranking Outliers Using Symmetric Neighborhood Relationship

Wen Jin[1], Anthony K.H. Tung[2], Jiawei Han[3], and Wei Wang[4]

[1] School of Computing Science, Simon Fraser University
wjin@cs.sfu.ca
[2] Department of Computer Science, National University of Singapore
atung@comp.nus.edu.sg
[3] Department of Computer Science, Univ. of Illinois at Urbana-Champaign
hanj@cs.uiuc.edu
[4] Department of Computer Science, Fudan University
weiwang1@fudan.edu.cn

Abstract. Mining outliers in database is to find exceptional objects that deviate from the rest of the data set. Besides classical outlier analysis algorithms, recent studies have focused on mining **local outliers**, i.e., the outliers that have density distribution significantly different from their neighborhood. The estimation of density distribution at the location of an object has so far been based on the density distribution of its k-nearest neighbors [2, 11]. However, when outliers are in the location where the density distributions in the neighborhood are significantly different, for example, in the case of objects from a sparse cluster close to a denser cluster, this may result in wrong estimation. To avoid this problem, here we propose a simple but effective measure on local outliers based on a symmetric neighborhood relationship. The proposed measure considers both neighbors and reverse neighbors of an object when estimating its density distribution. As a result, outliers so discovered are more meaningful. To compute such local outliers efficiently, several mining algorithms are developed that detects top-n outliers based on our definition. A comprehensive performance evaluation and analysis shows that our methods are not only efficient in the computation but also more effective in ranking outliers.

1 Introduction

From a knowledge discovery standpoint, outliers are often more interesting than the common ones since they contain useful information underlying the abnormal behavior. Basically, an outlier is defined as an exceptional object that deviates much from the rest of the dataset by some measure. Outlier detection has many important applications in fraud detection, intrusion discovery, video surveillance, pharmaceutical test and weather prediction. Various data mining algorithms [1, 2, 3, 8, 10, 11, 12, 13, 15, 18, 17, 20, 21, 23, 24, 25] for outlier detection were proposed. The outlierness of an object typically appears to be more outstanding with respect to its local neighborhood. For example, a network intrusion might cause a significant spike in the number of network events within a low traffic period, but this spike might be insignificant when a period of high network traffic is also included in the comparison. In view of this, recent work on outlier detection has been focused on finding **local outliers**, which are essentially objects that have significantly lower density [1] than its local neighborhood [2]. As an objective measure, the degree of outlierness of an object p is defined to be **the ratio of its density and the average density of its neighboring objects** [2]. To quantify what are p's neighboring objects, users must specify a value k, and neighboring objects are defined as objects which are not further from p than p's k^{th} nearest objects [2]. As an example, let us look at Figure 1 in which k is given a value of 3. In this case, the three neighboring objects of p will have higher density than p and thus p will have a high degree of outlierness according to the definition in [2]. This is obviously correct based on our intuition.

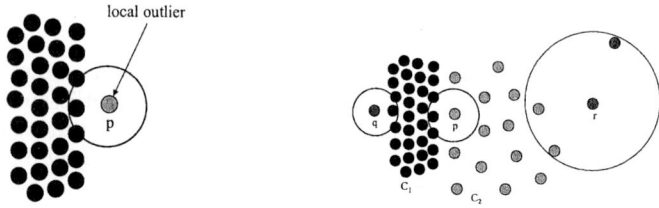

Fig. 1. A local outlier, p **Fig. 2.** Comparing the outlierness of p, q, r

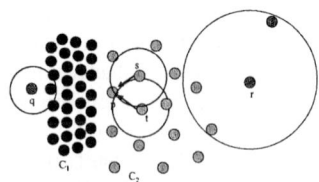

Fig. 3. Taking RNNs of p into account

Unfortunately, the same cannot hold in more complex situation. Let us look at the following example.

Example 1: We consider Figure 2 in which p is in fact part of a sparse cluster C_2 which is near the dense cluster C_1. Compared to objects q and r, p obviously displays less outlierness. However, if we use the measure proposed in [2], p could be mistakenly regarded to having stronger outlierness in the following two cases:
Case I: The densities of the nearest neighboring objects for both p and q are the same, but q is slightly closer to cluster C_1 than p. In this case, p will have a stronger outlierness measure than q, which is obviously wrong.

---
[1] The density of an object p is defined as $1/k_{dist}(p)$ where k is a user-supplied parameter and $k_{dist}(p)$ is the distance of the kth nearest object to p.
[2] Note that p's k^{th} nearest neighbor might not be unique and thus p could have more than k neighboring objects.

Case II: Although the density of r is lower than p, the average density of its neighboring objects (consisting of 2 objects from C_2 and an outlier) is less than those of p. Thus, when the proposed measure is computed, p could turn out to have a stronger outlierness measure than r, which again is wrong.

Note that the two cases we described are not only applicable to p but also to the two objects above and below p. In general, any member of C_2 that is lying near the border between the two clusters could have been misclassified as showing stronger outlierness than q and r. □

From these examples, we can see that existing outlierness measure is not easily applicable to complex situation in which the dataset contains multiple clusters with very different density distribution. The reason for the above problem lies in the inaccurate estimation for the density distribution of an object's neighborhood. In Figure 2, although p belongs to cluster C_2, it is closer to cluster C_1, and thus the estimation of p's neighborhood density distribution is derived from C_1 instead of C_2.

To get a better estimation of the neighborhood's density distribution, we propose to take both the nearest neighbors (**NNs**) and reverse nearest neighbors (**RNNs**) [14] into account. The RNNs of an object p are essentially objects that have p as one of their k nearest neighbors. By considering the symmetric neighborhood relationship of both NN and RNN, the space of an object influenced by other objects is well determined, the densities of its neighborhood will be reasonably estimated, and thus the outliers found will be more meaningful. As a simple illustration in Figure 3 which depicts the same situation as Figure 2, we show that p has two RNNs: s and t. This distinguishes it from q which has no RNNs, and r which has only an outlier as its RNNs. Later on in this paper, we will show how such an observation can be incorporated to ensure that the outlierness measure for p will indicate that it is a weaker outlier than both q and r. We now summarize our contributions in this paper:

(1) We propose the mining of outliers based on a symmetric neighborhood relationship. The proposed method considers the *influenced space* considering both neighbors and reverse neighbors of an object when estimating its neighborhood density distribution. To the best of our knowledge, previous work of outlier detection has not considered the effect of RNN. Such a symmetric relationship between NNs and RNNs will make the outlierness measurement more robust and semantically correct comparing to the existing method.

(2) We assign each object of database the degree of being **INFLuenced Outlierness(INFLO)**. The higher **INFLO** is, the more likely that this object is an outlier. The lower **INFLO** is, the more likely that this object is a member of a cluster. Specifically, $INFLO \approx 1$ means the object locates in the core part of a cluster.

(3) We present several efficient algorithms to mining top-n outliers based on **INFLO**. To reduce the expensive cost incurred by a large number of KNN and RNN search, a two-way search method is developed by dynamically pruning those objects with value $INFLO \approx 1$ during the search process. Furthermore, we take advantage of the micro-cluster [11] technique to compress dataset for efficient symmetric queries, and use two-phase pruning method to prune out those objects which will never be among the top-n outliers.

(4) Last but not the least, we give a comprehensive performance evaluation and analysis on synthetic and real data sets. It shows that our method is not only efficient and scalable in performance, but also effective in ranking meaningful outliers.

The rest of this paper is organized as follows. In section 2, we formally define a new outlier measurement using symmetric neighborhood relationship and discuss some of its important properties. In section 3, we propose efficient methods for mining and ranking outliers in databases. In section 4, a comprehensive performance evaluation is made and the results are analyzed. Related work is discussed in section 5 and section 6 concludes the paper.

2 Influential Measure of Outlierness by Symmetric Relationship

In this section, we will introduce our new measure and related properties. The following notations will be used in the remaining of the paper. Let D be a database of size N, let p, q and o be some objects in D, and let k be a positive integer. We use $d(p,q)$ to denote the **Euclidean distance** between objects p and q.

Definition 1 (k-distance and nearest neighborhood of p). *The k-distance of p, denoted as $k_{dist}(p)$, is the distance $d(p,o)$ between p and o in D, such that: (1) at least for k objects $o' \in D$ it holds that $d(p,o') \leq d(p,o)$, and (2) at most for $(k-1)$ objects $o' \in D$ it holds that $d(p,o') < d(p,o)$. The k-nearest neighborhood of p, $NN_k(p)$ is a set of objects X in D with $d(p,X) \leq k_{dist}(p)$: $NN_k(p) = \{X \in D \setminus \{p\} | \ d(p,X) \leq k_{dist}(p))\}$.* □

Definition 2 (local density of p). *The density of p, denoted as $den(p)$, is the inverse of the k-distance of p, i.e., $den(p) = 1/k_{dist}(p)$.* □

Although the k-nearest neighbor of p may not be unique, $k_{dist}(p)$ is unique. Hence, the density of p is also unique. The nearest neighbor relation is not symmetric. For a given p, the nearest neighbors of p may not have p as one of their own nearest neighbors. As we discussed in Section 1, these neighbors should also be taken into account when the outlierness of p is computed. Therefore, we introduce the concept of reverse nearest neighbors [14] as follows.

Definition 3 (reverse nearest neighborhood of p). *The reverse k-nearest neighborhood RNN is an inverse relation which can be defined as: $RNN_k(p) = \{q | q \in D, p \in NN_k(q)\}$.* □

For any object $p \in D$, NN_k search always returns at least k results, while the RNN can be empty, or have one or more elements. By combining $NN_k(p)$ and

$RNN_k(p)$ together in a novel way, we form a local neighborhood space which will be used to estimate the density distribution around p. We call this neighborhood space the **k-influence space** for p, denoted as $IS_k(\mathrm{p})$.

Example 2: Figure 4 gives a simple description of how to obtain RNN in $\{p, q_1, q_2, q_3, q_4, q_5\}$ when $k = 3$. $NN_k(q_1) = \{p, q_2, q_4\}$, $NN_k(q_2) = \{p, q_1, q_3\}$, $NN_k(q_3) = \{q_1, q_2, q_5\}$, $NN_k(q_4) = \{p, q_1, q_2, q_5\}$, $NN_k(q_5) = \{q_1, q_2, q_3\}$. During the search of k-nearest neighbors of p, q_1, q_2, q_3, q_4 and q_5, $RNN_k(p) = \{q_1, q_2, q_4\}$ is incrementally built. Similarly, $RNN_k(q_1)$, $RNN_k(q_2)$, $RNN_k(q_3)$, $RNN_k(q_4)$ and $RNN_k(q_5)$ are found. Note that $NN_k(p) = \{q_1, q_2, q_4\} = RNN_k(p)$ (here $IS_3(p) = \{q_1, q_2, q_4\}$). If the value of k changes, $RNN_k(p)$ may not be equal to $NN_k(p)$, or totally different.

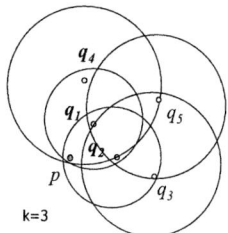

Fig. 4. RNN and Influence Space

Unlike the nearest neighborhood, the influence space for an object p contains influential objects affecting p, more precisely estimating density around p's neighborhood w.r.t. these objects.

Definition 4 (influenced outlierness of p). *The influenced outlierness is defined as:* $INFLO_k(p) = \frac{den_{avg}(IS_k(p))}{den(p)}$ *where* $den_{avg}(IS_k(p)) = \frac{\sum_{o \in IS_k(p)} den(o)}{|IS_k(p)|}$.

$INFLO$ is the ratio of the average density of objects in $IS_k(p)$ to p's local density. p's $INFLO$ will be very high if its density is much lower than those of its influence space objects. In this sense, p will be an outlier. We can assert p is a local outlier if $INFLO_k(p) > t$ where $t \gg 1$. On the other hand, objects with density very close to those in their influence space will have $INFLO \approx 1$. Without loss of generality, we assume that for any local outlier object q ($INFLO(q) > t$), we have $|RNN_k(q)| < j$(a value $< k$), and any non-local outlier p cannot belong to $RNN_k(q)$.

Lemma 1. *Given any object* $p, q \in D$, *if* $\max_{p' \in IS_k(p)} k_{dist}(p') < \min_{q' \in IS_k(q)} k_{dist}(q')$ *then* $den_{avg}(IS_k(p)) > den_{avg}(IS_k(q))$.

Proof. $den_{avg}(IS_k(p)) = \frac{\sum_{p' \in IS_k(p)} den(p')}{|IS_k(p)|} > \frac{|IS_k(p)| \cdot 1/\max_{p' \in IS_k(p)} k_{dist}(p')}{|IS_k(p)|} >$
$\frac{|IS_k(p)| \cdot 1/\min_{q' \in IS_k(q)} k_{dist}(q')}{|IS_k(p)|} = \frac{|IS_k(q)| \cdot 1/\min_{q' \in IS_k(q)} k_{dist}(q')}{|IS_k(q)|} \geq \frac{\sum_{q' \in IS_k(q)} den(q')}{|IS_k(q)|} =$
$den_{avg}(IS_k(q))$ □

Lemma 2. *For* $p \in D$, *if* $\frac{k_{dist}(p)}{max_{q' \in IS_k(p)} k_{dist}(q')} > t$, *then* p *is a local outlier.*

Proof. $INFLO_k(p) = \frac{den_{avg}(IS_k(p))}{den(p)} = \frac{\sum_{p' \in IS_k(p)} den(p')}{|IS_k(p)| \cdot den(p)} \geq$
$\frac{|IS_k(p)| \cdot 1/\max_{p' \in IS_k(p)} k_{dist}(p')}{|IS_k(p)| \cdot den(p)} = \frac{k_{dist}(p)}{max_{q' \in IS_k(p)} k_{dist}(q')} > t.$ □

Lemma 3. *For $p \in D$, if there exists $r \in RNN_k(p)$ such that $k_{dist}(p) \leq k_{dist}(r) \leq k_{dist}(q)$ where $q \in NN_k(RNN_k(p))$, $r \neq q$ and $\frac{den_{avg}(IS_k(q))}{k_{dist}(p)} > t$, then p is a local outlier.*

Proof. Since $k_{dist}(p) \leq k_{dist}(q)$, so $q \in NN_k(p) \cap RNN_k(p)$, thus $\max_{p' \in IS_k(p)} k_{dist}(p') = \max_{p' \in NN_k(p) \cup RNN_k(p)} k_{dist}(p') = \max_{p' \in RNN_k(p)} k_{dist}(p') \leq k_{dist}(r) = \min_{q' \in NN_k(q) \cup RNN_k(q)} k_{dist}(q') = \min_{q' \in IS_k(q)} k_{dist}(q')$. Based on Lemma 1, $den_{avg}(IS_k(p)) > den_{avg}(IS_k(q))$, so $INFLO_k(p) = \frac{den_{avg}(IS_k(p))}{den(p)} = den_{avg}(IS_k(p)) \cdot k_{dist}(p) > den_{avg}(IS_k(q)) \cdot k_{dist}(p) = \frac{den_{avg}(IS_k(q))}{k_{dist}(p)} > t$. So p is a local outlier. □

Lemma 4. *For $p \in D$, the value of $\frac{RNN_k(p) \cap NN_k(p)}{NN_k(p)}$ is proportional to the density value of p.*

Proof. Because the size of any cluster should be larger than k (usually $k = MinPts$ [2]), the higher the above ratio, the more influence for the local neighborhood to the object, and the higher density for this object. □

3 Mining Algorithms for Influence Outliers Using Symmetric Relationship

Essentially, mining influenced outliers is based on the problem of finding the influence space of objects, which is in KNN and RNN. In this section, we provide several techniques for finding influenced outliers, including the naive index-based method, the two-way search method and the micro-cluster method.

3.1 A Naive Index-Based Method

Finding influence outliers requires the operations of KNN and RNN for each object in the database, so the search cost is huge. If we maintain all the points in a spatial index like R-tree, the cost of range queries can be greatly reduced by the state-of-the-art pruning technique [19]. Suppose that we have computed the temporary $k_{dist}(p)$ by checking a subset of the objects, the value that we have is clearly an upper bound for the actual $k_{dist}(p)$. If the minimum distance between p and the MBR [3] of a node in the R-tree (called MinDist(p, MBR)) is greater than the $k_{dist}(p)$ value that we currently have, none of the objects in the subtree rooted under the node will be among the k-nearest neighbors of p. This optimization can prune entire sub-tree containing points irrelevant to the KNN search for p. Along with the search of KNN, the RNN of each object can be dynamically maintained in R-tree [14]. After building the index of KNN and RNN, the outlier influence degree can be calculated and ranked. The following algorithm is to mining top-n $INFLO$ by building KNN and RNN index within R-tree.

---

[3] Minimum bounding rectangle.

Algorithm 1 Index-based method.
Input: k, D, n, the *root* of R-tree.
Output: Top-n $INFLO$ of D.
Method:

1. FOR each object $p \in D$ DO
2. $MBRList = root$; $k_{dist}(p) = \infty$; $heap = 0$;
3.   WHILE (MBRList) != empty DO
4.   Delete 1st MBR from $MBRList$;
5.   IF ($1stMBR$ is a leaf) THEN
6.   FOR each object q in $1stMBR$ DO
7.   IF $(d(p,q) < k_{dist}(p))$ AND $(heap.size < k)$ THEN
8.   heap.insert(q);
9. $k_{dist}(p) = d(p, heap.top)$;
10.   ELSE
11.   Append MBR's children to MBRList;
12.   Sort nodeList by MinDist;
13.   FOR each MBR in MBRList DO
14.   IF $(k_{dist}(p) \leq MinDist(p, MBR))$ THEN
15.   Remove Node from MBRList;
16. FOR each object q in heap DO
17.   Add q into $NN_k(p)$, add p into $RNN_k(q)$;
18. FOR each object $p \in D$ DO
19.   Ascending sort top-n INFLO from KNN and RNN;

Here MBRs are stored in ascending order based on MinDist(p, MBR), as lines 11-12. The algorithm searches KNN_p only in those MBRs with MinDist smaller than the temporary $k_{dist}(p)$, otherwise these MBRs are pruned (lines 13-15). If any nearer object is located (lines 6-7), it will be inserted into the heap and the current $k_{dist}(p)$ will be updated (lines 8-9). Whenever $NN_k(p)$ are found, they are stored as p's nearest neighbors. Meanwhile, it need store p as a reverse nearest neighbor (lines 16-17). Finally, $INFLO$ is calculated based on KNN and RNN index.

3.2 A Two-Way Search Method

Two major factors hamper the efficiency of the previous algorithm. First, for any object p, RNN space cannot be determined unless all the other objects have finished nearest neighbor search. Second, large amount of extra storage is required on R-tree, where each object at least stores k pointers of its KNN, and stores m pointers (m varies from 0 to $o(k)$) for its RNN. The total space cost will be prohibitive. Therefore, we need reduce the computation cost for RNN and corresponding storage cost. By analyzing the characteristics of $INFLO$, it is clear that any object as a member of a cluster must have $INFLO \approx 1$ even without $INFLO$ calculation. So we can prune off these cluster objects, saving not only the computation cost but also the extra storage space.

Theorem 1. *For $p \in D$, if for each object $q \in NN_k(p)$, it always exists $p \in NN_k(q)$, then $INFLO_k(p) \approx 1$.*

Proof. Because for each $q \in NN_k(p)$, $p \in NN_k(q)$, p and its nearest neighbors are close to each other. They are actually in a mostly mutual-influenced neighborhood. Since k is potentially the number of objects forming a cluster, under this circumstance, p resides in core part of a cluster. □

To apply this theorem, we will first search p's k-nearest neighbor, then dynamically find the NN_k for each of these nearest neighbors. If $NN_k(NN_k(p))$ still contains p, which shows p is in a closely influenced space and is a core object of a cluster ($INFLO_k(p) \approx 1$), we can prune p immediately without searching corresponding RNN. Such a *early pruning* technique will improve the performance significantly. The two-way search algorithm is given as follows:

Algorithm 2 A Two-way search method.
Input: k, D, n, the *root* of R-tree, a threshold M.
Output: Top-n $INFLO$ of D
Method:

1. FOR each $p \in D$ DO
2. $count = |RNN_k(p)|$;
3.   IF *unvisited*(p) THEN
4. $S = getKNN(p)$; //*search k-nearest neighbors*
5.   *unvisited*(p) = $FALSE$;
6.   ELSE
7. $S = KNN(p)$; //*get nearest neighbors directly*
8.   FOR each object $q \in S$ DO
9.   IF *unvisited*(q) THEN
10. $T = getKNN(q)$; *unvisited*(q) = $FALSE$;
11.   IF $p \in T$ THEN
12.   Add q into $RNN_k(p)$;
13.   Add p into $RNN_k(q)$;
14. $count++$;
15.   IF $count \geq |S| * M$ THEN //*M is a threshold*
16.   Label p pruned mark;
17. FOR each object $p \in D'$ DO //*D' is unpruned database*
18.   Ascending sort top-n INFLO from KNN and RNN;

The algorithm aims to search and prune objects that are likely to have low $INFLO$, thus avoid unnecessary RNN search. The $|RNN_k(p)|$ is initialized to 0 for p. Search process is taken two directions, that is, from one object to its nearest neighbors, then to the new nearest neighbors (lines 8-14). If for p's nearest neighbors, their nearest neighbors' spaces contain p, or most of them contain p, p is a core object of a cluster and cannot be ranked as top-n outliers, and can be pruned (lines 15-16). Finally, top-n $INFLOs$ are calculated (lines 17-18).

3.3 A Micro-Cluster-Based Method

In order to further reduce the cost of distance computation, we introduce micro-cluster to represent close objects [11] so that the number of k-nearest neighbor search will be greatly reduced. The upper and lower bound of k-distance for each micro-cluster can be estimated in influenced space. Under the guidance of the two-way search, those micro-clusters which actually are "core parts" of clusters can be pruned and top-n outliers are ranked in the remaining dataset.

Definition 5. (MicroCluster) *The MicroCluster C for a d-dimensional dataset X is defined as the $(3 \cdot d + 2)$-tuple $(n, \overline{CF1(C)}, \overline{CF2(C)}, \overline{CF3(C)}, r)$, where $\overline{CF1}$ and $\overline{CF2}$ each corresponds to the linear sum and the sum of the squares of the data values for each dimension respectively. The number of data points $|C|$ is maintained in n, the centroid of $\overline{X_1} \ldots \overline{X_n}$ is $\overline{CF3(C)} = \frac{\overline{CF1(C)}}{n}$. The radius of the MicroCluster is $r = \max_{j=1}^{n} \sqrt{(\overline{X_j} - \overline{CF3(C)})^2}$.* □

[26] introduced an efficient clustering algorithm, BIRCH, with good linear scalability to the size of database, we borrow its basic idea to partitioning the database into micro-clusters. The detailed procedure can be referenced in [11]. The following theorem [11] can be used to estimate the lower and upper bound of k-distance of any object.

Theorem 2. *Let $p \in MC(n, c, r)$ and $MC_1(n_1, c_1, r_1), \ldots, MC_l(n_l, c_l, r_l)$ be a set of micro-clusters that could **potentially** contain the k-nearest neighbors of p. Each object o_i is treated as a micro-cluster $MC_i(1, o_i, 0)$. Thus we will now have $l + n - 1$ micro-clusters.*

1. *Let $\{d_{Min}(p, MC_1), \ldots, d_{Min}(p, MC_{l+n-1})\}$ be sorted in increasing order, then a lower bound on the k-distance of p, denoted as $\min k_{dist(p)}$ will be $d_{Min}(p, MC_i)$ such that $n_1 + \ldots + n_i \geq k$ and $n_1 + \ldots + n_{i-1} < k$*
2. *Let $\{d_{Max}(p, MC_1), \ldots, d_{Max}(p, MC_{l+n-1})\}$ be sorted in increasing order, then an upper bound on the k-distance of p, denoted as $\max k_{dist(p)}$ will be $d_{Max}(p, MC_i)$ such that $n_1 + \ldots + n_i \geq k$ and $n_1 + \ldots + n_{i-1} < k$.* □

The following is the micro-cluster based algorithm for mining top-n local outliers.

Algorithm 3 Micro-cluster method.
Input: A set of micro-clusters $MC_1, \ldots, MC_l, M$.
Output: Top-n $INFLO$ of D.
Method:

1. FOR each micro-cluster MC_i DO
2.   FOR each $p \in MC_i$ Do
3.   Get Max/Min of $k_{dist(p)}$; // based on theorem 2
4.   IF Min $k_{dist}(p) <$ Min$k_{dist}(MC_i)$ THEN
5.   Min $k_{dist}(MC_i) = $ Min$k_{dist}(p)$;
6.   IF Max $k_{dist}(p) >$ Max$k_{dist}(MC_i)$ THEN

7.   Max $k_{dist}(MC_i)$ = Max$k_{dist}(p)$;
8. FOR each micro-cluster MC_i DO
9. $count = |RNN_k(MC_i)|$;
10.   IF $unvisited(MC_i)$ THEN
11. $S = getKNN(MC_i)$; //search k-nearest micro-clusters
12. $unvisited(MC_i) = FALSE$;
13.   ELSE
14. $S = KNN(MC_i)$; //get nearest micro-clusters directly
15.   FOR each micro-cluster $q \in S$ DO
16.   IF $unvisited(q)$ THEN
17. $T = getKNN(q)$; $unvisited(q) = FALSE$;
18.   IF Min $k_{dist}(q) \geq$ Max $k_{dist}(MC_i)$ THEN
19.   Add q into $RNN_k(MC_i)$;
20.   Add MC_i into $RNN_k(q)$;
21. $count++$;
22.   IF $count \geq |S|*M$ THEN //M is a threshold
23.   Label MC_i pruned mark;
24 FOR each object $p \in$ unpruned micro-clusters MC' DO
25. Ascending sort top-n INFLO from KNN and RNN;

After building micro-clusters, the process of finding outliers is similar to the two-way search method. We simply treat each micro-cluster as a single object to search KNN. As the number of micro-clusters is much less than that of database objects, the computational cost will be saved a lot. The $|RNN_k(MC_i)|$ is initialized to 0 for each micro-cluster MC_i, and the lower/upper bound of k-distance of each MC_i is derived (lines 1-7) based on theorem 3.2. Then irrelevant objects in micro-clusters which cannot become top-n outliers are pruned if most of the k-nearest micro-clusters of a micro-cluster MC contain MC in their k-nearest micro-clusters as well, then MC will be located in the core part of clusters (lines 20-22) and could be removed. If the lower bound of k-distance for any MC's neighboring micro-cluster q is bigger than the upper bound of that for MC, then q belongs to MC's RNN (lines 18-21). By combining the two-way search and the micro-cluster technique, it achieves a significant improvement in performance.

4 Performance Evaluation

In this section, we will perform a comprehensive experimental evaluation on the efficiency and the effectiveness of our mining algorithm. We will compare our methods with the LOF method in [2] and show that our methods not only achieve a good performance but also identify more meaningful outliers than LOF. We perform tests on both real life data and synthetic data. Our real life dataset is the statistics archive of 2000-2002 National Hockey League (NHL), totally 22180 records with 12 dimensions[4]. Our synthetic datasets are generated

---
[4] http://www.usatoday.com/sports/hockey/stats/

based on multiple-gaussian distribution, where the cardinality varies from 1,000 to 1,000,000 tuples and the dimensionality varies from 2 to 18. The tests are run on 1.3GHZ AMD processor, with 512MB of main memory, under Windows 2000 advanced-server operating system. All algorithms are implemented by Microsoft Visual C++ 6.0.

Experiments on Effectiveness. To achieve a comprehensive understanding on the effectiveness of the **INFLO** measure, it is necessary to test on a series of datasets with different sizes and dimensions. We generate our dataset with complex density distribution by a mixture of Gaussian distribution. Most outliers detected by our methods are meaningful with good explanations, and some of them cannot be found by *LOF*. For easily illustrating, we just pick up a portion of 2-dimensional dataset containing a low density cluster A and a high density cluster B in Figure 5. The top-6 outliers are listed by *INFLO* and *LOF* respectively in Table 1.

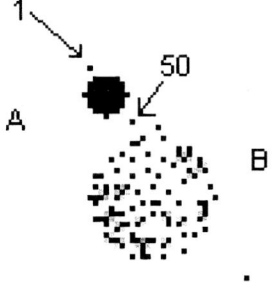

Fig. 5. A dataset

Table 1. Outliers Ranking

Rank	Index	LOF	Index	INFLO
1	147	3.47	147	17.34
2	101	2.80	101	8.899
3	146	2.56	146	8.81
4	50	1.74	1	4.50
5	65	1.57	50	3.52
6	4	1.45	16	3.03

Table 2. Outliers Ranking(INFLO)

Rank	INFLO	Player	Games	Goals	Shoot %
1	25.95	Nurminen	2	1	100
2	12.66	Lemieux	43	35	20.5
3	7.60	Holmstrom	76	16	21.6
4	7.25	Blake	67	19	7.1
5	7.03	MacInnis	59	12	5.5

Table 3. Outliers Ranking(LOF)

Rank	LOF	Player	Games	Goals	Shoot %
1	5.19	Nurminen	2	1	100
2	2.47	Jagr	81	52	16.4
3	2.61	Lemieux	43	35	20.5
4	2.31	McDonald	16	1	4.8
5	2.31	Skalde	19	1	4.2

Due to the limitation of space, we only show two instances. Table 1 lists the top 6 outliers based on the sample dataset in Figure 5, by both *LOF* and *INFLO* measures. The most outstanding outliers can be recognized by either measure. In this sample, 50 percentage of the top 6 outliers are the same points by both measures. When n is increased, *INFLO* will find even more different top outliers from *LOF*. By visual comparison, the top 6 outliers found by *INFLO* is more meaningful. Even for the same objects appeared in top-n lists of both measure, their position could be different and $INFLO$-based results are obviously more reasonable. In addition, $INFLO$ can detect outliers which can be overlooked by LOF. For instance, the 50th object and the 4th object have inversely ranking orders by different measure. LOF only considers nearest neighborhood as a density estimation space, and the NN of both the 1st and the 50th objects are in cluster A. Since the distance between the 50th object and A is larger than that of the 1st object and A, so the 50th object with low density is ranked as a higher outlier than the 1st with a high density. While $INFLO$ measure considers both NN and RNN, some objects of B will influence the 50th object, and thus make it being less outlierness than 1st object. It is clear that using $INFLO$ as outlierness measure preserves more semantics than using LOF. Another interesting phenomena in experiments is that $INFLO$ measure gives more rational indication for the outlier degree assignment. As an example, LOF value that are assigned to those bordering objects of a cluster has only a tiny difference with those in the core of a cluster. By $INFLO$, however, the bordering objects will have significantly larger $INFLO$ values than the core part of the same cluster while the value differences are smaller than objects in different cluster. Figure 6 presents such value differences curve by LOF and $INFLO$, in which the difference is evaluated by cluster bordering objects and cluster mean center.

In the following experiments, we run our proposed algorithms with NHL 2000-2002 playoff data (22180 tuples) to rank top-n exceptional players in NHL. The results are compared with those computed from LOF. We varied k from 10 to 50. Projection is done on dataset by randomly selecting dimensions, and the outlierness of hockey players is evaluated. For example, we focus on the statistics data in 3-dimensional subspace of Games played, Goals and Shooting percentage. Due to the limitation of space,

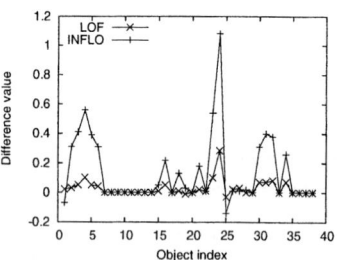

Fig. 6. LOF and INFLO

we only list top-5 players in Table 2 and Table 4. Lots of interesting and useful information can be found in our examination. For example, there are two players who are listed in both tables as top-5 outliers. Nurminen is the strongest outlier. Although he only took two games and got one point, his 100% shooting percentage dominated other two statistics numbers in comparison. As it happens in the synthetic dataset, we can still find some surprising outliers which cannot be identified by LOF. For example, Rob Blake ranks 4th in our method but is only ranked as the 31th outlier using LOF. Our reasoning for such surprising result is as follows. The variation of shooting percentage is usually small, since only a very few of players can become excellent shooter. Comparing to those players who have similar statistics number in Games Played and Goals dimensions, although Blake's shooting percentage is rather low, Blake is still not too far away from other player when viewed in term of distance. Thus based on LOF measure, Blake's could not be ranked in the top players. But the reason for him being a

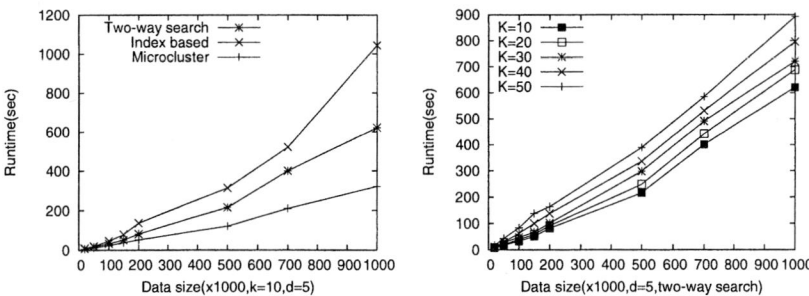

Fig. 7. Runtime vs datasize   **Fig. 8.** Effects of k

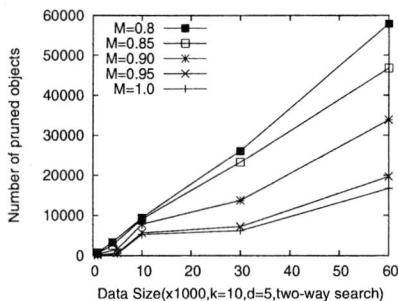

Fig. 9. Pruning results (1)

most exceptional player by $INFLO$ is that there is no such type of player whose Shooting Percentage is so low while having so many Goals. Actually, Blake is the only defence whose number of goals scored is over 12. He must have shot too many times in the games without getting goals.

Another interesting example is Jaromir Jagr, who scores in the 3rd position and ranks as the second outlier in LOF, but the 24th in our measure. The reason is that even though Jagr has a strong goaling capability and a big fame, there are over twenty players who have higher statistics than him in Shooting Percentage and Games. So objectively, he is not ranked as the most exceptional player during 2000-2002 seasons. *Note that we treat all the hockey data equal in the analysis not like hockey fans who always weigh goals much higher than other factors.*

Efficiency Issues of Experiments. We evaluate the efficiency of the proposed mining algorithms by varying the data size, dimension number, k and pruning parameter accordingly. Figure 7 shows the performance curves of different methods, along with the runtime (include CPU time and I/O time) corresponding to different size of dataset with 5 dimensions. It shows that the run time of three methods are similar when the number of tuple is less than 100k. When the data size increases to 200k or so, micro-cluster-based method is the best and the two-way search is better than index-based method. When the size of the load is near

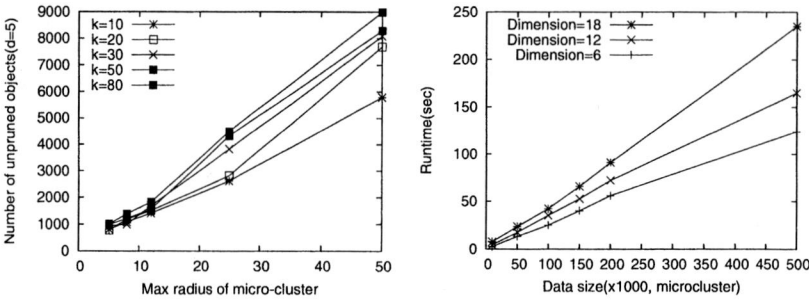

Fig. 10. Pruning results (2)   **Fig. 11.** Effects of dimensionality(1)

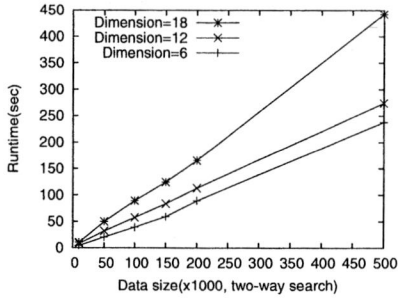

Fig. 12. Effects of dimensionality(2)

to 1000k, swapping operation between R-tree and disk will happen frequently. As such, the performance of index-based method starts to degrade. On the other hand, since the two-way search method does early pruning in the search process, it reduces the total computation cost greatly and saves much time. Micro-cluster method achieves best performance because it not only uses the similar pruning technique as the two-way search, but also reduces the huge number of the nearest neighbor search. So it takes the least time to finish mining outliers in each dataset and scales well to large databases. Unavoidably, this advantage in performance is done by sacrificing some precision in KNN approximation. However, if we adjust the micro-cluster to a suitable size, good quality mining results can still be obtained. Figure 9 shows the pruning results under the different values of threshold M (see the two-way search part in section 3). It can be seen that when M increases, more objects in the database remain unpruned, but the possibility of objects misses to be pruned will be reduced. If M decreases, more objects will be removed, and the cost of future computation will be reduced. It is particularly suitable for top-n case in which only a few objects can become the outlier candidates. Figure 10 shows the pruning results under the different radius of micro-cluster. We can see that when the radius increases, more objects will be inside the micro-clusters, and the difference between lower and upper bound of micro-clusters's k-distance will be larger. As a result, more micro-clusters will not be pruned. Figure 8 presents different performance results of the two-way search method when k varies from 10 to 50. If k is less than 30, the scalability is good with the support of R-tree. When k is over 30, the cost for the nearest search is rather expensive, more MBRs will be searched to compute the distance between the objects and the query object. Thus the running time would increase drastically with the increased number of distance computation.

We also studied the relationship between performance of our algorithm and the number of dimensions, and Figures 11 and 12 show the runtime of our algorithm with different dimensions and varying database with respect to the microcluster-based method and the two-way search method respectively. From the experiment results, we know that the algorithms on smaller dimensionality and data size always have shorter running time. Specifically, when dimensionality is larger than 12, the running time will be increased drastically, thus seriously hindering the efficiency of the algorithms.

5 Related Work

Knorr and Ng [12] initialized the concept of distance-based outlier, which defines an object o being an outlier, if at most p objects are within distance d of o. A cell-based outlier detection approach that partitions the dataset into cells is also presented. The time complexity of this cell-based algorithm is $O(N + c^k)$ where k is dimension number, N is dataset size, c is a number inversely proportional to d. For very large databases, this method achieves better performance than depth-based method, but still exponential to the number of dimensions. Ramaswamy et al. extended the notion of distance-based outliers by using the distance to the k-nearest neighbor to rank the outliers. An efficient algorithm to compute the top-n global outliers is given, but their notion of an outlier is still distance-based [20].

Some clustering algorithms like CLARANS [16], DBSCAN [6], BIRCH [26], and CURE [7] consider outliers, but only to the point of ensuring that they do not interfere with the clustering process. Further, outliers are only by-products of clustering algorithms, and these algorithms **cannot rank the priority of outliers**.

The concept of local outlier, which assigns each data a local outlier factor **LOF** of being an outlier depending on their neighborhood, was introduced by Breunig et al. [2]. This outlier factor can be used to rank the objects regarding their outlierness. To compute LOF for all objects in a database, O(n*runtime of a KNN query) is needed. The outlier factors can be computed efficiently if OPTICS is used to analyze the clustering structure. A top-n based local outliers mining algorithm which uses distance bound of micro-cluster to estimate the density, was presented in [11].

There are several recent studies on local outlier detection. In [5], [4], three enhancement schemes over **LOF** are introduced, namely LOF' and LOF" and GridLOF, and [22] introduces a connectivity-based outlier factor (COF) scheme that improves the effectiveness of an existing local outlier factor **LOF** scheme when a pattern itself has similar neighborhood density as an outlier. They extensively study the reason of missed outliers by **LOF**, and focus on finding those outliers which are close to some non-outliers with similar densities. While our measure based on the symmetric relationship is not only compatible with their improved measures, but also identifies more meaningful outliers. LOCI [17] addresses the difficulty of choosing values for MinPts in the LOF technique by using statistical values derived from the data itself.

6 Conclusion

In this paper, we discuss the problem with existing local outlier measure and proposed a new measure $INFLO$ which is based on a symmetric neighborhood relationship. We proposed various methods for computing $INFLO$ including the naive-index based method, the two-way pruning method and the micro-cluster based method. Extensive experiments are conducted showing that our proposed methods are efficient and effective on both synthetic and real life datasets.

References

1. C. Aggarwal and P. Yu: Outlier Detection for High Dimensional Data. *SIGMOD* 2001
2. M. M. Breunig, H.P. Kriegel, R.T. Ng, and J.Sander: LOF: Identifying Density-based Local Outliers. *SIGMOD* 2000
3. D. Chakrabarti: AutoPart: Parameter-Free Graph Partitioning and Outlier Detection. *PKDD 2004*
4. Z. X. Chen, A. W. Fu, J. Tang: On Complementarity of Cluster and Outlier Detection Schemes. *DaWaK* 2003
5. A. L. Chiu, A. W. Fu: Enhancements on Local Outlier Detection. *IDEAS* 2003
6. M. Ester, H. P. Kriegel et al.: A Density-based Algorithm for Discovering Clusters in Large Spatial Databases. *KDD* 1996
7. S. Guha, R. Rastogi, and K.Shim: Cure: An Efficient Clustering Algorithm for Large Databases. *SIGMOD* 1998
8. V. Hautamki, I. Krkkinen and P. Frnti: Outlier Detection Using k-nearest Neighbour Graph, *ICPR* 2004.
9. J. W. Han, M. Kamber: Data Mining: Concepts and Techniques. In *Morgan Kaufmann* Publishers.
10. H. Jagadish, N. Koudas, and S. Muthukrishnan: Mining Deviants in a Time Series Database. *VLDB* 1999
11. W. Jin, K. H. Tung and J. W. Han: Mining Top-n Local Outliers in Large Databases. *KDD* 2001
12. E. Knorr, R. Ng: Algorithms for Mining Distance-Based Outliers in Large Datasets. *VLDB* 1998
13. E. Knorr and R. Ng: Finding Intensional Knowledge of Distance-Based Outliers. *VLDB* 1999
14. F. Korn and S. Muthukrishnan: Influence Sets Based on Reverse Nearest Neighbor Queries. *SIGMOD* 2000
15. S. Muthukrishnan, R. Shah, J. S. Vitter: Mining Deviants in Time Series Data Streams. *SSDBM* 2004

16. R. Ng and J. W. Han: Efficient and Effective Clustering Method for Spatial Data Mining. *VLDB* 1994
17. S. Papadimitriou, H. Kitagawa et al. LOCI: Fast Outlier Detection Using the Local Correlation Integral. *ICDE* 2003
18. S. Papadimitriou, C. Faloutsos: Cross-Outlier Detection. *SSTD* 2003
19. N. Roussopoulos, S. Kelley and F. Vincent: Nearest neighbor queries. *SIGMOD* 1995
20. S. Ramaswamy, R. Rastogi, K. Shim: Efficient Algorithms for Mining Outliers from Large Data Sets. *SIGMOD* 2000
21. S. Shekhar, C. T. Lu, P. S. Zhang: Detecting Graph-based Spatial Outliers. *KDD* 2001
22. J. Tang, Z. X. Chen et al.: Enhancing Effectiveness of Outlier Detections for Low Density Patterns. *PAKDD* 2002
23. W. K. Wong, A. W. Moore et al.: Rule-Based Anomaly Pattern Detection for Detecting Disease Outbreaks. *AAAI* 2002
24. M. L. Yiu, N. Mamoulis: Clustering Objects on a Spatial Network. *SIGMOD* 2004
25. M. L. Yiu et al.: Aggregate Nearest Neighbor Queries in Road Networks. *IEEE Trans. Knowl. Data Eng.* 17(6), 2005
26. T. Zhang et al.: BIRCH: An Efficient Data Clustering Method for Very Large Databases. *SIGMOD* 1996

Construction of Finite Automata for Intrusion Detection from System Call Sequences by Genetic Algorithms

Kyubum Wee and Sinjae Kim

Ajou University, Suwon, 443-749, S. Korea
{kbwee, venddol}@ajou.ac.kr

Abstract. Intrusion detection systems protect normal users and system resources from information security threats. Anomaly detection is an approach of intrusion detection that constructs models of normal behavior of users or systems and detects the behaviors that deviate from the model. Monitoring the sequences of system calls generated during the execution of privileged programs has been known to be an effective means of anomaly detection. Finite automata have been recognized as an appropriate device to model normal behaviors of system call sequences. However, there have been several technical difficulties in constructing finite automata from sequences of system calls. We present our study on how to construct finite automata from system call sequences using genetic algorithms. The resulting system is shown to be very effective in detecting intrusions through various experiments.

1 Introduction

Intrusion means any behavior that damages the integrity, confidentiality, or availability of computer systems or networks by exploiting the vulnerabilities of the system or network resources. These days any computer or network user is exposed to threats of intrusions, and damages caused by intrusions are ever more increasing. Intrusion detection system tries to protect the normal users and system resources by detecting intrusions.

Approaches to detecting intrusions can be classified into misuse detection and anomaly detection. Misuse detection maintains patterns of intrusions and tries to detect those patterns. Anomaly detection establishes models of normal behavior and tries to detect behaviors that deviate from the normal ones. An advantage of anomaly detection is that it can learn models of normal behavior and, as a result, can detect intrusion patterns that are similar to known intrusion patterns. There have been many studies on how to model normal behavior including statistical approaches[3, 6, 8, 9, 12], neural networks[2], and finite state automata[7, 11, 13, 15].

Finite state automata monitor sequences of system calls generated by execution of privileged programs, and accept the normal sequences and reject the abnormal sequences. Forrest et al. introduced the use of system call sequences to model program behaviors [4, 5, 14]. They extract the short sequences of fixed length from the long sequence of system calls generated by processes, called N-grams, and maintain the database of such sequences. When a program is monitored, the N-grams are extracted and matched against the ones in the database of normal sequences. If the miss ratio exceeds the given threshold, a warning is issued. The use of system call sequences has been proven to be a very effective way of intrusion detection. However, since short sequences of fixed length are used, the intruder can dodge detection by carefully inserting spurious system calls within the fixed window size in order not to exceed the threshold.

Sekar et al. examine the program counters where the system calls are made [11]. States and edges of finite automata are labeled by program counters and system calls, respectively. This approach facilitates the construction of finite automata, but has difficulty in stack traversal and dealing with fork/exec to trace program counters.

Kosoresow et al. substitute macros for frequently occurring substrings of the system call sequence, and then construct finite automata recognizing the system call sequences generated by processes [7]. By introducing macros the system call sequences become shorter, and consequently the resulting automaton becomes smaller. This approach has neither the weakness of the N-gram method nor the difficulty of tracing program counters. However, the procedure of selecting macros and constructing finite automata are carried out manually using human insight and intuition. Wee et al. provided automatic ways to select macros and using suffix trees and to construct automata using multiple sequence alignment [15].

In this paper, we present our study on how to construct intrusion detection automata using genetic algorithms.

2 Construction of Finite State Automata

Execution of each privileged program generates many processes, and each process generates a long sequence of system calls. For each privileged program, we evolve a finite state automaton that can accept the sequences generated by the normal execution of the program and reject the sequences generated by the execution of the program while the system is being intruded.

2.1 Representation of Finite State Automata

A finite state automaton is represented as a matrix where the column indices correspond to states and the row indices input symbols. In our study input symbols are system calls. Table 1 is the matrix representation of the FSA in Figure 1. For our genetic algorithms, a matrix is encoded as a string that is obtained by splicing columns of the matrix in order. Figure 2 is the encoded string of the matrix in Table1. The entry –1 indicates that there is no transition from the state by the input symbol.

In our genetic algorithm each member of the initial population has –1 at about half the entries of the matrix. The number of states in FSA is initially set to the number of distinct system calls appearing in the execution of the given privileged program.

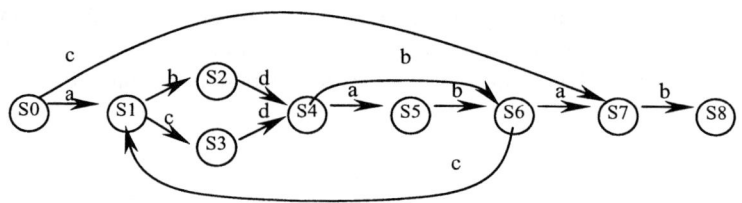

Fig. 1. a finite state automaton

Table 1. matrix representation of a finite automaton in Fig. 1

	S0	S1	S2	S3	S4	S5	S6	S7	S8
A	S1	-1	-1	-1	S5	-1	S7	-1	-1
B	-1	S2	-1	-1	S6	S6	-1	S8	-1
C	S7	S3	-1	-1	-1	-1	S1	-1	-1
D	-1	-1	S4	S4	-1	-1	-1	-1	-1

S1, -1, S7, -1, -1, S2, S3, -1, -1, -1, -1, S4, -1, -1, -1, S4 S5, S6, -1, -1, -1, S6, -1, -1, S7, -1, S1, -1, -1, S8, -1, -1, -1, -1, -1, -1

Fig. 2. encoding of the FSA in Table 1

2.2 Population, Selection, Crossover, and Mutation

We set the population size to 1000 and placed the individuals on a 20 × 50 grid. Each cell of the grid is assumed to have 8 neighbors: the cells immediately east, west, north, south, northeast, southeast, southwest, and northwest to the given cell, i.e. the Moore neighborhood. The cells at the edges of the grid find its neighbors in the wrap-around fashion. Hence the grid is actually a torus.

Crossover is performed on an individual and one of its neighbors. We randomly select an individual A and select as its mate the one that has the highest fitness among the 8 neighbors of A. The newly generated individual N, through crossover and mutation, replaces the individual L of the lowest fitness in the current population, if N is fitter than L. Otherwise the population stays the same, and the selection of two new mates starts over again.

We employ uniform crossover. Before the crossover is performed, the length of the child is determined first. The length of the child is randomly selected in the range of $P_{max}-2$ to $P_{max}+2$, where P_{max} is the length of the longer one between two parents. While the uniform crossover is performed, the extra portion of the longer parent which does not have corresponding positions in the shorter parent, is directly copied to the child. When the child's length is selected to be longer than P_{max}, the extra portion of the child's chromosome is set to arbitrary values.

When the uniform crossover is performed, the genotype of each parent is inherited to the child proportionately to their respective fitness. For example, if one parent's fitness is twice higher than the other's, then the child gets the fitter parent's genes twice more often than the less fit parent. Figure 3 illustrates the crossover operation.

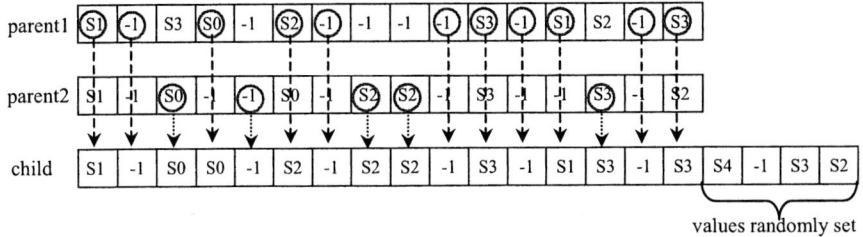

Fig. 3. crossover

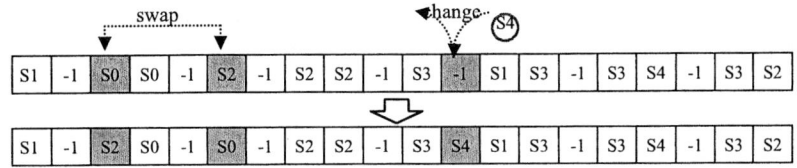

Fig. 4. mutation

Two kinds of mutation operation are employed. One is the change of the allele at a locus to a random allele. Another is the swap of alleles at two loci. The rate of each mutation operation is set to 2.5%. Figure 4 illustrates mutation operations.

The way the selection, crossover, and mutation are performed in our study is based on the scheme suggested in the Belz et al's work on induction of finite state automata with application to phonotactics [1].

2.3 Fitness

Fitness represents the degree of how well a given finite automaton deals with the sequences generated during execution of a given privileged program. Values of fitness range from 0 to 1.

There are three factors in evaluating fitness of finite automata: consistency, compactness, and distinction. Consistency is the rate of accepting normal sequences. Compactness is the size of finite automata. Distinction is the rate of rejecting abnormal sequences. Compactness is measured as the number of system calls occurred in the sequences divided by the number of states in the automaton. The smaller the number of states, the higher the compactness of the finite automaton.

Three factors of fitness are assigned their weights in evaluating the fitness. The weights are represented as percentages, and the sum of weights is 100%. The set of

number of states: 4
system calls: 1, 2, 3, 4
state transition table: S1, -1, S3, S0, -1, S2, -1, -1, -1, -1, S3, -1, S1, S2, -1, S3
weight of consistency: 40%
weight of compactness: 20%
weight of distinction: 40%
fitness = 0.75*40% + 1.00*20%
   +0.75*40% = 0.8

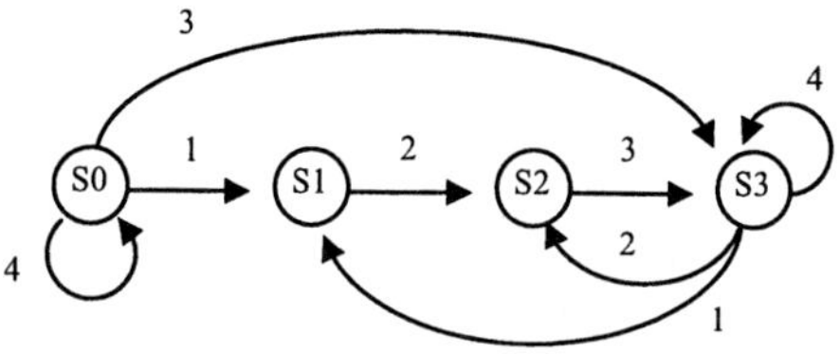

System call sequences	
normal sequences	abnormal sequences
(1) 1, 2, 3, 4	(1) 1, 4, 4, 3
(2) 3, 4, 2, 1	(2) 1, 4, 2, 3
(3) 3, 1, 2, 3	(3) 1, 4, 1, 3
(4) 3, 4, 2, 3	(4) 4, 1, 2, 3

Fig. 5. fitness of a finite automaton

weights can vary across privileged programs. For example, the set of weights for the program *sendmail* can be different from the set of weights for *ftp*. The following toy examples in Figure 5 illustrate how the fitness of automata is computed.

In Figure 5, the automaton has 4 states. State S0 is the starting state, and S3 is the final state. Weights of consistency, compactness, and distinction are 40%, 20%, and 40%, respectively. Consistency of the automaton is 0.75, since it accepts 3 sequences out of 4 normal sequences. The number of system calls occurring in the sequences is 4. Hence the compactness is 1.00. Distinction is 0.75, since it rejects 3 sequences out of 4 abnormal sequences. Hence the fitness is 0.75*40% + 1.00*20% + 0.75*40% = 0.8.

3 Experiments

We used Forrest et al's data sets for our experiments [16]. These data sets include the system call sequences of the privileged programs *ps, named, login, xclock, lpr, ftp,* and *sendmail*. An execution of a program generates many processes. Each sequence is obtained by collecting the system calls from the execution of a process. Table 2 describes the details of these data sets.

For example, the execution of the program *ps* under the normal condition generated 19 sequences. There are longer ones and shorter ones. If we add up all the lengths of these sequences, it comes to 6145. The alphabet of these sequences consists of 22 system calls. Each system call is represented by a unique number. Hence a sequence of system calls is actually a sequence of integers. Out of 19 sequences 11 are used as training data and the remaining 8 are used as test data.

Table 2. system call sequences data

program	normal sequences					abnormal sequences				
	number of seq.	sum of lengths of seq.	number of system calls	number of training seq.	number of test seq.	number of seq.	sum of lengths of seq.	number of system calls	number of training seq.	number of test seq.
Ps	19	6145	22	11	8	21	6969	22	13	8
named	26	44481	45	16	10	3	615	45	2	1
login	24	8907	46	14	10	9	4854	46	5	4
xlock	69	338898	40	40	29	2	950	36	1	1
Lpr	9	2399	37	5	4	8	1345	37	5	3
ftp	8	180316	50	5	3	5	1364	45	3	2
sendmail	49	414740	53	25	24	23	6505	50	13	10

Finite automata are evolved for each program using training data. The size of the population is set to 1000, crossover rate 90%, mutation rate 5%, and finite automata are evolved for 10,000 generations.

The best automaton thus obtained is run on normal sequences and abnormal sequences to evaluate the scores of the sequences and to determine the threshold value that can distinguish the normal sequences from the abnormal sequences. Every time the finite automaton makes a successful transition from the current state by the current symbol - actually a number representing a system call – in the sequence, the sequence gets one point. If there is no transition specified by the current symbol, the automaton goes back to the starting state instead of getting one point. The total points amassed by the sequence until the automaton consumes all the symbols in the sequence are the score of the sequence.

Every normal and abnormal sequence is evaluated and assigned its score by the automaton. Then the threshold score is determined that can best distinguish the normal sequences form the abnormal ones. The whole procedure is repeated several times with different sets of weights for consistency, compactness, and distinction factors of the fitness function in order to obtain the best combination of these three factors. The results are shown in Table 3.

The automaton along with the threshold value is evaluated depending on how well it can determine normal sequences and abnormal sequences. Detection rate represents

Table 3. the best automaton obtained for each program

Program	best automaton obtained		
	number of states	threshold	consistency/compactness/distinction
Ps	19	35	45/10/45
Named	45	50	70/10/20
Login	46	91	70/10/20
Xlock	37	92	70/10/20
Lpr	37	90	70/10/20
ftp	45	90	60/10/30
Sendmail	50	86	60/10/30

how well it can detect abnormal sequences. False positive rate represents how often it misreports a normal sequence as an abnormal one. Naturally high detection rate and low false positive rate are desirable. Table 4 shows the performance measures of intrusion detection. Table 5 and Table 6 show the performance of intrusion detection on the training data and testing data.

Table 4. performance measures of intrusion detection

TP (True Positive) : the number of cases where abnormal behavior is detected
TN (True Negative) : the number of cases where normal behavior is recognized
FP (False Positive) : the number of cases where normal behavior is misinterpreted as abnormal
FN (False Negative) : the number of cases where abnormal behavior is not detected
detection rate = TP / (TP + FN)
false positive rate = FP / (TN + FP)

Table 5. intrusion detection performance of the resulting finite automaton on training data

program	training TP	training TN	training FN	training FP	detection rate	false positive rate
ps	13/13	10/11	0/13	1/11	100%	9.1%
named	2/2	16/16	0/2	0/16	100%	0%
login	5/5	14/14	0/5	0/14	100%	0%
xlock	1/1	40/40	0/1	0/40	100%	0%
lpr	5/5	5/5	0/5	0/5	100%	0%
ftp	3/3	5/5	0/3	0/5	100%	0%
sendmail	12/13	24/25	1/13	1/25	92.3%	4.0%

Table 6. intrusion detection performance of the resulting finite automaton on test data

program	test TP	test TN	test FN	test FP	detection rate	false positive rate
ps	8/8	8/8	0/8	0/8	100%	0%
named	1/1	10/10	0/1	0/10	100%	0%
login	4/4	10/10	0/4	0/10	100%	0%
xlock	1/1	29/29	0/1	0/29	100%	0%
lpr	3/3	4/4	0/3	0/4	100%	0%
ftp	2/2	3/3	0/2	0/3	100%	0%
sendmail	10/10	24/24	0/10	0/24	100%	0%

As can be seen in Table 5 and Table 6, some of the programs have very scanty data. For such programs the generated automaton is not likely to be reliable. Hence we tried all the sequences as training data. The result is shown in Table 7 and Table 8.

The program *ps* has very heterogeneous sequences, and it seems that the behaviors of these sequences cannot be captured by finite automata. For the other programs, the

Table 7. the best automaton obtained for each program when all the sequences are used as training data

program	best automaton obtained		
	number of states	threshold	consistency/compactness/distinction
Ps	15	62	45/10/45
Named	40	70	70/10/20
Login	46	88	70/10/20
Xlock	37	85	70/10/20
Lpr	34	90	70/10/20
ftp	47	90	60/10/30
sendmail	52	69	60/10/30

Table 8. intrusion detection performance of the resulting finite automaton when all the sequences are used as training data

Program	TP	TN	FN	FP	detection rate	false positive rate
ps	21/21	18/19	0/21	1/19	100%	5.26%
named	3/3	26/26	0/3	0/26	100%	0%
login	9/9	24/24	0/9	0/24	100%	0%
xlock	2/2	69/69	0/2	0/69	100%	0%
lpr	8/8	9/9	0/8	0/9	100%	0%
ftp	5/5	8/8	0/5	0/8	100%	0%
sendmail	23/23	49/49	0/23	0/49	100%	0%

results show that our approach to construction of automata is a very effective means of intrusion detection.

4 Conclusions

Information security, including computer security and network security, is a serious matter these days. Intrusion detection is an indispensable ingredient of computer and network security. Use of system call sequences and finite automata have been recognized as an effective way of intrusion detection. However, previous techniques of constructing finite automata that can detect intrusions have drawbacks such as requiring additional information other than system call sequences or difficulty in handling sequences of significantly varying lengths.

In this paper we present a technique of constructing finite automata that can detect intrusions using genetic algorithms. It can construct finite automata from system call sequences without requiring additional information, and it can deal with sequences of varying lengths. Experiments show that the resulting system's detection rate is very high, and false positive rate is satisfactorily low. Our study also demonstrates the practical usefulness of machine learning by showing that induction of finite automata, which perform critical computation such as intrusion detection, can be effectively carried out through evolutionary computation.

Our future study plan includes testing the performance of our system by trying extensive data sets of system call sequences and improving the algorithm of constructing finite automata so that the resulting finite automata can handle such severely heterogeneous sequences as the ones generated by the program *ps*.

References

1. Belz A. and Eskikaya B.: A Genetic algorithm for Finite Automata Induction with an Application to Phonotactics. Proceedings of the European Summer School in Logic, Language, and Information Workshop on Automated Acquisition of Syntax and Parsing (1998) 9-17
2. Debar H., Becker M., and Siboni D.: A Neural Network Component for an Intrusion Detection System. Proceedings of the IEEE Symposium on Security and Privacy (1992) 240-250
3. Denning D.: An Intrusion Detection Model. Proceedings of the IEEE Symposium on Security and Privacy (1986) 119-131
4. Forrest S., Hofmeyr S., Somayaji A., and Longstaff T.: A Sense of Self for Unix Processes. IEEE Symposium on Security and Privacy (1996) 120-128
5. Hofmeyr S., Forrest S., and Somayaji A.: Intrusion Detection using Sequence of System Calls. Journal of Computer Security, Vol. 6 (1998) 151-180
6. Javitz H. and Valdes A.: The SRI IDES Statistical Anomaly Detector. Proceedings of the IEEE Symposium on Security and Privacy, Oakland, CA (1991)
7. Kosoresow A.: Intrusion Detection via System Call Traces. IEEE Software, Vol. 14, No.5 (1997) 35-42
8. Lankewicz L. and Benard M.: Real Time Anomaly Detection using a Nonparametric Pattern Recognition Approach. Proceedings of the Seventh Annual Computer Security Applications Conference, San Antonio, TX (1991)
9. Lunt T., Tamaru A., and Gilham F.: IDES: A Progress Report. Proceedings of the Sixth Annual Computer Security Applications Conference, Tucson, AZ (1990)
10. Me L.: GASSATA: A Genetic Algorithm as an Alternative Tool or Security Audit Trails Analysis. First International Workshop on the Recent Advances in Intrusion Detection, Louvain-la-Neuve, Belgium (1998)
11. Sekar R. and Bendre M.: A Fast Automaton-Based Method for Detecting Anomalous Program Behaviors. Proceeding of the 2001 IEEE Symposium on Security and Privacy (2001) 144-155
12. Smaha S.: Haystack: An Intrusion Detection System. Proceedings of the Fourth IEEE Aerospace Computer Security Applications Conference, Orlando, FL (1988)
13. Wagner D.: Intrusion Detection via Static Analysis. Proceedings of the 2001 IEEE Symposium on Security and Privacy (2001) 156-159
14. Warrender C., Forrest S., and Pearlmutter B.: Detecting Intrusions using System Calls: Alternative Data Models. Proceedings of the 20th IEEE Symposium on Security and Privacy (1999)
15. Wee K. and Moon B.: Automatic Generation of Finite Automata for Detecting Intrusions using System Call Sequences. Proceedings of International Workshop on Mathematical Methods, Models, and Architectures for Computer Network Security, St. Petersburg, Russia, Lecture Notes in Computer Science 2776 (2003) 206-216
16. http://www.cs.unm.edu/~immsec/systemcalls.htm

An Adaptive Intrusion Detection Algorithm Based on Clustering and Kernel-Method

Hansung Lee, Yongwha Chung, and Daihee Park*

Korea Univ. Dept. of Computer & Information Science,
{mohan, ychungy, dhpark}@korea.ac.kr

Abstract. An adaptive intrusion detection algorithm which combines the Adaptive Resonance Theory(ART) with the Concept Vector and the Mecer-Kernel is presented. Compared to the supervised- and the clustering-based Intrusion Detection Systems(IDSs), our algorithm can detect unknown types of intrusions in on-line by generating clusters incrementally.

Keywords: intrusion detection, ART, mercer kernel, concept vector.

1 Introduction

In the traditional *signature-based* IDSs, the rule-base has to be manually revised whenever each new type of attack is discovered. To solve this *manual revision problem*, some of the *machine learning* algorithms have been applied to the IDS[1][2]. However, most of these machine learning approaches are based on *supervised* learning, and have following problems: 1) a large volume of training data should be collected and classified manually; 2) the performance of the IDS depends on the quality of the training data; 3) a training phase with the huge data is computationally expensive and can not be performed in an incremental manner; 4) it is difficult to detect new intrusions which are not trained.

Recently, the *clustering* algorithms based on *unsupervised* learning have been proposed for IDS to overcome these problems[3][4][5]. However, the number of new intrusion types is increased rapidly and the volume of the information is too large. Thus, the general-purpose clustering algorithms used in artificial intelligence need to be modified to satisfy the following IDS requirements: 1) each event data should be processed as soon as it is received and clusters are generated adaptively without fixing the number of clusters; 2) clustering the huge volume of event data needs to be completed in few seconds; 3) the result of clustering needs to be insensitive to the order of input data since the sequence of event data is arbitrary in general.

In this paper, we propose a clustering-based intrusion detection algorithm which can satisfy all the requirements. First, we choose an on-line and incremental clustering algorithm, called *Adaptive Resonance Theory*(ART). In addition to that, we employ both *Concept Vector*[6] and *Mercer-Kernel*[7] to classify a high

---
* Corresponding author.

dimensional sparse pattern effectively and improve the separability, respectively. These two techniques can improve the detection rates of new intrusions because most of the information source for intrusion detection is high dimensional and very similar to each other. Based on the experimental results, our algorithm can provide superior performance by generating clusters incrementally and subdividing the patterns in detail.

The organization of this paper is as follows. Section 2 explains the data representation and the similarity measure, and Section 3 describes the proposed intrusion detection algorithm. The experimental results are given in Section 4, and conclusions are made in Section 5.

2 Data Representation and Similarity Measure

In this section, we define input dataset and similarity measure in order to evaluate the real world problems; the input patterns are represented by a mixture of variable types. For a given set of n input patterns $\mathbf{x} = \{\underline{x}_i\}_{i=1}^n$, we assume that the input pattern $\underline{x}_i$ consists of k numeric attributes and m symbolic attributes. Let R^k and R^m denote the k-dimensional numeric space and m-dimensional symbolic space, respectively. Then, $\mathbf{x}$ can be represented as follows:

$$\mathbf{x} = \{\underline{x}_i\}_{i=1}^n; \quad \underline{x}_i = \underline{x}_i^R + \underline{x}_i^S; \quad \underline{x}_i^R \in R^k, \underline{x}_i^S \in S^m \tag{1}$$

To avoid bias toward some features over other features, we perform L2 normalization on numeric attributes to have unit Euclidean norm.

$$\underline{x}_i^R = \frac{\underline{x}_i^R}{\|\underline{x}_i^R\|}; \quad \|\underline{x}_i\| = \sqrt{\sum_{j=1}^k x_{ij}^2} \tag{2}$$

Also, a similarity measure which computes the similarity between objects of mixed variable types is defined as follows: Let m and $\lambda \in [0,1]$ denote the dimension of the symbolic space and an adjustable parameter in order to weight the attribute types, respectively. Then,

$$S(\underline{x}_i, \underline{x}_j) = \lambda \cdot <\underline{x}_i^R, \underline{x}_j^R> + (1-\lambda) \cdot \frac{\sum_{l=1}^m \delta(x_{il}^S, x_{jl}^S)}{m} \tag{3}$$

where the delta function $\delta(\cdot)$ is defined as follows:

$$\delta(x_{il}^S, x_{jl}^S) = \begin{pmatrix} 1, & if \ x_{il}^S = x_{jl}^S \\ 0, & otherwise \end{pmatrix} \tag{4}$$

Since numeric attributes are normalized to be unit vectors, the cosine measure is obtained by the inner product of two vectors.

$$<\underline{x}_i^R, \underline{x}_j^R> = \|\underline{x}_i^R\| \cdot \|\underline{x}_j^R\| \cdot COS(\theta(\underline{x}_i^R, \underline{x}_j^R)) = COS(\theta(\underline{x}_i^R, \underline{x}_j^R)) \tag{5}$$

3 Adaptive Intrusion Detection Algorithm

An intrusion detection algorithm proposed in this paper is an "adaptive" algorithm which combines the on-line and incremental clustering algorithm ART with Concept Vector and Mercer-Kernel. By employing the Concept Vector, a weight vector of each cluster is normalized to the mean vector of each clusters. Thus, we need not consider the learning rate parameter in updating the weight vectors and can improve the speed of the execution. Also, we can improve the separability by mapping the input pattern to a feature space with Mercer-Kernel. Details of the proposed algorithm, called *Kernel-ART*, can be described as follows:

Initialization: The number of clusters is set to one initially, and the first input pattern is assigned to its initial weight vector as follows:

$$\underline{w}_1 = \underline{x}_1 = \underline{w}_1^R + \underline{w}_1^S = \underline{x}_1^R + \underline{x}_1^S \tag{6}$$

Then, the matching value, computed by the activation function between the initial weight vector and the first input pattern, is set to one. This ensures that the first input pattern is assigned to the first cluster for any vigilance parameter $\rho \in [0,1]$.

Activation Function: The basic idea of the Mercer-Kernel is to perform a nonlinear data transformation into some high dimensional dot-product space, called *feature space*, to increase the probability of the linear separability of the patterns within the transformed space[7]. By replacing the inner product in the similarity measure of equation (3) with the RBF kernel function $K(\underline{x}_i, \underline{x}_j) = exp\{-\frac{1}{c}\|\underline{x}_i - \underline{x}_j\|^2\}$, we can obtain a similarity measure function in the feature space. Then, the activation function is defined by the similarity measure in the feature space as follows:

$$AF(\underline{x}_i, \widehat{\underline{w}}_j) = \lambda \cdot exp\left\{-\frac{1}{c}\|\underline{x}_i^R - \widehat{\underline{w}}_j^R\|^2\right\} + (1-\lambda) \cdot \frac{\sum_{l=1}^m \delta(x_{il}^S, w_{jl}^S)}{m} \tag{7}$$

where $\widehat{\underline{w}}_j^R = \frac{\underline{w}_j^R}{\|\underline{w}_j^R\|}$ is the Concept Vector of a cluster j. The Concept Vector is the mean vector of the cluster normalized to the unit Euclidean norm. Since the Concept Vectors(i.e., clusters) are localized in the high dimensional sparse space, the clusters can represent the class structure of the dataset. That is, the clusters can represent each types of attacks individually.

Matching Function: If the activation function $AF(\cdot)$ and the matching function $MF(\cdot)$ are chosen as

$$MF(\underline{x}_i, \widehat{\underline{w}}_1) > MF(\underline{x}_i, \widehat{\underline{w}}_2) \Leftrightarrow AF(\underline{x}_i, \widehat{\underline{w}}_1) > AF(\underline{x}_i, \widehat{\underline{w}}_2), \tag{8}$$

then the mismatch reset condition and the template matching process of the original ART can be eliminated for the resonance domain[8]. The most simple way to define the activation and the matching functions under the condition of equation (8) is to set the activation function to equal to the matching function. By this setting, we can make the algorithm simple and improve the speed of execution.

Resonance Condition: According to the simple setting of the matching function, the resonance unit is selected as follows:

$$AF(\underline{x}_i, \widehat{\underline{w}}_{j^*}) \geq \rho; \quad j^* = arg\,max_{j=1,...,c}\{AF(\underline{x}_i, \widehat{\underline{w}}_j)\} \tag{9}$$

When the best-matching template does not satisfy the vigilance criterion, a new cluster unit can be created and the input pattern is assigned to it. This condition can speed-up the execution time of the algorithm further.

Update Weight Vector: When a cluster j^* is selected by equation (9), the input pattern is assigned to the cluster j^* and the weight vector is updated as follows:

$$\underline{w}_{j^*}^{R(t)} = \underline{w}_{j^*}^{R(t-1)} + \underline{x}_i^R \tag{10}$$
$$\underline{w}_{j^*}^{S(t)} = MostFrequentSymbol$$

Normalize input pattern with L2 norm and Initialize Weights:
$$\underline{w}_1 = \underline{x}_1 = \underline{w}_1^R + \underline{w}_1^S = \underline{x}_1^R + \underline{x}_1^S$$

While Stopping Condition is false
    For each input data
        Set activation of all F_2 to zero
        Compute Activation Function:
$$AF(\underline{x}_i, \widehat{\underline{w}}_j) = \lambda \cdot exp\left\{-\tfrac{1}{c}\|\underline{x}_i^R - \widehat{\underline{w}}_j^R\|^2\right\} + (1-\lambda) \cdot \tfrac{\sum_{l=1}^m \delta(x_{il}^S, w_{jl}^S)}{m}$$

        Find j^* with max activation
$$j^* = arg\,max_{j=1,...,c}\{AF(\underline{x}_i, \widehat{\underline{w}}_j)\}$$

        Test for reset:
            If $AF(\underline{x}_i, \widehat{\underline{w}}_{j^*}) \geq \rho$ then
$$\underline{w}_{j^*}^{R(t)} = \underline{w}_{j^*}^{R(t-1)} + \underline{x}_i^R$$
$$\underline{w}_{j^*}^{S(t)} = \{MostFrequentSymbol\}$$
        else
            allocation: c = c + 1
$$\underline{w}_c^{R(t)} = \underline{w}_c^{R(t)} + \underline{x}_i^R$$
$$\underline{w}_c^{S(t)} = \underline{w}_c^{S(t)}$$

Fig. 1. Outline of the Kernel-ART algorithm

The weight vector of the cluster j^* is defined by sum of input patterns that are assigned to the cluster j^*. Thus, we need not consider the learning rate parameter in updating the weight vectors, and our algorithm is less sensitive to the order of input patterns than that of previous clustering such as Fuzzy ART. This is because, in *Kernel-ART*, the weight vectors memorize the normalized mean vector of the input patterns assigned to each clusters. This intrusion detection algorithm, called *Kernel-ART*, is summarized in Fig. 1.

4 Experimental Results

To evaluate the effectiveness of *Kernel-ART*, KDD CUP 99 data[9] were used for the experiments. In order to make accurate analysis on the experiment result, we used only the Corrected-labeled dataset among KDD CUP 99 data. It was collected through the simulation on the U.S. military network by 1998 DARPA Intrusion Detection Evaluation Program, aiming at obtaining the benchmark dataset in the field of intrusion detection. The size of data is 311,029 and it consists of 9 symbolic attributes and 32 numeric attributes. The data is mainly divided into four types of attack: DOS, R2L, U2R and PROBING. In *Kernel-ART*, ρ is the vigilance parameter which affects the support of clusters. $\lambda \in [0, 1]$ and c denote the weight of the similarity measure function and the RBF kernel-width parameter of *Kernel-ART*, respectively. We set ρ to 0.93, λ to 0.5 and c to 1.

4.1 Comparison with Other Intrusion Detection Algorithms

Because many research results of intrusion detection have been reported recently, we compare our performance with those supervised and unsupervised(clustering) learning algorithms. Table 1 shows the classification capability of each research for normal data and four types of attack. The *Kernel-ART* proposed in this paper can provide good classification capability as a whole, as shown in Table 1. Most of previous methods except IDBGC[5] show considerable inferior performance only at the classification capability as to R2L and U2R. Note that R2L and U2R are host-based attacks which exploit vulnerabilities of the operating systems, not of the network protocol. Therefore, these are very similar to the "normal" data

Table 1. Comparison with Other Intrusion Detection Algorithms

		Supervised Learning			Unsupervised Learning	
		Bernhard [10]	KayAcik [11]	Ambwani [12]	IDBGC[5] Sampled	Proposed Kernel-ART
Normal		99.5%	95.4%	99.6%	-	97.1%
Attack	DOS	97.1%	95.1%	96.8%	56.0%	99.9%
	U2R	13.2%	10.0%	4.2%	78.0%	61.4%
	R2L	8.4%	9.9%	5.3%	66.0%	33.1%
	PROBING	83.3%	64.3%	75.0%	44.0%	95.5%

in the KDD CUP 99 data collected from network packets. However, our method can provide superior performance in separating these two patterns. It can be said, therefore, that the strategy of Kernel-Method employed in this paper, that is to improve the separability(see Eq. 7), is turned out to be very efficient. Note that results reported in [5] were average detection ratios with small number of "sampled" instance from KDD CUP 99 data, which are sampled with similar number of each attack type. The comparisons indicates that our method is not only comparable to [5] in general but also outperformed in DOS and PROBING, in particular.

4.2 Clustering Results of Each Subsidiary Types of Attack

To show the efficiency of *Kernel-ART*, we summarized the clustering results of each subsidiary types of attack in Table 2. In general, reasonable detection ratio with large number of instance. Depending on type of attack, the size of data available for the testing is quite different. The size of attack instance that pertained to U2R and R2L is much smaller than that of other types of attack. Therefore, some of the attacks in those two classes show low detection ratio. In case of PROBING, Saint is a network probing tool modeled after Satan. So, saint and Satan are clustered together. According to the experimental results of Table 2, detailed separation capability of *Kernel-ART* is relatively good. As in section 4.1, this result matches well with the strategy of Kernel-Method employed in this paper.

Table 2. Experimental Results of each Subsidiary Types of Attack

Type	Attacks	No. of instance	Detection Ratio	Attacks	No. of instance	Detection Ratio
DOS	land	9	100.0%	mailbomb	5000	100.0%
	processtable	759	100.0%	smurf	164091	100.0%
	neptune	58001	99.8%	apache2	794	99.6%
	back	1098	99.5%	pod	87	88.5%
	teardrop	12	0.0%	udpstorm	2	0.0%
U2R	multihop	18	38.9%	buffer overflow	22	27.3%
	ps	16	25.0%	perl	2	0.0%
	rootkit	13	0.0%	loadmodule	2	0.0%
	sqlattack	2	0.0%	xterm	13	0.0%
R2L	imap	1	100.0%	guess passwd	4367	99.6%
	httptunnel	158	65.8%	warezmaster	1602	54.8%
	xsnoop	4	50.0%	named	17	47.1%
	ftp write	3	33.3%	snmpgetattack	7741	0.5%
	snmpguess	2460	0.1%	phf	2	0.0%
	worm	2	0.0%	xlock	9	0.0%
	sendmail	17	0.0%			
PROBING	nmap	84	100.0%	mscan	1053	96.6%
	satan	1633	95.8%	portsweep	354	93.5%
	ipsweep	306	83.0%	saint	736	14.3%

Table 3. Input Parameters of Experiments: α and β are learning rates of Fuzzy Art, and ρ is the vigilance parameter. λ and c denote the weight of the similarity measure function and the RBF kernel-width parameter of Kernel-ART, respectively.

K-means	# of cluster = 39, repeat 30 experiments, using min-max normalization
Fuzzy ART	$\alpha = 0.00001$, $\beta = 1.0$, varying ρ from 0.35 to 0.95
Kernel-ART	$\lambda = 0.5$, c from 0.01 to 0.1, varying ρ from 0.35 to 0.95

Table 4. Comparison with Other Clustering Algorithms

		K-means	Fuzzy ART	Proposed Kernel-ART
Normal		75.6%	82.4%	**96.6%**
Attack	DOS	64.2%	93.8%	**93.2%**
	U2R	81.8%	84.1%	**87.5%**
	R2L	33.0%	62.3%	**73.9%**
	PROBING	96.6%	99.4%	**100.0%**

4.3 Comparison with Other Clustering Algorithms

To evaluate the clustering characteristics of *Kernel-ART*, we compared our method with typical clustering algorithms such as K-means and Fuzzy ART. Among the labeled 311,029 data instances, we sampled 880 data instances such as 176 normal instances, 176 DOS attacks, 176 R2L attacks, 176 U2R attacks, and 176 PROBING instances. The conditions of this experiment are summarized in Table 3, and the results of the experiment are shown in Table 4. These results show that *Kernel-ART* can provide better performance in classifying both "normal" and "attack" than the typical clustering methods.

5 Conclusions

In this paper, we have presented a robust and efficient intrusion detection algorithm which can detect various types of unknown intrusions in on-line by generating clusters incrementally. The Concept Vector and the Mercer-Kernel can classify a high dimensional sparse pattern effectively and improve the separability. Based on the experimental results, the proposed *Kernel-ART* can classify individual attacks more accurately than the previous methods. This classifying information can be exploited further for developing different responses to different attacks and several intrusion prevention strategies. Because our algorithm has no training phase and does not require periodical renewals of discovered attacks, the cost of system maintenance can also be reduced significantly. We believe our algorithm is very practical and can be employed in future IDSs because of its computational efficiency and ability in detecting new intrusions.

References

1. Lee, W., Stolfo, S., and Mok, K.: 'A Data Mining Framework for Building Intrusion Detection Models', Proceedings of the 1999 IEEE Symposium on Security and Privacy, pp. 120-132, 1999.
2. Hu, W., Liao, Y., and Vemuri, V. : 'Robust Support Vector Machines for Anomaly Detection in Computer Security', Proceedings of the International Conference on Machine Learning and Applications, pp.168-174, 2003.
3. Portnoy, L., Eskin, E., and Stolfo, S.: 'Intrusion Detection with Unlabeled Data using Clustering', Proceedings of the ACM Workshop on Data Mining Applied to Security, 2001.
4. Ye, N. and Li, X.: 'A Scalable Clustering Technique for Intrusion Signature Recognition', Proceedings of the IEEE Man, Systems and Cybernetics Information Assurance Workshop, 2001.
5. Liu, Y., Chen, K., Liao, X., and Zhang, W.,: 'A Genetic Clustering Method for Intrusion Detection', Pattern Recognition, Vol. 37, Issue 5, pp. 927-942. 2004.
6. Dhillon, I. and Modha, D.: 'Concept Decomposition for Large Sparse Text Data using Clustering', Technical Report RJ 10147(95022), IBM Almaden Research Center, 1999.
7. Girolami, M.: 'Mercer Kernel-based Clustering in Feature Space', IEEE Transaction on Neural Networks, Vol. 13, Issue 3, pp. 780-784, 2002.
8. Baraldi, A. and Chang, E.: 'Simplified ART: A New Class of ART Algorithms', International Computer Science Institute, TR 98-004, 1998.
9. KDD Cup 1999 Data, Available in http://kdd.ics.uci.edu/databases/kddcup99/kddcup99.html.
10. Results of the KDD 1999 Classifier Learning Contest, Available in http://www-cse.ucsd.edu/users/elkan/clresults.html.
11. Kayacik, H., Zincir-Heywood, A., and Heywood, M.: 'On the capability of an SOM based intrusion detection system', Proceedings of the International Joint Conference on Neural Networks, Vol. 3, pp. 1808-1813, 2003.
12. Ambwani, T.: 'Multi class support vector machine implementation to intrusion detection', Proceedings of the International Joint Conference on Neural Networks, Vol. 3, pp. 2300-2305, 2003.

Weighted Intra-transactional Rule Mining for Database Intrusion Detection

Abhinav Srivastava[1], Shamik Sural[1], and A.K. Majumdar[2]

[1] School of Information Technology,
[2] Department of Computer Science & Engineering,
Indian Institute of Technology, Kharagpur, India
abhinavs@sit.iitkgp.ernet.in, shamik@sit.iitkgp.ernet.in,
akmj@cse.iitkgp.ernet.in

Abstract. Data mining is the non-trivial process of identifying novel, potentially useful and understandable patterns in data. With most of the organizations starting on-line operations, the threat of security breaches is increasing. Since a database stores a lot of valuable information, its security has become paramount. One mechanism to safeguard the information in these databases is to use an intrusion detection system(IDS). In every database, there are a few attributes or columns that are more important to be tracked or sensed for malicious modifications as compared to the other attributes. In this paper, we propose an intrusion detection algorithm named weighted data dependency rule miner (WDDRM) for finding dependencies among the data items. The transactions that do not follow the extracted data dependency rules are marked as malicious. We show that $WDDRM$ handles the modification of sensitive attributes quite accurately.

Keywords: Data dependency, Weighted rule mining, Read-Write sequence, Intrusion detection.

1 Introduction

Data mining has attracted a great deal of attention in the industry in recent years due to the wide availability of huge volume of data and the imminent need for turning such data into useful information and knowledge [1]. Data mining generally refers to the process of extracting models or determining patterns from large observed data [2]. It involves an integration of techniques from multiple disciplines such as database technology, statistics, machine learning, high-performance computing, spatial data analysis, neural network and others.

Recently, researchers have started using data mining techniques in the emerging field of information and system security and specially in intrusion detection systems. An intrusion is defined as any set of actions that attempt to compromise the integrity, confidentiality or availability of a resource. Intrusion Detection is the process of monitoring the events occurring in a computer system or network and analyzing them for signs of intrusions [3].

Intrusion detection has been discussed in public research since the beginning of the 1980s. In the last few years, it became an active area of research and commercial IDSs started emerging [4]. Several research works also have been proposed that apply data mining for intrusion detection. Lee et al [5] have suggested data mining techniques for network intrusion detection. They consider several categories of data mining algorithms, namely, classification, link analysis and sequential analysis along with their applicability in the field of intrusion detection. Barbara et al [6] have built a testbed using data mining techniques to detect network intrusions . Though intrusion detection is a well researched area, only a few researches have focused on database intrusion detection. Chung et al [7] use the idea of "working scope" to find the frequent itemsets referenced together and used this information for anomaly detection. Lee et al [8] propose an intrusion detection system in real-time databases using time signatures. Lee et al [9] have suggested a method for fingerprinting the access patterns of legitimate database transactions and using them to identify database intrusions. Barbara et al [10] use hidden markov model (HMM) and time series to find malicious corruption of data. They use HMM to build database behavioral models that capture the changing behavior over time, and uses them to recognize malicious patterns. Zhong et al [11] have proposed an algorithm to mine user profiles based on the queries submitted by the user. Hu et al [12] have proposed an idea of determining dependency among data items in databases. The transactions that do not follow the mined data dependencies are identified as malicious transactions.

In this paper, we propose an algorithm for database intrusion detection using a data mining technique, which takes the sensitivity of the attributes into consideration. Sensitivity of an attribute signifies how important the attribute is for tracking against malicious modifications. This approach mines dependency among attributes in a database. The transactions that do not follow these dependencies are marked as malicious transactions.

The rest of the paper is organized as follows. In Section 2, we describe weighted data dependency rule mining $(WDDRM)$ algorithm with an example. We present details of our experiments and provide results in Section 3. Finally, we conclude the paper with some discussions.

2 Weighted Data Dependency Rule Mining

2.1 Intuition

Databases are increasing in size in two ways: the number N of records, or objects in the database, and the number d of fields, or attributes, per object. Databases containing of the order of $N = 10^9$ objects are increasingly common nowadays. The number d of attributes can easily be of the order of 10^2 or even 10^3 in various applications [2]. With the number of attributes increasing at such a high rate, it is very difficult for administrators to keep track of attributes whether they are accessed or modified correctly or not. By dividing the attributes into different categories based on their relative importance or sensitivity, it is comparatively easier to track only those attributes whose unintended modification can have the largest impact on the application or the system.

Practitioners as well as researchers have observed that IDS can easily trigger thousands of alarms per day, a number of which are triggered incorrectly by benign events [13]. Categorization of attributes helps the administrator to check only those alarms, which are generated due to malicious modification of sensitive data instead of checking all the attributes. Since the main objective of a database intrusion detection system is to minimize the loss suffered by the owner of the database, it is important to track high sensitive attributes with more accuracy.

If sensitive attributes are to be tracked for malicious modifications then we need to generate data dependency rules for these attributes. Unless there is a rule for an attribute, the attribute cannot be checked. If high sensitive attributes are accessed less frequently, then there may not be any rule generated for these attributes. The motivation for dividing attributes in different sensitivity groups and assigning weights to each group is to bring out the dependency rules for possibly less frequent but more important attributes. Once we have rules for these sensitive attributes, we can check them in each transaction and if any transaction does not follow the mined rules, it will be marked as malicious.

We discuss the main components of an IDS in the following subsections.

2.2 Security Sensitive Sequence Mining

The problem of finding sequences among the attributes along with the operations {read,write} is similar to the problem of mining sequential patterns. Mining sequences from large sets of data is a known problem. Agrawal et al [14] have proposed an algorithm for finding sequential patterns from data. In this algorithm, all the data items are considered at the same level without any weightage. We modify an existing sequential mining algorithm and make it security sensitive sequential mining by introducing weights for each attribute based on the sensitivity group. Higher the sensitivity of an attribute, higher is its weight. We have categorized the attributes in three sets : High Sensitivity (HS) attribute set, Medium Sensitivity (MS) attribute set and Low Sensitivity (LS) attribute set. The sensitivity of an attribute is dependent on the particular database application. Also, modification of the sensitive attributes are more important than reading those attributes from the point of view of integrity. For the same attribute say x, if $x \in HS$ then $W(x_w) > W(x_r)$, where W is a weight function, x_w denotes writing or modifying attribute x and x_r denotes reading of attribute x.

Given a schema, we categorize all the attributes into the above mentioned three sets based on their sensitivities and assign numerical weights to each set. Let $w_1, w_2, w_3 \in \mathbf{R}$, where $\mathbf{R}$ is the set of real numbers and $w_3 \leq w_2 \leq w_1$ are the weights of HS, MS and LS, respectively. Let $d_1, d_2, d_3 \in \mathbf{R}$ be the additional weights of the write operations for each category such that $d_3 \leq d_2 \leq d_1$. Let $x \in HS$ be an attribute which is accessed in a read operation. Then the weight given to x is w_1. If it is accessed in write operation then the weight given to x is $w_1 + d_1$.

TID	Attribute access sequence
1	$11_r, 13_w, 4_r, 8_r, 2_r, 16_r, 17_r, 14_r$
2	$7_r, 2_r, 7_r, 2_r, 14_r, 15_w$
3	$16_r, 17_r, 14_r, 14_r, 15_w, 17_w, 2_r, 7_w$
4	$11_r, 12_w, 2_r, 4_w, 16_r, 17_r, 14_r$
5	$2_r, 4_w, 2_r, 7_w, 7_r, 8_r, 2_r$
6	$11_r, 13_w, 4_r, 8_r, 2_r, 2_r, 4_w$
7	$14_r, 15_w, 4_r, 8_r, 2_r, 8_r, 2_r$
8	$7_r, 8_r, 2_r, 2_r, 2_r, 8_w, 5_w, 2_r, 4_w$
9	$8_r, 2_r, 14_r, 15_w, 7_r, 2_r$
10	$14_r, 15_w, 16_r, 17_r, 14_r, 14_r, 15_w, 17_w$

Fig. 1. Example transactions for the Sequence Mining Algorithm

Table Name	Column Name
Customer	$Name, Customer\_id, Address, Phone\_no$
Account	$Account\_id, Customer\_id, Status, Open\_dt, Close\_dt, Balance$
Account_type	$Account\_type, Max\_tran\_per\_month, Description$

Fig. 2. Bank database schema

For security sensitive sequence mining, we assign weights to each sequence based on the sensitivity groups of the attributes present in the sequence. The weight assigned to a sequence is the same as the weight of the most sensitive attribute present in that sequence. The weight assigned to each sequence also depends on the operation applied on the attributes.

The weights assigned to all the sequences are used in the second pruning step which calculates the support of each sequence in the transaction. If support value for any sequence is above the minimum support, the sequence is considered to be a frequent sequence. Let us assume that there is a sequence s with weight w_s. Let N be the total number transactions. If s is present in n transactions out of N transactions, then the support of sequence s would be:

$$Support(s) = (n * w_s) \; / \; N \qquad (1)$$

The effect of this weighted approach on sequence mining algorithm is significant. With this approach, sequences containing high sensitive attributes but accessed less in the transactions can become frequent sequences because each such sequence's count is enhanced by multiplying with its weight. The weighted support can now exceed the minimum support.

Consider the example transactions shown in Figure 1. There are 10 transactions. These transactions are generated from the bank database schema shown in Figure 2 with attributes encoded into integers. In Figure 3, the weight of each attribute is shown. These attributes are categorized into HS, MS and LS groups depending upon the sensitivity. First, these transactions are given input to a sequential pattern mining algorithm [14] for extracting the sequences using normal

Sensitivity Group	Attributes	Weights	Write Weights	Normalized Weights
HS	$7, 8, 13$	3	.25	.48
MS	$5, 16$	2	.25	.33
LS	$2, 4, 11, 12, 14, 15, 17$	1	.25	.19

Fig. 3. Weight table for the attributes used in the bank database

Sequence using Non-weighted Method	Sequence using Weighted Method
$< 4_r, 8_r, 2_r >, < 14_r, 15_w, 2_r >,$	$< 16_r, 17_r, 14_r, 15_w, 17_w, 7_w >,$
$< 2_r, 4_w >, < 2_r, 7_r >, < 2_r, 14_r >,$	$< 2_r, 4_w, 7_w, 7_r, 8_r >, < 7_r, 8_r, 8_w, 2_r, 4_w >,$
$< 16_r, 17_r, 14_r >,$	$< 7_r, 8_r, 2_r, 8_w, 4_w >, < 8_r, 2_r, 14_r, 15_w, 7_r >,$
$< 7_r, 2_r >, < 11_r, 2_r >$	$< 8_r, 14_r, 15_w, 7_r, 2_r >, < 11_r, 13_w, 8_r, 2_r, 4_w >,$
	$< 11_r, 13_w, 8_r, 2_r, 16_r >, < 13_w, 4_r, 8_r, 2_r >,$
	$< 7_w, 7_r, 8_r, 2_r >, < 2_r, 7_r, 14_r, 15_w >,$
	$< 7_r, 2_r, 14_r, 15_w >, < 14_r, 15_w, 2_r, 7_w >,$
	$< 14_r, 15_w, 2_r, 8_r >, < 14_r, 15_w, 8_r, 2_r >,$
	$< 4_r, 2_r, 8_r >, < 13_w, 8_r, 16_r, 17_r, 14_r >,$
	$< 13_w, 8_r, 2_r, 16_r, 14_r >$

Fig. 4. Mined sequence using Minimum Support value 25%

definition of support. These transactions and weights are also given as input to the proposed weighted sequential mining algorithm. Here, support values of the sequences are calculated using equation (1). In both the cases, minimum support is set to 25%. The sequences generated from the two algorithms are shown in Figure 4.

2.3 Read-Write Sequence Generation

In this subsection, we first define some of the terminologies used in the rest of the paper.

Definition 1. *A read sequence denoted as ReadSeq of attribute a_j is the sequence of the form $< a_{1_r}, a_{2_r}, a_{3_r}, ..., a_{k_r}, a_{j_w} >$, which is the sequence of attributes a_1 to a_k that are read before attribute a_j is written. All such sequences form a set named as read sequence set denoted by ReadSeqSet.*

Definition 2. *A write sequence denoted as WriteSeq of attribute a_j is the sequence of the form $< a_{j_w}, a_{1_w}, a_{2_w}, a_{3_w}, ..., a_{k_w} >$, which is the sequence of attributes a_1 to a_k that are written after attribute a_j is written. All such sequences form a set named as write sequence set denoted by WriteSeqSet.*

The sequences shown in Figure 4 are next used to generate read and write sequences. As per the definitions, *ReadSeq* and *WriteSeq* must contain at least one write operation. All the sequences that do not have any attribute with write operation, are not used for read and write sequence generation. A sequence that contains a single attribute does not contribute to the generation of dependency rules and hence will be ignored too. The read-write sequences are generated as follows.

Non-weighted Method		Weighted Method		
Read Set	Write Set	Read Set		Write Set
$< 14_r, 15_w >,$		$< 16_r, 17_r, 14_r, 15_w >, < 16_r, 17_r, 14_r, 17_w >,$		$< 15_w, 17_w, 7_w >,$
$< 2_r, 4_w >$		$< 16_r, 17_r, 14_r, 7_w >, < 2_r, 4_w >, < 2_r, 7_w >,$		$< 8_w, 4_w >,$
		$< 7_r, 8_r, 8_w >, < 7_r, 8_r, 2_r, 4_w >,$		$< 15_w, 7_w >,$
		$< 8_r, 2_r, 14_r, 15_w >, < 8_r, 14_r, 15_w >,$		$< 4_w, 7_w >,$
		$< 11_r, 13_w >, < 11_r, 8_r, 2_r, 4_w >,$		$< 13_w, 4_w >$
		$< 2_r, 7_r, 14_r, 15_w >, < 7_r, 2_r, 14_r, 15_w, >$		
		$< 14_r, 15_w >, < 14_r, 2_r, 7_w >, < 7_r, 8_r, 2_r, 8_r >$		

Fig. 5. Read Sequences and Write Sequences

For each write operation a_{j_w} in a sequence, add $< a_{1_r}, a_{2_r}...a_{k_r}, a_{j_w} >$ to *ReadSeqSet* where $a_{1_r}, a_{2_r}...a_{k_r}$ are the read operations on attributes a_1 to a_k before the write operation on attribute a_j. To generate write sequences, for each write operation a_{j_w} in a sequence, add $< a_{j_w}, a_{1_w}, a_{2_w}, ...a_{k_w} >$ to *WriteSeqSet* where $a_{1_w}, a_{2_w}, ...a_{k_w}$ are write operations on attributes a_1 to a_k after the write operation on attribute a_j. The read-write sequences generated from the mined sequences of Figure 4 are shown in Figure 5.

2.4 Weighted Data Dependency Rule Generation

There are two types of data dependency rules, namely, read rules and write rules. A read rule of the form $a_{j_w} \rightarrow a_{1_r}, a_{2_r}..., a_{k_r}$ implies that attributes a_1 to a_k are read in order to write attribute a_j. Write rule of the form $a_{j_w} \rightarrow a_{1_w}, a_{2_w}, ...a_{k_w}$ implies that after writing attribute a_{j_w}, attributes $a_{1_w}, a_{2_w}, ...a_{k_w}$ are modified. These rules are generated from the read and write sequences. Weighted data dependency rule generation uses weighted confidence. The confidence of the read and write rules are calculated by the following method.

Let R be a read rule of the form $a_{j_w} \rightarrow a_{1_r}, a_{2_r}, ...a_{k_r}$, generated from the read sequence $rs \in ReadSeqSet$. Let Count(a_{j_w}) and Count(rs) be the total count of the attribute a_{j_w} and that of rs among the total number of transactions. The weighted confidence of the rule R is defined as:

$$Confidence(C_R) = Count(rs) \ / \ Count(a_{j_w}) \qquad (2)$$

Count(a_{j_w}) is defined as follows:

$$Count(a_{j_w}) = \sum_{\forall \, Transaction \ T, \ a_{j_w} \in T \ and \ rs \notin T} (w_3 + d_3)$$
$$= \sum_{\forall \, Transaction \ T, \ rs \in T} max(W(rs)) \qquad (3)$$

Count(rs) is defined as:

$$Count(rs) = \sum_{\forall \, Transaction \ T, \ rs \in T} max(W(rs)) \qquad (4)$$

Non-weighted Method		Weighted Method	
Read Rules	Write Rules	Read Rules	Write Rules
$< 15_w \rightarrow 14_r >,$ $< 4_w \rightarrow 2_r >$		$< 17_w \rightarrow 16_r, 17_r, 14_r >, < 4_w \rightarrow 2_r >,$ $< 8_w \rightarrow 7_r, 8_r >, < 13_w \rightarrow 11_r >,$ $< 8_w \rightarrow 7_r, 8_r, 2_r >, < 15_w \rightarrow 14_r >,$ $< 7_w \rightarrow 2_r >$	$< 8_w \rightarrow 4_w >$

Fig. 6. Read and Write Dependency Rules with Confidence value 70%

ALGORITHM WDDRM:

Initialize two sets $ReadSeqSet = \{\Phi\}$, $WriteSeqSet = \{\Phi\}$ for storing read and write sequences respectively.

Initialize two sets $ReadRuleSet = \{\Phi\}$, $WriteRuleSet = \{\Phi\}$ for storing read and write rules respectively. Create a set weighted data dependency rules
  $WDDR = \{ReadRuleSet, WriteRuleSet\}$.

Execute sequential mining algorithm with minimum support $minSup$. At each step, calculate support of the sequences using equation (1).

For each sequential pattern P_i, where P_i contains at least one write operation
  IF $(a_{1_r}, a_{2_r}, ..., a_{k_r}, a_{j_w}) \in P_i$ and $a_{1_r}, a_{2_r}, ..., a_{k_r} \neq \emptyset)$ where a_{1_r} to a_{k_r} are all the read operation on attributes a_1 to a_k before a_{j_w}, the write operation on attribute a_j
    For each write operation a_{j_w}
      Generate read sequence $< a_{1_r}, a_{2_r}, ..., a_{k_r}, a_{j_w} >$ and Add to $ReadSeqSet$

  IF $(a_{j_w}, a_{1_w}, a_{2_w}, ...a_{k_w}) \in P_i$ and $a_{1_w}, a_{2_w}, ..., a_{k_w} \neq \emptyset)$ where a_{1_w} to a_{k_w} are all the write operation on attributes a_1 to a_k after a_{j_w}, the write operation on attribute a_j
    Generate write sequence $< a_{j_w}, a_{1_w}, a_{2_w}, ...a_{k_w} >$ and add to $WriteSeqSet$

For each read sequence rs of the form $a_{1_r}, a_{2_r}, ..., a_{k_r}, a_{j_w} \in ReadSeqSet$
  Construct read rule rr of the form $a_{j_w} \rightarrow a_{1_r}, a_{2_r}, ..., a_{k_r}$
  Calculate the confidence C of rr using equation (2)
  IF $(C \geq minConf)$ Add $rr \in ReadRuleSet$

For each write sequence ws of the form $a_{j_w}, a_{1_w}, a_{2_w}, ...a_{k_w} \in WriteSeqSet$
  Construct write rule wr of the form $a_{j_w} \rightarrow a_{1_w}, a_{2_w}, ..., a_{k_w}$
  Calculate the confidence C of wr using equation (2)
  IF $(C \geq minConf)$ Add $wr \in WriteRuleSet$

$Return\ WDDR = \{ReadRuleSet, WriteRuleSet\}$

Fig. 7. Weighted Data Dependency Rule Miner Algorithm

The rules generated from the read-write sequences are shown in Figure 6.

After the rules are generated, they are used to verify whether the incoming transactions are malicious or not. If an incoming transaction has a write operation, it is checked whether there are any corresponding read or write rules. If the write operation violates these rules, it is marked as malicious and an alarm is generated. Otherwise, normal operation proceeds. The complete algorithm for the weighted data dependency rule mining is shown in the Figure 7.

3 Experimental Results

We have carried out several experiments to show the efficacy of the developed method. The system has been developed using Java as front end and MS SQL 2000 Server as the back end database. We have used the bank database of Figure 2 for our experiments. Volunteers from our institute were invited to interact with the system and make malicious transactions. This was beneficial because the interaction by the volunteers helped us to capture real data that would be expected in a normal application. They were provided the schema and the information on sensitive attributes. The volunteers tried novel ways of committing malicious transactions since it was announced that scores would be awarded based on the total weight of attributes they could modify.

In the learning phase, we have generated a number of sets of training data with each set of size 10,000 transactions having different distributions. In one experiment, we have used the following distributions. Insert/Update=90%, Select=10%. We also choose the number of transactions containing most sensitive attributes in the training data as a parameter. We used 20% of the transactions with highly sensitive attributes in the training data. All these parameters are varied for different experiments. The support and confidence values are .25 and .70, respectively. Once the transactions are generated, we have run the non-weighted algorithm to generate the data dependency rules. After that, we have used WDDRM algorithm on the training data with weight ratios 1:2:3 for LS, MS and HS groups, respectively. In the experiments, we have taken additional weight of write operation as 0.25 for all the three categories. We used weights for different groups as another parameter. Dependency rules for each set of weights were finally generated.

In order to study relative performance, we have compared our work with the non-weighted dependency rule mining approach. We call this method as DDRM and use it for comparison. Figure 8(a) shows a comparison of $DDRM$ and $WDDRM$. The percentage of malicious transactions detected is plotted against the sensitivity ratio. When the weights of all three groups are equal, then $WDDRM$ reduces to $DDRM$. However, when distinct weights are assigned to the three groups, $WDDRM$ detects higher percentage of malicious transactions. $DDRM$ cannot be effectively applied in this situation. In Figure 8(b), comparative performances is shown for each sensitivity group. It is seen that $WDDRM$ outperforms $DDRM$ for more sensitive attributes.

Figure 9(a) shows the effect of the number of write operations on the performance of the intrusion detection system. As the number of write operations increases, the effectiveness of the system also increases. This is because write operations are required to generate the data dependency rules. If there are more write operations on attributes in the transactions, more rules are generated.

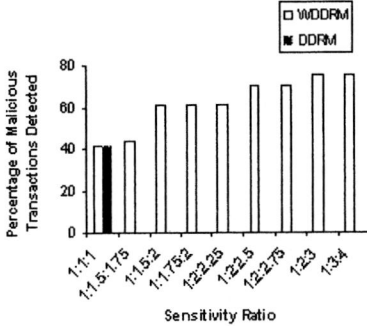 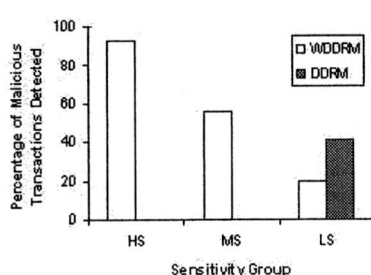

(a) Comparison of DDRM and WDDRM with different sensitivity ratio

(b) Comparison of DDRM and WDDRM for different sensitivity groups

Fig. 8.

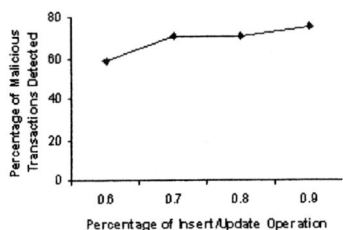 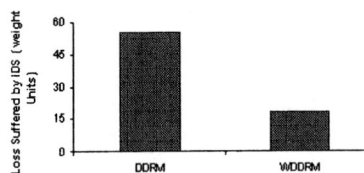

(a) Performance of $WDDRM$ algorithm with number of write operations

(b) Comparison of $DDRM$ and $WDDRM$ in terms of Loss Suffered by IDS

Fig. 9.

Hence, detection rate increases if more insert/update statements are present in the transactions. Figure 9(b) shows the loss suffered by the intrusion detection system in terms of weight unit using both the approaches. The ratio of weights used for the experiment is 3:2:1 for HS, MS and LS, respectively and distribution of Insert/Update=90% and Select=10%. Loss is computed by adding the weights of all the attributes whose malicious modifications are not detected by the IDS. It is evident from the figure that $WDDRM$ outperforms $DDRM$. This is because $WDDRM$ tracks the sensitive attributes in a much better way than $DDRM$ and hence overall loss is minimized.

4 Conclusions and Discussions

In this paper, we have identified some of the limitations of the existing data mining based intrusion detection systems, in particular, their incapability in treating database attributes at different levels of sensitivity. We proposed a novel weighted data dependency rule mining algorithm that considers the sensitivity of the attributes while mining the dependency rules. Experimental results show that our proposed algorithm performs better than some of the previous work done in this area. The sensitivity levels can be syntactically captured during data modeling through the E-R diagram notations.

Acknowledgements

This work is partially supported by a research grant from the Department of Information Technology, Ministry of Communication and Information Technology, Government of India, under Grant No. 12(34)/04-IRSD dated 07/12/2004.

References

1. J. Han, M. Kamber, *Data Mining: Concepts and Techniques*, Morgan Kaufmann Publishers (2001).
2. U. Fayyad, G. P. Shapiro, P. Smyth, *The KDD Process for Extracting Useful Knowledge from Volumes of Data*, Communications of the ACM, pages 27-34 (1996).
3. R. Bace, P. Mell, *Intrusion Detection System*, NIST Special Publication on Intrusion Detection System (2001).
4. E. Lundin, E. Jonsson, *Survey of Intrusion Detection Research*, Technical Report Chalmers University of Technology, (2002).
5. W. Lee, S.J. Stolfo, *Data Mining Approaches for Intrusion Detection*, Proceedings of the USENIX Security Symposium, pages 79-94 (1998).
6. D. Barbara, J. Couto, S. Jajodia, N. Wu, *ADAM: A Testbed forExploring the Use of Data Mining in Intrusion Detection*, ACM SIGMOD, pages 15-24 (2001).
7. C. Y. Chung, M. Gertz, K. Levitt, *DEMIDS: A Misuse Detection System for Database Systems*, IFIP TC-11 WG 11.5 Working Conference on Integrity and Internal Control in Information System, pages 159-178 (1999).
8. V.C.S. Lee, J.A. Stankovic, S.H. Son, *Intrusion Detection in Real-time Database Systems Via Time Signatures*, Proceedings of the Real Time Technology and Application Symposium, pages 124-133 (2000).
9. S.Y. Lee, W.L. Low, P.Y. Wong, *Learning Fingerprints for a Database Intrusion Detection System*, Proceedings of the European Symposium on Research in Computer Security, pages 264-280 (2002).
10. D. Barbara, R. Goel, S. Jajodia, *Mining Malicious Data Corruption with Hidden Markov Models*, IFIP WG 11.3 Working Conference on Data and Application Security, pages 175-189 (2002).
11. Y. Zhong, X. Qin, *Research on Algorithm of User Query Frequent Itemsets Mining*, Proceedings of the Machine Learning and Cybernetics, pages 1671-1676 (2004).
12. Y. Hu, B. Panda, *A Data Mining Approach for Database Intru sion Detection*, Proceedings of the ACM Symposium on Applied Computing, pages 711-716 (2004).
13. K. Julisch, M. Dacier, *Mining Intrusion Detection Alarms for Actionable Knowledge*, Proceedings of the ACM SIGKDD Conference on Knowledge Discovery and Data Mining, pages 366-375 (2002).
14. R. Agrawal, R. Srikant, *Mining Sequential Patterns*, Proceedings of the International Conference on Data Engineering, pages 3-14 (1995).

On Robust and Effective K-Anonymity in Large Databases

Wen Jin[1], Rong Ge[1], and Weining Qian[2]

[1] School of Computing Science, Simon Fraser University
{wjin, rge}@cs.sfu.ca
[2] Department of Computer Science, Fudan University
wnqian@fudan.edu.cn

Abstract. The challenge of privacy-preserving data mining lies in respecting privacy requirements while discovering the original interesting patterns or structures. Existing methods loose the correlations among attributes by transforming the different attributes independently, or cannot guarantee the minimum abstraction level required by legal policies. In this paper, we propose a novel privacy-preserving transformation framework for distance-based mining operations based on the concept of privacy-preserving MicroClusters that satisfy a privacy constraint as well as a significance constraint. Our framework well extends the robustness of the state-of-the-art k-anonymity model by introducing a privacy constraint (minimum radius) while keeping its effectiveness by a significance constraint (minimum number of corresponding data records). The privacy-preserving MicroClusters are made public for data mining purposes, but the original data records are kept private. We present efficient methods for generating and maintaining privacy-preserving MicroClusters and show that data mining operations such as clustering can easily be adapted to the public data represented by MicroClusters instead of the private data records. The experiment demonstrates that the proposed methods achieve accurate clusterings results while preserving the privacy.

1 Introduction

With the rapidly increasing amounts of data stored in electronic formats, the concerns about privacy of personal information have emerged globally. For example, bank databases with transactional information about every aspect of business which are now measured in gigabytes and terabytes, contain much sensitive information such as address, account balance, credit card number etc. Although data mining is a useful tool for such large databases and can discover valuable, non-obvious information, in the absence of adequate safeguards it can also jeopardize information privacy [16].

These privacy concerns have triggered regulations and laws protecting privacy in data collection and publishing. In particular, health-related data are very sensitive and most countries have established corresponding rules. For example, according to the United States Health Insurance Portability and Accountability Act (HIPAA), health records have to be de-identified before they can be released to some third party. For example, ZIP codes have to be generalized to their first three digits or replaced by 000 for units with 20,000 or fewer persons [8]. The HIPPA requirements correspond to different range constraints for the individual attributes such as "the first 3 digits" minimum range constraint(range threshold) on ZIP codes. Alternatively, a single global range constraint can be set by choosing the maximum of the minimum range thresholds over all dimensions. The HIPPA requirements also imply a significance constraint, for example, "20,000" is the minimum number of de-identified patients in a published group.

The topic of privacy-preserving data mining, has been initialized by [3], where a randomized approach [1, 2, 14] is proposed by adding random noise to all data records to protect the privacy and reconstructing the original data distribution to generate new dataset for data mining tasks such as classification. Another approach is by data generalization which hides individual record via generalized values. A typical model in this approach is k-anonymity [4, 20, 21] which generalizes the attribute values of the records such that for any record, there are at least k other records in the dataset from which it cannot be distinguished. The randomized approach meets the minimum range constraint required in privacy policies, however attributes are assumed to be *independently distributed* and thus the reconstructed data distribution may not accurately reflect the *correlations* among multiple attributes. The k-anonymity model, on the other hand, specifies a significance constraint, but no range constraint. Thus, the original data can be estimated very accurately from the anonymized data in the case of k similar or even identical records.

Figure 1 depicts a small sample database where the k-anonymity method will output the three dashed circles with radius r. The k-anonymity method condenses the k most similar (or even the same) objects into an anonymized group, so that the attacker can estimate the private attribute values with confidence interval of r which can become arbitrarily small. In addition, if the attacker has prior-knowledge of some original record, other information can be estimated step by step. For example if he knows one attribute x of a point is 20, from the published information of k-anonymity, he can infer another attribute y of this point with 100% confidence within a range of r.

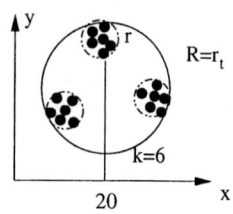

Fig. 1. A dataset

On the other hand, a model with significance and range constraints outputs one big solid circle, where the radius R is equal to r_t. This model on average has a better protection for the privacy. Even if the attacker knows that the x value of a point is 20, he can only guess the other attribute y of this point with 100% confidence within a range of R $(>> r)$.

In this paper, we propose a novel privacy-preserving transformation scheme which meets the constraints of privacy policies and is robust to different kinds of privacy attacks. Our framework, which well extends the k-anonymity model, is based on so-called privacy-preserving MicroClusters (*PPMicroClusters*) that satisfy a privacy constraint (minimum radius) as well as a significance constraint (minimum number of corresponding data records). Within the constraints given by these requirements, we try to minimize the overall cost of the *PPMicroClusters* resulting in maximum accuracy of subsequent analysis. These sphere-like MicroClusters are made public for data mining purposes, but the original data records are kept private.

The contributions of this paper are as follows:

(1) We propose the novel concept of PPMicroClusters that satisfy a privacy constraint (minimum radius) as well as a significance constraint (minimum number of corresponding data records).
(2) We develop an efficient local-search-based method for generating and maintaining PPMicroClusters in a dynamic database with updates.
(3) Our experimental evaluation on synthetic and real data sets demonstrates that the proposed methods achieve accurate clustering results while preserving the privacy.

The rest of the paper is organized as follows. Section 2 surveys related work. Section 3 introduces the new model. Section 4 and Section 5 present the algorithms. We give experimental results in Section 6 and conclude the paper in Section 7.

2 Related Work

The research on privacy preserving data mining was initialized by [3] and the current approaches to enforce privacy-protection in data mining applications can be categorized as follows:

●**Randomized Approaches:** These methods deliberately introduce noise in the data that hides data of the individual records. However, patterns summarizing the trends in the dataset are the information most relevant to data analysts that does not really need to access individual records. Novel reconstruction techniques to accurately reconstruct the distribution of original data values are presented in [3] and [1]. A perturbation mechanism was proposed in [2], where the model parameters are themselves characterized as random variables, and demonstrate that this feature provides improvements in privacy at a very marginal cost in accuracy. However, this approach only assumes data distributes independently in each attributes.
●**Cryptographic Techniques:** [18] introduces the problem of distributed data mining between more parties such that confidential information of any party is not disclosed. The method uses the results in [24] which forms the basis for applying cryptographic techniques to privacy-preserving in data mining. The challenge of this approach is the cost of computation/communication in cryption/decryption for large databases.
●**Data Generalization:** The major idea in this category is to transform the values of certain attributes that are marked for having the potential to breach

the privacy of a dataset and are called Quasi-Identifiers in [20]. The k-anonymity problem is discussed in [20] and [21], where a data transformation technique is described which makes use of concept hierarchies to bring about generalizations in the *Quasi-Identifiers* such that at least k or more tuples satisfy any given value combination for the *Quasi-Identifiers*. [17] presents a method based on a set of geometric data transformation primitives, to maintain the correlations. But the original attribute values can still be easily estimated with small variance.

•**Miscellaneous:** [9] discusses a novel genetic algorithm based method to achieve privacy protection in datasets. It uses a metric to evaluate the information loss due to hierarchy-based generalization. Schloer [22] develops a matrix based method for multidimensional data transformation. The authors in [3] use a measure that defines privacy in terms of variance, while in [1] a privacy measure based on the differential entropy of the generated distribution is proposed. The work of [14] applies a slightly different measure that considers the privacy of the model with respect to the actual objects in the dataset. Some recent work [4,5,6,12] extend k-anonymity in performance or the quality of preserved privacy. While *none of them considers the minimum radius requirement of the anonymized data, which means the precise estimate of the original data can still be easily reached if the anonymized data are bound in very small dense space.* The recent work on privacy-preserving clustering incudes *distribution-based method* where the data for a single entity are vertically split across multiple sites [23]; or the data are distributed horizontally [14], [10], among the sites. Another approach in privacy-preserving clustering is *centralization-based method* including [17] which proposes a target-oriented transformation method based on geometric transformations of digital images.

3 Privacy-Preserving MicroClustering

In this section, we introduce a framework based on privacy-preserving Micro-Clusters (*PPMicroClusters*) that satisfy a privacy constraint (minimum radius) as well as a significance constraint (minimum number of corresponding data records). Given a set of d-dimensional records, $X = \{\overline{X_1} \ldots \overline{X_k} \ldots, \overline{X_n}\}$, and each $\overline{X_i}$ is a record containing d dimensions which are denoted by $\overline{X_i} = (x_i^1 \ldots x_i^d)$. In order to perform a robust data transformation for distance-based data mining, we propose to use MicroClusters to approximate the subsets of any dataset with an user-specified privacy threshold.

Definition 1 (MicroCluster). *The MicroCluster C for a d-dimensional dataset X is defined as the $(3 \cdot d + 2)$-tuple $(n, \overline{CF1(C)}, \overline{CF2(C)}, \overline{CF3(C)}, r)$, where $\overline{CF1}$ and $\overline{CF2}$ each corresponds to the linear sum and the sum of the squares of the data values for each dimension respectively. The number of data points $|C|$ is maintained in n, the centroid of $\overline{X_1} \ldots \overline{X_n}$ is $\overline{CF3(C)} = \frac{\overline{CF1(C)}}{n}$. The radius of the MicroCluster is $r = \max_{j=1}^{n} \sqrt{(\overline{X_j} - \overline{CF3(C)})^2}$.* □

Note any point p which is within the radius from the center of a MicroCluster, will be assigned to this MicroCluster so that p may belong to multiple MicroClusters.

Definition 2 (Privacy Preserving MicroCluster). *Given a radius r_t (privacy threshold) and a number of objects n_t (statistical significance threshold). We call a MicroCluster $C = (n, \overline{CF1(C)}, \overline{CF2(C)}, \overline{CF3(C)}, r)$ a Privacy-Preserving MicroCluster, (PPMicroCluster) if the constraints: (1) $r \geq r_t$ (2) $n \geq n_t$ are satisfied.* □

PPMicroClusters are "small clusters" of points with sufficient statistics and privacy protection that can well represent the overall original data distribution including the inter-attribute correlations. If users wish to improve the privacy, they can increase the threshold r_t. Similarly, users can improve the significance by increasing the threshold n_t. However, the radius threshold should not become too large, because otherwise the accuracy of a subsequent data mining method will deteriorate. This tradeoff motivates the following definition of a privacy-preserving MicroClustering that best reflects the original cluster structure.

Definition 3 (Privacy Preserving MicroClustering). *Given a d-dimensional database X, the radius threshold r_t and the number of objects threshold n_t, the task of Privacy-Preserving MicroClustering (PPMicroClustering) is to find a set of PPMicroClusters $C_1, \ldots, C_m$ satisfying r_t and n_t such that the cost of $\sum_{j=1}^{m} \sum_{i=1, \overline{X_i} \in C_j}^{N} dist^2(\overline{X_i}, C_j)$ is minimized.* □

The smaller the sum of the squared distances of points to the center, the more PPMicroClusters, which means the smaller the size of each PPMicroCluster, and the less overlap among them. This motivates the objective of minimizing the sum of squared distances.

Theorem 1. *PPMicroClustering is an NP-hard problem.*

Proof. Since the special case of PPMicroCluster, where parameters n_t, r_t, m are set to 0, 0 and K respectively, is equivalent to the classic K-clustering problem [7], PPMicroClustering is an NP-hard problem. □

We use the cases in Figure 2 to illustrate the differences of our framework compared to the k-anonymity model. The *circle* indicates *the minimum radius constraint* r_t in our model and both models have the minimum points constraint

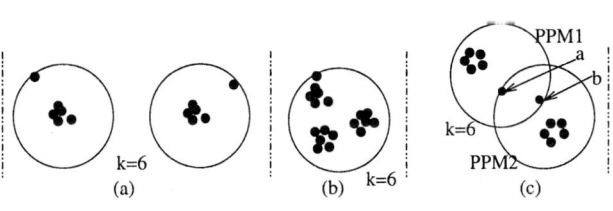

Fig. 2. (a) Case 1 (b) Case 2 (c) Case 3

$n_t = k = 6$. In case 1 (Figure 2(a)), when both models generate exactly the same two far-apart PPMicroClusters of 6 points, both achieve the same degree of privacy protection. In case 2 (Figure 2(b)) where points are distributed densely. The k-anonymity model will return 3 MicroClusters, while our model generates only one PPMicroCluster, leading to a much wider confidence interval when an adversary tries to estimate the private attributes. In case 3 (Figure 2(c)), k-anonymity assigns the points in the overlap area uniquely to the closest PPMicroCluster, i.e., "a" is assigned to PPM_1 and "b" belongs to PPM_2. Notice that now the real radius of PPM_1 or PPM_2 is less than the minimum radius constraint (i.e., the distance between the center and the farthest point in each PPMicroCluster after assigning "a" and "b" becomes less than r_t). However, we assign "a" and "b" to both PPM_1 and PPM_2, both PPMicroClusters will be "safer" in privacy since it would be harder for attacker to guess the individual record under the user-specified privacy standard. So in general, our model provides better privacy protection than k-anonymity.

Note that the constraint n_t in both our model is conceptually different from the constraint k in k-anonymity model. The value of k determines the privacy degree of the microcluster, while the value of n_t determines the statistical significance (accuracy) of the microcluster. To reach the expected privacy degree obtained by our model, k-anonymity needs to increase the value of k for the data in dense area, but meantime it has more difficulty to identify desired microclusters with good accuracy in sparse area. The generated PPMicroCluster can be used published for query processing and data mining applications with several forms: (1) Represent a PPMicroCluster by its center, radius and number of points contained. (2) Represent a PPMicroCluster by the CF-value obtained from aggregating all points located in a PPMicroCluster. (3) Randomly generate points in each PPMicroCluster.

4 A Local-Search-Based Algorithm

In this section, we introduce an efficient local search based algorithm to generate PPMicroclusters. The algorithm proceeds in the following three steps:

> **While** the dataset contains unmarked data points **Do**
> Step 1. Pick an unmarked data point p
> Step 2. Find a PPMicroCluster for p
> Step 3. Mark all data points in this PPMicroCluster

Fig. 3. Algorithm Sketch

Initially, the algorithm picks an unmarked data point randomly in step 1, and generates a PPMicroCluster for the point which is explained in section 4.1. Then it picks next seed considering the new generated PPMicroCluster. The details on how to pick next seed are elaborated in section 4.2.

4.1 Forming PPMicroCluster

The major step in this algorithm is the second step. We have the following observations to guide our algorithm design. Before going into details we define several notations to facilitate our explanation. Let denote point p's $k-1$ nearest neighborhood as kNN_p (here $k = n_t$, p is included), denote the radius of kNN_p as r_{kNN_p}. The mean vector of points in kNN_p is denoted as μ_p. If p is one of k nearest neighbors of μ_p, we call this neighboring region kNN_{μ_p} accordingly.

Lemma 1. *The sum of squared distance of kNN_p is greater than or equal to the sum of squared distance of kNN_{μ_p}*

Proof. For k points $p_1, p_2, \ldots, p_k$ which are members of kNN_p, and $\bar{p}$ is the means (μ_p) of such p points, i.e., $\bar{p} = \frac{\sum_{i=1}^{k} p_i}{k}$. Without loss of generality, suppose $p = p_1$, $\sum_{i=1}^{k}(p_i - \bar{p})^2 = \sum_{i=1}^{k} p_i^2 + k \cdot \bar{p}^2 - 2\bar{p} \cdot \sum_{i=1}^{k} p_i = \sum_{i=1}^{k} p_i^2 + k \cdot \frac{(p_1+\ldots+p_k)^2}{k^2} - 2 \cdot \frac{p_1+\ldots+p_k}{k} \cdot (p_1+\ldots+p_k) = \sum_{i=1}^{k} p_i^2 - \frac{(p_1+\ldots+p_k)^2}{k}$. So $\sum_{i=1}^{k}(p_i - p)^2 - \sum_{i=1}^{k}(p_i - \bar{p})^2 = \sum_{i=1}^{k} p_i^2 + k \cdot p^2 - 2p \sum_{i=1}^{k} p_i - (\sum_{i=1}^{k} p_i^2 - \frac{(p_1+\ldots+p_k)^2}{k}) = k \cdot p_1^2 - 2p_1 \cdot (p_1+\ldots+p_k) + \frac{(p_1+\ldots+p_k)^2}{k} = \frac{k \cdot p_1 + (p_1+\ldots+p_k)^2}{k} \geq 0$. □

If there is a point p' which is in kNN_p but not in kNN_{μ_p}, it shows that the distance between p' and μ_p is greater than the radius of kNN_{μ_p}. Thus, we conclude that the sum of squared distance of k points in kNN_{μ_p} to μ_p is smaller than or at least equal to the sum of squared distance of k points in kNN_p to p. In another word, kNN_{μ_p} is more compact. Since our objective function is the sum of square distance between any point and the center, the compacter the PPMicroClusters are, the smaller the overall sum of squared distance is. Hence we can base on this observation by replacing p by a "better" μ_p as a center to form a compact PPMicroCluster.

We start from finding the kNN_p for the data point p which is picked in step 1 and choose the mean vector of those points as a potential PPMicroCluster center. The PPMicroCluster for p is denoted by PPM_p. Since we aim to find a PPMicroCluster to cover p as much compact as it can, so we still need consider every point q in kNN_p as candidates and find corresponding potential PPMicroclusters for p in the following manner: find kNN_q, denote the mean vector of points in kNN_q as μ_q. If any kNN_{μ_q} exists (i.e., k-nearest neighbors of q include p), then determine potential PPMicroCluster PPM_q from center μ_q by extending radius $r_{kNN_{\mu_q}}$ of kNN_{μ_q} if r_t is

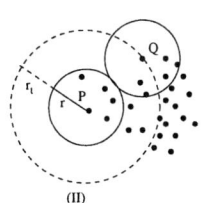

Fig. 4. Pick Q

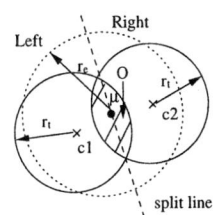

Fig. 5. Split

not satisfied. Furthermore, choosing a PPMicroCluster PPM' of these potential PPMicroClusters with a minimal sum of squared distance, denote its center as **mcenter**, which is different from the mean vector of all the points in PPM' since PPM' often has more than $k = n_t$ objects to meet r_t constraint. All the points including p in PPM' are then marked. Due to the different cardinality of potential PPMicroclusters, we make the decision based on the average squared distance of each PPMicroCluster.

In general, the strategy described above is trying to assign p to a better PPMicroCluster from possible PPMicroClusters which are not too far away from p and compact enough.

Remark. The k nearest neighbor search function considered in this section is slightly different from the general kNN search in that it needs not only consider the distance of the k-th neighbor but also retrieve the distance of the k-th neighbor when only unmarked points are issued. The details of obtaining guarantee of r_t is presented in section 4.2.

One minor issue left in step 2 is when the number of unmarked points is small, the r_{kNN_p} could be far less than the k nearest neighbor distance for only unmarked points (refereed to kNN'_p). Thus such a criterion holds: when $2 \times r_{kNN_p} \leq r_{kNN'_p}$, p is assigned to the closest PPMicroCluster generated before.

4.2 Picking the Next Seed

After forming a PPMicroCluster PPM_p for point p, it is very important to choose a "good" point as a "seed" to start another round of forming PPMicroClusters. As illustrated in Figure 4, we employ a range query on μ_p with radius r_{PPM_p} (bold line) plus the radius constraint r_t (dashed line), then picking the point which is located within radius $[r_{PPM_p}, r_{PPM_p} + r_t]$ and unmarked as the seed. Figure 4 also shows point Q is valid candidate for the next seed. When proceeding the seed searching, we consider points returned from the range query in the order of farthest point w.r.t the query center to the closest one. If no candidate is available, a randomly picked point will serve as the next seed.

The strategy we use in the seed selecting step follows the observation that the forming two touch PPMicroClusters is the most efficient way to cover all the corresponding points.

4.3 Handling Outliers

In the proposed framework, outliers will lead to MicroClusters that either have a too large radius or do not meet the significance threshold, and may consequently have more overlaps with existing MicroClusters. Note that we may apply different measures from the literature to define outliers, for example, the distance-based outlier definition [11,19]. In this paper, we measure the outlier at the abstraction level of MicroClusters which is more appropriate in our context than a definition at the abstraction level of individual records.

Definition 4 Outlying MicroCluster *Given any MicroCluster C of r and n, a percentage threshold of p, $0 \leq p \leq 1$, if $n \leq p \cdot n_t$, C is called an Outlying MicroCluster.* □

Concerning the treatment of outliers when refining the MicroClusters, we propose the following three methods:

- Method A: No special treatment of outliers, i.e., according to the algorithm presented in the previous subsection, outliers would be merged with the nearest PPMicroCluster until they meet the privacy constraints. The drawback is the radius of the merged PPMicroClusters may become (too) large.
- Method B: Detect which MicroClusters are outliers and remove them. The rationale is the MicroClusters will be used as input for a clustering algorithm, and for that purpose outliers are not of crucial importance. The drawback is the transformed, public dataset loses some information of the original, private data set.
- Method C: Detect which MicroClusters are outliers, increase their radius to minimum radius and add enough "fake objects" to these MicroClusters until they satisfy our minimum number of objects requirement (i.e., become PPMicroClusters). The rationale is it does not blow up other PPMicroClusters (which would imply a loss of clustering accuracy) while preserving the privacy of outlying MicroClusters. The drawback is increasing the number of PPMicroClusters.

5 An Incremental Local-Search Algorithm

In each PPMicroCluster, we keep information as follows:

1. PPMicroCluster **mcenter** (which is denoted in 4.1) c, mean μ and standard deviation δ of points in this PPMicroCluster.
2. The number of points located in the radius of r_t around c denoted as *core region*, where r_t is the radius constraint.
3. The number of points located outside of the core region.

When a new data point p_{new} comes, it is assigned to *mcenter* c of the closest PPMicroCluster. If p_{new} is located in c's core region. The number of data points in the core region, mean value vector and standard deviation, are updated accordingly. If a PPMicroCluster contains less than $2k$ points, we do not split it. However, when the number of points in a PPMicroCluster exceeds $2k$, a split operation is not necessarily triggered, but depending on the tradeoff between keeping all the points in this PPMicroCluster and splitting it into two or more PPMicroClusters. Obviously, the benefit of splitting a MicroCluster aims to decrease the sum of squared error. On the other hand, if the new split PPMicroClusters are highly overlapped, the points located inside the overlapping region contribute twice to the objective function, and the resulting qualities in accuracy are not desired. Thus, we introduce two criteria to evaluate the necessity of a split operation.

- **The loss-and-gain criterion:** evaluate the cost difference in the unsplit PPMicroCluster and the split PPMicroClusters in terms of the sum of squared error. If the gain wins over the loss, the split occurs; otherwise no split operation on this PPMicroCluster.
- **Effective radius criterion:** evaluate the "effective radius" r_e ($r_e \leq r$) of PPMicroCluster, which refers to the radius covering a region which a high percentage of points are included. If r_e is small, which means most of points in this PPMicroCluster locates densely around the center, so there is no necessary to split, just keep it as before. For the sake of quantifying the tradeoff in gain-and-loss criterion, we first estimate the possible overlap. To simplify the analysis, assuming points are uniformed distributed in the PPMicroCluster (note that this assumption is reasonable for estimating the contribution of points to the new split clusters, and other non-uniformed distributions have less trade-off to consider split).

The example depicted in Figure 5 shows that the original PPMicroCluster with radius r_e is cut in the middle and form two core regions on each side. The centers of two core regions are label c_1 and c_2 which are also the means on each half. Since the radius of a core region is the predefined r_t, we can estimate the size of the overlapping region marked as O in the figure as well as the sum of squared error. Thus, we have the following equation for quantifying the loss: $loss = (\frac{dist(c1,c2)}{2})^2 * num$, where num is the number of points in the overlapping region O and $dist(c1, c2)$ is the distance between two new centers. It is easy to see that $dist(c1, c2) = r_e$. Besides, the number of points in O is proportional to the width of O, i.e., $num = m * [(2r_t - r_e)/2r_e]$ where m is the total number of data points in a PPMicroCluster. Finally, the simplified equation becomes $loss = (\frac{r_e}{2})^2 * \frac{2r_t-r_e}{2r_e} * m$.

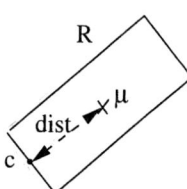

Fig. 6. Gain after split

Next we quantify the gain of moving the center of a MicroCluster to the mean of the data points located in the MicroCluster based on the following observation.

Lemma 2. *Given a set of points $P = \{p_1, \ldots, p_m\}$, and mean of them μ. The difference between the sum of squared distance of points in P to a center c and to the mean μ is $m \times dist(c, \mu)^2$.*

Proof. This observation can be proved easily. Consider the case showing in Figure 6, the mcenter of the split PPMicroCluster is labeled c and the mean of the data points in region R is μ. Based on the triangle inequality, the corresponding difference of sum of squared distance is at most $dist(c, \mu)^2 * num = dist(c, \mu)^2 * m$. □

Now we compare the difference between of loss and gain, if gain wins over loss, then the split operation can be employed; otherwise no split occurs in the current

PPMicroCluster. Based on lemma 2, we conclude that the gain of splitting a PPMicroCluster into two MicroClusters is at least $m \times (\frac{dist(c1,c2)}{2})^2$, which is also $m \times (\frac{r_e}{2})^2$.

In order to estimate the "effective" radius r_e ($r_e \leq r$) of current PPMicroCluster, we first define r_e to be the radius of a region which includes a high percentage (say 95%) of points in the PPMicroCluster. Afterwards, if the mean and the standard deviation of a random variable are labeled X with respect to μ_x and δ_x respectively, based on the Chebyshev's inequality theorem [15], for any positive constant t, $Pr(|X - \mu_x| < t) \leq 1 - \frac{\delta_x^2}{t^2}$.

Therefore, estimating the radius of a PPMicroCluster is equivalent to find the t such that $1 - \delta^2/t^2 = 95\%$. If t is less than r_t, the estimated radius, r_e, is set to r_t. Otherwise, the estimated radius is t.

In general, these two criteria help to provide a better maintenance of PPMicroClusters in a dynamic environment.

6 Experimental Evaluation

We have conducted an experimental evaluation of our method using a synthetic dataset and a real dataset. The synthetic dataset consists of 100,000, five dimensional records with 10 clusters generated from Gaussian distributions. The real dataset is a set of 50,000, seven dimensional health records. The number of "natural" clusters (10) in this dataset was determined by clustering with a series of different K-values and choosing the clustering, i.e., K-value, with the highest silhouette coefficient [13]. To evaluate the quality of the PPMicroClusters, we apply K-means clustering algorithm on the microclusters generated by our method and Charu's condensation method [4] respectively, and compare the results measured by (1) the accuracy of the clustering based on the PPMicroClusters (2) the degree of the privacy achieved. Both measures are analyzed with respect to the two parameters of PPMicroClustering, minimum radius threshold r_t and minimum number of objects threshold n_t.

We have implemented our methods in C++. All the experiments were conducted on an Intel 1GHZ processor with 512M RAM, 40G hard disk, running Windows XP.

Clustering Accuracy: The quality of clusters is evaluated by comparing the results of our modified K-means method on PPMicroClusters with the results of the traditional K-means method on the original database. We evaluate both the external quality and the internal quality of the clustering results. (1) The external quality will be measured by the entropy with respect to the given "true" cluster labels in the original dataset. For both the synthetic data and the real data, we take the results of K-means method on the original dataset as a substitute for the true class labels (their entropies are zero), and investigate the influence of changing the values of minimum radius r_t and n_t to the entropies. For the modified K-means clustering results of PPMicroClusters, the entropy of each cluster is measured based on the ratio of the number of objects with "true" label in the current cluster to the number of all objects in this cluster. (2) The

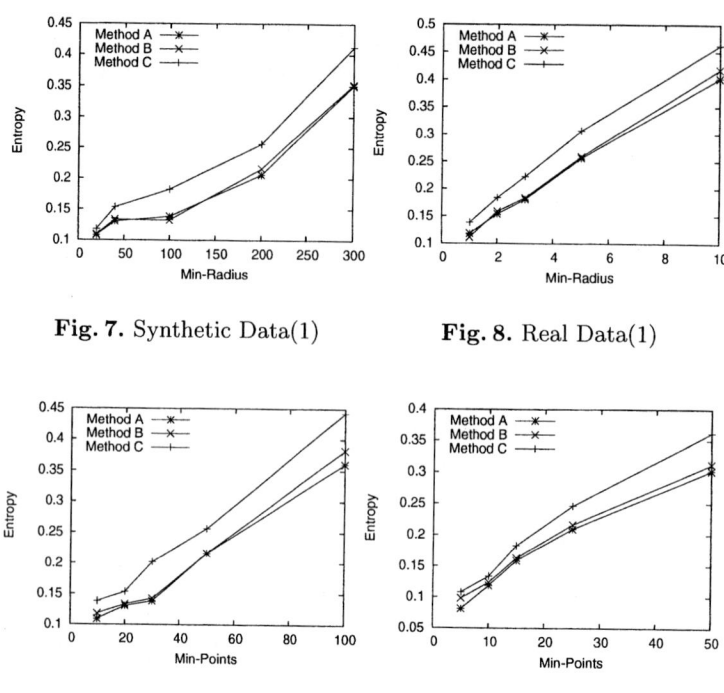

Fig. 7. Synthetic Data(1) **Fig. 8.** Real Data(1)

Fig. 9. Synthetic Data(2) **Fig. 10.** Real Data(2)

internal quality is measured by the sum-of-squares of the K clusters with respect to the closest centroids of clusters. That is, after K centroids are obtained by applying our modified K-means method over PPMicroClusters, the total sum of distances of all objects to the closest cluster centroids is derived and can be compared with the cost of baseline K-means method over the original dataset. We investigate the impact of changing values of r_t to the clustering cost.

The entropy evaluations with respect to the minimum radius r_t are shown in Figures 7 and 8 where $n_t = 10$. For both datasets, the entropies increase approximately linearly with increasing r_t values. PPMicroClusters can be obtained with three different ways of handling outliers (introduced in subsection 4.3 as method A, method B and method C). All three methods achieve relative good accuracy (entropies < 0.4). The best one is PPMicroClustering method A (i.e., outliers are assigned to the nearest PPMicroCluster) which achieves overall the lowest entropy values. We argue that this is due to the fact that the objects in the same PPMicroCluster always belong to the same cluster, and also outliers are always identified in the same (nearest) cluster as the traditional K-means method does. The PPMicroClustering method B (with removing outliers) obtains the second lowest entropies, and the PPMicroClustering method C (with "fake objects") yields clusters with larger entropies than the above two methods.

We also performed experiments to evaluate the clustering accuracy with respect to increasing n_t in Figures 9 and 10 where $r_t = 20$. For both datasets, all three methods achieve relatively low entropies, and the entropies increase

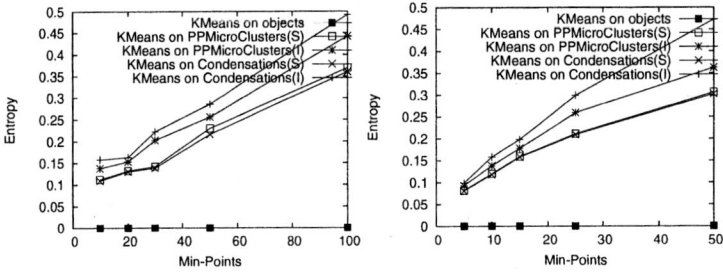

Fig. 11. Synthetic Data(3)   **Fig. 12.** Real Data(3)

steadily with increasing n_t values. The PPMicroClustering method A is slightly lower than the PPMicroClustering method B in entropy based on the similar reason as for the changing values of r_t (the increase of r_t leads to an increase of n_t and vice versa). The PPMicroClustering method C again ranks third w.r.t. entropy. Since method A outperformed the other outlier handling methods, we use method A in all the following experiments. We compare the clustering accuracy on PPMicroClusters generated by PPMicroClustering with that by Charu's Condensation method in both static and incremental ways (as shown in Figures 11 and 12). PPMicroClusters generated by PPMicroClustering static method are slightly less accurate than Condensation static method, meanwhile privacy constraint r_t are guaranteed. Since our incremental method evaluates split criteria of loss-and-gain and "effective" radius, it always prevents a microcluster from unnecessarily splitting into several "bad" microclusters. So the generated microclusters are always more accurate than those generated by Charu's incremental method.

The cost evaluations with respect to the minimum radius n_t are shown in Figures 13 and 14. Here $r_t = 20$ and we report K-means results on microclusters by PPMicroClustering and Charu's Condensation method in both static and incremental methods. The cost increases moderately with increasing minimum points n_t. The clustering cost on microclusters generated by static PPMicroClustering is almost the same as by Charu's static Condensation, while PPMicroClusters have privacy guarantee r_t. As indicated by the costs, the clustering cost on microclusters generated by both incremental methods are higher than by their static methods. Similar to the previous reason, the clustering cost on microclusters generated by our incremental method is slightly lower than by Charu's incremental method.

Privacy Measure. Since the generation of the PPMicroClusters is generalizing all attributes simultaneously, the level of privacy can be measured by as follows: (1) the variance of the radius of the PPMicroClusters; (2) the average radius of the PPMicroClusters. We analyze the influence of the changing value of r_t on these two measures. The variance of radius r is measured by $Var(r) = E[(r-r_t)^2] = \frac{\sum_{i=1}^{m}(r_i-r_t)^2}{m}$, where $r_1, \ldots, r_m$ are the radii of m PPMicroClusters. Here, to examine the privacy degree of PPMicroClusters, our radius

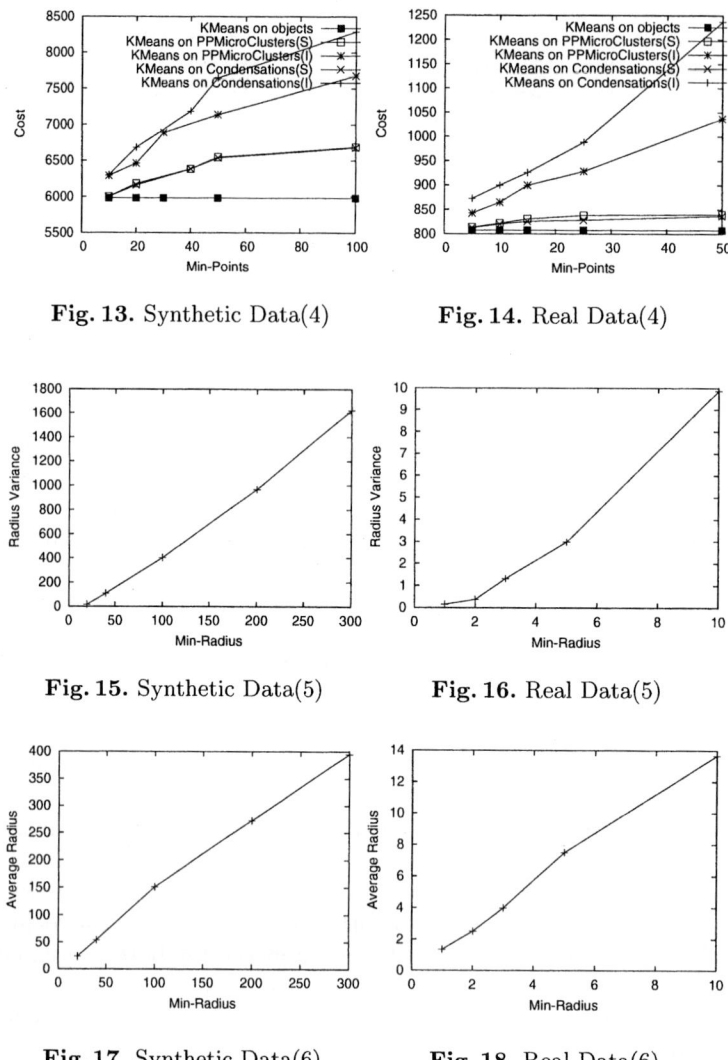

Fig. 13. Synthetic Data(4)

Fig. 14. Real Data(4)

Fig. 15. Synthetic Data(5)

Fig. 16. Real Data(5)

Fig. 17. Synthetic Data(6)

Fig. 18. Real Data(6)

variance is computed as the average squared deviation of each PPMicroCluster's radius from its radius (privacy) threshold r_t. Figures 15 and 16 show that the impact of changing numbers of r_t to the radius variance of PPMicroClusters. For both datasets, the smaller the minimum privacy threshold r_t, the smaller variance of the radius of each PPMicroCluster, which means the lower degree of privacy. On the other hand, the variance of radius of PPMicroClusters is larger for larger values of r_t (for example, when $r_t = 300$, the variance is 5 times of r_t), which means the higher degree of privacy. The average radius of PPMicroClusters is defined by $\frac{1}{m}\sum_{i=1}^{m} r_i$ where $r_1, \ldots, r_m$ are the radii of m PPMicroClusters. We examine the average radius of PPMicroClusters with respect to the changing values of r_t. Figures 17 and 18 depict the phenomena that for both datasets, the smaller the minimum radius threshold, the smaller average radius of PPMicroClusters (i.e., $r_t = 1.5$ in read data, the average radius of PPMicroClusters is 1.5), which means less privacy guarantee; while the average radius increases with increasing values of r_t, which generally means more privacy will be obtained. The figures on the accuracy and the privacy measures show the fundamental trade-off of privacy-preserving clustering: with increasing minimum radius (or minimum numbers), the accuracy decreases while privacy increases.

7 Conclusion

In this paper, we propose a novel robust transformation for privacy-preserving data mining based on the concept of privacy-preserving MicroClusters that satisfy a range constraint as well as a significance constraint. The privacy-preserving MicroClusters are made public for data mining purposes, but the original data records are kept private. We present methods for generating privacy-preserving MicroClusters and show that standard clustering algorithms can easily be adapted to cluster the public MicroClusters instead of the private data records. Our experimental evaluation on synthetic and real data sets demonstrates that the proposed methods achieve accurate clusterings while preserving the privacy.

Acknowledgement. We would like to thank Dr. Jiawei Han, University of Illinois at Urbana-Champaign and Dr. Martin Ester, Simon Fraser University for their valuable suggestions on the previous drafts.

References

1. D. Agrawal et al On the design and quantification of privacy preserving data mining algorithms. *PODS* 2001.
2. S. Agrawal, J. R. Haritsa. A Framework for High-Accuracy Privacy-Preserving Mining. *ICDE* 2005
3. R. Agrawal and R. Srikant. Privacy Preserving Data Mining. *SIGMOD* 2000
4. C. C. Aggarwal, P. S. Yu. A Condensation Approach to Privacy Preserving Data Mining. *EDBT 2004*
5. R. J. Bayardo, R. Agrawal. Data Privacy through Optimal k-Anonymization. *ICDE* 2005
6. B. M. Fung et al Top-Down Specialization for Information and Privacy Preservation. *ICDE* 2005
7. M. R. Garey and D.S.Johnson. *Computers and Intractability*. W.H.Freeman 1979
8. US Department of Health and Human Services. http://www.hhs.gov/ocr/hipaa/
9. V. S. Iyengar. Transforming Data to Satisfy Privacy Constraints. *KDD*, 2002
10. M.Klusch, S. Lodi et al. Distributed Clustering Based on Sampling Local Density Estimates. *IJCAI* 2003.
11. E. M. Knorr, R. T. Ng. Algorithms for Mining Distance-Based Outliers in Large Datasets. *VLDB* 1998

12. K. LeFevre, D. J. DeWitt, R.Ramakrishnan. Incognito: Efficient Full-Domain K-Anonymity. *SIGMOD* 2005
13. Kaufman L. et al. Finding Groups in Data: An Introduction to Cluster Analysis. John Wiley & Sons, 1990.
14. S. Merugu and J. Ghosh. Privacy-preserving Distributed Clustering using Generative Models. *ICDM* 2003
15. R. Motwani, P. Raghavan. Randomized Algorithms. *Cambridge University Press* 1995
16. Data Mining: Staking a Claim on Your Privacy. *Office of the Information and Privacy Commissioner.* 1/1998
17. S. R. Oliveira, O. R. Zaiane. Privacy Preserving Clustering By Data Transformation. *SBBD* 2003
18. B. Pinaks. Cryptographic Techniques for Privacy Preserving Data Mining. *SIGKDD Explorations* 2000,Vol 4,2
19. S. Ramaswamy et al. Efficient Algorithms for Mining Outliers from Large Data Sets. *SIGMOD* 2000
20. L. Sweeney. k-anonymity: A Model for Protecting Privacy. *IJUFKS* 2002
21. L. Sweeney. Achieving k-Anonymity Privacy Protection Using Generalization and Suppression. *IJUFKS* 2002
22. J. Schlrer. Security of Statistical Databases: Multidimensional Transformation. *TODS*, v.6 n.1, Mar. 1981
23. J. Vaidya, C. Clifton. Privacy-Preserving K-Means Clustering over Vertically Partitioned Data. *KDD* 2003
24. A. C. Yao. How to Generate and Exchange Secrets. *FOCS*, IEEE, 1986, pp. 162-167.
25. T. Zhang et al. BIRCH: An Efficient Data Clustering Method for Very Large Databases. *SIGMOD* 1996

Achieving Private Recommendations Using Randomized Response Techniques[*]

Huseyin Polat and Wenliang Du

Department of Electrical Engineering and Computer Science,
Syracuse University, CST 3-114, Syracuse, NY 13244-1240, USA
{hpolat, wedu}@ecs.syr.edu

Abstract. Collaborative filtering (CF) systems are receiving increasing attention. Data collected from users is needed for CF; however, many users do not feel comfortable to disclose data due to privacy risks. They sometimes refuse to provide information or might decide to give false data. By introducing privacy measures, it is more likely to increase users' confidence to contribute their data and to provide more truthful data. In this paper, we investigate achieving referrals using item-based algorithms on binary ratings without greatly exposing users' privacy. We propose to use randomized response techniques (RRT) to perturb users' data. We conduct experiments to evaluate the accuracy of our scheme and to show how different parameters affect our results using real data sets.

1 Introduction

Collaborative filtering (CF) is a recent technique for filtering and recommendation purposes. It has many important applications [1, 2] in E-commerce, direct recommendations, and search engines. With the help of CF, users can get recommendations about many of their daily activities. Using other users' data, CF systems try to predict how well an *active user* (*a*) will like an item that he/she did not buy before. The key idea is that *a* will prefer those items that like-minded users prefer, or that dissimilar users do not. Users might express their opinions about items they bought before or showed interest as *like (1)* or *dislike (0)*.

CF systems are advantageous; however, they fail to protect individual user's privacy and they have a number of disadvantages [1, 2]. The most important is that they are a serious threat to individual privacy. They pose various privacy risks [3] like unsolicited marketing, price discrimination, and being subject to government surveillance. Many users sometimes refuse to give data or might contribute false information. Customer data is a valuable asset and it has been sold when E-companies suffered bankruptcy [2]. By providing privacy measures, users feel comfortable to contribute more truthful information. The challenge is *how can people contribute their preferences about products for CF purposes without greatly compromising their privacy?*

---

[*] This work was supported by Grants ISS-0219560 and ISS-0312366 from the United States National Science Foundation.

In this paper, we propose a new scheme to achieve private recommendations using item-based algorithms on binary ratings. In our scheme, before sending data to the server, each user perturbs data in such a way that the server is not able to learn the true ratings. However, the perturbing scheme should still be able to allow the server to produce accurate referrals. We propose to use *randomized response techniques (RRT)* for data disguising. CF is based on aggregate values rather than individual data items. We hypothesize that it is possible to combine CF algorithms with the RRT to achieve users' privacy while still producing accurate referrals, because the aggregate data can be estimated with decent accuracy from disguised data if we have significantly large data. To verify this hypothesis, we implemented RRT for an item-based algorithm [11]. Using two existing data sets, we performed a series of experiments to evaluate the scheme's performance and to show how our results change with varying parameters.

2 Related Work

Canny [1, 2] proposes two schemes for privacy-preserving collaborative filtering (PPCF). A community of users can compute a public "aggregate" of their data without disclosing individual users' data. While his works focus on the peer-to-peer framework, in which users actively participate in the CF, our work focuses on another framework, in which users send their preferences to a server, which creates a model and provides referrals based on it. Polat and Du [8] apply randomized perturbation techniques (RPT) for PPCF. Although their schemes are based on numerical ratings and RPT, our work focuses on binary ratings and we employ RRT for data disguising. While they investigate how to provide private predictions using CF algorithms on user-user similarities, we focus on achieving private recommendations using item-based CF algorithms.

RRT were first introduced by Warner [12] as a technique to estimate the percentage of people in a population that has attribute A. The interviewer asks each respondent two related questions, the answers to which are opposite to each other. Using a randomizing device, respondents choose the first question with probability θ and the second question with probability $1-\theta$, to answer. The interviewer learns responses but does not know which question was answered. Sarwar et. al [11] propose item-based CF algorithms and present a model-based approach to pre-compute item-item similarity scores. After computing similarities, the best l similarities are retained (l is called the model size). The prediction is then computed by taking a weighted average of a's ratings on similar items.

3 Providing Private Predictions Using RRT

The algorithm proposed by [11] might be extended to provide referrals on binary ratings. In[7], referrals are provided on market basket data, where Tanimoto coefficient is used to find similarities between users. Polat and Du [9] extend Tanimoto coefficient to find user-user similarities on binary ratings. We propose the following scheme to provide predictions on binary ratings using item-based algorithms. The server finds item-item similarities using $W_{j_1 j_2} = [t(V_s) - t(V_d)]/t(V)$, where $t(V_s)$ and $t(V_d)$ are the number of similar and dissimilar votes for items j_1 and j_2, respectively and $t(V)$ is the number of co-ratings. Similarities range from -1 to 1. If $W_{j_1 j_2} > 0$, items are similar; otherwise they are dissimilar. When $W_{j_1 j_2} = 0$, they are not correlated. The server selects the best l similarities for each item and stores them. It receives a's known ratings and the target item q and finds a's ratings for those items that are the best similar items to q. It calculates the number of 1s (l_j) and 0s (d_j) among those ratings after it reverses the dissimilar items' ratings. It then computes $ld_j = l_j - d_j$. If $ld_j > 0$, then the item will be recommended as *like*, otherwise it will be predicted as *dislike*.

To further improve accuracy, significance weighting (SW) can be applied [6]. Herlocker et. al [6] add a correlation factor that can devalue similarity weights that are based on a small number of co-rated items. Since similarities are found between items, we apply SW based on number of users who rated such items. We applied $2c/n$ as a correlation factor, where c is the number of co-ratings and n is the number of users. If two items have less than $n/2$ co-ratings, we applied a significance weight of $2c/n$, otherwise a significance weight of 1 was applied.

A typical ratings vector includes the votes and empty cells for unrated items. An example of a ratings vector for user u is $V_u = (11 \perp 00 \perp 101)$, where $\perp$ means not rated. To perturb V_u, u generates a random number (r_u) using uniform distribution over the range [0,1]. If $r_u \leq \theta$, then u sends the true data, V_u. Otherwise, he/she sends the false data (exact opposite of the ratings vector), which is $\overline{V}_u = (00 \perp 11 \perp 010)$, where $\overline{V}_u$ is the vector that reverses the 1's in V_u to 0's and 0's to 1's; we call $\overline{V}_u$ the opposite of V_u. With probability θ, true data is sent while false data is sent with probability 1-θ. Although the server has the ratings vectors, it does not know whether they are true or false data.

3.1 Finding Private Predictions Using RRT

Without privacy as a concern, users send true ratings to the server, which can provide referrals as explained before. However, with privacy as a concern, the server should not be able to learn the true ratings of users including a. Users might send false data to accomplish perfect privacy but producing accurate predictions is impossible from this data. If they send actual data, finding accurate recommendations is possible but privacy is not preserved. CF systems should provide referrals efficiently. To achieve a good balance between accuracy, privacy, and efficiency, we propose to use both one-group and multi-group schemes.

One-Group Scheme. There are different RRT for data disguising. The one-group scheme [4], in which all ratings are put into the same group and all of them are either reversed together or keep the same values. The server cannot know whether users tell the truth or lie because the random numbers are only known by the users. In this scheme, we achieve the same accuracy on perturbed data with original scheme. The model, which includes item-item similarities, created from perturbed data is the same with the one created from original data because all ratings are either reversed together or keep the same values.

Although we achieve decent accuracy in this scheme, the privacy level is very low. If the server somehow learns the true rating for only one item, it can obtain true votes for all items. To further improve privacy, we propose to use multi-group schemes, in which the set of the items is partitioned into a number of groups; then the RRT are used to perturb each group *independently*. We can partition m items into m groups (m-group scheme), with each group containing only one item. For each group, users randomly decide whether to disclose its true rating (with probability θ) or to disclose the false rating (with probability $1 - \theta$). The users repeat this process for all groups; the random decisions are independent for each group. The m-group scheme is very secure. If the server can figure out one rating for any item, other ratings are still hidden. However, accuracy might become very low. A compromise between the one-group scheme and the m-group scheme is to partition the items into M groups, where $1 < M < m$. The decision is the same for all items in the same group, but the decisions for different groups are independent.

Multi-Group Schemes. Each user first groups the items in the same way. They then disguise their ratings in each group *independently*. Since the ratings in different groups are perturbed *independently*, even if the server knows information about one group, it will not be able to derive information about other groups. We improve privacy level by introducing multi-group schemes, while with increasing M, accuracy decreases because we add more randomness. Users disguise their ratings in each group as they do for one-group scheme based on random numbers and θ. They then send perturbed data to the server that needs to create a model by estimating item-item similarities from disguised data and storing the best l of them for each item. After model estimation, based on a's query, the server sends the estimated similarities for q to a who can compute predictions for q using them as explained previously. Since active users only need to send a query rather than their ratings, the server will not learn true preferences. To form the model on disguised data, the server should find a way to estimate item-item similarities from perturbed data.

The server does not know whether the collected data is true or not because users disguise their ratings based on the relation between random numbers, which are only known by them, and θ. However, it is possible to estimate the item-item similarities because the server is able to estimate the probabilities of having true or false data given disguised data. If we call the perturbed data Y_k, the true data X_k, and $\overline{X}_k$ represents the exact opposite of X_k or false data, where $k = 1, 2, \ldots, M$ and k shows the group name, the server needs to find $p(X_k|Y_k = X_k)$ and $p(\overline{X}_k|Y_k = X_k)$ for each group, where $p(X_k|Y_k = X_k) + p(\overline{X}_k|Y_k = X_k) = 1$. $p(X_k|Y_k = X_k)$ can be calculated using the Bayes' rule as follows:

$$p(X_k|Y_k = X_k) = \frac{p(Y_k = X_k|X_k)p(X_k)}{p(Y_k = X_k)} \quad (1)$$

where $p(Y_k = X_k|X_k)$ is θ. The value of $p(Y_k = X_k)$ can be calculated from disguised data while the value of $p(X_k)$ can be computed as follows using the facts that $p(Y_k = X_k|X_k) = \theta$ and $p(Y_k = \overline{X}_k|X_k) = 1 - \theta$:

$$p(Y_k = X_k) = p(Y_k = X_k|X_k)p(X_k) + p(Y_k = \overline{X}_k|X_k)p(\overline{X}_k)$$
$$p(Y_k = X_k) = \theta p(X_k) + (1-\theta)p(\overline{X}_k) \quad (2)$$

Eq. (2) can be solved for $p(X_k)$ as follows using the fact that $p(X_k) + p(\overline{X}_k) = 1$:

$$p(X_k) = \frac{p(Y_k = X_k) + \theta - 1}{2\theta - 1} \quad (3)$$

We get the following after replacing $p(X_k)$ with its equivalent in Eq. (1):

$$p(X_k|Y_k = X_k) = \frac{\theta^2 + \theta p(Y_k = X_k) - \theta}{2\theta p(Y_k = X_k) - p(Y_k = X_k)} \quad (4)$$

Note that X_k and Y_k are ratings vectors of original and disguised data, respectively. Therefore, to find $p(Y_k = X_k)$, the server finds posterior probabilities for all items in each group k, selects the best one, uses it as $p(Y_k = X_k)$, and computes $p(X_k|Y_k = X_k)$ and $p(\overline{X}_k|Y_k = X_k)$ values for each group. It then can estimate item-item similarities. Since all items in the same group are either reversed or keep the same values, similarities for them will be same with the ones computed from true data, while similarities computed from perturbed data for those items in different groups will be different from those calculated from true data due to RRT. However, the server can estimate similarities for them. Note that there are four possible situations because the received data might be true or false in each group. The server first computes four similarity values by considering all four situations. Then, it finds the probabilities of having such situations. Finally, it multiplies those weights with the corresponding probability values, sums the results, and finds the similarity values between items in different groups. After estimating similarities, the server forms the model by selecting the best l similarities for each item and starts providing filtering services.

It is still possible to achieve decent accuracy using multi-group schemes in addition to one-group scheme. Although the number of true similarities decrease with increasing M, there are still $(m^2 - Mm)/(2M)$ true similarities because similarity weights for items in the same groups will be identical to the ones computed from original data. In addition, since such weights between items are computed over all users who commonly rated them, the aggregate values can be estimated from disguised data when there are enough users' data.

Our scheme can also be extended to provide private top-N recommendations. To find top-N recommendations, the server computes ld_j values for all a's unrated items, sorts them, and provides first N items to a as top-N recommendations. Since online computation cost is critical, instead of finding referrals for all unrated items, a sends a query stating he/she is looking for recommendations for N_a items, where $N < N_a < m - m_{at}$ and m_{at} is the number of rated items by a. The server then sends item similarities for those N_a items to a who can compute ld_j values for them and find top-N recommendations.

3.2 Providing Private Recommendations with Full Privacy

In addition to preventing the server from learning the ratings, it should not be able to learn items rated by the users. We can extend our scheme to achieve this goal. Before they disguise their data, users conduct the followings to prevent the server from learning the rated items. First, each user u finds the number of rated items (m_{ut}) and randomly creates a uniform integer m_{ur} from the range $(1, m_{ut})$. They then randomly select m_{ur} unrated items and fill randomly selected $\lfloor m_{ur}/2 \rfloor$ items' cells with 1 and the remaining unrated items' cells with 0.

We can still provide accurate referrals while achieving full privacy because users fill empty cells with equal numbers of 1s and 0s. When there are enough users, the contributions of appended ratings to similarities will be close to zero. The changes in the number of similarly or dissimilarly rated items will be close to each other. The server does not know the rated items due to appended ratings. However, it can guess the randomly selected unrated items. The probability of guessing the number of randomly selected unrated items (m_{ur}) for the server is 1 out of $m'/2$, where m' represents the number of 1s or 0s, depending on which one is less, including the fake ratings for randomly selected unrated items. After guessing m_{ur}, the probability of guessing the $m_{ur}/2$ randomly selected unrated items filled with 1s is 1 out of $C_{m_{ur}/2}^{m'_1}$ and the probability of guessing the $m_{ur}/2$ randomly selected unrated items filled with 0s is 1 out of $C_{m_{ur}/2}^{m'_0}$, where m'_1 and m'_0 represent the number of 1s and 0s, respectively. Therefore, the probability of guessing the fake ratings is 1 out of $((m'/2)(C_{m_{ur}/2}^{m'_1})(C_{m_{ur}/2}^{m'_0}))$.

4 Overhead Costs and Privacy Analysis

Our scheme does not introduce additional communication and storage costs due to privacy concerns. Model creation is done off-line while referrals are computed and provided online. Our scheme's online component does not introduce overhead computation costs while off-line computation costs increase. However, off-line computation costs are not critical to the overall performance. Privacy can be measured with respect to the reconstruction probability (p) with which the server can obtain the true ratings vector of a user given his/her disguised data. Privacy level (PL) can be defined in terms of p as follows [10]: $PL = (1-p) \times 100$. With increasing p, privacy level decreases. To decrease p, the randomness should be increased, which makes accuracy worse because privacy and accuracy conflict each other. We can define p in terms of $p(X_k|Y_k = X_k)$ and M as follows:

$$p = \left[p(X_k|Y_k = X_k)\right]^M = \left[\frac{\theta^2 + \theta Y - \theta}{2\theta Y - Y}\right]^M \quad (5)$$

where Y represents $p(Y_k = X_k)$. With increasing M, p decreases while PL increases. The value of p depends on θ, M, and the value of Y or X, where X represents $p(X_k)$. Since randomization process is conducted independently for different groups, PL increases with increasing M. When $\theta = 1.0$ or $\theta = 0.0$, we disclose everything about the original data. However, when θ is away from 1.0 or 0.0 and approaches to 0.50, PL increases because we add more randomness. We calculated PLs and showed them in Fig. 1. We varied θ from 0.51 to 1.0 because the complementary θ values achieve the same PL and found PLs for X being 0.3, 0.4, and 0.5, where we fixed M at 3. We then varied M from 1 to 5, fixed X at 0.3, and found PLs for θ being 0.51, 0.6, and 0.7. As expected, PLs increase with decreasing θ from 1 to 0.51 and increasing M values.

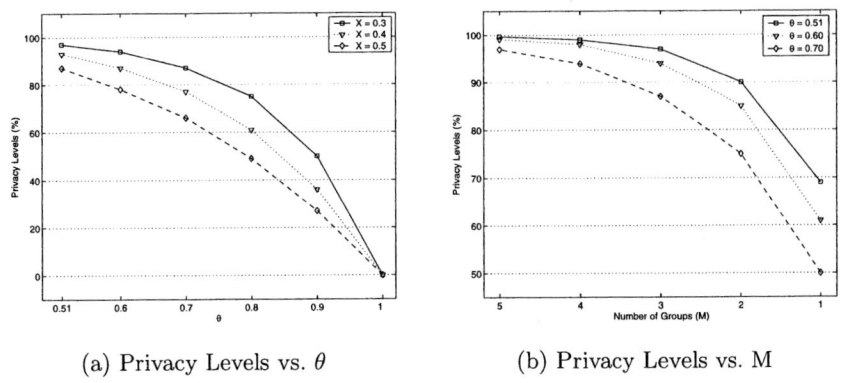

(a) Privacy Levels vs. θ (b) Privacy Levels vs. M

Fig. 1. Privacy Levels With Varying Parameters

5 Experiments

We used Jester and MovieLens (MLP) real data sets in our experiments. Jester data set [5] has 100 jokes and records of 17,988 users. The ratings range from -10 to +10, and the scale is continuous. MLP data was collected by the GroupLens Research Project (www.cs.umn.edu/research/Grouplens). It consists of ratings for 1,682 movies made by 943 users. Ratings are made on a 5-star scale. As evaluation criteria, we employed classification accuracy (CA), which is the ratio of the number of correct classifications to the number of classifications and F-measure (FM), where FM = (2 × precision × recall) / (precision + recall).

We transformed numerical ratings into two labels. We labeled items as *like* if numerical ratings were bigger than 3, or *dislike* otherwise in MLP. We labeled them as *like* if numerical ratings were above 2.0, or *dislike* otherwise in Jester. We randomly selected 2,000 and 800 training and 500 and 143 test users from Jester and MLP, respectively. Although we used the same test users throughout our trials, we used different numbers of training users based on various experiment settings. For each test user, we withheld 5 rated items' ratings, tried to find referrals for them, and compared them with true ratings. We ran data disguising 500 times and compute predictions on scrambled data. We compared referrals on perturbed data using our scheme with true ratings, calculated CAs and FMs for all test users, and displayed final values.

We hypothesize that accuracy and privacy depend on several factors including the significance weighting (SW), the model size (l), the number of users (n) and groups (M), the value of θ, and the number of appended ratings. We first performed testings to show how SW affects our results. We conducted trials with varying n values. We discarded such similarities calculated on fewer than 3 co-ratings. We found out that recommendation qualities become better when SW is applied for all n values because by applying it, we devalued such similarity weights calculated on limited number of co-ratings. CA is increased by 0.0045 with SW when n is 800 for MLP. We applied SW in the following experiments.

To find the optimum l, we performed trials using MLP while varying l values. For each l value, we ran the algorithm, found referrals for test items, compared them with true ratings, and calculated CAs and FMs. We ran this procedure 50 times and computed overall averages. With increasing l up to 500, the results are becoming better. However, after 500 best similarities, the results slightly become worse with increasing l. Therefore, we selected 500 as l for MLP. In the following experiments, we set l being 100 and 500 for Jester and MLP, respectively.

To show how data partition or various M values affect our results, we performed testings using both data sets with varying group schemes. We used randomly selected 500 and 200 training users from Jester and MLP, respectively, where we set θ at 0.65. We performed our experiments for up to five-group scheme, calculated CAs and FMs, and displayed the results in Table 1. As we expected, the results are becoming worse with increasing M because the bigger the M we use, the more randomness we add. For MLP, the loss in CA is 0.0338 when we changed M from 1 to 5 while it is 0.0188 for Jester.

Table 1. Recommendation Qualities With Varying M Values

	Jester				MLP			
M	1	2	3	5	1	2	3	5
CA	0.7095	0.6974	0.6962	0.6907	0.6813	0.6698	0.6580	0.6475
FM	0.6668	0.6550	0.6524	0.6455	0.7470	0.7391	0.7279	0.7158

Accuracy will be different for varying θ values because randomness differs with varying θ. We performed testings using both data sets, where we used 500 and 200 training users from Jester and MLP, respectively. We only showed results for three-group scheme, where we varied θ from 0.51 to 1.00, because complementary θ values give the same results. We compared the referrals estimated from disguised data with the ones computed from original data. We displayed CAs and FMs in Fig. 2. As expected, changes in accuracy are becoming worse while θ values are converging to 0.51. With decreasing θ values from 1 to 0.51, the randomness becomes bigger and causes losses in accuracy. Note that since all users send true data when $\theta = 1.0$, the changes in accuracy will be 0.

We hypothesize that we can still provide accurate recommendations with full privacy if we have enough users' data. Since false ratings are inserted into ratings vectors, referral qualities will decrease. To show how accuracy changes with

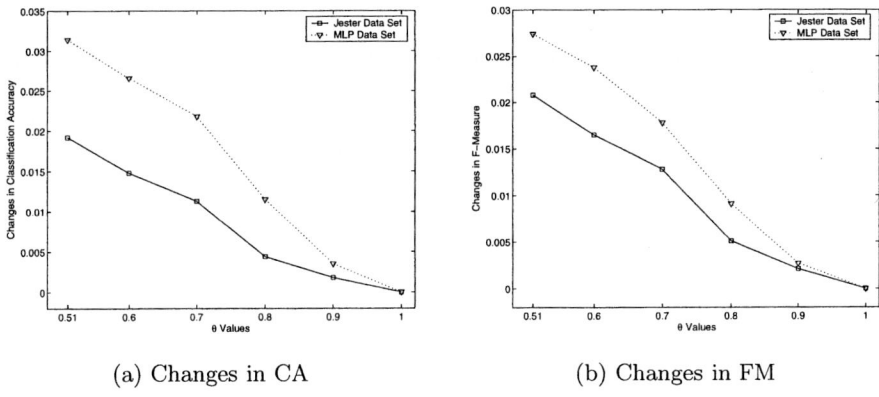

(a) Changes in CA   (b) Changes in FM

Fig. 2. Changes in Recommendation Qualities With Varying θ Values

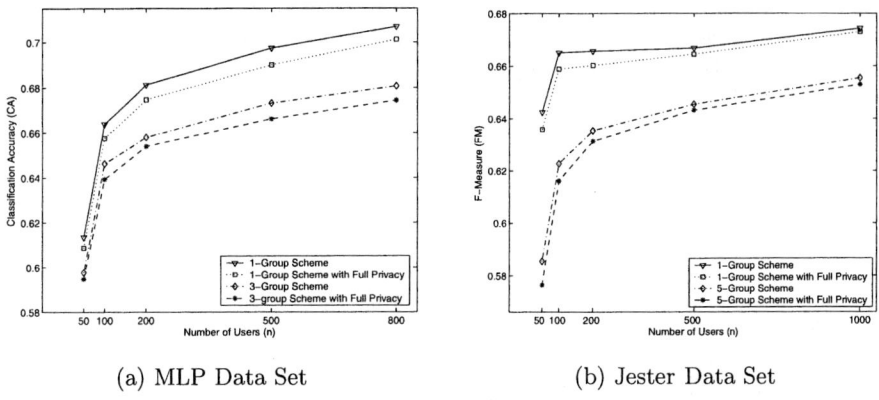

(a) MLP Data Set   (b) Jester Data Set

Fig. 3. Recommendation Qualities With Varying n Values and Full Privacy

appended ratings and varying n values, we performed experiments using both data sets. We appended ratings as explained in Section 3.2. We disguised users' ratings after inserting false ratings using θ being 0.65. We calculated CAs and FMs, and showed them in Fig. 3. As expected, recommendation qualities increase with increasing n. When there are more than 200 users, improvements become steady. However, recommendation qualities rapidly increases with increasing n when there are limited number of users. When ratings are appended, the results become worse compared to results without full privacy issues. However, the decrease in accuracy due to full privacy is small. For Jester, with increasing n, FM values based on appended ratings are becoming closer to the ones computed without appended ratings. For MLP data, when one-group scheme is used with n being 800, the accuracy lost is only 0.0057 due to appended ratings. It is 0.0067 for three-group scheme. Therefore, our scheme can be easily extended to achieve full privacy while still providing accurate referrals.

6 Conclusions and Future Work

We presented a solution to achieve binary ratings-based private referrals on item-based algorithms using RRT. We showed that it is possible to provide accurate recommendations on perturbed data. Besides one-group scheme, we introduced multi-group schemes to achieve an equilibrium between accuracy, privacy, and efficiency. We will investigate whether we can achieve accurate referrals based on inconsistently disguised data. We will study how to extend our scheme to other recommendation algorithms.

References

1. J. Canny. Collaborative filtering with privacy. In *Proceedings of the IEEE Symposium on Security and Privacy*, pages 45–57, Oakland, CA, USA, May 2002.
2. J. Canny. Collaborative filtering with privacy via factor analysis. In *Proceedings of the 25th ACM SIGIR Conference*, pages 238–245, Tampere, Finland, August 2002.
3. L. F. Cranor. 'I didn't buy it for myself' privacy and ecommerce personalization. In *Proceedings of the 2003 ACM Workshop on Privacy in the Electronic Society*, pages 111–117, Washington, DC, USA, 2003.
4. W. Du and Z. Zhan. Using randomized response techniques for privacy-preserving data mining. In *Proceedings of the 9th International ACM SIGKDD Conference*, Washington, DC, USA, August 2003.
5. D. Gupta, M. Digiovanni, H. Narita, and K. Goldberg. Jester 2.0: A new linear-time collaborative filtering algorithm applied to jokes. In *Proceedings of the Workshop on Recommender Systems, ACM SIGIR'99*, Berkeley, CA, USA, August 1999.
6. J. L. Herlocker, J. A. Konstan, A. Borchers, and J. T. Riedl. An algorithmic framework for performing collaborative filtering. In *Proceedings of the 22nd Annual International ACM SIGIR Conference*, Berkeley, CA, USA, August 1999.
7. A. Mild and T. Reutterer. Collaborative filtering methods for binary market basket data analysis. *Lecture Notes in Computer Science*, 2252:302–313, 2001.
8. H. Polat and W. Du. Privacy-preserving collaborative filtering. *International Journal of Electronic Commerce*, 9(4):9–36, 2005.
9. H. Polat and W. Du. Privacy-preserving top-N recommendation on horizontally partitioned data. In *Proceedings of the 2005 IEEE/WIC/ACM International Conference on Web Intelligence (WI'05)*, Paris, France, September 19–22 2005.
10. S. J. Rizvi and J. R. Haritsa. Maintaining data privacy in association rule mining. In *Proceedings of the 28th VLDB Conference*, Hong Kong, China, 2002.
11. B. M. Sarwar, G. Karypis, J. A. Konstan, and J. T. Riedl. Item-based collaborative filtering recommendation algorithms. In *Proceedings of the 10th International World Wide Web Conference (WWW10)*, pages 285–295, Hong Kong, May 2001.
12. S. L. Warner. Randomized response: A survey technique for eliminating evasive answer bias. *Journal of the American Statistical Association*, 60(309):63–69, 1965.

Privacy-Preserving SVM Classification on Vertically Partitioned Data*

Hwanjo Yu[1], Jaideep Vaidya[2], and Xiaoqian Jiang[1]

[1] University of Iowa, Iowa City IA 08544, USA
{hwanjoyu, xjia}@cs.uiowa.edu
[2] Rutgers University, Newark NJ 07102, USA
jsvaidya@rbs.rutgers.edu

Abstract. Classical data mining algorithms implicitly assume complete access to all data, either in centralized or federated form. However, privacy and security concerns often prevent sharing of data, thus derailing data mining projects. Recently, there has been growing focus on finding solutions to this problem. Several algorithms have been proposed that do distributed knowledge discovery, while providing guarantees on the non-disclosure of data. Classification is an important data mining problem applicable in many diverse domains. The goal of classification is to build a model which can predict an attribute (binary attribute in this work) based on the rest of attributes. We propose an efficient and secure privacy-preserving algorithm for support vector machine (SVM) classification over vertically partitioned data.

1 Introduction

The goal of data mining is to efficiently analyze large quantities of data to find interesting patterns and/or summarize the data in novel ways. Classification is one of the most common applications found in the real world. The goal of classification is to build a model which can predict the value of one variable, based on the values of the other variables. For example, based on financial, criminal and travel data, one may want to classify passengers as security risks. In the financial sector, categorizing the credit risk of customers, as well as detecting fraudulent transactions are both classification problems. Numerous such problems abound.

There is considerable research on different classification algorithms. Indeed, several different solutions are commonly used in the real world. A basic assumption is that complete access to data is available, either in centralized or federated form. However, privacy and security concerns restrict access to data. Sharing of data may not be possible due to either legal or commercial reasons. For example, due to HIPAA laws [1], medical data cannot be released for any purpose without appropriate anonymization. Similar constraints arise in many applications. European Community legal restrictions apply to disclosure of any individual data. Customer data, process data, etc., is often a valuable business asset for corporations. For example, complete manufacturing processes are

---

* This research was supported in part by a Faculty Research Grant from Rutgers Business School - Newark and New Brunswick.

trade secrets (although individual techniques may be commonly known). All of these cases require distributed knowledge discovery, without the disclosure of data. (Section 5 discusses related work in this area of Privacy-Preserving Data Mining.)

We assume vertically partitioned data with at least three participating parties, *i.e.*, three or more parties that collect different information about the same set of entities. For instance, a bank, health insurance company and auto insurance company collect different information about the same people. A bank has customer information like average monthly deposit, account balance. The health insurance company has access to medical information and other policy information. The car insurance company has access to information such as car type, accident claims, etc. Together, they might evaluate if the person is a credit risk for life insurance.

Support Vector Machine (SVM) classification is one of the most actively developed methodologies in data mining. SVM has proven to be effective in many real-world applications [2]. Like other classifiers, the accuracy of an SVM classifier crucially depends on having access to the correct set of data. Data collected from different sites is useful in most cases, since it provides a better estimation of the population than the data collected at a single site.

In this paper, we propose a privacy-preserving SVM (support vector machine) classification method on vertically partitioned data, PP-SVMV for short, such that each party (e.g., bank, insurance company) need not disclose its data or general information to other parties while still acquiring the same SVM classification accuracy as when the data is centralized. Our algorithm is efficient and secure. We first overview SVM (Section 2) and develop our PP-SVM technique (Section 3). We empirically show the practicality of our method in Section 4. Finally, related work is discussed in Section 5.

2 SVM Overview

We first describe the notation to overview SVM. All vectors are column vectors unless transposed to a row vector by a prime superscript $'$. The scalar (inner) product of two vectors x and y in the n-dimensional real space R^n is denoted by $x'y$ and the 2-norm of x is denoted by $||x||$. An $m \times n$ matrix $\mathcal{A}$ represents m data points in a n-dimensional input space. An $m \times m$ diagonal matrix $\mathcal{D}$ contains the corresponding labels (*i.e.*, +1 or -1) of the data points in $\mathcal{A}$. (A class label $\mathcal{D}_{ii}$, or d_i for short, corresponds to the i-th data point x_i in $\mathcal{A}$.) A column vector of ones of arbitrary dimension is denoted by e. The identity matrix of arbitrary dimension is denoted by $\mathcal{I}$.

First, consider a linear binary classification task, as depicted in Figure 1. For this problem, SVM finds the separating hyperplane ($w \cdot x = \gamma$) that maximizes the *margin*, denoting the distance between the hyperplane and closest data points (*i.e.*, support vectors). In practice, we use the "soft" margin to deal with noise, in which the distance from the boundary to each support vector could be different. The "hard" margin is formulated as $\frac{1}{||w||}$, as illustrated in Figure 1. To maximize the margin while minimizing the error, the standard SVM solution is formulated into the following primal program [2, 3]:

$$\min_{w,y} \quad \tfrac{1}{2}w'w + \nu e'y \qquad (1)$$

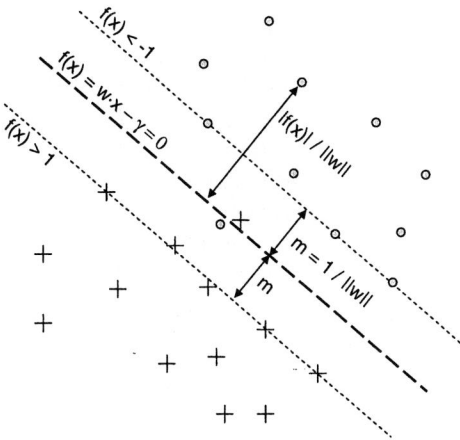

Fig. 1. The separating hyperplane that maximizes the margin. ('+' is a positive data point, i.e., $f('+') > 0$, and 'o' is a negative data point, i.e., $f('o') < 0$.)

$$\text{s.t.}\ \mathcal{D}(\mathcal{A}w - e\gamma) + y \geq e \ \text{and}\ y \geq 0 \tag{2}$$

which minimizes the reciprocal of the margin (i.e., $w'w$) and the error (i.e., $e'y$). By having the slack variable y in the constraint (2), SVM allows error or the soft margin. The slack or error is minimized in the objective function (1) and it will be larger than zero when the point is on the wrong side or within the margin area. The soft margin parameter ν (a user parameter) is tuned to balance the margin size and the error. The weight vector w and the bias γ will be computed by this optimization problem. Once w and γ are computed, we can determine the class of a new data object x by $f(x) = w'x - \gamma$, where the class is *positive* if $f(x) > 0$, or else *negative*.

In order to reduce the number of variables in the objective function and also be able to apply the kernel trick, we transform the primal problem to the following dual problem by applying the Largrange multipliers:

$$\min_{\alpha}\ \tfrac{1}{2}\alpha'Q\alpha - e'\alpha \tag{3}$$

$$\text{s.t.}\ 0 \leq \alpha_i \leq \nu\ \text{and}\ \sum_i d_i\alpha_i = 0,\ i = 0,...,m \tag{4}$$

where d_i (i.e., $\mathcal{D}_{ii}$) and α_i are the class label and the coefficient respectively for a data vector x_i. The coefficients α are to be computed from this dual problem. An $m \times m$ matrix Q is computed by the scalar product of every data pair, i.e., $Q_{ij} = K(x_i, x_j)d_i d_j$ where $K(x_i, x_j) = x_i \cdot x_j$ for linear SVM. The support vectors are the data vectors $\{x_i\}$ such that the corresponding coefficients $\alpha_i > 0$. The weight vector $w = \sum \alpha_i d_i x_i$ and thus the classification function $f(x) = \sum \alpha_i d_i x_i \cdot x - \gamma$ for linear SVM. For nonlinear SVMs, $f(x) = \sum \alpha_i d_i K(x_i, x) - \gamma$, where we can apply a nonlinear kernel for $K(x_i, x)$ (e.g., $K(x_i, x) = \exp(-\frac{\|x_i - x\|^2}{g})$) for RBF kernel, $K(x_i, x) = (x_i \cdot x + 1)^p$ for polynomial kernel,). [2] provides further details on SVM.

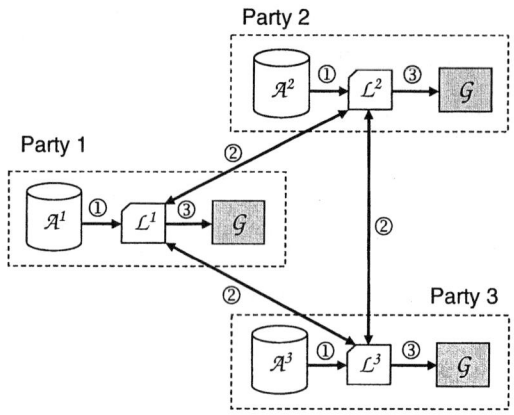

Fig. 2. PP-SVM: Framework for privacy-preserving SVM

3 Privacy-Preserving SVM

To generate the global SVM model (*i.e.*, the SVM model constructed from the data from multiple parties) without sharing any data among the parties, (1) the framework must be able to generate the global model only from models locally constructed by parties on their own data, without seeing others' data. We call this requirement *data privacy*. To prevent disclosing the general classification information on each party, (2) the local model must not be disclosed when jointly generating the global model. We call this requirement *model privacy*.

These two requirements lead to design our PP-SVM framework illustrated in Figure 2. (Figure 2 involves only three parties but can be generalized to more.) Each party builds a local model $\mathcal{L}$ from its own data $\mathcal{A}$ (①), and each party *securely* merges its model with others (②), in order to generate the global model $\mathcal{G}$ (③). The global model $\mathcal{G}$ will be the same for every party, which will be used for classifying new data objects. Assuming that the merge of the local models is done securely, this framework keeps private the local models (*i.e.*, $\mathcal{L}^1, \mathcal{L}^2, \mathcal{L}^3$) as well as the data of each party (*i.e.*, $\mathcal{A}^1, \mathcal{A}^2, \mathcal{A}^3$). This section presents techniques that implement the framework. First, we discuss the choice for the local model $\mathcal{L}$. Then, we present a method to securely merge the local models.

3.1 Local Model

As we see from the last paragraph of Section 2, an SVM model is represented by the bias γ, and a list of support vectors, their labels, and coefficients $\{(x_i, d_i, \alpha_i)\}$ such that $\alpha_i > 0$. That is, the global model $\mathcal{G}$ is composed of γ and $\{(x_i, d_i, \alpha_i)\}$ which are computed from the dual problem in Section 2.

Given vertically partitioned data over multiple parties, we cannot use a local SVM model (*i.e.*, computed only over local data) for our local model $\mathcal{L}$ in the framework (Figure 2), because the global SVM model $\mathcal{G}$ cannot be built only from local SVM models;

The globally optimal coefficients (computed by the dual problem) will be different from the locally optimal coefficients computed on local data. Since each party has the data of an attribute subset, the dual problem on the attribute subset will not generate the globally optimal coefficients. Thus, in our framework, the local model $\mathcal{L}$ needs to go beyond the standard SVM model.

To solve the dual problem globally, we need the $m \times m$ matrix $\mathcal{Q} = K(x_i, x_j) d_i d_j$ in Eq.(3) which is computed over the data of all the attributes. The diagonal matrix $\mathcal{D}$ for d_i is given as class labels, thus we only need to compute the global kernel matrix $\mathcal{K} = K(x_i, x_j)$. For linear kernel where $K(x_i, x_j) = x_i \cdot x_j$, the global matrix $\mathcal{K}$ can be directly computed from local matrices because $\mathcal{K}$ is a gram matrix and a gram matrix can be merged from gram matrices of vertically partitioned data, as Lemma 1 proves.

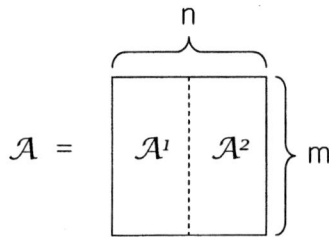

Fig. 3. Vertically partitioned matrix $\mathcal{A}$

Lemma 1. *Suppose the $m \times n$ data matrix $\mathcal{A}$ is vertically partitioned into $\mathcal{A}^1$ and $\mathcal{A}^2$ as Figure 3 illustrates. Let $\mathcal{K}^1$ and $\mathcal{K}^2$ be the $m \times m$ gram matrices of matrices $\mathcal{A}^1$ and $\mathcal{A}^2$ respectively. That is, $\mathcal{K}^1 = \mathcal{A}^1 \mathcal{A}^{1'}$ and $\mathcal{K}^2 = \mathcal{A}^2 \mathcal{A}^{2'}$. Then, $\mathcal{K}$, the gram matrix of $\mathcal{A}$, can be computed as follows:*

$$\mathcal{K} = \mathcal{K}^1 + \mathcal{K}^2 = \mathcal{A}^1 \mathcal{A}^{1'} + \mathcal{A}^2 \mathcal{A}^{2'} \qquad (5)$$

Proof. An $(i,j)^{th}$ element of $\mathcal{K}$ is $x_i \cdot x_j$, where x_i and x_j are i^{th} and j^{th} data vectors in $\mathcal{A}$. Let x_i^1 and x_i^2 be vertically partitioned vectors of x_i, which are the parts from $\mathcal{A}^1$ and $\mathcal{A}^2$ respectively. Then,

$$x_i \cdot x_j = x_i^1 \cdot x_j^1 + x_i^2 \cdot x_j^2 \qquad (6)$$

From Eq.(6), each element in $\mathcal{K}$ is equal to the sum of the elements in $\mathcal{K}^1$ and $\mathcal{K}^2$. Thus $\mathcal{K} = \mathcal{K}^1 + \mathcal{K}^2$.

Lemma 1 proves that local gram matrices are sufficient to build the global gram matrix which is the kernel matrix $\mathcal{K}$ for linear kernel. Some popular nonlinear kernel matrices can also be computed from the gram matrix: The polynomial kernel is represented by a dot product of data vectors (i.e., $K(x_i, x_j) = (x_i \cdot x_j + 1)^p$). The RBF kernel can also be represented by dot products (i.e., $K(x_i, x_j) = \exp(-\frac{||x_i - x_j||^2}{g}) = \exp(-\frac{|x_i \cdot x_i - 2 x_i \cdot x_j + x_j \cdot x_j|}{g})$). Thus, the local gram matrix from each party is sufficient to construct the global kernel matrix $\mathcal{K}$ for nonlinear kernels such as polynomial and RBF which can be represented by dot products.

Thus, we use the local gram matrix as the local model $\mathcal{L}$ in our framework. Section 3.2 discusses how to merge $\mathcal{L}$ securely from each party to securely build the global gram matrix. Once the global gram matrix is built, each party can run a quadratic programming solver to compute the global SVM model $\mathcal{G}$, which will be the same for every party.

3.2 Secure Merge of Local Models

To keep both data and model privacy, it is necessary to securely merge the local models which are the $m \times m$ local gram matrices. A *secure addition* mechanism for $m \times m$ matrices is required. For $k \geq 3$ parties, we developed such a method based on simple secure addition of scalars.

We first describe a simple method to securely calculate the sum of integers from individual sites under the assumption that there are at least three parties and the parties do not collude. We then extend the method so as to seamlessly merge the local models with high efficiency and privacy.

Secure Sum of Integers: Formally, we assume $k \geq 3$ parties, $P_0, \ldots, P_{k-1}$, with party P_i holding value v_i. Together they want to compute the sum $v = \sum_{i=0}^{k-1} v_i$. Assume that the sum v is known to lie in a field $\mathcal{F}$.

The parties also randomly order themselves into a ring. The ordering can be selected by one of the parties, or by a third party. If the parties cannot decide on a suitable order and no third party can be found, then a protocol developed by Sweeney and Shamos can be used [4] to fix upon a random ordering. The protocol developed by Sweeney and Shamos is quite efficient and requires only $O(k)$ communication. For this paper, to simplify the presentation, without loss of generality, we assume that this order is the canonical order $P_0, \ldots, P_{k-1}$. In general, any order can be decided on. The protocol proceeds as follows:

P_0 randomly chooses a number R, from a uniform distribution over $\mathcal{F}$. P_0 adds this to its local value v_0, and sends the sum $R + v_0 \mod |\mathcal{F}|$ to site P_1. For the remaining sites $P_i, i = 1, \ldots, k-1$, the algorithm proceeds as follows:
P_i receives

$$V = R + \sum_{j=0}^{i-1} v_j \mod |\mathcal{F}|.$$

P_i then computes

$$R + \sum_{j=1}^{i} v_i \mod |\mathcal{F}| = (v_i + V) \mod |\mathcal{F}|$$

and passes it to site $P_{i+1 \pmod{k}}$. Finally, P_0, subtracts R from the final message it gets (i.e., adds $-R \pmod{|\mathcal{F}|}$) to compute the actual result.

Clearly, the above protocol correctly calculates the required sum. In order to evaluate the security of the protocol, it is necessary to have a definition of *what* is meant by security. The area of Secure Multi-Party Computation (SMC) provides a theoretical framework for defining and evaluating secure computation. This protocol can be proven to be completely secure under our assumptions in the SMC framework. A complete proof of security is presented in our technical report [5].

Secure Sum of Matrices: We can extend the secure addition of scalars to securely adding matrices. The key idea is as follows. Suppose a master party wants to merge (*i.e.*, add) its local matrix with those in other slave parties. We assume that the parties have arranged themselves in some sequence and the master initiates the protocol.

1. The master party creates a random matrix of the same size as its local matrix. (The random matrix is hidden from the other parties.)
2. The master party merges (adds) the random matrix with its local matrix, sends the merged matrix to the following slave party.
3. Each slave party, receives the perturbed matrix, merges it with its local matrix and passes it to the following party (the last slave party sends the matrix back to the master).
4. The master subtracts the random matrix from the received matrix, which results in the matrix that adds the matrices of all the parties, without disclosing their local matrices to each other.

All addition is done in a closed field, and subtraction refers to addition of the complement. This secure addition mechanism is proven to be secure and efficient [5]. The extra computation required by the first party is the generation of the random matrix and the final subtraction. In terms of communication overhead, k rounds are required for every party to acquire the summed matrix, where k is the number of participating parties. One problem with the matrix summation method is that it is vulnerable to collusion. The parties preceding and following a party, can collude to recover its local matrix. However, the technique can easily be made collusion resistant to q parties by splitting up the local matrices into q random parts and carrying out the addition protocol q times. The sum of the final matrices from all q rounds gives the real global matrix. As long as the parties are ordered differently in each run, recovery of a local matrix is only possible if collusion occurs in all q rounds. Further details can be found in [5].

3.3 Security

Our method preserves "data privacy", since only the original party gets to exactly see the data; The local model is directly computed from the local data. However, to ensure "model privacy," we need at least three participating parties; Each party gets the final global model, which is simply the sum of local models. Thus, with only two parties participating, the other party's local matrix could be found simply by subtracting the local model from the global model. What is revealed, is the sum of the local models of the other parties. Since the SVM requires knowing the global matrix, this is always possible from the final result as well, and so is unavoidable.

We still need to analyze the effects of knowing the sum of gram matrices computed over the attributes of other parties. In general, the number and type of attributes of the other parties are still assumed to be unknown. As such, the summed matrix does not disclose any attribute values. If the exact number and types of attributes of the other parties are known, a number of quadratic equations will be revealed in the attribute values; Every cell of the gram matrix corresponds to a dot product – thus the quadratic equation. Since the matrix is symmetric, there are a total of $m(m+1)/2$ distinct equations (where m is the number of data objects). If the number of total variables (i.e., the sum of all the attributes of other parties) is larger than $m(m+1)/2$, it is impossible to recover the exact attribute values. Knowing that the matrix is symmetric and positive semidefinite does not disclose further information. While this does reveal more information than strictly necessary, this is a trade off in the favor of efficiency. If complete security is required, the summed matrix could be kept randomly split between two of the parties, and an oblivious protocol run to compute the global model using the generic circuit evaluation technique developed for secure multiparty computation [6, 7].

4 Experiment

The goal of the experiments is simply to demonstrate the scalability of our PP-SVMV. The accuracy will be exactly the same as that of SVM when the data is centralized. We revised the sequential minimal optimization (SMO) source[1] to implement the PP-SVMV. We used the Tic-Tac-Toe data set included in the SMO package for our experiment. We sampled around 958 data objects (m) and extracted around 27 features (n). PP-SVMV generates above 99% with an RBF kernel which is the same as that of the original SVM when the data is centralized.

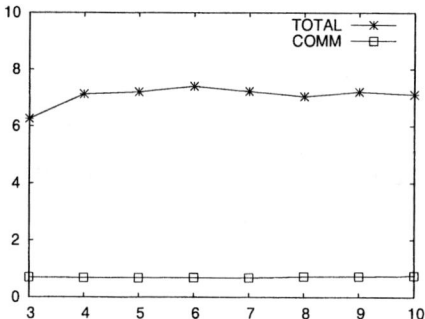

Fig. 4. X-axis:# parties; Y-axis: time (sec.); COMM: communication time; TOTAL: total training time

To check the scalability of the PP-SVMV on an increasing number of participating parties, we vary the number of parties from three to ten. We divide the 27 features about equally between the participating parties. For instance, when ten parties participate, three parties have two features, and the other seven parties have three features. Figure 4 shows results of our experiments: The total training time (including the parallel local computations) hardly changes; SVM is sensitive to the number of data objects more than the features, and the change on the number of features are not visibly influential to

---

[1] http://www.datalab.uci.edu/people/xge/svm

the total training time. The difference of the communication time is also not visible due to the dominant computation time. The results are averaged over ten runs.

5 Related Work

Recently, there has been significant interest in the area of Privacy-Preserving Data Mining. We briefly cover some of the relevant work. Several solution approaches have been suggested. One approach is to perturb the local data (by adding "noise") before the data mining process, and mitigate the impact of the noise from the data mining results by using reconstruction techniques [8]. However, there is some debate about the security properties of such algorithms [9, 10]. The alternative approach of using cryptographic techniques to protect privacy was first utilized for the construction of decision trees [11]. Our work follows the same approach. A good overview of prior work in this area can be found in [12]. Recently, some alternative techniques such as condensation[13] and transformation [14] have also been proposed.

In terms of data mining problems, work addressed includes association rule mining [15], clustering [16, 17], classification [18], and regression [19, 20]. All of the cryptographic work falls under the theoretical framework of Secure Multiparty Computation. Yao first postulated the two-party comparison problem (Yao's Millionaire Protocol) and developed a provably secure solution [6]. This was extended to multiparty computations by Goldreich et al. [7]. The key result in this field is that *any* function can be computed securely. Thus, the generic circuit evaluation technique can be used to solve our current problem. However, the key issue in privacy-preserving data mining is one of efficiency. The generic technique is simply not efficient enough for large quantities of data. This paper proposes an efficient technique to solve the problem.

Yu and Vaidya [21] developed a privacy-preserving SVM classification on *horizontally partitioned* data. Since their method is based on the trick of the proximal SVM [3], it is limited to linear classification. Our PP-SVMV is the first one proposing a secure SVM classification on *vertically partitioned* data, which uses the techniques of the secure matrix addition [5] and distributed SVM [22].

6 Conclusion

We propose a scalable solution for privacy-preserving SVM classification on vertically partitioned data (PP-SVMV). With three or more participating parties, our method PP-SVMV securely computes the global SVM model, without disclosing the data or classification information of each party to the others (*i.e.*, keeping the *model privacy* as well as the *data privacy*). Future work may address the idea of efficiently achieving complete security by keeping the global model split between parties as well.

References

1. "Standard for privacy of individually identifiable health information," *Federal Register*, vol. 66, no. 40, Feb. 28 2001.
2. V. N. Vapnik, *Statistical Learning Theory*, John Wiley and Sons, 1998.

3. G. Fung and O. L. Mangasarian, "Proximal support vector machine classifiers," in *Proc. ACM SIGKDD Int. Conf. Knowledge Discovery and Data Mining (KDD'01)*, 2001.
4. L. Sweeney and M. Shamos, "A multiparty computation for randomly ordering players and making random selections," Tech. Rep. CMU-ISRI-04-126, Carnegie Mellon University, 2004.
5. H. Yu and J. Vaidya, "Secure matrix addition," Tech. Rep., UIOWA Technical Report UIOWA-CS-04-04, http://hwanjoyu.org/paper/techreport04-04.pdf, 2004.
6. A. C. Yao, "How to generate and exchange secrets," in *Proceedings of the 27th IEEE Symposium on Foundations of Computer Science*. IEEE, 1986, pp. 162–167.
7. O. Goldreich, S. Micali, and A. Wigderson, "How to play any mental game - a completeness theorem for protocols with honest majority," in *ACM Symp. on the Theory of Computing*, 1987.
8. R. Agrawal and R. Srikant, "Privacy-preserving data mining," in *Proceedings of the 2000 ACM SIGMOD Conference on Management of Data*, 2000.
9. H. Kargupta, S. Datta, Q. Wang, and K. Sivakumar, "On the privacy preserving properties of random data perturbation techniques," in *Proceedings of the Third IEEE International Conference on Data Mining (ICDM'03)*, 2003.
10. Z. Huang, W. Du, and B. Chen, "Deriving private information from randomized data," in *Proc. of ACM SIGMOD Int. Conf. Management of data*, 2005.
11. Yehuda Lindell and Benny Pinkas, "Privacy preserving data mining," *Journal of Cryptology*, vol. 15, no. 3, pp. 177–206, 2002.
12. V. S. Verykios, E. Bertino, I. N. Fovino, L. P. Provenza, and Y. Saygin, "State-of-the-art in privacy preserving data mining," *SIGMOD Record*, vol. 33, no. 1, pp. 50–57, Mar. 2004.
13. Charu C. Aggarwal and Philip S. Yu, "A condensation approach to privacy preserving data mining.," in *EDBT*, 2004, pp. 183–199.
14. Stanley R. M. Oliveira and Osmar R. Zaiane, "Privacy preserving clustering by data transformation," in *SBBD*, 2004.
15. Jaideep Vaidya and Chris Clifton, "Secure set intersection cardinality with application to association rule mining," *Journal of Computer Security*, to appear.
16. Xiaodong Lin, Chris Clifton, and Michael Zhu, "Privacy preserving clustering with distributed EM mixture modeling," *Knowledge and Information Systems*, to appear 2004.
17. J. Vaidya and C. Clifton, "Privacy-preserving k-means clustering over vertically partitioned data," in *ACM SIGKDD Int. Conf. on Knowledge Discovery and Data Mining*, 2003.
18. J. Vaidya and C. Clifton, "Privacy preserving naïve bayes classifier for vertically partitioned data," in *2004 SIAM International Conference on Data Mining*, 2004.
19. A. F. Karr, X. Lin, A. P. Sanil, and Jerry P. Reiter, "Secure regressions on distributed databases," *Journal of Computational and Graphical Statistics*, 2005.
20. A. P. Sanil, A. F. Karr, X. Lin, and J. P. Reiter, "Privacy preserving regression modelling via distributed computation," in *ACM SIGKDD Int. Conf. Knowledge discovery and data mining*, 2004.
21. H. Yu, X. Jiang, and J. Vaidya, "Privacy-preserving svm using nonlinear kernels on horizontally partitioned data," in *Proc. ACM SAC Conf. Data Mining Track*, 2006.
22. F. Poulet, "Multi-way distributed SVM," in *Proc. European Conf. Machine Learning (ECML'03)*, 2003.

Data Mining Using Relational Database Management Systems*

Beibei Zou[1], Xuesong Ma[1], Bettina Kemme[1], Glen Newton[2], and Doina Precup[1]

[1] McGill University, Montreal, Canada
[2] National Research Council, Canada

Abstract. Software packages providing a whole set of data mining and machine learning algorithms are attractive because they allow experimentation with many kinds of algorithms in an easy setup. However, these packages are often based on main-memory data structures, limiting the amount of data they can handle. In this paper we use a relational database as secondary storage in order to eliminate this limitation. Unlike existing approaches, which often focus on optimizing a single algorithm to work with a database backend, we propose a general approach, which provides a database interface for several algorithms at once. We have taken a popular machine learning software package, Weka, and added a relational storage manager as back-tier to the system. The extension is transparent to the algorithms implemented in Weka, since it is hidden behind Weka's standard main-memory data structure interface. Furthermore, some general mining tasks are transfered into the database system to speed up execution. We tested the extended system, refered to as WekaDB, and our results show that it achieves a much higher scalability than Weka, while providing the same output and maintaining good computation time.

1 Introduction

Machine learning and mining algorithms face the critical issue of scalability in the presence of huge amounts of data . Typical approaches to address this problem are to select a subset of the data [4, 3], to adjust a particular algorithm to work incrementally (processing small batches of data at a time), or to change the algorithms such that they use data structures and access methods that are aware of the secondary storage. For instance, [8, 2] propose algorithms for decision tree construction that access special relational tables on secondary storage. Agarwal et al [7] develop database implementations of Apriori, a well-known algorithm for association rule mining, and show that some very specific implementation details can have a big impact on performance. Their approach achieves scalability by rearranging fundamental steps of the algorithm. Both of these pieces of work require the developer of the mining algorithm to be very familiar with database technology, implementing stored procedures, user defined functions, or choosing the best SQL statements. Machine learning researchers, however, are often not familiar enough with database technology to be aware of all optimization possibilities.

The goal of our research is to provide a general solution to scalability that can be applied to existing algorithms, ideally without modifying them, and that can be used

---
* Supported by NSERC, CFI, NRC.

by machine learning researchers to implement new algorithms without the need to be database experts. For that purpose, we have taken a very popular open-source package of machine learning algorithms, Weka [9], which can only be used on data sets that can fit into main memory, and extended it to be able to use a database as backend. In the extended system, WekaDB, a storage manager interface is defined with two implementations. One implementation is the original main-memory representation of data, the other uses a relational database management system (DBMS). All algorithms implemented in Weka can run in WekaDB without changes, and can use either of the two storage implementations depending on the data set size. Also, new algorithms can be added to the package without developers being required to know SQL.

Our basic approach couples Weka and the database rather loosely. The basic model uses the DBMS as a simple storage with the facility to retrieve records individually from the database, perform all computation in main memory, and write any necessary changes back to the database. However, accessing records individually is expensive, so WekaDB also implements several generally applicable optimizations. First, data is transferred in chunks between the database and WekaDB instead of one record at a time whenever possible. Second, many of the storage manager interface methods are implemented using advanced SQL statements; in particular, we take advantage of aggregate functionality (like sum, avg) provided by the DBMS. Third, some popular libraries (e.g., pre-processing filters) that were originally implemented on top of the storage interface, have been reimplemented to take advantage of DBMS functionality. Furthermore, even though WekaDB itself eases data size limitations, the implementations of the machine learning algorithms can create large internal data structures, imposing indirect limitations. In order to address this issue WekaDB provides database implementations for typical main memory data structures, like arrays. The algorithms can access these data structures as if they were implemented in main-memory.

We present an empirical evaluation of WekaDB on both synthetic and real data, using several machine learning algorithms from the original Weka package. The results show significant improvement in terms of scalability, while still providing reasonable execution time. For one of the algorithms, k-means clustering, we also compare with an implementation developed specifically for using a database backend [10]. In some situations, WekaDB's implementation even outperforms this specialized solutuion. In general, our approach is a practical solution providing scalability of data mining algorithms without requiring machine learning developers to be database experts.

2 Weka

Weka [9] is a popular, open source, machine learning software package implementing many state-of-the-art machine learning algorithms. These algorithms all access the data through one well-defined data-structure `core`. The data is represented by two main-memory data structures defined in `core`. A `Dataset` is a set of `Datarecord` objects. Each data record in a dataset consists of the same number of attribute/value pairs and represents one unit of data, e.g., information about a single customer. Additionally, the records have *weight* attributes, which are used by some learning algorithms.

`Dataset` keeps attribute and type information, and maintains a `DR vector` pointing to individual `Datarecord` objects. At the start of any algorithm, an initial `Dataset`

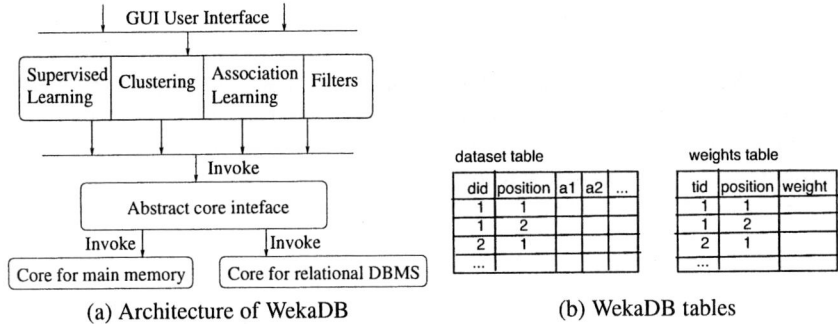

Fig. 1. WekaDB

object *DS* is created. Then, data records are loaded from an input file into individual `Datarecord` objects. For each object, a pointer is inserted into the `DR vector` of *DS*. During the computation, a copy *DS'* of a `Dataset` object *DS* could be made. Copying is lazy. Initially *DS'* shares the `DR vector` with *DS*. Only when a `Datarecord` object *o* in *DS'* needs to be modified, a new copy of the `DR vector` is created. All pointers in this vector still point to the old `Datarecord` objects. Then, a new copy of *o* is created, and the corresponding entry in the vector is adjusted accordingly.

Furthermore, `Dataset` provides methods to access and manipulate the data records: `enumerateRecords()` allows to iteratively retrieve data records, while `Record(index)` allows access to a record based on its index in the `DR vector`. General information about the data set can be returned by methods like `numRecords()`. There are also `delete/add` methods, which remove/add the corresponding `Datarecord` pointer from the `DR vector`. The `sort()` method sorts the `DR vector` based on an attribute. Summary statistics are provided by methods such as `sumofWeights()`.

3 Data in WekaDB

Fig. 1(a) shows the the redesigned system architecture. Based on the `core` interface from Weka, we defined a general data structure interface. Any data source that implements this interface can be plugged into Weka. Our new storage manager uses a relational database (currently DB2). `Dataset` and `Datarecord` have been modified to access the database. In `Dataset`, the `DR vector` was replaced by a `P-vector`. Instead of pointing to a `Datarecord`, each entry of the vector contains a `position` integer representing a record in the database. A `Datarecord` object is created (and the record loaded to main memory), whenever it is accessed by the learning algorithm.

3.1 Database Design

Data records reside in the database. Two extreme design alternatives are as follows.

- *Full table copy*: For each `Dataset` object *DS* there is one table *TDS* containing all records to which the `P-vector` of *DS* points. Making a copy *DS'* of *DS* leads to the creation of a table *TDS'*, and records in *TDS* are copied to *TDS'*.

- *Lazy approach*: There is only one Dataset table with a special attribute did indicating to which Dataset object a record belongs. Initially, all records have the same value *IDS* for did. When a copy *DS′* is made from an existing Dataset object *DS*, *DS′* shares the records with *DS*. Only if *DS′* changes a record for computation purposes, a copy of the original record is inserted into the Dataset table with the did attribute set to *IDS′*. This new record will be updated. Each Dataset object has to keep track of the set of did values with which its records might be labeled.

The full table copy approach is very time consuming if the machine learning algorithm performs many Dataset copy operations. However, it might be necessary if an algorithm adds or replaces attributes, i.e., changes the schema information. The lazy approach mirrors the lazy instantiation of new Datarecord objects in the main-memory implementation of core. Since most machine learning algorithms do not change attribute values, this seems to be the most efficient approach. However, many algorithms do change the weight attributes associated with the records. If this happens, basically all Dataset objects will have their own set of records in the Dataset table.

Therefore, we store the more static attribute information in a Dataset table and the frequently changing weight information in a weight table (Figure 1(b)). The Dataset objects share the same records in the Dataset table unless they change attribute values. If an algorithm never changes attribute values there is one set of data records in Dataset with did= *IDS*. In contrast, the weight table contains, for each existing P-vector (Dataset objects might share P-vectors) its own set of weight records. This set contains as many weight records as there are entries in the P-vector. Note that a P-vector might have fewer entries than the total number of records with did = *IDS* (e.g., in decision tree construction).

The Dataset table has one attribute for each attribute of the original data, a did attribute as described above, and a position attribute that links the record to the P-vector of the Dataset object. If a record *dr* has $did = IDS$ and $position = X$, then *DS*'s P-vector has one entry with value *X*. The weight table has attributes tid and position similar to the did and position attributes in the Dataset table, and a weight. In order to match the weights with the corresponding records in the Dataset table, we have to join over the position and match did/tid attributes. Both Dataset and weight tables have several indices (clustered and unclustered) on position and did/tid attributes in order to speed up the most typical data access.

Some algorithms change the structure, i.e., they remove or add attributes. This is usually only done in the preprocessing phase. In such cases, full table copies are made. When preprocessing is completed, a filtereddataset table is created, which will then be used instead of the original dataset table.

3.2 Main Memory Data Structures

Figure 2 shows how the main memory data structures are adjusted in order to allow for the main memory and database storage implementation to co-exist. The abstract class AbstractDataset implements variables and methods used in both storage implementations. MMDataset is the original Dataset implementation in Weka, and DBDataset is the abstract class for our relational database implementation. It contains commonly two subclasses. Recall that the Dataset object might have to keep track of several did.

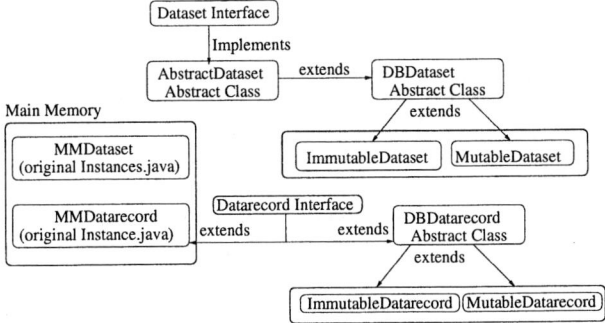

Fig. 2. WekaDB: Dataset and Datarecord

However, if an algorithm never changes attribute values (except the weights), there will be only one did value for all records. Hence, we allow algorithms to declare this fact in advance, and then use a simpler implementation which can ignore did. The class MutableDataset supports all the functions that allow algorithms to change attribute values, while ImmutableDataset does not support those functions (only the weights are allowed to change). The same class structure is used for data records.

4 Database Access

WekaDB accesses the database using a standard JDBC API. For space reasons, we only outline here how the ImmutableDataset class accesses the database. A special load interface allows the transfer of records from a file in ARFF format (used by Weka) to the database, creating dataset and weight tables and the corresponding records.

When an algorithm starts, an initial ImmutableDataset object is created with a corresponding P-vector based on the information in the dataset table. No data records are loaded. The P-vector is the only memory-based data structure that grows linearly with the size of the data set. It is needed because algorithms can reorder the records during the computation, for instance by sorting and resampling. In WekaDB, this is done by reordering the entries in the P-vector. This vector represents the data in the correct order for a specific ImmutableDataset object while the records in the dataset and weight table are unordered.

If a copy DS' is made from an ImmutableDataset object DS, it can share the P-vector and tid value with DS, or it can create its own P-vector, and receive a new tid value. In the latter case, it must call the add method for each record to which it wants to refer. This method adds the position of the record to the P-vector and inserts a weight record into the weight table, with the same position and the new tid value. No records are added into the dataset table for ImmutableDataset objects. The new "copy" of the data set is represented by the new P-vector.

Data records are accessed via the enumerateRecords() and Record(index) methods of the ImmutableDataset class. We only describe the latter here. Record(index) on object DS loads the record with position p if DS's P-vector V has $V[index] = p$.

The (slightly simplified) SQL statement is

```
SELECT * FROM dataset, weight
WHERE weight.tid = IDS
AND dataset.position = p
AND weight.position = p
```

If Record(index) is called in a loop accessing all records one by one, the statement is executed repeatedly, and data transfer takes place for each record. Looping is very common in data mining algorithms, which makes this type of data access very expensive. Hence, retrieval will be faster if we load a whole set of records with one query, buffer them within core and then provide the records to the user on request within the loop. Hence, we implemented a buffered version of Record(index), which retrieves B records at a time. B is an adjustable parameter. In the buffered implementation, when a record with position p (determined as above) is requested, we first check if the record is already in main memory. If so, the record is returned right away. Otherwise, we use the following (slightly simplified) SQL statement to retrieve B consecutive records, starting at position p:

```
SELECT *   FROM dataset, weight
WHERE weight.tid = IDS
AND dataset.position =  weight.position
AND dataset.position >= p
AND dataset.position < (p+B)
```

The B retrieved records are stored in a JDBC ResultSet. For data mining algorithms that access the data sequentially and do not perform any sorting, buffering can dramatically decrease the number of database accessed. If an algorithm had sorted the entries in the P-vector, the benefits of buffering are limited.

5 Using More Database Functionality

The loose-coupling approach discussed so far performs all computation on data records in main memory. It might be more efficient to perform some computation within the database by applying advanced SQL functionality. This leads to a semi-tight coupling between Weka and database. For that purpose we modified several core methods. As a simple example, our implementation of sumOfWeights of the ImmutableDataset class uses the SQL aggregate function sum to perform the operation within the database. Another example is sort(), which orders the data based on the values of one attribute. A main memory implementation requires retrieving records possibly multiple times from the database. In contrast, we use an SQL statement with an order by clause.

Data preprocessing [5] is a common step in many algorithms. It is used to clean data, and perform data transformation. Weka provides a set of filters using the filters interface. The implementation itself is built on top of core. Weka's main memory implementation of the filters accesses records one by one, and stores all the filtered data in a queue. This adds considerable overhead, and reduces scalability due to the queue data structure. We reimplemented the filters using a database oriented approach that does not require loading any records into main memory. For instance, for the filter that replaces all missing attribute values with the modes/means observed in the data, we precompute modes and means with SQL aggregate functions and use `update` SQL statements to replace missing values with these modes and means.

Since the machine learning algorithms are developed without considering space limitations they might create their own data structures that limit scalability. For instance, the logistic regression algorithm implemented in Weka normalizes the input data and stores it in a 2-dimensional array (one dimension represents the records, the other the normalized attributes). This is done by a pass through the data set using `Record(index)` calls. This array is, in fact, as large as the entire data set. Our approach is to provide adequate support in order to help developers eliminate such limitations. Normalizing records and then accessing the normalized data in a systematic way seem to be standard steps usable in various algorithms. Therefore, we offer extra normalization methods as part of the `Dataset` class. The methods use SQL queries to perform the normalization and store the normalized values in the database. The normalized data can be retrieved through an interface that provides the standard array representation. It can be used without knowing that a database implementation is used. Whenever a normalized record is accessed through the array interface, a corresponding SQL query retrieves the record from the database.

6 Optimizing JDBC Applications

Our implementation uses several standard mechanisms to speed up the JDBC application. First, our system uses a single database connection for all database access to optimize connection management. Since transaction management is expensive, we bundle related operations in a single transaction in order to keep the number of transactions small. Third, since the system runs in single user mode, we run our transactions in the lowest isolation mode provided by the database to minimize the concurrency control overhead. Finally, we use JDBC's `PreparedStatement` objects as much as possible, since these statements are parsed and compiled by the database system only once, and later calls use the compiled statements, improving performance significantly.

7 Empirical Evaluation

We evaluate WekaDB using the logistic regression algorithm, the Naive Bayes algorithm and the K-means clustering algorithm, on both synthetic and real data sets. The first two algorithms are used for classification problems, while the last one is an unsupervised learning algorithm. We note that at the moment, all algorithms in the Weka package, except the decision tree construction algorithm, work seamlessly with WekaDB. Once logistic regression and k-means clustering were fully functional, the other algorithms worked with WekaDB without any further tweaking.

The synthetic data sets were generated using the program of [1]. We generated training data sets with 10,000 to 1,000,000 records, and one testing data set with 5000 records. Each data set has 10 numerical attributes and 1 class attribute, without missing values. We also run tests with filters for replacing missing values and for discretizing continuous attributes. The results were very similar to the ones reported here.

The real data set is based on the AVIRIS (Airborne Visible/Infrared Imaging Spectrometer) data set, originally created at JPL (Jet Propulsion Laboratory, California Institute Technology) and extensively corrected by CCRS (Canadian Center for Remote Sensing, Natural Resources Canada). It contains hyperspectral data that was captured by the NASA/JPL AVIRIS sensor over Cuprite, Nevada on June 12, 1996 (19:31UT) (see [6] for more information). The original data set contains 314,368 records and 170 attributes. For the purpose of the experiments, we generated four different data sets, containing 12669, 19712, 35055 and 78592 records respectively, and one testing data set containing 3224 records. Each data set has 168 numeric attributes and 1 nominal class attribute without missing values.

We restricted the memory size to be used by Weka/WekaDB to 64MB in order to avoid long running times and be able to run many experiments. All experiments use the default values for all the algorithms, unless otherwise specified.

In all the experiments we measure the runtime of the algorithms when we increase the size of the training data set. In all cases, the main-memory implementation of Weka is significantly faster (between 2-3 orders of magnitude). However, the maximum number of instances that it can handle is below 40000. WekaDB, on the other hand, can handle up to 700000 - 800000 instances, which is a 20-fold improvement.

Figure 3 illustrates the running time on the synthetic data set for WekaDB with a buffer (size 10000) and without using a buffer for Naive Bayes (left figure) and logistic regression (right figure). As illustrated, the computation time increases linearly with the number of instances, which should be expected, as both algorithms have to loop through the data. A similar linear increase is observed in the original Weka implementation, but with a much smaller slope, of course. Using the buffer yields a 5-fold improvement in speed. Weka's computation time is 11 seconds at 28000 records (not shown in the figure), compared to 368 seconds for WekaDB with a buffer. Hence, Weka's computation is roughly 30 times faster. For logistic regression, the time difference is more pronounced, with Weka finishing in 0.9 seconds, compared to 251 seconds for WekaDB. However, Weka is showing a significant space limitation running out of memory at 29000 instances. At 700000 and 800000 records respectively, WekaDB also runs out of memory, because at this point the position vector becomes a memory constraint (since it grows linearly with the size of the data set). Recall that we use only 64MB. If we

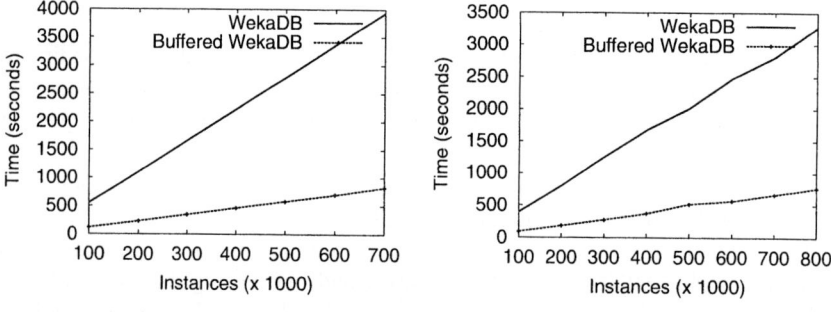

Fig. 3. WekaDB for Naive Bayes (left) and Logistic regression (right) on synthetic data

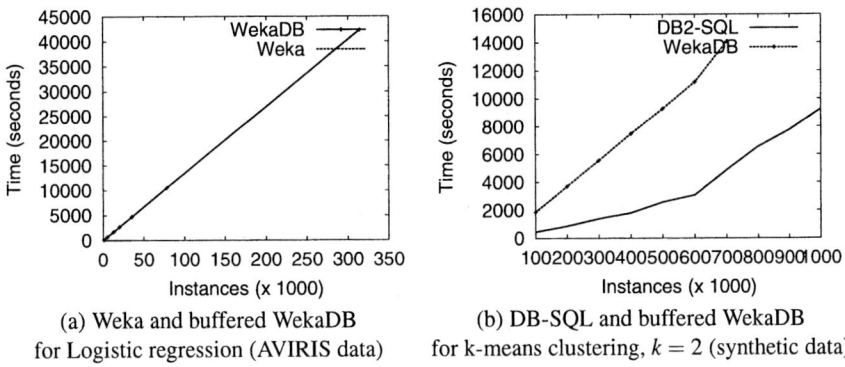

(a) Weka and buffered WekaDB for Logistic regression (AVIRIS data)

(b) DB-SQL and buffered WekaDB for k-means clustering, $k = 2$ (synthetic data)

Fig. 4. Performance Comparisons

assume that 1GB of main memory is available, we can expect WekaDB to handle on the order of 10,000,000 records while Weka will handle less than 500,000.

Figure 4(a) presents the performance of the original (main-memory) Weka implementation compared to WekaDB with a buffer on the AVIRIS dataset. The results are consistent with those on the synthetic data: Weka can only handle 35000 instances, roughly 10% of the size of the original data set. WekaDB is roughly 1000 times slower at 35000 instances, but can handle the entire data set successfully. Performance scales linearly with the number of instances. The reason for the large discrepancy in running time is the number of attributes, which is much larger than for the synthetic data. Very similar results are obtained with Naive Bayes.

For k-means clustering, the comparison of Weka, WekaDB and buffered WekaDB is very similar to those presented before, so we omit it here. Instead, Fig.4(b) compares WekaDB using a buffer to a k-means algorithm proposed by Ordonez [10], which also stores the data in a database and re-implements the computation in the database. This algorithm takes particular care to take advantage of any database-specific optimizations. We refer to it as DB2-SQL. As expected, the optimized algorithm can scale better (since there are no limiting memory data structures), and is faster for the particular number of clusters requested ($k = 2$). However, further analysis of the algorithms

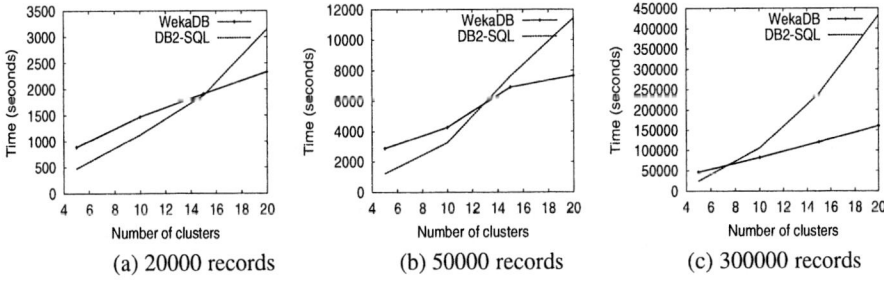

(a) 20000 records (b) 50000 records (c) 300000 records

Fig. 5. DB-SQL and buffered WekaDB for k-means clustering on different dataset sizes (synthetic data)

shows an interesting trend. Fig. 5 shows the behavior of the two algorithms as we vary the number of clusters between 5 and 20, for different dataset sizes. As the number of clusters increases, the computation time of WekaDB grows linearly, while that of DB2-SQL grows super-linearly. Hence, when there are many clusters, the simple k-means algorithm in WekaDB outperforms the specialized DB2-SQL implementation. As the number of instances also increases, WekaDB has even better performance. This is due to the fact that in the DB2-SQL implementation, the cluster centers and all auxiliary memory structures are in the database. As instances have to be compared to the cluster centers in order to decide where they belong, the computation time degrades. Also, larger numbers of clusters typically require more iterations of the algorithms. In WekaDB, since the number of clusters is still relatively small, a lot of the processing is done in main memory, which makes it much faster.

8 Conclusions

This paper presented an approach to the integration of learning algorithms with relational databases. We built an extension of the well-known Weka data mining library, WekaDB, which allows the data used by the learning algorithms to reside on secondary storage. This change is transparent to the developers of machine learning algorithms. Our empirical results show that this approach provides scalability up to very large data sets. From the point of view of evaluating empirically the performance of new data mining algorithms, we believe that WekaDB provides an interesting benchmark, since it provides a faithful implementation and execution of the learning algorithms. Other approaches, such as resampling and special-purpose algorithms, can be compared to it in terms of accuracy, as well as scalability and computation time. Also, in principle, WekaDB allows any new algorithms that are added to the Weka package to be able to work immediately on data stored in a database, without any further modifications. We currently work on removing the remaining memory limitation for WekaDB, by eliminating the need to have an internal memory data structure linear in the size of the data set. The idea is to store the position as additional attribute in the dataset table.

References

1. R. Agrawal, T. Imielinski, and A. Swami. Database mining: A performance perspective. *IEEE Transactions on Knowledge and Data Engieering*, 5(6), 1993.
2. J. Gehrke, R. Ramakrishnan, and V. Ganti. Rainforest: A framework for fast decision tree construction of large datasets. *Int. Conf. on Very Large Data Bases*, 1998.
3. A. W. Moore and M. Lee. Cached sufficient statistics for efficient machine learning with large data sets. *Journal of Artificial Intelligence Research*, 8, 1998.
4. W. Du Mouchel, C. Volinsky, T. Johson, C. Cortes, and D. Pregibon. Squashing flat files flatter. *ACM Int. Conf. on Knowledge Discovery and Data Mining*, 1999.
5. D. Pyle. *Data Preparation for Data Mining*. Morgan Kaufmann Publishers, 1999.
6. B. J. Ross, A. G. Gualtieri, F. Fueten, and P. Budkewitsch. Hyperspectral image analysis using genetic programmming. *The Genetic and Evolutionary Computation Conf.*, 2002.
7. S. Sarawagi, S. Thomas, and R. Agrawal. Integrating association rule mining with relational database systems: alternatives and implications. *ACM SIGMOD Int. Conf. on Management of Data*, 1998.

8. J. Shafer, R. Agrawal, and M. Mehta. SPRINT: A scalable parallel classifier for data mining. *Int. Conf. on Very Large Data Bases*, 1996.
9. I. H. Witten and E. Frank. Data mining software in Java. http://www.cs.waikato.ac.nz/ml/weka/.
10. Carlos Ordonez. Programming the K-means Clustering Alogrithm in SQL. *ACM Int. Conf. on Knowledge Discovery and Data Mining*, 2004.

Bias-Free Hypothesis Evaluation in Multirelational Domains

Christine Körner[1] and Stefan Wrobel[1,2]

[1] Fraunhofer Institut Autonome Intelligente Systeme, Germany
{christine.koerner, stefan.wrobel}@ais.fraunhofer.de
[2] Dept. of Computer Science III, University of Bonn, Germany

Abstract. In propositional domains using a separate test set via random sampling or cross validation is generally considered to be an unbiased estimator of true error. In multirelational domains previous work has already noted that linkage of objects may cause these procedures to be biased and has proposed corrected sampling procedures. However, as we show in this paper, the existing procedures only address one particular case of bias introduced by linkage. In this paper we therefore introduce *generalized subgraph sampling*, a sampling procedure based on bin packing, which ensures that test sets are properly chosen to match the probability of reencountering previously seen objects and which includes previous approaches as a special case. Experiments with data from the Internet Movie Database illustrate the performance of our algorithm.

1 Introduction

In machine learning one typically assumes that the true classification of an object depends only on the object itself and, given the object, is independent of the classification of other objects. In this case, the observed sample error on a sufficiently large and randomly chosen independent test set is an unbiased estimator of true error. However, many applications rely on relational data where the label of an object may probabilistically depend on the labels and/or attributes of other related objects or shared parts of objects ("autocorrelation"). As pointed out by [1], whenever there is autocorrelation, the observed error on a randomly chosen test set may not be an unbiased estimator anymore. In [1] this issue is addressed using *subgraph sampling*, which however completely eliminates the dependency between training and test sets and thus is applicable only to problem settings where future data *never* link to previously seen data.

In this paper we therefore propose *generalized subgraph sampling* (GSS), a sampling procedure based on bin packing, which ensures that test sets are properly chosen to match the probability p_S^{kn} of reencountering previously seen objects and which includes subgraph sampling as a special case for $p_S^{kn} = 0$. In the following section we first introduce the issues associated with autocorrelation in more detail and then present the GSS algorithm in Section 3. We experimentally compare two variants of our sampling algorithm with random sampling in Section 4. The paper concludes with a summary and further challenges for error estimation using multirelational data.

2 Linkage Bias in Multirelational Domains

In multirelational domains the assumption of independent instances cannot be taken for granted. Let us consider an example that clearly shows the dependencies between objects due to linkage and autocorrelation [1]. The Internet Movie Database[1] (IMDb) stores information on over 450,000 movies, including actors, producers, studios and box office receipts. We regard the learning task to predict whether a movie has box office receipts of more than \$2 million given information about the studio that made the movie. More formally, the application consists of two kinds of objects, movies X and studios A. Figure 1 shows the relevant structure of the movie data set. We will refer to a studio $a \in A$ as a *neighbor* of a movie $x \in X$ if the studio produced that movie. The set of movies sharing the same studio a forms the *neighborhood* of a. The *degree* δ_a specifies the number of movies produced by studio a.

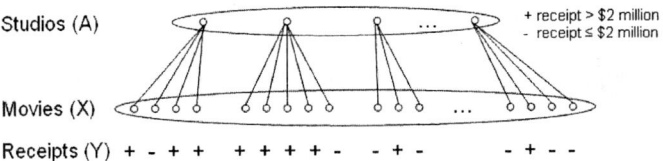

Fig. 1. Internal structure of the movie data

If we proceed as usual to estimate the error of a hypothesis and divide the movie data into a training and test set randomly (according to some split proportion), we will very likely assign movies of the same studio to both resulting sets. A dependency between the training and test set arises as the sets share some of their neighbors. In fact, when using random splits the relative frequency of common neighbors increases with the chosen split ratio between training and test set [2]. How does this dependency influence the error estimate? Since the labels of movies produced by the same studio are correlated (it is plausible that big studios will make many movies with big box office receipts), any learner capable of exploiting relations (in particular probabilistic relational learners [3, 4, 5]) will form a hypothesis that exploits the known objects, and thus will make fewer errors predicting the label of movies from known studios than of movies from unknown studios. Thus, the more objects with known neighbors are in the test set, the lower the estimated error will be even though the hypothesis is the same. This has led [1] to postulate that the dependency between the training and test set should be removed. They present a procedure, subgraph sampling, which ensures that any information (in this case studios) shared between different objects is included in either the training or the test set and thus eliminates the above mentioned bias in error estimation.

---

[1] http://www.imdb.com

This approach, however, considers only the application setting in which the studios of all future movies have never been seen before. Yet, in many applications future objects that we need to classify with our induced hypothesis will actually have *known* neighbors. In the movie domain it is quite likely that a new movie will be produced by one of the already existing studios. Therefore, the error will be overestimated if the links between training and test set are removed completely. Instead, the test set should reflect the probability that a randomly drawn (future) object is linked to known neighbors. We call this probability the *known neighbor probability*.

Definition 1 (Known Neighbor Probability). *For an instance space X, a distribution D_X, a sample S, a set A of neighbors and a function $nb : X \to A$ assigning to each instance $x \in X$ its neighbor $a \in A$, the known neighbor probability for an instance x randomly drawn according to D_X is defined as $p_S^{kn} := P\left(nb(x) \in \bigcup_{s \in S} nb(s)\right)$.*

Given a sample S, the known neighbor probability p_S^{kn} is a domain property. In a transductive learning setting or a context where the distribution over the complete instance space were known, its computation would be straightforward. In most cases, however, the known neighbor probability must be supplied by the user based on application considerations.

3 Generalized Subgraph Sampling

In order to arrive at an unbiased estimate, the fraction of objects in the test set with neighbors also present in the training set should match the known neighbor probability. How can this goal be achieved? Above we already remarked that random sampling is incapable of establishing the known neighbor frequency, as the amount of related objects in the test set varies with the chosen training/test set split proportion[2]. Therefore, we propose *generalized subgraph sampling* (GSS), which is a sampling procedure based on the known neighbor probability. It ensures that for a given data sample S, known neighbor probability p_S^{kn} and a chosen split proportion p_{train} the resulting test set contains the same proportion of objects with known neighbors with respect to the training set as specified by the known neighbor probability. GSS includes subgraph sampling as proposed by [1] as a special case for $p_S^{kn} = 0$.

The task to install the known neighbor probability into the test set can be considered as a bin packing problem. In general, bin packing requires to pack a set of items into a number of bins such that their total weight does not exceed some maximum value. More specific, GSS needs to fill three bins. The first bin, S_{train}, contains the training instances. The second and third bin, $S_{test,rel}$ and

---

[2] In certain domains, natural temporal splits can be used to form test sets. Alternatively, if we do not require an error estimate for a single user presentable hypothesis, it would be possible to induce and evaluate two hypotheses (on completely known vs. on completely unknown test data) and then average according to p_S^{kn} [6].

$S_{test,ind}$, contain the test instances which are either *rel*ated to or *ind*ependent of instances in the training set respectively. We designed two versions of GSS. The first version (Simple) prefers neighbors with a small degree in order to sustain the specified bin sizes and is allowed to adjust the chosen split proportion if necessary. The second version (Modified) chooses all objects randomly, yet may discard data tuples to preserve the known neighbor probability as well as the chosen split ratio. Algorithm 1 depicts Part 1 and 2 of GSS Simple, which satisfy the specification if the sample contains a sufficient number of neighbors with degree $\delta_a > 1$. We indicate a subset by adding a subscript to the name of the originating set, e.g. S_{test} denotes the test set created from sample S and S_a denotes the neighborhood of neighbor a. The sizes of S, $S_{test,rel}$ and $S_{test,ind}$ are denoted by n, $n_{test,rel}$ and $n_{test,ind}$ respectively. For further details see [2].

Algorithm 1 GSS Simple (Part 1 and 2).

Input: sample S, split proportion p_{train}, known neighbor probability p_S^{kn}
Output: S_{train}, S_{test}
1: $S_{train} = S_{test} = S_{test,kn} = S_{test,ind} = \emptyset$
2: $n_{test,rel} = |S| \cdot (1 - p_{train}) \cdot p_S^{kn}$
3: $n_{test,ind} = |S| \cdot (1 - p_{train})(1 - p_S^{kn})$
4: # Part 1: fill bin $S_{test,ind}$
5: compute set A of neighbors and neighborhood S_a for each $a \in A$
6: **while** $|S_{test,ind}| < n_{test,ind}$ **do**
7:     choose $a \in A$ with smallest δ_a
8:     $S_{test,ind} = S_{test,ind} \cup S_a$
9:     $S = S \backslash S_a$
10:    $A = A \backslash \{a\}$
11: **end while**
12: # Part 2: fill bin $S_{test,rel}$
13: $n_{test,rel} = (|S_{test,ind}|/(1 - p_S^{kn})) \cdot p_S^{kn}$
14: **while** $|S_{test,rel}| < n_{test,rel}$ and $|S| > 0$ **do**
15:     choose $a \in A$ randomly
16:     **if** $|S_a| \geq 2$ **then**
17:        choose two objects $s_1, s_2 \in S_a$
18:        $S_{test,rel} = S_{test} \cup \{s_1\}$
19:        $S_{train} = S_{train} \cup \{s_2\}$
20:        $S = S \backslash \{s_1, s_2\}$
21:     **else**
22:        $S_{train} = S_{train} \cup S_a$
23:        $S = S \backslash S_a$
24:        $A = A \backslash \{a\}$
25:     **end if**
26: **end while**
27: $S_{train} = S_{train} \cup S$
28: $S_{test} = S_{test,ind} \cup S_{test,rel}$

4 Experiments

We evaluated both variants of our algorithm on data from the IMDb and compared their performance against random sampling. Table 1 on top shows the achieved known neighbor probabilities for three chosen split proportions. As can be seen, both versions of our algorithm are successful in ensuring the required known neighbor probability of 0.45 regardless of the split proportion. As expected, the known neighbor probability obtained by random sampling varies as the chosen split proportion changes. The bottom of Table 1 shows that both algorithms produce exactly the required sizes at a split proportion of 0.7 and 0.9. For a chosen split proportion of 0.5 both algorithms yield significantly enlarged training sets, which results from an unexpected large number of neighbors with $\delta_a = 1$.

Table 1. Top: average obtained known neighbor probability for a target $p_S^{kn} = 0.45$; bottom: average obtained split proportion and percentage of unused objects

algorithm	KNP, split prop. = 0.5	KNP, split prop. = 0.7	KNP, split prop. = 0.9
Random	0.3184 ± 0.0182	0.3669 ± 0.0247	0.4014 ± 0.0467
Simple	0.4501 ± 0.0000	0.4502 ± 0.0000	0.4499 ± 0.0000
Modified	0.4500 ± 0.0026	0.4498 ± 0.0005	0.4494 ± 0.0009

	split prop. = 0.5		split prop. = 0.7		split prop. = 0.9	
algorithm	resultant split	unused	resultant split	unused	resultant split	unused
Random	0.5002 ± 0.0000	0.0000	0.7000 ± 0.0000	0.0000	0.9001 ± 0.0000	0.0000
Simple	0.5779 ± 0.0000	0.0000	0.6998 ± 0.0000	0.0000	0.8999 ± 0.0000	0.0000
Modified	0.6465 ± 0.0119	0.0014	0.7014 ± 0.0007	0.0013	0.9000 ± 0.0002	0.0011

5 Conclusion and Future Work

In relational domains it is well known that high linkage and autocorrelation cause a bias in test procedures. Therefore, sampling procedures must be adjusted to provide for an unbiased error estimate. Present approaches only address the special case where no further dependencies between the data sample and randomly drawn future objects are expected. We propose a sampling procedure that controls the amount of dependent objects in the test set. Our evaluation shows that GSS is an effective sampling procedure that guarantees to partition a sample according to a given known neighbor probability.

So far our procedure relies on the user to provide the known neighbor probability. It is a topic of future research to investigate whether certain conditions allow to estimate the known neighbor probability directly from the data sample.

References

1. Jensen, D., Neville, J.: Autocorrelation and linkage cause bias in evaluation of relational learners. In: Proc. of the 12th International Conference on Inductive Logic Programming, Springer-Verlag (2002)
2. Körner, C., Wrobel, S.: Bias-free hypothesis evaluation in multirelational domains. Technical report, Fraunhofer Institut Autonome Intelligente Systeme (2005) http://www.ais.fraunhofer.de/~ckoerner.
3. Getoor, L., Friedman, N., Koller, D., Pfeffer, A.: Relational data mining. In Dzeroski, S., Lavrac, N., eds.: Learning Probabilistic Relational Models. Springer-Verlag, Berlin (2001) 307–335
4. Taskar, B., Abbeel, P., Koller, D.: Discriminative probabilistic models for relational data. In: Proc. of the 18th Conference on Uncertainty in Artificial Intelligence. (2002)
5. Neville, J., Jensen, D.: Collective classification with relational dependency networks. In: Proc. of the 2nd Multi-Relational Data Mining Workshop, 9th ACM SIGKDD International Conference on Knowledge Discovery and Data Mining. (2003)
6. Fürnkranz, J. (personal communication)

Enhanced DB-Subdue: Supporting Subtle Aspects of Graph Mining Using a Relational Approach*

Ramanathan Balachandran, Srihari Padmanabhan, and Sharma Chakravarthy

The University of Texas at Arlington, Arlington TX 76019, USA

Abstract. This paper addresses subtle aspects of graph mining using an SQL-based approach. The enhancements addressed in this paper include detection of cycles, effect of overlapping substructures on compression, and development of a minimum description length for the relational approach. Extensive performance evaluation has been conducted to evaluate the extensions.

1 Introduction

Database mining has been a topic of research for quite some time [1-4]. Graph mining uses the natural structure of the application domain and mines directly over that structure. Graphs can be used to represent structural relationships in many domains. Subdue [5] is a mining approach that works directly on graph representation.

Subdue identifies interesting and repetitive substructures within the structural data. Subdue uses the principle of minimum description length [6] (or MDL) to evaluate the substructures. The major drawback of main memory algorithms is their scalability to larger problems. The DBMS version of Subdue is called DB-Subdue [7]. The input (which is a graph) is represented using relations and operations use joins (or other relational operations) for mining repetitive substructures. This paper extends the DB-Subdue (termed EDB-Subdue [8]). This paper proposes an approach to handle cycles and overlaps in a graph. This paper also addresses a new technique for evaluating the substructures, which is both scalable and capable of distinguishing the best substructure among substructures of equal size and frequency.

The rest of this paper is organized as follows. Section 2 discusses the related work briefly. Section 3 discusses the design issues for the various enhancements that have been added to DB-Subdue. Section 4 presents performance evaluation including comparison with the Subdue algorithm. Section 5 has conclusions.

2 Related Work

Related work include AGM (Apriori-based Graph Mining) [9], gSpan (graph-based Substructure pattern mining) [10], FSG (Frequent SubGraph discovery) [11] and

---

* This work was supported, in part, by NSF (grants IIS-0097517, IIS-0326505, and EIA-0216500).

Subdue [5]. AGM is a mathematical graph theory based approach which mines a complete set of subgraphs mainly using support measure. gSpan is a depth first search based canonical labeling approach, that uses a canonical tree representation of each graph instead of the adjacency matrix. FSG aims at discovering subgraphs which occur frequently over the entire set of graphs. Also the scope of all the above frequent item set mining algorithms differs entirely from the scope of EDB-Subdue which aims at discovering the best pattern within a forest of graphs as opposed to discovering frequent patterns in a large database of graphs.

3 Extensions to DB-Subdue (EDB-Subdue)

Database Minimum description length principle (DMDL) is a heuristic based on the minimum description length principle (MDL) [6]. Although the MDL principle is accurate, it cannot be applied directly to database representations (based on the number of 1's in the adjacency matrix) and is computationally expensive. On the other hand, DB-Subdue's frequency heuristic scales well for large datasets. DB-Subdue cannot distinguish between substructures that have the same number of vertices and edges and having the same frequency of occurrence. These attributes form the **signature** of a substructure. The adjacency matrix in the MDL principle plays a vital role in distinguishing between two substructures with the same signature. For example, in the graph shown in Fig. 1, there are two substructures, each appearing twice in the graph.

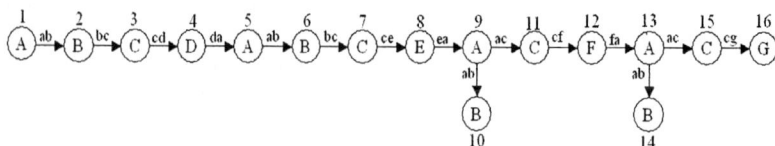

Fig. 1. Input Graph Example

Fig. 2 (graph1) and Fig. 3 (graph2) show two sub graphs with the same signature substructures. The adjacency matrices for the two graphs are different even though they have the same number of vertices and edges. The number of bits needed to encode *Fig 1* is less than the bits required to encode Fig 2. This is because for Fig 2 there is only one row in which there are 1s. But for Fug 3, both the first and second rows have 1s. As a result, two rows have to be represented in the second case whereas only one row has to be represented in the first case. For the input graph in Fig. 1, if graph compression is performed using the first substructure the resultant MDL value is 1.14539 and if the

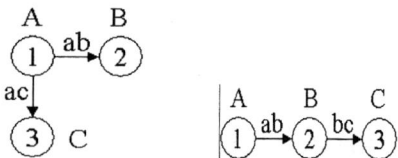

Fig. 2. Sample Graph1 **Fig. 3.** Sample Graph2

graph is compressed using the second substructure the MDL value obtained is 1.12849. Therefore, the MDL principle ranks *graph1* higher than *graph2*.

$$MDL = DL(G) / DL(S) + DL(G|S)$$

In the above formula, the lesser the number of bits needed to represent DL (S) + DL (G|S), the better is the substructure. The DMDL principle uses this representation difference for distinguishing same signature substructures. The DMDL value is calculated using a formula that helps us achieve the goal of differentiating same signature substructures.

$$DMDL = Value(G) / Value(S) + Value(G|S)$$

In the above formula, 'G' represents the entire graph, 'S' represents the substructure and 'G|S' represents the graph after it has been compressed using the substructure S.

Value(G) = graph_vertices + graph_edges
Value(S) = sub_vertices + uniquesub_edges
Value(G|S) = (graph_vertices − sub_vertices * count + count) +
(graph_edges − sub_edges * count)

Value (G) in the above formula represents the value of the entire graph. Value (S) represents the value of the substructure. Value (G|S) represents the compressed graph (replacing all the instances of the substructure in the graph). The parameter uniquesub_edges is calculated as the number of unique extensions in the substructure. This is because extensions are the only way we can determine how vertices are connected within a substructure. For the graph in Fig. 2 the extensions are 1, 1, as the second vertex is extended from the first vertex and the third vertex is also extended from the first vertex. Therefore the uniquesub_edges value is 1. The value 1 indicates that only one row of 1's are present in the corresponding adjacency matrix. For the graph shown in Fig. 3, the extensions are 1, 2 as the second vertex is extended from the first vertex and the third vertex is extended from the second vertex. The uniquesub_edges value is 2 indicating that there are two rows of 1's present in the corresponding adjacency matrix. Therefore, without computing the adjacency matrix for the substructure the same effect is obtained in the DMDL value. The DMDL value for the graph of Fig. 2 is 1.1481 and the value for graph of Fig. 3 is 1.1071. Hence, the first substructure is a better substructure than the second substructure as it has a higher value. In general, for higher edge substructures having the same signature, substructures with vertices having a higher out-degree are better substructures.

Detecting Cycles: Detecting cycles is important when dealing with graph mining algorithms. The main problem with the DB-Subdue algorithm is the fact that it loops within a cycle after detecting the same. In DB-Subdue, the substructure in Fig 4 will be extended to the substructure shown in Fig 5. However, this extended substructure does not exist in the input graph. Therefore this extension needs to be detected and avoided.

Fig. 4. Cycle   **Fig. 5.** Cycle extension

This might also affect the best substructure that is discovered. Cycles are detected by checking if the vertex number of any vertex in the new substructure formed is already present in the substructure. The conditions to prevent cycles are formalized as follows:

In the subgraph $V_1, V_2...V_j, V_{j+1}...V_{j+k}$, V_j there is a cycle as the vertex V_j appears twice. We need to eliminate expansion from the second occurrence of V_j, as it is the repetition of a vertex already present in the substructure. This is done by making the second occurrence of V_j different from the first occurrence of V_j (by modifying the vertex number). This prevents expansion from the second occurrence of V_j as its new value is not present in the input graph. The basic idea is to prevent that vertex from expanding further. The computation will not be affected in any manner, as the main objective is to substitute the repeated vertex with a vertex not present in the input graph. For example, in Fig 4 and Fig 5, for the extension from 3 → 3, vertex 3 repeats again and the presence of a cycle can be deduced. As the cycle is now detected, extension from the second occurrence of vertex 3 should be prevented. This is achieved by changing the vertex number to a value that is not present in the graph. This prevents the extension from the second occurrence of vertex 3, as the modified vertex value will not match any vertex in the input graph.

In general, each time a new vertex is added, it is checked to see if that vertex is already present in the substructure. If so, then the vertex number is changed to prevent future expansions from that vertex.

Overlapping Substructures: Two or more instances of a substructure are said to overlap if they have a common substructure between them.

We will first explain how overlap is avoided in a two-vertex substructure and then generalize it to higher-vertex overlap. An example of a two-edge substructure overlap is shown in Fig 6. In this example there are two instances of the substructure 'CDE' that are overlapping and the overlapping substructure among them is 'CD'. Since both the first and second vertex is overlapping, only one of the overlapping instances of the substructure must be considered for counting the number of instances of the substructure 'CD'. This problem is solved by only counting one instance; the instance with the greatest third vertex value while the count is computed. In the example, the substructure 'CDE' will have count of one rather than two, as the substructure 'CD' is overlapping.

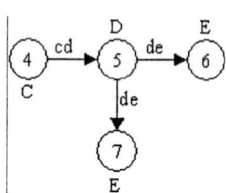

Fig. 6. Two edge substructure overlap

In general, for a higher-edge substructure, if there is an overlap on any of the vertices other than the last vertex, then only the overlapping instance that has the greatest last vertex value is included for computing the count. In case of an overlap on the last vertex, each vertex is checked starting from the vertex previous to the last one until the first vertex and the non-overlapping vertex is determined. Only the overlapping instance that has the greatest non-overlapping vertex value is included for calculating the count. Therefore when overlap is avoided the count of the substructure as computed as follows: No. of instances of the substructure = No. of non-overlapping instances + one overlapping instance.

4 Performance Evaluation

This section discusses the performance comparison between Enhanced DB-Subdue and Subdue main memory algorithm for various datasets. We used Oracle 9i, running on Linux, and Intel Xeon dual processor with 2GB of RAM. The set of experiments that were performed included datasets that had cycles. The substructures that were embedded are shown in Fig 7 and 8 respectively. The experiments were performed only between Enhanced DB-Subdue and Subdue main memory, as DB-Subdue cannot handle cycles. The enhanced DB-Subdue includes all the additional functionality (DMDL, Cycles and overlap). As it can be seen from the comparisons, the Enhanced DB-Subdue performs better than the main memory algorithm for large datasets. The running times using a beam value of 4 is shown in Fig. 9. Experiments were also performed for a beam value of 7 and 10 respectively. Fig. 9 clearly indicates that after incorporating all the features of Subdue into EDB-Subdue, it still outperforms Subdue. In addition, the main-memory Subdue would not handle data sets larger than 20K vertices and 40K edges (hence there are no performance values beyond 20K data sets) where as the EDB-Subdue could easily handle 800K vertices and 1600K edges. The graph is plotted on a log scale.

Fig. 7. Substructure1 with cycles   **Fig. 8.** Substructure2 with cycles

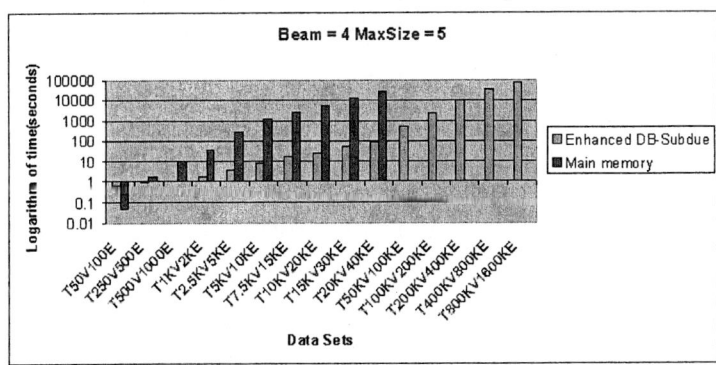

Fig. 9. Graphical Comparison of the approaches

5 Conclusions

The enhancements addressed in this paper include detection of cycles and handling overlapping substructures. All the enhancements were implemented using SQL. The experiments clearly demonstrate the scalability of the algorithm even after adding all the functionality.

References

1. Agrawal, R. and R. Srikant. *Fast Algorithms for Mining Association Rules*. in *Proceedings 20th International Conference Very Large Databases, VLDB*. 1994. Chile.
2. Sarawagi, S., S. Thomas, and R. Agrawal. *Integrating Mining with Relational Database Systems: Alternatives and Implications*. in *SIGMOD*. 1998. Seattle.
3. Mishra, P. and S. Chakravarthy. *Performance Evaluation and Analysis of SQL-92 Approaches for Association Rule Mining*. in *BNCOD Proceedings*. 2003.
4. Mishra, P. and S. Chakravarthy. *Performance Evaluation of SQL-OR Variants for Association Rule Mining*. in *Dawak (Data Warehousing and Knowledge Discovery)*. 2003. Prague.
5. Cook, D. and L. Holder, *Graph-Based Data Mining*. IEEE Intelligent Systems, 2000. 15(2): p. 32-41.
6. Quinlan, J.R. and R.L. Rivest, *Inferring decision trees using the minimum description length principle*. Information and Computation, 1989. 80: p. 227-248.
7. S. Chakravarthy, R. Beera, and R.Balachandran, *Database Approach to Graph Mining*, in Proc. of PAKDD Conference, 2004, Sydney, Australia.
8. R. Balachandran, *Relational Approach to Modeling and Implementing Subtle Aspects of Graph Mining*, MS Thesis, Fall 2003. http://www.cse.uta.edu/Research/Publications/Downloads/CSE-2003-41.pdf
9. A. Inokuchi, T.Washio, and H. Motoda. Complete mining of frequent patterns from graphs: mining graph data. Machine Learning, 50:p. 321-354, 2003.
10. X. Yan and J. Han. gSpan: Graph-based substructure pattern mining. In ICDM'02: 2nd IEEE Conf. Data Mining, pages 721-724, 2002.
11. M. Kuramochi and G. Karypis. Frequent subgraph discovery. In 1st IEEE Conference on Data Mining, 2001. http://citeseer.ist.psu.edu/kuramochi01frequent.html

Multimedia Semantics Integration Using Linguistic Model

Bo Yang and Ali R. Hurson

Department of Computer Science and Engineering,
The Pennsylvania State University, University Park, PA 16802, USA
{byang, hurson}@cse.psu.edu

Abstract. The integration of multimedia semantics is challenging due to the feature-based representation of multimedia data and the heterogeneity among data sources. From human viewpoint, multimedia data objects are often considered as perceptions of the real world, and therefore can be represented at a semantic-entity level in the linguistic domain. This paper proposes a paradigm that facilitates the integration of multimedia semantics in heterogeneous distributed database environments with the help of linguistic analysis. Specifically, we derive a closed set of logic-based form expressions for the efficient computation of multimedia semantic contents, which include conceptual attributes and linguistic relationships into the consideration. In the expression set, the logic terms give a convenient way to describe semantic contents concisely and precisely, providing a representation of multimedia data that is closer to human perception. The space utilization is also improved through the collective representation of similar semantic contents and feature values. In addition, the optimization can be easily performed on logic expressions using mathematical analysis. By replacing long terms with equivalent terms of shorter lengths, the image representation can be automatically optimized. Using a heterogeneous database infrastructure, the proposed method has been simulated and analyzed.

1 Introduction

In recent years, the rapid expansion of multimedia applications, partly due to the exponential growth of distributed and portable computing devices, has proliferated over the daily life of computer users. Consequently, research on multimedia technologies is of increasing importance in computer society. In contrast with the traditional text-based systems, multimedia applications usually incorporate much more powerful descriptions of human thought – video, audio, and images. Moreover, the large collections of data in multimedia systems make it possible to resolve more complex data operations such as imprecise query or content-based retrieval. However, the conveniences of multimedia applications come with challenges to the existing data management schemes:

First, multimedia applications generally require more resources; however, the storage space and processing power are limited in many practical systems, e.g., mobile devices and wireless networks [1,10]. Due to the large data volume and complicated operations of multimedia applications, new methods are needed to facilitate efficient representation, accessing, and processing of multimedia data while considering the technical constraints.

In addition, there is a gap between user perception of multimedia entities and physical represent-and-access mechanism of multimedia data. Users often browse and desire to access multimedia data at the object level ("entities" such as human beings, animals, or buildings). However, the existing multimedia retrieval systems tend to represent multimedia data based on their lower-level features ("characteristics" such as color patterns and textures), with little regard to combining these features into objects [2]. This representation gap often leads to unexpected retrieval results. The representation of multimedia data according to human's perspective is one of the focuses in recent research activities; however, few existing systems provide automated identification or classification of objects from general multimedia data collections.

Moreover, the collections of multimedia data are often diverse and poorly indexed [3]. In a distributed environment, due to the autonomy and heterogeneity of data sources, multimedia objects are often represented in heterogeneous formats [4]. The difference in data formats further leads to the difficulty of incorporating multimedia objects under a unique indexing framework.

Last but not the least, the present research on content-based multimedia retrieval is based on feature vectors. These features are extracted from the audio/video streams or image pixels, with the empirical or heuristic selection, and then combined into vectors according to the application criteria. Due to the application-specific multimedia data formats, this paradigm of multimedia data management lacks scalability, accuracy, efficiency, and robustness [3].

Motivated by the aforementioned challenges, in this work we introduce a semantic-aware paradigm that organizes the multimedia data objects based on concise and abstract description of data contents, and summarizes the data contents from different data sources as semantically equivalent and globally recognizable terms. To show the feasibility and effectiveness of the proposed paradigm, a simulator was developed to compare and contrast our method against several content-based searching methods as proposed in the literature.

The remaining part of this paper is organized into three sections: Section 2 briefly overviews the related work and background materials. Section 3 addresses the semantic representation of multimedia data objects and introduces an automated linguistic approach of integrating multimedia semantics. Finally, section 4 draws the paper into conclusions.

2 Preliminaries

2.1 Content Analysis

Due to the aforementioned importance of multimedia content processing, recent research work focused on organizing multimedia data based on their contents [1]. Content-based retrieval systems have attempted to provide solutions to multimedia searching based on specific features [3]. Most of these systems support searches on low-level features such as colors, textures, or shapes of images. However, in most practical cases, the semantic contents of multimedia objects may not be indicated as low-level features. As pointed out by Wang et al. [5], human beings tend to view images as whole objects. This object-oriented view on multimedia content processing has led to the research on two tracks: single-modal content processing and cross-modal content processing.

The single-modal processing focuses on content information within a single modality, such as image or audio. Most previous research topics in this area were highly domain-specific applications (such as face recognition [6]), with less emphasis on general-purpose object detection methods [7]. Moreover, the probabilistic schemes (Bayesian and etc.) employed in the earlier research needed large training data and sometimes user feedback to classify and detect objects of interest.

The cross-modal processing obtains content information by fusing visual-audio data and retrieving camera operations such as zooming or tracking. The recent research has focused on context-extraction models (Latent Semantic Index, Canonical Correlation Analysis, and etc.) to obtain semantic contents from video frames [8]. However, the extraction of semantic contents is a time-consuming task that includes complex matrix computations.

However, multimedia applications usually generate large volume of data. As a result, instead of direct manipulation of raw multimedia data, it is more practical to sketch the multimedia content using some concise representation. Hence, there is a need for a set of quantitative parameters that can be used for multimedia content representation as well as content-based operations such as similarity comparisons.

In most content-representation systems that have advanced in the literature [1], the content of each multimedia object is described as a combination of features. If we consider each feature as a dimension, a multimedia object can also be considered as a vertex in the high-dimensional feature space. Different multimedia objects are characterized by different feature values, therefore scattering in separated regions of the high-dimensional feature space. The boundaries between these regions distinguish the content differences between multimedia objects. Generally, the boundaries cannot be described by linear discriminant equations [7]; hence, some refined description methods (e.g. Gaussian mixtures, neural networks, and principle component analysis) are employed for classifying multimedia objects based on their features [11]. The feature extraction is the process of mapping multimedia objects into vertices in feature space. The extracted features are usually domain-specific characteristics that can distinguish the "object of interest" from a large number of multimedia objects.

There are two common types of features in multimedia retrieval systems: granule-level features and object-level features. The granule-level features are those characteristics that are derived directly or indirectly from the original format of multimedia storage — i.e., the pixels, such as hue, textures, and saturation. The object-level features, in contrast, are obtained from the recognition of the higher-level understanding of the multimedia data — the semantic topics of the multimedia data.

Most existing multimedia retrieval systems tend to use granule-level features for content representation. This is mainly due to the difficulty of obtaining object-level semantic concepts from multimedia raw data [7]. However, human's perception of multimedia data is always at an object-level, instead of focusing on granule-level features such as color histogram or texture. Hence, research on object detection is becoming important in recent years, and many models were proposed for efficient extraction of objects from multimedia data.

Object detection is a pattern recognition problem with two classes of involved patterns: possible objects versus "non-objects", where "non-objects" refer to the class of background patterns. The major challenge in object detection is the definition of distinctions between objects and non-objects. An object detector must cope with both the variation within the object category and with the diversity of visual components that exist in the world at large. For instance, human faces vary in color, aging, facial expression, and in small disguises such as the facial hair, glasses, or cosmetics. The illumination conditions, viewing directions or poses may also affect the appearance of human faces. Moreover, the detector for "faces" must also distinguish human faces from all other visual patterns that may occur in the world, such as similar looking objects.

Based on domain-specific knowledge or empirical observations, several experimental systems were built for detection of certain types of multimedia objects, such as face identification from images or video segments [12]. However, it is much more complex to build an object-detection system for general-purpose recognition of objects, such as analyzing the animals in a given image. Normally, most existing object-detection systems do not guarantee complete accuracy in recognizing visual objects; instead they provide approximate recognition for given objects. As mentioned before, a multimedia system usually maps multimedia data objects to vertices in a high-dimensional space of granule-level features. A given multimedia data object may reside in small region, surrounded by other objects that are semantically similar to it. Bigger distance between vertices in the feature space means smaller content similarity between multimedia data objects.

2.2 Logic-Based Semantic Description

To represent the contents of multimedia data objects in a computer-friendly structural fashion, we now describe a way of organizing the data objects into layers according to their semantic contents.

Definition 1: The Elementary Entities

The elementary entities are those data entities that semantically represent basic objects (objects that cannot be divided further). Let $E = f_1 \wedge f_2 \wedge \ldots \wedge f_n$, where $f_i = p_{i1} \vee p_{i2} \vee \ldots \vee p_{im}$ is the disjunction of some logic predicates (true/false values) and $p_{i1} \ldots p_{im}$ form a logic predicate set F_i. The semantic content of an elementary entity can then be defined as:

$$E = \bigwedge_{i=1}^{n} (\bigvee_{j=1}^{m} p_{ij}), \quad \text{for every } p_{ij} \in F_i \quad (1)$$

Definition 2: The Multimedia Data Objects

A multimedia data object is the combination of a series of elementary entities. Given the above definition of elementary entities $E_1, E_2, \ldots, E_k$, the content of a multimedia data object can be defined as:

$$S = \bigcup_{i=1}^{k} E_i. \quad (2)$$

As noted in definition 2, a multimedia data object is considered as a combination of logic terms, whose value represents the semantic content. The analysis of semantic contents is then converted to the evaluation of logic terms and their combinations. This content representation approach has at least the following advantages:

1. The logic terms provide a convenient way to describe semantic contents concisely and precisely. Easy and consistent representation of the elementary entities based on their features simplifies the semantic content representation of complex data objects using logic computations. As a result, the similarity between data objects can be considered as the equivalence of their corresponding logic terms.
2. This logic representation of multimedia content is often more concise than feature vector. In a specific multimedia database system, the feature vector is often fixed sized to facilitate the computation and representation. However, some features may be null in many cases. Although these null features do not contribute to the semantic contents of multimedia data objects, they still occupy space in the feature vectors — hence, lower storage utilization. In contrast, the logic representation can improve storage utilization by eliminating the null features from logic terms.
3. Compared with feature vectors, the logic terms provide an understanding of multimedia contents that is closer to human perception.
4. Optimization can be easily performed on logic terms using mathematical analysis. By replacing long terms with mathematically equivalent terms of shorter lengths, the multimedia data representation can be automatically and systematically optimized.
5. Based on the equivalence of logic terms, the semantically similar objects can be easily found and grouped into same clusters. This organization facilitates the content-based nearest-neighbor retrieval, and at the same time reduces overlapping and redundancy, resulting in efficient search and storage utilization.

3 Semantics Integration

3.1 Logic Expression Set

Given the set of multimedia data objects that are represented as combination of elemental logic expressions, the semantic relationships between these data objects can be described using a closed set of logic formula: Let $I_S = \{S_1, ..., S_n\}$ denote a collection of objects obtained from the complete set of multimedia data objects, and $I_Q = \{Q_1, ..., Q_n\}$ denote the list of content-based queries submitted during a period of time. Because any practical multimedia database only consists of limited number of data objects, the corresponding data object set should also be of limited cardinality. And the semantic contents of any data object in this database can be represented as the combination of several objects from the object set I_S.

Definition 3: The Instance-Of Relationship
The semantics of a given multimedia data object could be annotated as a word ω. Then an on-line thesaurus ψ (e.g. Roget's thesaurus or Wordnet) can be used to define the inter-relationship between the data objects. For two given data objects S_i and S_j, if

ω_i describes a generic concept that includes ω_j, then S_j contains a hyponym of S_i, or S_j is an instance of S_i, denoted as $S_i \xleftrightarrow{\succ} S_j$.

Definition 4: The Is-A Relationship
The is-a relationship is the controversial of instance-of relationship. In the above definitions, S_i contains a hypernym of S_j, or S_i is an extended concept of S_j, denoted as $S_i \xleftrightarrow{\prec} S_j$.

It can be proven that the instance-of/is-a (or hypernym/hyponym) relationships are a pair of partial order relations, which shows the "inclusion" relationship between semantic contents. Based on the hypernym/hyponym relationships, a Hasse diagram shaped hierarchy can be constructed to indicate the routes of searching semantic concepts within a closed linguistic system.

Definition 5: The Polysemy Relationship
Some multimedia data objects may contain words that can be interpreted as ambiguous semantic meanings, which are collectively defined as polysemy. The probability of polysemy in a multimedia dataset I_S, denoted as $\xi(I_S)$, can be defined as the percentage in the on-line thesaurus ψ of semantic lemmas that include non-tree-shaped branches in the Hasse diagram hierarchy, because the tree-shaped hierarchies are defined as single-ancestor relationships and therefore are free from polysemy. In other terms, the polysemy probability can be formalized as:

$$\xi(I_S) = 1 - \frac{\sum_{x \in I_S} |enclosure(x, I_S)|}{\sum_{|X|=0}^{|I_S|} |X|}, \text{ for } \forall X \subseteq I_S. \tag{3}$$

where $enclosure(x, I_S)$ returns all possible subsets of I_S that completely cover the semantic content of x and not shared with other subsets.

Definition 6: The Homology Relationship
In a multimedia database, the homology relationships exist among the semantically similar data objects in different data formats, same-format data objects in different physical locations, and/or similar data objects in different data formats at different physical locations. Given two multimedia data objects S_i and S_j, their homology relationship can be defined as:

$$S_i \xleftrightarrow{\equiv} S_j \quad iff (S_i \xleftrightarrow{\succ} S_j \wedge S_i \xleftrightarrow{\prec} S_j) \text{ is satisfied} \tag{4}$$

Definition 7: The Heterology Relationship
The heterology relationship shows the opposite semantic features of two multimedia data objects (or linguistic concepts). For these two data objects, no matter in which linguistic domain, their semantic contents cannot have homology, instance-of, or is-a relationships. For instance, "in" and "out", "up" and "down", "move" and "stop" are concepts that can be represented as the annotations or semantic contents of images, and this type of concepts form the relationship of heterology denoted as:

$$S_i \xleftrightarrow{\neq} S_j \quad iff \neg (S_i \xleftrightarrow{\succ} S_j \vee S_i \xleftrightarrow{\prec} S_j) \text{ is satisfied} \tag{5}$$

Definition 8: The Share-With Relationship

Given a multimedia data object and the content-relationship function as defined above, the share-with relationship returns a set of data objects that possess overlapping contents with the semantics of the given data object. Formally, the set returned by share-with relationship can be defined as:

$$S_i \xleftarrow{\cap} I_S = \{ S_k \mid \xi(I_S) \cap S_i \neq \emptyset \} \tag{6}$$

3.2 Content Integration

We have two major goals in the content integration process as follows:

- Specify the hidden semantic relationships among the multimedia data objects, and
- Minimize the logic-based representation of the exported terms.

These objectives allow higher QoS and performance, respectively. Inspired by the formation of Karnaugh Maps [9], we designed a combinatorial optimization table to shorten the complex combinations of features into condensed logic terms. The optimization table is a k-dimensional table where k is the size of the feature sets representing the underlying multimedia data objects. As a result each table entry represents a multimedia data object. The semantics integration process is performed on this combinatorial optimization table in the following phases:

Phase 1: Semantic domain partitioning

The partitioning of semantic domain provides a means of representing and organizing multimedia data objects based on their semantic contents. Given a collection of semantically similar data objects, one can collectively represent them using the description of their common semantic features.

Table 1. Notations related to semantic domain partitioning

Symbols	Notations
I	The set of terms
G	The set of semantic groups
$a(g)$	The cardinality of a group g
$t(G)$	The function of selecting a term from I with the minimum overlapping with elements in Γ
$S(x, G)$	The function selecting a group from Γ most related to x

```
Algorithm 1: Partitioning semantic domain
1.  G = ∅
2.  x = t(I, G)
3.  I = I - {x}
4.  WHILE |I| > 0 DO
5.     g = s(x, G)
6.     IF (distance of x and g ≤ threshold)
7.        g = g ∪ {x}
```

```
        8.   create Boolean set with size log(a(g))
        9.   ELSE
       10.        G = G U {{x}}
       11. RETURN the Boolean variable sets
```

The semantic terms are first translated into binary codes for convenience of processing – the terms are translated into a collection of Boolean variable sets. Assuming the data object set has $n_1, n_2, ..., n_k$ distinct semantic groups, respectively, we need $\lceil log_2(\alpha(n_1)) \rceil + \lceil log_2(\alpha(n_2)) \rceil + ... + \lceil log_2(\alpha(n_k)) \rceil$ Boolean variables to represent the semantic groups.

Phase 2: Similar content clustering

A combinatorial optimization table is constructed. Each cell is labeled with a combination of Boolean variables, either in the original form or in the complement form. As the indication of semantic content, the cells are filled with "1"s, "0"s, or "*"s. The "*"s indicate the non-applicable cases. Adjacent cells set to "1"s indicates the multimedia entities share some common features. Hence, we can cluster the "neighboring" entities with the common features as a semantically similar group. The clustering process is performed as indicated in the following rules:

- Each cluster contains 2^k adjacent 1s in a rectangular region in the combinatorial optimization table (*k* is any non-negative integer).
- The clusters with over 50% overlapping are merged into a larger cluster, which shows the share-with relationship as defined in section 3.1.
- For the adjacent orthogonal clusters, check with the on-line thesaurus ψ and determine whether they are under the same instance-of relationship. Merge the clusters with same hypernym into a larger cluster which is labeled as their hypernym.

Phase 3: Content-based retrieval

Table 2. Notations related to content-based retrieval

Symbols	Notations
t	The semantic distance threshold
R	The set of returned query result
Q	The query multimedia data object

```
Algorithm 2: Content-based retrieval
1. initialize t
2. R = ∅
3. IF (content clustering is finished)
4. THEN convert Q into query terms
5. FOR each term T in Q DO
6.      compute the semantic distance with T
7.      IF the semantic distance ≤ threshold t
8.      THEN   IF no further sub cluster
9.             THEN put the similar data objects in R
10.                 IF the objects is not enough
```

```
11.                    THEN increase the threshold t
                           pop a cluster and go to line 6
12.                    ELSE  go to line 17
13.             ELSE push current cluster in stack,
                           decrease the threshold t,
                           send query to sub clusters,
                           and go to line 6
14.     ELSE IF the current cluster is inaccessible
15.                THEN increase the threshold t
                           pop a cluster and go to line 6
16.                ELSE push the current cluster in stack,
                           send the query term to larger
                           super cluster and go to line 5
17. order the data objects in R and output
```

Lemma 1: If each cluster in the Hasse diagram hierarchy has no less than m sub clusters, then the height of the hierarchy is $O(\log_m n)$, where n is the number of clusters.

Theorem 1: If each node in the Hasse diagram hierarchy has no less than m sub clusters, then the insertion, deletion and normal retrieval in this hierarchy are $O(\log_m n)$.

Proof. The insertion, deletion and normal retrieval are proved separately.

1) The insertion of multimedia data objects starts at a non-dividable cluster and can be achieved in $O(1)$ time. However, it may cause the continuous modification of super clusters and make them rebuild cycles of homology. The modification of clusters and elimination of old homology relationship links are $O(1)$. If the proper hashing algorithms are employed, the searching of proper new homology cycles also has the $O(1)$ time complexity. Since the height of the Hasse diagram hierarchy is $O(\log_m n)$, there are at most $O(\log_m n)$ clusters need to be modified. Consequently, the total complexity of insertion is $O(\log_m n)$.

2) The processing of deletion is generally similar as that of insertion. But it also needs to remove the useless homology relationships. This removal is also proportional to the height of the hierarchy. As a result, the deletion is also achieved in $O(\log_m n)$ time.

3) As noted before, the normal retrieval is restricted in a sub branch of the Hasse diagram hierarchy. So the query processor examines at most $2 * \log_m n$ clusters. At each cluster, the time cost is $O(1)$. Thus the normal retrieval is also an $O(\log_m n)$ process.

Theorem 2: If each cluster in the Hasse diagram hierarchy has no less than m sub clusters, and the longest homology cycle has k clusters, then the nearest neighbor retrieval in this hierarchy is $O(\log_m n + \log_2 k)$.

Proof. As noted before, when processing the nearest neighbor retrieval, the system finds the semantically most similar data object and uses its homology cycle to find other data objects. The first step takes $O(\log_m n)$ time. Since the system needs to order the data objects according to their semantic similarities, the second step takes $O(\log_2 k)$ time. Consequently, the total time is $O(\log_m n + \log_2 k)$.

4 Conclusions

We proposed a novel content-aware retrieval model for multimedia data objects in heterogeneous distributed database environment. In contrast with the traditional feature-based indexing models, the proposed model employs a concise descriptive term to represent the semantic contents of multimedia objects. In short, the proposed model offers the following advantages: (1) the concise descriptions accurately represent the semantic contents of multimedia data objects using optimized logic terms; (2) the descriptive terms enable the search engine with capability of handling imprecise queries. Our future work would include improvements of the proposed model, such as more efficient search strategies and adaptation to cross-modal multimedia data.

References

1. Hsu, W., Chua, T. S., Pung, H. K.: Approximating Content-Based Object-Level Image Retrieval. Multimedia Tools and Applications. 12(2000) 59-79
2. Kim, J. B., Kim, H. J.: Unsupervised Moving Object Segmentation and Recognition Using Clustering and A Neural Network. Proc of the Intl Joint Conf on Neural Networks. 2(2002) 1240-1245
3. Huang, Y. P., Chang, T. W., Huang, C.-Z.: A Fuzzy Feature Clustering with Relevance Feedback Approach to Content-Based Image Retrieval. Proc of the IEEE Symposium on Virtual Environments, Human-Computer Interfaces and Measurement Systems.(2003) 57-62
4. Kwon, T., Choi, Y., Bisdikian, C., Naghshineh, M.: QoS Provisioning in Wireless/Mobile Multimedia Networks Using An Adaptive Framework. Wireless Networks. (2003) 51-59
5. Wang, J. Z., Li, J.: Learning-Based Linguistic Indexing of Pictures with 2-d Mhmms. Proceeding of ACM Multimedia. (2002) 436-445
6. Pentland, A.: View-Based and Modular Eigenspaces for Face Recognition, Proc of the IEEE Conf. on Computer Vision & Pattern Recognition, Seattle, WA, (1994)
7. Naphade, M. R.: Detecting Semantic Concepts Using Context and Audiovisual Features. IEEE Workshop on Detection and Recognition of Events in Video. (2001) 92-98
8. Li, D., Dimitrova, N., Li, M., Sethi, I. K.: Multimedia Content Processing through Cross-Modal Association. Proc of the ACM Conference on Multimedia. (2003) 604-611
9. Karnaugh, M.: The Map Method for Synthesis of Combinational Logic Circuits. Trans. AIEE. Part I. 9(1953) 593-599
10. Westermann, U., Klas, W.: An Analysis of XML Database Solutions for Management of MPEG-7 Media Descriptions. ACM Computing Surveys. (2003) 331-373
11. Naphade, M. R., Huang, T. S.: Recognizing High-Level Audio-Visual Concepts Using Context. Proc of the IEEE Intl Conf on Image Processing. (2001) 46-49
12. Li, M., Li, D., Dimitrova, N., Sethi, I. K.: Audio-Visual Talking Face Detection. Proc of IEEE Intl Conf on Multimedia and Expo. (2003) 473-476

A Novel Indexing Approach for Efficient and Fast Similarity Search of Captured Motions

Chuanjun Li and B. Prabhakaran

Department of Computer Science,
University of Texas at Dallas, Richardson, TX 75083
{chuanjun, praba}@utdallas.edu

Abstract. Indexing of motion data is important for quickly searching similar motions for sign language recognition and gait analysis and rehabilitation. This paper proposes a simple and efficient tree structure for indexing motion data with dozens of attributes. Feature vectors are extracted for indexing by using singular value decomposition (SVD) properties of motion data matrices. By having similar motions with large variations indexed together, searching for similar motions of a query needs only one node traversal at each tree level, and only one feature needs to be considered at one tree level. Experiments show that the majority of irrelevant motions can be pruned while retrieving all similar motions, and one traversal of the indexing tree takes only several microseconds with the existence of motion variations.

1 Introduction

Continuous motion data can be generated by many real-time and off-line applications in life sciences and animations, and can be employed for gesture recognition, gait analysis and rehabilitation, sports performance, film and video games [8]. To decide whether a motion segment in a motion stream is a known motion in a large motion database, or to recognize motions in the continuous motion data, not only is a motion similarity measure needed [5], but also an efficient and fast pruning algorithm is necessary. The pruning algorithm should prune most impossible motions in a large database for a motion query in real time. To prune motions efficiently and fast needs to address several challenges:

- Datasets of motions have multiple attributes. Each attribute describes the angular values or coordinates of a joint of the motion *subject*, and dozens of attributes are needed to capture a complete subject motion.
- Datasets of motions are high dimensional and even similar motions can have different dimensions. One dimension is for one sampling of all attributes, and every motion can have different durations and thus different dimensions.

Due to these issues, direct indexing of motion data is difficult and inefficient.

This paper proposes a new method for indexing motion data with dozens of attributes. The feature vectors are extracted by obtaining the equal-length dominating vectors from singular value decompositions (SVD) of motion data and by reducing vector dimensionalities. Corresponding feature values of all motion patterns are partitioned into several intervals. Motion or feature vector IDs are inserted into a tree of feature intervals by using the corresponding feature values. To take into consideration motion variations, a feature ID is allowed to be inserted into multiple neighboring feature intervals. Hence a feature vector ID can be in multiple leaf nodes instead of in only one leaf node. Searching for possible similar motions of a query needs only one node traversal at each tree level and takes only several microseconds.

2 Related Work

Equal length multi-attribute sequences are considered in [2]. A CS-Index structure is proposed for shift and scale transformations. In [4], multi-attribute sequences are partitioned into subsequences, each of which is contained in a Minimum Bounding Rectangle (MBR). Every MBR is indexed and stored into a database by using an R-tree or any of its variants.

Dynamic time warping (DTW) and longest common subsequence (LCSS) are extended for similarity measures of multi-attribute data in [9]. Before the exact LCSS or DTW is performed, sequences are segmented into MBRs to be stored in an R-tree. Based on the MBR intersections, similarity estimates are computed to prune irrelevant sequences.

Attributes of the data indexed in the previous work are less than ten. In contrast, our proposed indexing structure can handle dozens or hundreds of data attributes without loss of good performances. This work proposes a novel indexing approach which is different from that in [6], making it possible to search the indexing tree for similar motions in only several microseconds.

3 Geometric Structures Revealed by SVD

In this section, we give the definition and geometric interpolation of SVD for its application to the indexing of multi-attribute motion data.

SVD exposes the geometric structure of a matrix A. If the multi-dimensional row vectors or points in A have different variances along different directions and columns of A have zero means, the SVD of matrix A can find the direction with the largest variance. If columns of A do not have zero means, the direction along which row vector projections have the largest 2-norm or Euclidean length can be revealed by SVD. Figure 1 illustrates the data in an 18×2 matrix. The 18 points in the 18×2 matrix have different variances along different directions, hence data have the largest variance along v_1 as shown in Figure 1.

Along the direction of the first right singular vector, the projections of row vectors in A have the largest 2-norm, and along the second right singular vector direction, the projection 2-norm is the second largest, and so on. The singular values reflect the Euclidean lengths or 2-norms of the projections along the corresponding right singular vectors.

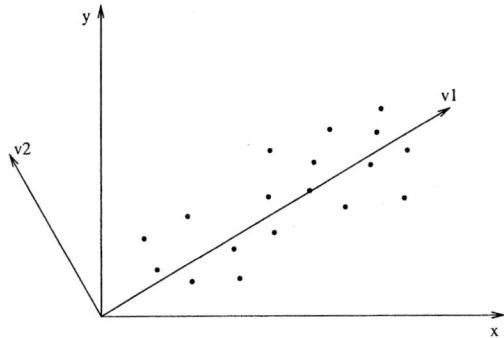

Fig. 1. Geometric structure of matrix exposed by its SVD

As shown in [1], any real $m \times n$ matrix A can be decomposed into $A = U\Sigma V^T$, where $U = [u_1, u_2, \ldots, u_m] \in R^{m \times m}$ and $V = [v_1, v_2, \ldots, v_n] \in R^{n \times n}$ are two orthogonal matrices, and Σ is a diagonal matrix with diagonal entries being the singular values of A: $\sigma_1 \geq \sigma_2 \geq \ldots \geq \sigma_{\min(m,n)} \geq 0$. Column vectors u_i and v_i are unit vectors and are the ith left and right singular vectors of A, respectively.

For similar motions with different lengths, their left singular vectors are of different lengths, but their right singular vectors are of the equal length. The singular values of matrix A are unique, and the singular vectors corresponding to distinct singular values are uniquely determined up to the sign, or a singular vector can have opposite signs [7]. For convenience, we will refer to the right singular vectors as singular vectors.

4 Feature Vector Extraction for Indexing

Motion matrices should have similar geometric structures if the corresponding motions are similar. Since the geometric similarity of matrix data can be captured by SVD, we propose to exploit SVD to generate representative vectors or feature vectors for motion matrices, and use these feature vectors for indexing the multi-attribute motion data.

As Figure 2 shows, the first singular values are the dominating ones among all singular values. Since the singular values reflect lengths or magnitudes of the row vector projections along their corresponding singular vectors, we can say that the first singular vectors are the dominating vectors. If two motions are similar, their corresponding first singular vectors u_1 and v_1 should be mostly parallel to each other geometrically, so that $|u_1 \cdot v_1| = |u_1||v_1||cos(\theta)| \doteq |u_1||v_1|$ $= 1$, where θ is the angle between the two right singular vectors u_1 and v_1, and $|u_1| = |v_1| = 1$ by the definition of SVD. Similarly, the first singular vectors are also very likely to be different from each other when two motions are different. Other corresponding singular vectors may not be close to each other even if two motions are similar as shown in Figure 3. This suggests that the first right singular vectors can be used to index multi-attribute motions for pruning the majority of different motions.

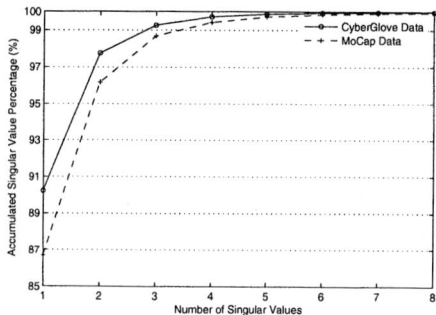

Fig. 2. Accumulated singular value percentages in singular value sums for CyberGlove data and captured human body motion data. There are 22 singular values for the CyberGlove data and 54 singular values for the captured motion data. The first singular values are more than 85% of the corresponding singular value sums.

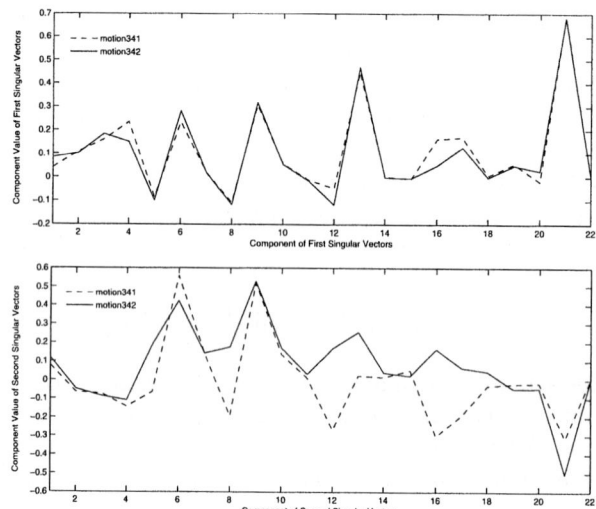

Fig. 3. Singular vectors of similar motions. The first singular vectors are similar to each other, while other singular vectors, such as the second vectors as shown at the bottom, can be quite different.

It is worth noting that for motions to be similar, other singular vectors and singular values should also be considered as shown in [5]. Although being necessary conditions for similarity measure, similar first singular vectors are sufficient for indexing purpose as to be demonstrated in Section 6.

Since the lengths or dimensions of the first singular vectors of multi-attribute motion data are usually larger than 15, dimensionality reduction needs to be performed on them first in order to avoid the so-called "curse of dimensionality." We use SVD further to reduce the dimensionality of the first singular vectors

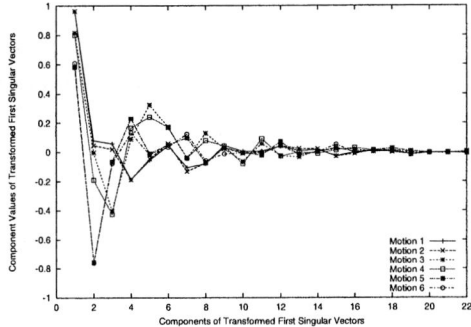

Fig. 4. Component distributions of the transformed first singular vectors

to be indexed. Let A be the matrix composing the first singular vectors of the motions to be indexed, and

$$A = W \Sigma Z^T$$

then $AZ = W\Sigma$ gives the projected/transformed first singular vectors of motion patterns in the coordinate system spanned by the column vectors of Z [3], and for a singular vector u_1 of a query motion, $u_1 Z$ gives a corresponding transformed singular vector of u_1 in the system spanned by the column vectors of Z.

Due to singular value decomposition, the component variations of the transformed first singular vectors are the largest along direction z_1, and decreases along directions $z_2, \ldots, z_n$ as shown in Figure 4. The differences among the first singular vectors are optimally reflected in the first several dimensions of the transformed first singular vectors, hence we can index the first singular vectors by indexing only the first several components of the transformed singular vectors. Differences among all the other corresponding components are small even if motions are different, so the other components can thus be truncated and the dimensionalities are reduced to the first several ones. We refer to the transformed singular vectors after dimensionality reduction as the *feature vectors* of the motions. If the first component of a feature vector is negative, all components of this vector are negated to obtain a consistent sign for feature vectors of similar motions [6].

5 Index Tree Construction

Let r be the dimension of the feature vectors, $r < n$. We designate one level of the index tree to each of the r dimensions. Let level 1 be the root node, level i includes nodes for dimension i, $i = 1, 2, \ldots, r$, and level $r + 1$ contains leaf nodes. Leaf nodes contain motion identifiers P_k, and non-leaf nodes contain entries of the form

$$(I_i, cp)$$

where I_i is a closed interval $[a, b]$ describing the component value ranges of the feature vectors at level i, $-1 \leq a, b < 1$. Each entry has the address of one child node, and cp is the address of the child node in the tree.

The width and boundary of interval I_i depend on the distribution of ith component values of feature vectors and the possible variations of the ith feature vector components of similar motions. Let δ_i be the maximum difference of the ith feature vector components of any similar motions, let x_i and y_i be the respective minimum and maximum values of the ith components of all feature vectors, and let ϵ be the entry interval factor for adjusting entry intervals. Then the width of entry intervals at the ith level is $\epsilon\delta_i$, and the number of entries of a node at level i is $\lceil (y_i - x_i)/(\epsilon\delta_i) \rceil$, limited by maximum number of entries per node allowed.

5.1 Insertion and Searching

Let the root node of the tree be T. The unique ID of a feature vector is inserted into the tree by comparing the ith component c_i of the feature vector and the entry interval $[a, b]$ of the node traversed and can be inserted into multiple neighboring intervals:

- **Subtree Insertion:** If T is a non-leaf node, find all entries whose I_i's overlap with $[c_i - \epsilon\delta_i, c_i + \epsilon\delta_i]$. For each overlapping entry, find the subtree whose root node T is pointed to by cp of the overlapping entry.
- **Leaf Node Insertion:** If T is a leaf node, insert the motion pattern identifier P_k of the feature vector in T.

Figure 5 illustrates how to insert an example feature vector into the first three levels of an example index tree. Root node at level 1 has four entries, each of which has a child node at level 2. Each node at level 2 and level 3 has three entries, and each of which has a child node at one lower level. Given a feature vector $f = (0.65, 0.15, -0.1, \ldots)$, and let $\delta_1 = 0.04$, $\delta_i = 0.08$ for $i \geq 2$, and $\epsilon = 1.0$. Entries at the root node are checked with $[0.65 - 0.04, 0.65 + 0.04] = [0.61, 0.69]$. Only the third entry overlaps with it, hence the vector f is forwarded only to node n_3 of level 2. At level 2, the feature vector covering range is $[0.15-0.08, 0.15+0.08]$ or $[0.07, 0.23]$. The second and third entries of node n_3 overlap with the feature vector covering range $[0.07, 0.23]$, hence the feature vector will be

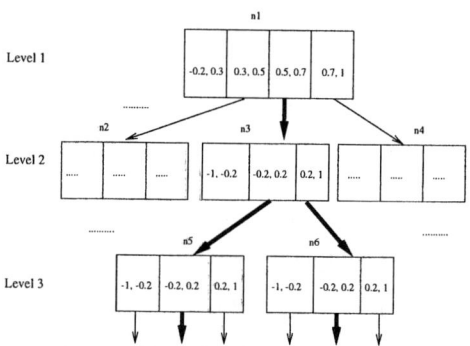

Fig. 5. An index tree example showing three non-leaf levels. Bold lines show where a feature vector is to be inserted.

forwarded to node n_5 and to node n_6 at level 3. At level 3, the feature vector covering range is [-0.1-0.08, -0.1+0.08] or [-0.16, -0.02]. Only the second entries of nodes n_5 and n_6 overlap with this range, so the nodes pointed by the second entries of nodes n_5 and n_6 will be traversed for insertion. This process goes on until the leaf nodes are traversed for holding P_k of the feature vector f.

A query searching can be very simple: find the entry of the node whose interval $[a, b]$ covers the ith component c_i of the query feature vector and traverse to the corresponding chile node pointed by the entry. When a leaf node is reached, all the motion identifiers included in that leaf node are returned for the query. Since a node entry contains all possible similar motions in neighboring entries of the same node, only one entry is needed to be traversed for a search at each level of the tree, rather than multiple entries to be traversed as in [6].

5.2 Similarity Computation

After the index tree has been searched for a query, the majority of irrelevant motions should have been pruned, and similar motions and a small number of irrelevant motions are returned as the result of the query. To find out the motion most similar to the query, a similarity measure shown below as defined in [5] can be used to compute the similarity of the query and all the returned motions, and the motion with the highest similarity is the one most similar to the query.

$$\Psi(Q, P) = \frac{1}{2} \sum_{i=1}^{k} ((\sigma_i / \sum_{i=1}^{n} \sigma_i + \lambda_i / \sum_{i=1}^{n} \lambda_i) |u_i \cdot v_i|)$$

where σ_i and λ_i are the ith singular values corresponding to the ith right singular vectors u_i and v_i of square matrices of Q and P, respectively, and $1 < k < n$. Integer k determines how many singular vectors are considered and depends on the number of attributes n of motion matrices. Experiments with hand gesture motions ($n = 22$) and human body motions ($n = 54$) show that $k = 6$ is large enough without loss of pattern recognition accuracy in streams.

6 Performance Evaluation

Let N_{pr} be the number of irrelevant motions pruned for a query by the index tree, and N_{ir} be the total number of irrelevant motions in the database. We define the pruning rate $\mathcal{P}$ as

$$\mathcal{P} = \frac{N_{pr}}{N_{ir}} \times 100\%$$

6.1 Motion Data Generation

Motion data was generated for hand gestures by using CyberGlove and for dances and other human motions captured by using 16 Vicon cameras. There are 22 attributes for the CyberGlove data, and each attribute is for the angular values of one joint of the glove. There are 54 attributes for the motion capture data, and each attribute is for the positional values of one joint of a moving subject. The captured motion data had been transformed so that similar motions performed at different locations, following different paths, or at different orientations have "similar" data matrices. One hundred and ten different hand gestures were generated, and each one was repeated for 3 times, resulting in 330 data matrices of 22 columns. Sixty two different motions, including Taiqi and dances were performed, and each one was repeated for 5 times, resulting in 310 data matrices of 54 columns.

6.2 Index Struction Building

We experimented with different tree configurations for CyberGlove data and motion capture (MoCap) data. For CyberGlove data of 22 attributes, feature vectors have 5 to 10 components, or trees of 5 to 10 levels were tested. For MoCap data of 54 attributes, trees of 5 to 12 levels were tested. The entry interval factors ϵ we tested were 1.5, 1.2, 1.0, 0.9, 0.8, 0.7, 0.6 and 0.5. The smaller the entry

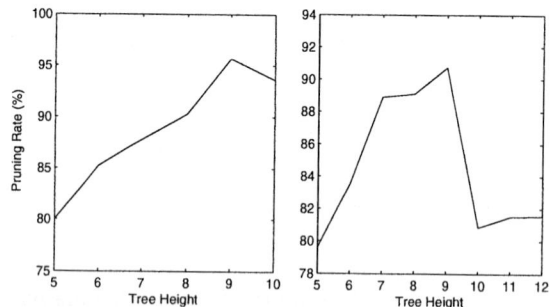

Fig. 6. Pruning rates of trees with different levels when all similar motions are to be retrieved. Left: CyberGlove data with $\epsilon = 1.0$; Right: MoCap data with $\epsilon = 0.8$.

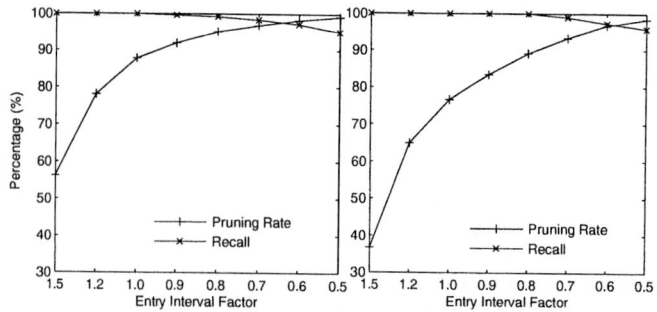

Fig. 7. Recalls and pruning rates for trees with height of 7 and different entry interval factors ϵ. Left: CyberGlove data; Right: MoCap data.

interval factors, the smaller the entry intervals, and the more the number of entries in a node at all levels.

6.3 Pruning Efficiency

We issued one query for every one of the 330 CyberGlove motions and the 310 MoCap motions. Figure 6 shows that when all similar motions were retrieved and the feature vectors have 9 features, 95.7% irrelevant CyberGlove motions and 91% irrelevant MoCap motions could be pruned. When the entry interval factor ϵ is no less than 1.0, all similar motions can be retrieved, and when ϵ is less than 1.0, the most similar motions can still be retrieved and only a small number of less similar motions can be pruned as indicated by the high recalls as shown in Figure 7.

6.4 Computational Efficiency

We tested the average CPU time taken by a query using different tree configurations. All experiments are performed on one 3.0 GHz Intel processor of a GenuineIntel Linux box.

The search time of a query by using the proposed index structure takes less than 3 μs as shown in Figure 8. As a comparison, the search time of a query

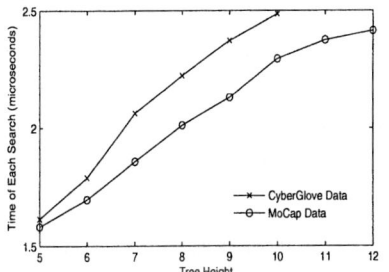

Fig. 8. Search time for one query by the proposed indexing approach

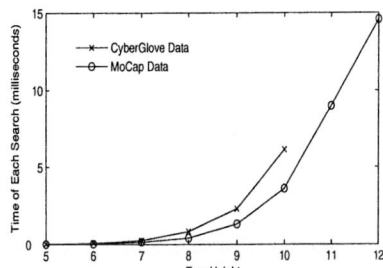

Fig. 9. Search time for one query by the index structure as proposed in [6]

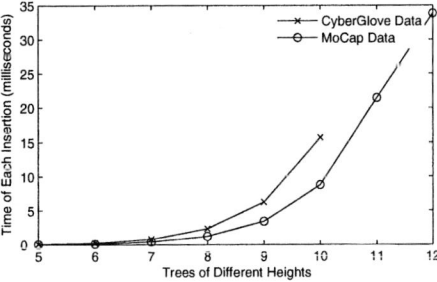

Fig. 10. Time taken for inserting a new motion ID in the indexing tree

by the algorithm in [6] can take several milliseconds as shown in Figure 9. As a tradeoff, the proposed approach in this paper takes a little longer for inserting feature vectors. Nevertheless, each insertion still takes less than 35 milliseconds as shown in Figure 10 and is usually done off-line.

7 Conclusions

This paper has proposed a novel approach for indexing multi-attribute motion data of different lengths. Feature vectors are extracted from motion data matrices by using SVD properties, and an interval-based tree structure is proposed for indexing the feature vectors. Feature vector IDs can be inserted into multiple neighboring feature value intervals to cope with motion variations and can be in multiple leaf nodes. As an advantage of this design, search of similar motions can be done in several microseconds by traversing only one node once at each tree level, and up to 95.7 % different CyberGlove motions and 91% captured human motions can be pruned.

References

1. G. H. Golub and C. F. V. Loan. *Matrix Computations*. The Johns Hopkins University Press, Baltimore, Maryland, 1996.
2. T. Kahveci, A. Singh, and A. Gurel. Similarity searching for multi-attribute sequences. In *Proceedings. of 14th Int'l Conference on Scientific and Statistical Database Management*, pages 175 – 184, July 2002.
3. F. Korn, H. V. Jagadish, and C. Faloutsos. Efficiently supporting ad hoc queries in large datasets of time sequences. In *SIGMOD*, pages 289–300, May 1997.
4. S.-L. Lee, S.-J. Chun, D.-H. Kim, J.-H. Lee, and C.-W. Chung. Similarity search for multidimensional data sequences. In *Proceedings. of 16th Int'l Conference on Data Engineering*, pages 599 – 608, Feb./Mar. 2000.
5. C. Li and B. Prabhakaran. A similarity measure for motion stream segmentation and recognition. In *Proceedings of the Sixth International Workshop on Multimedia Data Mining*, pages 89–94, August 2005.
6. C. Li, G. Pradhan, S. Zheng, and B. Prabhakaran. Indexing of variable length multi-attribute motion data. In *Proceedings of the Second ACM International Workshop on Multimedia Databases 2004*, pages 75–84, November 2004.
7. B. D. Schutter and B. D. Moor. The singular value decomposition in the extended max algebra. *Linear Algebra and Its Applications*, 250:143–176, 1997.
8. Online Vicon products introduction, http://www.vicon.com/jsp/products/products.jsp.
9. M. Vlachos, M. Hadjieleftheriou, D. Gunopulos, and E. Keogh. Indexing multi-dimensional time-series with support for multiple distance measures. In *SIGMOD*, pages 216–225, August 2003.

Mining Frequent Spatial Patterns in Image Databases

Wei-Ta Chen, Yi-Ling Chen, and Ming-Syan Chen

Department of Electrical Engineering,
National Taiwan University, Taipei, Taiwan, ROC
{weita, ylchen}@arbor.ee.ntu.edu.tw, mschen@cc.ee.ntu.edu.tw

Abstract. Mining useful patterns in image databases can not only reveal useful information to users but also help the task of data management. In this paper, we propose an image mining framework, Frequent Spatial Pattern mining in images (FSP), to mine frequent patterns located in a pair of spatial locations of images. A pattern in the FSP is associated with a pair of spatial locations and refers to the occurrence of the same image content in a set of images. This framework is designed to be general so as to accept different levels of representations of image content and different layout forms of spatial representations.

Index Terms: Image mining, spatial pattern.

1 Introduction

Data mining has attracted a significant amount of research attention due to its usefulness in many applications, including selective marketing, decision support, business management, and user profile analysis, to name a few [1]. However, most work focuses on extracting information from data stored in alphanumeric databases. Recently, advances in digital technologies have led to tremendous growth in the number of image repositories. A lot of studies have addressed the problem of content-based data management [2]. It is known that mining useful patterns from these image databases can not only reveal useful information to users but also help the task of data management [3]. As a result, image mining has emerged as an important research topic.

Compared with the traditional data mining, many new challenges arise in image mining due to the unique properties exhibited by images. We summarize these challenges into the following three issues: (1) Image content can be presented in several concept levels, ranging from pixel values, low-level features [4], visual thesaurus (category of features) [5] to objects [6], [7]. Which are the features to be used in the mining process? (2) The spatial information of image content conveys important messages to present an image [8]. How do we reflect the spatial information of image content? (3) What are the applications of image mining? In other words, how can we justify the usefulness of the mining results?

In our opinion, the third issue imposes the key challenge. The reason is that the specific features that should be used to represent image content vary with individual cases, and so do the layout forms used to partition the spatial locations of images. Only when we have an intended application in mind, can we select the suitable features and determine the ways of spatial partitions so as to start the mining process.

The main contribution of this paper is that we propose a general framework to mine frequent spatial patterns, which are of the most interest in the image mining context [4], [5], [6], [7]. The framework we propose, called Frequent Spatial Pattern mining in images (FSP), aims to mine frequent patterns located in specific spatial locations of images. A *pattern* in the FSP is associated with a pair of spatial locations and refers to the occurrence of the same image content in a set of images. In order to make this framework general enough to support various applications, we allow FSP to accept different representations of image content and different layout forms of representations of spatial locations. In FSP, an *item* is used to refer to an *abstract object* which represents the extracted feature from image content. Associated with each item is a region label which reflects the spatial location of this item. In this paper, when there is no confusion, *spatial location* and *region* are interchangeably used to mean the spatial property of an item.

The mining algorithm, FSP-Mining, in FSP can be decomposed into two phases: (1) Generation of the frequent itemsets in each of the partitioned regions that are covered by S. (2) Generation of the frequent spatial patterns through the frequent itemsets. Let S_c be the set of regions covered by S. In the 1st phase of FSP-Mining, the Apriori algorithm [9] or its extensions such as FP-tree [10] or DHP [11], can be used to generate the frequent itemsets in each region R_i, where $R_i \in S_c$. In the 2nd phase, we first use the frequent itemsets generated in the 1st phase to generate the set of candidate patterns. After one scan of the image set, we can obtain the support count of each candidate pattern. Consequently, we are able to generate all the frequent patterns corresponding to the set S of pairs of locations.

A significant amount of research effort has been elaborated upon addressing the problem of image mining. In [4], the authors mine patterns in global image features and associate these features with a class label. In contrast to this work, the objective of our work is to design a general image mining framework. Moreover, in our framework, we take the spatial locations of features into account rather than capture the global features. In [5], the authors partition images into regions as well and the regions are labelled using a visual thesaurus. However, in their paper, the patterns are not associated with regions, and the location information is only used to determine region labels. That is, they do not take into account the specific locations to mine the frequent itemsets. In [6] and [7], the authors propose mining objects in images. They both conduct their experiments on a synthetic data set composed of basic shapes. It is not clear, however, whether the results can be generalized to real images and be evaluated for their usefulness.

In the remainder of this paper, we present the design of the framework in Section 2 and conclude this paper with Section 3.

2 Design of FSP

In Section 2.1, we formally define the problem of mining frequent spatial patterns in images. In Section 2.2, we introduce the FSP-Mining algorithm in the proposed framework, Frequent Spatial Pattern mining in images (FSP).

2.1 Problem Formulation

Let $D = \{I_1, ..., I_n\}$ be a set of images, where I_i, $i \in [1..n]$, denotes an image. In the framework of FSP, we assume that each image I_i has been partitioned to a set $\Re = \{R_1, ..., R_r\}$ of regions (spatial locations) and that the abstract objects in each region have been extracted. Therefore, an image I can be represented as $\langle O_1, ..., O_r \rangle$, where O_i, $i \in [1..r]$, denotes the set of objects extracted from the region R_i. Let S be the set of pairs of locations in which we are interested, i.e., $S \subseteq \{(R_i, R_j) \mid 1 \leqslant i < j \leqslant r\}$.

Definition 1: A pattern p_{ij}, which is associated with regions R_i and R_j, is defined to be of the form (A, B), where $A \subseteq O_i$ and $B \subseteq O_j$ denote the itemsets located in R_i and R_j, respectively.

Note that in this paper when there is no confusion, we use the notation p, omitting the subscripts i and j, to refer to a pattern whose associated spatial pair could be any of those interesting location pairs in S.

Definition 2: A pattern p is called frequent if the ratio of the number of images contain this pattern to the total number of images is no less than the minimum support threshold min_sup. That is, p is frequent if $\frac{|\{I|I \text{ contains } p\}|}{|D|} \geqslant min\_sup$.

The problem of mining spatial patterns can be defined as follows: Given a set D of images, a set S of pairs of spatial locations and a minimum support threshold min_sup, we aim to mine the frequent patterns corresponding to each pair of locations in S.

2.2 Algorithm FSP-Mining

The mining algorithm, FSP-Mining, in FSP can be decomposed into two phases: (1) Generation of the frequent itemsets in each of the partitioned regions that are covered by S. (2) Generation of the frequent spatial patterns through the frequent itemsets. Let S_c be the set of regions covered by S. For instance, if $S = \{(R_1, R_2), (R_1, R_3), (R_2, R_4)\}$, $S_c = \{R_1, R_2, R_3, R_4\}$. In the 1st phase of FSP-Mining, the Apriori algorithm [9] or its extensions such as FP-tree [10] or DHP [11], can be used to generate the frequent itemsets in each region R_i, where $R_i \in S_c$. Let $L^i = \cup_k L_k^i$ be the set of all frequent itemsets mined in R_i, where L_k^i denotes the set of frequent k-itemsets found in R_i. To generate L^i in R_i, we only have to concern those items found in R_i. By restricting that only those items found in R_i are counted in the mining process, the well known Apriori algorithm or its extensions can be directly applied to the generation of frequent itemsets in R_i.

```
Algorithm: FSP-Mining(D, S, min_sup)
Input: a set D of images, a set S of region pairs,
       and a minimum support threshold min_sup;
Output: the set L of all frequent spatial patterns;
1.  Let S_c be the set of regions covered by S;
2.  C ← Φ;
3.  for each R_i ∈ S_c {
4.      D_i ← Φ;
5.      for each I = <O_1, ..., O_r> ∈ D {
6.          t ← <O_i>;
7.          D_i ← D_i ∪ t; }
8.      L^i ← gen_freqent_itemset(D_i, min_sup);}
9.  for each pair (R_i, R_j) ∈ S {
10.     C^ij ← { p_ij = (A, B) | (A, B) ∈ L^i×L^j};
11.     C ← C ∪ C^ij; }
12. L ← gen_frequent_pattern(D, C, min_sup);
13. return L;
```

```
Procedure: gen_frequent_pattern(D, C, min_sup)
Input: a set D of images, a set C of candidate patterns,
       and a minimum support threshold min_sup;
Output: the set L of all frequent spatial patterns;
1.  for each image I=<O_1, ..., O_r> ∈ D {
2.      for each candidate p_ij = (A, B) ∈ C {
3.          if (A⊆O_i ∧ B⊆O_j)
4.              p_ij.count++; } }
5.  L ← { p_ij ∈ C | p_ij.count ≥ min_sup};
6.  return L;
```

Fig. 1. Algorithm FSP-Mining

In the 2nd phase, we first use L^i generated in the 1st phase to generate the set of candidate patterns. Then, we discover the frequent spatial patterns by counting the support for these candidate patterns. Let C be the set of candidate patterns for S and C^{ij} the set of candidate patterns corresponding to the pair of locations (R_i, R_j), where $(R_i, R_j) \in S$. We can use L^i and L^j to generate C^{ij} as follows: $C^{ij} = \{(A, B) \mid (A, B) \in L^i \times L^j \}$. In other words, C^{ij} is set as the Cartesian product of L^i and L^j because a pattern is frequent only if the items in this pattern are frequent in their corresponding regions as well. Then, FSP-Mining combines all C^{ij} as C, i.e., $C = \cup_{ij} C^{ij}$, where $(R_i, R_j) \in S$. After one scan of the image set, we obtain the support count of each candidate pattern in C. Consequently, we are able to generate all the frequent patterns corresponding to the set S of pairs of locations.

Encoding Scheme. For the implementation of the 2nd phase, we devise an encoding scheme to utilize the hash-tree data structure [9] to speed up the counting process of generating frequent patterns. The devised encoding scheme is to associate an item with a region label of which this item appears in. For instance, if $C^{13} = \{(\{a\}, \{b\}), (\{b\}, \{a, c\})\}$ (i.e., there are two patterns in C^{13}), it can be encoded to $C^{13'} = \{\{a_1, b_3\}, \{b_1, a_3, c_3\}\}$, where a_1 stands for item a in R_1 and b_3 stands for item b in R_3 and so forth. Similarly, for an image $I = \langle \{a, b\}, \{b, c\}, \{a\}\rangle$, it can be encoded to $I' = \langle \{a_1, b_1, b_2, c_2, a_3\}\rangle$. With this encoding scheme, the hash-tree data structure can be directly applied to counting the support of a candidate pattern.

Algorithm FSP-Mining is outlined in Figure 1. We omit the pseudo code of gen_frequent_itemset() because the Apriori algorithm or its extensions, such as FP-tree or DHP, can be used to implement this procedure. Note that for ease of presentation, we explore in this paper patterns located in a pair of spatial locations. However, FSP-Mining can be easily extended to mine frequent patterns located in a more diversified combination of spatial locations. To mine patterns other than corresponding to a pair of spatial locations, we can modify both the candidate set C and Procedure $gen\_frequent\_pattern()$ accordingly to reflect the new form of patterns.

3 Conclusions and Future Work

Many new challenges arise in image mining due to the unique properties exhibited by images. To cope with these challenges, we have proposed in this paper a frequent spatial pattern mining framework for images. The mining algorithm in this framework is able to discover frequent spatial patterns in a set of images. This framework is designed to be general because the specific features used to represent image content vary with individual cases, and so do the layout forms used to present represent spatial locations. In our future work, we will evaluate the usefulness of the mining results by various image applications.

Acknowledgement

The work was supported in part by the National Science Council of Taiwan, R.O.C., under Contracts NSC93-2752-E-002-006-PAE.

References

1. Chen, M.S., Han, J., Yu, P.S.: Data mining: An overview from database perspective. IEEE TKDE **5** (1996) 866–883
2. Antani, S., Kasturi, R., Jain, R.: A survey on the use of pattern recognition methods for abstraction, indexing, and retrieval of images and video. Pattern Recognition **35** (2002) 945–965
3. Hsu, W., Lee, M.L., Zhang, J.: Image mining: Trends and developments. Journal of Intelligent Information Systems **19** (2002) 7–23
4. Qamra, A., Chang, E.Y.: Using feature patterns to assist automatic image categorization. In: IEEE International Conference on Multimedia and Expo. (2004)
5. Tesic, J., Newsam, S., Manjunath, B.S.: Mining image datasets using perceptual association rules. In: SIAM Sixth Workshop on Mining Scientific and Engineering Datasets. (2003)
6. Zaiane, O.R., Han, J., Zhu, H.: Mining recurrent items in multimedia with progressive resolution refinement. In: Proc. of ICDE. (2000) 461
7. Ordonez, C., Omiecinski, E.: Discovering association rules based on image content. In: Proc. of the IEEE Advances in Digital Libraries Conference. (1999)
8. Chua, T.S., Tan, K.L., Ooi, B.C.: Fast signature-based color-spatial image retrieval. In: ICMCS. (1997) 362–369
9. Agrawal, R., Srikant, R.: Fast algorithms for mining association rules in large databases. In: Proc. of Int'l Conf. on Very Large Data Bases. (1994) 487–499
10. Han, J., Pei, J., Yin, Y.: Mining frequent patterns without candidate generation. In: Proc. of the ACM SIGMOD. (2000) 1–12
11. Park, J.S., Chen, M.S., Yu, P.S.: Using a hash-based method with transaction trimming for mining association rules. IEEE TKDE **9** (1997) 813–825

Image Classification Via LZ78 Based String Kernel: A Comparative Study

Ming Li and Yanong Zhu

School of Computing Sciences, University of East Anglia,
Norwich, NR4 7TJ, UK
{mli, yz}@cmp.uea.ac.uk

Abstract. Normalized Information Distance (NID) [1] is a general-purpose similarity metric based on the concept of Kolmogorov Complexity. We have developed this notion into a valid kernel distance, called LZ78-based string kernel [2] and have shown that it can be used effectively for a variety of 1D sequence classification tasks [3]. In this paper, we further demonstrate its applicability on 2D images. We report experiments with our technique on two real datasets: (i) a collection of real-life photographs and (ii) a collection of medical diagnostic images from Magnetic Resonance (MR) data. The classification results are compared with those of the original similarity metric (i.e. NID) and several conventional classification algorithms. In all cases, the proposed kernel approach demonstrates better or equivalent performance when compared with other candidate methods but with lower computational overhead.

1 Introduction

Defining a similarity measure between two objects, without explicitly modelling of their task-specific statistical behaviour, is a fundamental problem with many important applications in areas like information retrieval and classification. A broad spectrum approach to this problem is to use the compression-based techniques as a tool for measuring the information redundancy among the objects. Informally, the more information shared between two objects, the more likely they are similar. Based on the concept of (conditional) Kolmogorov complexity [4], authors in [1] formalized such an idea into a similarity metric called Normalized Information Distance (NID). It is illustrated in Fig. 1. Although the Kolmogorov complexity $KC(\cdot)$ is not computable, any compression algorithm gives an upper bound and this can be taken as an estimate of $KC(\cdot)$. Some earlier studies on this approach can be found in [5, 6, 1], which were mainly focused on 1D sequences.

Recently, researchers start exploring the applicability of this compression-based technique (i.e. NID) in the field of two-dimensional images. The key to its feasibility lies in the fact that the raster-scanned version of the raw image preserves enough regularity in both dimensions for the compression algorithm to discover. The authors in [7] tested the technique on the task of handwritten digit recognition and reported an accuracy around 87% which is close to state-of-the-art performance (90% accuracy). However, this image collection is relatively simple due to the binary image representation (i.e. '#' for a black pixel and '.' for a white pixel) and consistent object scale. In [8], the authors tackled a more challenging task i.e. object identification from real-life photographs where images are of unknown and varying scale of scene. The best accuracy over the raw images is around 84% and it is reported to be better than conventional intensity-histogram based techniques. Technically, in both studies, the kernel of Support Vector Machine (SVM) was simply replaced with the similarity (i.e. NID) approximated by compressed length. The potential risk is that, compression-based similarity may result in a non positive-semi-definite (PSD) kernel matrix and thus the optimization problem is no longer convex; SVM learning with SMO-type implementation [9] could converge but the global optimality might not be guaranteed.

To avoid the above problem, we developed the notion of NID into a valid kernel distance, called LZ78-based string kernel, which is suitable for use with SVM classifier [2]. Essentially, it is based on the mapping of example input to a high-dimensional feature space that is indexed by all the phrases identified by a LZ78 parse of the input examples. Further comparisons with other state-of-the-art algorithms yield improved results for a variety of distinct tasks e.g. the classification of music genre, spoken words, and text documents [3].

In this paper, we investigate the applicability of the LZ78-based kernel on 2D image data and empirically demonstrated its advantage over pure compression-based techniques and conventional classification algorithms. The outline of the paper is as follows. In Section 2, we recall the fundamental tools used in this work: the concept of normalized information distance and Lempel-Ziv type compression algorithm. In Section 3, we describe our LZ78-based scheme for image classification. Section 4 presents the application of the proposed method to two practical image classification tasks and experimental results are presented and compared. Finally, conclusions are given in Section 5.

2 Background in Normalized Information Distance and Lempel-Ziv Coding

2.1 Approximation of Normalized Information Distance

The Normalized Information Distance (NID) as proposed in [1] is a similarity metric based on the concept of Kolmogorov Complexity. Informally, it is the ratio of the information shared by the two objects to the total information content of the pair of

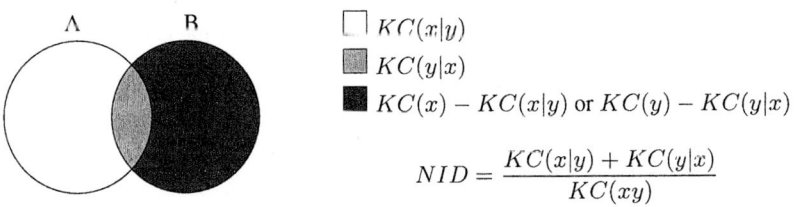

Fig. 1. Illustration of normalized information distance where circle A represents the Kolmogorov complexity $KC(x)$ of object x, circle B represents $KC(y)$ and the total area of two circles (A+B) is $KC(xy)$, i.e. the Kolmogorov complexity of the combination of the objects x and y

objects. This is illustrated in Fig. 1. Two identical objects will have NID=0, whilst two objects with no common information content will have NID=1. Given an object encoded as a binary string x, its Kolmogorov complexity $KC(x)$ is the minimum number of bits into which the string can be compressed without losing information [4]. Intuitively, Kolmogorov complexity indicates the descriptive complexity contained in an object. A random string has relatively high complexity since no structural pattern can be recognized to help reduce the size of program. Strings like structured texts and musical melodies should have lower complexity due to repeated terms and musical structure. Kolmogorov complexity is only an idealized notion be-cause it is not computable. However, any compression algorithm (e.g. LZ78 [10] and PPMZ [11]) gives an upper bound and this can be taken as an estimate of the Kolmogorov complexity. As a result, the theoretical elegant NID amounts to a normalized compression distance [7] in practice.

2.2 Lempel-Zip (LZ) Type Coding

As mentioned before, NID can be approximated by many compression algorithms. In this work, we select a compressor from LZ family [10] i.e. LZ78, which is simple and extremely fast. Moreover, LZ78 is driven by a dictionary-based coding scheme, which can be easily developed into a valid string kernel.

```
/* simplified LZ78 coding scheme */
clear dictionary;
w = λ;¹
while (more input)
   C = next symbol;
   pattern = wC;
   if(pattern in dictionary)
      w = wC;
   else
      add pattern to dictionary;
      w = λ;
   endif
endwhile
return dictionary;
```

The figure above captures the essence of LZ78, which works by identifying patterns, called phrases, of the data and stores them in a dictionary (i.e. encoding table) that defines shorter "entries" that can be used in their stead. In other words, it segments a sequence into several distinct phrases such that each phrase is the shortest subsequence that is not a previously parsed phrase. For example, given a sequence $x =$'abcabcabc', LZ78 parsing yields (a, b, c, ab, ca, bc), namely, $KC(x) = 6$.

3 Image Classification Scheme

In this work, image classification implies to be able to measure the similarity between the strings obtained by scanning the images in raster row-major order. As mentioned

---
[1] λ represents the empty string.

previously, such a raster-scanned version of the image retains enough regularity in both dimensions.

Based on the coding scheme mentioned in Section 2.2, two alternatives for the calculation of the similarity are: *using compressed length (i.e. dictionary size)* or *using compressed patterns (i.e. entries within the dictionary)*.

3.1 Using Compressed Length

In this way, compressed length is used to approximate the normalized information distance between two images. Following the works in [7] and [8], we choice a variant of NID to calculate the pairwise similarity (see Equation 1). Furthermore, to avoid the risk of finding local optimality because of the non-PSD problem, we convert the image into a vector form where the ith element corresponds to the NID score between current image and the ith image in the data, so that the standard kernel function (e.g. RBF kernel) can be applied. Note that with this technique the dimension of feature space is set by the number of examples.

$$NID(x,y) = \frac{KC(xy) - \min(KC(x), KC(y))}{\max(KC(x), KC(y))} \quad (1)$$

3.2 Using Compressed Patterns

The second way to calculate the similarity is based on the patterns (i.e. phrases) identified during compression. In our case, the image (i.e. raster-scanned version) is represented by the set of all the features (i.e. Φ_{lz78}) identified by our modified LZ78 parsing and the pairwise similarity is then defined as the inner product of the weighted[2] feature vectors:

$$K_{lz78}(s,t) = \langle \Phi_{lz78}(s), \Phi_{lz78}(t) \rangle \quad (2)$$

As illustrated in Fig. 1, it is natural to normalize the similarity score in order to take account of object size. In the kernel method, this effect can be achieved by normalizing the feature vectors in the feature space:

$$K_{lz78}^{norm}(s,t) = \frac{K_{lz78}(s,t)}{\sqrt{K_{lz78}(s,s) K_{lz78}(t,t)}} \quad (3)$$

4 Experiments

In this study, we take an empirical approach to evaluate the performance of our proposed scheme for image classification. More specifically, we are concerned about two issues: (i) how does our LZ78 kernel compare with other similarities approximated by compressed length? (ii) is our approach competitive with conventional classification algorithms applied in medical image classification? To conduct the evaluation, the proposed approach is applied to two distinct applications. The first is the classification of

---

[2] Feature importance is indicated by its relative frequency (i.e. tf) within the sting. The logarithm of tf is used to amend unfavorable linearity.

the photograph dataset used in [8] that contains two classes, each of which consists of 761 grayscale images with 247 × 165 pixel resolution[3]. The second is the detection of extracapsular extension (ECE) [12] of prostate cancer using a collection of 18 prostate MR images taken from 10 prostate cancer patients, among whom 6 have histological confirmed ECE and the others are proven to have organ confined tumors. All experiments are carried out using the *libsvm*[4] package, which guarantees the convergence to stationary points for non-PSD kernels.

4.1 Experiment on Photograph Collection

Some images about this collection are shown in Fig. 2. Based on a simple subset of the whole collection, the authors in [8] pointed out that the two classes of this dataset are separable on mean intensity alone. They therefore built a benchmark classifier based on comparing intensity histograms. The results showed that compression-based classifier outperforms the intensity-based classifier by 25% in accuracy, which demonstrates a desirable characteristic of the compression-based techniques-they can automatically identify discriminative regularities (i.e. patterns) from the training dataset. Further, a more complicated image dataset, containing all 1492 images, was used in [8] to test their technique.

Fig. 2. Photograph images showing either a battery-case or a coin-purse. Variation in the scene was introduced by altering the location and rotation of the target objects, changing lighting condition, presenting noise objects (e.g. a blue napkin, screws and a nut) together with complicated background (e.g. magazines and newspapers).

For comparison, we evaluate our method on the same image dataset and re-implement the alternative technique mentioned in [8] which uses partial matching compression technique (PPM) [11] to approximate the NID between two images. The performance is evaluated by a standard ten-fold stratified cross validation (CV). No sieve transformation [13] or ground truth data[5] are used in this experiment since we

---

[3] The original photographs are of 2470 × 1650 pixel resolution. Before processing, the images are down-sampled by a factor of ten using bicubic interpolation.

[4] Available at http://www.csie.ntu.edu.tw/ cjlin/libsvmtools

[5] In [8], the foreground (i.e. target object) of a subset of images (204 in total) was hand-labeled to produce the ground truth and a significant improvement (around 15% in classification accuracy) was observed.

Table 1. Accuracy obtained with various strategies for kernel matrix construction. RBF-Independent means to replace the kernel matrix by the compression-based similarity matrix directly; RBF-Dependent means that each image is firstly converted into a vector form in which the ith element represents the similarity between this image and the ith image in the data and then the distance between two such vectors is calculated via RBF kernel function.

Similarity Calculation	Compression Scheme	Kernel Matrix Construction	
		RBF Independent	RBF Dependent
Using compressed length (NID-based)	PPM over binary encodings	54.1685 (3.97229)	75.6036 (1.46874)
	LZ78 over descriptive symbols	80.2324 (1.75635)	90.7559 (2.65936)
Using patterns discovered by compression algorithm (LZ78-Kernel-based)	LZ78 over descriptive symbols (smoothed)[6]	89.1414 (3.42132)	91.6270 (2.17633)

are interested in comparing the performance made by the compression kernels rather than the improvement made by other techniques (e.g. feature transformation and image segmentation). Note that, when no ground truth data are utilized, the best performance in Lan's work (84%) is achieved by using sieve transformation. As shown later, our approach could achieve even better performance with raw images only.

Table 1 displays the accuracy of SVM classifiers with various strategies for kernel matrix construction. Lan's experiments in [8] are repeated and the performance of their SVM-based method on raw images (74% accuracy) is confirmed in our experiment: note that it occurred only when RBF kernel is applied (i.e. using RBF function to re-calculate the distance between two vectors representing the corresponding im-ages). The bottom row of Table 1 shows that, based on smoothed greyscale images, our LZ78 kernel could achieve better performance but without such extra computational overhead.

It is interesting to note that PPM generally performs better than LZ-type coding in terms of the compress ratio. However, comparing the results shown in the first and second rows, LZ78-based approach performs better in terms of classification accuracy. Two points should be noted. Firstly, in our implementation, PPM acts on binary-encodings of the image string while LZ-based algorithm directly acts on descriptive symbol sequences (i.e. greyscale values); the results may imply the fact that, the *format-similarity*[7] [14] introduced by the low-level (e.g. binary) encodings may deteriorate the complexity approximation, especially when the sequences are relatively short. Secondly, as mentioned before, compression-based similarity may result in a non-PSD kernel matrix; although a SMO-type implementation could converge but no global optimality is guaranteed, which is empirically confirmed in our experiments.

5 Experiment on Medical Diagnosis Image

Computer-aided diagnosis of diseases using medical images is an active research domain. We have previously described the use of a few classification algorithms to detect

---

[6] Obtained by mapping greyscale pixel values into the nearest multiples of five.

[7] It comes from either noisy or the duplication of the symbols used to encode the data.

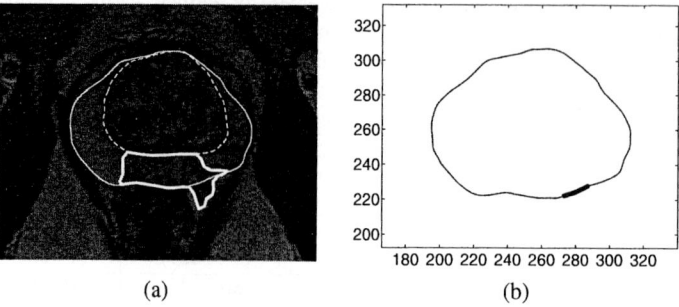

Fig. 3. (a) An example MR image slice (clipped), with the prostate boundary (slim solid line), central zone (dashed line) and a region of ECE (thick solid line) manually annotated. (b) Profile extraction positions on the slice, where large black dots indicate ECE positions according to the manual annotation.

the extra-capsular extension (ECE) of prostate cancer using Magnetic Resonance (MR) images [12]. Basically, each pixel along the prostate boundary is represented by a grey-level intensity profile extracted orthogonal to the prostate boundary and centered on the pixel. A classification model is trained to predict the ECE probability of the pixels (i.e. intensity profiles) along the prostate boundary. Then, a probability filtering process is applied to calculate the overall ECE probability of the image. The classifiers used in this study are k Nearest Neighbor (kNN) and Parzen classifier (PZC).

In this study, we present the application of our LZ78-based SVM to ECE detection. It begins with a sequence of intensity profiles representing a particular MR image, and then applies the following steps: (i) a string example is constructed by concatenating all the intensify profiles of the image in row-major order; (ii) the distances from the example to the decision boundary of positive class $c1$ and negative class $c2$ are calculated according to Equation 4; (iii) the final ECE probability of the image, $P(x)$, is given by Equation 5.

$$d_{ci}(x, sv^{(ci)}, \alpha^{(ci)}) = \left| \sum_{i=1}^{l} \alpha_i^{(ci)} K_{lz78}(sv_i^{(ci)}, x) \right| \quad (4)$$

$$P(x) = \frac{d_{c1}}{d_{c1} + d_{c2}} \quad (5)$$

where $sv^{(c)} = \{sv_i^{(c)}\}$ is the support vector from class c, $\alpha^{(c)} = \{\alpha_i^{(c)}\}$ is the combination coefficient and $K_{lz78}(\cdot)$ is the kernel function (see Equation 2) to perform the similarity calculation.

The MR image data set used in this study consists of 18 MR images from 10 patients, among which seven patients have histologically confirmed ECE, and the other three are proven to have organ confined (non-ECE) prostate cancer. Two mid-gland MRI slices are used for each of the non-ECE patients and five of the seven ECE patients, and one from each of the other two ECE patients. Each slice from the ECE patients includes at least one ECE region.

Since the main aim of this experiment is to evaluate the ability of a trained model to predict the ECE risk of unseen MR images, to avoid the effects of possible variations in

Table 2. Comparison of classification results in terms of AUROC, Sensitivity/Specificity and overall accuracy

Classifier	AUROC (0.0 − 1.0)	Sensitivity (at 66.7%specificity)	Accuracy (threshold=0.5)
kNN	0.736	75.0%	72.2%
PZC	0.764	83.3%	77.8%
LZ78-SVM	0.764	100.0%	88.9%

automatic boundary localization and to provide an objective evaluation of the classification models, manual annotations of the prostate and the ECE regions are used as the ground truth. The annotations are provided by an expert radiologist, and subsequently verified by a second expert to avoid inter-observer variation and ensure the accuracy and consistency. A typical example slice, with a region of ECE annotated can be found in Fig. 3 (a). Fig. 3 (b) shows the profile extraction positions on the example slice, as well as the determination of profile labels according to the manual annotation.

To conduct a statistical evaluation of the methods, the area under the receiver operator characteristic curve (AUROC) of the classification results is computed and compared. The results are obtained by using leave-one-image-out testing. As shown in Table 2, the LZ78-SVM classifier correctly classified 16 of all 18 images, with only 2 images from a non-ECE patient detected as false positive, and hence correctly identified the ECE status of 9 of 10 patients. Overall, based on the intensity profile features, LZ78-SVM offers the best detection results when compared to other classification methods.

6 Conclusion

This paper has described the application of the LZ78-based kernel technique to the classification of 2D images. Instead of following the commonly adopted strategy which approximates the NID by the compressed length of the input data, the technique uses a modified LZ78 compression algorithm as the heuristic for feature extraction and then builds a valid string kernel for SVM. Experiments based on two image collections show that, this method yields better performance when compared with previously proposed approaches. This implies a promising efficiency and wide applicability of the presented method.

Several issues would be considered for future work. Firstly, so far, we merely consider the raw image information (i.e., intensity values or intensity profiles). The performance of the proposed method can be further enhanced by incorporating more sophisticated image representation techniques, e.g., morphological operators and content-based descriptors. This should provide more compact and descriptive representation of the images, and hence better classification outcome. Second, the experiment on MR image data for prostate cancer diagnosis is based on a relatively small date set. Experiments on larger data sets are expected to conduct a more comprehensive evaluation, and will be performed in due time when such data are available. Furthermore, the use of other compression algorithms as the feature extraction techniques, such as the block-sorting algorithm [15] and the PPM family [11], will be investigated.

Acknowledgements

We wish to thank anonymous reviewers for their helpful comments and Yuxuan Lan for providing us the photograph image dataset. We also thank Prof. Ronan Sleep and Dr. Richard Harvey for interesting and useful discussion.

References

1. Li, M., Chen, X., Ma, B., Vitanyi, P.: The similarity metric. In: Proceedings of the 14th ACM-SIAM Symposium on Discrete Algorithms. (2003) 863–872
2. Li, M., Sleep, R.M.: A LZ78-based string kernel. In: Lecture Notes in Artificial Intelligence, Proceedings of the First Interna-tional Conference on Advanced Data Mining and Applications, Wuhan, China. Volume 3584. (2005) 678–689
3. Li, M., Sleep, R.M.: A robust approach to sequence classification. In: Proceedings of the 17th IEEE Conference on Tools with Artificial Intelligence, Hong Kong, China. (2005)
4. Li, M., Vitanyi, P.: An Introduction to Kolmogorov Complexity and Its Applications. Springer-Verlag, Berlin Heidelberg New York (1997)
5. Teahan, W.J., Harper, D.J.: Using compression-based language models for text categorization. In: Workshop on Language Modeling and Information Retrieval, Carnegie Mellon University. (2001) 83–88
6. Benedetto, D., Caglioti, E., Loreto, V.: Language trees and zipping. Physical Review Letters **88** (2000)
7. Cilibrasi, R., Vitanyi, P.: Clustering by compression. IEEE Transactions on Information Theory **51** (2005) 1523–1545
8. Lan, Y., Harvey, R.: Image classification using compression distance. In: Proceedings of the 2nd International Conference on Vision, Video and Graphics, Edinburgh. (2005)
9. Platt, J.: Sequential minimal optimization: A fast algorithm for training support vector machines. Microsoft Research Technical Report MSR-TR-98-14 (1998) Available at http://research.microsoft.com/users/jplatt/smo.html.
10. Ziv, J., Lempel, A.: Compression of individual sequences via variable-rate coding. IEEE Transactions on Information Theory **24** (1978) 530–536
11. Cleary, J., Witten, I.: Data compression using adaptive coding and partial string matching. IEEE Transactions on Communication **COM-32** (1984) 396–402
12. Zhu, Y., Williams, S., Fisher, M., Zwiggelaar, R.: The use of grey-level profiles for detection of extracapsular extension of prostate cancer from MRI. In: Proceedings of Medical Image Understanding and Analysis. (2005) 215–218
13. Bangham, A.J., Harvey, R., Ling, P., Aldridge, R.: Morphological scale-space preserving transforms in many dimensions. Journal of Electronic Imaging **5** (1996) 283–299
14. Keogh, E., Lonardi, S., Rtanamahatana, C.A.: Toward parameter free data mining. In: Proceeding of the 10th ACM SIGKDD, Seattle, Washington, USA. (2004) 206–215
15. Burrows, M., Wheeler, D.J.: A blocksorting lossless data compression algorithm. SRC Research Report 124 (1994)

Distributed Pattern Discovery in Multiple Streams

Jimeng Sun[1], Spiros Papadimitriou[2], and Christos Faloutsos[1]

[1] Carnegie Mellon University
{jimeng, christos}@cs.cmu.edu
[2] IBM Watson Research Center
spapadim@us.ibm.com

Abstract. Given m groups of streams which consist of $n_1, \ldots, n_m$ co-evolving streams in each group, we want to: (i) incrementally find local patterns within a single group, (ii) efficiently obtain global patterns across groups, and more importantly, (iii) efficiently do that in real time while limiting shared information across groups. In this paper, we present a distributed, hierarchical algorithm addressing these problems. Our experimental case study confirms that the proposed method can perform hierarchical correlation detection efficiently and effectively.[1]

1 Introduction

Streams are often inherently correlated and it is possible to reduce hundreds of numerical streams into just a handful of *patterns* that compactly describe the key trends and dramatically reduce the complexity of further data processing. Multiple co-evolving streams often arise in a large distributed system, such as computer networks and sensor networks. Centralized approaches usually will not work in this setting. The reasons are: **(i) Communication constraint**; it is too expensive to transfer all data to a central node for processing and mining. **(ii) Power consumption**; in a wireless sensor network, minimizing information exchange is crucial because many sensors have very limited power. **(iii) Robustness concerns**; centralized approaches always suffer from single point of failure. **(iv) Privacy concerns**; in any network connecting multiple autonomous systems (e.g., multiple companies forming a collaborative network), no system is willing to share all the information, while they all want to know the global patterns. To sum up, a **distributed online algorithm** is highly needed to address all the above concerns.

To address this problem, we propose a hierarchical framework that intuitively works as follows:1) Each autonomous system first finds its local patterns and shares them with other groups. 2) Global patterns are discovered based on the shared local patterns. 3) From the global patterns, each autonomous system further refines/verifies their local patterns.

---
[1] The technical report [6] is a longer version of this work.

2 Problem Formalization and Framework

Given m groups of streams which consist of $\{n_1, \ldots, n_m\}$ co-evolving numeric streams, respectively, we want to solve the following two problems: (i) incrementally find patterns within a single group (*local pattern monitoring*), and (ii) efficiently obtain global patterns from all the local patterns (*global pattern detection*).

More specifically, *local pattern monitoring* can be modelled as a function,

$$F_L : (S_i(t+1,:), G(t,:)) \to L_i(t+1,:), \qquad (1)$$

where the inputs are 1) the new input point $S_i(t+1,:)$ at time $t+1$ and the current global pattern $G(t,:)$ and the output is the local pattern $L_i(t+1,:)$ at time $t+1$. Details on constructing such a function will be explained in section 3. Likewise, *global pattern detection* is modelled as another function,

$$F_G : (L_1(t+1,:), \ldots, L_m(t+1,:)) \to G(t+1,:), \qquad (2)$$

where the inputs are local patterns $L_i(t+1,:)$ from all groups at time $t+1$ and the output is the new global pattern $G(t+1,:)$.

Now we introduce the general framework for distributed mining. More specifically, we present the meta-algorithm to show the overall flow, using F_L (*local patterns monitoring*) and F_G (*global patterns detection*) as black boxes.

Intuitively, it is natural that global patterns are computed based on all local patterns from m groups. On the other hand, it might be a surprise that the local patterns of group i take as input both the stream measurements of group i and the global patterns. Stream measurements are a natural set of inputs, since local patterns are their summary. However, we also need global patterns as another input so that local patterns can be represented consistently across all groups. This is important at the next stage, when constructing global patterns out of the local patterns; we elaborate on this later. The meta-algorithm is the following:

Algorithm DISTRIBUTEDMINING
0. (*Initialization*) At $t = 0$, set $G(t,:) \leftarrow$ null
1. For all $t > 1$
   (*Update local patterns*) For $i \leftarrow 1$ to m, set $L_i(t,:) := F_L(S_i(t,:), G(t-1,:))$
   (*update global patterns*) Set $G(t,:) := F_G(L_1, \ldots, L_m)$

3 Pattern Monitoring

Tracking Local Patterns. We now present the method for discovering patterns within a stream group. More specifically, we explain the details of function F_L (Equation 1). We first describe the intuition behind the algorithm and then present the algorithm formally. Finally we discuss how to determine the number of local patterns k_i.

The goal of F_L is to find the low dimensional projection $L_i(t,:)$ and the participation weights $W_{i,t}$ so as to guarantee that the reconstruction error $\|S_i(t,:) - \hat{S}_i(t,:)\|^2$ over time is predictably small.

The first step is, for a given k_i, to incrementally update the $k \times n_i$ participation weight matrix $W_{i,t}$, which serves as a basis of the low-dimensional projection for $S_i(t,:)$. Later in this section, we describe the method for choosing k_i. For the moment, assume that the number of patterns k_i is given.

The main idea behind the algorithm is to read the new values $S_i(t+1,:) \equiv [S_i(t+1,1), \ldots, S_i(t+1,n_i)]$ from the n_i streams of group i at time $t+1$, and perform three steps: (1) Compute the low dimensional projection $y_j, 1 \le j \le k_i$, based on the *current* weights $W_{i,t}$, by projecting $S_i(t+1,:)$ onto these.(2) Estimate the reconstruction error (e_j below) and the energy.(3) Compute $W_{i,t+1}$ and output the *actual* local pattern $L_i(t+1,:)$.

The term λ is a forgetting factor between 0 and 1, which helps adapt to more recent behavior. For instance, $\lambda = 1$ means putting equal weights on all historical data, while smaller λ means putting higher weight on more recent data.

In practice, we do not know the number k_i of local patterns. We propose to estimate k_i on the fly, so that we maintain a high percentage $f_{i,E}$ of the *energy* $E_{i,t}$. For each group, we have a low-energy and a high-energy threshold, $f_{i,E}$ and $F_{i,E}$, respectively. We keep enough local patterns k_i, so the retained energy is within the range $[f_{i,E} \cdot E_{i,t}, F_{i,E} \cdot E_{i,t}]$.

Algorithm F_L

Input: new vector $S_i(t+1,:)$, old global patterns $G(t,:)$
Output: local patterns (k_i-dimensional projection) $L_i(t+1,:)$
1. Initialize $x_1 := S_i(t+1,:)$.
2. For $1 \le j \le k$, we perform the following in order:

   $y_j := x_j W_{i,t}(j,:)^T$ (y_j = projection onto $W_{i,t}(j,:)$)
   If $G(t,:) = $ null, then $G(t,j) := y_j$ (handling boundary case)
   $d_j \leftarrow \lambda d_j + y_j^2$ (local energy, determining update magnitude)
   $e := x_j - G(t,j)W_{i,t}(j,:)$ (error, $e \perp W_{i,t}(j,:)$)
   $W_{i,t+1}(j,:) \leftarrow W_{i,t}(j,:) + \frac{1}{d_j}G(t,j)e$ (update participation weight)
   $x_{j+1} := x_j - G(t,j)W_{i,t+1}(j,:)$ (repeat with remainder of x).

3. Compute the new projection $L_i(t+1,:) := S_i(t+1,:)W_{i,t+1}^T$

Tracking Global Patterns. We now present the method for obtaining global patterns over all groups. More specifically, we explain the details of function F_G.

First of all, what is a global pattern? Similar to local pattern, global pattern is low dimensional projections of the streams from all groups. Loosely speaking, assume only one global group exists which consists of all streams, the global patterns are the local patterns obtained by applying F_L on the global group— this is essentially the centralized approach. In other words, we want to obtain the result of the centralized approach without centralized computation.

The algorithm exactly follows the lemma above. The j-th global pattern is the sum of all the j-th local patterns from m groups.

Algorithm F_G
Input: all local patterns $L_1(t,:), \ldots, L_m(t,:)$
Output: global patterns $G(t,:)$
0. Set $k := max(k_i)$ for $1 \leq i \leq m$
1. For $1 \leq j \leq k$, set $G(t,j) := \sum_{i=1}^{m} L_i(t,j)$ (if $j > k_i$ then $L_i(t,j) \equiv 0$)

4 Experimental Case Study

The Motes dataset consists of 4 groups of sensor measurements (i.e., light intensity, humidity, temperature, battery voltages) collected using 48 Berkeley Mote sensors at different locations in a lab, over a period of a month.

The main characteristics (see the blue curves in Figure 1) are: (1) Light measurements exhibit a clear global periodic pattern (daily cycle) with occasional big spikes from some sensors (outliers), (2) Temperature shows a weak daily cycle and a lot of bursts. (3) Humidity does not have any regular pattern. (4) Voltage is almost flat with a small downward trend.

The reconstruction is very good (see the red curves in Figure 1(a)), with relative error below 6%. Furthermore, the local patterns from different groups are correlated well with the original measurements (see Figure 2). The global patterns (in Figure 3) are combinations of different patterns from all groups and reveal the overall behavior of all the groups.

The relative reconstruction error as the evaluation metric. The best performance is obtained when all groups exchange up-to-date local/global patterns at every timestamp, which is prohibitively expensive. One efficient way to deal with this problem is to increase the communication period, which is the number of

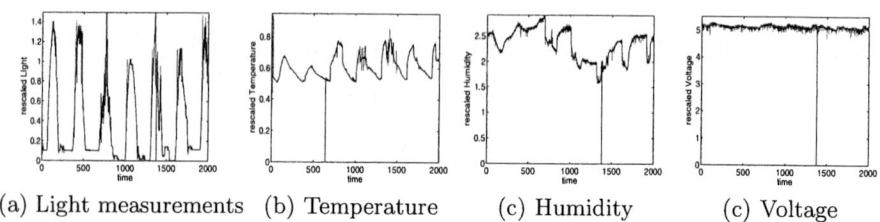

(a) Light measurements (b) Temperature (c) Humidity (c) Voltage

Fig. 1. original measurements (blue) and reconstruction (red) are very close

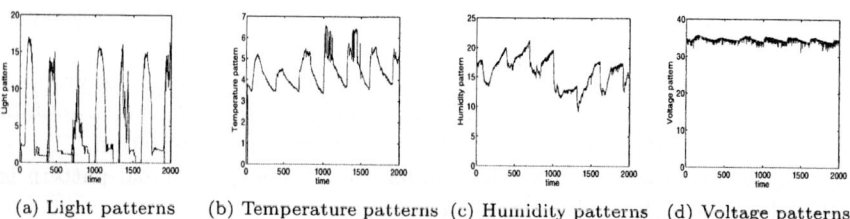

(a) Light patterns (b) Temperature patterns (c) Humidity patterns (d) Voltage patterns

Fig. 2. Local patterns

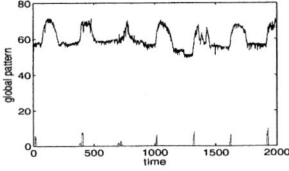 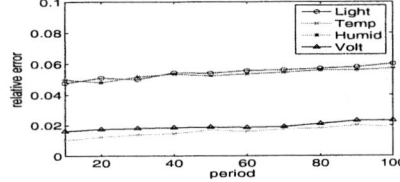

Fig. 3. Global patterns   **Fig. 4.** Error increases slowly

timestamps between successive local/global pattern transmissions. Overall, the relative error rate increases very slowly as the communication period increases (see Figure 4). This implies that we can dramatically reduce communication with minimal sacrifice of accuracy.

5 Related Work

Distributed Data Mining. Most of works on distributed data mining focus on extending classic (centralized) data mining algorithms into distributed environment, such as association rules mining [3], frequent item sets [5]. Web is a popular distributed environment. Several techniques are proposed specifically for that, for example, distributed top-k query [2] But our focus are on finding numeric patterns, which is different.

Privacy Preserving Data Mining. The most related discussion here is on how much privacy can be protected using subspace projection method [1,4]. Liu et al. [4] discuss the subspace projection method and propose a possible method to breach the protection using Independent component analysis(ICA). All the method provides a good insight on the issues on privacy protection. Our method focuses more on incremental online computation of subspace projection.

6 Conclusion and Acknowledgement

We focus on finding patterns in a large number of distributed streams. More specifically, we first find local patterns within each group, where the number of local patterns is automatically determined based on reconstruction error. Next, global patterns are identified, based on the local patterns from all groups. We evaluated our method on several datasets, where it indeed discovered the patterns. We gain significant communication savings, with small accuracy loss.

Work partially supported by the NSF under Grants No. IIS-0209107 IIS-0205224 INT-0318547 SENSOR-0329549 IIS-0326322 and the Pennsylvania Infrastructure Technology Alliance (PITA) This publication only reflects the authors views.

References

1. C. Agrawal and P. Yu. A condensation approach to privacy preserving data mining. In *EDBT*, 2004.
2. B. Babcock and C. Olston. Distributed Top-K Monitoring. In *SIGMOD*, 2003.
3. D. W. Cheung, V. T. Ng, A. W. Fu, and Y. Fu. Efficient Mining of Association Rules in Distributed Databases. *TKDE*, 8:911–922, 1996.
4. K. Liu, H. Kargupta, and J. Ryan. Multiplicative noise, random projection, and privacy preserving data mining from distributed multi-party data. In *TKDE*, 2005.
5. K. K. Loo, I. Tong, B. Kao, and D. Cheung. Online Algorithms for Mining Inter-Stream Associations From Large Sensor Networks. In *PAKDD*, 2005.
6. Jimeng Sun, Spiros Papadimitriou, and Christos Faloutsos. Distributed pattern discovery in multiple streams. Technical Report CMU-CS-06-100, Carnegie Mellon Univ., 2005.

COMET: Event-Driven Clustering over Multiple Evolving Streams

Mi-Yen Yeh, Bi-Ru Dai, and Ming-Syan Chen

Department of Electrical Engineering,
National Taiwan University,
Taipei, Taiwan, ROC
{miyen, brdai}@arbor.ee.ntu.edu.tw, mschen@cc.ee.ntu.edu.tw

Abstract. In this paper, we present a framework for event-driven Clustering Over Multiple Evolving sTreams, which, abbreviated as COMET, monitors the distribution of clusters on multiple data streams and online reports the results. This information is valuable to support corresponding online decisions. Note that as time advances, the data streams are evolving and the clusters they belong to will change. Instead of directly clustering the multiple data streams periodically, COMET applies an efficient cluster adjustment procedure only when it is required. The signal of requiring to do cluster adjustments is defined as an "event." We design a mechanism of event detection which employs piecewise linear approximation as the key technique. The piecewise linear approximation is advantageous in that it can not only be performed in real time as the data comes in, but also be able to capture the trend of data. When an event occurs, through split and merge operations we can report the latest clustering results effectively with high clustering quality.

1 Introduction

Research about mining in the data stream environment is flourishing in these years [1][2][3][4][5][6][7][8]. In addition to those on considering a data stream at a time, more and more emerging applications involve in monitoring multiple data streams concurrently. Such applications include online stock market data analysis, call detail records in telecommunication, sensor network, ATM operation in banks, etc. We are able to find out interesting and useful knowledge by analyzing the relationship between these multiple data streams. Therefore, mining multiple data streams has attracted an increasing amount of attention from related researchers. To discover the cross-relationship between streams, one way is to calculate the correlation between streams and report the stream pairs with high correlation [9][10][11][12]. Another one is to do similarity pattern query between multiple data streams [9][13]. Last but not least, some works are reported on applying the clustering technique to multiple data streams [14][15][16].

Among multiple evolving data streams, we want to trace not only those streams becoming similar to one another but also those becoming dissimilar along with the growing of streams. Clustering is a mining technique which puts the similar objects together and separates dissimilar ones into different clusters. As a result, by clustering the streams dynamically, we can achieve the goal of monitoring the evolution of stream clusters. From observing the clusters evolution we are able to get the useful information for decision making or data management in various applications. For example, in the stock market, the price of each stock may vary from time to time and some stocks tend to rise and fall concurrently in some time intervals. The stock monitoring system aims to find the streams which are in the same group and have similar behavior. From such evolving streams, the investors would like to buy a proper set of streams to maximize the profit.

In [14], an online data summarization framework is designed for offline clustering on multiple data streams when users submit requests. In contrast, we want to provide in this paper a more real-time and automatic system which performs online clustering. The system will report the revolution of clusters as time advances. To achieve this goal, one intuitive solution is to cluster these data streams periodically. We can just update each stream and apply an existing clustering algorithm on these streams at the pre-determined time point. However, due to the large stream number and the huge data volume, the distance update between each stream is very costly. Furthermore, periodical clustering is not able to cope with the data streams with different evolving speeds. If the values of data streams are relatively steady, most of the clustering tasks are unnecessary since the resulting clusters are likely to remain the same. On the other hand, if the values of data streams are relatively fluctuant, we may lose some cluster information when the fixed time period is too long. Concluding from above issues, we need a solution which is able to perform clustering whenever it is necessary. Consequently, a framework named *event-driven Clustering Over Multiple Evolving sTreams*, abbreviated as *COMET*, is proposed in this paper.

For generality, we consider the data on the numerical domain. Our work can be easily extended to the applications with categorical data via proper data transformation. Initially, the streams are divided into several clusters by applying any traditional clustering method. In fact, we can also apply our merge operation, which will be introduced later, to obtain initial clusters. Due to the evolving feature of data streams, a group of streams may be similar at this moment but become dissimilar to one another later. In order to capture the significant changes of each stream, we use continuous piecewise linear line segments to approximate the original data stream. Explicitly, the piecewise linear approximation can not only be performed in real time as the data comes in, but also be able to capture the trend of data. Two line segments with different slopes are connected by an end point. The end point represents the significant trend change point of the stream data. If a stream in a cluster has a significant change, it is possible to cause the split of this cluster. As a result, we can regard each end point of a stream as a "trigger" of the cluster evolution, and call the stream which has a newly encountered end point as "trigger-stream." When a trigger occurs, the distances between trigger-streams and other streams in the same cluster are then updated incrementally. If the distance of any stream pair in a cluster exceeds a given threshold, we say an "event" is detected. An event is a signal for the system to make necessary cluster modifications. Similar "event-driven" idea can be found in [13], but it has different definition and usage. When an event is found via the event detection mechanism, we perform necessary cluster split. Then, a procedure for checking whether there exist clusters being close enough to be merged together is activated. Since the split and merge processes are very efficient, the event processing procedure is able to handle thousands of streams concurrently.

2 Preliminaries

2.1 Problem Model

Given an integer n, an n-stream set is denoted as $\Gamma = \{S_1, S_2, ..., S_n\}$ where S_i is the ith stream. A data stream S_i can be represented as $S_i[t_1, ..., t_k, ...]$ where $S_i[t_k]$ is the data value of stream S_i arriving at time t_k. The objective of this paper is that given a set of data streams Γ and the threshold parameters, the summary of each stream S_i, which is denoted as $\widehat{S_i}$, is online maintained and the event detection mechanism is built. When events occur, cluster modifications will be performed instead of re-clustering all streams and the latest clustering results are reported.

2.2 Piecewise Linear Data Summarization

In COMET, piecewise linear approximation is adopting to detect the significant trend changes of the data streams. The end points between line segments are regarded as triggers of clustering evolution. The work in [17] describes the basic concept of online segmenting time series. Many variations are conceivable to adapt different types of data. For example, [13] provides a three-tiered online segmentation and pruning strategy for financial data streams. Base on sliding window techniques, a stream S_i is summarized as : $\widehat{S_i} = \{(S_i[t_{v1}], t_{v1})(S_i[t_{v2}], t_{v2}), ..., (S_i[t_{vk}], t_{vk})\}$.

3 Distance Measurement

We now discuss the distance measurement between two stream summaries. Since streams may vary at different level, instead of directly using Euclidean distance, we apply the distance measure in [18] and did some modification. Originally, it projects end points of one time series to another one, and then calculate the variance of length of these projected lines. The more similar these two series are, the smaller variance will be got. To avoid the cross of two series, [18] add some constant to separate them. In our case, due to the feature of streaming data, we cannot know how big the constant should be added in advance. As a result, the sign of the subtraction of two streaming data value is taken into consideration. Moreover, we accumulated all difference value at each time point. When an end point is met, we update the distance in an incremental fashion.

4 Event Detection and Clustering

In essence, a cluster is the set of summarized streams and all the clusters become a cluster set. Each cluster has a center which is simply the average of every member in that cluster. Consequently, the center of a cluster is also a sequence of end points. As data points come in, each stream is online doing piecewise linear approximation. For streams which have a new end point, we first find out the clusters that these streams belong to, and then only the distances between the trigger-streams and the rest of the streams in the same cluster will be updated. The stream pair distance is updated in an incremental manner as mentioned in Section 3.

4.1 Split of a Cluster

When the distance between the trigger-stream and other streams in the same cluster is updated, for each trigger-stream $\widehat{S}_i$, a list containing all $\widehat{S}_j$ in the same cluster C_k whose distance to $\widehat{S}_i$ exceeds the threshold δ_a is kept. If the size of the list is larger than a specific proportion, we regard $\widehat{S}_i$ itself as being very different from the original cluster. As a result, $\widehat{S}_i$ is required to be split out from the cluster C_k. On the other hand, if the size of this list is not larger than the specific amount, we consider that the streams *inside* the list could become quite different from the original cluster. Therefore the members of the list become the candidate streams to be moved out.

After splitting, we need to update corresponding cluster centers. Then, the inter-cluster distances are updated in the same way as updating distances between summarized streams.

4.2 Merge Clusters

The COMET framework checks whether there are clusters being close enough to be merged after splitting and updating the inter-cluster distances of each cluster pair. How close can two clusters be merged is defined by a user given threshold δ_e. If the inter-cluster distance between any two clusters is under the threshold δ_e, these two clusters are merged. Note that we can apply any agglomerative hierarchical clustering method in this merge process by setting the stop criteria as the threshold δ_e. The cluster number is relatively small compared to the original number of streams, and thus the execution time is relative low.

5 Conclusion

In this paper, we proposed the COMET framework for online monitoring clusters over multiple data streams. By using the piecewise linear approximation for data summarization, we can regard each end point of the line segment as a trigger point. At each trigger point, we update the distances between streams in the same cluster. Whenever an event happens, i.e., any distance between two streams in a cluster exceeds the pre-defined threshold, the clusters are modified by the split and merge processes. The COMET framework is efficient and of good scalability while producing cluster results of good quality.

Acknowledgements

The work was supported in part by the National Science Council of Taiwan, R.O.C., under Contracts NSC93-2752-E-002-006-PAE.

References

1. Babcock, B., Babu, S., Datar, M., Motwani, R., Widom, J.: Models and issues in data stream systems. In: Proc. of PODS. (2002)
2. Bulut, A., Singh, A.K.: SWAT: Hierarchical stream summarization in large networks. In: Proc. of ICDE. (2003)
3. Domingos, P., Hulten, G.: Mining high-speed data streams. In: Proc. of ACM SIGKDD. (2000)
4. Gaber, M., Krishnaswamy, S., Zaslavsky, A.: Cost-efficient mining techniques for data streams. In: Proc. of DMWI. (2004)
5. Ganti, V., Gehrke, J., Ramakrishnan, R.: DEMON: Mining and monitoring evolving data. Knowledge and Data Engineering **13** (2001)
6. Guha, S., Mishra, N., Motwani, R., O'Callaghan, L.: Clustering data streams. In: the Annual Symposium on Foundations of Computer Science. (2000)
7. Hulten, G., Spencer, L., Domingos, P.: Mining time-changing data streams. In: Proc. of ACM SIGKDD. (2001)
8. O'Callaghan, L., Mishra, N., Meyerson, A., Guha, S., Motwani, R.: Streaming-data algorithms for high-quality clustering. In: Proc. of ICDE. (2002)
9. Bulut, A., Singh, A.K.: A unified framework for monitoring data streams in real time. In: Proc. of ICDE. (2005)
10. Liu, X., Ferhatosmanoglu, H.: Efficient k-nn search on streaming data series. In: Proc. of SSTD. (2003)
11. Zhu, Y., Shasha, D.: Statstream: Statistical monitoring of thousands of data streams in real time. In: Proc. of VLDB. (2002)
12. Yi, B.K., Sidiropoulos, N., J., T., Jagadish, H.V., Faloutsos, C., Biliris, A.: Online data mining for co-evolving time sequences. In: Proc. of ICDE. (2000)
13. H. Wu, B. Salzberg, D.Z.: Online event-driven subsequence matching over financial data streams. In: Proc. of ACM SIGMOD. (2004)
14. Dai, B.R., Huang, J.W., Yeh, M.Y., Chen, M.S.: Clustering on demand for multiple data streams. In: Proc. of ICDM. (2004)
15. Rodrigues, P., Gama, J., Pedroso, J.P.: Hierarchical time-series clustering for data streams. In: Proc. of Int'l Workshop on Knowledge Discovery in Data Streams in conjunction with 15th European Conference on Machine Learning. (2004)
16. Yang, J.: Dynamic clustering of evolving streams with a single pass. In: Proc. of ICDE. (2003) 695–697
17. Keogh, E.J., Chu, S., Hart, D., Pazzani, M.J.: An online algorithm for segmenting time series. In: Proc. of ICDM. (2001)
18. Keogh, E.J.: A fast and robust method for pattern matching in time series databases. In: Proc. of ICTAI. (1997)

Variable Support Mining of Frequent Itemsets over Data Streams Using Synopsis Vectors

Ming-Yen Lin[1], Sue-Chen Hsueh[2], and Sheng-Kun Hwang[1]

[1] Department of Information Engineering and Computer Science,
Feng-Chia University, Taiwan
linmy@fcu.edu.tw, m9305966@webmail.fcu.edu.tw
[2] Department of Information Management,
Chaoyang University of Technology, Taiwan
schsueh@mail.cyut.edu.tw

Abstract. Mining frequent itemsets over data streams is an emergent research topic in recent years. Previous approaches generally use a fixed support threshold to discover the patterns in the stream. However, the threshold will be changed to cope with the needs of the users and the characteristics of the incoming data in reality. Changing the threshold implies a re-mining of the whole transactions in a non-streaming environment. Nevertheless, the "look-once" feature of the streaming data cannot provide the discarded transactions so that a re-mining on the stream is impossible. Therefore, we propose a method for variable support mining of frequent itemsets over the data stream. A synopsis vector is constructed for maintaining statistics of past transactions and is invoked only when necessary. The conducted experimental results show that our approach is efficient and scalable for variable support mining in data streams.

1 Introduction

Many data-intensive applications continuously generate an unbounded sequence of data items at a high rate in real time nowadays. These transient data streams cannot be modeled as persistent relations so that traditional database management systems are becoming inadequate in supporting the functionalities of modeling this new class of data [2]. The unbounded nature of data streams disallows the holding of the entire stream in the memory, and often incurs a high call-back cost even if the past data can be stored in external media. Any algorithm designed for streaming data processing would generally be restricted to scan the data items only once. Consequently, algorithms such as stream mining algorithms can present merely approximate results rather than accurate results because some data items will be inevitably discarded.

The discovery of frequent items and frequent itemsets has been studied extensively in the data mining community, with many algorithms proposed and implemented [1, 5, 9]. The 'one-pass' constraint, however, inhibits the direct application of these algorithms over data streams. The mining of frequent items/itemsets in a data stream has been addressed recently. An algorithm in [10] uses the Buffer-Trie-SetGen to mine frequent itemsets in a transactional data stream. The FP-stream algorithm [4] incrementally maintains tilted-time windows for frequent itemsets at multiple time granularities. The DSM-FI algorithm [7] uses a FP-tree [5] like forest and estimated supports for the mining. In addition, the Moment algorithm [3] employs a 'closed enumeration tree' for fast discovery of closed frequent itemsets in a data stream.

Note that the above approaches for mining frequent itemsets over data streams accept only one minimum support in the mining. The minimum support cannot be changed during the mining for these approaches. In reality, the minimum support is not a fixed value for the entire stream of transactions. The user may specify a threshold in the beginning, adjust the threshold after evaluating the discovered result, or change the threshold after a period of time after receiving volumes of transactions. The minimum support threshold therefore should be variable to suit the need of the user. In contrast to frequent itemset mining with a fixed support, the mining with respect to a changeable support is referred to as variable support mining. Although online association rule mining and interactive mining [8] may have changeable support thresholds, both algorithms are inapplicable to the stream data because a scanning of entire transactions is required.

In this paper, we formulate the problem of variable support mining in a data stream and propose the VSMDS (Variable Support Mining of Data Streams) algorithm for efficient variable mining of frequent itemsets in a stream of transactions. The VSMDS algorithm uses a compact structure (called PFI-tree) to maintain the set of potential frequent itemsets and update their support counts. A summary structure, called synopsis vector, is designed to approximate past transactions with a flexible distance threshold. The comprehensive experiments conducted show that VSMDS is highly efficient and linearly scalable.

2 Problem Statement

Let $\Psi = \{\alpha_1, \alpha_2, ..., \alpha_r\}$ be a set of literals, called *items*. A *data stream* DS = $\{t_1, t_2, ..., t_c, ...\}$ is an infinite sequence of incoming transactions, where each transaction t_i is an item-set associated with a unique transaction identifier. Let t_c be the latest incoming transaction, called *current transaction*. The *current length* of the data stream is the number of transactions seen so far. A transaction t_i *contains* an item-set e if $e \subseteq t_i$. The *support* of an item-set e, denoted by $sup(e)$, is the number of transactions containing e divided by the current length in DS.

The user specified a *minimum support* threshold $ms \in (0,1]$ in the beginning of the data stream. At any point of time, along with the incoming of transactions, the user may change the minimum support threshold so that the thresholds form a series of minimum supports. Let ms_c, called *current minimum support*, be the minimum support when we saw t_c. An item-set e is a *frequent itemset* if $sup(e) \geq ms_c$. The objective is to discover all the frequent itemsets in the data stream, with respect to current minimum support. Since the specified minimum support is not a fixed value, such a mining is called *variable support mining* over the data stream. In contrast, previous mining with only one unchangeable minimum support is called *fixed support mining*. The goal is to use the up-to-update minimum support ms_c and consider all the transactions, including the discarded ones, for the discovery of frequent itemsets.

3 VSMDS: Variable Support Mining for Data Streams

We process the stream, in a bucket-by-bucket basis, by grouping $|B|$ (called *bucket size*) incoming transactions into a bucket. A *potential frequent itemset tree* (called PFI-tree) is designed to maintain the set of potential frequent itemsets. To provide the user with the up-to-date result reflecting a newly specified minimum support, the proposed algorithm effectively compresses the discarded transactions into a summary structure called *synopsis vector* (abbreviated as SYV). Consequently, we may use the SYV to update the PFI-tree with respect to current minimum support. We use an idea similar to Proximus [6] for compressing the transactions but carry out a structure updating for more accurate results.

The series of minimum supports specified by the user is collectively referred to as the *support sequence* (ms_1, ms_2, ..., ms_λ), where ms_i indicates the minimum support used when DS has B_i buckets. In the following, The PFI_i is the PFI-tree and SYV_i is the *SYV* on seeing bucket B_i. Additionally, the ms_{PFI} denotes the minimum support threshold used in the PFI-tree.

Fig. 1 depicts the overall concept of the proposed VSMDS algorithm. On seeing a new bucket B_i, VSMDS updates the PFI_{i-1} and compresses B_i with SYV_{i-1} into SYV_i. The PFI_i is used to output the desired patterns to the user. The SYV_{i-1} is used to build PFI_i only when the PFI_i cannot provide the up-to-date results, that is, when $ms_i < ms_{PFI}$. The PFI_{i-1} keeps all the itemsets having supports at least ms_{PFI}, considering buckets up to bucket B_{i-1}, during the process. If $ms_i \geq ms_{PFI}$, the user are querying frequent itemsets that have higher supports. These itemsets can be located from PFI_{i-1} and VSMDS replies to the user without the participation of the SYV. If $ms_i < ms_{PFI}$, those itemsets having supports greater than or equal to ms_i but smaller than ms_{PFI}, thus being excluded in PFI_{i-1}, become frequent. Hence, VSMDS will use the SYV_{i-1} to build PFI_i for the mining of these itemsets at this moment. VSMDS utilizes the lexicographic property of consecutive item-comparisons [9] in PFI-tree for fast mining and updating of potential frequent itemsets. The SYV is a list of (delegate, cardinality) pairs. The cardinality indicates the number of occurrences of the delegate; the delegate represents a group of *approximated* itemsets. A delegate dg is said to *approximate* to an itemset e if the distance (eg. the number of different items between dg and e) is no more than certain distance threshold (defined by the user).

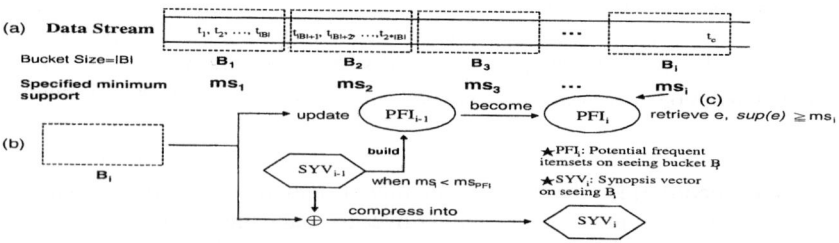

Fig. 1. Overall concept of the VSMDS algorithm: (a) bucketed transactions (b) update and compress operations on seeing a bucket B_i (c) retrieving the frequent itemsets from the PFI_i

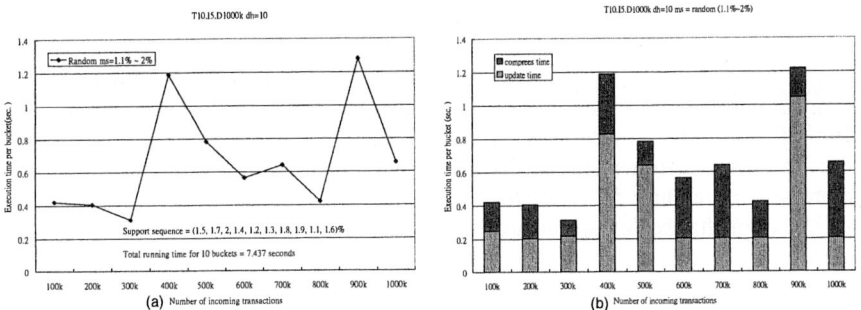

Fig. 2. (a) Mining the data stream with a support sequence of random thresholds (b) the breakdown of the processing time

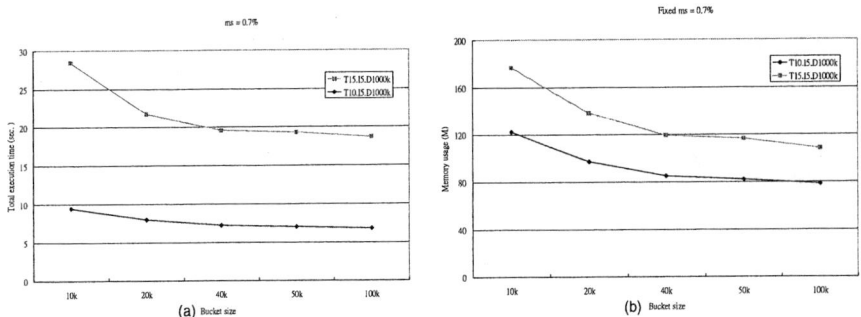

Fig. 3. (a) Effect on various bucket size (b) working memory size

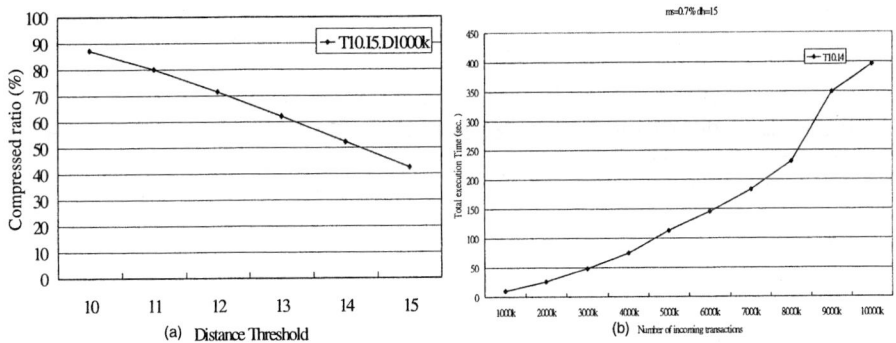

Fig. 4. (a) varying distance threshold (b) scalability evaluation. 1000k to 10000k

4 Experimental Results

We have conducted extensive experiments to evaluate the algorithm. The experiments were performed on an AMD Sempron 2400+ PC with 1GB memory, running the Windows XP, using data-sets generated from [1]. Due to space limit, we only report the results on dataset T10I5D1000k. The distance threshold is 10 and |B|=10.

Fig. 2(a) shows the performance of VSMDS algorithm with respect to a support sequence of random values ranging from 1.1% to 2%, the breakdown of execution time is shown in Fig. 2(b). The performance with respect to various bucket sizes is shown in Fig. 3(a), and the working memory sizes for the experiment are depicted in Fig. 3(b). Let the compression ratio be the size of the synopsis vector divided by that of the original transactions. Fig. 4(a) confirms that a distance threshold of 15 compresses more than 50% of the transactions in size. Fig. 4(b) indicates that VSMDS algorithm scales up linearly with respect to the dataset size (from 1000k to 10000k).

5 Conclusion

In this paper, we propose the VSMDS algorithm for mining frequent itemsets over a data stream with changeable support threshold. VSMDS utilizes the PFI-tree and the synopsis vector for the mining. The extensive experiments confirm that VSMDS efficiently mines frequent patterns with respect to variable supports, and has good linear scalability.

References

1. Agrawal, R. and Srikant, R.: Fast Algorithm for Mining Association Rules. In Proc. of the 20th International Conference on Very Large Databases (VLDB'94), pages 487-499, 1994.
2. Babcock, B., Babu, S., Datar, M., Motwani, R., and Widom, J.: Models and Issues in data stream systems. In Proc. of the 2002 ACM Symposium on Principles of Database Systems (PODS 2002), ACM Press, 2002.
3. Chi, Y. and Wang, H.: Moment: Maintaining Closed Frequent Itemsets over a Stream Sliding Window. In Proc. of the Fourth IEEE International Conference on Data Mining (ICDM'04), pages 59-66, Brighton, United Kingdom, 01-04 November 2004.
4. Giannella, C., Han, J., Pei, J., Yan, X., and Yu, P. S.: Mining Frequent Patterns in Data Streams at Multiple Time Granularities. In Proc. of the NSF Workshop on Next Generation Data Mining, 2002.
5. Han, J., Pei, J., and Yin, Y.: Mining Frequent Patterns without Candidate Generation. In Proc. of the 2000 ACM SIGMOD International Conference on Management of Data, Vol. 9, Issue 2, pages 1-12, 1999.
6. Koyuturk, M., Grama, A., and Ramakrishnan, N.: Compression, clustering and pattern discovery in very high dimensional discrete-attribute datasets. IEEE Transactions on Knowledge and Data Engineering, Vol. 17, no. 5, pages 447-461, 2005.
7. Li, H. F., Lee, S. Y., and Shan, M. K.: An Efficient Algorithm for Mining Frequent Itemsets over the Entire History of Data Streams. In Proc. of the First International Workshop on Knowledge Discovery in Data Streams, pages 20-24, Pisa, Italy, September 2004.
8. Lin, M. Y. and Lee, S. Y.: Interactive Sequence Discovery by Incremental Mining. Information Sciences: An International Journal, Vol. 165, Issue 3-4, pages 187-205, 2004.
9. Lin, M. Y. and Lee, S. Y.: A Fast Lexicographic Algorithm for Association Rule Mining in Web Applications. In Proc. of the ICDCS Workshop on Knowledge Discovery and Data Mining in the World-Wide Web, pages F7-F14, Taipei, Taiwan, R.O.C., 2000.
10. Manku, G. S., Motwani, R.: Approximate Frequency Counts over Data Streams. In Proc. of the 28th VLDB Conference, pages 346-357, Hong Kong, China, August 2002.

Hardware Enhanced Mining for Association Rules

Wei-Chuan Liu, Ken-Hao Liu, and Ming-Syan Chen

Department of Electrical Engineering,
National Taiwan University,
Taipei, Taiwan, ROC
mschen@cc.ee.ntu.edu.tw,
{kenliu, weichuan}@arbor.ee.ntu.edu.tw

Abstract. In this paper, we propose a hardware-enhanced mining framework to cope with many challenging data mining tasks in a data stream environment. In this framework, hardware enhancements are implemented in commercial Field Programmable Gate Array (FPGA) devices, which have been growing rapidly in terms of density and speed. By exploiting the parallelism in hardware, many data mining primitive subtasks can be executed with high throughput, thus increasing the performance of the overall data mining tasks. Simple operations like counting, which take a major portion of conventional mining execution time, can in fact be executed on the hardware enhancements very efficiently. Subtask modules that are used repetitively can also be replaced with the equivalent hardware enhancements. Specifically, we realize an Apriori-like algorithm with our proposed hardware-enhanced mining framework to mine frequent temporal patterns from data streams. The frequent counts of 1-itemsets and 2-itemsets are obtained after one pass of scanning the datasets with our hardware implementation. It is empirically shown that the hardware enhancements provide the scalability by mapping the high complexity operations such as subset itemsets counting to the hardware. Our approach achieve considerably higher throughput than traditional database architectures with pure software implementation. With fast increase in applications of mobile devices where power consumption is a concern and complicated software executions are prohibited, it is envisioned that hardware enhanced mining is an important direction to explore.

Keywords: Hardware enhanced mining, association rules.

1 Introduction

In several emerging applications, data is in the form of continuous *data streams*, as opposed to finite stored databases. Examples include stock tickers, network traffic measurements, web logs, click streams, data captured from sensor networks and call records. Specifically, a data stream is a massive unbounded sequence of data elements continuously generated at a rapid rate. It is recognized that the data stream processing has to satisfy the following requirements. First, each data point should be examined at most once when analyzing the data stream. Second, the storage cost of related data structures should be bounded. Third, newly generated data points should be processed as fast as possible to accomplish real-time computing, i.e., the processing rate should be at least the same as the data arrival rate. Finally, the up-to-date analysis results of a data stream should be instantly available when requested.

Note that traditional database architectures that focus solely on I/O optimization are not designed to utilize the continued evolution of hardware infrastructure resources, especially those on mobile devices, efficiently to meet the demand for high-speed data stream processing. Due to the dynamic and time-sensitive nature of most data stream applications, data stream processors need to be capable of handling huge amount of data in a limited length of time window with bounded memory space. To achieve this goal, we need to exploit the characteristics of modern hardware technologies to design efficient hardware framework to maximize the performance of data mining algorithms. In this paper, we propose a novel paradigm that comprises a hardware-enhanced framework, which exploits the massive parallelism in custom hardware to solve many high complexity problems in data mining tasks and to further increase the throughput and decrease the response time of the existing data mining systems. With fast increase in applications of mobile devices where power consumption is a concern and complicated software executions are prohibited, it is envisioned that hardware enhanced mining is an important direction to explore.

The novelty of our hardware-enhanced approach is that we transform the item transactions in a data stream into a matrix structure and efficiently map operations for discovering frequent itemsets to highly efficient hardware processing units. The matrix structure and the corresponding operations are optimally implemented as a hardware enhancement to the existing database architectures. Our approach finds the balance of the hardware and software design to solve the high complexity issues such as the level-2 itemset counting to enable high performance data stream processing systems that are not attainable with traditional architectures. Specifically, we realize Apriori-like algorithm within our proposed hardware-enhanced mining framework to mine frequent temporal patterns from data streams. Even with the quadratic increase of the size of 2-itemsets, the counts of frequent 1-itemsets and 2-itemsets are obtained after one pass of the datasets through our hardware implementation. The throughput obtained with our proposed hardware enhanced framework is two orders of magnitude larger than that attainable by reference software implementation. It is empirically shown that the hardware enhancements provide the necessary scalability to many high complexity operations such as subset itemsets counting and achieve considerably higher throughput than traditional database architectures with pure software implementation.

Many sequential algorithms to discover association rules are studies extensively [1][3][4][6]. Parallel and distributed schemes based on the sequential Apriori algorithm can be found in [2][5][10]. However, they did not focus on the scalability issues of the high complexity operations. To deal with the bottleneck of the Apriori-like algorithms, i.e., finding all frequent 2-itemsets of transaction, to mine for frequent itemsets in data streams, FTP-DS algorithm [13] utilized the delayed pattern recognition approach to address the time and the space constraints in a data stream environment. In [8][9], even though approximation approaches are employed, it still needs excessive time to scan all 2-itemsets of transaction.

We mention in passing that active storage which takes advantage of processing power on individual disk drives to run application-level code is proposed in [12]. As the number of hard disk drives increases, I/O-bounded scans are benefited by the partition of the data among the large number of disks and the reduction in bandwidth by filtering. However, [12] relies on storage parallelism, i.e. the number of physical hard disks, which does not scale up with the vast amount of data. The reduction of I/O traffic by filtering will also affect the accuracies of the data mining tasks. A commercial FPGA coprocessor board is used to accelerate the processing of queries on a relational database that contains texts and images in [7]. This approach is not directly applicable to the data mining tasks. [11] builds a model to parameterize the communication overhead between processor and programmable logic interface and logic delays in the programmable logic device to evaluate the speedup of the addition of programmable logic to RISC machine. To our knowledge, there was no prior work either designing hardware stream processor or balancing task partitions among hardware and software, let alone conducting the corresponding performance analysis. This feature distinguished our work from others.

The rest of the paper is organized as follows. The preliminaries of discovering frequent patterns over data streams are explored in Section 2. Hardware enhanced framework is described in Section 3. Performance analysis to evaluate the advantages of exploiting the application specific hardware for data mining tasks is conducted in Section 4. Empirical studies are showed in Section 5. This paper concludes with Section 6.

2 Preliminaries

By following the concept of general support framework [13], we briefly describe the determination of frequent temporal patterns as follows. A typical market-basket application is used here for illustrative purposes. The transaction flow in such an application is shown in Figure 1 where items a to h stand for items purchased by customers. For example, the third customer bought item c during time t=[0, 1), items c, e and g during t=[2, 3), and item g during t=[4, 5). With the sliding window model, the support of a temporal pattern is defined as follows.

Definition 1. *The support or the occurrence frequency of a temporal pattern X at a specific time t is denoted by the ratio of the number of customers having pattern X in the current time window to the total number of customers.*

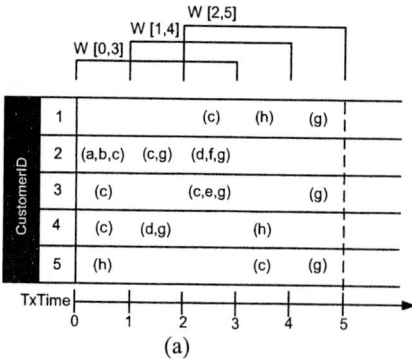

Fig. 1. (a) An example of online transaction flows. (b) The support values of the inter-transaction itemset {c,g}.

For example, given the window size N=3, three sliding windows, i.e. w[0,3], w[1,4], and w[2,5], are shown in Figure 1(a) for the transaction flows. According to above definition, supports of the inter-transaction itemset {c,g} from TxTime t=1 to t=5 are obtained as in Figure 1(b).

3 Hardware Enhanced Data Stream Processing

Because of the limited amount of instruction level parallelism (ILP) present in most of the data mining tasks [4][14], high speed data streams cannot be processed in time either by the multi-process or parallel systems to match their arrival rates. Many emerging data mining environments, such as data streams, sensor networks, and etc., demand higher throughputs and shorter response time than those attainable by traditional data mining infrastructures.

Modern VLSI technology makes it possible to pack millions of transistors in a single chip. Commercial FPGA devices provide millions of gates and also hundreds of thousands of logic elements integrated with large memory and high speed I/O interfaces. The hardware building blocks can be exploited for data mining tasks. Mining algorithms partitionable into independent subtasks can be executed in the hardware in a parallel fashion. Simple and frequently used routines are implemented in hardware redundantly to process incoming data simultaneously. Special purpose circuits can be implemented on field programmable gate array (FPGA) devices and interfaced to the host data mining system as an array processors. Similar architectures are used in the design of processors for digital signal processing applications which are characterized by intensive computations and real-time requirements. Similar coprocessors or accelerators for multimedia and networking applications have already been widely used in computing nowadays.

To achieve the throughput required in today's high speed data streams, high complexity operations in data mining tasks have to be executed within a relatively short period of time. The time required by most of the high complexity

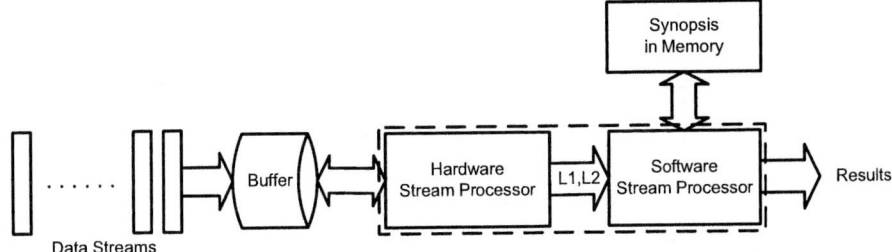

Fig. 2. Computation model with hardware enhancement for data streams

operations, such as 2-itemset enumeration and counting in the discovery of frequent patterns, becomes impractical as the size of the data and the data arrival rate increase. From our performance model described later in section 4, we explore a novel direction, a hardware enhanced framework, which is to exploit the massive number of parallel processing elements dedicated as an infrastructure for data mining tasks.

3.1 Stream Processor

The computation model of our hardware enhanced mining framwork for data streams is shown in Figure 2. There are various ways to partitions a data mining task into hardware and software components depending on the nature of the task. For the problem of finding frequent temporal patterns in data streams, since the computation of L1- and L2-itemsets is the most time-consuming task in our algorithm, we can offload this operation to the hardware to enhance performance. Subsequent rule generations can be processed in software implementations for flexibility.

Let a transaction $I = \{i_1, i_2, \ldots, i_N\}$ be a set of items, where N is the number of items and each item belongs to $\{0,1\}$. Each item stands for an event according to its position in a transaction. The first item indicates the event A and the second item stands for the event B, and so on. We use a bit to represent the occurrence of the event, i.e., the event occurred if the bit is set to one. Note that the number of items is pre-defined as part of the system specification. For example, in Figure 3, each bit arrives in an interval of one time unit. The third customer bought items $\{0,0,1,0,1,0,1,0\}$ in order during time t=[16, 24), where N is 8. Three bits are set to one to represent the occurrence of event C, E, and G, respectively.

Figure 4 shows the architecture of hardware stream processor. As the input to the stream processor, we have C customers and N distinct items that may appear in a transaction. There are four function blocks in this processor, namely, a serial/parallel converter, a sliding window buffer, a 2-itemset generator, and a frequent decision maker. N items are grouped as a transaction in a parallel form by the serial/parallel converter. The sliding window buffers the inputs of the most recent N time units. The 2-itemset generator enumerates all the combinations of 2-itemsets. Each frequent decision maker determines whether its

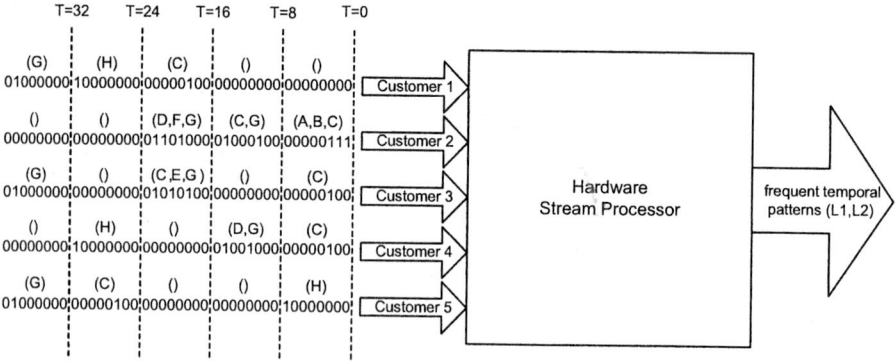

Fig. 3. The input and output of the hardware stream processor

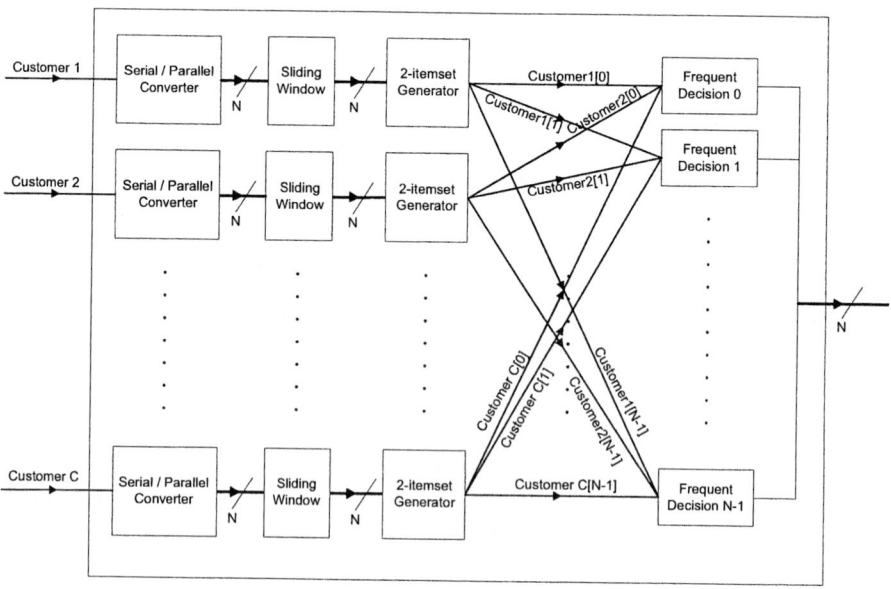

Fig. 4. The architecture of hardware stream processor

corresponding 2-itemset is frequent in the current sliding window. Here we use parallel adders and comparators to make the frequent decisions in real time. According to Definition 1 in previous section, an itemset is frequent if the number of occurrences in all customers exceeds in the number of user-specified threshold.

4 Performance Model

The characteristics of run-time behaviors are very different in hardware and software. The pipeline of hardware is achieved in register level while there is only limited instruction level pipelining in software implementations on traditional CPU-based framework. Only one task can be executed in any moment, such as calculating, scanning, and sorting. Specifically, most CPU has only one ALU to execute addition, shifting, comparison, and so on. Our specialized hardware design can process all tasks simultaneously, including 2-itemset enumeration, occurrence counting, etc. through the massive array of simple components. The basic characteristics of functions suitable for hardware enhancements are that they take up a significant portion of overall execution time, execute in a first-in-first-out manner with minimal state memory, and exhibit simple and regular structure. In this section we develop a simple model for the performance of the hardware enhancements to illustrate the limitation of traditional framework and the benefits of the proposed hardware enhanced framework.

Suppose that the operation i has N units of work. Each operation takes w_{cpu} clock cycles to complete in traditional architecture and w_{fpga} clock cycles to complete in the hardware enhancement. The CPU clock rate is f_{cpu} and the hardware enhancement clock rate is f_{fpga}. The hardware enhancement has L parallel units of processing elements. To keep the model simple, we assume that the overhead in communication for each unit of work takes a constant w_{comm} cycles. In traditional architecture, the execution time is

$$t_{op} = \frac{N * w_{cpu}}{f_{cpu}}$$

and the throughput is

$$throughput_{op} = \frac{N}{t_{op}} = \frac{f_{cpu}}{w_{cpu}}.$$

The throughput is limited by clock rate, f_{cpu}. Latest CPU operates at several GHz, beyond which the clock rate are not scalable. The execution clock cycle needed per operation, w_{cpu}, is constant for a given algorithm.

Now consider our hardware enhanced framework. For hardware enhancements, the execution time is

$$t'_{op,enhanced} = \frac{N * w_{fpga}}{L * f_{fpga}} + \frac{N * w_{comm}}{f_{fpga}}$$

and the throughput is

$$throughput'_{op,enhanced} = \frac{N}{t'_{op,enhanced}} = \frac{L * f_{fpga}}{w_{fpga} + L * w_{comm}}.$$

The throughput can be increased by increasing the number of parallel processing elements L, decreasing the clock cycles needed per operation w_{fpga}, or minimizing the communication delay w_{comm}. The density of processing elements packed into commercial FPGA devices is growing almost exponentially and thus provides tremendous room for optimization of the throughput.

Example 1: Consider the algorithm in [13] for the discovery of frequent patterns over data streams. The throughput of the algorithm is defined as the number of transactions that are processed every unit of time interval. Suppose that N is the average number of items per transaction and C is the number of customer. The amount of transaction items that our stream process can process in one unit of time is N. Note that the throughput is independent of the number of customers because the proposed hardware infrastructure can deal with all customer streams in a parallel fashion. The maximal throughput of hardware enhancement scales linearly with N. The bottleneck for the software implementation of Apriori algorithm is identified as the phase during which $N \cdot C$ comparisons are required to check if any of the $C_2^{|L1|}$ candidates is frequent, where $|L1|$ is the size of the large 1-itemset. Therefore, the maximal throughput of a reference software implementation scales with $\frac{1}{N \cdot C \cdot |L1|^2}$. For typical values of N, we observe that the throughput in our proposed hardware enhanced framework is many orders of magnitudes higher that that attainable with software implementation used in traditional database architectures.

5 Experiments

The hardware is implemented and verified with Altera's design software QuartusII and executes on the Altera's Stratix device. Software implementation of the algorithm is also executed on the same device, with a NiosII 50MHz CPU and 16MB of SDRAM. Transaction data sets are synthesized in a similar way to those in [3].

5.1 Performance and Scalability

Our experiments are conducted with synthetic data sets. In order to show the scalability of the proposed hardware enhanced framework, we measure the number of clock cycles needed to obtain the frequent patterns. The results are shown

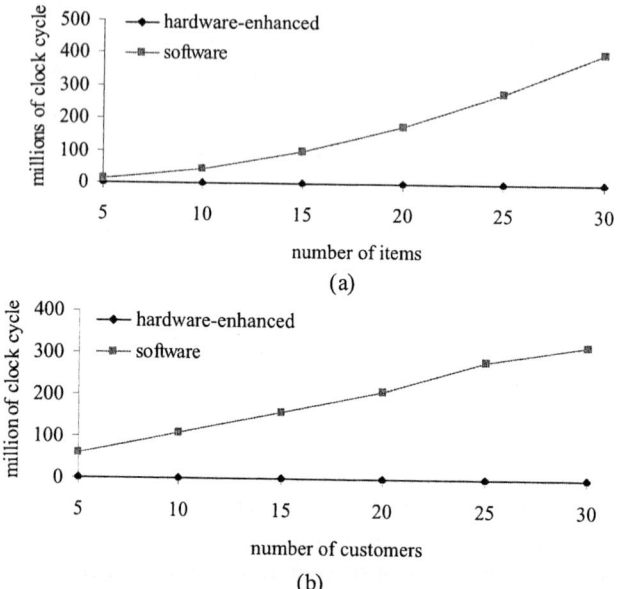

Fig. 5. The number of clock cycles needed for different number of (a) customers and (b) items

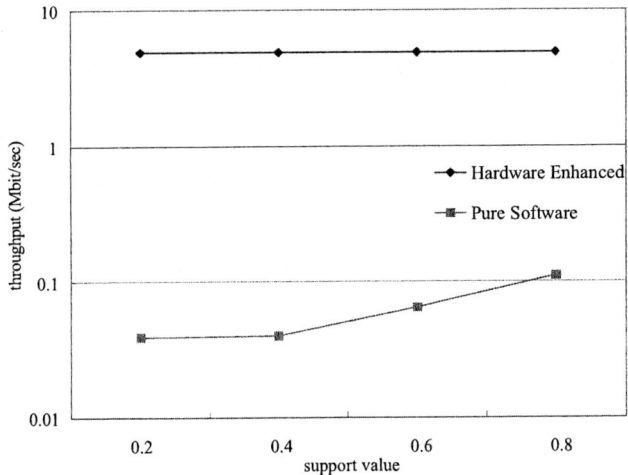

Fig. 6. Scalability with different support values

in Figure 5(a) and 5(b). The hardware enhanced stream processor offers throughput that is two order of magnitudes larger than its software couterpart does. We obtain similar results when we scale the support values as shown in Figure 6. The results are consistent with our previous analysis. The throughput of the hardware enhancement remains at constant level with different parameters, such as number of items, support value, density of data while the software couterpart scales poorly. Our hardware design scales linearly with both the number of items and the number of customers, i.e., data streams. The througput of the hardware enhanced data stream processing system remains constant while the throughput of the reference software implementation reduces exponentially as the number of items or customers increases.

6 Conclusion

The feasibility of our paradigm is shown by the implementation of hardware enhancements in commercial FPGA devices. The hardware enhanced mining framework is a promising new approach to boost the performance of many data mining algorithms and cope with many of their inherent high complexity issues. Specifically, our approach finds the balance of the hardware and software design to solve the level-2 itemset counting in Apriori algorithms. We also point out many applications that will benefit from the new paradigm. This promising problem we have addressed here is an unexplored territory in the field of data mining research. This paper is among the very first to explore this new direction.

Acknowledgements

The work was supported in part by the National Science Council of Taiwan, R.O.C., under Contracts NSC93-2752-E-002-006-PAE.

References

1. R. Agrawal, H. Mannila, R. Srikant, H. Toivonen, and A. I. Verkamo. Fast discovery of association rules. In *Advances in Knowledge Discovery and Data Mining*, pages 307–328. AAAI Press, 1996.
2. Rakesh Agrawal and John C. Shafer. Parallel mining of association rules. *IEEE Trans. On Knowledge And Data Engineering*, 8(6):962–969, 1996.
3. Rakesh Agrawal and Ramakrishnan Srikant. Fast algorithms for mining association rules. In *Proc. 20th Int. Conf. Very Large Data Bases*, pages 487–499, 1994.
4. Ming-Syan Chen, Jiawei Han, and Philip S. Yu. Data mining: an overview from a database perspective. *IEEE Trans. On Knowledge And Data Engineering*, 8:866–883, 1996.
5. Eui-Hong Han, George Karypis, and Vipin Kumar. Scalable parallel data mining for association rules. In *ACM SIGMOD Conf. on Management of Data*, pages 277–288, 1997.
6. Jiawei Han and Micheline Kamber. *Data Mining: Concepts and Techniques*. Morgan Kaufmann, 2000.
7. Jack S.N. Jean, Guozhu Dong, Hwa Zhang, Xinzhong Guo, and Baifeng Zhang. Query processing with an fpga coprocessor board. In *Proc. 1st Int. Conf. Engineering of Reconfigurable Systems and Algorithms*, 2001.
8. Ruoming Jin and Gagan Agrawal. An algorithm for in-core frequent itemset mining on streaming data. In *Proc. 5th IEEE Int. Conf. Data Mining*, 2005.
9. Richard M. Karp and Scott Shenker. A simple algorithm for finding frequent elements in streams and bags. In *ACM Trans. on Database Systems*, 2003.
10. Jong Soo Park, Ming-Syan Chen, and Philip S. Yu. Efficient parallel data mining for association rules. In *Proc. 4th Int. Conf. Information and Knowledge Management*, pages 31–36. ACM Press, 1995.
11. S. Rajamani and P. Viswanath. A quantitative analysis of processor - programmable logic interface. In *IEEE Symposium on FPGAs for Custom Computing Machines*, pages 226–234, 1996.
12. Erik Riedel, Christos Faloutsos, Garth A. Gibson, and David Nagle. Active disks for large-scale data processing. *IEEE Computer*, 34:68–74, 2001.
13. Wei-Guang Teng, Ming-Syan Chen, and Philips S. Yu. A regression-based temporal pattern mining scheme for data streams. In *Proc. 29th Int. Conf. Very Large Data Bases*, 2003.
14. Mohammed J. Zaki. Parallel and distributed association mining: A survey. *IEEE Concurrency*, 7(4):14–25, 1999.

A Single Index Approach for Time-Series Subsequence Matching That Supports Moving Average Transform of Arbitrary Order

Yang-Sae Moon and Jinho Kim

Department of Computer Science, Kangwon National University,
192-1 Hyoja Dong 2, Chunchon, Kangwon, Korea
{ysmoon, jhkim}@kangwon.ac.kr

Abstract. Moving average transform is known to reduce the effect of noise and has been used in many areas such as econometrics. Previous subsequence matching methods with moving average transform, however, would incur index overhead both in storage space and in update maintenance since the methods should build multiple indexes for supporting arbitrary orders. To solve this problem, we propose a single index approach for subsequence matching that supports moving average transform of arbitrary order. For a single index approach, we first provide the notion of *poly-order moving average transform* by generalizing the original definition of moving average transform. We then formally prove correctness of the poly-order transform-based subsequence matching. By using the poly-order transform, we also propose two different subsequence matching methods that support moving average transform of arbitrary order. Experimental results for real stock data show that our methods improve average performance significantly, by 22.4 ~ 33.8 times, over the sequential scan.

1 Introduction

Time-series data are the sequences of real numbers representing values at specific points in time. Typical examples of time-series data include stock prices, exchange rates, biomedical measurements, and financial data [1, 4, 11]. The time-series data stored in a database are called *data sequences*, and those given by users are called *query sequences*. And, finding data sequences similar to the given query sequence from the database is called *similar sequence matching* [1, 4, 8]. In many similar sequence matching models, two sequences $X = \{X[1], X[2], ..., X[n]\}$ and $Y = \{Y[1], Y[2], ..., Y[n]\}$ of the same length n are said to be *similar* if the distance $D(X, Y)$ is less than or equal to the user-specified *tolerance* ϵ [1, 4]. In this paper, we use the Euclidean distance, which has been widely used in [1, 4, 6, 7, 8, 9], as the distance function $D(X, Y)$, and define that X and Y are in ϵ-*match* if $D(X, Y)$ is less than or equal to ϵ.

In this paper we focus on the subsequence matching that supports moving average transform. Here, the *subsequence matching* [4, 8] is the problem of finding

Table 1. Summary of notation

Symbols	Definitions
$S[i:j]$	Subsequence of S, including entries from the i-th one to the j-th
$S^{(k)}$	k-moving average transformed sequence of S ($S^{(k)}[i] = \frac{1}{k}\sum_{j=i}^{i+k-1} S[j]$)
$S^{(k)}[i:j]$	Subsequence of $S^{(k)}$, including entries from the i-th one to the j-th
s_i	The i-th disjoint window of S ($= S[(i-1)*\omega+1 : i*\omega], i \geq 1$)
$s_i^{(k)}$	The i-th disjoint window of $S^{(k)}$ ($= S^{(k)}[(i-1)*\omega+1 : i*\omega], i \geq 1$)

subsequences, contained in data sequences, similar to a query sequence of arbitrary length. And, moving average transform [3, 10], which has been widely used in econometrics [3], converts a given sequence into a new sequence consisting of the averages of k consecutive values in the sequence, where k is called the *moving average order* or simply the *order* [6]. The moving average transform is very useful for finding the trend of the time-series data by reducing the effect of noise, and has been used in various applications [3]. Since the users want to control the degree of the noise reduction depending on the characteristics of data sequences to be analyzed [5], efficient support of arbitrary orders is also necessary. Table 1 summarizes the notation to be used throughout the paper.

In this paper we focus on the subsequence matching model that supports moving average transform of arbitrary order [6]. That is, the focused model uses the distance between two k-moving average transformed sequences $Q^{(k)}$ and $S^{(k)}[i:j]$, instead of the distance between two original sequences Q and $S[i:j]$, to determine whether the two sequences are in ϵ-match or not. We call this similarity model as *moving average transformed subsequence matching*. Previous research results [4, 6, 8], however, do not provide an efficient solution because of incurring index overhead.

In this paper we propose a single index approach for the moving average transformed subsequence matching. To explain our approach, we first provide the notion of *poly-order moving average transform* (or simply *poly-order transform*) by generalizing the original definition of moving average transform. The poly-order transform is different from the original moving average transform in a manner of using moving average orders. While the original transform uses only one specific order, the poly-order transform uses a set of moving average orders. That is, while the original transform makes a transformed sequence from an original sequence, the poly-order transform generates a set of transformed sequences from an original sequence. In this paper we show that, if constructing a single index using the poly-order transform and using the index, we are able to perform the moving average transformed subsequence matching correctly.

By applying the poly-order transform to both Faloutsos et al.'s method [4] (called *FRM* for convenience) and DualMatch [8], we propose two different moving average transformed subsequence matching methods. The first one is an FRM-based algorithm, which divides data sequences into sliding windows and a query sequence into disjoint windows. The second one is a DualMatch-based algorithm, which divides data sequences into disjoint windows and a query sequence into sliding windows. Experimental results show that two proposed matching methods improve performance significantly over the sequential scan regardless of selectivity ranges and query lengths.

The rest of this paper is organized as follows. Section 2 describes related work. Section 3 presents the concept of poly-order transform and explains the proposed matching algorithms. Section 4 presents the results of performance evaluation. Section 5 concludes the paper.

2 Related Work

We first review Agrawal et al.'s whole matching solution [1]. The whole matching solution consists of index building and similar sequence matching algorithms. In the index building algorithm, each data sequences of length n is transformed into f-dimensional points ($f \ll n$), and the transformed points are stored into an R^*-tree [2]. In the similar sequence matching algorithm, a query sequence is similarly transformed to an f-dimensional point, and a range query is constructed using the point and the tolerance ϵ. Then, by evaluating the range query using the index, the *candidates* are identified. This method guarantees there be no *false dismissal*, but may cause *false alarms* because it uses only f features instead of n. Thus, it performs the *post-processing step* that eliminates false alarms by accessing the actual data sequences from the disk [1].

Faloutsos et al. have proposed a subsequence matching solution (FRM) as a generalization of the whole matching[4]. FRM uses the window construction method of dividing data sequences into sliding windows and a query sequence into disjoint windows. In the index building algorithm, FRM divides data sequences into sliding windows and transforms each window to an f-dimensional point. However, dividing data sequences into sliding windows causes a serious problem of generating too many points to be stored into the index [4, 8]. To solve this problem, FRM does not store individual points directly into the R^*-tree, but stores only MBRs (minimum bounding rectangles) that contains hundreds or thousands of the f-dimensional points. In the subsequence matching algorithm, FRM performs subsequence matching based on the following Lemma 1 [4].

Lemma 1. *If two sequences S and Q are in ϵ-match, then at least one of the disjoint window pairs (s_i, q_i) are in $\epsilon/\sqrt{p}$ ($p = \lfloor Len(Q)/\omega \rfloor$). That is, the following Eq. (1) holds:*

$$D(S,Q) \leq \epsilon \implies \bigvee_{i=1}^{p} D(s_i, q_i) \leq \epsilon/\sqrt{p} \qquad (1)$$

According to Lemma 1, FRM divides a query sequence into disjoint windows, transforms each window to an f-dimensional point, makes a range query using the point and the tolerance ϵ, and constructs a candidate set by searching the R^*-tree. Finally, it performs the post-processing step to eliminate false alarms.

DualMatch [8] and GeneralMatch [9] have improved performance significantly in subsequence matching by using different window construction methods from FRM. In constructing windows, DualMatch is a dual approach of FRM, and GeneralMatch is a generalized approach of FRM and DualMatch. Except difference in window construction mechanism, index building and subsequence matching algorithms of DualMatch and GeneralMatch are similar to those of FRM.

Loh and Kim [6] have first proposed a subsequence matching method that supports moving average transform of arbitrary order. In the index building algorithm, the method builds an *m-index* by performing m-moving average transform on data sequences, by dividing the transformed sequences into windows, and by mapping the windows into lower-dimensional points. In the subsequence matching algorithm, given the order k that may or may not be equal to m, the method uses the m-index to perform k-order moving average transform. However, the method has a serious drawback that it is necessary to modify existing algorithms and node structures used in the R*-tree. Also, Loh and Kim have proposed the *index interpolation* [7] that constructs multiple m-indexes for arbitrary orders. However, this index interpolation causes another critical drawback that, as the number of m-indexes increases, much more space would be required for the indexes, and index maintenance overhead would be increased to maintain multiple indexes.

3 The Proposed Single Index Approach

3.1 The Concept

The motivation of the research is on how we can use Lemma 1, which has been used for a theoretical basis in many subsequence matching methods. If using Lemma 1, we can perform subsequence matching efficiently since we can reduce the index search range from ϵ to $\epsilon/\sqrt{p}$. To do this, we first derive the following Lemma 2 by applying k-moving average transform to Lemma 1.

Lemma 2. *If two k-order moving average transformed sequences $S^{(k)}$ and $Q^{(k)}$ are in ϵ-match, then at least one of the pairs $(s_i^{(k)}, q_i^{(k)})$ are in $\epsilon/\sqrt{p}\,(p = \lfloor Len(Q^{(k)})/\omega \rfloor)$. That is, the following Eq. (2) holds:*

$$D(S^{(k)}, Q^{(k)}) \leq \epsilon \implies \bigvee_{i=1}^{p} D(s_i^{(k)}, q_i^{(k)}) \leq \epsilon/\sqrt{p} \qquad (2)$$

PROOF: We omit the proof since it can be easily done using Lemma 1. □

To use Lemma 2 in moving average transformed subsequence matching without any modification, however, we have to build a lot of indexes since we require each index for every possible order k. To solve this problem, we propose an efficient approach that uses only one index rather than multiple indexes.

To support moving average transform of arbitrary order in FRM and DualMatch without incurring the problem of multiple indexes, we generalize the definition of moving average transform as the following Definition 1.

Definition 1. *Given a window $S[a:b]$ contained in a sequence S and a set $\mathbb{K}$ of orders, $k_1, k_2, \cdots, k_m$, the poly-order moving average transformed window set, $S^{(\mathbb{K})}[a:b]$, of $S[a:b]$ on $\mathbb{K}$ is defined as follows:*

$$S^{(\mathbb{K})}[a:b] = \{S^{(k_i)}[a:b] \mid k_i \in \mathbb{K},\ 1 \leq i \leq m\} \qquad (3)$$

To represent an area of containing multiple windows, we now rewrite the definition of MBR using a set of windows as follows.

Definition 2. *Given a set $\mathbb{W}$ of windows, $W_1, W_2, \cdots, W_m$, of the same size ω, an MBR of the set $\mathbb{W}$, $\mathbb{MBR}(\mathbb{W})$, is defined as an ω-dimensional MBR that contains every ω-dimensional point W_i in $\mathbb{W}$.*

According to Definitions 1 and 2, the poly-order transformed window set of s_i on $\mathbb{K}$ is denoted by $s_i^{(\mathbb{K})}$, and the MBR of containing all windows in $s_i^{(\mathbb{K})}$ is denoted by $\mathbb{MBR}(s_i^{(\mathbb{K})})$.

If using the poly-order transform, we can perform the moving average transformed subsequence matching correctly, i.e., we do not incur any false dismissal. To explain the correctness, we present Lemma 3 that represents the relationship between k-order transform and the poly-order transform on $\mathbb{K}$ containing k.

Lemma 3. *When $k \in \mathbb{K}$, if $q_i^{(k)}$ is in ϵ-match with $s_i^{(k)}$, then $q_i^{(k)}$ is also in ϵ-match with $\mathbb{MBR}(s_i^{(\mathbb{K})})$. That is, the following Eq. (4) holds:*

$$D(q_i^{(k)}, s_i^{(k)}) \leq \epsilon \implies D(q_i^{(k)}, \mathbb{MBR}(s_i^{(\mathbb{K})})) \leq \epsilon \qquad (4)$$

PROOF: We omit the proof due to space limitation. □

Based on Lemmas 2 and 3, we now derive Theorem 1, which provides a theoretical basis of the algorithms to be proposed.

Theorem 1. *When $k \in \mathbb{K}$, if $Q^{(k)}$ is in ϵ-match with $S^{(k)}[a:b]$, then at least one $q_i^{(k)}$ is in $\epsilon/\sqrt{p}$-match with $\mathbb{MBR}(S^{(\mathbb{K})}[a+(i-1) \cdot \omega : a+i \cdot \omega - 1])$. That is, the following Eq. (5) holds:*

$$D(Q_i^{(k)}, S^{(k)}[a:b]) \leq \epsilon$$
$$\implies \bigvee_{i=1}^{p} D(q_i^{(k)}, \mathbb{MBR}(S^{(\mathbb{K})}[a+(i-1) \cdot \omega : a+i \cdot \omega - 1])) \leq \epsilon/\sqrt{p}, \qquad (5)$$

where $p = \lfloor Len(Q^{(k)})/\omega \rfloor$, and $Len(S^{(k)}[a:b]) = Len(Q^{(k)})$.

PROOF: We can prove the theorem using Lemmas 2 and 3. We omit the detailed proof due to space limitation. □

Theorem 1 guarantees that the candidate set consisting of the subsequences $S^{(k)}[a:b]$ such that $q_i^{(k)}$ and $\mathbb{MBR}(S^{(\mathbb{K})}[a+(i-1) \cdot \omega : a+i \cdot \omega - 1])$ are in $\epsilon/\sqrt{p}$-match (i.e., satisfying the necessary condition of Eq. (5)) contains no false dismissal.

To use Theorem 1 for the FRM-based (or DualMatch-based) moving average transformed subsequence matching method, we need to construct a set of windows for each window of data sequences. That is, the methods to be proposed first construct an MBR that contains multiple poly-order transformed windows. The methods then transform the MBR to a lower-dimensional MBR, and finally build an index by storing the MBR.

3.2 FRM-MAT: FRM with Moving Average Transform

In this subsection we explain FRM-MAT, the moving average transformed subsequence matching method that is derived from FRM [4] by using the poly-order moving average transform.

Figure 1 shows the index building algorithm of FRM-MAT. In Step (1), we divides a data sequence S into sliding windows of length ω. In Steps (2) ∼ (6), for each sliding window, we construct an MBR and store the MBR into the multidimensional index. First, in Step (3), we make a set of transformed windows from a sliding window by performing the poly-order transform on a given set of orders. Next, in Step (4), we construct an f-dimensional MBR by using the lower-dimensional transformation on the set of windows. Last, in Step (5), we store the MBR into the multidimensional index with the starting offset of the corresponding sliding window.

Procedure *FRM-MAT-BuildIndex*(Data Sequence S, Window size ω, Set of orders $\mathbb{K}$)
(1) Divide S into sliding windows of length ω;
(2) **for** each sliding window $S[a:b]$ **do**
(3) Make a set of windows $S^{(\mathbb{K})}[a:b]$ by using the *poly-order moving average transform* on $\mathbb{K}$;
(4) Construct an *f*-dimensional MBR *f-D MBR* by using the lower-dimensional transformation on $S^{(\mathbb{K})}[a:b]$;
(5) Make a record < *f-D MBR, offset=a>*, and store it into the index;
(6) **endfor**

Fig. 1. The index building algorithm of FRM-MAT

Like FRM, however, FRM-MAT has a problem of generating a lot of MBRs to be stored in the index since it divides data sequences into sliding windows. To solve this problem, FRM has constructed an MBR that contains multiple points corresponding to multiple sliding windows [4]. Thus, in FRM-MAT, we also construct an MBR that contains multiple MBRs corresponding to multiple sliding windows. That is, in the index building algorithm, we construct an MBR that represents multiple consecutive sliding windows and store the MBR with the starting offsets of the first and the last windows. For easy explanation and understanding, however, we describe the algorithm in Figure 1 as that FRM-MAT stores an individual MBR for each sliding window directly.

Next, Figure 2 shows the subsequence matching algorithm of FRM-MAT. In Steps (1) and (2), for a give query sequence Q, we obtain p disjoint windows $q_i^{(k)}$ from the k-order moving average transformed sequence $Q^{(k)}$. In Steps (3) ∼ (8), for each disjoint window $q_i^{(k)}$, we find candidate subsequences by searching the index using the window and the given tolerance ϵ. First, in Step (4), we transform the corresponding window to an f-dimensional point using lower-dimensional transformation. Second, in Step (5), we construct a range query using the point and $\epsilon/\sqrt{p}$. Third, in Step (6), we search the multidimensional index using the range query and find the MBRs that are in $\epsilon/\sqrt{p}$-match with the point. Last, in Step (7), we obtain candidate subsequences using *offset*, which is stored in the record with the MBR as the starting position of the sliding window.

Procedure *FRM-MAT-SubsequenceMatching* (Query Sequence Q, Tolerance ε, Window size ω, Order k)
(1)   Make $Q^{(k)}$ from Q by using k-order moving average transform;
(2)   Divide $Q^{(k)}$ into disjoint windows $q_i^{(k)}$ ($1 \le i \le p$, $p = \lfloor Len(Q^{(k)})/\omega \rfloor$) of length ω;
(3)   **for** each window $q_i^{(k)}$ **do**
(4)   Transform the window to an f-dimensional point by using the lower-dimensional transformation;
(5)   Construct a range query using the point and $\varepsilon/\sqrt{p}$;
(6)   Search the index and find the records of the form <f-D MBR, offset>;
(7)   Include in the candidate set the subsequences $S[offset-(i-1)\cdot\omega : offset-(i-1)\cdot\omega + Len(Q^{(k)})-1]$;
(8)   **endfor**
(9)   Do the post-processing step;

Fig. 2. The subsequence matching algorithm of FRM-MAT

Finally, in Step (9), the post-processing step, we select only similar subsequences by discarding false alarms from the candidate set.

3.3 DM-MAT: DualMatch with Moving Average Transform

DM-MAT can also be derived from DualMatch [8] by using the poly-order transform. Since algorithms of DM-MAT are similar to those of FRM-MAT except difference in constructing windows, we omit the detailed algorithms of DM-MAT.

4 Performance Evaluation

4.1 Experimental Data and Environment

We have performed extensive experiments using two types of data sets. A data set consists of a long data sequence and has the same effect as the one consisting of multiple data sequences [4, 8, 9]. The first data set, a real stock data set used in FRM [4] and DualMatch [8], consists of 329,112 entries. We call this data set *STOCK-DATA*. The second data set, also used in FRM and DualMatch, contains random walk synthetic data consisting of one million enties: the first entry is set to 1.5, and subsequent entries are obtained by adding a random value in the range (-0.001,0.001) to the previous one. We call this data set *WALK-DATA*.

We have performed experiments on the following five matching methods:

- *SEQ-SCAN*: As a sequential scan solution, we find similar subsequences by fully scanning the entire database once.
- *FRM-MAT*: The FRM-based solution proposed in Section 3.2.
- *FRM-ORG*: As a simple solution obtained from FRM, we build each index for all orders and use algorithms of FRM with slight modification.
- *DM-MAT*: The DualMatch-based solution proposed in Section 3.3.
- *DM-ORG*: As a simple solution obtained from DualMatch, we build each index for all orders and use algorithms of DualMatch with slight modification.

The hardware platform for the experiment is a PC equipped with an Intel Pentium IV 2.80 GHz CPU, 512 MB RAM, and a 70.0GB hard disk. The operating system is GNU/Linux Version 2.6.6. We use the R*-tree [2] as a multidimensional index and extract six features[4, 8] from a window using Discrete Fourier Transform [4]. We use 256 as the minimum query length to be given, and accordingly, we set the window sizes for FRM-MAT and FRM-ORG to 256 [4], and those for DM-MAT and DM-ORG to 128 [8]. Next, we let $\mathbb{K} = \{2, 4, 8, 16, 32, 64, 128\}$. Therefore, we build only one index for FRM-MAT (or for DM-MAT) on $\mathbb{K}$, but seven indexes for FRM-ORG (or for DM-ORG) for each order in $\mathbb{K}$.

For the experimental results, we measure the elapsed time as the metric of efficiency and the storage space as the metric of overhead for the indexes. To avoid effects of noise, we experiment with 10 different query sequences of the same length and use the average as the result. We obtain the desired selectivity[4, 8] by controlling the tolerance ϵ.

4.2 Experimental Results

We conduct three different experiments: Experiment 1) measures the elapsed time by changing order k on different selectivities; Experiment 2) measures the elapsed time by changing order k on different query lengths; and Experiment 3) shows each index storage space required for the methods.

Experiment 1) The elapsed times on different selectivities

Figure 3 shows the experimental results for STOCK-DATA while changing order k on each selectivity of 0.0001, 0.001, and 0.01. Here, we use 512 as the query length. As shown in the figure, the proposed methods, both FRM-MAT and DM-MAT, reduce the elapsed time significantly over the sequential scan regardless of selectivity ranges. In summary, comparing with the sequential scan, FRM-MAT reduces the elapsed time by 22.4 times on the average, and DM-MAT by 33.8 times on the average. However, the elapsed times of FRM-MAT and DM-MAT are slighlty longer than those of FRM-ORG and DM-ORG respectively. It is because the sizes of MBRs stored in the index in FRM-MAT and DM-MAT, which build only one index for all orders, is relatively larger than those in FRM-ORG and DM-ORG, which build each index for all orders. And, in the figure, DM-MAT shows better performance than FRM-MAT. It is because DM-MAT like DualMatch can use the index-level filtering but FRM-MAT like FRM cannot [8]. Experimental results for WALK-DATA are very similar to those of STOCK-DATA. We omit the results due to space limitation.

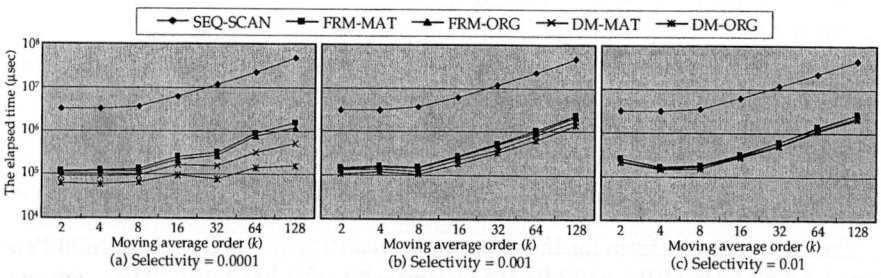

Fig. 3. The elapsed times for STOCK-DATA on different selectivities

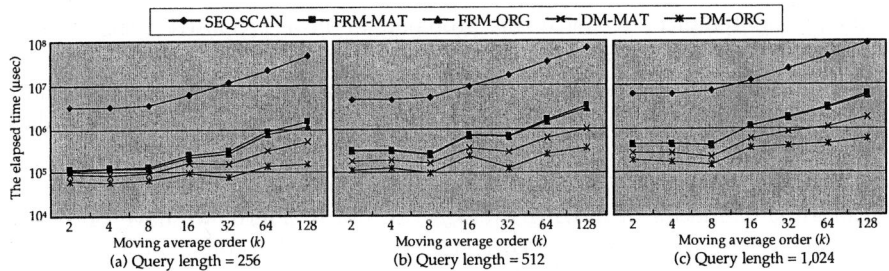

Fig. 4. The elapsed times for STOCK-DATA on different query lengths

Experiment 2) The elapsed times on different query lengths

Figure 4 shows the experimental results for STOCK-DATA while changing order k on each query length of 256, 512, and 1024. Here, we use 0.0001 as the selectivity. Figure 4 (a) is the case of using 256 as the query length, (b) for 512, and (c) for 1024. As shown in the figure, the proposed FRM-MAT and DM-MAT also reduce the elapsed time significantly over the sequential scan regardless query lengths. In summary of the results, FRM-MAT and DM-MAT reduce the average elapsed time by 20.2 ~ 42.6 times compared with the sequential scan. And, the results of WALK-DATA are also similar to those of STOCK-DATA.

Experiment 3) Storage space required for multidimensional indexes

Table 2 shows index storage spaces of five matching methods. The space for the sequential scan is 0 since it does not use any index. As shown in the table, FRM-ORG needs about seven times more storage space than FRM-MAT since it builds total seven indexes while FRM-MAT builds only one index. Similarly, DM-ORG needs about seven times more storage space than DM-MAT. Likewise, the number of indexes, i.e., index storage space, required for FRM-MAT (or DM-MAT) is only $1/|\mathbb{K}|$ of that for FRM-ORG (or DM-ORG). And accordingly, our methods can also reduce the index maintenance overhead, which are required to support insertion, deletion, and update of data sequences, by up to $1/|\mathbb{K}|$.

Table 2. Storage space comparison of the five matching methods for indexes

Data types	SEQ-SCAN	FRM-based approaches			DualMatch-based approaches		
		FRM-MAT	FRM-ORG	$\frac{\text{FRM-ORG}}{\text{FRM-MAT}}$	DM-MAT	DM-ORG	$\frac{\text{DM-ORG}}{\text{DM-MAT}}$
STOCK-DATA	0 KB	618 KB	1,526 KB	7.0	198 KB	1,434 KB	7.2
WALK-DATA	0 KB	618 KB	4,254 KB	6.9	562 KB	4,158 KB	7.4

5 Conclusions

Moving average transform is known to reduce the effect of noise and has been used in many areas such as econometrics since it is useful in finding overall trends. The previous researches on subsequence matching with moving average transform, however, would incur a critical overhead both in index space and in index maintenance. To solve this problem, we have proposed a single index approach for the moving average transformed subsequence matching.

The contribution of the paper can be summarized as follows. First, we have analyzed the problems that occur when we apply the previous matching algorithms to the moving average transformed subsequence matching. Second, we have formally defined the *poly-order moving average transform* by generalizing the original definition of moving average transform. Third, we have presented a related theorem to guarantee correctness of the poly-order transform-based subsequence matching and formally proven the theorem. Fourth, we have proposed two different moving average transformed subsequence matching algorithms, FRM-MAT and DM-MAT. Last, we have empirically shown superiority of the proposed methods through the extensive experiments on various data types, selectivity ranges, and query lengths. Experimental results for real-stock data show that our approach improves average performance by 22.4 ~ 33.8 times over the sequential scan. And, when comparing with the cases of building each index for all orders, our approach reduces the storage space and maintenance effort required for indexes significantly by sacrificing only a little performance degradation.

Acknowledgements

This work was supported by the Ministry of Science and Technology (MOST)/ Korea Science and Engineering Foundation (KOSEF) through the Advanced Information Technology Research Center (AITrc).

References

1. Agrawal, R., Faloutsos, C., and Swami, A., "Efficient Similarity Search in Sequence Databases," In *Proc. the 4th Int'l Conf. on Foundations of Data Organization and Algorithms*, Chicago, Illinois, pp. 69-84, Oct. 1993.
2. Beckmann, N., Kriegel, H.-P., Schneider, R., and Seeger, B., "The R*-tree: An Efficient and Robust Access Method for Points and Rectangles," In *Proc. Int'l Conf. on Management of Data*, ACM SIGMOD, Atlantic City, NJ, pp. 322-331, May 1990.
3. Chatfield, C., The Analysis of Time Series: An Introduction, 3rd Ed., Chapman and Hall, 1984.
4. Faloutsos, C., Ranganathan, M., and Manolopoulos, Y., "Fast Subsequence Matching in Time-Series Databases," In *Proc. Int'l Conf. on Management of Data*, ACM SIGMOD, Minneapolis, MN, pp. 419-429, May 1994.
5. Kendall, M., Time-Series, 2nd Ed., Charles Griffin and Company, 1976.
6. Loh, W.-K. and Kim, S.-W., "A Subsequence Matching Algorithm Supporting Moving Average Transform of Arbitrary Order in Time-Series Databases Using Index Interpolation," In *Proc. of the 12th Australasian Database Conference (ADC2001)*, Queensland, Australia, pp. 37-44, Jan., 2001.
7. Loh, W.-K., Kim, S.-W., and Whang, K.-Y., "A Subsequence Matching Algorithm that Supports Normalization Transform in Time-Series Databases," *Data Mining and Knowledge Discovery*, Vol. 9, No. 1, pp. 5-28, July 2004.

8. Moon, Y.-S., Whang, K.-Y., and Loh, W.-K., "Duality-Based Subsequence Matching in Time-Series Databases," In *Proc. the 17th Int'l Conf. on Data Engineering (ICDE)*, IEEE, Heidelberg, Germany, pp. 263-272, April 2001.
9. Moon, Y.-S., Whang, K.-Y., and Han, W.-S., "General Match: A Subsequence Matching Method in Time-Series Databases Based on Generalized Windows," In *Proc. Int'l Conf. on Management of Data*, ACM SIGMOD, Madison, WI, pp. 382-393, June 2002.
10. Rafiei, D. and Mendelzon, A. O., "Querying Time Series Data Based on Similarity," *IEEE Trans. on Knowledge and Data Engineering*, Vol. 12, No. 5, pp. 675-693, Sept./Oct. 2000.
11. Wu, H., Salzberg, B., and Zhang, D., "Online Event-driven Subsequence Matching Over Financial Data Streams," In *Proc. of Int'l Conf. on Management of Data*, ACM SIGMOD, Paris, France, pp. 23-34, June 2004.

Efficient Mining of Emerging Events in a Dynamic Spatiotemporal Environment*

Yu Meng and Margaret H. Dunham

Department of Computer Science and Engineering,
Southern Methodist University,
Dallas, Texas 75275-0122
ymeng(mhd)@engr.smu.edu

Abstract. This paper presents an efficient data mining technique for modeling multidimensional time variant data series and its suitability for mining emerging events in a spatiotemporal environment. The data is modeled using a data structure that interleaves a clustering method with a dynamic Markov chain. Novel operations are used for deleting obsolete states, and finding emerging events based on a scoring scheme. The model is incremental, scalable, adaptive, and suitable for online processing. Algorithm analysis and experiments demonstrate the efficiency and effectiveness of the proposed technique.

1 Introduction

We present an efficient data mining technique for modeling multidimensional time variant data series and its suitability for mining emerging events in a spatiotemporal environment. Given an ordered time series or a data stream that is composed of a (large) set of data points (events) collected by a real-world application, we are interested in many cases in finding those events that are relatively new but potentially have significant impact on the system. The data mining technique is desired to model the dynamically changing profile and provide capabilities to accommodate new trend and to forget obsolete profile.

The significance of mining emerging events rests on detecting them dynamically at an early stage. Thus we aim at finding them when they are rare but new in occurrence in a soft real time manner. The rarity of emerging events makes it related to identifying patterns of rarity [2-9]. However previous work does not address this problem in a dynamic spatiotemporal environment. First, existing algorithms require that the entire dataset be accessed at one time [5, 8, 10] or mine within a data window [3, 4, 9]. Mining with the entire dataset implicitly assumes stationarity and therefore losses the dynamically changing nature of the dataset. On the other hand, mining within a time window has made an assumption that the history prior to the window does not influence current behavior and is totally forgettable. The second issue is that existing algorithms either keep temporal information of the datasets without examining spatial

---

* This material is based upon work supported by the National Science Foundation under Grant No. IIS-0208741.

relationship among data points [4, 5] or otherwise focus on spatial clustering but ignore temporal dependency of data [3, 10]. In the practical examples such as computer network traffic, highway traffic and electric power demand management, both the spatial relationship of data points and their temporal dependency are important.

Therefore previous related techniques can be viewed at three different levels. The first level work (outlier detection, anomaly detection, and rare event detection) is to detect those events which our deviate from the majority in the whole dataset. The second level work (surprising patterns, concept drifting) takes a time-variant statistical distribution of the data profile into consideration. The third level work (emerging events) seeks those events which are rare but with a support larger than a threshold. Moreover, mining of rarity can be either spatial or temporal or both. Our work represents this new fourth level.

The proposed technique is built based on the Extensible Markov model (EMM), a spatiotemporal modeling technique proposed by the authors [1]. EMM interleaves a clustering algorithm with a dynamic Markov chain to model spatiotemporal data. In this paper, modules for adding and deleting states of Markov chain are used in modeling. To extract emerging events, an *aging score of occurrences* is proposed to reflect decay of importance. Both emerging events and obsolete events are judged using functions of the score and thus the proposed technique is able to continuously model the change of the data profile. The proposed technique inherits the traits of EMM and therefore is efficient, scalable, incremental and thus suitable for unsupervised online processing.

2 Methodology

In this section we present new EMM techniques to be used in the identification of emerging events. An additional labeling element, the *aging score of occurrence* (or the *score* in short), is introduced to each node (or cluster or state) and each link (or transition) in EMM. We first define the score, and then investigate its properties and present how the score scheme is applied to identifying obsolete events and emerging events.

The score of node or link is built using an *indicator function*:

$$I_{Ek}(\xi_t) = \begin{cases} 1 & E_k = \xi_t \\ 0 & E_k \neq \xi_t \end{cases}$$

Here E_k is an EMM component (either node or link) and ξ_t is the current component of the same type at time t. We may eliminate the subscript k for simplicity.

Definition 1: *(Aging Score of Occurrence for an EMM component).* At time t, the aging score of occurrence *for an EMM component E* is defined by

$$S_t^{(E)} = \sum_{i=1}^{t}(I_E(\xi_i) \cdot (1-\alpha)^{t-i}),$$

where $0 < \alpha < 1$ is an aging coefficient, t is current time.

A *Cluster Feature (CF)* refers to a labeling presentation of a cluster or a state of EMM. A cluster feature introduced by BIRCH [10] is defined using a vector of three attributes which denote the count of occurrence, CN_j, the first moment, LS_j, and the second moment, SS_j, of data points in a cluster. To use the score of the EMM components, we extend the labeling schemes of both with a score S_t and a time t, as defined in Definitions 2 and 3. It is easy to see that the CF, CL_{ij}, and S_t are additive, and thus computation of e-CF and e-CL is as efficient as the adopted clustering method such as BIRCH.

Definition 2: Extended Cluster Feature (e-CF) is a vector with five attributes to summarize the information of a cluster or a node at time t, defined by:

$$e\text{-}CF_t = <CN_t, \vec{LS}_t, SS_t, S_t^N, t>.$$

Definition 3: Extended EMM Transition Labeling e-CL is defined by:

$$e\text{-}CL_{ij} = <CL_{ij}, S_t^L, t>.$$

The t indicates the last time that e-CF was updated. This makes us not have to update all EMM components at every time but only update current coponents.

In addition to the decay of importance, we use a sliding window to achieve the stationary approximation. The idea is to examine whether a node of EMM has been visited (to be current) in the window w. If a node is visited, then it is in active use; otherwise the node is considered an obsolete node and will drop from EMM along with associated links. Note that not all links occurred within the sliding window.

Definition 4: Obsolete Events. Assume current time is t. If an EMM node N_o is not seen in a window $[t-w, t]$, it is recognized as an obsolete node. All links in and out of obsolete node N_o are obsolete and are removed from EMM.

Definition 5: Emerging Event. Assume a transition L_{ij} between two EMM nodes N_i and N_j occurs at time t. The transition L_{ij} and the absorbing node N_j is considered as an emerging events if

$$R_t^{(L)} = S_t^{(L)}/CL_t > 1 - \varepsilon, \text{ or}$$

$$R_t^{(N)} = S_t^{(N)}/CN_t > 1 - \varepsilon,$$

where ε is a predefined threshold.

Definition 5 discerns the EMM components with majority of occurrences introduced in the recent history and thus considered to be associated with developing trends. Scores of a node and a link are computed incrementally. In addition, two comparisons are needed to determine an emerging event. Thus for an EMM with m nodes, the time for these computations has $O(1)$ complexity. Thus the proposed technique inherits the efficiency of the EMM framework.

3 Experiments

In this section, we briefly report the performance experiments in terms of efficiency and effectiveness of the proposed technique. Experiments were performed on the VoIP traffic data provided by Cisco Systems which represents 1.5 million logged

VoIP CDRs in their Richardson, Texas facility from Sep 22 to Nov 17. The site related traffics comprising the calls of internal to internal, internal to local, internal to national, internal to international, as well as those in opposite directions were selected for investigation. Statistics were measured every 15 minutes. After preprocessing, there are 5422 points and the data format at each time point is:

$$V_t = <D_t, T_t, S_{1t}, S_{2t}, ... S_{7t}>,$$

where D_t denotes type of day, T_t time of the day, and S_{it} the value of statistic volume found at that call direction i, at time t. BIRCH and nearest neighbor clustering algorithms are used. Euclidean distance is used as the dissimilarity measure. Four parameters, namely clustering threshold th, window size w, decay coefficient α and threshold of score/count ratio r (note that $r=1-\varepsilon$), are used throughout the experiments. Default values, i.e. $th = 30$, $w = 1000$, $\alpha = 0.01$ and $r = 0.9$ are used unless the parameter is a variable of investigation. The th is calibrated using EMM prediction [1].

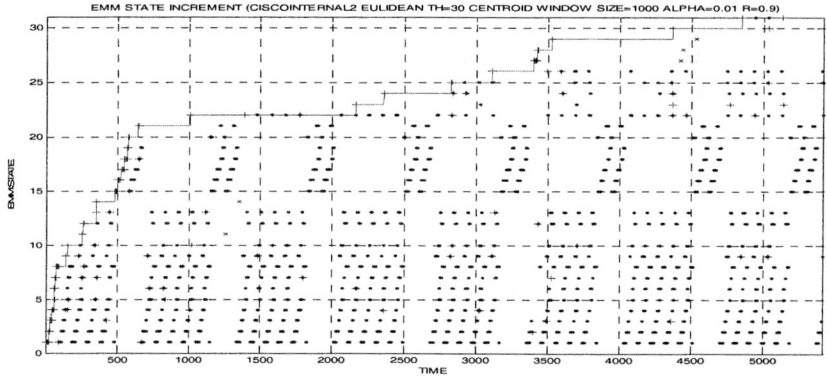

Fig. 1. Node increment, node deletion, and emerging event finding with EMM

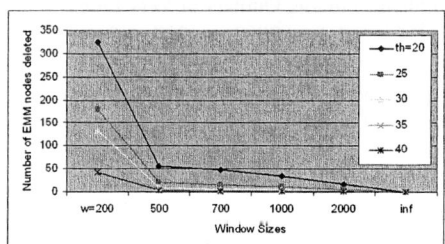

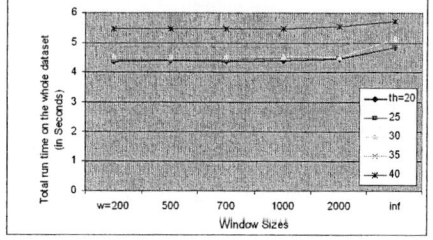

Fig. 2. Number of EMM Nodes deleted   **Fig. 3.** Run time to process the dataset

Fig. 1 illustrates the modeling process with increment of nodes (the envelope), deletions of obsolete nodes (x) and detections of emerging nodes (+) of EMM on the fly. The growth rate is sublinear and decreases as the time goes. The growth rate is less than 0.6% at the end of the modeling process and thus is efficient in terms of space usage. Fig. 2 shows that the number of deletions dramatically changes with different window sizes. However the total run time of the application at hand does not change much as illustrated in Fig. 3. Further experiments show that examinations of both spatial and temporal relationships are important. Temporal dependency gives more conservative judgments. Plateaus exist for parameters such as window size, score/count ratio and decay coefficient, with which appropriate ranges of parameters that reflect the dynamic profile of the data can be determined.

4 Conclusions

In this paper, we have presented an efficient data mining modeling technique suitable for finding emerging events in a spatiotemporal environment. The model accommodates anomaly, concept drifting, support, and temporality in one framework without losing time and space efficiency. Novel operations are proposed for deleting obsolete states, and finding emerging events based on a scoring scheme. Selection of parameters for appropriate capturing the dynamic data profile is found to have a range. Offline mining of the results generated by this model is possible for more complex patterns.

References

1. Margaret Dunham, Yu Meng, and Jie Huang, "Extensible Markov Model", ICDM, 2004, pp. 371-374.
2. Aggarwal, Han et al., "A Framework for Clustering Evolving Data Streams", VLDB 2003.
3. Kevin B.Pratt and Gleb Tschapek, "Visualizing Concept Drift", SIGKDD 2003.
4. N.Ye and X.Li, "A Markov Chain Model of Temporal Behavior for Anomaly Detection", *Proc. IEEE Systems, Man, and Cybernetics Information Assurance and Security Workshop*, 2000.
5. E. Keogh et al., "Finding Surprising Patterns in a Time Series Database in Linear Time and Space", SIGKDD, 2002, pp. 550-556.
6. D. Yu, G. Sheikholeslami, and A. Zhang, "FindOut: Finding Outliers in Very Large Datasets", *Knowledge and Information Systems*, vol. 4, no. 4, Oct. 2002, pp. 387–412.
7. G.M.Weiss and H.Hirsh, "Learning to Predict Extremely Rare Events", *AAAI Workshop Learning from Imbalanced Data Sets*, 2000, pp. 64-68.
8. P. Domingos and G. Hulten, "Mining High-speed Data Streams", *Knowledge Discovery and Data Mining*, pages 71-80, 2000.
9. Gerhard Widmer and Miroslav Kubat, "Learning in the Presence of Concept Drift and Hidden Contexts", *Machine Learning*, 23, 69-101 (1996).
10. T. Zhang, R. Ramakrishnan, and M. Livny, "BIRCH: A New Data Clustering Algorithm and Its Applications", *Data Mining and Knowledge Discovery*, 1(2): 141-182, 1997.

A Multi-Hierarchical Representation for Similarity Measurement of Time Series

Xinqiang Zuo and Xiaoming Jin

School of Software,
Tsinghua University, Beijing, 100084, China
zuoxq04@mails.tsinghua.edu.cn
xmjin@mail.tsinghua.edu.cn

Abstract. In a large time series database, similarity searching is a frequent subroutine to find the similar time series of the given one. In the process, the performance of similarity measurement directly effects the usability of the searching results. The proposed methods mostly use the sum of the distances between the values on the time points, e.g. Euclidean Distance, dynamic time warping (DTW) etc. However, in measuring, they do not consider the hierarchy of each point in time series according to importance. This causes that they cannot accurately and efficiently measure similarity of time series. In the paper, we propose a Multi-Hierarchical Representation (MHR) to replace the original one based on the opinion that the points of one time series should be compared with the ones of another with the same importance in measuring. MHR gives the hierarchies of the points, and then the original one can be represented by the Multi-Hierarchical subseries, which consist of points in the same hierarchy. The distance between the representations can be computed as the measuring result. Finally, the synthetic and real data sets were used in the effectiveness experiments comparing ours with other major methods. And the comparison of their efficiencies was also performed on the real data set. All the results showed the superiority of ours in terms of effectiveness and efficiency.

1 Introduction

Time series has been a ubiquitous data in the real-world, e.g. daily temperature, stock prices, various sensor data etc. There have been a lot of research works in searching and mining time series. Similarity measurement is a frequent subroutine in many applications. Due to the variety of different data, it is difficult to design a direct distance function to obtain better performance. As a preprocessing step, representation, which transforms time series into more meaningful and usable format, is a suitable solution.

Many popular representations are based on the segmentation of time domain, e.g. Symbolic Representation, Piecewise Linear Representation (PLR). Generally, the time series is segmented into many subseries. Then the original one is represented by the resulting segmentation based on the content of each subseries or the distribution of the values using various techniques, e.g. clustering or statistic etc. And other methods give the new representation using frequency transforms, e.g. Fourier Transforms [1], Wavelets [2]. But they are not enough to solve the similarity problem of time series, due to their ignoring on the hierarchy of each point in time series according to importance.

The accurate and suitable representation problem has brought challenges in the research on time series. So, it is by no means trivial to study this problem and put forward an effective solution that represents time series accurately and usably for similarity measurement. In this paper, we propose a novel approach, called Multi-Hierarchical Representation (MHR), to fill this gap. Our strategy is to partition the points into different hierarchies using Fast Fourier Transform (FFT). Then the time series is represented by the Multi-Hierarchical subseries, which consist of the points in the same hierarchy. The distance between the new representations can be computed by the sum of distances of the subseries in different hierarchies. We used best match searching and clustering experiments on both synthetic and real data sets to evaluate the effectiveness of MHR comparing with other methods. And the comparison of their efficiencies was also performed on the real data set. All the results showed the superiority of ours in terms of effectiveness and efficiency.

The rest of the paper is organized as follows. Sect. 2 provide some background materials. Sect. 3 introduces MHR and its similarity measurement. In Sect. 4, we give the exhaustive performance comparisons between ours and other methods. Finally, in Sect. 5 we offer some conclusion remarks.

2 Background

2.1 Related Work

Many representations of time series have been proposed in searching or mining applications. Symbolic representation is a popular method to transform the numerical series into symbolic sequence. The simple is to automatically cluster all the subseries in a fixed-window into some classes, and then use the symbols standing for the classes, to replace each subseries [3]. The method may be disabled due to the inaccuracy of the interval boundaries, e.g. a whole shape (or content) is segmented. Recently in [4], the Symbolic Approximation (SAX) has been proposed with an approximate distance function that lower bounds the Euclidean Distance. In [5], a new method for meaningful unsupervised symbolization of time series, called Persist, was proposed utilizing incorporation of temporal information. Clipped representation has attracted much interest [6,7], and it has superior space benefits due to only saving 0 and 1. In [8], a multiresolution symbolic representation was proposed, and Hierarchical Histogram Model was used as the distance function. It integrated the segmental results of several fixed-window. The multiresolution seems similar to MHR, but, actually, we have the essential difference that we emphasize the corresponding hierarchical relation of the points in similarity measure, and they used the multiresolution segmental windows to solve the inaccuracy using only one fixed-window.

Dimensionality reduction is also one kind of representation, representing the time series with a multidimensional vector. In [1], the Discrete Fourier Transform (DFT) was utilized to perform the dimensionality reduction, and other techniques have been suggested, including Singular Value Decomposition (SVD) [9] and the Discrete Wavelet Transform (DWT) [2]. In [10], an extended representation of time series using piece-wise linear segments was proposed, as well as a weight vector that contains the relative importance of each individual linear segment, which allows fast and accurate classification, clustering and relevance feedback. In [11, 12], Piecewise Aggregate Approximation (PAA) was proposed, and in [13], a more effective method Adaptive Piecewise Constant Approximation (APCA) was proposed with segments of varying lengths of each time series.

All the methods above is to find a new representation that can represent the time series accurately with simple format or high level content to obtain the preferable results in the relevant post-process, e.g. similarity measurement, searching, clustering etc. Most of them are based on the local information. In this paper, we focus on the accurate and suitable representation based on the global consideration.

2.2 Preliminaries

In this subsection, we start with some basic definitions. A time series $X = x_1, x_2, ..., x_n$ is a sequence of real values in which each value corresponds to a time point. x_i (or $X[i]$) stands for the value at i-th sampling time. The value might be of various dimensions. $|X| = n$ denotes the length of X. We give the definitions of two kinds of subseries for our approach as follows:

Definition 1. Sequential Subseries: A sequential subseries from time point s to e is defined as $X[s,e] = x_s, x_{s+1}, ..., x_e$ with the length of $|X[s,e]| = e - s + 1$.

Definition 2. Unsequential Subseries: An unsequential subseries is formalized by $X(S)$ where $S = S(1), S(2), ..., S(m)$ is the ordered subset of natural number with the restriction $S(1) < S(2) < ... < S(m)$. Each value in $X(S)$ can be got by the formula $X(S)_i = X[S(i)]$.

Then we introduce DFT, which transforms a time series from time domain into frequency domain. The contrary process is named Inverse Discrete Fourier Transform (IDFT). Formally, the DFT of X is defined to a sequence $X_f = X_1, X_2, ...X_n$ consists of n complex numbers. And FFT and Inverse Fast Fourier Transform (IFFT) are the faster algorithms with $O(nlog(n))$ time performance corresponding to DFT and IDFT with $O(n^2)$.

3 MHR Approach

The hierarchical strategy is adopted in our approach. Firstly, the time points in time series are partitioned into different hierarchies. In each hierarchy, an unsequential subseries defined in Definition 2 is formed by the values on the time points belong to the hierarchy. Then a time series can be represented by

Table 1. Symbols and their Explanations

Symbol	Explanation
X	Time series, $X = x_1, x_2, ..., x_n$
S_{xi}	The set contains the points of X in i-th hierarchy
$X(S_{xi})$	The unsequential subseries of X in i-th hierarchy
X'_i	The reconstructed with a linear combination of the first i Fourier waves
n_x	Length of time series X
n_{xi}	Length of $X(S_{xi})$
h	Number of the hierarchies

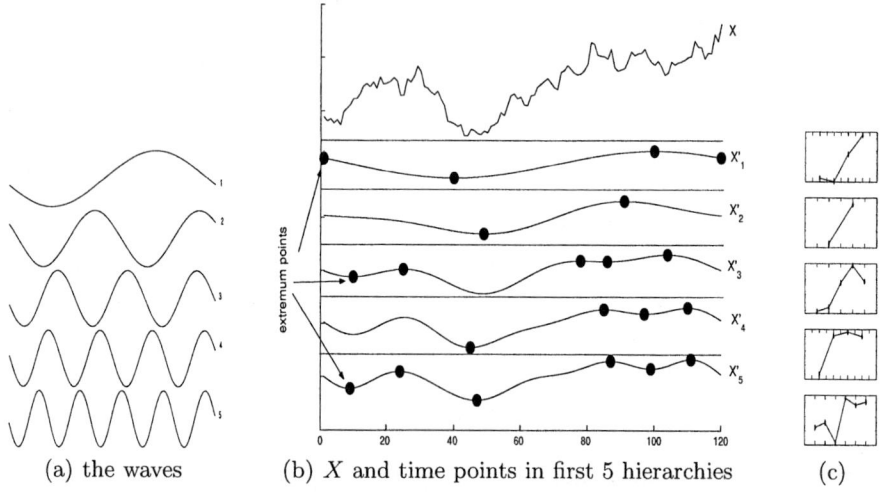

(a) the waves   (b) X and time points in first 5 hierarchies   (c)

Fig. 1. The original time series and its unsequential subseries of first 5 hierarchies (c)

the unsequential subseries of all the hierarchies. Table 1 shows the main symbols used in this paper. Then we introduce our approach detailedly in the following.

3.1 Hierarchical Representation

Given a time series X with length n and a hierarchy parameter h ($h < n$), we use FFT to generate h waves as Fig. 1(a). Then we combined linearly the first i waves, labelled as X'_i with the same length as X, like the five curves in Fig. 1(b). The larger i is, the more similar X'_i is to the original X, which can be found from Fig. 1(b).

Then we can get the unsequential subseries of each hierarchy. In i-th ($1 \leq i \leq h$) hierarchy, S_{xi} consists of the time points with extremum in X'_i, i.e. the black points in Fig. 1(b). The unsequential subseries can be formed according to Definition 2 as shown in Fig. 1(c). In addition, if a time point has been in the higher hierarchy, it will not be added in the lower, even if it is also the extremum point in any lower hierarchy. The first and last time points are initialized as the points in first hierarchy. Choosing the extremum points is due to their importance

```
Input: X, h //Time series X, hierarchy parameter h
Output: S_x1, S_x2, ..., S_xh  //S_xi is the set of the points in i-th hierarchy
1.  X_f = FFT(X)
2.  generate the first h waves series.
3.  for i=1:h
4.      X'_i=combination of the first i waves
5.      construct S_xi with the extremum points set of X'_i
6.  end
7.  return S_x1, S_x2, ..., S_xh
```

Fig. 2. The algorithm generating each S_{xi} in MHR

in the whole series. The algorithm generating each S_{xi} is illustrated formally in Fig 2, and then MHR of X can be formed using Definition 2.

3.2 Distance Measurement

In this subsection, we give the distance measurement of MHR. After hierarchical partition of two time series, we can compute the distance between the unsequential subseries in each hierarchy using a distance function that allows time warping, e.g. DTW. And then the sum of distances of all h hierarchies is calculated as the distance of the two time series. Given two time series X and Y with lengths n_x and n_y respectively, the hierarchy parameter h ($h < n_x, n_y$), and a warping distance function D_w. Formally, the MHR Distance (MHRD) can be defined as follows: $\text{MHRD}(X, Y, h) = \sum_{i=1}^{h} D_w(X(S_{xi}), Y(S_{yi}))$.

We take DTW as the distance function to analysis the time performance of MHRD. The time consumption of calculating the distance of two time series is $O(\sum_{i=1}^{h} n_{xi} n_{yi})$ with obvious superiority over DTW with $O(n_x n_y)$ ($n_{xi} \ll n_x$ and $\sum_{i=1}^{h} n_{xi} < n_x$). Mostly, it is more accurate with larger h, but with more time consumption. So h can be selected according to the particular requirement. We performed the experiment to show the time performances of ours and others.

3.3 Additional and the Expansion

In our approach, there might be the case that $n_{xi} = 0$ but $n_{yi} \neq 0$, i.e. the extremum points in i-th hierarchy of X are also with extremum in k-th hierarchy ($k < i$). We use the unsequential subseries in the $(i-1)$ hierarchy of X instead of that of the i-th in this case. This is the important addition for our approach.

Our approach can be expanded in the following directions: ordering the waves, and giving a gradual segmentation approach using the time points in each hierarchy. We can reset the order of the waves with decreasing amplitude of the corresponding Fourier coefficients of X_f. Then X'_i can be got by the linear combination of the first waves with the largest i Fourier coefficients. Then the following process is as same as the description above. Using the extremum points in each hierarchy, we can segment the time series gradually. Then it can be used in many methods, which are based on segmentation of time series.

4 Experimental Evaluation

In this section, we used best match searching and clustering experiments to test the accuracies of ours and others on both synthetic and real data sets. In the experiments, DTW was selected as the warping distance function D_w in our approach (MHRD), because it is widely used in many applications. The competitors included Euclidean Distance and DTW. Finally, we give the efficiency comparison in Sect. 4.4.

4.1 Datasets

The synthetic data set is Synthetic Control Chart Time Series (SYNDATA) data set which was downloaded from the UCI KDD archive[1]. It contains 600 examples of synthetic control charts belong to 6 different classes, and each class consists of 100 time series. The length of each time series is equal to 60.

The real data set is the Standard and Poor 500 index (S&P) historical stock data from Mar. 27, 2004 to Mar 26, 2005 [2]. We chose the opening price as our experimental data. Each stock data is a series of length d, where $d \leq 252$ (d might be smaller if the company is removed from the Index). We only used the stocks whose length is 252. Based on the official S&P clustering information, we divided the stock data into the classes. Finally, 50 classes contain 442 stock data were used by removing the classes which contain only one stock.

4.2 Best Match Searching

Evaluation method. Best match searching is the process to find the time series whose distance with the given one is below a predefined threshold, or the most similar k matches in database. Because the threshold is difficult to set for different measurements with different value ranges, we used KNN searching for the experiment. The accuracy evaluation is following, which was also adopted in [8]: The standard (right) searching result of a query Q, labelled as std(Q), is the set of time series which belong to the same class as Q, which can be got by the apriori classification information, and the results by different methods are marked knn(Q). The accuracy (precision) related to Q is defined as follows:

$$\text{Accuracy}(Q) = \frac{|\text{knn}(Q) \cap \text{std}(Q)|}{k} \quad (1)$$

In our experiment, we set the number of time series belong to the same class as the query as the value of k. In the experiment, each time series is treated as a query. The average of the accuracies is calculated as the final result.

---
[1] The UCI KDD Archive, http://kdd.ics.uci.edu
[2] S&P500, http://kumo.swcp.com/stocks/

Method	h	Accuracy
MHRD	5	0.6324
	7	0.6531
	9	0.6952
	11	0.7451
	13	0.7725
	15	0.8156
	17	0.8616

Method	$window$	Accuracy
Euclidean	5	0.2380
	7	0.2364
	9	0.2779
	11	0.2792
	13	0.3194
	15	0.3406
	17	0.3724
	60	0.5112
DTW		0.8207

Fig. 3. Matching accuracy on SYNDATA

Method	h	Accuracy
MHRD	5	0.3722
	7	0.3925
	9	0.4067
	11	0.4532
	13	0.4623
	15	0.4511
	17	0.4632

Method	$window$	Accuracy
Euclidean	5	0.3637
	7	0.4514
	9	0.4164
	11	0.3792
	13	0.3657
	15	0.3564
	17	0.3478
	252	0.2186
DTW		0.1538

Fig. 4. Matching accuracy on S&P

Experiment on SYNDATA. We show the matching accuracies of the methods on SYNDATA in Fig. 3. In the experiment, we also realized the piecewise normalization in Euclidean Distance, which can get better clustering results than plain Euclidean Distance on stock data proofed in [14]. When $window$ is set to 60 (252) in Fig. 3 (Fig. 4), it is the plain Euclidean Distance. The results of ours with different h are given, as well as that of Euclidean distance and DTW. From the results, we can get that DTW obtains a much better result. And the results of Euclidean Distance are the worst, so we can conclude that the data are not sensitive to time warping. Our results are approximative to that of DTW, but more efficient. When h in ours is set as 17, we can get the best result of all.

Experiment on S&P. Fig. 4 show the matching accuracies of the methods on S&P data. The results display that ours is the best of all, though ours is inferior to the Euclidean Distance only in a few cases, e.g. $h = 5$ or 7. The result of DTW is the worst, so we can conclude that the stock data is sensitive to time warping. The results also demonstrated that Euclidean Distance with the piecewise normalization is an accepted method on stock data.

4.3 Clustering Experiment

Evaluation method. We used Hierarchical Agglomerative Clustering (HAC) to realize the clustering experiment. The clustering result can be taken with the predefined clustering number. We used the complete distance to compute the distance between two classes. We computed the clustering accuracy using the method, which is adopted in many applications [8, 14]. Given the standard clustering result $C = C_1, C_2, ..., C_k$ from the apriori classification information

Method	h	Accuracy
MHRD	5	0.5993
	7	0.6012
	9	0.6211
	11	0.7100
	13	0.7521
	15	0.7912
	17	0.8455

Method	window	Accuracy
Euclidean	5	0.2724
	7	0.2979
	9	0.3551
	11	0.2980
	13	0.3967
	15	0.4088
	17	0.4177
	60	0.4996
DTW		0.6850

Fig. 5. Clustering accuracy on SYNDATA

Method	h	Accuracy
MHRD	5	0.2735
	7	0.3006
	9	0.3454
	11	0.3844
	13	0.4026
	15	0.4255
	17	0.4923

Method	window	Accuracy
Euclidean	5	0.3404
	7	0.4507
	9	0.3934
	11	0.2709
	13	0.3091
	15	0.2447
	17	0.2651
	252	0.2443
DTW		0.2113

Fig. 6. Clustering accuracy on S&P

and the clustering result using each method $C' = C'_1, C'_2, ..., C'_k$, compute the accuracy by the following formulas:

$$\text{Accuracy} = \frac{\text{sim}(C, C') + \text{sim}(C', C)}{2} \quad (2)$$

$$\text{sim}(C, C') = (\sum_i \max_j \text{sim}(C_i, C'_j))/k; \quad \text{sim}(C_i, C'_j) = 2\frac{|C_i \cap C'_j|}{|C_i| + |C'_j|} \quad (3)$$

$\text{sim}(C', C)$ above can be calculated similarly as $\text{sim}(C, C')$ in Eq 3. We computed both $\text{sim}(C', C)$ and $\text{sim}(C, C')$, because they are not symmetric. The clustering numbers used in HAC were set to 6 and 50 on SYNDATA and S&P data set respectively as same as their class numbers.

Experiment on SYNDATA. The clustering results on SYNDATA are given in the Fig. 5. From the figure, we can observe that the best result is also obtained by ours as best match searching experiment, when $h = 17$. The results of Euclidean Distance are also the worst, and DTW can get much better results. Our approach can be seen a "unsequential piecewise" DTW algorithm, and it have better accuracy and efficiency than DTW.

Experiment on S&P. The experimental results on S&P are listed in Fig. 6. From the figure, we can find that Euclidean Distance with piecewise normalization is superior to DTW. It is the further proof to confirm that the stock data is sensitive to time warping. But ours with time warping, when $h = 17$, is the best due to its hierarchical strategy. And the results of ours with other parameters also gain the advantages over other methods.

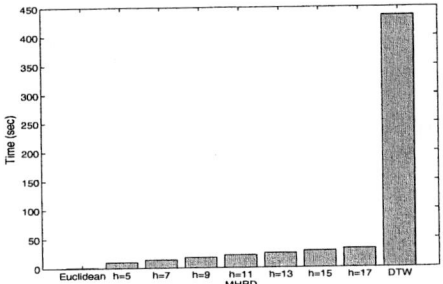

Fig. 7. Time performances of the methods in the experiment

4.4 Time Performance Comparison Experiment

In this subsection, we tested time performances of the methods using 1-NN queries with sequential scan in the S&P data set. The time consumption of each method was the time in 442 times queries corresponds to each time series, and we assume that each time series had been preprocessed for each method. The experiment were conducted on the machine with CPU of Celeron 1.70Ghz and 512 MB of physical memory, running Microsoft Windows Server 2003. We only counted the time consumption of distance calculation for exact comparison except accessing the disk. In Fig. 7, we give the experimental results, which show that DTW is much shower than others. The time of MHRD increases linearly along with h. Though Euclidean Distance is faster than ours, considering the accuracy and efficiency, our approach is superior to it.

5 Discussion and Conclusion

In this paper, we introduce a new representation of time series that can be used in similarity measurement with better effectiveness and efficiency, named Multi-Hierarchical Representation (MHR). Our idea is based on the opinion that the points of one time series should be compared with the ones of another with the same importance in similarity measurement. The unsequential subseries is defined in time series different from the general sequential subseries. In our approach, firstly, we partition the time points into different hierarchies. Then time series is represented by the Multi-Hierarchical subseries consisting of points in each hierarchy. The sum of distances between subseries in all hierarchy can be taken as the measuring result using a distance function that allows time warping.

We used best matching searching and clustering experiments on both SYN-DATA and S&P to evaluate the effectiveness of our approach comparing with other methods. The results showed the superiority of ours on accuracy over others. There are some cases that the effectiveness of ours is little inferior to the others. But in most cases, MHR is a more accurate representation for measuring the similarity of time series. And we also tested the time performances of ours and other competitors. The results demonstrated that ours is slower than Euclidean Distance but much faster than DTW. In the future, we will continue to research the extensive problems. And in our approach, the weight of the distance in each hierarchy is same. We will also study further in this aspect.

Acknowledgement

The work was supported by the NSFC 60403021 and the 973 Program 2004CB-719400. We thank the anonymous reviewers for their helpful comments.

References

1. Rakesh Agrawal, Christos Faloutsos, and Arun N. Swami. Efficient similarity search in sequence databases. In *FODO '93*, pages 69–84.
2. Kin pong Chan and Ada Wai-Chee Fu. Efficient time series matching by wavelets. In *ICDE '99*, pages 126–133.
3. Gautam Das, King-Ip Lin, Heikki Mannila, Gopal Renganathan, and Padhraic Smyth. Rule discovery from time series. In *KDD '98*, pages 16–22.
4. Jessica Lin, Eamonn Keogh, Stefano Lonardi, and Bill Chiu. A symbolic representation of time series, with implications for streaming algorithms. In *DMKD '03*, pages 2–11.
5. Fabian Mörchen and Alfred Ultsch. Optimizing time series discretization for knowledge discovery. In *KDD '05*, pages 660–665.
6. A. J. Bagnall and G. J. Janacek. Clustering time series from arma models with clipped data. In *KDD '04*, pages 49–58.
7. Chotirat (Ann) Ratanamahatana, Eamonn Keogh, Anthony J. Bagnall, and Stefano Lonardi. A novel bit level time series representation with implication of similarity search and clustering. In *PAKDD '05*, pages 771–777.
8. Vasileios Megalooikonomou, Qiang Wang, Guo Li, and Christos Faloutsos. A multiresolution symbolic representation of time series. In *ICDE '05*, pages 668–679.
9. Flip Korn, H. V. Jagadish, and Christos Faloutsos. Efficiently supporting ad hoc queries in large datasets of time sequences. In *SIGMOD '97*, pages 289–300.
10. Eamonn Keogh and M. Pazzani. An enhanced representation of time series which allows fast and accurate classification, clustering and relevance feedback. In *KDD '98*, pages 239–241.
11. Byoung-Kee Yi and Christos Faloutsos. Fast time sequence indexing for arbitrary lp norms. In *VLDB '00*, pages 385–394.
12. Eamonn Keogh and Michael J. Pazzani. Scaling up dynamic time warping for datamining applications. In *KDD '00*, pages 285–289.
13. Eamonn Keogh, Kaushik Chakrabarti, Michael Pazzani, and Sharad Mehrotra. Locally adaptive dimensionality reduction for indexing large time series databases. In *SIGMOD '01*, pages 151–162.
14. Martin Gavrilov, Dragomir Anguelov, Piotr Indyk, and Rajeev Motwani. Mining the stock market: which measure is best? In *KDD '00*, pages 487–496.

Multistep-Ahead Time Series Prediction

Haibin Cheng, Pang-Ning Tan, Jing Gao, and Jerry Scripps

Department of Computer Science and Engineering,
Michigan State University
{chenghai, ptan, gaojing2, Scripps}@msu.edu

Abstract. Multistep-ahead prediction is the task of predicting a sequence of values in a time series. A typical approach, known as multi-stage prediction, is to apply a predictive model step-by-step and use the predicted value of the current time step to determine its value in the next time step. This paper examines two alternative approaches known as independent value prediction and parameter prediction. The first approach builds a separate model for each prediction step using the values observed in the past. The second approach fits a parametric function to the time series and builds models to predict the parameters of the function. We perform a comparative study on the three approaches using multiple linear regression, recurrent neural networks, and a hybrid of hidden Markov model with multiple linear regression. The advantages and disadvantages of each approach are analyzed in terms of their error accumulation, smoothness of prediction, and learning difficulty.

1 Introduction

Many time series problems involve the task of predicting a sequence of future values using only the values observed in the past. Examples of this task, which is known as *multistep-ahead time series prediction* [1], include predicting the time series for crop yield, stock prices, traffic volume, and electrical power consumption. By knowing the sequence of future values, we may derive interesting properties of the time series such as its projected amplitude, variability, onset period, and frequency of abnormally high or low values. For example, multistep-ahead time series prediction allows us to forecast the growing period of corn for next year, the maximum and minimum temperature for next month, the frequency of El-Nino events in the next decade, etc.

A typical approach to solve this problem is to construct a single model from historical values of the time series and then applies the model step by step to predict its future values. This approach is known as *multi-stage prediction*. Since it uses the predicted values from the past, it can be shown empirically that multi-stage prediction is susceptible to the error accumulation problem, i.e., errors committed in the past are propagated into future predictions.

This paper considers two alternative approaches for multistep-ahead time series prediction. The first approach, known as *independent value prediction*, builds a separate model for each prediction step using only its past observations. The second approach, known as *parameter prediction*, fits a parametric function to the time series and builds regression models to predict the parameters of the function.

We implement all three prediction approaches using multiple linear regression [2], recurrent neural networks [3], and a hybrid of hidden Markov model with multiple linear regression [7] as the underlying regression methods. The advantages and disadvantages of each prediction approach are analyzed in terms of their error accumulation, smoothness of prediction, and learning difficulty.

2 Methodology

A time series is a sequence of observations in which each observation x_t is recorded at a particular timestamp t. A time series of length t can be represented as a sequence $X = [x_1, x_2, ..., x_t]$. We use the notation X_{t-p}^{t} to denote a segment of the time series $[x_{t-p}, x_{t-p+1}, ..., x_t]$. Multistep-ahead prediction is the task of predicting a sequence of h future values, X_{t+1}^{t+h}, given its p past observations, X_{t-p+1}^{t}.

2.1 Regression Methods

This section presents the regression methods used for modeling the time series.

2.1.1 Multiple Linear Regression (MLR)

The MLR model, which is also called the AR model, is given by the equation:

$$f(X_{t-p+1}^{t}) = \sum_{i=1}^{p} a_i x_{t-i+1} + \varepsilon_t$$

, where ε_t corresponds to a random noise term with zero mean and variance σ^2. The coefficient vector $[a_1, a_2, ..., a_p]^T$ is estimated using the least square method by minimizing the sum of squared error, SSE, of the training data. The variance is estimated using SSE/h, where h is the size of the prediction window.

2.1.2 Recurrent Neural Networks (RNN)

RNN has been successfully applied to noisy and non-stationary time series prediction. In RNN, the temporal relationship of the time series is explicitly modeled using feedback connections [3] to the internal nodes (known as hidden units). An RNN model is trained by presenting the past values of the time series to the input layer of the Elman back propagation network [4]. The weights of the network are then adjusted based on the error between the true output and the output predicted by the network until the algorithm converges. Before the network is trained, the user must specify the number of hidden units in the network and the stopping criteria of the learning algorithm.

2.1.3 Hybrid HMM/MLR Model

Hybrid HMM/MLR model is an extension of traditional hidden Markov model applied to regression analysis [7]. This method is an effective way for modeling piecewise stationary time series, where the observed values are assumed to be generated by a finite number of hidden states. Let (Z_t) denote the Markov chain on the state space $S = \{s_1, s_2, ..., s_N\}$. The initial probability for a given state s is denoted as π_s, while the transition from one state to another is characterized by the transition matrix $A = (a_{ij})$, where $P(Z_{t+1} = s_j | Z_t = s_j) = a_{ij}$. At time t, the observed value x_t depends only on the current state Z_t : $x_t = f_{z_t}(X_{t-p}^{t-1}) + e(0, \sigma_{z_t})$ where $f_{z_t} \in \{f_{s_1}, f_{s_2}, ..., f_{s_N}\}$ is the corresponding regression function and $e(0, \sigma_s)$ is a noise term with mean zero and a variance σ_s^2 that depends on the current state, s. We use the regression function produced by MLR in our experiments. The hybrid HMM/MLR model is trained by maximizing the following likelihood function:

$$L_\theta(X_1^t) = \sum_Z P(X_1^t; Z) = \sum_Z \pi_{z_1} \prod_{i=2}^t P(z_{i+1} | z_i) \Phi\left(X_i - f_{z_i}(X_{i-p}^{i-1})\right) \quad (1)$$

A brute force method for maximizing the likelihood function requires a complexity of $O(N^T)$ operations. However, an efficient approach called the forward-backward procedure can reduce the complexity of the computation down to $O(N^2 T)$. This procedure is based on the well-known expectation-maximization (EM) algorithm.

2.2 Prediction Approaches

We investigate three approaches for predicting the sequence of future values X_{t+1}^{t+h} from a given time series X_1^t. A training set D is initially created from the time series using a sliding window of length $p+h$ (see Figure 1). Each instance of the sliding window corresponds to a record in the training set, as shown in Table 1. The input X corresponds to the first p values of the window while the output Y corresponds to the remaining h values of the window. For example, the first record of the training set D contains $X = [x_1, x_2, ..., x_p]$ as its input variables and $Y = [x_{p+1}, x_{p+2}, ..., x_{p+h}]$ as its output variables. Similarly, the second record contains $X = [x_2, x_3, ..., x_{p+1}]$ as its input variables and $Y = [x_{p+2}, x_{p+3}, ..., x_{p+h+1}]$ as its output variables, while the last record contains $X = [x_{t-h-p+1}, x_{t-h-p+2}, ..., x_{t-h}]$ as its input variables and $Y = [x_{t-h+1}, x_{t-h+2}, ..., x_t]$ as its output variables. For notational convenience, we use $Y(i)$ to refer to all the values in the ith column of Y in D. For example, $Y(3) = [x_{p+3}, x_{p+4}, ..., x_{t-h+3}]^T$.

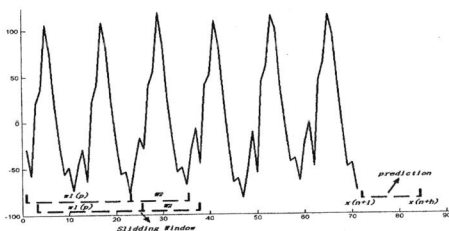

Fig. 1. A sliding window is used to create the regression training set D=X'+Y

Table 1. Traning Set $D = X \times Y$

$X = [X(1), ..., X(p)]$	$Y=[Y(1), ..., Y(h)]$
$[x_1, x_2, ..., x_p]$	$[x_{p+1}, x_{p+2}, ... x_{p+h}]$
$[x_2, x_3, ..., x_{p+1}]$	$[x_{p+2}, x_{p+3}, ... x_{p+h+1}]$
⋮	⋮

2.2.1 Multi-stage Prediction

Multi-stage prediction predicts the future values of a time series in a step by step manner. We first predict x_{t+1} using the previous p values, $x_{t+1-p},...,x_{t-1},x_t$. We then predict x_{t+2} based on its previous p values, which includes the predicted value for x_{t+1}. The procedure is repeated until the last value, x_{t+h}, has been estimated. In this approach, it is sufficient to construct a single model for making the prediction.

2.2.2 Independent Value Prediction

Independent value prediction predicts the value at each time step using a separate model. Given the initial data set shown in Table 1, we first create h training sets, each of which has the same input X, but different output Y. We use Y(1) as the output variable for the first training set, Y(2) as the output variable for the second training set, and so on. By learning each training set independently, we obtain h regression models f_i ($i = 1,2,..., h$). The models are then used to predict the next h values as follows: $\hat{x}_{t+i} = f_i(X)$, $i = 1,2,...,h$.

2.2.3 Parameter Prediction

Parameter prediction transforms the problem of predicting h output values into an equivalent problem of predicting $(d+1)$ parameters. For each record in Table 1, we fit a parametric function g to the output vector Y. Let ($c_0, c_1,..., c_d$) denote the parameters of the function g. We then replace the original output vector Y=[Y(1),Y(2),...,Y(h)] with a modified output vector Y'=[$c_0, c_1, ..., c_d$]. We now construct $(d+1)$ regression models f_i ($i = 0,1,2,...,d$), one for each output column Y'. The models are then applied to predict the $(d+1)$ parameters of a test sequence. The test sequence is reconstructed by substituting the predicted parameters into the parametric function g. While this methodology is generally applicable to any family of parametric functions, we use polynomial functions in our experiments.

2.3 Model Selection

The parameters for our prediction approaches include the order of regression model p, the size of prediction window h, and the degree of polynomial fit d (for parameter prediction). The size of the prediction window h is domain dependent and depends on the nature of the application. We use Akaike's final prediction error (FPE) [8] as our criterion for determining the right order for p in the MLR model.

$$FPE = \hat{\delta}^2 \frac{t+p}{t-p} \text{ where } \hat{\delta}^2 = \frac{\sum_{j=1}^{t-p-h}(y_{j1} - \hat{y}_{j1})^2}{t-p-h} \quad (2)$$

The same criterion is applicable to estimate the degree of the polynomial function used in parameter prediction. To determine the correct order for RNN, we employ the method described by Kennel in [5]. Let X_p denote as an instance of the training data and $X_p^{(n)}$ denote its nearest neighbor. The pair is declared as false nearest neighbors if $\left| \frac{d(X_p, X_p^{(n)}) - d(X_{p+1}, X_{p+1}^{(n)})}{d(X_p, X_p^{(n)})} \right|$ exceeds a user-specified threshold (where d refers to the distance between a pair of observations). Our goal is to choose a value for p such that the number of false nearest neighbors is close to zero.

3 Experiments and Discussions

We perform a comparative study on the three prediction approaches using both real and synthetic datasets. The real datasets are obtained from the UCI Machine Learning Repository [9] and the Time Series Data Library [6]. Our experiments were conducted on a Pentium 4 machine with 3GHz CPU and 1GB of RAM.

3.1 Evaluation Metric

The estimation error of a prediction approach is evaluated based on the following measure: $RMSE = \sum_i (y_i - \hat{y}_i)^2 / \sum_i (y_i - \bar{y})^2$, where y_i is the true value, $\hat{y}_i$ is the predicted value, and $\bar{y}$ is the average value of the time series. The RMSE values recorded in our experimental results are obtained using ten-fold cross validation.

A **Win-Draw-Loss Table** is created to compare the relative performance between two prediction approaches when applied to n data sets. We use the criterion of 0.01 difference in RMSE to determine whether one approach wins or loses against another approach. For a stricter evaluation, we also apply the **paired t significance test** to determine whether the observed difference in RMSE is statistically significant. To do this, we first calculate the difference (d) in the RMSE obtained from two prediction approaches on each data set. The mean $\bar{d}$ and standard deviation s_d of the observed differences are also calculated. To determine whether the differences are significant, we compute their T-statistic: $t = \bar{d}\sqrt{n}/s_d$ which follows a t-distribution with $n-1$ degrees of freedom. Under the null hypothesis that the two prediction approaches are comparable in performance, we expect the value of t should be close to zero. From the computed value for t, we estimate the p-value of the difference, which corresponds to the probability of rejecting the null hypothesis. We say the difference in performance is statistically significant if $p<0.05$ and highly statistically significant if $p < 0.001$.

3.2 Error Accumulation

Error accumulation refers to the propagation of past prediction errors into future predictions. To gain a better insight into this problem, we employ the bias-variance decomposition for squared loss functions. Consider a time series generated by the model $x_{t+1} = f(X_{t-p}^t) + e(0, \sigma^2)$. Let $(y_1, y_2, \cdots, y_h)$ denote the observed values of the time series in a prediction window of length h, i.e., $y_1 = x_{t+1}, y_2 = x_{t+2}, \ldots, y_h = x_{t+h}$. Furthermore, let $(y_1^*, y_2^*, \cdots, y_h^*)$ be the corresponding values generated by the deterministic model f. In other words, $y_i = y_i^* + e(0, \sigma^2)$. We use the notation $(v_1, v_2, \cdots, v_h)$ to denote the values predicted by a regression model, g. The mean squared error (MSE) at each prediction step j is defined as: $MSE(j) = E[(y_j - v_j)^2]$. The MSE at each step can be decomposed into the following three components [10]:

$$MSE(j) = (E(v_j) - y_j^*)^2 + E[(v_j - E(v_j))^2] + E[(y_j - y_j^*)^2] \tag{3}$$

The first term represents the squared bias of the model; the second term represents the variance of the model; while the third term represents the variability due to noise. The next example illustrates the error accumulation problem for the noise term.

Example 1: *Consider the following AR(2) model:* $x_{t+1} = a_1 x_t + a_2 x_{t-1} + \varepsilon$, *where* ε *has mean zero and variance* σ^2. *Suppose we were able to model accurately the coefficients* a_1 *and* a_2 *using MLR. For the multi-stage approach, we can show that:*

$$y_1 = a_1 x_t + a_2 x_{t-1} + \varepsilon_1 = v_1 + \varepsilon_1$$
$$\therefore MSE\ (1) = E\left[(y_1 - v_1)^2\right] = E(\varepsilon_1^2) \propto \sigma^2.$$
$$y_2 = a_1 y_1 + a_2 x_t + \varepsilon_2 = a_1 v_1 + a_2 x_t + (a_1 \varepsilon_1 + \varepsilon_2) = v_2 + (a_1 \varepsilon_1 + \varepsilon_2)$$
$$\therefore MSE\ (2) \propto (a_1^2 + 1)\sigma^2.$$
$$y_3 = a_1 y_2 + a_2 y_1 + \varepsilon_3 = a_1 v_2 + a_2 v_1 + (a_1^2 \varepsilon_1 + a_1 \varepsilon_2 + a_2 \varepsilon_1 + \varepsilon_3)$$
$$= v_3 + (a_1^2 \varepsilon_1 + a_1 \varepsilon_2 + a_2 \varepsilon_1 + \varepsilon_3)$$
$$\therefore MSE\ (3) \propto (a_1^4 + a_1^2 + a_2^2 + 1)\sigma^2.$$

The preceding formula shows the accumulation of errors due to noise for multi-stage prediction as the prediction step increases. For independent value prediction:

$$y_1 = a_1 x_t + a_2 x_{t-1} + \varepsilon_1$$
$$y_2 = (a_1^2 + a_2) x_t + (a_1 a_2) x_{t-1} + (a_1 \varepsilon_1 + \varepsilon_2)$$
$$y_3 = (a_1^3 + 2 a_1 a_2) x_t + (a_1^2 + a_1^2 a_2) x_{t-1} + (a_1^2 \varepsilon_1 + a_1 \varepsilon_2 + a_2 \varepsilon_1 + \varepsilon_3)$$

Assuming that the coefficients for x_t *and* x_{t-1} *can be accurately estimated by MLR, the noise terms for multi-stage and independent value prediction are identical.*

The preceding example illustrates that error accumulation due to noise is unavoidable, regardless of the prediction approach. To analyze the error accumulation due to the bias and variance of a model, we generate the following time series: $X_t = 0.418 X_{t-1} + 0.634 X_{t-2} + \varepsilon$, where ε is a Gaussian noise with mean zero and variance $\sigma^2 = 0.1$. The length of the time series is set to 1000 and the prediction window is $h = 50$. To ensure there is sufficient bias in the model, we set $p = 1$. The bias and variance of the models are estimated by generating 500 bootstrap replicates of the training set D and inducing a model g from each bootstrap replicate. The models are then applied to the test sequence to obtain 500 estimated values (v_j) for each prediction step j. The empirical bias is computed by taking the average value of the 500 predictions ($\bar{v}_j$) and subtracting it from the value predicted using the deterministic model. The variance of the models can also be estimated as follows: $\text{var}(j) = E[(v_j - \bar{v}_j)^2]$. Figures 2 and 3 illustrate the bias and variance for multi-stage and independent value predictions (using MLR and a hybrid HMM/MLR as the underlying regression methods). Both figures show that the bias and variance for multi-stage prediction grows steadily with increasing time steps, whereas the bias and variance for independent value prediction do not appear to be propagated into future predictions.

Therefore, error accumulation is a major problem in multi-stage prediction, irrespective of the choice of regression methods. However, it is not a problem for parameter prediction because the models for predicting different parameters are built independently (similar to independent value prediction).

Fig. 2. Bias and variance for MLR

Fig. 3. Bias and variance for HMM/MLR

3.3 Learning Difficulty

Multi-stage prediction builds a single model to fit the entire time series. In contrast, we need to build h models for independent value prediction and $(d+1)$ models for parameter prediction. Model building is therefore more expensive for independent value and parameter prediction approaches compared to the multi-stage approach.

The model to be learnt by independent value prediction also becomes more complex with increasing time steps. To illustrate this, let f denote the true model that generates the data, i.e., $X_t = f(X_{t-p}^{t-1})$. For simplicity, let $(x_1, x_2, \ldots, x_p)$ denote the input variables and $y_i = x_{p+i}$ denote the h output variables:

$$y_1 = f(x_1, x_2, \ldots, x_p) = f_1(x_1, x_2, \ldots, x_p)$$
$$y_2 = f(x_2, \ldots x_p, y_1) = f(x_2, \ldots x_p, f_1)$$
$$= f_2(x_1, x_2, \ldots, x_p)$$
$$\vdots$$
$$y_h = f(\ldots, y_{h-2}, y_{h-1}) = f(\ldots, f_{h-2}, f_{h-1})$$

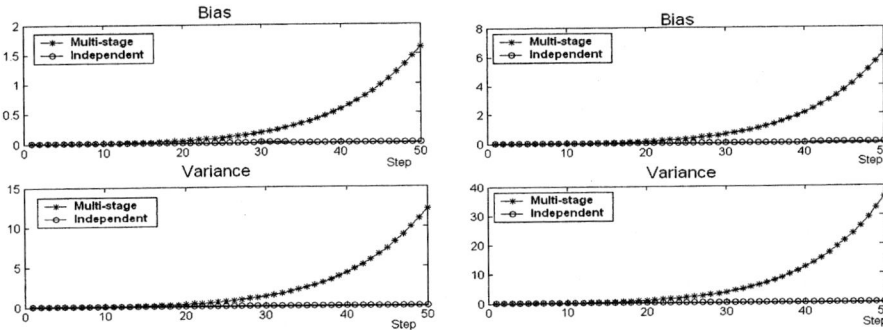

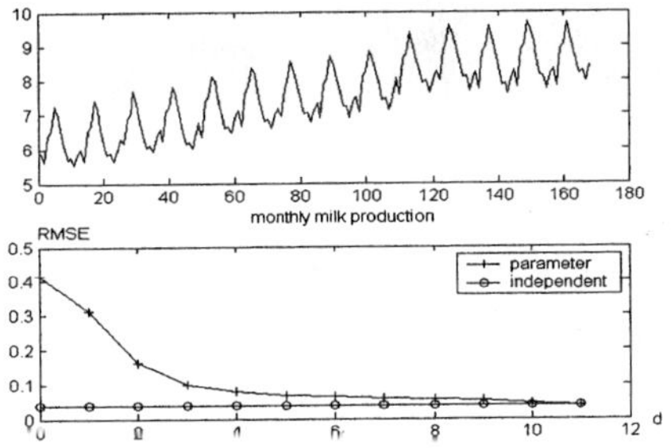

Fig. 4. Prediction Results (p=12,h=12)

If f is a linear function, then all the f_k's constructed by the independent value prediction approach are also linear functions. If f is non-linear, then the f_k's become increasingly complex functions of the input variables $(x_1, x_2, \ldots, x_p)$. Unless the regression method is very flexible, learning the appropriate model for each time step can be quite a challenging task. For parameter prediction, the learning difficulty depends on how easy it is to find the appropriate function that fits the output vector.

To compare parameter prediction against independent value prediction, we apply both methods to the monthly *milk production data* (see the top diagram in Figure 4) using $h=12$ and $p=12$. For parameter prediction, we use a polynomial function to fit the output vector and vary the degree of the polynomial from 0 to 11. We then employ MLR to predict the parameters of the polynomial. The bottom diagram of Figure 4 shows a comparison between the RMSE of parameter prediction against independent value prediction as the degree of the polynomial function is varied. Observe that the RMSE for parameter prediction drops dramatically when the polynomial degree increases to 3 and decreases slowly thereafter. This result suggests that it is sufficient to fit a polynomial of degree 4 to the output vector and achieves comparable accuracy as independent value prediction (which must construct 12 regression models).

3.4 Smoothness of Prediction

Another factor to consider is the influence of noise on the prediction approaches. To do this, we conduct an experiment using a simple, stationary time series, i.e., white noise, as shown in Figure 5. Figure 6 shows that multi-stage prediction tends to smooth out the time series to its mean value after p time steps. Such smoothing effect is not present in independent value prediction, which makes spurious predictions around the mean, because the prediction at each time step is modeled independently. This method may suffer from overfitting as it tries to capture the fluctuations of the noise time series. For parameter prediction, the best fit model of the data is found to be a polynomial of degree zero. Even though the parameters are predicted independently, the smoothness of the time series is guaranteed by the parametric function used to fit the output vector.

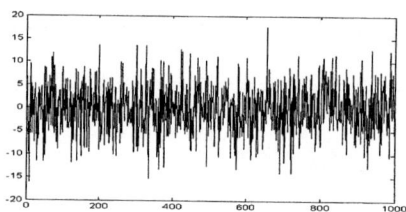

Fig. 5. White Noise WN(0,0.5)

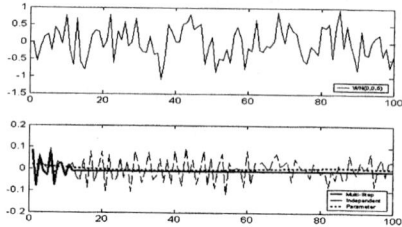

Fig. 6. Predicting Results (p=12,h=100,d=6)

3.5 A General Comparison

Finally, we apply the three prediction approaches to 21 real data sets to compare their relative performance. The RMSE value for each data set is obtained by 10-fold cross validation. The size of the prediction window is set to $h=24$. Table 2 summarizes the RMSE for the three prediction approaches using MLR as the underlying regression method. Their relative performance is summarized in Table 3 in terms of the number of wins, draws and losses. We also test the significance of the difference using paired t-significance test. The result shows that the observed difference between the RMSE of multi-stage and independent value prediction is not that significant. However, the performance of parameter prediction is significantly worse than independent value prediction. This is because MLR may not be suitable to fit the parameters of the function, which have nonlinear relationships with the time series values.

Table 2. Multiple Linear Regression

	Multi-stage	Independent	Parameter
milk	0.0733	0.0705	0.0918
Temp.	0.2776	0.2959	0.2936
PET	0.0419	0.0414	0.0619
PREC	0.0310	0.0317	0.0534
Solar	0.1218	0.1236	0.2132
appb	0.2974	0.3804	0.3803
appd	0.2152	0.2395	0.2766
appf	0.3147	0.2445	0.2991
appg	0.8642	0.9343	0.9218
deaths	0.7309	0.5560	0.5633
lead	0.4195	0.4207	0.4206
sales	0.3187	0.3637	0.3637
wine	0.2738	0.2902	0.3209
seriesc	0.9359	0.9845	0.9845
odono	0.4226	0.4731	0.4712
qbirth	0.5450	0.4793	0.5231
Bond2	0.5226	0.5886	0.5884
Daily	0.2006	0.2137	0.2137
food	0.1995	0.1929	0.1950
treerin	0.8929	0.8807	0.8818
pork	0.9462	0.7948	0.7918

Table 4. RNN

	Multi-stage	Independent	Parameter
Milk	0.0733	0.0705	0.0918
Temp.	0.2776	0.2959	0.2936
PET	0.0419	0.0414	0.0619
PREC	0.0310	0.0317	0.0534
Solar	0.1218	0.1236	0.2132
Appb	0.2974	0.3804	0.3803
Appd	0.2152	0.2395	0.2766
Appf	0.3147	0.2445	0.2991
Appg	0.8642	0.9343	0.9218
Deaths	0.7309	0.5560	0.5633
Lead	0.4195	0.4207	0.4206
Sales	0.3187	0.3637	0.3637
Wine	0.2738	0.2902	0.3209
Seriesc	0.9359	0.9845	0.9845
Odono	0.4226	0.4731	0.4712
Qbirth	0.5450	0.4793	0.5231
Bond2	0.5226	0.5886	0.5884
Daily	0.2006	0.2137	0.2137
Food	0.1995	0.1929	0.1950
Treerin	0.8929	0.8807	0.8818
Pork	0.9462	0.7948	0.7918

Table 3. Win-Draw-Loss results for MLP

	Multi vs Indep	Multi vs Param	Indep vs Param
0.01diff	10-6-5	14-2-5	8-12-1
T value	0.1496	0.8761	2.7299
P value	0.8826	0.3914	0.0129

Table 5. Win-Draw-Loss results for RNN

	Multi vs Indep	Multi vs Param	Indep vs Param
0.01 diff	7-0-14	5-2-14	13-1-7
T value	2.6396	3.3884	0.3012
P value	0.0157	0.0029	0.7664

Tables 4 and 5 show the results using RNN as the underlying regression method. Observe that the independent value and parameter prediction approaches perform significantly better than multi-stage prediction ($p < 0.05$). For multi-stage prediction, the RMSE for RNN is higher than the RMSE of MLR in 10 out of 21 data sets, which suggests the possibility of model overfitting when using a flexible regression method such as RNN. Nevertheless, we still find 17 data sets in which independent value prediction with RNN outperforms all the prediction approaches using MLR and 12 data sets in which parameter prediction with RNN outperforms all the prediction approaches using MLR. This result suggests that, using nonlinear regression methods such as RNN, the independent value and parameter prediction approaches may achieve better performance than multi-stage prediction. Moreover, for parameter prediction, most of the data sets require $d < 5$, which makes it more efficient to build compared to independent value prediction (which requires building $h = 24$ models).

4 Conclusions

In this paper, we conduct an empirical study on three prediction approaches for solving multistep-ahead time series prediction problems. The tradeoffs among these approaches are studied using both real and synthetic data sets. Our experimental results show that multi-stage prediction tends to suffer from error accumulation problems when the prediction period is long. This is because the bias and variance from previous time steps are propagated into future predictions. Independent value prediction is less susceptible to this problem because its predictions are made independently at each time step. However, it has difficulty in learning the true model because the function to be modeled becomes more complex with increasing time steps. This approach also does not smooth out the effect of noise unlike multi-stage prediction. Parameter prediction handles noisy data by fitting a function over the entire output sequence while alleviating the error accumulation problem by making independent predictions. It also tends to be more efficient than independent value prediction when the number of parameters to be fitted is small. However, finding the appropriate parameter function to fit the time series can be quite a challenging task. Finally, we observe successful applications of both independent value and parameter prediction approaches when applied to real data sets using RNN.

References

1. Gershenfeld N. A. and Weigend A. S.: The Future of Time Series. In "Time Series Prediction: Forecasting the Future and Understanding the Past",(1993) 1-70.
2. Jones R. H.: Maximum likelihood fitting of ARMA models to time series with missing observations. Technometrics 20, (1980) 389–395.
3. Giles C.L., Lawrence S. and Tsoi A.C.: Noisy Time Series Prediction using a Recurrent Neural Network and Grammatical Inference. Machine Learning, 44(1-2) (2001) 161-183.
4. Elman J.L.: Distributed Representations, Simple Recurrent Networks, and Grammatical Structure. Machine Learning, 7 (2/3) (1991) 195–226.
5. Kennel M. B., Brown R., and Abarbanel H. D. I.: Determining embedding dimension for phase-space reconstruction using a geometrical construction. Phys. Rev. A 45 (1992) 3403.
6. Hyndman R., Time Series Data Library. http://www-personal.buseco.monash.edu.au/~hyndman/ TSDL/
7. Joseph Rynkiewicz: Hybrid HMM/MLP models for time series prediction. Proc of the European Symposium on Artificial Neural Networks Brugges, Belgium,(1999). 455-462.
8. Akaike H.: Fitting autoregressive models for prediction. Annals of the Institute of Statistical Mathematics, 21 (1969) 243-247
9. UCI Machine Learning Repository http://www.ics.uci.edu/~mlearn/MLRepository.html
10. Y. Le Borgne.: Bias variance trade-off characterization in a classification. What differences with regression? Technical Report N°534, ULB, (2005).

Sequential Pattern Mining with Time Intervals

Yu Hirate and Hayato Yamana

Dept. of Computer Science, Waseda University,
3-4-1 Okubo Shinjuku-ku Tokyo, Japan
{hirate, yamana}@yama.info.waseda.ac.jp

Abstract. Sequential pattern mining can be used to extract frequent sequences maintaining their transaction order. As conventional sequential pattern mining methods do not consider transaction occurrence time intervals, it is impossible to predict the time intervals of any two transactions extracted as frequent sequences. Thus, from extracted sequential patterns, although users are able to predict what events will occur, they are not able to predict when the events will occur. Here, we propose a new sequential pattern mining method that considers time intervals. Using Japanese earthquake data, we confirmed that our method is able to extract new types of frequent sequences that are not extracted by conventional sequential pattern mining methods.

1 Introduction

Sequential pattern mining methods[1, 2] extract frequent sequences with their transaction occurrence order, but without time intervals between their occurrence time. Thus, it is impossible to identify the time intervals between any two transactions extracted as frequent sequential patterns. It is useful to be able to understand what events will occur, and moreover when these events will occur. To distinguish time intervals from extracted sequences using conventional sequential pattern mining methods, it is necessary to add constraints of time infomation to extracted sequences[3, 4]. However, in order to distinguish sequences which consist of the same items with different time intervals, constraint approach algorithms need to be re-executed with changing their constraints.

To solve this problem, we propose a new sequential pattern mining with time intervals. The proposed method enables sequential patterns to be distinguished with any time interval that are multiples of a user-defined base interval.

2 Related Work

Modified PrefixSpan[5], which is an alternative sequential pattern mining, defines sequences with the same items and different item gaps as different sequences. Item Gaps are defined as the number of items between any two items. Modified PrefixSpan works well when applied to datasets whose item intervals are defined as the number of items, such as DNA sequences. However, it does not work well when applied to datasets whose items have their own occurrence time.

3 Sequential Pattern Mining with Time Intervals

3.1 Problem Definition

Let $I = \{i_1, i_2, \cdots, i_n\}$ be a set of items. A transaction t is a list of items sorted alphabetically. A transaction t has a time-stamp infomation, denoted as $t.time$. The time interval extended sequence, denoted as ts, is a list of transactions with time intervals, and is defined as follows:

$$ts = <-D(t_1,t_1) - t_1, -D(t_1,t_2) - t_2, -D(t_1,t_3) - t_3, \cdots, -D(t_1,t_m) - t_m >$$
$$\wedge t_1.time \leq t_2.time \leq t_3.time \leq \cdots \leq t_m.time \tag{1}$$

Here, $D(t_\alpha, t_\beta)$ is the time interval between t_α and t_β, and is defined by the following expression:

$$D(t_\alpha, t_\beta) = d_n \mid n = \lfloor \frac{t_\beta.time - t_\alpha.time}{\Delta t} \rfloor \tag{2}$$

Note that Δt is a user-defined parameter, and determines the unit of time interval partition. For example, when Δt is set to 1 day, $D(t_\alpha, t_\beta)$ is defined as follows:

$$D(t_\alpha, t_\beta) = \begin{cases} d_0 & (t_\beta.time - t_\alpha.time = 0) \\ d_1 & (0 < t_\beta.time - t_\alpha.time \leq 1 day) \\ d_2 & (1 day < t_\beta.time - t_\alpha.time \leq 2 days) \\ \vdots & \end{cases} \tag{3}$$

Note that when ts starts with $< -d_0 - t_1, \cdots >$, it is possible to omit "$-d_0-$" and to represent as $< t_1, \cdots >$.

The time interval extended sequential database, $TSDB$, which is a target of pattern extraction, is a set of time intervals for the extended sequence ts, and is defined as $TSDB = \{ts_1, ts_2, \cdots, ts_t\}$. When two time interval extended sequences, $ts1 =< -D(t_1,t_1) - t_1, \cdots, -D(t_1,t_m) - t_m >$ and $ts2 =< -D(t'_1,t'_1) - t'_1, \cdots, -D(t'_1,t'_m) - t'_m, \cdots, -D(t'_1,t'_n) - t'_n > (m \leq n)$, are given, we say that $ts2$ includes $ts1$ iff $t_i \subset t'_i$ for all $\{i | 1 \leq i \leq m\}$ and $D(t_1, t_k) = D(t'_1, t'_k)$ for all $\{k | 1 \leq k \leq m\}$. The support of time interval extended sequence ts in $TSDB$, denoted as $sup_{TSDB}(ts)$, is the percentage of time interval extended sequences that include ts. A frequent time interval extended sequence is defined as the time interval extended sequence whose support is higher than $min\_sup (0 \leq min\_sup \leq 1)$. Given $TSDB$ and min_sup, sequential pattern mining with time intervals extracts all the frequent time interval extended sequences.

3.2 Proposed Method

To extract frequent time interval extended sequences, we extended the sequential database projection operation in the PrefixSpan algorithm[2]. Similar to PrefixSpan, our algorithm extracts frequent time interval extended sequences with a depth-first search by executing the projection operation recursively. In this section, we describe our projection operation in detail.

Definition1. **Projection Level**
Projection level is the number of items included in a projection sequence. For example, let $ts1$ be a time interval extended sequence with l items. Generating $TSDB|ts1$ is level l projection.

Definition2. **Prefix and Postfix of Time Interval Extended Sequences**
Let $ts = < -D(t_1,t_1) - t_1, \cdots, -D(t_1,t_m) - t_m >$ be a time interval extended sequence, and t_α be any transaction. When there exists an integer $j (1 \leq j \leq m)$ satisfying $t_\alpha \in t_j$ and $D(t_1, t_\alpha) = D(t_1, t_j)$, we define the time interval extended sequence $< -D(t_1,t_1) - t_1, \cdots, -D(t_1,t_j) - t_\alpha >$ as a prefix with regard to t_α and $D(t_1, t_\alpha)$, denoted as $prefix(ts, t_\alpha, D(t_1, t_\alpha))$, and the remaining time interval extended sequence $< -D(t_j, t_j) - t'_j, \cdots, -D(t_j, t_m) - t_m >$ as a postfix with regard to t_α and $D(t_1, t_\alpha)$, denoted as $postfix(ts, t_\alpha, D(t_1, t_\alpha))$, where t'_j is the subset of t_j from which items also including in t_α are excluded. When $t'_j = \phi$, $postfix(ts, t_\alpha, D(t_1, t_\alpha))$ becomes $< -D(t_j, t_{j+1}) - t_{j+1}, \cdots, -D(t_j, t_m) - t_m >$. On the other hand, when there exists no integer j, both $prefix(ts, t_\alpha, D(t_1, t_\alpha))$ and $postfix(ts, t_\alpha, D(t_1, t_\alpha))$ are defined as ϕ.

Our projection operation performs differently between Level 1 projection and Level 2 or later projection.

Level 1 Projection. In the case of Level 1 Projection, as it is impossible to define time intervals with a single transaction, our algorithm scans $TSDB$ and checks all items' supports, similar to PrefixSpan. For every item i_α whose support is higher than the minimum support, generate $TSDB|i_\alpha$ then execute the level 2 projection operation. Note that when item i_α appears more than once in the same ts, our algorithm generates multiple prefixes and postfixes at each item i_α, and then treats them as different sequences. For example, let $ts1 = < -d_0 - a, -d_1 - (abc), -d_2 - (ac) >$, then the projection result whose projection transaction is $< a >$ becomes 3 time interval extended sequences, $< -d_1 - (abc), -d_2 - (ac) >$, $< -d_0 - (bc), -d_2 - (ac) >$, and $< -d_0 - c >$.

Level 2 or later Projection. In the case of Level 2 or later projection, our algorithm scans projected $TSDB$ and counts pairs of items included in projected sequences and their time intervals. Our algorithm projects $TSDB$ in the following way. Let $ts_\alpha = < -D(t_1,t_1) - t_1, \cdots, -D(t_1, t_{\alpha-1}) - t_{\alpha-1}, -D(t_1, t_\alpha) - t_\alpha >$, $ts_{\alpha-1} = prefix(ts_\alpha, t_{\alpha-1}, D(t_1, t_{\alpha-1}))$ ($\alpha \geq 1$). ts_α-projected database, denoted as $TSDB|ts_\alpha$, is defined as follows:

$$TSDB|ts_\alpha = \Big\{ ts' \Big| ts \in TSDB|ts_{\alpha-1} \land ts' = postfix(ts, t_{\alpha-1}, D(t_{\alpha-1}, t_\alpha))$$
$$\land sup_{TSDB|ts_{\alpha-1}}(t_\alpha, D(t_{\alpha-1}, t_\alpha)) \geq min\_sup \land ts' \neq \phi \Big\} \quad (4)$$

ts_α-projected database is a collection of projected time interval extended sequences, which are postfixes of time interval extended sequences included in $ts_{\alpha-1}$-projected database, with regard to t_α and $D(t_{\alpha-1}, t_\alpha)$, and ts_α is called a projection time interval extended sequence of $TSDB|ts_\alpha$.

4 Evaluation

In this evaluation, a Japanese earthquake dataset was used. The dataset is distributed via K-net[6] provided by the National Research Institute of Earth Science and Disaster Prevention and includes data from 3,296 earthquakes that occurred from May 1995 to December 2003.

A sequence of this dataset is defined as a list of earthquakes in the same grid. A grid is defined as 1-degree latitude and 1-degree longitude square. And all earthquake data are itemized to the same item referring to depth of the epicenter and magnitude on the Richter scale.

Extracted sequence quality. Table 1 shows a part of the frequent time interval extended sequential patterns extracted by proposed method with $min\_sup = 0.05$ and $\Delta t = 1 day$. Calculating the confidence based on the support of patterns shown in Table 1 yielded the following knowledge. Once item-A occurred, item-A will occur again:

- within 1 day with probability of $\frac{0.222}{0.723} \times 100 = 30\%$.
- within 1 to 2 days with probability of $\frac{0.101}{0.723} \times 100 = 13\%$.
- within 2 to 3 days with probability of $\frac{0.081}{0.723} \times 100 = 11\%$.

On the other hand, there is no time interval information in extracted patterns using conventional sequential pattern mining. Thus, users are not able to predict how long after item-A the event will occur again. These results indicate that the patterns extracted by the proposed method are more useful than those extracted by the conventional sequential pattern mining algorithm.

Performance evaluation. In the next section, the running times of both PrefixSpan and the proposed Algorithm are compared. Figure 1 shows the relation between the number of extracted frequent sequential patterns and min_sup. Figure 2 shows the relation between execution time and min_sup.

Figure 1 and Figure 2 shows the following. Using PrefixSpan, as the number of extracted sequences increased exponentially as min_sup decreased, thus execution time increase exponentially as min_sup decreased. On the other hand,

Table 1. Partial results extracted using the proposed method

Extended Time Interval Extended Sequential Patterns	Support
$< A >$	0.723
$< A, -d_0 - A >$	0.222
$< A, -d_1 - A >$	0.101
$< A, -d_2 - A >$	0.081

item-A: Earthquake of magnitude >4.0 and <6.0, and with an epicenter depth from 10 km to 100 km.

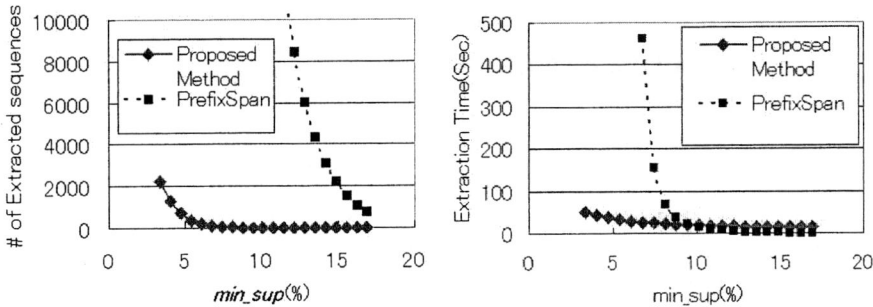

Fig. 1. # of Patterns vs. min_sup   **Fig. 2.** Execution time vs. min_sup

since the proposed method distinguishes between sequences consisting of the same transactions with different time intervals, proposed method can prevent the huge increase in both the number of extracted sequences and running time. These results indicate that users can analyze more sensitive patterns in a short time using the proposed method.

5 Conclusions

Here, we propose a new type of sequential pattern mining method based on time intervals. Evaluations confirmed that the sequential pattern mining based on time intervals is able to extract more useful patterns for feedback to the real world than conventional sequential pattern mining, and that users can also analyze more sensitive patterns in a shorter time in comparison with using conventional sequential pattern mining.

Acknowledgments. This research was funded in part by both "e-Society" and "21-century COE Programs: ICT Productive Academia" of MEXT in Japan.

References

1. R. Agrawal and R. Srikant, "Mining Sequential Patterns," In Proc. of ICDE'95, pp. 3-14, 1995.
2. J. Pei, J. Han, B. Mortazavi-Asl, H. Pinto, Q. Chen, U. Dayal, and -C. M. Hsu, "PrefixSpan: Mining Sequential Patterns Efficiently by Prefix-Projected Pattern Growth," In Proc of ICDE'01, pp.215-224, 2001.
3. M.J. Zaki, "Sequence Mining in Categorical Domains: Incorporating Constraints," In Proc. of CIKM'00, pp. 422–429, 2000.
4. J. Pei, J. Han and W. Wang, "Mining Sequential Pattern with Constraints in Large Databases," In Proc. of CIKM'02, pp. 18–25, 2002.
5. H. Kitakami, T. Kanbara, Y. Mori, S. Kuroki, and Y. Yamazaki, "Modified PrefixSpan Method for Motif Discovery in Sequence Databases," In Proc. of PRICAI2002, pp.482-491, 2002.
6. K-NET Kyoshin Network, http://www.k-net.bosai.go.jp

A Wavelet Analysis Based Data Processing for Time Series of Data Mining Predicting*

Weimin Tong, Yijun Li, and Qiang Ye

School of Management, Harbin Institute of Technology,
Harbin, 150001, China
tongweimin@hit.edu.cn

Abstract. This paper presents wavelet method for time series in business-field forecasting. An autoregressive moving average (ARMA) model is used, it can model the near-periodicity, nonstationarity and nonlinearity existed in business short-term time series. According to the wavelet denoising, wavelet decomposition and wavelet reconstruction, the hidden period and the nonstationarity existed in time series are extracted and separated by wavelet transformation. The characteristic of wavelet decomposition series is applied to BP networks and an autoregressive moving average (ARMA) model. It shows that the proposed method can provide more accurate results than the conventional techniques, like those only using BP networks or autoregressive moving average (ARMA) models.

1 Introduction

The reliable, secure and economical commercial forecasting plays an important role in the business field. Therefore forecasting methods have been studied deeply and can be divided into two categories, namely classical methods and intellectual technologies [1]-[3]. The classical methods mainly include many models based on various statistical theories, whereas intellectual technologies include artificial neural network methods and expert system approaches [4]-[11]. In practice all these methods have been applied in the research of time series to certain degrees. However, because of time series' near-periodicity, non-stationarity and nonlinearity, difficulties do occur when using these methods to solve practical problems [4], [6]. Firstly, Classical time series analysis methods mainly depend on linear time model and linear spectral estimation. Although these methods are based on the simple theories and are convenient to be applied in practice, when forecasting precision needs to be enhanced so as to amplify the model's scale, the forecasting precision becomes lower and the forecasting speed becomes slower due to time series' essence of non-linearity. Secondly, expert system techniques, utilizing the knowledge and analogical reasoning of experienced human operators, have been investigated [12]. Although expert system can synthesize many influencing factors, its knowledge base is very difficult to describe and build and its parameters can't be adjusted flexibly. As a result, in

---

* Sponsored by National Natural Science Foundation of China (Grant No. 70501009).

practice, the application of expert system is much limited. In addition, several research groups have studied the use of artificial neural networks (ANN) [13], [14] for ANN techniques are excellent tools to describe nonlinear relation [5]. However when the dependence of ANN learns represents relatively strong nonlinearity, it requires a large numbers of input cells, so the neural network training can often run into local minimum and the convergence speed is extremely low, and finally a great quantity of cyber-resources and times have been consumed. This proves that ANN techniques are inappropriate to solve practical problems [15]-[17]. In this paper, we intend to apply wavelet methods into data mining in the business field, and elucidate in details the procedure and effect of this new method applied in the time series prediction, through analyzing the real data of one supermarket.

2 Wavelet Analysis and ARMA Model

In this paper the data processing flow can be divided into four phases. The first phase is wavelet denoising. Its aim is to eliminate the noise of signal. The second phase is the wavelet decomposition. It aims to attain the wavelet decomposition series and the last approximation series of original time series at each scale domains. The third is modeling and forecasting. Its role is modeling and forecasting for these wavelet decomposition series at each transformation domains by using BP neural net, and for the final scale decomposition series by using ARMA model. The fourth is the wavelet reconstruction. Here the wavelet reconstructions have been achieved with purpose of synthesizing those forecast series obtained at each transformation domains, to system final forecasting, utilizing the wavelet reconstruction technique.

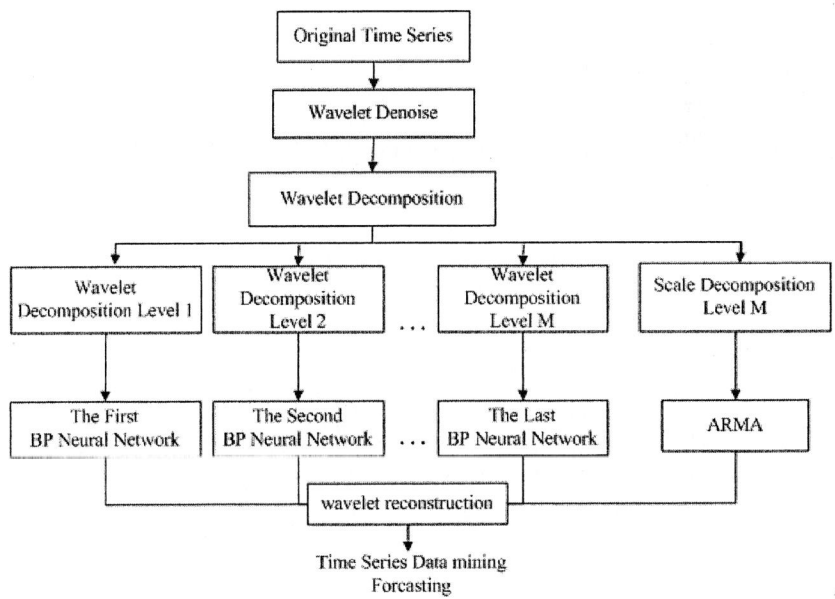

Fig. 1. The architecture of wavelet-neural network-ARMA model

In figure 1, original system is any time series needs to be analyzed. Wavelet denoising level 1, Wavelet decomposition level 2, wavelet decomposition level 3… wavelet decomposition level M are M wavelet decomposition series of original time series at M levels (decomposition layer) and their notations are d_1, d_2, …, d_M, respectively. Scale decomposition level M is M^{th} scale decomposition series, its notation is a_M.

The final phase of these forecasting models is to synthesize forecasting series at various transformation domains in order to obtain final forecasting series by using wavelet reconstruction technology.

3 Time Series-Based Data Predict

There will be an analysis of 2688 sample datum (4 weeks) of one supermarket, and make forecasting.

3.1 Noise Elimination of Original Data by Wavelet Analysis

Because the original data we collected contain big noise, the original data have obvious false data and can't be used directly. Aiming to this problem, noise elimination processing using the fault soft threshold due to the lack of enough data and correlative empirical equation. The fault threshold we assigned is 32.259. The low frequency coefficient hasn't quantization of threshold. The wavelet we chose is biorthogonal wavelet, and its parameter is (4,4) and level number is 5. The results comparison between original data and de-noised results are shown in figure 2 and figure 3.

In the following text, the time series of load we will analyze are the de-noised time series except for special illustration.

3.2 Wavelet Decompositions

According to the characteristic of the wavelet transformation, the length of decomposition series is decreased doubly with scale increased doubly. Since the scale

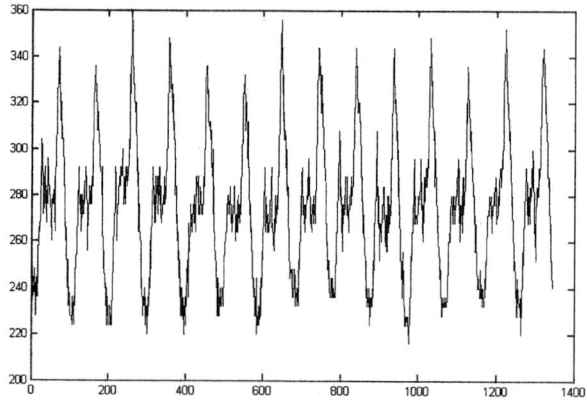

Fig. 2. The sample time series of original system load

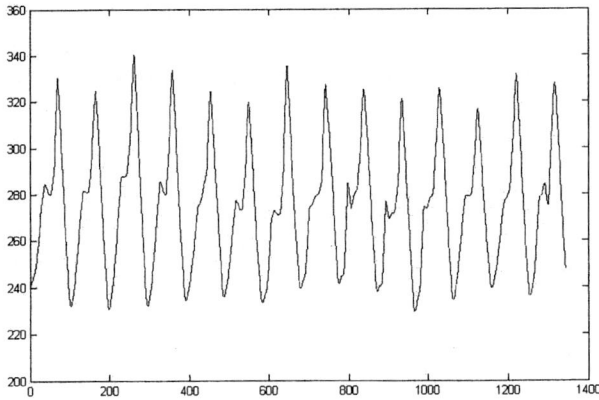

Fig. 3. The de-noised signal by using the default threshold

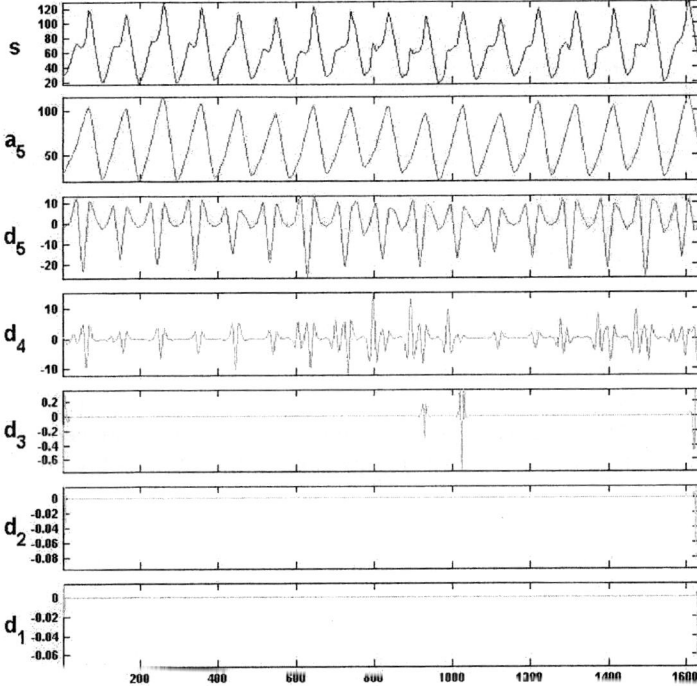

Fig. 4. Time series and its redundant wavelet and scale transforms

of the new series is different on each scale, the intervals of samples from neighboring data of series are different too. In order to facilitate modeling, it can make the new series and original series have the same length and sample, and name them as redundant wavelet transformation and redundant scale transformation. The coefficient of Low-pass filter in Biorthogonal wavelets bior(4,4) is (0, 0.0378, -0.0238, -0.1106, 0.3774, 0.8527, 0.3774, -0.1106, -0.0238, 0.0378), the coefficient of High-pass filter is (0, -0.0645, 0.0407, 0.4181, -0.7885, 0.0407, 0.4181, -0.0645, 0, 0), they also have 10 coefficient.

In figure 4, on the top is the primitive time series of 1344 data, its notation is S. In addition to it, from the top to the bottom the scale transformation time series of the 5th scale level is a_5. The wavelet transformation time series of 5th, 4th, 3rd, 2nd, 1st scale level is d_5, d_4, d_3, d_2, d_1. By all appearances, there is obvious near-periodicity in series S, a_5 and d_5, and this near- periodicity is one day, it reflects for 96 continuous points in the figure.

As for scale level, in fact 1st scale level corresponds to scale 2, afterwards 4, 8, and 16, 32 in turn. When scale is short, wavelet transformation series express strong change, embodying that the reliance relation of the data is not good, namely, the reliance time is short, but change scope is little (compared with original data). All these show that this component have no great impact on change trend of original time series, it is just a kind of part influence factor, so the prediction data of this time series has little impact on total predict of original time series. With the increasing scale, wavelet time series becomes more and more smooth. It becomes plainness in the 5th scale, showing that it has stable impact on original time series, and effect time is much longer. In addition, the scale transformations in 5th scale has also become plain and keep the same trend with the original time series, furthermore its value is close to the original time series. This shows that the scale transformation series has the essential impact on original time series. The characters of the time series make wavelet-neural networks-ARMA model available to predict and analysis the original time series.

3.3 Modeling and Forecasting in Scale

When BP neutral nets are built at five wavelet transformation domains, every net model includes three layers, namely, an input layer, a hidden layer and a output layer, and there is only one output cell in every output layer. Design parameters of the five net models are shown in the table 1. In this table, the notation BP_1 is the BP neutral net model, which is built for d_1.

Table 1. Design parameters of BP networks at every scale

Model Parameter	BP_1	BP_2	BP_3	BP_4	BP_5
Input	7	7	7	7	13
Output	14	28	28	56	112

Because the characteristic of these 5 wavelet transformation series is that the range changed is very small relative to original series, and d_5 which range changed is a little greater than other fours has near-periodicity, every input cell of this 5 models is not data on continuous position of time series, but is a vector that contains 96 datum in succession. For example, there are 13 input cells in model BP_5, and X={$x_1, x_2... x_{13}$}

is the input vector, then x_1 is a vector which consists of the first 96 data of series d_5, and x_2 is a vector which consists of the second 96 data of series d_5, and so on.

For conveniently describing, $x_1, x_2...x_n$ represent each value in the processed 5th scale decomposition series a_5 received zero equalization transfers, then the problem of model building for the 5th scale decomposition series a_5 is equal to how to make the series $(x_1, x_2...x_n)$ fit into ARMA (p, q) model, the model is:

$$x_t = \alpha_1 x_{t-1} + \alpha_2 x_{t-2} + \cdots + \alpha_p x_{t-p} + \varepsilon_t - \beta_1 \varepsilon_{t-1} - \beta_2 \varepsilon_{t-2} - \cdots - \beta_q \varepsilon_{t-q}.$$
$$t = p+1, p+2, \cdots, n \quad (1)$$

$\{\varepsilon_t\}$ is white noise series, $E\varepsilon_t^2 = \sigma^2$, $E\varepsilon_t x_s = 0 \ (s < t)$

Firstly the parameter evaluation is processed. The regression approximation method is applied in this paper and consists of two steps. The first step is to make the series $(x_1, x_2...x_n)$ fit the regression model AR (p). And properly large positive integer is empirically selected as order P $(>> p, q)$. Base on the premise that assure precision of evaluation and reduce calculation amount, we select $p= 300$ in this paper. The fitted model is:

$$x_t = \hat{\alpha}_1 x_{t-1} + \hat{\alpha}_2 x_{t-2} + \cdots + \hat{\alpha}_P x_{t-P} + \hat{\varepsilon}_t. \quad (2)$$

The second step is to calculate fitted residual error through the formula (2).

$$\hat{\varepsilon}_t = x_t - \hat{\alpha}_1 x_{t-1} - \hat{\alpha}_2 x_{t-2} - \cdots - \hat{\alpha}_P x_{t-P} \,.\, t = P+1, P+2, \cdots, n \quad (3)$$

This fitted residual error series, $\hat{\varepsilon}_{P+1}, \hat{\varepsilon}_{P+2}, \cdots, \hat{\varepsilon}_n$, can be seen as the sample values of $\{\varepsilon_t\}$ series in the formula (1). Then, the formula (1) can be near written as

$$x_t = \alpha_1 x_{t-1} + \alpha_2 x_{t-2} + \cdots + \alpha_p x_{t-p} - \beta_1 \hat{\varepsilon}_{t-1} - \beta_2 \hat{\varepsilon}_{t-2} - \cdots - \beta_q \hat{\varepsilon}_{t-q} + \varepsilon_t.$$
$$t = P+1, P+2, \cdots, n \quad (4)$$

Besides $\{\varepsilon_t\}$ in the formula (4), $\{x_t\}$ and $\{\hat{\varepsilon}_t\}$ have sample values or near-sample values in the formula $t = P+1, P+2, \cdots, n$. Thereby the formula (4) has matrix form:

$$x = (X\hat{E})\begin{pmatrix} \alpha \\ -\beta \end{pmatrix} + \varepsilon. \quad (5)$$

Due to only value of data series $x_1, x_2...x_n$ are known, the order is also unknown when we fit into ARMA (p, q) model. Therefore it is essential to evaluate the values of p and q when the formula (1) is fitted, AIC order-confirmation criterion is employed to get the ARMA (2, 1) model in this paper. Next this model will forecast the scale transformation series of 5th scale as follows:

Because $p=2, q=1$, it can be deduced that

$$\hat{\varepsilon}_k = x_k - \alpha_1 x_{k-1} - \alpha_2 x_{k-2} + \beta_1 \hat{\varepsilon}_{k-1} \ k = 1, 2, \cdots, n \quad (6)$$

After $\hat{\varepsilon}_k$ values have been confirmed, each x_{n+k} can be ascertained via confirmation of x_{n+k} conditional expectation, namely

$$\begin{aligned} x_{n+1} &= E(x_{n+1}) \\ &= E(\alpha_1 x_n + \alpha_2 x_{n-1} + \varepsilon_{n+1} - \beta \hat{\varepsilon}_n) \\ &= \alpha_1 x_n + \alpha_2 x_{n-1} - \beta \hat{\varepsilon}_n \end{aligned} \quad (7)$$

When $k>1$, moving average part disappear,

$$x_{n+k} = \alpha_1 x_{n+k-1} + \alpha_2 x_{n+k-2} \quad (8)$$

x_{n+k} near-forecasting value can be calculated combining with known $\hat{\varepsilon}_k$, then the forecasting series can be gotten by each value of series $\{x_{n+k}\}$ plus scale transformation series mean value finally.

3.4 Short Term Time Series Forecasting

According to the wavelet method, in order to get the systematic short-term forecast of 15- minute interval, wavelet rebuild technology is used to combine the six forecasting series in turn and finally get the prediction data.

In order to compare with other methods (for example, only using BP neural networks or ARMA for model building), $MSE = \sum_{t=1}^{N}(z_t - \hat{z}_t)^2 / N$ is defined and mean square error is used to describe index of forecasting performance. z_t stands for the actual data of time series, and $\hat{z}_t$ stands for the forecasting data. After using 1334 datum of two weeks predicts 96 data in the next day through three different methods, the indexes of forecasting performance of three methods, which are wavelet-NN-ARMA method, BP-NN method and ARMA method, are shown in the table 2. These results prove that the performance of wavelet-NN-ARMA is more ideal than the other two methods'.

Table 2. The comparison of mean-square error of the three methods

	Method of This Paper	BP	ARMA
Forecasting Data	5.6228	115.9997	19.4109

In order to show forecasting performance further there are statistic result by relative error in the table 3, 4 and 5. According to this comparison, the advantage of wavelet-NN-ARMA method is even apparent. The statistical results of relative error analysis of load forecast on the next day forecasted by above-mentioned three forecast methods are shown respectively in the table 3, 4 and 5. These comparison results demonstrate the superiority of the method in this paper adequately.

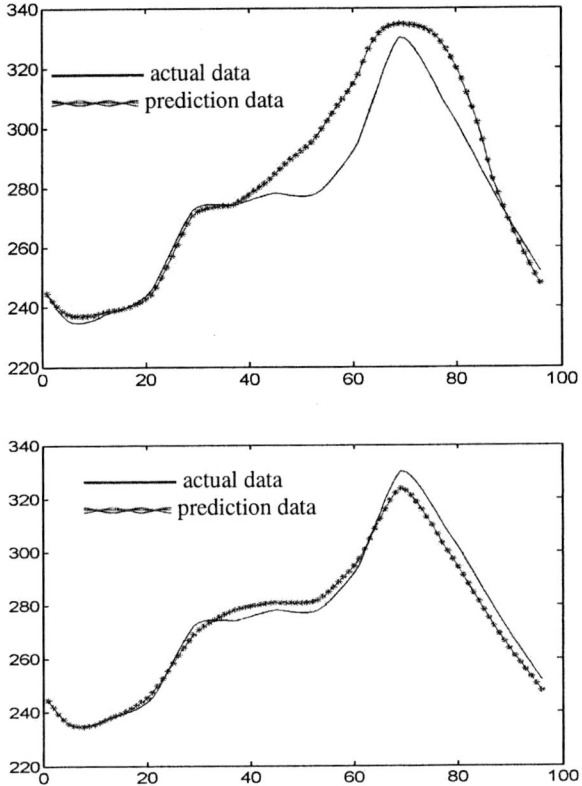

Fig. 5. The forecasting effects of BP network and ARMA

Table 3. The relative error accuracy analysis of the method in this paper

Statistic	RE<1%	RE<2%	RE<3%	RE>3%
Points	60	96	96	0
Percentage	62.5%	100%	100%	0%
Maximal RE	1.80%	Average RE	0.69%	

Table 4. The relative error accuracy analysis of BP network forecasting method

Statistic	RE<1%	RE<2%	RE<3%	RE>3%
Points	46	57	62	34
Percentage	47.92%	59.38%	64.58%	35.42%
Maximal RE	7.80%	Average RE	2.60%	

Table 5. The relative error accuracy analysis of ARMA forecasting method

Statistic	RE<1%	RE<2%	RE<3%	RE>3%
Points	40	73	96	0
Percentage	41.67%	76.04%	100%	0%
Maximal RE	2.64%	Average RE	1.22%	

4 Conclusions

Wavelet denoising is emphasized in this paper. Combining wavelet, ARMA model and neural networks in data mining time series forecast can not only separate all kinds of hidden periods and describe them effectively, but also well depict its essential nonlinearity, thus increase the forecast precision of the data mining time series. The analysis of the examples in practice proves that, the forecast method mentioned in this paper can be applied successfully and effectively in the business field.

References

1. C. C. Aggarwal./Yu, P. S.: Data Mining Techniques for Associations, Clustering and Classification. In: Zhong, N./Zhou L. Z., (eds.), Methodologies for Knowledge Discovery and Data Mining, PAKDD-99, Springer, Berlin, 1999, P. 13-23.
2. D. Agrawal and C. C. Aggarwal. On the design and quantification of privacy preserving data mining algorithms. In PODS, pages 247--255, 2001.
3. Christopher J.C. Burges. A Tutorial on Support Vector Machines for Pattern Recognition. Data Mining and Knowledge Discovery,1998.
4. G. P. Nason, and R. V. Sachs. (1999) Wavelets in time- series analysis. Phil. Trans. R. Soc. Lond. A, 357 (1760), 2511-2526.
5. A. J.Campbell and Murtagh, F.(1997) Combining neural networks forecasts on wavelet transformed time series. Connection Sci., vol. 9, 113–121.
6. Wang Jianze, Ran Qiwen, Ji Yaochao, Liu Zhuo(1998) Frequency Domain Analysis of Wavelet Transform in Harmonics Detection. AEPS. 1998, 22(7):40-43
7. S. Rahman and R. Bhatnagar(1988) An expert system based algorithm for short term load forecast. IEEE Trans. Power Systems, 3(2): 392–399.
8. Song Aiguo and Lu Jiren(1998). Evolving Gaussian RBF network for nonlinear time series modelling and prediction. Electro- nics Lett., 34(12): 1241-1243.
9. S. Chen(1995) Nonlinear time series mo- delling and prediction using Gaussian RBF network with enhanced clustering and RLS learning. Electron Lett. 31(2): 117-118.
10. H.T.Yang and C.M.Huang. A new short term load forecasting approach using self organizing fuzzy ARMAX models. IEEE Trans.Power Systems, 1998, 13(1): 217-225.
11. B.Geva. Scale Net-Multiscale neural network architecture for time series prediction. IEEE Trans. Neural Networks, 1998, Vol.9:1471-1482.
12. Moghram and S. Rahman. Analysis and evalua- tion of five short term load forecasting techniques. IEEE Trans.Power Systems,1989, 5(4):1484-1491.

13. A. Campbell and F. Murtagh. Combining neural networks forecasts on wavelet transformed time series. Connection Sci., 1997, vol. 9, pp:113–121.
14. N. Amjady and M. Ehsan. transient stability assessment of power systems by a new estimating neural network. Can. J. Elect. & Comp. Eng., 1997, 22(3): 131–137.
15. Zhao Hongwei, Ren Zhen, Huang Weiying. A Short Load Forecasting Method Based on PAR Model. Proceedings of the CSEE., 1997, 17(5): 348-351
16. S. Rahman. Generalized knowledge-based short- term load forecasting technique. IEEE Trans. Power Systems, 1993, 8(2): 508–514.
17. S. G. Mallat. A theory for multiresolution signal decomposition: the wavelet representation. IEEE Trans. PAMI, 1989, Vol. 11, pp.674–693.

Intelligent Particle Swarm Optimization in Multi-objective Problems

Shinn-Jang Ho[1], Wen-Yuan Ku[2], Jun-Wun Jou[2], Ming-Hao Hung[2], and Shinn-Ying Ho[3]

[1] Department of Automation Engineering, National Formosa University,
Huwei, Yunlin 632, Taiwan
`sjho@nfu.edu.tw`
[2] Department of Information Engineering and Computer Science,
Feng Chia University, Taichung 407, Taiwan
[3] Department of Biological Science and Technology and Institute of Bioinformatics,
National Chiao Tung University, Hsinchu 300, Taiwan
`syho@mail.nctu.edu.tw`

Abstract. In this paper, we proposes a novel intelligent multi-objective particle swarm optimization (IMOPSO) to solve multi-objective optimization problems. High performance of IMOPSO mainly arises from two parts: one is using generalized Pareto-based scale-independent fitness function (GPSISF) can efficiently given all candidate solutions a score, and then decided candidate solutions level. The other one is replacing the conventional particle move process of PSO with an intelligent move mechanism (IMM) based on orthogonal experimental design to enhance the search ability. IMM can evenly sample and analyze from the best experience of an individual particle and group particles by using a systematic reasoning method, and then efficiently generate a good candidate solution for the next move of the particle. Some benchmark functions are used to evaluate the performance of IMOPSO, and compared with some existing multi-objective evolution algorithms. According to experimental results and analysis, they show that IMOPSO performs well.

1 Introduction

Multi-objective optimization is an important research topic for both scientists and engineers. The use of evolutionary algorithms for multi-objective optimization problem (MOOP) has significantly grown in the last few years. This gives rise to a wide variety of new algorithms [1]-[7]. Particle swarm optimization (PSO) is one of the evolutionary computation techniques. Kennedy and Eberhart, inspired by the choreography of a bird flock, proposed PSO [8]. PSO is characterized by its simplicity and straightforward applicability, and it has proved to be efficient for a lot of problems in science and engineering. New variants of the method more suitable for such problems have been developed [9]-[12].

In this paper, we propose an intelligent multi-objective particle swarm optimization IMOPSO using a novel intelligent move mechanism (IMM) and a novel generalized

Pareto-based scale-independent fitness function (GPSISF) to solve multi-objective optimization problem. Based on orthogonal experimental design (OED) [13]-[15], IMM uses a divide-and-conquer approach to efficiently determine the next move of a particle. Generally, a D-dimensional vector of the next move is divided into N ($\leq D$) partial vectors. IMM spends at most $2N$ objective function evaluations to find a potentially good solution consisting of N good partial vectors to be the next move. IMOPSO with IMM has both the advantages of global exploration and local exploitation by focusing on accuracy and computation time.

OED with both orthogonal array (OA) and factor analysis is a typical method in quality control, also an efficient technique in the Taguchi method [16]. Tsai et al. [17] proposed a hybrid Taguchi-genetic algorithm for global numerical optimization therein the Taguchi method is inserted between crossover and mutation operations of a traditional GA. Tanaka proposed an orthogonal design algorithm (ODA) for a comparison with genetic algorithm (GA) searching mechanisms [18]. ODA uses GA-encoding and OED, but uses no recombination or mutation. OED can also be incorporated into the recombination operation of GA. Zhang and Leung proposed an orthogonal genetic algorithm (OGA) [19]. Leung and Wang proposed an improved OGA with quantization (OGA/Q) using an OA-based initial population for global numerical optimization [20]. Both OGA and OGA/Q use OA, but use no factor analysis. Ho et al. proposed population-based evolutionary algorithms with an OED-based recombination for efficiently solving large parameter optimization problems [7], [21]-[23]. In addition, OED also performs well when cooperating with simulated annealing (SA). A point-based orthogonal simulated annealing (OSA) algorithm with OED-based generation mechanism for efficiently determining the next move of SA is recently proposed by Ho et al. [24], [26].

2 Description of the IMOPSO

IMM is the main phase in adjusting the particles' velocities of IMOPSO. Let the particle be a D-dimensional vector $X = [x_1, x_2, ..., X_D]^T$. The major concerns of efficiently using IMM are 1) how to encode system parameters into the particle X and 2) how to effectively divide X into N partial vectors where each partial vector is treated as factor of OED.

2.1 Particle Representation

A suitable way of encoding system parameters into the particle X is important in efficiently using IMM. Problem-specific particle representation and specialized operations are generally more efficient in solving constrained problems than any other methods using penalty approaches [25]. It is shown that maintaining feasibility can make the OED-based operation more efficient than the penalty approach [26]. An illustrative example using an effective parameter transformation to reduce the degree of epitasis and confine searches within feasible regions for designing genetic-fuzzy systems can be referred to in [21].

2.2 IMM

For each particle $X_i(t) = [x_{i1}, x_{i2}, ..., x_{iD}]^T$, IMM generates two temporary moves $C=[C_1, ..., C_D]^T$ and $S = [S_1, ..., S_D]^T$ corresponding to cognitive and social parts respectively:

$$C = x_i(t) + wV_i(t) + c_1\vec{r_1}(P_i(t) - X_i(t)), \qquad (1)$$

$$S = x_i(t) + wV_i(t) + c_2\vec{r_2}(P_g(t) - X_i(t)), \qquad (2)$$

IMM aims at efficiently combining good partial vectors of C and S to generate the next move $X_i(t+1)$, as described below.

Divide the D-dimensional vector of X_i into N non-overlapping partial vectors with sizes l_i, $i=1, ..., N$, using the same division scheme for C and S such that

$$\sum_{i=1}^{N} l_i = D. \qquad (3)$$

The proper value of N is problem-dependent. The larger the value of N, the more efficient the IMM if the interactions among partial vectors are weak. Considering the tradeoff, an efficient division criterion is to minimize the interactions among partial vectors while maximizing the value of N. to efficiently use all columns of OA excluding the study of intractable interactions, the used OA is $L_{N+1}(2^N)$ and largest value of N is equal to $2^{\lfloor \log_2(D+1) \rfloor} - 1$ where the bracket represents a lower ceiling operation. For example, a coarse-to-fine strategy using a variable value of N is sometimes more efficient [22]. In this study, IMOPSO uses a constant value of N.

How to perform an IMM operation using the particle $X_i(t)$ of a D-dimensional vector with an objective function f is described as follows:

Step 1: Generate two temporary moves C and S for $X_i(t)$ using Eq. (1) and Eq. (2).
Step 2: Divide each of C and S into N partial vectors where each partial vector is treated as a factor.
Step 3: Use the first N columns of an OA $L_M(2^{M-1})$, where $M = 2^{\lfloor \log_2(N+1) \rfloor}$.
Step 4: Let levels 1 and 2 of factor i represent the ith partial vector of C and S, respectively.
Step 5: Compute the objective function value f_t of the combination t, where $t=1,...,M$. Note that f_1 is the value of $f(C)$.
Step 6: Compute the main effect S_{jk} where $j=1, ..., N$ and $k=1,2$.
Step 7: Determine the better one of two levels of each factor based on the main effect.
Step 8: The next move $X_i(t+1)$ is formed using the combination of the better partial vectors.
Step 9: Verify that $X_i(t+1)$ is better than the $X_i(t)$ and the M sample solutions derived from OA combinations. If it is not true, let $X_i(t+1)$ be the best one of $X_i(t)$ and these M sample solutions.

OA specifies a small number M of representative combinations that are uniformly distributed over the neighborhood of the particle $X_i(t)$. The number of objective function evaluation is $M+1$ per IMM operation, which includes M evaluations in Step 5 and one in Step 9. IMM spends $M+1$ function evaluations while PSO spends one function evaluation using the generate-and-go method to determine the next move. However, the M sampling solutions and factor analysis make IMM more efficient in obtaining the next move. If interactions among partial vectors are weak, $X_i(t+1)$ is a potentially good approximation to the best one of all the 2^N combinations in the neighborhood of $X_i(t)$.

2.3 Fitness Function GPSISF

GPSISF considers the quantitative fitness performances in the objective space for both dominated and non-dominated individuals and makes the best use of Pareto dominance relationship to evaluate individuals using a single measure of performance. Let the fitness value of an individual X be a tournament-like score obtained from all participating individuals according to the following function:

$$\text{GPSISF}(X) = p - q + c, \qquad (4)$$

where p is the number of individuals which can be dominated by X, and q is the number of individuals which can dominate X in the objective space. Generally, a constant c can be optionally added in the fitness function to make fitness values positive. In this study, c is the number of all participant individuals.

First GPSISF uses a pure Pareto-ranking fitness assignment strategy, which differs from the traditional Pareto-ranking methods, such as non-dominated sorting and Zitzler and Thiele's method, in two aspects. First GPSISF can assign discriminative fitness values to not only non-dominated individuals but also dominated ones. IMM can take advantage of this assignment strategy to accurately estimate the main effect of factors. It is less efficient for IMM to use Zitzler and Thiele's method where the fitness values of dominated individuals in cluster are always identical. Fig. 1(a) shows an example illustrating the fitness values using GPSISF for a bi-objective minimization problem. For

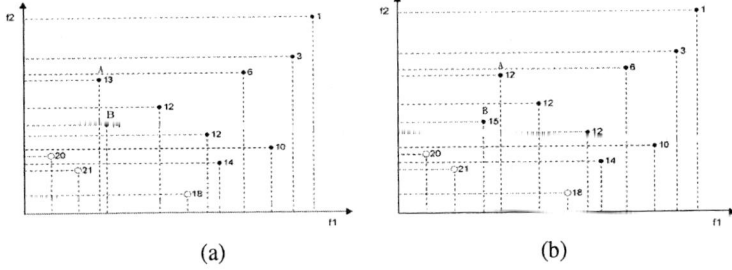

Fig. 1. The fitness values of 12 participant individuals in the objective space of a bi-objective minimization problem. The fitness values of the same dominated individual A in (a) and (b) using GPSISF are different depending on the Pareto dominance relationship.

example, three individuals are dominated by A ($p=3$) and two individuals dominate A ($q=2$). Therefore, the fitness value of A is 3-2+12=13. It can be found that one individual has a larger fitness value if it dominates more individuals. On the contrary, one individual has a smaller fitness value if more individuals dominate it.

Second, GPSISF has an implicit niching mechanism based on the Pareto dominance relationship. Fig. 1(b) is derived from Fig. 1(a) by moving the individual A, such that B dominates A. Considering the rectangle formed by A, the value of q is increased from 2 to 3 and thus the fitness value of A is decreased from 13 to 12. This scenario reveals that the same individual dominated by crowed individuals has a smaller fitness value than that dominated by sparse individuals. Therefore, no additional techniques such as fitness sharing are needed to achieve the niching effect.

2.4 IMOPSO

Since it has been recognized that the incorporation of elitism may be useful in maintaining diversity and improving the performance of multi-objective PSOs, IMOPSO uses an elite set E with capacity N_{Emax} to maintain the best non-dominated individuals generated so far. The simple IMOPSO can be written as follows:

Step 1: (Initialization) Randomly generate an initial population N_{POP} and create an empty elite set E and an empty temporary elite set E'.
    1) The position of the i-th particle, X_i are initialized with random real number within the specified decision variable range;
    2) The velocities of the i-th particle, V_i are initialized to 0.

Step 2: (Evaluation) Compute all objective function values of each individual particle in the population. Assign each individual a fitness value by using GPSISF.

Step 3: (Update elite sets) Add the non-dominated individuals in both the population and E' to E, and empty E'. Remove the dominated individuals in E. If the number N_E of non-dominated individuals in E is larger than N_{Emax}, randomly discard excess individuals.

Step 4: (Selection *gbest*) Select a global best P_g for the i-th particle from E.

Step 5: (Update position and velocity) Apply IMM to generate next position $X_i(t+1)$ and velocity $V_i(t+1)$.

Step 6: (Update *pbest*) If new solution dominates the current *pbest* then *pbest* updated.

Step 7: (Termination test) If a stopping condition is satisfied, stop the algorithm. Otherwise, go to Step 2.

3 Performance Comparisons of IMOPSO

High performance of IMOPSO is demonstrated by showing its superiority over the following multi-objective PSOs and EAs, such as AMOPSO[9], [10], DNPSO[11], mDNPSO[12], SPEA[5], SPEA2[6], NSGA[2], NSGA2[3], and IMOEA[7].

Zitzler et al. [4] constructed six test problems and investigated the performance of various popular multi-objective EAs. Each of the test problems is structured in the same manner and consists of three functions f_1, g, h:

$$\begin{aligned}\text{min. } & T(X) = (f_1(x_1), f_2(X)), \\ \text{s. t. } & f_2(X) = g(x_2, \ldots, x_m)h(f_1(x_1), g(x_2, \ldots, x_m)),\end{aligned} \quad (5)$$

where $X = [x_1, x_2, \ldots, x_m]^T$, f_1 is a function consisted of the first decision variable x_1 only, g is a function of the remaining m-1 parameters, and the two parameters of the function h are the function values of f_1 and g. In this paper, we were used five test problems ZDT_1 and ZDT_2 to evaluate the performance of IMOPSO and compared with some existing algorithms. Those test problems can be retrieved from [4].

3.1 Small-Scale Problem

The non-dominated solutions merged from 30 runs of NSGA, SPEA, DNPSO, mDNPSO, AMOPSO and IMOPSO in the objective space are reported, as shown in Figs. 2 and 3, and the curve in each figure is the Pareto-optimal front of each test problem. In our experiments, benchmark of NSGA and SPEA were downloaded from [27], source code of AMOPSO was downloaded from [27]. DNPSO and mDNPSO were written by ourselves. For test problems ZDT_1 and ZDT_2, IMOPSO, AMOPSO, DNPSO and mDNPSO can obtain well-distributed Pareto fronts, and the Pareto fronts for each test function are very close to the Pareto-optimal fonts.

As shown in Figs. 2 and 3, the quality of non-dominated solutions obtained by IMOPSO is superior to those of all compared algorithms in terms of the number of non-dominated solutions, the distance between the obtained Pareto front and Pareto-optimal front, and the distribution of non-dominated solutions.

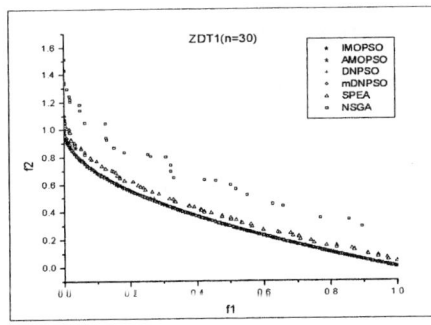

Fig. 2. Convex test ZDT_1 (n=30)

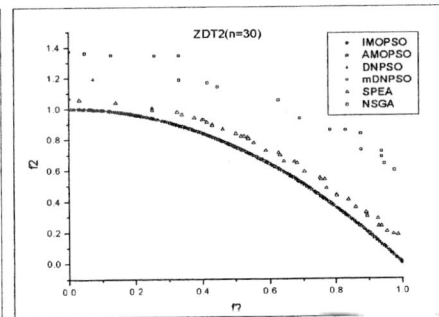

Fig. 3. Non-convex test ZDT_2 (n=30)

3.2 Large-Scale Problem

Ho et al.[7] extended test problems with a large number of parameters (n=63) and proposed IMOEA to solving large MOOPs. For shown IMOPSO have high performance at large scale problem. We apply IMOPSO to solving large scale problem and compare the performance with IMOEA and other algorithms.

The non-dominated solutions merged from 30 runs of NSGA2, SPEA, SPEA2, IMOEA, AMOPSO and IMOPSO in the objective space are reported, as shown in Figs. 4 and 5. As shown in Figs. 4 and 5, the quality of non-dominated solutions obtained by IMOPSO is superior to those of all compared algorithms in terms of the number of non-dominated solutions, the distance between the obtained Pareto front and Pareto-optimal front, and the distribution of non-dominated solutions.

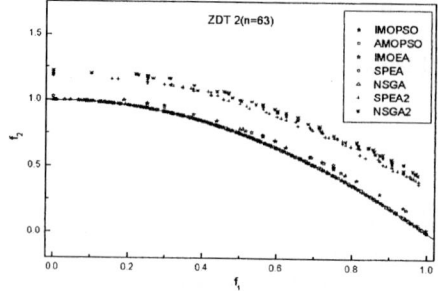

Fig. 4. Convex test ZDT_1 ($n=63$)   **Fig. 5.** Non-convex test ZDT_2 ($n=63$)

We also used three issues are normally taken into consideration, describe as below:

1) Minimize the distance of the Pareto front produced by our algorithm with respect to the global Pareto front
2) Maximize the spread of solutions found, so that we can have a distribution of vectors as smooth and uniform as possible.
3) Maximize the number of elements of the Pareto optimal set found.

Based on this notion, we adopted one metric to evaluate each of three aspects previously indicated.

1) **Generational distance** (GD): The concept of generational distance was introduced by Van Veldhuizen and Lamont as a way of estimating how far the elements are in the set of non-dominated vector found so far from those in the Pareto optimal set and is defined as

$$GD = \sqrt{\sum_{i=1}^{n} d_i^2} \,/\, n$$

where n is the number of vectors in the set of non-dominated solutions found so far and d_i is the Euclidean distance (measured in objective space) between each of these and the nearest member of the Pareto optimal set.

2) **Spacing** (SP): Here, one desires to measure the spread of vectors throughout the non-dominated vectors found so far. Schott proposed such a metric measuring the range variance of neighboring vectors in the non-dominated vectors found so far. This metric is defined as

$$S \triangleq \sqrt{\frac{1}{n-1}\sum_{i=1}^{n}(\bar{d}-d_i)^2}$$

where $d_i = \min_j \left(\left| f_1^i(\vec{x}) - f_1^j(\vec{x}) \right| + \left| f_2^i(\vec{x}) - f_2^j(\vec{x}) \right| \right)$, $i, j = 1, \ldots, n$, $\bar{d}$ is the mean of all d_i, and n is the number of non-dominated vectors found so far.

3) **Error ratio** (ER): This metric was proposed by Van Veldhuizen to indicate the percentage of solutions that are non members of the true Pareto optimal set

$$ER = \sum_{i=1}^{n} e_i \,/\, n$$

where n is the number of vectors in the current set of non-dominated vectors available, $e_i = 0$ if vector i is a member of the Pareto optimal set, and $e_i = 1$, otherwise.

A. ZDT$_1$: Fig 4 show the graphical results produced by our IMOPSO, the AMOPSO, the IMOEA, the SPEA, the SPEA2 and the NSGA2. The true Pareto front of the problems is shown as a continuous line. Tables 1-3 show the comparison of results among the six algorithms considering the metrics previously described. It can be seen that the average performance of IMOPSO and AMOPSO are the best compared with other algorithms.

B. ZDT$_2$: Fig 5 show the graphical results produced by our IMOPSO, the AMOPSO, the IMOEA, the SPEA, the SPEA2 and the NSGA2. The true Pareto front of the problems is shown as a continuous line. Tables 4-6 show the comparison of results among the six algorithms considering the metrics previously described. It can be seen that the average performance of IMOPSO is the best than other algorithms.

Table 1. Results of the error ratio metric for zdt1

ER	IMOPSO	AMOPSO	IMOEA	NSGA2	SPEA2	SPEA
Worst	0.1061	0.0002	1.0000	1.0000	1.0000	1.0000
Best	0.0000	0.0000	1.0000	1.0000	1.0000	1.0000
Average	0.0320	0.0000	1.0000	1.0000	1.0000	1.0000
Median	0.0253	0.0000	1.0000	1.0000	1.0000	1.0000
Std. Dev.	0.0253	0.0000	0.0000	0.0000	0.0000	0.0000

Table 2. Results of the generational distance metric for zdt1

GD	IMOPSO	AMOPSO	IMOEA	NSGA2	SPEA2	SPEA
Worst	0.0018	0.0000	0.0347	0.0353	0.0285	0.0503
Best	0.0001	0.0000	0.0138	0.0224	0.0212	0.0308
Average	0.0004	0.0000	0.0200	0.0296	0.0246	0.0392
Median	0.0003	0.0000	0.0194	0.0295	0.0248	0.0387
Std. Dev.	0.0003	0.0000	0.0041	0.0027	0.0021	0.0042

Table 3. Results of the spacing metric for zdt1

SP	IMOPSO	AMOPSO	IMOEA	NSGA2	SPEA2	SPEA
Worst	0.0252	0.0002	0.0658	0.0574	0.0538	0.0899
Best	0.0009	0.0000	0.0162	0.0382	0.0313	0.0415
Average	0.0033	0.0000	0.0302	0.0464	0.0400	0.0586
Median	0.0025	0.0000	0.0281	0.0471	0.0401	0.0566
Std. Dev.	0.0042	0.0000	0.0127	0.0052	0.0049	0.0091

Table 4. Results of the error ratio metric for zdt2

ER	IMOPSO	AMOPSO	IMOEA	NSGA2	SPEA2	SPEA
Worst	0.9444	1.0000	1.0000	1.0000	1.0000	1.0000
Best	0.0302	0.0204	1.0000	1.0000	1.0000	1.0000
Average	0.4107	0.7257	1.0000	1.0000	1.0000	1.0000
Median	0.4350	0.8245	1.0000	1.0000	1.0000	1.0000
Std. Dev.	0.2512	0.2938	0.0000	0.0000	0.0000	0.0000

Table 5. Results of the generational distance metric for zdt2

GD	IMOPSO	AMOPSO	IMOEA	NSGA2	SPEA2	SPEA
Worst	0.0002	0.0131	0.0559	0.0696	0.0595	0.1691
Best	0.0000	0.0000	0.0159	0.0501	0.0424	0.0657
Average	0.0001	0.0011	0.0287	0.0581	0.0499	0.0920
Median	0.0001	0.0001	0.0274	0.0581	0.0497	0.0885
Std. Dev.	0.0000	0.0033	0.0097	0.0049	0.0042	0.0207

Table 6. Results of the spacing metric for zdt2

SP	IMOPSO	AMOPSO	IMOEA	NSGA2	SPEA2	SPEA
Worst	0.0027	0.0683	0.1034	0.0744	0.0639	0.0972
Best	0.0001	0.0000	0.0060	0.0462	0.0382	0.0437
Average	**0.0010**	0.0063	0.0287	0.0617	0.0548	0.0756
Median	0.0010	0.0007	0.0221	0.0617	0.0556	0.0758
Std. Dev.	0.0005	0.0173	0.0181	0.0068	0.0057	0.0134

4 Conclusion

We have proposed an intelligent multi-objective particle swarm optimization IMOPSO using a novel intelligent move mechanism IMM and generalized Pareto-based scale-independent scoring function (GPSISF) to solve multi-objective optimization problems (MOOPs). Since OED is advantageous for problems with weak interactions among parameters and IMOPSO works without using linkage identification, it is essential to encode parameters into particle such that the degree of epitasis can be minimized. IMOPSO is powerful based on the abilities of the proposed GPSISF and IMM. We believe that the auxiliary techniques, which can improve performance of conventional MOPSOs, can also improve performances of IMOPSO. Due to its simplicity, theoretical elegance, generality, and superiority, IMOPSO can be most widely used for solving real-world applications of MOOPs.

References

1. K. Deb, "Multi-objective Genetic Algorithms: Problem Difficulties and Construction of Test Problems," in *MIT Evolutionary Computation*, Vol. 7, No. 3, pp. 205-230, Fall 1999.
2. N. Srinivas and K. Deb, "Multiobjective optimization using non-dominated sorting in genetic algorithms," in *Evolutionary Computation*, Vol. 2, No. 3, pp. 221-248, 1994.
3. K. Deb, A. Pratap, S. Agarwal, and T. Meyarivan, "A fast and elitist multiobjective genetic algorithms: NSGA-II," IEEE Trans. Evol. Comput., vol. 6, no. 2, pp. 182-197, April 2002.
4. E. Zitzler, K. Deb and L. Thiele, "Comparison of Multiobjective Evolutionary Algorithms: Empirical Results," in *MIT Evolutionary Computation*, Vol. 8, No. 2, pp. 173-195, Summer 2000.
5. E. Zitzler and L. Thiele, "Multiobjective evolutionary algorithms: a comparative case study and strengthen Pareto approach," in *IEEE Trans. on Evolutionary Computation*, Vol. 3, No. 4, pp. 257-271, 1999.
6. E. Zitzler, M. Laumanns, and L. Thiele, "SPEA2: Improving the strength Pareto evolutionary algorithm," *Technical Report 103*, Computer Engineering and Communication Networks Lab (TIK), Swiss Federal Institute of Technology (ETH) Zurich, Gloriastrasse 35, CH-8092 Zurich, May 2001.
7. S.-Y. Ho, L.-S. Shu, J.-H. Chen, "Intelligent Evolutionary Algorithms for Large Parameter Optimization Problems," in *IEEE Trans. Evolutionary Computation*, Vol. 8, No. 6, pp. 522-541. December 2004.
8. J. Kennedy and R. C. Eberhart, "Particle swarm optimization," in *Proc. IEEE Conf. Neural Networks IV*, Vol. 4, pp. 1942-1948, 1995.
9. C. A. Coello Coello and M. S. Lechuga, "MOPSO: A Proposal for Multiple Objective Particle Swarm Optimization," in *Proc. Congress on Evolutionary Computation(CEC)*, pp. 1051-1056, May 2002.
10. C. A. Coello Coello, G. T. Pulido and M. S. Lechuga, "Handling Multiple Objectives With Particle Swarm Optimization," in *IEEE Trans. Evolutionary Computation*, Vol. 8, No. 3, pp. 256-279, June 2004.
11. X. Hu and R. C. Eberhart, "Multiobjective Optimization Using Dynamic Neighborhood Particle Swarm Optimization," in *Proc. Congress on Evolutionary Computation(CEC)*, pp. 1677-1681, May 2002.
12. X. Hu, R. C. Eberhart and Y. Shi, "Particle Swarm with Extended Memory for Multiobjective Optimization," in *Proc. of IEEE International Conference on Swarm Intelligence Symposium(SIS)*, pp. 193-197. April 2003.
13. Q. Wu, "On the optimality of orthogonal experimental design," in *Acta Math. Appl. Sinica*, vol. 1, no. 4, pp. 283-299, 1978.
14. A. Dey, *Orthogonal Fractional Factorial Designs*. New York:Wiley, 1985.
15. A. S. Hedayat, N. J. A. Sloane, and J. Stufken, *Orthogonal Arrays: Theory and Applications*. New York: Springer-Verlag, 1999.
16. T.-P. Bagchi, *Taguchi Methods Explained: Practical Steps to Robust Design*. Prentice-Hall, 1993.

17. J. -T. Tsai, T. -K. Liu, and J. -H. Chou, "Hybrid aguchi-genetic algorithm for bloal numerical optimization," in *IEEE Trans. Evolutionary Computation*, Vol. 8, no. 4, pp. 365-377, August 2004.
18. H. Tanaka, "A comparative study of GA and orthogonal experimental design," in Proc. IEEE Int. Conf. Evolutionary Computation, 1997, pp. 143-146.
19. Q. Zhang and Y. -W. Leung, "An orthogonal genetic algorithm for multimedia multicast routing, in *IEEE Trans. Evolutionary Computation*, Vol. 3, no. 1, pp. 53-62, April 1999.
20. Y.-W. Leung and Y. Wang, "An orthogonal genetic algorithm with quantization for global numerical optimization," *IEEE Trans. Evolutionary Computation*, vol. 5, no. 1, pp. 41-53, Feb. 2001.
21. S.-Y. Ho, H.-M. Chen, S.-J. Chen, "Design of accurate classifiers with a compact fuzzy-rule base using an evolutionary scatter partition of feature space," in *IEEE Trans. Systems, Man, and Cybernetics, Part B: Cybernetics*, vol. 34, no. 2, pp. 1031-1044, April 2004.
22. H.-L. Huang and S.-Y. Ho, "Mesh optimization for surface approximation using an efficient coarse-to-fine evolutionary algorithm," *Pattern Recognition*, vol. 36, no. 5, pp. 1065-1081, 2003.
23. S.-Y. Ho, S.-J. Ho, Y.-K. Lin, and W.-C. Chu, "An orthogonal simulated annealing algorithm for large floorplanning problems," in *IEEE trans. VLSI system*, Vol. 12, no. 8, pp.874-86, August 2004.
24. S.-J. Ho, S.-Y. Ho and L.-S. Shu, "OSA: Orthogonal Simulated Annealing Algorithm and Its Application to Designing Mixed H2/H Optimal Controllers, " in *IEEE Trans. System, Man, and Cybernetics-Part A*, Vol. 34, No. 5, pp. 588-600, September 2004.
25. Z. Michalewicz, D. Dasgupta, R. G. Le Riche, and M. Schoenauer, "Evolutionary algorithms for constrained engineering problems," *Computers & Industrial Engineering*, vol. 30, no. 4, pp. 851-870, Sept. 1996.
26. S.-Y. Ho and Y.-C. Chen, "An efficient evolutionary algorithm for accurate polygonal approximation," *Pattern Recognition*, vol. 34, no. 12, pp. 2305-2317, 2001.
27. Test Problems and Test Data for Multiobjective Optimizers URL: http://www.tik.ee.ethz.ch/~zitzler/testdata.html/

Hidden Space Principal Component Analysis*

Weida Zhou, Li Zhang, and Licheng Jiao

Institute of Intelligence Information Processing, Xidian University, Xi'an 710071, China
wdzhou@mail.xidian.edu.cn, zhangli@mail.xidian.edu.cn,
lchjiao@mail.xidian.edu.cn

Abstract. A new nonlinear principle component analysis (PCA) method, hidden space principal component analysis (HSPCA) is presented in this paper. Firstly, the data in the input space is mapped into a high hidden space by a nonlinear function whose role is similar to that of hidden neurons in Artificial Neural Networks. Then the goal of features extraction and data compression will be implemented by performing PCA on the mapped data in the hidden space. Compared with linear PCA method, our algorithm is a nonlinear PCA one essentially and can extract the data features more efficiently. While compared with kernel PCA method presented recently, the mapped samples are exactly known and the conditions satisfied by nonlinear mapping functions are more relaxed. The unique condition is symmetry for kernel function in HSPCA. Finally, experimental results on artificial and real-world data show the feasibility and validity of HSPCA.

1 Introduction

Linear PCA (LPCA) is an important method of the statistical analysis of data. LPCA is a powerful tool for extracting linear structures from possibly high-dimensional data sets. It directly deals with the covariance matrix of the data to extract second-order uncorrelated principal components of the data, and it can implement features extraction and data compress of the data. Kernel PCA (KPCA) presented by Schölkopf is a generalization of LPCA [1]. In KPCA, the data in an input space firstly is mapped into a high (possibly infinite) dimensional feature space by a nonlinear kernel function. And then LPCA is performed on the mapped data in the feature space, which performs a nonlinear PCA in the input space essentially. It is necessary for kernel functions adopted in KPCA to satisfy Mercer's condition. The nonlinear mapping functions never need to be known explicitly. Since it is the mapping functions that are unknown, the mapped data is unknown, it is impossible to compute of the statistics of the data such as one-order or high-order statistics, except for two-order ones. It constrains the further analysis and processing on the mapped data to a certain extent.

A new nonlinear PCA method, hidden space principal component analysis (HSPCA) is presented in this paper. Firstly, the data is mapped into a high-dimensional hidden

---

* This work was supported in part by the Shaanxi Province Natural Science Foundation of China under grant 2004F1.

space by a nonlinear function whose role is similar to that of hidden neurons in Artificial Neural Networks. Then linear PCA is performed on the mapped data to implement the goal of feature extraction and data compression. Compared with linear PCA, HSPCA is an intrinsic nonlinear method. It can extract the data features more efficiently than LPCA does. Compared with kernel PCA presented recently, the mapped data is known explicitly in HSPCA. Therefore we can handle these mapped data in the feature space directly. Moreover the nonlinear kernel functions adopted in HSPCA need not to satisfy Mercer's condition and only satisfy the symmetry condition which is the basic condition for kernel function.

2 Hidden Space Principle Analysis (HSPCA)

As we know, the introduction of kernel functions is to implement the nonlinear mapping in SVMs. Before SVMs were presents, the nonlinear mapping had been successfully applied to some other methods, such as the hidden function mapping in forward neural networks (FNNs) and radial basis function networks (RBFNs). It has been proven that the separability of patterns will be increased greatly if patterns in a low dimensional space are mapped into a high-dimensional space [2]. SVMs for classification and regression estimation in hidden space have been presented in [3]. Note that the unique condition for the hidden kernel functions is the symmetry, which will extend the set of usable Mercer kernel functions in SVMs [3]. Some useful symmetric hidden kernel functions were given in [3]. In the following, we present the PCA method in hidden space.

Let the set of i.i.d. patterns be $X = \{\mathbf{x}_1, \mathbf{x}_2 \cdots, \mathbf{x}_N | \mathbf{x}_i \in \mathbb{R}^d\}$ and kernel function be $k(\mathbf{x}, \mathbf{y})$. The set of the mapped patterns by $k(\mathbf{x}, \mathbf{y})$ in the hidden space

$$\{\mathbf{z}_i = [k(\mathbf{x}_i, \mathbf{x}_1), k(\mathbf{x}_i, \mathbf{x}_2), \cdots, k(\mathbf{x}_i, \mathbf{x}_N)]^T, i = 1, \cdots, N\} \tag{1}$$

Since the mapped patterns in the feature space or the hidden space are known definitely shown in Eq. (1), we can simply compute the statistics of the patterns in the hidden space, which is impossible for KPCA. For the sake of simplicity, we have made the assumption that the data is centered. If not, we can remove the following mean from the mapped data simply

$$\mathbf{m} = \frac{1}{N} \sum_{i=1}^{N} \mathbf{z}_i$$

Let

$$K = [\mathbf{z}_1, \mathbf{z}_2, \cdots, \mathbf{z}_N] \tag{2}$$

Obviously the matrix K is a real-valued symmetric matrix of order $N \times N$.

The covariance matrix of the patterns in the hidden space can be expressed as

$$C = \frac{1}{N} \sum_{i=1}^{N} \mathbf{z}_i \mathbf{z}_i^T = \frac{1}{N} K K^T \tag{3}$$

Implement eigen-decomposition of the real-valued symmetric matrix (3), we have

$$C\mathbf{v} = \lambda \mathbf{v} \qquad (4)$$

which can be rewritten as

$$C\mathbf{v} = \frac{1}{N} K K^T \mathbf{v} = K\mathbf{u} = \lambda \mathbf{v} \qquad (5)$$

where $\mathbf{u} = \frac{1}{N} K^T \mathbf{v} \in \mathbb{R}^N$. Hence $\mathbf{v} \in span\{\mathbf{z}_1, \mathbf{z}_2, \cdots, \mathbf{z}_N\}$. Now suppose

$$\mathbf{v} = \sum_{i=1}^{N} \alpha_i \mathbf{z}_i = K\alpha \qquad (6)$$

where $\alpha = [\alpha_1, \alpha_1, \cdots, \alpha_N]^T$. Combining Eq. (6) and Eq. (4), we get

$$\frac{1}{N} K K^T K \alpha = \lambda K \alpha \qquad (7)$$

The coefficient $1/N$ in Eq. (7) can be ignored because its role is only to multiply the eigenvalues by N. Eliminating K in both sides in Eq. (7), we obtain

$$K^T K \alpha = \lambda \alpha \qquad (8)$$

With respect to the positive semi-definite, symmetrical and real-valued matrix KK^T, we can obtain N normalized eigenvectors $\alpha^1, \alpha^2, \cdots, \alpha^N$ where $\|\alpha^i\| = 1, i = 1, \cdots, N$ and the corresponding eigenvalues $\lambda^1, \lambda^2, \cdots, \lambda^N$. But the work is not complete, our goal is to obtain the eigenvector $\mathbf{v}$ in the hidden space. Normalizing $\|\mathbf{v}^i\| = 1, i = 1, \cdots, N$, from Eq. (6) we have

$$\mathbf{v}^i = K\beta^i, \beta^i = \frac{\alpha^i}{\sqrt{\lambda^i}}, i = 1, \cdots, N \qquad (9)$$

Thus we select the n larger eigenvalues from N eigenvalues $\lambda^1, \lambda^2, \cdots, \lambda^N$ and the corresponding n eigenvectors, which compose the transform matrix of HSPCA as follows

$$V = \left[\mathbf{v}^{i_1}, \mathbf{v}^{i_2}, \cdots, \mathbf{v}^{i_n}\right], i_1, i_2, \cdots, i_n \in \{1, \cdots, N\} \qquad (10)$$

For each pattern, the extracted principle components are

$$\hat{\mathbf{z}}_i = V^T \mathbf{z}_i = \left[\beta^{i_1}, \beta^{i_2}, \cdots, \beta^{i_n}\right]^T K\mathbf{z}_i$$
$$= \left[\frac{\alpha^{i_1}}{\sqrt{\lambda_{i_1}}}, \frac{\alpha^{i_2}}{\sqrt{\lambda_{i_2}}}, \cdots, \frac{\alpha^{i_n}}{\sqrt{\lambda_{i_n}}}\right]^T K\mathbf{z}_i, i = 1, \cdots, N \qquad (11)$$

Now we have finished the derivation of extracting principle components in the hidden space as above.

3 Simulation

In order to evaluate the performance of HSPCA, we performed experiments on two-spiral data. We used two methods to perform our simulation. One method is nonlinear SVMs. Other is a mixed method of HSPCA and linear SVM. We used the HSPCA to extract the principal components of data that was the training and testing patterns of linear SVM.

The two-spiral problem is referred to as "touchstone" to test the performance of a learning algorithm in the field of pattern recognition [4]. In this experiment, the two-spiral data can be expressed as

$$X_1 = \left\{(x,y) \middle| \begin{array}{l} x = (0.05\theta)\sin\theta + \sigma_x n_x, \quad n_x \sim N(0,1), \quad \sigma_x = 0.04 \\ y = (0.05\theta)\cos\theta + \sigma_y n_y, \quad n_y \sim N(0,1), \quad \sigma_y = 0.04 \end{array} \right\}$$

and

$$X_2 = \left\{(x,y) \middle| \begin{array}{l} x = (0.05\theta + 0.2)\sin\theta + \sigma_x n_x, \quad n_x \sim N(0,1), \quad \sigma_x = 0.04 \\ y = (0.05\theta + 0.2)\cos\theta + \sigma_y n_y, \quad n_y \sim N(0,1), \quad \sigma_y = 0.04 \end{array} \right\}$$

We sampled 32 patterns as the training patterns and 126 ones on $\theta \in [0, 2\pi]$ as the test patterns in each spiral randomly. We performed 30 runs totally and adopted two kinds of kernel function: Gaussian RBF kernel $k(\mathbf{x},\mathbf{y}) = \exp\left(-\|\mathbf{x}-\mathbf{y}\|^2 / 2p^2\right), p \in \mathbb{R}$ and generalized two-quadrics kernel $k(\mathbf{x},\mathbf{y}) = \left(1 + \|\mathbf{x}-\mathbf{y}\|^2\right)^p, p \notin \mathbb{Z}$ (which is Mercer kernel if and only if $p < 0$). Let $C = 100$. Table 1 shows the average results of

Table 1. Comparison of HSPCA+LSVMs and NSVMs on two-spiral data

Kernel function	Optimal parameter	Method	# PC extracted by HSPCA	Test error (%)
Gaussian RBF	p=0.07	NSVMs	—	2.7
	p=0.11	HSPCA+LSVMs	32	3.5
	p=0.11	HSPCA+LSVMs	16	4.1
Generalized Multi-quadrics	p= -7.5	NSVMs	—	3.8
	p= -9.9	HSPCA+LSVMs	15	5.0
	p= -9.5	HSPCA+LSVMs	7	6.0

HSPCA+LSVMs and NSVMs over 30 runs. The recognition performance of the mixed method is lightly lower than that of NSVMs. It is natural. HSPCA only extract a part of principal components of the data and delete other minor components. So the mixed method only used a part of features of the data, while NSVMs used all features of the data. It is acceptable that the recognition error does not increased greatly as the reducing of the number of principal components.

4 Conclusion

A new nonlinear PCA method, hidden space principal component analysis (HSPCA) is presented in this paper. Firstly, the data is mapped in a high-dimensional hidden space by a nonlinear function whose role is similar to that of hidden neurons in Artificial Neural Networks. Then LPCA is performed on the mapped data to implement the goal of features extraction and data compression. Compared with linear PCA, HSPCA is a nonlinear method essentially. It can extract data features more efficiently than LPCA does. Compared with kernel PCA presented recently, the mapped data is known explicitly in HSPCA. Therefore we can handle these mapped data in the feature space directly and at any means. Moreover the nonlinear kernel functions adopted in HSPCA need not to satisfy Mercer's condition and only satisfy the symmetry condition, the most relaxed condition for kernel function.

Since the inverse mapping of kernel mapping adopted in HSPCA and KPCA almost does not exist. It is almost impossible for HSPCA and KPCA to perform data construction or de-noising. It is worth studying how to implement data reconstruction or de-noising base on nonlinear PCA.

References

[1] B. Schölkopf, A.J. Smola and K.-R. Müller, Nonlinear component analysis as a kernel eigenvalue problem. *Neural Computation*, 10:1299-1319, 1998.
[2] T.M. Cover, Geometrical and statistical properties of systems of linear inequalities with applications in pattern recognition. *IEEE Transactions on Electronic Computers*, vol.EC-14, pp.326-334, 1965.
[3] L. Zhang, W. Zhou, L. Jiao. Hidden space support vector machines. *IEEE Trans. NNs.* 2004,15 (6): 1424-1434.
[4] K.J. Lang and M.J. Witbrock, Learning to tell two spirals apart. In Proc.1989 Connectionist Models Summer School, pp.52-61, 1989.

Neighbor Line-Based Locally Linear Embedding

De-Chuan Zhan and Zhi-Hua Zhou

National Laboratory for Novel Software Technology,
Nanjing University, Nanjing 210093, China
{zhandc, zhouzh}@lamda.nju.edu.cn

Abstract. Locally linear embedding (LLE) is a powerful approach for mapping high-dimensional data nonlinearly to a lower-dimensional space. However, when the training examples are not densely sampled, LLE often returns invalid results. In this paper, the NL^3E (Neighbor Line-based LLE) approach is proposed, which generates some virtual examples with the help of *neighbor line* such that the LLE learning can be executed on an enriched training set. Experiments show that NL^3E outperforms LLE in visualization.

1 Introduction

Many real-world problems suffer from a large amount features [5]. Therefore, dimensionality reduction techniques are needed. Popular linear dimensionality reduction methods such as PCA [7] and MDS [2] are easy to implement. The main idea of PCA is to find the projection direction with the largest possible variance and then project the original data onto that direction, while that of MDS is to find the low-dimensional embeddings which best restore the pair-wised distances between the original samples. Although these linear methods have achieved some success, most real-world data are non-linearly distributed and therefore, these methods can hardly work well.

Recently, a number of non-linear dimensionality reduction methods have been proposed, e.g. the manifold learning methods LLE [9], ISOMAP [10], etc. LLE preserves the information of local distance between the concerned data and its neighbors, while ISOMAP preserves the pairwise *geodesic distances* between the original samples. Both LLE and ISOMAP have been applied to data visualization [6][9][10], and encouraging results have been reported when the data are densely sampled and there is no serious noise in the data.

De Silva and Tenenbaum [4] proposed two improved ISOMAP algorithms, namely, C-ISOMAP and L-ISOMAP. C-ISOMAP has the ability to invert conformal maps while L-ISOMAP attempts to reduce the computational load. Unfortunately, similar to ISOMAP, C-ISOMAP performs pool when the training data are not densely sampled [4][10]. Even more worse, C-ISOMAP requires more samples than ISOMAP [4]. L-ISOMAP reduces computational complexity by mapping only the landmark points. Unfortunately, it is more unstable than ISOMAP since the landmarks may be not densely sampled [4]. Like ISOMAP series of algorithms [10], given sufficient data, LLE is guaranteed asymptotically to recover the geometric structure [9]. Recently, LLE and LDA have been combined as new classification algorithms [3][12], which work well only with dense samples either. Since in real-world tasks it is hard to guarantee that the data is densely sampled, the performance of these manifold learning algorithms are often not satisfying.

In this paper, the NL³E (Neighbor Line-based LLE) method is proposed. Through generating virtual samples with the help of the *neighbor line*, NL³E can work well in some cases where the data are not densely sampled.

The rest part of this paper is organized as follows. In section 2, the LLE and NNL algorithms and some works utilizing virtual samples are introduced. In section 3, the NL³E method is proposed. In section 4, experiments are reported. Finally, in section 5, conclusions are drawn.

2 Background

2.1 LLE

LLE [9] maps a data set $\mathbf{X} = \{x_1, x_2, \cdots, x_n\}$, $x_i \in \mathcal{R}^d$, to a data set $\mathbf{Z} = \{z_1, z_2, \cdots, z_n\}$, $z_i \in \mathcal{R}^m$, where $d > m$. It assumes the data lie on a low-dimensional manifold which can be approximated linearly in a local area of the high-dimensional space. Roughly speaking, LLE firstly fits hyperplanes around each sample x_i, based on its k nearest neighbors, and then calculates the reconstruction weights. After that, it finds the lower-dimensional coordinates z_i for each x_i, which preserve those reconstruction weights as good as possible.

Formally, the k nearest neighbors of x_i are identified according to Euclidean distance at first. Then the neighboring points are used to reconstruct x_i, and the total reconstruction error over all the x_i's is defined as Eq. 1, where x_{ij} is the jth neighbor of x_i, and w_{ij} encodes the contribution of x_{ij} to the reconstruction of x_i. By minimizing Eq. 1, w_{ij} can be determined.

$$\varepsilon(\mathbf{W}) = \sum_i \left| x_i - \sum_j w_{ij} x_{ij} \right|^2, \qquad (1)$$

Then, the weights w's are fixed and the corresponding z_i's are sought through minimizing Eq. 2. Like Eq. 1, Eq. 2 is based on local linear reconstruction errors, but here the weights w's are fixed while the coordinates z_i's are optimized.

$$\varepsilon(\mathbf{Z}) = \sum_i \left| z_i - \sum_j w_{ij} z_{ij} \right|^2. \qquad (2)$$

The LLE algorithm has been applied to visualization and achieved some success [9]. It is noteworthy that the original LLE algorithm was mainly designed for visualization, which does not take into account the label information. However, the working scheme of LLE can be modified to utilize the label information, and therefore it can also be used in classification [3][12]. Nevertheless, as mentioned before, like other existing manifold learning algorithms, LLE can hardly work well when the data are not densely sampled.

2.2 NNL

The NFL (Nearest Feature Line) method was originally proposed for face recognition [8]. In NFL, *feature line* is defined as the line passing through two points from a same class. The distance between an unseen point to the line is regarded as a measurement of the strength of the point belonging to the concerned class. Besides suffering from large computational costs, NFL often fails when the query point, i.e. the unseen data point to be classified, is far from the prototype points because in this case, unreliable extrapolated points may be queried for classifying the unseen data. In order to reduce the influence of this problem, the NNL (Nearest Neighbor Line) method, which is a modified version of NFL, was proposed, where only the neighbors of the query point instead of all the possible feature lines are used [13].

Formally, let $\{x_j^i\}$ $(i = 1, \cdots, c; j = 1, \cdots, N_i)$ denote the training set, where c is the number of classes, x_j^i is the jth sample of the ith class, and N_i is the number of the samples belonging to the ith class. Let x denote the query sample. Suppose x_a^i and x_b^i denote the two nearest neighbors of x in the ith class. Then, the straight line $x_a^i x_b^i$ passing through x_a^i and x_b^i is called the *neighbor line* of x in the ith class. The *neighbor line distance* between x and $x_a^i x_b^i$ is given by $dist(x, x_a^i x_b^i) = \|x - \mathbf{I}_{x_a^i x_b^i}\|$, where $\|\cdot\|$ stands for the Euclidean distance, and $\mathbf{I}_{x_a^i x_b^i}$ is the image of x projected onto the neighbor line, or equally, the plumb root. Then, x is classified as belonging to the class corresponding to its nearest neighbor line, that is,

$$label(x) = \arg\min_{i \in \{1,\cdots,c\}} dist(x, x_a^i x_b^i). \tag{3}$$

2.3 Virtual Samples

In pattern recognition, much effort has been devoted to tackling the *small sample problem*, where the utilization of virtual samples is an effective scheme. For example, in the $(PC)^2A$ method designed for face recognition with one training image per person [11], the horizontal and vertical projections of the original face image are used to help create some virtual face images such that the intra-class differences can be computed in PCA.

In machine learning, virtual samples have been used in *comprehensible learning*. For example, virtual samples were generated to help extract symbolic rules from complicated learning systems such as neural network ensembles [16]. In the *twice-learning* paradigm [14][15], a learner with strong generalization ability is used to generate virtual samples which are then given to a learner with good comprehensibility, such that the learning results are with high accuracy as well as good comprehensibility.

Virtual Samples are also useful in learning with imbalanced data sets. For example, in the SMOTE algorithm [1], virtual samples of the minority class are generated such that the number of minority training samples is increased. Here the virtual samples are generated through interpolating between each minority class point and its k nearest neighboring minority class points, which looks somewhat like the interpolating scheme used in NFL [8] and NNL [13]. In fact, the NFL and NNL algorithms have implicitly utilized virtual samples since they use a virtual point instead of a real data point to help compute the distance between a data point and a class.

3 NL³E

Since many real-world data sets are not densely sampled, the performance of manifold learning algorithms are often not satisfying. It can be anticipated that if more samples with helpful information are available, the learning results could be better. As introduced in section 2.3, virtual samples are useful in many areas. However, almost all the existing techniques for generating virtual samples were developed in fields other than manifold learning. In order to design suitable methods for manifold learning, the characteristics of manifold learning algorithms must be taken into account. Here only LLE is considered.

The principal idea of LLE is to keep the local relationship between the samples during the mapping process. Therefore, in order to keep the local relationship, virtual samples must be created in local areas. Thus, the neighbor line used in NNL seems helpful.

Here the neighbor line method is generalized. In its original form, the interpolated points on only the nearest neighbor line could be used as potential virtual samples. Here the interpolated points on a number of neighbor lines can be used as virtual samples. It is anticipated that through generating more virtual samples, the data set will become densely distributed meanwhile the underlying distribution is reserved.

In LLE, there is a *neighbor selection* parameter, k. When the input samples are not densely sampled, if k is set to a large value, LLE may return invalid answer due to the lose of locality; but if k is set to a small value, LLE can hardly get sufficient information. In NL³E, the k-nearest neighboring points of the concerned data point will be identified, as that in LLE. But rather than using only these k neighboring points as LLE does, NL³E can obtain more data points to use because a number of virtual samples on the neighbor lines corresponding to the identified k neighboring points will be generated. Therefore, with the same setting of k, the samples used by NL³E can cover the local area better than that used by LLE. Fig. 1 gives an illustration.

In Fig. 1, the concerned point is x_i, and its four nearest neighbors, i.e. x_{ij} ($j = 1, \cdots, 4$), have been identified. Assume the circle around x_i specifies the underlying locality of x_i. Thus, x_i has only one neighbor, i.e. x_{i1}, locating in the real local area. It is obvious that x_i can hardly be faithfully reconstructed when the local information is too little. Fortunately, NL³E can use virtual samples to enrich the local information. As Fig. 1 shows, there are six virtual samples created with the help of the neighbor lines. In this case, if k is set to 1 in the original data set to find neighborhood area, then after virtual sample creation, 7 points should be selected in order to get a neighbor area with similar size. In general, in order to obtain the local area with similar size, the neighbor selection parameter used after the virtual sample generation process should be bigger than

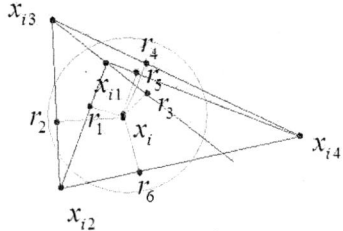

Fig. 1. An illustration of the virtual samples generated in NL³E

k. Let p denote the number of virtual examples generated when k is 1. Then, the neighbor selection parameter used after the virtual selection process could be determined according to $k' = (1 + p) \times k$ because roughly p virtual samples will be generated based on every points of the k nearest neighbors.

Note that the neighbor area used for generating virtual examples is not needed to be the same as that used for reconstructing the concerned data point x_i. Actually, NL³E identifies a big neighbor area through consulting the l ($l \geq k$) nearest neighbors of x_i, in which the virtual examples are generated. Then, on the enriched training set, k' ($k' = (1 + p) \times k$) nearest neighbors of x_i will be used to reconstruct x_i according to Eq. 1.

When a big l value is used, a lot of virtual samples will be created. Considering that in a d-dimensional space $(d+1)$ neighbors are sufficient for reconstructing a data point, in NL³E the number of virtual samples to be generated is restricted to $(d + 1)$. That is, only $(d + 1)$ number of virtual samples is really generated among the C_l^2 possible virtual samples.

The pseudo-code describing the NL³E algorithm is shown in Table 1. In contrast to LLE, NL³E has only one more parameter to set, that is, l.

4 Experiments

For visualization, the goal is to map the original data set into a two- or three-dimensional space that preserves as much as possible the intrinsic structure. In many previous works on visualization, the results are mainly compared through examining the figures to point out which looks better. To compare the results more impersonally, it was suggested to use the variance fraction to measure the visualization effect [10]. However, variance fraction in fact measures the relationship between the reconstructed pairwise geodesic distances and the lower-dimensional distances, not the structure inflexibility. In another work, the correlation coefficient between the distance vectors, i.e. the vectors that comprises the distances between all pairs of the true structure and that of the recovered structure, was used [6]. It has been shown that this method provides a good measurement of the validity of the visualization [6]. Suppose the distance vector of the true structure is **DV** and that of the recovered structure is **DV'**, then the correlation coefficient between **DV** and **DV'** is computed by

Table 1. Pseudo-code describing the NL^3E approach

NL^3E (**X**, l, k, m)

Input:
    X: original samples $\{x_1, x_2, \cdots, x_n\}$, $\mathbf{x} \in \mathcal{R}^d$
    l: the neighbor selection parameter used in virtual sample generation
    k: the neighbor selection parameter used in LLE
    m: the dimensionality of the output coordinates

Process:
1. $V = \emptyset$
2.   **For** $i = 1, 2, \cdots, n$ **do**
3.   identify x_i's l-nearest neighbors according to Euclidean distance
4.   **For** $j = 1, 2, \cdots, d$ **do**
5.   Select a pair of neighbors randomly and non-repeatedly, and
           assume they are x_{iR_1} and x_{iR_2} ($1 \leq R_1 < R_2 \leq l$)
6. $r_{ij} = \text{getPlumbRoot}(x_i, x_{iR_1} x_{iR_2})$
           % r_{ij} is the plumb root of x_i on the neighbor line $x_{iR_1} x_{iR_2}$
7.   **If** r_{ij} is identical to x_i % $x_{iR_1} x_{iR_2}$ passes through x_i
8. $j = j - 1$
9.   **else**
10. $V = V \bigcup \{r_{ij}\}$
11.   **End If**
12.   **End For**
13.   **End For**
14. $\mathbf{X}' = \mathbf{X} \bigcup V$
15. $\mathbf{Z}' = \text{LLE}(\mathbf{X}', (d+1) \times k, m)$
      % $(d+1) \times k$ is the neighbor selection parameter used by LLE
16. $\mathbf{Z} = \text{getXcor}(\mathbf{Z}', \mathbf{X})$ % Get **X**'s lower-dimensional coordinates

Output: **Z**

$$\rho = \frac{(\mathbf{DV} \cdot \mathbf{DV}') - \overline{\mathbf{DV}} \cdot \overline{\mathbf{DV}'}}{\sigma(\mathbf{DV})\sigma(\mathbf{DV}')}, \qquad (4)$$

where $(\mathbf{A} \cdot \mathbf{B})$ is the inner product of **A** and **B**, $\overline{\mathbf{U}}$ returns the average value of **U** and $\sigma(\mathbf{U})$ is the standard deviation of **U**. Generally, the larger the ρ, the better the performance.

Several synthetic data sets are used in the experiments. First, a two-dimensional rectangle is selected as the basic structure, and then 200, 300, or 400 points are randomly sampled from the structure. After that, the points are separately embedded onto "Scurve" (SC) or "Swiss roll" (SW). So there are 6 data sets, i.e. SC-200, SC-300, SC-400, SW-200, SW-300 and SW-400. SC-400 and SW-400 are shown in Fig. 2. The colors reveal the structure of each data set.

NL^3E is used to map these data sets onto two-dimensional space, and then the visualization effect is evaluated. The performance of NL^3E is compared with that of LLE according to the correlation coefficient. Since the data sets are generated through embedding some two-dimensional samples onto higher-dimensional

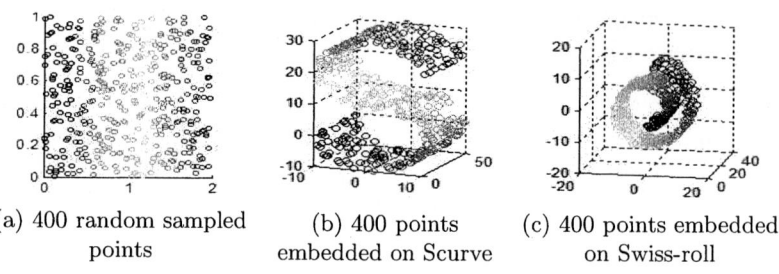

(a) 400 random sampled points   (b) 400 points embedded on Scurve   (c) 400 points embedded on Swiss-roll

Fig. 2. Embedded Samples

space, the intrinsic dimension of these data sets are all two. The parameter l of NL^{3}E is set from 6 to 10. The experiments are repeated for 5 times under each configuration, and the average value of ρ which is denoted by $\bar{\rho}$ is recorded. The parameter k of both NL^{3}E and LLE is set from 3 to 7.

In Table 2, the performance measured by ρ is reported ($k = 6$, $l = 8$), where R_i ($i = 1, 2, \cdots, 5$) denotes the ith run of the experiment. The table shows that NL^{3}E outperforms LLE in most situations. Only on SW-300, the average value of $\bar{\rho}$ of NL^{3}E is worse than that of LLE.

Table 2. ρ value of each data set ($k = 6$, $l = 8$)

	NL^{3}E						LLE
	R_1	R_2	R_3	R_4	R_5	AVG.	
SC-200	0.740	0.771	0.865	0.829	0.831	**0.807**	0.747
SC-300	0.832	0.843	0.808	0.812	0.852	**0.829**	0.714
SC-400	0.828	0.716	0.751	0.823	0.763	**0.776**	0.576
SW-200	0.381	0.341	0.367	0.355	0.372	**0.363**	0.305
SW-300	0.610	0.659	0.620	0.726	0.615	0.646	**0.694**
SW-400	0.781	0.707	0.716	0.711	0.450	**0.673**	0.588

Fig. 3 shows the visualization results of NL^{3}E and LLE on the SC series data sets when $k = 6$ and $l = 8$. Note that under each configuration the experiment has been run for 5 times, and Fig. 3 shows the situation where the ρ value is the median in these 5 runs. Colors reveal the structure of embedding samples. It is obvious that LLE's performance is poor while the result of NL^{3}E are quite well.

Fig. 4 shows the visualization results of NL^{3}E and LLE on the SW series data sets when $k = 6$ and $l = 8$. It can be found that the performance of NL^{3}E is not so good as that in Fig. 3. This can also be observed in Table 2, where the $\bar{\rho}$ value of NL^{3}E on the SW series data sets are lower than these on the SC series data sets. Nevertheless, it is obvious that NL^{3}E still performs better than LLE.

For studying how the parameter l affecting the performance of NL^{3}E, more experiments are conducted. The results are shown in Fig. 5, where $k = 6$ while

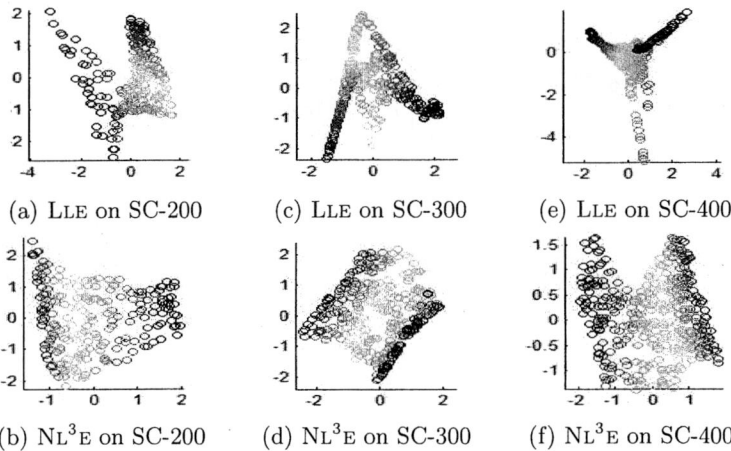

Fig. 3. Visualization results on SC series data sets

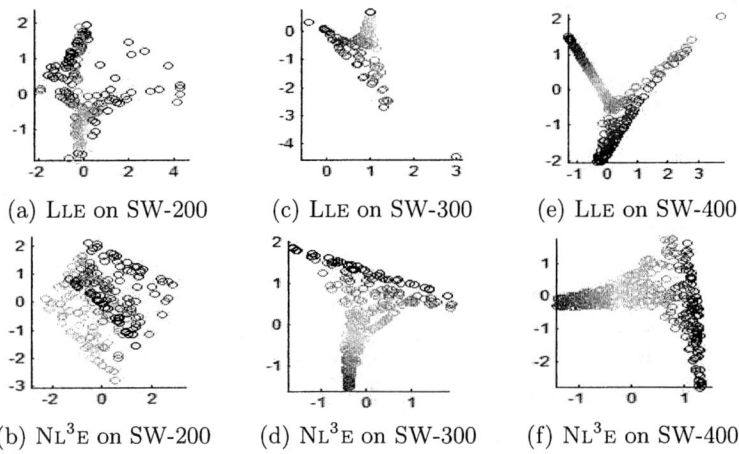

Fig. 4. Visualization results on SW series data sets

l changes from 6 to 10. It's obvious that no matter which value l takes, except on SW-300, the performance of NL^3E is better than that of LLE in most cases.

In order to explore the influence of the parameter k on the performance of NL^3E, further more experiments are performed. The results are shown in Fig. 6, where $l = 8$ and k changes from 3 to 7. As this figure tells, when k increases, firstly LLE's performance becomes better and NL^3E's becomes better too. If k continuously increases, LLE's performance may decrease. This can be observed clearly on SC-200, SW-300 and SW-400. Although NL^3E's effect may decrease either, it is more slowly. Almost in every point, NL^3E is better than LLE. In fact, no matter which value is set to k, the performance of NL^3E remains quite well.

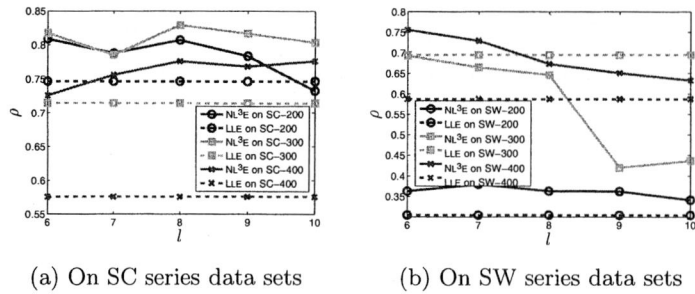

(a) On SC series data sets (b) On SW series data sets

Fig. 5. The influence of the parameter l on the performance of NL^3E

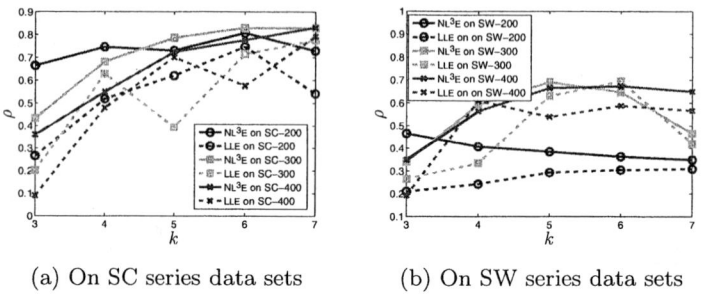

(a) On SC series data sets (b) On SW series data sets

Fig. 6. The influence of the parameter k on the performance of NL^3E

5 Conclusion

Many manifold learning algorithms often return invalid results when the data is not densely sampled. This paper proposes the NL^3E algorithm, which is a variant of LLE but can work well in some cases where the data is not densely sampled. The reason lies in the fact that using virtual samples, NL^3E can use more information. Experiments on synthetic data sets show that the performance of NL^3E is better than that of LLE. The performance of NL^3E on real-world data will be evaluated in the future.

In this paper, the virtual samples generated with the help of neighbor lines are all plumb roots in a local area, therefore the local information is enriched while the locality is kept. It is evident that such kind of virtual samples can also be used by other manifold learning algorithms such as ISOMAP, C-ISOMAP, etc. to relax the requirement of dense samples. This will be studied in the future.

Note that in order to enrich the local information, the computational cost of NL^3E is bigger than that of LLE. Fortunately, the computational cost of NL^3E can be reduced by using smaller number of virtual samples. Nevertheless, designing efficient virtual sample utilization scheme is also an important future work.

Acknowledgements

This work was supported by the Foundation for the Author of National Excellent Doctoral Dissertation of China under the Grant No. 200343, the National Science Fund for Distinguished Young Scholars of China under the Grant No. 60325207, and the Fok Ying Tung Education Foundation under the Grant No. 91067.

References

1. Chawla, N.V., Bowyer, K.W., Hall, L.O., Kegelmeyer, W.P.: SMOTE: Synthetic minority over-sampling technique. Journal of Artificial Intelligence Research **16** (2002) 321–357
2. Cox, T., Cox, M.: Multidimensional Scaling. Chapman and Hall, London (1994)
3. de Ridder, D., Loog, M., Reinders, M.J.T.: Local fisher embedding. In: Proceddings of the 17th International Conference on Pattern Recognition. Cambridge, UK (2004) 295–298
4. de Silva, V., Tenenbaum, J.B.: Global versus local methods in nonlinear dimensionality reduction. In: Becker, S., Thrun, S., Overmayer, K. (eds.): Advances in Neural Information Processing Systems 15. MIT Press, Cambridge, MA (2002) 705–712
5. Duda, R.O., Hart, P.E., Stork, D.G.: Pattern Classification, 2nd edition. Wiley, New York, NY (2004)
6. Geng, X., Zhan, D.-C., Zhou, Z.-H.: Supervised nonlinear dimensionality reduction for visualization and classification. IEEE Transactions on System, Man and Cybernetics-Part B: Cybernetics **35** (2005) 1098-1107
7. Jolliffe, Z.T.: Principal Component Analysis. Springer, New York, NY (1986)
8. Li, S.Z., Lu. J.: Face recognition using the nearest feature line method. IEEE Transactions on Neural networks **10** (1999) 439-443
9. Roweis, S.T., Saul, L.K.: Nonlinear dimensionality reduction by local linear embedding. Science **290** (2000) 2323-2326
10. Tenenbaum, J.B., de Silva, V., Langford, J.C.: A global geometric framework for nonlinear dimensionality reduction. Science **290** (2000) 2319-2323
11. Wu, J., Zhou, Z.-H.: Face recognition with one training image per person. Pattern Recognition Letters **23** (2002) 1711–1719
12. Zhang, J., Shen, H., Zhou, Z.-H.: Unified locally linear embedding and linear discriminant analysis algorithm (ULLELDA) for face recognition. In: Li, S.Z., Lai, J., Tan, T., Feng, G., Wang, Y. (eds.): Lecture Notes in Computer Science 3338. Springer, Berlin (2004) 296–304
13. Zheng, W., Zhao, L., Zou, C.: Locally nearest neighbor classifiers for pattern classification. Pattern Recognition **37** (2004) 1307-1309.
14. Zhou, Z.-H., Jiang, Y.: Medical diagnosis with C4.5 rule preceded by artificial neural network ensemble. IEEE Transactions on Information Technology in Biomedicine **7** (2003) 37–42
15. Zhou, Z.-H., Jiang, Y.: NeC4.5: neural ensemble based C4.5. IEEE Transactions on Knowledge and Data Engineering **16** (2004) 770–773.
16. Zhou, Z.-H., Jiang, Y., Chen, S.-F.: Extracing symbolic rules from trained neural network ensembles. AI Communications **16** (2003) 3–15

Predicting Rare Extreme Values

Luis Torgo and Rita Ribeiro

LIACC-FEP, University of Porto, R. de Ceuta, 118, 6., Porto 4050-190, Portugal
{ltorgo, rita}@liacc.up.pt
http://www.liacc.up.pt/~[ltorgo,rita]

Abstract. Modelling extreme data is very important in several application domains, like for instance finance, meteorology, ecology, etc.. This paper addresses the problem of predicting extreme values of a continuous variable. The main distinguishing feature of our target applications resides on the fact that these values are rare. Any prediction model is obtained by some sort of search process guided by a pre-specified evaluation criterion. In this work we argue against the use of standard criteria for evaluating regression models in the context of our target applications. We propose a new predictive performance metric for this class of problems that our experiments show to perform better in distinguishing models that are more accurate at rare extreme values. This new evaluation metric could be used as the basis for developing better models in terms of rare extreme values prediction.

1 Introduction

In several applications the main focus of interest is a small proportion of the available data. These unusual cases have a large importance, and as such, anticipating them is a critical task for these domains. An example of such applications is the prediction of the future returns of a stock. Unusually high (low) returns are rare, but they are the most interesting values for investors and thus they should be the target of any financial prediction model.

A related problem has been receiving great attention in the data mining community: the construction of classification models based on samples with unbalanced class distributions (e.g. [1]). Predicting extreme values of a continuous variable can be handled through a classification approach by means of a discretization process (e.g. [2]) off the continuous target variable. This would have the advantage of using all work that has been around in the areas of unbalanced classification problems and evaluation under differentiated misclassification costs. However, this approach would require to establish the number of classes and, moreover, would lead to an undesirable crisp division between what is an extreme and what is a "normal" case. These are some of the major drawbacks of handling regression as a classification problem[1].

The problem of predicting rare extreme values is a particular case of multiple regression where a target continuous variable Y is being modelled using a set

---

[1] More details on this argument can be found on [4],an extended version of this work.

of predictor or input variables $X_1, X_2, \cdots, X_p$. Any modelling method tries to find the model parameters that minimise an error function over the training sample. Standard functions used in regression setups are the Mean Squared Error, $MSE = \frac{1}{n} \sum_{i=1}^{n} (y_i - \hat{y}_i)^2$, or the Mean Absolute Deviation, $MAD = \frac{1}{n} \sum_{i=1}^{n} |y_i - \hat{y}_i|$. Both these measures take all errors equally (with the same cost) and thus can be regarded as less adequate for our target applications, where errors on extreme values are more important.

One possible method for giving more weight to the errors on extreme values is to use case weights. Some algorithms allow the user to attach a weight to each case of the training sample. Model parameters can then be obtained by minimising a criterion that takes into account these weights. Using case weights that depend on the respective Y value being an extreme allows us to bias the obtained model to correctly predict these extreme cases. The main drawback of this approach is that it only sees one side of the problem, the true values. In effect, this method does not try to avoid (or penalise) the cases where an extreme value is predicted by the model, but the truth value is "normal", i.e. false positives according to the classification terminology. This drawback stems from the fact that the weights are dependent solely on the true value of the cases, y_i, instead of being dependent on both y_i and $\hat{y}_i$. Our proposal builds upon this idea by trying to eliminate this drawback through the use of a weight function that depends on both y_i and $\hat{y}_i$.

2 Our Proposal

The overall goal of this work is to have an evaluation metric that is biased towards valuating more the predictions of rare extreme values. Our proposal was developed with the following requirements in mind: i) the cost of a prediction error should depend on both the predicted and the true values, i.e. we should penalise both false positives and false negatives; ii) the cost of the errors should vary smoothly (no crisp divisions between extremes and non-extremes); iii) the method should have reasonable default costs (according to the overall goal) for applications where knowledge about the costs is not available.

We propose an evaluation metric that is basically a weighted average of the errors. Our key contribution lies on the form of calculating the weights. We use a weight function that depends on both the true and predicted values. We propose to use a smooth cost surface, $w(Y, \hat{Y})$, that can be seen as a continuous version of cost matrixes used in classification tasks. Summarising, our proposed Rare Extremes Error metric is defined as,

$$RExE = \frac{1}{n} \sum_{i=1}^{n} w(y_i, \hat{y}_i) \times L(y_i, \hat{y}_i) \qquad (1)$$

where $L(y_i, \hat{y}_i)$ can be any loss function, e.g. the squared error.

In order to make the use of smooth cost surfaces practical we need to devise an easy way of specifying them. Our proposal consists of requiring the specification of the cost values at a small set of properly selected points and then using a function approximation method to interpolate the complete surface. The axes of the surface are the true, Y, and predicted, $\hat{Y}$, values of the target variable. These range from low extreme values ($extr_L$) to high extreme values ($extr_H$). The points selected for specifying the cost surface should be related to the most relevant areas of the surface. These are the areas of lower cost (the model accurately predicts and extreme as such), and of the worse performance (the model predicts an extreme high for a true extreme low, or vice versa).

For applications where no cost information is available but still extremes are more important, we need to describe means to setup the costs for the key points used for surface approximation. The critical question is to define what is a rare extreme value. We use the same definition as in Torgo and Ribeiro [3]. This means that we set $extr_L = adj_L$ and $extr_H = adj_H$, where $adj_L(adj_H)$ is the smallest observation that is greater or equal to the 1st quartile minus $1.5r$, with r being the interquartile range. After having defined these two extreme values we artificially create n grid points by diving the interval $extr_H - extr_L$ in n equally spaced bins. This means that we will have a $(n+2)\mathrm{x}(n+2)$ matrix to fill in with costs. We use an arithmetic progression to setup the costs from the lowest to the highest cost. Full details and illustrative examples can be found in [4].

3 An Experimental Evaluation of the Proposal

We have carried out a series of experiments with the goal of checking the validity of our proposed metric in the task of identifying the models that are better from the perspective of being more accurate at rare extreme values. With this purpose we have designed the following experimental setup for each data set:

1. Draw a stratified test sample with 50% of the cases;
2. Randomly generate a set of prediction errors with the same size as the test sample. The errors are drawn from a normal distribution. We then pick the n largest errors, where n is the number of extreme values of the distribution of Y, and increase these errors by a constant k. The overall objective of this step is to obtain a set of credible prediction errors for a standard model when making predictions for a problem with some extremes. For this type of problems we expect (we have confirmed this experimentally using several modelling techniques and several real world data sets), the models to achieve a performance of this type: normal-shape distribution of the error with some extreme errors typically occurring on test cases with extreme values of Y.
3. We then artificially allocate this set of generated errors to each case on the test set in two different ways, leading to the "artificial performance" of models A and B. For model A, the smallest errors are allocated to the extremes in the test set, thus leading to what could be considered to be an ideal model for our target applications. On the contrary, model B has the largest errors on the extreme values of the target, in what could be considered a "normal" behaviour of a model in this type of tasks.

A performance metric that is biased towards accurate predictions on extremes, should clearly indicate that the performance of Model A is better than the

Table 1. The results in terms of percentage difference between Models A and B

Data Set	SigMetric (avg±sd)	NRExE (avg±sd)	Data Set	SigMetric (avg±sd)	NRExE (avg±sd)
algae1	52.6±4.8	**82.9±1**	deltaAilerons	**55.2±0.6**	5±0.4
algae2	55±4.6	**79.3±1.6**	ibm	**71.9±0.3**	7.4±0.4
algae3	71.1±2.4	**88.1±1**	abalone	**70.5±1.1**	6.5±0.5
algae4	73.8±14.1	**87.3±5.1**	cpuSmall	63±1	**81.1±0.4**
algae5	56.1±4.6	**83.9±1.5**	servo	74.4±5.8	**85.2±0.9**
algae6	84.2±0.8	**91.1±0.5**	cwDrag	**57.4±1.6**	2.8±2.6
algae7	52.4±11.1	**82.7±2.3**	co2Emission	**58.4±0.6**	17.8±6.5
Boston	**65±1.6**	21±25.3	availablePower	69.7±1.5	**71.5±0.7**
machineCpu	76.5±3.4	**77.9±1**	china	68.9±2.8	**71.8±1.4**
bank8FM	55.3±0.5	**63.6±0.8**	add	**56.6±0.3**	5.5±0.7

performance of Model B. Notice that, given that the errors of the two models are exactly the same (only occurring at different test cases), metrics like the MSE or the MAD will show both models as having exactly the same score.

As we are testing on a large set of domains with a quite different range of target variable values, we have used a normalised version of our performance statistic to allow comparisons across domains,

$$NRExE = \frac{\sum_{i=1}^{n_{test}} w(y_i, \hat{y}_i) \cdot |y_i - \hat{y}_i|}{\sum_{i=1}^{n_{test}} w(y_i, \widetilde{Y}) \cdot |y_i - \widetilde{Y}|} \qquad (2)$$

where $\widetilde{Y}$ is the sample median.

The goal of our experiments is to assert the score difference between models A and B, when evaluating them using our proposed metric and an alternative measure. With this purpose we have measured the percentual difference of scores for all data sets. Positive values of this difference indicate that our metric is able to identify Model A as performing better than Model B. We obviously want the difference to be as high as possible, as Model A has an "ideal" performance. We have compared our proposed metric, $NRExE$, against the score obtained by the most similar alternative, an error measure using case weights as mentioned in Section 1. For this competitor we have setup the case weights such that more weight is given to cases with extreme values of the target (details on [4]). Notice that contrary to our approach the weights of this measure only consider the true value of the target, thus not taking into account the predictions of the models.

For each data set we have repeated the experiment outlined above 10 times. The results shown on Table 1 are the average and standard deviation of the observed percentual differences between Model A and B, when using $NRExE$ and the metric with sigmoid-based case weights. The best scores for each data set are indicated in bold. The used datasets are real world problems with a diverse set of rare extreme values types. For instance, some include both low and high extremes, while others include only one type of extremes. Due to space reasons we are not able to present the full characteristics of these problems.

The results reported in Table 1 show the advantages of our proposed metric for domains where the main objective is to be accurate at rare extreme values. In effect, in most problems our metric correctly signals model A as being significantly better than model B, in spite of being compared against a competitor metric that also take extremes into account. Notice that standard measures, like MSE, would signal both models as being equal (difference equal to zero).

4 Conclusions

In this paper we have described the particular features of a class of problems with high practical importance: the prediction of rare extreme values. We claim that existing metrics for evaluating the performance of different models have several drawbacks and perform poorly on identifying the best models in terms of predictive accuracy on the most important cases for these applications. We have presented a new metric that is particularly suited for these applications.

In a set of experiments using real world data we have shown that this measure is able to identify the best model in terms of accuracy on the rare extreme values, even on the most difficult scenario where both models have exactly the same error distribution and thus have the same score in "standard" metrics like MSE.

One of the main impacts of the results of this work is that our metric can be used to compare different existing models on tasks where the main goal is the accuracy on rare extreme values. The use of our metric should provide better information concerning the merits of alternative models for these important tasks. Another important side effect of this work is the possibility of using the described metric in the search process of any modelling technique, so as to develop models that are built for maximising the predictive performance on extreme values.

Acknowledgements

This work was supported by FCT project MODAL (POSI/SRI/40949/2001) cofinanced by POSI and by the European fund FEDER and by a PhD scholarship of FCT (SFRH/BD/1711/2004) to Rita Ribeiro.

References

1. G.Weiss and F. Provost. Learning when training data are costly: The effect of class distribution on tree induction. *JAIR*, 19:315–354, 2003.
2. L. Torgo and J. Gama. Regression using classification algorithms. *Intelligent Data Analysis*, 1(4), 1997.
3. L. Torgo and R. Ribeiro. Predicting outliers. In N. Lavrac, D. Gamberger, L. Todorovski, and H. Blockeel, editors, *Proceedings of Principles of Data Mining and Knowledge Discovery (PKDD'03)*, number 2838 in LNAI, pages 447–458. Springer, 2003.
4. L. Torgo and R. Ribeiro. Predicting rare extreme values. Technical Report 2006-01, LIACC-NIAAD. University of Porto, 2006.
   (http://www.liacc.up.pt/~ltorgo/Papers/PREVext.pdf).

Domain-Driven Actionable Knowledge Discovery in the Real World[*]

Longbing Cao and Chengqi Zhang

Faculty of Information Technology, University of Technology Sydney, Australia

Abstract. Actionable knowledge discovery is one of Grand Challenges in KDD. To this end, many methodologies have been developed. However, they either view data mining as an autonomous data-driven trial-and-error process, or only analyze the issues in an isolated and case-by-case manner. As a result, the knowledge discovered is often not actionable to constrained business. This paper proposes a practical perspective, referred to as *domain-driven in-depth pattern discovery* (DDID-PD). It presents a domain-driven view of discovering knowledge satisfying real business needs. Its main ideas include constraint mining, in-depth mining, human-cooperated mining, and loop-closed mining. We demonstrate its deployment in mining actionable trading strategies in Australian Stock Exchange data.

1 Introduction

Actionable knowledge discovery can afford important grounds to business decision makers. In the panel discussions of SIGKDD 2002 and 2003 [2, 7], it was highlighted by panelists as one of the Grand Challenges for extant and future data mining. This situation partly results from the scenario that extant data mining is a data-driven trial-and-error process [2] where data mining algorithms extract patterns from converted data via some predefined models based on experts' hypothesis. Data mining is presumed as an automated process producing automatic algorithms and tools without human involvement and the capability to adapt to external environment constraints.

However, data mining in the real world, for instance financial data mining, is highly constraint-based [8, 11]. Constraints involve technical, economic and social aspects. The real-world business problems and requirements are often tightly embedded in domain-specific business rules and process with expertise (*domain constraint*). Patterns actionable to business are often hidden in large quantities of data with complex structures, dynamics and source distribution (*data constraint*). Often mined patterns are not actionable to business even though they are interesting to research. There exist big interestingness gaps between academia and business (*interestingness constraint*). Furthermore, interesting patterns often cannot be deployed to real life if they are not integrated with business rules, regulations and processes (*deployment constraint*). There could be other types of constraints such as knowledge constraint, dimension/level constraint and rule constraint [8].

---

[*] This work is sponsored by UTS Chancellor research fund and ECRG Fund.

To discover actionable knowledge from data embedded in the above constraints, it is essential to slough off the superficial and captures the essential information from the data mining. However, this is a non-trivial task. Tricks may not only include how to find a right pattern with a right algorithm in a right manner, they also involve a suitable process-centric support with a suitable deliverable to business. Even many methodologies are studied, they either view data mining as an automated process, or deal with the constraints in a case-by-case manner. Our experience [3] and lessons learned in data mining in capital markets [6] show that the involvement of domain knowledge and experts, the consideration of constraints, and the development of in-depth patterns are essential for filtering subtle concerns while capturing incisive issues. Combining these aspects together, it can advise the process of real-world data mining in a manner more actionable and reliable to business. These are our motivation to develop a practical framework, called *domain-driven in-depth pattern discovery* (DDID-PD), for the discovery of actionable knowledge from the real world.

DDID-PD views actionable knowledge discovery as an iteratively interactive in-depth pattern mining process in domain-specific context. It exploits key components including (i) constraint mining, (ii) incorporating domain knowledge through human-mining-cooperation, (iii) in-depth mining, and (iv) loop-closed mining. Mining constraint-based context requests to develop workable mechanisms to deal with comprehensive constraints. The involvement of domain experts and their knowledge can reduce the complexity of the knowledge discovery process in the constrained world. In-depth pattern mining discovers actionable patterns. A system following the DDID-PD framework can embed effective supports for domain knowledge and experts' feedback, and refines the lifecycle of data mining in an iterative manner.

Taking financial data mining as an example, this paper introduces some case studies deploying the DDID-PD framework to mine actionable trading strategies for improving trading performance and costs. It shows that the DDID-PD can benefit the actionable knowledge mining in a more realistic and reliable manner than data-driven methodology such as CRISP-DM [13].

2 Domain-Driven In-Depth Pattern Discovery

The existing data mining methodology, for instance CRISP, generally supports autonomous pattern discovery from data. The DDID-PD, on the other hand, highlights a process that discovers in-depth patterns from constraint-based context with the involvement of domain experts/knowledge. This section outlines key ideas and relevant research issues of the DDID-PD.

2.1 Pattern Actionability

Let $I = \{i_1, i_2, \ldots, i_m\}$ be a set of items, DB be a database consisting of a set of transactions, x is an itemset in DB. Let P be a pattern discovered in DB through a model M. In DDID-PD [4], the following concepts measure pattern actionability, i.e., whether or not, or to what extent, P can be used to answer real business needs.

DEFINITION 1. Technical Interestingness – The technical interestingness $tech\_int()$ measures how interesting the pattern is from technical perspective. It is measured through certain technical metrics specified for a data mining method. For instance, the following logic formula indicates that an association rule P is technically interesting if it satisfies a user-defined *min_support* and *min_confidence*.

$$\forall x \in I, \exists P: x.min\_support(P) \wedge x.min\_confidence(P) \rightarrow x.tech\_int(P)$$

DEFINITION 2. Business Interestingness – The business interestingness *biz_int()* of a pattern is determined by some domain-oriented social and/or economic criteria accepted by real users. In stock data mining, a stock price predictor P is interesting to trading if it satisfies the *profit* and *roi* (*return on investment*) requests.

$$\forall x \in I, \exists P: x.profit(P) \wedge x.roi(P) \rightarrow x.biz\_int(P)$$

DEFINITION 3. Actionability of a pattern – The actionability of a pattern *act()* indicates to what degree it satisfies both technical and business interestingness. If both technical and business interestingness or a hybrid interestingness measure integrating both aspects are satisfied, it is called an *actionable* pattern. Such kind of patterns are not only interesting to data miners, but generally interesting to decision-makers.

$$\forall x \in I, \exists P: x.tech\_int(P) \wedge x.biz\_int(P) \rightarrow x.act(P)$$

2.2 Actionable Knowledge Discovery Process

The components of the DDID-PD are shown in Figure 1, where we highlight those processes specific to DDID-PD in thicken boxes. The lifecycle of DDID-PD is as follows, but be aware that the sequence is not rigid, some phases may be bypassed or moved back and forth in a real problem. Every step of the DDID-PD process may involve domain knowledge and the interaction with real users or domain experts.

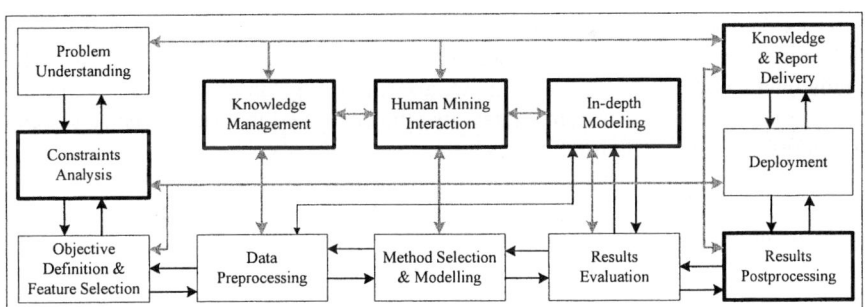

Fig. 1. DDID-PD process model

P1. Problem understanding;
P2. Constraints analysis;
P3. Analytical objective definition, feature construction;
P4. Data preprocessing;
P5. Method selection and modeling; or
P5'. In-depth modeling;
P6. Initial generic results analysis and evaluation;

P7. *It is quite possible that each phase from P1 may be iteratively reviewed through analyzing constraints and interaction with domain experts in a back-and-forth manner; or*
P7': *In-depth mining on the initial generic results where applicable;*
P8. *Results post-processing;*
P9. *Reviewing phases from P1 may be required;*
P10. *Deployment;*
P11. *Knowledge delivery and report synthesis for smart decision making.*

The DDID-PD process highlights four highly correlated procedures that are critical for the success of data mining in the real world. They are (i) *constraint mining*, (ii) *in-depth mining*, (iii) *human-cooperated mining*, and (iv) *loop-closed mining*. The following sections discuss them respectively.

2.3 Constraint Mining

Specifically, in Section 1, we list several types of constraints, which play significant roles in a process effectively discovering knowledge actionable to business. In practice, many other aspects such as data stream and the scalability and efficiency of algorithms may be enumerated. In DDID-PD, constraints are domain-specific, functional, nonfunctional and environmental. These ubiquitous constraints form a *constrained context* for actionable knowledge discovery. All the above constraints must, to varying degrees, be considered in relevant phases of DDID-PD. In this case, the analysis is called *constraint mining* [8, 11].

Some major aspects of domain constraints include the domain and characteristics of a problem, domain terminology, specific business process, policies and regulations, user profiling and favorite deliverables. Potential matters to satisfy or react on domain constraints could consist of building domain model, domain metadata, semantics and ontologies [5], supporting human involvement, human-machine interaction, qualitative and quantitative hypotheses and conditions, merging with business processes and enterprise information infrastructure, fitting regulatory measures, conducting user profile analysis and modeling, etc. Relevant hot research areas include interactive mining, guided mining, and knowledge and human involvement.

Constraints on particular domain data may be embodied in terms of aspects such as very large volume, ill-structure, multimedia, diversity, high dimensions, high frequency and density, distribution and privacy, etc. Data constraints seriously affect the development of and performance requirements on mining algorithms and systems, and constitute some grand challenges to data mining. As a result, some popular researches on data constraints-oriented issues are emerging such as stream data mining, link mining, multi-relational mining, structure-based mining, privacy mining, multimedia mining and temporal mining.

What makes this rule, pattern and finding more interesting than the other? In the real world, simply emphasizing technical interestingness such as objective statistical measures of validity and surprise is not adequate. In DDID-PD, social and economic interestingness such as user preferences and domain knowledge are also considered in assessing whether a pattern is actionable or not.

Furthermore, DDID-PD advocates the delivery of an interesting pattern integrated with the domain environment such as business rules, process, information flow, presentation, etc. In addition, many other realistic issues are considered. For instance, a software infrastructure may be established to support the full lifecycle of data mining; the infrastructure needs to integrate with existing enterprise information systems and workflow; parallel KDD [10] with parallel supports are implemented on multiple sources, parallel I/O, parallel algorithms and memory storage; visualization, privacy and security should receive much-deserved attention.

2.4 In-Depth Mining

In general, data mining publications tend to push the use of specific algorithms rather than answer real business needs. As a result, patterns interesting to data miners often can not achieve business benefits when deployed. We call them *generic* patterns. Such situations have hindered the deployment and adoption of data mining in real applications. Therefore it is essential to evaluate the actionability of a pattern and focus on discovering actionable patterns satisfying both *tech_int(P)* and *biz_int(P)* to support realistic and reliable decision-making. This is *in-depth pattern mining*. Its objective is not to push the use of a specific algorithm, rather try to answer real business needs in a workable manner.

In-depth patterns mining targets to improve both technical (*tech_int()*) and business (*biz_int()*) interestingness in the above constraint-based context. Technically, it could be through enhancing or generating more effective interestingness measures [12]. It could also be through developing alternative models for discovering deeper patterns. Some other options include rule reduction, model refinement or parameter tuning by optimizing *generic* pattern set. Additionally, techniques can be developed to deeply understand, select and refine the target data set.

In in-depth mining, more attention should be paid to business requirements, objectives, domain knowledge and qualitative intelligence of domain experts for their impact on mining deep patterns. This could be through selecting and adding business features, considering domain and background knowledge in modeling, supporting interaction with domain experts, fine tuning parameters and data set by domain experts, optimizing models and parameters, adding factors into technical interestingness measures or building business measures, improving result evaluation mechanism through embedding domain knowledge and human involvement, etc.

2.5 Human Cooperated Mining

The real requirements for discovering actionable knowledge in constraint-based context determine that real data mining is more likely to be human involved rather than automated. Human involvement is embodied through cooperation between humans (including users and business analysts, mainly domain experts) and data mining system. This is achieved through the compensation between human qualitative intelligence such as domain knowledge and field supervision, and mining quantitative intelligence like computational capability. Therefore, real-world data mining likely presents as a human-machine-cooperated interactive knowledge discovery process.

In DDID-PD, the role of human (mainly domain users and experts) could be embodied in the full period of data mining from business and data understanding, problem definition, data integration and sampling, feature selection, hypothesis proposal, business modeling and learning to the evaluation, refinement and interpretation of algorithms and resulting outcomes. For instance, experience, metaknowledge and imaginary thinking of domain experts can guide or assist with the selection of features and models, adding business factors into the modeling, creating high quality hypotheses, designing interestingness measures by injecting business concerns, and quickly evaluate mining results. This assistance may largely improve the effectiveness and efficiency of mining actionable knowledge.

In general, human often serve on the feature selection and result evaluation. DDID-PD views that human could be an essential constituent of or the centre of data mining system. The complexity of discovering actionable knowledge in constraint-based context determines to what extent human must be involved. As a result, human mining cooperation could be, to varying degrees, human-centred or guided mining [2, 8], or human-supported or assisted mining, etc.

To support human involvement, human mining interaction, or in a sense presented as interactive mining [1, 2], is absolutely necessary. Interaction often takes explicit form, for instance, setting up direct interaction interfaces to fine tune parameters. Interaction interfaces may take various forms as well, such as visual interfaces, virtual reality technique, multi-modal, mobile agents, etc. On the other hand, it could also go through implicit mechanisms, for example accessing a knowledge base or communicating with a user assistant agent. Interaction quality relies on performance such as user-friendliness, flexibility, run-time capability and understandability.

2.6 Loop-Closed Mining

Actionable knowledge discovery in a constraint-based context is likely to be a closed rather than open process. It encloses iterative feedback to varying stages such as sampling, hypothesis, feature selection, modeling, evaluation and interpretation in a human-involved manner. On the other hand, real-world mining process is highly iterative because the evaluation and refinement of features, models and outcomes cannot be completed once, rather is based on iterative feedback and interaction before reaching the final stage of knowledge and decision-support report delivery.

The above key points of the DDID-PD indicate that real-world data mining cannot be dealt just with an algorithm, rather it is really necessary to build a proper data mining infrastructure to discover actionable knowledge from constraint-based scenarios in a loop-closed iterative manner. To this end, agent-based data mining infrastructure [14] presents good facilities since it provides good supports for both autonomous problem-solving and user modeling and user agent interaction.

3 Case Study: Developing Actionable Trading Strategies

In stock data mining [6], we deploy the DDID-PD to mine actionable trading patterns. Our objective is to develop actionable trading strategies which can not only trigger profitable trading signals, but also result in the proper measurement and support of market dynamics. For space limit, we only illustrate two case studies here. There is other work under development in terms of domain-driven perspective such as trading strategy-stock correlation analysis, broker-based association analysis, and so forth.

3.1 Designing Actionable Trading Strategy

A quality trading strategy can be designed from scratch via analyzing market dynamics and microstructure. For instance, in real trading, traders often trade multiple stocks to manage risk. DDID-PD based technology extracts evidences about what stocks are correlated with others, and discover trading patterns effective on multiple instruments. A typical example is the pairs trading strategy, which is based on the correlation analysis between stocks. We find that an effective pairs trading strategy is not only dependent on correlations but also considering constraints and domain knowledge such as relevant market factors.

The design of an actionable strategy is a human-mining interaction process supporting iterative development, back-testing, refinement and optimization of trading strategies. To this end, we built a financial trading rule automated development and evaluation system called F-Trade[1]. Figure 2 shows signals for a stock pair in ASX market. Figure 3 further shows the impact of business factors – distance and weight on return and the number of triggered signals.

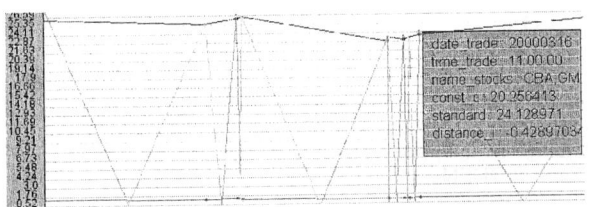

Fig. 2. Simulated trading via pairs trading strategy in F-Trade (ASX intraday data from 1 Jan 2000 to 20 Jun 2000)

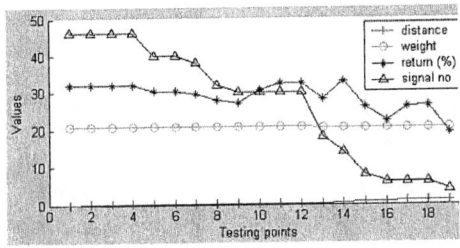

Fig. 3. Relation between d_0, *weight*, *return* and *signal number*

The exercise in testing ASX Top 32 stocks from January 1997 to June 2002 shows that DDID-PD based strategy has led to some interesting emergence beyond the normal mining algorithm design and domain expectation.

- Pair relationship between stocks and the combination of the above four factors interesting to trading cannot just be determined by technical measures such as coefficient ρ. They are also highly affected by stock movement such as volatility and liquidity. High volatility improves return while high liquidity balances the market impact on return.
- All 13 correlated stocks mined in Top 32 ASX come from different sectors. This finding means that pairs are not necessary from the same sector as presumed by financial researchers.

---

[1] F-Trade: accessible from http://www.f-trade.info with authorization.

3.2 Mining Actionable Trading Strategy

There exist many generic trading strategies in the literature and trading houses. Let's take the very common moving average strategy MA(*sr*, *lr*) as an instance. It actually indicates a generic correlated trading pattern between indicators *short-run moving average* (*sr*) and *long-run moving average* (*lr*).

ALGORITHM 1: A generic strategy MA(*sr*, *lr*)
IF *sr* > *lr* THEN *Buy*
IF *sr* < *lr* THEN *Sell*

Generally speaking, this pattern doesn't work in market trading. The actionability of a MA instance is determined in terms of the performance in real data, traders' interestingness and market dynamics such as transaction cost. To this end, the involvement of domain knowledge is quite significant for finding actionable rules. Using the DDID-PD ideas, we improve the generic MA and design an in-depth rule MA(t, sr, lr, δ_x, δ_y, h, d) as follows.

ALGORITHM 2: A revised MA(t, sr, lr, δ_x, δ_y, h, d)
IF $sr*(1-\delta_x)$ >= lr, triggering '*buy*' signal
    $t = t+h$; holding 'h' transactions or days
    IF $sr*(1-\delta_x)$ >= lr THEN
        Buy, '*buy*' signal is steady
        $t = t+d$; delaying 'd' transactions or days
IF $sr*(1+\delta_x)$ <= lr, triggering '*sell*' signal
    $t = t+h$; holding 'h' transactions or days
    IF $sr*(1+\delta_x)$ <= lr THEN
        Sell, '*sell*' signal is steady
        $t = t+d$; delaying 'd' transactions or days

This in-depth rule considers the following constraints and background knowledge, which make it more adaptable to market dynamics compared with MA(*sr*, *lr*).

- More filters are imposed on the generic MA to assist in filtering out false trading signals which would result in losses, for instance, fixed percentage band filter δ, time delay filter d, and time hold filter h;
- The fixed band filter δ_x (or δ_y) requires the buy or sell signal to exceed sr or lr by a fixed multiplicative band δ_x (or δ_y);
- The time delay filter d requires the buy or sell signal to remain valid for a prespecified number of transactions or days d before action is taken;
- The time hold filter h requires the buy or sell signal to hold the long or short position for a prespecified number of transactions or days h to effectively ignore all other signals generated during that time;
- In practice, note that only one filter is imposed at a given time.

Furthermore, we built interaction interfaces to support the definition and refinement of both technical and business parameters. Figure 4 illustrates some of such interfaces for the revised MA(t, sr, lr, δ_x, δ_y, h, d). Through the interfaces, users can trigger the process in terms of either Automated execution or Interactive mode with the involvement of users. In Interactive mode, technical analysts can advise the above process as well as refining technical factors for setting data mining process and tuning algorithm parameters. Business analysts can supervise the construction of features, fine tune the parameters, and set evaluation criteria for the business concerns. For instance, the measure *sharpe_ratio* is used for evaluating the business actionability of an identified rule.

$$sharpe\_ratio = (r_P - r_R) / \sigma_P$$

where r_P is expected portfolio return, r_R is risk free rate, and σ_P is portfolio standard deviation. Higher *sharpe_ratio* means more return with lower risk. Additionally, the system supports ad-hoc execution. Users can tune the parameters and interestingness measure at run time to evaluate the strategy.

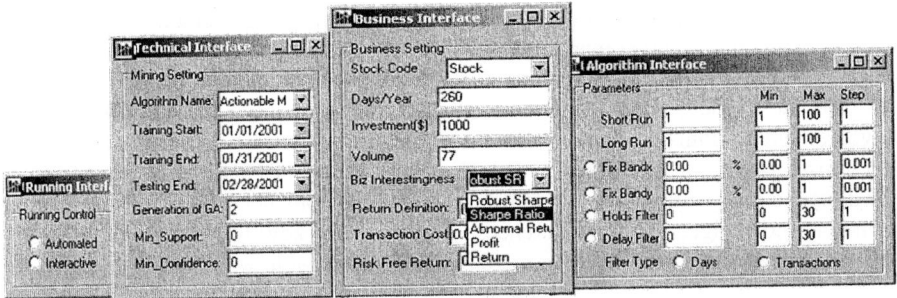

Fig. 4. Interfaces supporting human-mining system interaction

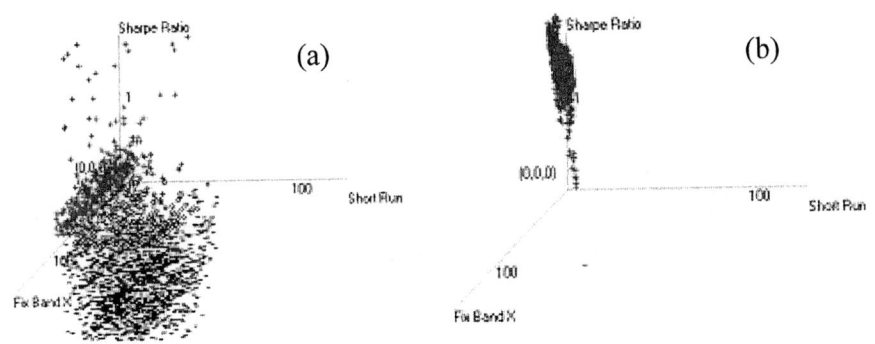

Fig. 5. Improved business interestingness by actionable rules

DDID-PD assists us in finding a collection of actionable rules. For instance, in ASX interday data, MA(4, 19, 0.033) could be an actionable rule using training data from 1 January 2000 to 31 December 2000 and testing set between 1 January 2001 and 31 December 2001. The number of trading signals generated by this rule is much bigger with better *sharpe_ratio* than other possible rules. Figure 5 (b) shows that its *sharpe_ratio* has a greatly improved positive scope compared with (a) the results of a generic MA rule. This demonstrates that DDID-PD driven strategy mining can improve strategy actionability.

4 Conclusions and Future Work

Actionable knowledge discovery is significant and also very challenging. It is nominated as one of Grand Challenges of KDD in the next 10 years. The research on this issue may change the existing situation where a great number of rules are mined while few of them are interesting to business, and promote the widely deployment of data mining into business. This paper has developed a new data mining framework, referred to as Domain-Driven In-Depth Pattern Discovery (DDID-PD). It provides a systematic overview of the issues in discovering actionable knowledge, and advocates the methodology of mining actionable knowledge in constraint-based context through human-mining system cooperation in a loop-closed iterative refinement manner.

The main phases and components of the DDID-PD include almost all phases of the CRISP-DM. It has enclosed some big differences from the CRISP-DM. For instance, (i) some new essential components, such as constraint mining, in-depth mining, the involvement of domain experts and knowledge, are taken into the lifecycle of KDD for consideration, (ii) in the DDID-PD, the normal steps of CRISP-DM are enhanced by dynamic cooperation with domain experts and the consideration of constraints and domain knowledge. These differences actually play key roles in improving the existing knowledge discovery in a more realistic and reliable way.

References

[1] Aggarwal, C., Towards effective and interpretable data mining by visual interaction, *ACM SIGKDD Explorations Newsletter*, 3(2): 11-22, 2002.
[2] Ankerst, M., Report on the SIGKDD-2002 panel the perfect data mining tool: interactive or automated? *ACM SIGKDD Explorations Newsletter*, 4(2):110-111, 2002.
[3] Cao, L., Dai., R., Human-Computer Cooperated Intelligent Information System Based on Multi-Agents, *ACTA AUTOMATICA SINICA*, 29(1):86-94, 2003.
[4] Cao, L., et al., Domain-driven in-depth pattern discovery: a practical perspective. Proceeding of AusDM, 101-114, 2005.
[5] Cao, L., et al., Ontology-Based Integration of Business Intelligence. *Int. J. on Web Intelligence and Agent Systems*, Vol.4 No 4, 2006.
[6] Financial data mining program: http://datamining.it.uts.edu.au/.
[7] Fayyad, U., Shapiro G., Uthurusamy R., Summary from the KDD-03 panel – Data mining: the next 10 years. *ACM SIGKDD Explorations Newsletter*, 5(2): 191-196, 2003.
[8] Han, J., Towards Human-Centered, Constraint-Based, Multi-Dimensional Data Mining. *An invited talk* at Univ. Minnesota, Minneapolis, Minnesota, Nov. 1999.
[9] Tan, P., Kumar, V., Srivastava, J., Selecting the Right Interestingness Measure for Association Patterns, SIGKDD'02, pp32-41.
[10] Manlatty,M., etc. Systems support for scalable data mining, *SIGKDD Explorations*, 2(2):56-65, 2000.
[11] J-F. Boulicaut, B. Jeudy. Constraint-based data mining. The Data Mining and Knowledge Discovery Handbook, O. Maimon and L. Rokach (Eds.), Springer, pp. 399-416, 2005.
[12] Omiecinski, E., Alternative Interest Measures for Mining Associations. *IEEE Transactions on Knowledge and Data Engineering*, 15:57-69, 2003.
[13] http://www.crisp-dm.org.
[14] Zhang, C., Zhang, Z., Cao, L., Agents and Data Mining: Mutual Enhancement by Integration, *LNCS 3505*, 50-61, 2005.

Evaluation of Attribute-Aware Recommender System Algorithms on Data with Varying Characteristics

Karen H.L. Tso and Lars Schmidt-Thieme

Computer-based New Media Group (CGNM),
Department of Computer Science, University of Freiburg,
Georges-Koehler-Allee 51, Freiburg 79110, Germany
{tso, lst}@informatik.uni-freiburg.de

Abstract. The growth of Internet commerce has provoked the use of Recommender Systems (RS). Adequate datasets of users and products have always been demanding to better evaluate RS algorithms. Yet, the amount of public data, especially data containing content information (attributes) is limited. In addition, the performance of RS is highly dependent on various characteristics of the datasets. Thus, few others have conducted studies on synthetically generated datasets to mimic the user-product relationship. Evaluating algorithms based on only one or two datasets is often not sufficient. A more thorough analysis can be conducted by applying systematic changes to data, which cannot be done with real data. However, synthetic datasets that include attributes are rarely investigated. In this paper, we review synthetic datasets applied in RS and present our synthetic data generation methodology that considers attributes. Furthermore, we conduct empirical evaluations on existing hybrid recommendation algorithms and other state-of-the-art algorithms using these variable synthetic data and observe their behavior as the characteristic of data varies. In addition, we also introduce the use of entropy to control the randomness of the generated data.

1 Introduction

Recommender systems use collaborative filtering to generate recommendations by predicting what users might be interested in, given some user's profile. Several prominent online commercial sites (e.g. amazon.com and ebay.com) offer this kind of recommendation services.

There are two different recommendation tasks typically considered: (i) predicting the ratings, i.e. how much a given user will like a particular item, and (ii) predicting the items, i.e. which N items a user will rate, buy or visit next (topN).

For RSs, nearest-neighbor methods, called collaborative filtering (CF ; [7]), is the prevalent method in practice. On the other hand, methods that rely only on attributes and disregard the rating information of other users, are commonly called the Content-Based Filtering (CBF). They have shown to perform very poorly. Yet, attributes usually contain valuable information; hence it makes it desirable to include attribute information in CF models – so called hybrid collaborative/content-based filtering methods.

There are many proposals on how to integrate attributes in CF for ratings. For instance, few others attempt linear combination of recommendation of CBF and CF predictions [5, 8, 10, 16]. There also exists methods that apply a CBF and a CF model sequentially, i.e. predict ratings by means of CBF and then re-estimate them from the completed rating matrix by means of CF [13]. There are also further proposals on how to integrate attributes when the problem is viewed as a classification problem [3, 4, 19]. As we lose the simplicity of CF, we do not consider those more complex methods here. We have selected three basic methods that predict items and try to keep the simplicity of CF, but still should improve prediction results.

When evaluating these recommendation algorithms, suitable datasets of users and items have always been demanding, especially when diversity of public data is limited. To compare the recommendation quality of different algorithms, it is not enough to evaluate the algorithms on just one or two datasets. Instead, one should investigate the behavior of the algorithms as systematic changes are applied to the data. Although there are already few attempts in generating synthetic data for the use in RS, to our best knowledge, there is no prior approach in generating synthetic data for evaluating recommender algorithms that incorporate attributes.

In this paper, we will make the following contributions: (i) we will propose our Synthetic Data Generator which produces user-item and user/item-attribute datasets and introduce the use of entropy to measure the randomness in the artificial data, (ii) we will survey some of the existing hybrid methods that consider attribute information in CF for predicting items. In addition, (iii) we will conduct empirical evaluations on three existing hybrid recommendation algorithms and other state-of-the-art algorithms using the generated synthetic data and observe their behavior when the characteristic of attribute data varies.

2 Related Works

One of the most widely known Synthetic Data Generators (SDG) in data mining is the one provided by the IBM Quest group [2]. It mimics the "real" world transactions in the retailing environment. It generates data with a structure and was originally intended for evaluating association rule algorithms. Later on, Deshpande and Karypis used this SDG for evaluating their item-based top-N recommendation algorithm [6]. Popescul *et.al* have proposed a simple approach by assigning a fixed number of users and items into clusters evenly and draw a uniform probability for each user and item in each cluster [17]. A similar attempt has been done for Usenet News [11, 14] as well as Aggarwal *et.al* for their horting approach [1]. Traupman and Wilensky tried to reproduce data by introducing skewed data to the synthetic data similar to a real dataset [20]. Another approach is to produce datasets by first sampling a complete dataset and re-sample the data again by missing data effect [12].

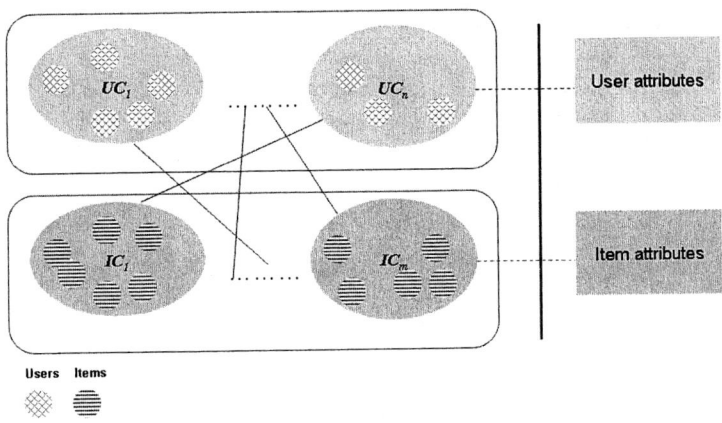

Fig. 1. Overview structure of synthetic data

The focus of this paper is to investigate SDG for CF algorithms which consider attributes. To the best of our knowledge, there is no prior attempt in examining SDGs for hybrid RS algorithms.

3 Synthetic Data Generator

The SDG can be divided into two phases: drawing distributions and sampling data. In the first phase, it draws distribution of User Cluster (UC) and Item Cluster (IC), next it affiliates UC or IC with user/item attribute respectively as well as to associate the UC and IC. Using these generated correlations, the users, items, ratings and item/user-attribute datasets can then be produced in the second phase. Fig. 1 presents an overview of how the artificial data are generally structured.

3.1 Drawing Distributions

To create the ratings and attributes datasets, we generate five random distributions models:

- $P(UC)$, how users are distributed in N number of UC.
- $P(IC)$, how items are distributed in M number of IC.
- $P(A|UC) \ \forall \ UC$, how user attributes (A) are distributed in UC.
- $P(B|IC) \ \forall \ IC$, how item attributes (B) are distributed in IC.
- $P(UC|IC) \ \forall \ IC$, how UC are distributed in IC.
- q be the probability that an item in IC_i is assigned to UC_j

The SDG first draws $P(UC)$ and $P(IC)$ from a Dirichlet distribution (with parameters set to 1). This asserts that the sum of $P(UC)$ or $P(IC)$ forms to one. $P(B|IC)$ shows the affiliation of item attributes with the item clusters by drawing from a special Chi-square distribution rejecting values greater than

Algorithm 1. Drawing distribution

Input: $|A|, |B|, N, M, \epsilon_A, \epsilon_B, \epsilon_C$
Output: $P(UC), P(IC), P(A|UC), P(B|IC), P(UC|IC)$
$h = 0$
$P(UC) \sim Dir_{a_1, a_2 \ldots, a_N}$
$P(IC) \sim Dir_{b_1, b_2 \ldots, b_M}$
repeat
$\quad P(B|IC)_h = S\chi^2 ED(|B|, M, h, \epsilon_B)$
$\quad P(UC|IC)_h = S\chi^2 ED(N, M, h, \epsilon_{IC})$
$\quad P(A|UC)_h = S\chi^2 ED(|A|, N, h, \epsilon_A)$
$\quad h = h + 0.1$
until $h < 1$

Algorithm 2. Drawing Special χ^2 distribution with specified entropy values

$S\chi^2 ED(n, m, H_{XY}, \epsilon_{XY}) :$
$d = 1$
repeat
$\quad P(X_i|Y_j) \sim \chi^2_d|_{[0,1]} \quad \forall i = 1 \ldots n, \forall j = 1 \ldots m$
$\quad d = d + 1$
until $|H(X|Y) - H_{XY}| < \epsilon_{XY}$
$P(X|Y)$

---

1. Likewise, the correlation between UC and IC, $P(UC|IC)$, as well as the correlation between user attributes and user clusters, $P(A|UC)$, are done with similar manner. However, the attribute-aware CF algorithms we discuss in this paper do not take user-attributes into account. The overall drawing distributions process is summarized in (Algo. 1.).

By virtue of the randomness in those generated models, it is necessary to control or to measure the informativeness of these random data. Hence, we apply the Information Entropy and compute the average normalized entropy of the models.

$$H(X) = -\sum_{x \in \text{dom}(X)} \frac{P(x) \log_2 P(x)}{\log_2 |\text{dom}(X)|}. \quad (1)$$

The conditional entropy for the item-attribute data therefore is:

$$H(B_i|IC) = -\sum_{b=0}^{1} \sum_{j \in \text{dom } IC} \frac{P(B_i = b, IC = j) \cdot \log_2 P(B_i = b|IC = j)}{\log_2 |\text{dom } IC|} \quad (2)$$

In our experiment, $P(B|IC)$ is sampled eleven times for eleven different entropy values from 0 to 1 with 0.1 interval. By rejection sampling, $P(B \mid IC)$ is drawn iteratively with various Chi-square degrees of freedom until $H(B|IC)$ reaches desired entropies (Algo. 2.). Other types of distribution have also been examined, yet, Chi-square distribution has shown to give the most diverse en-

Algorithm 3. Sampling data

$uc_u \sim P(UC)$ user cluster of user u
$ic_i \sim P(IC)$ item cluster of item i
$oc_{l,k} \sim P(UC_l|IC_k)$ user of cluster l who prefer item of cluster k
$o_{u,i} \sim binom(q)$ $\forall u,i : oc_{uc_u,ic_i} = 1$ occurrence of user of uc_u prefers item of ic_i
$o_{u,i} = 0$ else
$b_{i,t} \sim P(B_t|IC = ic_i)$ item i contains attribute t

---

tropy range. We expect that as the entropy increases, which implies the data is less structured, the recommendation quality should decrease.

3.2 Sampling Data

Once these distributions have been drawn, users, items, ratings and item-attributes data are then sampled accordingly to those distributions. Firstly, users are assigned to user clusters by random sampling from $P(UC)$. Similar procedure, applies for sampling items. The user-item(ratings) data is generated by first sample $P(UC_l|IC_k)$ of users belonging to UC_l who prefer items in IC_k, then sample q portion of items of IC_k to these sampled users. The affiliation between items and attributes is done by sampling $P(B|IC)$ of items which contain attribute B. The same procedure can be applied to generate the user-attributes datasets. The overall sampling data process is summarized in (Algo. 3.).

4 Hybrid Attribute-Aware CF Methods

Here, we discuss three existing hybrid methods [21], which will be evaluated using the data generated from the SDG.

1. Sequential CBF and CF (Adapted Content-Boosted CF),
2. Joint Weighting of CF and CBF, and
3. Attribute-Aware Item-Based CF.

Sequential CBF and CF is the adapted version of an existing hybrid approach, Content-Boosted CF, originally proposed by [13] for predicting ratings. This method has been conformed to the predicting items problem here. It first uses CBF to predict ratings for unrated items and then filters out ratings with lower scores (i.e. keeping ratings above 4 on a 5-point scale) and applies CF to recommend topN items.

Joint Weighting of CF and CBF (Joint-Weighting CF-CBF), first applies CBF on attribute-dependent data to infer the fondness of users for attributes. In parallel, user-based CF is used to predict topN items with ratings-dependent data. Both predictions are joint by computing their geometric mean.

Attribute-Aware Item-Based CF (Attr-Item-based CF) extends item-based CF [6]. It exploits the content/attribute information by computing the similarities between items using attributes thereupon combining it with the similarities between items using ratings-dependent data.

All three approaches recommend items that contain the highest frequency of their neighboring items. For the last two algorithms, λ is used as a weighting factor to vary the significance applied to CF or CBF.

5 Evaluation and Experimental Results

In this section, we present the evaluation of the selected attributes-aware CF algorithms using artificial data generated by SDG discussed in Section 3 and compare their performances with their corresponding non-hybrid base models, which do not integrate attributes, i.e. user-based and item-based CF, as well as to observe the behavior of the algorithms after supplement of attributes.

Metrics. Our paper focuses on the item prediction problem, which is to predict a fixed number of top recommendations and not the ratings. Suitable evaluation metrics are Precision, Recall and F1.

Similar to Sarwar et al. [18], our evaluations consider any item in the recommendation set that matches any item in the testing set as a "hit". F1 measure is then used to combine Precision and Recall into a single metric.

$$\text{Precision} = \frac{\text{Number of hits}}{\text{Number of recommendations}}$$

$$\text{Recall} = \frac{\text{Number of hits}}{\text{Number of items in test set}}$$

$$F1 = \frac{2 \cdot \text{Precision} \cdot \text{Recall}}{\text{Precision} + \text{Recall}}$$

Parameters. Due to the nature of collaborative filtering, the size of neighborhood has significant impact on the recommendation quality [9]. Thus, each of the randomly generated data should have an assorted neighborhood sizes for each method. In our experiments, we have selected optimal neighborhood sizes and λ parameters for the hybrid methods by means of a grid search. See Table 1. Lambda is used to weight the contribution of attribute-dependent and rating-dependent models. Threshold and max, for the Sequential CBF-CF are set to 50 and 2 accordingly as chosen in the original model [13]. For more detail explanation of the parameters used in those algorithms, please refer to [21] and [13].

Table 1. The parameters chosen for the respective algorithms

Method	Neighborhood Size	λ
user-based CF	35-50	–
item-based CF	40-60	–
joint weighting CF–CBF	35-50	0.15
attr-aware item-based CF	40-60	0.15

Table 2. The parameters settings for the synthetic data generator

Description	Symbol	Value
Number of users	n	250
Number of items	m	500
Number of User Clusters	N	5
Number of Item Clusters	M	10
Number of Item Attributes	$|B|$	50
Probability of i in IC assigned to a UC	q	0.2

As our algorithms do not consider user attributes, our SDG only generates models for item attributes. The parameter settings for our experiments are summarized in Table 2.

Experimental Results. In our experiments, we have generated five different trials. For each trial, we produce one dataset of user-item (ratings) and eleven different item-attributes datasets with increasing entropy from 0-1 with 0.1 interval, by rejection sampling. In addition, to reduce the complexity of the experiment, it is assumed that the correlation between the user and item clusters to be fairly well-structure and have a constant entropy of 0.05. The results of the average of five random trials where only item-attributes with entropy of 0.05 are presented in Fig. 2.

As shown in Fig. 2, Joint-Weighting CF-CBF achieves the highest Recall value by around 4% difference w.r.t. its base method. On the other hand, Attr-Item-based CF does not seem to be effective at all as attributes are appended to its base model. It also has a very high standard deivation. This suggests that the algorithms to be rather unstable and unreliable. Although Melville *et al.* [13] reported that Content-Boosted CF performed better than user-based and pure CBF for ratings, it fails to provide quality top-N recommendations for items

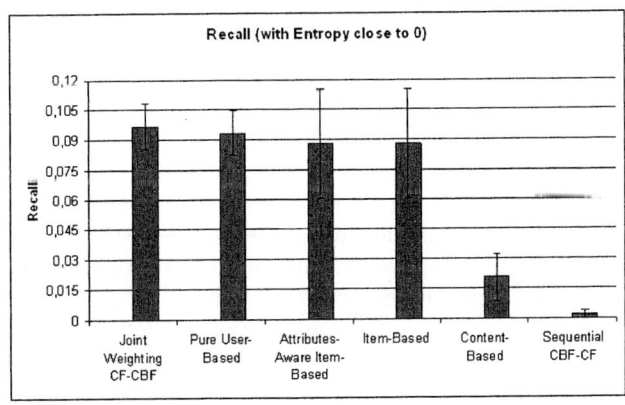

Fig. 2. Recall by selecting item-attributes with entropy ≤ 0.05

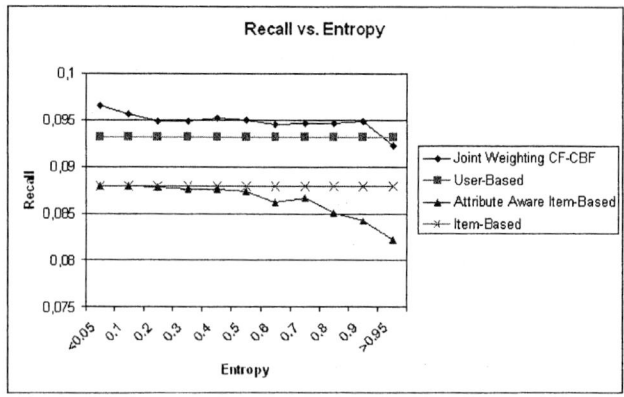

Fig. 3. Recall vs. Entropy ranging from 0-1

in our experiments. Therefore, we will focus our evaluation on the other two algorithms in the rest of the paper.

As the aim of the paper is to examine the behavior of the models as the characteristic of data varies, what is more important is to observe the performance as entropy varies. As anticipated, the recommendation quality increases, when there exists more structure in the data. The results of an average of five random trials of item-attribute datasets with eleven various entropies are presented in Fig. 3.

We can see that for both Attr-Item-based CF and Joint-Weighting CF-CBF algorithms, the quality of recommendation reaches its peaks when the entropy approaches zero and it gradually decreases as entropy increases. As for Attr-Item-based CF, although it carries the right entropy trend, its peak does not surpass its base model and the quality drops gradually below its base model, which does not make use of attributes. On the other hand, for Joint-Weighting CF-CBF, the value of recall descends gradually as the entropy raises, still the recall maintain above its base-model until entropy approaches 1 where recall plummets to below its base-line score.

6 Conclusions and Future Works

The aim of this paper is to conduct an empirical evaluation on three existing hybrid recommendation models and other state-of-the-art algorithms with data generated by the SDG presented in this paper. All algorithms discussed here focus on the predicting items problem. Joint-Weighting CF-CBF, appears to enhance recommendations quality when reasonable amount of informative attributes are presented. The other algorithms do not seem to be sensitive to attributes. Yet, we expect the outcomes could be ameliorated by adding more structural dependency between clusters. In addition, currently the data are only controlled by the entropy of item-attribute datasets; however, other distributions such as the user-item data should also be investigated when various entropies are considered. Furthermore, more extensive experiments should be done to examine the effect of varying other parameters settings and to conduct an empirical evaluation with models that predict ratings.

References

1. Aggarwal, C. C., Wolf, J. L., Wu, K.-L. and Yu, P. S.: Horting hatches an egg: A new graph-theoretic approach to collaborative filtering. In Proceedings of ACM-SIGKDD International Conference on Knowledge Discovery and Data Mining. ACM, New York (1999)
2. Agrawl, R. and Srikant, R.: Fast algorithms for mining association rules. In Proceedings of the 20th International Conference on Very Large Data Bases (VLDB). Morgan Kaufmann (1994) 487-499
3. Basilico, J. and Hofmann, T.: Unifying collaborative and content-based filtering. In Proceedings of the 21st International Conference on Machine Learning. Banff, Canada (2004)
4. Basu, C., Hirsh H., and Cohen, W.: Recommendation as classification: Using social and content-based information in recommendation. In Proceedings of the 1998 Workshop on Recommender Systems. AAAI Press, Reston, Va. 11-15 (1998)
5. Claypool, M., Gokhale, A. and Miranda T.: Combining content-based and collaborative filters in an online newspaper. In Proceedings of the SIGIR-99 Workshop on Recommender Systems: Algorithms and Evaluation (1999)
6. Deshpande, M. and Karypis, G.: Item-based top-N recommendation algorithms, ACM Transactions on Information Systems **22/1** (2004) 143–177
7. Goldberg, D., Nichols, D., Oki, B. M. and Terry, D.: Using collaborative filtering to weave an information tapestry. Commun. ACM **35** (1992) 61–70
8. Good, N., Schafer, J.B., Konstan, J., Borchers, A., Sarwar, B., Herlocker, J., and Riedl, J.: Combining Collaborative Filtering with Personal Agents for Better Recommendations. In Proceedings of the 1999 Conference of the American Association of Artificial Intelligence (AAAI) (1999) 439-446
9. Herlocker, J., Konstan, J., Borchers, A., and Riedl, J.: An Algorithmic Framework for Performing Collaborative Filtering. In Proceedings of ACM SIGIR'99. ACM press (1999)
10. Li, Q. and Kim, M.: An Approach for Combining Content-based and Collaborative Filters. In Proceedings of the Sixth International Workshop on Information Retrieval with Asian Languages (ACL) (2003) 17–24
11. Konstan, J. A., Miller, B. N. , Maltz D., Herlocker, J. L., Gordon, L. R. and Riedl, J.: Group-Lens: Applying collaborative filtering to usenet news. Commun. ACM **40** (1997) 77-87
12. Marlin, B., Roweis, S. and Zemel, R.: Unsupervised Learning with Non-ignorable Missing Data. In Proceedings of the 10th International Workshop on Artificial Intelligence and Statistics (AISTATS) (2005) 222–229
13. Melville, P., Mooney, R. J. and Nagarajan, R.: Content-Boosted Collaborative Filtering for Improved Recommendations. In Proceedings of the Eighth National Conference on Artificial Intelligence(AAAI-2002). Edmonton, Canada (2002) 187–192
14. Miller, B. N., Riedl, J. and Konstan, J. A.: Experiences with GroupLens: Making Usenet useful again. In Proceedings of the 1997 USENIX Technical Conference (1997)

15. MovieLens: Available at http://www.grouplens.org/data (2003)
16. Pazzani, M. J.: A framework for collaborative, content-based and demographic filtering. Artificial Intelligence Review 13(5-6):393408 (1999)
17. Popescul, A., L.H. Ungar, D.M. Pennock, and S. Lawrence: Probabilistic models for unified collaborative and content-based recommendation in sparse-data environments. In Proceedings of the Seventeenth Conference on Uncertainty in Artificial Intelligence (2001) 437–444
18. Sarwar, B. M., Karypis, G., Konstan, J. A. and Riedl, J.: Analysis of recommendation algorithms for E-commerce. In Proceedings of the 2nd ACM Conference on Electronic Commerce (EC). ACM, New York (2000) 285–295
19. Schmidt-Thieme, L.: Compound Classification Models for Recommender Systems. In Proceedings of the IEEE International Conference on Data Mining (ICDM). New Orleans, USA (2005) 559–570
20. Traupman, J. and Wilensky, R.: Collaborative Quality Filtering: Establishing Consensus or Recovering Ground Truth?. In Proceedings of WebKDD 2004: KDD Workshop on Web Mining and Web Usage Analysis, in conjunction with the 10th ACM SIGKDD International Conference on Knowledge Discovery and Data Mining (KDD 2004), August 22-25 2004. Seattle, WA (2004)
21. Tso, H. L. K., Schmidt-Thieme, L.: Attribute-Aware Collaborative Filtering. In Proceedings of the 29th Annual Conference of the German Classification Society 2005. Magdeburg, Germany (2005)

An Intelligent System Based on Kernel Methods for Crop Yield Prediction

A. Majid Awan and Mohd. Noor Md. Sap

Faculty of Computer Sci. & Information Systems, University Technology Malaysia,
Skudai 81310, Johor, Malaysia
awanmajid@hotmail.com, mohdnoor@fsksm.utm.my

Abstract. This paper presents work on developing a software system for predicting crop yield from climate and plantation data. At the core of this system is a method for unsupervised partitioning of data for finding spatio-temporal patterns in climate data using kernel methods which offer strength to deal with complex data. For this purpose, a robust weighted kernel k-means algorithm incorporating spatial constraints is presented. The algorithm can effectively handle noise, outliers and auto-correlation in the spatial data, for effective and efficient data analysis, and thus can be used for predicting oil-palm yield by analyzing various factors affecting the yield.

1 Introduction

Clustering is a useful machine learning technique that can capture meaningful patterns in the agro-hydrological data. Finding good quality clusters in spatial data (eg, temperature, precipitation, pressure, etc) is more challenging because of its peculiar characteristics such as auto-correlation, non-linear separability, outliers, noise, high-dimensionality, and when the data has clusters of differing shapes and sizes [10, 15, 18]. The popular clustering algorithms, like k-means, have some limitations for this type of data [16, 18]. Therefore, we present a weighted kernel k-means clustering algorithm incorporating spatial constraints bearing spatial neighborhood information in order to handle spatial auto-correlation, outliers and noise in the spatial data.

A number of kernel-based learning methods have been proposed in recent years [3, 4, 7-9, 11, 13]. Generally speaking, a kernel function implicitly defines a non-linear transformation that maps the data from their original space to a high dimensional space where the data are expected to be more separable.

2 Application Area and Methods

A simplified view of the problem domain is shown in Figure 1. The data consists of a sequence of snapshots of the earth areas consisting of measurement values for variables like temperature, pressure, precipitation, crop yield, etc.

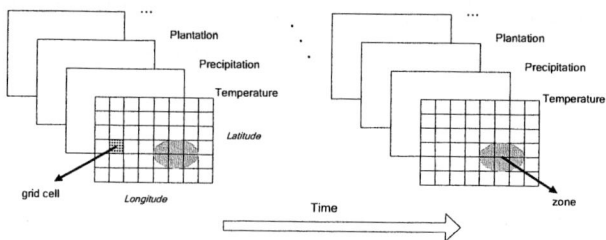

Fig. 1. A simplified view of the problem domain

This work uses clustering to divide areas of the land into disjoint regions in an automatic but meaningful way that enables us to identify regions of the land whose constituent points have similar short-term and long-term characteristics. The spatial and temporal nature of the target data poses a number of challenges. For instance, such type of data is noisy. In addition, such data displays autocorrelation (i.e., measured values that are close in time and space tend to be highly correlated, or similar), high dimensionality, clusters of non-convex shapes, outliers.

If we apply a clustering algorithm to cluster time series associated with points on the land, we obtain clusters that represent land regions with relatively homogeneous behavior. We can then identify how various parameters influence the climate and oil-palm produce of different areas using correlation. A simplified architecture of the agro-hydrological system is shown in Figure 2:

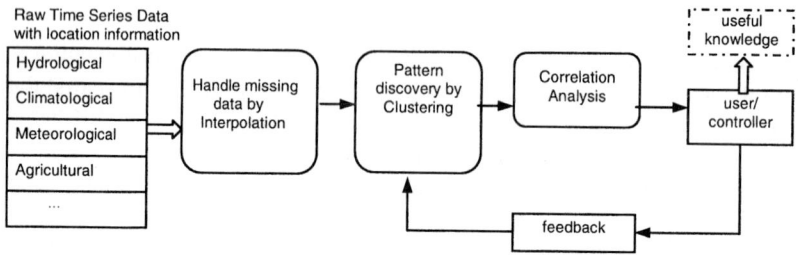

Fig. 2. A simplified architecture for the system

3 Kernel-Based Methods

The kernel methods are among the most researched subjects within machine-learning community in recent years and have been widely applied to pattern recognition and function approximation [2,5,6,12,14,17]. There are instances where a linear hyperplane cannot separate classes without misclassification, an instance relevant to our problem domain. However, those classes can be separated by a nonlinear separating hyperplane. This concept is based on Cover's theorem on the separability of patterns.

Let a nonlinear transformation function ϕ maps the data into a higher dimensional space. Suppose there exists a function K, called a kernel function, such that,

$$K(x_i, x_j) = \phi(x_i) \cdot \phi(x_j)$$

A kernel function is substituted for the dot product of the transformed vectors. So, kernels allow large non-linear feature spaces to be explored while avoiding curse of dimensionality. Further, the use of the kernel function is less computationally intensive. The formulation of the kernel function from the dot product is a special case of Mercer's theorem [13].

4 Weighted Kernel K-Means with Spatial Constraints (SWK-Means)

Let $X = \{x_i\}_{i=1,...,n}$ be a data set with $x_i \in R^N$. We call codebook the set $W = \{w_j\}_{j=1,...,k}$ with $w_j \in R^N$ and $k \ll n$. The k-means clustering algorithm can be enhanced by the use of a kernel function. The kernel k-means algorithm can be generalized by introducing a weight for each point x, denoted by $u(x)$, as:

$$E(W) = \sum_{j=1}^{k} \sum_{x_i \in V_j} u(x_i) \|\phi(x_i) - w_j\|^2 \quad (1)$$

where,
$$w_j = \frac{\sum_{x_j \in V_j} u(x_j)\phi(x_j)}{\sum_{x_j \in V_j} u(x_j)} \quad (2)$$

The Euclidean distance from $\phi(x)$ to center w_j is given by the following eq.

$$\left\|\phi(x_i) - \frac{\sum_{x_j \in V_j} u(x_j)\phi(x_j)}{\sum_{x_j \in V_j} u(x_j)}\right\|^2 = K(x_i, x_i) - 2\frac{\sum_{x_j \in V_j} u(x_j)K(x_i, x_j)}{\sum_{x_j \in V_j} u(x_j)} + \frac{\sum_{x_j, x_l \in V_j} u(x_j)u(x_l)K(x_j, x_l)}{(\sum_{x_j \in V_j} u(x_j))^2} \quad (3)$$

If we adopt Guassian radial basis function (RBF), viz., $K(x_i, x_j) = \exp(-\|x_i - x_j\|^2 / 2\sigma^2)$, then $K(x,x)=1$. And, writing the last term in eq. (3) as C_k. we can get:

$$\left\|\phi(x_i) - \frac{\sum_{x_j \in V_j} u(x_j)\phi(x_j)}{\sum_{x_j \in V_j} u(x_j)}\right\|^2 = 1 - 2\frac{\sum_{x_j \in V_j} u(x_j)K(x_i, x_j)}{\sum_{x_j \in V_j} u(x_j)} + C_k \quad (4)$$

For increasing the robustness of fuzzy c-means to noise, an approach is proposed in [1]. Here we propose a modification to the weighted kernel k-means to increase the robustness to noise and to account for spatial autocorrelation in the spatial data. It can be achieved by a modification to eq. (1) by introducing a penalty term containing spatial neighborhood information, as:

$$E(W) = \sum_{j=1}^{k} \sum_{x_i \in V_j} u(x_i) \|\phi(x_i) - w_j\|^2 + \frac{\gamma}{N_R} \sum_{j=1}^{k} \sum_{x_i \in V_j} u(x_i) \sum_{r \in N_k} \|\phi(x_r) - w_j\|^2 \quad (5)$$

where N_k stands for the set of neighbors that exist in a window around x_i and N_R is the cardinality of N_k. The parameter γ controls the effect of the penalty term. The distance in the last term of eq. (5), can be calculated as

$$\left\|\phi(x_r) - \frac{\sum_{x_j \in V_j} u(x_j)\phi(x_j)}{\sum_{x_j \in V_j} u(x_j)}\right\|^2 = 1 - 2\frac{\sum_{x_j \in V_j} u(x_j)K(x_r, x_j)}{\sum_{x_j \in V_j} u(x_j)} + C_k = 1 - \beta_r + C_k \quad (6)$$

The expression for effective minimum distance from each point to every cluster representative can be obtained from eq. (5) using eq. (4) and (6):

$$-2\frac{\sum_{x_j \in V_j} u(x_j)K(x_i,x_j)}{\sum_{x_j \in V_j} u(x_j)} + C_k + \frac{\gamma}{N_R}\sum_{r \in N_k}(\beta_r + C_k) \qquad (7)$$

As $K(x_i,x_j)$ measures similarity between x_i and x_j, and when x_i is an outlier, then $K(x_i,x_j)$ will be very small. So, the second term in the above expression will get very low value. The total expression will get higher value and hence results in robustness by not assigning the point to the cluster. For detail about the algorithm, pls. see [11].

5 Experimental Results

The system is implemented in C++. We get results regarding analyzing various factors impacting oil-palm yield. However, because of space constraints, here we briefly describe the clustering results of the SWK-means algorithm.

For experimentation we selected 24 rainfall stations. A 12-month moving average is used for removing seasonality from the data. For monthly rainfall values for 5 years, we get a data matrix of 24×60. SWK-means partitioned it into 2 clusters. For visualization of results, we also applied the algorithm to the monthly average rainfall values of this period. Its results are shown in Figure 3. For the next five year periods of time for the selected 24 rainfall stations we get data matrices as 48×60, 72×60 and so on. The algorithm proportionally partitioned the data into two clusters. The corresponding results are given in table 1 (a record represents 5-year monthly rainfall values taken at a station). It also validates the proper working of the algorithm.

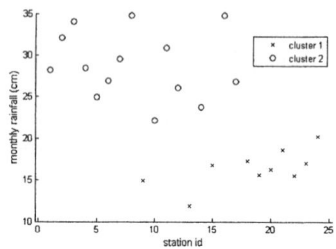

Fig. 3. Clustering results of SWK-means algorithm showing two clusters of monthly rainfall (average) of 24 stations

Table 1. Results of SWK-means on rainfall data at 24 stations for 5, 10, 15, 20, 25, 30 years

No. of Records	No. of records in cluster 1	No. of records in cluster 2
24	10	14
48	20	28
72	30	42
96	40	56
120	50	70
144	60	84

For the overall system, the information about the landcover areas of oil palm plantation is gathered. The analysis of these and other time series (e.g., precipitation, temperature, pressure, etc) is conducted using clustering. We can then identify how various parameters, such as precipitation, temperature etc, influence the climate and oil-palm produce of different areas using correlation. Our initial study shows that the rainfall patterns alone affect oil-palm yield after 6-7 months. This way we are able to predict oil-palm yield for the next 1-3 quarters on the basis of analysis of present plantation and environmental data.

6 Conclusions

Computational machine learning techniques like clustering can be effectively used in analyzing the impacts of various hydrological and meteorological factors on vegetation. Kernel methods are helpful for clustering complex and high dimensional data that is non-linearly separable in input space. Consequently for developing a system for oil-palm yield prediction, an algorithm, weighted kernel k-means incorporating spatial constraints, is presented which is a central part of the system. We get promising results on our test data sets. It is hoped that the algorithm would prove to be robust and effective for spatial (climate) data analysis, and it would be very useful for oil-palm yield prediction.

References

1. M.N. Ahmed, S.M. Yamany, et al. A modified fuzzy C-means algorithm for bias field estimation and segmentation of MRI data. IEEE Trans. on Medical Imaging, 21 (2002).
2. A. Ben-Hur, D. Horn, H. Siegelman, V. Vapnik. Support Vector Clustering. JMLR, 2(2001).
3. F. Camastra. Kernel Methods for Unsupervised Learning. PhD thesis, Uni of Genova (2004).
4. F. Camastra, A. Verri. A Novel Kernel Method for Clustering. IEEE Trans. PAMI, 27(2005)
5. J. H. Chen, C. S. Chen. Fuzzy kernel perceptron. IEEE Trans. NN, 13 (2002) 1364–1373.
6. N. Cristianini, J.S.Taylor. An Intr. to Support Vector Machines. Cambridge Ac Press (2000).
7. I.S. Dhillon, Y. Guan, B. Kulis. Kernel kmeans, Spectral Clustering and Normalized Cuts. KDD 2004.
8. C. Ding, X. He. K-means Clustering via Principal Component Analysis. Proc. ICML 2004.
9. M. Girolami. Mercer Kernel Based Clustering in Feature Space. IEEE Trans. NN, 13 (2002).
10. J. Han, M. Kamber, K.H. Tung. Spatial Clustering Methods in Data Mining: A Survey. In: H.J. Miller, J. Han (eds.): Geographic Data Mining & Knowledge Discovery, T & F (2001).
11. M.N. Md. Sap, A. Majid Awan. Finding Spatio-Temporal Patterns in Climate Data using Clustering. Proc. Int. Conf. on Cyberworlds (CW'05), Singapore (2005).
12. V. Roth, V. Steinhage. Nonlinear discriminant analysis using kernel functions. NIPS 12.

13. B. Scholkopf, A. Smola. Learning with Kernels: Support Vector Machines, Regularization, Optimization, and Beyond. MIT Press (2002).
14. B. Scholkopf, A. Smola, K. R. Müller. Nonlinear component analysis as a kernel eigenvalue problem. Neural Computation, 10 (1998), 1299–1319.
15. S. Shekhar, et al. Trends in Spatial Data Mining. In Data Mining: Next Generation Challenges and Future Directions, H. Kargupta, et al. (eds.), MIT Press, 2003.
16. M. Steinbach, P-N. Tan, V. Kumar, S. Klooster, C. Potter. Data Mining for the Discovery of Ocean Climate Indices. Proc. 5th Workshop on Scientific Data Mining at ICDM, 2002.
17. V.N. Vapnik. Statistical Learning Theory. John Wiley & Sons, 1998.
18. P. Zhang, et al. Discovery of Patterns of Earth Science Data Using Data Mining. In Next Generation of Data Mining Applications, J. Zurada, M. Kantardzic (eds.), IEEE Press, 2003.

A Machine Learning Application for Human Resource Data Mining Problem[*]

Zhen Xu[1] and Binheng Song[2]

[1] School of Software, Tsinghua University, Beijing, 100084, P.R. China
z-xu03@mails.tsinghua.edu.cn
[2] Dept. Math, Tsinghua University, 100084, P.R. China
bsong@math.tsinghua.edu.cn

Abstract. Apply machine learning methods to data mining domain can be more helpful to extract useful knowledge for problems with changing conditions. Human resource allocation is a kind of problem in data mining domain. It presents machine learning techniques to dissolve it. First, we construct a new model which optimizes the multi-objectives allocation problem by using fuzzy logic strategy. One of the most important problems in the model is how to get the precise individual capability matrixes. Machine learning method by being told is well used to settle the problem in this paper. In the model, appraisal values about employees are saved in knowledge warehouse. Before tasks allocation, machine learning approach provides the capability matrixes based on the existing data sets. Then Task-Arrange or Hungarian Algorithm provides the final solution with our proposed matrixes. After present tasks are finished, machine learning method by being told can update the matrixes according to the suggestions on employees' performance provided by the specialists. Useful knowledge can be well mined in cycles by learning approach. As a numerical example demonstrated, it is helpful to make a realistic decision on human resource allocation under a dynamic environment for organizations.

1 Introduction

As data sets are continuously growing in size and number, there is a need to extract useful information from collections of data. The disciplines of data mining and machine learning are concerned with the application of methods such as clustering, classification, rule induction, and others to potentially large data repositories in order to extract relevant information and eventually convert data and information into knowledge [1, 2].

Machine learning makes the computer be intelligent and deals with the issue of how to build programs that improve their performance at some task through experience. A key research area in machine learning is extracting conceptual description from data samples. Meanwhile, the key purpose of data mining is to search interesting patterns and important rules from large-scale databases. Therefore, many machine

---

[*] National 973 Project in China (Grant No. 2004CB719401).

learning methods could be used directly into data mining domain. In this paper, we will apply machine learning to human resource allocation, which is one of the problems in data mining domain.

Surveying the past research, there are two types of modeling about Human Resource Allocation. One is linear programming model [3]; the other is goal programming [4]. Linear programming is a single goal optimization technique; this is not the situation for a majority of firms with multiple goals. Goal programming can not deal with the organizational differentiation problems [6]. In addition, neither do the models studied take advantage of the machine learning method.

Currently many resource allocation algorithms have existed. It uses multi-processors to solve distributed resource allocation in [7], where the processor number is correlated with the task number. The approaches proposed by [8-9] provide algorithms on task scheduling problem. These algorithms do not consider the time constraints among tasks. It limits further application

There are great deals of data about every employee. Before tasks allocation, managers will provide appraisal values aimed to every employee. But the appraisal values each specialist gives are very subjective. Moreover, data is changing according to one's changing experience. Actually, Human resource allocation is a kind of problem in data mining domain. In the process of task allocation, it is crucial to identify and mine useful information from existing data sets. Here we bring forward a machine learning approach to dissolve the allocation problem.

In our paper, a new model is proposed. And we apply fuzzy logic concept into the model. Individual capability matrixes about employees are the data need to be mined. Machine learning methods can offer the matrixes not only based on the previous appraisal data sets the managers provide but also suited to changing conditions. Through the matrixes, we get the fuzzy synthesis appraisal matrixes finally. Then Task-Arrange or Hungarian Algorithm [12] can obtain the best assignment solution based on the matrixes. After the allocation, the specialists feed back the evaluation values which will be saved in the knowledge warehouse. The self-learning process revises the original data in the warehouse. The refreshed matrixes will be offered when next allocation comes. So we can mine useful knowledge based on existing data sets in cycles.

The layout of the paper is as follows: In section 2, we first present our basic model. Section 3 introduces the machine learning method by being told. And then we describe fuzzy concepts and define fuzzy membership sets in Section 4. We give the optimization target in Section 5. Section 6 briefly describes our algorithm which achieves the optimization solution, and a numerical example is also given in section 7 to show how the algorithm works. Section 8 gives the comparison with other algorithms. Finally, we conclude in section 9.

2 The Fuzzy Modeling

The management of human resource includes three aspects: human capital control, tasks control and time control.

2.1 Model on Human Capital Control

1. The follow parameters are aimed at the whole human resource of the firm:

- Matrix $S_{hr} = [s_{i,j} \mid i,j \in N, 1 \leq i \leq |A \text{ or } B|, 1 \leq j \leq |T|, 0 \leq s_{i,j} \leq 1]$

The individual capability matrix is in this form.
$s_{i,j}$ means the satisfaction degree certain employee doing all the jobs.
A or B are the satisfaction degree fuzzy sets of the tasks, $|T|$ is the number of tasks.

- Matrix $S_2 = [s_{i,j} \mid i,j \in N, 1 \leq i \leq |R|, 1 \leq j \leq |A \text{ or } B|, s_{i,j} = 0 \text{ or } 1]$

S_2 is the satisfaction degree appraisal matrix specialists give.
$s_{i,j}$ means the satisfaction degree employee i doing the assigned job.
A or B are the satisfaction degree fuzzy sets of the tasks.

- Matrix $S_3 = [s_{i,j} \mid i,j \in N, 1 \leq i \leq |R|, 1 \leq j \leq |T|, 0 \leq s_{i,j} \leq 1]$

In the above definition, S_3 is an evaluation matrix about the employees, the element of which shows the performance evaluation of employee i on task j.
$|R|$ represents the number of the employees; $|T|$ is the number of the tasks.
The final fuzzy synthesis appraisal matrixes needed in our algorithm is in the form.

- group $_r^m$

The parameter r is the group name in a firm, and m represents the number of employees in the group.

2. For concrete human resource, it needs other parameters:
hr{group, ifFree, freeTimes, tasks, cost}

- group: The name of technique group the employee belongs to.
- ifFree: Indicate whether or not the employee is free currently. The value concludes "1"(occupied) and "0"(free).
- freeTimes: Time during which the employee can keep free.
- tasks: Which task the employee has been assigned to.
- cost: The premium paid for accomplishing certain job.

2.2 Model on Tasks Control

1. For the whole set of tasks:

- Matrix $T = [t_{i,j} \mid i,j \in N, 1 \leq i,j \leq |T|]$

The matrix T represents the time orders among tasks. If $t_{i,j} = 1$, task i is earlier than task j; otherwise, $t_{i,j} = 0$ means task i is not earlier than task j or there is no time order between the two tasks. We can also use PERT for distinguishing distinctly.

2. For concrete task, the parameters are defined here:
task{state, endTime, totalTime, needGroup, penaltyFunc, budget}

- state: Indicate whether or not the task has begun. The value includes "1"(yes) and "0"(no).
- endTime: The end time of a task.
- totalTime: The total time of a task.

- needGroup: The group and the number of the employees needed in a task. Its data is in the form of g_r^m.
- penaltyFunc: The penalty function if the task is delayed.
- budget: The max cost of the task.

2.3 Model on Time Control

time{endTime, totalTime}

- endTime: The end time of a task.
- totalTime: The total time of a task.

3 Learning Method by Being Told

The basic model of machine learning can be pictured as follows.

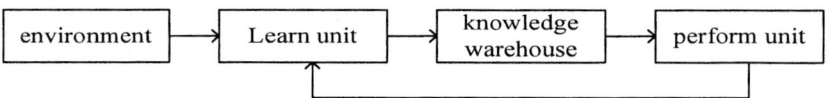

Machine learning in Artificial Intelligent (AI) kingdom is the core of data mining. There are many applications about collaborations of data mining and machine learning. Many machine learning methods could be used directly into data mining domain.

Induction learning is one of the machine learning methods. Learning by being told is one of the induction learning methods. In our paper, we will apply machine learning by being told method to mine useful data about human resources.

The learning process of learning method by being told can be described as follows:

1. **Machine asks appraisals from specialists**
   The manner may be simple or complex, maybe initiative or passive.
2. **Machine accepts and explains the appraisals.**
   Machine makes the suggestions be identified by the system.
3. **Machine converts the appraisals into concrete knowledge.**
   The suggestions can be operational and usable.
   Here the appraisals are in the matrix form.
   Matrix $S = [s_{i,j} \mid i,j \in N, 1 \leq i \leq |R|, 1 \leq j \leq |A \text{ or } B|, s_{i,j} = 0 \text{ or } 1]$
   $s_{i,j}$ means the satisfaction degree employee i doing the assigned job.
   The above matrixes will be converted into the individual capability matrixes, which is in the following form.
   Matrix $S_{hr} = [s_{i,j} \mid i,j \in N, 1 \leq i \leq |A \text{ or } B|, 1 \leq j \leq |T|, 0 \leq s_{i,j} \leq 1]$
   A or B are the satisfaction degree fuzzy sets of the tasks, $|T|$ is the number of the tasks.
   If the satisfaction degree implies that the employee is very capable of doing a certain task, his/her appraisal values in the individual capability matrix will be adjusted bigger; in the contrary, the appraisal values will be adjusted smaller if the employee is overrated, otherwise, the value will remain still.

4. **Machine adds the new knowledge into the knowledge warehouse.**
   The problems about knowledge redundancy and conflict must be considered. The old data is replaced by the new ones.
5. **System provides the refreshed results with the help of the knowledge.**

In fact, the whole process is rotated. Machine provides the results necessary for the allocation; the algorithm gets the allocation plan through the above results; when the tasks have been finished, specialists feed back appraisals which will be converted and added to the knowledge warehouse; the new data will be provided when the next allocation comes. The method will adjust the values from time to time according to the employees' changing experience. And the data will be closer to the individual capability.

4 Fuzzy Logic

Define the fuzzy membership sets we use as follows:

1. **fuzzy sets for degree of job satisfaction**
   Specialists will build the fuzzy sets for degree of job satisfaction. The values of the sets are different aiming at different goals.
2. **weight values of the fuzzy sets**
   The values of the elements in the weight sets are between "0" and "1". The sum in one set is usually "1".
3. **individual capability matrixes**
   The elements' values indicate the different satisfaction probability if certain employee occupy different tasks. For example:
4. **fuzzy synthesis appraisal matrixes**
   The matrixes are used in Fuzzy appraisal decision-making method which is an effective appraisal way to solve problems involving multi-factors. The matrixes are derived through the model $M(\cdot,+)$.

5 Optimization Goal of the Allocation Model

When directors implement the human resource allocation among tasks, they usually expect the optimized result in the following multi-objectives programming.

1. Minimize the time in executing the entire tasks

$$\text{Time}_{\text{Task}}(R^*) = T_{\text{Task}}(R^*) + T_{\text{Task}}(R^*)$$
$$\text{s.t: } 0 \leq \text{Time}_{\text{Task}}(R^*) \leq \text{Task.totalTime} \quad R^* \subset R$$

R is the set of the human resource, R^* is the subset of R. $T_{\text{Task}}(R^*)$ is the calculated time, which is necessary to perform task set. $T_{\text{Task}}(R^*)$ is the calculated waiting time, which appears only in the case of allocation of so-called busy resources. The time spent on the tasks must be less than the total time regulated beforehand.

2. Minimize the whole premium cost to the whole employees

$$\text{Cost}_{\text{Task}}(R^*) = T_{\text{Task}}(R^*) \sum_{r \in R^*} \text{cost}$$
$$\text{s.t: } 0 \leq \text{Cost}_{\text{Task}}(R^*) \leq \text{Task.budget} \quad R^* \subset R$$

R is the set of the human resource, R* is the subset of R. And the money cost by the employees must be less than the budget regulated beforehand.

To simplify the model, we only consider the human pay with no regard of cost of penaltyFunc of the tasks here.

6 Our Approach

The optimized solution of multi-objective-Function is to make multi-objective-Function be best satisfied. But these functions can not reach the best optimization at the same time. So we can only acquire the fuzzy optimization solution.

1. Definition 1:
If the number of employee certain task i needs is j, then dividing the task i into subtasks. And the number of the subtasks is j. The whole amount of the tasks will be added to j-1. The set of the subtasks replaces its parent task.

$$t_i = t_{i1} + t_{i2} + \ldots\ldots + t_{ij}$$

2. The algorithm can be described as follows:
Suppose Task = {$task_1, task_2 \ldots\ldots task_n$}, Hr = {$hr_1, hr_2 \ldots hr_m$}

Algorithm:
1. Scheduling the current set of tasks according to matrix T.
   If the value of the element in T is "1", schedule the two tasks. Otherwise, calculate the latest-begin time ($task_i$.endTIme − $task_i$.totalTime) of the other tasks. Scheduling them based on the time.
2. Calculating g_r^m = $task_i$.needGroup. Acquire the sets of subtasks, and replace their parent task. Changing the value of the number of the new tasks set.
3. Finding out the sets of the usable human resource.
   If hr_i.ifFree == true then labeling the employee.
4. Grouping the labeling employee according to the employee's attributes "group". The result is represented in the form of hr_i^j.
5. Building the fuzzy sets for degree of job satisfaction A_i based on factor time.
6. Acquiring A_i's weight Q_A and B_i's weight Q_B.
7. Building the fuzzy sets for degree of job satisfaction B_i based on factor cost.
8. The individual capability matrix R_{hr} considering the time factor is provided by the machine learning method.
   Matrix $R_{hr} = [s_{i,j} \mid i,j \in N, 1 \leq i \leq |T|, 1 \leq j \leq |A|, 0 \leq s_{i,j} \leq 1]$
   $|A|$ is the number of the elements in the satisfaction degree fuzzy set A.
9. The individual capability matrix R_{hr} considering the cost factor is provided by the machine learning method.
   Matrix $R_{hr} = [s_{i,j} \mid i,j \in N, 1 \leq i \leq |T|, 1 \leq j \leq |B|, 0 \leq s_{i,j} \leq 1]$
   $|B|$ is the number of the elements in the satisfaction degree fuzzy set B.
10. Calculating the fuzzy synthesis appraisal matrix C by Using the model M(·,+) in the fuzzy mathematics.
    Matrix $C_{ij} = [c_{i,j} \mid i,j \in N, 1 \leq i \leq |R|, 1 \leq j \leq |T|, 0 \leq c_{i,j} \leq 1]$
11. Using Task-Arrange or Hungarian Algorithm to obtain the optimization result. Refer to [12] for more detailed description of the algorithms.

12. After the present tasks have been finished, specialists feed back the satisfaction appraisal matrixes according to employees' performance.
    Matrix $V_{ij} = [v_{i,j} \mid i,j \in N, 1 \leq i \leq |R|, 1 \leq j \leq |A|, v_{i,j} = 0 \text{ or } 1]$
    Matrix $V_{ij}' = [v_{i,j}' \mid i,j \in N, 1 \leq i \leq |R|, 1 \leq j \leq |B|, v_{i,j}' = 0 \text{ or } 1]$
    $v_{i,j}$ or $v_{i,j}'$ means the satisfaction degree employee i doing the assigned job.
    $|A|$ or $|B|$ is the number of the elements in the satisfaction degree fuzzy set A or B.
13. If   the appraisal result is A1(shorter than expected),
    Then the element's value in individual capability matrix is added 0.01;
    If   the appraisal result is A2(approximately expected),
    Then the element's value in individual capability matrix is kept invariable;
    If   the appraisal result is A3(longer than expected),
    Then the element's value in individual capability matrix is subtracted 0.01;
14. If   the appraisal result is B1(less than expected),
    Then the element's value in individual capability matrix is added 0.01;
    If   the appraisal result is B2(approximately expected),
    Then the element's value in individual capability matrix is kept invariable;
    If   the appraisal result is B3(more than expected),
    Then the element's value in individual capability matrix is subtracted 0.01;
15. Updating the values of the employees' individual capability matrix in knowledge warehouse.

7 A Numerical Example

Conditions:
   Task = {task1, task2, task3}
   task1:$\{g_1^2, g_2^2\}$ task2:$\{g_1^1, g_3^2\}$ task3:$\{g_2^2, g_3^2\}$
   The set value in the machine learning method is 9/10
After the fifth step we can get the data:
   Task={$task_{11}, task_{12}, task_{13}, task_{14}, task_{21}, task_{22}, task_{23}, task_{31}, task_{32}, task_{33}, task_{34}$}
   usable hr = $\{hr_1^3, hr_2^4, hr_3^2\}$
Then we know:
   $\{hr_{11}, hr_{12}, hr_{13}\}$ are allocated among $\{task_{11}, task_{12}, task_{21}\}$
   $\{hr_{21}, hr_{22}, hr_{23}, hr_{24}\}$ are allocated among $\{task_{11}, task_{12}, task_{31}, task_{32}\}$
   $\{hr_{31}, hr_{32}\}$ are allocated among $\{task_{21}, task_{22}, task_{33}, task_{34}\}$
First we consider the first allocation.
1. Building the fuzzy sets for degree of job satisfaction due to the goal of time.
   A = {A1(shorter than expected), A2(approximately expected), A3(longer than expected)}, weight Q_A = (0.5, 0.3, 0.2)
2. Building the fuzzy sets for degree of job satisfaction due to the goal of cost.
   B = {B1(less than expected), B2(approximately expected), B3(more than expected) }, weight Q_B = (0.6, 03, 0.1)
3. The weight of the two goals Q = {0.6, 0.4}
4. The individual capability matrixes of $\{hr_{11}, hr_{12}, hr_{13}\}$ about $\{task_{11}, task_{12}, task_{21}\}$ aiming at A are provided by the machine learning method:

$$R1_{r11} = \begin{bmatrix} 0.6 & 0.6 & 0.7 \\ 0.2 & 0.3 & 0.2 \\ 0.2 & 0.1 & 0.1 \end{bmatrix} \quad R1_{r12} = \begin{bmatrix} 0.4 & 0.5 & 0.6 \\ 0.3 & 0.4 & 0.3 \\ 0.3 & 0.1 & 0.1 \end{bmatrix} \quad R1_{r13} = \begin{bmatrix} 0.5 & 0.4 & 0.5 \\ 0.2 & 0.4 & 0.3 \\ 0.3 & 0.2 & 0.2 \end{bmatrix}$$

5. The individual capability matrixes of $\{hr_{11}, hr_{12}, hr_{13}\}$ about $\{task_{11}, task_{12}, task_{21}\}$ aiming at B are:

$$R1_{r11} = \begin{bmatrix} 0.6 & 0.7 & 0.5 \\ 0.2 & 0.2 & 0.4 \\ 0.2 & 0.1 & 0.1 \end{bmatrix} \quad R1_{r12} = \begin{bmatrix} 0.4 & 0.5 & 0.4 \\ 0.3 & 0.2 & 0.3 \\ 0.3 & 0.2 & 0.3 \end{bmatrix} \quad R1_{r13} = \begin{bmatrix} 0.5 & 0.4 & 0.6 \\ 0.3 & 0.4 & 0.3 \\ 0.2 & 0.2 & 0.1 \end{bmatrix}$$

6. Using the model $M(\cdot,+)$ in the fuzzy mathematics, we can get the fuzzy synthesis appraisal matrix.

$$C = \begin{bmatrix} 0.236 & 0.2548 & 0.25 \\ 0.1736 & 0.2102 & 0.228 \\ 0.2042 & 0.182 & 0.223 \end{bmatrix}$$

7. Task-Arrange or Hungarian Algorithm gives the best allocation.
   $task_{11}$ is assigned to hr_{12}, $task_{12}$ is assigned to hr_{13}, $task_{21}$ is assigned to hr_{11}.
8. Also we can get the allocation result about other tasks and human resources.
9. After the tasks have been finished, some specialists will feed back the appraisals according to employees' performance. The appraisals are in matrix form.

    Matrix $V = [v_{i,j} \mid i,j \in N, 1 \leq i \leq |R|, 1 \leq j \leq |A|, v_{i,j} = 0 \text{ or } 1]$
    Matrix $V = [v_{i,j} \mid i,j \in N, 1 \leq i \leq |R|, 1 \leq j \leq |B|, v_{i,j} = 0 \text{ or } 1]$
    $|A|$ or $|B|$ is the number of the elements in the satisfaction degree fuzzy set A or B.

    Here we have three candidates in each degree fuzzy set. The columns of the matrix represent the members in the satisfaction degree sets. For example, if the satisfaction degree about hr_{11} doing $task_{21}$ the specialist think is A1(approximately expected), then we set $v_{12} = 1$. The rest may be deduced by analogy.

$$V11 = \begin{bmatrix} 0 & 1 & 0 \\ 1 & 0 & 0 \\ 0 & 1 & 0 \end{bmatrix} \quad V21 = \begin{bmatrix} 0 & 1 & 0 \\ 1 & 0 & 0 \\ 0 & 1 & 0 \end{bmatrix} \quad V31 = \begin{bmatrix} 1 & 0 & 0 \\ 0 & 1 & 0 \\ 0 & 1 & 0 \end{bmatrix}$$

The above matrixes are aimed to Time factor.

$$V12' = \begin{bmatrix} 0 & 0 & 1 \\ 0 & 0 & 1 \\ 1 & 0 & 0 \end{bmatrix} \quad V22' = \begin{bmatrix} 0 & 1 & 0 \\ 0 & 1 & 0 \\ 1 & 0 & 0 \end{bmatrix} \quad V32' = \begin{bmatrix} 1 & 0 & 0 \\ 0 & 0 & 1 \\ 1 & 0 & 0 \end{bmatrix}$$

The above matrixes are aimed to Cost factor.

10. Refresh the individual capability matrixes.
    If   the appraisal result is A1,
    Then the element's value in individual capability matrix is added 0.01;
    If   the appraisal result is A2,
    Then the element's value in individual capability matrix is kept invariable;
    If   the appraisal result is A3,
    Then the element's value in individual capability matrix is subtracted 0.01;
    If   the appraisal result is B1,
    Then the element's value in individual capability matrix is added 0.01;

If   the appraisal result is B2,
Then   the element's value in individual capability matrix is kept invariable;
If   the appraisal result is B3,
Then   the element's value in individual capability matrix is subtracted 0.01;
The new matrixes about factor Time are as follows.

$$R1_{r11} = \begin{bmatrix} 0.6 & 0.6 & 0.71 \\ 0.2 & 0.3 & 0.2 \\ 0.2 & 0.1 & 0.09 \end{bmatrix} \quad R1_{r12} = \begin{bmatrix} 0.42 & 0.5 & 0.6 \\ 0.3 & 0.4 & 0.3 \\ 0.28 & 0.1 & 0.1 \end{bmatrix} \quad R1_{r13} = \begin{bmatrix} 0.5 & 0.4 & 0.5 \\ 0.2 & 0.4 & 0.3 \\ 0.3 & 0.2 & 0.2 \end{bmatrix}$$

The new matrixes about factor Cost are as follows.

$$R1_{r11} = \begin{bmatrix} 0.6 & 0.7 & 0.5 \\ 0.2 & 0.2 & 0.4 \\ 0.2 & 0.1 & 0.1 \end{bmatrix} \quad R1_{r12} = \begin{bmatrix} 0.38 & 0.5 & 0.4 \\ 0.3 & 0.2 & 0.3 \\ 0.32 & 0.2 & 0.3 \end{bmatrix} \quad R1_{r13} = \begin{bmatrix} 0.5 & 0.43 & 0.6 \\ 0.3 & 0.4 & 0.3 \\ 0.2 & 0.17 & 0.1 \end{bmatrix}$$

11. These weight values will be added to the knowledge warehouse and the individual capability matrixes will be refreshed in the warehouse.
12. When the next allocation comes, the new evaluation matrixes above will be provided to the managers.

8 Comparison with Other Models and Methods

- Comparison with linear programming model
    1. Linear programming is a single goal optimization technique; this is not the situation for a majority of firms with multiple goals.
    2. Our new model can deal with allocation problems with two optimization goals.
- Comparison with goal programming model
    1. Goal programming can not deal with the organizational differentiation problems.
    2. Though our new model, managers can get concrete information about every employee. So it could be more flexible and accurate when planning.
- Comparison with distributed resource allocation algorithm [7]
    1. Distributed algorithm is one of the algorithms to solve the allocation problems. It uses multi-processors to solve distributed resource allocation, and the processor number is correlated with the task number. In our model, there are eleven tasks and nine employees to be allocated. The number of the processor is $(m+1)(n-1)+2 = 10*10 + 2 = 102$. It is so big that the algorithm is not suitable to dissolve the allocation problem involved many tasks and employees. And the procedure takes time $O(nm^2)$.
    2. The complexity of the algorithm is $O(nm)$.

9 Conclusions

Human resource allocation is a kind of problem in data mining domain. How to get useful information from so much existing data is crucial to solve the problem. First we construct a human resource fuzzy allocation model. In the model it uses machine learning method by being told to get the individual capability evaluation matrixes. It is suitable for the multi-objective assignment problem under a dynamic environment. Machine learning method and fuzzy logic concepts provide better solutions in planning problems than the previous approaches. In the solution, the individual capability matrixes in the knowledge warehouse are updated in time through the process of the self-learning. Machine learning method by being told will adjust the elements' values in the matrixes to be closer to the reality from time to time. Based on the values, we can get the final synthesis evaluation matrix via fuzzy logic. Machine learning and fuzzy set theories are proper to be applied to human decision-making problems.

References

1. Hand, D., Mannila, H., Smyth, P.: Principles of Data Mining. The MIT Press. (2001)
2. Mitchell, T.: Machine Learning. McGraw Hill. (1997)
3. Do Zhang.: Machine Learning and Software Engineering. Software Quality Journal. 11(2): (2003) 87–119
4. Grant, E. W, Jr., and Hendon, F. N., Jr.: An application of linear programming in hospital resource allocation. J. Health Care Market. 1: (1987) 69–72
5. Welling, P.: A Goal Programming Model for Human Resource Accounting in a CPA Firm. Accounting, Organizations and Society. 2(4): (1977) 307–316
6. Wikil Kwak.: Human Resource Allocation in a CPA Firm: A Fuzzy Set Approach. Review of Quantitative Finance and Accounting. 20: (2003) 277–290
7. CaiFen Wang.: The Distributed Algorithms of Optimum Distribution for the Limited Resource. Journal of Northwest Normal University. 30(1): (1994) 26–30
8. Keyser Thomas K, Davis Robert.: Distributed computing approaches toward manufacturing scheduling problems. J. IIE Transactions. 30 (4): (1998) 379–390
9. Vigo Daniele, Maniezzo Vittorio1.: A genetic/ tabu thresholding hybrid algorithm for the process allocation problem .J. Journal of Heuristics. 3 (2): (1997) 91–110
10. Omer, K, Andre de Korvin.: Utilizing Fuzzy Logic in Decision-Making: New Frontiers. Application of Fuzzy Sets and Theory of Evidence to Accounting II, Stamford, Connecticut: JAI Press Inc. (1998) 3–14
11. Ian Cloete, and Jacobus van Zyl.: A Machine Learning Framework for Fuzzy Set Covering Algorithms. 2004 IEEE International Conference on Systems, Man and Cybernetics. (2004) 3199–3203
12. KaiCheng Lu.: Single-objective, multi-objectives and Integer programming. Beijing, Tsinghua University Press. (1999) 171–180
13. Liu, Y. H. and Y. Shi.: A Fuzzy Programming Approach for Solving a Multiple Criteria and Multiple Constraint Level Linear Programming Problem. Fuzzy Sets and Systems. 65: (1994) 117–124
14. Zebda, A.: Fuzzy Set Theory and Behavioral Models for Decision Making under Ambiguity. Application of Fuzzy Sets and the Theory of Evidence to Accounting II, Stamford, Connecticut: JAI Press Inc. (1998) 15–27
15. Shaw M J.: Knowledge based scheduling in flexible manufacturing systems. TI Tech. J. Winter, 1987, 54-61
16. K. Srinivasan and D. Fisher.: Machine learning approaches to estimating software development effort.: *IEEE Trans. SE.* 21(2): (1995) 126–137.
17. Chen H, Schuffels C, Orwig R.Internet Categorization and Serach: A Self-Organizing Approach. Journal of Visual Communication and Image Representation. 7(1): (1996) 88–102

Towards Automated Design of Large-Scale Circuits by Combining Evolutionary Design with Data Mining*

Shuguang Zhao[1], Mingying Zhao[2], Jun Zhao[1], and Licheng Jiao[1]

[1] School of Electronic Engineering, Xidian University
[2] School of Mechanic-Electronic Engineering, Xidian University,
Xi'an 710071, P.R. China
Sgzhao@xidian.edu.cn

Abstract. As an important branch of evolvable hardware, evolutionary design of circuit (EDC) is a promising way to realize automated design of complex electronic circuits. To improve EDC in efficiency, scalability and capability of optimization, a novel technique was developed. It features an adaptive multi-objective genetic algorithm and interactions between EDC and data mining. It was validated by the experiments on arithmetic circuits, showing some exciting results. Some circuits evolved are the best ones ever reported in terms of gate count and operating speed. Moreover, some novel knowledge, e.g., efficient and scalable design formulae and generalized transform rules have been discovered by mining the data and results of EDC, which are easy to verify but difficult to dig out by human experts with existing knowledge.

1 Introduction

With ever-growing scales and complexities of electronic circuits, both automated design and knowledge discovery of them become even more challenging tasks for artificial intelligence. As an important branch of Evolvable Hardware (EHW) [1,2], an emerging field, evolutionary design of circuits (EDC) [2] refers to applications of artificial-evolution based techniques especially Genetic Algorithms (GAs)[3] to circuit design to seek optimal or feasible solutions (i.e. structures and parameters of circuits) according to design targets. It is theoretically capable of automatic searching out optimized solutions to problems of circuit design even without a priori knowledge and human interventions, although the reality is far from ideal.

As to logic or digital circuits, EDC is mainly implemented at two distinct abstract-levels, gate-level and function-level. In contrast with a function-level evolution [1,2,4] that usually employs larger scale modules with more complex functions and directly evaluates fitness on hardware (e.g. programmable logic devices), a gate-level evolution [5-7] usually takes logic gates as basic units and evaluates individual's fitness through software simulation, which endow it with some advantages such as applicability and analyzability. So far, most works reported in gate-level evolution are concentrated on combinational circuits especially arithmetic circuits, mainly for the

---

* This work was partially supported by National Natural Science Foundation of China (under grants No. 60133010 and No. 60374063) and China Postdoctoral Science Foundation.

sake of seeking novel or efficient building-blocks. But they seldom managed to deal with multiple objectives or design requirements involving function, gate count and operating speed of logic circuits [5-7, 10].

We have developed a novel approach to gate-level multi-objective evolution of larger scale logic circuits. The approach features an adaptive multi-objective genetic algorithm and a mechanism of two-way enhancements of EDC and knowledge discovery, as illustrated in Fig. 1. The remainder of this paper was contributed to introduction of main ideas, algorithms and experimental results of it.

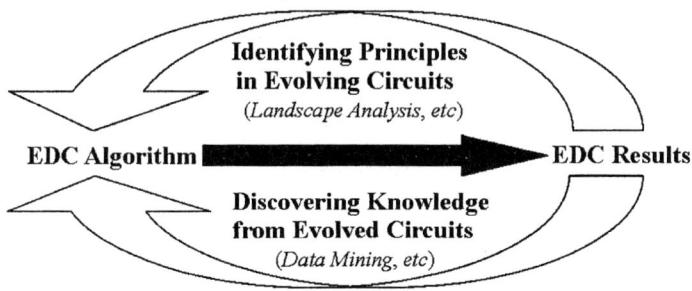

Fig. 1. An illustration of two-way enhancements between EDC and knowledge discovery

2 Circuit Model and Representation Scheme

As depicted in Fig. 2, the gate-level abstract model adopted in this paper take the form of a rectangle array comprising logic units of C rows by L columns, each of which has K inputs, M outputs and N kinds of configurable functions. For all logic units, let the input set is $CI=\{CI_{i,j} | 1 \leq i \leq C, 1 \leq j \leq L\}$ and the output set is $CO=\{CO_{i,j} | 1 \leq i \leq C, 1 \leq j \leq L\}$, while for the array the external input set is $S=\{S_i | 1 \leq i \leq p\}$ and the output is $O=\{O_i | 1 \leq i \leq q\}$. Then, if each unit is allowed to input from the arrays' external inputs, units' outcomes and logic constant $LC=\{0,1\}$ and to feed its outcomes to other units and the array (i.e., its outputs), or formally $CI \subset S \cup CO \cup LC$, $O \subset CO$, feedbacks (i.e. resultant signals are subsequently taken as sources) are permitted which consequently result in

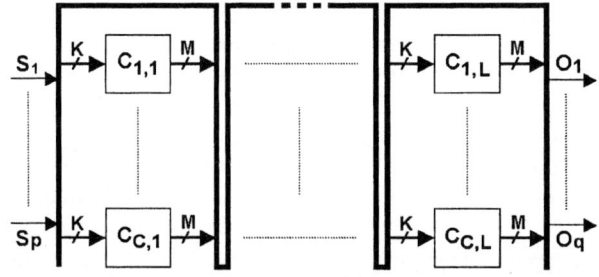

Fig. 2. Abstract model for gate-level evolution

a sequential circuit. Otherwise, if each unit is not allowed to input from the units located in the subsequent (dextral) columns, or formally $CI=S \cup LC \cup \{CO_{i,j} | 1 \leq i \leq C, 1 \leq j \leq x\}$ for a unit located in column x, feedbacks are actually prohibited and the array is restricted to represent combinational circuits. Thus, such an array model is universal and convenient for representation of both sequential and combinational circuits.

To encode the array, all units including the virtual units for external inputs are numbered orderly from rows to columns. A unit located in row i and column j, $1 \leq i \leq C$, $0 \leq j \leq L$, is assigned a sequence number $CN=i+j \cdot C$ and encoded as a character string $[IS_{1,CN}, IS_{2,CN}, \ldots, IS_{K,CN}, TS_{CN}]$, where $\{IS_{j,CN} | 1 \leq j \leq K\}$ indicate its input sources that equal to the sequence numbers of the K units that output to it respectively, and TS_{CN} corresponds to its current function selected. By linking all units' encoding strings and an encoding string indicates sources of the array's outputs, $[OS_1,\ldots,OS_q]$, the array's encoding, i.e. chromosome of a potential circuit looks as follows

$$[IS_{1,1}, IS_{2,1}, \cdots IS_{K,1}, TS_1] \cdots [IS_{1,G}, IS_{2,G} \cdots IS_{K,G}, TS_G][OS_1, \cdots, OS_q] \quad (1)$$

For the binary encoding adopted in this paper, the string length of ISx_i and OS_i is $LS=\lfloor \log_2 (C \cdot (L+1)) \rfloor$, and that of TS_i is $LT=\lfloor \log_2 N \rfloor$ ($\lfloor x \rfloor$ denotes upper bound of x). Thus, the chromosome length of a circuit candidate is $CL=C \cdot L \cdot (K \cdot LS+LT)+q \cdot LS$. To decrease the problem scale and to improve efficiency, it is helpful to define as compact as possible a subset of logic functions, considering research intention of and the design task. For most EDC tasks of combinational circuits, it is feasible to adopt just two-input logic gates with 4 configurable functions, AND2, OR2, NOT and XOR (eXclusive-OR), which are most often used in conventional designs and form a complete logic set. As to sequential circuits, it is usually necessary to include additional memory-units, e.g., registers or latches in the subset.

3 Dynamic Multi-objective Evaluation

To design circuits is unavoidable to meet multiple targets or specifications, which make it a typically difficult multi-objective optimization problem expressed as

$$\text{Maximize} \quad f(X) = (f_1(X), f_2(X), \cdots f_n(X)), \quad X \in \Omega \quad (2)$$

To solve the problem with efficiency and simplicity, a well-known fitness function in the form of 'sum of weighted objective functions' was adopted with modifications, which converts the problem into its single-objective equivalent that a GA is good at

$$\text{Maximize} \quad Fit(X) = \sum_{i=1}^{n} w_i \bullet Fit_i(X) \quad (3)$$

Where $Fit_i(X)$ denotes the normalized objective function corresponding to object function $f_i(X)$; w_i denotes the relevant weight factor for $i=1\ldots N$, which satisfies

$$\sum_{i=1}^{n} w_i = 1 \quad (4)$$

To let the genetic search process pay attention to all objectives, the weight factors $\{w_i\}$ are designed to change dynamically in a way similar to that of the back-propagation learning algorithm commonly used in artificial neural networks

$$w_i(t+1) = \alpha \bullet w_i(t) + (1-\alpha) \bullet [2/N - \overline{Fit_i(t)} / \sum_{j=1}^{N} \overline{Fit_j(t)}] \qquad (5)$$

Where $\alpha \in [0, 1]$ is a constant suggested 0.8 or so by experiments; $\overline{Fit_i(t)}$ denotes the average fitness of all individuals in the population. Provided that $\sum w_i(0)=1$, Equation (4) will hold for all t as expected. Then, the more an objective optimized, the less the relevant weight factor and resultant optimizing pressure on the objective would be, while an less optimized objective and its weight factor will be treated in a reverse direction. Meanwhile, initial values of the weight factors are still meaningful to express user-preferences, although it is usually set as $w_i(0)=1/n$, $1 \leq i \leq n$.

For gate-level EDC, design targets mainly include expected functions or behaviors, efficiency of resource usage and operating speed of circuits. It is vital for an evolved circuit to behave 100% correctly according to the expectation, which is usually described with a truth table for combinational circuit or a state table for a sequential one. So, it is natural to express such a design object with a ratio of Number of Matched Operations (*NMO*) to Total Number of defined Operations (*TNO*)

$$Fit_1 = NMO/TNO \qquad (6)$$

Each operation corresponds to a specified combination of (input, output) or (input, current state, output, next state), and it is scored 1 for 'correct' and 0 for 'incorrect' by a simulation subprogram designed to figure out every unit's logic level and *active time* for every operation. The *active time* is estimated by the unit's location in a signal path, based on the assumption that each unit's propagation-delay is independent of its logic function. To get a smoother landscape that is consequently easier to search, each output variable is counted respectively when computing *NMO* and *TNO*.

As a less gate-number is usually preferable, it is also natural to measure efficiency of resource usage of a circuit in terms of gate count. While every candidate circuit seemingly occupies all units in the array, it is feasible to express the objective with Number of *Unused Gates* (*NUG*), which are gates that have no effective effect on the candidate's behavior, e.g. even numbers of *NOT* gates linked in serial, a gate with its output not referred, etc. By identifying all *Unused Gates* in a circuit with their predefined features and simulation results, the relevant objective function can be computed as *NUG* divided by *TNG* (i.e., Total Number of Gates comprised)

$$Fit_2 = NUG/TNG \qquad (7)$$

Operating speed of a candidate circuit can be estimated with its Maximal Propagation-Delay (*MPD*) from external inputs to array outputs, using simulation results regarding *active time*. The objective function can be computed as

$$Fit_3 = 1/(1 + k_1 \bullet MPD) \qquad (8)$$

Where, k_1 is a user-defined parameter. With the normalized objective functions to be maximized, Fit_1, Fit_2 and Fit_3, the fitness function to synthetically evaluate a candidate circuit can be defined as

$$Fit(t) = \sum_{i=1}^{3} w_i(t) \bullet Fit_i(t) \qquad (9)$$

4 Adaptation of Genetic Parameters

Some GA parameters especially crossover probability P_c and mutation probability P_m have great effects on performances of the GA, and their optimal values are hard to be predefined because they vary with conditions. To improve the GA's performances, P_c and P_m are ordered to self-adapt to genetic-procedure and individuals' diversity. The former is estimated by *relative generation number*, a ratio of the current generation number t to the maximal generation number defined t_{max}. The latter is measured with concentration degree of individuals in the population, which is estimated as

$$f_d(t) = \overline{f}(t) / [\varepsilon + f_{max}(t) - f_{min}(t)] \tag{10}$$

Where $f_{max}(t)$, $f_{min}(t)$ and $\overline{f}(t)$ are maximal, minimum and average fitness of all individuals in the population respectively. Obviously, the index $f_d(t)$ satisfies $0 < f_d(t) < +\infty$, and it is closely related to the distribution or diversity of individuals, or more exactly, it will simultaneously vary with the latter but in a reverse direction. On these bases, P_c and P_m are ordered to adapt themselves in the following manners

$$P_c = \begin{cases} P_{c0} / f_d(t) & t < t_0 \\ P_{c0} \cdot e^{-k_3 \cdot (t-t_0)/t_{max}} / f_d(t) & t_0 \leq t \leq t_1 \\ P_{c0} \cdot e^{-k_3 \cdot (t_1-t_0)/t_{max}} / f_d(t) & t_1 \leq t \leq t_{max} \end{cases} \tag{11}$$

$$P_m = \begin{cases} P_{m0} \cdot f_d(t) & t < t_0 \\ P_{m0} \cdot e^{-k_4 \cdot (t-t_0)/t_{max}} \cdot f_d(t) & t_0 \leq t \leq t_1 \\ P_{m0} \cdot e^{-k_4 \cdot (t_1-t_0)/t_{max}} \cdot f_d(t) & t_1 \leq t \leq t_{max} \end{cases} \tag{12}$$

Where P_{c0} and P_{m0} are initial values of P_c and P_m respectively, and it is usually feasible to let $P_{c0} = 0.8$ and $P_{m0} = 0.1$; t_0 and t_1 are user-defined parameters that satisfy the inequality, $0 \leq t_0 \leq t_1 \leq t_{max}$, and it is suggested to let $t_0 = 0.2 t_{max}$ and $t_1 = 0.8 t_{max}$; k_3 and k_4 are constant parameters that can be assigned as $k_3 = k_4 = 3$. In this way, P_c and P_m will slowly decrease in the evolution process, meanwhile they will respond to the changes of individuals' diversity. With such a self-adaptation mechanism simulating some principles of bionomics and developmental biology [8-9], P_c and P_m will be probably suitable for a whole evolution process, e.g., a higher P_c and a higher P_m to speed the genetic search at the first stage of evolution ($t < t_0$), a lower P_c and a higher P_m to stop the coming premature (local) convergence implied by an increasing $f_d(t)$ during the whole process of evolution search, a lower P_c and a lower P_m to improve the quality of the 'optimal solution' at the final stage of evolution ($t_0 \leq t \leq t_{max}$), etc.

5 CBR-Based Extraction and Reuse of Principles

Although EDC has succeeded in obtaining efficient or novel solutions, it is weak in solving large-scale and complex problems mainly due to its computation-expensive nature. Therefore, the idea of *divide and conquer* featured by conventional design methodologies is also useful to EDC, that is, to decompose a problem into several simpler sub-problems and solve them (top-down), or conversely, to build a circuit by connecting modules of several types together (bottom-up). As shown in Fig. 1, these involve extracting meaningful sub-circuits or design principles from the evolved solutions and reusing them by integrating EDC with Case-Based Reasoning (CBR), so as to solve the scaling problem and to understand the evolved circuits.

CBR is an artificial intelligence technique that solves new problems by using or adapting past solutions. In CBR systems knowledge is embodied in a library of past cases (i.e., case-base), rather than being encoded in classical rules. Each case typically contains a description of the problem, plus a solution containing implicit knowledge and/or the outcome. To solve a current problem with CBR, it is matched against the cases in the case-base, and similar cases are retrieved and used to suggest a solution which is reused and tested for success and then revised if necessary. Finally the current problem and the final solution are retained as part of a new case. Therefore, CBR quite suits EDC requirements of extraction and reuse of principles contained in EDC results, as demonstrated by some preliminary results [11, 12].

The primary difficulties herein exist in building a case-base. While CBR relies on cases that have known structure, e.g. attribute value pairs, evolved circuits lack any *understanding* incorporated in their structures. As a result, all knowledge beyond their functionality must be identified before building a useful case-base. Instead of using a GA as a *knowledge lean* method to generate knowledge for a case-base, we employed the GA with efficient and flexible genotype-phenotype mapping to produce efficient solutions and to consequently ease the creation of a case-base.

The first step to create a case-base is to remove redundant information and duplicate circuits in EDC results, and to compress (in terms of the spaces left in the genotypes) and normalize (in terms of the cells' order) the resultant circuits so as to facilitate the CBR functions of matching, retrieval and adaptation. The second step involves splitting the normalized circuits into sub-circuits and calculation of their structure, behavior and functionality. It is followed by separating the sub-circuits into perfect and imperfect solution elements for the given requirements. Finally, the circuits are indexed according to their function, behavior, structure, and sub-circuits, by using a case-based indexing mechanism. Each circuit or case in the case-base stores its own information on its similarity to all other cases, which is represented by four indexes correspond to its function, behavior, structure and sub-circuits respectively. These indexes need only be calculated once, and additional cases can be indexed in linear time proportional to the size of the case-base.

With the case-base built, matching of cases is achieved efficiently by using the *Nearest Neighbor Matching* function that ranks the cases, giving rise to applications such as discovery of equivalent circuits or implementations of logic functions, transform formulas and optimal circuits in terms of gate count, speed, etc. Moreover, imitating the way that human designs manage to solve relatively large scale problems mainly by assembling verified build-blocks, appropriate cases or circuits in the case-base can be retrieved, assembled and tested automatically towards larger and more complex circuits in accordance with the expected behaviors and performances of circuits. In this way, novel knowledge for circuit design could be discovered from EDC results, and in return the knowledge discovered will help understand the evolved circuits and partly solve the scaling problem of EDC.

6 Experimental Results of EDC

By integrating the ideas discussed above into an EGA framework with roulette-wheel selection, one-point crossover and simple mutation, which is theoretically proved able to converge in a probability 1, the approach for evolutionary discovery is created. With this approach, evolutionary design experiments on some benchmark problems for gate-level evolution [6,7,10], including even-parity checkers and multipliers, have been implemented successfully. Thanks to the adaptation measures that are validated to improve robustness of GA, the experiments worked very well with a set of parameters. Besides the parameters given, initial values of weight factors are set as $w_1(0)=w_2(0)=w_3(0)=1/3$, whereas population size and maximal generation number were specified according to the problem scale.

An n-bit digital multiplier is a combinational circuit that *output the product of two groups of n-bit binary numbers*. A series of experiments were carried out on multipliers of increasing scales. For each scale from 2-bit to 4-bit, 20 runs of the program were performed. A 2-bit multiplier evolved is depicted in Fig. 3, which is as good as the best one ever known. A 3-bit multiplier evolved is depicted in Fig. 4, where GN is short for gate number and DN for delay number. It is 10.7% more efficient and 20% faster than the best one designed by a human expert, which features

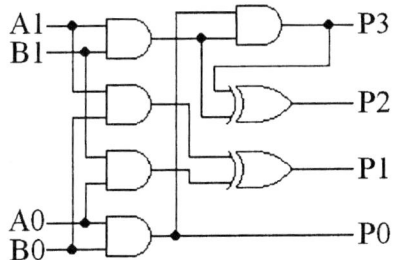

Fig. 3. A 2-bit multiplier evolved

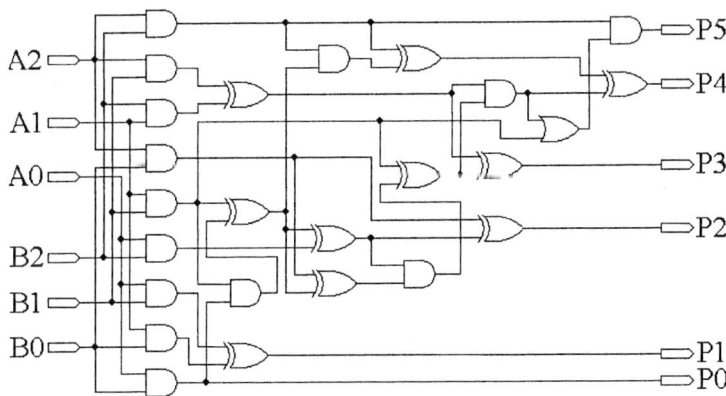

Fig. 4. A 3-bit multiplier evolved (GN=25, DN=9)

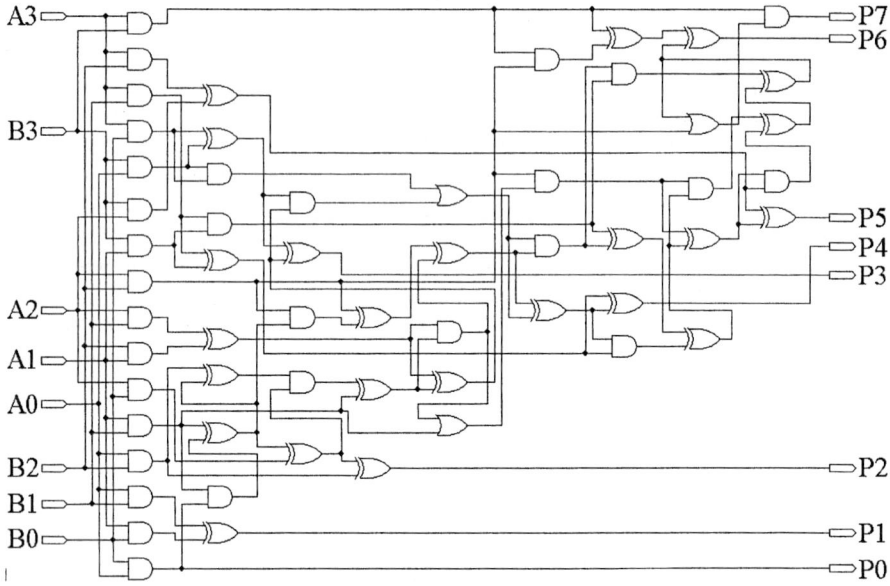

Fig. 5. Standard implementations of the AND gate with one input inverted, as A⊕AB=AB'

Fig. 6. A 4-bit multiplier evolved (GN=58, DN=18. With input-inverted AND gates, GN=55)

GN=28 and DN=10. Meanwhile, it is in fact as good as the most efficient one evolved by Miller *et al* [6, 10], which consists of 21 standard gates and 2 nonstandard AND gates with one input inverted. One such nonstandard AND gate is actually equivalent to 2 gates, i.e., a AND gate in conjunction with a XOR gate as emerged in our results or a Not gate in series with a AND gate, as shown in Fig. 5. As to a 4-bit multiplier that is rather difficult even for human experts to design, our result evolved from scratch is depicted in Fig. 6, which is much better than that of a human expert (with GN=64, DN=24) and that evolved by Miller *et al* [6, 10] from a conventional design result, which consists of 57 gates including 10 nonstandard AND gates and features DN=20. All these evolved circuits feature wondrous reuses of inner outcomes. They imply that our approach is effective and it surpassed its congeners and human experts, especially with larger circuits.

7 Experimental Results of Data Mining

With the techniques discussed in section 5, some experiments were carried out on evolved circuits. For a binary addition, the core of a binary multiplication, following expressions that derive the carry from the addends have been identified

$$CF_{n+1}=CF_n\oplus[I_{n+1}\bullet(I_1\oplus\ldots\oplus I_n)] \quad (13)$$

$$CS_n = \begin{cases} 0 & n<4 \\ CS_4 = I_1I_2I_3I_4 & n=4 \\ CS_5 = CS_4 \oplus \{I_5[I_1I_2(I_3\oplus I_4)\oplus I_3I_4(I_1\oplus I_2)]\} & n=5 \\ CS_6 = CS_5 \oplus \{I_6\{I_1I_2(I_3\oplus I_4)\oplus I_3I_4(I_1\oplus I_2)\oplus I_5\{I_1I_2\oplus I_3I_4 \oplus[(I_1\oplus I_2)\bullet(I_3\oplus I_4)]\}\}\} & n=6 \end{cases} \quad (14)$$

Where CF_n and CS_n denote respectively the least-bit and the secondary-least-bit of the carry derived from n bit-addends, $I_1,\ldots,I_n$; CF_{n+1} denotes the successor of CF_n with an additional input I_{n+1}, $n\geq 1$. As $CF_1=0$, it is easy to derive and efficiently implement CF_n for binary additions of arbitrary bits by iteratively using Equation (13). Although Equations (14) have complex forms and limited applicability, they are also useful for efficient implementations of multipliers.

By analyzing the experimental results of even-parity checkers of increasing inputs, which *output '1' iff inputs contain nonzero even numbers of '1'*, a universal design formula for even-parity checkers of arbitrary bits was also found as

$$F_{n+1}=I_{n+1}\bullet(I_1+\ldots+I_n)\oplus F_n \quad (15)$$

Where F_n denotes the output of a n-bit even-parity checker with n inputs, $I_1,\ldots,I_n$; F_{n+1} denotes the successor of F_n updated by I_{n+1} joining; $F_1=0$. In addition, some novel transform formulas regarding exclusive-OR logic were also obtained as

$$A\bullet(B\oplus C)=(A\bullet B)\oplus(A\bullet C) \quad (16)$$

$$A\oplus B+A\bullet B=A+B \quad (17)$$

$$A\oplus B=(A+B)\bullet(A\bullet B)'=(A+B)\oplus(A\bullet B) \quad (18)$$

$$(A+B)\bullet(A+C)=B\oplus C\oplus(A\bullet B)\oplus(A\bullet C) \quad (19)$$

Although the above extracted principles are easy to prove with knowledge of Boolean algebra, they are rather difficult for human experts to dig out with conventional approaches. These results imply that the approach is useful to acquisition and discovery of relevant knowledge.

8 Conclusions

In this paper, a novel EDC approach involving CBR-based data mining was proposed and verified. By virtue of a series of measures cooperating together, this approach was proved capable of automated design of multi-objective optimized logic circuits, and it was shown very helpful in discovery of novel and optimized building-blocks along with new principles and rules. In future we will study the ways to enhance EDC and knowledge discovery more efficiently and interactively so as to solve large-scale problems and acquire new knowledge.

References

1. Yao X., Higuichi T.: Promises and Challenges of Evolvable Hardware. IEEE Trans. On Systems Man and Cybernetics-Part C. 1 (1999) 87–97
2. Zhao S. G.: Study of the Evolutionary Design Methods of Electronic Circuits. Ph.D. dissertation (in Chinese), Xidian University, China (2003)
3. Goldberg D. E.: Genetic Algorithms in Search, Optimization and Machine Learning. Addison-Wesley, Reading, MA (1989)
4. Zhao S. G., Yang W. H.: Intrinsic Hardware Evolution Based on a Prototype of Function Level FPGA. Chinese Journal of Computers. 6 (2002) 666–669
5. Koza J. R.: Genetic Programming: On the Programming of Computers by Means of Natural Selection. MIT Press, Cambridge, MA (1992)
6. Vassilev V. K., Job D., Miller J. F.: Towards the Automatic Design of More Efficient Digital Circuits. In: Proceedings of EH'00. IEEE, PaloAlto (2000) 151–160
7. Coello Coello A. C., *et al.*: Use of Evolutionary Techniques to Automate the Design of Combinational Circuits. Inter. J. of Smart Engineering System Design. 4 (2000) 299–314
8. Shang Y. C., Cai X. M.: General Bionomics. Beijing University Press, Beijing (1992)
9. Gilbert S. F., Developmental Biology. 6th edn. Sinauer Associates Inc., Sunderland (2000)
10. Miller J. F., Job D., Vassilev V. K.: Principles in the Evolutionary Design of Digital Circuits: Part I. J. of Genetic Programming and Evolvable Machines. 1/2 (2000) 7–35
11. Miller J. F., Job D., Vassilev V. K.: Principles in the Evolutionary Design of Digital Circuits: Part II. J. of Genetic Programming and Evolvable Machines. 3 (2000) 259–288
12. Zhao S. G., Jiao L.C., Tang M.: Automated Design and Knowledge Discovery of Logic Circuits Using a Multi-objective Adaptive GA. In: S. Zhang and R. Jarvis (Eds.): AI 2005, LNAI 3809. Springer-Verlag, Berlin Heidelberg (2005) 997–1000

Mining Unexpected Associations for Signalling Potential Adverse Drug Reactions from Administrative Health Databases

Huidong Jin[1], Jie Chen[1], Chris Kelman[2], Hongxing He[1], Damien McAullay[1], and Christine M. O'Keefe[1]

[1] CSIRO Health Informatics, GPO Box 664, Canberra ACT, 2601, Australia
{Warren.Jin, Jie.Chen, Hongxing.He, Damien.McAullay, Christine.OKeefe}@csiro.au
[2] NCEPH, The Australian National University, Canberra ACT, 0200, Australia
Chris.Kelman@anu.edu.au

Abstract. Adverse reactions to drugs are a leading cause of hospitalisation and death worldwide. Most post-marketing Adverse Drug Reaction (ADR) detection techniques analyse spontaneous ADR reports which underestimate ADRs significantly. This paper aims to signal ADRs from administrative health databases in which data are collected routinely and are readily available. We introduce a new knowledge representation, Unexpected Temporal Association Rules (UTARs), to describe patterns characteristic of ADRs. Due to their *unexpectedness* and *infrequency*, existing techniques cannot perform effectively. To handle this unexpectedness we introduce a new interestingness measure, *unexpected-leverage*, and give a user-based exclusion technique for its calculation. Combining it with an event-oriented data preparation technique to handle infrequency, we develop a new algorithm, MUTARA, for mining simple UTARs. MUTARA effectively short-lists some known ADRs such as the disease **esophagitis** unexpectedly associated with the drug **alendronate**. Similarly, MUTARA signals **atorvastatin** followed by **nizatidine** or **dicloxacillin** which may be prescribed to treat its side effects **stomach ulcer** or **urinary tract infection**, respectively. Compared with association mining techniques, MUTARA signals potential ADRs more effectively.

1 Introduction

Adverse Drug Reactions (ADRs) represent causal relationships between adverse events and use of medicines, such as **alendronate**[1] causing **esophagitis** [9]. Adverse events contribute to about 5% of all hospital admissions, and are about the 5th commonest cause of death in hospital [4]. Among them, 30% to 60% are preventable/avoidable by careful prescribing and monitoring [1]. With ADRs like drug-symptom/diagnosis patterns, computerised systems can search health

---

[1] Alendronate is an antiresorptive agent, approved for treatment of osteoporosis [7]. Its adverse reactions or side effects include **esophagitis**, **diarrhoea**, **vomiting**, etc [7,9].

records to monitor adverse events [1]. Such patterns may be used to find at-risk patient groups [5], or to help practitioners ameliorate their diagnoses and prescriptions [3]. Therefore, systematically signalling and validating ADRs is of financial and social importance.

Because of limited trial size and duration, pre-market drug testing cannot recognise all ADRs [8]. There exist several post-marketing ADR detection techniques, known as signal detection in pharmacovigilance, like EBGM (Empirical Bayes Geometric Mean) and BCPNN (Bayesian confidence propagation neural network) [3]. They work mainly on spontaneous ADR reports in which drugs reportedly cause symptoms/diagnoses [3]. However, when based only on these spontaneous ADR reports, the frequency of adverse reactions is underestimated, typically by a factor of about 20 [1]. Adverse reactions may go unnoticed until lots of drug users have been affected, e.g., recent experience with Vioxx [7, 9]. In contrast, administrative health databases routinely record health events such as medical services, diagnoses, and drug prescriptions for, say, subsidy purposes. They cover quite extensive users and are readily available. Signalling ADRs from these databases would complement existing techniques. This data mining work is the first and preliminary attempt on this new direction.

For each patient in an administrative health database, an event sequence can be generated using event timestamps. Among these sequences, ADRs as patterns are normally unexpected and infrequent due to rigorous pre-market drug testing. It is inappropriate to signal ADRs by looking for frequent patterns/association from the event sequences, as is done in current temporal data mining [2]. Another difficulty is that a drug is strongly associated with certain diagnoses due to treatment/prevention. Thus, new techniques are essential for signaling ADRs from the sequences.

We signal ADRs by finding patterns where an event pattern C occurs unexpectedly in a T-sized period after another event pattern A. For simplicity, we assume that both A and C comprise a set of event types. These patterns are denoted by $A \stackrel{T}{\hookrightarrow} C$, called Unexpected Temporal Association Rules (UTARs). T, a period length, constrains the temporal relation between the **antecedent** A and the **consequent** C, and so ensures the UTARs' plausibility. To handle the unexpectedness, we introduce a new interestingness measure, *unexpected-leverage* and give a user-based exclusion technique for its calculation. The basic idea is to exclude expected events in a single T-constrained subsequence and then aggregate unexpectedness over all the remaining T-constrained subsequences. In contrast to [11], it need not compare new rules with existing knowledge rules during the mining procedure in order to find unexpected rules. We also use an event-oriented data preparation technique to handle infrequency. We develop a new data mining algorithm, MUTARA (Mining UTAR given the Antecedent), to signal simple ADRs where a drug causes a symptom. The technique is easily extended for longer patterns such as adverse events induced by drug interaction. Its performance is demonstrated on linked administrative health databases by short-listing ADRs. Compared with OPUS_AR^t, extended from OPUS_AR (OPUS Association Rule) [12], MUTARA can short-list ADRs more effectively.

The rest of the paper is organised as follows. We define UTARs in Section 2, and propose its mining algorithm MUTARA in Section 3. Results and comparisons are presented in Section 4, followed by conclusion in Section 5.

2 Problem Formulation

We first introduce Temporal Association Rules (TARs) as a representation for signalling ADRs from a set of health event sequences Ω. The effect of an event is usually short-acting, e.g., less than 6 months for acute or sub-acute ADRs [8]. By embedding a temporal constraint into association rules $A \rightarrow C$, we get a category of TARs $A \xrightarrow{T} C$. The notation $\xrightarrow{T}$ is used to indicate explicitly that the **antecedent** A and the **consequent** C occur within subsequences constrained by a time window of length T. For simplicity, we choose such a T-constrained subsequence from each sequence in Ω. As illustrated in Fig. 1, e.g., the subsequences for drug User 1 and drug Nonuser 1 are $\{A_1, A_5, A_6, C_1\text{-}C_3\}$ and $\{A_3\text{-}A_5, A_8, C_2\}$ within the hazard and the control periods respectively. The **support** of a TAR, $supp(A \xrightarrow{T} C)$, is the proportion of subsequences where A occurs before C at least once, among all the T-constrained subsequences. Thus, the support describes how often A followed by C occurs in the subsequences. For the two subsequences in Fig. 1, e.g., $supp(A_6 \xrightarrow{T} C_3) = \frac{1}{2}$. Similar to association rules, we have **confidence** of a TAR, $conf(A \xrightarrow{T} C) = \frac{supp(A \xrightarrow{T} C)}{supp(A \xrightarrow{T})}$ where $supp(A \xrightarrow{T})$ indicates the proportion of T-constrained subsequences that contain A. As another measure of the association strength [12], **leverage** is defined as $leverage(A \xrightarrow{T} C) = supp(A \xrightarrow{T} C) - supp(A \xrightarrow{T}) \times supp(\xrightarrow{T} C)$. For the two subsequences in Fig. 1, $leverage(A_6 \xrightarrow{T} C_1) = \frac{1}{2} - \frac{1}{2} \times \frac{1}{2} = \frac{1}{4}$, and $leverage(A_6 \xrightarrow{T} C_3) = \frac{1}{2} - \frac{1}{2} \times \frac{1}{2} = \frac{1}{4}$.

Adverse events are infrequent in administrative health databases because (1) all drugs are screened before marketing, and (2) some post-marketed drugs that strongly led to adverse events are removed from the market. As a result, if we want to signal ADRs by finding valid TARs whose support and confidence exceed pre-specified thresholds θ_s and θ_c respectively, two problems emerge. (1) The thresholds θ_s and θ_c should be set pretty small [2]. This leads to innumerable valid TARs and makes mining algorithms unmanageable [12]. (2) It is hard to set problem-specific thresholds properly. The same difficulties exist in other temporal data mining models, e.g., sequential patterns [2].

One solution to signal ADRs from innumerable TARs is to short-list those most interesting rules [12]. For example, we may simply apply OPUS_AR [12] on the T-constrained subsequences. We call this OPUS_AR^t. It can return a pre-specified number of TARs that maximise a rule quality measure.

The existing interestingness measures, including risk ratio, odds ratio, and leverage, are not suitable for ranking TARs to signal ADRs. Risk ratio and odds ratio are two appealing measures of effect to medical experts [6, 8]. For example,

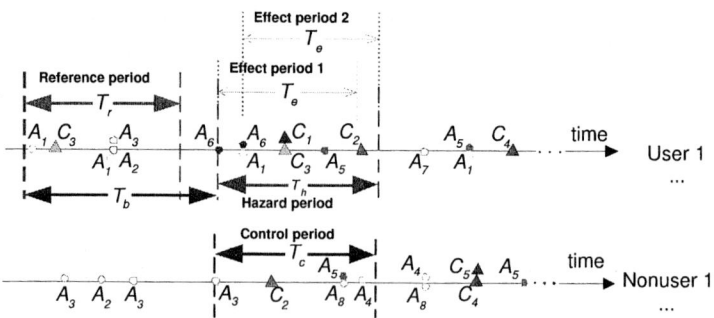

Fig. 1. Illustration of two event sequences and concepts of MUTARA given the antecedent A_6. A_1-A_8 are prescribed drugs. C_1-C_5 are, e.g., diagnoses. T-constrained subsequences for User 1 and Nonuser 1 include event types within the hazard and the control periods respectively. $T_h \doteq T_c = T$. In MUTARA, a hazard period unites the first 2 effect periods around A_6, or a control period is set randomly for each sequence.

risk ratio is $RR(A \xrightarrow{T} C) = \frac{supp(A \xrightarrow{T} C)/supp(A \xrightarrow{T})}{supp(\neg A \xrightarrow{T} C)/supp(\neg A \xrightarrow{T})}$ where $\neg A$ indicates 'no A'. It describes to which degree A increases the occurrence rate of C with the T-constraint. Because these two measures are commonly used in pre-market testing, their values for ADRs in administrative health databases are relatively low. Leverage can not indicate the unexpectedness between drugs and adverse events for effectively signalling ADRs. Based on leverage, e.g., hypertension NOS (last row in Tables 1 and 2) is strongly associated with alendronate[1]. This condition is highlighted due to its high frequency among alendronate users, even before taking the drug. It is not evidentially an adverse reaction.

Thus we need a new knowledge representation for ADRs. Our strategy is to embed unexpectedness into rules directly. To clearly indicate temporal unexpectedness, we introduce an **Unexpected Temporal Association Rule** (UTAR), denoted by $A \xrightarrow{T} C$, which means that the **consequent** C occurs unexpectedly within a T-sized period after the **antecedent** A. Rather than defining unexpectedness explicitly, we aggregate it from individual sequences.

Definition 1. *The **support** of the UTAR $supp(A \xrightarrow{T} C)$ is the proportion of the T-constrained subsequences that contain A unexpectedly followed by C, among all of the T-constrained subsequences.*

In other words, only the subsequences that contain A and then unexpectedly contain C will favor $A \xrightarrow{T} C$. Within a single sequence, we bypass the problem of determining whether event types unexpectedly follow A and only exclude expected event types following A. In the first sequence in Fig. 1, e.g., C_3 occurs both after and before the first A_6. For its subsequence within the hazard period, C_3 is expectable after A. That is, this subsequence does not favor the unexpected temporal association between A_6 and C_3. Thus $supp(A_6 \xrightarrow{T} C_3) = 0$. Sequence information outside of subsequences can be used to prune expected event types, and remaining subsequences may be aggregated together to express the unexpectedness of UTARs. A simple user-based exclusion operation will be given in Section 3.

Having defined the support for UTARs, we can now introduce a new measure.

Definition 2. *The* **unexpected-leverage** *of the UTAR $A \xrightarrow{T} C$ is the proportion, among all the T-constrained subsequences, of the subsequences that exhibit the unexpected association in excess of those that would be supposed if A and unexpected C were independent of each other. That is,*

$$unexlev(A \xrightarrow{T} C) = supp(A \xrightarrow{T} C) - supp(A \xrightarrow{T}) \times supp(\xrightarrow{T} C) \quad (1)$$

where $supp(\xrightarrow{T} C)$ is the proportion of subsequences that unexpectedly contain C.

Thus, only the subsequences which unexpectedly contain C will contribute to unexpected-leverage. For example, for the two subsequences in Fig. 1, we have $unexlev(A_6 \xrightarrow{T} C_1) = \frac{1}{4} > unexlev(A_6 \xrightarrow{T} C_3) = 0$. The measure successfully distinguishes C_1 from the others.

To avoid the problem of proper threshold settings, we generate a pre-specified number of, say 10, UTARs with high unexpected-leverage for signalling ADRs. Another reason to rank UTARs only based on unexpected-leverage is that large unexpected-leverage also indicates large support and confidence.

3 Mining UTARs Given the Antecedent (MUTARA)

In this section, we develop a simple but effective algorithm to search for interesting UTARs when the antecedent, say a drug, is specified in advance. We concentrate on simple UTARs corresponding to an ADR in which the use of a drug causes a diagnosis, i.e., drug $A \xrightarrow{T}$ diagnosis C. These UTARs are of great practical value, and the success on them may also pave the way for mining sophisticated UTARs in future. The mining algorithm, MUTARA, is outlined in Algorithm 1. We exemplify it on the two health event sequences as illustrated in Fig. 1.

In Step 1, we initialise parameters as explained below.

- The antecedent A is specified to restrict the search space, e.g., A_6. The sequences having A are called **user sequences**, and otherwise **nonuser sequences**.
- Event types of interest determine the possible candidates for the consequent C, e.g., diagnoses C_1-C_5.
- A study period is specified by $[t_S, t_E]$ according to the antecedent A. User sequences that do not contain A in this period are ignored.
- The time lengths T_e, T_r, T_b, and T_c indicate lengths of, respectively, *the effect period, the reference period, the period between the first A and the starting point of the reference period,* and *the control period* as illustrated in Fig. 1.

Algorithm 1 Mining UTARs given the Antecedent (MUTARA).

1. *Initialise parameters, such as the antecedent A, event types of interest, the study period $[t_S, t_E]$, time period lengths $T_e, T_r, T_b,$ and T_c.*
2. *Prepare user subsequences from user sequences which have A during the study period: choose event types from the hazard period, and exclude some of them based on the user-based exclusion with respect to the antecedent A;*
3. *Choose nonuser subsequences from the control period from nonuser sequences;*
4. *Calculate supports and unexpected-leverage of each event type of interest;*
5. *Rank the event types in the descending order of unexpected-leverage, and return the top 10 UTARs with high unexpected-leverage.*

- In order to offset low frequency of adverse reactions, a hazard period may cover several effect periods in a single user sequence. Effect periods starting with A ensures there exists A within a T_e-size period before any event in the hazard period. Based on empirical results, the hazard period is set as the union of the first two effect periods as illustrated in Fig. 1.

Clearly, event types within the hazard period may associate with the antecedent A. Among them, some are not likely to be unexpectedly induced by A. For each user sequence, we can use the event types which occurred before the first occurrence of A to infer event types which are not unexpectedly induced by A. We can disregard these expected ones in the hazard period, and infer that the remaining ones are more likely caused by A unexpectedly. An underlying reason is that most event types are more common in certain patient groups, say, the condition **hypertension** for the elderly [6] no matter whether A occurs or not. To simplify this exclusion operation for Step 2, we borrow the concept of the reference period from case-crossover studies [6]. It is a T_r-sized period which is a T_b-sized interval before the first occurrence of A as illustrated in Fig. 1. If the event type (e.g., suffering a disease) occurs in the reference period, the patient (or his/her doctors) would be not surprised to see its occurrence after A, especially if A is likely used to treat/prevent that event type. The event types within the reference period are probably not unexpected to the user with respect to the antecedent A, and they can be excluded for mining simple UTARs. For User 1 in Fig. 1, e.g., C_3 is in the reference period, and C_1, C_2, C_3 in the hazard period. Then, C_3 is excluded and $\{C_1, C_2\}$ are left for this subsequence. This exclusion operation is carried out only based on a single user sequence, and is termed as the **user-based exclusion**. It is basically designed to exclude some expected event types from a single drug user's viewpoint.

In Step 3 of MUTARA, for each nonuser sequence, we may randomly choose the control period within $[t_S, t_E + T_c]$. In order to avoid other confounding factors like age and gender, all nonusers are chosen from the same age-gender group as drug users. The event types within the control period comprise the nonuser subsequence.

In Step 4, putting the user subsequences after exclusion and the nonuser subsequences together, we calculate supports for each event type with respect to A. We compute unexpected-leverage of each event type according to Eq. (1).

Unexpected-leverage then aggregates unexpectedness among the subsequences. Finally, MUTARA outputs the top 10 UTARs with highest unexpected-leverage.

In MUTARA, the hazard, the effect and the control periods are set according to the antecedent A, and it restricts the event types in the calculation of supports. This **event-oriented data preparation** makes it possible to signal the usually infrequent ADRs. The user-based exclusion in Step 2 is simple and easily implemented, but it is crucial in MUTARA. In order to make a clear comparison, the only difference of our implemented OPUS_AR^t from MUTARA is without using the user-based exclusion.

4 Experimental Results on Administrative Databases

The CSIRO, through its Division of Mathematical and Information Sciences, was commissioned by the now Australian Government Department of Health and Ageing (DoHA) in August 2002 to analyse a linked data set produced from MBS, PBS and Queensland Hospital morbidity data, more commonly referred to as the Queensland Linked Data Set (QLDS). The objective was to provide a demonstration of the utility of data mining on de-identified administrative health data to investigate patterns of utilisation, adverse events and other health outcomes.

The QLDS was made available to CSIRO under a negotiated agreement between DoHA and Queensland Health. The data set contained de-identified and confidentially linked patient level hospital separation data (1 July 1995 to 30 June 1999), Medicare Benefits Scheme (MBS) data and Pharmaceutical Benefits Scheme (PBS) data (both 1 January 1995 to 31 December 1999). All data were de-identified, and actual dates of service were removed, so that time sequences were indicated only by time from first admission. This process provided strong privacy protection, consistent with the requirements of the relevant Federal and State legislations. CSIRO held the QLDS in a secure computer environment and limited access to authorised staff directly involved in the data analysis.

Each record in the hospital separation data may have several inpatient diagnoses coded in the ICD9 system. There are 2020 different diagnoses. Each record in the PBS data corresponds to one prescription supplied to one patient, and there are 758 distinct codes in the WHO ATC system [10]. The ATC codes for alendronate and atorvastatin, e.g., are M05BA04 and C10AA05 respectively [10]. For convenience, we refer to 1 January 1995 as Day 1 hereinafter. Thus the time period is [1, 1826] for all the health event sequences.

Like other data mining results, it is unrealistic to expect every interesting UTAR generated to be of value to domain experts. There may several reasons for UTARs being mined from the QLDS. (1) The QLDS only contains data for hospitalised patients which may not be representative of the general population. (2) The QLDS contains incomplete health data for each patient. Thus, similar to signal ADRs from spontaneous ADR reports [3], a practical goal is to short-list the unexpected associations between adverse events and use of medicines among the most interesting UTARs. These associations would have to be further evalu-

Table 1. Comparison results on inpatient diagnoses for older females given alendronate (4341 patients have used alendronate during [672, 1465], 121962 nonusers, and totally there are $N = 4341 + 121962 = 126303$ subsequences)

Rank based on Unexlev	Rank based on Leverage	ICD9 code	Diagnosis name	Unexlev	$supp(\xrightarrow{T} C) \times N$	UTAR support	Leverage	$supp(\xrightarrow{T} C) \times N$	TAR support
1	1	73300	OSTEOPOROSIS NOS	1.13E-03	954	175	2.02E-03	1071	292
2	3	73313	PATH FX VERTEBRAE	3.57E-04	260	54	5.63E-04	287	81
3	6	—	—	1.90E-04	118	28	3.50E-04	139	49
4	13	53011	REFLUX ESOPHAGITIS	1.59E-04	579	40	2.20E-04	587	48
5	11	—	—	1.52E-04	344	31	2.28E-04	354	41
6	24	—	—	1.35E-04	85	20	1.43E-04	86	21
7	5	—	—	1.35E-04	1396	65	3.64E-04	1426	95
8	9	—	—	1.31E-04	682	40	2.69E-04	700	58
9	23	—	—	1.13E-04	225	22	1.43E-04	229	26
10	15	—	—	1.12E-04	489	31	1.89E-04	499	41
...						...			...
1163	2	4019	HYPERTENSION NOS	-4.61E-04	6757	174	7.24E-04	6912	329

ated by causality analysis [3], and other health aspects taken into consideration in interpretation on any findings. Because this isn't yet to be done for our results, only those results corresponding to known ADRs are reported. If a drug is prescribed to treat a condition, it can be taken as a proxy for side effects associated with the condition. Prescription of nizatidine, e.g., may be a proxy for ulcers [7]. Some ADRs then may be signalled from unexpected associations between drugs probably used to treat adverse reactions and a given drug.

We report some preliminary results generated by MUTARA, in comparison with OPUS_AR^t. We concentrate on two drugs, alendronate[1] and atorvastatin[2]. Since atorvastatin is well-tolerated and its adverse drug reactions rarely lead to hospitalisation, we use prescribed drugs as proxies for signalling its side effects.

Due to space limitation, we only report some results on two cohorts, 'older (≥ 60) females' and 'older (≥ 60) males'. The parameters are set as follows. $T_e = T_c = 180$ and $T_c = T_b = 6 \times T_e$ in days. For the study period $[t_S, t_E]$, in order to leave reasonable room for hazard periods, $t_E = 1645 - T_e$ for inpatient diagnoses and $t_E = 1826 - T_e$ for prescribed drugs. t_S is set as the drug introduction day, i.e., 672 and 1114 for alendronate and atorvastatin respectively.

Table 1 lists some inpatient diagnoses in the descending order of unexpected-leverage for older females given alendronate generated by MUTARA. They are compared with results generated by OPUS_AR^t. In Table 1, reflux esophagitis (53011) is ranked as No 4 based on its unexpected-leverage value of 1.59E-04 (Column 5) among 2020 different diagnoses. It is worth noting that 48 (Column 10) patients suffer reflux esophagitis within 180 days after taking alendronate. Among them, about 40 (Column 7) drug users start suffering reflux esophagitis after drug usage. That is, only 8 patients suffered reflux esophagitis in their hazard and reference periods. As a comparison, reflux esophagitis is ranked as No 13 (Column 2) based on leverage (Column 8). It is interesting to observe that the first two diagnoses in Table 1 are closely related with osteoporosis which

---

[2] Atorvastatin is used to reduce the amount of cholesterol and certain fatty substances in the blood. Its side effects include **stomach ulcer, urinary tract infection**, etc [7].

Table 2. Comparison results on inpatient diagnoses for older males given alendronate (1027 alendronate users during [672,1465], 101304 nonusers, and $N=102331$)

Rank based on		ICD9 code	Diagnosis name	Unexlev	$supp(\xrightarrow{T} C)$ $\times N$	UTAR support	Leverage	$supp(\xrightarrow{T} C)$ $\times N$	TAR support
Unexlev	Leverage								
1	1	—	—	2.13E-04	119	23	3.78E-04	136	40
2	11	53081	ESOPHAGEAL REFLUX	4.85E-05	402	9	5.82E-05	403	10
3	4	—	—	4.83E-05	305	8	9.66E-05	310	13
4	13	—	—	4.31E-05	358	8	5.27E-05	359	9
5	25	—	—	3.41E-05	250	6	3.41E-05	250	6
...			...			...			...
495	2	4019	HYPERTENSION NOS	-7.87E-05	4390	36	2.02E-04	4419	65

Table 3. Top 10 drugs unexpectedly associated with atorvastatin for older males (6236 atorvastatin users during [1114, 1646], 78800 non-users, and $N=85036$)

Rank based on		ATC code	Drug name	Unexlev	$supp(\xrightarrow{T} C)$ $\times N$	UTAR support	Leverage	$supp(\xrightarrow{T} C)$ $\times N$	TAR support
Unexlev	Leverage								
1	8	—	—	4.01E-03	3463	595	5.33E-03	3584	716
2	24	—	—	2.59E-03	2588	410	2.75E-03	2603	425
3	46	—	—	1.09E-03	1490	202	1.65E-03	1541	253
4	56	A02BA04	Nizatidine	7.39E-04	1461	170	1.40E-03	1522	231
5	19	—	—	6.73E-04	2247	222	2.94E-03	2455	430
6	98	J01CF01	Dicloxacillin	6.25E-04	993	126	7.67E-04	1006	139
7	103	—	—	5.51E-04	466	81	7.25E-04	482	97
8	90	—	—	5.45E-04	1100	127	9.04E-04	1133	160
9	125	—	—	5.01E-04	237	60	5.23E-04	239	62
10	63	—	—	4.85E-04	1388	143	1.25E-03	1458	213

alendronate[1] is used to treat/prevent. We attribute these results to 'therapeutic failures' and the incomplete data. If the health event sequences contained all the diagnoses/symptoms before alendronate prescription, the user-based exclusion would decrease the unexpected-leverage values of these failures. This would further highlight ADRs like alendronate $\xrightarrow{T}$ reflux esophigitis in Table 1. Though hypertension NOS is ranked as No 2 by OPUS_AR^t, it is ranked as low as 1163 by MUTARA. Similar results can be found in Table 2. On the other hand, MUTARA and OPUS_AR^t run about 63.1 and 64.8 seconds on an Intel Pentium 4 (3.2GHz)/Linux computer respectively. Both run comparably fast.

Table 2 shows inpatient diagnoses with highest unexlev values for older males given alendronate. Among the 10 drug users suffering esophageal reflux, only one suffers it in its reference period. Based on its unexlev value of 4.85E-5, esophageal reflux which belongs to esophagitis[1] is ranked as No 2 by MUTARA. It is No 11 based on leverage by OPUS_AR^t.

Table 3 lists top 10 drugs with highest unexlev values for older males when atorvastatin is specified. It is interesting to see that No 4 and 6 drugs may be prescribed to treat side effects of atorvastatin[2]. Nizatidine (A02BA04) is used to treat the recurrence of ulcers and to treat other conditions where the stomach produces too much acid [7]. Nizatidine is presumably prescribed to treat side effects like stomach ulcer related to atorvastatin. Similarly, dicloxacillin (J01CF01) is presumably prescribed for urinary tract infection[2]. As a comparison, these two drugs are ranked as low as 56 and 98 by OPUS_AR^t respectively.

5 Conclusion and Future Work

In this paper, we have proposed the new knowledge representation, Unexpected Temporal Association Rule (UTAR), and its interestingness measure, unexpected-leverage, in the context of signalling unexpected and infrequent Adverse Drug Reactions (ADRs) from administrative health databases. Based on our user-based exclusion and event-oriented data preparation techniques, we have developed a simple mining algorithm MUTARA to signal simple ADRs.

From the QLDS, MUTARA has short-listed some known ADRs such as alendronate $\stackrel{T}{\hookrightarrow}$ esophagitis, and atorvastatin $\stackrel{T}{\hookrightarrow}$ stomach ulcer or urinary tract infection using drugs as proxies for diagnoses. Considering data biases and incompleteness in the QLDS, these shortlists are quite promising to help medical experts identify ADRs efficiently and effectively. It demonstrates the usefulness of this research. We are currently extending the proposed techniques to signal more complicated ADRs. In addition, we are interested in mining ADRs from large administrative health databases without specifying the drug exposure or outcome event.

Acknowledgements. The authors acknowledge the Australian Government Department of Health and Ageing (DoHA) and the Queensland Department of Health for providing data for this research. The authors thank K. Mackay, P. Graham, and anonymous reviewers for their comments and suggestions.

References

1. D. Bates and *et al.* Detecting adverse events using information technology. *Journal of American Medical Informatics Association*, 10(2):115–128, 2003.
2. J. Chen, H. He, G. Williams, and H. Jin. Temporal sequence associations for rare events. In *Proceedings of PAKDD'04*, pages 235–239, 2004.
3. M. Hauben and X. Zhou. Quantitative methods in pharmacovigilance: focus on signal detection. *Drug Safety*, 26(3):159–186, 2003.
4. J. Lazarou, B. Pomeranz, and P. Corey. Incidence of adverse drug reactions in hospitalized patients: a meta-analysis of prospective studies. *The Journal of the American Medical Association*, 279(15):1200–1205, 1998.
5. J. Li and *et al.* Mining risk patterns in medical data. *SIGKDD'05*, pp.770–5, 2005.
6. M. Maclure. The case-crossover design: a method for studying transient effects on the risk of acute events. *American Journal of Epidemiology*, 133(2):144–153, 1991.
7. MedlinePlus. http://medlineplus.gov/, 2005.
8. M. Stephens, J. Talbot, and P. Routledge, editors. *Detection of New Adverse Drug Reactions*. Macmillan Reference Ltd, London, United Kingdom, 4th edition, 1998.
9. The Adverse Drug Reactions Advisory Committee (ADRAC). Australian ADR bulletin. DoHA, Australia, 2005. http://www.tga.gov.au/adr/aadrb.htm.
10. The Drug Utilisation Sub-Committee (DUSC). Australian statistics on medicines. DoHA, Australia, 2000.
11. K. Wang, Y. Jiang, and L. V. Lakshmanan. Mining unexpected rules by pushing user dynamics. In *Proceedings of SIGKDD'03*, pages 246–255, 2003.
12. G. I. Webb. Efficient search for association rules. In *SIGKDD'00*, pp.99–107, 2000.

Author Index

Abe, Hidenao 70, 509
Achtert, Elke 119, 174
An, Aijun 107
Appan, Preetha 371
Awan, A. Majid 841

Baig, Abdul Rauf 504
Balachandran, Ramanathan 673
Bashir, Shariq 504
Becker, Karin 346
Bell, G.D. 285
Benatallah, Boualem 250
Berti-Équille, Laure 440
Bhowmick, Sourav S. 323, 351, 425
Böhm, Christian 119
Brecheisen, Stefan 179

Cao, Longbing 821
Chakravarthy, Sharma 673
Chan, Shing-Kit 334
Chang, Elizabeth 450
Chau, Michael 199
Chen, Chien-Yu 530
Chen, Hung-Ming 520
Chen, Jie 867
Chen, Kuan-Wei 520
Chen, Ling 425
Chen, Ming-Syan 699, 719, 729
Chen, Wei-Ta 699
Chen, Xiaojun 189
Chen, Yi-Ling 699
Cheng, Haibin 765
Cheng, James 462
Cheng, Reynold 199
Cho, Sungzoon 215
Chung, Yongwha 603

Dai, Bi-Ru 719
de la Iglesia, B. 285
Deng, Shengchun 567
Dillon, Tharam S. 450
Driessens, Kurt 60
Du, Wenliang 637
Dunham, Margaret H. 750

Faloutsos, Christos 713
Fan, Hongjian 91
Fan, Hongqin 557
Fan, Ming 91
Feng, Guang 313
Feng, Ling 450
Foss, Andrew 557
Frank, Eibe 97

Galloway, John 410
Gao, Jing 765
Gao, Li 240
Ge, Rong 621
Goh, John 415
Gruenwald, Le 323

Hadzic, Fedja 450
Han, Jiawei 577
Hand, David J. 1
Hassan, Mahbub 250
He, Hongxing 867
He, Qinming 15
He, Zengyou 567
Hilario, Melanie 546
Hirate, Yu 775
Ho, Shinn-Jang 520, 790
Ho, Shinn-Ying 520, 790
Ho, Tu Bao 265, 493
Honavar, Vasant 45, 55
Hong, Tzung-Pei 483
Hsieh, Chih-Hung 520
Hsu, Chen-Ming 530
Hsu, Ching-Chi 530
Hsueh, Sue-Chen 724
Hu, Xiaohua 303
Huang, Hui-Ling 520
Huang, Joshua Zhexue 189, 567
Huang, Xiangji 107
Huang, Yaochun 552
Hung, Ming-Hao 790
Hurson, Ali R. 679
Hwang, Sheng-Kun 724

Jaroszewicz, Szymon 35
Jia, Weijia 361

Jiang, Lizheng 149
Jiang, Xiaoqian 647
Jiao, Licheng 801, 857
Jin, Huidong 867
Jin, Wen 577, 621
Jin, Xiaoming 755
Jou, Jun-Wun 790

Kalousis, Alexandros 546
Kang, Bo-Yeong 129
Kang, Dae-Ki 45, 55
Kao, Ben 199
Ke, Yiping 462
Kelman, Chris 867
Kemme, Bettina 657
Kim, Dae-Won 129
Kim, Dongil 215
Kim, Jinho 739
Kim, Sinjae 594
Kleinberg, Jon 380
Koh, Yun Sing 473
Körner, Christine 668
Kriegel, Hans-Peter 139, 174, 179
Kröger, Peer 119
Kryszkiewicz, Marzena 468
Ku, Wen-Yuan 790

Lam, Wai 334
Le, Hoai Minh 160
Le, Minh-Hoang 265
Le, Si Quang 493
Le Thi, Hoai An 160
Lee, Hansung 603
Leschi, Claire 60
Leskovec, Jure 380
Li, Chuanjun 689
Li, Chunping 400
Li, Jinyan 425
Li, Ming 704
Li, Xiaoming 361
Li, Yijun 780
Liao, Lingzhi 20
Lim, Ee-Peng 250, 415
Lin, Ming-Yen 724
Ling, Ping 225
Liu, Baw-Jhiune 530
Liu, Fei Tony 81
Liu, Ken-Hao 729
Liu, Mengxu 91
Liu, Rey-Long 255

Liu, Tie-Yan 240, 313
Liu, Wei-Chuan 729
Liu, Yang 107
Lu, Jingli 499
Lu, Yumao 205
Luo, Siwei 20

Ma, Wei-Ying 240, 275, 313
Ma, Xiuli 149
Ma, Xuesong 657
Majumdar, A.K. 611
McAullay, Damien 867
Meng, Yu 750
Mogi, Akira 390
Moon, Yang-Sae 739
Motoda, Hiroshi 390

Nakamori, Yoshiteru 265
Nayak, Richi 292
Newton, Glen 657
Ng, Jackey 199
Ng, Wilfred 462
Ng, Yen Kaow 540
Nguyen, Phu Chien 390
Nichele, Caren Moraes 346

O'Keefe, Christine M. 867
O'Keefe, Richard 473
Ohara, Kouzou 390
Ohsaki, Miho 509

Padmanabhan, Srihari 673
Papadimitriou, Spiros 713
Park, Cheong Hee 30
Park, Daihee 603
Pfahringer, Bernhard 60, 97
Pfeifle, Martin 179
Polat, Huseyin 637
Prabhakaran, B. 689
Precup, Doina 657
Pryakhin, Alexey 139, 174

Qian, Weining 621
Qiang, Qi 15

Raghavan, Prabhakar 11
Ramamohanarao, Kotagiri 91
Reutemann, Peter 60
Ribeiro, Rita 816
Rountree, Nathan 473
Roychowdhury, Vwani 205

Author Index

Saad, F.H.   285
Sap, Mohd. Noor Md.   841
Schmidt-Thieme, Lars   831
Schubert, Matthias   139, 174
Scripps, Jerry   765
Shinohara, Takeshi   540
Silvescu, Adrian   45, 55
Simoff, Simeon J.   410
Simovici, Dan A.   35
Singh, Ajit   380
Song, Binheng   847
Srivastava, Abhinav   611
Sun, Aixin   250, 351
Sun, Jimeng   713
Sundaram, Hari   371
Sung, Sam Y.   552
Sural, Shamik   611

Tan, Henry   450
Tan, Pang-Ning   765
Tang, Shiwei   149
Taniar, David   415
Tao, Pham Dinh   160
Thuraisingham, Bhavani M.   12
Ting, Kai Ming   81
Tong, Weimin   780
Torgo, Luis   816
Tseng, Belle   371
Tseng, Shian-Shyong   483
Tso, Karen H.L.   831
Tsumoto, Shusaku   509
Tung, Anthony K.H.   577

Vaidya, Jaideep   647
Vasile, Flavian   55
Vinh, Le Sy   493

Wang, Ching-Yao   483
Wang, Wei   577
Wang, Yan   225
Wang, Zhihai   20
Washio, Takashi   390
Wee, Kyubum   594
Wen, Ji-Rong   275
Williams, Graham   189
Wong, Tak-Lam   334
Woźnica, Adam   546

Wrobel, Stefan   668
Wu, Chienwen   435
Wu, Junfeng   557
Wu, Weili   552

Xiong, Hui   552
Xu, Sumei   292
Xu, Xiaofei   189, 567
Xu, Zhen   847

Yajima, Yasutoshi   230
Yamaguchi, Takahira   70, 509
Yamana, Hayato   775
Yang, Bo   679
Yang, Dongqing   149
Yang, Huai-Yuan   240
Ye, Qiang   780
Ye, Shaozhi   275
Ye, Yunming   189
Yeh, Mi-Yen   719
Yoo, Illhoi   303
Yu, Hwanjo   647
Yu, Jeffrey Xu   499

Zaïane, Osmar R.   557
Zhan, De-Chuan   806
Zhang, Chengqi   499, 821
Zhang, Dehui   149
Zhang, Li   801
Zhang, Shichao   499
Zhang, Xu-Dong   313
Zhang, Zhigang   361
Zhao, Jun   857
Zhao, Lianwei   20
Zhao, Mingying   857
Zhao, Qiankun   323, 351
Zhao, Shuguang   857
Zhao, Yanchang   20
Zhou, Chun-Guang   225
Zhou, Shuigeng   189
Zhou, Weida   801
Zhou, Xueyuan   400
Zhou, Zhi-Hua   806
Zhu, Yanong   704
Zou, Beibei   657
Zuo, Xinqiang   755

Lecture Notes in Artificial Intelligence (LNAI)

Vol. 3918: W.K. Ng, M. Kitsuregawa, J. Li, K. Chang (Eds.), Advances in Knowledge Discovery and Data Mining. XXIV, 879 pages. 2006.

Vol. 3910: S.A. Brueckner, G.D.M. Serugendo, D. Hales, F. Zambonelli (Eds.), Engineering Self-Organising Systems. XII, 245 pages. 2006.

Vol. 3904: M. Baldoni, U. Endriss, A. Omicini, P. Torroni (Eds.), Declarative Agent Languages and Technologies III. XII, 245 pages. 2006.

Vol. 3899: S. Frintrop, VOCUS: A Visual Attention System for Object Detection and Goal-Directed Search. XIV, 216 pages. 2006.

Vol. 3890: S.G. Thompson, R. Ghanea-Hercock (Eds.), Defence Applications of Multi-Agent Systems. XII, 141 pages. 2006.

Vol. 3885: V. Torra, Y. Narukawa, A. Valls, J. Domingo-Ferrer (Eds.), Modeling Decisions for Artificial Intelligence. XII, 374 pages. 2006.

Vol. 3881: S. Gibet, N. Courty, J.-F. Kamp (Eds.), Gesture in Human-Computer Interaction and Simulation. XIII, 344 pages. 2006.

Vol. 3874: R. Missaoui, J. Schmidt (Eds.), Formal Concept Analysis. X, 309 pages. 2006.

Vol. 3873: L. Maicher, J. Park (Eds.), Charting the Topic Maps Research and Applications Landscape. VIII, 281 pages. 2006.

Vol. 3863: M. Kohlhase (Ed.), Mathematical Knowledge Management. XI, 405 pages. 2006.

Vol. 3862: R.H. Bordini, M. Dastani, J. Dix, A.E.F. Seghrouchni (Eds.), Programming Multi-Agent Systems. XIV, 267 pages. 2006.

Vol. 3849: I. Bloch, A. Petrosino, A.G.B. Tettamanzi (Eds.), Fuzzy Logic and Applications. XIV, 438 pages. 2006.

Vol. 3848: J.-F. Boulicaut, L. De Raedt, H. Mannila (Eds.), Constraint-Based Mining and Inductive Databases. X, 401 pages. 2006.

Vol. 3847: K.P. Jantke, A. Lunzer, N. Spyratos, Y. Tanaka (Eds.), Federation over the Web. X, 215 pages. 2006.

Vol. 3835: G. Sutcliffe, A. Voronkov (Eds.), Logic for Programming, Artificial Intelligence, and Reasoning. XIV, 744 pages. 2005.

Vol. 3830: D. Weyns, H. V.D. Parunak, F. Michel (Eds.), Environments for Multi-Agent Systems II. VIII, 291 pages. 2006.

Vol. 3817: M. Faundez-Zanuy, L. Janer, A. Esposito, A. Satue-Villar, J. Roure, V. Espinosa-Duro (Eds.), Nonlinear Analyses and Algorithms for Speech Processing. XII, 380 pages. 2006.

Vol. 3814: M. Maybury, O. Stock, W. Wahlster (Eds.), Intelligent Technologies for Interactive Entertainment. XV, 342 pages. 2005.

Vol. 3809: S. Zhang, R. Jarvis (Eds.), AI 2005: Advances in Artificial Intelligence. XXVII, 1344 pages. 2005.

Vol. 3808: C. Bento, A. Cardoso, G. Dias (Eds.), Progress in Artificial Intelligence. XVIII, 704 pages. 2005.

Vol. 3802: Y. Hao, J. Liu, Y.-P. Wang, Y.-m. Cheung, H. Yin, L. Jiao, J. Ma, Y.-C. Jiao (Eds.), Computational Intelligence and Security, Part II. XLII, 1166 pages. 2005.

Vol. 3801: Y. Hao, J. Liu, Y.-P. Wang, Y.-m. Cheung, H. Yin, L. Jiao, J. Ma, Y.-C. Jiao (Eds.), Computational Intelligence and Security, Part I. XLI, 1122 pages. 2005.

Vol. 3789: A. Gelbukh, Á. de Albornoz, H. Terashima-Marín (Eds.), MICAI 2005: Advances in Artificial Intelligence. XXVI, 1198 pages. 2005.

Vol. 3782: K.-D. Althoff, A. Dengel, R. Bergmann, M. Nick, T.R. Roth-Berghofer (Eds.), Professional Knowledge Management. XXIII, 739 pages. 2005.

Vol. 3763: H. Hong, D. Wang (Eds.), Automated Deduction in Geometry. X, 213 pages. 2006.

Vol. 3755: G.J. Williams, S.J. Simoff (Eds.), Data Mining. XI, 331 pages. 2006.

Vol. 3735: A. Hoffmann, H. Motoda, T. Scheffer (Eds.), Discovery Science. XVI, 400 pages. 2005.

Vol. 3734: S. Jain, H.U. Simon, E. Tomita (Eds.), Algorithmic Learning Theory. XII, 490 pages. 2005.

Vol. 3721: A.M. Jorge, L. Torgo, P.B. Brazdil, R. Camacho, J. Gama (Eds.), Knowledge Discovery in Databases: PKDD 2005. XXIII, 719 pages. 2005.

Vol. 3720: J. Gama, R. Camacho, P.B. Brazdil, A.M. Jorge, L. Torgo (Eds.), Machine Learning: ECML 2005. XXIII, 769 pages. 2005.

Vol. 3717: B. Gramlich (Ed.), Frontiers of Combining Systems. X, 321 pages. 2005.

Vol. 3702: B. Beckert (Ed.), Automated Reasoning with Analytic Tableaux and Related Methods. XIII, 343 pages. 2005.

Vol. 3698: U. Furbach (Ed.), KI 2005: Advances in Artificial Intelligence. XIII, 409 pages. 2005.

Vol. 3690: M. Pěchouček, P. Petta, L.Z. Varga (Eds.), Multi-Agent Systems and Applications IV. XVII, 667 pages. 2005.

Vol. 3684: R. Khosla, R.J. Howlett, L.C. Jain (Eds.), Knowledge-Based Intelligent Information and Engineering Systems, Part IV. LXXIX, 933 pages. 2005.

Vol. 3683: R. Khosla, R.J. Howlett, L.C. Jain (Eds.), Knowledge-Based Intelligent Information and Engineering Systems, Part III. LXXX, 1397 pages. 2005.

Vol. 3682: R. Khosla, R.J. Howlett, L.C. Jain (Eds.), Knowledge-Based Intelligent Information and Engineering Systems, Part II. LXXIX, 1371 pages. 2005.

Vol. 3681: R. Khosla, R.J. Howlett, L.C. Jain (Eds.), Knowledge-Based Intelligent Information and Engineering Systems, Part I. LXXX, 1319 pages. 2005.

Vol. 3673: S. Bandini, S. Manzoni (Eds.), AI*IA 2005: Advances in Artificial Intelligence. XIV, 614 pages. 2005.

Vol. 3662: C. Baral, G. Greco, N. Leone, G. Terracina (Eds.), Logic Programming and Nonmonotonic Reasoning. XIII, 454 pages. 2005.

Vol. 3661: T. Panayiotopoulos, J. Gratch, R.S. Aylett, D. Ballin, P. Olivier, T. Rist (Eds.), Intelligent Virtual Agents. XIII, 506 pages. 2005.

Vol. 3658: V. Matoušek, P. Mautner, T. Pavelka (Eds.), Text, Speech and Dialogue. XV, 460 pages. 2005.

Vol. 3651: R. Dale, K.-F. Wong, J. Su, O.Y. Kwong (Eds.), Natural Language Processing – IJCNLP 2005. XXI, 1031 pages. 2005.

Vol. 3642: D. Ślęzak, J. Yao, J.F. Peters, W. Ziarko, X. Hu (Eds.), Rough Sets, Fuzzy Sets, Data Mining, and Granular Computing, Part II. XXIII, 738 pages. 2005.

Vol. 3641: D. Ślęzak, G. Wang, M. Szczuka, I. Düntsch, Y. Yao (Eds.), Rough Sets, Fuzzy Sets, Data Mining, and Granular Computing, Part I. XXIV, 742 pages. 2005.

Vol. 3635: J.R. Winkler, M. Niranjan, N.D. Lawrence (Eds.), Deterministic and Statistical Methods in Machine Learning. VIII, 341 pages. 2005.

Vol. 3632: R. Nieuwenhuis (Ed.), Automated Deduction – CADE-20. XIII, 459 pages. 2005.

Vol. 3630: M.S. Capcarrère, A.A. Freitas, P.J. Bentley, C.G. Johnson, J. Timmis (Eds.), Advances in Artificial Life. XIX, 949 pages. 2005.

Vol. 3626: B. Ganter, G. Stumme, R. Wille (Eds.), Formal Concept Analysis. X, 349 pages. 2005.

Vol. 3625: S. Kramer, B. Pfahringer (Eds.), Inductive Logic Programming. XIII, 427 pages. 2005.

Vol. 3620: H. Muñoz-Ávila, F. Ricci (Eds.), Case-Based Reasoning Research and Development. XV, 654 pages. 2005.

Vol. 3614: L. Wang, Y. Jin (Eds.), Fuzzy Systems and Knowledge Discovery, Part II. XLI, 1314 pages. 2005.

Vol. 3613: L. Wang, Y. Jin (Eds.), Fuzzy Systems and Knowledge Discovery, Part I. XLI, 1334 pages. 2005.

Vol. 3607: J.-D. Zucker, L. Saitta (Eds.), Abstraction, Reformulation and Approximation. XII, 376 pages. 2005.

Vol. 3601: G. Moro, S. Bergamaschi, K. Aberer (Eds.), Agents and Peer-to-Peer Computing. XII, 245 pages. 2005.

Vol. 3600: F. Wiedijk (Ed.), The Seventeen Provers of the World. XVI, 159 pages. 2006.

Vol. 3596: F. Dau, M.-L. Mugnier, G. Stumme (Eds.), Conceptual Structures: Common Semantics for Sharing Knowledge. XI, 467 pages. 2005.

Vol. 3593: V. Mařík, R. W. Brennan, M. Pěchouček (Eds.), Holonic and Multi-Agent Systems for Manufacturing. XI, 269 pages. 2005.

Vol. 3587: P. Perner, A. Imiya (Eds.), Machine Learning and Data Mining in Pattern Recognition. XVII, 695 pages. 2005.

Vol. 3584: X. Li, S. Wang, Z.Y. Dong (Eds.), Advanced Data Mining and Applications. XIX, 835 pages. 2005.

Vol. 3581: S. Miksch, J. Hunter, E.T. Keravnou (Eds.), Artificial Intelligence in Medicine. XVII, 547 pages. 2005.

Vol. 3577: R. Falcone, S. Barber, J. Sabater-Mir, M.P. Singh (Eds.), Trusting Agents for Trusting Electronic Societies. VIII, 235 pages. 2005.

Vol. 3575: S. Wermter, G. Palm, M. Elshaw (Eds.), Biomimetic Neural Learning for Intelligent Robots. IX, 383 pages. 2005.

Vol. 3571: L. Godo (Ed.), Symbolic and Quantitative Approaches to Reasoning with Uncertainty. XVI, 1028 pages. 2005.

Vol. 3559: P. Auer, R. Meir (Eds.), Learning Theory. XI, 692 pages. 2005.

Vol. 3558: V. Torra, Y. Narukawa, S. Miyamoto (Eds.), Modeling Decisions for Artificial Intelligence. XII, 470 pages. 2005.

Vol. 3554: A.K. Dey, B. Kokinov, D.B. Leake, R. Turner (Eds.), Modeling and Using Context. XIV, 572 pages. 2005.

Vol. 3550: T. Eymann, F. Klügl, W. Lamersdorf, M. Klusch, M.N. Huhns (Eds.), Multiagent System Technologies. XI, 246 pages. 2005.

Vol. 3539: K. Morik, J.-F. Boulicaut, A. Siebes (Eds.), Local Pattern Detection. XI, 233 pages. 2005.

Vol. 3538: L. Ardissono, P. Brna, A. Mitrović (Eds.), User Modeling 2005. XVI, 533 pages. 2005.

Vol. 3533: M. Ali, F. Esposito (Eds.), Innovations in Applied Artificial Intelligence. XX, 858 pages. 2005.

Vol. 3528: P.S. Szczepaniak, J. Kacprzyk, A. Niewiadomski (Eds.), Advances in Web Intelligence. XVII, 513 pages. 2005.

Vol. 3518: T.-B. Ho, D. Cheung, H. Liu (Eds.), Advances in Knowledge Discovery and Data Mining. XXI, 864 pages. 2005.

Vol. 3508: P. Bresciani, P. Giorgini, B. Henderson-Sellers, G. Low, M. Winikoff (Eds.), Agent-Oriented Information Systems II. X, 227 pages. 2005.

Vol. 3505: V. Gorodetsky, J. Liu, V.A. Skormin (Eds.), Autonomous Intelligent Systems: Agents and Data Mining. XIII, 303 pages. 2005.

Vol. 3501: B. Kégl, G. Lapalme (Eds.), Advances in Artificial Intelligence. XV, 458 pages. 2005.

Vol. 3492: P. Blache, E.P. Stabler, J.V. Busquets, R. Moot (Eds.), Logical Aspects of Computational Linguistics. X, 363 pages. 2005.

Vol. 3490: L. Bolc, Z. Michalewicz, T. Nishida (Eds.), Intelligent Media Technology for Communicative Intelligence. X, 259 pages. 2005.

Vol. 3488: M.-S. Hacid, N.V. Murray, Z.W. Raś, S. Tsumoto (Eds.), Foundations of Intelligent Systems. XIII, 700 pages. 2005.

Vol. 3487: J.A. Leite, P. Torroni (Eds.), Computational Logic in Multi-Agent Systems. XII, 281 pages. 2005.